LANDSLIDES in PRACTICE

LANDSLIDES in PRACTICE

Investigations, Analysis, and
Remedial/Preventative Options in Soils

DEREK H. CORNFORTH, PhD, PE

WILEY

JOHN WILEY & SONS, INC.

This book is printed on acid-free paper. ♾

Published by John Wiley & Sons, Inc., Hoboken, New Jersey
Published simultaneously in Canada

For general information on our other products and services or for technical support, please contact our Customer Care Department within the United States at 800-762-2974, outside the United States at (317) 572-3993 or fax (317) 572-4002.

Wiley also publishes its books in a variety of electronic formats. Some content that appears in print may not be available in electronic books.

Library of Congress Cataloging-in-Publication Data:

Cornforth, Derek H.
Landslides in practice : investigation, analysis and remedial / preventative options in soils / Derek H. Cornforth
Includes bibliographical references and index.
ISBN 0-471-67816-3 (cloth)
1. Soil mechanics. 2. Landslide hazard analysis. 3. Landslides I. Title.
TA710.C65 2004
624.1'51363—dc22

2004007921

Printed in the United States of America

10 9 8 7 6 5 4 3 2 1

Contents

Relative Cost of Construction

When comparing remedial/preventative options, the cost of construction is one of the more important considerations. The cost, together with the likely effectiveness and specific site conditions, determines the preferred option.

The table shown below provides geotechnical practitioners with a quick guide to relative construction costs for the 45 remedial/preventative techniques presented in Part B of this book. It is based on the author's experiences in the Pacific Northwest region of the United States and should be useful in preliminary landslide studies to reduce potential options to a manageable number, and also to avoid overlooking less frequently used techniques that may be appropriate for a particular site. See pages 267–269 for an expanded discussion of this topic.

	Relative Construction Contract Cost				
	L		M		H
Slope regrading	—	—			
External buttress		—	—	—	
Infill buttress		—			
Shear key		—	—		
Filter systems	—				
Riprap/grouted riprap			—		
Gabion mattresses		—			
Shotcrete	—	—			
Chunam plaster	—				
Bioremediation		—			
Horizontal drains		—	—		
Trench drains			—		
French drains	—				
Drainage blanket	—	—			
Deep wells			—	—	
Wellpoints, ejectors		—	—		
Relief wells	—	—			
Vertical gravity drain	—	—			
Tunnels, adits				—	—
Vertical shaft/drain array					—
Surface water control	—	—			
Foundation consolidating berms		—	—		
Prefabricated vertical drains		—			

	Relative Construction Contract Cost				
	L		M		H
Slurry trench cutoff wall				—	—
Slope liners			—		
Grout curtains			—	—	—
Tied-back soldier pile wall			—	—	
Concrete shear pile wall				—	—
Tied-back slurry trench wall					—
Masonry or concrete gravity wall		—	—	—	
Concrete cantilever wall			—	—	
Concrete crib or bin wall			—		
Gabion wall	—	—			
Soil nailing		—	—		
Mechanically stabilized earth wall			—	—	
Reinforced soil slope		—	—		
Reticulated micropiles		—	—		
Compaction grouting		—	—	—	
Dynamic compaction			—	—	
Vibro-compaction			—	—	
Stone columns			—	—	
Excavation and replacement		—	—		
Deep soil mixing			—	—	
Shear piles				—	—
Tied-back element wall			—	—	

L – Low M – Medium H – High

Metric Conversions

Length	1 inch	= 25.4 mm	1 mm	= 0.0394 inch
	1 foot	= 0.3048 m	1 m	= 3.281 feet
	1 mile	= 1.6093 km	1 km	= 0.621 mile
Area	1 sq. in.	= 6.452 cm^2	1 cm^2	= 0.155 sq. in.
	1 sq. ft.	= 0.0929 m^2	1 m^2	= 10.764 sq. ft
	1 acre	= 0.4047 hectare	1 hectare	= 2.47 acre
	1 sq. mile	= 2.59 km^2	1 km^2	= 0.386 sq. mile
Volume	1 cu. in.	= 16.387 cm^3 (cc)	1 cm^2	= 0.061 cu. in.
	1 cu. ft.	= 0.0283 m^3	1 m^3	= 35.31 cu. ft.
	1 cu. yd.	= 0.7646 m^3	1 m^3	= 1.308 cu. yd.
	1 U.S. gallon	= 3.785 liters	1 liter	= 0.264 U.S. gallon
Mass	1 lb.	= 0.4536 kg	1 kg	= 2.205 lb.
Force	1 lb.	= 4.448 N	1 N	= 0.225 lb.
	1 ton	= 8.896 kN	1 kN	= 0.1124 U.S. ton
Density	1 lb./cu. ft.	= 16.019 kg/m^3	1 kg/m^3	= 0.0624 lb./cu. ft.
		= 0.1571 kN/m^3	1 kN/m^3	= 6.365 lb./cu. ft.
Pressure/Stress	1 lb./sq. in.	= 0.0703 kg/cm^2	1 kg/cm^2	= 14.22 lb./sq. in.
		(= 6.895 kPa)	1 kPa	= 0.145 lb./sq. in.
	1 lb./sq. ft.	= 4.882 kg/cm^2	1 kg./cm^2	= 0.2048 lb./sq. ft
		(= 0.04788 kPa)	1 kPa	= 20.886 lb./sq. ft
	1 U.S. ton/sq. ft.	= 95.76 kPa	1 kPa	= 0.01044 US ton/sq. ft
	[Note: 1 kPa	= 1 kN/m^2]		
Flow Velocity	1 gal./min.	= 6.309x10^{-5} m^3/sec	1 m^3/sec	= 15,850 gpm
	1 ft./sec.	= 0.3048 m/sec	1 m/sec	= 3.28 ft./sec
Coefft. of compressibility m_v:				
	1 sq. ft./U.S. ton	= 0.0104 m^2/kN		
	1 sq. in./lb.	= 14.22 cm^2/kg		
Coefft. of consolidation c_v:				
	1 sq. ft./year	= 0.0929 m^2/year	1 m^2/year	10.76 sq. ft./year
		(= 0.002946 mm^2/sec)	1 mm^2/sec	= 339.4 sq. ft./year
Moment	1 lb.-ft.	= 0.1383 kg-m	1 kg-m	= 7.23 lb.-ft
		(= 1.3558 Nm)	1 N-m	= 0.7376 lb.-ft
Speed	1 mile/hour	= 1.609 km/hour	1 km/hour	= 0.622 mile/hour
		(= 0.447 m/sec)	1 m/sec	= 2.237 mph
	1 foot/sec	= 0.3048 m/sec	1 m/sec	= 3.281 feet/sec

U.S. Conversions

	1 U.S. gallon	=	0.1337 cu. ft.	1 cu. ft.	=	7.48 U.S. gallons
	1 acre	=	43,560 sq. ft.			
	1 U.S. gallon/min	=	0.002228 cu. ft./sec.	1 cu. ft./sec.	=	449 U.S. gal./min.
	1 kip	=	1,000 lb.			
	1 lb./sq. in.	=	2.307 feet of water			
c_v:	1 sq. in./min.	=	3,652 sq. ft./year			
k:	1 cm/sec	=	1,035,000 ft./year			
m_v:	1 sq. in./lb.	=	13.89 sq. ft./U.S. ton			
γ:	1 gm/cc	=	62.43 lb/cu. ft.			

Abbreviations

AASHTO	American Association of State Highway and Transportation Officials, Washington D.C.
ASCE	American Society of Civil Engineers
ASTM	American Society for Testing and Materials, Philadelphia, Pennsylvania
BGS	British Geotechnical Society
BSI	British Standards Institution, London, England
CIRIA	Construction Industry Research and Information Association, London, England
EERC	Earthquake Engineering Research Center, Univ. of California
ENR	Engineering News-Record Magazine
FHWA	Federal Highway Administration, Washington, D.C.
IAEG	International Association of Engineering Geologists
ICE	The Institution of Civil Engineers, London, England
ISSMFE	International Society of Soil Mechanics & Foundation Engineering
NAVFAC	Naval Engineering Facilities Engineering Command, Alexandria, Virginia
PTI	Post Tensioning Institute, Phoenix, Arizona
USCE	U.S. Army Corps of Engineers

About This Book

This book has taken many years of preparation and reflects the author's knowledge gained from almost 50 years of practice as a geotechnical landslide consultant. It is written for graduate students and practicing engineers/geologists in geotechnical engineering and engineering geology. The book first describes landslide analyses and investigations, but the primary focus is on available remedial and preventative measures for landslides in soils and weathered rocks. Numerous case histories are described throughout the book to illustrate the application of these measures in the real world. Selected case histories are given fuller treatment in the final section of the book. The writing style is easy to follow. For each remedial technique, information is provided on (i) how it is used, (ii) engineering design, (iii) construction procedure, (iv) relative cost, and (v) practical examples.

The book includes the author's technical contributions on appropriate factors of safety, original profile analysis, Cohesive Index for classifying silts and clays, method of estimating the drained strength of sands and gravels, analytical wedge analyses, and thought-provoking comments on laboratory undrained shear tests and back analysis of landslides.

About the Author

Derek Cornforth obtained his M.S. degree (as a Fulbright Scholar) at Northwestern University and subsequently earned a Ph.D. in geotechnical engineering from London University's Imperial College of Science & Technology. After working for several years as a geotechnical specialist in Britain, he moved to the United States and was a vice president with Shannon & Wilson for nine years. In 1983, he founded Landslide Technology in Portland, Oregon. Through his technical leadership, this firm has studied and remediated hundreds of landslides at sites in the western United States, Alaska, and abroad. It has received many engineering excellence awards for providing innovative landslide repairs. Dr. Cornforth has written 25 published technical papers, hundreds of geotechnical engineering reports, and has made numerous presentations to universities and technical societies on landslides and other geotechnical subjects. He also has provided expert witness and arbitration services, and has served on national committees for the American Society of Civil Engineers and the British Geotechnical Society. During his career, Dr. Cornforth was a chartered engineer in the U.K., and a registered professional engineer in several states.

Preface

The book is divided into three parts:

PART A describes site investigations, soil shear strength properties, and slope stability analyses. These are the technical tools needed to study landslides.

PART B describes 45 methods for stabilizing landslides or preventing instability from occurring in soils. The intent is to make the reader knowledgeable about these techniques so that appropriate choices can be made, depending on site conditions. In many cases, there are entire books devoted to some of these techniques, and such works can provide more details where a particular technique is of interest.

PART C describes 12 selected case histories in which multiple technical points relevant to landslide practice are illustrated. These points are listed at the beginning of each case history and are cross-referenced throughout Parts A and B. It should be mentioned that there are many other case histories (or examples) within Parts A and B where briefer explanations are provided.

Equations are numbered consecutively in *each chapter,* but are not numbered consecutively from chapter to chapter. U.S. agencies are sometimes named with "U.S." omitted. After the first reference within a chapter, agencies are often referred to by letters; e.g., U.S. Federal Highway Administration is designated as Federal Highway Administration and FHWA. The U.S. Army Corps of Engineers is referred to as Corps of Engineers or USCE.

Acknowledgments

This book would not have been possible without the active support and encouragement from Landslide Technology. Many current and former staff members have contributed to the technical input and chapter reviews (in alphabetical order): Darren Beckstrand, Brent Black, Chris Carpenter, Charles Hammond, Jonathan Harris, Kristie Hartfeil, Gerry Heslin, Randy Hill, George Machan, Erica Meyer, Mike Meyer, Larry Pierson, president Ermel Quevedo, John Sager, Will Shallenberger, Andy Vessely, and Kenji Yamasaki. Carole Schmidt typed the draft manuscript and Lauren O'Neal, assisted by Michael Tardif, produced the hundreds of drawings within the book. Greg Westergaard prepared the photographs.

Other contributions have been made by Gordon Green and Erik Mikkelsen (field instrumentation), Frank Fujitani (Shannon & Wilson), Lee Schroeder (Oregon State University), Greg McInnis (Geo-Tech Explorations), Ron Chapman (Schnabel Foundation Engineering), Mark Koelling (Hayward Baker), George Gross (Spencer Gross Surveying), Gary Reynolds (Portland General Electric), and Jim Lowell (Alaska DOT).

The help of all these technical resources has been invaluable and much appreciated.

LANDSLIDES in PRACTICE

PART A Investigations and Analyses

The coverage of each subject in Part A is deliberately uneven, with more information being provided on topics that engineers and geologists typically have difficulties in understanding. For example, a detailed presentation is given on the shear strength of stiff clays and the failure of such slopes many years after construction (delayed failure). Other techniques, such as cone penetration testing, are only mentioned briefly because they are rarely used on landslides. These choices have been made to control the length of Part A in deference to Part B, the main focus of the book. Nevertheless, engineers and geologists should find much useful information on site investigations and stability analyses for landslides in Part A.

Many textbooks are currently available on these subjects. Three of them are especially recommended by the author as being complementary and supplementary to the present volume.

Soil Mechanics in Engineering Practice, 3rd ed., by Karl Terzaghi, Ralph B. Peck, and Gholamreza Mesri. John Wiley & Sons, Inc. Third Edition published 1996, 548 pp.

Geotechnical Instrumentation for Monitoring Field Performance, by John Dunnicliff. John Wiley & Sons, Inc., published 1988, 575 pp.

Landslides Investigation and Mitigation, Special Report 247. Multiple authors; edited by A. Keith Turner and Robert L. Schuster. National Academy Press, published 1996, 675 pp.

CHAPTER

1

Landslides

1.1 SCOPE OF THE BOOK

Many books have been written about the identification, analysis, and treatment of landslides. There are specific books on laboratory testing, field instrumentation, and methods of slope stability analysis. Professional societies with interests in landslides sponsor seminars, conferences, and research publications. In many parts of the world, the economic losses caused by landslides are large, and there is significant loss of life.

This book is written primarily for the benefit of professional civil engineers and engineering geologists who are involved in civil engineering works. It is intended to be a practical guide for such practitioners. Other professionals who may benefit from knowledge in the book include structural engineers, highway and railroad engineers, mining engineers, planners, educators, construction attorneys, civil engineering contractors, dam and power generator engineers, public works engineers, water supply engineers, and others in related fields.

The remedial and preventative treatment of landslides is the principal focus of the book. The intent of this section is to provide the reader with a reasonably thorough understanding about each technique that is sufficient to allow the professional to choose between alternatives. For each treatment, the provided information generally covers the following:

- Appropriate (and inappropriate) use
- Principle of effectiveness for landslide remediation/prevention
- Design method
- Construction/installation procedure and field equipment
- Examples or case histories taken from the author's experiences or published materials
- Construction costs in the United States using the Engineering News-Record Construction Cost Index (ENRCCI) as a guide to converting historical data to current cost

Practical issues are stressed. References are given that can provide more detailed information on each subject. It should be understood that remedial/preventative techniques are continually changing as new technologies become available.

The opinions expressed in the book are those of the author, and other knowledgeable persons may disagree on specific points. In this regard, the author generally mentions dissenting opinions where appropriate.

The length of the book has been a concern, and this has required choices to be made on what topics to include or exclude from the book. It was decided that landslides in hard rocks would be excluded. These are landslides in which failure is controlled by discontinuities such as joints, faults, or bedding planes. However, slopes of weathered rock, where slippage occurs through soil or broken rock (and respond as soil mechanics), are included. Another excluded group of landslides are those involving cold region engineering such as permafrost or seasonally frozen ground.

The book contains numerous case histories or examples to illustrate the practical uses of remedial/preventative techniques. In addition, 12 case histories are presented at the end of the book that demonstrate technical points discussed in the main body of the book. These are cross-referenced.

Regarding individual topics, the author has chosen to be (i) mainstream; i.e., describing methods more commonly used or followed, and (ii) selective in coverage; i.e., writing more extensively on subjects for which many engineers experience difficulty, such as strength of clays, and briefly on subjects that are considered common knowledge. Thus, coverage of topics is deliberately uneven.

1.2 LANDSLIDE DESCRIPTIONS

The term *landslide* is sometimes felt to be inadequate because many types of slope movement do not involve sliding. The geomorphologist term *mass wasting* is little improvement. Cruden (1991) has suggested a simple definition of landslide: the movement of a mass of rock, debris, or earth down a slope. For this book, the term *landslide* covers all slope movements that occur from natural or manmade causes except ground subsidence.

The dimensions and geometry of a landslide have been described by Varnes (1978) using the cutaway drawing shown on Figure 1.1. Subsequently, the International Association of Engineering Geologists (IAEG) created a Commission on Landslides that has produced the section and definitions shown on Figure 1.2 (IAEG, 1990). There are many variations from these terms in common use. In this book, alternative terms are used:

IAEG TERM	BOOK TERM
main scarp	headscarp
minor scarp	secondary scarp
surface of rupture	slip surface (occasionally slip plane if planar)

The author also uses the following terms:

Length—horizontal distance from top of headscarp to toe (upslope to downslope). The *horizontal* distance for slope length is chosen because it is a quick method of determining size from plan drawings.

Width—generally, the widest dimension across the slope.

Depth—usually described as "up to" the maximum depth below the existing ground surface.

Slope—average slope in degrees from horizontal, or an average gradient of horizontal : vertical.

Landslide Size

There is no standard for describing landslides by size, but it is useful to provide some reference. Table 1.1 has been used as the guide to describing landslide size throughout this book.

Table 1.1 Grouping Landslides by Area in Plan

Descriptor	Area, sq. ft.	Area, sq. m.
Very small	<2000	<200
Small	2,000–20,000	200–2,000
Medium	20,000–200,000	2,000–20,000
Large	200,000–2,000,000	20,000–200,000
Very large	2,000,000–20,000,000	200,000–2,000,000
Huge	>20,000,000	>2,000,000

Note: 1 sq. mile = 27,878,400 sq. ft. 1 acre = 43,560 sq. ft. Length is measured horizontally, not along the slope. If the size is near the border between categories, both sizes are mentioned (e.g., landslide of around 20,000 sq. ft. is described as a small-to-medium landslide). For slopes steeper than 45° to the horizontal, it is recommended that vertical height replace horizontal length in the area calculation. For flow slides, it is recommended that the area be based on the eroded bowl at the initiation site, ignoring any further erosion further downslope in the valley below.

1.3 LANDSLIDE CLASSIFICATION

In the original work by Varnes (1978) which has been updated and partly revised by Cruden and Varnes (1996) slope movements have been subdivided into six categories:

1. Falls
2. Topples
3. Slides—rotational and translational
4. Lateral spreads
5. Flows
6. Composites—combination of types

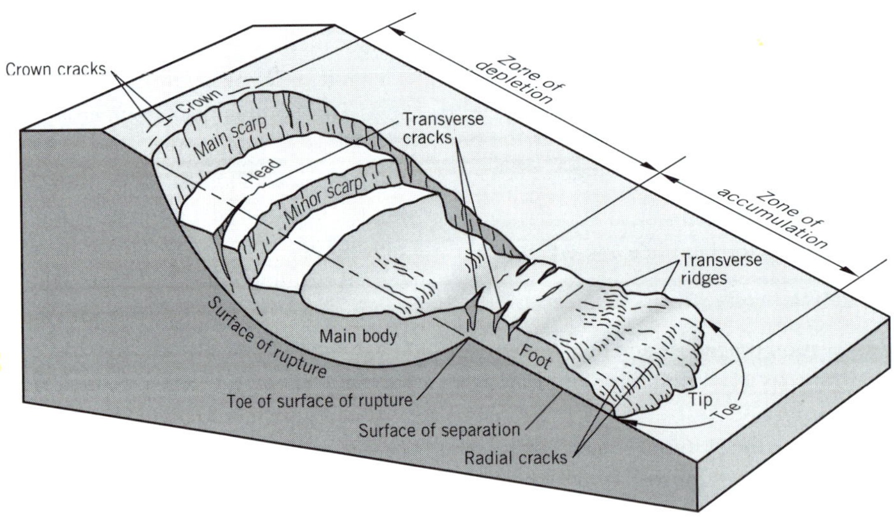

Figure 1.1 Terminology for describing landslide features (modified from Varnes, 1978).

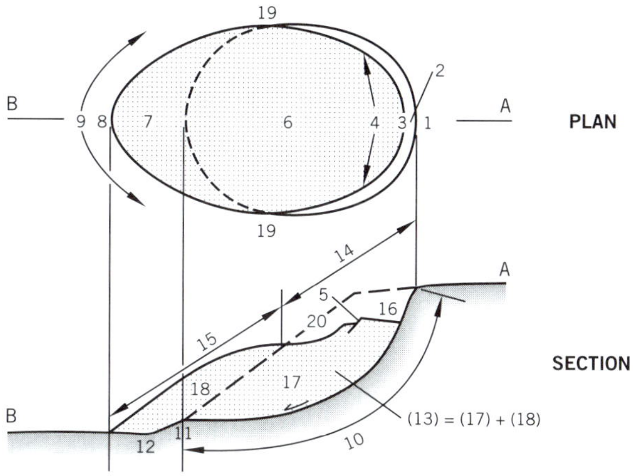

No.	Name	Definition
1	Crown	Practically undisplaced material adjacent to highest parts of main scarp
2	Main scarp	Steep surface on undisturbed ground at upper edge of landslide caused by movement of displaced material (13, stippled area) away from undisturbed ground; it is a visible part of surface of rupture (10)
3	Top	Highest point of contact between displaced material (13) and main scarp (2)
4	Head	Upper parts of landslide along contact between displaced material and main scarp (2)
5	Minor scarp	Steep surface on displaced material of landslide produced by differential movements within displaced material
6	Main body	Part of displaced material of landslide that overlies surface of rupture between main scarp (2) and toe of surface of rupture (11)
7	Foot	Portion of landslide that has moved beyond toe of surface of rupture (11) and overlies original ground surface (20)
8	Tip	Point on toe (9) farthest from top (3) of landslide
9	Toe	Lower, usually curved margin of displaced material of a landslide, most distant from main scarp (2)
10	Surface of rupture	Surface that forms (or that has formed) lower boundary of displaced material (13) below original ground surface (20); also termed *slip surface* or *shear surface*; if planar, can be termed *slip plane* or *shear plane*
11	Toe of surface of rupture	Intersection (usually buried) between lower part of surface of rupture (10) of a landslide and original ground surface (20)
12	Surface of separation	Part of original ground surface (20) now overlain by foot (7) of landslide
13	Displaced material	Material displaced from its original position on slope by movement in landslide; comprises both depleted mass (17) and accumulation (18)
14	Zone of depletion	Area of landslide within which displaced material lies below original ground surface (20)
15	Zone of accumulation	Area of landslide within which displaced material (13) lies above original ground surface (20)
16	Depletion	Volume bounded by main scarp (2), depleted mass (17), and original ground surface (20)
17	Depleted mass	Volume of displaced material (13) that overlies surface of rupture (10) but underlies original ground surface (20)
18	Accumulation	Volume of displaced material (13) that lies above original ground surface (20)
19	Flank	Undisplaced material adjacent to sides of surface of rupture; if left and right are used, they refer to flanks as viewed from crown; otherwise use compass directions
20	Original ground surface	Surface of slope that existed before landslide took place

Figure 1.2 Description of landslide parts (based on UNESCO Working Party, 1993, with minor modifications).

For each of these subdivisions, the materials are grouped as either (1) **rock**, (2) predominantly coarse material (**debris**), and (3) predominantly fine material (**earth**). Predominantly coarse is defined as having 20–80% of particles in the gravel/boulder size (>2mm).

Sketches of various failure types (excluding composites) from Varnes (1978) are given on Figure 1.3 with abbreviated comments. For more detailed information, the reader is referred to the original publication.

Skempton and Hutchinson (1969) provide another classification system for landslide types. Their system is useful for engineering work but is less widely accepted than the Varnes system.

1.4 PREVENTION OF LANDSLIDES

Prevention rather than remediation is desirable where slope failure is likely to be rapid and there is a high risk of damage

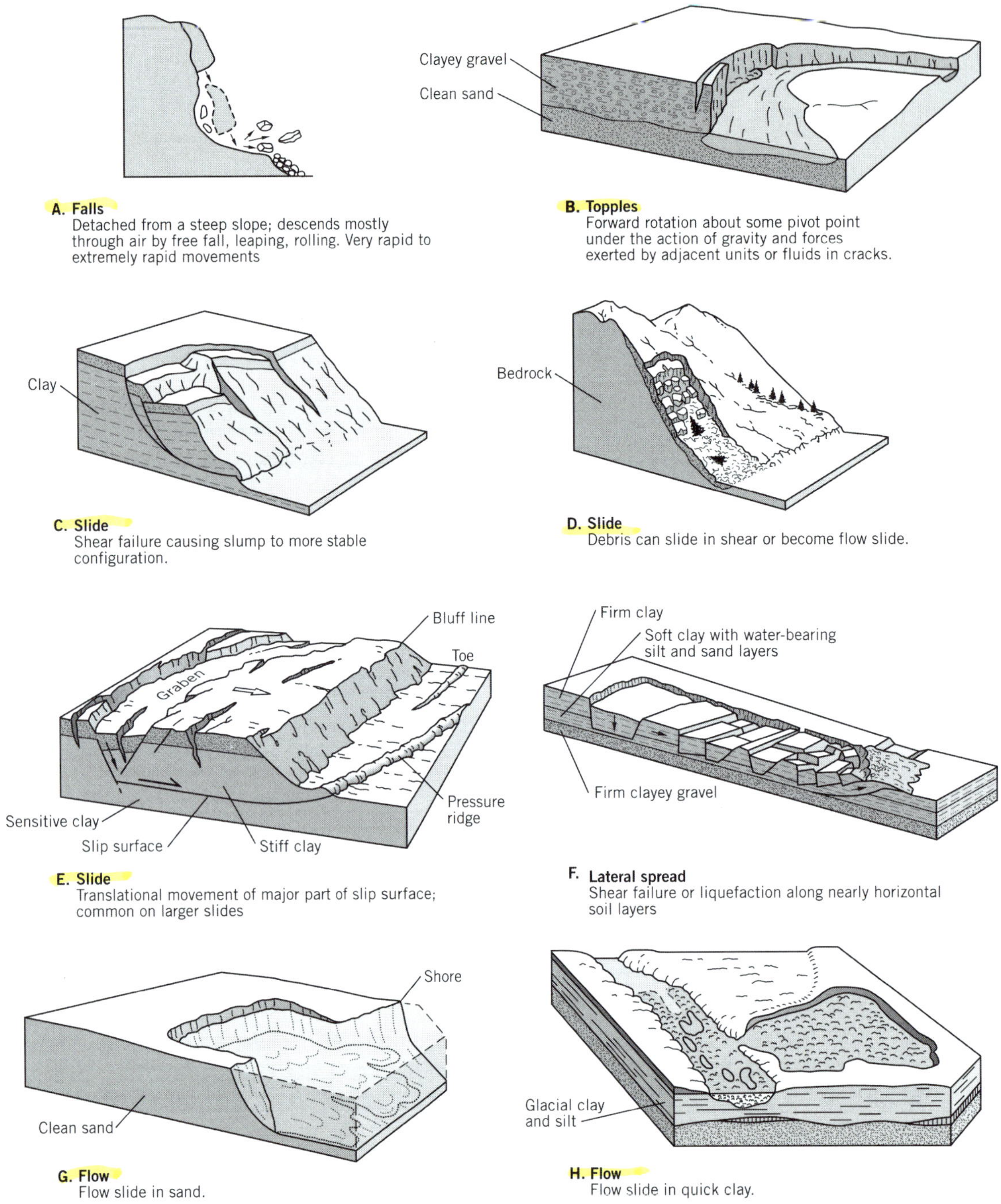

Figure 1.3 Examples of landslide occurrences (after Varnes, 1978).

and injuries. These failures include flow slides, earthquake-induced slides, and rock slides.

Potential landslides due to these causes can be reduced in a cost-effective way by taking these actions:

- Identifying landslide risk through hazard mapping and past experience, then implementing a plan of hazard reduction on a prioritized basis
- Periodic inspections of facilities that are vulnerable to landslides to observe any early signs of distress and, if appropriate, take preventative action to avert a landslide
- Maintaining and improving drainage measures in areas vulnerable to landslides
- Protecting lifeline facilities, buildings, and other places of public access from earthquake-induced landslides

1.5 REMEDIATION OF LANDSLIDES

Two basic questions need to be addressed:

1. *What is the cause of the landslide?*
2. *What is the amount of remediation needed to maintain stability for reasonably foreseeable future conditions?*

The two-step approach recommended for landslide remediation avoids the common pitfalls of developing a "one solution fits all" mentality (e.g., install horizontal drains or construct support berms) and blindly applying a "design" factor of safety of 1.3 or 1.5 to a landslide, when the amount of remediation needed to answer Question 2 can be rationally determined from an understanding of causation.

Question 1 is often the focus of a legal claim over responsibility for economic losses or loss of life. However, the technical reasons for determining causation are that this understanding often can lead to the most appropriate treatment and help to determine the amount of treatment needed. A few simple examples will illustrate the relationship between causation and treatment:

- A leaking canal triggers instability of a nearby slope.
  Causation: Raised groundwater levels due to leaking water
  Possible treatments: (i) line the canal, (ii) create a seepage barrier downslope, (iii) install a trench drain to lower groundwater level back to pre-canal levels.
- Road widening project causes cut slope failure.
  Causation: Unloading toe of slope
  Possible treatments: (i) build structural wall, (ii) raise road level to reinstate ground, or (iii) relocate road, plus other options. However, the knowledge that the slope was stable before the road-widening cut allows the designer to calculate how much force is needed to reinstate the prior stability. This approach is described as Original Profile Analysis in Chapter 10 and can be applied in other innovative ways for slope stabilization.
- Landslide develops from erosion in a cut slope.
  Causation: Loss of ground undermining the slope
  Possible treatments: (i) control seepage with a filter blanket, (ii) install horizontal drains to reduce seepage at the face, or (iii) reinstate original slope face, including drainage sufficient to prevent future erosion.

Question 2 generally requires an extra margin of safety to be put into remediation to allow for reasonably foreseeable future conditions. For example, it is improbable that the storm that triggered a landslide will be the heaviest likely to be experienced during the service life of the facility. Therefore, the remedial design should assume that somewhat higher (but reasonable) groundwater levels will occur in the future.

If stability calculations are involved in remedial studies, an extra margin of safety allows for uncertainties in analysis. Here the geotechnical designer of a *landslide* remediation has distinct advantages over a *conventional slope stability* analysis. The advantages are: (i) the factor of safety is known to be 1.00 in a landslide at the onset of failure; (ii) the shape of the slip surface can be measured by inclinometers; and (iii) groundwater pressures at the actual slip surface can be measured. Thus, a back analysis of the landslide at failure can be modeled with good accuracy, allowing parametric analyses (variations of shear strength and groundwater) to be examined. Since remedial treatments and back analysis are performed on the same cross-sections they are comparative ("before" and "after") studies. Any errors in the assumptions of the back analysis are carried forward into the remedial analysis. Thus, there is a very high probability that stability will be achieved. Accordingly, a lower factor of safety F is permissible for landslide remediation than for general slope stability design because the potential errors of the analysis are lower (see Chapter 10 for a fuller discussion of this issue).

CHAPTER

2

Landslide Occurrences

This chapter discusses the causes of landslides. It should be emphasized that there can be several contributory causes, but there is usually one dominant cause. In legal terms, it is the *proximate cause*. For example, a homeowner puts a steep cut up to a property boundary and it remains stable for many years; the adjoining owner then places a fill above the cut slope and in the following winter a landslide occurs, affecting both properties. The owner who placed the fill is likely to have the legal responsibility due to proximate cause (i.e., if the fill had not been constructed, the landslide would not have occurred). Even when no legal issue is involved, the geotechnical consultant should look for the dominant cause of the landslide as part of the technical investigation to ensure that the causation has been corrected by the proposed treatment.

2.1 RAINFALL

The well-known relationship between heavy rainfall and landslides does not need further elaboration. Chapter 5, Section 5.1, describes the buildup of groundwater levels in clay slopes during the winter season and the spikes or peaking of groundwater pressures during severe storms. It is likely that continuous monitoring of piezometers and rainfall will become more common on major landslides now that the technology is readily available (see Chapter 4, Section 4.3).

Nowhere in the world is there a higher interest in the relationship between landslides and rainfall than in landslide-prone Hong Kong, where about 6 million people live in an area of 400 sq. miles (Brand, 1985). Much of the highly populated area has very steep hills of granite and rhyolitic volcanic rocks, which are mantled by residual soils and colluvium derived from the underlying rock. These produce a highly variable transition from rock to soil and are characterized by shallow slides, especially in cut slopes. Deaths and casualties are common during severe storms.

Between 1978 and 1984, the Geotechnical Control Office installed 46 automatic rain gauges throughout Hong Kong. Rainfall data are continuously recorded and transmitted by phone lines to a central computer.

The Geotechnical Control Office has studied the rainfall data over a 20-year period (1963–1983) during which there were 13 major storm events. The data showed that the number of significant landslides was related to the *short-term intensity* of rainfall rather than the *cumulative total* rainfall of a storm. Consider the following two examples:

- The June 1966 storm (Figure 2.1a) caused 29 significant landslides that killed or injured 35 people and required 8,500 others to be temporarily evacuated. The hourly rates of rainfall measured at the Royal Observatory and the maximum rates measured elsewhere have similar patterns of intensity. The cumulative rainfall was about 460 mm (18 inches). The large majority of the slides (23) occurred near the peak intensity of 150 mm (6 inches) per hour.
- The October 1978 storm (Figure 2.1b) had almost the same 24-hour cumulative rainfall as the June 1966 event. However, the maximum intensity of rainfall was only 38 mm (1.5 inch) per hour, one-quarter of the rate in 1966. There was only one significant landslide and only one casualty.

These results show the importance of short-term, high-intensity rainfall in triggering major landslides. However, the 24-hour reading has value because it often includes a short-duration, high-intensity rainfall period within it. The other conclusions of the Hong Kong study are specific to the condi-

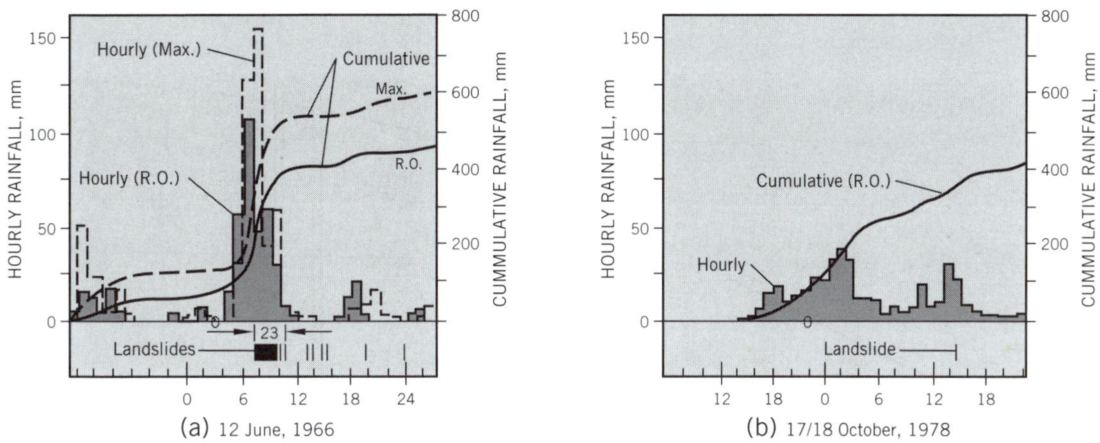

R.O. – Royal Observatory, Hong Kong
Max. – Maximum observed at all measuring stations

Figure 2.1 Effect of rainfall intensity on landslides in Hong Kong storms of (a) June 12, 1966 (b) October 17–18, 1978 (after Brand, 1985).

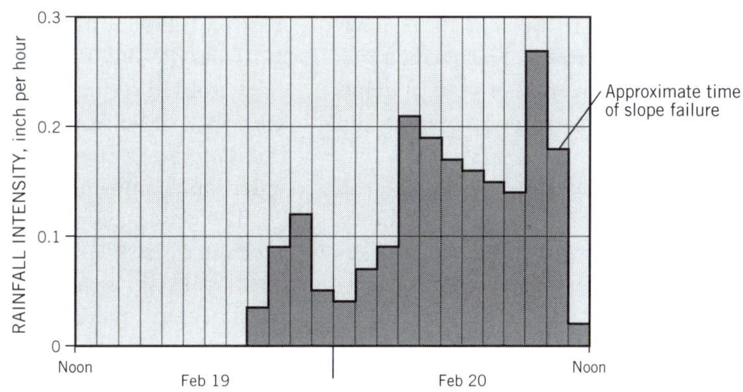

Figure 2.2 Failure of a vertical slope near the peak intensity of the storm on February 20, 1982, Portland, Oregon.

tions of that region. It is of interest that the study concluded that antecedent rainfall could cause minor slides but is not a significant factor in major slide occurrences.

Western Oregon has similar shallow slides in residual soils. On the west slopes of Portland there is a transitional zone of clay-to-weathered rock above the basalt bedrock. At higher elevations, a thin mantle of loess is involved in many of the landslides caused by storms, including the storm of February 1996 when more than 400 landslides were recorded within the city limits. In this storm, 17 homes were destroyed and 64 others were damaged to the level that the occupants had to be relocated (Burns, 1998).

Rainfall is recorded hourly at Portland International Airport. At a site within 7 miles of the airport, a storm on February 20, 1982, caused a landslide 50 feet wide on an 11-foot vertical cut slope. The rainfall on that day was 2.08 inches, the highest daily total for 8 years, and the slope failure occurred very close to the peak intensity of 0.27 inch per hour (Figure 2.2). It supports the conclusions from Hong Kong that high-intensity rainfall of short duration is a major triggering factor for landslides in residual soils and other shallow landslides.

2.2 SPRINGS AND SEEPAGE

Where waterbearing sands and gravels overlie more impermeable soils (or bedrock), the groundwater is usually *perched;* i.e., the water head A in the permeable soil is higher than the head B in the stratum below (Figure 2.3). Rainfall seeps into the ground by gravity until it reaches the less-permeable stratum. Groundwater builds up above the impermeable stratum and flows laterally to a slope face where it emerges (C) as a line of seepage or spring. Perched groundwater occurs commonly in layered strata, such as alluvium, colluvium and fills built from various source materials.

Under natural conditions, a spring may erode the more permeable soils at point C and undermine the slope above. The undercut slope face then sloughs into the eroded void and, with time, a zone of very loose soil will develop at this location. A severe storm may then raise the groundwater table quickly and cause a flow slide.

The same train of events happens after a cut has been made through similar geological conditions to those shown on Figure 2.3. Such cuts may be made during dry weather, and instability may not occur until the following winter.

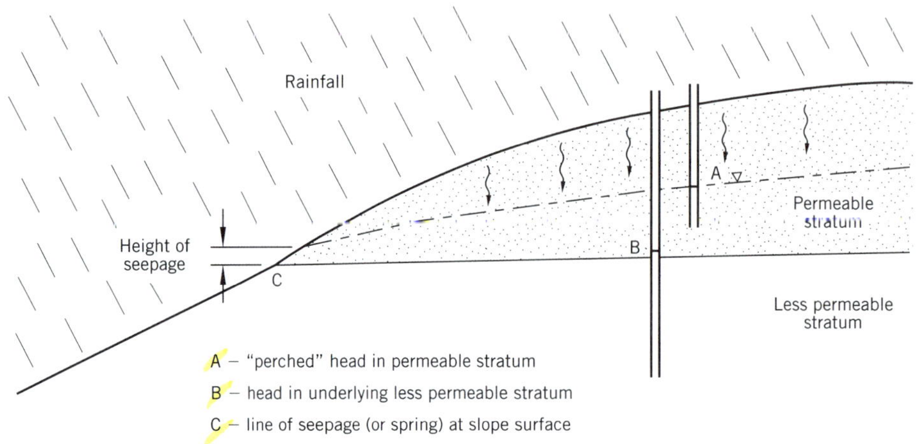

Figure 2.3 Perched groundwater and seepage.

Landslides caused by aggressive erosion from seepage are common. They can be prevented or remediated by several techniques, including (i) construction of a reverse filter at the ground surface, or (ii) lowering the groundwater level within the slope. The reverse filter (Chapter 16, Section 16.2) allows the groundwater to emerge without eroding fine-grained soil. Lowering the groundwater with horizontal or vertical drains may prevent groundwater levels from rising in winter and causing erosion at the surface.

Where natural springs and seepage lines occur, the soil above the spring line is typically at marginal stability. Therefore, any fills that encroach over the spring area are likely to become unstable. Spring flows may be seasonal and only active during the wet season.

Example

Bull Run Dam No. 2, within the watershed that supplies drinking water to Portland, Oregon, is built across an ancient landslide. The designers put the main spillway downstream of the dam and cut a canal through ancient landslide debris to reach it.

The landslide debris (Figure 2.4) comprises large blocks of broken rock, gravel, and clay in a heterogeneous mixture that includes nests of broken rock. Since construction of the dam in 1961, seepage from the spillway approach canal has raised piezometric levels by up to 40 feet, apparently due to progressive leaching of the fines within the ancient landslide debris. This probably created subsurface flow channels between the canal and the Bull Run River. Over time, what was originally a few isolated springs on the river bank became a broad seepage area. A 1995 storm caused a flow slide on the bank of the Bull Run River (Figure 2.4) that broke two of the city's main water supply pipelines below the slope.

The erosion bowl was infilled with rockfill immediately after the flow slide. To reduce seepage from the canal to the river bank, the approach canal has been partly lined with a plastic liner covered with 2 feet of crushed rockfill (see Chapter 18, Section 18.2).

2.3 IRRIGATION AND AQUEDUCTS

Irrigation of the arid lands of the United States puts large amounts of water into the ground. In a typical situation, 36 inches of water will be applied during the growing season. Of this amount, about one-third is lost to evapo-transpiration,

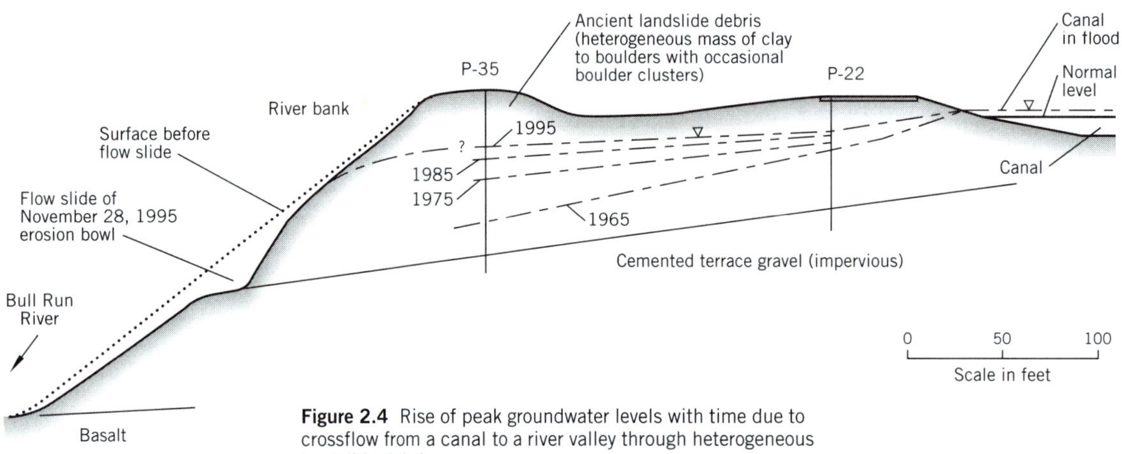

Figure 2.4 Rise of peak groundwater levels with time due to crossflow from a canal to a river valley through heterogeneous landslide debris.

one-third is used to grow plants, and one-third passes into the groundwater. The latter fraction is important because it prevents the build-up of salts at the surface that would make the ground sterile over time.

Once it enters the groundwater regime, the irrigation water flows by gravity. If it reaches a barrier to downward movement, it builds up head and may flow laterally to a riverbank or cut slope. The Columbia River in Washington has spectacular large slides that have been caused by irrigation of the uplands adjacent to the river. Similar types of landslides occur in many other areas where irrigation is close to slopes.

Irrigation ditches, water supply canals, and hydroelectric canals are aqueducts that often follow the contours of hillsides. They are rarely lined, and there is often extensive leakage into the ground. Many landslides have been caused by the lateral flow from these canals that raises the groundwater levels in the surrounding terrain.

2.4 WEATHERING

Weathering is the natural process of breaking down rocks to soil through freeze-thaw, wet-dry and the slow action of physical and chemical deterioration. Some of the ways in which weathering leads to landsliding are described in this section.

Residual Soils

Exposed rock faces weather from the outside, changing from unweathered rock to soil. In the case of fine-grained rocks, such as basalt, the depth of weathering is relatively shallow, resulting in surficial sliding generally triggered by heavy rainfall. The location of such relatively small and shallow sliding is unpredictable. In the case of coarser-grained rock, such as granite, the rock is more permeable and weathering penetrates deeper. In tropical climates with heavy rainfall, the depth of the residual weathering can be large. Residual soils can be as deep as 100 feet in Hong Kong (Brand, 1985).

A residual soil profile is very heterogeneous, making it difficult to sample and test. Based on their extensive experience with this type of landsliding, the Hong Kong Geotechnical Control Office (1984) developed a method for logging the ground profile in terms of six material grades and four zones (Tables 2.1 and 2.2). The categories are based on the weathering of granite but appear to be suitable for most residual soils. In the *Grade* classification, Grades IV to VI are treated as soils and Grade I to III as rock; for the *Zone* classification, Zones A, B are soils and C, D are rock.

A further complication in working with slopes of residual soils is that they are usually partly saturated and develop negative pore pressures. Conventional stability analyses are

Table 2.1 Weathering *Grade* Classification System Recommended by the Geotechnical Control Office (1984) for Use in Hong Kong (1)

Grade		Typical Characteristics (2)
VI	Residual soil	Soil formed by weathering in place but with original texture of rock completely destroyed.
V	Completely decomposed rock	Rock wholly decomposed but rock texture preserved. No rebound from Schmidt hammer (3). Slakes readily in water (4). Geological pick easily indents surface when pushed.
IV	Highly decomposed rock	Rock weakened—large pieces can be broken by hand. Positive N Schmidt rebound value up to 25. Does not slake readily in water. Geological pick cannot be pushed into surface. Hand penetrometer strength index >5000 psf (5). Individual grains may be plucked from surface.
III	Moderately decomposed rock	Completely discolored. Considerably weathered but possessing strength such that pieces 2 inches diameter cannot be broken by hand. N Schmidt rebound value 25 to 45. Rock material not friable.
II	Slightly decomposed rock	Discolored along discontinuities. Strength approaches that of fresh rock. N Schmidt rebound value greater than 45. More than one blow of the hammer needed to break specimen.
I	Fresh rock	No visible signs of weathering; not discolored.

Note: (1) This table is based on Moye (1955) and Hencher & Martin (1982).

(2) These characteristics may be affected by moisture content and microfracturing.

(3) N Schmidt rebound hammer values are for the hammer held horizontally. After "seating" blows, take the average of the highest five of ten blows at the same location. Only record as zero if there is no rebound. This index test is not applicable for the description of drillcore.

(4) Samples that are already saturated are less likely to slake.

(5) The presence of residual quartz in decomposed coarse-grained rocks can result in wide variation of hand penetrometer values. Take an average of ten values, avoiding disturbed or friable areas, and divide by two to arrive at the strength index.

Table 2.2 Weathering *Zone* Classification System Recommended by the Geotechnical Control Office (1984) for Use in Hong Kong

Zone	Zones of Weathering (based on Ruxton & Berry, 1957)
A	Structureless sand, silt and clay. May have boulder concentration at the surface.
B	Residual material with corestones. Rock percentage is less than 50% and corestones are rounded and not interlocked.
C	Corestones with residual material. Rock percentage is 50 to 90%, and corestones are rectangular and interlocked.
D	More than 90% rock. Minor residual material along major structural discontinuities which may be considerably iron stained.

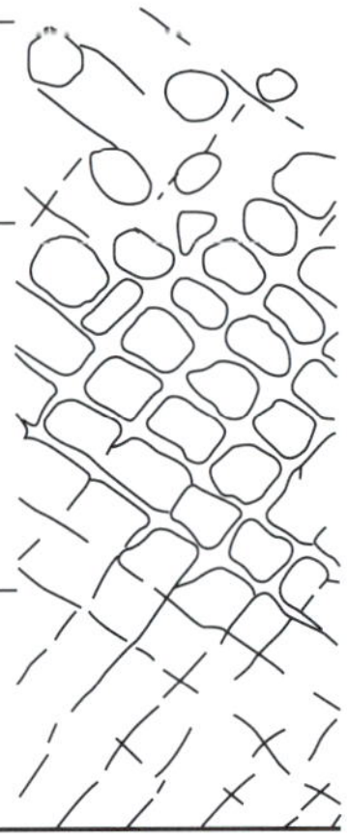

difficult to perform due to the combination of unreliable strength data, negative pore pressures, and shallow depths to the shear surface. Brand (1985) recommends using the observational approach during construction (see Chapter 10) as a supplement to site investigations and office studies.

Differential Weathering

Cut slopes made through sedimentary rocks may pass through jointed hard rock, partly to fully cemented conglomerates, indurated sand, clay shale, etc. Over a period of years, differential weathering may erode the less resistant rock layers, undermining the slope above. Seepage, occurring between two layers of different permeabilities, can accelerate the weathering and erosion at the face.

Many landslides are the result of differential weathering. Case History 12 concerns differential weathering of a steep rock cliff comprised of hard and soft rocks. Failures can be prevented by support, infilling, and other procedures.

Local Loss of Ground

Large blocks of rock are sometimes barely stable and rely on friction provided by broken rock debris in cracks or bedding planes. Should sufficient numbers of these stones fall out of the face by weathering, a large rock failure can occur. Frequent inspection of critical slopes needs to be performed by an engineering geologist to prevent future failures. Rock slopes that are found to be at risk can receive preventative treatment such as rock fences, netting, fallout ditches, rock bolting, breakup of a large rock piece at a dangerous location, infilling of eroded layers, or other preventative measures. Such treatments are outside the scope of this book.

2.5 FILLS

Fills cause landslides through increasing the load on the ground and/or by artificially steepening an existing slope. Some common examples are shown on Figure 2.5. On Figure 2.5(a) an embankment fill is placed over a flood plain. Alluvium may have weak, near-horizontal layers of clayey silt or clay within it and can induce a failure by *lateral spreading*. A major portion of the slip surface passes through a clay layer with significant lateral continuity.

Another common type of failure is a fill placed over the edge of a moderate slope to provide level ground for a structure (Figure 2.5b). Construction occurs in the summer when groundwater levels are seasonally low. In the following winter, or sometimes several years later, the groundwater level will rise high enough to cause slope failure. The direct cause (or "trigger") for the landslide is the rise of the groundwater which lowers the effective stresses in the native ground, thereby reducing the soil strength. However, the underlying cause is the weight of the fill and its steep outer slope, which has locally oversteepened the natural slope. If the fill had not been built, it is probable that the natural slope would have remained stable.

Roads are often built along the contour of a slope, as shown on Figure 2.5(c). A large fill on the outer part of the slope, sometimes used to reduce the horizontal curvature on sharp bends, may cause a landslide in the slope below the road. Such failures are more likely during the winter due to seasonal rises in groundwater levels within the natural slope. Sidehill construction is usually a "balanced" cut and fill. However, if there is very little cut at a particular section, as in the example of Figure 2.5(c), the fill acts as a stabilizing buttress to the slope above the road. On the other hand, an approximately equal cut-and-fill cross-section can (i) undermine the upper slope, causing it to fail, (ii) overload the downhill slope, causing it to fail, or (iii) cause the entire slope to become an active landslide. These types of failure are relatively common in colluvium and other slopes that are marginally stable in their natural condition. The type of failure is affected by site-specific conditions, such as the depth of sliding, groundwater levels, and presence of ancient slip surfaces. These issues are discussed in Sections 2.14 and 2.15.

Several examples of fills causing landslides are described in the case histories (Part C) at the end of the book:

CH-6: Railroad embankment fill at top of slope producing a delayed failure

CH-7: Large stockpile of riprap causing slope failure at extreme low tide

CH-8: Delayed failure landslide due to fill placed on upper part of slope

CH-10, CH-11: Landslide caused by constructing road fills across ancient landslide terrain

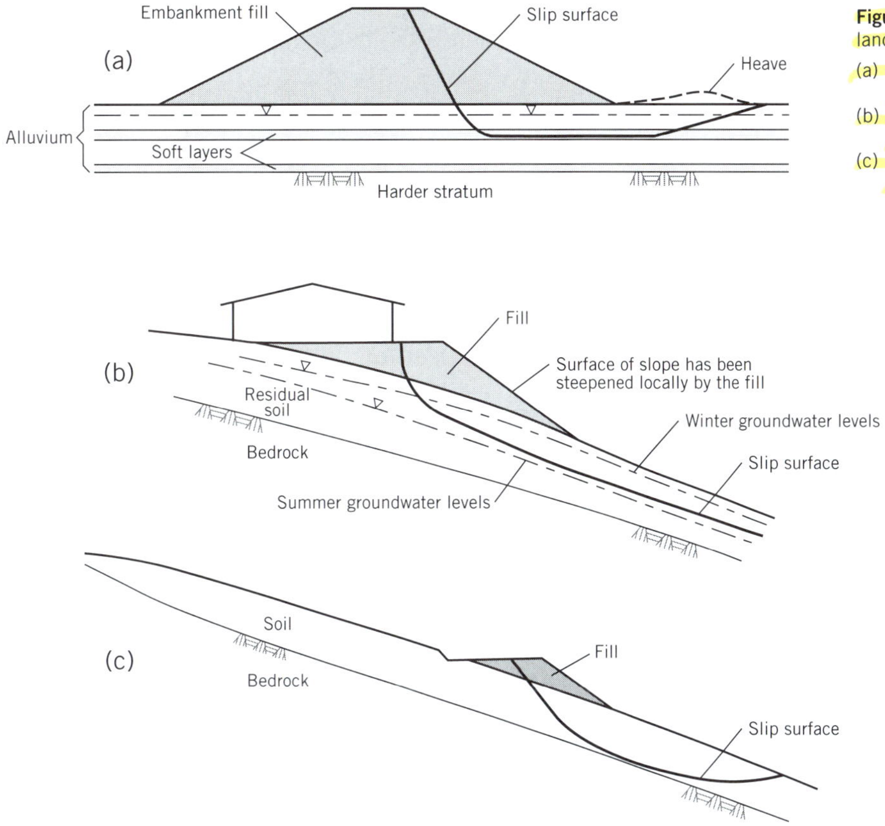

Figure 2.5 Examples of landslides caused by fills:
(a) lateral spreading of weak foundation
(b) fill overloading top of slope
(c) fill placed partway down a slope

2.6 EARTH DAMS AND RESERVOIRS

Earth dams are engineered fills that have to be capable of performing a dual function as a water barrier and gravity-retaining structure. The factor of safety for normal static operating conditions is designed to be at least 1.50; for pseudo-static seismic loading, the minimum design factor of safety is 1.15. However, the potential loss of strength due to temporarily elevated groundwater pressures (liquefaction) during strong earthquakes is a major concern. Much research is being directed to achieve rational design procedures for understanding the liquefaction phenomenon since the near-catastrophic failure of the Lower San Fernando Dam, California, in 1971 (Seed, 1979).

Excluding the issue of earthquake loading, the principal stability concerns after construction have been (i) the potential for internal erosion within the dam and foundation, (ii) lateral movements in stiff clay and clay shale foundations, (iii) development of artesian pressures below the dam caused by the reservoir, (iv) potential failure of the upstream slope of the dam and reservoir shoreline by rapid drawdown, and (v) instabilities in the slopes around the reservoir due to the raising of groundwater levels by impoundment.

In the United States, earth dams for hydroelectric projects are routinely inspected every five years and developing stability problems are usually addressed in a timely manner. However, many privately owned dams are not routinely inspected.

Many earth dams have been built below old landslide slopes. In most cases, the landslide had narrowed the valley profile, making the site topographically attractive for dam construction. The presence of an old landslide in a dam abutment generally is not harmful because the embankment dam provides a buttress to support the landslide. However, where recognized, owners generally install field instrumentation (inclinometers, piezometers, survey hubs, etc.) to monitor the long-term behavior of the landslide.

There have been several spectacular failures of dams (e.g., Carsington Dam, England; Vaiont Dam, Italy; Teton Dam, Idaho), due to some of the causes just listed. These failures have been well documented in the literature.

2.7 CUTS

Cuts remove support from a slope. If the slope is originally in a marginally stable condition—examples would be talus slopes, colluvium, or ancient landslide—a cut can often result in a landslide. Other causes of failure in cut slopes include (i) removing the soil overlying an artesian pressure condition—see Section 2.8, (ii) intercepting a perched groundwater, causing a line of seepage or springs on the slope face, (iii) over-

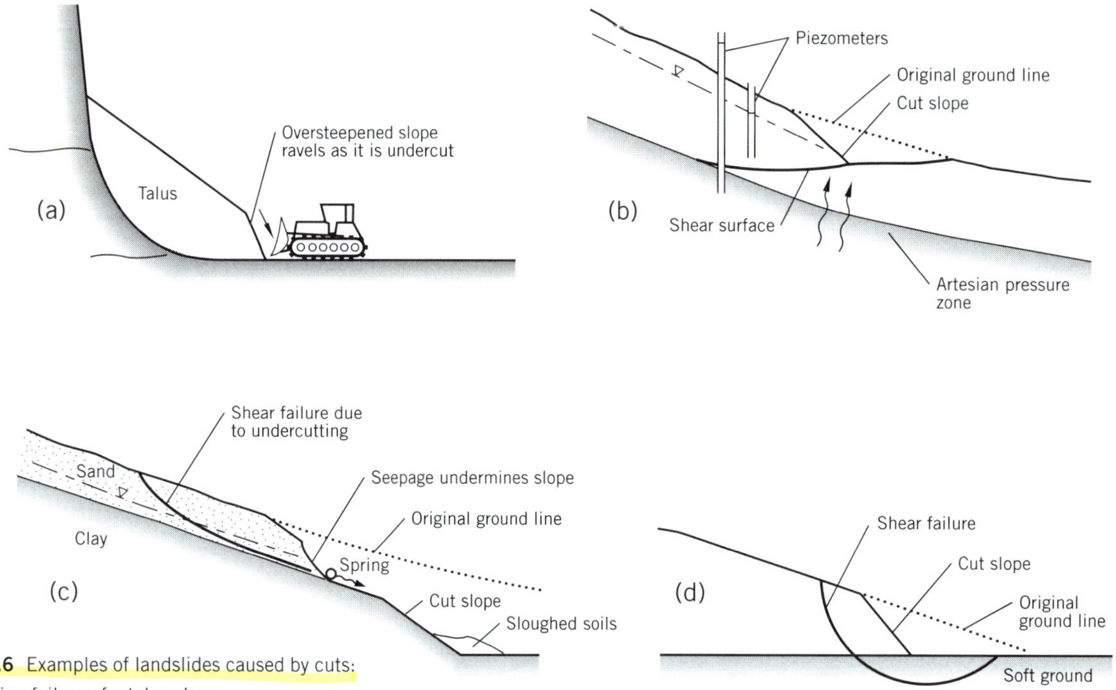

Figure 2.6 Examples of landslides caused by cuts:
(a) raveling failure of a talus slope
(b) cut made toward an artesian condition
(c) intercepting a perched groundwater
(d) base failure below cut slope

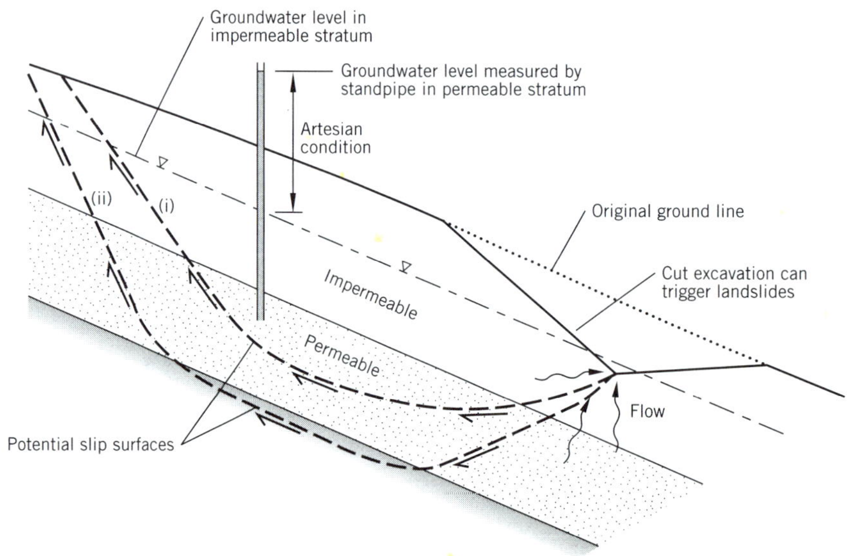

Figure 2.7 Example of an artesian pressure causing slope failure above a cut.

steepening the base of a slope on a soft foundation, causing the foundation to fail, or (iv) delayed failure of a cut slope made into stiff clay. The first three examples are illustrated on Figure 2.6. Delayed failure is discussed in Section 2.15.

Two examples of cut slopes causing landslides are discussed in the case histories of Part C:

CH-1: Excavation into ancient landslide terrain to construct a reservoir at the base of a slope

CH-3: Temporary excavation into an old landslide with artesian conditions

2.8. ARTESIAN PRESSURES

An *artesian pressure* occurs when the groundwater head in soil or rock is higher than the head in the overlying ground. In some cases, the artesian head rises above the ground surface. Artesian conditions develop when groundwater, fed by an upslope source, becomes trapped by a less permeable layer above it.

An example of artesian groundwater is shown on Figure 2.7 and is a fairly common occurrence within a layered strata, such as colluvium. The buried pressure head is stable under the existing natural conditions but becomes unstable when a cut is made that removes part or all of the less permeable soil above it. The hydraulic gradient from the artesian layer to the ground surface increases, causing the excavation to heave and/or erupt as a spring. In many cases, the ground undermining produces a shear slide or mudflow.

Case History 3 (Part C) describes a highway construction project in Idaho that found artesian conditions in one boring near the center of a ravine. When soft soils within the ravine were excavated to provide a firm foundation for a new highway fill, the ground became unstable and triggered a major flow slide that regressed several hundred feet upslope from the work area. The flow slide caused extensive damage to a railroad and another highway below the construction project.

Case History 7 (Part C) describes a site in Alaska where seepage within bedrock and talus is trapped by more impermeable harbor sediments (Figure 2.8). Similar conditions have been encountered in the Norwegian fjords (Bjerrum, 1971).

Artesian conditions encountered during a project site investigation should serve as a warning flag for any proposed earthworks. Since the ground is often marginally stable, any increase in shear stresses from a cut or fill can trigger instability unless steps are taken to relieve the excess pressures in advance of construction. Cut slopes are particularly vulnerable to instability if the cut approaches an artesian zone.

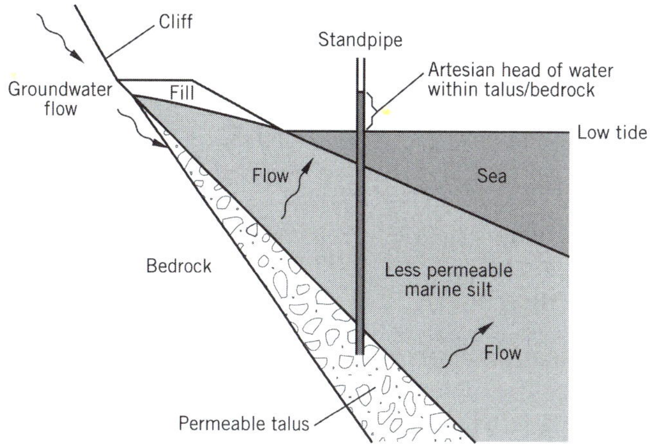

Figure 2.8 Schematic representation of artesian head condition at Skagway, Alaska.

2.9 CONCENTRATED WATER SOURCES

One of the most common causes of landslides, especially in urban areas, is water entering the ground from a concentrated water source during a storm. Some examples:

- Overflowing ditch on a highway that allows flood water to cross at a low point, causing failure of the outside slope (flow slide or slump)
- Broken culvert under fill causing internal erosion (usually a flow slide)
- Blocked culvert at the upstream end, causing ponding of water and blow out (usually flow slide or slump below the outlet end)
- Residential gutters and/or downspouts pouring water onto the ground (erosion and gullies, failure of nearby slopes)
- Broken water pipes and storm sewers, especially if placed on the outer edge of a slope (erosion and failure of nearby slopes)
- Untreated natural springs and lines of seepage (slumps, flow slides)
- Blockage of natural ravines by brush or garbage (flow slides)
- Discharge of surface water collected from urban developments into a ravine (high erosion, slumps downstream of the pipe)

All these causes are preventable. Concentrated flows of water should not be permitted to discharge onto slopes that are at marginal stability or susceptible to erosion. Ditches and culverts should be routinely maintained and adequately sized to handle the flow of major storms.

2.10 RIVER EROSION

Riverbank erosion is a natural geological process by which rivers are widened and deepened. Most erosion occurs during flood conditions when the water is high, rapidly flowing, and turbulent. Floods also raise groundwater in the riverbanks. Failure of the banks may occur as the river level recedes after the peak flow.

Rivers erode into the bank on the outside of a bend. Undercutting erodes the slope at and below stream level, and the slope above the river eventually slumps down. Downed trees at a river edge can disturb soils in the bank; the loosened soil may enlarge into a slope failure during a later flood.

Stream velocity may be increased by manmade developments. Higher flow and increased turbulence above the preexisting natural environment may cause slope instability. Examples would be (i) discharge of surface water from a recently built development into a ravine and (ii) release of water from a dam into the river below the dam.

Riverbanks are improved by riprap, gravel blankets, slope flattening, retaining walls, or bioremediation techniques (plantings, erosion mats, etc.). If very turbulent water is present, a grouted riprap can be used; the grout prevents large rocks from being plucked out.

The case histories (Part C) include two examples of riverbank instability:

CH-2: Beaver shoreline erosion

CH-9: Goat Lick Slide

2.11 COASTAL EROSION

Heavy storms lashing against a coastline can cause extensive loss of ground to the sea. In England, the north Yorkshire coast, the Isle of Wight at Ventnor, and the Dorsetshire cliffs at Lyme Regis have been extensively studied in recent years and are being protected by concrete walls and heavy rock. Similarly, coastlines in the United States have areas that are continually being lost to sea erosion.

The rocks that are more susceptible to wave damage are sedimentary units (siltstone, sandstones, shales) and metamorphic rocks such as schists and slates. In Oregon, the loss of weak sedimentary rocks has left many stacks of harder volcanic rocks projecting out of the ocean offshore from the retreating coastline. Weak zones, such as exposed faults, experience faster rates of erosion than adjacent intact rocks.

Coastal erosion is also affected by lateral drift of sandy beaches. Loss of sand can undermine a slope and trigger landsliding upslope. The weather patterns (El Nino, La Nina) influence the rate of lateral movements of beaches.

Sand spits are created by lateral drifts across a coastal inlet. They can be breached by a winter storm that is followed by rapid erosion and loss of land. Houses built on spits are at high risk unless there have been geotechnical studies and protective measures put in place to retain the existing spit.

Protective treatments of a coastline include reinforced concrete seawalls, heavy riprap, or artificial riprap (proprietary interlocking concrete devices). Large riprap needs to be backed by two filter layers to protect the underlying soils; otherwise, it is likely that the native soils will be pulled through the riprap by wave action and undertow.

Cliffs are eroded at the base by the sea, and a later collapse creates a steep scarp. The scarp is likely to recede back over time to a more stable configuration. Also, springs exposed on a cliff slope should be treated to prevent erosion undermining the slope above.

2.12 SUBAERIAL SUBMARINE FLOW SLIDES

Deep Submarine Slides

There is increasing interest in the subject of very deep submarine flow slides. On the commercial side, it has been spurred by concern over underwater slides affecting cables and pipelines on the ocean floor and the safety of overwater oil-drilling platforms. For geologists and other marine scientists, recent improvements in geophysical methods of exploration have allowed the ocean floor to be seen in considerable detail.

The huge seafloor landslides are of primary interest to geomorphologists. However, landslides combining both above and below water (subaerial and submarine) are of interest to geotechnical engineers and engineering geologists. Many such slides are initiated by manmade construction activities such as fill placements and dredging for port facilities, land reclamation, airports, and refineries.

Flow Slides in Norwegian Fjords

Bjerrum (1971) summarized Norwegian experience with six submarine landslides in the fjords around Trondheim. They occurred in postglacial deltas and estuaries where rivers flow into the fjord. The deltas are continuously growing from accumulation of sand and silt being carried in by the rivers.

Bjerrum explains that the slides begin with an *initial shear slide*. In four of his six case studies, the initial slide was caused by manmade fills. The initial slide failure frequently coincided with low tide, when the slope has the maximum degree of instability due to loss of lateral support from the sea. In other cases the initial slide occurred in such a large depth of water that little was known about what might have initiated the first (shear) slide. He speculated that it probably was caused by the ground becoming oversteepened from accumulation of sediments carried into the fjords by rivers. These initial slides could have been predicted by a slope stability analysis.

The initial shear slide is followed by a massive *flow slide*. The flow slide develops retrogressively and has a tendency to widen in loose soils. The dimensions of the flow slide are disproportionately large compared with the size of the initial shear slide.

Bjerrum noted that when a liquid sand mass flows downhill over a subaqueous sand or silt slope, it will erode a deep canyon, growing in volume as new sand is collected by erosion. If the fjord is wide enough and the slope gradient is small, the velocity will slow down and eventually halt. He also noted that artesian pressures, originating from high water pressures in the fissures of the bedrock beneath the soft deposits, may have been a contributing factor to the instabilities.

Bjerrum's descriptions of the individual slides in his study show that secondary slides along the path of the flow slide are common occurrences. In one case, slide A at Orkdalsfjorden in 1930 (Figure 2.9), there were two large secondary slides (B, C) that followed the onset of the original flow slide, one being 2 km (1.2 miles) away. Shortly afterward, two underwater telephone cables were broken, one of these being 18 km (11 miles) from the site at a depth of more than 400 m (1,300 feet) below water. Other slides (Follafjorden, 1952; Finnvika, 1940) caused significant waves to be generated from the rapidity of the slide movements.

Bjerrum's explanation of a flow slide mechanism closely fits the events that caused a flow slide in the fjord at Skagway, Alaska, in 1994. A large pile of riprap placed on top of the slope for a construction project caused an initial slide in the soft marine silts below, which rapidly developed into a flow slide. Most of the 1,300-foot long dock was destroyed, one worker was killed, and a large wave caused by the slide severely damaged a floating ferry terminal. Four construction workers provided eyewitness accounts of the slope failure that occurred at extreme low tide (see Case History 7 in Part C).

Susceptibility of Loose Sands to Initiate Flow Slides

For many years, the banks of the Mississippi River have experienced flow slides that have destroyed parts of the levee system. These flow slides are examples of river erosion undermining loose sand deposits at bar points of the river. Typically, the loose

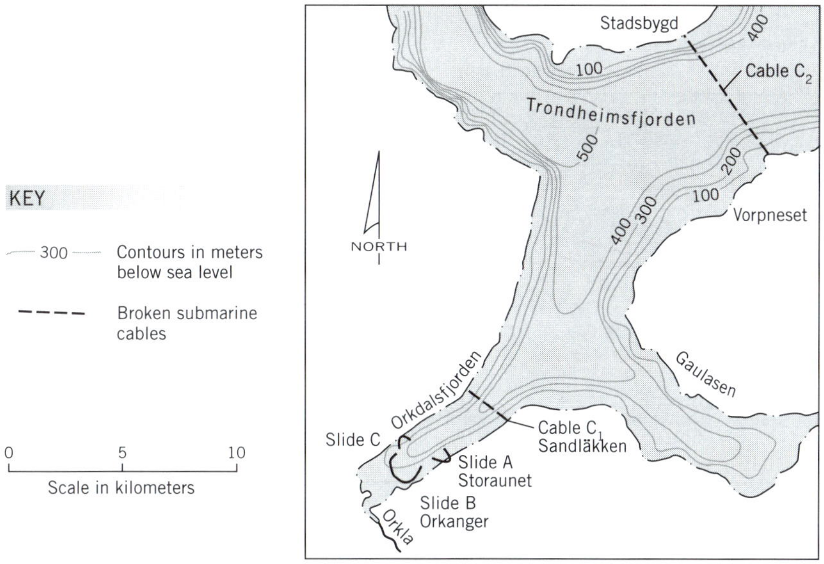

Figure 2.9 Map of flow slide at Orkdalsfjorden on May 2, 1930 (after Terzaghi, 1956; Bjerrum, 1971).

sand stratum overlies a much denser sand stratum. The dense sand gradually erodes and eventually undermines the loose sand, resulting in a flow slide of loose sand.

Seed (1983) wrote a summary of liquefaction-type landslides affecting coastal deposits and fills. His research of case histories concluded that the soils susceptible to failure included:

1. Loose sands and silty sands with relative densities less than 45%,
2. Loose non-plastic silts with water contents of about 33% or higher, and
3. Highly sensitive quick clays with water contents equal to or greater than (liquid limit + 5%)

He identified three principal triggering mechanisms: (i) earthquake shaking and possibly low-level vibrations, (ii) relatively rapid changes in shear stresses on a slope due to natural erosion, deposition, or construction (fills, dredging), and (iii) low tidal levels, especially extreme low tides.

The ease with which relatively small triggering mechanisms can destabilize a slope is explained by the behavior of loose cohesionless soils in undrained conditions (Figure 2.10).

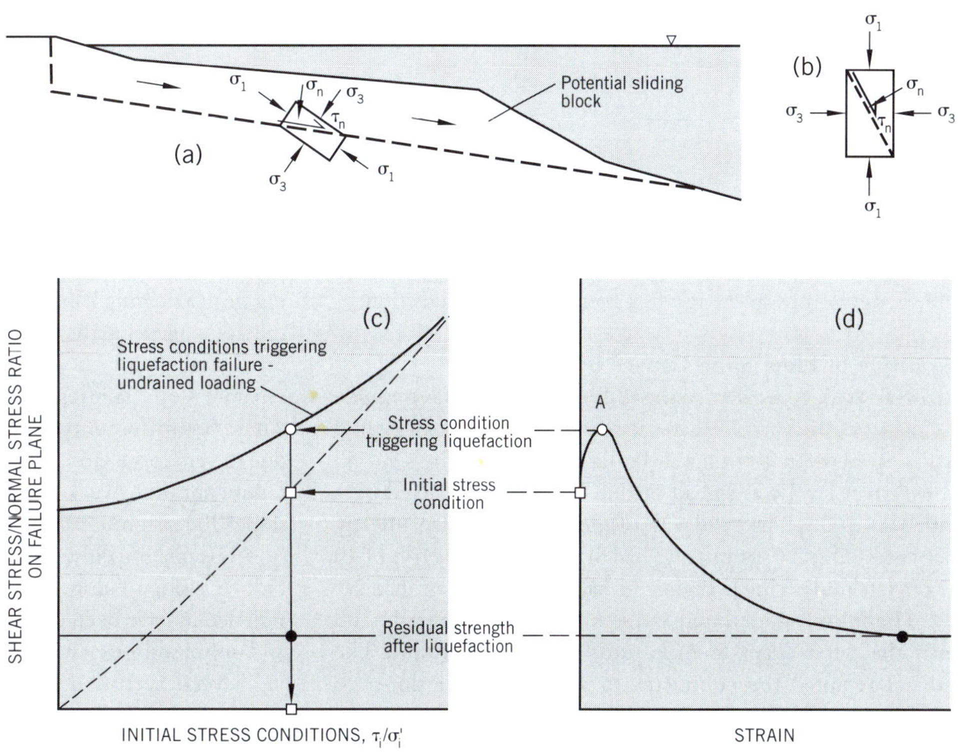

Figure 2.10 Explanation of weak reserve strength of loose sand under undrained conditions:
(a) stresses on a soil element on a potential failure surface
(b) equivalent stress conditions on triaxial test specimen
(c), (d) typical results of undrained tests on specimen of loose sand (after Seed, 1983)

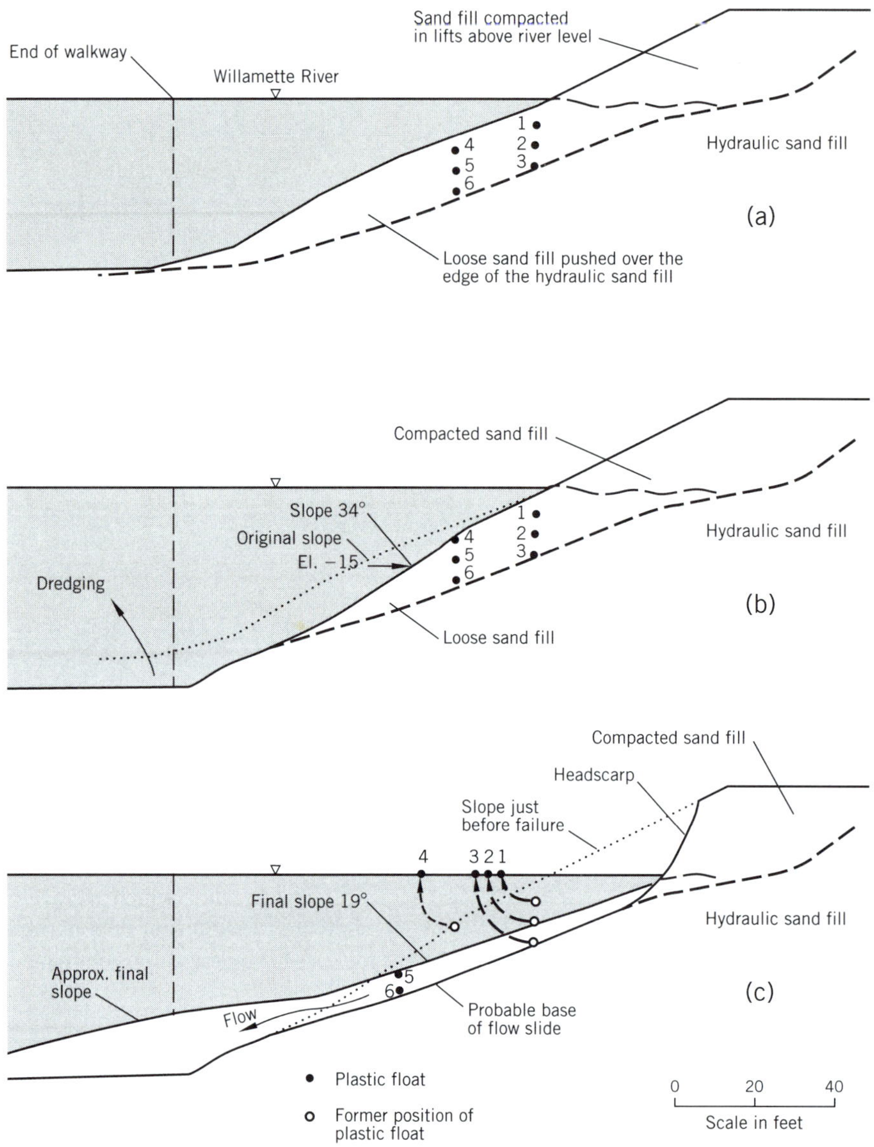

Figure 2.11 Terminal 2 Flow Slide, Portland, Section A-A:
(a) original conditions
(b) conditions immediately before the flow slide
(c) conditions after the flow slide

With an existing deviator stress acting on the slip plane of a potential sliding block, a small increase can exceed the available reserve of strength (point A on Figure 2.10), after which the strength drops to the residual strength after liquefaction.

Example of Flow Slide Caused by Dredging

In 1966, dredged sand from the Willamette River was pumped onto an area of about 250 acres to reconstruct an existing wharf and ship berth at Terminal 2, Portland, Oregon. The sand fill above river level was spread in thin lifts and compacted by a vibratory roller. The sand was of medium grain size (mostly between 0.2 and 0.8 mm) and had SPT blow counts (N) of 40 to 60. The underlying hydraulic fill had N values of less than 20.

The fill was underbuilt, and it was necessary to push sand over the outer edges to make up the required ground. The project required the contractor to dress the outer slopes to 2 horizontal : 1 vertical before constructing an anchored bulkhead partway up the slope. The height of the slope was 68 feet, with about 20 feet being above river level at the time of construction (Figure 2.11).

The slopes were excavated by a crane and clamshell operating from a floating barge. The sand was dug out from the base of the slope, allowing the sand higher up the slope to ravel into the excavation ("glory hole" technique). Within a short time, four separate slope failures occurred around the edge of the fill. There was uncertainty about the cause: Was it sand liquefaction due to very loose sand fill around the perimeter, or was it due to failure through weak silt layers of river alluvium below the slope? To learn more about the nature of the problem, the decision was taken to instrument a section of the slope and try to induce a slope failure using the same excavation techniques. Full details have been described earlier (Cornforth et al., 1974) and only a summary will be provided here.

Several overwater borings were put down adjacent to the earlier slope failures, using a barge-mounted rotary boring rig with mud flush. SPTs were performed and 3-inch diameter thin-wall Shelby tube samples were recovered by hydraulic jacking. Soil properties measured in the sand fill slope were:

SPT blow count	0 to 2 (upper 12 feet dozed fill)
	4 to 12 (underlying hydraulic fill)
Relative dry density	20 to 40%
Static angle of repose	34 degrees

For the field experiment, float balls were buried in the sand fill; these are labeled 1 to 6 in Figure 2.11(a). An observation barge was anchored alongside the slope test section. A floating walkway alongside the barge allowed the slope profile to be measured continually during the experiment. The contractor started to excavate a trench about 120 feet long, parallel to the shoreline. During the experimental dredging, the depth of the trench never exceeded the design depth of the project by more than 3 feet.

As excavation proceeded, there were short periods of bubbling immediately upslope of the dredge, indicating that surface raveling and slope readjustments were occurring. Two minor slumps occurred, accompanied by water turbulence and bubbling. In each case, the lower part of the slope had reached approximately 34 degrees (the angle of repose) just before the slumps.

A flow slide occurred after about 7 hours of dredging. At the start of failure, violent agitation and bubbling were observed near the shoreline. Dredging was stopped. Within 5 minutes, below-water loss of ground had reached the river surface, and the slope above the river began to break away in thin cupped slices. The steep headscarp eventually stopped growing about 4 feet in front of the slope top (at the center), and formed a concave scarp about 150 feet wide at the waterline. There was a thick scum of aerated silt at the water surface. The main slide activity lasted 20 minutes, although minor sloughing continued for another 20 minutes. The ground below the outermost end of the walkway (at the approximate design slope toe) had risen by 18 feet (Figure 2.11c).

The slope profile *immediately before* the flow slide is shown on Figure 2.11(b). The lower part of the loose sand fill was standing at an angle of 34° to the horizontal, the angle of repose of the sand.

During the flow slide, plastic floats 1, 2, 3, and 4 all rose to the water surface showing that the sand at their buried positions had liquefied (flowed). Floats 5 and 6 remained buried.

The soil profile *after* the flow slide (Figure 2.11c) shows that the slope below the headscarp was at an angle of 19° to the horizontal and approximately parallel to the interface separating hydraulic fill from the loose sand fill dozed over the edge. Since float 3 had liquefied and risen to the water surface, it is probable that the remaining thin layer is material that accumulated on the flatter slope toward the end of the flow slide due to caving from the headscarp.

The plan view of the failure (Figure 2.12) was measured 10 weeks after the experimental slope failure. It shows that the sand flowed through a fairly narrow opening near the mid-height of the slope that was at the angle of repose just before the flow slide. The hourglass shape of the slide in plan view has an upper bowl and a debris fan below it. The change from bowl to fan occurred near elev. −15 feet.

The initial liquefaction probably occurred at the mid-height of the slope where the neck is located on Figure 2.12 (i.e., elev. −15 feet approx). Above this point, sand fill was lost and below it the sand accumulated in a debris fan. As shown on Figure 2.11(b), elev. −15 is where the slope had steepened to 34° to the horizontal just before the flow slide. As the sand face collapsed and flowed downslope, the newly exposed face peeled off in turn, causing retrogressive movement upslope. After about 5 minutes, the headscarp reached the river surface, by which time floats 1, 2, and 3 had floated out of the liquefied sand to reach the surface. This shows that liquefaction occurred as deep as the contact between the loose bulldozed sand and the underlying hydraulic sand fill. The after slide profile is shown on Figure 2.11(c).

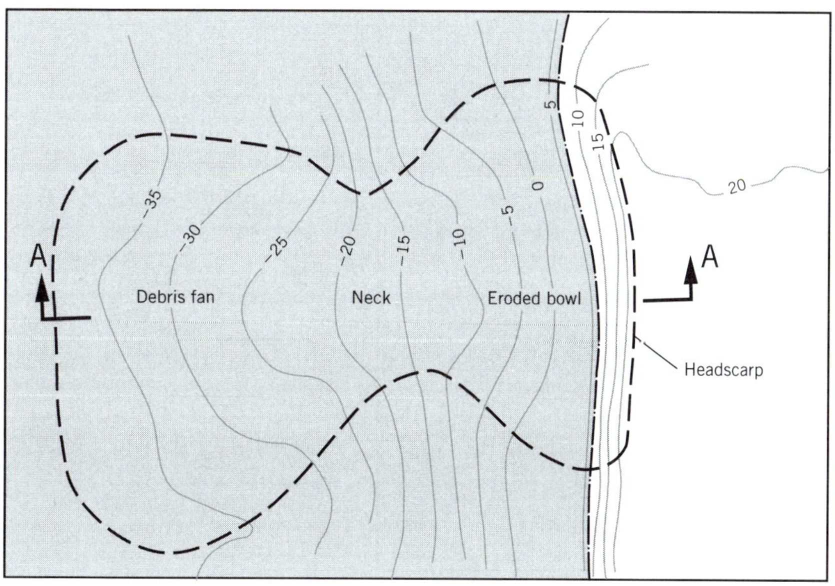

Figure 2.12 Terminal 2 Flow Slide, Portland. After failure contours.

These slope failures could have been avoided by initially *overbuilding* the hydraulic fill and then cutting back to the design 2:1 slopes. The hydraulic sand fill had a significantly higher relative density than sand bulldozed over the edge. Clearly, bulldozing sand into a river over the edge of a slope is a poor construction technique. It trapped air in the sand (as observed from the bubbles coming to the surface during the slope failure) and it is likely that the bulldozed sand was at, or below, the minimum intergranular density (see Chapter 7), making it highly susceptible to collapse and liquefaction during excavation.

2.13 DEBRIS FLOWS

Mountainous regions of the world are very susceptible to debris flow slides that can cause extensive damage along their paths. Many deaths and injuries have occurred. These very destructive landslides affect roads, railways, rivers, canals, and residences, and occur during high-intensity storms.

A natural debris flow is essentially a normal geological process by which young, steep valleys are widened and deepened. It is likely to recur at periodic intervals.

The initiation site for a debris flow typically occurs in a narrow ravine with slopes steeper than the angle of repose (approx. 70% grade). The soil is either weathered in-place rock or colluvium and may be anchored to the slope (before failure) by tree roots. Many initiation sites are in hollows of small ravines where runoff converges under high-intensity rainfall. Some initiation sites have springs.

The mechanism of these flow slides appears to be as follows: The soil at the site is either loose or becomes loose during preliminary movements. When the saturated loose soil breaks away, the slide scours into the valley floor and picks up additional soil, trees, and undergrowth as it flows downslope. The runout does not stop until relatively flat ground is reached. Here it can create a logjam within the valley or form a debris fan on the flood plain of a larger river. In extreme cases, the flow slide may temporarily dam a river downslope.

Squier and Harvey (2000) describe two flow slides from the winter of 1996 in which five persons were killed. Significantly, four of these deaths occurred at a residence that had been built on the remnants of a previous debris flow within the ravine. These deaths stimulated research into (i) ways in which the public can be protected from flow slides, and (ii) the influence of clearcutting forests on the incidence of flow slides.

One slide is shown in plan and profile on Figures 2.13 and 2.14. At the initiation site, the natural slope is around 90 percent. At the time of the flow slide, the rainfall was 5 inches over a 24-hour period. The sand and gravel overburden had 10–30% fines and was less than 3 feet (0.9 m) thick. The bedrock was sandstone with thin beds of mudstone and siltstone.

The photograph on Figure 2.15 shows a huge boulder that was carried downslope during the Rock Creek debris flow. Debris flows destroy almost everything in their path and cause massive erosion of streambanks and vegetation.

Figure 2.13 Site plan, 1996 Rock Creek Flow Slide, Oregon (after Squier & Harvey, 2000).

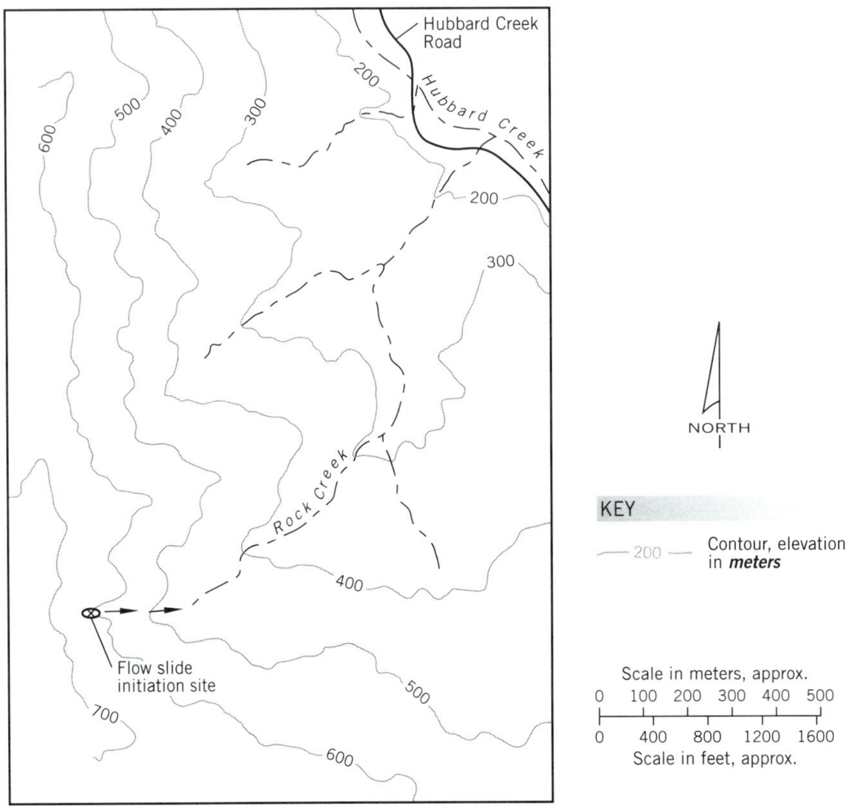

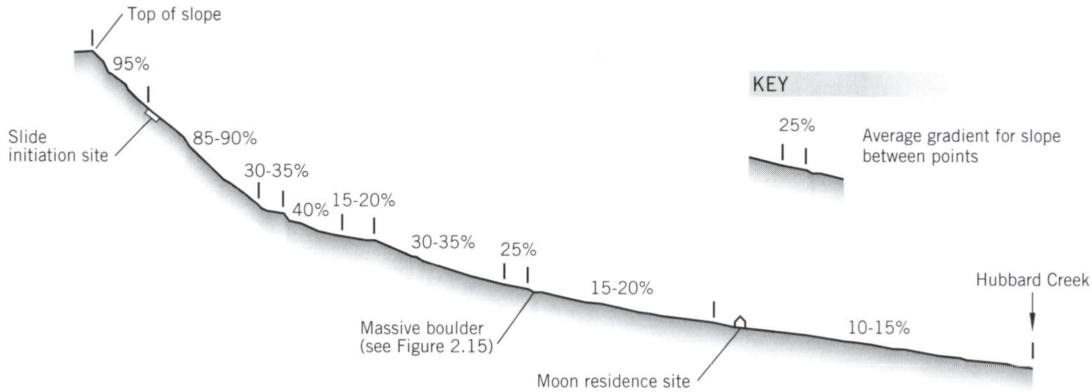

Figure 2.14 Valley profile, Rock Creek Flow Slide, 1996 (after Squier & Harvey, 2000).

Figure 2.15 Massive boulder, 18 feet across, lodged in a narrow ravine during the Rock Creek Flow Slide, 1996.

Effect of Clearcutting

An issue of considerable interest to the logging community, attorneys specializing in wrongful death lawsuits, federal and state agencies, and the general public is whether clearcutting forest slopes leads to an increase in the incidence of debris flow landslides. As previously stated, such slides occur naturally. Since they occur during high-intensity rainstorms, it is difficult to separate a natural climatological landslide from one that would not have occurred at that time except for man-made intervention (i.e., clearcutting). Intensities of storms can vary significantly over short distances in mountainous terrain, and there are usually no rain gauges close to a landslide site.

There are some technical arguments supporting a decrease in slope stability due to clearcutting:

- Clearcutting reduces evapotranspiration so the ground after clearcutting remains wetter prior to a storm.
- Ground disturbance increases the rate of infiltration of water into the ground during a storm.
- Root strength is lost.

On the latter issue, data suggest that root decay takes 3 to 5 years to occur and most failures in clearcuts have occurred within 12 years of harvest. The time for maximum loss of root support is about 7 years after logging; thereafter, new vegetative root mass returns to the slope. Other research is quantifying the root contribution to slope stability according to species.

Greenway (1987) lists the effects of trees on slope stability, as shown on Figure 2.16. Obviously, some factors have more effect than others, depending on the circumstances. For example, tree roots can anchor a thin mantle of soil overlying bedrock on a steep slope (Factor 6, Figure 2.16) but the thickness of the soil and the type of root system are critical elements. According to Gray and Sotir (1996), the mechanical reinforcing or restraining influence of roots on a slope is limited to the zone up to about 5 feet from the surface.

On steep slopes with rock close to the surface, the reinforcement of the soil provided by the roots is probably the key element. After the ground is clearcut, the roots then rot out

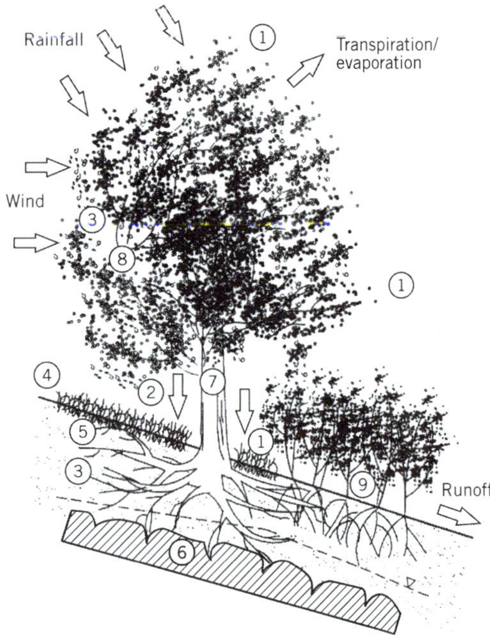

KEY

Hydrological mechanisms		Influence
1	Foliage intercepts rainfall, causing absorptive and evaporative losses that reduce rainfall available for infiltration.	Beneficial
2	Roots and stems increase the roughness of the ground surface and permeability of the soil, leading to increased infiltration capacity.	Adverse
3	Roots extract moisture from the soil, moisture that is lost to the atmosphere via transpiration, leading to lower pore water pressures.	Beneficial
4	Depletion of soil moisture may accentuate dessication cracking in the soil, resulting in higher infiltration capacity.	Adverse
Mechanical mechanisms		
5	Roots reinforce the soil, increasing shear strength.	Beneficial
6	Tree roots may anchor into firm strata, providing support to the upslope soil mantle through buttressing and arching.	Beneficial
7	Weight of trees surcharges the slope, increasing normal and downhill force components.	Adverse/Beneficial
8	Vegetation exposed to wind transmits dynamic forces into slope.	Adverse
9	Roots bind soil particles at the ground surface and increase surface roughness, thereby reducing susceptibility to erosion.	Beneficial

Figure 2.16 Effect of vegetation on slope stability (after Greenway, 1987).

and the slope becomes vulnerable to landslides within the soil mantle prior to re-establishment of a new root system.

The effect of trees and other vegetation on landslide occurrences are very complex and site-specific to the soils, vegetation, climate, slopes, etc. The two references cited earlier provide research summaries.

Avoidance of Debris Flows

Since debris flows are initiated on very steep and relatively inaccessible sites, it is difficult to identify and treat such sites with preventative measures. However, ravines with a history of debris flows can be mapped, including debris fans at the outlet from such ravines. Some possible actions:

- Ban logging on very steep slopes.
- Ban residential construction on foundations of old debris flow soils in ravines or flood plains.
- Post cautionary signs where ravines with flow debris remnants cross highways, canals, etc., to alert the public to possible danger during heavy rainfall.

2.14 ANCIENT LANDSLIDE REACTIVATION

Geology

During the Pleistocene era, which ended about 8,000 years ago, the large ice masses advanced and retreated from the polar regions and created many of the surface soils present in northern Europe and North America. Sea levels were several hundred feet lower than today, causing deep erosion in river valleys. At the periphery of the ice advance, huge slope failures probably occurred from high groundwater levels, slope erosion, etc. These ancient landslide areas, characterized by uneven hummocky ground and shallow depressions, cover extensive areas. In North America, they are termed *ancient landslide terrain*. Geologically, these slopes are colluvium and sometimes are extensive enough to be classified by geologists as a separate formation.

Ancient landslide terrain can be recognized by an experienced engineering geologist on a field reconnaissance. On large translational slides, the top may have a graben feature that is a marsh or lake. This is where unstable ground has moved away from stable ground, causing the head of the unstable slope to sink into the gap and form a *graben*. There may be no overt signs of active shear movement; the landslide may be dormant or moving at a slow creep.

The landslide debris within ancient landslide terrain is often a heterogeneous mixture of the soil and rock units formerly higher up the slope. Thus, clays may contain organics, scattered pebbles, or broken rock fragments. There may be waterbearing pockets of broken rock surrounded and isolated by impermeable soils. Such pockets often can be recharged from the surface and feed water continuously into the landslide mass. If they can be identified by site explorations, and relieved by drains, it may be possible to improve landslide stability.

A surprising aspect of ancient landslide terrain is the relatively high stiffness of clay soils within the slide mass. It might be expected that the heavy intermixing of soil types would have occurred under very turbulent or flow slide conditions. Overriding glaciation from later ice advances may explain the stiffness.

The geological origin of ancient landslide terrain can be more diverse than the common assumption that they occurred during the Pleistocene era. They conceivably could be much older within the Pliocene. However, it is of interest that a camel's tooth attributed to the Pleistocene era was unearthed in the ancient landslide affecting Washington Park in Portland, Oregon (Case History 1, Part C).

Discrete Shear Zones

The slip surface of an ancient landslide is significantly weaker than the soil above and below it. The author has observed no exceptions. Such landslides usually have a relatively thin zone of shear displacement, termed a *discrete shear zone*. The slip surface, on examination, often appears to have undergone physical and chemical deterioration of the soil. The discrete shear zone is practically impossible to sample for laboratory undisturbed shear tests.

In many stiff clays (London Clay, for example) a set of three reversals in a direct shear test is usually sufficient to obtain a close estimate of the residual strength. On the weathered siltstones of western Oregon, however, residual tests performed in the general overburden of ancient landslide terrain may show no drop-off in shear strength through three shear box reversals (Cornforth and Fujitani, 1991). Instead, it requires (i) the test sample to be taken at, or very close to, the slip surface; and (ii) many stress reversals in a direct shear machine during which the strength progressively declines. Case History 10: Hagg Lake, Slides 4 and 3 (Part C), is a good example of this effect.

Measurements of residual strength within a discrete shear zone can be obtained in one of the following ways:

1. Expose the discrete shear zone in a test pit or adit and perform an in-situ direct shear test. This is costly, time-consuming, and measures residual strength at only one location.
2. Determine the depth of the discrete shear zone by inclinometers, then put down a second hole alongside to take an undisturbed sample at the ancient shear zone (plus other samples above and below). Perform a stress reversal direct shear test or ring shear test on or very close to the shear zone. If this is impossible (the zone is often highly fractured), use material from the shear zone for a remolded ring shear test.
3. Determine as in 2, but use disturbed soils (SPT sample or cuttings) to perform a remolded ring shear test

Of these options, the first is of considerable interest to research-level studies but is otherwise limited by cost to major landslide investigations. The second provides the investigator with a visual picture of the shear zone and soils immediately above and below it. The last is the simplest and least costly.

An alternative procedure is to calculate the residual strength by back analysis. For landslide analysis, it can be assumed that the static factor of safety is 1.00 for very slow-moving landslides. By measuring the other parameters in the landslide model, the average residual strength on the slip surface can be obtained. In most cases, this is a better method of performing a stability analysis. If there is uncertainty of groundwater levels at failure, a parametric analysis can be undertaken and the residual strengths obtained in this way can be carried forward into remediation analyses for the same assumed conditions.

Stability of Ancient Landslide Terrain under Natural Conditions

The significance of ancient landslide terrain is that, in temperate zones of relatively high precipitation, the ground is often in a state of marginal stability during the wet season. Whether or not the ground moves is usually a combination of the physical conditions of the site and the intensity of rainfall during winter storms.

Examples of physical conditions affecting ancient landslide activity could be rockfalls onto the top of the slope or river erosion at the slope base. These would increase activity. Examples of a more stable environment are an ancient landslide that bears onto a rock outcrop or a beach at the slope base.

The stability of ancient landslide terrain under natural conditions can range between (i) completely stable at all times; (ii) usually stable, but vulnerable to creep movements intermittently from intense winter storms or earthquake; (iii) moves each winter but by different amounts, depending on storm patterns; and (iv) moves year round due to unfavorable site conditions.

Effects of Construction Activities on Ancient Landslide Terrain

Manmade construction activities on ancient landslide terrain can reactivate the slide or significantly increase the rate of movement from pre-existing creep conditions. When the City of Portland excavated into a ravine to build two water storage reservoirs at the base of an ancient landslide in 1893–94, major movements exceeding 12 inches per year occurred over the next three years. An extensive investigation, described in Case History 1 (Part C), showed that excavation for the two reservoirs (100,000 cu. yd.) was only about 3% of the landslide volume of 3,400,000 cu. yd. Prior to the reservoir construction, the slope was either dormant or undergoing very slow creep. In the lawsuit brought by property owners against the city, it was successfully argued that such a small percent change in the landslide volume could not have caused such a major landslide. It was a plausible but wrong argument.

The construction of these two reservoirs reactivated a pre-existing ancient landslide, and the increased rate of movement caused extensive damage. More recent calculations show that the calculated factor of safety dropped by about 5% as a result of the excavations. Pumping from a shaft near the middle of the site restored stability temporarily, and a series of drainage tunnels eventually reduced activity back to creep levels.

This example is one of many sites where small earthworks have caused large reactivated movements of a preexisting ancient landslide. They frequently become lawsuits.

A typical scenario is as follows: The presence of ancient landslide conditions is overlooked. Site investigation borings find unusual mixtures of soils. Laboratory tests of the overburden give relatively high shear strengths. The calculated factor of safety is satisfactory, exceeding 1.50. It is erroneously concluded that an adequate safety factor for stability is available.

The project is built during dry summer months. In the following winter, or possibly several winters later (if delayed failure is involved), movements begin and worsen in each succeeding year. The cost of remediation is large. After the lawyers become involved, no expense is spared to investigate the situation. The cost of studies and litigation after the landslide can be many times the relatively modest cost of the original site investigation.

Partial reactivation of a large expanse of ancient landslide terrain is fairly common. The hypothetical examples shown on Figure 2.17 include breakouts at intermediate locations between the head and toe of the ancient landslide. The shear strength mobilized in these breakout points involve "first time" sliding in which the strength is usually much higher than the residual strength at the preexisting ancient slip surface. Thus, the total shear resistance is a combination of first-time sliding strengths at the ingress/egress points and residual strength along the ancient slip surface.

The hypothetical ancient landslide reactivation illustrated on Figure 2.17 is a long translational slide on a modest slope. A new highway is built near the center of the slide involving a deep cut. This seemingly innocuous "notch" in the slope can cause reactivation of the whole or parts of the ancient landslide. In Figure 2.17(a), the whole length of the slope moves because the highway excavation has removed soils above the water table (see Infinite Slope Analysis Chapter 9, Section 9.6). Other reactivation possibilities shown on Figure 2.17 include: (b) a breakout into the base of the cut slope with a short length of first-time sliding (FTS) near the cut, (c) a shorter slide into the cut with a new headscarp of FTS (or tension crack), or (d) a slide developing from above the cut to the base of the slope.

A geotechnical specialist (engineer or geologist) faced with a construction project on ancient landslide terrain may not be able to reliably determine the existing status of slide activity. In some cases, it may be possible to install inclinometers and monitor the slope over one or two wet seasons. In other cases, a short construction schedule may preclude such observations. Local information may be available and should be sought.

In temperate wet climates, it should be assumed that the slope is marginally stable. If a geological reconnaissance has not revealed any evidence of recent landslide activity, *the minimum requirement is that the stability of the slopes during and after development should not be worse than it was under preexisting natural conditions.* In other words, any effect of the development to lower slope stability should be offset by stabilizing measures. This is a *minimum* requirement. It is recommended that stability be improved over the preexisting conditions to offset uncertainties in stability analyses.

For the hypothetical example of Figure 2.17, some preventative treatment possibilities are to support the road with shear piles, tieback anchors, dewatering, a buttress at the slope base, and other possible treatments as discussed in Part B.

If the landslide mass is a stiff overconsolidated clay, an increase of shear stresses will cause a temporary reduction in pore pressures due to dilatancy (see Section 2.15 following). There can be a delay of many years before failure occurs in such slopes. Therefore, failure may not be immediate, despite the construction activity being the "cause" of the failure.

Site conditions involving ancient landslide terrain are described in the following case histories (Part C): CH-1, CH-3, CH-4, CH-10, and CH-11.

2.15 DELAYED FAILURE

Development of Delayed Failures

This is an important category of landslides affecting slopes of stiff to hard clays, clayey silts, or rocks that have weathered in-place to a stiff clay matrix. In delayed failure, the slope failure

Figure 2.17 Examples of potential instability of ancient landslide terrain due to construction of a deep cut at mid-slope.

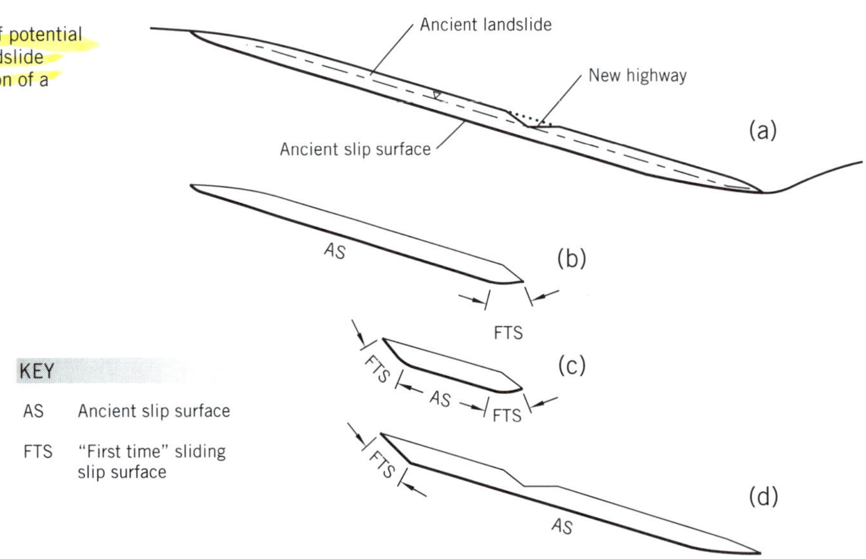

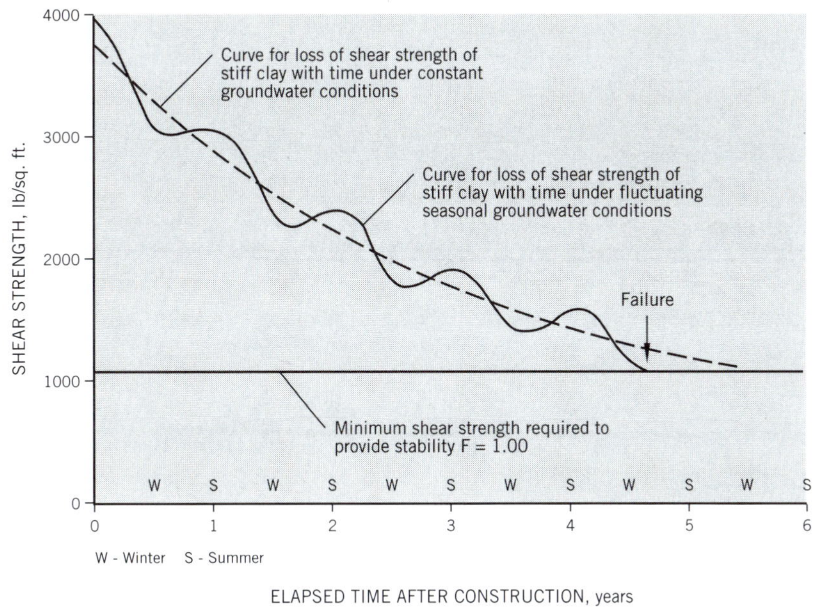

Figure 2.18 Hypothetical example to explain the phenomenon of delayed failure in stiff clay slopes.

does not occur concurrently with the causative factor, which may be a cut, fill, or erosion. The reason for the delayed failure is that overconsolidated clays dilate when sheared and develop negative changes in pore water pressures. The induced negative changes in pore pressures temporarily increase the shear strength and provide slope stability. As the pore pressures rise toward the long-term equilibrium levels within the slope, the shear strength of the soil decreases with elapsed time and may cause a landslide years or decades after construction.

A technical explanation of delayed failure, using the Skempton pore pressure parameter A, is presented in Chapter 8, Section 8.12. A simpler understanding can be demonstrated by the graph on Figure 2.18, which is a hypothetical example. At the time of construction, the stiff clay has a high undrained shear strength that easily exceeds the strength (shown by a horizontal line on the figure) needed to preserve slope stability. With time, the clay shear strength drops as the initial negative change in pore pressures slowly rise towards the long-term equilibrium levels within the slope. If natural groundwater levels are constant, the decreasing strength may follow a path similar to the broken line graph on Figure 2.18 and the slope would fail if the broken line reaches the horizontal line representing a factor of safety F = 1.00.

A more realistic scenario would be a natural groundwater table that fluctuates between summer and winter. In this case, it can be anticipated that the clay will lose strength in the winter months at a faster than average rate because the differential between the depressed pore water pressure and the higher winter groundwater level will be greater than average. Conversely, the clay will lose strength at a slower rate than average in the summer when the groundwater is lower than in winter. The roller-coaster curve overlaid on the broken curve is a simplified representation of this effect. With a fluctuating groundwater regime, it is likely that a delayed failure will occur in winter when a high groundwater spike (or just seasonally high groundwater) triggers the slope failure as the decreasing strength approaches the horizontal line on Figure 2.18.

As a stiff clay slope loses strength with time, the rate of movement will increase. In cases where delayed failure takes decades to occur, the rate of movement in the early years may be imperceptible. An example of the rate of movement with time is given on Figure 2.19(a) for the Faraday Slide, which is described in Case History 8. It can be seen that the movements accelerated after they were first observed in 1976. This movement was stopped by a toe buttress to prevent a more catastrophic landslide. Similar plots of increasing rates of movement were measured at Percy Slide, Leaburg, Oregon (not shown) and a closely monitored retaining wall (which failed) at Kensal Green, London (Skempton, 1964).

Monitoring of a slow-moving landslide by survey hubs or inclinometers can show over a period of years whether the landslide is developing into a delayed failure. If so, preventative measures can be taken to stabilize the slope.

A delayed failure landslide has three characteristics:

1. The soil in the slope is an overconsolidated clay, clayey silt, or silt (i.e., slow draining, with dilatant shear behavior).
2. An event occurs that reduces the stability of the slope and increases the shear stresses in the soil; examples are manmade cuts and fills, and erosion.
3. The velocity of the shear movements within the discrete shear zone progressively increase with time.

Delayed failures may explain the statement occasionally heard from geotechnical practitioners that landslides can occur for no apparent reason. It is probable that many such instances are the result of delayed failures, which are poorly understood within the profession.

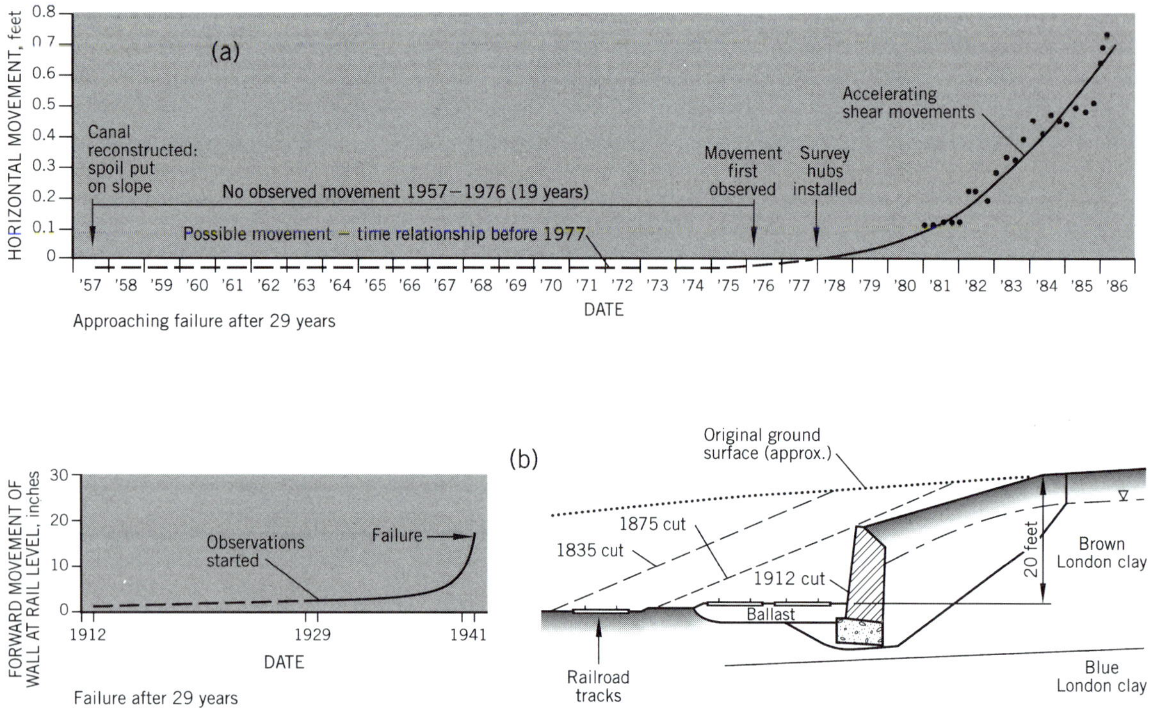

Figure 2.19 Increasing velocity of landslide movement in a developing delayed failure:
(a) Faraday Slide (see Case History 8)
(b) Kensal Green, London railway cut (after Skempton, 1964)

Examples of Delayed Failures in London Clay

London Clay is a stiff, fissured clay and is typical of many similar stiff clays in southern England (e.g., Oxford Clay, Kimmeridge Clay, Lias Clay). The phenomenon of delayed failure has been extensively studied by A. W. Skempton and other British research workers. Here are five examples (Skempton, 1964) for London Clay in the London metropolitan area:

1. *Wembley Hill.* Slope 3:1 and 40 feet high with a 25-foot high retaining wall at the base (total 65 feet). Built 1905; failed 1918, 13 years after construction.
2. *Kensal Green* (Figure 2.19b). Cutting in the London-Birmingham railway. Widened in 1912 involving a 20-foot high excavation and 12-foot high retaining wall. Failed in 1941, 29 years after construction. The wall was surveyed for 12 years before failure, initially moving outward at only ¼ inch per year. The horizontal movement at failure was 18 inches.
3. *Sudbury Hill.* Railway cutting 23 feet deep with side slopes of 3:1. Excavated in 1900, failed in 1949, 49 years after construction.
4. *Wood Green.* Slope 21 feet high with 3:1 side slopes above a 16-foot high retaining wall (total height 37 feet). Built in 1893; failed in 1948, 55 years after construction.
5. *Holloway.* 18-foot high slope. Built 1870; failed 1951, 81 years after construction.

Typical properties for London Clay are LL 81, PL 29, PI 52, w 31%.

Examples of Delayed Failures in Oregon

The Pacific Coast region of the United States has mountainous terrain with exposures of stiff tuffaceous clays that can fail many years after construction. Although geotechnical engineers and geologists usually attribute these failures to rainfall, it is probable that many, if not most, of them are delayed failures in which the final triggering event is the seasonal rise of groundwater and/or a major storm.

Three examples from Oregon are given below:

Pelton Upper Slide, Madras

This slide is described in Case History 6 (Part C) and Cornforth and Vessely (1992a). In brief: A railroad fill was built across the top of a slope in about 1910. Rapid failure of the slope occurred in 1975, 65 years after construction. This was a double wedge failure caused by a fill built over the upper (driving) wedge. The loss of shear strength over time occurred within the very stiff soil below the horizontal lower wedge. Back-calculated properties: $c_r' = 0$, $\phi_r' = 8\frac{1}{2}°$ in a three-dimensional stability analysis. Average properties of the soil: LL 82, PL 57, PI 25, w 54%.

Faraday Slide, Estacada

Described in Case History 8. In brief: Fill placed on the top of a slope during canal construction in 1957 caused a circular arc failure that was almost entirely within a very stiff sandy silty clay (decomposed tuff breccia). The average soil properties were LL 70, PL 39, PI 31, w 41%. The rate of movement became noticeable to the owner in 1976 and accelerated over the next 10 years (Figure 2.19a). To prevent an impending catastrophic slide, which could have released canal water and caused other harmful effects downstream, the slide was stabilized by a buttress in 1989, 32 years after the causative fill construction of 1957.

Silver Point Landslide, Cannon Beach

This landslide will be explained in more detail because of the legal implications of a delayed failure. This large landslide (Figures 2.20 and 2.21) is 1,200 feet long and about 600 feet wide on average. It occurred overnight on February 3, 1974. Four properties on the landslide were destroyed.

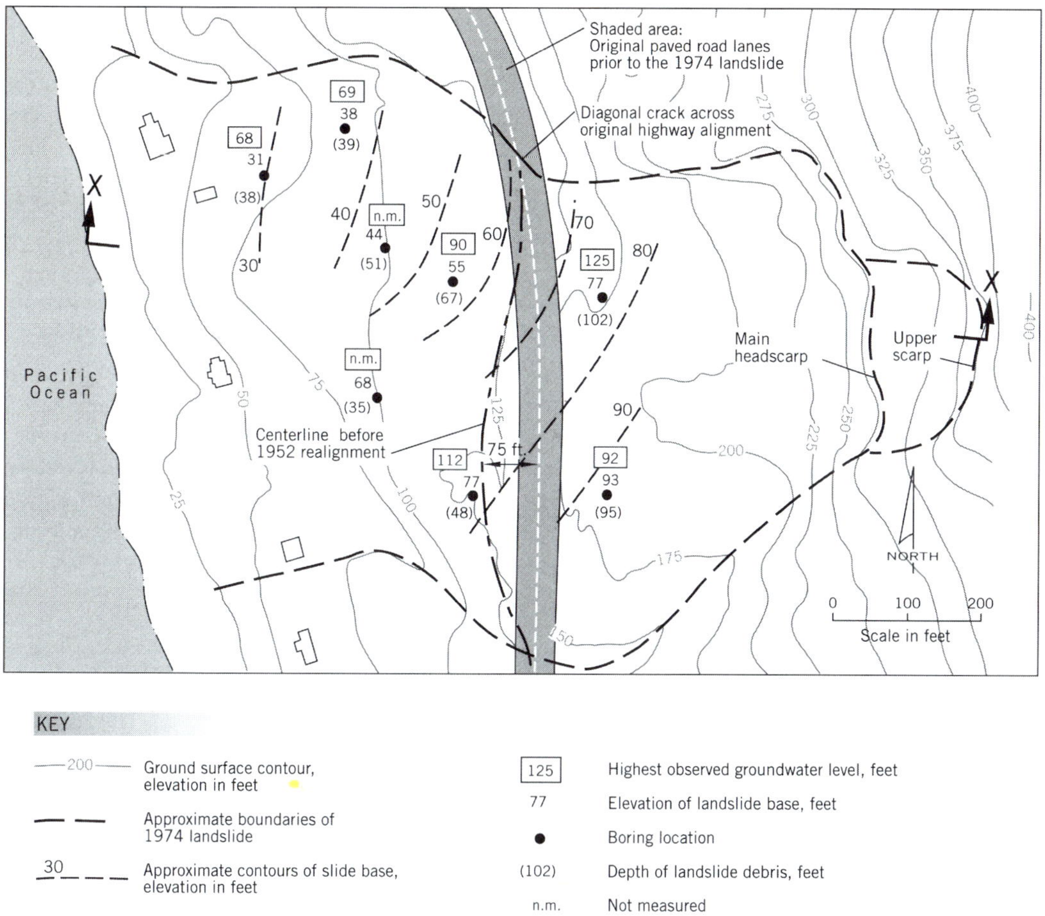

Figure 2.20 Silver Point Slide. Site plan showing ground contours after the February 1974 landslide.

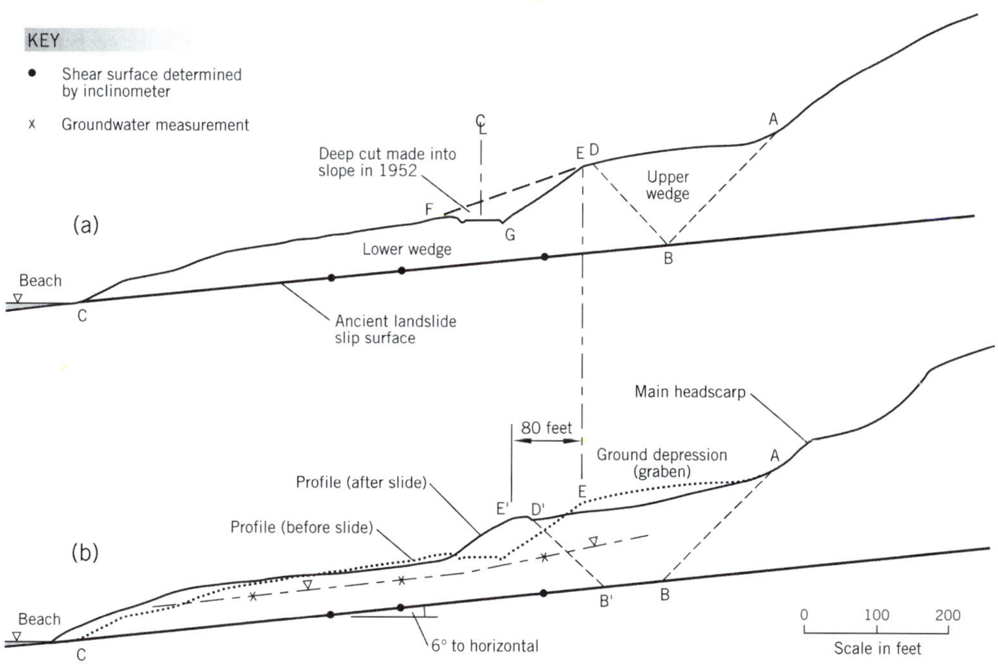

Figure 2.21 Silver Point Slide. Section X-X near center of slide:

(a) before landslide, June 1973 topography

(b) after landslide, February 1974 topography

The coast road was realigned in 1952 to replace an S-curve with a single curve (Figure 2.20). As-built sections show that a 20-foot wide two-lane road was replaced with a 48-foot wide four-lane road. The maximum horizontal cut into the slope was 86 feet.

It is calculated that approximately 100,000 cu. yds. of soil were excavated from the 700-foot road length affected by the 1952 slide. There is an estimated 1.8 million cu. yd. of debris in the landslide mass. Therefore, the excavated quantity is approximately 5.5% of the landslide mass.

After the realignment of 1952, the highway had no maintenance problems until 1967, when a diagonal crack appeared that subsequently became the north edge of the 1974 slide. From 1971 to 1973, records showed that maintenance crews placed 1,200 cu. yds. of crushed rock and asphalt in this area to maintain the road grade. In the winter of 1972–73, it was observed that, south of the diagonal crack, the highway centerline had moved downslope over a length of several hundred feet. The slide started to move more rapidly after January 14, 1974. The road was closed to traffic on February 3, the evening before the overnight slide.

Very detailed photogrammetry analyses showed that the slide area moved significantly during the 1967–1973 period. Furthermore, the 1973 stereo pairs showed a developing headscarp above the road that became the headscarp of the 1974 catastrophic slide.

The overburden generally was described as a stiff silty clay and clayey silt with occasional buried wood fragments, logs, and scattered basalt and sandstone fragments. The average properties of the landslide debris were LL 61, PL 27, PI 34, w 35%. A direct shear test on a clay specimen with precut shear planes gave residual strength parameters of $c_r' = 0$, $\phi_r' = 12.8°$.

Double wedge analyses for the "before cut" and "after cut" cross-section of Figure 2.21(a) showed that the 1952 excavation lowered the factor of safety of the eventual landslide by about 6 percent. It can be seen from the section that the excavation removes soil from the lower wedge that provides resistance to the upper (driving) wedge. Thus, the net resistance to stability is lowered.

A value of $\phi' = 34°$ was used for "first time" sliding of the upper wedge slip surface AB. With groundwater as measured in the winter of 1974–75, and assigning $\phi_r' = 13°$ along the ancient slip surface, the calculated factor of safety of the entire landslide is close to unity.

The author's interpretation of the Silver Point landslide is that the upper wedge area ABD was stressed by the loss of lateral support caused by the excavation area EFG. (Note that Figure 2.21(a) represents the *average* cut made through the slide area.) The strength of the stiff clay on plane AB slowly decreased with time leading to increasing movements downslope of it and the need for increasing amounts of maintenance from 1967 onward to maintain the road grade. By the winter of 1973–74, slope movements were moving fairly quickly and a modest storm in early February 1974 caused the large 70-foot landslide movement. It is a classical delayed failure meeting all three criteria stated earlier. The loss of shear strength with time occurs in the plane AB of the upper wedge and not in the plane BC of the ancient landslide that preexists at residual strength.

The probable movements of the double wedges at the end of the catastrophic landslide are shown on Figure 2.21(b). Points B and E, before the slide, move to B' and E' respectively after the slide. Surface AE sinks to form a depressed (graben) feature. The "before" profile has been superimposed on the section to demonstrate this effect.

Four properties were destroyed by this landslide and the state acquired the land through inverse condemnation, offering the owners valuations that were based on "after slide" conditions, i.e., essentially worthless. The property owners sued the state, alleging that the construction work in 1952 caused the later landslide. Four separate trials were held in which three out of four verdicts supported the state's position that the Silver Point landslide was a natural event (act of God). One verdict supported the delayed failure explanation. Jurors interviewed after the trials stated that they did not form an opinion about conflicting technical testimony but could not accept that the state should be held responsible for construction work that occurred 22 years before the massive landslide. This case illustrates the difficulty of successfully pursuing a claim for damages due to a delayed failure after a significant passage of time.

2.16 EARTHQUAKES

Earthquake Types

Strong motion earthquakes can occur in many parts of the world, especially along the "ring of fire" where the surface crustal plates collide. Four types of earthquake source are recognized in earthquake engineering:

1. *Subduction–Interface* earthquake, caused by slip between two adjoining plates.
2. *Subduction–Intraslab* earthquake, caused by a breakup of a subducting plate as it is reabsorbed into the molten mass of the earth's core
3. *Crustal* earthquake, caused by rupture along a preexisting active fault within the earth's crust.
4. *Random Crustal* earthquake, having the same cause as a crustal earthquake but occurring where no active fault has been identified by geological mapping; this usually occurs where bedrock is mantled by overburden soils such that the presence of the fault is unknown.

A cross-section through northwest Oregon (Figure 2.22) illustrates the earthquake types described above. The power (release of energy) of an earthquake is measured on the Richter scale. Each increase of 1.0 on the Richter scale represents a 30-fold increase in the energy. Subduction-Interface ruptures produce the largest earthquakes, sometimes exceeding M 8.

Keefer (1984) studied 30 historical earthquakes and compiled the data plotted on Figure 2.23 to show the areas affect-

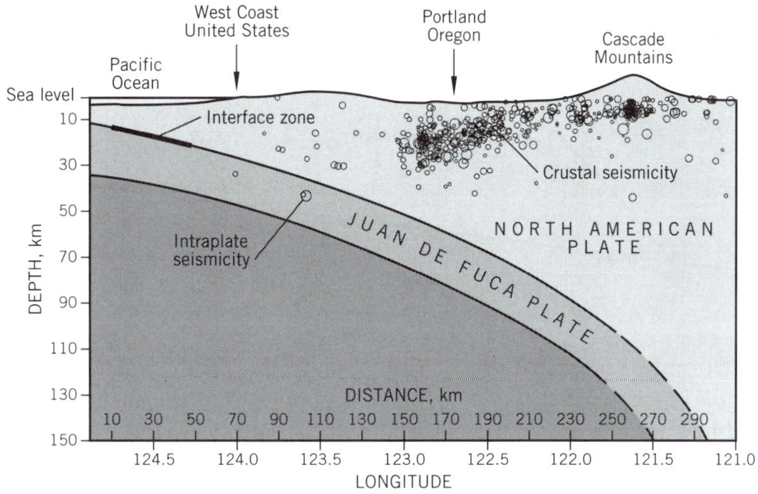

Figure 2.22 Northwest Oregon Seismicity: Cross-section centered on latitude 45.5° with inferred location of the subducted portion of Juan de Fuca Plate.

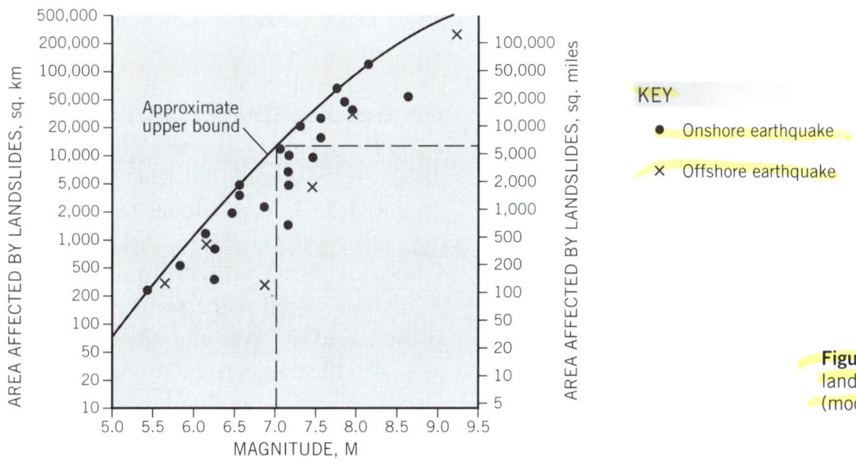

Figure 2.23 Area affected by landslides due to earthquakes (modified from Keefer, 1984).

ed by major landslides. Thus, for an earthquake of magnitude 7.0, landslides could extend over an area as much as 6,000 sq. miles. Keefer noted that the greatest losses of human life were due to rock avalanches, rapid soil flows, and rock falls. On a more comprehensive database of 300 earthquakes within the United States, the smallest earthquake reported to have caused landslides had a magnitude 4.0. The minimum magnitude that causes lateral spread and flows was 5.0, and this value is consistent with the research of others on soil liquefaction.

Earthquake-Induced Landslide Groups

Excluding consideration of rock slopes, earthquake-induced landslides can be grouped as follows:

- Failure of marginally stable slopes
- Translational-slide movements in clay soils
- Liquefaction of cohesionless soils

Landslides in Marginally Stable Soils

Examples of marginally stable slopes include: ancient landslide terrain, ocean cliffs, actively eroding river banks, manmade cuts and fills on steep terrain, talus slopes, weathered rock faces, and stratified volcano slopes.

When marginally stable slopes are subjected to the horizontal forces from a strong motion earthquake, failure can occur. Usually these failures are local and fairly small (Chleborad and Schuster, 1989). However, the Olympia earthquake of 1949 (Magnitude 7.1) produced a slide in a 300-foot high cliff into the Tacoma Narrows near Fort Nisqually (Noson et al., 1988). The main body of the slide occurred in the Esperance Sand stratum in a slope averaging about 32° to the horizontal. A photograph of the slide indicates a surficial break typical of sands. Although some reviewers have suggested that liquefaction may have been partly responsible for the failure, the fact that failure was delayed until three days after the earthquake, and the type of failure, suggest that it can be classified as failure of a marginally stable sand slope. Approximately 50 slope failures were caused by the Olympia earthquake.

Seed and Goodman (1964) and Goodman and Seed (1966) discuss the analysis of movements during earthquakes in slopes of cohesionless soils with marginal stability. These

Figure 2.24 Translational landslide caused by an earthquake.

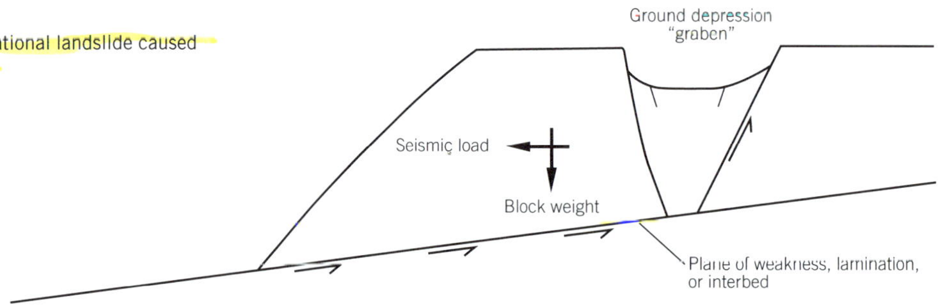

analyses assume that the slopes are above the water table and will not be subjected to liquefaction during the earthquake.

In clay slopes of ancient landslide terrain, significant movements could damage structures located at the margins of the landslide. Although this specific issue has not been studied extensively, Makdisi and Seed (1978) provide an approximate method for estimating ground movements in clay slopes during a major earthquake. Briefly, the method calculates the horizontal *yield acceleration,* k_y, needed to bring the factor of safety of the slope to 1.00. The slope is assumed to move during the part of the earthquake-induced acceleration-time graph that exceeds the calculated "yield acceleration." Thus, significant total movements occur during a large magnitude earthquake in which the duration of strong motions is high. The calculated "yield acceleration" is low for slopes with marginal stability, and is likely to be exceeded for longer periods of time (during the earthquake) than in a slope with a higher static factor of safety. For more information on this subject, see Chapter 12, Section 12.3.

Translational Landslides in Clay Slopes

These slides occur in clay slopes that have adequate stability under normal static conditions but become unstable when subjected to horizontal forces from a large earthquake. This type of failure is likely to occur where the ground has a plane of weakness in the near-horizontal direction (Figure 2.24). Several major slides of this type occurred in Anchorage, Alaska, during the 1964 Good Friday earthquake (Richter M_L 8.4–8.6; M_W 9.2). The clay stratum that sheared along near-horizontal surfaces during approximately 5 minutes of strong motions is the Bootlegger Cove clay, a blue-gray plastic clay, 200 to 300 feet thick, that is sensitive to remolding and loses strength under cyclic loading (Seed and Wilson, 1967). Silty and sandy beds occur within the clay, especially near the surface of the stratum, and liquefaction pore water pressures within these more permeable beds may have contributed to the failures. However, the slippage at Anchorage appears to have occurred within the sensitive clay and thus needs to be distinguished from failures that result from loss of strength within loose sand layers.

The failure of soft sensitive lacustrine clays caused spectacular movements and resultant damage. In Anchorage, the L Street slide moved about 12 to 15 feet horizontally, and the Fourth Avenue slide moved about 10 feet (Long and George, 1967). In each case, the block movement created a ground depression (graben) at the head of the slide where the unstable block separated from the stable ground (Figure 2.24). The length of the Fourth Avenue slide, from headscarp to toe, of about 600 feet suggests that the half wavelength of the seismic waves may control the breakaway point and has provided one method of making a pseudostatic analysis of the slide (Long and George, 1967). These two slides were stabilized against future major earthquakes by construction of rockfill buttresses at the toe of the block slide.

The Turnagain Heights landslide in Anchorage covered a very large area: 8,500 feet wide along the coastline and up to 1,200 feet inland (130 acres). The slide moved into the sea for distances of up to 1,200 feet. The slide mass broke up and destroyed 75 homes. Model tests of the Turnagain Heights slide were performed on a shaking table at the University of California (Seed and Wilson, 1967). The results, Figure 2.25(a), showed that the failure was retrogressive (i.e., started at the toe of the slide and moved backward) and the broken up soil in the model had a strong resemblance to the geologic section observed in the detailed site investigations, Figure 2.25(b). These results indicate that the extent of such landslides depend on the *duration* of strong ground motions.

A landslide on the San Pedro River near Lake Rinihue, Chile, during the earthquake of 1960 caused shear failure in lacustrine clay with considerable breakup of the ground surface (Figure 2.26). This failure extended about 1,700 feet behind the original cliff (Davis and Karzulovic, 1961) and involved 30 million cu. yd. of slide materials.

Liquefaction-Induced Landslides

Temporary liquefaction of sands during strong ground motions causes the third group of earthquake-induced landslides. Sands (and other cohesionless soils, including gravels and coarse silts) have point-to-point contact between the grains. When the soil structure is disturbed by earthquake shaking, the sand grains may go into a fluid state, depending on the compactness of the sand, weight of overburden, duration and severity of shaking, etc. When liquefaction occurs, the soil loses most of its shear strength; sand "boils" often erupt at the ground surface. The surface itself may be broken up, but more often is subjected to extensive cracking as the ground shifts laterally.

Landslides caused by liquefaction are reported in virtually every major earthquake. To liquefy, the sand deposits have to be relatively loose and below groundwater. Areas at particular

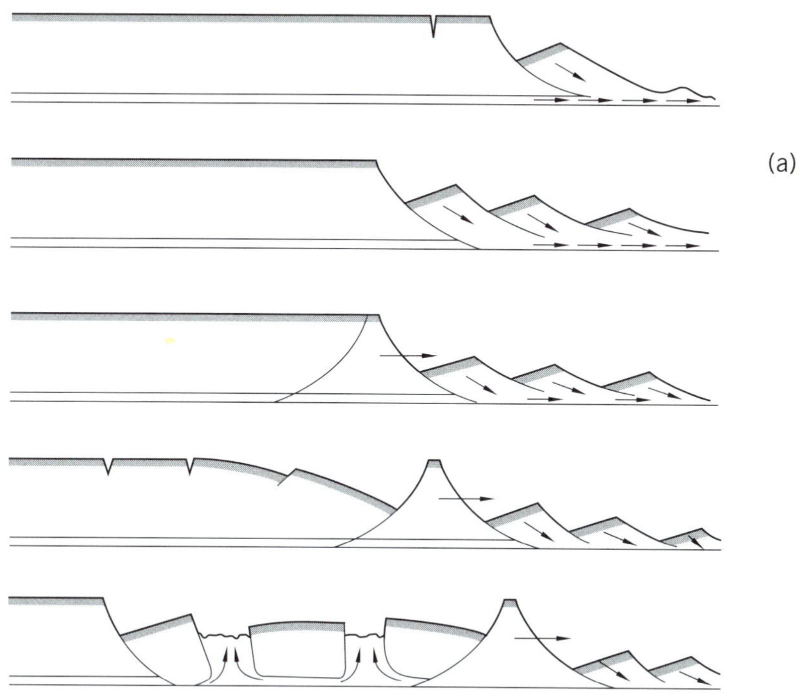

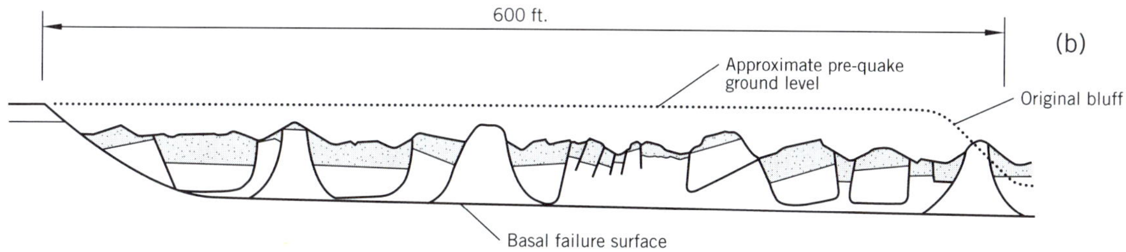

Figure 2.25 Turnagain Heights Landslide, Anchorage, Alaska:
(a) progressive failure mechanism observed in model tests
(b) actual soil profile mapped by site investigations (after Seed & Wilson, 1967)

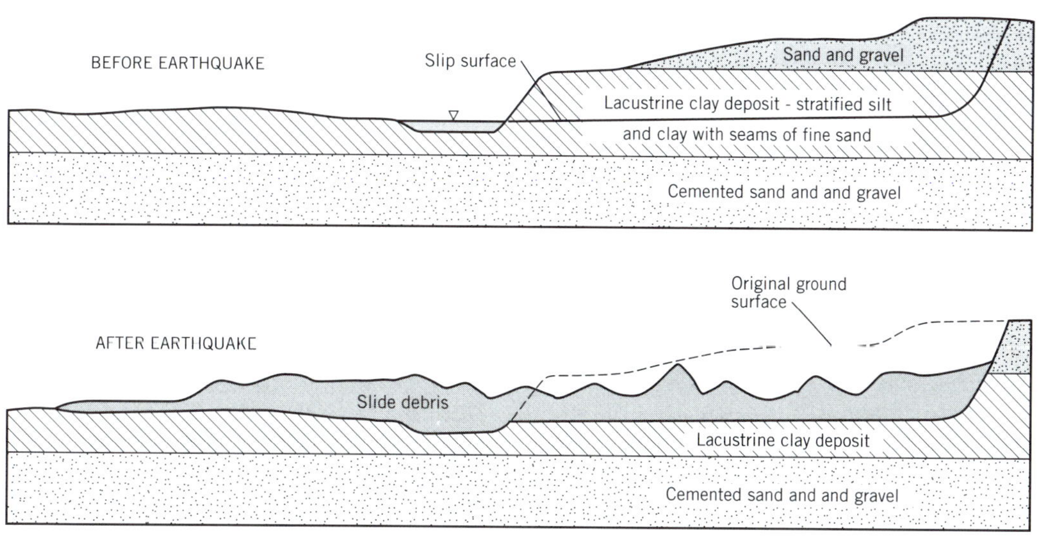

Figure 2.26 Large translational slide near Lake Rinihue, Chile (1960) (after Davis & Karzulovic, 1961).

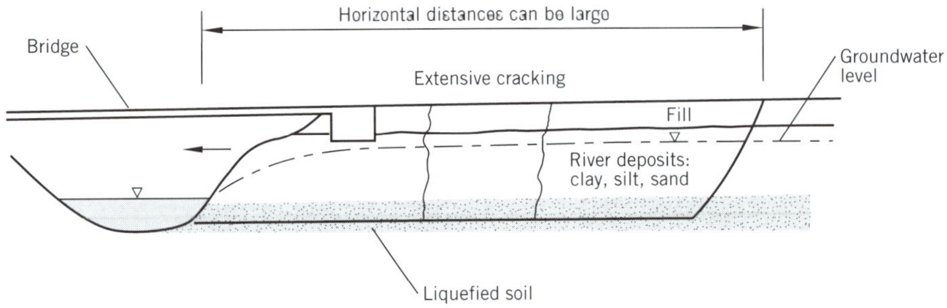

Figure 2.27 Liquefaction causing lateral slide movements on flood plains.

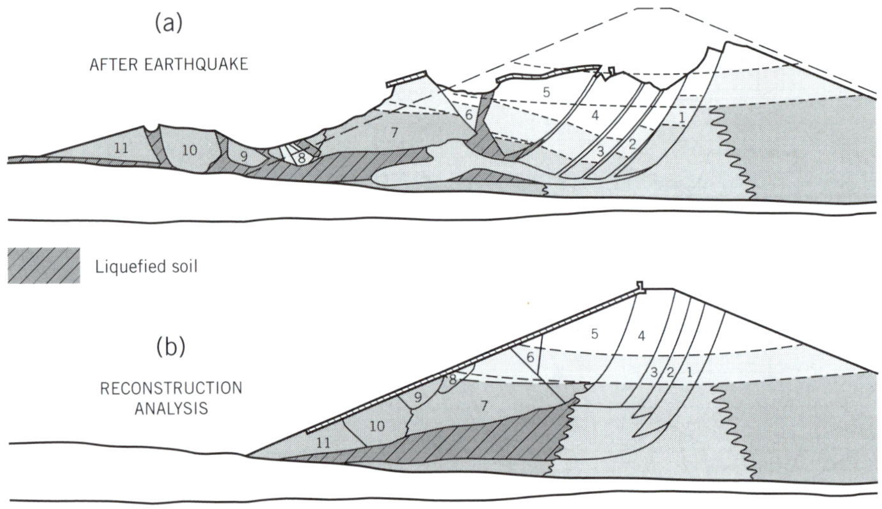

Figure 2.28 Near-failure of the Lower San Fernando Dam, 1971. Sections through the dam (after Seed & Harder, 1990).

risk of liquefaction slope failure include flood plains of rivers, deltas, estuaries, and loosely placed sand fills (including hydraulic sand fills). Highways or railroad fills may fail and bridges may be damaged by ground moving toward the river from one or both sides (Figure 2.27). Buried pipelines may be broken or float to the surface.

Flow slides may develop. At Valdez, Alaska, extensive sections of the waterfront were destroyed by flow slides during the earthquake of 1964 (Coulter and Migliaccio, 1966). Shannon (1966) reported a similar waterfront flow slide at Seward.

It is estimated that 270 bridges were severely damaged by the 1964 Alaska earthquake, with movements of up to 6½ feet towards the rivers being observed at some locations (McCulloch and Bonilla, 1967). The damage was greatest on the deltas at Whittier and Seward.

Dames and Moore (1989) examined damage in the San Francisco area immediately after the Loma Prieta earthquake (Magnitude 7.1). Although the 10-second duration of strong shaking was relatively short, they noted extensive damage to uncontrolled fills. By contrast, engineered sand fills placed under careful control survived the shaking extremely well.

There has been a considerable advance in knowledge of soil liquefaction since 1970. The impetus to research came after the near-failure of the Lower San Fernando Dam in California in 1971 (Magnitude 6.6). The 30-foot drop in the crest of this dam, which had been declared safe in an inspection only five years earlier, was attributed to liquefaction of hydraulic sand fill placed in the shoulder of the dam during construction in 1915. The "before" and "after" sections through the dam (Figure 2.28) show the liquefied zone reconstructed from a detailed study after the event (Seed and Harder, 1990).

A liquefaction analysis is a fairly complicated study, but many geotechnical specialists can perform such studies. The basic approach is described in Chapter 12.

Many facilities, such as dams and lifeline structures, can be made earthquake-resistant by ground compaction, in-situ treatment, drainage, or slope support (see Chapter 21). In some cases, relocation of the facilities is an option. For new facilities, preventative treatment of liquefaction-prone soils can be undertaken before construction. Given a choice, structures in earthquake regions should avoid the high-risk ground conditions that have experienced severe damage in past earthquakes.

2.17 ROCK SLOPES

Rocks are broadly separated into two categories: hard rocks and soft rocks. In practice, there is a transitional change from hard *rocks* to hard *soils*, but the two broad categories are sufficient for a brief discussion of landslides in rocks.

Hard rocks are capable of standing at very steep slopes. The key factor in their stability is not the rock itself but the spacing and orientation of discontinuities such as joints, faults, bedding planes, and interbeds. If the principal sets of joints or bedding are dipping *into* the slope, stability generally is excellent and steep cuts can be made with minimal maintenance. Conversely, should the discontinuities be sloping *toward* the slope, which often controls topographical development in nature, the situation is adverse for stability and ground treatment may be needed for cut slopes.

Hard rock slopes exposed to weathering have rock debris within the joints. If additional weathering, springs, or heavy rainfall remove the debris, slippage along a detrimental contact angle may occur. Therefore, when such conditions have been identified by a geological survey of the slope, monitoring, removal, or preventative treatment may be required to safeguard people and property below the hazard.

Another stability issue for hard rocks is the possibility of toppling failure in which an unjointed block of rock may rotate about its base. This occurs where the joints or weathering separate the rock in a near-vertical orientation. If there is a slope on one side of the vertically separated rock, slow erosion can eventually undermine one side of the rock block, causing it to topple. Adjacent rock columns can also fail or rotate outward due to loss of lateral support.

Soft rocks include most weakly cemented sedimentary units of conglomerate, sandstone, siltstone, and various tuffs and breccias. If much of the cementation has been lost, these rocks behave like very hard soils. The principal hazard at most sites is erosion from weathering or wave action, causing differential erosion, as discussed in Section 2.4. However, when overstressed by a fill or cut, the soil may behave as an overconsolidated clay and be subject to delayed failure.

Many lawsuits in the United States are brought about by rocks falling down cliffs or slopes onto highways and other public areas. Boulders of all shapes drop, bounce, and roll off steep slopes with considerable velocity. The prediction of how this occurs is the subject of ongoing research. However, some slopes have a history of such rockfalls, so good recordkeeping by highway maintenance personnel and other owners can identify the more vulnerable slopes. Prevention is the key need to reduce the incidence of deaths, injuries, and property damage. Preventative techniques include: (i) altering the cut slope angle (steeper slopes being somewhat safer because loose rocks drop rather than roll), (ii) use of wire mesh nets over the slope face, (iii) adequate fallout ditches (with soft landing areas), (iv) protective fences, (v) periodic field inspections by geologists, (vi) periodic scaling, (vii) reinstating undermined slopes, and other techniques that may be appropriate at a particular slope. The services of an experienced engineering geologist is essential to reaching an optimum treatment.

Public and private agencies generally have insufficient construction and maintenance budgets to be able to deal with all potential slope failures. Therefore, priorities have to be established. One rational procedure for ranking the potential danger is described in Chapter 23, Section 23.2, and an example of its use is presented in Case History 12: Crown Point Highway Rock Slide.

The mechanics of rock slope instability and rock slope treatments are not covered in this textbook due to space limitations. Good reference texts include Brawner (1994), Wyllie and Norrish (1996), and Norrish and Wyllie (1996).

2.18 LOESS SLOPES

Loess Distribution and Characteristics

Loess is windblown silt from the late Pleistocene epoch. The melting glaciers deposited large quantities of silt into drainages. The silt was subsequently eroded and blown by winds to be deposited as relatively uniform strata. It has been estimated that about 17 percent of the continental United States is covered by loess at the surface (Figure 2.29). The greatest concentration of loess is in China, but it occurs in many other countries that bordered the Pleistocene glaciers.

Loess deposits are composed of silt and fine sand that are coated with a clay binder. The clay cements the particles into a loose structure with a high cohesive strength in a natural unsaturated condition. When loess has contact with abundant water, the soil structure collapses; it is easily eroded and produces gullies, sinkholes, etc.

Holtz and Gibbs (1951) suggested that loess be subdivided into three categories according to their grain size distribution (Figure 2.30), namely: sandy loess, silty loess, and clayey loess. The large majority of loess deposits fall into the silty loess category. Data collected by Holtz and Gibbs (1951) showed most loess has liquid limits ranging from 28 to 34 percent and plasticity indices ranging from 5 to 12 percent.

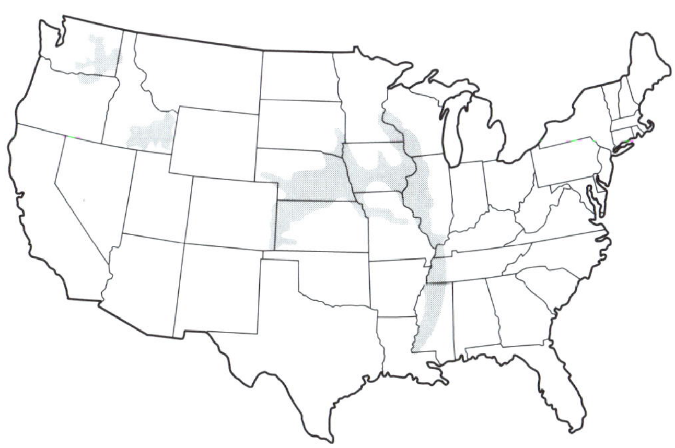

Figure 2.29 Loess deposits in the United States (after Turnbull, 1965).

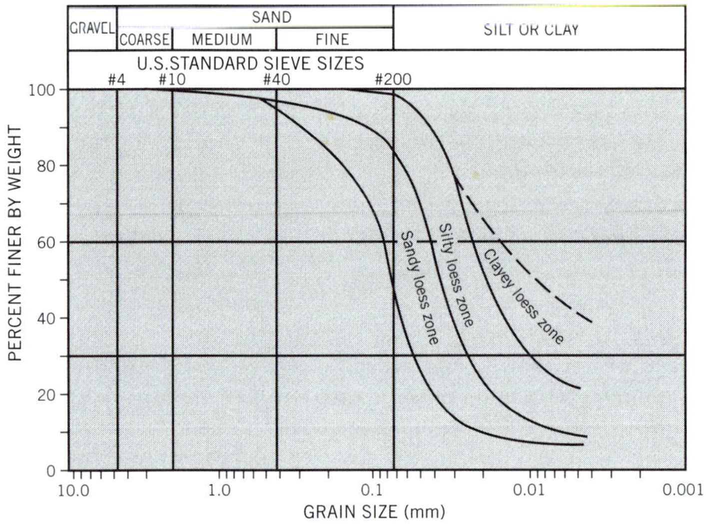

Figure 2.30 Categories of loess (after Holtz & Gibbs, 1951).

Kane (1968) introduced the concept of *critical water content* as the water content at which the clay binder in loess becomes saturated. It typically ranges from 17 to 20 percent (Higgins and Fragaszy, 1988). Below the critical water content, the clay binder shrinks and develops increasingly negative pore water pressures, thereby increasing the undrained shear strength. A good summary of laboratory research results on loess is presented by Higgins and Modeer (1996).

The fragile structure of loess makes it difficult to sample for laboratory strength tests. Block sampling from pits or slopes is the best technique, and is feasible because loess has comparatively uniform properties with depth. An alternative is Pitcher sampling. Transportation of samples from site to laboratory also has to be handled with extreme care.

Landslides in Loess Slopes

Landslide experience in the United States is that loess causes relatively shallow shear failures. Many of the slope movements are caused by erosion. Therefore, prevention or remediation of landslides in cut slopes is focused on methods of eliminating or treating erosion. Excluding surface erosion, the following types of failures are mentioned by Higgins and Modeer (1996):

- Failures due to shrinkage cracks in clayey loess where vertical cuts fail as blocks or slabs of soils.
- Freezing damage in vertical cuts of silty loess is caused by the formation of an ice lens at a shallow depth within the slope, followed by thawing and failure along the resulting surface of weakness.
- Thaw in late winter/early spring, causing an upper layer of loess to become saturated over frozen or unsaturated soil; the thawed soil then slides or flows downhill.
- Cleaning out roadside ditches can locally undercut the adjacent slope, causing slumps of the oversteepened loess.
- Cuts that intercept the groundwater table.

Once a shallow failure occurs and is left untreated, erosion gullies and sinkholes can increase the volume of disturbed ground fairly rapidly.

Although loess formations have the same geologic origins throughout the world, the composition and thickness varies from place to place, and the local climates differ. It is known, for example, that there have been major catastrophes in China due to the failure of loess soils. Nevertheless, the same principles of prevention and treatment should apply everywhere, namely (i) prevention of erosion and (ii) prevention of saturation/collapse from water acting on unsaturated loess below the critical water content. The latter is undoubtedly the key requirement in the avoidance of most major landslides and mudflows involving loess soils. Drainage measures to keep water away from the slope are the principal ways of combating both surface erosion and saturation/collapse.

2.19 HIGHLY SENSITIVE SILT AND CLAY

Highly sensitive soils are known as *"quick"* clays and occur in Canada, Norway, and Sweden. Their remolded shear strength is a small fraction of their undisturbed shear strength and has caused spectacular large flow slides.

Original studies of these clays were performed at the Norwegian Geotechnical Institute and were first described by Bjerrum (1954). Norway was covered by large ice masses during the Pleistocene era. These ice masses depressed the ground and scoured out the residual soil. When the glaciers retreated, the soil and crushed rock were sedimented on the sea floor as *marine* silt and clay. The land uplifted and brought the marine clays above present-day sea level. However, the upper soils around Oslo were deposited in *fresh* water. Groundwater fluctuations produced a stiffer crust at and near the ground surface.

After being uplifted, the Norwegian marine clays were subjected to leaching from groundwater, which has reduced the salt concentration in the pore water. This decreases both the liquid limit (LL) and plasticity index (PI) of the soils. The clays are relatively "inactive" i.e. the plasticity index is low relative to the clay fraction (defined as less than 0.002 mm parti-

Table 2.3 Average Properties of Some Norwegian Marine Clays*

Salt Concentration	Number of Sites	W	LL	PL	PI
> 10 gm/liter	3	43	53	24	29
< 10 gm/liter	5	47	33	20	13

*Bjerrum, 1954

cle size), the latter typically ranging from 45 to 65 percent. This decrease in "activity" appears to be directly attributable to the change in salt content from 35 gm per liter in the sea to as little as 3 to 5 gm per liter in some leached clays. The index properties do not significantly change until the salt content falls below about 10 gm per liter. The average properties from various sites studied by Bjerrum (1954) are given on Table 2.3.

The principal clay mineral is illite, which ranges from 25–40% of the sample and 45–100% of the clay fraction. Minor constituents are quartz and feldspar. Illite is also the clay mineral in Canadian quick clays (Quigley, 1980).

It can be seen from Table 2.3 that marine clays containing less than 10 gm of salt per liter have a much lower average PI than clays with a higher salt concentration. Moreover, the average natural water content of the salt-deficient clays is well above the average LL, unlike the high salt clays.

Laboratory studies have shown that leaching significantly reduces the *remolded* shear strength but creates only a small reduction in the *undisturbed* shear strength. Thus, the sensitivity (i.e., ratio of undisturbed/remolded strength) increases dramatically from around 6 to more than 60. Some quick clays have sensitivities of up to 500. The very low remolded strength accounts for the flow conditions that are characteristic of slope failures in quick clay soil.

Landslides in quick clay often begin as a modest cut or excavation through the upper crust of clay. The initial slope failure can quickly enlarge by retrogressive sliding as the flowing mud passes through the opening created by the initial failure. A "bottleneck" opening is sometimes created with a large bowl behind it (Chapter 1, Figure 1.3).

Some very large landslides spurred the need to fully understand their causation. Movies taken of the mud "river," often with houses riding on the more rigid crust, have been seen by many geotechnical consultants. Some of the more damaging slides include:

- Lemieux Slide, Ontario, Canada, 1993. Involved 3 to 4 million cu. yds. Debris flowed about 1 mile upstream and 1.1 mile downstream in the South Nation Valley (Brooks et al., 1994).
- Vaerdalen Slide, near Trondheim, Norway, 1893. Involved 72 million cu. yd and about 1.1 sq. mile of land; destroyed 22 farms, killing 111 people (Holmsen, 1953). The principal mass went out like "rumbling thunder"; the sliding went to depths of up to 110 feet and movements continued for about 30 minutes.
- Gauldalen Valley Slide, near Trondheim, Norway, 1345. Probably the largest flow slide ever recorded in Norway. The slide dammed the Gaula River and flooded the valley. After the clay debris dam failed, 48 farms were destroyed and 250 people lost their lives (Helland, 1894).

Other quick clay slides occur as a single event with no retrogression or follow-on failures. An example is the Bekkelagel Slide near Oslo of October 17, 1953. An area of 4 acres and about 110,000 cu. yd of slide mass moved out in about 15 to 20 seconds, according to eye-witness statements. It moved the main road and a railway almost 500 feet downslope. In the critical parts of the subsurface profile, the salt content was only 2 to 3 gm per liter.

There is no apparent correlation between heavy rainfall and landslide occurrences (Holmsen, 1953). However, river erosion and manmade cuts and fills have been responsible for the initial slope failure that precedes the flow slide phase.

CHAPTER

3

Field Investigations

3.1 SCOPE OF SITE INVESTIGATIONS

The scope of a site investigation depends on the size and complexity of the landslide under study. For a small landslide on which the cause is easily understood, the scope may require only a site inspection followed by a brief letter report. At the other end of the scale (e.g., a large landslide controlled by geological conditions at depth), the scope could include multiple borings, instrumentation, test pits, adits, etc. in determining the cause, nature, and most appropriate treatment of the landslide. Site investigation borings, laboratory shear tests, and analyses are expensive to perform. In many cases, landslide studies can exceed $100,000 (year 2003 costs in United States) and be a relatively high percentage of the construction cost to remediate the landslide. Therefore, landslide studies should not be judged in terms of some fixed percentage of construction cost as, for example, is a common practice in using the services of an architect for a building design. Field investigations and laboratory tests are a necessary part of determining an appropriate remediation for a landslide.

The first consideration is to understand the cause of the landslide, then obtain sufficient data to properly model it, and finally determine the options that best suit the situation revealed by the data collection. These options are usually focused on different methods of remediating the landslide—that is, achieving stabilization with a sufficient allowance for uncertainties (usually measured by factor of safety) to be reasonably sure that it will remain permanently stable. Different stabilization methods have different levels of assurance of success. For example, a tied-back soldier pile wall can be built to provide a high factor of safety and carries a high probability of permanent stabilization. It also carries a relatively high construction cost. By contrast, a mitigation scheme using horizontal drains requires a more modest investment. Horizontal drains depend on their ability to draw down the groundwater profile, and provide a modest increase in factor of safety. If horizontal drains are unable to adequately lower the groundwater, they have been wasted money, and this may not become known until many drains have been installed. Therefore, in terms of assurance of success prior to construction, horizontal drains are a riskier proposition than a soldier pile wall.

The geotechnical engineer's mission is to provide the most appropriate solution to the owner's landslide based on technical benefits, cost, constructibility, environmental constraints, property ownership, and public safety. All these factors should be considered before selecting a remedial treatment. Other alternatives to full remediation are appropriate for particular instances: maintenance, avoidance, selective stabilization, and marginal stabilization. These alternatives are discussed in Chapter 14.

Site investigations provide the facts on which the technical solutions and costs can be quantified. These in turn allow appropriate decisions to be made.

For a medium to large landslide, site investigations can be subdivided into three phases: preliminary, general, and specific.

Preliminary. This phase includes the site visit, information collection, and planning of a study program.

General. This is usually the key study, and involves the collection of data to understand and model the landslide. The most important requirement is to put a line of borings through the center of the landslide to determine the geological conditions, measure the depth of slippage, measure groundwater levels near the slip surface, and collect samples for laboratory testing of classification and strength. Other borings can expand this information to other parts of the landslide and occasionally outside the landslide area. In wide landslides, one or two additional lines of borings may be needed to properly model the landslide.

At the conclusion of the general site investigation, the causation should be understood. Various remedial options can be examined together with conceptual construction costs of the preferred choices.

Specific. A third phase of field investigations may be required to provide more details of the subsurface conditions for the selected remedial solution. As examples, this could be (i) a line of borings where a wall is to be built, (ii) a program to confirm high groundwater levels in part of the landslide, or (iii) confirming the lateral extent of a soil or rock formation that plays a key role in the proposed remediation. The third phase site investigation also can provide additional data to improve the second phase objectives, such as special sampling, additional groundwater data, or extra borings to complete the landslide model.

3.2 PRELIMINARY SITE INVESTIGATION

The preliminary work consists of collecting any existing data about the landslide area and a field visit (site reconnaissance).

Data Collection

The following information should be obtained, if it is available: previous site investigation reports; past construction activities on the property; topographic plans of the site; aerial photos (stereo pairs); oblique photos of the slide from the air; relevant photos of site development or landslide event; regional geology maps; correspondence pertaining to the landslide site, including prior landslide and maintenance records; and newspaper accounts of the landslide event.

Photographs are useful, especially if the landslide debris has been removed. Old site investigation reports need to be complete, including pullout maps, oversize drawings, tables, boring logs, and laboratory data. The text by itself has limited value.

Site Reconnaissance

The site visit is the most important part of the preliminary work because it is the earliest visit to the landslide site (often within hours of the event) and provides the opportunity to obtain photographs and descriptions of the conditions before they are changed. Even if the landslide occurred months or years earlier, it is still the first opportunity to assess the situation and should be used to collect as much information as possible.

The author has reviewed many files of other geotechnical practitioners and is often surprised by the casual manner in which the preliminary site reconnaissance of a landslide has been conducted. In many instances, a group of engineers and geologists will visit the site, write down some observations, take a few photographs, and discuss possible ways of stabilizing the ground. This information is then summarized by an internal memorandum of about one page, sometimes accompanied by sketches. Most of the information that could have been obtained is never taken.

To fully observe the conditions, engineers and geologists should spend at least two to three hours at the site of a small landslide and up to eight hours at a large landslide. In general, two people should participate in the site visit, one being the project engineer or geologist for the study. Two people can take field measurements more accurately than one person, and it provides a measure of safety on steep and/or dangerous situations. On small landslides, a single person may perform the work.

Landslides occur in a wide variety of circumstances, and it is difficult to provide advice that is germane to all landslides. However, the more important requirements of a site reconnaissance are:

- Make a plan of the landslide.
- Make a section through the approximate center of the landslide.
- Take many photographs from different viewpoints at the site.
- Quantify observations as much as possible.
- Consider access for future explorations.

Table 3.1 is a suggested checklist for a site reconnaissance.

Site Plan

First, establish a baseline on the landslide, preferably on or adjacent to a road, railroad, canal, or river. The baseline should be linear and can be marked with survey stationing at 25- to 50-foot intervals by plastic cones, wood stakes, stones, or chalk marks. The location of features in the baseline vicinity are obtained by tape measurements from one of the baseline stations or by offsets from any point on the baseline. Ground

Table 3.1 Checklist for Landslide Site Reconnaissance

Suitable clothing, including raingear, gloves, and jacket with multiple pockets
Lightweight, waterproof boots with good grips on the soles
"Write in the rain" paper, graph paper, pencils, pens, erasers, etc.
200-foot nonmetallic tape measure
16-foot pocket tape measure
Survey rod, collapsible and lightweight
Machete to cut through undergrowth
Towels, for drying equipment and hands
Clinometer, for measuring angles and percent slope (or Brunton compass)
Hand level
Distance-measuring binoculars (Bushnell)
Small orange cones (or wood stakes) to mark the baseline or key points on traverses
Brightly colored survey marking tape
35mm camera, with 4 rolls of film 200 ASA, or digital camera
Zoom lens: 28–105mm for a 35mm camera
Geology pick
Sample bags and labels, and sampling shovel
Altimeter on high slides (or GPS)
Change of clothes and plastic containers for soiled clothes and boots
Business cards (to gain access to private property)

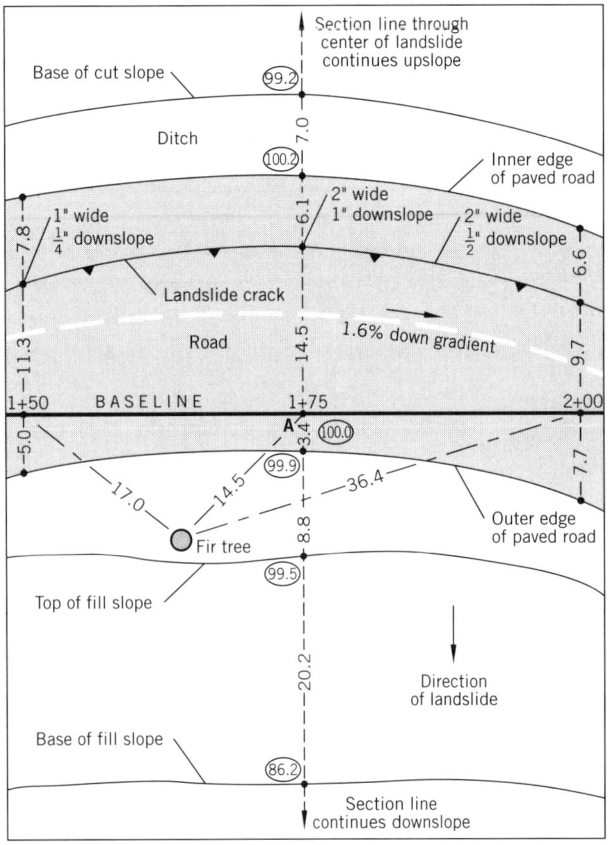

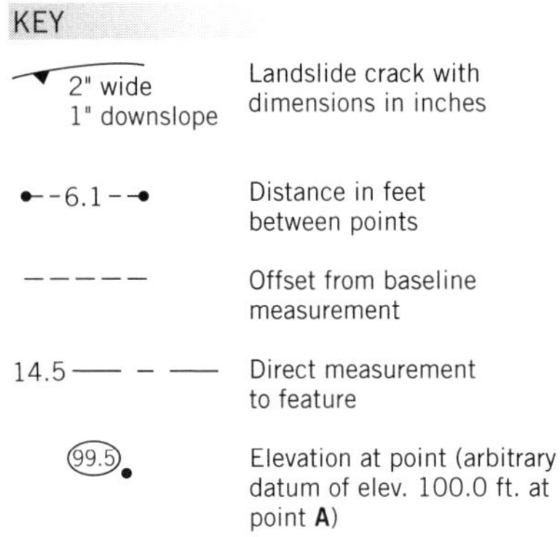

Figure 3.1 Partial field reconnaissance mapping of a landslide using a reference baseline on a road.

level readings can be taken by (i) a hand level, (ii) theodolite and survey rod, or (iii) tape and clinometer reading of grade (as described later). These levels can be related to an arbitrary datum set at some point on the baseline.

These methods are illustrated on Figure 3.1, which shows *part* of a site plan for a landslide affecting a road. The landslide is 300 feet wide and the baseline is made 350 feet long so that it extends past the edges of the landslide (not shown in the figure). Stations are marked along the baseline at 25-foot intervals, and an arbitrary datum level of elev. 100.0 feet is assigned to point A at the mid-length of the baseline. Offset measurements from baseline stations are used to map the curvature of the road in plan. A large fir tree is located by direct taping from three of the station points. Details are also collected for a section passing upslope/downslope through the center of the landslide at A (Sta. 1+75). Landslide cracks in the road surface are mapped, and the width/vertical offsets are noted. Cut and fill limits are also measured.

It usually takes many sketch sheets, similar to Figure 3.1, to complete a site reconnaissance map for a landslide. Back at the office, the field data are compiled into a plan and (at least) one section through the landslide. It is, or course, recognized that the accuracy is insufficient for later construction drawings, but the plan and section provides the information needed for determining causation and planning site explorations.

Field measurements are time-consuming but valuable because the road, railroad, canal, or river is often the facility requiring protection from the landslide movements. Many landslide practitioners do not make these measurements because they (or the owners) will eventually hire a surveyor. However, surveyors may omit details that are significant to a landslide study. Furthermore, it may be weeks before the survey is completed. For this reason, it is strongly recommended that project engineers/geologists prepare a preliminary plan and section. It provides them with firsthand knowledge of the landslide conditions and allows the information to be obtained in a timely manner. A site plan prepared by a licensed surveyor will be needed for preparing a remediation contract. There is usually very good agreement between the surveyor's map and the more crudely measured plan obtained from a site reconnaissance.

Measuring Height and Distance from a Baseline

The calculation of a cut height or cliff on the uphill side of a *level baseline* (road, railroad, etc.) can be determined by the methods shown on Figure 3.2. The percent slope method is the easiest, but some observers may prefer to measure angles to the horizontal. A handheld clinometer measures both "angles" and "percent slope." Note: The tangent of the slope angle is the same as percent slope.

The results can be obtained by graphical plots or calculated by formulas. It is recommended that the horizontal distance be taped rather than paced. If distance x (Figure 3.2) is wide enough, sets of readings taken at three locations are preferable to only two locations. However, standing in the middle of a road requires care (i.e., a second person to watch for traffic).

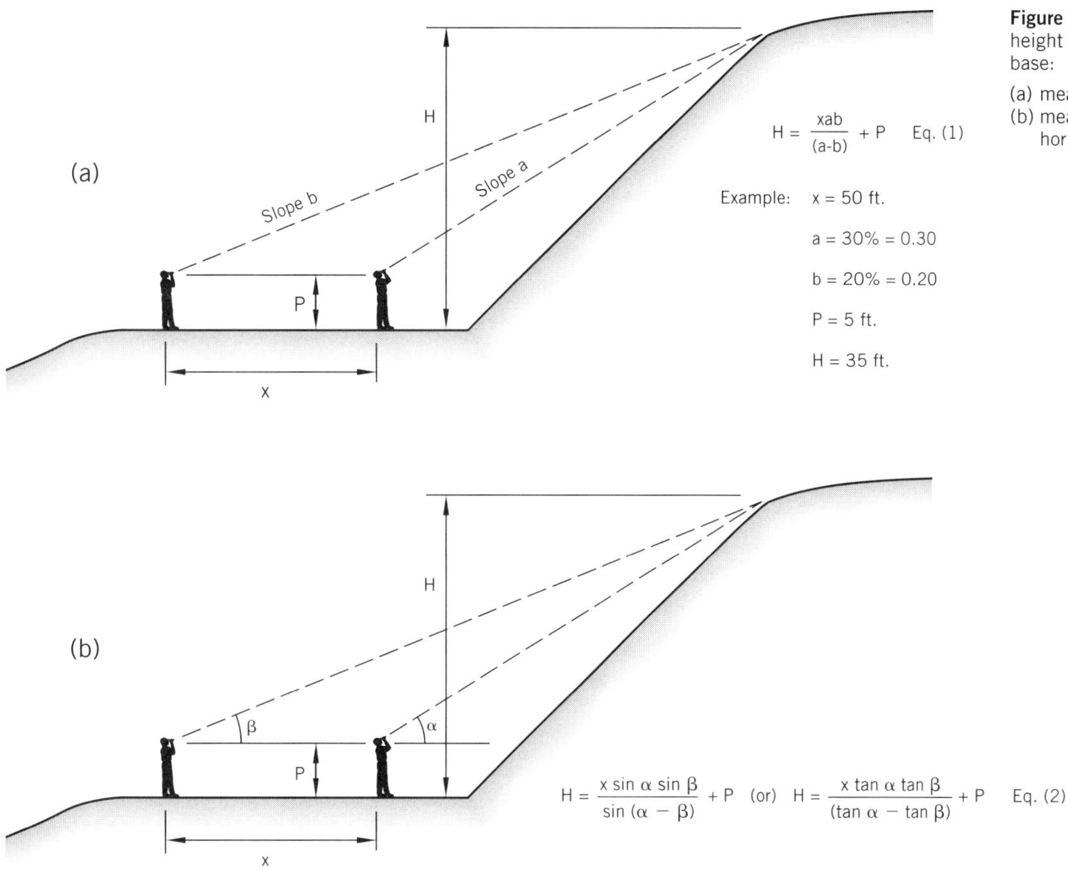

Figure 3.2 Determining the height of slopes from a *level* base:

(a) measuring slope percent
(b) measuring angles to the horizontal

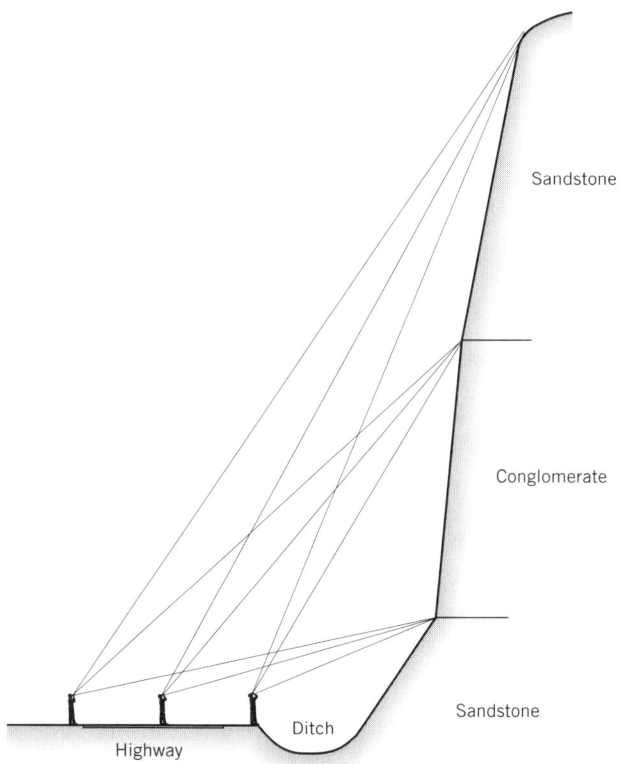

Figure 3.3 Clinometer mapping of a steep rock cliff.

The actual data measured by clinometer readings at the steep cliff of Case History 12 are shown on Figure 3.3. The three observer positions give good agreement on the contacts between the bedrock units. At this section, the cliff is 122 feet high above the road. Before making the calculations, the engineer's guess was that the total height of the slope was only about 70 feet. This significant underestimate of the actual height of a steep cliff at close range has occurred at other sites, confirming the desirability of measuring (rather than estimating) the height of slopes.

The calculation of a cut height or cliff on the uphill side of *sloping ground* can be determined by the method shown in Figure 3.4. In this case, clinometer readings are taken of both the top and bottom of the steep slope. In general, the line of sight to the bottom of the slope will be down (i.e., below the horizontal). However, the set of readings taken furthest from the slope may require lines of sight that are above horizontal for both the top and bottom of the slope, as shown in the example of Figure 3.4; in this case, the measured angle (or percent slope) of the reading to the bottom of the slope is negative in the equations (3) and (4).

The height measurement can be performed graphically. The accuracy of graphical construction decreases as the angle A' DB' (Figure 3.4) becomes more acute.

The height of the observer's eye is not required in the equation to calculate slope height. The base slope angle θ is measured by taking an eye-to-eye slope measurement on the

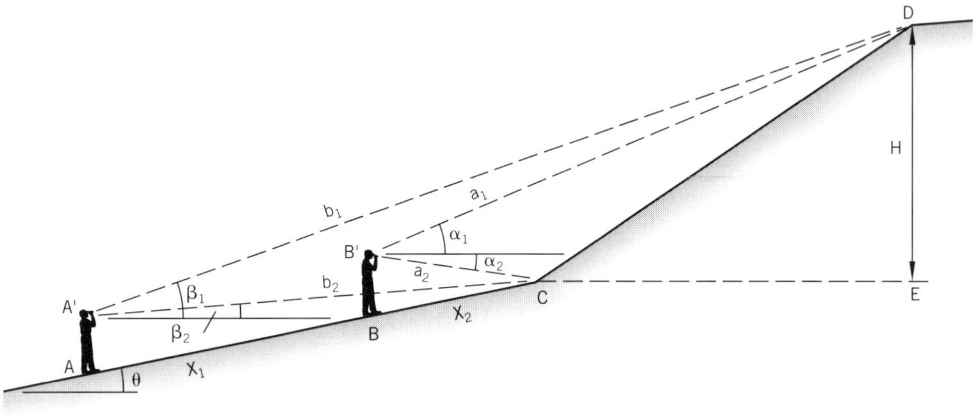

Clinometer can measure both "angles" and "percent slope".

Observer stands on slope inclined at θ to the horizontal. Observer reads top and bottom of slope using clinometer.

Observer locations A, B; $AB = X_1$; $BC = X_2$ measured along the slope

Required vertical height of slope H = DE

If reading angles, read $\alpha_1\ \alpha_2\ \beta_1\ \beta_2$

If reading percent slope, read $a_1\ a_2\ b_1\ b_2$

Note: If observer at A is reading uphill to the bottom of the slope C (as shown in figure), enter negative values in the equations for b_2 or tan β_2.

Angles to horizontal method

$$H = \frac{\cos\theta\,\{(X_1 + X_2)(\tan\beta_1 + \tan\beta_2)\tan\alpha_1 - X_2(\tan\alpha_1 + \tan\alpha_2)\tan\beta_1\}}{(\tan\alpha_1 - \tan\beta_1)} \qquad \text{Eq. (3)}$$

Percent slope method (use fractions in calculation)

$$H = \frac{\cos\theta\,\{a_1(b_1 + b_2)(X_1 + X_2) - b_1(a_1 + a_2)X_2\}}{(a_1 - b_1)} \qquad \text{Eq. (4)}$$

Example $a_1 = 0.47$; $a_2 = 0.19$; $b_1 = 0.35$; $b_2 = -0.02$;
$X_1 = 30$ feet; $X_2 = 16.3$ feet; $\theta = 7°$ to horizontal

$$H = \frac{(0.993)(0.47)(0.33)(46.3) - (0.35)(0.66)(16.3)}{(0.12)}$$

= 28.3 feet (28 feet)

The slope height can also be obtained graphically.

Note: For angle θ of less than 10° (the usual case), cos θ is approximately 1.0. Assuming cos θ = 1 in the above example, the calculated H is 28.5 feet; in such cases, cos θ can be omitted from the equations.

Figure 3.4 Determining the height of slopes from a *sloping* base.

second person of the field crew (allowing for personal height difference).

Finally, there is a redundancy in the calculation because ground slope AC must intersect with the two lower lines of sight made by the observer at locations A and B. If it is impossible to reach the base of the slope or cliff due to a ditch, fence, or dangerous situation, the length X_2 can be determined from the intersection of the two lower lines of sight taken from locations A and B.

Other ways of measuring slope and distance from a baseline area are illustrated in Figure 3.5. The slope of a fill or the slope of the ground above a cut can often be measured by standing in line with the extension of the slopes (at D and U on Figure 3.5). This provides a direct measure of slope angle.

A recently-developed tool for measuring slopes is the hand-held laser ranging instrument. An example is the Impulse LR instrument manufactured by Laser Technology, Inc., Englewood, Colorado. This device measures slope distance, horizontal and vertical distances, cumulative and difference distances, heights, and inclination (degrees). The Impulse LR can measure distances of up to 1880 feet (575 m) and has resolution of 0.05 feet. The stated accuracy is +/− 0.1 degree.

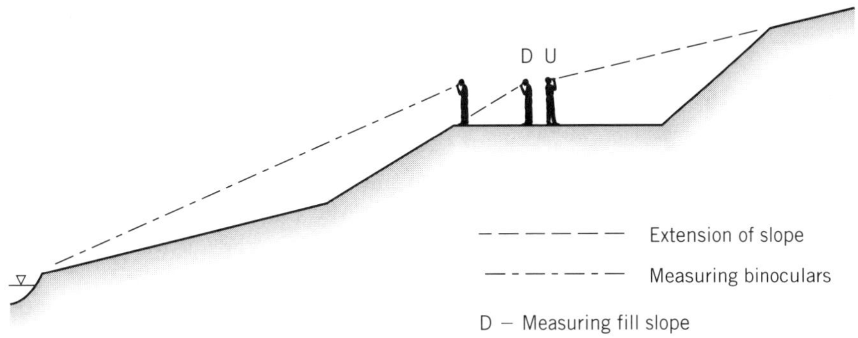

Figure 3.5 Other methods of measuring slope and distance.

The cost (year 2004) is around $ 2,500. The instrument can be mounted on a tripod.

Another hand-held laser ranging instrument has been developed by binocular manufacturer Bushnell specifically for golfers. It is much simpler and less expensive than the Impulse LR described above. It measures distances of 20-400 yards, measuring to the nearest yard. It can be used in conjunction with a clinometer.

Using these instruments, a geotechnical practitioner can measure the distance to breaks in slope, trees, houses, or other objects on a landslide. These measurements can later be tied into a cross-section prepared from an uphill/downhill traverse. It is desirable to take readings of the distances to the top and base of the landslide (such as a riverbank or street).

Once the baseline area has been mapped, the landslide plan can be compiled from available site topography, aerial photographs, and lateral offsets taken from one or more traverses going up and down the landslide. In the case of a long narrow landslide, one traverse through the center should be sufficient. For wide landslides, several traverses may be needed to provide a good site plan.

Table 3.2 relates slope angle, percent slope, and the horizontal : vertical ratio. This table is useful for a variety of purposes in landslide analysis. For example, if a soil has an angle of repose of 35°, the outer slope will be at 1.43 horizontal : 1 vertical. Conversely, measuring the top and base of a dry talus slope of granular soils by the methods shown on Figures 3.2 or 3.4 can be used to calculate the angle of repose of the talus.

Section through the Landslide

Measuring a cross-section along the direction of landsliding can be dangerous work. The acceptable level of personal risk has to be based on the judgment of the field personnel. However, a measured cross-section is valuable for accurately observing ground cracks, scarps, seeps, etc. and provides the project engineer/geologist with a "feel" for the site that is difficult to experience when there are only topographic maps available.

Table 3.2 Relationships Between Slope Angle, Slope Percent and Ratio of Horizontal : Vertical (H : V)

Slope Angle	Slope %	H : V	Slope Angle	Slope %	H : V	Slope Angle	Slope %	H : V
5	8.7	11.43	22	40.4	2.48	39	81.0	1.24
6	10.5	9.51	23	42.4	2.36	40	83.9	1.19
7	12.3	8.14	24	44.5	2.25	41	86.9	1.15
8	14.1	7.12	25	46.6	2.15	42	90.0	1.11
9	15.8	6.31	26	48.8	2.05	43	93.3	1.07
10	17.6	5.67	27	51.0	1.96	44	96.6	1.04
11	19.4	5.15	28	53.2	1.88	45	100.0	1.00
12	21.3	4.71	29	55.4	1.80	46	103.6	0.97
13	23.1	4.33	30	57.7	1.73	47	107.2	0.93
14	24.9	4.01	31	60.1	1.66	48	111.1	0.90
15	26.8	3.73	32	62.5	1.60	49	115.0	0.87
16	28.7	3.49	33	64.9	1.54	50	119.2	0.84
17	30.6	3.27	34	67.5	1.48	55	142.8	0.70
18	32.5	3.08	35	70.0	1.43	60	173.2	0.58
19	34.4	2.90	36	72.7	1.38	65	214.5	0.47
20	36.4	2.75	37	75.4	1.33	70	274.7	0.36
21	38.4	2.61	38	78.1	1.28	75	373.2	0.27

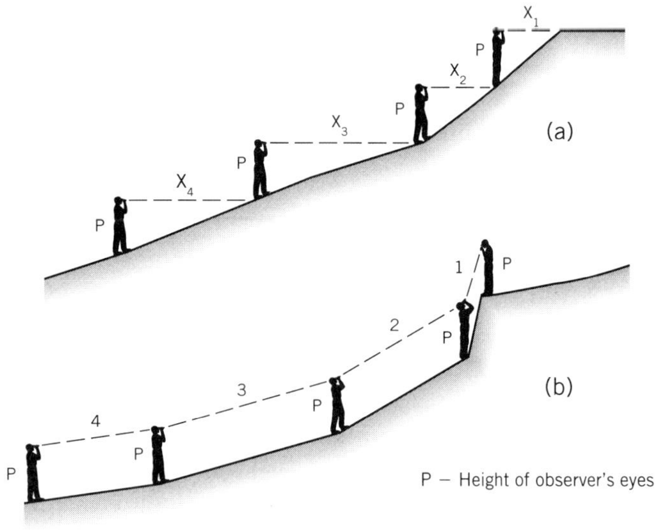

Figure 3.6 Slope traverses:
(a) one-person step measurement;
(b) two persons

Measuring Height and Distance on a Slope

Two common methods of obtaining a slope profile are illustrated on Figure 3.6. A one-person survey is done by a series of steps using a hand level (Figure 3.6a). Starting at the top of the slope, a stake can be driven into the ground and one end of a cloth measuring tape is attached to it. The person goes down the slope until the ground level of the stake is at eye level. The horizontal distance is measured and a second stake is installed at the observer's feet. The first stake is later removed. The process is repeated down the slope.

There are many variations of the procedure. A lightweight surveyor's rod can be used to measure the horizontal distance, which avoids the necessity of continual staking. It is also possible to lay the tape along the slope and measure angles to breaks in the slope with a clinometer; stakes are used only at the slope breaks. However, undergrowth usually prevents this method from being used on landslides in wet climates. The one-person step measurements are slow and tedious, and the method is only practical for small landslides.

A two-person team usually profiles a slope by the method illustrated on Figure 3.6(b). A cloth tape is held at hand or chest level by each person (with allowance for personal height differences). The two-person team moves up or down the hill, taking measurements at slope breaks, or places of interest such as ground cracks, seepage areas, or landmarks (e.g., large trees). One person takes percent slope or angles readings with a clinometer. The results are plotted on graph paper later.

Although these methods appear to be rather crude, they have frequently been checked by professional surveying and generally show very good agreement. Any landslide features are noted during the traverse.

Traverses often have to pass through dense undergrowth. If this type of ground is anticipated, it is advisable to set out the traverse lines in advance and have workers cut a path down the landslide.

Global Positioning System

Global Positioning System (GPS) handheld instruments are now widely available and have very good accuracy. They are especially useful for larger landslides. A sketch plan can be made of the landslide, marking key points from which GPS measurements are taken.

Photographs

Photographs should be taken from several viewpoints. Close-ups should be taken of important landslide features plus typical exposures of soil and rock.

Landslides can be difficult to photograph because a panorama of the slide may cover an arc of 120° or more from a single vantage point. If a very wide angle lens is used, the observed features become very distorted, especially at the edges of the photograph. To obtain a panoramic view, it is preferable to take multiple photographs using a lens that is comparable to the human eye for field of view (i.e., 50mm lens for a 35mm camera). By photographing in the portrait position (camera turned through 90°) and allowing about 15% overlap at the edges, a panoramic photograph can be produced by overlapping the photographs with acceptable fit at the edges. Where it is difficult to achieve a good overlap fit, such as a head-on view of a nonvertical slope, the better fit should be made on the more important parts of the photograph. It is generally better to produce a rectangular panoramic photograph from a series of individual photographs than a curved photograph that is obtained by trying to get good fit over the entire edge length of individual photographs.

Photographs should be mounted on heavy stock paper (110 lb.—199 gm/sq.m or heavier). Information should include: project (usually name and job number), date, direction of view, and brief description of the photograph. It is recommended that all sets of negatives be kept in a central file (such as a series of index file boxes) with a sequential number for each set of negatives, date taken, name of project, and job number. A summary file of the films provides a rapid means of recovering negatives when needed.

Digital cameras are replacing cameras with film. Digital photographs have the advantages that (i) the results can be viewed on site, (ii) the photo can be relayed to the client or office engineer/geologist by e-mail shortly after they have been taken, and (iii) panoramic photos, which are very useful for landslide work, can be manipulated by computer software to provide a nearly seamless contact between adjoining photo images. A potential disadvantage of digital photography is that the images could be lost unless they are routinely transferred to a company storage device in a systematic manner and are properly indexed to the project.

Oblique aerial photographs can be very useful for large landslides or landslides covered by dense trees or undergrowth. The author has flown over numerous landslides, including many featured in the case history section of this

Table 3.3 Checklist for Landslide Observations

Headscarp: height, angle of ingress, soil exposed
Cracks on the slide: location, length, vertical offsets, horizontal widths (note reverse scarps and cracks above the headscarp)
Toe of slide: describe heaving ground or other disturbance due to landslide
Springs: location, estimate flow
Seepage areas: quantify the affected area
Manmade features: measure roads, houses, power poles, culverts, walls, fills and cuts, pipelines, ditches
Rock and soil types exposed on the landslide: bedding and joint planes (if relevant to the slide); description of soils, including stiffness
Presence of ravines, streams within or adjacent to the slide
Trees: distressed orientations, or curved conifer trunks
Slope gradients above and below the landslide
Tilt of power poles and tautness of overhead lines
Structural distress around houses: separation cracks; tilting; cracks in brickwork, concrete walls, patio; distortion of doors, windows, and overhead porches
Information from local residents and property owners
Check access for drill rigs including utility line locations and overhead lines, other obstructions
Determine north direction of the site

book, to observe and photograph landslides from the air. Most of these flights have been made in a small, single-engine plane with a professional pilot. The site should be first evaluated to ensure that it is photogenic and reasonably accessible to filming from the air. North-facing landslides are difficult to photograph, as are slides within narrow ravines.

On reaching the site, the plane is slowed down and a side window is opened to provide unobstructed shots. Several circles are made at different heights. In general, the better angles range from about 20° to 45° to the horizontal with the sun behind the camera. Cloudy or overcast days should be avoided; clouds cast shadows to produce a spotty effect and overcast skies produce dull color. It is advisable to check the orientation of the landslide so that the flight occurs during a favorable sun position. Haze and polarizing filters are also useful accessories for flight photography.

Helicopters also can be used for aerial photography. They vibrate more than a fixed-wing aircraft and are a greater safety risk, especially when hovering.

Observations

The suggested checklist on Table 3.3 is a useful guide to ensure that information is not overlooked during the site reconnaissance. Many of these items will not be relevant to a specific site. Except for the baseline measurements and cross-sections, most of the other information can be quantified by walking (paces); these include offset lines from the uphill/downhill traverses or the measurement in plan of a house within the landslide.

It can be argued that too much information is collected by these methods and much of it may be irrelevant. However, collecting too much information is better than collecting too little, especially when the site is remote and distant from the engineering office. It is also surprisingly common that seemingly small details become significant as the investigation progresses.

3.3 GEOLOGICAL MAPPING

On most large landslides and landslide complexes, the role of an engineering geologist is essential to achieving a successful outcome of the study. The series of landslides affecting the Cromwell Gorge in New Zealand is an excellent example; another is the coastal landslides of England. The geologist brings the broad framework of the geological environment into the study, allowing the landslide to be seen as part of this environment. In some cases, a geologist can help to set the probable boundaries of an active landslide within a much larger area of similar geology when the actual boundaries are not clearly defined by broken ground. In other cases, the work of the geologist can provide a rational interpretation of a complicated drill log and may, for example, determine whether bedrock has been reached or not reached.

Geological mapping generally differs from the mapping described earlier under Preliminary Site Investigation. The geologist's work almost always extends into a much broader area than the immediate landslide (or landslide-prone facility). This broader look is intended to gain a sufficient understanding of site geology.

For a large landslide, the geologist may spend two to three days in the field examining slopes and outcrops around and within the slide area. The geologist also would review available geology maps, dissertations, and other published and unpublished geological studies of the area. If available, aerial stereo pairs would be examined to see landforms. During the course of a study, the engineering geologist should work closely with the geotechnical engineer to ensure that the landslide interpretation is consistent with the site geology.

Another role of an experienced engineering geologist is the prevention and/or avoidance of landslides. This group of issues includes: (i) mapping new routes for highways, railroads, canals, and pipelines, (ii) designing new tunnels or expanding existing tunnels, (iii) mapping ancient and active landslides for cities to prevent development in unsuitable areas, (iv) identifying geological hazards on lifeline projects (such as water conduits) in earthquake-prone regions, and (v) examining the potential for rockfalls along highways.

3.4 TOPOGRAPHY

An accurate topographic map is essential to landslide analysis and preparation of documents for remediation. These maps can be obtained by land surveying, bathymetry (underwater surveys), or by photogrammetry (aerial surveys).

Land Surveys

In the United States, land surveying is provided by Professional Land Surveyors, registered by the state. You can be reasonably assured that changes in slopes and manmade features will be accurately recorded. However, it is important for landslide practitioners to make sure that the surveyor understands the landslide information that needs to be collected, such as the top and base of scarps, locations of cracks and scarps within the landslide mass, toe of slide, etc. The quality of the map depends on the skill and equipment of the surveyor.

Bathymetry Surveys

Underwater surveys are obtained by a recording fathometer on a boat. The fathometer sends out pulses that are reflected off the ocean or lake bottom and converted to depths. The position of the boat at each reading is determined by Global Positioning System (GPS). Time is recorded concurrently. The boat makes a grid of traverses across the water and the recording equipment shows the prior traverse lines so that the boat operator does not repeat earlier work. A transponder on land is needed to provide a high level of accuracy to the GPS data.

For ocean surveys, the basic data have to be corrected for tidal variations to obtain ground levels relative to a datum, such as mean low low water (MLLW). Corrections can be made for local tidal conditions using a nearby tide gage as the control.

It is likely that unusual underwater ground conditions will be found after a landslide. Some survey ships are equipped with sidescan sonar equipment that can produce a sonograph of the underwater ground surface.

Aerial Surveys

Photogrammetry provides overlapping vertical stereo pair photographs that allows ground contours to be mapped by special equipment. The photographs can also be used by geologists and engineers to examine the landslide. When viewed through stereoscopic glasses, the vertical height differences appear exaggerated, showing the scarps and slope features with excellent relief.

The aircraft flies at a speed of about 125 mph and takes photographs using a 6-inch (150mm) or 12-inch (300mm) focal length lens. The 6-inch lens is more commonly used for landslide mapping. The degree of detail on the photographs depends on the height of the plane above the ground. The lowest practical height is 1,000 feet above ground, which provides a scale of 1 : 2,000. A good rule of thumb is that the scale should be about 10 times the intended map scale. For example, if a landslide is to be mapped at 1 inch = 40 feet with 2-foot contour intervals, the photographs would be taken at a scale of 1 inch = 400 feet (1 : 4,800).

Bright sunshine provides the best contrast for mapping. However, if the site is covered with many trees, shadows cast by the sun make ground mapping difficult; for this situation, a light overcast day provides better mapping conditions.

Ground control is an important part of a photogrammetric survey, and typically accounts for about one-half of the cost. Ground control points (pre-marks) are established around the site using white spray paint or white plastic strips in the form of a + sign with a survey pin at the center of the cross. A surveyor measures the precise location and elevation of each pre-mark. If a landslide area is covered by a stereo pair (two photographs, each 9 inches square) a total of 5 pre-marks is desirable, comprising 4 pre-marks around the perimeter and 1 pre-mark near the center. The photogrammetrist only needs 3 control points to make a map, but the other 2 pre-marks provide redundancy and generally improves accuracy through crosschecking.

All pre-marks have to be clearly visible from the overhead camera. They can be placed on the sides of roads, driveways, open spaces, etc. with permission from owners. The plastic strips are removed after the flight.

Stereo photographs are relatively plentiful for all metropolitan areas, navigable waterways, major roads, rivers, and coastline within the United States. However, it is usually difficult to obtain existing stereo pairs for remote areas. Aerial photographs prior to 1940 are rare. Sources for photographs include the Corps of Engineers, state highway departments, and private firms specializing in aerial photography and/or surveying.

A major problem in photogrammetry is that the ground cannot be seen through dense trees and undergrowth. When these conditions are encountered, the photogrammetrist shows broken lines (form lines) to denote the loss of accuracy. Therefore, in a landslide covered by dense vegetation, ground traverses may be needed to supplement aerial photography. In general, however, a contoured site plan obtained from aerial photographs is the best method of surveying a medium to large size landslide.

New techniques of aerial photography involve Global Positioning System (GPS) techniques. On larger projects, a GPS receiver is connected to the camera and is triggered at the same time as each photograph, giving the precise coordinates of the plane's position at the center of the photograph. However, some ground control is still necessary. A more recent development is to include an Inertial Measuring Unit (IMU), a form of gyroscope that can measure the angle of the camera during flight. Although still not perfected, it offers the possibility of eliminating the need for ground control.

Yet another developing technology is LIDAR, a light detection and ranging device that has a pulsating laser capable of providing 4 to 15,000 pulses per second. The equipment is able to penetrate through a canopy of trees because some of the numerous pulses evade the obstructions and reach the ground.

Another type of aerial photograph uses infrared film, which colors moisture red. It can be used to detect high groundwater levels, seepage, or springs in a hillside. It is occasionally used in landslide studies.

The results of an aerial survey are usually provided to the engineer as a contour map. The data are often digitized and put onto disc for use in AutoCAD or similar computer software programs.

3.5 SURVEY MONITORING

Surface monuments, typically hubs or ⅝-inch dia. iron rods, can be installed on a landslide to supplement information obtained from inclinometers. The monuments can be set out as a line in the general direction of landslide movement or in two lines at right-angles (x and y components) to determine the precise direction of movement in the horizontal plane. The vertical movement (component z) is also monitored. The combination of the resultant horizontal component with the vertical component can provide a good indication of the shape of the failure surface below each survey point. Case Histories 6 and 8 (Part C) are good examples of such use.

To establish the position of the survey monuments, the surveyor establishes at least two control points outside the landslide area to use as a baseline. A light-reflecting prism on a tripod is positioned, in turn, over each of the points to be monitored. A total station survey instrument is set up on the baseline. After the baseline check has been made, angles and distances to the prisms are measured.

The data are read to the nearest 0.001 foot, and angles are read to the nearest second of arc. Although these measurements are within the accuracy of the surveying *instrument*, the accuracy of the *system* is less. Survey measurements probably are accurate to about 0.02 foot (0.25 inch). The accuracy of surveying has improved significantly since the advent of electronic distance-measuring instruments.

Readings need to be taken a sufficient number of times to establish trends. For a slow-moving slide, readings taken monthly or quarterly can be satisfactory for the intended purpose of monitoring the landslide. For fast-moving slides, readings may have to be taken twice daily.

The main disadvantage to using surface hubs is that they are prone to disturbances that are not related to landslide movements. The hubs can be damaged accidentally by construction equipment, animals, and people. They can move from surface creep unrelated to deep-seated movements, especially on clay slopes in winter. There is a risk of vandalism. Extreme weather may also affect the readings, especially frost heave. Perhaps the most important factor is that the readings must be referenced to a stationary benchmark. On steep slopes or very large landslides, it is often difficult to find a suitable reference point.

Survey monitoring is a good investigative tool, and some large property owners (e.g., utilities) have their own equipment and trained personnel to perform the measurements. Readings should be taken by the same team of surveyors to achieve consistency. Case Histories 1, 5, 6, and 8 are examples of how surface survey monitoring supplemented a landslide investigation.

3.6 DIFFICULT ACCESS

Many landslides occur on very difficult terrain at a time of inhospitable weather. Where solutions have to be obtained quickly, the site investigation team has to perform their work with these hardships and potential personal risk. Access to drilling sites is a key requirement.

Steep Soil Slopes

Talus and other steep soil slopes at marginal stability present very difficult access for drilling rigs because even small temporary cuts (to create a level area) may become locally unstable.

One practical solution to this dilemma is to move further upslope to more stable (less steep) ground and drill angled holes into the slope below. Some inclinometer probes can obtain readings at up to 30° to the vertical (see Case History 9: Goat Lick Slide and the Bull Run Access Road Slide described in Chapter 10, Section 10.2 Original Profile Analysis).

For flatter slopes with difficult access, a drill crew can dig out a small level area to erect a drilling platform on scaffolding. A relatively light drill rig can be lifted by helicopter onto the platform. The use of a helicopter significantly increases the cost of the site investigation. A light drill rig may have limited down thrust but holes of up to about 100 feet deep can be achieved. Another option is to build a platform on scaffolding and lower the drill rig down the slope on skids using a large dozer as an anchor at the top of the slope.

On many steep access roads, the drill rig may have to be brought to the site with the assistance of a dozer. The dozer is often kept at the site (on standby) to pull the drill rig out daily or after drilling is finished.

Another option is to use a "spider" backhoe. It has long, articulated outrigger legs that stabilize the backhoe on a steep slope. The backhoe can be used to dig a test pit or a trench up to 15 feet deep, depending on subsurface conditions and slope.

Overwater Drilling

As a rule of thumb, drilling over water typically costs 3 to 4 times the cost of drilling on land for the same amount of footage. Overwater drilling requires good advance planning and preparedness for unexpected ground conditions and equipment breakdown.

The three methods of supporting the drill rig are: (i) a platform on scaffolding; (ii) boat or barge; and (iii) pontoon.

A *platform* or *scaffolding* (Figure 3.7) is the least expensive option but has limited application. The scaffolding has to be erected on dry ground and the maximum distance from the shoreline is governed by tidal range, river stage or, in the case of a landslide within a reservoir, by low water of the reservoir. Typically, the scaffolding and platform is erected and the drill rig is winched down to it from above.

For a *boat* or *barge,* a drill platform is cantilevered over the edge of the boat and a safety railing is built around the outer perimeter. The hole is drilled through an opening within the platform.

A *pontoon* provides an ideal platform for overwater drilling. British practice is to use shallow draft pontoons originally developed by the British Army for rapid deployment. They are supplied in standard units that can be connected to

Figure 3.7 Shell and auger drilling from a scaffold-supported platform.

provide a larger platform. The units are small enough to be transported by road without a special convoy. Drilling is performed, for example, through the center of four units connected together. The pontoons are equipped with winches to anchor the pontoon in strong currents or allow it to move with the tide.

The boat, barge, or pontoon should be as large as possible for stability, although restrictions such as the width of the waterway, other environmental constraints, and cost may dictate the size. A small powerboat or rowing boat is needed for daily access. The position of the borehole can be fixed by taking either sextant readings from the floating platform or by intersecting sight lines taken from fixed locations on the shore. Readings of position can also be obtained by GPS methods. Redundant measurements should be taken to ensure that the position is accurately located while drilling is in progress.

Tidal changes create additional difficulties. The tidal level has to be checked continually during drilling and sampling and anchors require continual adjustments. Undisturbed sampling and field vane tests have to be performed under calm conditions and preferably during a slack water period near high or low tide. Skagway Marine Slide (Case History 7) required drilling through very deep water, large tidal variations, and adverse Alaska winter conditions.

3.7 OVERBURDEN DRILLING

Drilling practices vary widely, and a geotechnical consultant usually has to choose from equipment available in the locality. Most equipment is appropriate for the different soil and rock types likely to be encountered in a particular region. However, competitive bid pricing affects the standards of performance, and rigid contract specifications can stifle initiative and alternative choices. This is especially true in Britain where contracts usually allow no alternative to the ubiquitous shell and auger drilling, a system that is both inefficient and produces poor-quality samples.

Some of the more common drilling methods, and their use for landslides, are shown on Figure 3.8 and are briefly described below. Note that the terms *drillhole* and *borehole* are used interchangeably in this book; some engineers limit the word *drilling* to coring of hard rocks.

Rotary Drilling with Mud

Technique

See Figure 3.8(a). Power rotary drill rig rotates drill rods with a bit, usually a tri-cone, at the lower end. A short length of casing is used to start the hole and a T-junction directs the return fluid into a trough. The filtered mud is then returned to the drill head for recirculation in the drilling process. Bentonite drilling mud is most frequently used. "Revert" is an alternative drilling mud that breaks down after use and allows the hole to be used for piezometer installations.

To sample, the drill rods and bit are taken out of the hole and are replaced with drill rods and a sampler. The sampler is either driven into the ground (for SPT) or is jacked in by the hydraulic head of the power auger.

Record: (i) Loss of drill fluid return, (ii) any chatter of drill rods, (iii) slow rates of penetration

Speed of drilling: Relatively fast, typically about 60 feet per day with sampling at 5-foot intervals

Suitable soils: Soft clay, silt, sand

Advantages

1. Good for site investigations because the soils at the bottom of the hole receive only minor disturbance prior to sampling.
2. Simple to operate and relatively trouble-free drill rig.

Disadvantages

1. Cannot lift pebbles back to the surface with the drilling mud, and such stones may continue down with the hole and be recovered by a sampler at a different level than that at which they occurred.
2. No information is obtained on groundwater levels during the drilling operation.
3. Further penetration can be stopped by compact gravels or stones within landslide debris.
4. Cannot see the soils between sampling except by examining the mud return.

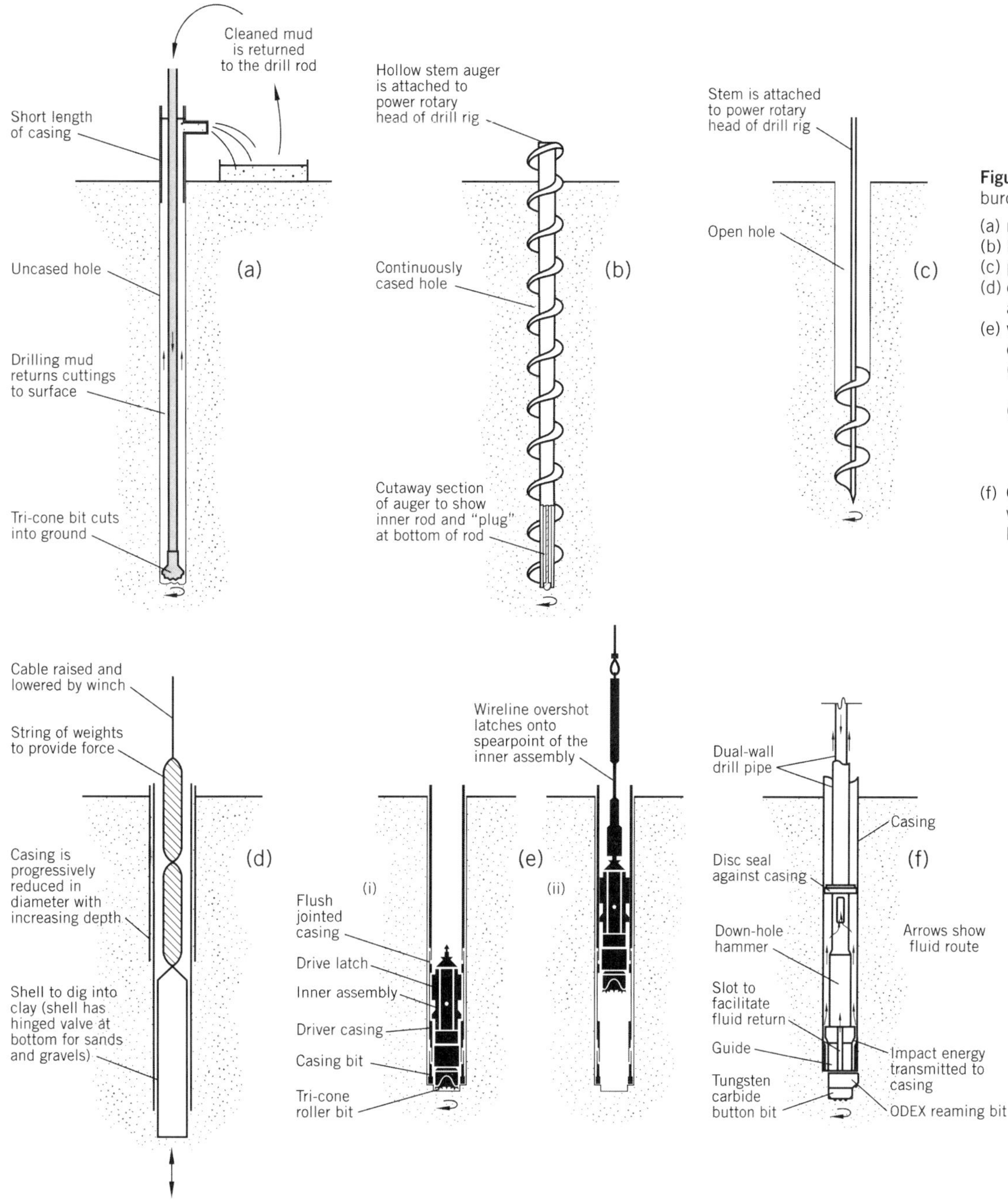

Figure 3.8 Common overburden drilling techniques:
(a) rotary with mud
(b) hollow stem auger
(c) power auger
(d) cable tool (Britain: shell and auger)
(e) wireline drilling using casing advancer
  (i) casing advancer during drilling
  (ii) lifting out the roller bit prior to sampling through the cased hole
(f) ODEX drilling method with down-the-hole hammer

Hollow Stem Auger

Technique

See Figure 3.8(b). Power rotary drill rig rotates continuous-flight hollow stem augers. The augers have a hollow pipe (the "stem" of the auger) with spiral flights welded to the outside. A plug at the lower end of a drill rod is used to prevent soil from entering the hollow stem as the hole is advanced between samples. The 5-foot long augers typically have stems of 8-inch outside diameter and 4-inch inside diameter; other sizes are available. Additional lengths of augers are added as the hole deepens. The hollow stem auger method provides a totally cased hole.

To sample, the central plug is removed and replaced with the sampler attached to drill rods. The sampler is either driven into the ground (for SPT) or is jacked in by the hydraulic head of the power auger.

Record: (i) Any heave of ground into the hollow stem prior to sampling, (ii) any chatter of the drill rods indicating rocks or gravel strata, and (iii) rates of drilling penetration if unusual; slow, fast or sudden drops (indicating voids)

Speed of drilling: Fast, typically about 80 to 90 feet per day with sampling at 5-foot intervals

Suitable soils: Clay, silt, sand; moderately stony soils if shallow in thickness

Advantages

1. Probably the fastest method of drilling for most soils.
2. There is no drilling mud to contaminate samples or groundwater level readings.
3. There is relatively minor disturbances of the ground below the augers if proper drilling techniques are used.

Disadvantages

1. Removal of the plug in water-bearing sands can disturb the sand in the SPT; to prevent this, water should be added inside the stem to create a reverse head of water.
2. A very powerful machine is needed to turn the augers; most hollow stem augers have a maximum range of about 120 to 150 feet and can be "stopped" at shorter depths by boulders or compact rocky soils. Access and leveling of the drill rig can also be problems for these drill rigs on landslide slopes.
3. The soils between samples are not observed.

Power Auger

Technique

See Figure 3.8(c). Power rotary drill rig rotates a drill rod with a short length of continuous flight auger at the lower end. The auger is screwed into the ground and lifted out for inspection of the soils. The cuttings are shaken into a heap beside the hole. Some solid shaft power augers have continuous flights for the full depth of the hole.

To sample, the flight auger is removed from the hole and is replaced with drill rods and a sampler. The sampler is either driven into the ground (for SPT) or is jacked in by the hydraulic head of the power auger.

Record: (i) Groundwater levels, (ii) depths of soil strata changes, and (iii) difficulty of drilling

Speed of drilling: Fast, typically about 80 to 90 feet per day with sampling at 5-foot intervals in appropriate soil formations

Suitable soils: Stiff clay, dense silt and sand, gravels with binder of fines

Advantages

1. Good for examining the full depth of soil conditions and for obtaining groundwater data during drilling.
2. Relatively lightweight drill rig and truck allows easier access to difficult sites.

Disadvantages

1. The drilling depth is usually limited by weak power or caving soils.
2. Caving of the sides of the hole can contaminate samples lower in the hole.

Shell and Auger (Cable Tool)

Widely used in Britain and many other countries with current or past ties to Britain. Although it is known as "shell and auger" boring, the auger is rarely used. It is termed "cable tool" in the United States where it is rarely used because of the slow rate of progress.

Technique

See Figure 3.8(d). A steel tube or shell at the end of a cable is raised and lowered to dig into the ground at the bottom of the hole. A series of jarring links (the "string of tools") above the shell adds weight to the percussive blows. The shell for clay soils is open-ended and the clay is pried out into a heap next to the hole. For sands or gravels, the shell is fitted with a hinged flap valve to retain the soil after it enters the shell. A heavy chisel is used to break up hard rock or boulders.

The casing is driven into the ground to support the sides of the hole. At various depths (typically 40 feet), the frictional resistance becomes too great, and a smaller diameter casing is placed inside the casing already in the ground so that the hole can be advanced. It is not uncommon for four sizes of casing to be used on deeper holes. The casing diameters typically range from 16 inches to 6 inches, and different size shells are needed as the diameter of the hole changes with depth.

To sample, the string of tools is removed from the hole and is replaced with drill rods and a sampler. The sampler is driven into the ground for both SPT samples and tube samples. The tube samples have to be thick-walled to withstand the blows of the hammer without buckling.

Record: (i) Groundwater levels, (ii) depth of soil strata changes, and (iii) difficulty of drilling

Speed of drilling: Very slow, typically about 20 feet per day with sampling at 5-foot intervals

Suitable soils: All types

Advantages

1. It can penetrate through very difficult soils, such as landslide debris containing hard rock fragments intermixed with clay.
2. It can examine the full depth of soil and groundwater conditions during drilling, but recovered soil is heavily disturbed.
3. It is light and maneuverable, and can be easily towed onto limited access sites with minimal environmental disturbance.

Disadvantages

1. There is considerable disturbance at the bottom of the hole due to the percussive blows of the drilling technique; this affects relative density measurements in cohesionless soils.
2. It is very slow, which adds to the cost of site investigation, especially if a field inspector is present throughout the exploration.
3. Undisturbed tube samples are of poor quality.
4. Field log is usually taken by the driller and is less reliable than that of an independent field inspector.

Wireline Drilling with Casing Advancer

Technique

See Figure 3.8(e). Power rotary drill rig rotates the outer casing and drill bit at the lower end of the casing. Uses flush-jointed casing of various sizes (typically 4.5 inch o.d., 4.0 inch i.d.). Drill water returns up the outside of the casing. Power is transmitted from the rotating casing to the tri-cone roller bit through drive latches that key into the inside of the driver casing wall. The tri-cone roller bit is slightly ahead of the casing shoe bit during drilling. To sample at the bottom of the hole, a wireline overshot device is lowered down the hole and latches onto the body (inner assembly) of the casing advancer. After the casing advancer has been removed, the sampler is lowered down the cased hole.

Record: (i) Slow rates of penetration, (ii) any obstruction encountered.

Speed of drilling: Fast, typically exceeding 60 feet per day with sampling at 5-foot intervals

Suitable soils: Soft clay, silt, sand

Advantages

1. Soils at the bottom of the hole receive only minor disturbance prior to sampling.
2. Very rapid drilling capacity.

Disadvantages

1. Gravel and pebbles cannot be lifted to the surface and may be pushed down to contaminate samples taken lower down.
2. No information is obtained on groundwater while drilling.
3. Compact gravel, boulders, or other obstructions can stop further penetration of the hole.
4. Cannot observe the soils encountered between the sampling interval.

Odex Method with Down-the-Hole Hammer

Technique

See Figure 3.8(f). Down-the-hole hammer imparts dynamic blows just above the bottom of the hole so that very little applied energy is lost at the bit. The Odex system (Figure 3.9) comprises a pilot bit (usually a button bit—see Figure 3.9b) with a reamer above it. The bit has a bottom discharge.

The bit breaks up the ground at the bottom of the hole and the cuttings are returned to the ground surface up the inside of the casing. Air flush, water flush, or foam can be used to pick up the cuttings and lift them to the surface.

When the hammer is operating, the guide rotates and forces the eccentric reamer to undercut the casing shoe, thus allowing the casing to follow the bit down the hole. The hole is cased at all times. When the drill is reversed, the reamer retreats and can be withdrawn up the casing. This occurs for sampling and at the end of drilling.

Record: (i) Loss of drill fluid return, (ii) rate of penetration during drilling

Suitable soils: Dense boulders, gravel, sand, rockfill

Figure 3.9 ODEX drilling:
(a) button bit showing button discharge holes and reamer above the bit
(b) from top to bottom: impact hammer base, guide, reamer, bit

Advantages

1. Very fast in compact hard soils, boulders, or weathered bedrock.
2. Penetrates into difficult and variable ground conditions where most other drill rigs would meet refusal, and it can get past obstructions, such as floater boulders in landslides.
3. It can drill a short hole into hard bedrock (without casing) to anchor inclinometers without the need for rotary coring equipment

Disadvantages

1. It can be stopped by cohesive strata.
2. The vibrating hammer invalidates SPT's at the bottom of the hole.
3. Only operates efficiently in relatively hard materials, so it is less versatile than most other drilling methods and is best utilized as a backup procedure for hard layers encountered in landslides.

3.8 STANDARD PENETRATION TEST

The Standard Penetration Test, commonly referred to as the SPT, is a simple and relatively inexpensive way of obtaining:

- A sample that can be described and classified by water content, Atterberg limits and gradation, and
- An indirect measure of the relative density of sandy soils through the SPT blow count N

The Standard Penetration Test is anything but "standard" in terms of field procedures, and needs both careful field observations and good interpretation to provide acceptable information. However, SPT sampling is routinely undertaken for intermittent sampling of boreholes, and is a very useful procedure because almost all drill rigs are equipped to perform the test.

Field Procedure

The SPT sampler is shown on Figure 3.10; it is known as a split-spoon sampler because the main sampling barrel can be divided into two halves when the sampler is unscrewed from the drill rods and the cutting shoe is unscrewed from the lower end of the barrel. The external diameter of the barrel is 2 inches and the internal diameter is 1.375 inches. The thick-walled sampler can withstand repeated pounding from the hammer. The top end of the sampler has a ball-valve that allows water or air to escape as the sample is being driven but closes when the sampler is withdrawn, helping to retain the sample.

The SPT sampling procedure begins by cleaning out the hole of loose deposits to the depth at which a sample is to be taken. The split-spoon sampler (attached to the drill rods) is lowered down the hole by a rope that passes over a pulley at the masthead and is wrapped around a revolving "cathead." At the top of the drill string, above ground, a special rod is attached that has a circular anvil at the lower end. Once the sampler, drill rods, and anvil rod are in position, a 140 lb. hammer is attached to the rope and is placed over the anvil rod. The driller then marks (with chalk) the drill rod at three 6-inch intervals above the point where the drill rods emerge from the casing. The hammer is lifted 30 inches above the anvil using friction between the spinning cathead and the rope to lift the hammer. Typically, the hammer is released by rapidly slackening the rope so that the hammer can fall onto the anvil with a free-fall action. The number of blows required to drive the sampler through each of three 6-inch intervals is recorded. The first 6-inch blow count is disregarded on the assumption that the first 6 inches of penetration passes through soil disturbed by drilling. The number of hammer blows needed to drive the subsequent 12 inches of penetration into the soil is recorded as the SPT blow count N. The blow count is usually terminated when there is less than 6 inches of penetration after 50 blows of the hammer; in these cases, the SPT blow count can be recorded as >100 or, for example, 50/2½". Stopping the blow count at 50 with less than 6 inches of penetration reduces wear and tear on the equipment, and signifies that a hard stratum or obstruction has been encountered.

The 140 lb. hammers used in the SPT are of three types (two are shown on Figure 3.11):

- *Donut hammer*—a large steel ring that is lifted by a rope or steel cable from eyebolts attached diametrically across from each other on top of the ring.
- *Safety hammer*—a cylindrical sleeve, which loosely fits around the shaft of a drill rod with an anvil at the upper end, or
- *Automatic trip hammer*—a hammer that is lifted by a clasping device and released by a trip mechanism so that the free fall is exactly 30 inches. This equipment is often specified for SPT measurements in Japan, China, Britain, and other countries.

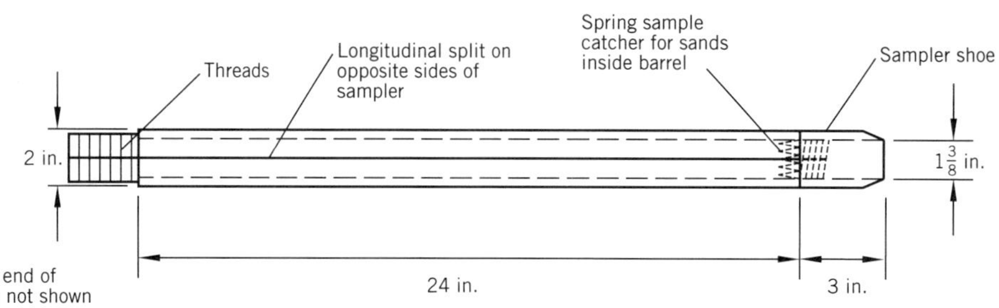

Figure 3.10 Standard Penetration Test sampler.

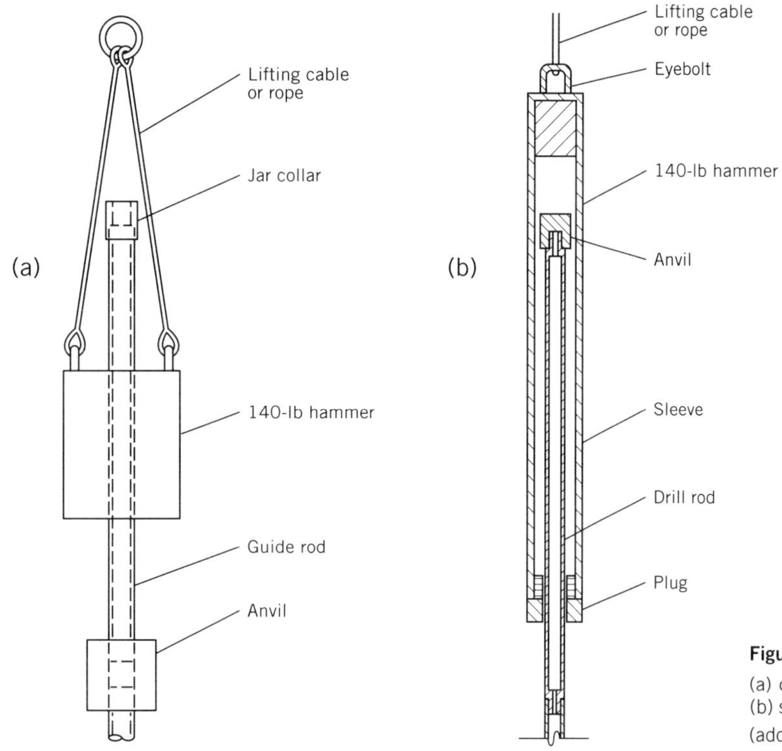

Figure 3.11 SPT hammer types:
(a) donut hammer
(b) safety hammer
(adopted from Steinberg, 1981)

After SPT sampling, the hammer and anvil rod are removed, and the sampler is pulled up to the surface. The sampler is unscrewed from the drill rods, the cutting shoe is removed, and the split-tube is opened for examination. A geologist, geotechnical engineer, or technician describes the soil and preserves pieces for later examination. All samples should be labeled at the site before being sent to the soils laboratory. The labeling information should include: site, job number, boring number, depth interval of sample, blow counts at 6-inch intervals (including the first 6 inches), length of sample recovered, and a brief soil description.

Technical Concerns with Field SPT Sampling

Loss of Sample

A sample occasionally falls out of the sampler as it is being lifted out of the hole, especially when sampling loose to medium dense sands and silts below the water table. Certain devices can be inserted between the cutting shoe and the sampler barrel to prevent the soil from slipping out as the sample is pulled out. The most common "sample catcher" is a spring of thin metal or plastic (Figure 3.10) that allows the sample to enter the barrel but grips the soil if the sample starts to move out.

A lost sample should be recorded on the field log. On the summary boring log, an asterisk (*) or other appropriate note should signify that no sample was recovered in the SPT test. A good practice is to clean out the hole to the base of the lost SPT immediately; sample and repeat the SPT; otherwise, a sampling program based on 5-foot intervals would produce no information over a 10-foot depth.

Soil Disturbance by the Boring Technique

In open-hole boring, the casing usually penetrates below the groundwater table. Groundwater level outside the casing may be higher than the water level in the borehole, and the resulting hydraulic gradient can disturb the sands or silts to be sampled. This is a fairly common cause of low blow counts being measured below the groundwater table in hollow stem auger, shell and auger, posthole auger, and other open-hole methods of boring. To avoid this problem, the casing should be filled with water so that a reverse head is created (i.e., water within the casing is higher than the groundwater level).

A second cause of disturbance specific to the hollow stem auger is the suction caused by pulling out the plug from the hollow stem prior to sampling. The suction can pull sand up into the hollow stem and thus disturb the ground immediately below the hole. Although the first 6 inches of penetration is not counted in the SPT, it is better to avoid deliberate disturbance of the sand below the auger. By twisting and very slowly withdrawing the plug for the first 1 foot, the problem generally can be avoided. Where the SPT blow count data are critical, the inside of the hollow stem auger can be probed to find out if soil has risen inside the hollow stem. The probe result should be noted on the field log.

The third potential cause of disturbance during boring is the heavy pounding of a casing to support the sides of the hole. This is especially critical in loose sands. In general, driving the casing must be avoided when trying to determine the relative density of cohesionless soils. Rotary drilling with mud is the preferred technique but a hollow stem auger is acceptable, subject to the precautions stated in the previous paragraphs.

Gravel in the Sample

If pebbles are within a sandy stratum, a stone is likely to be caught by the edge of the SPT cutting shoe or become wedged in the sampler opening and be pushed into the soil below. The SPT blow count N will be increased by the larger cross-sectional area being driven into the soil.

Some practitioners disregard blow counts recorded in sands with pebbles. However, the results can be used with discretion. For example, a low blow count must indicate that the sampler is passing through soil of low relative density; similarly, a very high blow count is likely to be within a very compact stratum. Intermediate blow counts will be more difficult to interpret, and relative density interpretations should be lowered from those for pure sandy soils. In sands with occasional scattered pebbles, a series of SPT measurements can be used to obtain a general indication of relative density within a stratum because it is unlikely that pebbles will impede the penetration blow count in all tests.

Length of Drill Rods

Research at the University of Florida (Schmertmann and Smith, 1977) indicates that short drill rods (less than 10 feet long) give resistance measurements that are too high. The effect is attributed to reflection of energy that reduces the energy available for driving the sampler into the ground. Seed et al. (1983) recommend that the blow counts for depths of up to 10 feet be multiplied by a factor of 0.75 to correct this problem.

Table 3.5 Correction Factors C_N for Effective Overburden Pressure σ'_v

Effective Overburden Pressure ton/sq. ft.	Correction Factor C_N			
	Peck and Bazaraa (1969)	Peck, Hanson, and Thornburn (1974)	Seed, Arango, and Chan (1975)	Average
0.1	2.86	—	2.25	2.55
0.2	2.22	1.54	1.87	1.88
0.3	1.82	1.40	1.65	1.62
0.4	1.54	1.31	1.50	1.45
0.5	1.33	1.23	1.38	1.31
0.6	1.18	1.17	1.28	1.21
0.7	1.05	1.12	1.19	1.12
0.8	0.99	1.08	1.12	1.06
0.9	0.96	1.04	1.06	1.02
1.0	0.94	1.00	1.00	0.98
1.1	0.92	0.97	0.95	0.95
1.2	0.90	0.94	0.90	0.91
1.3	0.88	0.91	0.86	0.88
1.4	0.86	0.89	0.82	0.86
1.5	0.84	0.87	0.78	0.83
1.6	0.82	0.84	0.74	0.80
1.7	0.81	0.82	0.71	0.78
1.8	0.79	0.81	0.68	0.76
1.9	0.78	0.79	0.65	0.74
2.0	0.76	0.77	0.62	0.72

Relationship between SPT Blow Count N and Relative Density

Sands are difficult to sample, transport, and test in a relatively undisturbed condition. Geotechnical engineers therefore rely on *estimates* of relative density to determine strength and compressibility for engineering design. The most widely used relationship is that put forward by Terzaghi and Peck (1948) given on Table 3.4. The quantification of relative density in the third column of Table 3.4 has never been sanctioned by Terzaghi and Peck but was suggested by Gibbs and Holtz (1957).

As mentioned in Chapter 7, sand in the field can have a relative density greater than 100 percent because the upper density limit is determined by a laboratory procedure. Heavily indurated sand (e.g., samples from Mangla Dam, Pakistan) can exceed 100 percent relative density, but it is a fairly rare condition limited to natural deposits and not compacted fills. It is also possible, without significant error, to substitute relative dry density (RDD) for relative density D_r (see Section 7.3, Chapter 7). Relative dry density is a simpler calculation than relative density; it does not require the measurement or assumption of the specific gravity G of the sand grains.

Table 3.4 Relative Density of Sands According to Results of the Standard Penetration Test

Relative Density	SPT Blow Count N*	Relative Density,** D_r	$(N_1)_{60}$***
Very loose	0–4	0–15%	0–3
Loose	4–10	15%–35%	3–8
Medium dense	10–30	35%–65%	8–25
Dense	30–50	65%–85%	25–42
Very dense	greater than 50	85%–100%	greater than 42

* original Terzaghi and Peck (1948) correlation
** relative densities suggested by Gibbs and Holtz (1957)
*** recommended by Skempton (1986); for D_r = 50%, $(N_1)_{60}$ = 15; for D_r = 100%, $(N_1)_{60}$ = 58.

The Terzaghi and Peck relationships of Table 3.4 were developed for an effective overburden pressure of around 1 ton/sq. ft. (about 25 to 30 feet depth, depending on groundwater position). Sand deposits develop strength and resistance to penetration from the weight of overburden, so corrections have to be made to the measured blow count at other depths below the surface. Several such corrections have been offered, including Gibbs and Holtz (1957), Peck and Bazaraa (1969), Peck, Hanson, and Thornburn (1974), and Seed, Arango, and Chan (1975) (Table 3.5). The correction factors C_N generally are above 1 at effective overburden pressures of less than 1 ton/sq. ft. and below 1 at higher overburden pressures.

In practice, the three sets of correction factors are very similar. Therefore, a method of estimating relative densities is to average the three similar correlations and divide the average factors into the Terzaghi and Peck relationships to produce the graph shown on Figure 3.12. The practical benefit of this graph is that field SPT's can be plotted directly onto the graph to estimate relative density. However, subsequent to the described modifications of the Terzaghi and Peck relationship, there have

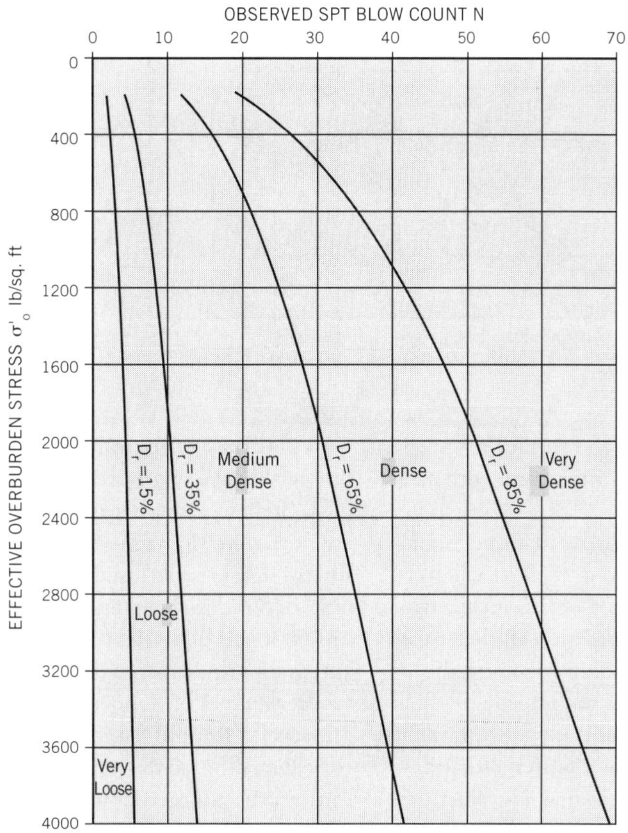

Figure 3.12 Observed SPT blow count vs. relative density as modified from the Terzaghi and Peck (1948) correlations.

This graph is appropriate for **donut** hammers

D_r – relative density

been efforts to "standardize" the Standard Penetration Test for seismic liquefaction analyses, and this requires an additional change to Figure 3.12, as described next.

Standardizing the Standard Penetration Test

The SPT is the best available method of measuring the in-place relative density of sands. As earthquake engineering has become increasingly important in many regions of the world, the SPT is being relied upon to estimate the susceptibility to liquefaction of sand foundations and sand fills. It is also being used in research work to check the response of sands to actual earthquakes. Therefore, much effort has been made to isolate the variables of the procedure so that the results can be made comparable (Seed et al., 1985).

One of the principal components of the SPT is the energy transmitted to the anvil by the hammer. This depends on the technique of the driller and the type of hammer being used. The long safety hammers are more efficient in transferring energy into the rods than are the more squat donut hammers. The driller affects the result through (i) the ability to fully release the hammer by rapidly slackening the rope, (ii) lifting the hammer to the precise 30-inch height above the anvil, and (iii) the rate at which the blows are delivered because rapid delivery can develop undissipated pore pressures between blows.

A study by Kovacs et al. (1983) was performed on 56 drill rigs and operators in the United States. In all cases, the SPT was performed by a procedure of wrapping the lifting rope twice around the cathead (rotating pulley). By comparison to the theoretical free-fall energy of a 140 lb. hammer falling 30 inches, the mean energy achieved was:

- Safety hammer: 61 % of theoretical energy
- Donut hammer: 45 % of theoretical energy

Based on these data, the donut hammer can be expected to produce SPT blow counts that are 30-35% higher than a safety hammer. Other field studies, although showing wide scatter, confirm this average relationship between the safety and donut hammers (Figure 3.13).

Automatic mechanical trip hammers have given similar results to the safety hammer. Three field studies (Decker, 1983; Douglas and Strutznsky, 1984; and Liang, 1983) gave energy ratios averaging 55–62% of the theoretical energy. (*Note:* A dissenting opinion has been put forward by Goble (1995). Citing the results of almost 40 energy calibration tests, Goble states that certain automatic trip hammers achieve 85–95% transfer efficiency and are significantly higher than cathead-and-rope operated safety hammers at 65–80% transfer efficiency. Both these results are higher than the values put forward by Seed. For purposes of this ongoing discussion, the Seed data, which have independent research backup, Figure 3.13, will be accepted.)

The Japanese obtain higher energy transfers to the drill rods. Yoshimi and Tokimatsu (1983) report an average efficiency of 78% for a mechanically-tripped donut hammer and 67% for a rope and pulley system using the special throw release used by Japanese drillers. This more efficient energy transfer

Figure 3.13 Comparison of SPT N-values measured with donut and safety hammers (after Seed et al., 1985).

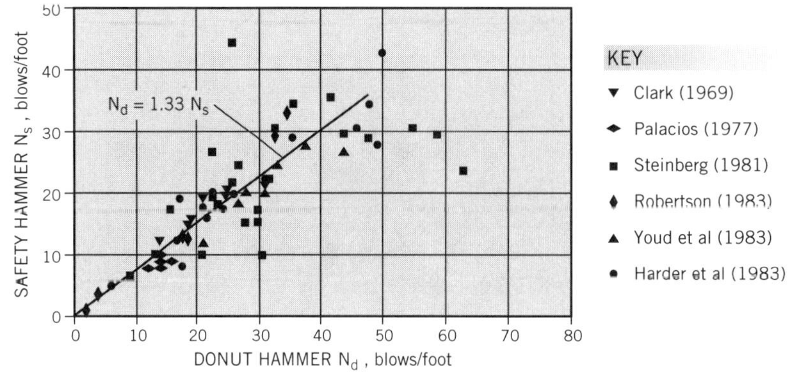

has been attributed to the smaller anvils and smaller diameter pulleys used in Japanese practice in comparison with the United States and elsewhere.

Seed et al. (1985) have proposed that the SPT technique be standardized at an energy transfer ratio of 60%, as achieved in the United States by a safety hammer or automatic trip hammer. In the case of the safety hammer, the lifting rope should be wrapped only twice around the revolving cathead (pulley) so that the release of the rope will not impede the free fall of the hammer. The SPT blow count at this energy transfer level is termed N_{60}. Any other energy transfer efficiency should be applied as a ratio to the measured N value; for example, if the energy transfer ratio is only 45%, the N value measured by that SPT test would be reduced by a factor of 45/60 (= 0.75) to obtain N_{60} values.

The field procedures recommended by Seed et al. (1985) are:

- Borehole 4 to 5 inches diameter
- A-size (1.62 inch o.d.) drill rods up to 50 feet, N-size (2.37 inch o.d.) rods below 50 feet
- Blows delivered at about 30 blows per minute (one blow every 2 seconds)
- Drill bit with upward deflection of drilling mud
- Straight-sided interior to sampler barrel

It is doubtful whether all these recommendations will be fully adopted because of the wide variety of boring techniques throughout the world. However, the intent to provide standardized techniques for the SPT is laudable, especially for investigations of earthquake-resistant design and research.

When the Terzaghi and Peck interpretation of SPT blow counts was published in 1948, safety hammers and automatic trip hammers did not exist. Therefore, the energy transfer ratio was that of a donut hammer of 45% rather than the 60% recommended by Seed in 1985 and now widely adopted by the geotechnical profession. Therefore, to obtain graphs representing the "standard" SPT of N_{60}, the modified Terzaghi and Peck correlation of Table 3.5 and Figure 3.12 has to be further changed by a factor of 45/60 = 0.75. The revised chart, Figure 3.14, allows SPTs obtained by an automatic trip hammer or a safety hammer (both N_{60} values) to be plotted directly on the graph to obtain relative density. If a donut hammer is used, Figure 3.12 can be used.

Calculation of Normalized SPT Blow Count $(N_1)_{60}$

For earthquake engineering, the normalized SPT blow count $(N_1)_{60}$ is being adopted as the standard. This refers to the blow count that would be measured at an effective overburden pressure of 1 ton/sq. ft. using a hammer with an energy transfer ratio of 60%. For practical purposes, a pressure of 1 kg/sq. cm. or 100 kPa is equivalent to 1 ton/sq. ft. If a safety hammer or automatic trip hammer is used, both of which have an energy transfer ratio of 60%, no adjustment is needed for the hammer. However, if a donut hammer is used, the energy transfer ratio is only about 45% so the number of blows recorded by the operator must be adjusted downwards to reflect the lower dynamic energy applied to the sampler. The observed blow counts are multiplied by 45/60 = 0.75 to obtain $(N)_{60}$ values.

To obtain $(N_1)_{60}$, the values of $(N)_{60}$ have to be multiplied by the overburden correction factor C_N, which is obtained from one of the correlations (or the average values in column 5) given in Table 3.5.

$$(N_1)_{60} = C_N (N)_{60} \qquad \text{Eq. (5)}$$

Additional Refinements for Standardization of the SPT

Skempton (1986) has attempted to relate some of the factors affecting the SPT into a quantifiable whole—factors such as aging, overconsolidation ratio, gradation, borehole size—and their effect on the normalized blow count $(N_1)_{60}$. The full technical approach cannot be presented here, but a selective summary will suffice.

Skempton first discusses the energy transfer ratio, which he terms the Rod Energy Ratio ER_r. Much of his research encompasses the same studies referred to by Seed et al. (1985) but also includes more data from China and Japan. This includes the Japanese Tombi trigger mechanism and the Japanese driller's practice of throwing the rope sideways off the cathead to obtain a free fall of the hammer.

He reaches similar conclusions to Seed on the energy transfer to the rods, as summarized on Table 3.6. Note that Skempton's Rod Energy Ratio ER_r is changed to Energy Transfer Ratio ETR (Seed's term), for consistency of presentation in this chapter. It should be noted that Skempton's interpretation for the safety hammer in the United States is a ETR/60 ratio of 0.92, compared to 1.00 by Seed. Ratios from

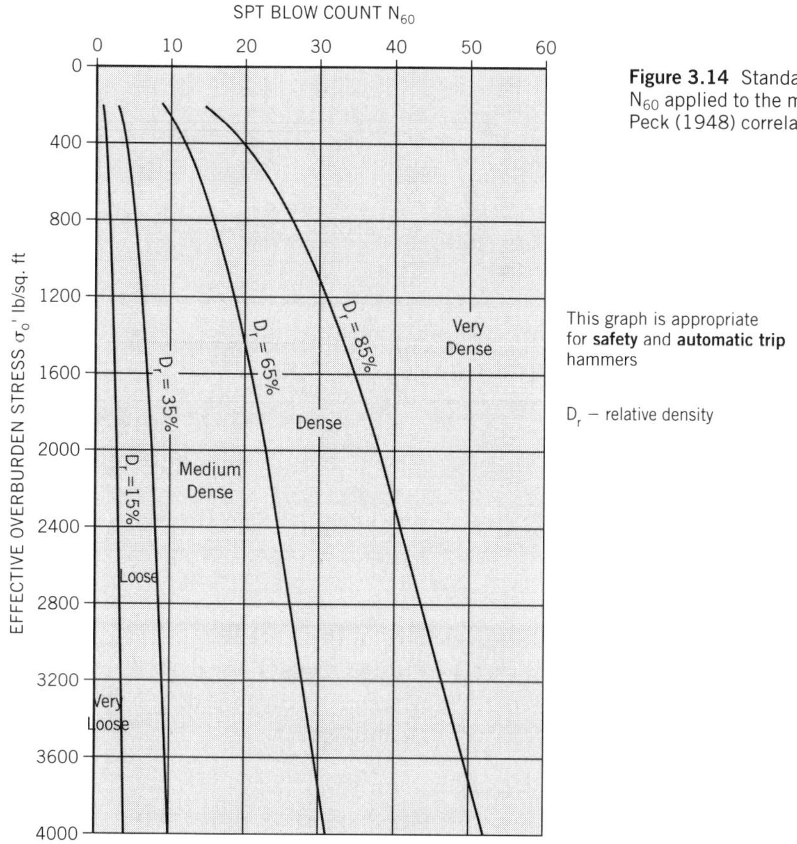

Figure 3.14 Standardized SPT blow count N_{60} applied to the modified Terzaghi and Peck (1948) correlation.

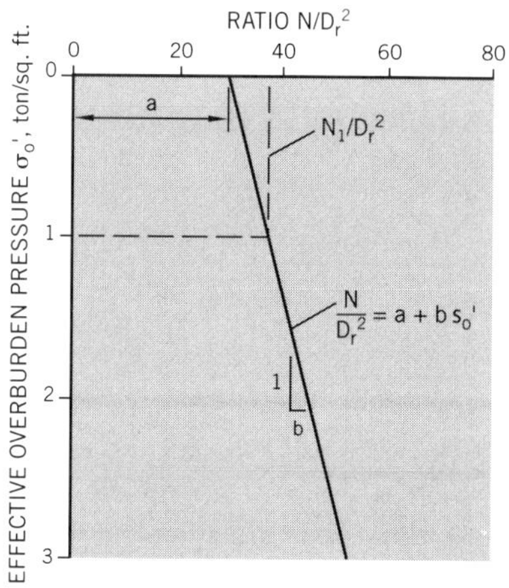

Figure 3.15 Approximate relationship between N/D_r^2 and σ_o' for sand (after Skempton, 1986).

other countries are the results of different equipment and field techniques. The results of field blow counts must be multiplied by the factor in the final column of Table 3.6 to calculate N_{60}.

Other correction factors recommended by Skempton are listed in Table 3.7. In brief: The effect of *rod length* is based on wave equation studies by Schmertmann and Palacios (1979) and is attributed to energy reflections; a correction for *sample liners* is due to the fact that some American samplers have them and they may be left out during a SPT. *Borehole diameter* effects have been estimated from several studies but still require more research. All these corrections are relatively minor and are subject to future revision. However, they may be worth adopting for liquefaction studies.

In examining the various other sand properties that affect the SPT penetration resistance N, Skempton notes that, for any given effective overburden pressure σ_v', the value of N increases approximately as the square of the relative density D_r. Thus, N/D_r^2 = constant and the relationship with depth is of the form $N/D_r^2 = a + b\,\sigma_v'$ as shown on Figure 3.15, where a and b are constant for a given site. This relationship was first suggested by Meyerhof (1957). The value of N_1/D_r^2 is obtained from the graph. In these formulas, D_r is a ratio, not a percentage. The values of N_{60}/D_r^2 can be obtained by making the appropriate correction for the type of hammer used (Table 3.6) and other test procedure variations (Table 3.7).

Table 3.6 Summary of Rod Energy Ratios (Energy Transfer Ratios) for Various Equipment*

	Hammer	Release	Energy Transfer Ratio	ETR/60
Japan	Donut	Tombi	78%	1.30
	Donut	2 turns of rope	65%	1.08
China	Pilcon type	Trip	60%	1.0
	Donut	Manual	55%	0.92
USA	Safety	2 turns of rope	55%	0.92
	Donut	2 turns of rope	45%	0.75
UK	Pilcon, Dando, old standard	Trip	60%	1.0
		2 turns of rope	50%	0.83

*after Skempton, 1986

Table 3.7 Recommended Corrections to Measured Blow Counts N for Rod Length, Samplers Without Liners, and Borehole Diameter*

Rod length:	>10 m	(> 33 feet)	1.0
	6–10 m	(20–33 feet)	0.95
	4–6 m	(13–20 feet)	0.85
	3–4 m	(10–13 feet)	0.75
Standard sampler			1.0
U.S. sampler without liners			1.2
Borehole diameter:	65–115 mm	(2.5–4.5 inches)	1.0
	150 mm	(6 inches)	1.05
	200 mm	(8 inches)	1.15

*Skempton, 1986

Skempton analyzed ten case histories using the relationship:

$$\frac{N_{60}}{D_r^2} = a + b\sigma_v' \qquad \text{Eq. (6)}$$

to study the effect of the intrinsic properties of sand on the blow count N_{60}. He identifies these properties as particle size (based on the D_{50} size), overconsolidation, and aging. For the ten case histories, the range of results fall within the following extremes: [insert EQ 6B art]

	$\frac{N_{60}}{D_r^2}$	$\frac{(N_1)_{60}}{D_r^2}$
Recent fine sand fill, Ogishima Island, Japan	$17 + 17\,\sigma_v'$	34
Heavily overconsolidated fine sand, Sizewell, U.K.	$38 + 50\,\sigma_v$	88

The parameters a and b remain approximately constant for a particular site for relative densities between 35% and 85% and for effective overburden pressures between ½ and 2½ ton/sq. ft.

Allowing for the lower energy transfer ratio of equipment 50 years ago, Skempton has modified the Terzaghi and Peck relationship to correspond with the $(N_1)_{60}/D_r^2 = 60$, which is about in the middle of the range for the ten case histories cited earlier, and is shown in the final column of Table 3.4.

Skempton recommends the following corrections to the measured N values for sand:

- *Particle size.* Fine sand: reduce N values by factor 0.92
  Coarse sand: increase N values by factor 1.08
  This observed effect is probably due to the higher ϕ values for coarser sands at a particular relative density.
- *Overconsolidation.* Increase parameter b by the multiplier C_{oc}

$$C_{oc} = \frac{1 + 2\,K_o}{1 + 2\,(K_o)_{NC}} \qquad \text{Eq. (7)}$$

  where $K_o = K_o$ at the overconsolidation ratio (OCR) of the sand
  $(K_o)_{NC} = K_o$ for the sand under normal consolidation
  $(K_o)_{NC}$ can be obtained from the relationship $(K_o)_{NC} = 1 - \sin\phi'$
  K_o can be estimated from $K_o = (K_o)_{NC}\,(OCR)^{\sin\phi'}$

  Values of K_o and C_{oc} are given in Table 3.8.
- *Aging.* Aging appears to increase resistance to deformation and increase parameter a. Some typical results for normally consolidated fine sands are given in Table 3.9.

Skempton provides no specific recommendations for dealing with overconsolidation and aging, although geotechnical engineers can take account of these factors in interpreting SPT data. With respect to aging, it can be assumed that the Terzaghi and Peck correlation relates to natural deposits. If these are disturbed by vibratory compaction, then the $(N_1)_{60}/D_r^2$ presumably drops from around 55–60 to 35–40 according to Table 3.9. This probably explains the limited increase in SPT blow counts observed immediately after vibrocompaction, even though the relative density has been increased by compaction (see Chapter 21, Section 21.2).

Table 3.8 Values of K_0 and C_{oc}*

Overconsolidation Ratio	$\phi' = 32°$		$\phi' = 36°$		$\phi' = 40°$	
	K_o	C_{oc}	K_o	C_{oc}	K_o	C_{oc}
1	0.47	1.00	0.41	1.00	0.36	1.00
2	0.68	1.22	0.62	1.23	0.56	1.23
3	0.84	1.38	0.78	1.41	0.73	1.43
4	0.98	1.53	0.93	1.57	0.87	1.59
6	1.21	1.76	1.17	1.84	1.14	1.91
10	1.59	2.15	1.58	2.28	1.58	2.42

*Skempton, 1986

Table 3.9 Effect of Aging*

	Age, years	Ratio $(N_1)_{60}/D_r^2$
Laboratory tests	0.01	35
Recent fills	10	40
Natural deposits	>100	55

*Skempton, 1986

Relevant Experiences of the Author

This author has been using the SPT for more than 40 years. Two relevant experiences are worth presenting as footnotes to the research work described earlier.

Skagway, Alaska

(Case History 7). AW drill rods (1.72 inch O.D.; 1.22 inch I.D.) were being used for deep water borings. In 88 feet of water, samples of soft to medium stiff, slightly clayey silt required 4 to 7 blows per foot to be sampled by SPT just below the mudline; in 175 feet of water, similar soils required 11 to 18 blows per foot. Although this example relates to a slightly clayey silt, rather than sand, the implication is that the measured SPT blow count rises in very long drill rods, apparently due to energy lost in the longitudinal whiplash along the rods. As previously mentioned, Seed recommends switching from A rods to N rods below 50 feet but there does not seem to have been any experimental study to quantify this effect. Thus, the SPT blow count N probably should be *reduced* at large depths where thin (small diameter: length ratio) drill rods are being used. This is the opposite of the rising correction with depth for drill rod length given in Table 3.7. More research on drill rod effects would be valuable to the profession.

Limassol, Cyprus

This was a lawsuit involving the dredging of sands in Limassol harbor, Cyprus (unpublished). Bidders were provided with boring logs that showed comparatively low SPT blow counts near the surface of the sand. The contractor brought to the site lightweight clamshells to excavate the sand and quickly found that the clamshells bounced off the bottom without penetrating into the sand. Larger, heavier equipment had to be brought in to complete the dredging and a claim was filed for "changed conditions."

The author's firm was engaged to perform a detailed site investigation including SPT and thin-wall tube sampling with a piston sampler. The tests were made from a large floating

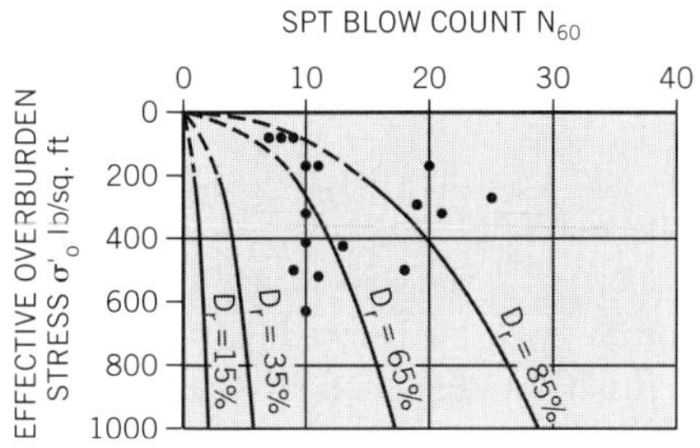

Figure 3.16 Results of SPT in Limassol, Cyprus, harbor as plotted on Figure 3.14.

pontoon in about 10 feet of water. A very experienced senior engineer was sent to the site to perform the measurements.

Several wash borings were made, paying careful attention to the position of the casing shoe relative to the start of SPT sampling. The hammer was raised by hand and dropped by a free fall, and thus was probably close to the N_{60} energy transfer. The drill rods were 1.65 inch in diameter. The hole diameter was 4.4 inches (112 mm). Submerged density of the sand was about 60 lb/cu. ft. The sand was gray uniform fine sand, almost all particles being between 0.06 and 0.2 mm size (D_{50} = 0.12 mm).

Assuming that the SPT methodology used at this site was at the N_{60} level, the results of the SPT program are plotted on Figure 3.16 (i.e., the upper part of the chart, Figure 3.14). It shows that the measured blow counts indicate relative densities of about 50 to 100%, median 72%.

For comparison, four undisturbed sand samples were taken by a Geonor fixed piston sampler using 3-inch o.d. by 33-inch long thin-wall tubes. The tubes were jacked into the sand using the two hydraulic rams of the drill rig. All the undisturbed samples were retrieved in good condition. Six other samples were rejected, including one no recovery, one disturbed, three short samples and one non-representative silt sample. In the laboratory, the undisturbed samples were cut into 8-inch long segments and density measurements were taken. Density limits were determined by the rapid tilt test (for minimum dry density) and Kango vibratory hammer at 2,000 impacts/minute (for maximum dry density).

The results for the four samples covered a narrow range of 66 to 77% relative density (average 71%). The average is comparable with the relative density estimated from the SPT results and gives confidence in the reliability of overburden corrections. It is unlikely that the sampling procedure increased the relative density of the specimens.

The rod lengths in the SPT ranged from 11 feet (3.4 m) to 23 feet (7.0 m) which require corrections ranging from 0.75 to 0.95 according to Skempton (Table 3.7). These would reduce the estimated relative densities of the SPT and make the comparison with the measured relative densities worse.

Relative Density of Gravels

Until about 1980, conventional wisdom held that gravels are not susceptible to liquefaction during strong earthquake motions. It was reasoned that gravels, because of their high permeability, would dissipate excess pore pressures almost as quickly as they would be generated. However, studies have shown that there have been sites where gravels liquefied during earthquakes, and that stones of up to 1-inch size have been brought to the surface by flowing water. Given the difficulty of trying to collect undisturbed gravel samples for laboratory tests, emphasis has been placed on performing an in-situ field test comparable with the Standard Penetration Test (SPT) to obtain the relative density of gravels.

The Becker Penetration Test (BPT) has been developed to explore gravels. By correlating the BPT with the SPT, the large body of data for evaluating the liquefaction of soils through the SPT can be applied to gravels. Knowledge of the relative density of gravels can also be used to estimate the peak shear strength of the gravel for static stability analyses (see Chapter 7, Section 7.6).

The Becker Hammer Drill drives a double-walled casing into the ground using a double-acting diesel hammer. Air is forced down the annulus of the casing to the bit and fractured pieces of gravel are carried up the inside of the casing by airflow to be collected in a cyclone. The reverse circulation technique is illustrated schematically in Figure 3.17.

The diesel hammer used on the Becker drill is an International Construction Equipment (ICE) Model 180 rated at a maximum energy of 8,100 ft.-lb. per blow. The hammer is closed at the top, and part of the energy during driving occurs from compression of the air trapped in the top of the hammer cylinder in each cycle. By measuring the trapped air pressure (known as the *bounce chamber pressure*), the driving energy for each blow can be estimated. Correlations between the hammer energy and the bounce chamber pressure have been developed by the manufacturer.

Only the outer pipe absorbs the impact of the hammer (the inner pipe floats, separation being provided by neoprene

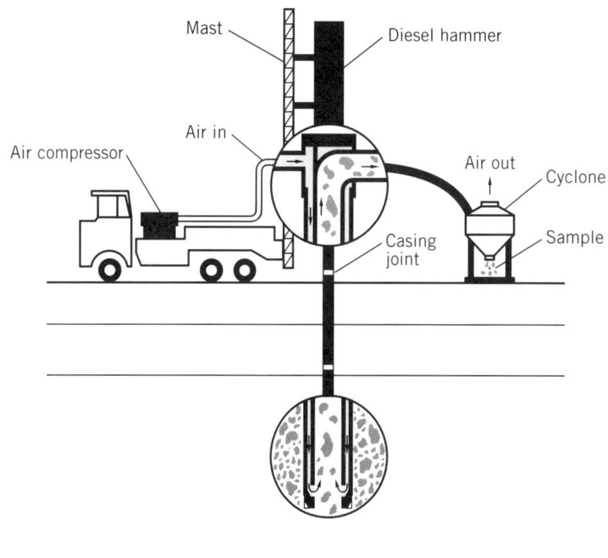

Figure 3.17 Schematic illustration of the Becker Drill sampling procedure (after Harder and Seed, 1986).

cushions). Casing pipes are available in three sizes, and drill bits have two basic shapes: crowd-in and crowd-out. Harder and Seed (1986) have developed recommended standardized procedures for the BPT as follows:

- Use an AP-1000 drill rig, ICE 180 hammer operating at full throttle with the rotary blower on.
- Use the 6.6 inch o.d. casing.
- Use a closed-end 6.6 inch o.d., 8 tooth crowd-out drill bit (the closed bit has a sealed end to prevent gravel entering the inner pipe).

During field drilling, the Becker blow counts and bounce chamber pressures are measured. The Becker Penetration Test (BPT) is the number of hammer blows required to drive the casing 1 foot into the ground. Therefore, by counting the hammer blows, a continuous record of penetration resistance can be obtained for the entire depth of the gravel stratum.

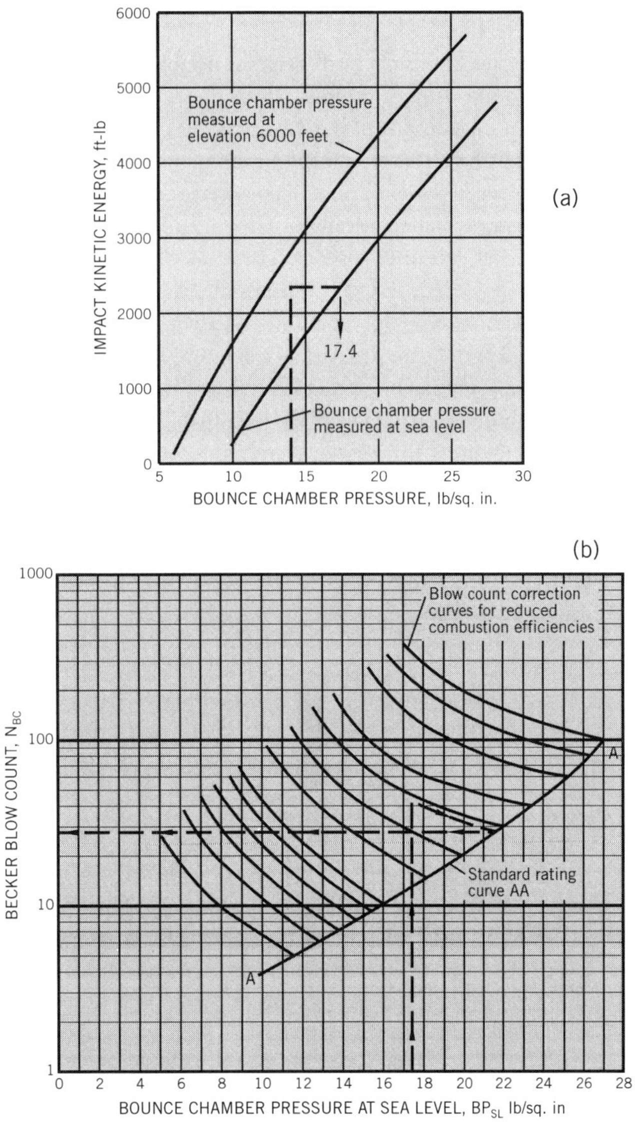

Figure 3.18 Becker Penetration Test:
(a) estimation of equivalent bounce chamber pressure at sea level from data obtained at other elevations
(b) determination of corrected blow count N_{BC} from equivalent bounce chamber pressure and uncorrected Becker blow count (after Harder and Seed, 1986)

Many firms and agencies prefer to use an open bit so that a sample is collected. Their objective is to obtain both a BPT and a boring log. However, experience has shown that the blow count for an open bit is roughly one-half of the blow count for a closed bit. Experiments by Harder and Seed (1986) showed that the recirculation process draws up excessive amounts of sand into the casing and out of the hole, thereby loosening the sand ahead of the bit and giving a low blow count. Therefore, the *closed-end bit* should be used for BPT measurements of relative density.

The energy delivered by the double-acting diesel hammer is not constant. The total potential energy is the weight of the ram (W) multiplied by the drop (H) plus the energy stored in the bounce chamber. This acts like a spring to build energy when the ram is lifted by the double-acting mechanism. When the ram is falling, it compresses the fuel and air in the combustion chamber; this acts as a cushion to slow down the falling ram. The amount of energy lost is a function of the atmospheric pressure and the dimensions of the combustion chamber.

Based on field experiments, Harder and Seed (1986) have "standardized" the Becker blow count N_{BC}. The procedure is as follows:

1. Measure the Becker blow counts and bounce pressures.
2. Correct the bounce pressures at high elevations to equivalent bounce pressure at sea level using Figure 3.18(a). The bounce pressure curve at elevations between sea level and 6,000 feet can be estimated by interpolation.
   Example: Measured bounce pressure at elev. 4,000 feet = 14 psi
   From Figure 3.18(a), equivalent bounce pressure at sea level = 17.4 psi
3. Using the equivalent bounce pressure and measured Becker blow count, use the curves on Figure 3.18(b) to obtain the Corrected Becker Blow Count, N_{BC}.
   Example: Using equivalent bounce pressure of 17.4 psi and a measured blow count of 43, follow the closest curve to terminal curve AA. For this example, $N_{BC} = 28$.
4. The correlation between N_{BC} and the SPT blow count N_{60} is shown on Figure 3.19. This graph is taken from Harder (1997) and includes more data than the original Harder and Seed (1986) paper. The graph shows a lot of scatter around the mean line, which is to be expected from such diverse procedures. However, the trend is that N_{60} and N_{BC} are roughly comparable up to blow counts of around 20, the critical region for most liquefaction studies. At higher blow counts, N_{60} is generally lower than N_{BC}.

A criticism of the Harder and Seed technique is that it ignores the casing friction as a variable yet there must be some casing friction incorporated in the Becker blow count. Youd and Idriss (2001) suggest that casing friction effects are a con-

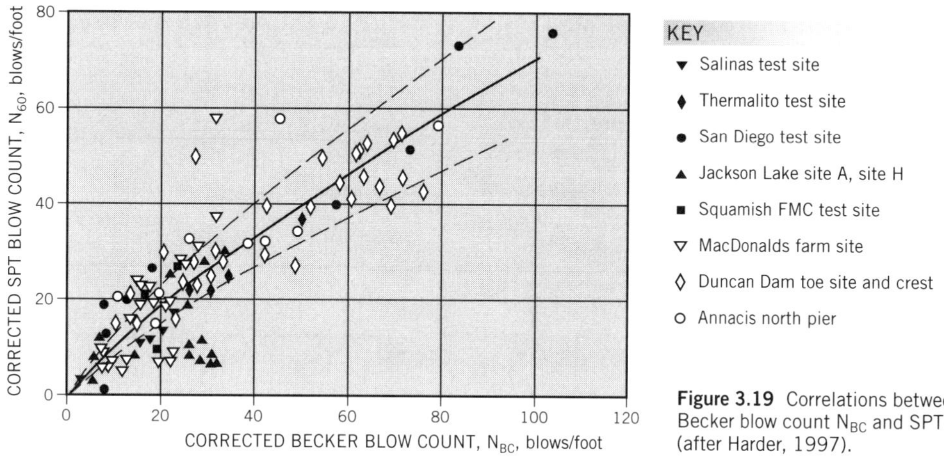

Figure 3.19 Correlations between corrected Becker blow count N_{BC} and SPT blow count N_{60} (after Harder, 1997).

cern for depths exceeding 100 feet and for measurements in soft soils underlying thick deposits of dense soil. For these conditions, they suggest that mud be used to reduce casing friction.

Another procedure to correct Becker blow counts to a reference level has been proposed by Sy and Campanella (1993). It is based on stress wave measurements to determine the energy transferred into the top of the Becker casing and convert it to a standard energy level, comparable with the procedure for converting SPT blow count N to N_{60}. It also takes into account casing friction. The Sy and Campanella technique requires two sets of force and acceleration transducers to be attached to a 2-foot length of Becker pipe. The transferred energy is determined from force and velocity measurements, similar to the procedure used in dynamic monitoring of pile driving. The measured Becker blow counts are corrected to a reference ENTHRU level of 30% of the manufacturer's rated energy for the ICE 180 diesel hammer. This technique, used in Canada, requires more stoppage time than the bounce pressure measurements but is a direct measurement of the energy applied to the Becker casing.

Static Cone Penetrometer

The cone penetrometer is popular in Europe and is being increasingly used in the United States. It is often compared with the standard penetration test. The instrument (Figure 3.20a) has a 60° cone at the end of a rod, and a friction sleeve immediately behind the cone. The cone can be pushed into the ground independently of the sleeve, or the sleeve and cone can be pushed together. The resistance to penetration is measured and the data are interpreted to determine the type of soil around the penetrometer and various soil properties. Numerous technical papers and textbooks have been devoted to the subject, and there are correlations with the SPT blow count.

The main advantages of a cone penetrometer are that a continuous record of the subsurface conditions is obtained and the data are measured statically. In the case of sands, the latter advantage is that the sand is not affected by the blow of the hammer hitting the rods. This can cause unconfined sand, between the sampler and borehole wall, to jump under dynamic blows.

A major disadvantage of the cone penetrometer is that no samples are recovered, so the soil profile record depends on interpretations by an engineer or geologist. This is very unsatisfactory for landslide work where soil layers may be intermixed. Therefore, the technique is not recommended for most landslides. In soft clays, the penetrometer can be an alternative (or supplement) to the field vane.

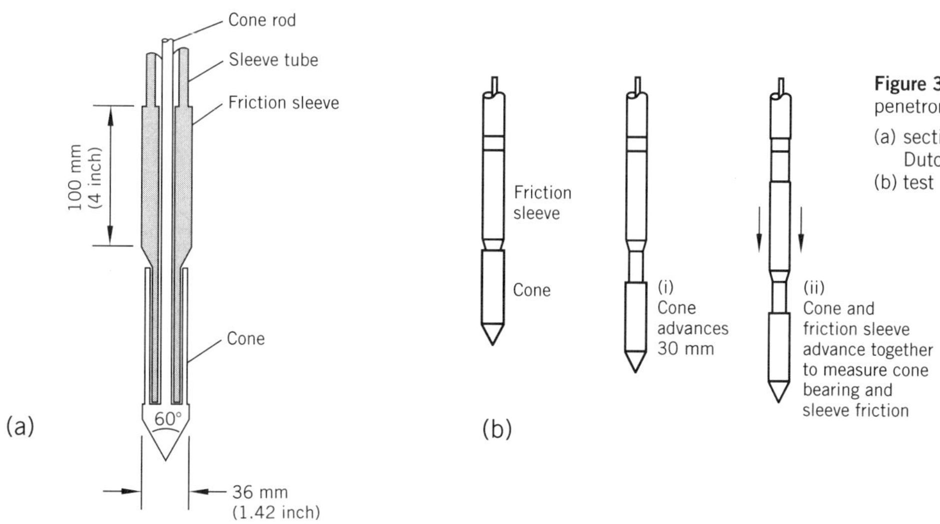

Figure 3.20 Static cone penetrometer
(a) section through the Dutch cone penetrometer
(b) test procedure

3.9 RELATIVELY UNDISTURBED SAMPLING

The methods of recovering soil samples have changed very little for several decades. It seems that the profession has accepted current practices as adequate. However, the quality of samples obtained for laboratory shear, consolidation, and permeability testing is one of the weakest links in the analysis of landslides. The reasons for this deficiency are the lack of improvements in sampling, and competitive pressures to keep costs down in site investigations.

Sampling in Landslides

"Relatively undisturbed" samples, taken at intervals within the boring, provide samples for more sophisticated laboratory tests such as shear strength. These samples are more time-consuming to handle, including the need to mark, wax the exposed soil faces at the ends of the sample, transport, store, etc.

Factors Affecting Undisturbed Sampling

The ideal method of sampling soils would be to use a tube with an infinitely thin wall. The tube would cut through the soil with no side friction between the sampler and the soil. After reaching full penetration, the soil would be sheared off at the base without disturbing the soil on the shear plane. The sample would be taken out of the ground, maintaining the stresses that were previously acting on it at the sampling depth. The process would recover a piece of the soil exactly as it was in the ground; i.e., "undisturbed" by the sampling technique.

This ideal sampling procedure can never be achieved in practice, but the concept is worth bearing in mind when comparing samples and sampling techniques. Hvorslev (1949), in his classic work on sampling of soils, examined numerous factors that influence the quality of samples. The three major factors are:

1. The end area ratio of the sampler tube
2. Friction between the soil and inside of the sampler tube
3. Sampling technique

End Area Ratio (EAR)

This may be the most important factor. It is the ratio of the volume of soil taken by the sampler to the volume of soil displaced by the sampling procedure. It can be represented by the End Area Ratio (EAR) as defined by:

$$EAR = \frac{d_0^2 - d_i^2}{d_i^2}$$

where d_0, d_i are the outside and inside diameter of the sampler tube, respectively.

If the sampler has a cutting shoe of a larger diameter than the sampling tube, the cutting shoe represents the outside diameter d_0.

Hvorslev recommended that the EAR be kept to the practicable minimum, preferably less than 10 percent. EAR values for commonly used samplers are listed on Table 3.10.

Table 3.10 End Area Ratios of Selected Soil Sampling Tubes

Sampling Tube	d_o	d_i	End Area Ratio percent
Thin-wall (Shelby) tube,	2.00	1.875	14
straight sides	3.00	2.875	9
Standard Penetration Test	2.00	1.375	112

d_o – outside diameter d_i – inside diameter

Friction between the Soil and Sampling Tube

Friction between the soil and the inside wall of the sampler causes drag disturbance at the edges of the soil sample. In extreme cases, it can disturb the entire sample. Hvorslev (1949) demonstrated this effect by photographing samples of varved clays split down the middle. The degree of bending of the individuals layers at the edges of the varved clay samples is a measure of the disturbance that occurred during sampling.

Hvorslev supported the concept of the inside diameter of the cutting edge (at the bottom of the sampling tube) being of a slightly smaller diameter than the inside of the tube itself. The intent was to reduce the frictional drag between the soil sample and the inside of the tube, and to allow the sample to expand slightly due to stress relief. For many years, tube manufacturers turned in the lower end of the tubes before reaming out the specified inside diameter at the cutting edge. This practice is no longer followed; it increases the end area ratio and adds to the cost of manufacturing thin-wall sampling tubes.

There is also friction between the *outside* of the sampling tube and the soil. A cutting shoe, such as that used in Britain on the U100 sampling tube, helps to reduce outside friction; however, a cutting shoe increases the end area ratio of the sampler.

After the sampler has penetrated into the soil, one practice is to slowly rotate the sampler so that the sample separates from the in-place soil below the end of the sampling tube. The sample then can be withdrawn without the risk of the sample being left behind in the hole. Twisting off the bottom of the sample relies upon having sufficient friction between the soil and inside of the sampling tube to resist the amount of twist needed to shear off the soil.

The author's opinion is that a thin-wall sampling tube should be straight-sided, both on the inside and outside, to minimize the end area ratio. Friction between the soil sample and inside of the sampling tube can be reduced by applying a thin coat of light oil or dry lubricant to the tube lining. However, this increases the risk that the sample will fall out during withdrawal of the sampler from the hole.

Sampling Technique

Hvorslev investigated different methods of taking samples. He concluded that piston samplers take the least disturbed samples in soft cohesive soils. More importantly, he recommended that thin-wall sampling tubes be *rapidly jacked* into the ground using the hydraulic rams mounted on a rotary drill rig. This combination is still the best method of recovering soft cohesive soils. However, open tubes are routinely used for thin-wall sampling and generally provide an acceptable result.

Hammering the samples into the ground is the worst technique. It usually takes numerous hammer blows to obtain a full tube of soil, and penetration occurs at a variable rate. The soil sample is subjected to the vibration passing along the drill rods attached to the sampler head.

Effect of Sampling Disturbance on Stiff Clay

The effect of sampling disturbance on *soft* clays is well-known, and no further comment is needed here. However, it is sometimes believed that stiff, overconsolidated clays are relatively insensitive to disturbance. This is incorrect, as demonstrated by samples of Keuper Marl taken at Hatton, England, in 1973. The results have not been previously published.

At this site, the Keuper Marl is a heavily overconsolidated, indurated, fissured, hard sandy clayey silt with occasional siltstone lithorelicts. At each of three locations, borings were made by a shell and auger drill rig to depths of up to 20 feet. Continuous 4-inch diameter open-drive samples were taken using the British U100 procedure of hammering thick-walled tubes with a cutting shoe (end area ratio 27 percent). For comparison of soil properties, seven block samples were taken from three test pits close to the borings. The pits were excavated by a backhoe to leave a "table" of marl in the center of the pit. The table was reduced to a cube with 10-inch sides by hand trimming, and wrapped in muslin fabric impregnated with paraffin wax to prevent moisture loss.

Table 3.11 Average Identification Properties of Block and Tube Samples, Hatton Site

	Block Specimens	U100 Tube Specimens
Liquid Limit	65	70
Plastic Limit	34	31
Plasticity Index	31	39
Natural water content	25.1%	26.9%
Cohesive Index (Chapter 8, Section 8.2)	0.91	1.26

U100 is U.K. sampling tube of nominal 100 mm (4 inches) inside diameter.

Average identification properties measured on the samples are given on Table 3.11. The results obtained on the block and tube specimens are similar, the tube samples apparently being slightly more clayey than the block samples. It can be noted that the average natural water contents are well below the plastic limits, which indicates a high degree of overconsolidation.

The other laboratory tests gave very significant differences in measured soil properties between the specimens cut from block samples and those extruded from tube samples, as summarized in Table 3.12. The block samples were hand trimmed to obtain 1.5 inch diameter triaxial test specimens. Fissures were excluded from all the test specimens.

From these laboratory statistics, it can be seen that, comparing specimens cut from block samples to specimens taken from sampling tubes:

- The median undrained strength is more than double.
- The median modulus E is almost quadruple.
- Compressibility is much lower.
- Coefficient of consolidation is increased by a factor of 10,000.

Average water content of the block samples was slightly lower than the average water content of the U100 tube samples.

As shown by the graphs on Figure 3.21, the soil properties at the same water contents are significantly different:

- *Undrained shear strength.* The results diverge at natural water contents below the plastic limit (Figure 3.21a). The tube samples have an approximately constant strength at all the tested water contents, whereas the block samples show an increasing strength with decreasing water content, as should be expected.
- *Consolidation tests.* The average curves show a higher compressibility in the tube samples (Figure 3.21b).
- *Modulus of elasticity.* The shape of the deviator stress vs. axial strain curves for typical specimens (Figure 3.21c) and the calculated modulus E vs. water content (Figure 3.21d), like the shear strength data, show a strong divergence between block and sampling tube specimens below the plastic limit.

Table 3.12 Median Values of Selected Soil Properties, Hatton Site (#)

	Block Specimens	U100 Tube Specimens
Median undrained shear strength*	8,700 psf	4,200 psf
Median modulus of elasticity*	9,000 psi	2,300 psi
Median coefficient of compressibility**	0.011 sq. ft./ton	0.016 sq. ft./ton
Median coefficient of consolidation***	c. 130,000 sq.ft./year	14 sq. ft./year

(#) all results converted from metric units
* undrained triaxial tests at cell pressures of 7.25 psi, modulus measured at 33% of failure stress
** oedometer tests in the stress range of approx. 2,000 to 4,000 psf
*** triaxial dissipation tests 4-inch long specimens
U100 is U.K. sampling tube of nominal 100 mm (4 inches) inside diameter

It is concluded that the U100 sampling procedure causes considerable disturbance to the very stiff Keuper Marl, especially at water contents below the plastic limit. The end area ratio of 27% for this sampler means that 100 area units of sampler tube is capturing approximately 79 area units of soil. In very stiff to hard soils, the intense hammering needed to obtain a sample (more than 100 blows from the jarring link) must cause a profoundly adverse change in the original "undisturbed" soil structure preexisting in the ground. The very high coefficient of consolidation of the block specimens (Table 3.12) suggests that the Keuper Marl may have porous root holes near the ground surface (see Rowe, 1972) that probably are destroyed by the U100 sampling procedure.

Recommended Sampling Techniques

Very Soft to Medium Stiff Clays, Silts

Use *thin-wall steel tube samplers* with an end area ratio of around 10 percent and 2-inch to 5-inch diameter. A higher quality sample is likely to be obtained from a larger diameter sampler. The sampler should be rapidly jacked into the soil using the hydraulic rams of the drill rig. Open driving of the samples is acceptable. *Piston sampling* provides some suction on withdrawal that helps to provide a higher quality sample, especially in soft clays.

Advantages of a piston sampler are (i) no slough enters the sampling tube prior to sampling, (ii) excess soil cannot enter during sampling, and (iii) sample quality and amount recov-

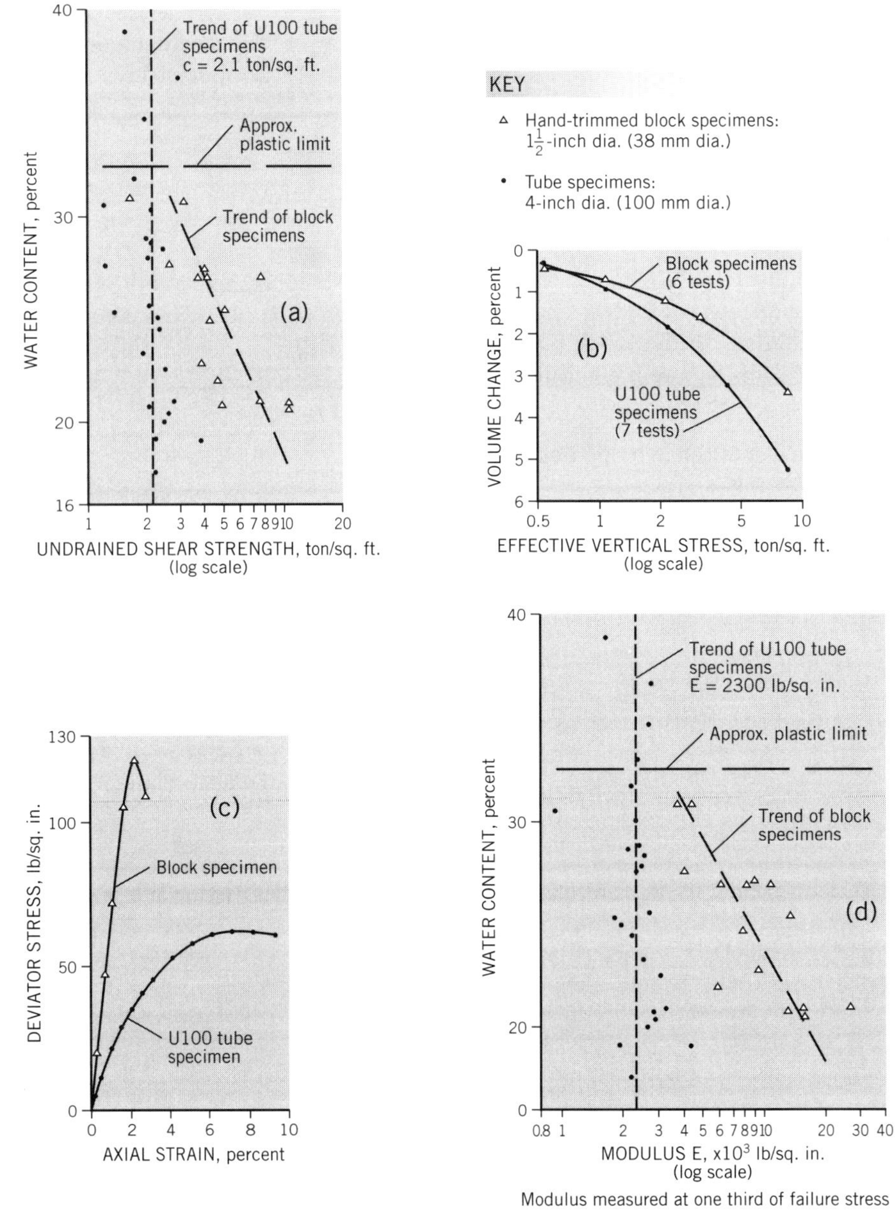

Figure 3.21 Effect of sample disturbance on measured soil properties:
(a) undrained strength vs. water content
(b) consolidation tests
(c) stress-strain curves in the undrained triaxial shear test
(d) modulus E vs. water content

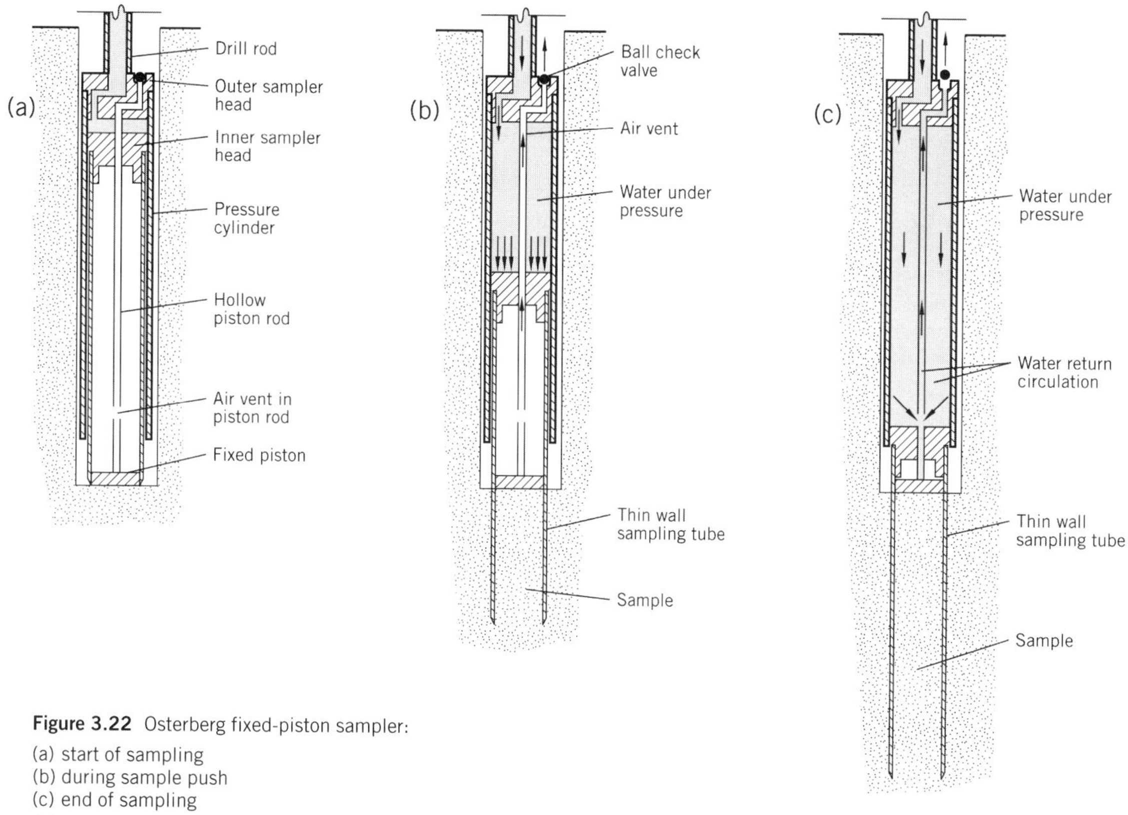

Figure 3.22 Osterberg fixed-piston sampler:
(a) start of sampling
(b) during sample push
(c) end of sampling
(after U.S. Dept. of Interior, 1974)

ered are improved. The main disadvantages are higher costs and longer field time. A fixed-piston sampler is preferred (Hvorslev, 1949) and, of these, the hydraulically activated Osterberg sampler is the easiest to use. The Osterberg sampler (Figure 3.22) is commercially available and uses 3-inch and 5-inch diameter sampling tubes. It is better suited to water flush than mud (U.S. Corps of Engineers, 1996) because suspended sand particles in mud can damage the O-ring seals.

In operation, the fixed piston sits level with the cutting edge of the sampler tube when it is lowered to the bottom of the borehole (Figure 3.22a). Drilling water is pumped through the drill rods to push the thin-wall tube past the fixed piston into the soil (Figure 3.22b). At the end of the stroke, a bypass hole in the hollow piston rod relieves the pressure and returns the drilling water back to the ground surface, thus preventing any further push on the tube (Figure 3.22c). The shaded areas on Figure 3.22 show only the water travel that pushes the sampler tube, but water is exiting through the hollow piston rod throughout the sampling process. When the drive is completed, the sampler is rotated to shear the bottom of the sample and lock the sampler for withdrawal. After the sampler has been recovered from the hole, the attaching screws are removed, and a small hole is drilled into the top of the sampler tube to release the vacuum.

Stiff to Very Stiff Clays, Silts

Use the *Pitcher sampler* (Figure 3.23) with a thin-wall steel tube of the type used for softer clays. The thin-wall sampling tube is attached to a spring-loaded head and does not rotate. In stiff clays, the spring compresses in response to the soil resistance and the thin-wall tube retracts into the outer rotating barrel, leaving only a short length of the sampler tube ahead of the outer shoe. As the cutting shoe reduces the soil resistance, it allows the thin-wall tube to take a sample in a stiff clay without buckling. For less stiff clays, the cutting edge of the sampler tube projects further out from the rotating shoe to minimize soil disturbance from the drilling fluid. Thus, the distance between the lower cutting edge of the sampler tube and the relieving bit of the outer shoe is variable, depending on the resistance (strength) of the clay. The drilling fluid passes between the stationary sampling tube and the rotating outer barrel. The quality of this sampling procedure is very high.

An alternative procedure is to cut *block samples* inside a vertical shaft or horizontal adit. The block samples have to be wrapped immediately to prevent drying. One technique is to wrap the samples with several layers of cheesecloth impregnated with microcrystalline wax (or paraffin wax: microcrystalline wax mixed in equal proportions). It requires very careful handling to avoid breaking the seal. At the laboratory, further sampling by jacking thin-wall samplers into the block, or hand trimming on a soil lathe (Head, 1988), provides specimens for triaxial testing.

Loose to Dense Fine-to-Coarse Sands

Sand provides many difficulties to undisturbed sampling and laboratory tests. Even if a relatively undisturbed sample is

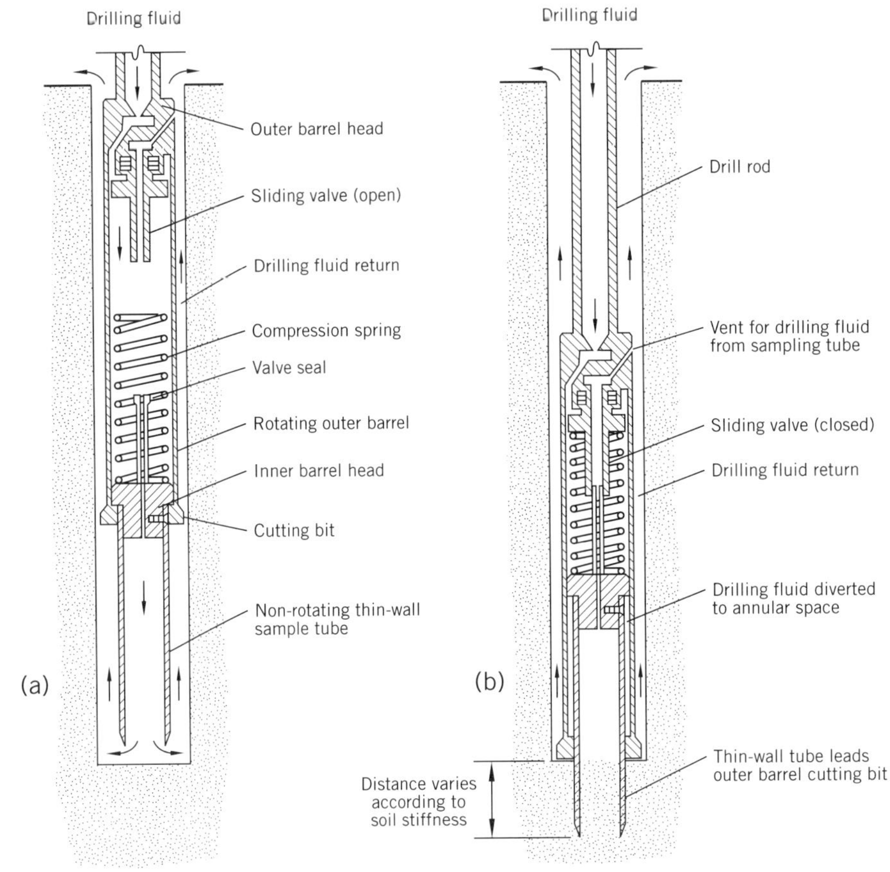

Figure 3.23 Pitcher sampler:
(a) lowering sampler to bottom of hole
(b) sampling
(after Winterkorn & Fang, 1975)

obtained, it is almost impossible to extrude it and cut a triaxial test specimen without disturbance unless there is a significant silt binder present. It is usually better to determine strength indirectly from relative density and angle of repose tests (Cornforth, 1973; see also Chapter 7). The most common method of obtaining the relative density is through Standard Penetration Tests (see Section 3.8).

Loose to dense sands can be sampled by a thin-wall sampler tube, preferably by fixed-piston sampling methods. For dense sand, a shorter stroke is needed (e.g., 12 inches) to avoid buckling the tube. For very loose sands, which often fall out of the sampler on being pulled to the surface, an option is to use a smaller diameter tube (e.g., 2-inch diameter instead of 3-inch) to provide more side friction between the sand and sampler tube per unit volume of sample.

On the issue of disturbance during thin-wall sampling of sands, many investigators believe that the amount of disturbance is unacceptable, with loose sands becoming denser and dense sands becoming looser during the sampling process (Marcuson and Franklin, 1979). It is difficult to prove one way or the other. The author believes that the disturbance during piston sampling is within acceptable limits for sands at relative densities around the critical levels for liquefaction studies.

The *Bishop compressed air sand sampler* (Bishop, 1948) has been used successfully to take loose to medium dense sand samples (Nixon, 1954). It has not been used extensively in recent years but remains a viable option for collecting relatively undisturbed sand samples for relative density measurements.

The principle of operation is shown on Figure 3.24. The brass sampling tube is 16 inches long, 2.375 inch i.d., and has an end area ratio of 11 percent. The sampler, inside an outer shell, is lowered to the bottom of a water-filled casing (Figure 3.24a). At the bottom of the hole, the sampler is jacked into the sand (by pulleys or hydraulic head of a drill rig) by means of a guide rod attached to the sampler head. The drill rods are removed and compressed air is fed into the guide rod to close the port in the sampler head and expel water from the outer shell (Figure 3.24b), creating an air chamber similar to a diving bell. The sampler is raised into the outer shell by a cable where it is held in place by the compressed air (Figure 3.24c). The outer shell and sampler are pulled to the ground surface by a cable. A bicycle pump is sufficient to provide the compressed air.

Another sampling technique is to *freeze the ground* and then core through it (Yoshimi et al., 1978; U.S. Corps of Engineers, 1996). This is an expensive and time-consuming procedure. It is unlikely to be used in landslide remediation or prevention except for research purposes.

Gravels, Boulder Clays

Soils containing gravel-size stones cannot be recovered in a relatively undisturbed condition by tube sampling. One option is to cut block samples from pits, shafts, or adits.

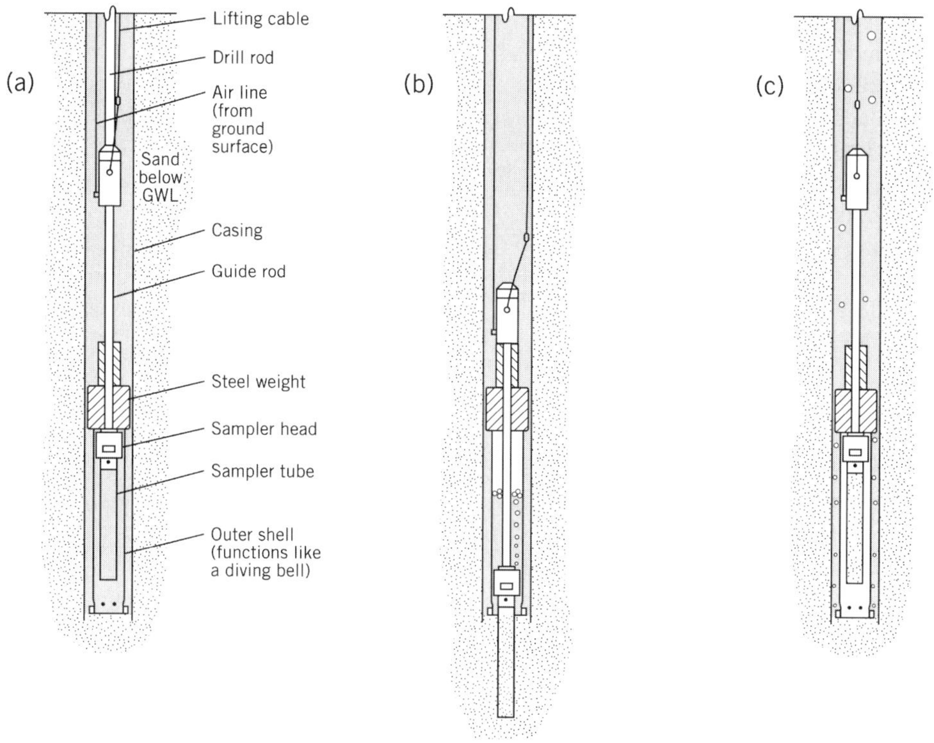

Figure 3.24 Bishop compressed air sand sampler (after Bishop, 1948)

Rocks

Cemented rocks or highly indurated soils can be recovered by rock coring. *Triple-tube core barrels* are recommended for high-quality samples and should be taken in relatively short runs (2 to 5 feet) especially around the slip surface of the landslide. The minimum core size should be HQ (2.50 inches diameter) but higher recovery ratios generally are obtained in larger size cores, such as PQ (3.35 inches diameter). The circulating medium can be water, drilling mud, or compressed air.

The triple-tube core barrel sampler has an outer tube with the cutting bit at the lower end and an inner stationary tube. The drilling fluid passes between the inner and outer barrels before discharging through ports in the bit. The third tube is a liner inside the sampler that collects the core and can be used for shipping.

Handling Samples

Handling undisturbed samples after recovery is well understood within the geotechnical community. The details are omitted here. The Corps of Engineers manual on Soil Sampling (EM 1110-1-1906) is a good reference on handling procedures.

3.10 TEST PITS, TRENCHES, SHAFTS, AND ADITS

Excavations into landslides can sometimes provide valuable information on ground conditions that may not be apparent (or present a confusing picture) from borings alone. Test pits and trenches can be dug to depths of about 12 feet by relatively inexpensive earthmoving equipment, and deeper excavations can be made by larger equipment. Safety is always an issue, especially if the ground is very loose, intermixed loose and compact soils, or waterbearing soils. In general, personnel should not enter an excavation with vertical sides that are more than 4 feet deep without shoring or some other form of protection against side collapse.

For landslide work, *test pits* and *trenches* can be used to examine and measure the ground breaks (headscarps, landslide side edges, toe eruption, etc.) that aid analysis. For example, the very steep headscarp at Pelton Upper Slide (Case History 6) was excavated below the level of the remedial trench to check that the shear surface continued to plunge steeply. This helped to validate the double wedge failure model and the assumed shape of the upper wedge failure surface.

The location of a landslide toe may be apparent from heaving ground, but is usually less visible than the headscarp. A toe trench excavated in the direction of sliding can provide a better understanding of the slip surface as it rises to the ground surface.

Test pits and trenches are not needed on the large majority of landslides. If they are needed, the consultant should use a contractor with extensive experience of this type of construction to excavate and support deep temporary excavations.

Shafts and *adits* are relatively costly to construct and usually can only be justified on large to very large landslides, unless it is the intention to use them for both site explorations and remediation (e.g., drainage tunnel). The most common other reasons are (i) to build a shaft or adit to examine a deep-seated failure surface, or (ii) to examine the subsurface conditions that are difficult to determine from borings (e.g., coarse gravels, boulder formations, large voids, weak interbeds, etc.).

Finally, all the above types of exploration can be used to conduct in-situ tests to include a larger quantity of soil than is possible from borehole samples. Such tests include direct shear tests along the slip surface, permeability tests (falling or constant head in a pit). In complex geology, the geologist can check the continuity and nature of weak strata, interbeds, faults, etc. Large block samples can be recovered for sophisticated laboratory tests.

3.11 GEOPHYSICAL EXPLORATIONS

Use of Geophysical Explorations for Landslides

Geophysical techniques measure properties of *contrast* between materials in the ground. In all types of geophysical explorations, one or more borings **must** be put down in the survey area to help the geophysicist interpret the field measurements.

Although most geophysical surveys take measurements at the ground surface, it is also possible to suspend seismic instruments through water to obtain sub-bottom profiles. Another technique is to place the instruments in adjacent boreholes to obtain crosshole surveys of materials between boreholes.

Field time is usually short and ranges from one to three days for most projects. A field crew of two to three persons is normal. Wet weather, frozen ground, strong winds, and stray electrical currents can interfere with the field measurements.

For landslide work, the following techniques are used:

- Seismic (sonic) methods, primarily *seismic refraction*. Measures the velocity of shock waves passing through the ground and their return to the ground surface. It can only measure boundaries between soils in which the seismic velocities are progressively higher in the underlying strata.
- *Resistivity.* Detects the effect of electric currents put into the ground. The main benefit of resistivity over seismic refraction is that it can detect weak soils of low resistivity (e.g., soft clay) underlying materials of high resistivity (e.g., gravels, bedrock). However, it is rarely used for landslide work.

In this section, only a summary of the seismic refraction method will be presented.

Seismic Refraction and Reflection Techniques

The principal use of seismic refraction for landslide studies is to determine the top of bedrock which, in many cases, is close to the line of slippage. Seismic lines can be an attractive alternative to borings when access is difficult and/or the landslide covers an extensive area.

To make a seismic refraction survey, a line of geophones is laid out in a straight line from the source ("shot point") of the seismic energy. A geophone is a sensor that converts ground shaking into a voltage response. The data from all the geophones are recorded on a seismograph.

Seismic energy is provided by detonating a small amount of explosive in a shallow hole, or manually by a hammer falling onto a metal plate. The explosives method generally is preferred, but requires permits and considerable caution in urban areas. The shock wave radiates from the point source into the ground. When it strikes the boundary of a soil or rock with different transmission properties, the wave is either reflected off the surface, refracted into the adjoining material, lost as heat, or changed to other wave types. The various waves reaching the geophones are recorded. The key wave is the compressional wave (known as the P-wave for primary or pressure), which is the first wave to arrive at the geophones.

On Figure 3.25(a), the P-wave passes through layer 1 at a slower velocity than through layer 2. Dense, hard strata such as bedrock have higher P-wave velocities than soils. The sequence of increasing P-wave velocity for successively deeper strata is essential in seismic refraction surveys. Table 3.13 lists typical P-wave velocities V_P for several types of soils and rocks.

For geophones close to the shot point, the path of the wave arriving first at the geophones will be through the slower-velocity layer 1. For more distant geophones, the first arrival wave will pass downward through layer 1 and then refract through the upper boundary of higher-velocity layer 2 before passing back through layer 1 to the surface (Figure 3.25a). It is refraction that allows the thickness of individual strata to be calculated.

A plot of first-arrival times for all the geophones in the survey line is shown on Figure 3.25(b). Note that time is on the ordinate rather than the absissa of the graph, so the lower velocity V_1 shows as a steeper gradient (contrary to conventional engineering graphs). Beyond a certain point, termed the crossover distance X_c, all the first-arrival waves pass through the deeper layer 2 before reaching a geophone. As can be seen on Figure 3.25(a), the distance traveled through layer 1 is the same for all geophones that refract back to the surface from layer 2. For these geophones, the time difference from one geophone to the next is the time required to pass along the higher velocity horizontal boundary of layer 2. Thus, velocity V_2 can be measured off the graph, Figure 3.25(b). The depth Z of layer 1 can be determined from Eq. (11) accompanying this graph.

When multiple strata and non-parallel surfaces are involved, the analysis becomes increasingly complicated and is

Table 3.13 Typical Compressional (P-Wave) Velocities V_P for Various Soils and Rocks*

Material	V_P (ft./sec.)	Material	V_P (ft./sec.)
Damp loam	1000–2500	Weathered rocks	1500–12,000
Dry sand	1500–3000	Shale	2600–12,000
Clay	3000–6000	Sandstone	7000–13,000
		Unweathered basalt	8500–14,000
Saturated loose sand	5000	Unweathered granite	15,500–22,000
Till	5500–7500		
Fresh water	4800		
Salt water	5250		

*from U.S. Corps of Engineers Design Guide No. 23

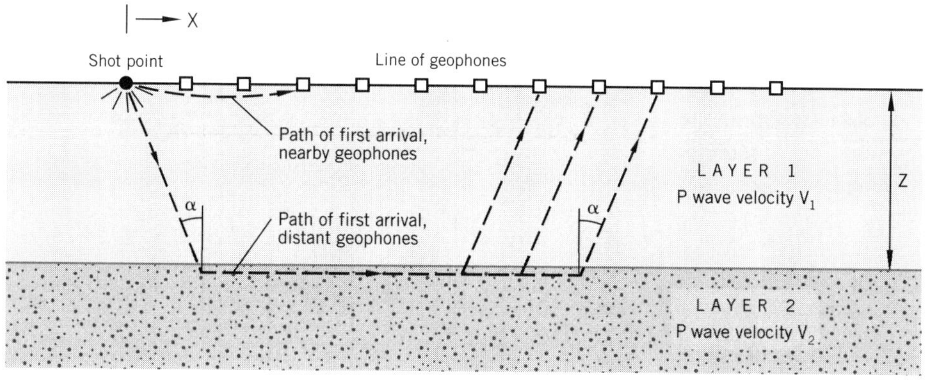

(a)

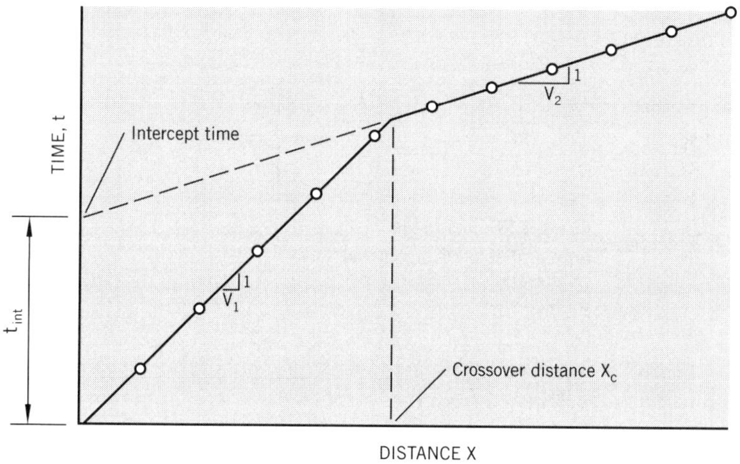

(b)

$$\text{Depth } Z = \frac{t_{int}}{2} \cdot \frac{V_1 \cdot V_2}{\sqrt{(V_2^2 - V_1^2)}} \qquad \text{Eq. (11)}$$

where V_1 = P wave velocity in Layer 1

V_2 = P wave velocity in Layer 2

t_{int} = intercept time

Figure 3.25 Seismic refraction:
(a) principle of seismic refraction in a two-layer system with parallel boundaries
(b) time-distance plot for first-arrival waves for a simple two layer system with parallel boundaries

outside the scope of this book. However, the principles of refraction apply to the paths taken by energy waves.

Seismic refraction surveys are attractive to consulting engineers and owners as a cost-effective way of extending subsurface information obtained from borings. Other advantages are that (i) they can be used to measure groundwater levels in very pervious sands, gravels, and boulders (seismic velocity is usually much higher in saturated soils compared to dry or damp soil), and (ii) the results are unaffected by scattered boulders that sometimes interfere with borings.

The main disadvantage of seismic refraction surveys is that they often do not perform well in showing a non-linear profile of bedrock between borings. The author has experienced this problem on several sites using different geophysical subcontractors. Such unreliability may lead to construction-related claims for differing quantities. The author has rarely used geophysical explorations, preferring instead to spend available funds on additional borings that provide factual rather than interpreted data.

Other disadvantages of seismic refraction surveys are: (i) they cannot measure the presence of lower velocity soils below higher velocity soils (as previously mentioned), (ii) they cannot separate many types of overburden soils (such as alluvium) below the water table because there is insufficient velocity difference (Table 3.13), (iii) they do not provide a reliable interpretation of the weathered zone thickness of bedrock, which is often thin and transitional in nature, and (iv) assumptions of homogeneity and isotropy may cause significant errors in interpretation.

Subbottom Profiling

This technique uses the principles of seismic reflection to provide a picture of the soft sediments below a body of water. It has been developed by the Waterways Experiment Station of the U.S. Army Corps of Engineers (Ballard et al., 1993). The method can replace or supplement overwater borings.

The seismic source depends on the required depth of penetration and resolution. As the energy level increases, penetration increases but resolution decreases. The seismic sources are referred to (in increasing strength) as pingers, boomers, sparkers, and airguns. The resolving accuracy ranges from <0.6 feet for a pinger to >3 feet for an airgun (U.S. Corps of Engineers, 1998).

The seismic source is towed on a sled or catamaran, which in turn tows a line of hydrophones. These measure rapid and continuous reflection soundings of the soils beneath the water. The position of the source is obtained from a GPS for the time of each firing along the tow path. The data are analyzed to obtain the P-wave velocities and depth of the subsurface layers. A continuous profile can be obtained for each tow path. At Skagway, Alaska (Case History 7), subbottom profiling provided dramatic pictures of the scoured edges of the flow slide.

3.12 FIELD VANE TEST

The field vane test is an excellent method of measuring the undrained shear strength of soft clays in-place. It avoids the necessity of trying to obtain high quality undisturbed samples in soils that are prone to disturbance in handling.

Equipment

The vane shear test consists of inserting a four-bladed vane in undisturbed soil and rotating it to measure the torsional force needed to shear a cylindrical surface of clay. The torsional force is then converted to average undrained shear strength.

Vanes can be rectangular or tapered, as illustrated in Figure 3.26. All field vanes have a height : diameter ratio of 2 and vane diameters range from 1.50 to 3.62 inches. The steel rod attached to the vane is 0.5 inch diameter, and blade thicknesses are 0.06 inch (for blades up to 2-inch diameter) and 0.125 inch (for the larger blades). Dimensions and specifications are different in countries outside the United States. All blades have a sharp cutting edge at the lower end. The field vane is manufactured as either (i) a vane on the end of a rod, or (ii) a vane enclosed by a low friction sheath.

Field Test Procedure

The vane is lowered to the bottom of the borehole, and the vane is pushed a distance of at least 5 hole diameters into the soft clay. When the vane has reached the desired test depth, the vane is rotated at a standard rate of 6° per minute until the maximum torque is reached. Readings are usually taken at 15-second intervals. Failure generally occurs 2 to 10 minutes after the start of the test, softer clays taking longer than stiffer clays. After failure, the vane is rotated rapidly for 10 revolutions, and a second test is conducted to measure the remolded strength. Tests are typically taken at 2.5- to 5-foot intervals within the clay stratum.

To consistently apply the constant rate of rotation, some vanes have a worm gear and hand-crank mechanism. Additionally, tests should be performed to measure the rod friction by removing the vane head and following the same procedure. The torque measured in the rod friction test is subtracted from the torque of the vane tests to obtain the net torque T applied to the vane. In the case of the sheathed vane, a similar test measures the friction in the bearings between the rod and the sheath.

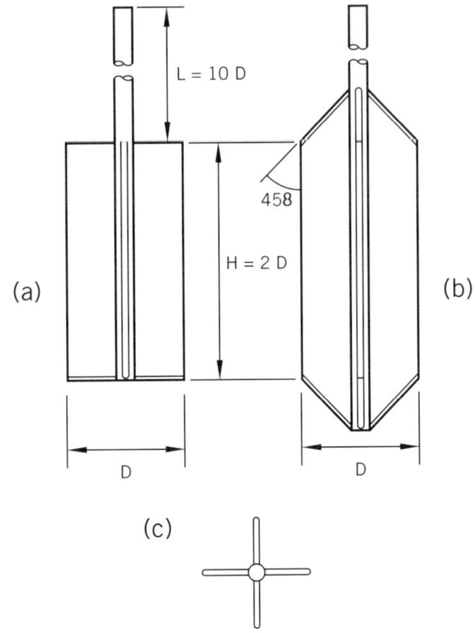

Figure 3.26 Field vane:
(a) elevation of rectangular vane
(b) elevation of tapered vane
(c) plan view

The undrained shear strength c of the clay (in lb./sq. ft.) is calculated from the formula:

$$c = \frac{T}{K} \qquad \text{Eq. (12)}$$

where T = net torque in lb.-ft.
K = vane constant in ft.^{3}, depending on the dimensions and shape of the vane

Values of 1/K are given in Table 3.14.

The usual interpretation of the vane test is that the soil is sheared vertically along the cylindrical perimeter and horizontally across the two ends of the vane. For the rectangular vane, for example, with a height of twice the diameter, it can be shown:

$$c = \frac{6T}{7\pi D^3} \qquad \text{Eq. (13)}$$

where D = vane diameter in feet

The field vane is usually limited to clays with undrained shear strengths of less than 1,000 lb/sq. ft. Although the test is

Table 3.14 Values of 1/K for Various Common Vanes with Length : Diameter Ratio of 2

	Value of 1/K in ft.$^{-3}$			
	D = 1.5 inch	D = 2.0 inches	D = 2.5 inches	D =3.625 inches
Rectangular vane	139.7	58.93	30.17	9.897
Tapered vane	132.3	55.65	28.47	9.333

D = vane diameter

strongly recommended for measuring the strength of very soft and soft clay strata, the main drawback is that the soil being tested is not seen. This drawback can be partly overcome by following a procedure of alternating vane tests with SPT or thin-wall tube sampling so that the intervening soil strata can be sampled. The test is unsuitable for sands and silts, which may occur as lenses or thin strata within alluvial or lacustrine sediments. Stones and shells, if present, also affect the test results. Any unusually high test results from the strength-depth profile should be ignored as likely involving sand or hard inclusions (e.g., shells) within the clay.

Interpretation of Field Vane Tests

The vane test does not precisely duplicate the conditions for mobilizing shear strength in landslides. The differences have been attributed to four factors:

- *Time to failure.* The field vane test is completed in a few minutes but slope failures on soft foundations can take days or weeks to mobilize the shear strength. It is well-known from laboratory studies that the undrained shear strength of a soft clay decreases with an increasing time to failure. Therefore, for this factor, the field vane test is likely to overestimate the actual field strength of the clay.
- *Clay anisotropy.* The field vane test primarily measures shear strength in the vertical plane; in most ground failures shear occurs along inclined or horizontal surfaces. Horizontal surfaces are often weaker than vertical surfaces due to bedding planes (e.g. alluvial flood plain or lacustrine deposits).
- *Progressive yielding.* Slope failures generally cannot mobilize the peak shear resistance simultaneously at all points along the slip surface. Yielding develops progressively from the more severely stressed zones of the potential slip surface to less severely stressed zones. Thus, the average shear strength mobilized along the slip surface is less than the peak. This issue affects clay soils with a peak in the stress-deformation curve, usually clays that have some degree of overconsolidation.
- *Soil disturbance.* Pushing the vane into the undisturbed soil below the borehole has the same effect as pushing a sampler tube into soil. Experiments by La Rochelle et al. (1973) with vanes of different blade thicknesses demonstrated that the measured vane strengths decreased with increasing blade thickness (Figure 3.27). This factor is likely to cause the measured vane strength to be lower than the actual strength. For a rectangular blade of 2 inch diameter, vane thickness of 0.062 inch, and rod of 0.5 inch diameter, the end area ratio (defined in Section 3.9) is about 12%.

The most practical way to overcome these concerns is to correlate field vane results with back-calculated strengths from actual slope or foundation failures. This was first done by Bjerrum (1972, 1973) in the form of a correction factor, and additional data have been put into Bjerrum's chart by others (Figure 3.28). As might be expected, the scatter is significant, but the trend can be clearly seen.

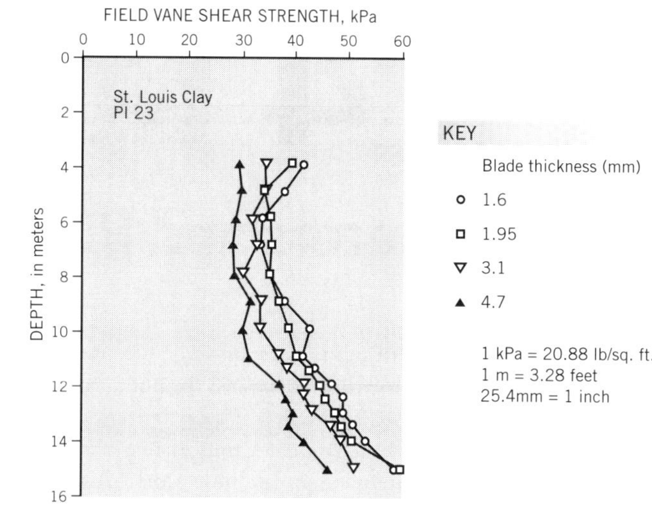

Figure 3.27 Effect of soil disturbance due to vane blade thickness on measured vane shear strength (after La Rochelle et al., 1973).

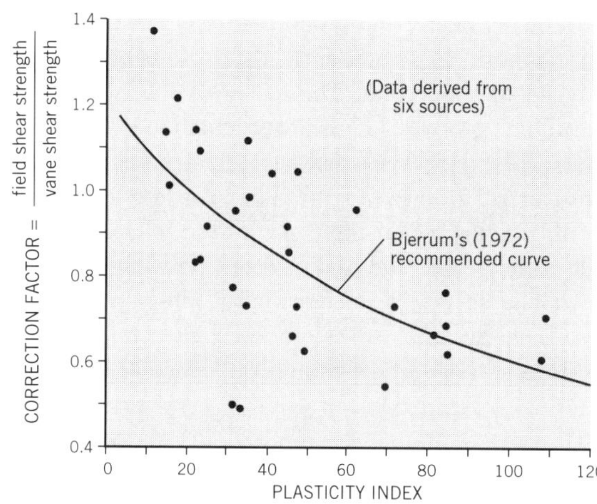

Figure 3.28 Correction factor for field vane shear test (after Bjerrum, 1972).

CHAPTER

4

Inclinometers and Piezometers

An essential part of modeling a landslide is to determine the location of the slip surface and the pore water pressure acting on it (Figure 4.1). These data are obtained by field instruments. Because of the importance attached to obtaining these measurements in landslide work, this chapter has been separated from Site Investigations. Comprehensive coverage of these and other field instruments is available in Dunnicliff (1988).

4.1 INCLINOMETERS

Summary of Technique

Lateral movements below the ground surface can be measured by an inclinometer system. First, a special casing (Figure 4.2) is installed in a borehole. The inside of the casing has four longitudinal grooves (at the four quadrants) and the inclinometer probe has wheels that track along a diametrically opposite pair of grooves (Figure 4.3a). An accelerometer within the probe, aligned in the plane of the wheels, measures the tilt of the probe (and casing) at any position along its length. By taking successive incremental readings (measurement interval, Figure 4.3b) as the probe is pulled up the casing, the in-ground shape of the casing is obtained. If landslide movements occur after the casing has been installed and initially read, the tilt of the casing in the shear zone of the landslide will change (Figure 4.4a). The depth and amount of shear movement is obtained by subtracting the initial set of tilt readings from the subsequent readings. In the example shown on Figure 4.4(b), a shear movement of 0.25 inch has occurred over a 4-foot thick shear zone 32 to 36 feet below ground surface. In landslides, the lower sections of the inclinometer casing should always be installed in stable ground and thus provide a fixed reference for movements in the casing above it. Embedment into stable ground should be 10 to 20 feet, wherever feasible.

Application of Inclinometers to Landslide Work

Inclinometers are probably the most valuable tool available to a landslide analyst and their use, whenever economically possible, cannot be overemphasized. Inclinometers provide information on:

- The *depth* of landslide movements; this is essential for modeling the landslide shape.
- The *thickness* of the shear zone (typically 1 foot to 5 foot deep zone, commonly referred to in this book as the *discrete* shear zone); this is especially needed for (i) shear pile

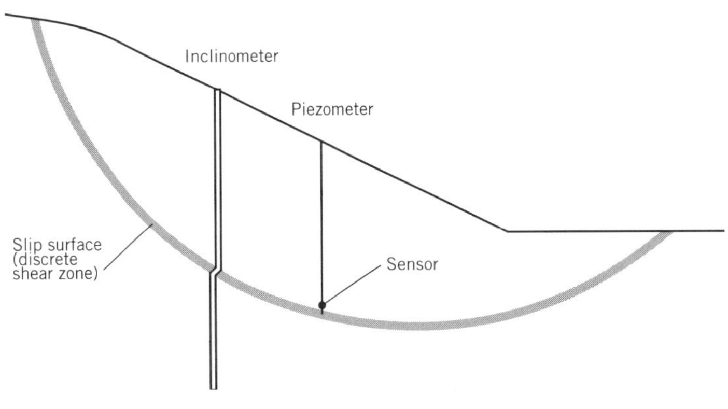

Figure 4.1 Monitoring landslides with inclinometers and piezometers.

design which is governed by the bending moments developed across the shear zone; (ii) selection of samples for laboratory shear tests, and (iii) the location for installing an in-place inclinometer, if needed.

- The *amount* of movement; this can be measured for a few inches of simple shear displacement at the discrete shear zone, the amount depending on the deflected shape of the casing, casing diameter, and the length of the probe. When the inclinometer probe is unable to pass through the shear zone, further movements can be measured, if needed, by surface survey hubs or by replacing the inclinometer casing.
- The *rate* of movement; this may be needed (i) to estimate the "static" factor of safety in back analysis (see Chapter 9, Section 9.4), (ii) to measure the variation in rate with rainfall or other causation factors, or (iii) to confirm the effectiveness of a landslide remediation and demonstrate that a landslide has been stabilized (i.e., rate is zero).
- The *direction* of movement; this may be obvious in most landslides, but directions can vary on large slides where differential movements may occur in parts of the site due to subsurface obstructions or change in slip surface characteristics.

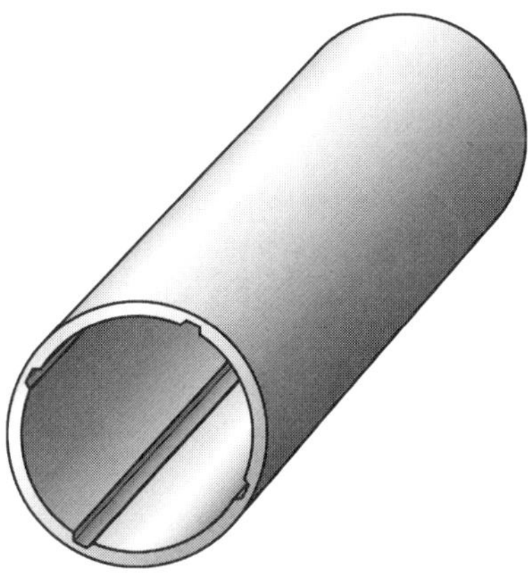

Figure 4.2 Isometric view of inclinometer casing showing internal longitudinal grooves.

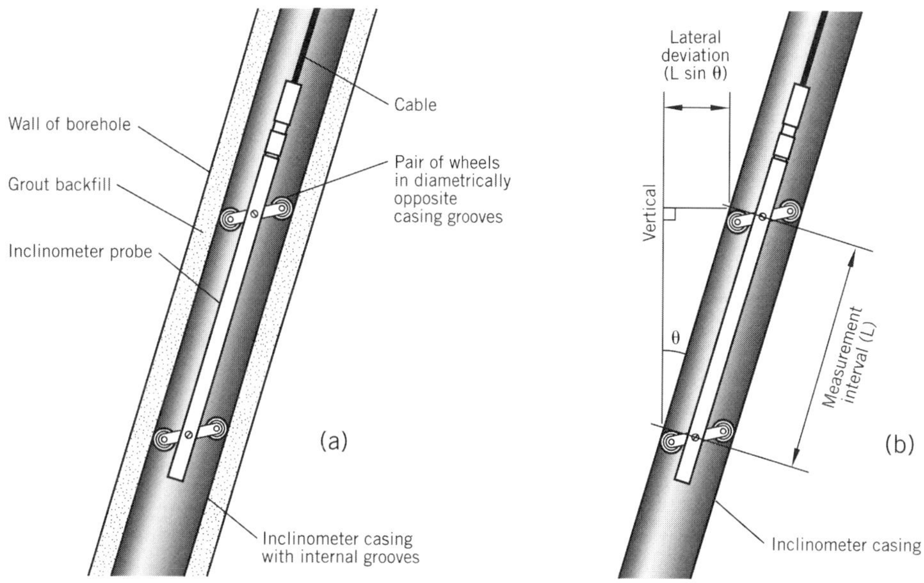

Figure 4.3 Inclinometer system:
(a) probe and casing within borehole
(b) measurement of tilt
(Slope Indicator Co.)

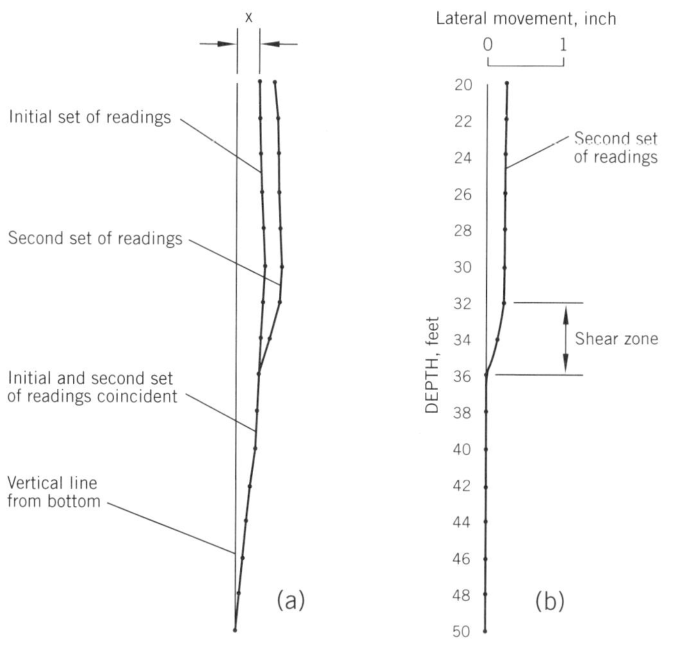

Figure 4.4 Example of inclinometer data:
(a) shape of casing in the ground for two sets of inclinometer readings
(b) determination of shear movement and depth of shear during time interval between reading sets

Note: This drawing is to illustrate the principle only. In practice, distance x is likely to be many inches (or possibly feet) out of vertical.

A useful (and underused) technique is to install inclined inclinometer casings to determine the depth of slippage on steep and/or inaccessible points of a landslide. A forested landslide in Montana (Figure 4.5) can serve as an example. The upper reach of the landslide is omitted from the figure. The landslide is about 2,000 feet long and 300 feet wide, and the depth of sliding is likely to be at least 100 feet below the surface. A road (termed "Middle Road" on the figure) crosses the landslide near the mid-length of the slide and provides easy access for drilling equipment. At this location, it can be seen that a vertical inclinometer casing, supplemented by two other casings at 30° to the vertical, would provide good information on the depth and slope of the slip surface near the mid-length of the slide. Since the top (headscarp) and toe of the landslide are known, the overall shape of the landslide slip surface can be obtained despite the slope being inhospitable for exploration equipment.

When using inclined casings, it should be recognized that the accuracy of the inclinometer measurements decreases with increasing deviation from vertical. This is rarely of concern where the principal objective is to measure the position of the slip surface. Examples of using inclined inclinometer casings on landslide studies are given in Case History 9 and in Chapter 10, Section 10.2.

In addition to these direct landslide applications, inclinometers can be used in special situations related to landslide work. For example, inclinometers can be installed within a concrete pile or attached to a steel H-pile to measure the deflected shape of shear piles used for landslide remediation. This application not only provides reassurance on satisfactory performance but can be used to determine bending moments as a check on the design calculations. In another application, inclinometers can be installed prior to construction to observe lateral movements caused by removal of support during construction. This use may be needed where landslide remediation is close to buried cables or pipelines. Yet another use of inclinometers is to provide an early warning of developing movements because the inclinometer can measure subsurface lateral deflections better than any other survey procedure.

Inclinometer System

The inclinometer system has two components: a portable instrument (probe) and a permanently installed casing. Although the inclinometer probe itself has high accuracy, it is the system as a whole that controls the accuracy of the field data. There are several manufacturers of inclinometer systems, but the system offered by Slope Indicator Company (SINCO), the original developers of the Wilson inclinometer, is described herein to explain the technology.

The inclinometer *casing* is made of high-impact ABS plastic which is suitable for long-term contact with soils, grout, and groundwater. It is manufactured in 10-foot (or 3 m) lengths and has glued and riveted couplings that seal the casing to prevent grout or soil entering the casing through the joints. In recent years, a coupling has been developed (available from Roctest) that allows the casing lengths to be rapidly joined together without glue or rivets. Telescopic couplings are available to allow for ground settlement after installation. These are rarely needed on landslides.

Inclinometer casing is available in three outside diameters ranging from 1.9 to 3.34 inches. A sufficient gap should be allowed between the inclinometer casing and borehole to ensure that the annular space can be completely backfilled with grout over the entire depth. It is preferable to use the largest diameter casing wherever possible; this allows more shear deformation to occur before the inclinometer probe is unable to travel past the distorted casing segment.

The inside of the inclinometer casing has four longitudinal grooves aligned at the four quadrants (Figure 4.2). The probe travels along the grooves to take measurements. The grooves have a spiral tolerance of less than ⅓° per 10-foot length to minimize spiral error. For most landslide work, casing spiral is not a problem, but very deep installations may require the use of a spiral sensor, available from the manufacturer, to measure the precise orientation of the grooves at the shear zone of the landslide. At the Clyde project in New Zealand, for example, inclinometer casings 500 to 700 feet long were installed in deep landslides, and casing spiral was an important consideration.

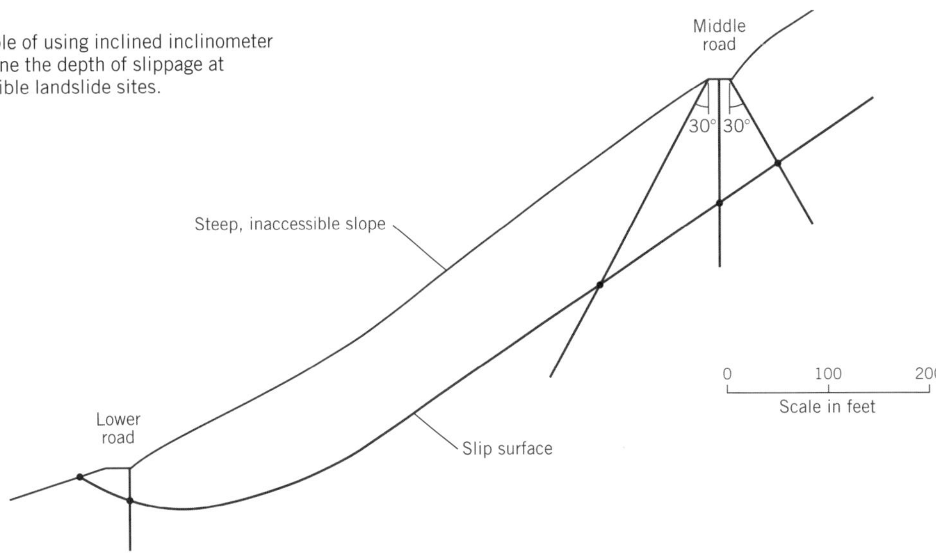

Figure 4.5 Example of using inclined inclinometer casings to determine the depth of slippage at relatively inaccessible landslide sites.

The Digitilt inclinometer *probe* incorporates two force-balanced servo-accelerometers to measure the tilt of the instrument. One accelerometer is set in the plane of the wheels that track in the grooves of the casing. The second accelerometer is set at right-angles to the first. The accelerometers measure the deviation of the probe from vertical (Figure 4.3b) and are housed in a stainless steel body, 1 inch diameter and about 32 inches long.

The wheel assemblies swivel to fit the size of casing. The wheels have sealed bearings and self-center within the grooves. The distance between the upper and lower wheels (gauge length) is 2 feet for U.S. models and 0.5 m for metric models.

The measurement range of the probe is ±35° from the vertical for the U.S. models. Metric probes have a range of ±53° from vertical. Accuracy is reduced with increasing initial tilt of the casing. The manufacturer's stated *system field accuracy,* for a borehole inclined within 3° of vertical, correctly installed casing, and proper reading techniques, is ±0.3 inch per 100 feet. However, correction of systematic errors can improve the accuracy of readings, as discussed later.

A specially-manufactured cable connects the probe to the readout unit at the ground surface. The cable has a steel core to minimize stretching, and the electrical wires are spaced around and bonded to the core. The outer jacket is neoprene and remains flexible in cold weather. The cable has markers to measure depth below ground. Measurements are related to the *middle* of the gauge length.

A pulley wheel assembly temporarily attaches to the top of the casing. It has a set of jaws so that the cable can be clamped at selected markers to obtain reproducible depth control. The pulley assembly should always be used because it not only provides consistent measurements of depth but also avoids the risk of excessively bending the cable at the top of the casing.

The readout unit (Digitilt DataMate) has electronics housed in a compact plastic case and is sealed against condensation and humidity. It stores a list of up to 40 inclinometer installations in memory. At a site, the operator selects the installation and the DataMate displays the starting depth, casing axis, etc. The readings at each depth are recorded via a hand switch. The readout unit also performs check-sums to validate the data. After screening the data, questionable readings can be retaken before leaving the site. At the office, the inclinometer data are downloaded to a computer for data processing.

Inclinometer Casing Installation

Inclinometer casings usually are installed in cased boreholes that have been sampled. It is very important that the bottom of the inclinometer casing is anchored at least 10 feet into bedrock or very stiff ground below the base of the landslide. Readings of ground movement are referenced to the stable bottom of the casing. If the investigator is in doubt about the location of the shear zone of the landslide, drilling should be continued. It is a better choice to risk wasting this additional footage than to risk missing information needed for a landslide analysis. Mikkelsen (1996) recommends an even deeper fixity of 20 feet into stable ground so that the bottom 10 inclinometer readings can provide calibration data to detect and quantify errors. However, for most routine landslide investigations, a fixity length of 10 feet should be sufficient. In very hard rock at the base of a landslide, end fixity may be limited to only a few feet due to drilling equipment limitations.

The first sections of inclinometer casing are bonded together by glue and rivets (where needed), and are suspended in the hole by a clamp at the top of the drill casing. Because inclinometer casing joints are sealed, water or drilling mud in the borehole may cause the inclinometer casing to float. If this occurs, clean water is poured into the casing to get the casing to sink.

Additional casing sections are attached to reach the desired length. If there is no standing water in the hole, glued joints can break under the weight of the suspended inclinometer casing, especially in holes deeper than 100 feet. To prevent separation, the installer may (i) rivet the joints for additional security and/or (ii) support the casing from the base via a wire rope (in this situation, the base should be reinforced with a steel plate or similar).

After the casing reaches the bottom of the hole, alignment is adjusted so that the plane of two opposing grooves is in the anticipated direction of sliding. The groove pointing downhill should be marked as the A_0 groove. The diametrically opposite groove is designated A_{180} and the imaginary plane connecting the two grooves is the A *axis* (Figure 4.6). The other two grooves (B_0, B_{180}) provide the B axis. In the past, the designations were A+ and A– forming the A *plane,* and these terms probably are still being used by many installers.

Next, the inclinometer casing should be filled with water to prevent grout from lifting it. The annular space between the inclinometer casing and sides of the borehole is grouted from the base upward to ensure that there are no unfilled gaps within the annular space. Grout can be delivered either (i) through a tremie pipe inside the inclinometer casing, or (ii) through an external tremie pipe outside the inclinometer casing.

For grouting *through* the inclinometer casing, a non-return valve (available from the inclinometer manufacturer) is attached to the base of the casing with a grout tube coming up

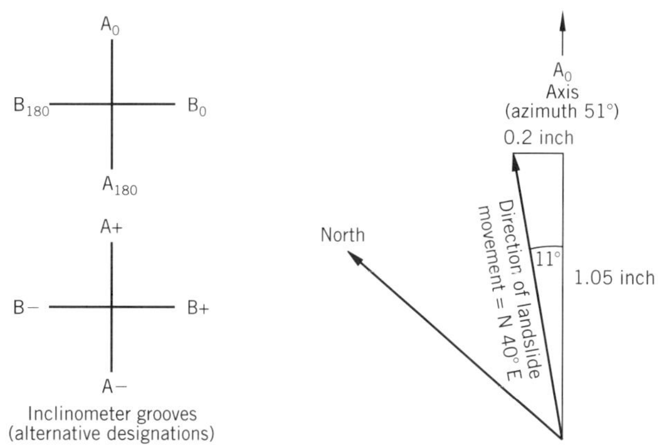

Figure 4.6 Groove designations and example of vector summation of movement.

to the surface. Grout is fed under a low pressure through the valve and slowly fills the annular space. When the grout level reaches the surface, the feed pipe is pulled off the valve and the non-return valve prevents grout from returning to the inside of the inclinometer casing. A small amount of grout within the feed pipe drops to the bottom of the inclinometer casing and is immediately flushed out with clean water. As grout backfills the hole annulus, the drill casing is carefully extracted.

The fluid pressure of grout may lift the inclinometer casing. One technique of counteracting the uplift is to leave the drill rods in the casing overnight to provide weight until the grout sets. Another choice is to externally grout in two or more stages. The method of casing restraint should be carefully evaluated, especially on deep installations, because a sudden rise of the inclinometer casing may cause serious injuries. Loading the inclinometer casing at the top is not recommended. This practice can distort the casing and increase the likelihood of measuring errors.

Grouting through the casing would seem to be an ideal way to backfill the inclinometer casing, but in practice it carries risk. The first risk is that the return valve may malfunction, allowing a backflow of grout into the inclinometer casing. This can be overcome by adding a second non-return valve in series with the first one; this adds to cost and requires a slightly deeper hole. The second risk is that some grout is left *inside* the inclinometer casing and could affect the reproducibility of readings. This second risk can be prevented by filling the grout tube with water before breaking the connection at the bottom of the hole; then additional water is flushed through the inside of the inclinometer casing to fully clean it.

The alternative grouting procedure is to grout through a tremie pipe *outside* the inclinometer casing. The grout tube, typically assembled from PVC pipe with threaded flush couplings, can be lightly attached to the casing as it connected together and lowered into the hole. Alternatively, the grout tube can be put in the hole after the inclinometer casing is in place. When the casing is at the full depth, grouting begins and the grout pipe is raised with the rise of the grout in the annular space, keeping the tip of the grout pipe a few feet below the grout surface. Some installers grout the entire annular space with the tip at the bottom of the hole and then withdraw the grout tube; others follow the same procedure but leave the grout pipe in the hole permanently.

The main advantage of grouting externally is that there is no possibility of grout coating any part of the inner grooves of the inclinometer casing. Also, if grouting has to be temporarily halted because of fluid losses into a highly permeable soil or rock stratum, the grouting can be continued later after sealing the borehole wall. Therefore, the external method of grouting is the recommended technique.

Some consultants try to match the stiffness of the grout to the stiffness of the soil in which the inclinometer casing is being placed. The intent is to improve conformance. This issue has received increasing attention in recent years. It is the author's opinion that the concern is overstated for the following reasons:

1. The main purpose for using grout as a backfill is to ensure that the inclinometer is fully supported over its entire length. In the past, sand- and gravel-filled annular spaces on deep (or unstable) holes have developed large voids where the backfill bridges the annular gap and subsequently produces erratic data.
2. At the depth of most landslides, the earth pressures are so high that the small area of grout around the casing offers insignificant extra resistance to the shearing soils. Since cement grout has low tensile strength, the backfill cracks and conforms to the ground movements irrespective of the grout strength.
3. As a practical matter, it is impossible to provide a single conforming grout for landslides due to significant variations in soil strength (modulus) with depth, the bedrock properties, and the changing properties of the grout as it hardens over time.
4. The principal impediment to soil/casing conformance is the very stiff plastic casing itself. This is a benefit on landslides with a very thin shear zone because a sharp bend produced by true soil/casing conformance would prevent the probe from passing through the shear zone after very little movement. Therefore, a high degree of conformance is undesirable in many landslides.

Grout backfill for inclinometers is usually made from a cement-bentonite-water mix. There are no set standards and mixes vary widely; usually, an attempt is made to match the grout to the soil type (Table 4.1). The bentonite provides some plasticity to the grout and prevents shrinkage on setting. A key benefit of bentonite in the mix is that it helps to suspend cement particles in a high water-cement ratio mix and minimizes bleed (Mikkelsen, 2002).

The *cement* is added to the water first and mixed. Sodium bentonite powder is added slowly and mixing continues until the slurry reaches the consistency of a thick cream. A field test is that it should slowly drop off an extended finger and make a small "crater" on hitting the slurry surface in the mixing tank. If the grout becomes too thick, it will be difficult to pump.

Loss of grout into the surrounding soil can present serious difficulties at some sites. Permeable gravels, openwork boul-

Table 4.1 Typical Mixes for Cement-Bentonite Grout in Inclinometer Installations

	Mikkelsen (2002)		Landslide Technology	
	Soft Soils	Medium to Hard Soils	Soft Soils	Stiff Soils
Water	75 gallons	30 gallons	35 gallons	35 gallons
Cement	94 lbs.	94 lbs.	94 lbs.	188 lbs.
Bentonite	39 lbs.*	25 lbs.*	25–37 lbs.*	12–15 lbs.*
28-day compressive strength	about 4 psi	about 50 psi	—	—

(*Note:* 1 sack of cement = 94 lbs.)

*approximate quantities as required to obtain correct grout consistency for pumping—see text

ders, a crossflow of groundwater, and fractured rocks are some of the causes. These require innovative techniques to complete the backfill. Openwork voids, for example, usually require a more viscous grout mix with inclusion of sand or fine gravel, and/or a quick-setting additive. Such choices depend on the size of the hole and require careful and sometimes time-consuming techniques to achieve success. In some cases, it may be appropriate to first grout the open hole and then redrill to install the inclinometer casing. The initial grouting seals the walls of the hole.

Inclinometer installers sometimes use uniform coarse sand as a backfill, and such sand gradations (e.g., No. 8/No. 12 or No. 10/No. 20) are readily available in the United States. In shallow inclinometer installations (e.g., less than 50 feet deep) the sand can be poured down the annular space in short depth increments and tamped into place by a threaded length of pipe. Plenty of water should be added to prevent the sand from bridging the annular space. Pea gravel is another backfill option for shallow holes. Sand or gravel backfill is commonly used in rapidly moving landslides (e.g., more than 0.5 inch of shear per day). They are also used on landslides where a grout mix operation is impractical due to location or other difficulties. In general, however, grout backfill should be the first choice for inclinometer casing installations.

When backfilling is completed, the top of the inclinometer casing is cut off either about 1 foot above the ground surface (in protected areas, so that it can be easily seen) or close to the ground surface (in areas where traffic or vandalism may damage an above-ground installation). The casing "stick-up" height should be noted. The inclinometer casing should be cut at an even footage (either 2, 4, 6, or 8 feet). The horizontal orientation (azimuth) of the A_0 groove should be measured by a compass and the A_0 groove marked for identification. A lockable protective monument should be added.

The installation of inclinometer casings should never be entrusted to inexperienced personnel. Many of the difficulties encountered in interpreting inclinometer data have resulted from poor or unknown installation procedures. If problems do occur during backfill, it is important for the geotechnical consultant to know how they occurred and what actions were taken. Therefore, the consultant's representative needs to be on site full time during this work.

Inclinometer Readings

After the inclinometer casing has been installed, the probe should be run down and up the casing to check that there are no obstructions. "Dummy" probes can be used for this purpose rather than to risk damage to the reading probe. Obstructions are a rarity. (A "dummy" probe also is useful after significant movements have occurred to check that a probe can pass through the shear displacement zone.)

Assuming the landslide is dormant at the time the inclinometer casing is installed, it is advisable to wait 1 to 3 days for the grout to set before initial readings are taken. The initial readings are of key importance because all subsequent readings are referenced to the changes from the initial readings. For this reason, it is good practice to take two sets of initial readings either on a single site visit or on successive days. The average of the two sets at each depth can be used as the reference. Should there be an obvious anomaly in one or more individual readings, these readings can be rechecked during the second set of initial readings. This procedure eliminates unusual kinks that sometimes persist in all future sets of readings when one of the initial readings is anomalous. Some computer programs cannot average two sets of initial readings. In this case, the engineer should carefully examine the two data sets and select one as the "initial" set.

Readings are taken at even numbers of feet; this avoids the undesirable practice of the probe wheels being at the casing joints. For example, at a depth of 20 feet (as measured to the center of the probe), the joint is at 20 feet and the wheels are at 19 and 21 feet, away from the joint.

The inclinometer probe is placed in the casing with the upper wheel in the A_0 groove. It is lowered to the bottom of the casing and the cable is clamped in the jaws of the pulley wheel assembly on the foot marker corresponding to the lowest depth reached in the initial reading set. The probe should be left in this position for approximately 10 minutes to allow the temperature of the probe to equalize with that of the water in the casing. This procedure prevents errors due to sensor warm-up drift. Next, the probe is raised to the surface in 2-foot increments (U.S. practice) with readings being taken on the A_0 and B_0 directions at each depth increment. The B_0 direction is at 90° clockwise from the A_0 and the B axis tilt is measured by the second sensor within the probe. After reaching the surface, the probe is turned through 180° and another set of readings are obtained in the A_{180} and B_{180} directions. This second set of readings should have the opposite tilt of the first set. The sum of the two sets, known as the *checksum,* at each depth should be close to zero. Small differences are measured, however, due to variations in casing grooves, the positioning of the probe, and some "zero-offset" in the probe. For the zero-offset alone, the checksums should be constant, as explained later.

SINCO literature states that checksums for the A axis should be within 10 units of the mean checksum for that axis. For example, if the mean checksum is 2, acceptable A axis checksums can range from −8 to +12. The suggested limit for the B axis is twice as high: 20 units range. These checks can be made in the field and, if necessary, repeat readings can be taken before leaving the site.

The quality of a data set can be validated against earlier sets of readings by statistics. The SINCO DataMate calculates the mean and standard deviation of the checksums. This is compared to the typical standard deviation obtained previously at the same site. If it is within 3 to 5 units of typical, the data set is acceptable. For example, if the typical standard deviation is 4, then subsequent reading sets can be accepted if they do not exceed 7 to 9.

For the data set of 4/17/2000 for the landslide in Weber County, Utah, shown on Figure 4.7, the *quality of data* checks are shown on Table 4.2.

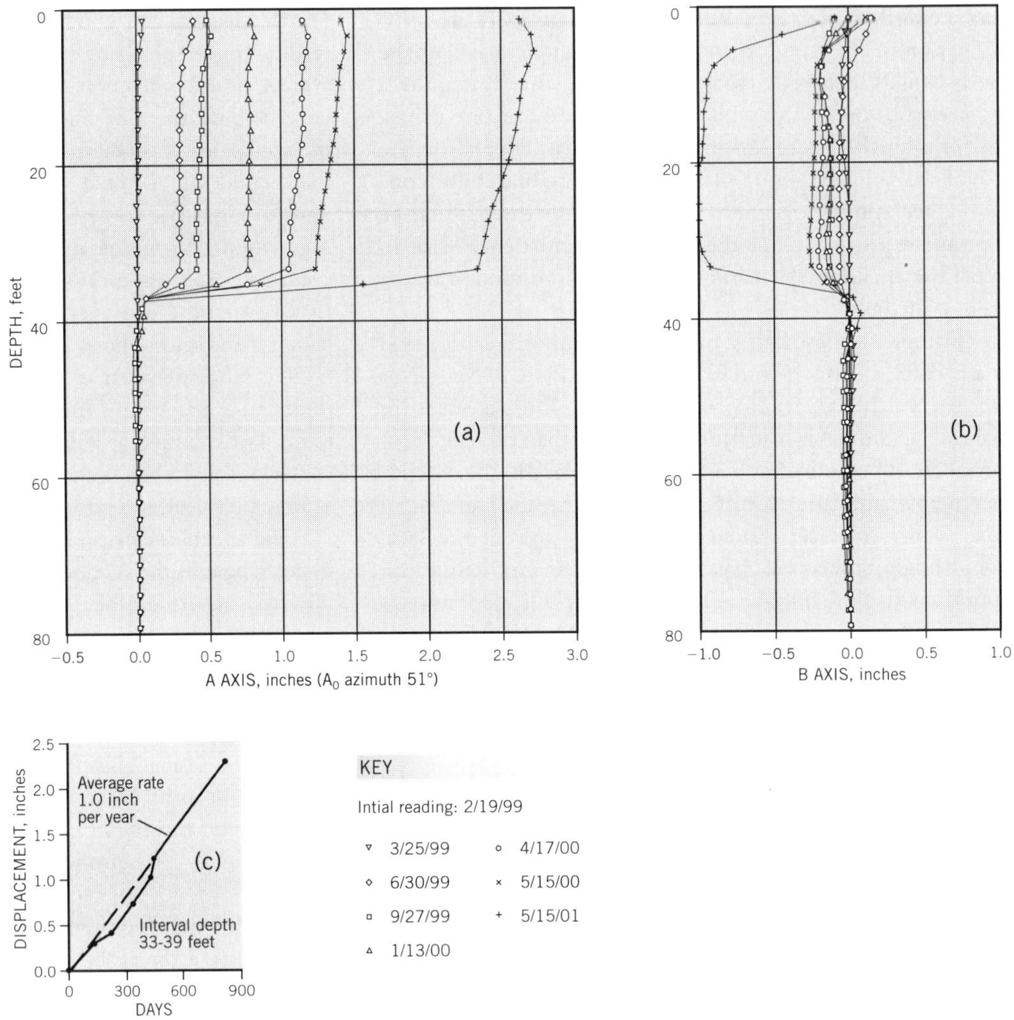

Figure 4.7 Example of inclinometer data plots for a landslide in Weber County, Utah, using Slope Indicator DigiPro Software; data corrected for bias shift error.

The following handling practices need to be observed:

- The probe should not be dropped onto a hard surface or the bottom of the inclinometer casing. A heavy jolt can damage or break the sensors. If it accidentally receives a heavy jolt, it should be checked carefully through test surveys in the field or be checked out by the manufacturer.
- The cable should be attached to both the probe and the readout box under completely dry conditions; this should occur within a vehicle or building at the site. Do not allow any moisture into these electrical connections.
- The cable should be transported as a large diameter coil (cable reels are available for very long cables). Use the pulley wheel assembly to prevent sharp bends in the cable when taking readings.
- The probe should be cleaned, dried, lightly oiled, and replaced in the carrying case before leaving the site. The carrying case should be placed on a seat and not the hard bed of a truck.
- Follow the manufacturer's recommendations for care and maintenance of the probe and cable.

Inclinometer probes from the same manufacturer should be interchangeable. However, it is advisable to use the same probe, same cable, and same operator on a project whenever possible. Changes of cable can be especially troublesome because of small variations in the length markings between cables.

Data Reduction and Interpretation

As shown on Figure 4.3(b), the inclinometer probe measures the slope (tilt) of the inclinometer casing. When the probe is raised by the gauge length of 2 feet, the lower set of wheels moves up to the position occupied by the upper set of wheels in the previous reading. Thus, the tilts measured at each read-

Table 4.2 Quality of Data Checks: Example for Weber County Data Set of 4/17/2000

Standard Deviation of Checksums:	
Typical value for installation	3
Value for current readings on 4/17/00.	2
Value of 2 is less than 3 (typical) + 3 to 5 = 6 to 8.	Acceptable* (√)
Individual Checksums:	
Mean value on A axis	23
Range of checksums for current readings on 4/17/00	17–28
Range is within ± 10 units of mean.	Acceptable* (√)

**Source:* Digitilt DataMate Appendix 1 (2003)

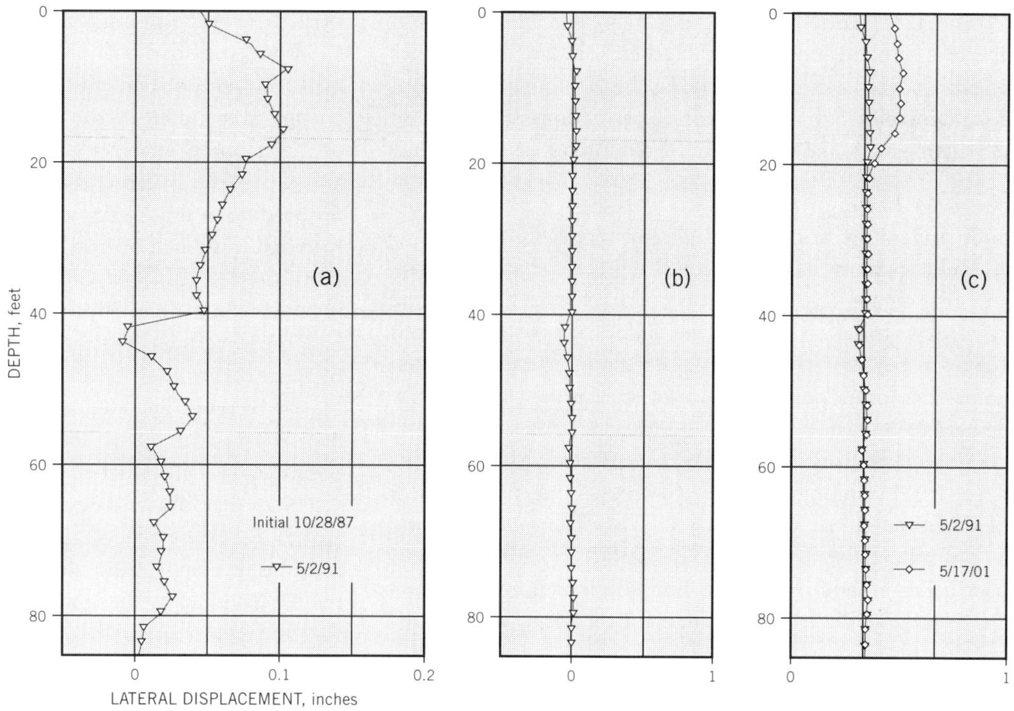

Figure 4.8 Effect of using highly exaggerated horizontal scales:
(a) horizontal scale 1 inch = 0.1 inch
(b) same data to a horizontal scale 1 inch = 1 inch with bias shift systematic error correction applied
(c) later data showing the actual shear movement at a depth of 15 to 21 feet (including bias shift error correction)

ing, when plotted on a graph, show the shape of the casing as it exists in the ground. In a poorly drilled boring, the base of the casing may be offset several feet from a vertical line below the top. A vertical or near-vertical hole alignment generally provides more accurate inclinometer data than a hole with greater deviation from vertical.

In landslide studies, the second and subsequent sets of inclinometer readings are compared to the initial set of readings, it being assumed that the bottom of the casing is anchored in stable ground. The data reduction is performed by a computer program and provides a plot of the lateral deflection against depth for the plane of the grooves. If the A grooves have been perfectly aligned in the plane of sliding, the entire shear movement will be shown in the A-axis plot and no movement will have occurred within the B axis. In practice, the major part of the shear movement takes place along the A axis and a minor amount occurs in the B axis at right-angles.

Normal practice is to plot the inclinometer data in the A and B axes as graphs of lateral deflection vs. depth (Figure 4.7). Scales of 1 inch = 1 inch horizontally and 1 inch = 10 feet vertically are recommended. The horizontal : vertical scale exaggeration is thus 120. Allowing for errors between sets of readings, a shear movement of 0.1 to 0.2 inch over a 5-foot depth generally can be relied upon as confirmation of actual movement. This assumes, of course, that a good casing installation was made and good reading practice was followed. A horizontal scale of 1 inch = 0.5 inch and a vertical scale of 1 inch = 20 feet are also common plots.

Some engineers plot the lateral displacements at highly exaggerated scales, either hoping to detect landslide movement at the earliest opportunity or through misunderstanding the accuracy of the inclinometer *system*. Such plots can be interpreted incorrectly. In most cases, these alleged small movements do not exist and can be attributed to either systematic errors of the instrument, poor backfill practices, or variances in field reading techniques.

An example is given on Figure 4.8. The graph on the left is drawn to 1 inch = 0.1 inch of lateral displacement. It can be seen that there are various apparent movements with depth. When the scale is reduced to 1 inch = 1 inch of lateral displacement and the bias shift error (see later) has been removed, the resulting graph Figure 4.8(b) shows no discernible movement. Readings taken 10 years later, Figure 4.8(c), show shear movements from 15 to 21 feet. The readings below 21 feet are practically identical to the readings taken 10 years earlier.

Care should be taken to avoid reaching interpretative conclusions too quickly. It is usually advisable to wait for an additional set of readings before acting on the evidence of very small lateral movements.

The actual amount and direction of movement are determined by vector summation of the two components measured in the A and B axes. For example, in the data of Figure 4.7, the measured shear movement in the A_0 axis on 4/17/00 is 1.05 inch and in the B_{180} axis is 0.20 inch. By vector summation (Figure 4.6), the actual movement is 1.07 inch at an angle 11° counterclockwise from the A_0 axis. The azimuth of this groove is 51° so the azimuth of the slide movement is 40°, i.e., N 40° E.

Available software is capable of extracting additional useful information from the inclinometer data. SINCO's DigiPro software, for example, can provide movement vs. time graphs for a selected depth interval, such as the discrete shear zone of a landslide. As shown by the insert graph Figure 4.7c, the rate of movement of a landslide in Weber County, Utah, is averaging 1.0 inch per year. The DigiPro software can also mathematically rotate the orientation of the two measuring axes into the axis (plane) of movement; then there is no need to plot the B plane component. Other software capabilities include adding a simplified boring log alongside the lateral movement plots, applying various corrections to eliminate systematic errors, and applying a correction for casing spiral.

Systematic Errors

Product literature from Slope Indicator Co. states that the system field accuracy is ±0.3 inch per 100 feet, subject to qualifiers. This figure is a combination of random and systematic errors. According to Mikkelsen (2003), the *random* error is 0.05 inch per 100 feet after all systematic errors have been removed. Another possible source of error is the care taken by the field technician.

Systematic errors can usually be separated from displacement data by mathematical techniques. Experienced users should be able to make some or all of these corrections using the available software. If help is needed, there are consultants specializing in field instrumentation. Equipment manufacturers can provide the names of consultants who are knowledgeable about their products.

Mikkelsen (2003) provides a detailed explanation of systematic errors and their correction. The following paragraphs provide a brief summary.

Three systematic errors to be aware of are: bias shift error, rotation error, and depth positioning error.

Bias Shift Error

Previously known as Zero Shift or Offset Error, this is the most common systematic error and the easiest to correct. It can be recognized by the "windshield wiper" effect in which the displacement graphs have a back and forth linear tilt from the vertical. The swing occurs around an apparent hinge point at the base of the casing (Figure 4.9).

The sensor bias is the reading of the probe when it is vertical. Although it is set close to zero in the factory, the bias will change over the life of the probe and can change during field use. In a probe with zero bias, the reading taken in the A_{180} axis should be numerically identical but of opposite sign to the reading in the A_0 axis at the same depth. For example, the readings could be 330 and −330. If the probe has a bias of 10 units, the A_0 reading would be 330 + 10 = 340 and the A_{180} reading would be −330 + 10 = −320. In a probe with zero bias, the checksum 330 + (−330) = 0; for a probe with a bias of 10, the checksum 340 + (−320) = 20 units, i.e., twice the bias of 10 units. The checksum calculation, usually performed by software, is the basic check for finding inconsistencies and poor data in inclinometer readings.

Now that bias has been explained, the *bias shift* can be discussed. Bias shift occurs *within each data set* between opposite readings, probably from slight jarring of the probe. It must remain constant with depth in a set of readings for the bias shift to be corrected. The magnitude of the bias, and the fact that it can also shift *between* data sets, is of no consequence. The main reason for taking readings in opposite grooves is to eliminate bias from the results of a particular data set.

The error caused by a constant bias shift in a Digitilt probe can be determined from the formula:

Bias Shift Error = b (0.0006) N units of inches
where b = the bias in units
N = number of vertical readings taken

For example in a 100-foot inclinometer casing, N would be 50. For a bias shift error (b) of 10 units

Error = 0.3 inch at top of inclinometer casing

Figure 4.9 "Windshield wiper" effect. Percy Slide, Oregon:
(a) data uncorrected
(b) data corrected for bias shift error

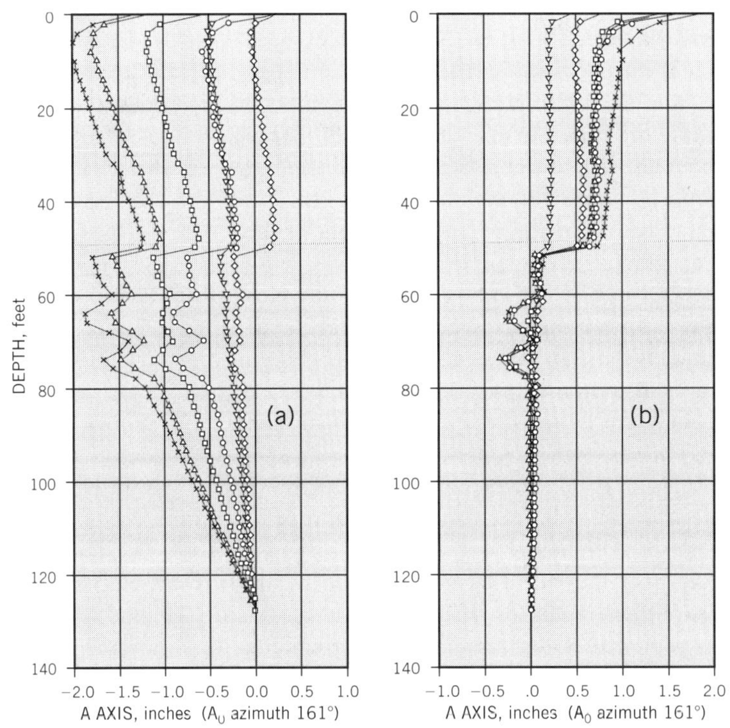

Table 4.3 Bias-Shift Analysis
Example: Weber County Landslide, Utah

(1)	(2)	(3)	(4) =(3)−(2)	(5)	(6)	(7) =(6)−(5)	(8) =(2)−(5)	(9) =(3)−(6)	(10) =(9)+(12)	(11) =(10)−(8)
Depth	A_0 groove			A_{180} groove			$A_0 - A_{180}$			Difference
	Initial 2/99	4/00	Differ.	Initial 2/99	4/00	Differ.	Initial 2/99	4/00	4/00 corrected	4/00 corrected −2/99
42	545	581	36	−558	−561	−3	1103	1142	1133	30
44	375	402	27	−390	−379	11	765	781	772	7
46	343	365	22	−356	−341	15	699	706	697	−2
48	342	364	22	−356	−341	15	698	705	696	−2
50	402	417	15	−411	−393	18	813	810	801	−12
52	444	465	21	−455	−440	15	899	905	896	−3
54	453	485	32	−474	−463	11	927	948	939	12
56	468	487	19	−477	−463	14	945	950	941	−4
58	475	494	19	−486	−471	15	961	965	956	−5
60	508	525	17	−517	−501	16	1025	1026	1017	−8
62	527	546	19	−537	−522	15	1064	1068	1059	−5
64	549	569	20	−557	−546	11	1106	1115	1106	0
66	559	576	17	−563	−553	10	1122	1129	1120	−2
68	559	579	20	−565	−555	10	1124	1134	1125	1
70	567	584	17	−572	−560	12	1139	1144	1135	−4
72	570	589	19	−575	−561	14	1145	1150	1141	−4
74	567	586	19	−573	−563	10	1140	1149	1140	0
76	550	570	20	−553	−545	8	1103	1115	1106	3
78	549	565	16	−551	−539	12	1100	1104	1095	−5
			Mean 21			Mean 12 (12) Correction −9				Mean 0

To eliminate bias shift error, a correction is applied to the A_0–A_{180} calculations that are used to determine lateral displacement. Using a data set from the Weber County, Utah, landslide (shown already corrected in Figure 4.7), analysis is made of the stable ground below 40 feet where it is known that there is no lateral displacement. The data from 40 to 80 feet for 4/17/00 are compared with the initial data of 2/19/99 on Table 4.3. The mean bias shifts are 21 (A_0 groove) and 12 (A_{180} groove), requiring a correction to the $A_0 - A_{180}$ data of 4/17/00 of −9 units (column 10, Table 4.3). The corrected differences between the initial and subsequent $A_0 - A_{180}$ are shown in column 11, Table 4.3. They are close to zero at all depths, which is the result that should be expected in stable ground.

The correction of −9 units is applied to all the $A_0 - A_{180}$ readings of the current data set from base to top of the inclinometer casings. The corrected data are then converted to lateral displacement using the probe multiplier (0.0006 inch/unit for the Digitilt probe).

The correction for bias shift error has been presented in some detail because it is the most common systematic error and is easy to correct by computer software. It allows the results to be presented graphically in a manner that can be easily assimilated and interpreted (Figure 4.7).

A more dramatic example of the "windshield wiper" effect is shown on Figure 4.9. Only the A_0 axis, with and without the correction for bias shift error, is shown on Figure 4.9.

From the standpoint of correcting the data, it is beneficial to have a significant length of inclinometer casing in stable ground. A depth of 10 to 20 feet into stable ground is needed to make the correction (Mikkelsen, 2003). Bias shift error correction tends to be more critical on deep landslides or landslides where shear movements are occurring at creep rates.

The bias shift changes for the Digitilt probe in the A axis should be less than 20 units. If it exceeds 20, the probe probably should be returned to the manufacturer for repair or replacement of the sensors.

Rotation Error

This error occurs in inclinometer casings which significantly deviate from vertical. The error is caused when a small shift or rotation of the accelerometer's sensing axis (e.g., the A axis) occurs toward the inclined plane at right-angles to it (e.g., the B axis). The error can occur for shifts of less than 1° (Mikkelsen, 2003). The Digitilt manufacturer's tolerance in the past has been ±0.5°, but has been tightened to ±0.25°.

The error can be detected by (i) knowing the inclinometer casing is severely out of vertical alignment by plotting the cumulative *deviation from vertical* (i.e., the shape of the casing in the ground, see Figure 4.4a) and (ii) observing that the *lateral displacement* graph on the other plane is similar in shape to the cumulative deviation graph.

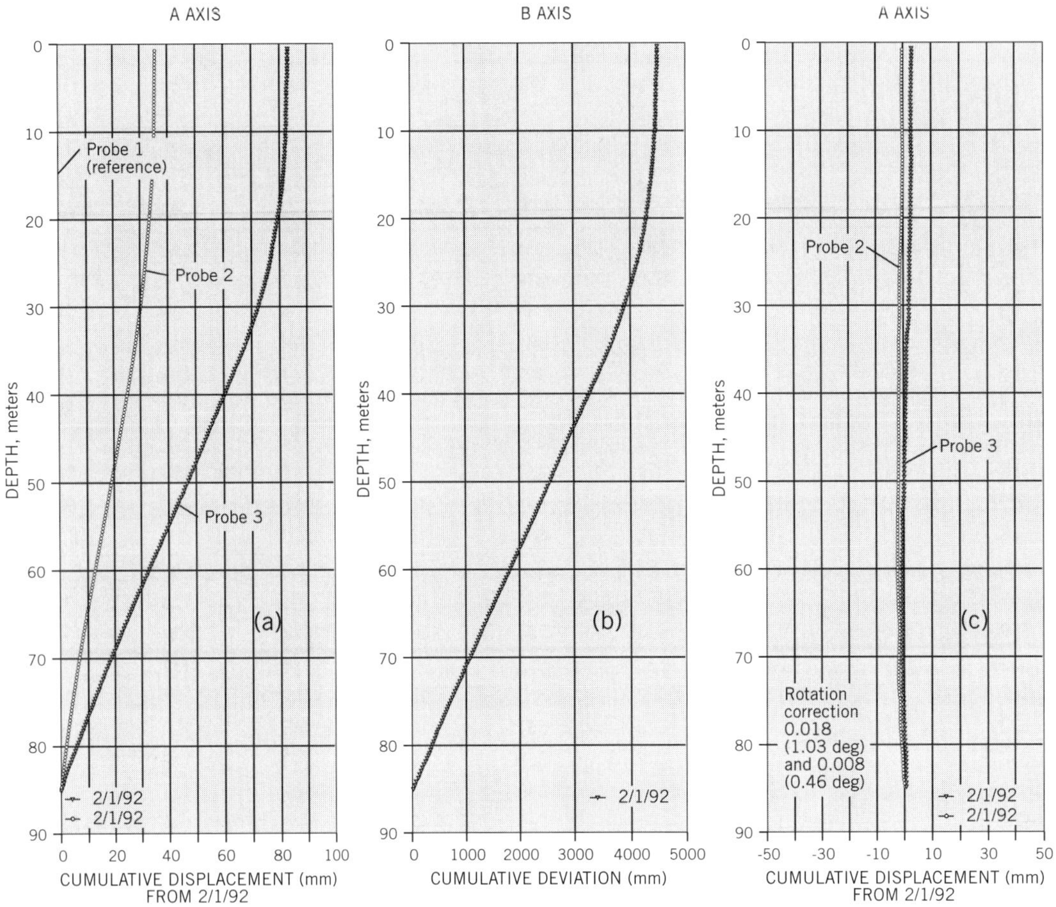

Figure 4.10 Example of rotation error and correction:
(a) *lateral displacement* on A axis of two probes compared with a third probe as reference
(b) *deviation from vertical* on the B axis
(c) *lateral displacement* on the A axis after correction for rotation error
(after Mikkelsen, 2003).

An example from Mikkelsen (2003) is shown on Figure 4.10. To demonstrate rotation error, three separate probes were used to take readings on the same day on the same inclined casing, one probe acting as the reference. On Figure 4.10(a) are the lateral displacement graphs in the A axis. The graph of deviation from vertical of the inclinometer casing in the B axis is shown on Figure 4.10(b).

The methodology is explained in Mikkelsen (2003). At any level in the casing:

Rotation error in plane A = sine A plane rotation error × B plane inclination

For example, at the ground surface in Figure 4.10, probe 2 has a displacement error of 35 mm in the A plane corresponding to a 4500 mm cumulative deviation in the B plane.

If Δ is the A plane rotation error, then:

$$\sin \Delta = \frac{\text{rotation error in A plane}}{\text{B plane inclination}} = \frac{35}{4,500}$$

$$\Delta = 0.45°$$

Similarly, for probe 3, the Δ correction is 1.05°.

Rotation error correction can be made by DigiPro software. The correction is entered as the sine of the rotation angle. Generally, the rotation angle is less than 1°.

When bias shift error and rotation error occur together, the bias shift error is removed first. At that point, the rotation error should be evident and the second (rotation) error correction can be made.

Rotation error is likely to occur when a replacement or different inclinometer probe is brought in to continue the inclinometer observations at a site. It may also occur if the original probe has undergone repair by the manufacturer. If possible, the same probe should be used throughout a monitoring program.

Depth Positioning Error

The error is caused by the sensor being positioned at different levels from the initial data set. It can be caused by compression or settlement of the inclinometer casing, change of cable (the replacement cable having depth markings slightly different from the original cable), and operator error (e.g., not using the wheel pulley during readings).

Depth positioning errors are not a common problem for most landslide studies. Unless the error is obvious, it can be very difficult and time-consuming to correct.

Data Quality Issues

A major difficulty in some landslide situations is to persuade the owner that experienced personnel need to be used in collecting and interpreting inclinometer data. The previous section has described in some detail the system errors that can be reduced or eliminated by knowledgeable people.

Some common problems noted from past experience are the following:

- *Data are not timely.* Information is collected according to a set schedule, but the data are not reduced and plotted for several days or weeks. When the data are analyzed, field errors are noted. It is then too late to correct the situation. In some cases, it is discovered that movements have occurred and the reading schedule should have been changed.

  It is good practice to reduce and plot the data on the same day or the next day after taking field measurements. Also, if the data have been collected by a technician, an engineer or geologist familiar with the project should review the results within the same period.
- *Plotting data to an inappropriate scale.* Already discussed.
- *Reaching conclusions that are unsupported by the data.* Already discussed. The most common scenario is to mistake the bias shift error for actual movement of the ground. In general, the data need to show a simple shear displacement of at least 0.15 inch. Movement of points above the shear zone should be consistently downslope from those below it.
- *Contracting practices.* In the United States and elsewhere, many agencies put field instrumentation installation and readings under the main contract of medium- to large-size projects. The contractor is chosen by a low bid procedure and the winning contractor often holds a "rebid" with subcontractors to obtain a lower price for the instrumentation work. Quality is compromised by such practices.

Field instrumentation of landslides should be considered a *design requirement* rather than a quality assurance issue. Monitoring should be placed under the control of the design team (or their specialist instrumentation consultant). Poor contracting practices often lead to near-useless data, claims, and other legal/technical wrangling.

In-Place Inclinometers

An in-place inclinometer (IPI) is a modified probe that is lowered down the inclinometer casing to a preselected depth and kept at that depth throughout the period of study (Figure 4.11). It has several advantages over the conventional inclinometer probe that is lowered and raised for each set of readings. The advantages are: (i) the wheels remain at exactly the same depth; (ii) the position of the wheels within the groove do not change; and (iii) there is no temperature drift within the probe. These benefits reduce all the random and systematic errors referred to earlier. However, the main advantage of the in-place inclinometer is that it can be positioned across a shear zone and monitored continuously by an automatic data acquisition system. It can function at remote locations or in highly sensitive site conditions and, if needed, can trigger an alarm if the shear displacement exceeds a predetermined threshold value. Another option is to place several IPIs in series so that they can observe deflections over an extended vertical range. This arrangement could be used for a deep shear zone, for deflections of a shear pile, etc.

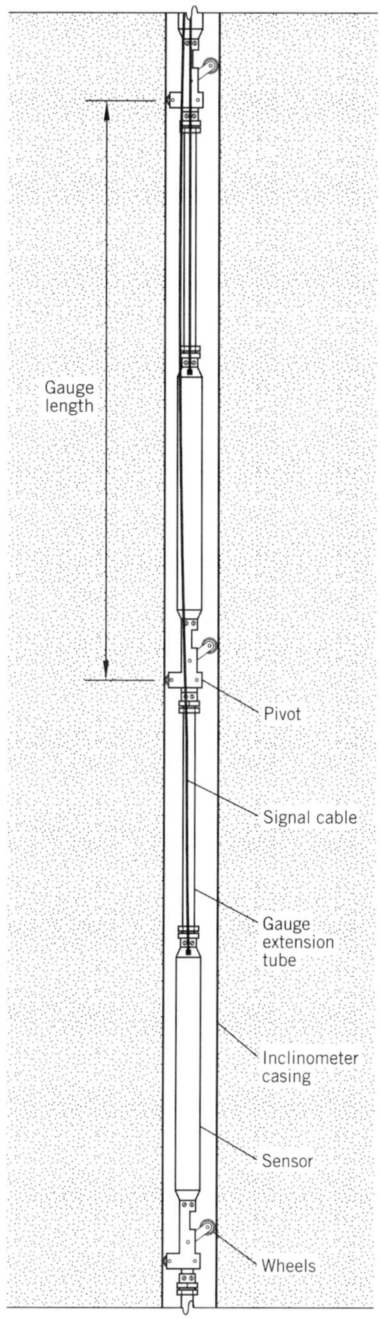

Figure 4.11 In-place inclinometer sensors (two probes coupled in series) (Slope Indicator Co.).

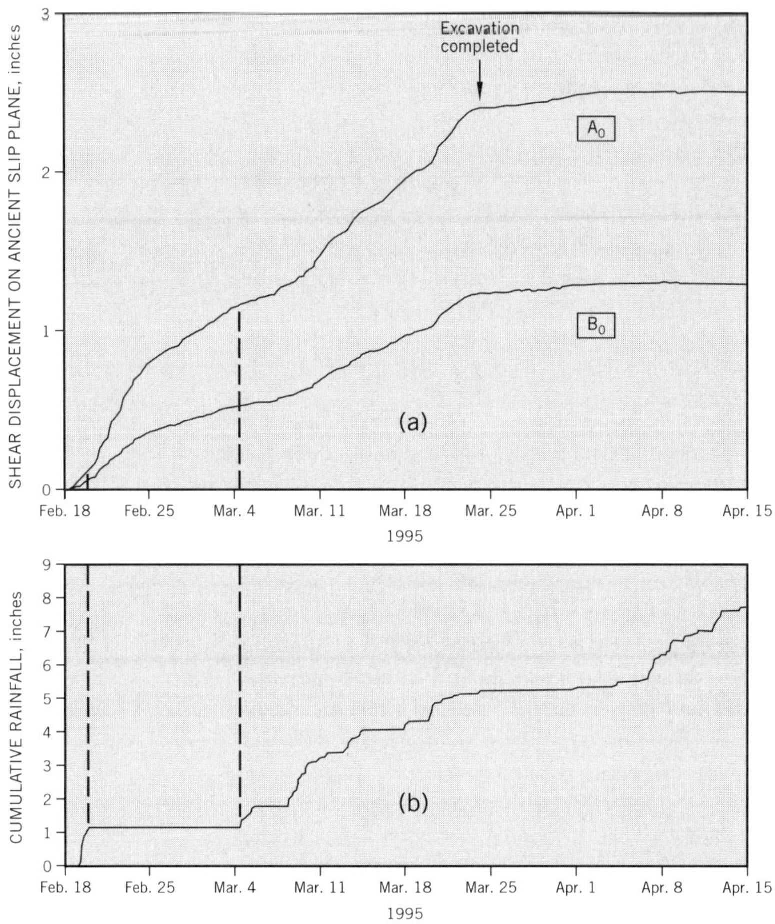

Figure 4.12 Continuous readings from a single in-place inclinometer placed across the slip surface of a deep-seated ancient landslide during a building excavation above it:
(a) readings of A and B axes
(b) cumulative rainfall during the period of observations

In monitoring landslides, the most common procedure is to install the inclinometer casing and take readings with the conventional probe until the location of a discrete shear zone has been established. Depending on the thickness of the shear zone, either one or two IPI sensors are placed across this depth to provide continuous readings of subsequent shear deformations. The precise depth of the shear zone can be determined by taking closely-spaced readings across it, as described in Chapter 22, Section 22.1. After the shear zone location has been determined, the upper wheel of the IPI is set above it and the lower wheel below it. Thus, the IPI straddles the shear zone.

The Slope Indicator Co. IPI has a biaxial electrolytic sensor including a built-in temperature sensor. It has a tilt range of ±10° from the vertical. Experience with this sensor is that it is up to 10 times more precise than the traversing digitilt inclinometer system.

The IPI can be briefly removed from the inclinometer casing to allow a traversing inclinometer probe to survey the entire depth of the casing or to confirm the observations of the IPI in the shear zone. Providing the IPI is replaced immediately, a new series of continuous readout measurements can begin and be added almost seamlessly to the earlier data.

The principal limitation of the IPI is the relatively small tilt range (±10° from vertical), and occasional problems with long-term signal drift. The IPI can also become stuck within the casing in a landslide with a very thin shear zone.

An example of the benefits of IPI data, taken at the Washington Park Station Slide (described in Case History 4), is shown on Figure 4.12. The in-place inclinometer was placed across the shear zone of an ancient landslide upslope of the proposed excavation for a transit station in Portland, Oregon, and was connected to an automatic data acquisition system. When excavation began, shear movements occurred in the shear zone more than 70 feet below the ground surface and continued until the 35-foot deep excavation was completed. The excavation took place in the winter months and a rain gauge at the site recorded cumulative rainfall, Figure 4.12(b). No rain fell during a two-week period (February 19–March 4), during which time the shear movements continued to increase as the depth of excavation increased (Figure 4.12a). In this case, the continuously monitored IPI clearly demonstrated that the measured shear movements were related to the building excavation project and not to winter rainfall.

4.2 PIEZOMETERS

Piezometers measure groundwater pressures. This information is needed for effective stress stability analyses of landslides and to observe the variation of pore pressures vertically in the landslide environment. Effective stress is discussed in Chapters 6 and 9.

Piezometers in Landslides

There are two principal requirements for piezometers installed in landslides:

- The tip (sensor) should be at, or very close to, the slip surface of the landslide.
- The piezometer should reliably measure the groundwater pressure at the sensor over an extended time period.

Many types of piezometers are available (Dunnicliff, 1988). In this chapter, only the standpipe, pneumatic, and vibrating wire piezometers will be presented. These are the principal types of piezometers used in landslide studies, excluding special situations.

Hvorslev (1951), Penman (1960), and others have studied the response time of various piezometers used in geotechnical practice. The vibrating wire and pneumatic piezometers each use diaphragms that undergo very small volume changes (movements of pore water) to register pressure changes. They are classified as *rapid response* piezometers and quickly measure the rise and fall of groundwater pressures. In contrast, the standpipe piezometer has a column of water that has to move up and down the riser tube; this can take hours or days to reach equilibrium within soils of low permeability. Thus, the standpipe piezometer cannot measure the peak groundwater level reached during storms when it is installed in clays.

Sensor Location

Piezometers are usually installed as part of the site investigation program, often before the depth of slippage has been measured by inclinometers. The geotechnical engineer then has to make an estimate of slip surface depth based on the observed soil and rock conditions. In many cases, this estimate may be within a few feet of the correct depth; if not, it may be necessary to revisit the site and install sensors at the correct depths. This decision may depend on the critical nature of groundwater to the investigation and the willingness of the parties to incur extra costs. Another option is to install two or more sensors at different selected levels in the site investigation boring and trust that at least one of them is positioned close to the slip surface.

In an ideal installation, the sand pack should enter but not pass through the shear zone so that the pressure in the sand reflects the pore water pressure on the slip surface of the landslide. The sensor itself should be just above the slip surface so that it is not damaged by landslide movements.

There are two difficulties with an incorrectly positioned piezometer sensor:

1. If the sensor is *too deep,* the sensor, standpipe, or other leads may be damaged or destroyed by the shear displacements of the landslide. Furthermore, the base of the sand pack intake may penetrate into bedrock or a deeper stratum with different groundwater pressures. Piezometers installed and isolated into bedrock below landslides frequently have lower groundwater pressures than the landslide materials above.
2. If the sensor is *too shallow,* the measured groundwater pressure may be incorrect. For demonstration purposes, consider an infinite slope-type landslide with the outer surface, slip surface, and bedrock surface at 3 : 1 (horizontal : vertical) as shown on Figure 4.13. Assume that the slip surface and bedrock surface are coincident, the bedrock is impermeable, and that steady state seepage flows parallel to the ground and bedrock surfaces. In this case, the flow lines are parallel to the ground surface and the lines of equipotential are at right angles to the ground surface.

   Piezometer sensor 1, correctly positioned at the slip surface, measures groundwater level h_1 in the vertical standpipe or pressure gauge. However, piezometer sensor 2, at a higher level in the overburden, is at a different equipotential line and measures a higher groundwater level h_2 above the slip surface. Since the *vertical* height is used in the slope stability analysis, piezometer sensor 2 is measuring the incorrect vertical head on the slip surface below it.

Piezometer Locations

In Landslide Technology's field practice, piezometers are placed in a separate borehole located about 10 to 20 feet away (along the contour) from an inclinometer casing. The incli-

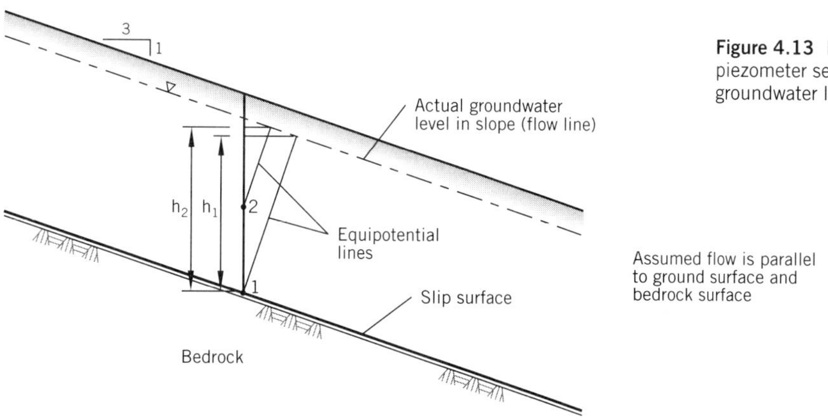

Figure 4.13 Effect of a badly positioned piezometer sensor on the measured groundwater level h.

nometer hole is used to explore the subsurface conditions: Samples are taken and drilling is continued into bedrock or stable stratum to a depth of at least 10 feet. The second hole for the piezometer is far enough away to be unaffected by the inclinometer hole and is terminated within the shear zone and not below it. Since no sampling is required, the piezometer hole can be rapidly drilled and does not significantly increase the cost of a site investigation.

Piezometer Reliability

A puzzling and often frustrating aspect of piezometer installations is that, despite careful field techniques, a small percentage of a group will fail to function properly. It applies to both standpipe piezometers and diaphragm piezometers. The typical occurrences are: (i) no reading of pressure (dry hole or zero response); (ii) readings that never change (irrespective of rainfall or season); and (iii) piezometer readings that are inconsistent with others in the group and with rational expectation (such as a low pressure when the landslide generally has high pressures). After a group of piezometers have been installed and apparently reached equilibrium, the performance of the group needs to be reviewed. If an anomalous set of readings is noted, the possible causes of an error should be identified and corrective action taken. However, in many cases, there is no apparent cause of the malfunction. A piezometer at a key location should be replaced as quickly as possible.

Standpipe Piezometer

This piezometer is also termed "open standpipe" and "Casagrande-type" piezometer, but will be referred to simply as a *standpipe piezometer* in this book. It is easy to install, the materials are relatively inexpensive, and it is the workhorse for most landslide projects. It can be used to measure in-situ permeability of the surrounding soil (see Chapter 5, Section 5.5). A disadvantage is the slow response to measuring pore pressures in soils of low permeability.

A standpipe piezometer has a porous measuring tip attached to a riser pipe that extends from the tip to the ground surface (Figure 4.14). The tip is embedded in a saturated sand pack to provide the groundwater pressure measuring zone, and is usually sealed above by a bentonite layer. The remainder of the borehole is backfilled with a relatively impervious cement-bentonite grout. A protective casing is provided at the ground surface to protect the standpipe from damage and vandalism.

Two types of tips are commonly used: (i) a porous stone made of aluminum oxide (68 μm pores) or made from polyethylene (60 μm pores); these are often referred to as Casagrande-type piezometer tips, (ii) a PVC slotted screen, 1 inch or 2 inches diameter, also used for horizontal drain applications; the slot widths are usually 0.01 inch. Some installers wrap the slotted screen with a geotextile.

The riser pipe generally is made of PVC and is glued together in 10-foot lengths. A cap with a vent hole is push-fitted over the top of the riser pipe. A protective cover or monument is held in place by a concrete plug (Figure 4.14). Surface water should be diverted away from the installation, an important detail.

The sand surrounding the sensor should be a uniform sand passing the No. 20 but retained by the No. 40 U.S. sieves. Other sand packs (No. 8 to No. 12 U.S. sieves, for example) are also used. However, the sand should meet the filter requirements for the adjoining soil; otherwise, fine-grained soils may clog the pores or slots of the sensor. The sand pack is typically 5 feet deep.

The bentonite seal can be made from bentonite chips or granules. Coarse bentonite chips are slower to hydrate than

Figure 4.14 Standpipe piezometer:
At top of borehole
(a) protective casing approximately flush with ground surface
(b) protective monument above the ground surface
At measurement zone
(c) porous tip
(d) slotted screen

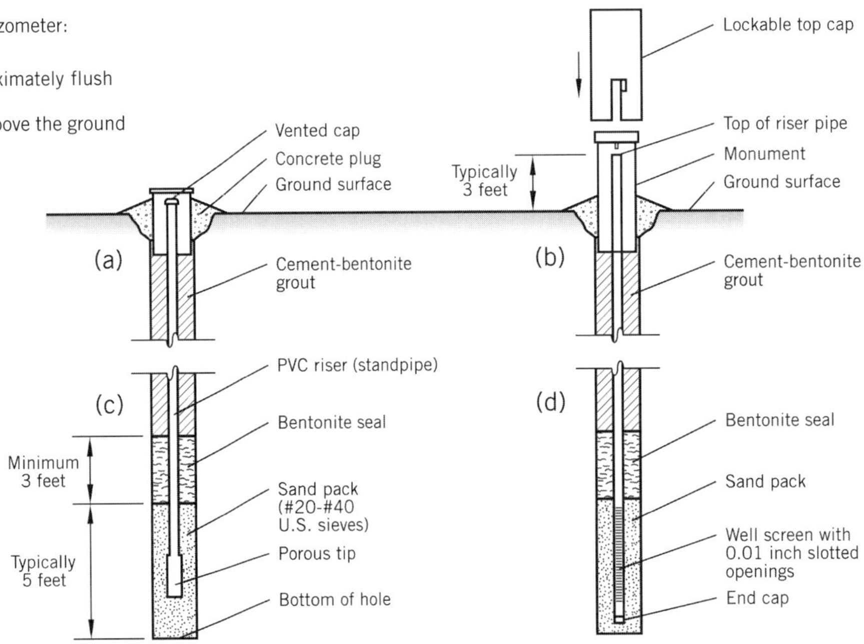

other bentonite products and can be dropped through standing water in the hole. A 50-lb. bag provides a seal 3 to 4 feet thick.

Cement-bentonite grout mixes have been previously discussed for inclinometer casing installation. For piezometer installations, Landslide Technology (year 2003) uses the same amount of bentonite as that used for inclinometers in soft soils (Table 4.1). The intent is to safeguard against shrinkage between the grout and borehole wall after construction. Consultants seem to use many different backfill grout mixes.

Some common drilling methods for installing all types of piezometers include power rotary with water flush (or Revert mud), hollow stem auger, or casing advancer. If a casing has been used to support the hole, the insertion and withdrawal of the casing may smear the borehole walls in the piezometer measuring zone. Before installing the piezometer, the bottom of the hole should be thoroughly flushed with clean water. For a hollow stem auger, the final 5 feet can be drilled by a rotary tricone bit through the bottom of the hollow stem auger using water flush. Bentonite drilling mud should not be used at the piezometer measuring depth. If necessary, bentonite (or Revert) drilling mud should be replaced with water flush for the final 5 feet, i.e., the sensing depth. (Note: Revert is a biodegradable drilling mud that breaks down over time and is supposed to be acceptable for piezometers. However, the author prefers to take no chances at the piezometer measuring depth.)

An apparently malfunctioning standpipe piezometer is usually difficult to correct. Some investigative procedures include:

- Filling up or drawing down the water level in the standpipe to find out if the water surface subsequently returns to the preexisting level
- Flushing the porous tip or slots to remove clogging (comparable to flushing out horizontal drains)
- Checking that surface water is not getting into the standpipe; this is a fairly common problem with riser pipes cut off at the ground surface and improperly protected from runoff water
- Checking that the cap has a hole in it to equalize pressure
- Probing the riser pipe to check for obstructions

The constructed cost of a typical 60-foot deep standpipe piezometer installation in the United States, including the consultant's representative, is around \$2,000 (year 2003). This includes an overnight stay, vehicle expense, preparing a summary log and a drillers' monitoring well report. Drilling and standpipe installation typically take 6 hours, including moving onto the hole from another part of the site. These costs exclude mobilization to the site and subsequent piezometer readings.

The main drawback of the standpipe piezometer is the slow response time in low-permeability soils. Groundwater has to rise up the standpipe after installation to reach equilibrium with the groundwater pressure at the porous tip. Once this level has been attained, the level has to move up and down in response to groundwater pressure changes, and this requires a significant movement of water into or out of the soil around the sensor.

The Hvorslev formula for a standpipe piezometer embedded in uniform soil is:

$$k_h = \frac{d^2 \log_e\left(\frac{2mL}{D}\right)}{8L\,(t_1 - t_0)} \log_e\left(\frac{h_0}{h_1}\right) \qquad \text{Eq. (1)}$$

where $m = \sqrt{k_h / k_v}$ and all other terms are defined on Figure 4.15

Rearranging Eq. (1), putting $t_0 = 0$, $t_1 = t$ and assuming $m = 1$:

$$t = \frac{d^2 \log_e\left(\frac{2L}{D}\right) \log_e\left(\frac{h_0}{h_1}\right)}{8Lk} \qquad \text{Eq. (2)}$$

For $r = 90\%$, $h_0/h_1 = 10$; for $r = 99\%$, $h_0/h_1 = 100$.

$\log_e(100) = 2\log_e(10)$

Therefore, the response time for 99% equalization takes twice as long as the response to 90%, as shown on Figure 4.15 for a rise pipe of 1 inch diameter. For a change in diameter, the response time varies as the square of the diameter (i.e., a 2-inch dia. riser pipe takes 4 times as long to respond as a 1 inch dia. pipe).

It can be seen from Figure 4.15 ($r = 99\%$, $d = 1$ inch) that soils with permeabilities below 10^{-6} cm/sec take more than 1 day to respond. This slow response makes the standpipe piezometer inaccurate for measuring peak or spike groundwater levels (due to heavy rainfall) in soils of low permeability. To make such observations accurately in silty clays and clays, a rapid response diaphragm-type piezometer with continuous readout is needed, as discussed later.

Existing standpipe piezometers can be retrofitted for automatic data acquisition systems by suspending a vibrating wire piezometer with the sensor below the lowest likely groundwater level. The pressures are recorded through a data logger and converted to head above sensor level.

A simple and inexpensive option for measuring the approximate peak of groundwater during a storm is to suspend a series of Halcrow "buckets" within a standpipe piezometer (Figure 4.16). These patented plastic devices are threaded onto a nylon line at a selected depth above the normal groundwater level in the standpipe. When groundwater level rises, it passes through the intake hole and is retained there until the line is pulled out of the pipe. The highest filled bucket is the approximate maximum level of the water rise. These devices do not change the slow response of the standpipe piezometer and thus may only measure a maximum that is lower than the actual maximum. However, it is a convenient way to obtain the approximate maximum head data without the expense of an automatic data acquisition system.

In the United States, environmental concerns about groundwater contamination have led to increasing regulatory control over piezometer installation. Most states have laws governing construction, maintenance, and abandonment. ASTM has issued a standard practice for design and installation of groundwater monitoring wells (D 5092).

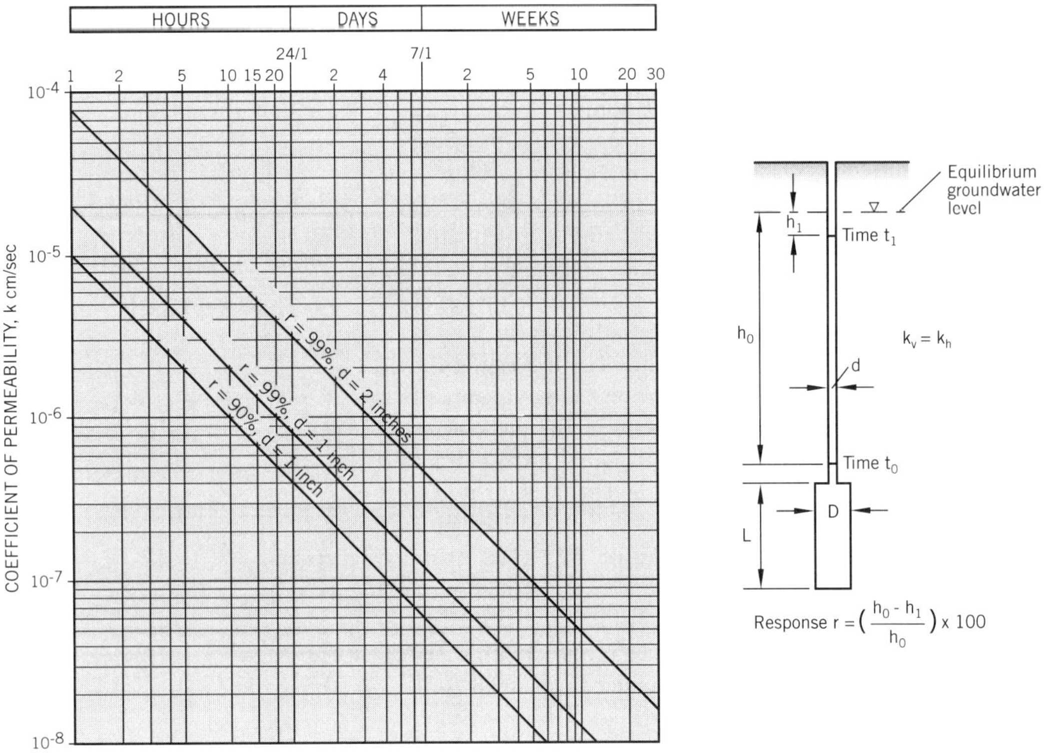

Figure 4.15 Response times for standpipe piezometers embedded in homogeneous, isotropic soil.

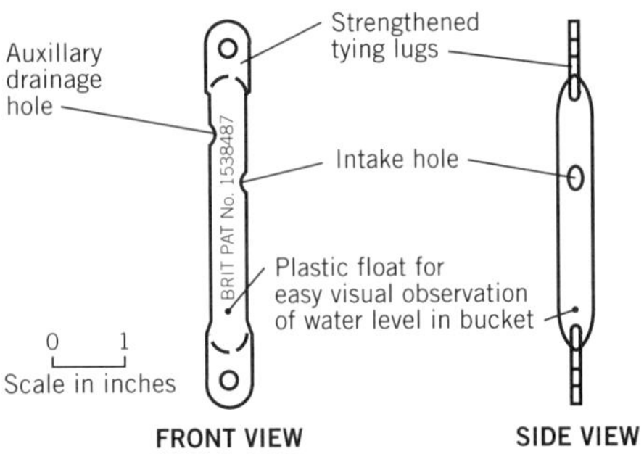

Figure 4.16 Halcrow plastic bucket for measuring the peak groundwater level in a standpipe piezometer.

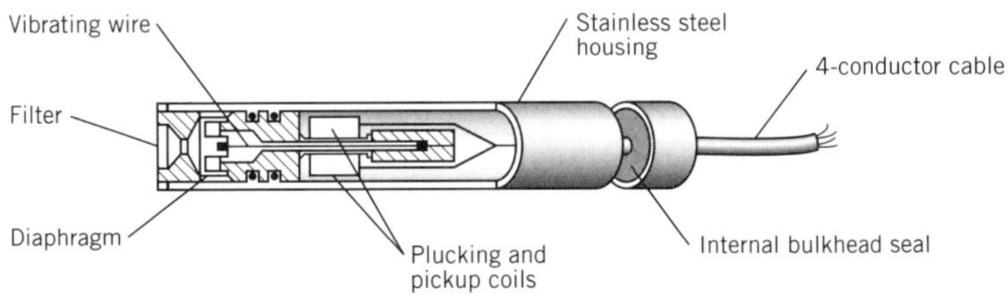

Figure 4.17 Vibrating wire piezometer: section through sensor.

Diaphragm Piezometers

These rapid response piezometers require very small displacements to measure the pore water pressure. There are two types in widespread use: The vibrating wire piezometer and the pneumatic piezometer.

Vibrating Wire Piezometer

This type has a pre-tensioned steel wire attached to the middle of a metallic diaphragm and the opposite end is clamped to the sensor body (Figure 4.17). The wire is "plucked" at its mid-length by a magnetic coil, causing the wire to vibrate at its natural frequency. Electrical leads transmit an induced frequency signal to a readout box at the ground surface.

When a groundwater pressure change occurs, the pressure applied to the outside of the diaphragm changes the natural frequency of the tensioned wire. By reading the frequency of the current induced in the coil by the vibrating wire, the axial strain of the wire is measured and the applied stress can be calculated (see Dunnicliff, 1988).

The housing around the tensioned wire is evacuated and hermetically sealed to prevent corrosion of the wire. Potential problems are creep of the wire under tension and slippage at the clamping points. These can produce zero drift that is unrelated to the measurement of groundwater pressure. Some manufacturers overcome these problems by stress-relieving the sensor under high temperature before attaching the electrical coil. The measured long-term creep of Geokon vibrating wire piezometers is less than 0.1% of the range per year.

The thickness of the diaphragm is varied to change the pressure range and sensitivity of the instrument. A removable porous stone tip (Figure 4.18) provides a cavity between the diaphragm and stone. The cavity protects the diaphragm from

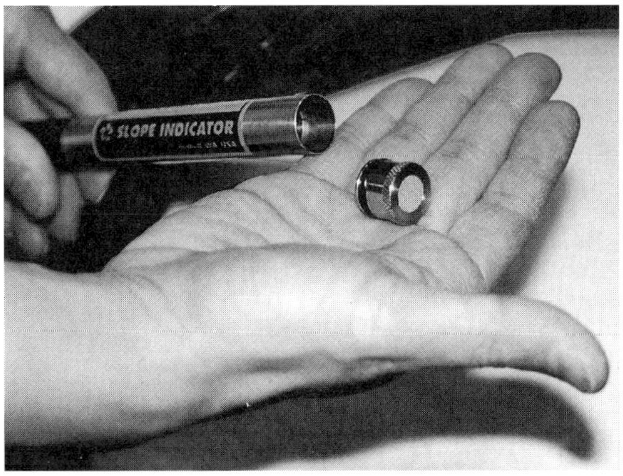

Figure 4.18 Vibrating wire piezometer. Removable porous stone tip with O-ring seal.

soil contact and can be filled with water prior to installation in the borehole.

The sensor is susceptible to damage from lightning. Lightning protection is placed in a sealed enclosure at the top of the borehole. Some manufacturers (e.g., Geokon) also put lightning protection inside the sensor. Another useful device is to include a temperature sensor within the piezometer housing. A temperature correction improves accuracy of the water pressure measurements especially just after installation.

The main benefit of the vibrating wire piezometer is that the signal can be transmitted without degradation over relatively long distances and it can be easily connected to an automatic data acquisition system. It is a rapid response piezometer that can measure the equilibrium groundwater pressure in clays shortly after installation and the short-term rises that can trigger landslide movements.

Pneumatic Piezometer

This piezometer has an external diaphragm in contact with the groundwater and two tubes behind the diaphragm to allow gas to circulate. The principle of operation is illustrated on Figure 4.19 for one of these systems. Before measuring groundwater pressure, the diaphragm seals the return tube. The gas pressure in the inlet tube is slowly increased. When it exceeds the groundwater pressure, the diaphragm deflects and gas circulates back to the ground surface via the return tube. A return flow indicator on the readout box alerts the operator that the gas is circulating. The gas flow is shut off and the pressure drops slowly until it is in balance with groundwater pressure again and seals the return tube. The pressure in the input tube then equals the groundwater pressure on the outside of the diaphragm.

The gas is either carbon dioxide or dry nitrogen. Both are preferred to air to avoid condensation of water in the tubing. The portable readout unit includes a small gas cylinder that can be recharged from an auxiliary large cylinder as needed. Dunnicliff (1988) recommends the following types of tubing for this application: unplasticized Nylon 11 with an outer sheath of polyethylene or polyurethane. A 3/16 inch (5 mm) diameter tubing is suitable and 1/8 inch (3 mm) can be used where lengths are less than 250 feet. These sizes refer to the outside diameter of the gas tubing and not the exterior sheath.

Careful control of the gas flow is essential to minimize diaphragm movement when the reading is taken. A precision flowmeter built into the readout unit helps the operator maintain a relatively constant flow.

The pneumatic piezometer has a rapid response time, is not affected by freezing, and is less expensive than other diaphragm piezometers. It is essential that the connections are leak-free, especially on the input tube. It also requires gas to be carried to the site, and good operator skills. Small amount of dirt, if they reach the diaphragm, can interfere with the closure. In general, however, it is a reliable piezometer that is preferred by some geotechnical consultants. However, it is impractical to automate the readings of a pneumatic piezometer.

Installation of Diaphragm Piezometers

Diaphragm piezometers are traditionally installed in a similar manner to the standpipe piezometer previously described i.e. using a sand intake covered by a bentonite seal and a cement-bentonite grout. Some different features are:

Vibrating Wire (VW) Piezometer. A simple method of ensuring that water is always in contact with the VW diaphragm is to install the sensor upside down, i.e., porous stone uppermost. This requires the cable to be bent through a small radius and taped in position as the piezometer is low-

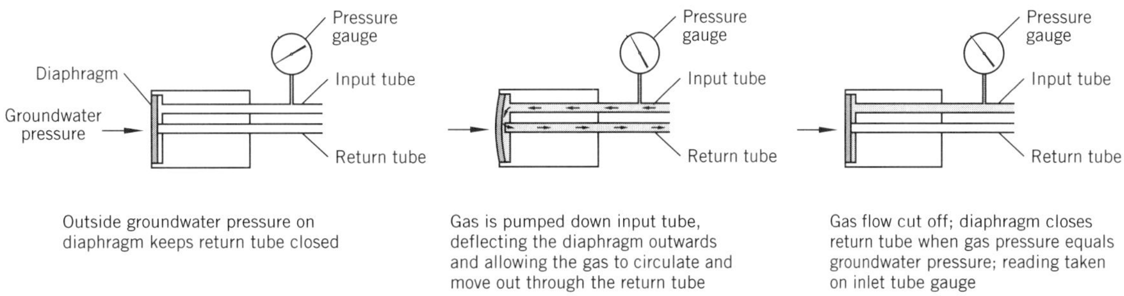

Figure 4.19 Operating principle of a pneumatic piezometer (Slope Indicator Co.).

ered down the hole. A short length of rebar can provide extra weight and prevent the piezometer tip from sticking to the borehole wall. Any casing in the hole must be withdrawn very carefully to avoid damaging the electrical leads and connections.

At the ground surface, the cables, junction boxes, etc., have to be protected from surface water, mechanical damage, animals, and vandalism; this often requires a trench to convey the leads to a protective monitoring station. The cable may be encased in an electrical conduit.

Pneumatic Piezometer. The tip should be immersed in a bucket of clean water for at least 15 minutes before installation. The sensor can be placed in a geotextile bag filled with saturated sand, and the weighted tip is then lowered down the hole. At the specified level, the bag is seated by adding more sand around the bag. Although an effort should be made to keep the porous stone saturated, a small amount of air is needed near the diaphragm to prevent dynamic loading when the diaphragm moves while taking a reading.

After lowering the piezometer to the required depth, it is advisable to check that it is functioning correctly before the sand pack and bentonite seal is added to the hole. It should be checked again after the installation is completed.

At the top of the hole, the tubes have to be terminated in a water-free space. This may require a monument box above the ground surface.

Fully Grouted Installation. An alternative and possibly better installation technique for diaphragm piezometers is to dispense with the sand pack and bentonite seal and fully grout the piezometer from the borehole base to the ground surface. The benefits: (i) saving in installation time and cost, (ii) ability to install multiple piezometers at different levels within the same hole, (iii) permits diaphragm piezometers and an inclinometer casing to be installed in the same hole, and (iv) from an environmental regulatory standpoint, a fully grouted piezometer installation is more desirable than other backfilling procedures in cutting off potential contamination of groundwater. Fully grouted pneumatic piezometers have been used since 1982 (Tofani, 2000). Tofani also describes projects where grout-in-place piezometer tips were installed alongside conventional open standpipes. Where steady state hydrostatic groundwater conditions were present, the two types of installation procedures gave similar results. McKenna (1995) also compared conventionally installed and fully grouted pneumatic piezometers and obtained similar piezometric levels and response times.

The fully grouted installation is explained in Mikkelsen and Green (2003) and is an outgrowth of earlier work by Vaughan (1969). Diaphragm piezometers require very small displacements to register pressure changes and thus can operate effectively in a relatively impermeable backfill (cement-bentonite grouts typically have permeabilities of around 10^{-7} cm/sec). The distance from the borehole wall to the diaphragm is so short that the sensor will measure the correct pore pressure at that level even if the grout backfill is up to 100 times more permeable than the soil stratum adjacent to it. Thus, leakage up or down the grouted annulus is not considered to be a problem by proponents of this technique.

Artesian Pressures

Installing a piezometer under artesian high flow conditions can be difficult due to the upwelling of water to the ground surface. One option is to raise the outer drill casing above the ground to the height of the artesian head and backfill the piezometer in the normal manner. Another option is to install an inflatable packer just above the bentonite seal and grout the packer in place.

At the surface, the uppermost length of riser pipe can be fitted to accept quick-connect couplings so that a pressure gauge can be added to measure the artesian pressure. For long-term use, the top of the pipe should be protected from freezing.

4.3 AUTOMATIC DATA ACQUISITION SYSTEMS

Landslide piezometers and in-place inclinometers can be continuously monitored using an automatic data acquisition system (ADAS). Continuous monitoring allows short-term groundwater rises to be observed and the information can be correlated with rainfall intensity by installing an automatic rain gage at the site. An example is given in Chapter 5, Section 5.1.

Deep-seated shear movements can be monitored continuously by placing an in-place inclinometer (IPI) across the shear zone. Small incremental shear movements are detected by such a system. If the IPI is within close proximity of piezometers, the correlation between groundwater pressures and shear movements can be observed.

A typical ADAS control unit comprises a data logger, power supply, and a surge (lightning) protector (Figure 4.20), all housed in a weatherproof box that can be attached to a post near a number of boreholes or at a protected location indoors. Cables from the instruments are enclosed in PVC conduits buried in shallow backfilled trenches to prevent damage. For piezometers, only vibrating wire piezometers are appropriate for this type of installation. The information can be downloaded to a computer and displayed in many formats.

The data can be stored on site and retrieved during periodic visits. It also can be accessed remotely by telephone by providing a modem at the site central unit. However, the more complex the system becomes the more likely it is to require special design considerations and fine-tuning during installation. The batteries typically have to be replaced every 2 to 6 months, depending on the number of instruments and reading intervals. Another option is to use large, deep discharge batteries recharged by solar power. There will undoubtedly be future improvements in this developing technology.

Another system uses small (1- to 3-channel) data loggers at the top of the hole where the instrument is installed. This eliminates the cost of trenching the wires to a central loca-

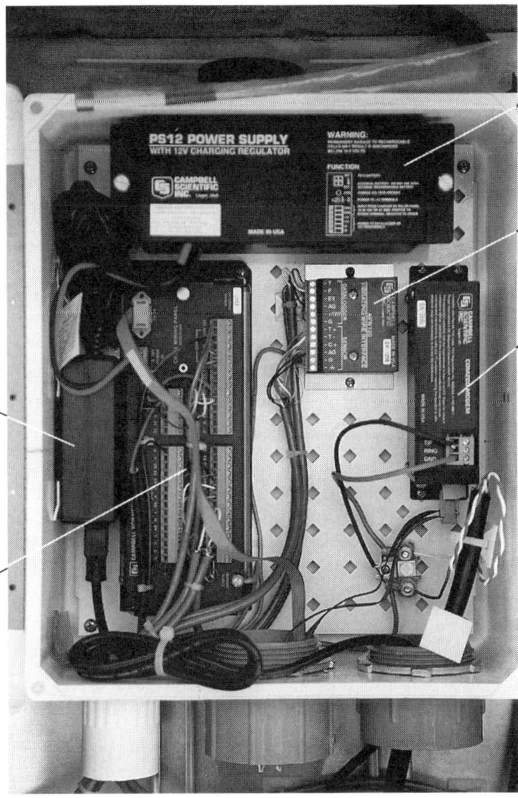

Figure 4.20 Automatic Data Acquisition System (ADAS). Typical control panel showing datalogger, power source, vibrating wire piezometer interface, modem.

tion and reduces the risk of damage from lightning and construction activities. However, it prevents remote access to the data (although wireless technology continues to improve and may overcome this obstacle). Vandalism is a serious concern and an appropriate protective housing has to be provided.

An automatic data acquisition system can provide: (i) readings as frequently as desired by the investigator to obtain a complete time record, (ii) accurate and reliable data with no human error, (iii) data at critical times such as floods or high intensity storms, (iv) quick and easy conversion of data to charts, (v) the possibility of taking readings remotely, especially useful in difficult terrain and winter conditions, and (vi) real time information and early warning system capability.

Some drawbacks to an automatic data acquisition system are: (i) it adds significantly to the costs of an investigation and generally takes longer to become operative than manual readings, (ii) there are generally fewer site visits to check for changes at the landslide such as spring activities, cracks, etc., (iii) maintenance costs after installation can be significant if problems occur; also it requires regular field checks to validate reliability and replace batteries, and (iv) the system can be susceptible to damage from construction work, animals, vandalism, and weather.

There is a significant learning curve for installing and running an ADAS efficiently. The services of an experienced field instrumentation specialist to expedite the process is strongly advised.

CHAPTER

5

Groundwater

Groundwater flow and seepage is a major subject covered by numerous textbooks (e.g., Driscoll, 1986; Fetter, 1988; Domenico and Schwartz, 1990). This chapter will discuss only a few aspects of groundwater that directly relate to landslide studies.

Due to the complex subsurface conditions within most landslides, modeling of groundwater is difficult. Much effort can be expended without achieving worthwhile practical results. In most cases, therefore, drawdown and similar calculations of ground treatment can serve as a guide but need to be supplemented by direct field observations (piezometric measurements) to determine effectiveness.

5.1 GROUNDWATER PROFILE

The groundwater regime in a landslide can be complex, and piezometer tips have to be carefully located to correctly measure pore water pressures. Some common variations are shown on Figure 5.1, as follows:

1. Piezometer tips *in the shear zone* measure the vertical head of water or pressure affecting the shear strength along the slip surface. By connecting these head levels, the groundwater profile needed for stability analyses is obtained.

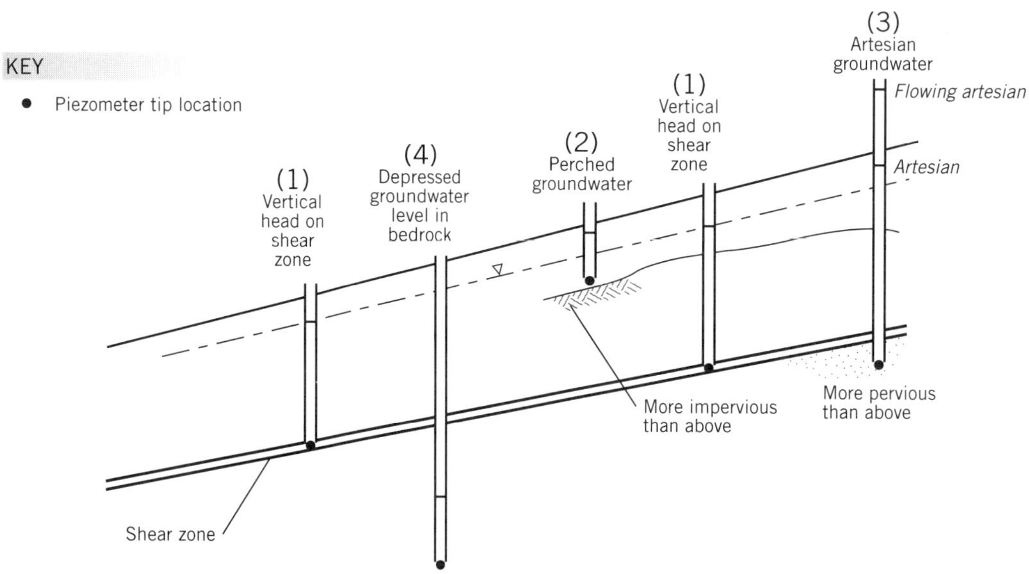

Figure 5.1 Groundwater regimes.

2. If more impervious soils are encountered in a boring, it is possible that a *perched* groundwater will be found, either permanently or intermittently during storms. Perched groundwater is higher than the groundwater table acting on the shear zone and may be feeding water toward the shear zone by vertical infiltration. Perched groundwater can be intercepted by drains.
3. Permeable ground that is overlain by less permeable soils may create an *artesian* condition in slopes, a common cause of landslides when an excavation is made into such a slope (see Chapter 2, Section 2.8). Geologists refer to groundwater level above the ground surface as a *flowing artesian* condition.
4. A *hard stratum* or *bedrock* below a shear zone often has a significantly different groundwater pressure than the shear zone above it. In many cases, bedrock is weathered at its surface and the underlying unweathered rock is more permeable than the overburden. Therefore, care needs to be taken to avoid putting piezometer tips through the shear zone into bedrock.

   The lower groundwater pressures sometimes measured in bedrock suggest the possibility of draining a landslide mass by putting vertical drains into the bedrock. In recent years, U.S. environmental laws have not allowed water to be drained from one aquifer to another, even in remote locations. Therefore, although technically feasible, it is not an option where such laws apply.

5.2 GROUNDWATER FLOW ALONG A SHEAR ZONE

An enduring myth among some geotechnical specialists is that the shear zone of a landslide has soils that are badly broken up by shear movements and thus are very permeable. Geologists are especially prone to this viewpoint because the fault zones of hard rocks do include highly broken rock pieces that can be permeable.

In clay soils, there is little direct evidence that the shear zone is any more permeable than elsewhere in the landslide debris. Stiff clay soils at depth are usually tight and no seepage is observed. Some examples:

- At the Washington Park Reservoirs Slide in Portland, Oregon, the slip zone in the stiff clay at the base of 22 vertical shafts was meticulously examined, and the excavations at each shaft were described by Clarke (1904)—see Case History 1. These excavations encountered zones of water-bearing broken rocks in most of the shafts but the slippage zone was always found within a clay seam. There is no mention of any seepage coming through the clay slip plane.
- At the Washington Park Station Slide in Portland, Oregon (Case History 4), large elevator shafts passed through the ancient landslide shear zone. No seepage was observed. The shear zone was estimated to be about 1½ feet thick.
- At the West Percy Slide, Leaburg, Oregon, a deep shear key intercepted the shear zone in highly weathered flow breccia. The shear zone appeared to be only a few inches thick. The clayey silt of the shear zone was softer than the surrounding stiff soils, presumably due to the shear zone being at residual strength. However, no seepage was observed at the slip surface exposures.
- At the Aldous Avenue Slide, Salem, Oregon, it was alleged in a lawsuit that stormwater collected in a detention basin at the top of the slope entered the headscarp of an ancient landslide, traveled 1,100 feet along the ancient shear zone, and caused a local landslide further downslope. When the detention basin was later filled with water, there was no change in the groundwater levels measured in the piezometers installed at the local landslide downslope.

These examples demonstrate that a shear zone in clay is not a conduit for rapid transmission of groundwater or groundwater pressures. The clay shear zone is usually tight and behaves as a clay, not as broken fragments.

5.3 EFFECT OF RAINFALL ON GROUNDWATER LEVELS

Groundwater levels in clay slopes generally rise during the wet season and rise further during and immediately after storms. With the advent of automatic data acquisition systems, it is possible to monitor piezometric levels continuously. This provides a better understanding of the lag between rainfall and groundwater response, the nature of the upward spikes during high intensity rainfall, the general level of seasonal rise during the wet season, etc.

Piezometer LT-21 Installation at Frog Lake Landslide
A good example of long-term groundwater observations has been the monitoring of piezometer LT-21 at Frog Lake landslide near Estacada, Oregon (Cornforth and Mikkelsen, 2000). The piezometer is located near the mid-length of an ancient landslide slope 8,000 feet long with steep cliffs of basalt and conglomerate at the upper end and the Clackamas River at the base of the slope (Figure 5.2). Part of the ancient landslide was reactivated by construction of Frog Lake reservoir 5,000 feet upslope from the river. This very large translational landslide is believed to be slipping at depths of up to 300 feet below the surface, probably through the Bull Creek Beds.

Piezometer LT-21 was installed in July 1995. The summary drillhole log (Figure 5.3) showed that all materials were landslide debris. The upper 71 feet of hole was drilled through sandy silt and sand by the ODEX technique using air feed. The lower 79 feet was drilled with HQ coring and passed through highly broken rock.

On removing the drilling rods, the lower part of the hole collapsed. A plastic standpipe piezometer, nominal 1 inch inside diameter, was installed to a depth of 90 feet in the

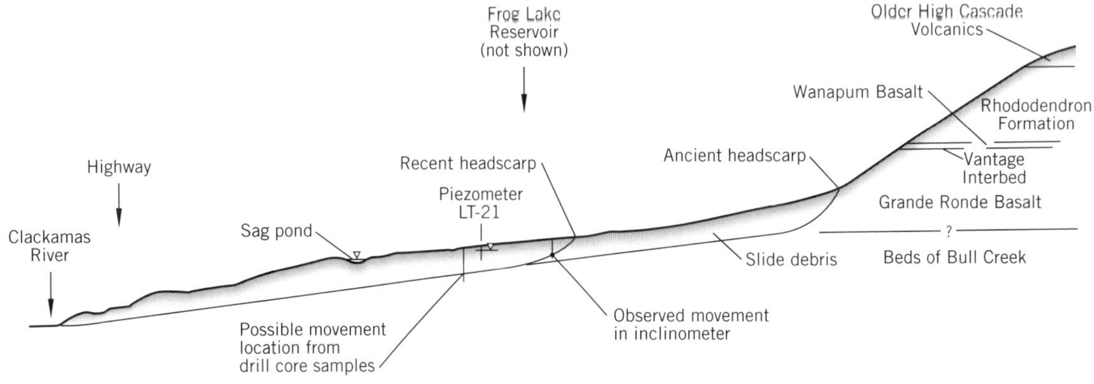

Figure 5.2 Geological section, Frog Lake Landslide, showing location of piezometer LT-21 near mid-slope.

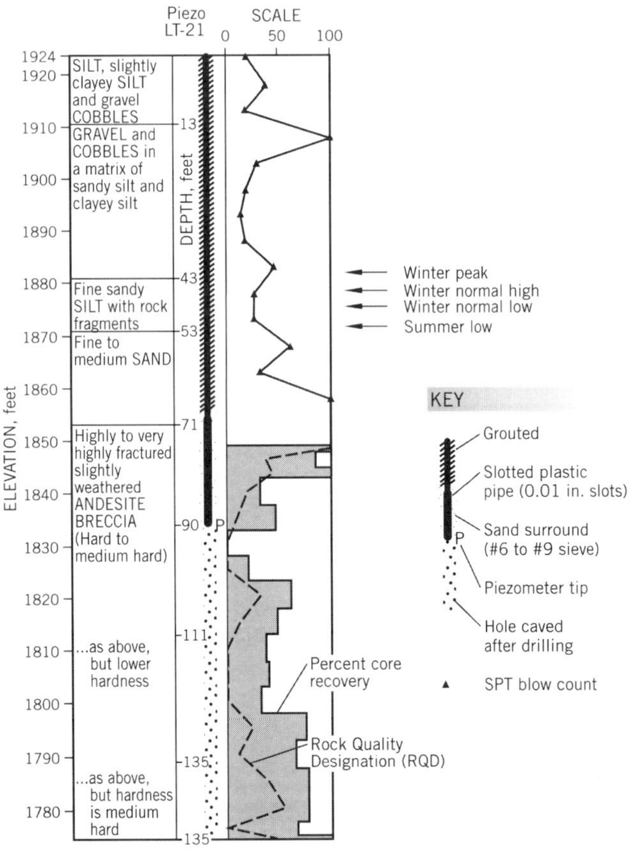

Figure 5.3 Summary log of subsurface conditions and piezometer LT-21.

remaining open hole. The lower 20 feet of the plastic pipe has 0.010-inch wide slots and is surrounded by clean sand (No. 6 to No. 9 U.S. sieve sizes). The upper 70 feet of the hole is grouted with a cement-bentonite (8:1 mix by weight) slurry.

The standpipe piezometer was automated in 1997. A ¾-inch diameter vibrating wire pressure sensor was lowered into the standpipe by a cable and suspended at the 90-foot depth. The measured water head is added to the stationary tip elevation to obtain the piezometric elevation.

The PVC riser is enclosed in a 6-inch diameter steel protective casing at the surface. Here, a small junction box was installed to terminate the sensor cable, provide lightning protection, and connect to the trenched-in telephone cable that extends from the hole to a control house. The wires were connected to a Campbell Scientific CR10X datalogger inside the building. The datalogger was programmed to collect data every hour, the data being vibrating-wire frequency readings, heads in feet, and elevations.

Rainfall and intensity was measured by a gauge placed on a steel pole next to the control house. Telephone wires carried the pulse signal from the rain gauge to the CR10X datalogger 200 feet away. A tipping bucket style rain gauge (Met-One) was installed. A heater for the gauge funnel activated below 40° Fahrenheit, melting any snow or ice.

The bucket in the gauge tipped for every 0.01 inch of accumulated rainfall. Each tip activated a magnet/reed switch and sent a pulse signal to the datalogger. The logger tracked the number of pulses and the program calculated the cumulative rainfall and intensity. The results were stored hourly.

Groundwater Response Analysis

At Frog Lake site, both the surface and hydraulic gradients are about 13 percent (Figure 5.2). Rainfall not only infiltrates vertically (approx. 50 feet to the average water level in piezometer LT-21) but moves downslope on reaching the water table. This, combined with the very heterogeneous nature of the landslide debris (Figure 5.3), controls the groundwater response to rainfall.

The groundwater response has been subdivided into three 13-week periods on Figures 5.4, 5.5, and 5.6. Although one must be cautious in giving too much significance to a single site record, Frog Lake piezometer LT-21 provides insights into groundwater response that probably apply to other sites. For

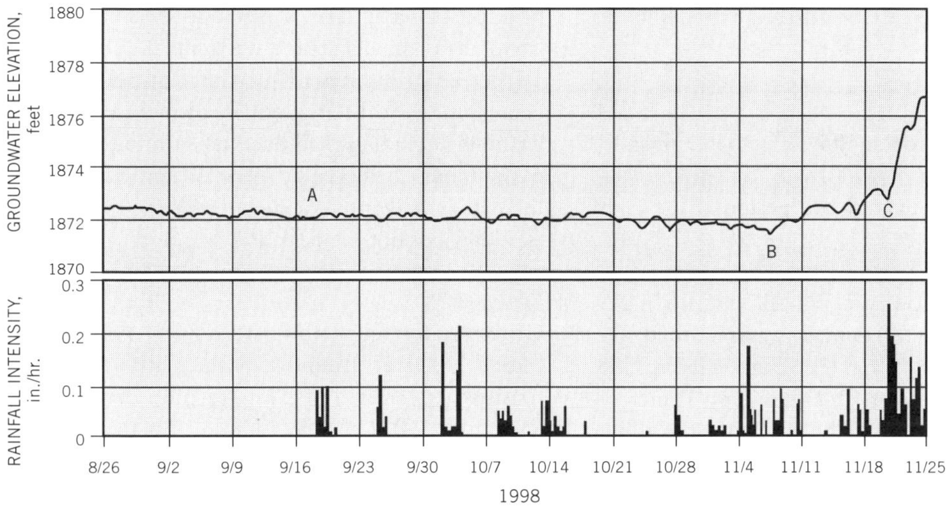

Figure 5.4 Piezometer LT-21: groundwater levels and rainfall intensity, 8/26/98 to 11/25/98.

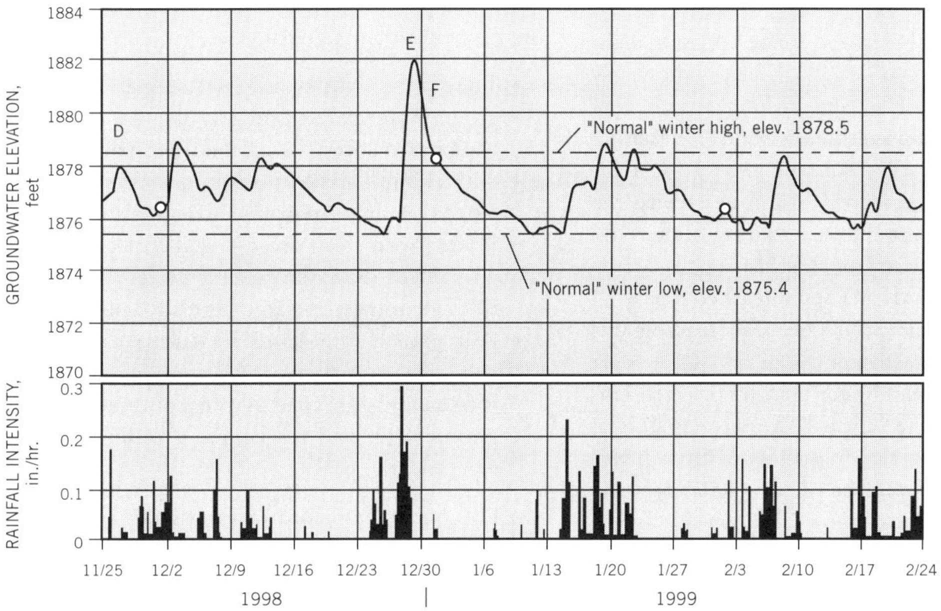

Figure 5.5 Piezometer LT-21: groundwater levels and rainfall intensity, 11/25/98 to 2/24/99.

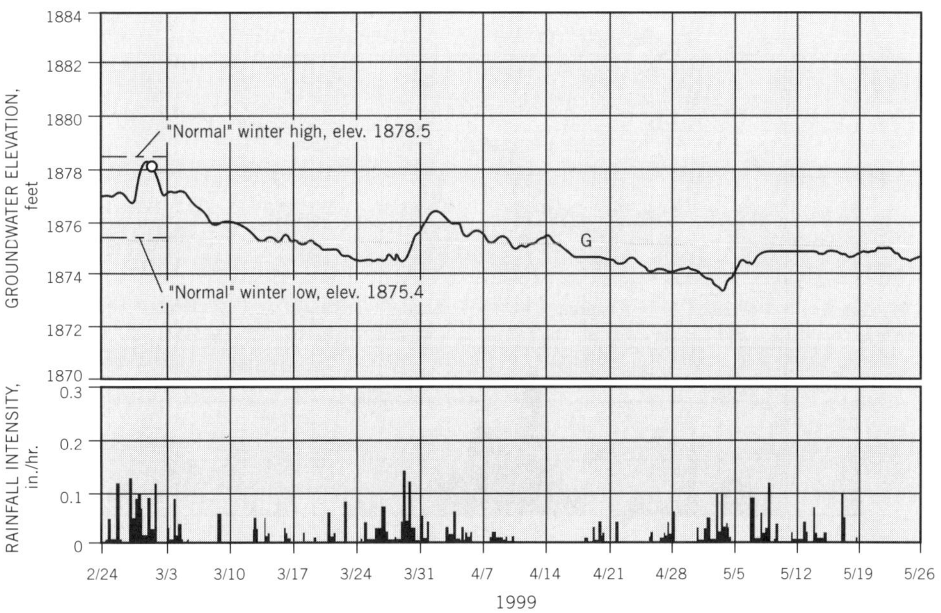

Figure 5.6 Piezometer LT-21: groundwater levels and rainfall intensity, 2/24/99 to 5/26/99.

simplicity, elevation data such as elev. 1,875.0 is simplified to elev. 75.0; the units are in feet.

Late Summer 1998

No rainfall occurred at the site from July 30 to September 18. The end of this period is shown by A on Figure 5.4. Groundwater levels fell throughout the period at a rate of about ¼ inch/day.

During the seven-week period from September 18 to November 10, there was sporadic rainfall but no major storms. A total of 7 inches of rainfall was recorded, but groundwater level *continued to decline,* reaching the annual low of elev. 71.5 near mid-November (B, Figure 5.4). Therefore, the rainfall infiltration was essentially rewetting the upper 53 feet of soil above the groundwater table after the dry summer months.

Winter 1998–99

The first winter storm produced 7.4 inches of rainfall in 6.6 days; the groundwater level rose by 5.1 feet, as shown by CD on Figures 5.4 and 5.5. Thereafter, winter storms kept groundwater levels generally within a band fluctuating between about elev. 75.4 and elev. 78.5, a range of 3.1 feet, for about 16 weeks. These limits (Figure 5.5) can be termed the "normal" highs and lows of groundwater during winter and are well above the late summer low of elev. 71.6.

In late December, a groundwater spike reached a peak of elev. 82.0 (E, Figures 5.5 and 5.7). The peak reading was caused by an intense storm with rainfall of 6.0 inches over 1.8 days which raised the groundwater level by 6.1 feet. This peak value is 3.5 feet above the "normal" winter high. It is of interest that the peak lasted for only two to three hours (9:00 p.m. to midnight on December 28, see Figure 5.7) and the groundwater level declined at almost 2 feet per day for the next two days (F, Figure 5.7).

The elapsed time (delay) from peak rainfall intensity to peak piezometric level was about 25 hours (Figure 5.7). In later storms, the delay time was generally about 19 hours.

Periodic readings of piezometers (the usual practice in geotechnical work) are unlikely to record the peak level. For example, if readings had been taken monthly on the first day of the month as shown by open circles on Figures 5.5 and 5.6, the data would approximately cover the normal winter highs and lows, but not the peak at E.

Spring 1999

Groundwater fell below the normal winter range in late March. However, groundwater fluctuations in spring are more sensitive to rainfall (G, Figure 5.6) than in late summer (B, Figure 5.4). This may indicate that the ground above the water table is closer to full saturation in spring than in fall.

Conclusions

Some key points emerge from the continuous monitoring of the piezometer and rain gauge:

- The first storms of the wet season have no effect on groundwater levels probably due to resaturation of the overburden above the groundwater table.
- Winter rainfalls keep groundwater levels in a general range that always remains above the late summer levels.
- Severe storms can create a spike of groundwater that is several feet above the general winter range and may cause landslide movements in marginally stable slopes.
- Peak groundwater levels are very transitory and are difficult to measure by periodic site visits; however, automatic data acquisition systems can continuously monitor the groundwater and record the groundwater fluctuations, especially the peaks.

The one-year record at Frog Lake covered the period August 1998–May 1999. Similar records have been obtained in subsequent years that verify the general conclusions just stated.

Figure 5.7 Piezometer LT-21: spike of groundwater in December 1998.

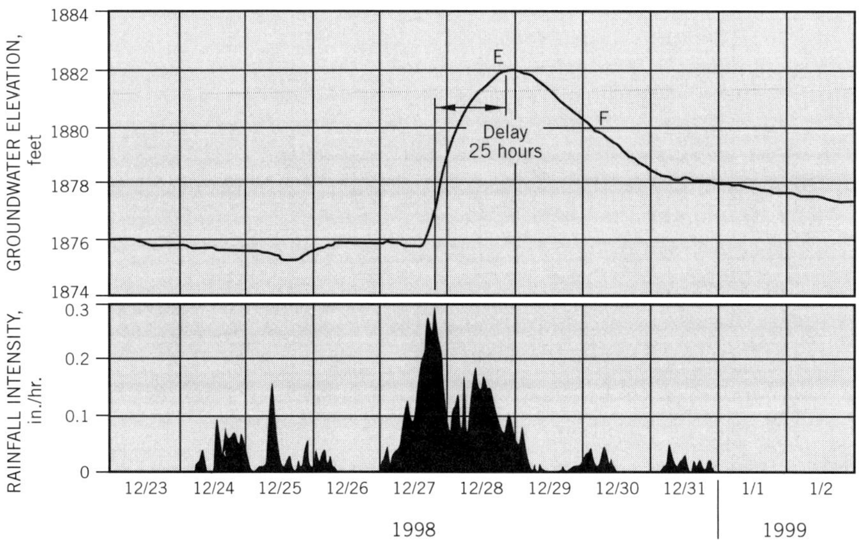

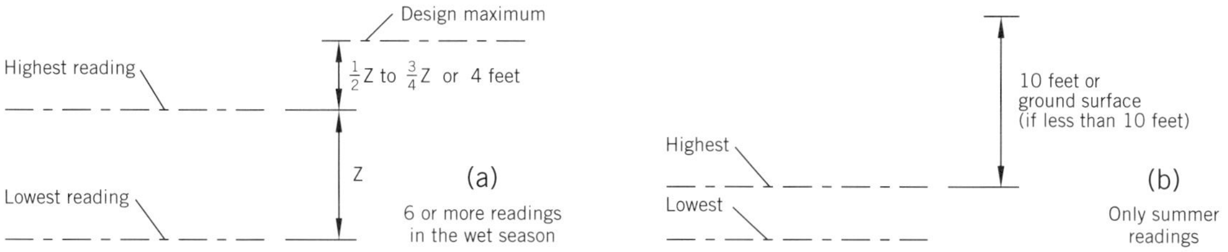

Figure 5.8 Suggested guidelines for estimating the maximum probable winter groundwater levels in clay slopes from limited sets of readings.

5.4 SELECTION OF GROUNDWATER LEVELS IN A STABILITY ANALYSIS

Estimation of Groundwater Levels Based on Experience

In many cases, a landslide occurs during an extreme storm event. After the storm subsides, a site investigation explores the site and installs piezometers to measure groundwater levels. These measurements may continue for an entire winter season before remedial action is taken. The question is: What groundwater levels should be used in stability analyses?

Much depends on the amount of groundwater data collected after the landslide has occurred. If piezometers are installed quickly and there is sufficient time left in the wet weather season, a good indication of the normal winter highs probably can be obtained from twice weekly or weekly readings over a two-month period with effort being made to obtain readings during or shortly after a period of heavy rainfall. If such data cannot be obtained, the consultant has to make estimates of groundwater highs and select higher factors of safety for remediation to allow for the lack of reliable groundwater information.

Assuming that the landslide was caused by very heavy rainfall, any groundwater levels measured after the event will be lower than the levels at the time of failure so some upward adjustment in the measured groundwater levels is needed. The amount that groundwater "spikes" during high intensity rainfall varies from site to site. As more research is conducted on this subject, using continuous readout piezometers, there should be better guidelines on how to estimate or predict such spikes. Based on limited information, the author suggests the following tentative procedures (Figure 5.8):

- If at least 6 readings have been taken at intervals during the wet season, take 50–75% of the difference between the high and low readings and add it to the highest reading to estimate a storm maximum reading. *Example:* Measured high-measured low = 7 feet. Select 5 feet above the measured high reading. Irrespective of this calculation, use at least 4 feet above the measured high reading.
- If the only readings are summer (dry season) readings, storm levels are likely to be at least 8 feet above summer levels. Select 10 feet or the ground surface, whichever is lower.
- Use experience from past records of sites with similar soil conditions and slopes.

Peak groundwater levels are used in back analysis to calculate the average shear strength along the slip surface of a landslide. The same data are carried forward into the stability analyses of the proposed remedial treatment. Therefore, some error in the forecast of groundwater levels generally is not critical. The selected safety factor should be sufficient to overcome such errors in what is a comparative set of calculations.

Analytical Estimation of Groundwater Rise

Lumb (1975) derived the following formula to calculate the rise Z of the "wetting front" (Figure 5.9) during a storm:

$$Z = \frac{kt}{n(S_f - S_o)} \qquad \text{Eq. (1)}$$

where k = saturated coefficient of permeability; t = elapsed time of rainfall; n = soil porosity; S_o, S_f = initial and final degrees of saturation.

This equation has been put into practice in Hong Kong where the height of the wetting front is added to the winter groundwater level to estimate the peak groundwater levels. For this purpose, a rainfall return period of 10 years is commonly used. Typically, the wetting band thickness is calculated to be about 2 m (6.6 feet) (Brand, 1985).

Direct Monitoring of Peak Groundwater Level

The best way to obtain peak groundwater levels for landslide analysis is to measure them directly over one or more winter seasons using vibrating wire piezometers and an automatic data acquisition system. These systems are described in Chapter 4. In practice, this opportunity is rarely available because (i) landslides usually have to be remediated before the next wet season, and (ii) cost. On larger landslides, however, sufficient time may be available.

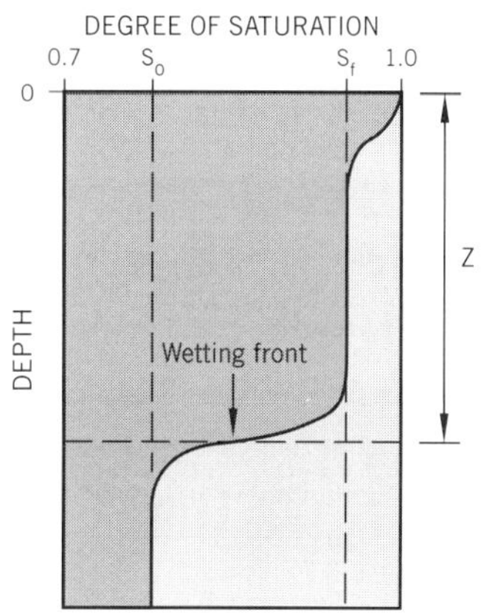

Figure 5.9 Advance of the wetting front due to rain infiltrating into soil (after Lumb, 1975).

5.5 MEASUREMENTS OF FIELD PERMEABILITY

Field measurements of permeability can be used for several purposes in landslide studies:

- To determine permeability for seepage calculations
- To identify permeable zones feeding water into a landslide
- To determine the capability of the soil or rock to accept remedial grouting

General

The permeability of soil or rock is a fundamental property that expresses the ease with which water can pass through the material. The controlling equation (Darcy's law) is:

$$Q = kiA \qquad \text{Eq. (2)}$$

where Q = flow
i = hydraulic gradient = water head difference : length of passage
k = coefficient of permeability
A = cross-sectional area through which flow occurs

Much has been written about the coefficient of permeability in basic textbooks on groundwater and will not be repeated here. There is a huge range in values of parameter k in soils of different gradation.

In layered soils and rocks, the permeability in the bedding plane direction can be much higher than in the direction normal to the bedding. For horizontal bedding planes, such as commonly occurs in flood plain deposits, the value of k_h is much greater than k_v, where subscripts h and v represent the horizontal and vertical directions. Flow net analyses make use of a transformation factor m defined by:

$$m = \sqrt{\frac{k_h}{k_v}} \qquad \text{Eq. (3)}$$

Factor m appears in many theoretical relationships for flow through soils.

Landslide deposits can exhibit anisotropy of $k_x > k_y$ where x, y are directions perpendicular to each other. However, many landslides are heterogeneous and contain pockets of boulders or gravel within fine-grained soils. The permeability differences are large, and perched water in the coarser soils may act as water feeders to the fine-grained soils, keeping groundwater levels generally high within the landslide. (Such landslides are good candidates for drainage by horizontal drains.)

Figure 5.10 Measurement of field permeability through the bottom of a casing:
(a) constant head
(b) falling head

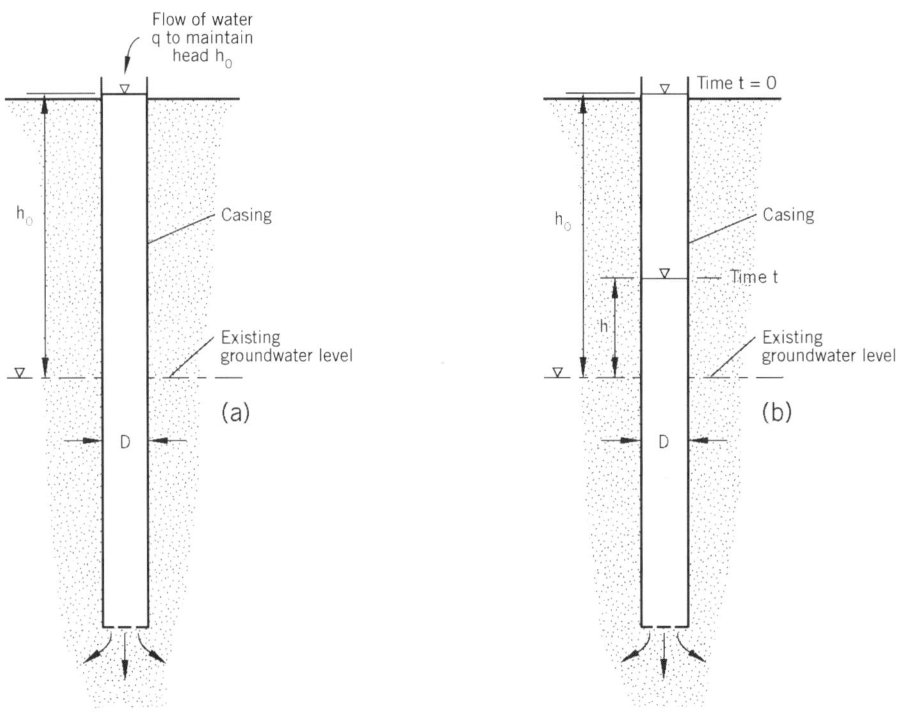

Constant head test:

$$k = \frac{q}{(2.75)\,D\,h_0}$$

q flow in cu. cm/sec
D inside diameter of casing, cm
h_0 constant head, cm
k in cm/sec

Source: U.S. Bureau of Reclamation, 1960

Falling head test:

$$k = \frac{0.657\,D}{t}\log_{10}\frac{h_0}{h}$$

h_0 initial head (t =0) in cm
h head after time t, cm

Using basic time lag T (see Figure 5.11c)

$$k = \frac{D}{3.5\,T}$$

T basic time lag, seconds

Due to the historical development of soil mechanics, the units of permeability are almost always expressed as cm/sec. The conversion to U.S. units is 1 cm/sec = 1.97 ft./min.

Field Tests

Three permeability tests that are commonly performed in the field are (a) falling head or constant head through bottom of casing, (b) rising head test in a piezometer, and (c) packer tests during drilling. All permeability calculations need to be carefully checked to ensure consistency between units of measurement and also to obtain the correct value of x in the multiplier 10^{-x}.

Tests through Bottom of Casing

This is a fairly crude test because only the soil at the bottom of the casing is being directly subjected to the water head. However, the test is useful in sands and gravels, which are difficult to sample and are too permeable for piezometer testing (i.e., the piezometer porous tip is less permeable than the gravel and negates the measurements).

The hole is cased through water-bearing sands and gravels, and cleaned out to the bottom of the casing. Groundwater is allowed to reach static equilibrium (which occurs quickly in gravels and cobbles but may require overnight in sands). The casing is filled to the top with water from a pump. After filling, the fall of the water level in the casing with time is recorded. Alternative test procedures are: (i) perform a constant head test by throttling the pump to maintain the head a few inches below the casing top and measure the pump's flow rate; and (ii) if the soil is too permeable for the pump to fill the casing to the top, begin the falling head or constant head tests at a lower level. Figure 5.10 illustrates the tests and formulas.

The drill bit can sometimes be advanced below the bottom of the casing to produce a cylindrical cavity similar to the shape around a piezometer (or the casing can be retracted by a few feet to produce the same effect). A rising head test can be performed using the formula shown on Figure 5.11.

Rising Head Test in Piezometer

In this relatively common test, water is sucked out of the piezometer standpipe to a predetermined depth below groundwater level and the rise of the water level with time is measured (Figure 5.11 a,b). First, lower the water level in the standpipe a few feet below the predetermined starting depth.

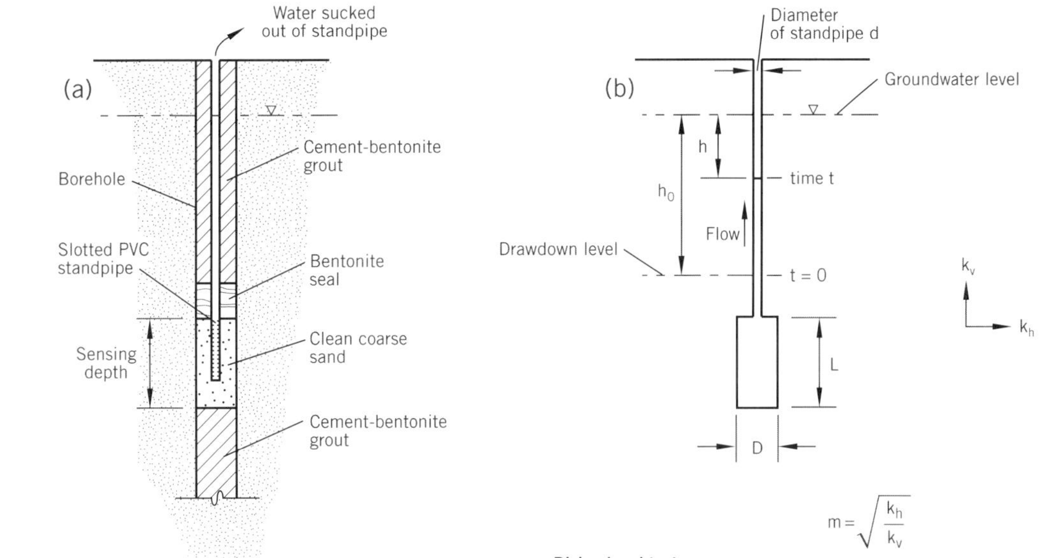

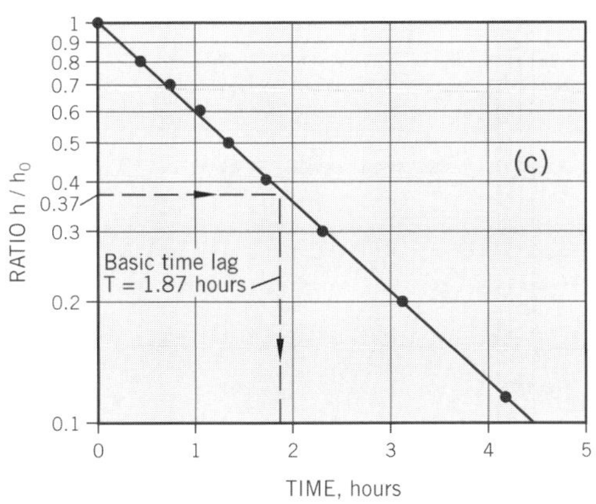

$$m = \sqrt{\frac{k_h}{k_v}}$$

Rising head test:

Full formula (Hvorslev, 1951)

$$k_h = \frac{d^2 \log_e \left[\frac{m\,L}{D} + \sqrt{1 + \left(\frac{m\,L}{D}\right)^2} \right]}{8\,L\,(t_2 - t_1)} \log_e \frac{h_1}{h_2}$$

For $\frac{m\,L}{D} > 4$, $m = 1$, and $h_1 = h_0$ at $t_1 = 0$, the formula becomes

$$k_h = \frac{d^2\,(2.3)^2 \log_{10}\left(\frac{2\,L}{D}\right)}{8\,L\,t} \log_{10}\left(\frac{h_0}{h}\right)$$

Using the basic time lag T obtained from plot (c)

$$k_h = \frac{d^2 \log_{10}\left(\frac{2\,L}{D}\right)}{3.48\,L\,T}$$

Parameters are defined on (b) and (c) above

Figure 5.11 Measurement of field permeability in piezometer rising head test:
(a) typical piezometer installation
(b) definitions of measurements for permeability calculations
(c) plotted data and time lag determination (example)

Next, lower the water level measuring device (usually referred to as a "stinger" because of the electrical buzzer or light that signifies when the tip has reached water) into the standpipe and set the tip at a predetermined depth. When the water in the standpipe rises to this level, the stinger buzzes and the stopwatch time is begun. The stinger is raised to a new level and the time at which the water level triggers the buzzer is again noted. The stinger continues to be lifted incrementally and provides data for the semi-log chart, Figure 5.11(c).

The formulas are provided on Figure 5.11. The easiest solution is to determine the basic time lag T. A typical calculation is given in Case History 11: Hagg Lake Slide 6.

Field permeability tests require careful drilling procedures to minimize disturbance and smear at the wall of the borehole where the tests are to be performed. Rotary drilling methods are preferred. In the past, a synthetic product (Revertex) was used in place of drilling mud, but its use is limited due to environmental concerns. A high initial head of water is now used with reasonable success to prevent hole caving.

Packer Tests

Packer tests can be performed in either rock drillholes or soil boreholes that do not cave when unsupported. The downhole assembly for a *double packer test* is illustrated on Figure 5.12. The upper and lower packers have expandable rubber sleeves that are inflated by compressed air or nitrogen to seal the packer against the hole wall. Between the packers, a perforated pipe section forces water under pressure into the surrounding soil or rock to measure permeability. This test is usually performed *after* the hole has been completed. The consultant selects the vertical interval to be tested based on the cores or boring log information. It is thus more selective than a *single packer test* that is usually performed *during* drilling. The single packer is placed at some predetermined height above the bottom of the hole. It can be sealed to the hole wall by air pressure (or nitrogen) or mechanically (screw arrangement) that squeezes a rubber plug tightly against the hole wall. The test zone extends from the packer to the bottom of the hole (not illustrated). Single packer tests often provide a continuous profile of permeability (average permeability over a selected vertical interval) for the full length of the exploratory hole or, more selectively, in an area of interest. The formulas for calculating permeability are given on Figure 5.12.

The field packer test has similar limitations to a laboratory permeability test. If the soil permeability is high, the measurements will be inaccurate, reflecting the inability of the pump and hoses to deliver water fast enough to the soil (typically 35 gpm maximum). This difficulty can be partly overcome by narrowing the test interval and lowering the applied water pressure. These changes can extend the test range by about one order of magnitude but still may be insufficient for openwork gravels and rocks. In permeable, fractured rocks, the packer may not be able to seal and prevent water from passing around the packer through the fractures. At the less permeable end of the scale, the flow of water into the ground ("take") can be very small. In this case, there may be uncertainty about whether the readings are genuine or result from slight leakage around an imperfect seal or creep expansion in the pressure

Figure 5.12 Measurement of field permeability in a double packer test.

$$k = \frac{(0.508)\, Q \log_e \left(\frac{L}{r}\right)}{2 \pi L H} \quad \text{for } L \geq 10r$$

where

Q = constant rate of flow into hole, cu. ft./min

L = length of test interval between packers, feet

r = radius of hole tested, feet

H = differential head between packer and outside groundwater level, feet

k = coefficient of permeability, cm/sec

(0.508 converts ft./min to cm/sec)

Source: U.S. Bureau of Reclamation, 1960

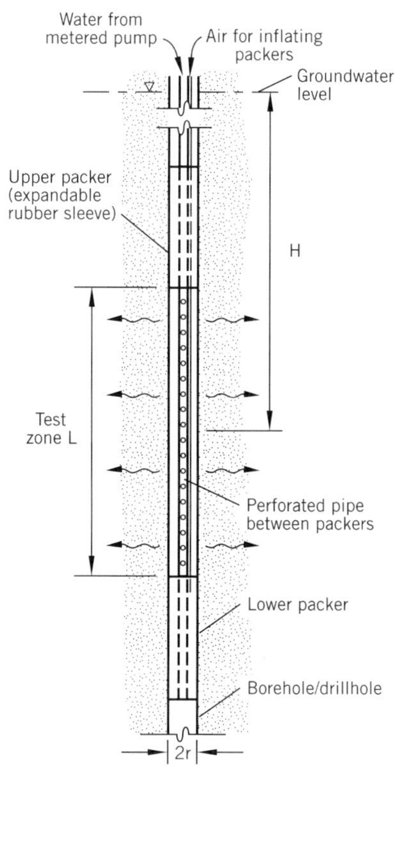

line. Field engineers are likely to increase the water pressure in low permeability soils. However, the water pressure should not exceed the overburden stress at the test section; otherwise, there is a risk of hydraulic fracturing that invalidates the results.

Permeability computations should be recorded to two significant figures. Further mathematical refinement is unwarranted.

An example of a permeability profile is given on Figure 5.13. These tests were made near the spillway approach canal at Bull Run Dam No. 2, Sandy, Oregon. The field technique was to drill and case through highly variable ancient landslide materials of broken rock and mixed clay-to-gravel soils. At a preselected depth, the casing was partly withdrawn to expose the test section. Single packer tests were made at constant water pressure, measuring the flow rate needed to maintain the head. Falling head tests supplemented the packer tests and gave good agreement over the same test intervals.

The test holes were 3.8-inch diameter and the typical test interval was 10 feet. For a measured coefficient of permeability k of 1×10^{-4} cm/sec, the quantity of water flowing into the ground over a 5-minute test at a pressure head of 50 feet is 1.1 cu. ft. Therefore, the lower limit of measurable permeability at this site condition is probably about 5×10^{-6} cm/sec. Slightly lower permeabilities can be measured by converting the test to falling head.

The maximum measurable permeability is governed by the water pump and/or the ability of the connecting pipes to transfer water into the test section of the hole. For these tests, the limit is probably around 2 to 3×10^{-3} cm/sec (for a pump rate of about 30 gpm). Thus the range of permeabilities that can be measured at this site is around 500:1 by the constant head method. The results on Figure 5.13 suggest that both limits were reached at this site.

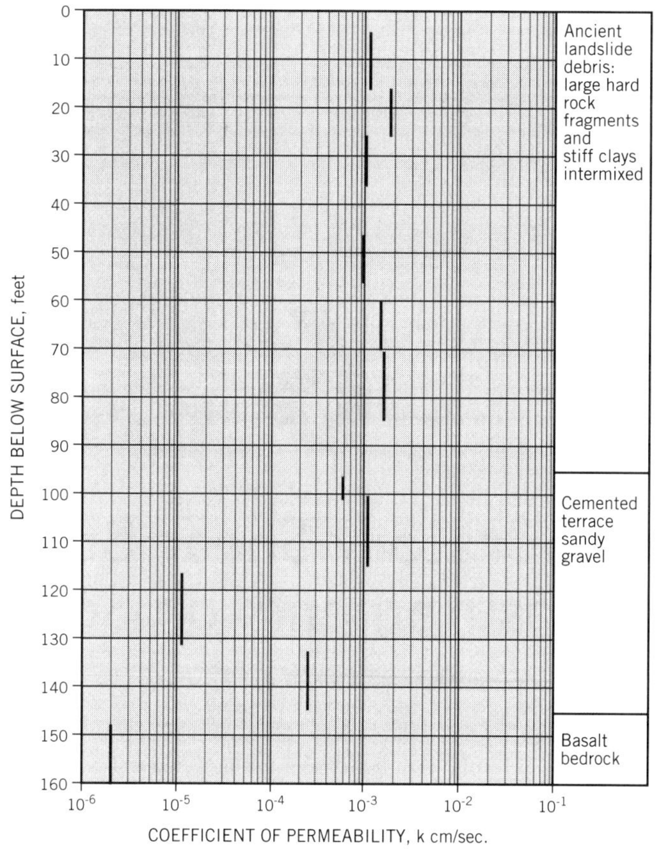

Figure 5.13 Example of packer test results.

CHAPTER

6

Laboratory Shear Strength Measurements on Soils

This chapter, and the two following chapters, discuss shear strength testing of soils for landslide analysis. For a more detailed description of soil testing, the reader should review the standard procedures set out by the American Society for Testing and Materials (ASTM) and the British Standards Institution (BSI). Another good source of information on laboratory testing is provided in the three-volume set written by K. H. Head (1986).

6.1 BASIC CONCEPTS

Soil can be a three-phase material: solid, liquid, and gas. However, except for arid areas or shallow slides in temperate climates, soils within landslides have the spaces between soil particles infilled with water. These spaces are termed *voids.*

The three-phase system can be separated into its components as shown on Figure 6.1.

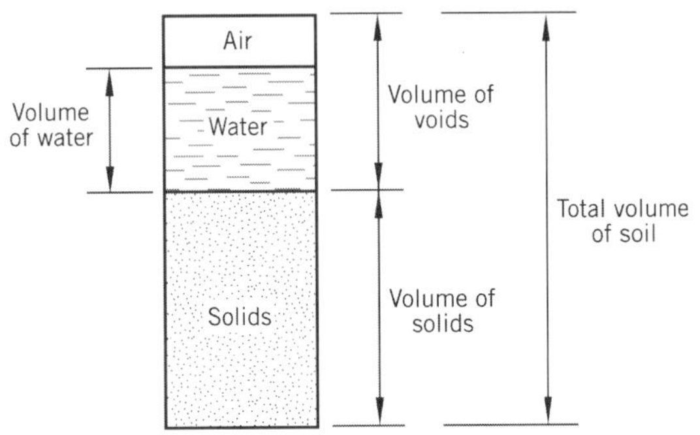

Figure 6.1 Soil components.

The *voids ratio* e is defined:

$$e = \frac{\text{Volume of voids}}{\text{Volume of solids}} \qquad \text{Eq. (1)}$$

e is usually written as a decimal.

The *porosity* n of soil is defined:

$$n = \frac{\text{Volume of voids}}{\text{Total volume of soil}} = \frac{e}{(1+e)} \qquad \text{Eq. (2)}$$

n is usually written as a percentage.

$$e = \frac{n}{(1-n)} \qquad \text{Eq. (3)}$$

Another measure of the compactness of soil particles is *dry density* γ_d given by:

$$\gamma_d = \frac{G\gamma_w}{(1+e)} \qquad \text{Eq. (4)}$$

where G = specific gravity of the soil

γ_w = density of water (62.35 lb./cu. ft. in fresh water)

The *degree of saturation* S of soil is defined:

$$S = \frac{\text{Volume of water}}{\text{Volume of voids}} \qquad \text{Eq. (5)}$$

S is usually written as a percentage

The *wet density* γ is given by:

$$\gamma = \frac{(G + eS)\gamma_w}{(1+e)} \qquad \text{Eq. (6)}$$

with G, e, S, γ_w defined previously.

In saturated soils, S = 100%; i.e., 1.00 in Eq. (6)

$$e = w\,G \text{ in saturated soil} \qquad \text{Eq. (7)}$$

where w = water content of the saturated soil expressed as a decimal.

Equation (7) is very useful because fine sands, silts, and clays are usually fully saturated, including the zone of capillary rise above the groundwater table. Therefore, the expression $e = wG$ can be used to calculate e and then γ in a landslide investigation. It is important, therefore, to obtain a water content profile with depth by measuring the water contents of all recovered samples. Most consulting engineers perform these measurements routinely.

The water within the voids of a soil is referred to as *pore water.* The pressure in the water, relative to atmospheric pressure, is termed *pore water pressure* (u) or simply *pore pressure.* It can be measured in pressure units (e.g., psi) or water head (e.g., feet). 1 psi = 2.31 feet of fresh water or 2.25 feet of sea water.

Pore pressures can become less than atmospheric in laboratory shear tests. When the pressure approaches absolute zero (i.e., −14.7 psi), any air within the soil expands rapidly and the soil cavitates. Cavitation can affect undrained shear tests on sands and silts due to the decrease in pore pressure during shear of dilatant soil.

A *principal stress* is a stress acting on a plane that has no shear stress. An element of soil, shown by the cube on Figure 6.2, has three principal stresses acting along axes mutually perpendicular to each other. Under static conditions, there are equal stresses on opposite sides of the cube, not shown.

Figure 6.2 Total principal stresses acting on an element of soil.

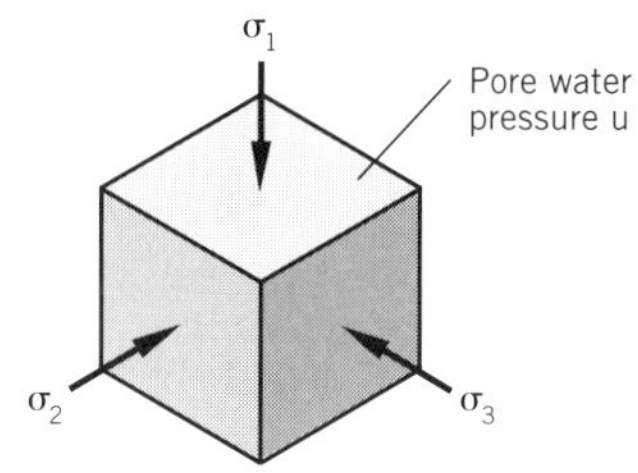

6.2 PRINCIPLE OF EFFECTIVE STRESS

The principle of effective stress states that the strength and deformation of a soil depend only on the effective stresses acting on it.

If the pore water pressure within the soil is u, the *effective principal stresses* are:

Major	$\sigma_1' = \sigma_1 - u$	Eq. (8)
Intermediate	$\sigma_2' = \sigma_2 - u$	Eq. (9)
Minor	$\sigma_3' = \sigma_3 - u$	Eq. (10)

Rearranging the equations, the *total principal stresses* acting on the soil cube are:

Major	$\sigma_1 = \sigma_1' + u$
Intermediate	$\sigma_2 = \sigma_2' + u$
Minor	$\sigma_3 = \sigma_3' + u$

For a soil element below an infinitely wide, horizontal ground surface (as illustrated on Figure 6.3) the major principal stress σ_1 is the vertical stress. The intermediate stress σ_2 and minor principal stress σ_3 are horizontal and of equal magnitude.

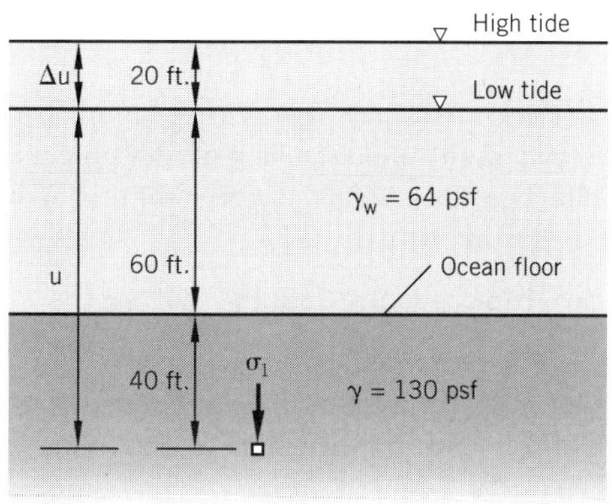

Figure 6.3 Element of soil below the ocean with a fluctuating tide.

The principle of effective stress can be illustrated by an element of soil below the ocean floor, as shown on Figure 6.3. There is 60 feet of water above the floor at low tide and a 20-foot tidal range. The element of soil is 40 feet below the surface of the ocean floor.

At low tide, $\sigma_1 = (130)(40) + (64)(60) = 9{,}040$ psf

At high tide $\sigma_1 = (130)(40) + (64)(80) = 10{,}320$ psf

At high tide, therefore, the total stress acting vertically on the soil element is 1,280 psf higher than at low tide.

The effective stresses are as follows:

Low tide
$u = (64)(100) = 6{,}400$ psf
$\sigma_1' = 9{,}040 - 6{,}400 = 2{,}640$ psf

High tide
$u = (64)(120) = 7{,}680$ psf
$\sigma_1' = 10{,}320 - 7{,}680 = 2{,}640$ psf

Thus, the vertical effective stresses are identical at high and low tide (or at any other tidal level above the ocean floor). The principle of effective stress applied to this situation indicates that the change of tide has no effect on the strength of the soil, and no deformation (settlement or heave) occurs. The soil simply does not react to the changing tide because it affects both the total stresses and pore water pressures equally within the soils, but does not change the effective stresses.

Mathematically, if the total stress is σ_1 and the pore pressure is u at low tide, and the change in pore pressure from low to high tide is Δu, then:

Low tide $\sigma_1' = \sigma_1 - u$

High tide $\sigma_1' = (\sigma_1 + \Delta u) - (u + \Delta u) = \sigma_1 - u$

6.3 PORE PRESSURE PARAMETERS A AND B

Skempton (1954) introduced two pore pressure parameters, A and B, to help in the understanding of how pore water pressure u is influenced by ambient and deviator principal stresses. Skempton's equation is:

$$\Delta u = B\{\Delta\sigma_3 + A(\Delta\sigma_1 - \Delta\sigma_3)\} \qquad \text{Eq. (11)}$$

where A, B are empirical parameters to allow for the deviation of real soils from elastic theory. Parameter B is used extensively in laboratory shear testing to determine when full saturation has been achieved during the pre-saturation phase of a triaxial test (as discussed in Section 6.4). Parameter A has been used by Skempton to explain the concept of delayed failure (Chapter 8, Section 8.12).

A brief description of the procedure followed to arrive at Eq. (11) is helpful to understanding the concept (Bishop and Henkel, 1957).

The elastic strains ε of soil subjected to changes in effective stresses σ' are:

$$\varepsilon_1 = \frac{1}{E}\{\Delta\sigma_1' - \nu(\Delta\sigma_2' + \Delta\sigma_3')\} \qquad \text{Eq. (12)}$$

where E = Young's modulus
ν = Poisson's ratio

Similar expressions to Eq. (12) can be written for ε_2 and ε_3. Total volumetric strain $\frac{\Delta V}{V}$ is:

$$\varepsilon_1 + \varepsilon_2 + \varepsilon_3 \; \frac{\Delta V}{V} = \frac{(1-2\nu)}{E}(\Delta\sigma_1' + \Delta\sigma_2' + \Delta\sigma_3') \qquad \text{Eq. (13)}$$

For an equal all-around (ambient) effective stress change of $\Delta\sigma'$, Eq. (13) becomes:

$$-\frac{\Delta V}{V} = \frac{3(1-2\nu)}{E}(\Delta\sigma')$$

$$\text{and } -\frac{\Delta V}{V} = c_s(\Delta\sigma') \qquad \text{Eq. (14)}$$

where c_s = compressibility coefficient of the soil skeleton

$$\text{i.e., } c_s = \frac{3(1-2\nu)}{E} \qquad \text{Eq. (15)}$$

The decrease in volume of the soil is almost entirely due to the decrease in volume of the void spaces. This volume change can be related to the pore water pressure change as follows:

$$-\Delta V = nV\, c_f\, \Delta u \qquad \text{Eq. (16)}$$

where n = porosity
c_f = compressibility of the fluid in the pore space

Substituting Eq. (13) in Eq. (16):

$$n\, c_f\, \Delta u = \frac{(1-2\nu)}{E}(\Delta\sigma_1' + \Delta\sigma_2' + \Delta\sigma_3') \qquad \text{Eq. (17)}$$

In a triaxial test, $\Delta\sigma_2' = \Delta\sigma_3'$. The test is usually performed by first consolidating the soil under an all-around stress change, followed by application of a deviator stress $\Delta\sigma_1' - \Delta\sigma_3'$. These two phases of the test can be separated through elastic theory.

From Eq. (17) and Eqs. (8), (9) and (10):

$$\Delta u = \frac{(1-2\nu)}{n\, c_f\, E}(\Delta\sigma_1 + 2\Delta\sigma_3 - 3\Delta u)$$

$$= \frac{3(1-2\nu)}{n\, c_f\, E}\{\tfrac{1}{3}(\Delta\sigma_1 - \Delta\sigma_3) + \Delta\sigma_3 - \Delta u\}$$

Substitute for Eq. (15):

$$\Delta u = \frac{c_s}{n\, c_f}\{\tfrac{1}{3}(\Delta\sigma_1 - \Delta\sigma_3) + \Delta\sigma_3 - \Delta u\}$$

$$\Delta u\left\{\frac{nc_f}{c_s} + 1\right\} = \{\Delta\sigma_3 + \tfrac{1}{3}(\Delta\sigma - \Delta\sigma\ )\}$$

$$\Delta u = \frac{1}{\left\{1 + \frac{nc_f}{c_s}\right\}}\{\Delta\sigma_3 + \tfrac{1}{3}(\Delta\sigma_1 - \Delta\sigma_3)\} \qquad \text{Eq. (18)}$$

It is now seen that Eq.(18) is in the same form as Eq.(11) with

$$A = \tfrac{1}{3} \text{ and } B = \left\{\frac{1}{1 + \frac{nc_f}{c_s}}\right\}$$

When the soil is fully saturated, the fluid in the soil is entirely water and $c_f = c_w$. The compressibility of water c_w is typically about 200 times lower than the compressibility of the soil skeleton c_s. Therefore, when soil is saturated B = 1.00.

Pore pressure parameter A at failure is often around +1 for normally consolidated soils. On overconsolidated clays, parameter A can be negative at failure due to dilatancy.

Note: The coefficient of compressibility of the soil structure c_s is often denoted by the symbol m in soil mechanics literature.

6.4 TRIAXIAL TESTS

The triaxial test is a versatile testing method for applying a variety of stress and drainage conditions to soils to measure strength, deformation, or permeability. Bishop and Henkel (1957) describe the tests in detail, and Head (1986) is a good source of information.

Triaxial Apparatus

The triaxial test layout is shown on Figure 6.4. The soil specimen can be prepared in several ways:

- Extruded directly from a relatively undisturbed thin-wall tube sample
- Taken from a block sample and trimmed to shape by a wire saw on a hand-turned lathe
- Compacted by a compaction hammer in a two- or three-way split mold
- Deposited, tamped, or vibrated in a split mold (sand)
- Remolded by hand in a split mold

Compaction can be performed to a preselected density or may be specified for a preselected water content and compactive effort (to replicate standard or modified compaction tests).

The soil specimen is placed on the lower pedestal of the triaxial cell and the top cap is placed on it. Coarse porous stones or low air entry porous stones are placed at the ends of

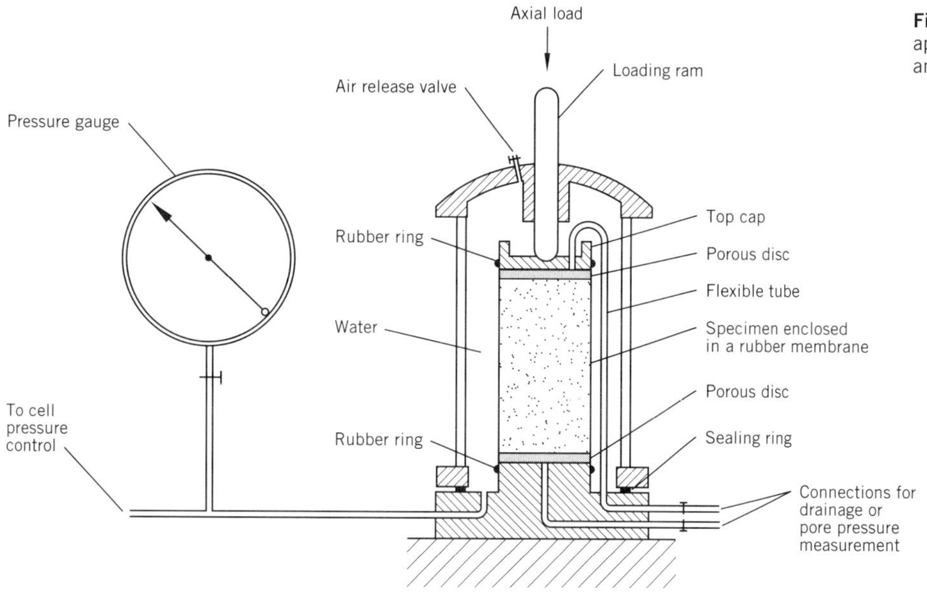

Figure 6.4 Triaxial test apparatus (from Bishop and Henkel, 1957).

the soil. A rubber membrane is placed around the test specimen and is sealed by rubber O-rings at the top and bottom.

The outer chamber is bolted onto the base and filled with de-aired water. The tubes leading to the top and base of the specimen are filled with de-aired water. The use of de-aired water, and connecting tubes that have very low permeability and expansion under pressure, are needed to provide reliable measurements of pore water pressures during tests that last for several hours or days.

Pressure gauges measure the cell pressure, back pressure (a pressure applied to the pore water), and changes in pore water pressure. Volume change devices that can measure volume changes under pressure are usually placed between the triaxial cell and the pressure gauges (not shown on Figure 6.4). A constant pressure source, typically an air compressor with an air/water interface, supplies pressure to the cell and back pressure to the test specimen. All pressure gauges need to be periodically checked for accuracy, and also be calibrated against each other.

The cell pressure provides the total all-around (ambient) stress. The pore water pressure u is measured from the top or bottom (usually the bottom) of the specimen. Drainage can be permitted from either or both ends of the specimen.

The triaxial cell is placed on the platen of a load frame, which is raised at a constant speed preselected for the test. A multispeed motor can be geared to reach soil failure over a time period ranging from minutes to several weeks.

The axial load, required to load the specimen to failure, can be applied through a piston passing through the top of the triaxial cell. An external proving ring or load cell measures the applied load. Piston friction, plus allowance for the cell pressure, can be measured at the start of the shear phase of the test by running the piston into the cell at the preselected strain rate without touching the specimen. Despite these precautions, piston friction can be significant, especially when a failure plane develops and pulls the piston sideways against the bushing. It is possible to use cells fitted with a rotating bushing that eliminates the vertical component of friction from the proving ring measurements. Another common alternative is to use a load cell placed within the cell directly above the specimen.

Back pressure is water pressure applied directly to the pore water within the soil specimen. Since the cell pressure is the total stress σ and the back pressure is the pore pressure u, the back pressure always has to be less than the cell pressure to obtain a positive effective stress on the test specimen. In many triaxial shear tests, the back pressure is raised to around 60 psi so that any entrapped air is dissolved into solution. In this way, the test specimen behaves as a fully saturated soil rather than a soil with air present. Thus, a soil tested with a cell pressure of 70 psi and a back pressure of 60 psi has an effective stress (after consolidation) of 10 psi and, according to the principle of effective stress, will have the same measured properties as a fully saturated soil tested with a cell pressure of 10 psi and zero pore pressure.

Pore water pressures are measured using a *null indicator*. The principle of measurement is illustrated on Figure 6.5. A low-expansive tube (such as mylar) filled with de-aired water leads from the triaxial cell to the null indicator. Within the null indicator, the de-aired water from the cell makes contact with an immiscible liquid (e.g., dyed alcohol or mercury) at Point A (Figure 6.5a). When the pore water pressure rises during a triaxial test, the increased pressure pushes down the contact between the two liquids. However, if an equal pressure is applied to the second liquid from an external source to keep the contact at point A, the applied external pressure equals the pore water pressures u within the soil specimen. Many devices have been made to measure pore pressures using the null principle. Figure 6.5(b) shows a null indicator using mercury within a lucite block. Twisting and locking the device at an angle moves the mercury into the larger chamber and allows de-aired water to be circulated prior to using the device in a test. In addition, the apparatus is calibrated to correct for the small expansions in the leads under pressure of the tube between null level A and the triaxial cell.

Another important laboratory device for performing triaxial tests is the *volume change indicator*, illustrated schematically

Figure 6.5 Null indicator used to measure pore water pressure in triaxial tests
(a) principle
(b) typical null indicator device

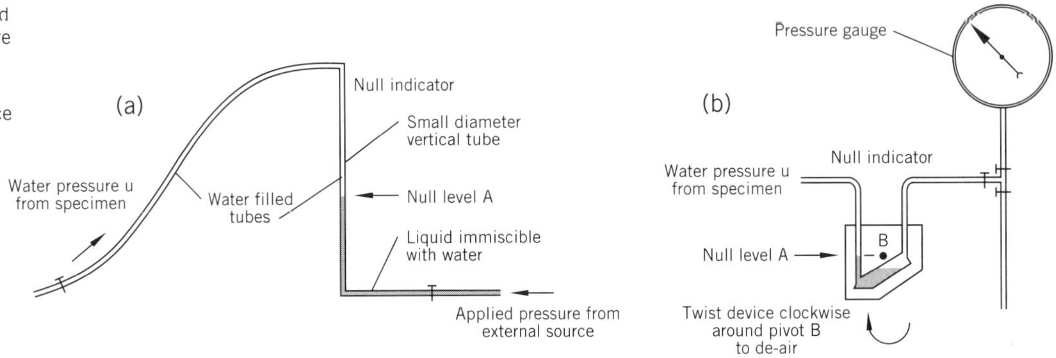

on Figure 6.6. It measures volume changes of the specimen and/or the triaxial cell during consolidation and shear when the water is at a pressure above atmospheric (if back pressures are not being used, an external burette measures volume changes). The measuring cylinder is enclosed within a larger diameter lucite cylinder (Figure 6.6a). A colored liquid, which is immiscible with water and has a slightly different specific gravity, is used to provide the measuring point C within the calibrated cylinder. When water from the test specimen enters the volume change indicator at point A, the contact C between the two liquids goes down and the volume change can be measured. If contact C nears the indicator base at B and water is continuing to enter at A, a set of switch-over valves (Figure 6.6b) reverses the direction of flow within the indicator. The inflow of water then comes into the apparatus at B instead of A and contact C goes up instead of down; this extends the range of volume change that can be measured while preserving a high level of accuracy in measurement. The measuring cylinder has the same pressure on the inside and outside except for an insignificant difference due to the small specific gravity difference between the two liquids.

Types of Triaxial Tests

The more common triaxial shear tests performed in commercial soil laboratories are described below. All the tests are compressive strength measurements in which the principal stresses in the horizontal plane are equal, i.e., $\sigma_2 = \sigma_3$.

1. ***Unconfined Undrained.*** The specimen is set up in the apparatus without a rubber membrane. No cell pressure is applied. The rate of strain is typically set at 2 percent of the axial length per minute. This test is suitable for intact clays or clayey silts. Gives the "immediate" undrained shear strength c (also known as S_u). The test should not be performed on fissured clays or weakly cohesive soils.
2. ***Confined Undrained.*** A cell pressure is applied to the specimen and no drainage is allowed throughout the test. The specimen is failed in the same manner as (1) at a relatively fast rate of strain. The cell pressure is usually set at the total vertical overburden stress acting on the soil in the ground. In the case of intact clays or clayey silts, the shear strength results of tests (1) and (2) should be the same; for fissured clays or weakly cohesive silts and silts, test method (2) gives a higher undrained strength. This is sometimes referred to as a Q test.
3. ***Consolidated Undrained with Pore Pressure Measurements.*** A cell pressure is applied to the specimen, and it is allowed to fully consolidate. After consolidation, no drainage is allowed and the specimen is sheared to failure at a constant rate of strain. This test is usually conducted at a fairly slow rate of strain of 2 to 3 percent axial strain per hour. Readings are taken at ½ to 1 percent intervals of axial strain or continuously. This is sometimes referred to as $\bar{R}$ test.

Figure 6.6 Volume change indicator for triaxial tests:
(a) measuring unit
(b) flow reversal system (switch-over valves)

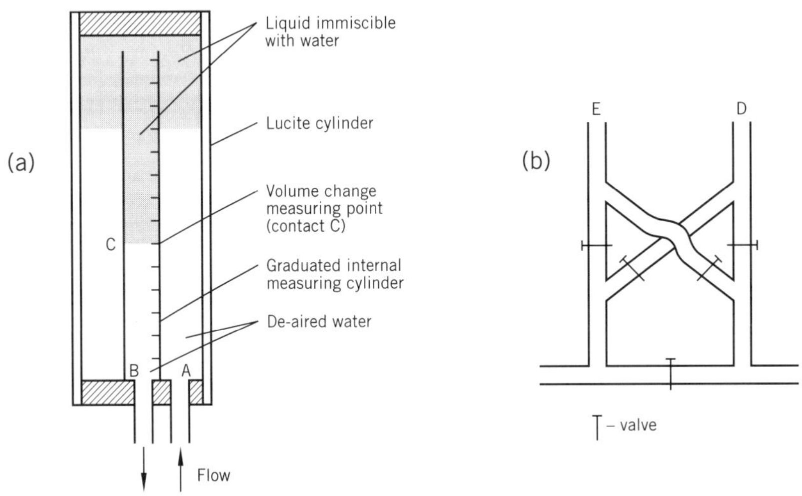

4. ***Consolidated Drained.*** A cell pressure is applied to the specimen and it is allowed to fully consolidate. After consolidation, the specimen is allowed to drain while it is sheared to failure at a constant rate of strain. The shear test is performed at a slow rate to allow at least 95% dissipation of excess pore pressures. This is sometimes referred to as S test.

Presaturation Phase

Air is always trapped between the test specimen and rubber membrane. A technique to eliminate the trapped air is to *presaturate the specimen by back pressures* immediately after setting up the specimen in the triaxial apparatus. A high back pressure puts the trapped air into solution so that pore water pressures or volume changes measured in the consolidation and shear phases of the triaxial test reflect the behavior of a fully saturated soil.

Presaturation has to be performed carefully to avoid overconsolidating the soil prior to the consolidation phase of the test. Therefore, the back pressures are applied in small increments, allowing time for the air to dissolve into the water after each increment. It normally takes from two hours to one day to complete the saturation process (depending on soil type), and the back pressure usually reaches 50 to 70 psi before saturation is achieved.

As previously shown in Eq. (18), pore pressure parameter B is defined by:

$$B = \frac{1}{\left\{1 + \frac{nc_f}{c_s}\right\}}$$

where n = porosity; c_f = compressibility of fluid; c_s = compressibility of the soil structure.

tur

When the soil is fully saturated, the fluid (water) is much less compressible than the soil skeleton c_s. The term nc_f/c_s is zero and B = 1.00. In the triaxial test, presaturation by back pressures is stopped when B >0.96. Soft compressible soils achieve B >0.96 at lower back pressures than stiffer soils.

The back pressure saturation procedure is as follows: one end (in this example, it will be assumed to be the top end) of the specimen is connected to a constant pressure device with a volume change indicator to measure the volume of water entering the specimen during the saturation phase. The opposite (lower) end of the specimen is connected to a pore pressure measuring device.

Suppose the objective is to test the specimen under a consolidation stress of 28.8 psi (2 U.S. ton./sq. ft.). The objective during presaturation should be not to exceed an effective stress of more than about 25 percent of the final consolidation stress at any time during presaturation; i.e., the effective stress should always be less than about 7.2 psi during the presaturation phase for this example.

The initial pore pressure is measured, after which the pore pressure valve is closed. A cell pressure (σ) of up to 10 psi is put into the triaxial cell, and the pore pressure is again measured. As shown on Table 6.1, the B value is calculated as $\Delta u/\Delta\sigma$. Next, a back pressure equal to the cell pressure minus 1 to 2 psi is introduced at the top of the specimen. With time, this back pressure approximately equalizes throughout the specimen and its progress is observed through the pore pressure device at the base of the specimen. When the pressures at the top and bottom of the specimen are nearly equal, the back pressure valve is shut off and the

Table 6.1 Example of Saturation by Back Pressures

	Cell Pressure		Back Pressure	Pore-Water Pressure		Parameter B	Effective Consolidation Stress
	σ psi	$\Delta\sigma$ psi	psi	u psi	Δu psi	$= \frac{\Delta u}{\Delta\sigma}$	σ' psi
	0			0.4(1)			
(ii)	10.0	10.0	(off)	3.6	3.2	0.32	6.4
(i)	10.0		9.0	5.0(2)			
(ii)	20.0	10.0	(off)	7.5	2.5	0.25	12.5
(i)	20.0		19.0	17.2(2)			
(ii)	30.0	10.0	(off)	22.2	5.0	0.50	7.8
(i)	30.0		29.0	28.3(2)			
(ii)	40.0	10.0	(off)	35.2	6.9	0.69	4.8
(i)	40.0		39.0	38.9(2)			
(ii)	50.0	10.0	(off)	47.3	8.4	0.84	2.7
(i)	50.0		49.0	49.0(2)			
(ii)	60.0	10.0	(off)	58.2	9.2	0.92	1.8
(i)	60.0		59.0	59.0			
(ii)	70.0	10.0	(off)	68.6	9.6	0.96	1.4
(iii)	87.8	27.8	(off)	85.8	26.8	0.96	2.0
(iv)	87.8		60.0	85.8			

(1) Initial, on setup (2) At the conclusion of backpressure increment

(i) Using backpressure to raise the specimen's pore-water pressure

(ii) Raising the cell pressure to check the response of pore-water pressure and measure B. Effective stress on soil σ' is highest at this part of the cycle.

(iii) Cell pressure increased to preconsolidation level with back pressure at 59.0 psi. Final check on parameter B.

(iv) Back pressure raised from 59.0 to 60.0 psi (for convenience only) before start of consolidation phase of triaxial test.

cell pressure is raised by another increment. The pore pressure is again read, and a new B value is calculated. The back pressure is raised again to within 1 psi of the cell pressure, and the process is repeated by adding increments of cell pressure and back pressure until B is 0.96 or higher. As the pressures rise, the B values increase such that larger increments of cell pressure can be added without overstressing the test specimen; i.e., maintaining $\sigma' < 7.2$ psi (in this example) throughout the presaturation phase.

In an actual example of a back pressure saturation, shown on Table 6.1, the effective stress σ' is shown in the right-hand column. It will be noted that the effective stress reached 12.5 psi on the second increment of presaturation which is 45 percent of the effective consolidation stress of 27.8 psi (87.8 – 60.0) set for this test. It exceeds the 25 percent desirable limit. This test probably should have been left for a longer period of time at a back pressure of 9.0 psi, which had not equalized throughout the specimen before the cell pressure was raised to 20 psi. Another choice would have been to raise the cell pressure by only 5 psi to 15 psi and include an extra cycle of saturating in the early stages of the test. The latter choice is recommended by the author; the increments of cell pressure should be smaller in the early stages at the test and increase later as the value of B approaches 1.00.

The main purpose of setting a limit on allowable prestress during presaturation is to provide a consolidation curve later that is not significantly affected by the prestress induced during presaturation. ASTM sets a very tight upper limit of 5 psi for σ' during the presaturation stage.

Presaturation by back pressures is performed almost routinely for effective stress triaxial tests on saturated soils to remove air trapped between the soil specimen and the rubber membrane. The need to keep the effective stress low during the presaturation phase is rarely questioned. However, a counterpoint argument to this procedure is that very stiff clay test specimens swell and soften at a low effective confining stress σ'. At the end of the presaturation phase, these soils have been artificially softened. The softening affects the subsequent consolidation curve and probably the peak effective stress strength parameters, especially the measured cohesion intercept c'. Therefore, it may be advisable on stiff clays to require a minimum (lower limit) confining stress σ' of, for example, 5 psi rather than 1 to 2 psi to prevent swelling during presaturation by back pressures.

Consolidation Phase

When the B parameter has reached 0.96, the cell pressure is raised to provide the required effective stress during the consolidation phase of the test. For example, if B = 0.96 at a back pressure of 60 psi, the cell pressure is raised to 87.8 psi for a desired consolidation effective stress of 27.8 psi.

To consolidate the soil, the valve to the constant head back pressure is opened, and water drains from the specimen to the back pressure device. Volume changes with time are recorded during consolidation. If the specimen is drained from one end only, and pore pressure dissipation with time is measured at the other end, soil permeability (k_v), compressibility (m_v) and consolidation coefficient (c_v) can be obtained (Figure 6.7). Alternatively, the specimen can be drained from both ends; this quadruples the rate of consolidation. Subscript v refers to the vertical direction of drainage.

On clays, silty clays, and very silty clays, it is common practice to add filter strips around the perimeter of the triaxial test specimen to speed up the rate of consolidation. There are theoretical relationships (Bishop and Henkel, 1957) that allow k, m_v and c_v to be calculated for this arrangement; in practice, the filter strips are imperfect drains and the calculated consolidation parameters are unreliable. The filter strips also complicate the back pressure saturation procedure because the back pressure can pass *around* rather than *through* the soil. They also trap more air between the specimen and the membrane. Based on laboratory data (Cornforth, 1961b), filter strips should be eliminated on all 1.5-inch diameter triaxial specimens except for the soil types mentioned at the start of this paragraph; i.e., soils with a coefficient of consolidation c_v greater than 1000 sq. ft./year and Cohesive Index (Chapter 8) of less than 0.40.

The consolidation phase of a triaxial test is usually carried out under an ambient effective consolidation stress; i.e., $\sigma'_c = \sigma'_1 = \sigma'_2 = \sigma'_3$. The soil consolidates in all directions and the length and cross-sectional area have to be adjusted to reflect the consolidation. Assuming isotropic consolidation:

$$L_c = L_o\left[1 - \frac{1}{3}\frac{\Delta V_c}{V_o}\right] \quad \text{Eq. (19)} \qquad A_c = A_o\left[1 - \frac{2}{3}\frac{\Delta V_c}{V_o}\right] \quad \text{Eq. (20)}$$

where L_o and A_o are the original length and area of the specimen, L_c and A_c are the length and area of the specimen after consolidation, ΔV_c is the volume change at the end of the consolidation phase, and V_o is the original volume.

Consolidation can also be performed under K_o conditions in which the horizontal cross-sectional area of the specimen remains constant during the consolidation phase of the triaxial test. This procedure simulates vertical one-dimensional consolidation of a stratum. K_o is defined by:

$$K_0 = \frac{\sigma'_3}{\sigma'_1} \text{ for } \varepsilon_2 = \varepsilon_3 = 0 \quad \text{Eq. (21)}$$

The value of K_0 for a particular soil is measured in a separate triaxial test. It is comparatively easy to perform the test on rapid-draining cohesionless soils but is much more difficult in clays due to slow pore pressure changes during consolidation. The apparatus is a *lateral strain indicator* developed by A. W. Bishop and described in Bishop and Henkel (1957). The device is placed around the circumference of a test specimen at the mid-height. As the triaxial cell pressure is slowly increased in small increments, a vertical load is applied to maintain the K_0 condition. A null indicator on the lateral strain indicator is used to keep a constant specimen diameter. K_0 test results for a sand are discussed in the next chapter (Chapter 7, Section 7.9). Some typical K_0 values are: loose sand, 0.45; dense sand, 0.33; normally consolidated soft clay, 0.6; overconsolidated clay, variable, depending on the degree of overconsolidation, but can be higher than 1 in-situ.

At the end of K_0 consolidation, the specimen is already loaded by the deviator stress required for K_0. Consequently, the pore pressure at failure in a consolidated-undrained triaxial test when consolidated under K_0 conditions will be significantly different than in a test consolidated under equal all-around stresses.

Permeability Tests

In addition to measuring permeability of the soil as part of the consolidation phase, it is also possible to conduct permeability tests by passing water through the test specimen from end to end. The test is typically performed after consolidation is completed. Back pressures are applied to both ends of the specimen and one is raised higher than the other to produce a flow gradient. For example, a specimen could have a cell pressure of 100 psi and back pressure of 70 psi at each end. One back pressure is raised to 75 psi, causing a differential pressure of 5 psi. The volume changes are measured by a volume change indicator.

There is a limited range of soil permeabilities for which this test can be undertaken. If the soil permeability is too low, the test takes too long and there is a likelihood of the results being affected by membrane leakage, volume changes due to creep of the leads, etc. If the soil is too permeable, the observations will reflect the time for water to pass through the tubes and porous stones rather than through the soil. The practical range varies with the apparatus and size of specimen. As a guide, the range at which the triaxial test is effective for small specimens has been measured at 2×10^{-7} cm/sec to 4×10^{-4} cm/sec. Therefore, the triaxial test is a good method for measuring the permeability and coefficient of consolidation c_v of silts (Figure 6.7).

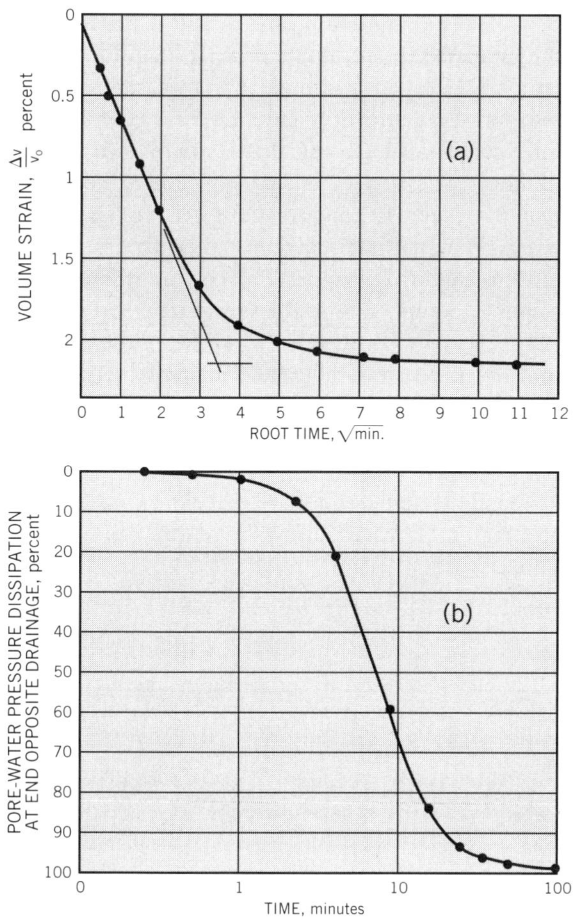

Start Date: 9/1/99 Clock Time: 12:44
Constant back pressure = 60.0 lb./sq. in. Initial excess pore pressure = 25.8 lb./sq. in.

Time		Volume			Pore-Water Pressure		
Elapsed min.	Root min.	Gauge	Change c.c.	Strain %	Gauge lb./sq. in.	Change lb./sq. in.	Dissip %
0	0	20	0	0	85.8	0	0
0.25	0.5	17.9	2.1	0.33	85.7	0.1	0.4
0.5	0.71	17.25	2.75	0.49	85.6	0.2	0.8
1	1	16.3	3.7	0.65	85.3	0.5	1.9
2.25	1.5	14.8	5.2	0.92	83.9	1.9	7.4
4	2	13.2	6.8	1.2	80.4	5.4	20.9
9	3	10.6	9.4	1.66	70.5	15.3	59
16	4	9.2	10.8	1.91	64.2	21.6	84
25	5	8.6	11.4	2.01	61.7	24.1	93
36	6	8.3	11.7	2.07	61.0	24.8	96
51	7.14	8.1	11.9	2.10	60.6	25.2	98
63	7.94	8.05	11.95	2.11	60.6	25.2	98
100	10	7.95	12.05	2.13	60.3	25.5	99
121	11	7.9	12.1	2.14	60.3	25.5	99
1170	39.21	7.9	12.1	2.14	60.0	25.8	100

Final volume strain = 2.14%

Zero volume strain (corrected) = 0.06%

Net volume strain $\frac{\Delta V}{V_0}$ = 2.08%

Change in effective stress Δu = 25.8 lb./sq. in.

Coefficient of compressibility $m_v = \frac{(0.1390)\left(\frac{\Delta V}{V_0}\right)}{\Delta u}$

V_0 = 566.6 ml. m_v = 0.0112 sq. ft./ton

From Work sheet { Initial specimen length = 5.615 in.; Consol. specimen length = 5.575 in. }

Average specimen length $\bar{h}$ = 5.595 in.

Single drainage $h_d = \bar{h}$

$\therefore (h_d)^2$ = 31.30 sq. in.

Coefficients of Consolidation and Permeability

(a) Root time curve (volume)

$\sqrt{t_{100}} = 3.5 \sqrt{\text{min.}}$ t_{100} = 12.25 min.

$C_v = 2865 \frac{(h_d)^2}{t_{100}}$ = 7320 sq. ft./yr.

$k = (0.301)(m_v)(C_v)(10^{-7}) = 2.5 \times 10^{-6}$ cm./sec.

(b) Log time curve (dissipation)

$C_v = 1388 \frac{(h_d)^2}{t_{50}}$ t_{50} = 7.4 min.

C_v = 5870 sq. ft./yr. $k = 2.0 \times 10^{-6}$ cm./sec.

Figure 6.7 Example of consolidation phase of a triaxial test.

Shear Phase

The rate of strain in a drained shear test has to be slow enough that excess pore pressures at failure are very low and have a negligible impact on the measured strength parameters. The theory of consolidation was applied to this issue by Gibson and Henkel (1954). They concluded that a pore pressure dissipation of 95% is sufficient. Based on this analysis, the time to failure t_f for a drained test can be determined as follows:

$$t_f = \frac{20h^2}{\eta c_v} \qquad \text{Eq. (22)}$$

where h = one-half of the specimen height
c_v = coefficient of consolidation
η = factor depending on the drainage conditions at the specimen boundaries

The factor η is 0.75 for single drainage, 3 for double drainage (both ends of specimen) and 40 for radial drainage combined with double drainage (Bishop and Henkel, 1957).

The coefficient of consolidation c_v can be calculated from the volume change vs. root time graph; a good example is shown in Figure 6.7(a). The theoretical relationship, based on Terzaghi's theory of consolidation, is shown on Figure 6.8(a). Note that the theoretical graph starts to diverge from the straight-line relationship after about 50% of primary consolidation has taken place. Therefore, in fitting the straight-line to actual laboratory consolidation test data, it is important that the straight line starts near the origin of the graph and fits the data up to about 50% of the final volume change. If the soil is more permeable than clay, the linear part of primary consolidation may have occurred before any laboratory readings have been taken. In this event, a curve similar to that shown on Figure 6.8(b) is obtained and t_{100} is incorrectly measured.

The consolidation phase of a triaxial test is usually concluded after 24 hours and the volume change that has occurred after 24 hours is taken to be 100% primary consolidation on the root time plot. The calculation of the minimum time to failure t_f during the shear phase of the triaxial tests is given on Table 6.2.

Table 6.2 Formulas for Relating Minimum Time to Failure t_f with t_{100}

Drainage of Specimen	t_{100}	t_f
Single drainage (one end)	$t_{100} = \frac{\pi h^2}{c_v}$	$t_f = 8.5\,t_{100}$
Double drainage (both ends)	$t_{100} = \frac{\pi h^2}{4c_v}$	$t_f = 8.5\,t_{100}$
Radial and double drainage	$t_{100} = \frac{\pi h^2}{100c_v}$	$t_f = 18\,t_{100}$

Notes: (1) h = one-half height of specimen
(2) radial drainage formula assumes height of specimen = 2 × diameter

Radial drains, which are used to increase the rate of consolidation and shear, are made from strips of filter paper. These compress under high pressure and their full effectiveness as drains is questioned. They also increase the measured deviator stress at failure and a correction has to be made (Bishop and Henkel, 1957).

When the minimum time to failure has been calculated, the estimated time and rate of axial strain for the test can be set. Stiff clays fail at around 6% axial strain in the triaxial test; soft clays usually require 12–20% axial strain. Suppose, for example, that the calculated minimum time to failure is 17 hours. The actual test can be set for 32 hours, i.e., 2 days plus overnight. In this situation, it is preferable to run the test at a constant rate of strain rather than to slow down the test overnight. Changing the rate of strain during a test causes a "bump" in both the stress–strain and volume change curves; it is also inadvisable to speed up the test on the second day just before failure. A laboratory technician may be able to take readings after normal working hours (e.g., 8 p.m. and 11 p.m.) to avoid large gaps in the stress–strain data.

Sands are tested at a speed that allows readings to be taken at a reasonable pace during the test. A typical test time is 2 hours.

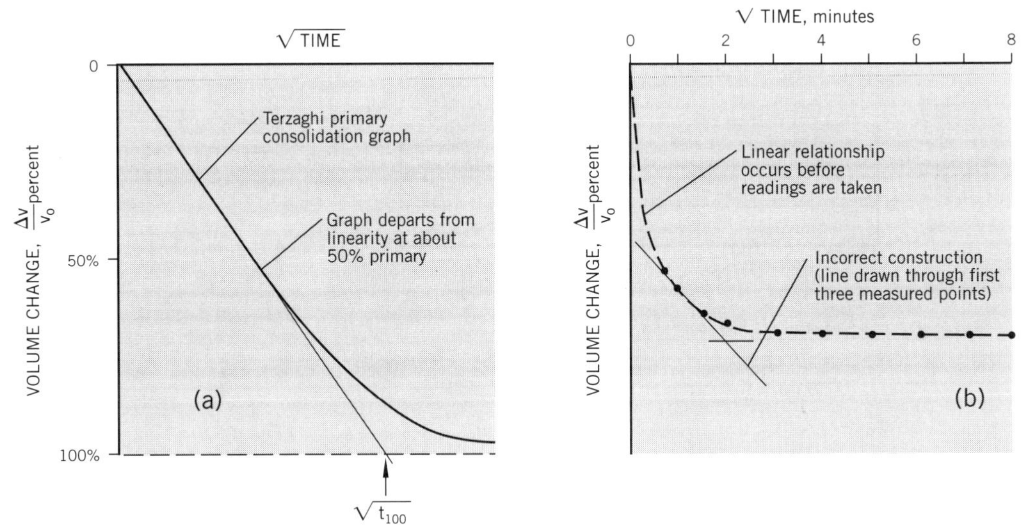

Figure 6.8 Interpretation of consolidation data in triaxial tests:
(a) Terzaghi theory of consolidation
(b) incorrect interpretation due to rapid consolidation

For routine commercial testing, the approximate corrections (reductions) to the measured deviator stress at failure are:

1½ inch dia. specimen	0.6 psi for membrane
	2.0 psi for membrane and filter drain combination
4 inch dia. specimen	0.3 psi for membrane
	0.8 psi for membrane and filter drain combination

Undrained Strength of Dilatant Soils

Dilatant soils, such as stiff overconsolidated clays and dense sands, undergo "strain softening" during shear. After the failure plane starts to develop, pore water migrates to the failure zone, causing a local increase in water content and decrease in density relative to other parts of the soil specimen. Although shearing occurs at *constant volume* in an undrained test on dilatant soils, pore water moves from nondilating areas to the dilating (failure) plane during the test.

It has important consequences in terms of what the undrained test is actually measuring. This topic is discussed in more detail in Chapters 7 and 8.

Consolidated-undrained triaxial tests on dilatant clays should be sheared at about the same rate of strain as the drained test to allow time for pore pressures to approximately equalize throughout the specimen.

Consolidated-Undrained Effective Stress Tests on Soft Clays

Normally consolidated clays have a stress–strain curve that rises to a maximum deviator stress and maintains the maximum stress for additional strain; pore pressures do not change appreciably near failure. Therefore, these soils can be sheared at slightly faster rates than is indicated by time-to-failure calculations.

Drained Strength of Soft Clays

The rate of strain in the shear test should be based on the calculations of Table 6.2. As a check that any excess pore pressures are at an acceptable level, one end of the specimen can be used for drainage and the other end can observe excess pore pressures. If the pore pressures near failure at the non-draining end are more than about 7% of the deviator stress, the specimen can be switched to double drainage or the shear rate can be slowed down.

6.5 SHEAR BOX TEST

The shear box creates a horizontal shear plane at the mid-height of the test specimen. The two halves (upper, lower) of the shear box split and separate horizontally by about 0.35 inch during the test. Although the shear box is used for measuring residual strength, the limited travel of the shear box makes the ring shear test a better alternative.

Test Equipment

The most common size of shear box tests a specimen 6 cm (2.36 inches) square and 2 cm (0.79 inch) thick. The apparatus is shown on Figure 6.9. The shear box consists of a square metal frame that is split into two parts at the mid-height. The soil specimen is placed in the box and porous stones are inserted above and below it. A top cap (loading pad) is put on the upper porous stone and a normal load is applied using a loading yoke. A vertical dial gauge measures the vertical settlement when the soil specimen is loaded through the yoke.

The square frame containing the soil specimen is placed into a bowl container. Water is added to the bowl to prevent the soil from drying out during the test. After consolidating the soil, the shear box, which is supported by ballbearing runners, is pushed horizontally at a constant speed by a loading jack. The container pushes on the lower half of the shear box frame. The upper half, above the mid-height split, bears against a horizontal proving ring. The soil is sheared across the mid-height of the specimen and the shear load is measured by the proving ring. The horizontal displacement of the specimen is calculated from the distance traveled by the loading jack

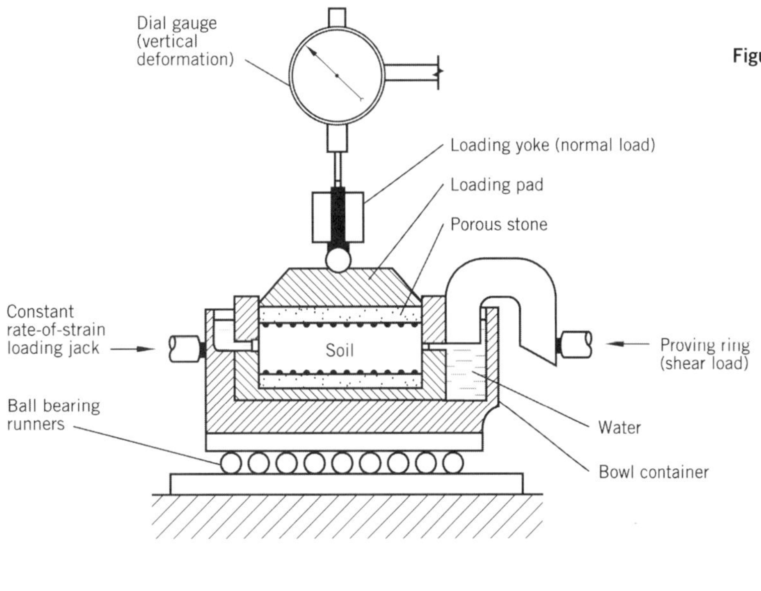

Figure 6.9 Shear box apparatus.

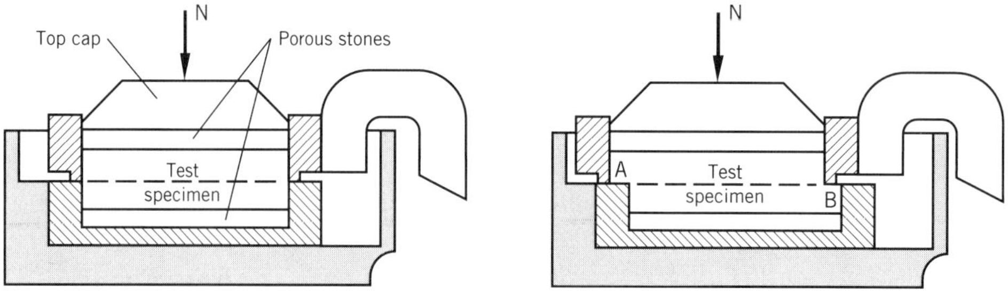

Figure 6.10 Separation of shear box halves during shear box test:
(a) before test
(b) during test

minus the deflection of a proving ring. The apparatus usually has a multi-speed gearbox that allows tests to be run over a wide range of time from a few minutes to several weeks.

The shear box test has significant shortcomings due to the separation of the two halves of the box during shear. At the start of the test, the normal force is above the center of the entire specimen and shear occurs across a soil interface (Figure 6.10a). During the test, the separation of the two halves of the box creates a soil-to-metal contact at A and B and a soil-to-air contact near point B (Figure 6.10b). Also, the normal force becomes eccentric over the lower half of the specimen. Despite these issues, the normal stress is assumed to remain constant throughout the test.

The standard shear box has been modified to accept other specimen sizes. One modification allows the use of cylindrical specimens obtained from small-diameter undisturbed samples. Larger shear boxes, 12 inches to 36 inches per side, have been built to test gravels and compacted fills. For example, at Muddy Run dam in Pennsylvania, a square shear box with sides of 6 feet was manufactured to measure the shear strength of a compacted schist rockfill.

The more common uses of the shear box are as follows:

- To measure the residual strength of clay using strain reversal
- To measure the strength of soils with large stones such as gravels (fairly rare use, but several laboratories have large 12-inch square shear boxes)
- To measure the adhesion between geotechnical fabrics or fabric/soil interfaces, especially for landfill liners and other environmental purposes

Residual Strength Measurements in the Shear Box

The residual strength of an overconsolidated clay is usually performed on a test specimen cut from an undisturbed sample. The specimen is first sheared to failure at a constant rate of strain to measure the peak strength. The two halves of the shear box are then returned to the starting point by using a faster rate of strain or hand cranking. The test is repeated (second cycle). Multiple cycles are made until the maximum shear stress is unchanged from the previous cycle, thus indicating that the residual strength has been reached. An example of a 5-cycle test is shown on Figure 6.11. On some clays, as many as 12 cycles are required to reach the residual strength condition. A minimum of three undisturbed specimens are tested at three different normal stresses (e.g., 1, 2, and 3 ton./sq. ft.) to obtain Mohr envelopes of peak and residual strengths.

A procedure in some commercial laboratories is to retest the *same* specimen at progressively higher (normal) stresses. The concept for a multistage test is that the residual shear plane has been established in the first set of tests. The main disadvantages of this approach are: (i) the peak envelope is not obtained; (ii) the same piece of soil is being tested, which is undesirable; and (iii) the specimen typically "sloughs" into the container bowl with each load cycle and becomes increasingly thinner; this effect, combined with additional consolidation settlement at the higher normal stresses, moves the slip surface below the mid-height split of the shear box. In general, it is preferable to test separate specimens at each normal stress, but a multistage test makes good sense where only a limited

Figure 6.11 Shear stress-horizontal displacement curves measured on a specimen from the slip surface at Hagg Lake Slide 4 (Case History 10).

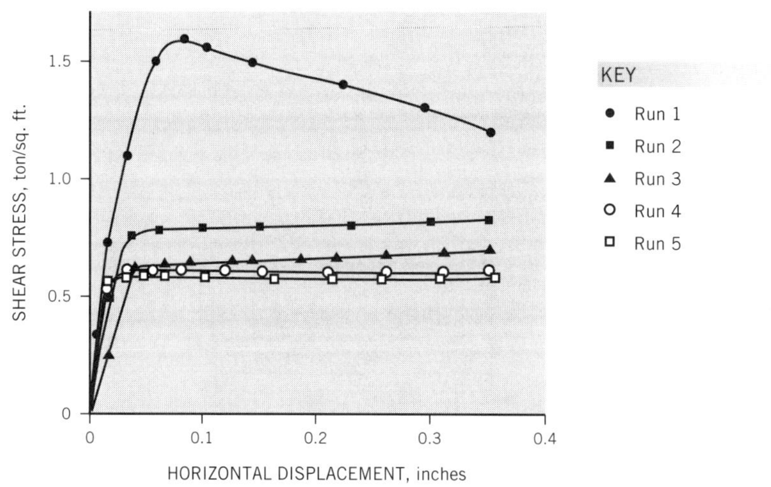

amount of material can be obtained from the actual shear zone of a landslide.

Two other procedures are used to speed up the development of a polished shear surface: (i) cut through the test specimen across the mid-height prior to the test; and (ii) apply up to six rapid stress reversals after completing the first run.

The interpretation of residual strength of clay is discussed in Chapter 8. The shear box has limitations as a means for measuring residual strength and the time required for large numbers of reversals makes it an expensive test in commercial laboratories.

Figure 6.12 Ring shear test concept.

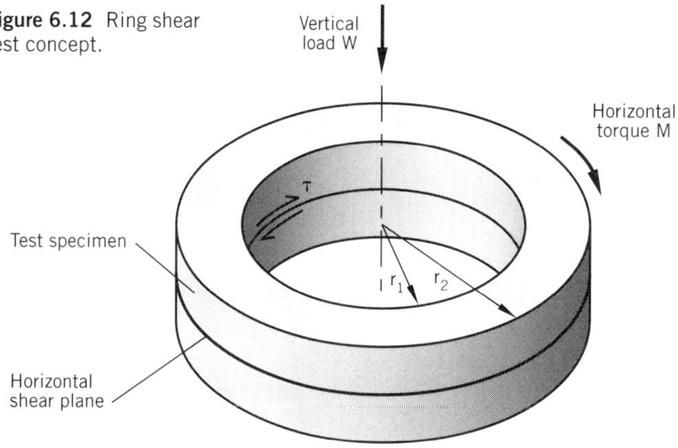

6.6 RING SHEAR TEST

In the ring shear test, an annulus of soil is sheared by torque. The relative motion, and development of a slip plane, occurs at the mid-height of the specimen (Figure 6.12). A major advantage of the test (compared to the shear box) is that shear movement is continuous, without reversals, to residual condition.

The concept of performing torsion tests on soils in the laboratory goes back to the 1930s, but the impetus to use the test to measure residual shear strength can be attributed to the work of Bishop et al. (1971). The only commercial unit currently available is the Bromhead apparatus manufactured in England.

Test Equipment

The Bromhead ring shear apparatus is illustrated on Figure 6.13. The test specimen is a thin, annular ring of soil with the following dimensions: outside diameter 3.937 inches (100 mm); inside diameter 2.756 inches (70 mm) and thickness 0.197 inch (5 mm). Thus, the annular width of the specimen is 0.59 inch (15 mm).

The specimen is set up within an annular groove on the base plate (Figure 6.13b). A loading platen and torque arm assembly is centered over the top of the base plate so that the annular porous stone in the loading platen is aligned over the

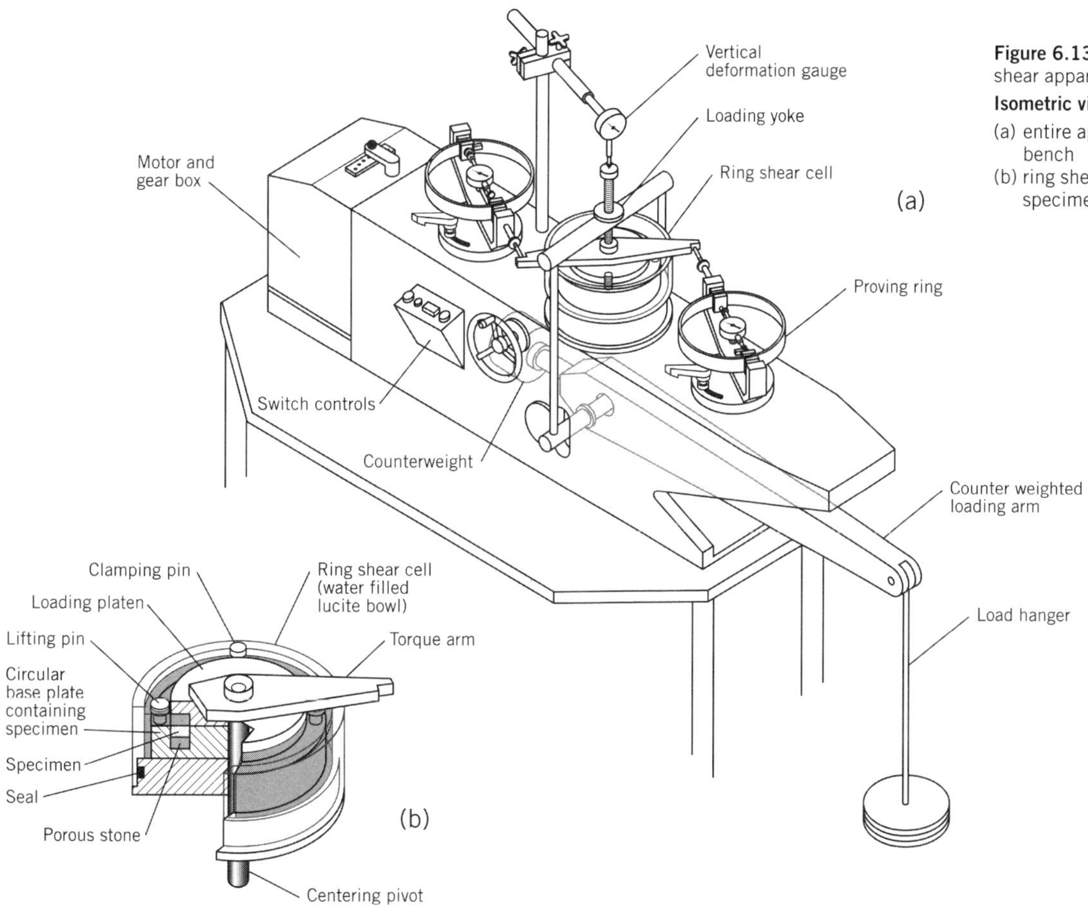

Figure 6.13 Bromhead ring shear apparatus

Isometric views:

(a) entire apparatus on a bench
(b) ring shear cell and test specimen details

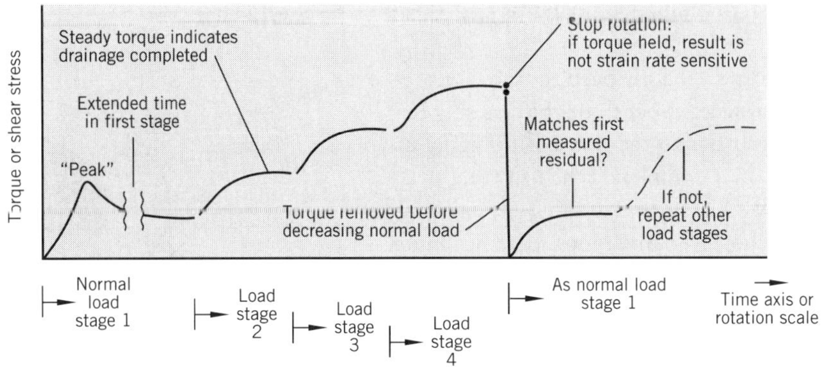

Figure 6.14 Multistage residual strength ring shear test on single specimen (after Bromhead, 1986).

specimen. Water is added to the lucite (perspex) bowl to prevent the soil from drying out during the test. A normal stress is applied by loading the soil through a loading yoke and lever load arm (Figure 6.13a). In the shear test, the base plate containing the soil specimen is steadily rotated by a multispeed motor located below it. This rotation is resisted by the soil specimen and the shear stress is determined by loads measured on two proving rings (Figure 6.13a). A failure plane develops just below the loading platen/soil contact.

The gearbox has a choice of 25 constant speeds ranging from 60° per minute to 0.024° per minute (105 inches per hour to 0.04 in./hr.). Early tests showed that it takes 5 to 20 inches of shear displacement to reach residual strength for London Clay and 40 inches for Weald Clay (Bishop et al., 1971).

Residual Strength Measurements in the Ring Shear Test

The test specimen is usually prepared from remolded soil, which is placed in the annular groove of the specimen container. It is kneaded into place using a wood dowel or similar implement, and the upper surface is made smooth and level. The loading yoke and torque arm assembly is placed over the specimen and it is consolidated under the selected normal load. The minimum time to failure can be calculated from Table 6.2, but since the specimen is very thin and in double drainage, dissipation of excess pore pressures will occur relatively quickly in all soil types. For example, a fat clay with a coefficient of consolidation c_v of 10 sq. ft./year when tested in the Bromhead ring shear device requires a minimum shear time of 23 minutes according to Table 6.2.

For the shear test, Bromhead (1986) recommends a multistage test procedure. After applying the first normal load, the specimen is strained until the residual strength appears to have been reached, then the next higher normal load is applied. When a full sequence of normal loads has been applied, the total load is reduced and the strength is measured again at the initial (lowest) normal load. Provided that the strength is comparable with the first run, it is inferred that the residual strength of the first run is acceptable. With this procedure, it may be necessary to check more than one point if the first check shows an unacceptable level of agreement. As a practical matter, the torque at the highest normal load of the series needs to be taken off before the lowest normal load (check) test is performed; otherwise, rapid deformation of the specimen will occur, which can lead to erratic results in the check test. This approach to ring shear testing is shown on Figure 6.14.

The ring shear test shares a problem with the shear box test. As the shearing continues, small amounts of soil squeeze out or slough from the test specimen into the surrounding water-filled bowl. Therefore, in a multistage test, the specimen becomes progressively thinner. In the case of the ring shear, this loss of soil during shear, plus earlier consolidation, causes the upper porous stone to move into the annular groove. According to Stark and Vettel (1992), the wall friction increases the measured residual strength. They recommend that settlement from both sources be limited to 0.75 mm (15% of the original height). This can be accomplished by adding soil and reconsolidating as needed. Prior to drained shearing, the specimen surface should be flush with the top of the baseplate. Subsequently, Stark and Eid (1993) developed an alternative specimen container that can be overconsolidated and precut prior to shearing with little or no wall friction. This equipment is not commercially available. However, to minimize the side friction issue in the Bromhead apparatus, it is advisable to use stiff clay to keep consolidation settlement low and reduce the quantity of clay being lost into the bowl during the shear test.

The reliable measurement of peak strength of undisturbed samples in the ring shear apparatus presents a difficulty because the stresses developed across the test specimen, from the inner to the outer radius, are nonuniform. (For residual strength, there is no problem—the entire shear surface is at the same stress.) Bishop et al. (1971) investigated this issue in some detail. However, the assumption that the normal stress and shear stress are uniformly distributed across the plane of relative rotary motion provides a reasonable interpretation of both the peak and residual strengths. The relevant formulas (see Figure 6.12) are:

$$\text{Normal stress } \sigma'_n = \frac{W}{\pi(r_2^2 - r_1^2)} \qquad \text{Eq. (23)}$$

$$\text{Shear stress } \tau = \frac{3M}{2\pi(r_2^3 - r_1^3)} \qquad \text{Eq. (24)}$$

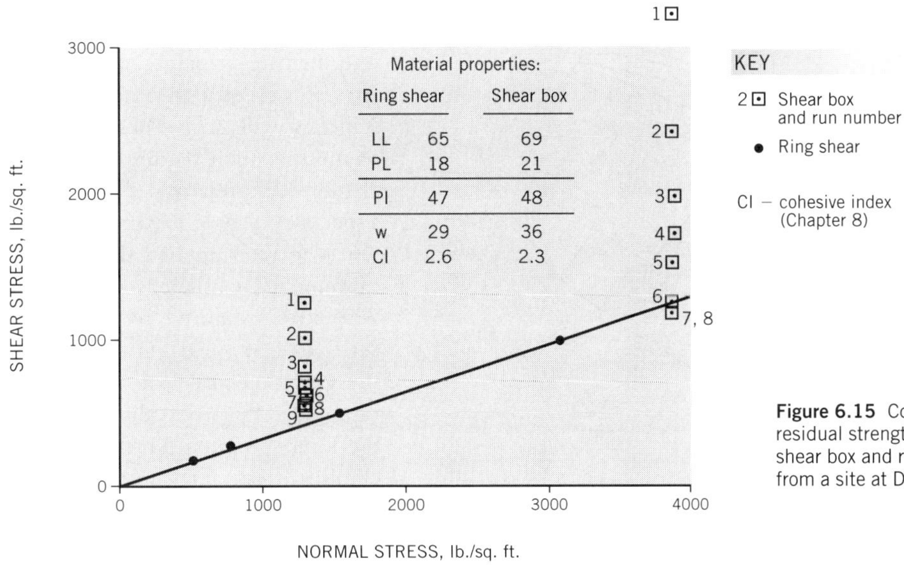

Figure 6.15 Comparison of residual strengths measured by shear box and ring shear tests from a site at Danville, California.

Torque $M = \frac{1}{2}(F_1 + F_2)L$ Eq. (25)

where W = vertical load on specimen

M = torque

r_1, r_2 = inner, outer radius respectively

F_1, F_2 = loads on the two proving rings

L = distance along torque arm between proving ring contacts

Another difficulty in testing undisturbed samples is to obtain a specimen with an annular shape. This difficulty can be overcome by cutting one-fourth of an annulus using specially made steel cutters (Vessely and Cornforth, 1998). The four segments are then put into the grooves of the baseplate. This procedure works satisfactorily.

The test procedure for undisturbed clays is to shear the specimen at a very slow rate of 0.024° to 0.048° per minute up to the peak strength. After passing the peak, the machine speed is significantly increased (typically 25 times the slow speed) to produce slickensided shear faces. The test then is slowed down to the original slow rate. After the shear stress levels off, the residual strength is measured. The cycle of speeding up and slowing down can be repeated to check the residual strength at a larger shear deformation.

Comparison of Shear Box and Ring Shear Results

Both the shear box and ring shear measure residual strength. Each test has limitations in accuracy as a result of theoretical considerations, the design of the apparatus, the testing techniques, the skills of the technician performing the test, interpretations of the test results, etc. It is almost impossible to isolate all these variables in making comparisons.

Bishop et al. (1971) compared the ring shear with other methods of measuring residual strength of brown London Clay using: (i) reversal shear box on undisturbed and slurried specimens, (ii) cut-plane shear box on undisturbed soil, (iii) direct shear on the slip surface, (iv) cut-plane triaxial tests, and (v) triaxial tests on samples taken from the slip surface. These methods all gave apparent residual strengths that were 2° to 6° higher than the ring shear. Tests on other clays showed the same trend, except for tests on Cucaracho Shale, where the shear box and ring shear gave similar results.

At Landslide Technology, limited experience from three sites indicates that if samples are taken from the slip surface observed by inclinometers, the shear box and ring shear give similar results, but the shear box requires numerous reversals. A typical set of results from a site at Danville, California, is shown on Figure 6.15. It is the author's opinion that the ring shear is inherently the better test of the two for measuring residual strength because it allows uninterrupted movements in one direction similar to failures in the field. The shear box can provide acceptable data if a sufficient number of reversals is made. According to Bishop et al. (1971), the shear box needs 10+ inches of total travel (about 30 reversals), although, in the author's opinion, fewer reversals generally will suffice in practice. Because the ring shear test can continue shearing day and night, it is the more practical test for commercial work.

6.7 PLANE STRAIN TEST

When landslides occur on long embankments or cuts, it frequently occurs that the width of the landslide, W, is large compared to the depth of slippage, D (Figure 6.16). In such cases, the direction of movement is at right angles to the axis of the embankment, and the soil has sheared in *plane strain*. Many landslides fall into this category. Furthermore, it is common practice for stability calculations to be performed on a two-dimensional cross-section (i.e., plane strain).

There is no commercially available test to measure the shear strength of soils in plane strain. The laboratory equipment to perform the test is too expensive to develop for general use, and the specimen shape is incompatible with borehole sampling. The plane strain apparatus developed by A. W. Bishop at Imperial College, London, is shown diagrammatically on Figure 6.17, and the author used it for several years on research.

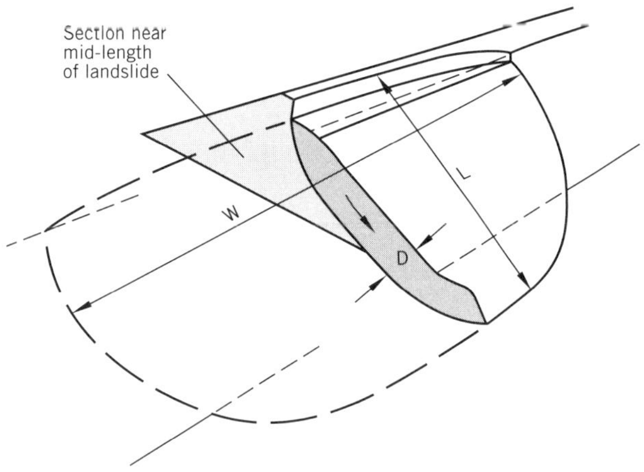

Figure 6.16 Wide landslide producing a two-dimensional (plane) failure.

In plane strain, the axial strain ε_2 in the direction of the intermediate principal stress σ_2 is zero. Bishop's plane strain apparatus tested a specimen 16 inches long, 4 inches high, and 2 inches wide. A special clamp was fitted across the two ends of the specimen (Figures 6.17 and 6.18). At one end of the clamp was a pressure cell connected to a null indicator. When a specimen tries to expand longitudinally during a plane strain test, a force is applied through the pressure cell to prevent movement (i.e., the end clamp maintains $\varepsilon_2 = 0$ at all times). The stress required to maintain zero longitudinal strain ($\varepsilon_2 = 0$) is $\sigma_2' - \sigma_3'$ from which the intermediate principal stress σ_2' can be calculated. The vertical deviator stress, $\sigma_1' - \sigma_3'$ is provided by two loading pistons set at the quarter points of the long specimen. These are fitted with rotating bushes to eliminate vertical friction. In all other respects, the equipment set-up is similar to a conventional triaxial test.

Figure 6.17 Bishop plane strain apparatus.

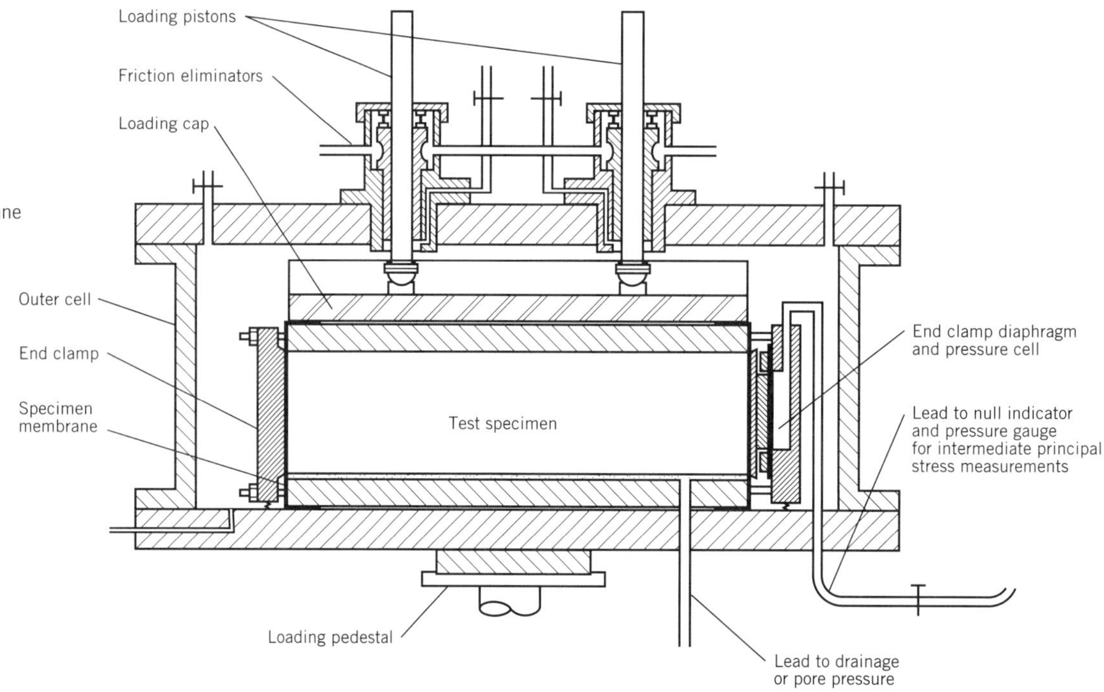

Figure 6.18 Plane strain test specimen with end clamp in position.

Some of the results obtained with the Bishop plane strain apparatus are presented in Chapter 7. They are included in this book to show the role played by strain on the strength of dense sands, and also to demonstrate the nature of undrained tests on dilatant soils.

6.8 MOHR DIAGRAM

The Mohr diagram is the most common method for presenting the results of shear tests on soils. Another method is the stress path, which will not be presented here. On the Mohr diagram, the ordinate is shear stress and the abscissa is normal stress. A principal stress is one in which there is no shear stress, so principal stresses are plotted on the bottom line of a Mohr diagram.

The results of triaxial shear tests are shown by Mohr circles and the line or curve representing soil failure for a group of tests is the Mohr *envelope*. A common practice is to show total stress circles by solid lines and effective stress circles by broken lines, as in Figure 6.19. When the pore pressures are positive at failure, the effective stress circle is to the left of the total stress circle; when pore pressures are negative, the effective stress circle is to the right of the total stress circle. The diameter of the circle does not change because $\sigma_1 - \sigma_3 = \sigma_1' - \sigma_3'$. The Mohr envelope is usually depicted as a straight-line making the best fit (statistically or by judgment) to the circles on the diagram.

Overconsolidated clays give very inconsistent shear test results. This is probably due to the poor sampling procedures still in use. The British U100 "undisturbed" sampler, hammered into stiff clay, gives especially poor results. By contrast, soft, normally consolidated clays generally show very consistent data on the Mohr envelope. They are usually sampled by thin-wall samplers and have less sampling disturbance. Normal practice is to perform a minimum of three shear tests to determine the Mohr envelope for a soil. Due to erratic results, it is often necessary to obtain four or five shear test results on overconsolidated clay samples to provide a reasonable interpretation of the shear strength parameters.

Cohesion and Cohesion Intercept

Many geotechnical practitioners are confused by the terms cohesion c and cohesion intercept c′. On Figure 6.20 are

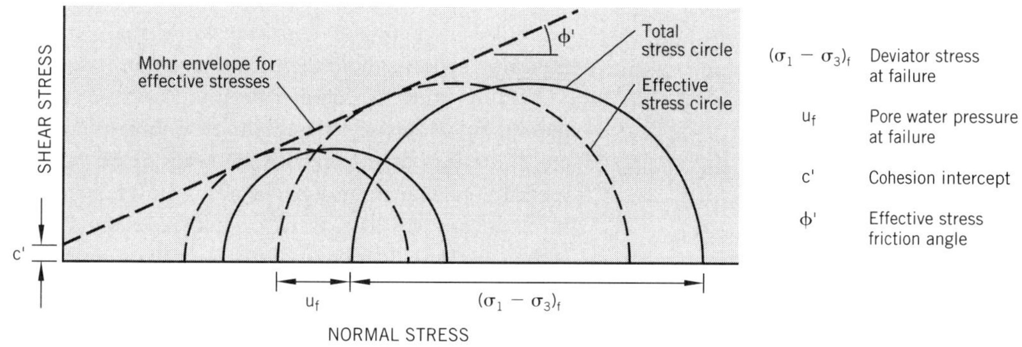

Figure 6.19 Mohr diagram for total and effective stresses.

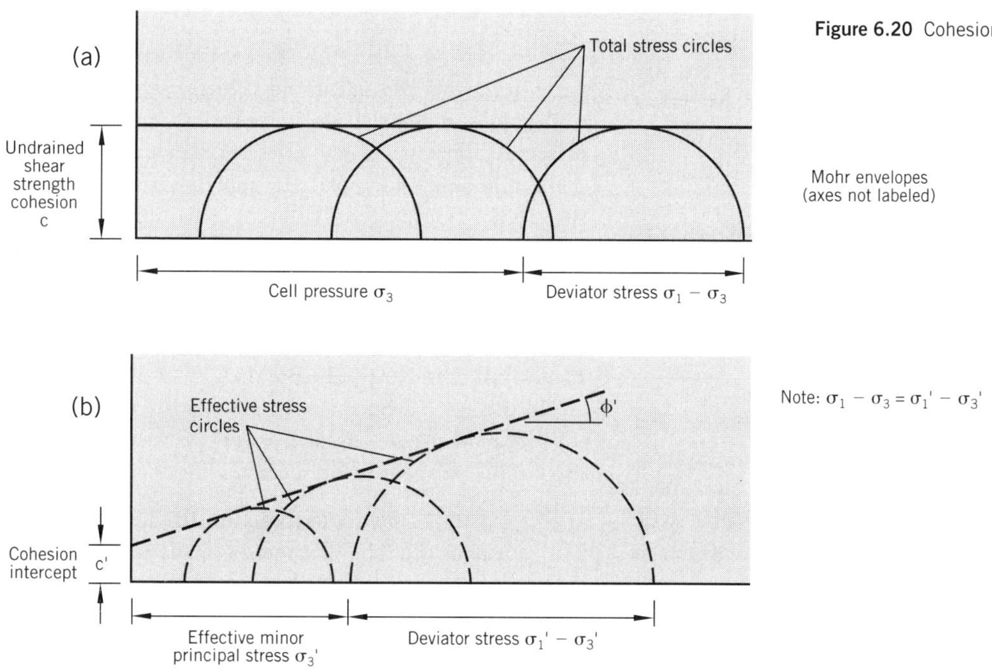

Figure 6.20 Cohesion c and cohesion intercept c'.

Mohr diagrams for a hypothetical medium stiff clay. On the upper diagram, Figure 6.20(a), are the result for undrained shear tests in which the specimens were placed in the triaxial cell, a cell pressure was applied, but no drainage was allowed. Because the specimens are not consolidated, and there is no change in effective stress, the measured undrained strengths should be approximately equal, independent of cell pressure. The Mohr envelope is a horizontal line, and one-half of the deviator stress is the undrained shear strength c. This is usually referred to as the *cohesion* of the clay, or *immediate shear strength*. The $\phi = 0$ analysis of slope stability is based on this total stress approach.

On the lower diagram, Figure 6.20(b), the results of consolidated-undrained shear tests (with pore pressure measurements) are plotted as effective stress circles. The envelope has a slope angle ϕ' and intercepts the ordinate axis to give a *cohesion intercept*, c'. There is no relationship between cohesion intercept c' and the cohesion (immediate strength) c.

Cohesion intercept c' typically is measured in effective stress tests on stiff overconsolidated clays. Soft, normally consolidated clays usually have a Mohr envelope that passes through the origin of the Mohr diagram in effective stress tests (but still has cohesion in undrained tests). Therefore, it is speculated that cohesion intercept is a result of the clay's past history under the forces of overconsolidation (Skempton, 1964). Another school of thought has been that the Mohr envelope is curved near the origin in these soils and cohesion intercept c' does not exist. As a practical matter, the linear c' and ϕ' effective stress parameters are the best choice for defining strength provided the laboratory shear tests cover the range of stresses acting on the slip surface of the landslide.

6.9 LIQUEFACTION TEST

The cyclic testing of a cylindrical specimen of saturated silt or sand is used to measure the susceptibility of the soil to liquefaction during an earthquake. The details of the test are described in American Society for Testing and Materials (ASTM) test procedure D 5311.

Testing Equipment

The equipment is similar to the conventional triaxial cell previously described except that the top of the cell has a larger piston that can transmit cyclic loading to the top of the test specimen. The loading piston is fitted with low friction bushings and is connected to the top platen so that extension as well as compression can occur during cyclic loading.

The equipment has to apply an uniform sinusoidal load at a frequency range of 0.1 to 2 cycles/second (Hz); the most common frequency used is 1 cycle/second. The equipment must be able to apply the cyclic load on top of an initial static load. An air pocket in the top of the cell allows the piston to oscillate in and out of the cell without creating pressure surges in the cell water.

Axial load measurements are obtained by an electronic load cell. Axial deformation is measured by a linear variable differential transformer (LVDT) or by potentiometer-type deformation transducers. The pore water pressures are monitored by a very stiff electronic pressure transducer accurate to about ±0.25 psi.

Specimen Preparation

Tests are performed on both undisturbed and remolded soils. Since original soil structure obviously affects the performance of soils in this test, the undisturbed condition is preferable. However, it is extremely difficult to obtain relatively undisturbed samples of sand or silt, especially those with a low relative density. Piston samples using 2-inch or 3-inch thin-wall sampler tubes combined with very careful drilling techniques (such as mud rotary) can sometimes achieve satisfactory undisturbed samples. Another technique is to freeze the ground. However, it is impossible to obtain zero disturbance during sampling.

After the sample has been taken, the handling and transport of the sample is another challenge. The ends of the tube have to be plugged or waxed to ensure that the sample is firmly held in the tube. Wax has to be removed later, and any remnants of wax left on the inside of the sampler can disturb the sand during extrusion. Cutting through the sampler tube to minimize the length of extrusion can damage the sample by the grip needed to hold the tube in position for cutting.

Transporting the sampler from the field is usually accomplished by wrapping the samples in bubble wrap, styrofoam, etc. and either hand-carrying the tubes or putting them in a special container in an upright position. Whatever method is used, the samples have to be carefully reexamined at the test laboratory for evidence of disturbance by vibrations or shock loads. Samples should not be transported in the back of a pickup truck or the cargo space of commercial aircraft.

Assuming that the sample has reached the laboratory in a condition deemed satisfactory for testing, the specimen has to be prepared for testing. Three-or four-way split molds can be used for this purpose, although extreme care has to be taken when removing the split mold segments because suction can easily pull out the side of the specimen. Once on the lower platen of the testing machine, the specimen is sealed by a membrane and held in place by applying a negative pore pressure to the pore water. The above difficulties apply to other types of triaxial tests on sand.

A procedure for preparing specimens of frozen sand for testing is described in ASTM D 5311. Essentially, the frozen ground is prepared like a rock core and is allowed to thaw in the triaxial chamber just before the test.

In the case of specimens of recompacted sand prepared in the laboratory, there are at least two procedures that can be used to prepare a saturated specimen for testing:

1. Set up a split-mold frame on the lower platen of the triaxial cell. The split mold is made airtight and a vacuum is applied through the side of the mold to pull the membrane tight against the inside of the mold (Figure 6.21a). De-aired water is put inside the membrane and

Figure 6.21 Preparation of triaxial specimens of recompacted sand in the laboratory:
(a) split mold in position
(b) mold filled with sand, top cap sealed
(c) mold removed, small negative pore pressure applied to the sand specimen

saturated sand (sand boiled with water, then cooled) is placed into the standing water within the split mold. Vibration by tamping or vibrator (with or without a surcharge weight) is used to densify the saturated sand to the desired level of relative density. This process can be done in one to four lifts, as selected by the laboratory experimenter. When the specimen has been formed, the top is carefully leveled; the top cap is put on and the membrane is sealed with an O-ring (Figure 6.21b). The volume change burette is lowered to put a negative pore pressure on the sand and the split mold is carefully removed (Figure 6.21c). (*Note:* The inside of the mold in contact with the membrane needs to be kept perfectly dry so that it can be easily removed without pulling on the membrane.) All measurements of the length and diameter of the specimen have to be taken with great care to obtain an accurate relative density at placement. The weight of dry sand in the specimen is obtained at the end of the test.

2. A split-mold specimen former is set up as described in the first procedure and damp sand is placed and compacted. The top cap is added and de-aired water is passed from bottom to top of the specimen to remove as much trapped air as possible. A negative pore pressure is applied to the specimen and the split mold is removed. The test equipment is assembled and saturation is completed by the application of back pressures (previously described in Section 6.4).

If the sand has sufficient fine sand and silt, it may be possible to compact the specimen in a split mold and add the membrane later. Another variation in technique is to draw de-aired water up through the sand using a suction at one end.

Cyclic Shear Test

The cyclic stress ratio (CSR) is defined by:

$$\text{CSR} = \frac{\tau}{\sigma'} \qquad \text{Eq. (26)}$$

where τ = shear stress
σ' = effective overburden stress

For laboratory tests:

$$\text{CSR} = \frac{\text{Peak cyclic deviator stress}}{2\sigma_c'} \qquad \text{Eq. (27)}$$

where σ_c' = effective consolidation pressure

A typical objective of the shear test program is to test several potentially liquefiable soil specimens at a range of CSR values and determine the number of cycles to produce either liquefaction ($\sigma_c' = 0$ during the cycle) or unacceptable levels of double amplitude cyclic strain; the latter can be set by the experimenter but typically ranges from 5 to 20%.

The specimen is first consolidated to a selected all-around stress σ_c'. The cyclic shear test is undrained. The piston is set to provide a pulsating load calculated for the selected CSR. The test is then run at a selected vibration in the range of 0.1 to 2 cycles/sec. until the failure criterion has been met or some upper limit (e.g., 500 cycles) has been reached for specimens that do not liquefy. After the test is completed, and without dissipating any excess pore pressure set up by the cyclic loading, a conventional triaxial test is performed to measure the undrained strength of the soil.

A typical set of results is given on Figure 6.22, based on tests at the University of California, Berkeley, by Dr. Michael Riemer. This relatively undisturbed silt sample from Forest Grove, Oregon, was consolidated under 75 kPa (10.88 psi) all-

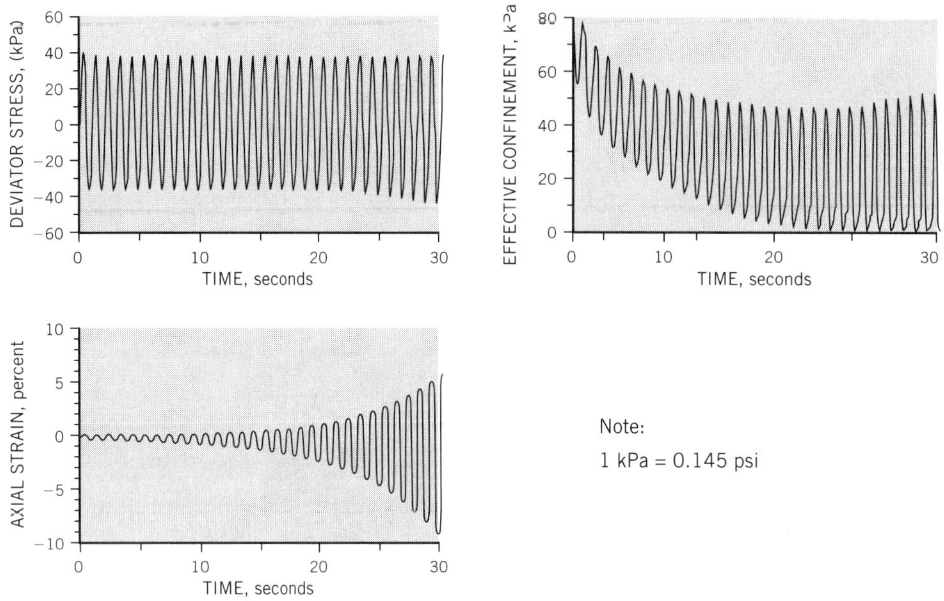

Figure 6.22 Typical data obtained from a cyclic shear stress test.

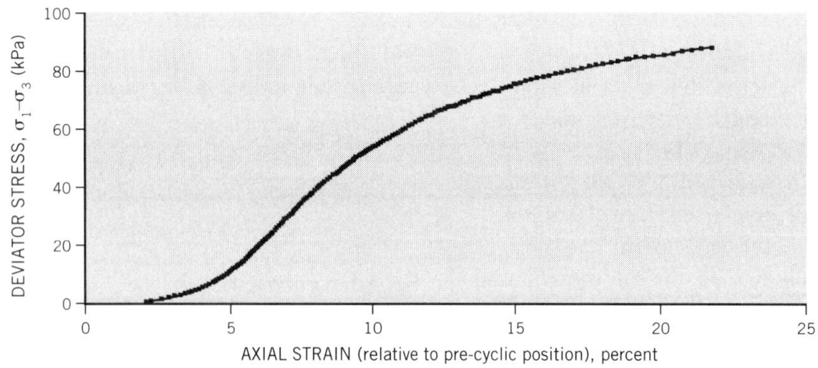

Figure 6.23 Undrained shear test performed on specimen after cyclic shear test shown on Figure 6.22.

around pressure and tested at 1 cycle/sec. under a vertical cyclic stress of 37.5 kPa (5.44 psi). Therefore, the CSR ratio was 0.25. The specimen reached liquefaction ($r_u = \Delta u/\Delta\sigma_3' = 100\%$) after 22 cycles and cyclic strains (double amplitude) of 5% and 10% at 23 and 28 cycles, respectively. The undrained shear test after the cyclic test is shown on Figure 6.23.

It can be argued that sampling disturbance will tend to densify a loose sand or silt. Therefore, the test results are likely to overestimate the resistance of the soil to strong ground shaking, and cyclic shear test data need to be treated conservatively. Of even greater concern is the difficulty of recovering very loose and loose saturated sands in a sampler when they are being pulled to the ground surface. The unrecovered (lost) samples are likely to be looser than the recovered samples, and thus are more prone to liquefaction. These concerns support the view that laboratory cyclic shear test results should be interpreted conservatively.

6.10 ADDITIONAL LABORATORY SHEAR STRENGTH TESTS

There are three tools in general use that directly measure the undrained shear strength c of clay soils: the pocket penetrometer, torvane, and laboratory vane.

Pocket Penetrometer

The pocket penetrometer is a spring-loaded piston, ¼ inch diameter, which is pushed ¼ inch into the surface of the clay. As the clay resists the penetration, a pointer on the piston barrel housing registers the undrained shear strength when the piston has penetrated to the ¼ inch depth. Some penetrometers are calibrated in terms of unconfined compressive strength (i.e., twice the shear strength c).

The penetration has to be done slowly, and more reliable results are obtained on medium stiff clays. On stiff, brittle

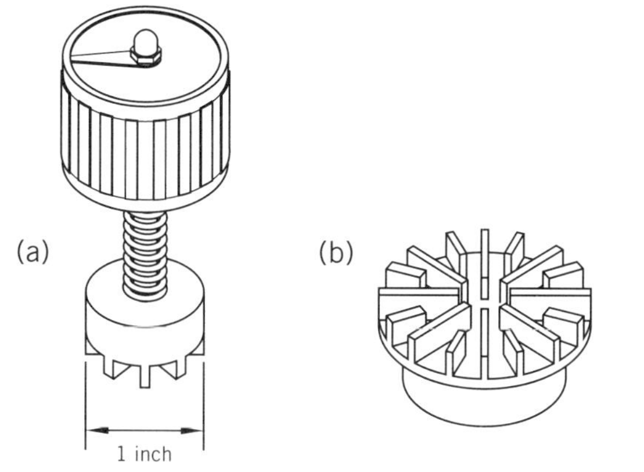

Figure 6.24 Torvane:
(a) with vane head suitable for medium stiff clay
(b) large diameter head suitable for soft clay

clays, the penetration can break up the clay around the piston tip. On soft clays, there is insufficient resistance to register an accurate reading on the spring-loaded pointer.

Pocket penetrometers are used in both the field and the laboratory. It is a relatively crude measuring tool, and variable results are obtained, depending on the operator. The measurements should not be used in stability calculations, but can provide a rough measure of clay stiffness for descriptive purposes.

Torvane

The torvane head has shallow vanes arranged radially around the circumference (Figure 6.24). The vanes are pushed into the clay up to the full depth of the vanes, and a torque is applied via a calibrated spring until the clay fails. The undrained shear strength c is read off the dial incorporated in the hand grip. The vanes have to be carefully inserted and the torque applied very slowly.

There are two other accessory vane heads: one has a larger diameter with 16 shallow vanes around the circumference (Figure 6.24b); it is used to measure the undrained strength of soft clays. The second accessory is a smaller head that is used for very stiff clay. The two accessory vane heads can be added to the basic unit by screws, and there are multiplication factors to correct the calibrated head of the instrument for these other vanes.

The larger torvane is the best measuring unit. It gives reproducible results in soft clays. The smallest vane head is unreliable. When pushed into stiff clays, the penetration of the vanes breaks up the clay locally and the test result is inaccurate.

Laboratory Vane

The laboratory vane can be used to measure the undrained shear strength of very soft and soft clays. A common procedure is to test the base of thin-wall tube samples shortly after the samples arrive at the laboratory. Vane tests provide a simple and reliable way to obtain undisturbed shear strengths at a low cost, thereby allowing numerous tests to be performed for a strength vs. depth soil profile of soft clay deposits.

The laboratory vane is similar to the field vane (Section 3.12), except that there is generally no outer sheath. There is also no standard size for the vanes, which can range from 0.5 inch to 1.3 inches diameter and have height : diameter ratios of 1:1 to 2:1.

Test procedures are given in ASTM Designation D 4648 and Head (1986), the latter being the British practice.

All vanes have four rectangular blades. The vane shaft is attached to a spring-loaded head, which measures the torque applied to the soil. Some laboratory vanes are handheld, but it is preferable that they be mounted on a stand so that the vane is lowered into the clay with minimal disturbance. A schematic drawing of a laboratory vane set up vertically is shown on Figure 6.25.

Procedures for this test vary widely. The clay sample should have a plane face so that the vane penetrates into it at right angles. The depth of embedment, e (Figure 6.25) ranges from h to 4h (h is the vane height). The author suggests h is sufficient, and appropriate to minimize disturbance and shaft adhesion affecting the results. ASTM recommends a motorized rate of 60° to 90° per minute for the speed of rotation. Serota and Jangle (1972) recommend a speed of 360° per minute. After a peak shear strength has been measured, the vane is rotated rapidly through 2 to 10 revolutions, then slowed down to the original speed to measure the remolded shear strength. The ratio of the peak: remolded strength is the *sensitivity* of the clay.

After testing, it is common practice to examine the clay specimen to check that there are no sand lenses or stones within the tested zone. Moisture contents and sometimes Atterberg limits are measured on the clay.

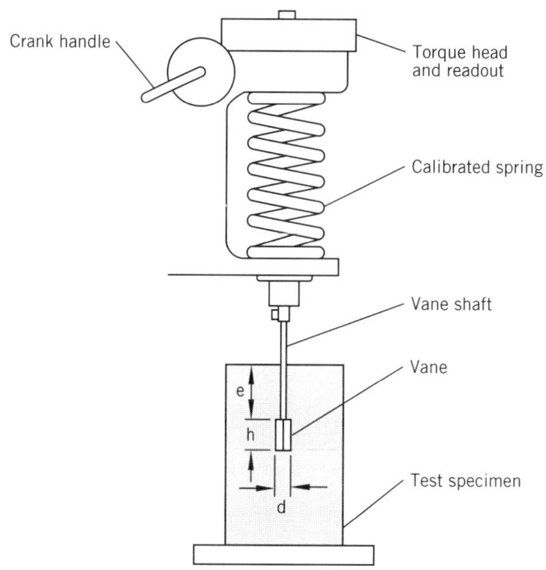

Figure 6.25 Laboratory vane test.

The metal blades of the vane disturb the clay as the vane is inserted into the soil. For the larger vane, the *disturbance ratio* or *vane area ratio* is about 14% and rises to 21% for the smaller vane. The disturbance ratio is the ratio of the volume of the vane blades to the cylindrical volume of soil involved in the test. ASTM recommends that the vane area ratio should be less than 15% to minimize soil disturbance from displacement by the vane.

$$\text{Vane area ratio} = \frac{4(R - r)t + \pi r^2}{\pi R^2} \qquad \text{Eq. (28)}$$

where R = radius of vane; r = radius of shaft; t = thickness of vane blade

Stiff soils are especially affected by the relatively hard push needed to insert the vane. These soils often show evidence of an induced prefailure when the vane is inserted. For this reason, vanes and penetrometers are not suitable for stiff clays. The upper limit appears to be a shear strength of around 1000 lb./sq. ft.

CHAPTER

7

Properties of Sands and Other Cohesionless Soils

This chapter describes the shear strength of sands and the effect of strain conditions on strength. Most landslides fail in plane strain, for which sand has a higher peak strength than the strength measured in a conventional triaxial test. However, since the strength peaks abruptly in plane strain, this reserve of strength has limited practical benefit to slope design, and no benefit after a landslide has occurred.

The extrusion and shear testing of undisturbed sand samples in the laboratory is almost impossible and is rarely attempted commercially. As an alternative to selecting a strength value from published tables, a method of estimating the drained strength of sands is presented. The required information is easy to obtain and consists of (i) an angle of repose test on dry sand, and (ii) the relative density of the sand, obtained either (a) from a field SPT or (b) by measuring natural density on undisturbed samples, followed by maximum and minimum density tests; most practitioners probably will use SPT data.

Finally, there is an important discussion on the nature of the undrained triaxial shear test on dilatant sands. The conclusion is that the test does not measure the strength of the soil at constant voids ratio, but instead is a constant volume test that masks the movement of pore water to the failure zone from the remainder (nonfailing) sand within the test specimen. This effect is also relevant to the interpretation of the "undrained" test on clays in Chapter 8.

The chapter does *not* discuss liquefaction of sandy soils. This property is associated with earthquake loading and, in loose saturated soils, can temporarily drop the available shear strength to a low value (termed "residual"). It is discussed in Chapter 12, and is estimated from back analysis of past ground movements associated with liquefaction.

7.1 CLASSIFICATION

Cohesionless soils are described by the percentage by weight of the various grain sizes within a sample. For sands and gravels, these percentages can be determined by sieve analyses. The procedures are well-known and are described in standard tests.

The results of a sieve analysis are presented on a gradation chart that shows the percent of the sample finer by weight for each sieve. An example of the U.S. chart is shown on Figure 7.1, and is based on the Unified Soil Classification System (USCS).

The British standard of soil classification, also known as the Massachusetts Institute of Technology (MIT) classification system, separates sands and gravels into three subcategories of *fine, medium,* and *coarse,* using grain size rather than sieve size as the control. The British classification is

- *Gravel.* Coarse, 60 mm–20 mm; medium, 20 mm–6 mm; fine, 6 mm–2 mm
- *Sand.* Coarse, 2 mm–0.6 mm; medium, 0.6 mm–0.2 mm; fine, 0.2 mm–0.06 mm.

The MIT and USCS systems have significant differences: In the USCS, fine sand covers a much wider range of grain sizes, and gravel begins at 4.8 mm, compared with 2.0 mm in the MIT system.

The USCS also requires calculation of two coefficients:

Coefficient of uniformity $C_u = D_{60}/D_{10}$ Eq. (1)

Coefficient of curvature $C_c = (D_{30})^2/(D_{10} \cdot D_{60})$ Eq. (2)

where D_{10}, D_{30}, and D_{60} are taken from the gradation curve as the grain size in mm corresponding to 10%, 30% and 60% finer by weight

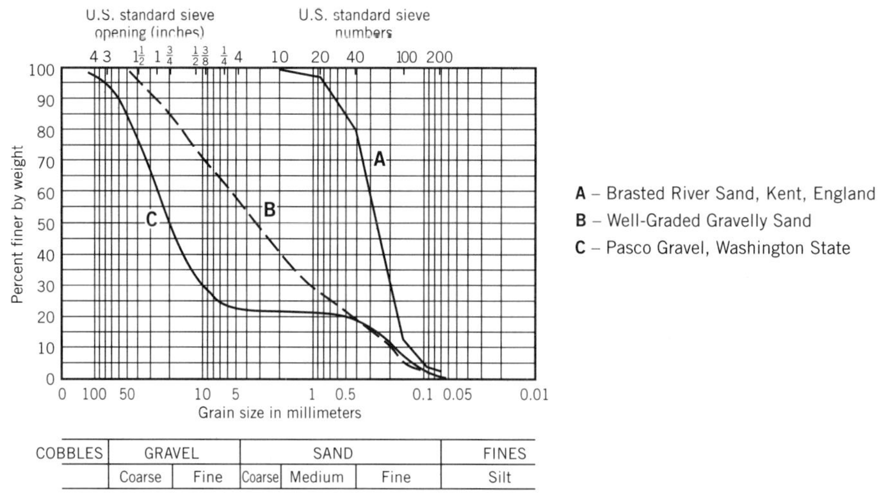

Figure 7.1 Example gradations.

Table 7.1 Classification Chart for Coarse-Grained Soils (USCS)

Criteria for Assigning Group Symbols and Group Names Using Laboratory Tests[A]				Soil Classification: Group Symbol	Soil Classification: Group Name[B]
COARSE-GRAINED SOILS More than 50% retained on No. 200 sieve	Gravels More than 50% of coarse fraction retained on No. 4 sieve	Clean Gravels Less than 5% fines[C]	$C_u \geq 4$ and $1 \leq C_c < 3$	GW	Well-graded gravel[F]
			$C_u < 4$ and/or $1 > C_c > 3$	GP	Poorly graded gravel[F]
		Gravels with Fines More than 12% fines[C]	Fines classify as ML or MH	GM	Silty gravel[F, G, H]
			Fines classify as CL or CH	GC	Clayey gravel[F, G, H]
	Sands 50% or more of coarse fraction passes No. 4 sieve	Clean Sands Less than 5% fines[D]	$C_u \geq 6$ and $1 \leq C_c \leq 3$	SW	Well-graded sand[I]
			$C_u < 6$ and/or $1 > C_c > 3$	SP	Poorly graded sand[I]
		Sands with Fines More than 12% fines[D]	Fines classify as ML or MH	SM	Silty sand[G, H, I]
			Fines classify as CL or CH	SC	Clayey sand[G, H, I]

Notes: [A]Based on the material passing the 3-inch (75-mm) sieve.

[B]If field sample contained cobbles or boulders, or both, add "with cobbles or boulders, or both" to group name.

[C]Gravels with 5 to 12% fines require dual symbols:
GW-GM well-graded gravel with silt
GW-GC well-graded gravel with clay
GP-GM poorly graded gravel with silt
GP-GC poorly graded gravel with clay

[D]Sands with 5 to 12% fines require dual symbols:
SW-SM well-graded sand with silt
SW-SC well-graded sand with clay
SP-SM poorly graded sand with silt
SP-SC poorly graded sand with clay

[F]If soil contains ≥ 15% sand, add "with sand" to group name.

[G]If fines classify as CL-ML, use dual symbol GC-GM, or SC-SM.

[H]If fines are organic, add "with organic fines" to group name.

[I]If soil contains ≥ 15% gravel, add "with gravel" to group name.

Using these tools, the classification system assigns a Group Symbol to the soil. Table 7.1 shows the classification table for coarse-grained soils, abstracted from ASTM Designation D 2487.

The Unified Soil Classification System is a cumbersome procedure and has received mixed acceptance by geotechnical practitioners. Most consultants prefer to use a descriptive system rather than a group symbol and include color, relative density, and other observations in the soil description. It is also helpful to provide the relative density and main soil constituent in capital letters so that this information can be quickly assimilated.

Examples of such descriptions are:

- MEDIUM DENSE, light brown, medium to fine SAND, trace silt (Curve A, Figure 7.1)
- DENSE, multicolored, gravelly, fine to coarse SAND; gravel rounded to subrounded (Curve B, Figure 7.1)

- VERY DENSE, dark gray, sandy, fine to coarse GRAVEL; gravel rounded, gap graded (Curve C, Figure 7.1)

The adjectives for the minor constituents are:

very	(silty, sandy)	30–50% by weight
	(silty, sandy)	10–30% by weight
slightly	(silty, sandy)	2–10% by weight
	trace	0–2% by weight

Classification is more comprehensive than is being covered here. However, certain terms used in this textbook are relevant:

- *Well-graded* means that there is a broad range of particle sizes over the entire gradation (e.g., curve B on Figure 7.1).
- *Uniform* gradation means that the soil particles are mostly of the same size (e.g., curve A on Figure 7.1).
- *Gap-graded* means that there is a broad range of soil particles, but the soil is missing a group of particle sizes within a broad range (e.g., curve C, Figure 7.1, is missing the soil sizes from ¼ inch to No. 30 sieve; i.e., there are no medium and coarse sand sizes within the soil; it comprises gravel with some fine sand).
- *Clean* sand and gravel means that there is essentially no fines (materials passing the No. 200 sieve) within the soil.

It is beneficial to also describe the shape of coarse sand and gravels as part of the description. The usual terms are:

- *Angular*—sharp edges
- *Subangular*—similar to angular but has slightly rounded edges
- *Subrounded*—well-rounded corners and edges
- *Rounded*—smoothly curved sides and no edges.

The describer should mention whether the individual stones of gravels or boulders are elongated or flat-sided, using the criteria:

- *Elongated*—stones having a length:width ratio of more than 3
- *Flat-sided*—stones having a width:thickness ratio greater than 3

Large size stones are usually described by their origin or purpose. Examples would be RIPRAP, angular, 6- to 24-inch size; or ROCKFILL, angular, very hard basalt, 8 inch to 1 inch.

There is a difference of terminology between engineers and geologists. Geologists describe a well-graded soil as "poorly sorted" and a uniform gradation as "well sorted." Engineers are describing the soil according to the gradation chart and geologists are relating to origins. Sands that have been deposited in rivers are generally uniform in gradation because of relatively uniform flow conditions (well sorted by the river). Poorly sorted soils usually have not been transported by water or wind (e.g., boulder clay, weathered in-place rock).

7.2 GRADATION AND ENGINEERING PROPERTIES

This section will briefly deal with the effects of gradation on the engineering behavior of cohesionless soils.

Free-draining Conditions

Free-draining is a subjective evaluation but it is intended to describe soil that does not hold water or excess water pressures in the void spaces for longer than a few minutes. Many landslide corrective measures require the use of a free-draining soil or rockfill. Coarse sand, gravel, and boulders are free draining; fine sand is *not* free-draining and medium sand is a transition soil that can be free-draining in some conditions but not in other situations.

If more than 5 to 10% by weight of fines (materials passing No. 200 sieve) is within a soil, it is unlikely to be free-draining. In specifying a free-draining soil, it is important to restrict the fines to about 3 to 5% by weight and further stipulate that the fines be non-plastic. Rock quarries often have a soil overburden above the hard rock face. If the quarry owner does not remove the overburden, but includes it in the blasting operations, the mixed soil and rock probably will be unsuitable for landslide remediation where a free-draining rockfill is required. Similarly, a well-graded sandy gravel from a gravel quarry may not be free-draining if the silt content is around 10% or higher by weight. This can be very deceptive because the gravel component is dominant in the material appearance. Such material is unsuitable for trench drains where the need to drain quickly during storms is important.

Some engineers and geologists assume that any soil described as sand is free-draining. However, fine sand drains slowly, especially in thick deposits. It should be anticipated that excess pore pressures will develop within a thick saturated sand deposit on loading with a large fill (i.e., it should not be assumed that an effective stress, fully drained, condition will exist immediately after construction).

The term "free-draining" as discussed above is referring to normal construction use of soils. Earthquakes provide very rapid loading that can cause liquefaction in gravels as well as in all sand groups.

Strength and Compressibility

Sands and gravels with *angular* particles have higher strengths than soils of similar gradation with *rounded* particles. Much of the particle-to-particle contact occurs at these sharp edges. As the stress level increases, the point contacts break down. These soils have a curved Mohr envelope that becomes less steep (lower ϕ') as the stress increases.

Angular soils generally have lower densities and higher compressibilities than similar soils with rounded particles. Therefore, a dense alluvial gravel with rounded stones generally is more dense, less compressible, but weaker than angular rockfill.

7.3 RELATIVE DENSITY

Sands, gravels, and other cohesionless soils have an intergranular *structure* that ranges from very loose to very dense. Adding an increased *static* stress to the soil structure has a minor effect on the structural packing of the soil particles. A loose sand will not become a dense sand. As an example, Brasted Sand, an alluvial deposit, with the gradation A on Figure 7.1, has the following changes when subjected to a static consolidation stress of about 6 ton/sq. ft.:

Relative density of 8% changes to relative density of 16%
Relative density of 73% changes to relative density of 77%

These are fairly minor increases. To densify sand, the soil grains have to be rearranged into a tighter packing by a combination of vibration and confinement. *Vibration* allows the sand grains to rearrange. *Confinement* prevents the energy of vibration from loosening the sand near an open surface. Persons familiar with field compaction of sand know that a vibratory roller will densify a loose sand below the surface, but the surface itself will remain loose. If a compaction test is made at the surface, the result often fails to meet the relative compaction requirement of the project. If the upper 6 inches of the layer is removed and a test is performed below this layer, the compaction achieved is almost always satisfactory. Over the years, there have been numerous site disputes (and occasionally delays, claims, and lawsuits) over failure to meet compaction specifications when the field technician has measured the relative compaction of a sand at the ground surface instead of just below it.

Clean sands and gravels do not perform consistently in the standard and modified compaction tests. The falling compaction hammer hits only a part of the soil surface in the mold, and the remaining unconfined surface jumps in response to the dynamic blow. When the next blow falls on another part of the soil surface, the previously compacted surface jumps and loosens. The net effect is a compaction curve that often does not resemble the familiar dry density vs. moisture content relationship of silts, clays, or sands/gravels with significant fines content.

A better measure of the compactness (structural packing) of clean sands and gravels is the *relative density* test.

The *minimum density* of sand or gravel is a fundamental property. It is the loosest condition at which the soil can maintain an *intergranular* structure; i.e., a structure in which the soil grains can move by sliding or rolling around each other without collapsing. This is dealt with in greater detail in the next section.

The *maximum density* is an arbitrary upper limit of density set by laboratory procedures. It requires vibration and overburden weight to provide confinement during vibration. However, it is not an absolute maximum density because nature can achieve even higher densities. Undisturbed samples of indurated sand from the Mangla Dam project in Pakistan had measured relative densities of around 140 percent.

The *relative density* D_r of sands and gravels (Terzaghi and Peck, 1948) is defined as follows:

$$D_r = \frac{e_{max} - e}{e_{max} - e_{min}} \qquad \text{Eq. (3)}$$

expressed in percent
where e = voids ratio of the soil being measured
e_{max} = voids ratio of the soil at the minimum density
e_{min} = voids ratio of the soil at the maximum density

It is unfortunate that this definition is the accepted standard worldwide. It requires measurement (or assumption) of the specific gravity G of the soil to calculate voids ratio e, an unnecessary step. Also, the value of *maximum e* corresponds to the *minimum density,* and vice versa, which is confusing.

A better method of measuring relative density is directly through density measurements. The dry density is a measure of the soil's compactness. The relative dry density (RDD) is defined as follows:

$$RDD = \frac{\gamma_d - \gamma_{d\,min}}{\gamma_{d\,max} - \gamma_{d\,min}} \qquad \text{Eq. (4)}$$

expressed in percent
where γ_d = dry density of the soil being measured
$\gamma_{d\,min}$ = dry density of the soil at the minimum density
$\gamma_{d\,max}$ = dry density of the soil at the maximum density

Relative density D_r and *relative dry density* (RDD) are identical at the two density limits but differ slightly between the two limits. The maximum divergence occurs near the midpoint of the density range with the relative density being slightly higher than the relative dry density. The relationships are:

$$RDD = D_r \left\{\frac{(1 + e_{min})}{(1 + e)}\right\} \qquad \text{Eq. (5)}$$

or

$$D_r = RDD \left\{\frac{\gamma_{d\,max}}{\gamma_d}\right\} \qquad \text{Eq. (6)}$$

For practical purposes, the two definitions of relative density are interchangeable. Relative dry density, using directly measured densities, is easier to calculate and a more logical choice. *Note:* ASTM refers to the relative dry density as the Density Index, I_d. Other useful relationships:

Relative compaction, R_c can also be stipulated:

$$R_c = \frac{\gamma_d}{\gamma_{d\,max}} \qquad \text{Eq. (7)}$$

expressed in percent

Dry densities can be calculated from the following relationships:

$$\gamma_d = \frac{G\gamma_w}{(1 + e)} \qquad \text{Eq. (8)}$$

$$\gamma_d = G\gamma_w (1 - n) \qquad \text{Eq. (9)}$$

where γ_w = density of water
G = specific gravity of soil
n = porosity, expressed as a decimal

U.S. Maximum Laboratory Density Test

The U.S. procedure for this test is ASTM Designation D 4253. An apparatus designed for the test is shown on Figure 7.2. There are two test molds, depending on the maximum particle size in the test sample, as shown in Table 7.2.

The test mold is rigidly attached to a vibrating table. In the case of the saturated method of testing, the mold is filled with saturated sand as the table vibrates. After filling the mold with sand, a guide sleeve is attached to the top of the mold and a lead-filled steel pipe, with a baseplate at the lower end, is lowered onto the sand surface to provide a surcharge load and confinement at the top of the specimen. The vibrating table applies a sinusoidal vertical displacement with a double amplitude of 0.013 inch for 8 minutes at 60 cycles/second. The maximum density is then measured on the compacted soil.

This ASTM standard reads more like recommendations than a "standard" procedure. The standard allows the soil to be tested dry or saturated, and with different types of vibrating tables. It also allows the operator to experiment with different vibration amplitudes to determine the optimum maximum value of density "in conjunction with design or special studies." These variations in procedure do not constitute a standard procedure. It is a method that searches for a maximum density rather than setting the procedure and accepting the maximum value produced by it. In practice, the maximum density is achieved on saturated sand (Kolbuszewski, 1948(a); Cornforth, 1973).

U.S. Minimum Laboratory Density Test

The U.S. test procedure is ASTM D 4254. It uses the same molds listed on Table 7.2, one of which is shown on Figure

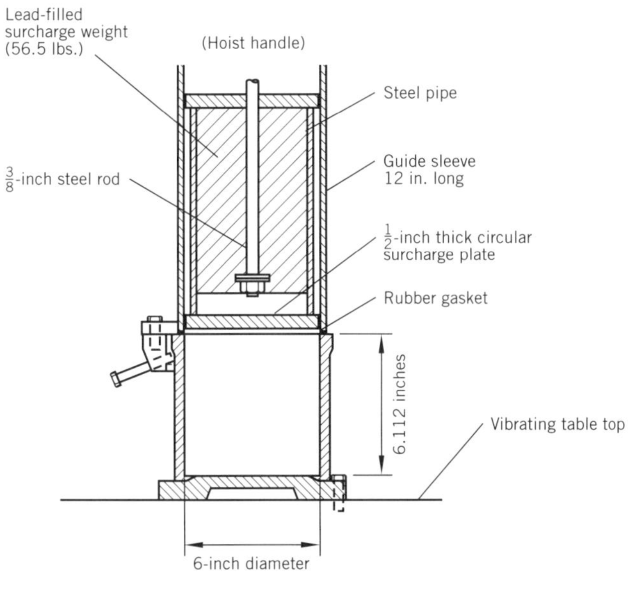

Figure 7.2 Apparatus for maximum density test: 0.1 cu. ft. mold and surcharge weight (ASTM D 4253, some details omitted).

Table 7.2 Test Mold Sizes for Maximum and Minimum Density Tests (ASTM)

	Mold Details			
Maximum Particle Size	Volume	Height	Diameter	Surcharge Load
3/4 inch	0.1 cu. ft.	6.112 in.	6.0 in.	56.5 lbs.
3 inch	0.5 cu. ft.	9.092 in.	11.0 in.	190 lbs.

7.2. Dry soil is poured through a funnel in a steady stream, holding the funnel vertical. The lower tip of the spout is continually adjusted to maintain a free fall of about ½ inch onto the top of already-deposited soil. The pouring device is moved in a spiral pattern from outside to center of the mold. After filling the mold to overtopping, the top of the mold is struck level with a straight edge. The weight of dry sand within the mold is measured to determine the minimum dry density.

An alternative test for sands only is the so-called "tilt" test. A 1,000 gm sample of oven-dried sand is put in a glass measuring cylinder, preferably one without a pouring lip. A stopper or membrane is placed over the top of the cylinder. In the test, the cylinder is turned upside down and then rapidly turned upright. The volume reading of the sand surface is measured, allowing the minimum dry density to be calculated.

ASTM recommends that the test be repeated until three consistent values are obtained. ASTM specifies a 2,000 ml graduated cylinder. The author's experience has been to use a 1,000 ml graduated cylinder with 10 ml gradations as specified for this test in British Standard 1377:1990. This allows the volume to be estimated to the nearest ml. Once the measuring cylinder has been turned upright, the sand surface can be brought level by a small amount of further tilt without affecting the results.

Both the maximum and minimum density tests can include up to 15 percent by weight of noncohesive fines (British practice sets the limit at 10 percent). The tests on gravel in the larger mold permit up to 30 percent by weight of 3 inch-to-1½ inch particle sizes.

British Standard Tests for Density Limits

British standards are different to ASTM standards, but they use the same principle (vibration with confinement) for measuring the maximum density (British Standard 1377:1990). The maximum density test uses an electric vibrating hammer with a power consumption of 600–700 watts operating at a frequency within the range 25–45 Hz (Figure 7.3). A special supporting frame can be used to hold the hammer, but is not essential. A steel tamper head, slightly smaller than the inside of the compaction mold, is attached to the vibrating hammer. A 4-inch diameter standard compaction mold is used in the British standard test.

The mold is placed on a concrete base and is filled to the one-third height with saturated sand. The layer is compacted for 60 seconds under water using a downward force of 70 to

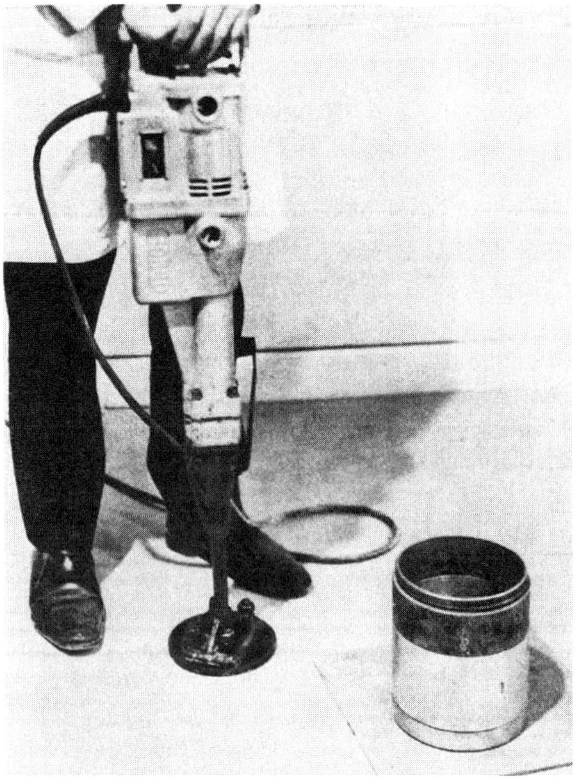

Figure 7.3 Vibrating hammer used to measure the maximum dry density of sands according to British Standard 1377:1990.

90 lbs. (which includes the weight of the hammer and tamper). Three levels of sand are placed and compacted. After the compacted sand has been weighed and dried, one or two repeat tests are performed to check the reliability of the measurement (agreement should be within 50 gm dry weight).

For particle sizes of up to 1½ inch, the British Standard uses a 6-inch diameter California Bearing Ratio (CBR) mold. A larger tamper is used; otherwise, the test procedure is essentially the same as for sands.

Based on a limited database of direct comparison, the U.S. and British methods of measuring minimum and maximum densities seem to give comparable results, but there is a need for further research and standardization of these tests worldwide. The relative density of sands and gravels is usually measured indirectly in the field by penetration tests, such as the Standard Penetration Test.

When compacting *gravels and rockfills* for remediation of landslides, a "methods" specification is generally used. The contractor is required to provide certain types of compaction equipment and make a specified minimum number of passes with the equipment over each lift of soil. The maximum allowable thickness of the uncompacted lift is usually specified. The methods specification has an expectation that a relative dry density of at least 70–80% will be achieved, but the site inspector relies on compliance with the method specification and a visual assessment.

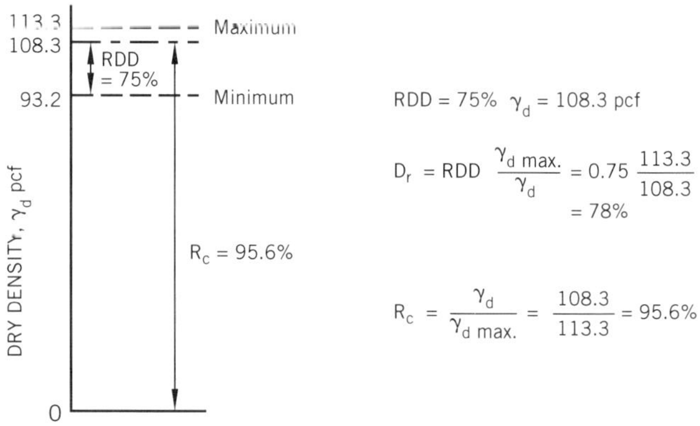

Figure 7.4 Example of relative dry density RDD, relative density D_r and relative compaction R_c for Brasted Sand.

If field densities are measured, the results may be based on relative compaction, R_c (Eq. 7). For Brasted Sand, for example, a relative dry density of 75% would correspond to a relative compaction R_c of 95.6% (Figure 7.4).

7.4 ANGLE OF REPOSE

The *minimum* strength of a sand, gravel, or rockfill with an intergranular structure can be measured by the static angle of repose.

Laboratory Measurement for Sand

A large glass plate is carefully leveled. Dry sand is deposited through a funnel into a conical heap approximately 3 inches high (Figure 7.5). The heap is undercut by scraping away sand from the bottom of the heap in *very small* quantities until sand grains ravel down the face of the slope. The equilibrium slope of the heap is then measured by some means as, for example, by using cardboard templates cut to different angles, the angles differing by 1 degree. Measurement should take place only after a surface raveling has occurred down the whole face of

Figure 7.5 Measurement of the angle of repose of dry sand.

the slope. However, the extreme top of the slope does not always move, and a slightly steeper slope angle at the top of the heap should be ignored. Similarly, flatter slopes resulting from more dynamic wave runs should not be measured.

The test is performed 10 to 20 times, and the average value is taken to be the static angle of repose.

Field Measurement for Gravel and Rockfill

A heap of material is tipped onto flat ground, where it sets up in a relatively loose condition. If some of the material is excavated from the base of the heap, the cohesionless rocks will start to ravel down the face of the slope. At equilibrium, the face is at the angle of repose. The angle of repose is a simple, practical way to measure the minimum strength of gravel and rockfills in the field.

7.5 LABORATORY DRAINED STRENGTH OF SAND

All the tests described in this section were performed by the author at Imperial College, London University in the late 1950s using sand from the Brasted River in Kent, England (Gradation A on Figure 7.1). The results have been previously reported (Cornforth 1961(a); 1964; 1973; 1974(a)). The information is abbreviated to focus on understanding shear strength relevant to landslide studies.

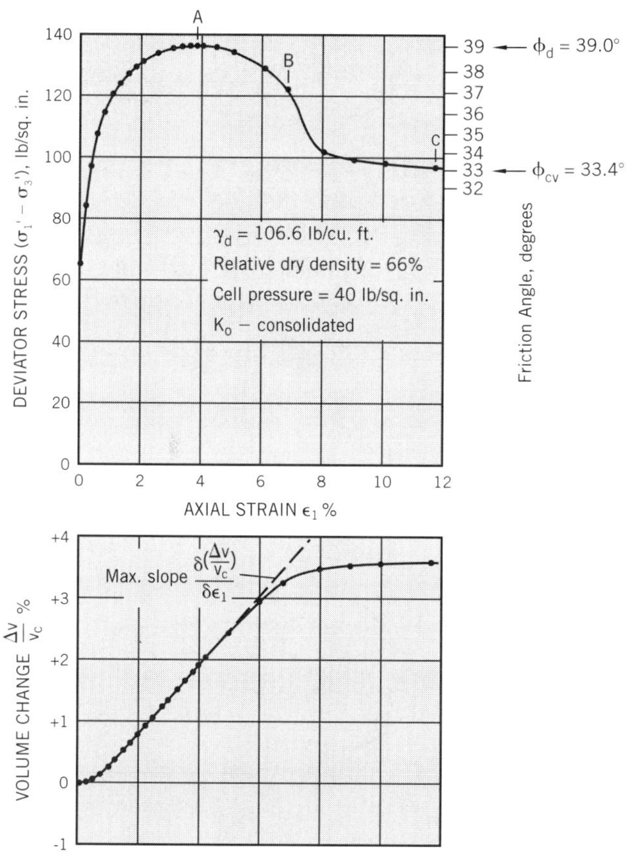

Figure 7.6 Stress-strain and volume change of a dense sand in a consolidated-drained triaxial compression test.

Table 7.3 Maximum/Minimum Densities for Brasted Sand (G = 2.684)

	Dry Density, γ_d lb./cu. ft.	Porosity, n %	Voids Ratio, e
Maximum density	113.5	32.2	0.475
Minimum density	93.4	44.2	0.792
Difference, max to min	20.1	12.0	0.317

Table 7.4 Relationship between $(\sigma_1' - \sigma_3')/\sigma_3'$ and ϕ'
(for an effective stress envelope passing through the origin of the Mohr diagram)

ϕ'	$\left(\frac{\sigma_1' - \sigma_3'}{\sigma_3'}\right)$	Difference	ϕ'	$\left(\frac{\sigma_1' - \sigma_3'}{\sigma_3'}\right)$	Difference
18	0.894		32	2.255	
		0.071			0.137
19	0.965		33	2.392	
		0.075			0.145
20	1.040		34	2.537	
		0.077			0.153
21	1.117		35	2.690	
		0.081			0.162
22	1.198		36	2.852	
		0.084			0.171
23	1.282		37	3.023	
		0.089			0.181
24	1.371		38	3.204	
		0.093			0.191
25	1.464		39	3.395	
		0.097			0.204
26	1.561		40	3.599	
		0.102			0.216
27	1.663		41	3.815	
		0.107			0.230
28	1.770		42	4.045	
		0.112			0.244
29	1.882		43	4.289	
		0.118			0.261
30	2.000		44	4.550	
		0.124			0.279
31	2.125		45	4.829	
		0.130			0.30

Fractions of a degree may be estimated by linear interpolation

The maximum and minimum density limits for Brasted Sand are summarized in Table 7.3.

Characteristics of a Drained Triaxial Compression Test on Dense Sand

A set of shear test results for dense sand is shown on Figure 7.6. The specimen was consolidated under K_o conditions, so there is an initial deviator stress acting on the specimen at the end of consolidation and beginning of shear. As shearing progresses, the specimen starts to dilate (expand) and the maximum rate of dilation, $\delta(\Delta V/V_c)/\delta\varepsilon_1$, occurs at or near the peak stress, point A on Figure 7.6. As shearing continues beyond the peak stress, the stress slowly decreases until a failure plane develops (point B). Once the slip plane develops, the stress rapidly decreases to an approximately constant value, commonly referred to as the ultimate strength (point C). At ultimate strength, the sand specimen shears at constant volume and the corresponding friction angle is termed ϕ_{cv}.

Assuming that the Mohr envelope for the sand passes through the origin of the Mohr diagram, the ratio of the peak deviator stress $(\sigma_1' - \sigma_3')$ to the all-around consolidation stress $(\sigma_3')$is related to the shear strength parameter ϕ' of the soil (Table 7.4). This relationship is useful for quickly determining ϕ' from the result of a triaxial compression test. For example, the peak deviator stress for the test specimen of Figure 7.6 is 136.0 psi at a confining cell pressure ϕ'_3 of 40.0 psi. The ratio is 3.40, which, from Table 7.4, indicates an angle of shearing resistance ϕ' (for $c' = 0$) of 39.0°. At the ultimate strength (point C on the deviator stress-strain curve of Figure 7.6), the ratio is 2.45 ($\phi_{cv} = 33.4°$, by interpolation on Table 7.4). For a drained test, ϕ' is sometimes referred to as the *drained angle of shearing resistance,* ϕ_d rather than ϕ'.

Effect of Relative Density on Strength

The results of four consolidated-drained *triaxial compression* tests on Brasted Sand at relative dry densities (RDD) ranging from 80% to 10% are shown on Figure 7.7. The relative dry density shown on this and all other figures in this chapter refer to the *initial* relative dry density at which the specimen was placed. The term "relative dry density" will be spelled out so that the reader will not be confused by the unfamiliar initials RDD. As previously stated, the terms "relatively dry density" and "relative density" are, for practical purposes, interchangeable.

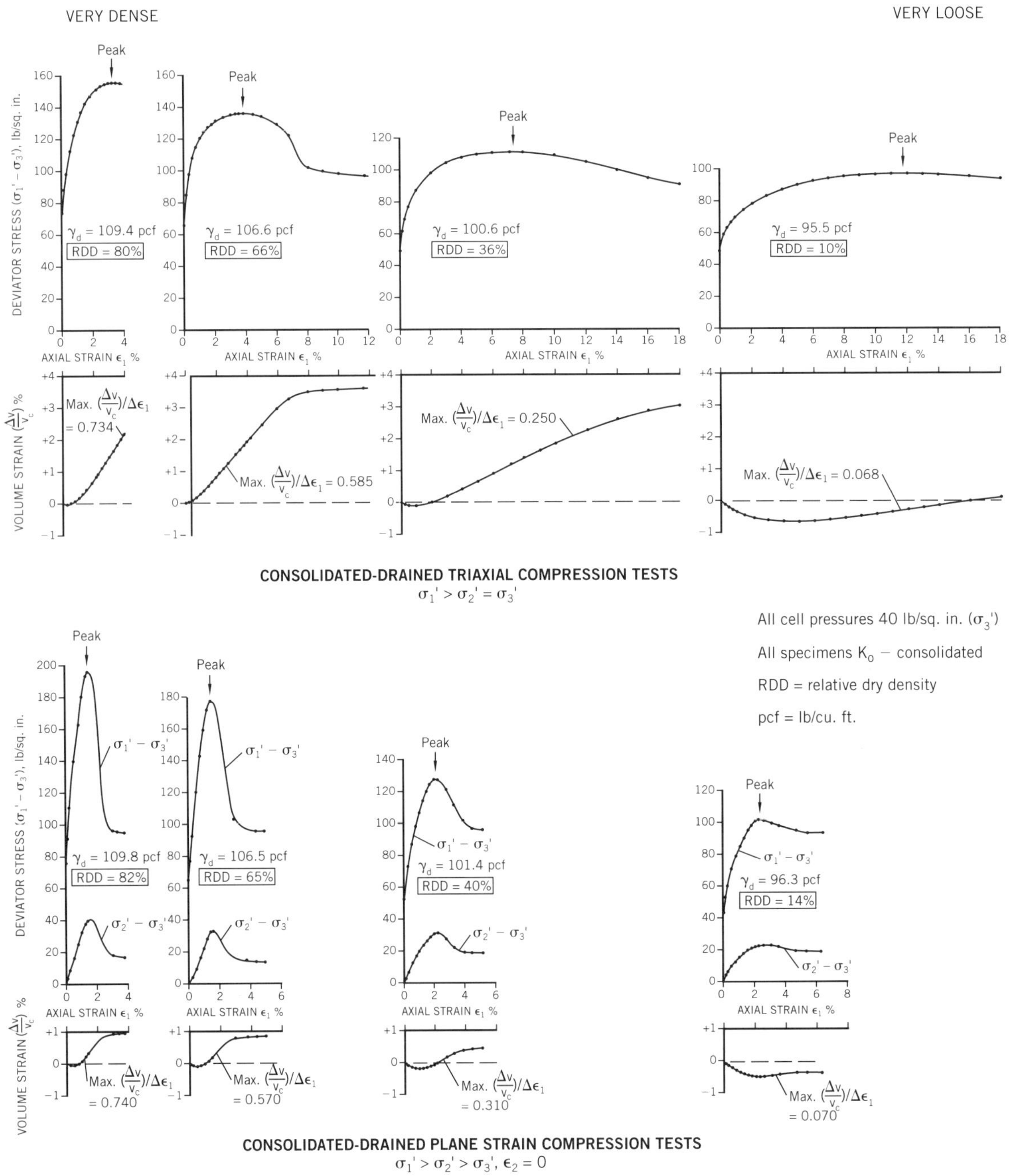

Figure 7.7 Stress-strain and volume change data for triaxial and plane strain compression tests performed at relative dry densities ranging from 10% to 80%.

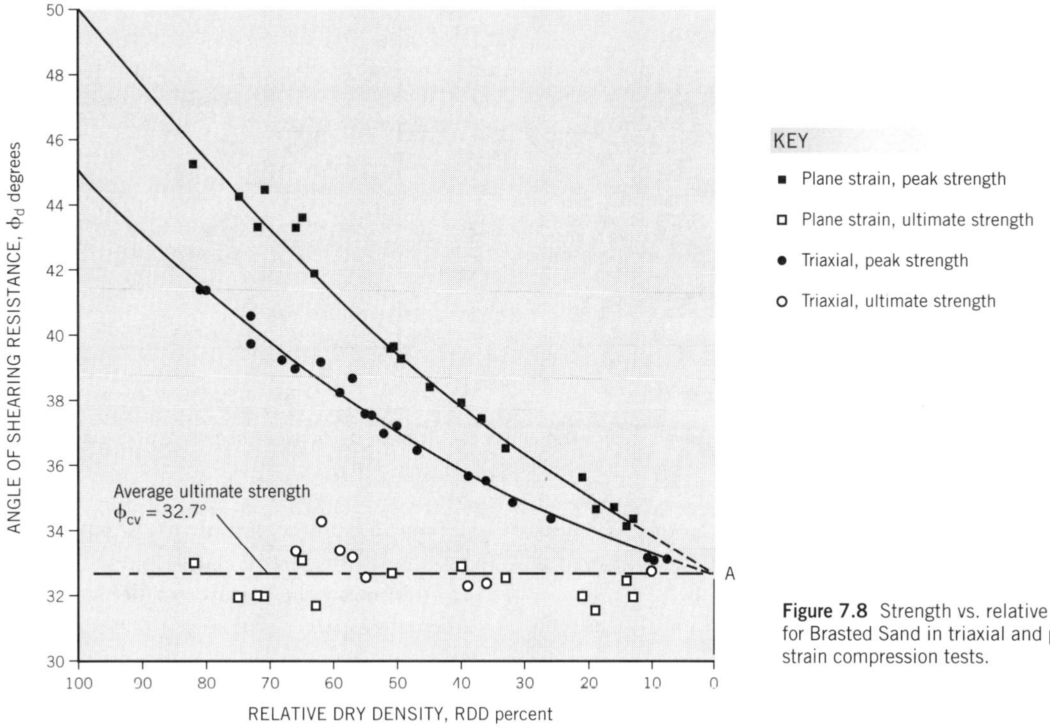

Figure 7.8 Strength vs. relative dry density for Brasted Sand in triaxial and plane strain compression tests.

On the four sets of triaxial compression test graphs on Figure 7.7, the initial density increases from very loose (right) to very dense (left). It can be seen that, as the initial density increases: (i) the peak strength increases, (ii) the maximum rate of dilation $\frac{\Delta V}{V_c}/\Delta \epsilon_1$ increases, and (iii) the ultimate strength remains approximately constant, irrespective of the initial relative dry density.

Plane strain tests covering the same range of relative dry densities and plotted to the same scales are shown below the triaxial tests on Figure 7.7. It can be seen that: (i) the strength of the sand is higher in plane strain than the strength in triaxial compression at the same relative dry density, and failure occurs at lower axial strains, (ii) the rate of dilation in plane strain increases with increasing relative dry density, and (iii) the ultimate strength in plane strain remains approximately constant, irrespective of the initial relative density.

The peak shear strengths of the entire program of triaxial and plane strain tests are plotted against relative dry densities on Figure 7.8. For more details about the test program, see Cornforth (1964).

From the data, the following conclusions can be drawn:

- The peak shear strength measured in plane strain is *always higher* than the peak shear strength measured in triaxial tests at the same initial relative dry density.
- The difference in strength *progressively increases* as the initial relative dry density increases.
- The ultimate strengths in both types of test are *approximately constant* over the entire range of initial relative dry densities. The average ultimate strength for the triaxial tests is ϕ_{cv} = 33.0°; for plane strain, the average result is 32.3°. On Figure 7.8 the average, ϕ_{cv} = 32.7°, is shown.

Minimum Intergranular Density Concept

The lowest relative dry densities in this test program were around 10%. It can be seen on Figure 7.7 that the stress–strain curves for the very loose specimens have peak strengths that are only slightly higher than the ultimate strength, and dilatancy is very weak. If it had been possible to set up the specimens at *zero* relative dry density, in either the triaxial test or the plane strain test, it could be expected that the stress–strain curve would reach a maximum directly at the ultimate strength and there would be no dilatancy, as illustrated hypothetically on Figure 7.9. The stress–strain curve and volume

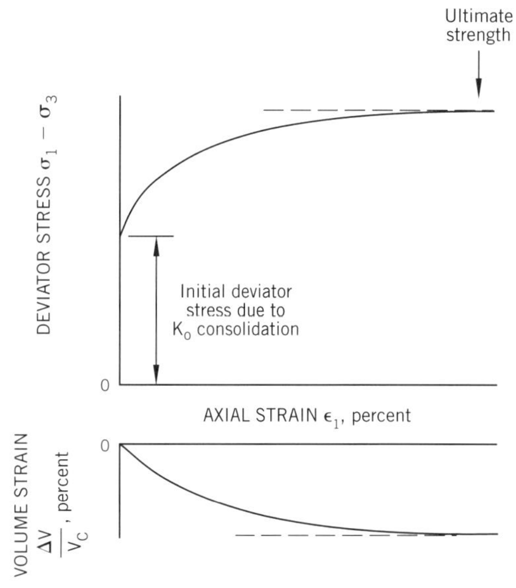

Figure 7.9 Stress-strain and volume change graphs for sand at zero relative dry density.

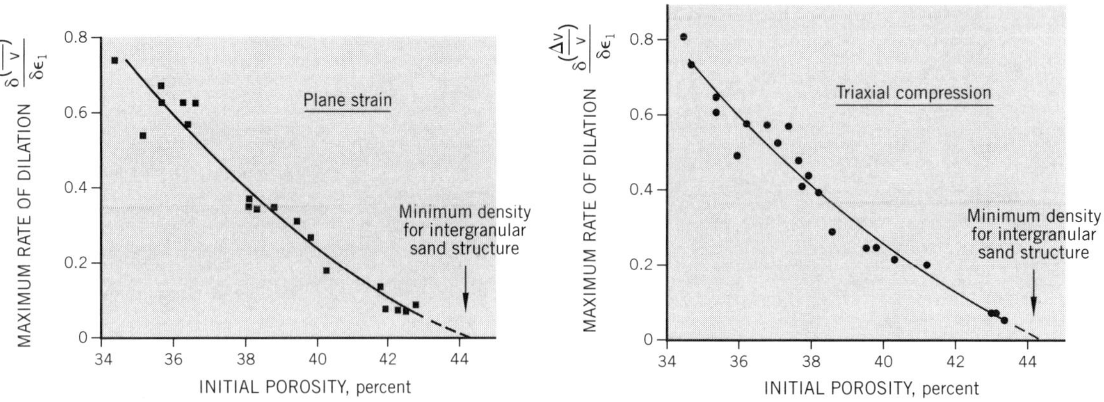

Figure 7.10 Determination of the minimum intergranular density (porosity) from rate of dilatancy measurements.

change curve would be mirror images of each other. The peak and ultimate strengths would intersect at Point A on the strength vs. relative dry density plot, Figure 7.8.

There are several ways of estimating the initial density at Point A:

- Graph the maximum rate of dilation against initial density and extrapolate to zero rate of dilation. Figure 7.10 shows the results for both triaxial tests and plane strain.
- Graph the difference between peak and ultimate strengths against initial density and extrapolate to zero difference.
- Graph the ratio of peak:ultimate strength against initial density and extrapolate to a ratio of 1.0.
- Graph volume changes from peak to ultimate strengths against initial density and extrapolate to zero volume change.

These methods show that point A is reached at porosities ranging from 44.1% to 44.3%. The minimum density test (rapid tilt) gave a porosity of 44.2% for Brasted Sand.

This unique placement density, at which the peak and ultimate strengths coincide for both symmetric (triaxial) strain and plane strain, and there is no dilation, has been termed the *minimum intergranular density* (Cornforth, 1964). It marks the transition point from an intergranular sand structure to a collapsible sand structure.

It is possible to set up laboratory specimens at the minimum intergranular density by modifying a Norwegian Geotechnical Institute (Bjerrum et al., 1961) technique. A three-part split mold is used to prepare the test specimen directly onto the base pedestal of the triaxial apparatus. Damp, very loose sand is placed inside the mold, and de-aired water is allowed to percolate slowly upwards from the base. The rising water causes the sand to slowly collapse, and more damp sand is added as necessary until the water level reaches the top of the mold. After sealing the top cap and applying a small negative pore-water pressure, the split mold is removed and the triaxial cell is assembled for the test. The test specimen is then fully saturated by the application of back pressures. This puts the test specimen density at, or very close to, the minimum intergranular density. In the subsequent triaxial shear test, the sand specimen shears directly to the ultimate strength without a peak in the stress–strain curve and without dilating.

In summary, the **minimum density** test measures a fundamental property—the lowest density at which the packing of the sand grains produces an intergranular structure. At looser densities, the sand is in an unstable structure that will collapse on saturation. The **maximum density** test measures an arbitrary upper limit set by a laboratory procedure that has no significance in nature but defines the relative density limits for geotechnical work.

Effect of the Strain Conditions on Strength

The test program on Brasted Sand included a series of triaxial *extension* tests. In the extension test, the cylindrical specimen is pulled vertically, decreasing the axial effective stress. The cell pressure is both the major principal stress σ_1' and the intermediate principal stress σ_2'. A special apparatus is needed to provide the pull (Cornforth, 1964). The cell pressures were 130 to 140 lb./sq. in.

One of the difficulties in performing triaxial extension tests is that the specimens fail by narrowing the diameter across the failure zone (necking). As soon as the maximum deviator stress had been reached, the tests were stopped and very careful measurements were taken by a traveling microscope. In the calculations, it was assumed that a failure plane would make an angle $(45 + \frac{1}{2}\phi_d)°$ to the vertical axis, passing symmetrically through the necked zone.

The triaxial extension test results are plotted on Figure 7.11. The data scatter around the curve representing the results of the triaxial compression results and are well below the plane strain curve. It is concluded that the strength of the sand in triaxial extension tests is approximately the same as the strength in triaxial compression tests.

This phenomenon can be explained as follows: During shear, individual sand grains slide, roll, or are fractured by adjacent sand grains. They follow directions of least resistance as strain occurs under an increasing deviator stress. In triaxial compression and extension, two of the three principal stresses are equal, creating a plane of equal stress. The sand grains have an equal opportunity to move in any direction within the

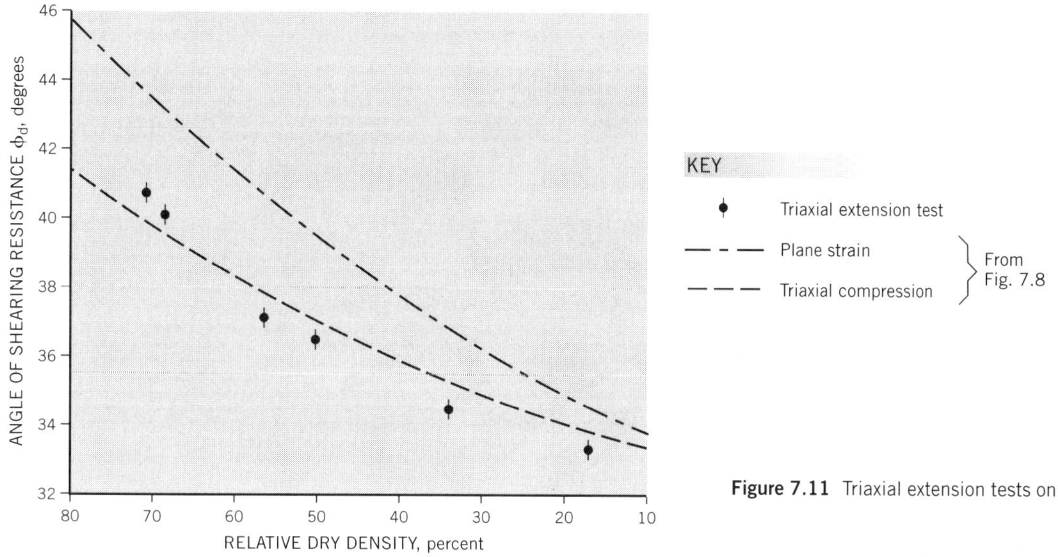

Figure 7.11 Triaxial extension tests on Brasted Sand.

equal plane and in mass, move radially within this circular plane. In plane strain, the individual grains are much more restricted and will only move with limited deviation from the two-dimensional movement of the specimen. This restriction mobilizes a higher resistance because the individual grains are being forced in a particular direction of shear by plane strain conditions. Therefore, for dilatant sands, *symmetric strain (triaxial compression and extension) produces the lowest mobilized strength for a particular sand structure (i.e., relative density) and plane strain produces the highest mobilized strength.* At the minimum intergranular density, at which the sand can shear at constant volume without dilatancy, the strength under all strain conditions is identical. The sand on the failure plane at the ultimate strength has a density that is a combination of the minimum initial intergranular density plus a small density increment due to the average principal stress acting on the sand.

The very loose condition of the sand on the failure plane is demonstrated on Figure 7.12. After removing the rubber membrane at the end of a test, a touch on the top cap caused the specimen to separate down the shear plane as shown in the photograph. This shows that the sand within the thin shear zone is much looser than elsewhere in the specimen. This dense sand specimen had been sheared to the ultimate condition.

Figure 7.12 Plane strain failure of dense sand. Specimen separates easily at the failure plane, demonstrating that the sand in the failure zone is very loose at the end of the shear test.

Investigations by Others

Comparisons of the strength of sand in triaxial compression, extension, and plane strain have been studied elsewhere. With respect to triaxial compression and extension tests, the conclusions have ranged from no difference over the entire relative density range (Eldin, 1951; Kirkpatrick, 1957; Taylor, 1941) to the extension tests giving lower results (Peltier, 1957; Habib, 1953). It is not known whether any correction was made for "necking" in these comparisons. Omission of a correction for necking will result in the extension tests having an apparently lower strength than compression tests. In comparing plane strain and triaxial compression tests, the research consistently shows that plane strain produces higher strengths than triaxial compression at the same placement densities (Bjerrum, et al., 1961; Rowe, 1969; Marachi et al., 1969).

7.6 DRAINED STRENGTH ESTIMATES

Background Information

A method for estimating the drained strength of sands evolved from the author's test program on Brasted Sand. Earlier, Hafiz (1950) had shown that the strength-relative density curves for five very different sands and gravels fell into a narrow range

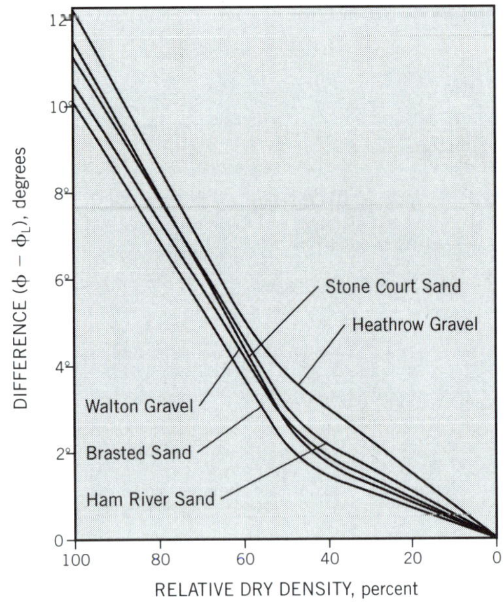

	Angle of repose	Loosest strength ϕ_L
Heathrow Gravel	36.9°	37.8°
Walton Gravel	34.8°	34.2°
Brasted Sand	32.2°	31.8°
Stone Court Sand	31.8°	31.4°
Ham River Sand	32.0°	31.7°
Average	33.5°	33.4°

Figure 7.13 Graph of ($\phi - \phi_L$) vs. relative dry density in large shear box (after Hafiz, 1950).

when plotted as the difference in friction angle (Figure 7.13). These tests were performed in a large shear box and the limiting densities were determined by the methods of Kolbuszewski (1948a,b), which are similar to current techniques.

The author's approach on Brasted Sand was to divide the peak shear strength by the ultimate strength. It was reasoned that the ultimate strength is a function of the original rock mineral, the grain shape and angularity, etc. These same factors also affect the measured strengths at other relative densities. Therefore, a dimensionless density factor R can be calculated:

$$\text{Density Factor } R = \left(\frac{\sigma_1' - \sigma_3'}{\sigma_3'}\right)_{\text{peak}} \Big/ \left(\frac{\sigma_1' - \sigma_3'}{\sigma_3'}\right)_{\text{ultimate}} \qquad \text{Eq. (10)}$$

The resulting graph is shown on Figure 7.14. R increases from 1.0 at the minimum intergranular density to higher values at higher relative densities. The density factor (i.e., increase in shear strength above the ultimate strength) is approximately 70 percent higher in plane strain than in triaxial compression at the same initial relative dry density. This allows the experimental data to be extrapolated up to 100% relative dry density.

The next step was to extend the data to other cohesionless soils by calculating the strengths for a range of ultimate strengths ranging from 28° to 36°. When these calculated strengths were converted back to shear angles ϕ_d, it was found that the difference ($\phi_d - \phi_{cv}$) was almost the same throughout the entire relative density range for all of the above ultimate strengths. Therefore, as a practical matter, one curve would suffice (Figure 7.15). The increase in shear angle due to increased density is termed the *density component of strength*, ϕ_{dc}.

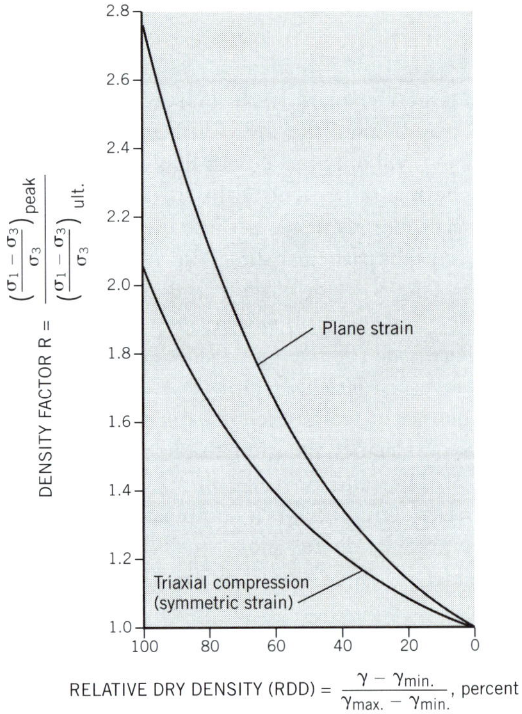

Figure 7.14 Density factors for Brasted Sand.

Estimation of the Maximum Drained Strength in Sand

Estimate of the *maximum* (peak) drained strength of sand ϕ_d is given by:

$$\phi_d = \phi_{cv} + \phi_{dc} \qquad \text{Eq. (11)}$$

where ϕ_{cv} = minimum intergranular density
ϕ_{dc} = density component of shear strength

The value of ϕ_{cv} can be obtained by performing a static angle of repose test on dry sand or a triaxial test continued to the ultimate strength. The density component can be obtained by estimating the relative dry density from SPT blow count and taking ϕ_{dc} from Figure 7.15.

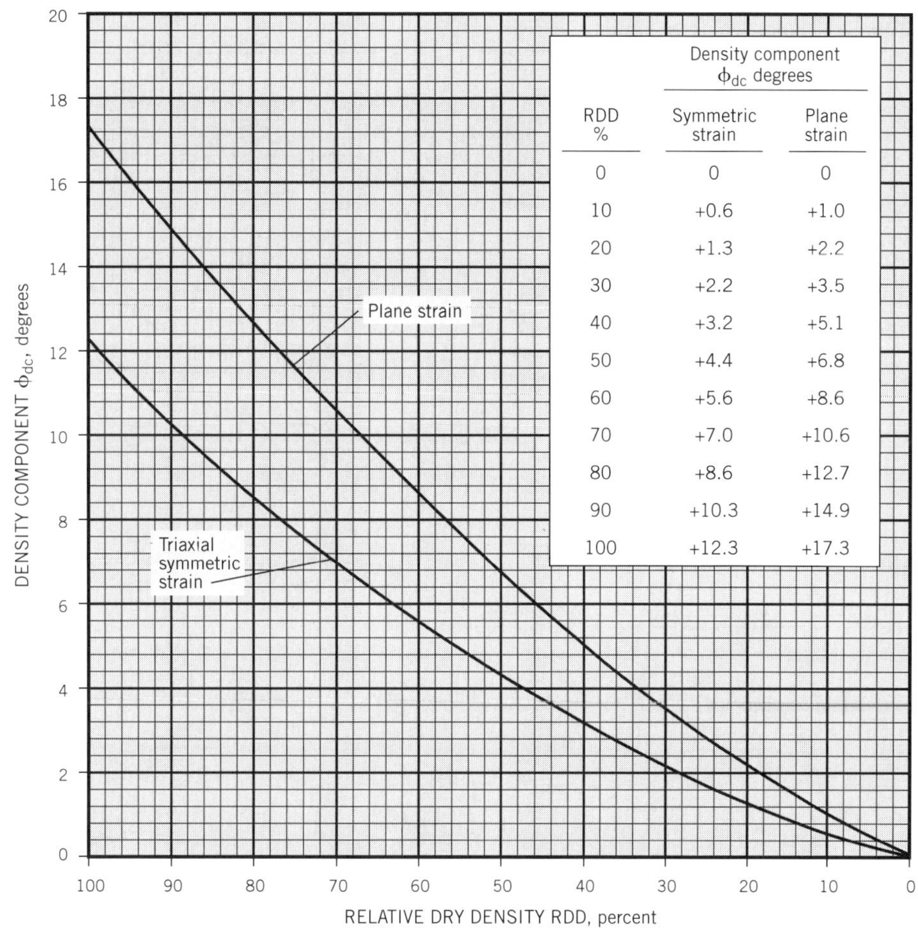

RDD %	Density component ϕ_{dc} degrees: Symmetric strain	Density component ϕ_{dc} degrees: Plane strain
0	0	0
10	+0.6	+1.0
20	+1.3	+2.2
30	+2.2	+3.5
40	+3.2	+5.1
50	+4.4	+6.8
60	+5.6	+8.6
70	+7.0	+10.6
80	+8.6	+12.7
90	+10.3	+14.9
100	+12.3	+17.3

Figure 7.15 Density component ϕ_{dc} vs. relative dry density.

Example:

SPT blow count N_{60} = 23 at an effective overburden stress of 1 ton/sq. ft.
From Figure 3.14, relative density ≈ relative dry density = 65%
Angle of repose = 33.5° (average of 20 tests) = ϕ_{cv}
Angle of shearing resistance $\phi_d = \phi_{cv} + \phi_{dc}$
In symmetric strain (triaxial test) ϕ_{dc} = + 6.2° at RDD of 65% (Figure 7.15)
$\therefore \quad \phi_d = 33.5° + 6.2°$
$= 39.7°$

Comparison with other soils

The strength estimate technique has been compared with three other sands that were tested under the author's supervision. The gradations of the three samples (with Brasted Sand for comparison) are shown on Figure 7.16, and density limits and test conditions are summarized on Table 7.5.

Comparisons between the sand strength measurements and the *predicted* strengths are shown on Figure 7.17. Despite the substantial difference between these three sands and Brasted Sand in terms of origin, gradation, and maximum/minimum densities, the predicted strengths (based on Figure 7.15) show very good agreement with the actual measurements.

There is only a limited amount of data on plane strain strengths. However, the results obtained by Rowe (1969) and Marachi et al. (1969) are also in generally good agreement with the predictive curves (Cornforth, 1973).

Selection of Drained Shear Strength of Sands for Stability Analyses

In the field, the shear stresses developed along a potential failure surface are nonuniform. Certain parts of the shear surface reach the strain corresponding to the peak shear strength more quickly than others. Since the peak strength of a dense sand is only sustained over a short interval of strain, it is evident that all points on the potential failure surface cannot concurrently attain their peak strength. A failure surface will achieve its maximum resistance to first-time sliding when various parts of the slip surface are below, at, and past the maximum failure strength of the sand. Therefore, the peak shear strength measured in the triaxial test should not be used as the design strength; a lower strength has to be chosen.

The shear strength that can be relied upon as a minimum is the ultimate strength at which there is no further decline in strength under further shear strain. The ultimate strength can be measured by continuing the triaxial drained shear test past the peak to the ultimate value, which is reached at about 12 to 20 percent axial strain, depending on the initial density. This

Table 7.5 Test Data and Density Limits for Three Sands

	Guinea Sand	Portland Sand	Limassol Sand
Test conditions	saturated	saturated	saturated
Specimen diameter, in.	4	4	4
Type of consolidation	ambient	ambient	ambient
Cell pressures, psi	20/40/60	25	30
Friction eliminators used*	yes	no	yes
Minimum density test	rapid tilt	rapid tilt	rapid tilt/ASTM
Maximum density test	Kango hammer	surcharged on vibrating table	Kango hammer
Test results:			
Minimum density, lb./ft.^{3}	87.8	81	78.2
Maximum density, lb./ft.^{3}	110.2	104	108.2
Static angle of repose, deg	33.3	35.0	33.2
Measured avg ϕ_{cv}-values, deg	...	36.1	34.4
Specific gravity, soil grains	2.68	2.72	2.73

*rotating bush mechanisms, which eliminate the vertical component of piston friction during shear test.

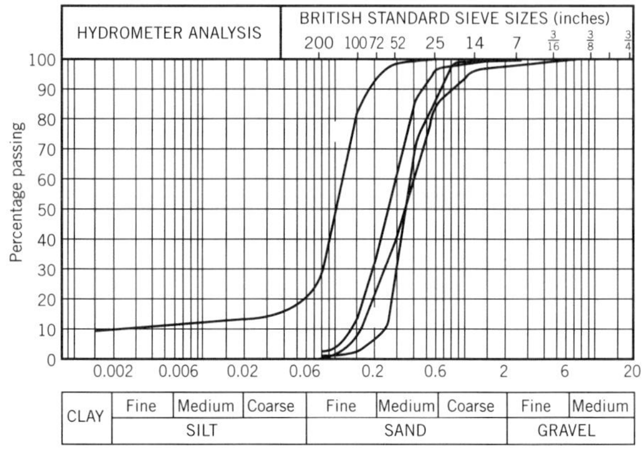

A – River sand from Brasted, Kent, England
B – Marine sand from Guinea, West Africa
C – River sand from Portland, Oregon (U.S. sieve sizes)
D – Marine sand from Limassol, Cyprus

Figure 7.16 Gradation curves of four sands used to compare actual and predicted drained strengths of sands.

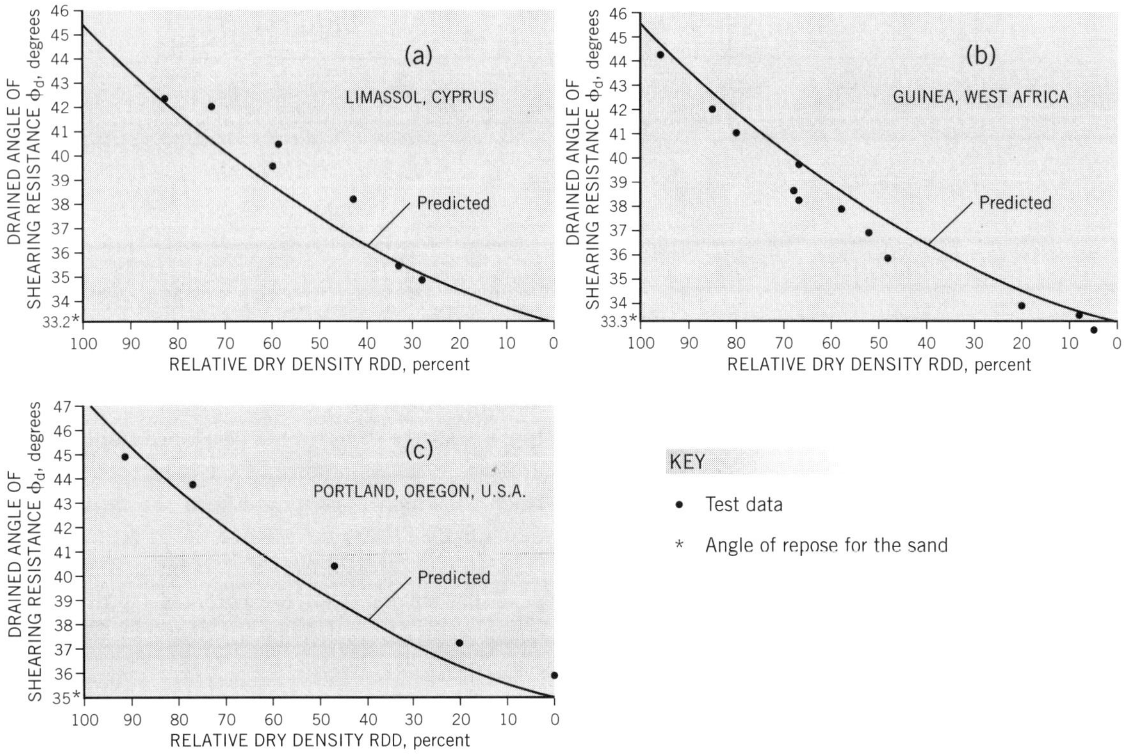

Figure 7.17 Comparison of predicted strength—relative dry density curves with data obtained from commercial laboratories:

(a) Limassol, Cyprus
(b) Guinea, West Africa
(c) Portland, Oregon, U.S.A.

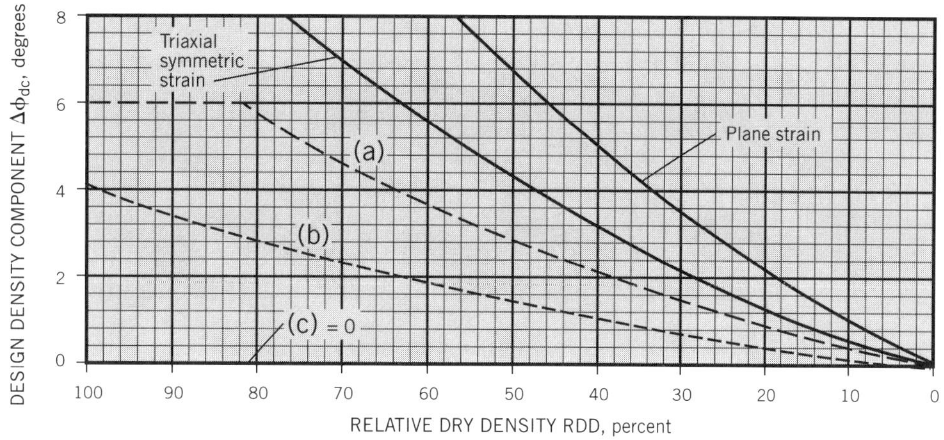

Figure 7.18 Suggested curves of $\Delta\phi_{dc}$ for stability analysis purposes:
(a) "realistic" average strength for first-time sliding
(b) general slope stability design
(c) landslide back analysis

procedure should be followed for all laboratory shear tests on sands.

The ultimate strength can also be measured by a static angle of repose test on dry sand. A comparable test can be performed in the field on cohesionless gravels or rockfills to obtain the ultimate strength, ϕ_{cv}.

The increment of strength that can be added to ϕ_{cv} to reflect the relative density component is a judgment decision for the geotechnical practitioner and will be influenced by the purpose of the study. If a realistic value of the strength prior to a landslide is needed in stability calculations, curve (a) on Figure 7.18 is suggested. This is probably close to the maximum resistance that can be attained in a slope before a failure occurs. If the objective is to design a stable slope (in a compacted fill, for example) the curve (b) may be sufficiently conservative when used in conjunction with a factor of safety in stability analyses. Finally, if the purpose of the strength estimate is a back analysis of an actual landslide, there should be *zero* additional density component because the slip surface will have already reached the ultimate condition.

Thus, for sands and other cohesionless soils, the suggested procedure is:

1. Measure the ultimate strength by a triaxial test and/or static angle of repose test.
2. Select an allowance for the density component of strength from curves (a), (b), or (c), depending on the purpose of the study.

The **design** strength of a sand can, therefore, be estimated:

$$\phi_{design} = \phi_{cv} + \Delta\phi_{dc} \qquad \text{Eq. (12)}$$

where ϕ_{cv} = minimum intergranular density ≈ static angle of repose
$\Delta\phi_{dc}$ = reduced density component of shear strength obtained from the appropriate curve (a) (b) or (c) on Figure 7.18

Example: $\phi_{cv} = 34.6°$

$\Delta\phi_{dc}$ for a realistic value at relative dry density of 50% from curve (a) is 2.9°.
$\phi_{design} = 34.6 + 2.9 = 37.5°$.

7.8 LABORATORY UNDRAINED STRENGTH OF SANDS

Note: Geotechnical engineers and geologists with less interest in issues of shear strength may choose to move directly from Section 7.7 to Section 7.9.

The undrained shear test attempts to measure the strength of soil when there is insufficient time for drainage to occur. As this section will demonstrate, this objective is not attained because pore water moves around within a triaxial test specimen. Therefore, the laboratory undrained triaxial test is a constant volume test rather than a test in which the soil is "undrained"; i.e., one in which there is no change in the density on the failure plane.

This section describes, on the basis of research on Brasted Sand, how the density on the failure plane of a dense sand loosens during shear in an undrained triaxial shear test in a similar manner to a drained test. This phenomenon can be observed over a very wide range of initial densities in plane strain tests because the volume of sand undergoing dilatancy is much smaller than in triaxial tests. These internal changes within the undrained shear test—the failure plane becoming looser and the rest of the test specimen becoming denser—has significant implications in the interpretation of the undrained test.

The discussion leads to the conclusion that "undrained" strength is a transient behavior. In laboratory triaxial tests, for

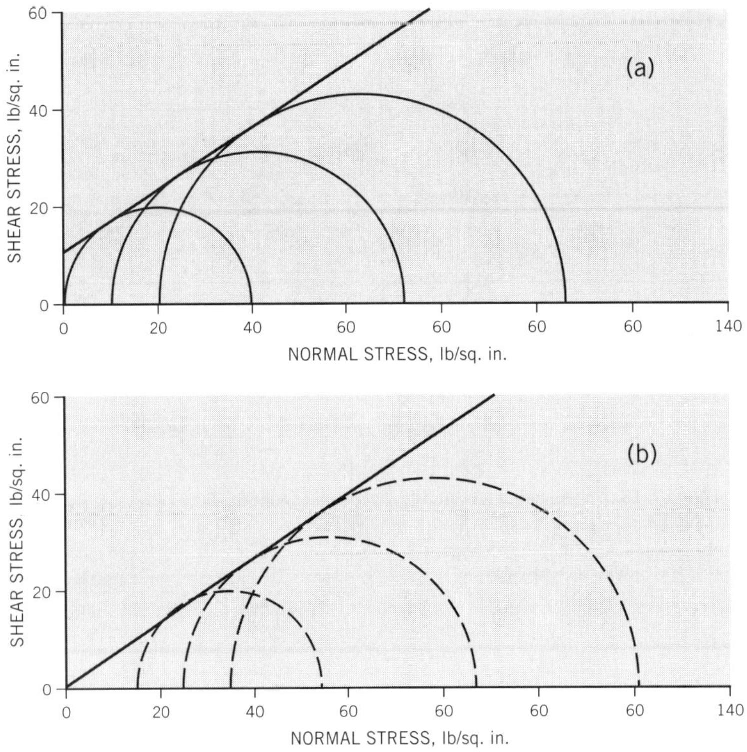

Figure 7.19 Effect of cavitation on the undrained strengths of dilatant cohesionless soils:
(a) Mohr envelope of measured undrained strengths
(b) Mohr envelope after adding 14.7 lb./sq. in. to cell pressure σ_3

example, faster rates of strain produce higher strengths (and lower pore pressures at failure) because there is a nonsteady state flow gradient toward the failure plane.

Cavitation Problem

When medium dense or dense sand is sheared in an undrained triaxial test, the sand dilates and develops negative pore pressures. If the pore pressures approach zero atmospheric pressure (approximately –14.7 psi), the air dissolved in the pore water comes out of solution and the soil cavitates. Therefore, undrained tests performed on saturated sand or silt often produce the results shown on Figure 7.19(a). The soil appears to be a $c-\phi$ material. However, the soil has cavitated, and therefore the pore pressure has dropped to around –14 to –14.7 psi and cannot go lower. The effective stresses for each Mohr circle can be obtained by adding 14.7 psi to each cell pressure σ_3 in the tests, producing an effective stress Mohr envelope shown on Figure 7.19(b). What was set up as a series of undrained tests using total stresses (and often reported as such by commercial soil laboratories) is, in fact, a set of effective stress tests.

To try to overcome the problem of cavitation, undrained tests can be performed using a high initial pore pressure in the test. If, for example, an initial pore pressure of 90 psi is selected (by applying a back pressure to the specimen), the pore pressure would have to fall by about 104 psi in the undrained test before the soil cavitates. Despite this technique, a medium dense or dense sand can be impossible to shear to maximum deviator stress in the laboratory without cavitating.

Experimental Data

An example of an undrained triaxial shear test on medium dense (RDD = 50%) specimen of Brasted Sand is shown on Figure 7.20(a). The initial pore pressure was set at +137 psi. During the shear test, the pore pressure plummets due to the dilation of the sand. The effective all-around pressure σ_3' rises correspondingly and the deviator stress $\sigma_1' - \sigma_3'$ rises at a fast rate throughout the test without ever reaching a peak value. However, the deviator stress ratio $\frac{\sigma_1' - \sigma_3'}{\sigma_3'}$ does reach a peak value.

A very loose specimen (RDD=14%) develops positive pore pressures at small axial strains and then shows a falling trend (Figure 7.20b). The test was stopped at an axial strain of 16%. The deviator stress $\sigma_1' - \sigma_3'$ has unusual characteristics. An early peak occurs and is followed by a dip and then a rising trend. When plotted in terms of deviator stress ratio, $\frac{\sigma_1' - \sigma_3'}{\sigma_3'}$ the curve has a conventional rise to a peak value similar to a drained test.

The probable explanation for the unusual stress-strain curves and the lack of any peak deviator stress $\sigma_1' - \sigma_3'$ in the undrained tests on Brasted Sand is that the sand is not shearing undrained at the original placement density. The specimen is dilating internally at constant volume. Because the specimen is unable to draw in water from a burette or back pressure device, it must produce water for the dilating sand of the failure zone from non-dilating parts of the specimen.

The very loose sand (RDD=14%) that was tested in the undrained shear test depicted on Figure 7.20(b) is shown in Figure 7.21 at the end of the test. The specimen has changed

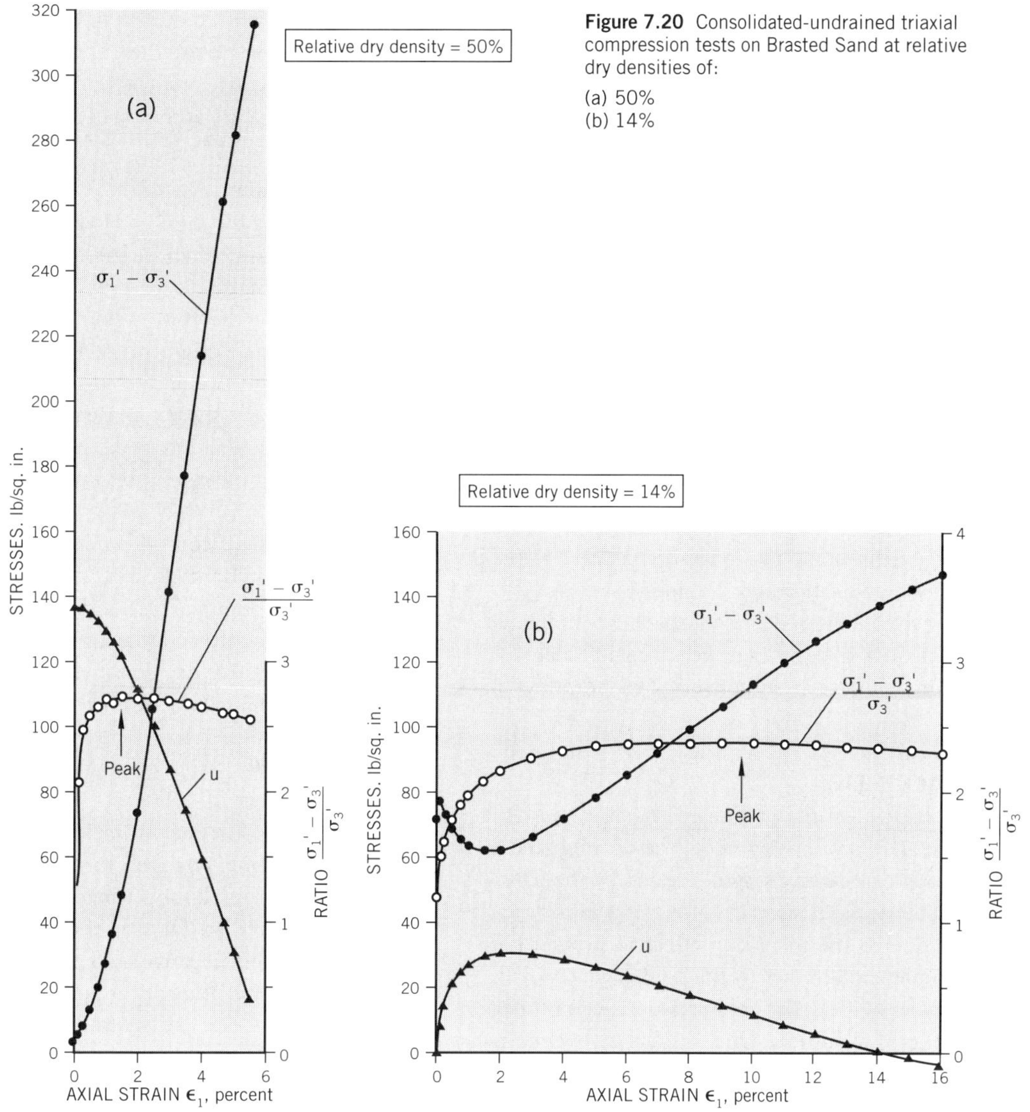

Figure 7.20 Consolidated-undrained triaxial compression tests on Brasted Sand at relative dry densities of:
(a) 50%
(b) 14%

Figure 7.21 Consolidated-undrained triaxial test specimen at the end of the shear test.

from its initial cylindrical shape to a barrel shape, which is typical for a triaxial test. It indicates that a high proportion of the sand is participating in the shear. (By contrast, if the top cap had been aligned eccentrically, a single failure plane would have developed, common on smaller-diameter specimens tested in commercial laboratories.) The only parts of the specimen shown on Figure 7.21 that are not dilating are the areas immediately below the two ends of the specimen, where friction between the sand and porous stones prevents the sand from moving radially.

The migration of water from the non-dilating areas of a specimen to the dilating areas can be demonstrated by setting up two identical triaxial test specimens. One is consolidated and sheared to failure, and moisture contents are taken at several levels. The second specimen is consolidated but not sheared, and moisture contents are taken at the same levels as the sheared specimen. Without describing all the details of these techniques, it can be stated that such comparison tests, both for triaxial and plane strain, show a consistent decrease in

Table 7.6 Porosity Changes in a Consolidated-Undrained Triaxial Compression Test on Medium Dense (RDD 50%) Brasted Sand

	Measured Porosities		
Position in Specimen	Initial (untested)	Final (after undrained test)	Difference
Top	37.1	35.2	− 1.9
Three-quarter height	38.2	39.4	+ 1.2
Mid-height	37.8	38.8	+ 1.0
One-quarter height	38.0	39.1	+ 1.1
Bottom	38.9	37.3	− 1.6
Average	38.0	38.0	

moisture content (i.e., higher density than initial) at the top and bottom of the specimen adjacent to the end platens, and an increase in moisture content (i.e., lower density than initial) at the mid-height where dilation is occurring.

The results for the medium dense consolidated-undrained triaxial test shown on Figure 7.20(a) are given in Table 7.6. The moisture contents have been converted to porosities in the table.

Interpretation of the Data

What appears to happen in the undrained triaxial test is that these non-dilating areas produce water for the dilating sand as a result of the increased effective stresses created by the falling pore pressure u. The deviator stress $\sigma_1' - \sigma_3'$ rises rapidly and continues to rise due to the increasing effective stresses. This relationship continues until either (i) the sand cavitates or (ii) in the case of very loose sand, the pore pressure levels off near the ultimate strength condition. Under these two circumstances, the deviator stress reaches a peak. Therefore, the peak deviator stress measured in an undrained triaxial test is *not* measuring the strength at *constant voids ratio* of the sand.

Consolidated-undrained tests on Brasted Sand were sheared at the same rates of strain as the consolidated-drained tests. This practice is commonly followed when pore pressures are being measured to avoid significant pore pressure gradients within the specimen (Bishop and Henkel, 1957). By using this procedure, the pore pressure measurements in the undrained test are measuring the effective stress increase required to provide water from non-dilating parts of the specimen to dilating parts.

To summarize this section, the intent of an undrained shear test is to measure the strength of soil at the existing voids ratio (density). In dilatant sands, this objective does not occur because the sand "drains" internally with water migrating to the shear zone as it loosens in shear. Since no external drainage is permitted in the test, the additional water needed in the failure zone is produced by negative pore pressures that have to occur during shear to consolidate the non-dilating volumes at the two ends of the specimen. By shearing the test very slowly, the test measures the negative pore pressure needed at a particular strain condition. In undrained triaxial tests on medium dense to dense sands, the test is never able to provide enough water to loosen the sand to the ultimate condition; the deviator stress continues to rise and the pore pressure continues to fall until cavitation occurs, as shown in Figure 7.20(a).

Plane Strain Data

The concept that the sand is "draining" internally during an undrained test can be demonstrated over a wide density range in plane strain tests. In these tests, most of the specimen does not dilate during the test, unlike the barrel-shaped triaxial specimen, but fails along a single failure plane (Figure 7.12). Shearing is largely confined to this thin shear zone, at least in the later stage of the shear test. Because a lower percentage of the sand is actively dilating in the plane strain test, the volume expansion during shear tests on dense sand is smaller than in triaxial tests.

A comparison of consolidated-drained and consolidated-undrained tests of Brasted Sand in plane strain is shown on Figure 7.22. On Figure 7.22(a), the results of the undrained tests are in good agreement with the drained tests over the entire range of relative dry densities when undrained failure is determined by the maximum principal stress ratio $\frac{\sigma_1'}{\sigma_3'}$.

Note: $$\frac{\sigma_1'}{\sigma_3'} = \frac{(\sigma_1' - \sigma_3')}{\sigma_3'} + 1 \qquad \text{Eq. (13)}$$

The pore pressures at the maximum principal stress ratio mirror the volume changes in the drained tests (Figure 7.22b). When the volume changes in the drained tests are positive (dilatant) at failure, the pore pressures in the undrained tests are negative, and vice versa in looser initial states. The crossover point from positive to negative occurs at relative dry densities of around 40%. For three of the undrained tests, the volume changes that would be expected to occur in drained tests due to the observed negative pore pressures are shown by x's on Figure 7.22(b). It is seen that these calculated volume changes are in excellent agreement with the curve of volume changes at failure measured in actual drained tests. This result supports the opinion that the negative pore pressures in the undrained tests provide the free water needed for the dilating areas of the specimen. In drained tests, that water would have been provided from an external burette. Therefore, the "undrained" test is really a drained test in which the water needed for dilation is produced *internally* rather than *externally*. The plane strain tests demonstrate that a slow undrained test on dilatant sand is simply a disguised form of drained test. More importantly, undrained tests do not measure the shear strength of the "unchanged" soil. Loosening of the sand on the failure plane occurs in a similar way to a drained test.

Undrained Tests as a Transient Phenomenon

The loosening of the soil on the failure plane in dilatant soils, with consequent migration of pore water, has been known for a long time, but the implications have received limited discussion.

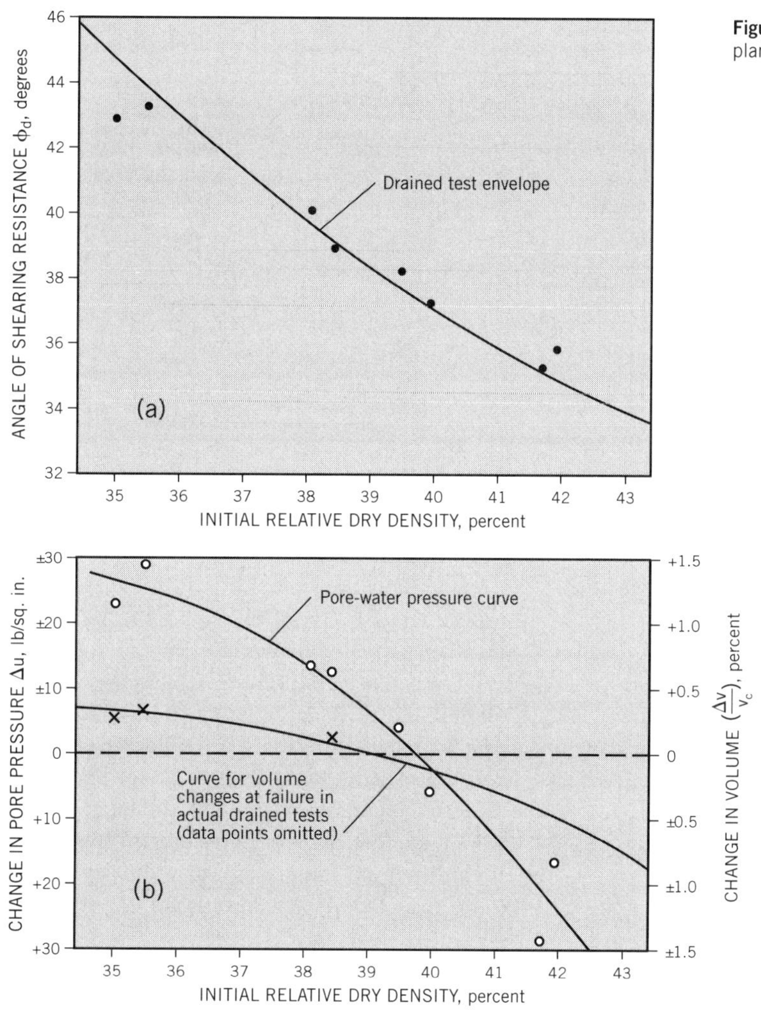

Figure 7.22 Comparison of consolidated-undrained plane strain tests on Brasted Sand with drained tests.

KEY

- ● Strength of consolidated-undrained tests at maximum principal stress ratio σ_1' / σ_3'

KEY

- o Pore-water pressures in consolidated-undrained tests at maximum principal stress ratio σ_1' / σ_3'
- × Denotes calculated volume change that would have occurred in a drained test (see text)

If the undrained test is run more quickly than a drained test, significant hydraulic gradients will occur within the specimen. To obtain the water needed for dilatant shear, the specimen has to develop higher negative pore pressures to produce more water from soil with a shorter drainage path to the failure plane. The result will be a higher shear strength. The only limitation will be the extent to which the pore pressure can go negative, which is governed by cavitation.

In terms of landslides, the construction of a fill at the top of a slope or the excavation of a cut at the base of a slope increase the shearing stresses within the slope. In dilatant soils, such as dense sand or stiff overconsolidated clay, the first effect is that a high strength will develop from the negative pore pressures induced in the dilatant soils. This soil strength will decrease as pore water migrates to the overstressed zone within the slope to loosen the soil there, and may cause failure if the reduced soil strength during the transition back to equilibrium groundwater conditions is insufficient to support the slope. How long it takes for a long term, steady state condition to be established depends on the rate of drainage. In clays, this transient condition can last for many years (see Delayed Failure Chapter 2); in fine sands, it may take hours or weeks.

To provide a design method that can reliably measure or predict the transient short-term strength changes of dilatant soils (in both sands and clays) is a remaining challenge in soil mechanics.

7.9 ACTIVE, PASSIVE, AND AT-REST EARTH PRESSURE COEFFICIENTS

For cohesionless soils, the active and passive pressures acting on a vertical wall with a horizontal backfill surface are shown on Figure 7.23. The passive coefficient K_p is the reciprocal of the active coefficient K_A. For a rigid, nonyielding vertical face AB, construction of successive horizontal layers of backfill can produce at-rest K_0 conditions in the underlying soils (see Chapter 19). Similarly, other types of consolidation with complete lateral restraint develop K_0-conditions at depth, such as the natural depositing of cohesionless soils by streams.

To develop reliable data on K_0 from laboratory experiments, it is essential to use large diameter (at least 4 inches) test specimens and very careful set-up procedures to measure the relative density of the soil. The Bishop lateral strain indicator

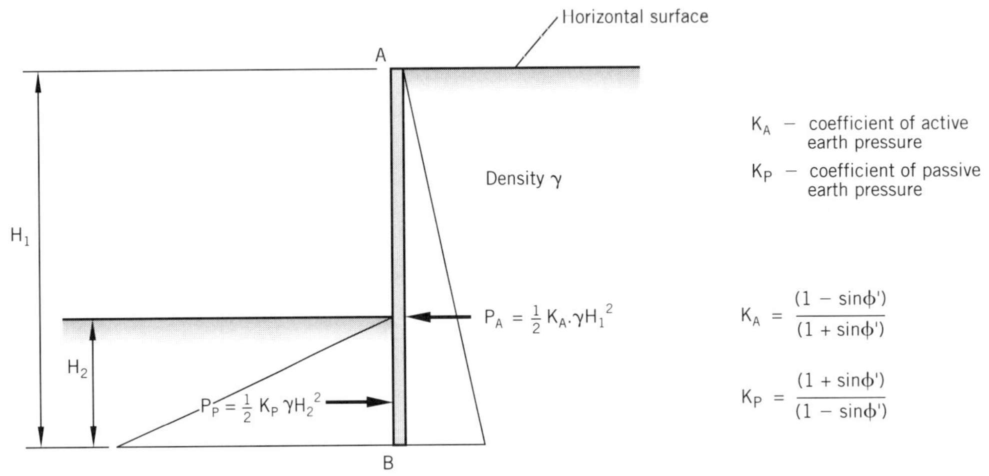

Figure 7.23 Active and passive pressures acting on a vertical retaining wall with a horizontal backfill surface.

(Bishop and Henkel, 1957) maintains the constant specimen diameter in a K_0 triaxial test.

Returning again to the laboratory experimental data on Brasted Sand, as discussed previously in this chapter, Fraser (1957) obtained the K_0 vs. density relationship shown on Figure 7.24(a). Since K_0 tests are generally not available in commercial soils laboratories, geotechnical designers have to make estimates. Two commonly used relationships are:

$K_0 \approx (1 - \sin\phi')$, based on theoretical work by Jaky (1948) Eq. (13)

$K_0 \approx 1.5\, K_A$, empirical Eq. (14)

Both equations provide reasonable approximations to the experimental K_0 range shown on Figure 7.24. The Jaky relationship is the slightly better fit.

The active earth pressures corresponding to the level backfill of Figure 7.23 are shown on Figure 7.24(b). For triaxial test data, K_A ranges from about 0.20 in dense sand to about 0.30 for very loose sand. The values of K_A in plane strain are lower, reflecting higher shear strengths, and range from around 0.16 to 0.30. Therefore, the active pressures on compacted cohesionless backfill behind long retaining walls are likely to be slightly lower than are calculated from triaxial test strengths.

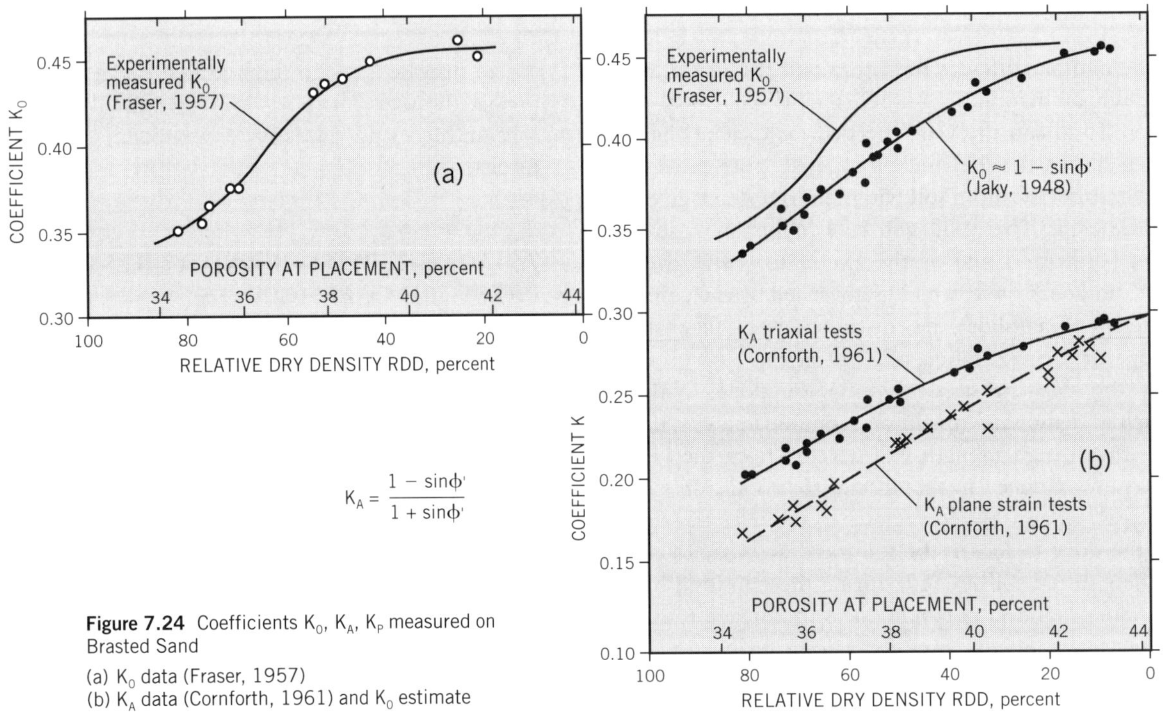

Figure 7.24 Coefficients K_0, K_A, K_P measured on Brasted Sand

(a) K_0 data (Fraser, 1957)

(b) K_A data (Cornforth, 1961) and K_0 estimate

7.10 FIELD BEHAVIOR OF SANDS AND OTHER COHESIONLESS SOILS

Stability

The most critical condition for stability of cohesionless soils is at or near the edge of a slope. It is a common practice to build temporary roads, or roads across steep sidehills, by spreading rockfill using a dozer. If the outer rockfill slope is disturbed in any way, the rock fragments ravel because the outer slope has been constructed at the angle of repose. However, traffic can move over the road with little risk; the interior of the fill has a higher stability than the edge.

The differences in stability are demonstrated by the slope stability analyses shown on Figure 7.25. The slope is 60 feet high, stands at 35° to the horizontal, and has shear strength parameters $c' = 0$, $\phi' = 35°$. For an infinite slope analysis, Chapter 9, Section 9.6, the factor of safety of the outer slope $F = 1.00$. By studying potential slip surfaces at various distances behind the crest of the slope, and passing through the toe, the effect of slip surface location on stability can be evaluated. If there is an external body of water, such as a river, sea, or lake, it is assumed that the free-draining soil in the slope has a groundwater table at the same elevation as the external water surface; i.e., height H_w above the toe of the slope.

The graphs on Figure 7.25(b) show the stability with either no external body of water or a fully submerged condition (water surface at or above the top of the slope). The shallow arc is slightly less stable than the wedge and it can be seen that the upper part of this potential failure is approximately parallel to the outer slope. The deep arc always has a higher factor of safety than the other two slip surfaces for the same horizontal distance behind the crest.

When an external body of water is present, the results shown on Figure 7.25(c) are obtained. To minimize confusion, only the data for the shallow arc (SA) are plotted on this graph. With the water surface part way up the slope, the stability falls below the graph for the dry ($H_w/H_s = 0$) and fully submerged ($H_w/H_s = 1$) cases. The lowest factors of safety occur when the external water is 50 to 75% above the base of the slope ($H_w/H_s = 0.5$ and 0.75). These two curves are practically coincident and have not been drawn separately; the most critical level may occur between these two levels. It is interesting that the factor of safety F reaches 1.00 when the top of the shallow arc is 6 feet from the crest. These data show that a free-draining cohesionless soil will be stable at an angle *slightly less* than the angle of repose when the slope lies within a tidal range or some other fluctuating water regime outside the slope. If the groundwater lags behind a falling tide, river after flood, or rapid drawdown of a reservoir, the stability is adversely impacted to an even greater extent.

Sand

When exposed by a cut slope, damp sand develops negative pore pressures, creating a condition of *apparent cohesion*. It allows sand to remain stable in steep or vertical cuts for

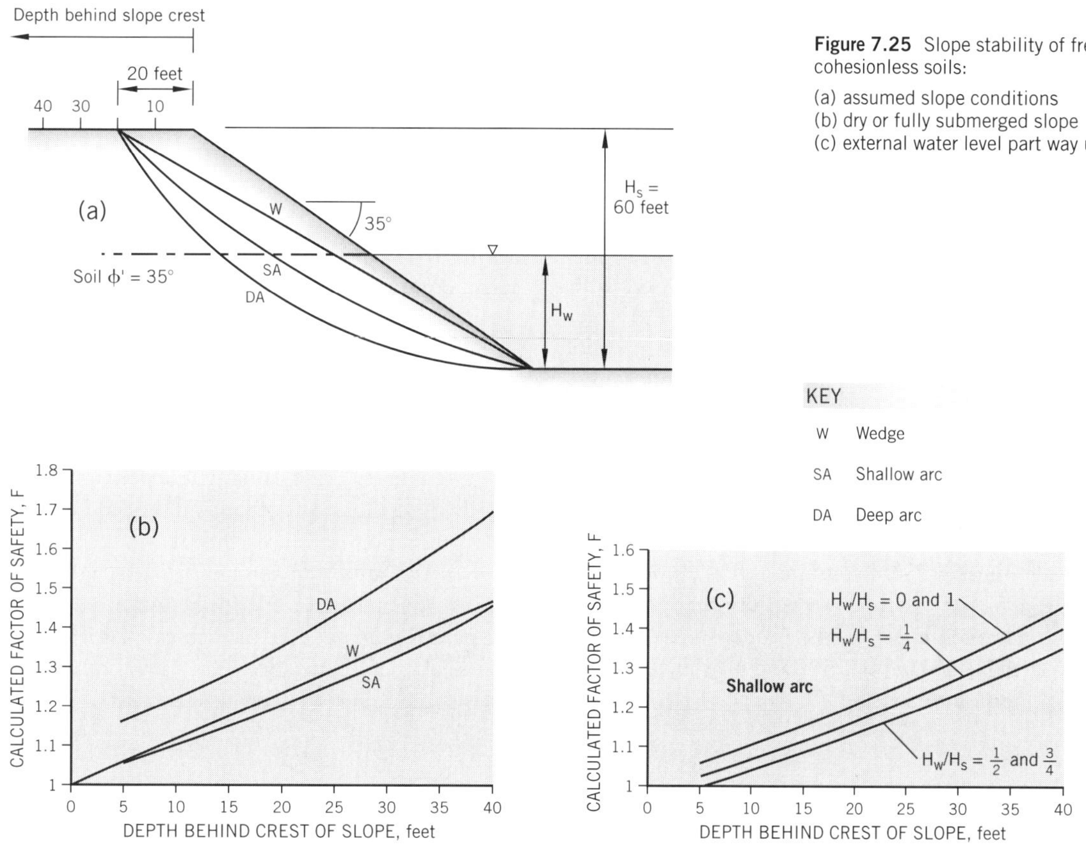

Figure 7.25 Slope stability of free-draining cohesionless soils:
(a) assumed slope conditions
(b) dry or fully submerged slope
(c) external water level part way up the slope

extended periods of time. Very compact fine sand in a damp condition can be left for weeks in a test pit with no sloughing. Should the sand dry out or the pit become waterlogged, local slope failure will occur.

Extremely loose, damp sand containing entrained air is referred to as *bulked sand*. It can remain stable for short periods but, if it is at a lower density than the minimum intergranular density, it will collapse when flooded by water. Water settlement is a popular way for U.S. contractors to infill trenches. Damp sand backfill is loosely thrown into the trench and is subsequently flooded with water from a hosepipe. The soil in the trench settles. Assuming that the sand backfill becomes fully saturated, it will remain in a very loose condition, close to zero relative dry density. This loose condition, plus any air voids remaining in the backfill, will cause longer-term ground settlement after construction. It is also a potential source of slope failure if a trench is dug through the same piece of ground later. Trench backfill should be performed in thin (6-inch loose measure) lifts and provided with at least 95 percent of standard (Proctor) compaction.

Loose, air-entrained sand fill caused flow slides around the edges of a sand fill at the Port of Portland, Oregon. The sand was pushed over the edges of a hydraulic fill to make up the specified widths of the filled area. In a loose bulked condition, it flowed out when undermined by dredging for dock construction (see Chapter 2, Section 2.12).

Stability of Trenches

Loose sand is a potentially dangerous material in the sides of trenches. Other cohesionless soils including gravel, rocks and construction rubble also can collapse into an open trench and cause death or injuries. If there is any concern about stability, the side slopes should be fully supported by a trench shield, sheetpiles, or strutted timbers. A short-term alternative is to provide the trench with side slopes no steeper than 1:1. Such slopes are flatter than the Rankine wedge failure slope and should avoid a rapid collapse, provided there is no seepage to undermine the trench sides. However, the long-term stability of a cohesionless soil slope is the angle of repose, typically requiring a relatively flat slope of about 1¾ horizontal : 1 vertical. Construction personnel should not enter a trench or pit that is deeper than 4 feet unless the sides are fully supported or laid back to a safe slope.

A groundwater table above the bottom of a trench is hazardous in sands and other cohesionless soils. In such cases, the excavation should be performed either completely under water or the area around the trench should be dewatered by deep wells, wellpoints, etc. Sump pumping should not be permitted because it simply removes water from the base of the trench but has a limited effect on the groundwater pressures acting on the trench sides and bottom, and may allow a heave or quicksand condition to develop.

Shot Rockfill

Many landslide remediation projects use shot rockfill as a construction material. Although it is relatively expensive to procure at most sites, shot rockfill has a high strength and excellent drainage characteristics. What is sometimes overlooked by practitioners is that shot rockfill has a low density because the voids ratio is high (due to the angular rock fragments) and there is very little water within the voids. A second issue is that the strength decreases for high rockfills due to breakage at the point-to-point rock contacts. This is not a problem on most landslide projects but the estimated rockfill strength used in stability calculations should be progressively reduced as the embankment height increases.

Rockfills are more compressible than most other fills. If a pipe or conduit crosses from recently constructed rockfill to undisturbed native ground, the possibility of a break occurring due to differential settlements needs to be evaluated.

The strength and drainage properties of rockfill can be adversely affected by fines. Clays and other fines can be intermixed by poor quarry practices, as discussed earlier in the chapter.

Aging of Sand

When a sand deposit stays in place for many years, the penetration resistance of the sand gradually increases without any significant change in the density. This phenomenon, termed *aging*, is thought to result from secondary compression (e.g., Terzaghi et al., 1996). The grains apparently move into a more stable arrangement with the surface roughness becoming more engaged. The aging effect has been borne out by various studies of sand deposits during large earthquakes where recent fills have been more prone to liquefaction than older deposits of similar relative density.

Laboratory tests by Daramola (1980) showed that sand specimens in the triaxial test became much stiffer over time. His tests on river sand (D_{50} = 0.3 mm) showed that the secant modulus increased by 60% after 30 days and 100% after 150 days compared to freshly prepared specimens. All tests were conducted at a relative density of about 67%. The drained shear strength was essentially unchanged. The change in deformation properties is attributed to improved bonding of the grains with time.

In terms of the field observations of penetration resistance, the initial SPT profile of a loose sand stratum may reflect aging that, in the case of natural deposits, can be measured in hundreds or thousands of years. If the sand density is increased (as a precaution against earthquake-induced liquefaction) the after-densification test results have zero aging and often produce a similar SPT profile to the untreated sand even though the sand is much denser. In such cases, it is advisable to wait a few weeks before measuring the "after" treatment SPT data to allow some aging to occur.

CHAPTER

8

Properties of Clays and Cohesive Soils

This chapter is mostly devoted to a detailed discussion of the shear strength of clay. It includes the recommended procedures for measuring and selecting the strength parameters for short-term and long-term stability of clay slopes for normally consolidated and overconsolidated clays. In the case of overconsolidated clays, the distinction is drawn between first-time sliding and reactivation of a preexisting landslide condition.

The chapter also includes the introduction of a new classification system for silts and clays—the *Cohesive Index* CI. This easy-to-use classification method relates directly to the visual description of silts and clays. A CI of 0 is a cohesionless silt and a CI of 1.0 or above is clay. Thus, a value between these limits instantly conveys the cohesiveness of the soil; for example, a CI of 0.2 indicates a weakly cohesive silt. The CI values can be correlated with certain soil properties, such as the coefficients of permeability (k) and consolidation (c_v).

The properties of clay under earthquake loading is not discussed in this chapter. Chapter 12 describes liquefaction (in silts) and silt-clay deformations during major earthquakes.

8.1 DESCRIPTION AND CLASSIFICATION OF SILTS AND CLAYS

Gradation Analysis

Soils that pass through the No. 200 U.S. sieve size are silts and/or clays. The individual grains cannot be discerned by eye except for coarse silt. In a gradation analysis, the usual test is to pretreat the soil with a dispersing agent (e.g., sodium hexametaphosphate) and perform a hydrometer analysis (sedimentation test) on the treated soil. Other pretreatments can be performed on soils containing organics or calcareous constituents. Details of the test procedures are described in ASTM D 422, BS 1377:2:1990, and Head (1992).

The silt/clay grain sizes are measured by a sedimentation test. A key issue with the sedimentation test is that the results are based on Stoke's Law for spheres falling freely through a fluid. Clay particles are of various shapes, including flakes that are a totally different shape from spheres. The test itself, carried out in a 1,000 ml glass cylinder with a bulbous hydrometer being raised and lowered within the soil during the test, interferes with the sedimentation process. This is a seriously flawed test based on assumptions that are at odds with the shapes of clay particles.

Despite these significant limitations, the hydrometer analysis is widely used. Based on the sedimentation graph, the clay fraction is determined as the percent by weight finer than 2 microns (0.002 mm) in British practice and 5 microns (0.005 mm) in U.S. practice. British practice sets the upper limit of silt at 0.06 mm compared to the U.S. limit of 0.074 mm. These differences are relatively minor.

If the soil is not properly dispersed before the hydrometer analysis, the particles will aggregate and a lower value of clay fraction will be measured. Some soils, notably the clay mineral kaolinite, remain in solution and do not sediment easily. This clay mineral registers a very high clay fraction yet exhibits the properties of a silt.

Plasticity Tests

The best method of describing and classifying silts and clays is through their *plasticity*. Silty soils have no or low plasticity and clayey soils have medium to high plasticity. The range of water contents over which the soil exhibits plastic behavior is quantified by the plasticity index (PI), the difference between the liquid limit (LL) and plastic limit (PL).

- *Liquid limit* is the water content at which the soil changes from a plastic material to a fluid.
- *Plastic limit* is the water content at which the soil becomes too dry to remain in a plastic condition.

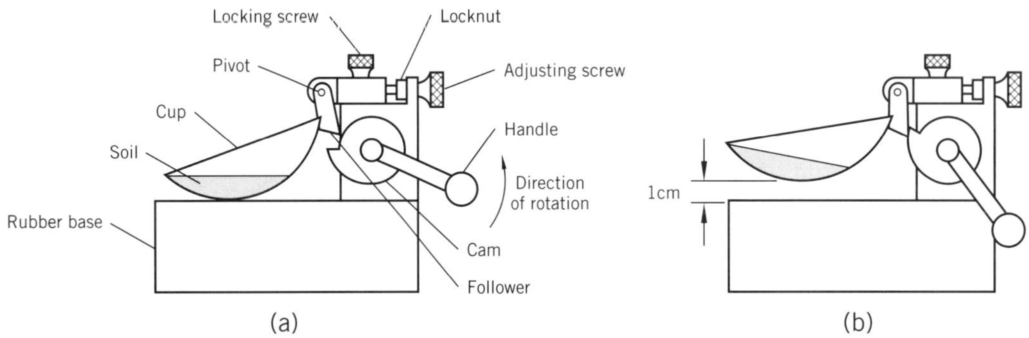

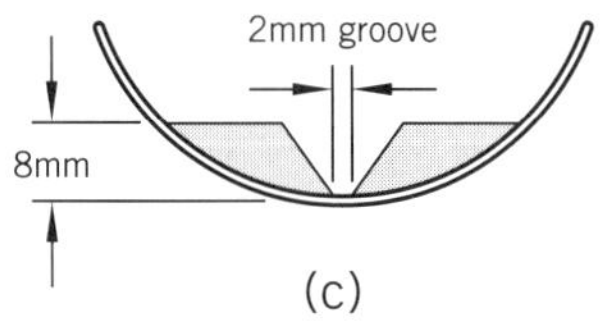

Figure 8.1 Casagrande apparatus for measuring liquid limit:
(a) cup on base (after Head, 1992)
(b) cup raised to the maximum height of 1 cm (after Head, 1992, 1992)
(c) cross-section of soil pat prior to test

The tests to measure these limits are described in detail in ASTM D 4318-95 (U.S. practice) and BS 1377: Part 2: 1990 (British practice).

Briefly, the test methods are as follows:

Liquid Limit

Very wet soil is placed in a special cup (Figure 8.1) and a groove is cut through the soil with a special tool. The cup is raised 1 cm by winding a cam (Figure 8.1b). This causes the cup to rise and freely fall onto a hard rubber base. The number of falls (blows) needed to close the groove in the clay over a length of 0.5 inch (13 mm) is recorded. The test is repeated at several water contents and the results are graphed as water content vs. log N (blows) to determine the water content corresponding to 25 blows. This water content is the liquid limit. (*Note:* Current British practice uses a cone penetrometer test to measure liquid limit.)

Plastic Limit

The soil is air-dried to a relatively low water content and rolled into a thread by hand (using a glass plate below the soil). The soil is kneaded and rolled out several times until the thread of soil starts to crumble at a diameter of ⅛ inch (3 mm). The test is repeated on another piece of soil.

Both tests need to be performed with great care to obtain consistent results. The plastic limit is the more difficult test to perform accurately. It should be noted that the liquid and plastic limits are carried out on *remolded* soil. In nature, clays can exist at water contents above the liquid limit and not be fluid. However, they can become fluid when disturbed and result in a flow slide. Similarly, many very stiff clays have water contents below the plastic limit. The plasticity index is the water content range for plastic behavior in remolded soil.

ASTM Classification

Classification of silts and clays is specified in ASTM D 2487-93 and D 2488-93 and has been developed from the Unified

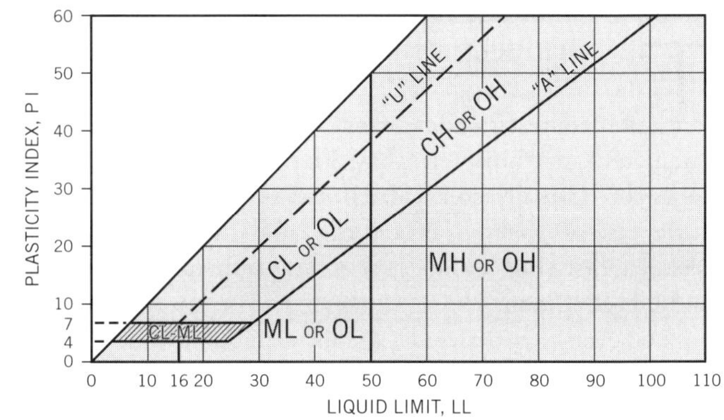

Figure 8.2 Casagrande classification chart for fine-grained soils (ASTM D 2487).

Table 8.1 Classification of Fine-Grained Soils: ASTM

(50% or more by weight passing No. 200 sieve)

	A-line	LL	PI
Clay (C)	Above		
Silt (M)	Below		
Lean clay (CL)	Above	<50	
Fat clay (CH)	Above	>50	
Silty clay (CL-ML)	Above		4–7
Inorganic silt	Below		<4
Silt (ML)	Below	<50	
Elastic silt (MH)	Below	>50	
Organic	[If LL after oven drying <75% LL before oven drying]		

If 15–30% by weight is retained on the No. 200 sieve, add with sand or with gravel (whichever is predominant) to the group name; e.g., silt with gravel, ML.

If 30% or more by weight is retained on the No. 200 sieve, add sandy or gravelly (whichever is predominant) to the group name; e.g., sandy lean clay, CL.

Table 8.2 Descriptions of Various Characteristics of Fine-Grained Soils*

Soil Characteristic	Description	Criteria
Moisture condition	Dry	Absence of moisture, dusty, dry to the touch.
	Moist	Damp but no visible water.
	Wet	Visible free water, usually soil is below water table.
Reaction to acid (HCl)	None	No visible reaction.
	Weak	Some reaction, with bubbles forming slowly.
	Strong	Violent reaction, with bubbles forming immediately.
Consistency	Very soft	Thumb will penetrate soil more than 1 in. (25 mm).
	Soft	Thumb will penetrate soil about 1 in. (25 mm).
	Firm	Thumb will indent soil about 1/4 in. (6 mm).
	Hard	Thumb will not indent soil but readily indented with thumbnail.
	Very hard	Thumbnail will not indent soil.
Cementation	Weak	Crumbles or breaks with handling or little finger pressure.
	Moderate	Crumbles or breaks with considerable finger pressure.
	Strong	Will not crumble or break with finger pressure.
Structure	Stratified	Alternating layers of varying material or color with layers at least 6 mm thick; note thickness.
	Laminated	Alternating layers of of varying material or color with the layers less than 6 mm thick; note thickness.
	Fissured	Breaks along definite planes of fracture with little resistance to fracture.
	Slickensided	Fracture planes appear polished or glossy, sometimes striated.
	Blocky	Cohesive soil can be broken down into small angular lumps which resist further breakdown.
	Lensed	Inclusion of small pockets of different soils, such as small lenses of sand scattered through a mass of clay; note thickness.
Dry strength	Homogenous	Same color and appearance throughout.
	None	The dry specimen crumbles into powder with mere pressure of handling.
	Low	The dry specimen crumbles into powder with some finger pressure.
	Medium	The dry specimen breaks into pieces or crumbles with considerable finger pressure.
	High	The dry specimen cannot be broken with finger pressure. Specimen will break into pieces between thumb and a hard surface.
	Very high	The dry specimen cannot be broken between the thumb and a hard surface.
Dilatancy	None	No visible change in the specimen.
	Slow	Water appears slowly on the surface of the specimen during shaking and does not disappear or disappears slowly upon squeezing.
	Rapid	Water appears quickly on the surface of the specimen during shaking and disappears quickly upon squeezing.
Toughness	Low	Only slight pressure is required to roll the thread near the plastic limit. The thread and the lump are weak and soft.
	Medium	Medium pressure is required to roll the thread to near the plastic limit. The thread and the lump have medium stiffness.
	High	Considerable pressure is required to roll the thread to near the plastic limit. The thread and the lump have very high stiffness.

*from ASTM D 2488-93

Soil Classification System (USCS). The soil is classified as fine-grained if 50% or more by dry weight passes the No. 200 sieve.

The results of a set of Atterberg limits are plotted on the Casagrande classification chart, Figure 8.2. This chart is well known. Clays plot on or above the A line; silts plot below the A line. Clays and silts of low plasticity have liquid limits of less than 50%; high plasticity clays and silts have liquid limits higher than 50%. Table 8.1 summarizes these criteria.

Abbreviated soil classification symbols combine the main constituent (as obtained from the plasticity chart) with lower case prefixes and suffixes. Example: g (ML) sc is gravelly silt with sand and cobbles. The U-line on the Casagrande chart (Figure 8.2) represents an approximate upper bound for soils.

ASTM D 2488 provides numerous tables for describing various characteristics of fine-grained soils. They have been collected together in Table 8.2.

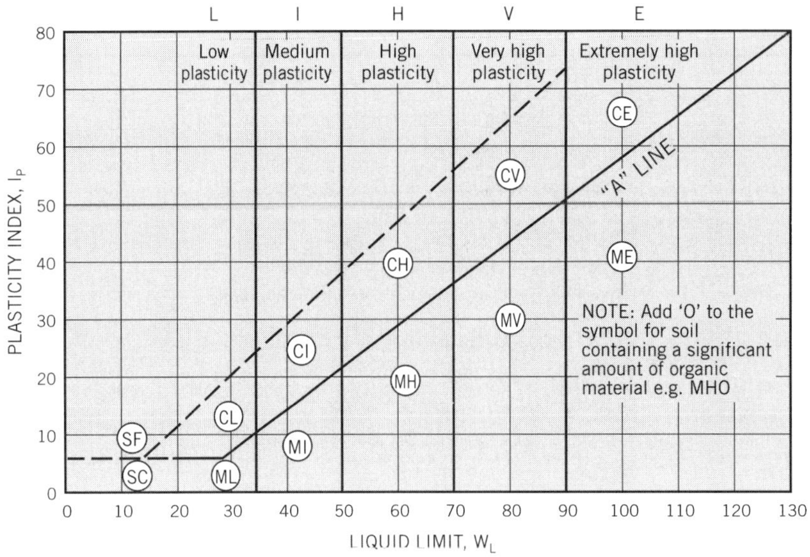

Figure 8.3 Plasticity chart used in British practice (after Head, 1992).

British Classification

In British practice, the Casagrande chart is further modified. The A-line is unchanged, but plasticity is separated for both silts and clays into several boundaries, depending on the liquid limit (Figure 8.3). Organic soils are noted by adding the symbol O after the main soil descriptor. Example: CVO—organic clay of very high plasticity. The following symbols are used for plasticity properties:

	BRITISH	U.S.
Liquid limit	W_L	LL
Plastic limit	W_P	PL
Plasticity index	I_P	PI

Note: U.S. plasticity symbols are used throughout this book.

8.2 SILT AND CLAY CLASSIFICATION USING COHESIVE INDEX

Cohesive Index

The author has developed an alternative method of describing and classifying silts and clays that has some advantages over the traditional classification systems previously described. The proposed method makes identification simpler and provides a parameter, Cohesive Index (CI), which can be used in practice as a means to describe the cohesive transition from silt to clay. It can also be used to estimate values of silt/clay permeability k and coefficient of consolidation c_v, as described later.

The *range* of plasticity in fine-grained soils is strongly influenced by the clay minerals or constituents. For example, if the soil contains organics, the organic content will affect both the plastic limit and liquid limit by increasing the water content of both. A second example is a soil containing the clay mineral halloysite (fairly common in volcanic rocks weathered to soil under certain ambient weathering conditions). Halloysite grains have water encapsulated in the soil grains and have relatively high water contents when oven-dried at 105° C because the water within the soil grain dries out. It is not, however, a highly plastic clay mineral.

These anomalies can be essentially neutralized by using a dimensionless parameter, Cohesive Index, defined by:

$$\text{Cohesive Index} \quad CI = \frac{PI}{PL}$$

Cohesive Index

The Cohesive Index concept has been in use by the author for about 25 years and appears to produce soil classifications that are consistent with field classification methods for silts and clays. The recommended procedure for visual examinations are given on Table 8.3.

The classification chart is shown on Figure 8.4. The plasticity index and plastic limit values are plotted on the chart to obtain the soil description corresponding to these parameters.

One of the advantages of this classification method is that cohesionless silt has CI = 0 and a fat clay (for which the adjective "silty" is dropped) is classified as clay and has CI $\geqslant$1.0. Fine-grained soils between silt and clay have CI values between 0 and 1, so the degree of "cohesiveness" can be quickly understood. For examples, soil with a CI of 0.2 is a weakly cohesive silt; soil with a CI of 0.8 is close to being a clay.

Knowing the CI value, a geotechnical practitioner can make an estimate of parameters k and c_v for the soil. Examples are provided later in this chapter.

Table 8.3 Cohesive Index and Visual Classification of Fine-Grained Soils

Cohesive Index (CI)	Soil Description	Observations during Visual Examination of Soil
0	SILT	• After rubbing between the fingers and thumb, washes off instantly without leaving stains. • Has feel of velvet. • Drier soil cannot be rolled into a thread without breaking up (toughness test). • No dry strength.
0–0.2	SILT, trace clay	• After rubbing soil between the fingers and thumb, washes off easily but leaves stain on fingers that has to be rubbed off. • A pat of moist soil on the first two fingers exhibits minimal but discernible cohesion when the fingers are separated.
0.2–0.4	slightly clayey SILT	• Free water rises quickly to the surface of a wet pat of soil when adjacent fingers are moved in a scissor action or the hand is shaken (dilatancy test). Surface of pat becomes dull when the free water is removed by the thumb. • Cohesive ball of moist soil crumbles when manipulated by the thumb and fingers. • Wet soil feels slightly greasy.
0.4–0.6	clayey SILT	• Free water rises to the surface of a wet pat of soil after fairly vigorous shaking or scissor-like (shearing) finger action (dilatancy test). • Under warm (20°C) conditions, soil dries out fairly quickly around the edge of a wet pat of soil, normally beginning within 1 minute of handling.
0.6–0.8	very silty CLAY	• A pat of wet soil does not produce free water at the surface, even with vigorous shaking (dilatancy test). • After shaking test, and putting dry thumb on surface, soil has a wet appearance with low sheen. • Under warm (20°C) conditions, some drying of the soil occurs around the edges of the wet soil during a more prolonged examination.
0.8–1.0	silty CLAY	• Cohesive ball of moist soil will not crumble on handling. • Soil has a glossy sheen. • Stiff soil requires moderate pressure and manipulation to soften with water.
> 1.0	CLAY	*Stiffer Soil:* • Considerable finger pressure needed to soften up soil with water. • Soil cut with difficulty by a knife and leaves a polished surface. • Dries to a hard, brick-like condition (dry strength test). *Softer Soil:* • Very greasy (soapy) feel and sticky when wetted up. • Washes off the hands with considerable difficulty. • Can be molded and rolled between the palms into 1/8 inch (3 mm) diameter threads several inches long without breaking (toughness test).

Comparison of Cohesive Index and USCS Classification Charts

Data from two landslide projects are plotted on Figure 8.5:

1. *Lake Pend Oreille, Idaho.* Lacustrine sediments, predominantly soft clay and clayey silt in a deep lake environment
2. *Skagway Harbor, Alaska.* Marine sediments of soft to very soft, sandy, slightly clayey silt in a fjord environment

In the Cohesive Index plot, Figure 8.5(a), the plotted data has good separation and classification is easily accomplished. On the ASTM Casagrande plasticity chart, Figure 8.5(b), the Pend Oreille sediments generally plot around the A line and fit into all four subcategories CL, CH, ML, and MH. The Skagway sediments cluster at the lower end of the A line. Several points for the Skagway soils plot as borderline clay. Any person examining these soils could not possibly classify any of these samples as clay, or borderline clay. They are weakly cohesive silt and are correctly classified by the author's chart (Figure 8.5a). In summary, the Cohesive Index classification system is easy to use and can be linked to the visual classification techniques of Table 8.3.

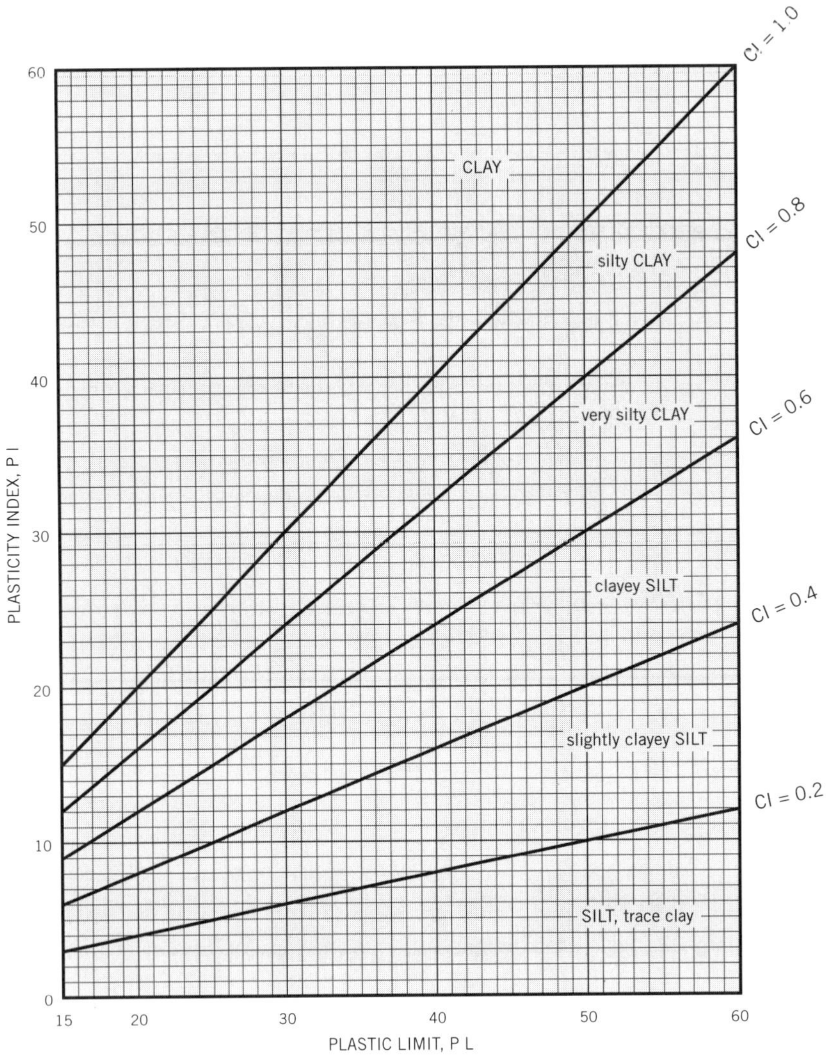

Figure 8.4 Cornforth classification of silts and clays based on Cohesive Index CI.

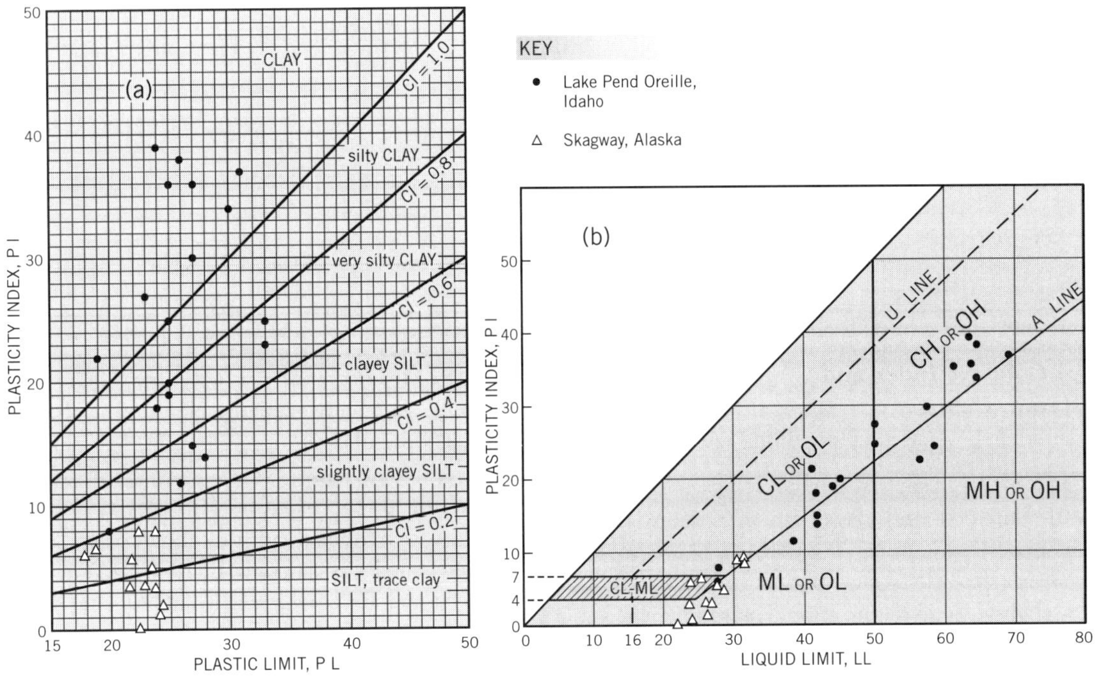

Figure 8.5 Example of Lake Pend Oreille lacustrine sediments and Skagway marine silts plotted on the Cornforth and Casagrande charts.

8.3 SILT AND CLAY CONSISTENCY

Clay

A good indication of the undrained shear strength of clay can be obtained from visual examination of undisturbed material. The U.S. practice is covered on Table 8.2 and Table 8.4. Note that there are some differences in the terms used to describe consistency between these two tables. Table 8.5 represents British practice. Practice in the United States uses lower shear strengths at each consistency level than in Britain (e.g., "soft" clay is 250–500 lb./sq. ft. in the U.S., and 400–800 lb./sq. ft. in Britain). Thus, care has to be exercised in reviewing clay descriptions from different countries. It should be noted that some published tables use unconfined compressive strength, which doubles the shear strength values shown in Tables 8.4 and 8.5.

For technical reports, it is recommended that the description of a fine-grained soil use consistency first (in capital letters) followed by the soil constituents, putting the main constituent in capitals.

> Example: "VERY SOFT, green-gray and brown-gray, silty CLAY, finely laminated; occasional fine sand and silt partings; bedding planes oriented at about 5 degrees to the horizontal."

This format allows the reader to quickly identify the principal soil type in the sample and the soil consistency.

Silt

The descriptions given to silt consistency are quite variable. The author suggests: very soft, soft, firm, very firm, hard. The word *stiff* is usually avoided for describing silty soils.

8.4 RATE OF CONSOLIDATION

When soft soil is first loaded by a fill, excess pore pressures are developed in the foundation. The rate at which the pore pressure dissipates back to preexisting levels depends on the rate of consolidation properties of the foundation.

In Terzaghi's theory of consolidation, the rate of consolidation in soil is controlled by the *coefficient of consolidation*, c_v.

Table 8.4: Identification of Clay Consistency (U.S. Practice*)

Consistency	Undrained shear strength c lb./sq. ft.	Handling Characteristics
Very soft	<250	See Table 8.2 for ASTM D 2488 tests for handling characteristics of consistency.
Soft	250–500	
Medium	500–1,000	
Stiff	1,000–2,000	
Very Stiff	2,000–4,000	
Hard	>4,000	

*Terzaghi, Peck, and Mesri (1996)

Table 8.5: Identification of Clay Consistency (British Practice*)

Consistency	Undrained shear strength c lb./sq. ft.	Handling Characteristics
Very soft	<400	Exudes between fingers when squeezed
Soft	400–800	Molded by light finger pressure
Soft to firm	800–1,000	
Firm	1,000–1,500	Molded by fairly strong finger pressure
Firm to stiff	1,500–2,000	
Stiff	2,000–3,000	Can be indented by thumb pressure, but cannot be molded by fingers
Very stiff or hard	>3,000	Can be indented by thumb nail

*Head (1992)

The relationship is:

$$T = \frac{c_v t}{H^2} \qquad \text{Eq. (1)}$$

where T = time factor
t = elapsed time after loading (assumed instantaneous)
H = drainage path

The drainage path is the maximum distance that pore water has to travel to reach a permeable boundary during the dissipation of excess pore pressures (Figure 8.6). A permeable boundary can be the ground surface (air or water interface) or

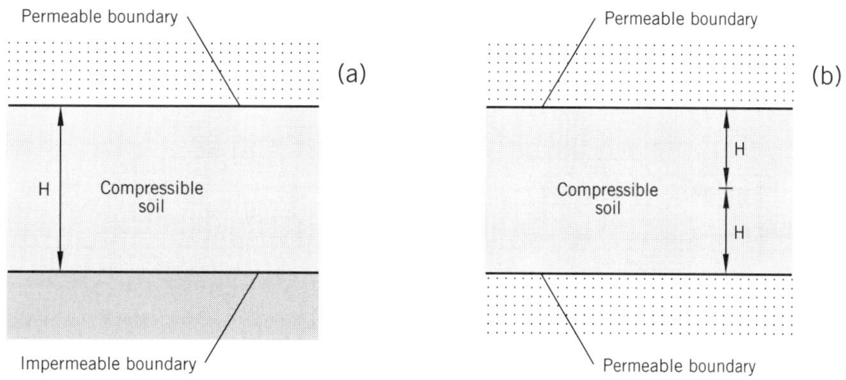

Figure 8.6
Drainage path H
(a) single drainage
(b) double drainage

a soil/rock boundary at least 100 times more permeable than the consolidating soil.

Eq. (1) can be rewritten:

$$t = \frac{TH^2}{c_v} \qquad \text{Eq. (2)}$$

Eq. (2) is the more useful expression and allows calculation of the time required to reach specified levels of consolidation.

Time Factor T

The time factor T depends on the type of loading. The most commonly used relationship is based on uniform loading. If U is the average degree of consolidation achieved in the compressible stratum under uniform surface loading, the relationship between U and T is given in Table 8.6.

Table 8.6 Relationship between U and T for Uniform Surface Loading

Average degree of consolidation, U percent	0	10	20	30	40	50
Time Factor T	0	0.008	0.031	0.071	0.126	0.196
Average degree of consolidation, U percent	60	70	80	90	95	100
Time Factor T	0.286	0.403	0.567	0.848	1.129	Infinity

The dissipation of pore pressures is always 100% at the drainage boundary. At the non-draining boundary, or at the center of a stratum with double drainage (Figure 8.6b), the soil is the slowest to drain. Using the values from Table 8.6, T is 0.196 when the average degree of consolidation (dissipation of pore pressures) is 50%. On Figure 8.7, which shows isochrones representing the degree of consolidation as a function of depth along the drainage path H, it can be seen from the graph for T = 0.20 that the degree of consolidation at the farthest point from the drainage boundary is only 23%

Table 8.7 Relationship between U_H and T for Uniform Surface Loading*

U_H	0	0.3	5.1	13.5	22.8	39.3	52.4	62.8
T	0	0.05	0.10	0.15	0.20	0.30	0.40	0.50
U_H	71.0	77.2	82.2	86.0	89.2	93.4	96.0	99.1
T	0.60	0.70	0.80	0.90	1.00	1.20	1.40	2.00

*after Bishop and Henkel, 1957

when the average for the whole stratum is 50%. Similarly, when the average degree of consolidation for the stratum reaches 90%, the degree of consolidation is about 84.8% (T = 0.848 curve on Figure 8.7) at the farthest point from the drainage boundary.

For settlement calculations, the average degree of consolidation is of most interest. For landslide work, the non-dissipated pore pressures at the farthest point from the drainage boundary is often more important than the average. The reason is that slip surfaces seek the weakest zone, which is often near bedrock. If H is the drainage path and U_H is the degree of consolidation at distance H from the drainage boundary, the relationship between U_H and T is given in Table 8.7. The relationships of Tables 8.6 and 8.7 are shown graphically on Figure 8.8.

Terzaghi's theory of consolidation assumes that the loading is applied instantaneously at time t = 0. This assumption is acceptable for most construction projects provided the time of interest is much larger than the construction time. If not, it is common practice to assume that t = 0 occurs at the middle of construction, assuming that the rate of construction is uniform. Thus, for a 3-month construction period of loading, time t = 0 is assumed to begin after 1½ months.

A more precise analytical approach is to use a construction time factor T_{vc} based on a linear build up of loading from t = 0 to t = t_c (Terzaghi, Peck and Mesri, 1996). The graphs are shown on Figure 8.9. If T_{vc} is close to zero (0.01 or less on

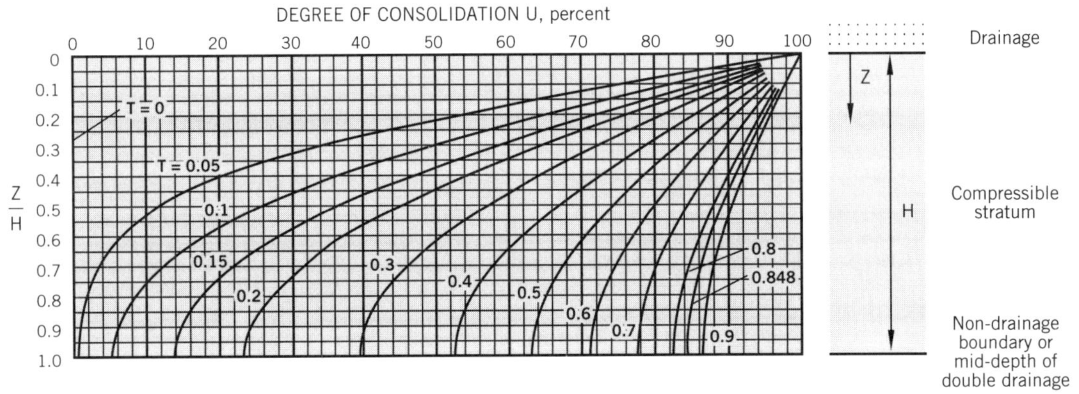

Figure 8.7 Degree of consolidation of a compressible half-stratum (modified from Lambe and Whitman, 1969).

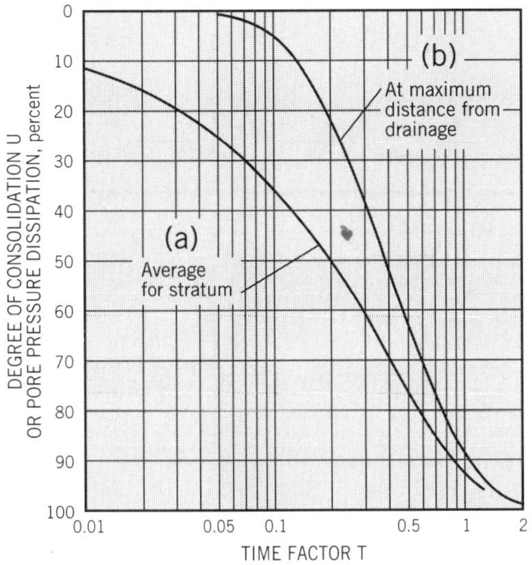

Figure 8.8 Degree of consolidation as a function of time factor T:
(a) average within compressible stratum
(b) at maximum distance from drainage layer

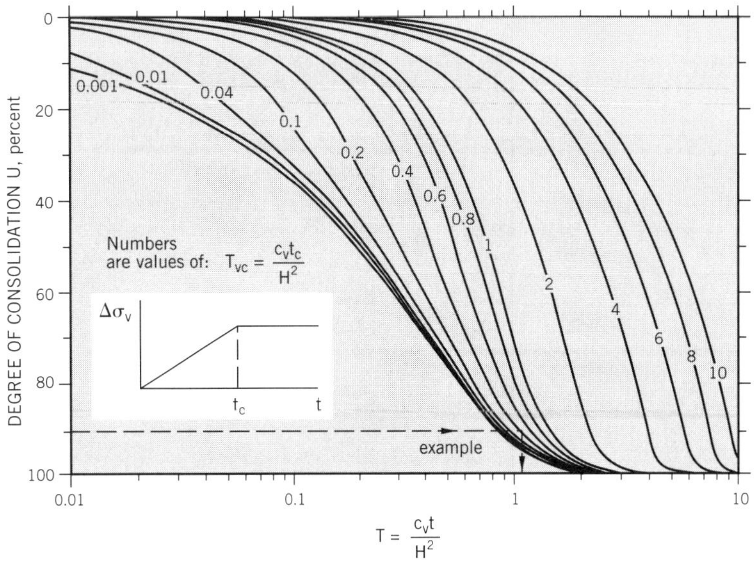

Figure 8.9 Degree of consolidation as a function of time factor T to account for a linear rate of loading (after Terzaghi, Peck, and Mesri, 1996).

Figure 8.9, the effect of loading is almost instantaneous and the curve is the same as curve (a) on Figure 8.8. The following example will illustrate the use of the graph.

t_c = 1 month = 0.083 year
H = 30 feet
c_v = 5,000 sq. ft./year
T_{vc} = 0.46

For 90% average consolidation and $T_{vc} = 0.46, T = 1.1$

t_{90} = 0.2 year = 2.4 months

The stratum will reach 90% average degree of consolidation 1.4 months after construction and 2.4 months from the start of construction.

Coefficient of Consolidation c_v

The coefficient of consolidation is a measure of the rate at which consolidation will occur within a soil stratum. It is the constant of proportionality in the theory of consolidation defined as:

$$c_v = \frac{k}{\gamma_w m} \qquad \text{Eq. (3)}$$

where k = coefficient of permeability

m = coefficient of compressibility $\left(= \frac{\Delta V}{V_0} \Big/ \Delta p'\right)$

γ_w = density of water

$\frac{\Delta V}{V_0}$ = volumetric strain

$\Delta p'$ = change in effective stress

In laboratory tests, the coefficient of consolidation generally decreases slightly with increasing effective stresses. The principal component affecting the value is the soil permeability k, which has a much greater range of variability than the coefficient of compressibility m.

In landslide studies, it is worthwhile to check the likely time for a stratum to consolidate under a fill load. An example might be a buttress that is placed above a clay stratum at the toe of a slope. If a slip surface passes partly (or completely) through the clay foundation below the buttress, it will develop a higher strength as foundation consolidation occurs.

The estimated time to reach an average of 90% primary consolidation in the foundation can be obtained from the chart, Figure 8.10. Thickness H is the drainage path, which is the full stratum thickness in single drainage and one-half of the stratum thickness for double drainage.

Example: H = 20 feet; c_v = 100 sq. ft./year.
From Figure 8.10, the t_{90} is 3.4 years.

If the requirement is to find t_{90ND} for the farthest point of the drainage path

t_{90ND} = 1.22 t_{90}
= 4.1 years

In performing an effective stress analysis of a landslide remediation, it may be necessary to calculate the excess pore pressures remaining in the foundation at the end-of-construction (especially if it involves a fill). In such calculations, the t_{90ND} may be the relevant input parameter for a potentially deep-seated failure surface.

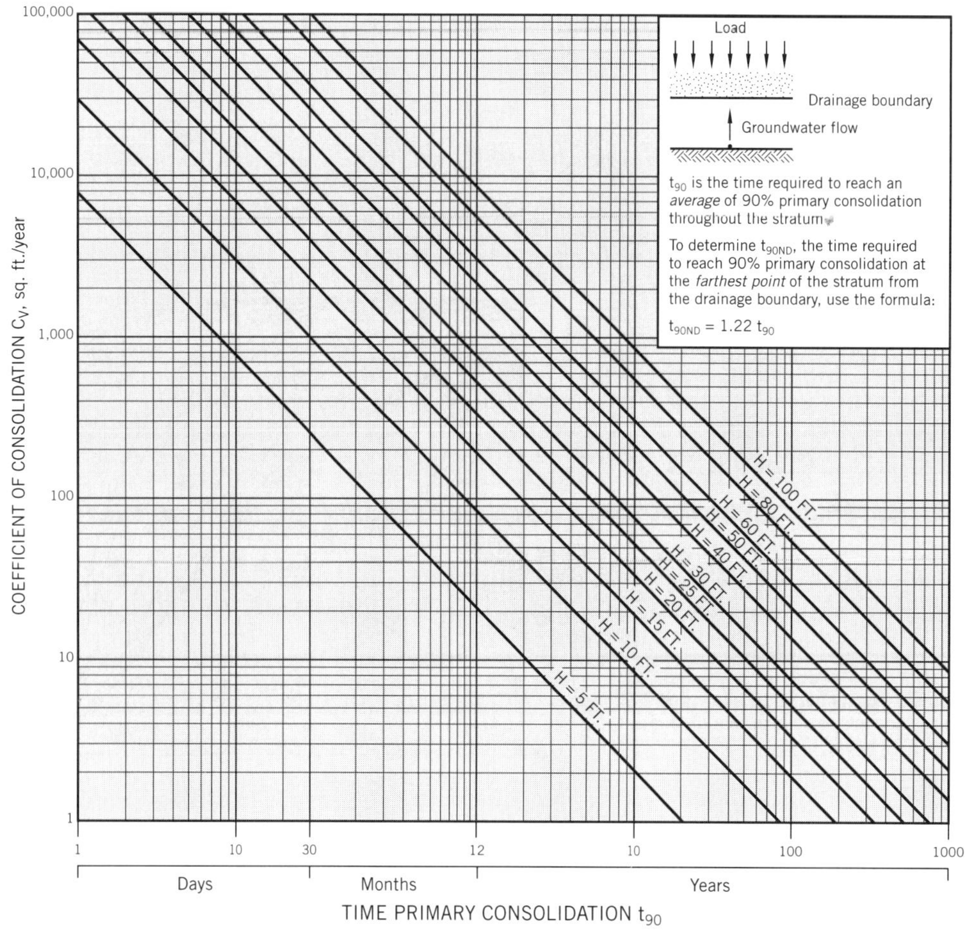

Figure 8.10 Graph of coefficient of consolidation c_v vs. time to 90% primary consolidation t_{90}.

The coefficient of consolidation has been correlated with cohesive index (described under Section 8.2) on the semi-log graph of Figure 8.11. Knowing the CI value of a soil, an estimate of c_v can be made. Remember that the rate of consolidation is considerably influenced by soil fabric (Rowe, 1972), especially near the ground surface. The parameter is affected to a lesser extent by stress level, tending to decrease with increasing effective stress. The relationship shown on Figure 8.11 should be used only as a general guide. On Figure 8.12, a similar graph has been prepared to correlate soil permeability with cohesive index CI.

A compressible stratum is often made up from sedimentary deposits with slightly different gradations, or has thin layers of different gradation (such as fine sand) within a more uniform soil unit. Since the principal component of the coefficient of consolidation c_v is the permeability k, small changes in gradation can provide significantly different values of c_v in laboratory tests. How can a *composite* value of c_v be determined for analysis?

One option is to test several soil samples and then assume that the results from each sample represent individual layers of equal thickness (Figure 8.13). It is also assumed that flow occurs normal to the bedding, with no lateral flow along the individual layers (i.e., flow is confined laterally). The hydraulics of flow (e.g., Das, 1994) is given by:

$$\frac{n}{k_c} = \frac{1}{k_1} + \frac{1}{k_2} + \frac{1}{k_3} + \ldots \frac{1}{k_n} \qquad \text{Eq. (4)}$$

where k_c = a composite value of permeability representing the behavior of the entire stratum

k_1, k_2, k_3, etc. = vertical permeabilities of the individual layers within the stratum

n = number of layers

Now $k = c_v\, m\, \gamma_w$, as defined previously in Eq. (3). γ_w is a constant and m has much less variation than k. Therefore, as a first approximation, the term c_v can be substituted for k in Eq. (4), as shown below:

$$\frac{n}{(c_v)_c} = \frac{1}{(c_v)_1} + \frac{1}{(c_v)_2} + \frac{1}{(c_v)_3} + \ldots \frac{1}{(c_v)_n} \qquad \text{Eq. (5)}$$

where $(c_v)_c$ = composite c_v for the entire stratum

$(c_v)_1, (c_v)_2, (c_v)_3, \ldots (c_v)_n$ = coefficients of consolidation measured in the individual layers.

n = number of layers

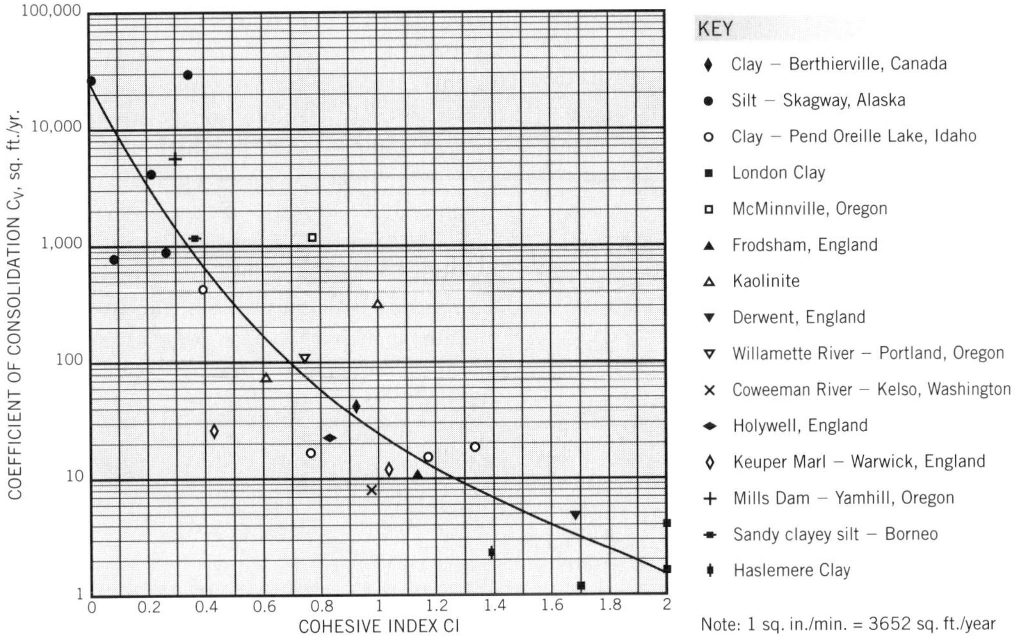

Figure 8.11 Graph of coefficient of consolidation c_v vs. cohesive index CI

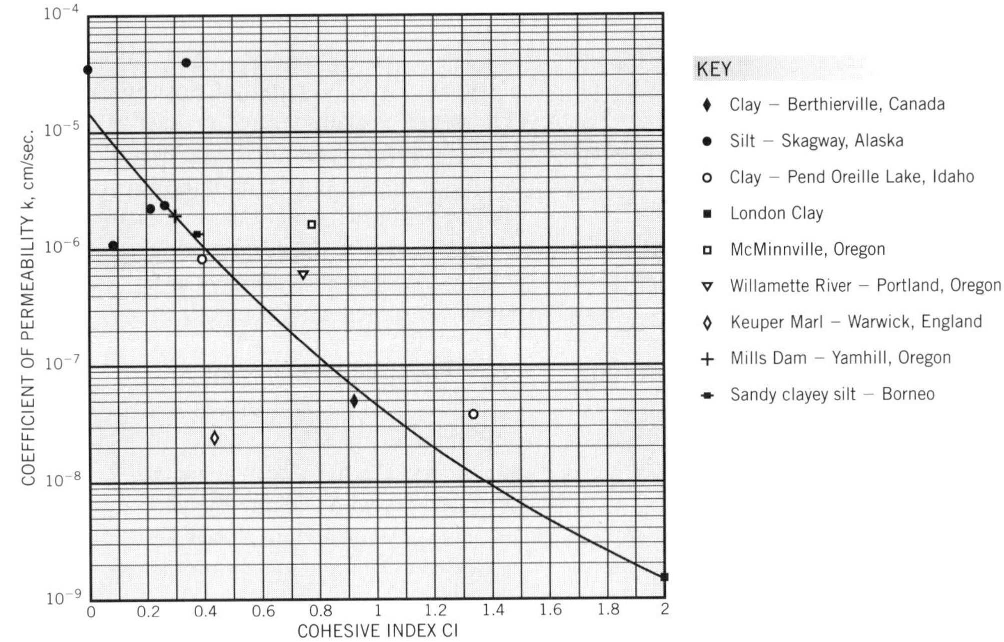

Figure 8.12 Graph of coefficient of permeability k vs. cohesive index CI.

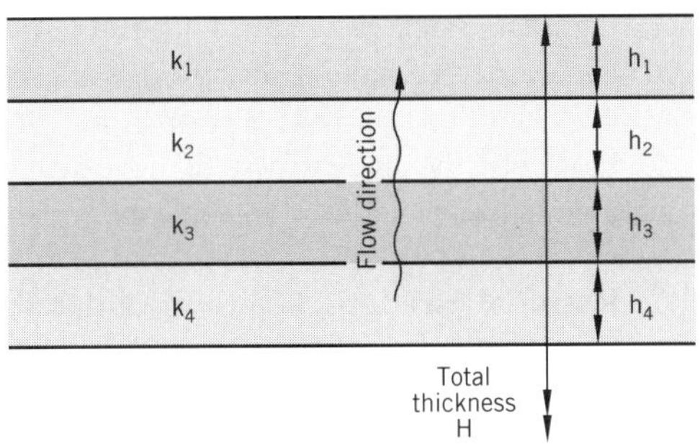

$$\frac{H}{k_c} = \frac{h_1}{k_1} + \frac{h_2}{k_2} + \frac{h_3}{k_3} + \ldots \frac{h_n}{k_n}$$

For layers of equal thickness h, H = nh

$$\frac{n}{k_c} = \frac{1}{k_1} + \frac{1}{k_2} + \frac{1}{k_3} + \ldots \frac{1}{k_n}$$

k_c = composite permeability

Figure 8.13 Determination of the composite coefficient of permeability for flow passing through numerous layers of different permeabilities.

At the Skagway flow slide (Case History 7), five samples were tested in the laboratory and gave the following average results: 790; 890; 4,130; 27,800; and 29,700 sq. ft./year. When analyzed by Eq. (5), the composite value of c_v for the entire stratum is calculated to be 1,850 sq. ft./year. The reason that the composite value is closer to the lower values of the individual measurements is that it takes much longer for water to pass through less permeable layers than more permeable layers. Therefore, in terms of time for water to pass through the entire stratum, the total time is mostly governed by the time taken to pass through less permeable soils.

8.5 NORMALLY CONSOLIDATED AND OVER-CONSOLIDATED CLAYS

Normally consolidated clay is clay that has never been subjected to a greater overburden pressure than currently exists on it. These clays commonly occur in lakes, river flood plains (below the lowest seasonal groundwater level), tailings ponds, etc. The water contents are usually close to the liquid limit.

Alluvial deposits usually have pockets, lenses, or thin strata of fine sand interspersed with the fine-grained silts and clays. In theory, sand strata act as drainage layers and will speed up the rate of consolidation. However, due to the nature of flood plain deposition, lateral continuity is likely to be poor, especially normal to the direction of river flow.

Lake deposits are also stratified but appear to be more uniform in gradation. Both river and lake deposits can have organics within the soils, either dispersed throughout the soil or as local lenses of peaty soils.

Overconsolidated clay is a clay that has been subjected to greater overburden pressures than presently exist. Overconsolidated clays occur from a variety of origins: (i) clay that was previously covered by a much thicker overburden that has since eroded; (ii) clay that has been overridden by glaciers; (iii) clay that has been subjected to desiccation hardening near the ground surface; and (iv) rock that has weathered to clay.

Where clay has lost overburden pressure due to erosion, the in-situ K_0 stresses can exceed 1. Relief of stress has probably produced the numerous slickensided shear surfaces seen throughout the London Clay stratum and similar overconsolidated clays in southern England. These slickensides should not be interpreted as evidence of current slippage.

Overconsolidated clays are mostly stiff to hard in consistency. The natural water content may be lower than the plastic limit in hard clays.

The optimum water content for (standard) compaction usually is close to the plastic limit (PL); therefore, heavily overconsolidated clays can be used for fills. In Britain, fills have been constructed from London Clay with water contents up to 1.2 x PL. At this upper limit of water content, the undrained shear strength was around 1,000 lb./sq. ft. Similar fill strength vs. water content relationships can be developed for other overconsolidated clays that are being considered for fills.

In California, clay fills are specified on the basis of the modified compaction test, which requires a higher compactive effort and has lower optimum water contents than the standard test. The clay fills are hard and may subsequently soften (swell) over many years causing vertical and lateral ground movements. This has produced lawsuits in residential developments built on clay fills due to ground movements. Therefore, a high degree of compaction can provide unintended consequences. Swelling clay fills have to be firm enough to provide an adequate foundation, but must avoid excessive swelling after construction. In this instance, the standard compaction test may be a better control than the modified compaction test.

8.6 LABORATORY DRAINED STRENGTH OF CLAYS AND SILTS

Since drained tests are effective stress tests, the shear strength parameters in this book are generally shown as c', ϕ' rather than c_d, ϕ_d, the latter sometimes being used to denote drained tests.

Peak Strength, Normally Consolidated

The results of a consolidated-drained triaxial compression test on a specimen of soft, slightly clayey silt is shown on Figure 8.14. The stress-strain curve of the normally consolidated lake silt rises to a peak at an axial strain of 16 percent. The volume change decreases, then levels off at failure without dilating. These characteristics are similar to sand at the minimum intergranular density (Chapter 7).

Peak Strength, Overconsolidated

A consolidated-drained triaxial test on a specimen of overconsolidated stiff clay is shown on Figure 8.15. The stress-strain curve reaches a peak at an axial strain of 8 percent and then declines with additional axial strain. The volume change initially decreases and then increases (dilates) as the specimen is sheared to failure. After failure, the rate of dilation decreases. These characteristics are similar to the behavior of sand with an initial density higher than the minimum intergranular density.

Stiff clays are difficult to sample and are "disturbed" by hammering a sampler into the ground and/or a sampler having a large end area ratio. This disturbance, and the scattered occurrence of natural slickensides, affects the test results and the circles usually have a poor fit to a straight line envelope. However, the line of best fit normally produces a cohesion intercept c' in addition to angle ϕ'.

Residual Strength, Overconsolidated

Consolidated-drained ring shear tests on specimens of stiff

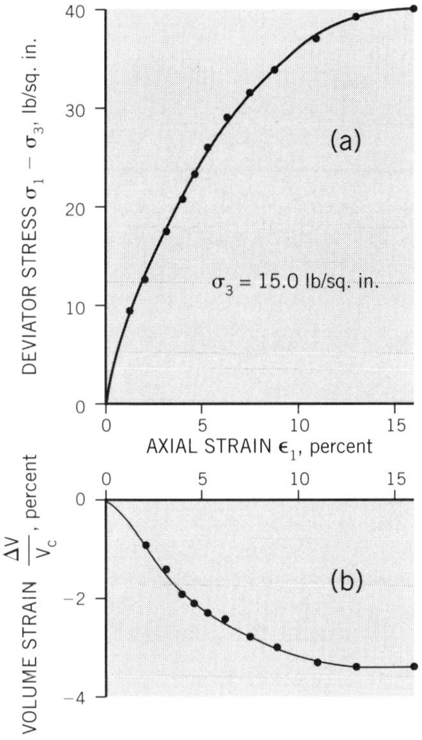

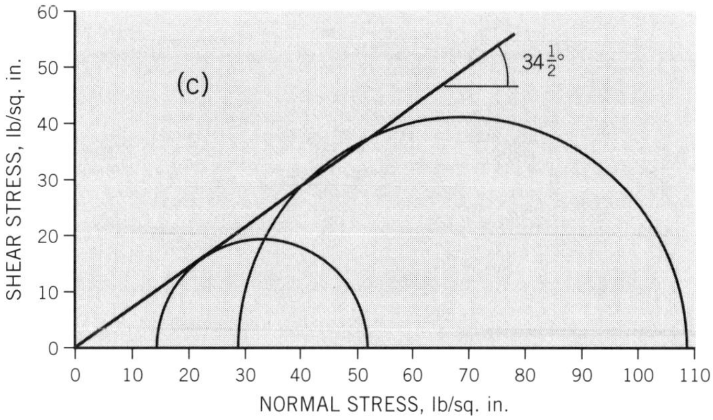

Figure 8.14 Consolidated-drained triaxial compression test on a normally consolidated clayey silt:

(a) stress–strain curve
(b) volume change curve
(c) Mohr envelope

Lake Pend Oreille lacustrine sediments

Properties:
LL 50, PL 36 w 49% CI 0.39
c_v = 420 sq. ft./year c = 660 lb/sq. ft.

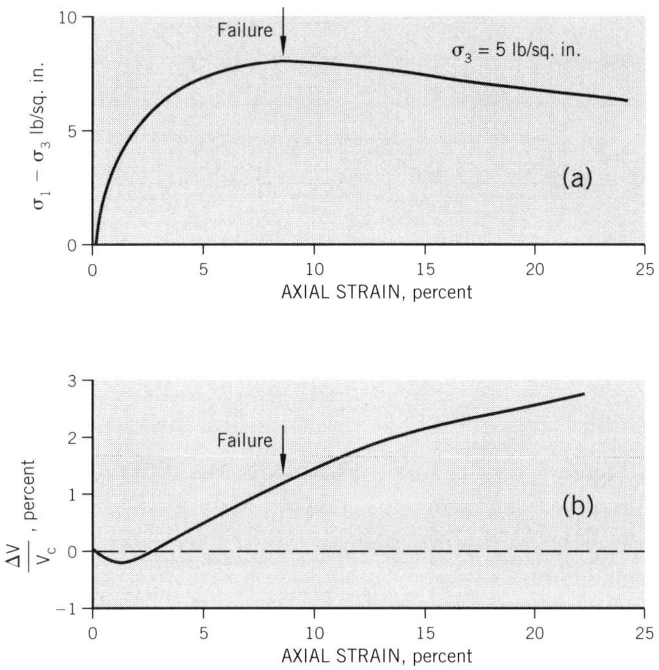

Figure 8.15 Consolidated-drained triaxial compression test on an overconsolidated clay:

(a) stress–strain curve
(b) volume change curve
(after Bishop and Henkel, 1957)

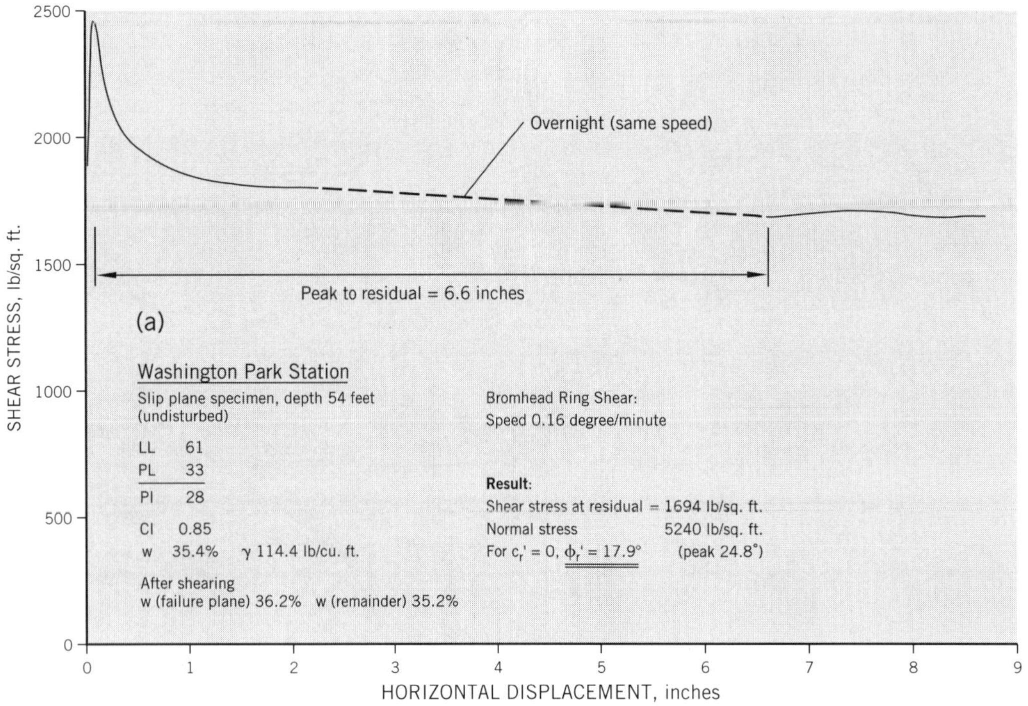

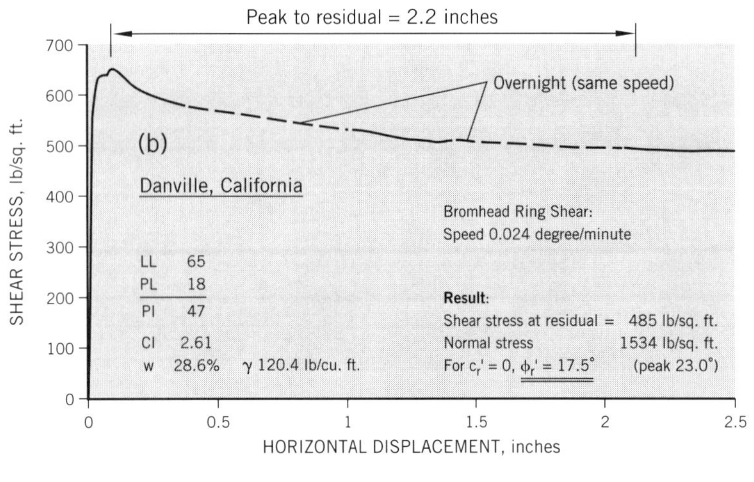

Figure 8.16 Two examples of consolidated-drained ring shear tests on stiff clay specimens:
(a) Washington Park Station, Portland, Oregon
(b) Danville, California

overconsolidated clay are shown on Figure 8.16. The ring shear test is performed to measure the residual strength of a clay at very large shear displacements. The stress-horizontal displacement curve has a peak value followed by a slow decrease to residual. In the two examples shown in Figure 8.16, the peak occurred at 0.1 inch of shear displacement.

For these two soils, it required 2 to 6 inches of relative displacement to reach the residual strength. Some experimenters continue the test longer, but the subsequent changes in shear stress are very small. The large relative displacements produce a shear plane with a glossy sheen (slickenside) due to the parallel orientation of the clay minerals in the plane of shear.

8.7 LABORATORY UNDRAINED STRENGTH OF CLAYS AND SILTS

There are two ways in which the undrained strength of clays and silts are measured: (i) at the existing density and (ii) consolidated to a different density, either to obtain an effective stress Mohr envelope (the more common reason) or to simulate the stress conditions acting on the clay in the ground under some future situation (such as a fill loading). The methods of measurement have been described in Chapter 6.

Undrained Strength at Existing Density

The undrained test on clay at the existing density measures the *immediate shear strength,* usually referred to as the cohesion

$\sigma_1 = \gamma H$

$\sigma_1' = \gamma H - \gamma_w h$

$K_o = \sigma_3'/\sigma_1'$

$\sigma_3' = K_o(\gamma H - \gamma_w h)$

$\sigma_3 = K_o(\gamma H - \gamma_w h) + \gamma_w h$

$\bar{\sigma} = \frac{1}{3}(\sigma_1 + \sigma_2 + \sigma_3)$

$= \frac{1}{3}(\gamma H + 2K_o(\gamma H - \gamma_w h) + 2\gamma_w h)$

$$\bar{\sigma} = \frac{1 + 2K_o}{3}(\gamma H - \gamma_w h) + \gamma_w h \qquad \text{(Eq. } i\text{)}$$

$$\left[\text{i.e. } \bar{\sigma} = \frac{1}{3}(\sigma_1' + \sigma_2' + \sigma_3') + \gamma_w h\right] \qquad \text{(Eq. } ii\text{)}$$

Use Eq. (*i*) or Eq. (*ii*) to calculate $\bar{\sigma}$

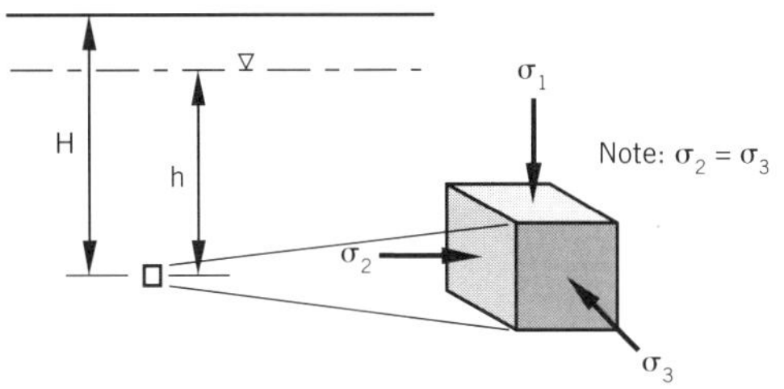

Example: Soft, normally consolidated clay

Assume $K_o = 0.6$ $\gamma = 125$ lb/cu. ft., H = 30 feet, h = 20 feet

$\sigma_1 = 3750$ lb/sq. ft. (26.0 psi)

$\bar{\sigma} = 0.733\,(3750 - 1248) + 1248$

$= 3083$ lb/sq. ft. (21.4 psi)

Figure 8.17 Calculation of the average total stress acting on an element of soil, assuming K_0.

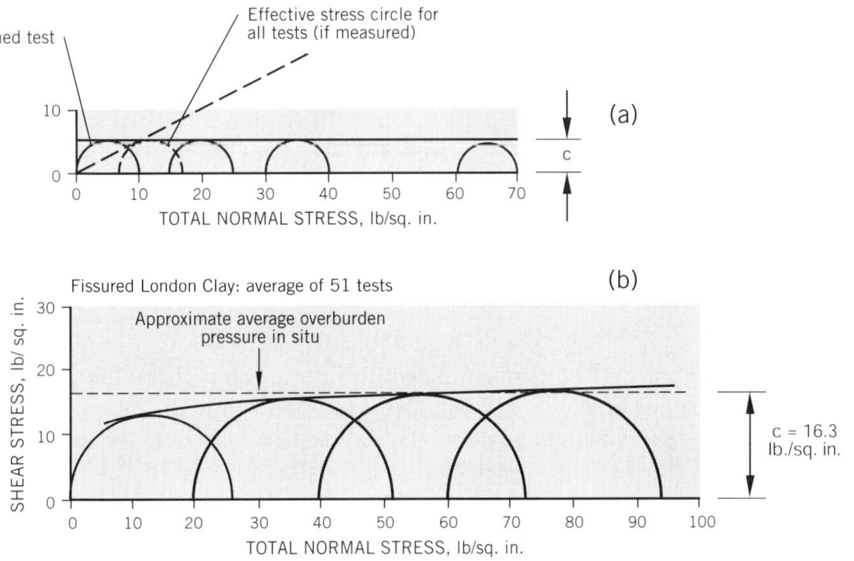

Figure 8.18 Mohr envelopes in total stresses for unconsolidated-undrained triaxial tests on saturated clays:

(a) soft, normally consolidated clay

(b) stiff, fissured, overconsolidated clay

(after Bishop and Henkel, 1962)

of the clay. The value of the cohesion determines the description of the clay consistency: very soft, soft, etc. (Tables 8.4, 8.5). The shear tests are performed without any consolidation, but there is usually a cell pressure equal to either the vertical total stress or the average total stress acting on the clay in the ground, determined from depth and density measurements, and an assumption of K_0 (see calculation on Figure 8.17).

If different cell pressures are used on several specimens from the same sample, the results shown on Figure 8.18(a) are obtained for soft soils. For stiff, overconsolidated clays, the shear strength slightly increases with increasing cell pressure (Bishop and Henkel, 1957) as shown on Figure 8.18(b).

Silts and weakly cohesive silts may cavitate during the undrained test in a similar manner to sands. This phenomenon has been discussed in some detail in Chapter 7, Section 7.8.

The undrained shear strength of very soft, soft, and firm clays develop positive pore pressures during a shear test. Therefore, they are weakest at the existing density and become stronger as the pore pressures dissipate and the clay consolidates and strengthens. Stability and landsliding is more likely to occur during or at the end of construction of man-made fills and cuts rather than longer term. An exception is in tidal areas (e.g., Case History 7) where extreme low water may decrease slope stability to a critical level shortly *after* construction before the soft soil has had sufficient time to increase in strength under the weight of a fill.

The slope stability of very soft to firm clays is studied by the $\phi = 0$ approach based on Figure 8.18. The clay does not have time to change effective stresses before failure occurs. It is a *total stress analysis.*

As the depth below ground increases, a progressive increase in immediate shear strength c can be expected with increasing overburden weight. If the cohesion c increases linearly with overburden effective stress p′, the value of c/p′ is approximately constant with depth. This is difficult to demonstrate in practice because most shear strength vs. depth profiles, whether obtained by laboratory measurements or in-situ vane tests, show significant scatter but generally have a *trend* of increasing shear strength with increasing depth.

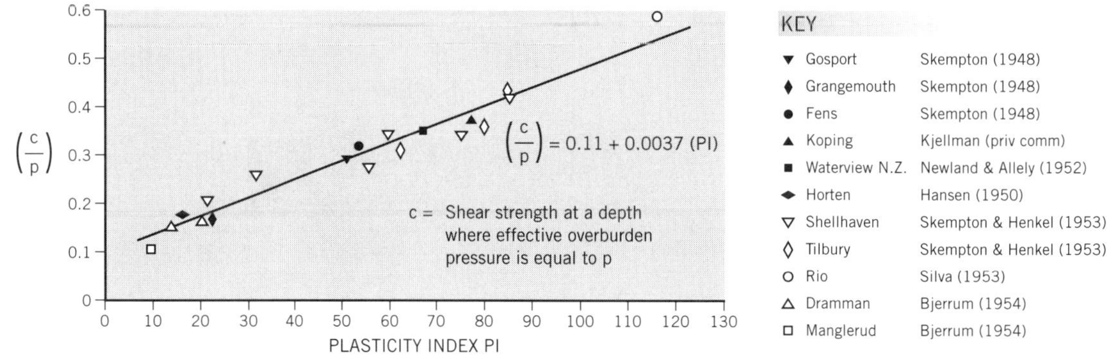

Figure 8.19 Relationship between c/p ratio and plasticity index PI for normally consolidated clay (after Skempton, 1957).

Skempton (1957) first showed an approximate relationship between c/p and plasticity index PI (Figure 8.19). A more recent compilation of vane shear strength data by Tavenas and Leroueil (1987) also plots c/p against the plasticity index (Figure 8.20). There is much scatter, and the relationship is of limited value.

Another property of normally consolidated clays of interest to landslide work is the *remolded* shear strength. If the clay is remolded at an existing natural water content near the liquid limit, it usually shows a very significant drop in undrained strength. This property can be measured in the laboratory by triaxial tests or torvane, and in the field by the vane test. Sensitivity (S) to remolding is defined by:

$$S = \frac{\text{Undisturbed shear strength}}{\text{Remolded shear strength}}$$

Sensitivities in normally consolidated clays are often in the range of 5 to 10, showing that they can lose 80 to 90% of their strength by remolding. Quick clays have even greater sensitivities. Thus, clays with natural water contents above the liquid limit can have reasonably good shear strengths in place but, once disturbed, rapidly lose strength and become a flow slide.

Undrained Strength at Changed Density

Triaxial tests that are performed consolidated-undrained with pore pressure measurements mirror the corresponding drained tests. Soft clays that are contractive (consolidate) during a drained test develop positive pore pressures during an undrained test. Stiff clays that have a peak in the stress–strain curve and dilate (soften, expand) on the failure plane during a drained test develop negative pore pressures at failure in an undrained test. The parameter A (Chapter 6, Section 6.3) is around +1.0 at failure in soft, normally consolidated clays and becomes negative as the overconsolidation ratio increases (Figure 8.21).

As previously stated in Chapter 6, the effective stress Mohr envelope for a normally consolidated soft clay almost always shows a very consistent linear relationship to the individual test circles and passes through the origin. The Mohr envelope for an overconsolidated clay typically has a cohesion intercept c′. The cohesion intercept is simply an intercept on the ordinate of the Mohr diagram and has no connection with cohesion (immediate shear strength) of clay. The cohesion intercept c′ is considered to be a result of the stress history of overconsolidation; as the clay rebounds from the higher stresses of its geological past, it retains some additional component of shear strength that exceeds that of a normally consolidated clay under the same stresses.

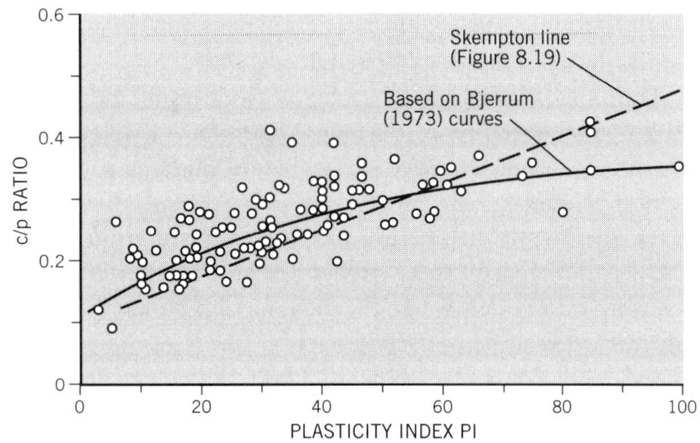

Figure 8.20 Relationship between c/p ratio and plasticity index PI for field vane tests on inorganic, normally consolidated clays (after Tavenas and Leroueil, 1996).

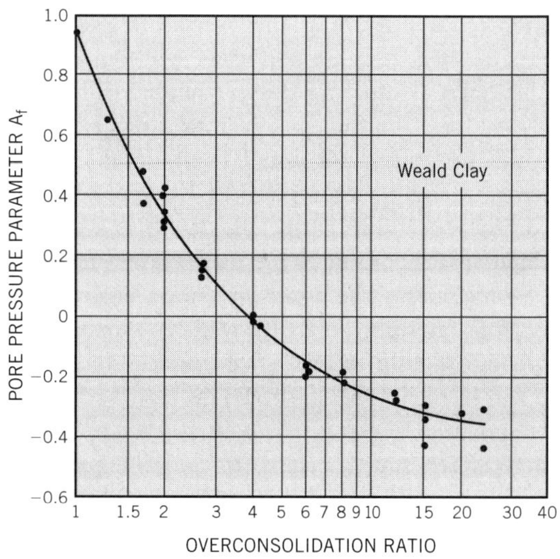

Figure 8.21 Pore pressure parameter A at failure for remolded Weald Clay (after Bishop and Henkel, 1957).

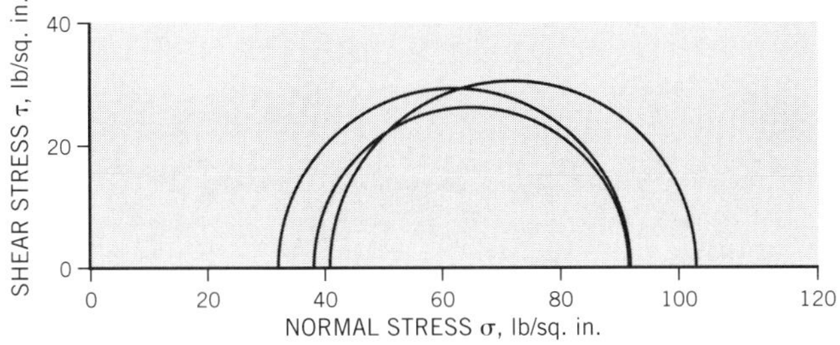

Figure 8.22 Effective stress Mohr circles for a set of three consolidated-undrained triaxial compression tests on a stiff clay. Despite the use of three significantly different effective consolidation stresses, the three effective stress circles overlap at failure.

Many geotechnical practitioners specify consolidated-undrained shear tests with pore pressure measurements rather than consolidated-drained shear tests in the mistaken belief that there is a considerable saving in the test time. As previously discussed for dilatant sands in Chapter 7, which will not be repeated in detail here, pore water in an undrained test migrates to the shear plane from other parts of the specimen in a dilatant (heavily overconsolidated) clay, and migrates away from the shear plane in a contractive (normally consolidated) clay. The pore water redistributes *internally* in the undrained test, and moves *externally* in the drained test. The drainage path is slightly shorter in the undrained test but not to a significant extent. Therefore, there is no real time saving by testing undrained (in which the test has to be run slow enough to equalize the pore water pressures) than in a drained test (in which the test has to be run slow enough to prevent pore pressure changes).

There are some disadvantages to selecting the consolidated-*undrained* test over the consolidated-*drained* test. In normally consolidated clays, the high pore pressures at failure crowd the effective stress circles into the left side of the Mohr diagram, making it difficult to measure parameter ϕ'. For heavily overconsolidated clays, the negative pore pressures at failure move the effective stress circles to the right of the Mohr diagram, making it difficult to accurately measure the cohesion intercept c'. Whether the clay is normally consolidated or overconsolidated, Mohr circles in a consolidated-undrained test series tend to be close together (Figure 8.22).

The best procedure for measuring the effective stress parameters of a clay is to perform consolidated-*drained* triaxial tests using side filters and draining from both ends of the specimen. The drained test spreads out the Mohr circles so that a more reliable determination of c' and ϕ' is achieved. The time to failure in the lab test can be determined from Table 6.2 of Chapter 6.

8.8 RESIDUAL STRENGTH OF CLAY

Residual strength is the constant strength of a clay at large shear displacements. It can be measured in a reversible shear box or ring shear apparatus, as presented in Chapter 6.

Dependency on Clay Fraction

As explained by Skempton (1985), the *peak strength* in the stress-displacement curve (ring shear tests) occurs at a shear displacement of 0.02 to 0.12 inch for stiff overconsolidated clays and at 0.12 to 0.25 inch for normally consolidated clays (Figure 8.23). For clays with a low clay fraction, overconsolidated clays lose strength after the peak as the clay weakens on

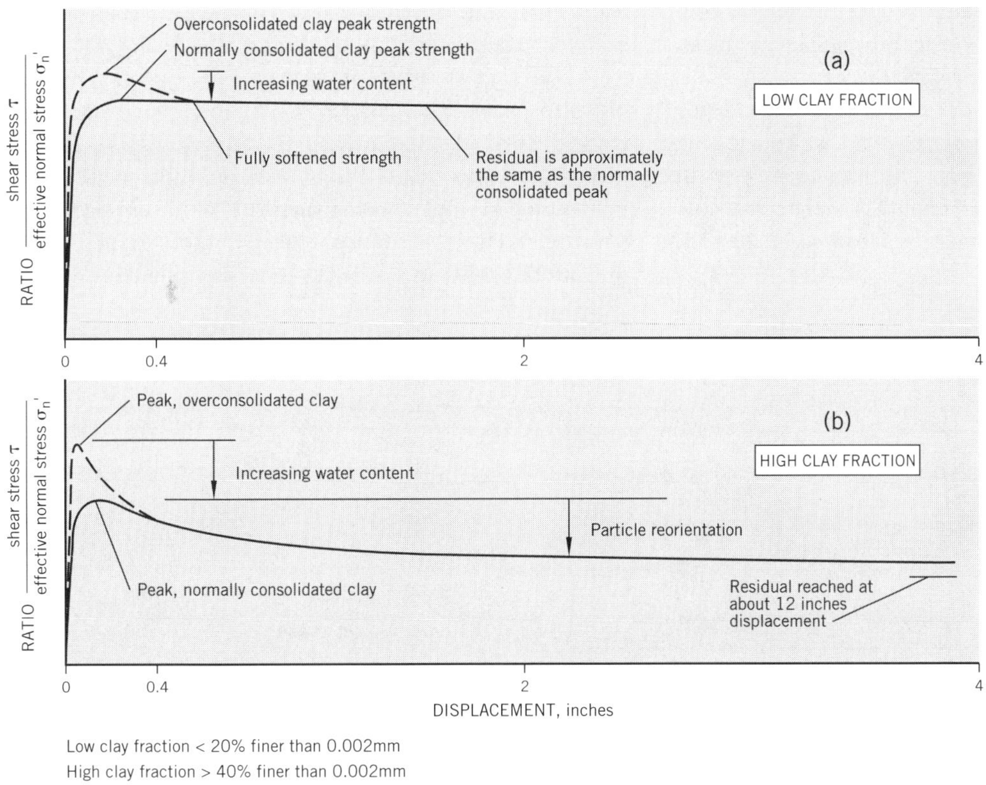

Figure 8.23 Ring shear tests on soils with
(a) low clay fraction
(b) high clay fraction
(adapted from Skempton, 1985)

Figure 8.24 Ring shear tests on sand–bentonite mixtures (after Lupini, Skinner and Vaughan, 1981 and Skempton, 1985).

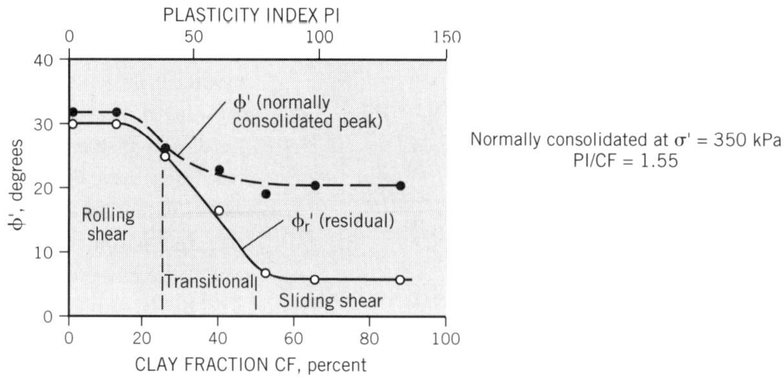

the failure plane i.e. increase in water content. This reduced strength is referred to as the *fully softened strength* where the clay shears at constant strength with no further volume change. It is reached after shear displacements of 0.16 to 0.4 inch.

For clays with a much higher clay fraction (e.g., >40% of particles less than 0.002 mm in the hydrometer test) particle reorientation occurs and the strength continues to decrease with increase in shear displacement until the *residual strength* ϕ_r is reached at 4 to 20 inches of shear displacement (Bishop et al., 1971).

The behavioral change from low to high clay fractions was studied experimentally by Lupini et al. (1981) using various mixtures of sand and bentonite. Their results are shown on Figure 8.24. For low clay fractions, there was little change from peak to residual strength because the mixture is controlled by the sand grains; the strength is mostly developed from rolling shear. At high clay fractions, above about 50%, there is a large difference between peak and residual strength and the clay fraction is dominant in causing sliding shear. Between these extremes is a transitional stage.

Skempton's research has indicated some stress dependency in residual strength tests; i.e., lower ϕ_r' values with increasing normal effective stress. Accordingly, he elected to present residual strength ϕ_r' values corresponding to an effective normal stress of 100 kPa (approximately 1 ton/sq. ft. in U.S. units).

The data from many other clays are shown on Figure 8.25. The shape of the curve agrees with the shape obtained from the sand–bentonite tests of Figure 8.24.

Effect of Deformation Rate on Residual Strength

An issue of considerable importance in earthquake engineering is the effect of rapid shaking on preexisting landslides with clay at residual strength. Laboratories have attempted to measure the residual strength of clays at high velocity and compare the results with the slow rate of shear displacement (0.0002 in./min. or 0.005 mm/min) of the usual test for residual strength. Data obtained by R. J. Petley and J. F. Lupini and reported by Skempton (1985) are shown on Figure 8.26. The change of strength is around 2.5% per log cycle of velocity. The highest velocities in these tests are well below the maximum velocities that occur during a major earthquake. However, these data are reassuring in that there are only minor changes in strength over the range of velocities relevant to slow-moving landslide practice and the laboratory test is valid for such studies.

Faster rates of shear were made for the earthquake design of Kalabagh Dam in Pakistan (Skempton, 1985). These tests were performed on remolded soil specimens, which were preconsolidated to about 9 ton/sq. ft. and subsequently tested in the ring shear at normal pressures of 2 ton/sq. ft. and 5 ton/sq. ft. The "slow" residual shear strength was measured at the rate of 0.0004 inch/min. to shear displacements of about

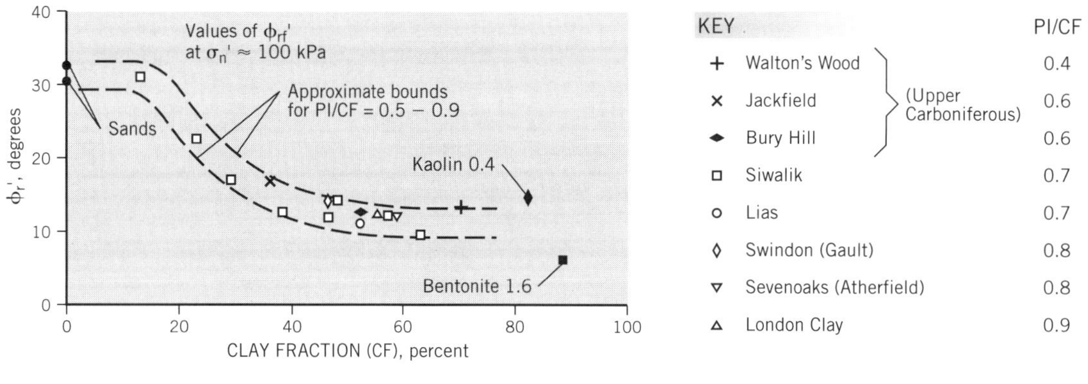

Figure 8.25 Residual strength ϕ_r' vs. clay fraction for numerous clays (after Skempton, 1985).

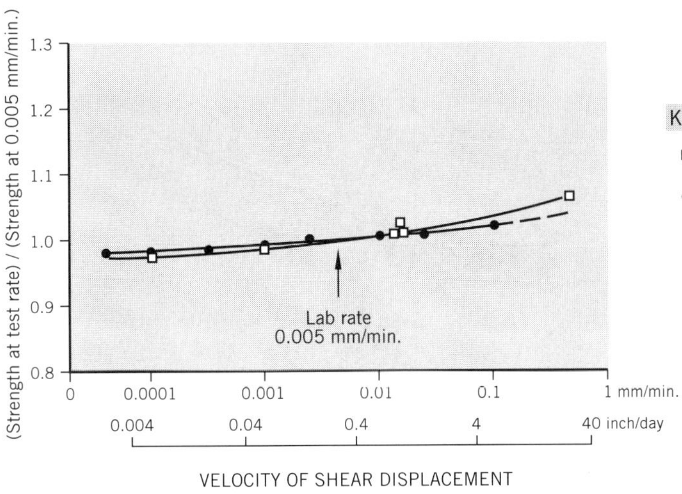

Figure 8.26 Variation of residual strength with shear velocity (after Skempton, 1985).

20 inches, after which the rate was increased and maintained until reaching steady conditions. After a pause to allow any pore pressures to dissipate, the slow rate was reimposed. This was followed by a still higher rate of shear, etc. until several rates had been measured, each followed by a drop back to the slow rate.

The results obtained on a clay with a clay fraction of 47% are shown on Figure 8.27(a). It can be noted that the residual strength dropped back to a τ/σ ratio of about 0.156 after each fast shear episode. At the higher velocities of shear, the τ/σ ratio rose rapidly to a peak and then dropped to a lower approximately constant value, which was higher than the slow rate of 0.156.

On Figure 8.27(b) the results from a second sample are shown with the peak being labeled "maximum" and the lower approximately constant value being labeled "minimum." A sharp increase in strength occurs at shear velocities above 4 inches/minute. This is attributed by Skempton to a change from sliding shear at the slower rates to a turbulent shear at the higher speeds. He suggests that the surface of shear is being increasingly disturbed at the higher speeds in contrast with sliding of particles oriented parallel to the plane of displacement in the slower tests. Other tests in this study showed more divergent results (Skempton, 1985), leaving some unresolved questions about clay behavior under these circumstances. However, all the published maximum (peak) ratios show a trend of increasing strength with increasing rate of shear. It is this ratio that is relevant to earthquake response.

There clearly needs to be more study of the behavior of residual strength clays during earthquakes. Another source of guidance on this topic is the extensive amount of data on the effect of strain rate on the undrained shear strength c of clays.

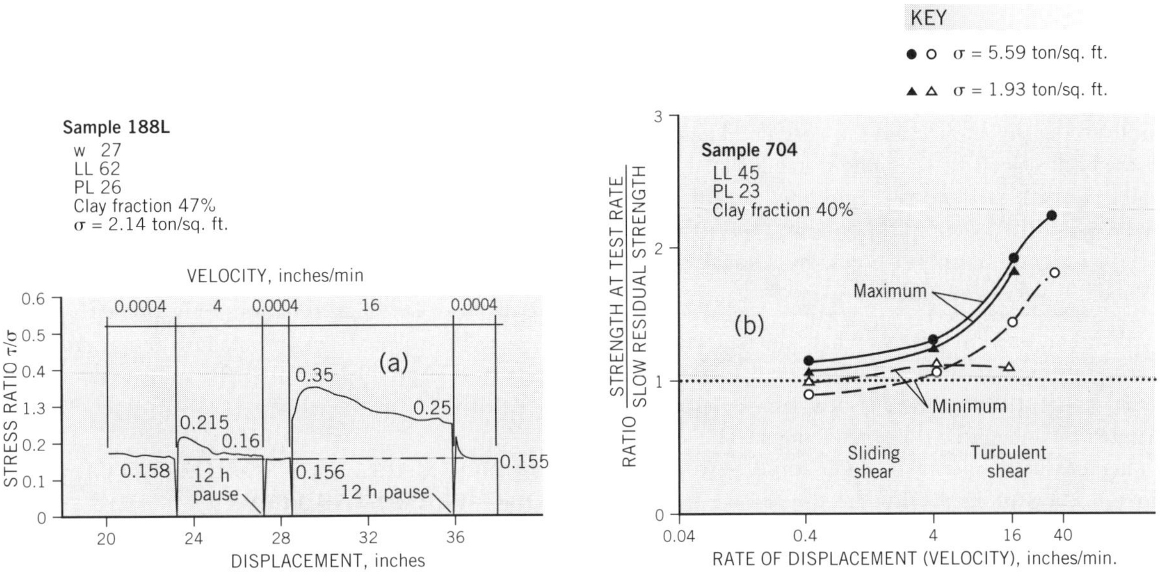

Figure 8.27 Effect of fast rates of shear on the residual strength of a clay with a high clay fraction (after Skempton, 1985).

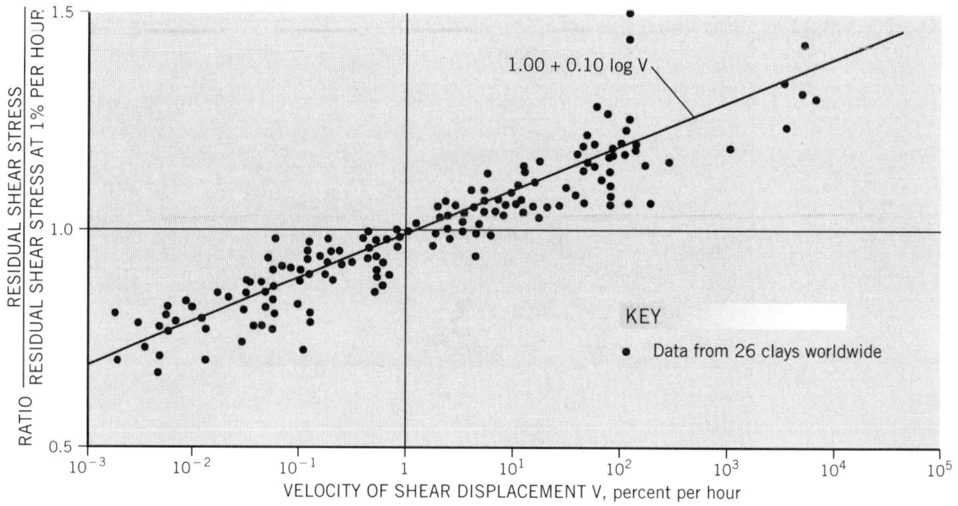

Figure 8.28 Effect of shear velocity on the undrained strength of clays (adapted from Kulhawy and Mayne, 1990).

A summary (Figure 8.28) has been published by Kulhawy and Mayne (1990), which indicates that the strength increases at a rate of about 10% per log cycle of shear rate. This collection of data cover 26 different clays from numerous sources. It has been used as the basis of a method of calculating the displacement of marginally stable clay slopes during earthquakes using a modified Newmark analysis (Vessely and Cornforth, 1998). The method is described in Chapter 12, Section 12.3.

Effect of Variable Weathering on Residual Strength

Landslides affecting the perimeter road at Hagg Lake reservoir, near Gaston, Oregon, were reactivations of ancient landslides by road fill construction. The soil is a weathered siltstone of volcanic origin. The unusual results of the residual strength tests are more fully described in Case History 10. However, the key points were:

- Samples taken from the general overburden above the slip surface showed virtually no change in strength on repeated shearing in a reversal shear box.
- Samples taken from the slip surface had large drops from the peak to residual at the two landslides studied.
- The measured laboratory residual strengths on samples taken from the slip surface were in very good agreement with the back-calculated strengths in stability analyses.
- Despite the fact that the two landslides were relatively close to each other, were in the same soil formation, and had similar identification properties, the residual strengths of 9° and 22° were very different.

As with other sites in Oregon where ancient landslides have occurred in highly weathered siltstones, the soil on and immediately adjacent to the discrete shear zone appears to be very disrupted by ancient landslide movements. The slip surface has a higher water content and softer consistency than the adjoining soils. It is probable that the slip surface has been altered by physical and chemical weathering. The degree to which weathering has occurred may account for very different residual strengths measured (and back-calculated) at the two Hagg Lake sites. Therefore, a single or narrow range of ϕ_r values may not be an accurate way of describing soils from such sites; this is contrary to the concept that residual strength ϕ_r has a single value for soil with a common origin, as in the case for clays such as London Clay.

8.9 NORMALLY CONSOLIDATED CLAY: SHORT TERM STABILITY

Normally consolidated clays develop positive pore pressures during shear; when these dissipate, the clay consolidates and becomes stronger. Therefore, slope stability is more critical in the short term, and a landslide usually occurs at this time.

The immediate strength, cohesion c, can be measured in one of two ways:

1. In-place vane shear tests, applying the Bjerrum (1972) correction (Figure 2.39) to reflect anisotropy and other factors (Chapter 3, Section 3.12).
2. Undisturbed samples taken by a piston sampler and tested unconsolidated-undrained in the laboratory to obtain undrained shear strength c (Chapter 3, Section 3.9). The cell pressure can be set at the approximate total overburden stress.

The slope stability analyses is usually termed a $\phi' = 0$ (total stress) analysis within the normally consolidated clay stratum. Any free-draining parts of the landslide slip surface, such as a gravel fill, have strengths calculated by effective stresses. In a landslide situation, only the actual slip surface needs to be analyzed. For a nonfailure design problem, stability charts (Chapter 9, Section 9.12) can be used.

8.10 NORMALLY CONSOLIDATED CLAY: LONG-TERM STABILITY

Since long-term stability of normally consolidated clays is always higher than short-term stability, the need to determine the factor of safety long term is usually only needed as information related to a remedial treatment. If a buttress, shear key, or groundwater lowering has been incorporated into the landslide section, consolidation and strengthening of the clay will occur with time until long-term equilibrium has been established. The long term (equilibrium) factor of safety can be calculated by an effective stress analysis.

8.11 OVERCONSOLIDATED CLAY: SHORT-TERM STABILITY

There is no reliable method of calculating the short-term stability of stiff clay. It is known, however, that the immediate strength of a stiff clay decreases with time, so short-term stability is time-dependent. The risk of a slope failure increases with elapsed time after construction of a cut or fill.

Construction of a Cut or Fill

In the absence of a rational design procedure for natural clay slopes, some possible courses of action are:

- Use local experience with the clay stratum, if available, to handle short-term stability, but with appropriate conservatism.
- Temporarily or permanently support the slope by, as examples, tied-back soldier pile walls, slurry trench walls, or soil nail walls.
- Use 1:1 temporary cut slopes.

Other options are available, depending on circumstances. However, the undrained shear strength c measured in a triaxial test on overconsolidated clay should not be used for slope stability analysis. The reason has been explained for dense dilatant sands in Chapter 7 and applies equally to dilatant clays. The use of such strengths in stability analyses are likely to significantly overestimate the actual clay shear strength available at the site except in the very short term.

8.12 OVERCONSOLIDATED CLAY: LONG-TERM STABILITY

Shear Strain Dependency

There are three types of effective stress strengths that can be recognized and each applies to different situations in landslide analyses. The three strengths are shown conceptually on the shear stress–shear displacement graph of Figure 8.29(a). This data can be obtained from a ring shear test or reversal shear box test on an undisturbed, intact specimen of overconsolidated clay. The three strengths are:

1. The **peak strength** (point A), reached at a relatively small displacement. Typically, soils in nature have some preexisting shear stress acting on them, unlike the conditions at the start of a triaxial test.
2. The **fully softened strength** (point B), which occurs after a slip surface has formed and the clay in the slip

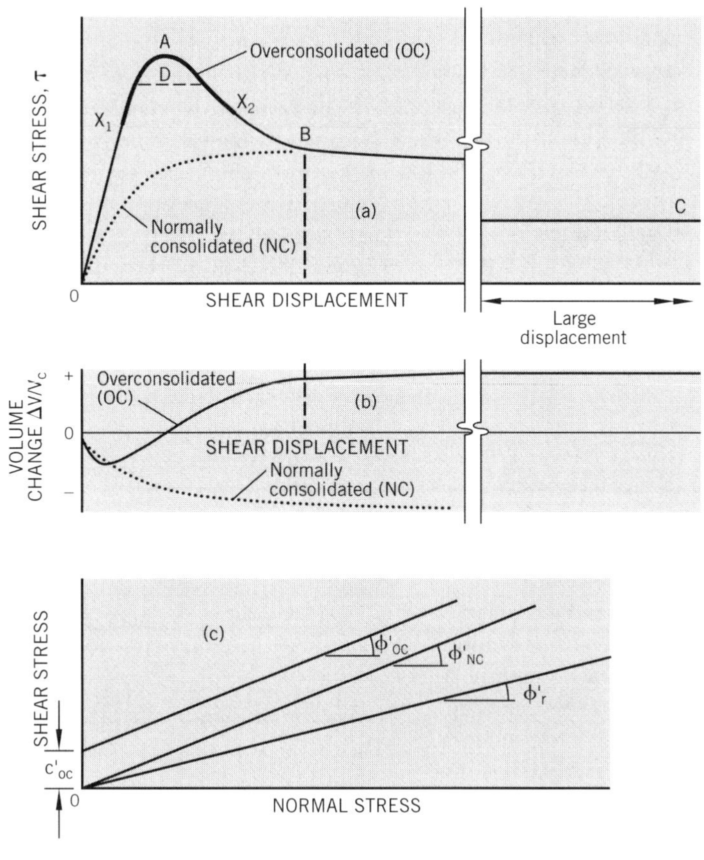

Figure 8.29 Conceptual explanation of strengths developed in an overconsolidated clay:
(a) shear stress—shear displacement curves
(b) volume change—shear displacement curves
(c) Mohr envelopes

zone has increased in water content and the volume change (Figure 8.29b) has essentially become constant. It can be noted that point B is equivalent to failure in a normally consolidated clay in which the shear stress rises to a maximum without a peak (dotted line on Figure 8.29a) and the volume change has leveled off to a constant value (dotted line in Figure 8.29b).

3. The **residual strength** (point C), which is reached at very large strains, orienting clay particles in the direction of the shear strain. In clays containing smectites, such as montmorillonite, the drop in strength from point B to point C is very significant and the shear surfaces typically are highly polished (slickensided).

Mohr Envelopes

The Mohr envelopes for these strengths are shown on Figure 8.29(c). For the peak strength A, a series of tests run at different effective normal stresses produces an envelope with a cohesion intercept c' and shear angle ϕ'. For discussion purposes, they will be labeled c'_{oc} and ϕ'_{oc}. If the clay is pre-softened before the tests, the angle ϕ'_{oc} remains about the same and c'_{oc} decreases toward zero. The concept that a fully softened condition will cause c'_{oc} to fall to zero gave rise to the so-called $c' = 0$ analysis for the long-term strength of overconsolidated clay slopes (Skempton, 1954). It is also known that remolded specimens of overconsolidated clay usually have a Mohr envelope that passes through the origin (i.e., $c' = 0$) and has approximately the same ϕ' as undisturbed specimens. Therefore, a fully softened strength for many clays can be approximated by the envelope $c'_{oc} = 0$ and ϕ'_{oc}. This result is similar to the $c' = 0, \phi' = \phi'_{nc}$ for a normally consolidated clay. The rationale is that overconsolidation produces the cohesion intercept c'. When this stress history is eradicated by shearing to the constant volume condition, and the water content on the slip surface has risen, the clay strength is comparable with a normally consolidated clay.

The Mohr envelope for residual strength has a ϕ_r' which is lower than ϕ'_{oc} and, usually, $c_r' = 0$. This condition is reached after very large shear displacements of about 2 to 4 feet in single discrete shear zones (Section 8.8) or even larger displacements where shear movements occur in parallel bands over a deep shear zone.

Progressive Shear Failure

A key concept in the understanding of mobilized strength in stiff overconsolidated clay is the role of *progressive failure.* This term refers to the progressive transfer of shear stress along a developing shear surface with increasing shear strain.

The slope section on Figure 8.30 illustrates the concept of progressive failure. The directions of the major and minor principal stresses change along the length of the slip surface. At different parts of the slip surface, the clay may be below the peak strength, at the peak strength, or past the peak strength, depending on the amount of shear deformation that has occurred at each location along the slip surface.

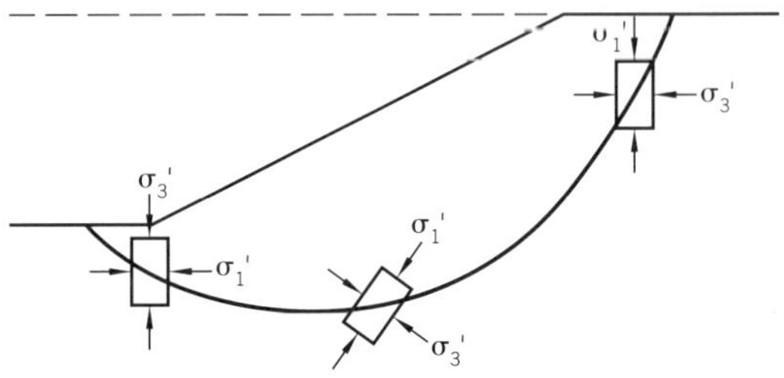

Figure 8.30 Rotation of the principal stresses along a potential circular arc slip surface in a cut slope (after Wu and Sangrey, 1978).

When a cut is first made into a stiff overconsolidated clay, shear movements occur to mobilize sufficient strength to provide stability. High negative pore pressures temporarily provide high strengths, limiting the amount of shear movement needed to mobilize the required resistance for stability. As pore pressures in the slope slowly rise and strains continue, some areas of the potential slip surface cross their peak resistance and become weaker with increasing strain. This loss of resistance is transferred to adjacent areas that are still able to mobilize increased resistance with increasing strain, i.e., have not yet reached the peak resistance. Thus, shear failure "progresses" along the potential slip surface and may result in failure of the entire block if the total resistance along the slip surface becomes less than the total shear stresses induced by gravity.

Variable strain rates within a slope ensure that the peak resistance of the clay cannot be mobilized simultaneously at all parts of the slip surface. Therefore, some reduction in the peak strength Mohr envelope is needed in slope stability and back analysis of landslides. The reduction should be greatest where the clay has a very sharp peak at low shear strain.

The concept of variable strengths along the slip surface is illustrated by the bold line X_1 to X_2 around the peak strength on Figure 8.29(a). The total shearing resistance that can be mobilized along the potential slip surface is a combination of shear strengths that have an average value of D. The value of D is always less than the peak strength A and is always more than the fully softened strength B.

The determination of the average maximum strength D for "first-time" sliding is a judgment decision. The reliability of the test data depends on many factors. Assuming that the data quality is reasonable good, it is likely that D can be selected as being within the range 25% to 60% of the difference between points A and B.

Selection of Strengths for Practical Applications

Based on the foregoing discussion, the following procedures are recommended to obtain the shear strength parameters for landslide studies in stiff overconsolidated clays:

1. Maximum Attainable Strength

This is the maximum attainable strength along a slip surface (point D, Figure 8.29), allowing for progressive failure development. It is *not* the peak strength, point A. It can be used for:

- The design strength in first-time sliding (FTS) with an appropriate factor of safety.
- Recreation of conditions at the onset of landsliding, for causation investigations.

Perform consolidated-drained triaxial tests on a minimum of four and preferably six test specimens at different effective stress cell pressures. Continue the test past failure point A to reach the fully softened condition, point B. This probably will require an axial strain of 12–20%. Use high quality field sampling procedures and slow rate of strain in the laboratory.

Overconsolidated clays typically give rather erratic results (especially fissured clays)—hence, the need for as many test results as practicable. The estimation of the average strength D due to progressive failure can be accommodated by reducing cohesion intercept c'_{oc} to about one-third to one-half of the measured value while keeping ϕ'_{oc} unchanged.

2. Fully Softened Strength

This is the reduced strength (point B, Figure 8.29) where the overconsolidated clay has passed the peak strength and starts to shear at constant volume. It has been termed the $c' = 0$ analysis. It can be used for:

- A more conservative design strength for maximum attainable strength, again using an appropriate design factor of safety
- Existing conditions in a landslide on a first-time sliding (FTS) slip surface after (i) very small movement in highly plastic clay (see Figure 8.23b) or (ii) after moderate movements in a low plastic clay (see Figure 8.23a)

Using the test program outlined for Maximum Strength, it is recommended that a strength of $c'_{oc} = 0$, $\phi' = \phi'_{oc}$ is suitable for this situation.

It should be borne in mind that many FTS shear surfaces are at the head and toe of reactivated landslides. The headscarp may have tension cracks where no strength is provided to landslide resistance.

3. Residual Strength

The residual strength (point C, Figure 8.29) should be used on all overconsolidated clay slip surfaces where present or past landslide shear movements exceed 1 foot (0.3 m), assuming that only a single shear surface exists. This group includes:

- Reactivation of ancient landslides, irrespective of the amount of movement
- First-time slides, after significant movement

It is recommended that residual strength be measured in the ring shear apparatus; remolded or disturbed soil can be used. A stress reversal shear box is also acceptable, but is less satisfactory than a ring shear test. The author does not recommend triaxial tests with precut shear plane due to the difficulty of analyzing the results.

In ancient landslide terrain, samples should be taken at, or very close to, the actual shear zone because the residual strength may be different from elsewhere in the overburden as a result of long-term deterioration (see Case History 10).

4. Strength Assessed Through Back Analysis

Back analysis of landslides uses the known factor of safety of 1.00 to calculate the shear strength properties along the slip surface. This subject is discussed in detail in Chapter 9, Section 9.4. The geotechnical practitioner can determine either (i) the shear strength at the onset of sliding or (ii) the shear strength when the sliding has just stopped moving. For landslide remediation analysis, choice (ii) may be more relevant and would be generally lower than choice (i) in first-time slides.

Although it is reassuring to obtain good agreement between laboratory data and the back-analyzed strength, it is the back analysis shear strength data that should be used in remedial studies. The reason for preferring the back-analyzed soil strength is that laboratory tests measure only an infinitesimally small part of the shearing surface, whereas back analysis reflects the strength of the entire landslide mass.

8.13 SHEAR MOVEMENTS AND FAILURE IN OVERCONSOLIDATED CLAY SLOPES

Shear movements in overconsolidated clay slopes may be reflecting the seasonal rise and fall of groundwater levels in a marginally stable landslide or the changing strength of the clay in response to some manmade construction. This section describes ways of interpreting the movements according to whether the velocities of movement are accelerating, constant, or decelerating.

Monitoring Slope Movements

Total movements, and velocities of movement, within the shear zone of a clay slope can be tracked directly by inclinometers or indirectly by surface survey hubs. Inclinometer surveys are described in Chapter 4, and ground surveys are more briefly covered in Chapter 3, Section 3.5. The benefits of the latter are described in three case histories—CH-1, CH-6, and CH-8.

In many situations, shear movements at the edges of a landslide are more damaging than total movements. Shear movements at the edges damage pipelines, utilities, or structures. By contrast, houses and other facilities within the landslide, but away from the edges, may sustain little or no damage. On large landslides, the occupants of houses within the confines of the slide may not know that they are living on sliding ground.

Velocity of Shear Movements Accelerating

In stiff clay, an increase in shearing stresses, caused by loading or unloading the clay, initially produces a decrease in the pore water pressures. This effect can be seen in a consolidated-undrained triaxial test with pore pressure measurements. At failure, the clay has a negative change in pore pressure. In the field, the negative pore pressure produced in stiff clays initially provides a high strength and shear resistance along a potential shear surface. With the passage of time, however, the groundwater reverts toward the long-term conditions of equilibrium;

i.e., groundwater and pore pressures rise, and the clay on the potential slip surface softens from its earlier stiffness. If the slowly reducing strength of the clay is insufficient to maintain stability of the slope, failure can occur months or decades after the causative factor that initiated the pore pressure changes. These slope failures have been termed *delayed failures* (Chapter 2, Section 2.15), and are often misunderstood by many geotechnical practitioners.

A. W. Skempton, in unpublished work, has explained the phenomenon of delayed failure in terms of changes in the pore pressure coefficient A (Chapter 6, Section 6.3). For illustrative purposes, assume the ground is initially level as shown on Figure 8.31(a). Consider an element of soil at point X in the ground at depth H below the surface. Original groundwater level is at 0.2H depth. Also assume that $\gamma = 2\,\gamma_w$.

Initial Conditions—Figure 8.31a

$$\sigma_1 = \gamma H$$
$$\text{Assume } \sigma_3 = \sigma_1 = \gamma H$$
$$u = 0.8\,\gamma_w H = 0.4\,\gamma H$$
$$\sigma_1' = 0.6\,\gamma H$$

Long-term Equilibrium Conditions—Figure 8.31b

A cut excavation is made. Based on an elastic analysis, the total stresses become:

$$\sigma_1 = 0.81\,\gamma H$$
$$\sigma_3 = 0.51\,\gamma H$$

These total stresses will not change, but the effective stresses change with changes in pore water pressures.

$$\text{Pore pressure } u = 0.54\,\gamma_w H = 0.27\,\gamma H$$
$$\sigma_1' = (0.81–0.27)\,\gamma H = 0.54\,\gamma H$$
$$\sigma_3' = (0.51–0.27)\,\gamma H = 0.24\,\gamma H$$

Now assume that the factor of safety F is exactly 1.00 under the long-term equilibrium conditions. For a Mohr circle passing through the origin (fully softened clay)

$$\sin \phi'_{LT} = (\sigma_1' - \sigma_3') / (\sigma_1' + \sigma_3') = 0.3/0.78 = 0.385$$
$$\phi'_{LT} = 22.6° \quad \text{(where LT = long term)}$$

End of Construction Conditions—Figure 8.31c

$$\Delta u = B\{\Delta\sigma_3 + A\,(\Delta\sigma_1 - \Delta\sigma_3)\} \quad \text{Chapter 6, Section 6.3, Eq. (11)}$$

Assume that pore pressure parameter A = 0 and B = 1.00

$$\Delta u = \Delta\sigma_3$$
$$\Delta u = (0.51 - 1)\,\gamma H$$
$$= -0.49\,\gamma H = -0.98\,\gamma_w H$$
$$u = 0.8\,\gamma_w H - 0.98\,\gamma_w H$$
$$= -0.18\,\gamma_w H\ (= -0.09\,\gamma H)$$
$$\sigma_1' = (0.81 + 0.09)\,\gamma H = 0.90\,\gamma H$$
$$\sigma_3' = (0.51 + 0.09)\,\gamma H = 0.60\,\gamma H$$

The ϕ'_{ST} corresponding to the effective principal stresses is $\sin \phi'_{ST} = 0.3/1.5 = 0.2$

$$\phi'_{ST} = 11.5° \quad \text{(where ST = short term)}$$

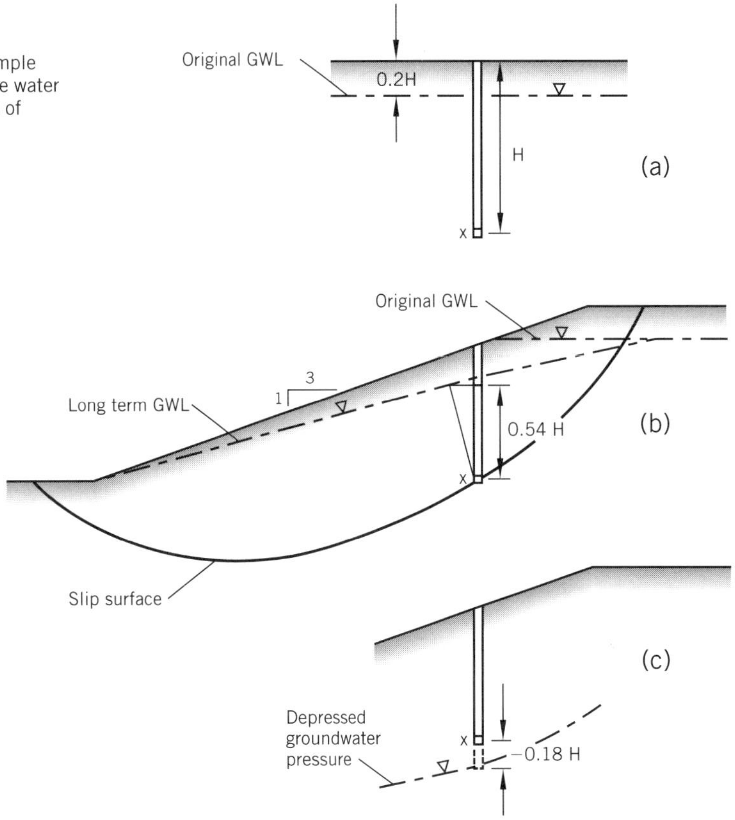

Figure 8.31 Illustrative example depicting the changes of pore water pressures behind a cut slope of overconsolidated clay:
(a) before construction
(b) long-term equilibrium
(c) end of construction

If the factor of safety F was exactly 1.00 in the long-term, the short-term F is tan ϕ'_{LT}/tan ϕ'_{ST} = 0.416/0.203 = 2.05.

The conclusion of this illustrative example, including all the simplifying assumptions, is that the factor of safety at element X has dropped from 2.05 at the end of construction to 1.00 long term. It is this loss of strength with time in overconsolidated clays—due to the initial drop in pore pressure subsequently rising to the equilibrium conditions—that accounts for delayed failure.

When discussing the principle of effective stress in Chapter 6, Section 6.2, it was pointed out that an element of soil below the sea floor is unaffected by the rise and fall of the tide because the total stresses and pore water pressure change by the same amount; i.e., the effective stress is the same at high tide, low tide, and intermediate tides. What is different in this discussion is that the total stresses in the cut slope remain unchanged from the end of construction to the long term, but the pore water pressures rise, decreasing the effective stresses. Thus, unlike the tide example, the clay in the cut slope swells and becomes weaker with time. It is common knowledge that the water content on the slip surface of failure in an overconsolidated clay is significantly higher than the clay elsewhere. This strain-softening phenomenon can be observed in laboratory test specimens and slip surface samples taken in the field.

Delayed failure can be attributed to the time required for groundwater to migrate through relatively impervious clay to soften the clay in the shearing zone, combined with the time for progressive failure to develop at the slip surface. These two factors occur concurrently as pore pressures change from those initially induced by a cut (or fill) toward the long-term equilibrium conditions. In the case of progressive failure, shear strains develop nonuniformly, and some parts of the slip surface reach their peak resistance while other parts are still building up resistance. Eventually, the maximum resistance along a slip surface is a combination of strengths below, at, and past the peak of the stress–strain curve of the clay, as discussed previously in Section 8.12.

For a delayed failure to occur, three conditions have to be present:

1. The soil is an overconsolidated clay or clayey silt; i.e., dilates on shearing to failure.
2. An event increases the shear stresses in the clay (e.g., erosion at base, construction of fills and cuts, etc.).
3. The rate of shear movement increases with time.

If these three conditions are met, observed movements may eventually result in a major catastrophic failure.

Case History 8 is an example of a slide that was monitored by surface hubs for many years. After 25 years from the triggering event (placing fill at the top of the slope), the progressively increasing velocity of movements were stopped by construction of a buttress at the toe of the slope. This prevented a potentially catastrophic landslide.

Case History 6 is an example of a delayed failure occurring 53 years after a fill was constructed across the top of the slide area. In this case, the slide area was on a high bluff and the major slide movement occurred without the owner being aware of shear movements leading up to it. Although it lacks direct evidence of condition (3) listed above, this landslide was almost certainly a delayed failure.

Percy slide on the Leaburg Canal, Oregon, is another example of long-term survey monitoring that showed accelerating landslide movements. A cut into the base of the slide area was the triggering event. It was stabilized by a buttress 30 years later after the increasing rates of movement threatened the operation of the hydropower project.

Velocity of Shear Movements Constant

A landslide in clay soil that is moving at an approximately constant velocity is likely to be shearing at residual strength on the slip surface provided that there are no significant conditions at the top and base of the slide that will change the velocity over time. Generally, the lower end of a slide is passing over a cliff, river bank, or reservoir slope such that continued movement does not build up resistance at the base.

The stability of a clay slope with an underlying slip surface at residual strength will change with the seasonal rise and fall of groundwater levels. During the rainy season the velocity will increase, and during the dry season the velocity will slow down or stop. Therefore, the "constant" velocity for these slopes will occur only if the groundwater conditions are constant.

On an annual basis, a landslide at residual strength will show a random (trendless) variation of movement that should have an approximately constant average velocity over a period of several years. Annual rainfall is quite variable, and departures from the long-term average can persist for many years. Fortunately, the combination of a known large shear displacement and a series of trendless annual movements is usually sufficient to show that a landslide is likely to continue moving at rates similar to the past, subject to variations caused by the weather patterns from year to year.

A good example of a landslide moving at a long-term constant velocity is the Washington Park Reservoir slide in Portland, Oregon. The landslide is described in Case History 1. This large slide is 2,100 feet long, 1,700 feet wide at the base, and approximately 70 feet deep on average. In 1896, the Portland Water Bureau created two storage reservoirs to serve the west side of Portland, Oregon. Soils were excavated from the base of the slope to create the reservoirs. The chosen site was a slope of ancient landslide terrain and the effect of the excavations was to create a massive, rapidly moving landslide. Extraordinary measures were taken to stabilize the slide, including a maze of drainage tunnels constructed along the clay/bedrock contact.

The extensive drainage measures succeeded in slowing down the rates of movement. A large number of survey points were set on the landslide and have been monitored for more than 100 years. Survey Line 20 crosses the landslide from side to side near the base. The average annual downslope movement along survey line 20 over a 25-year period (1930–1954) is shown on Figure 8.32, together with the annual rainfall over

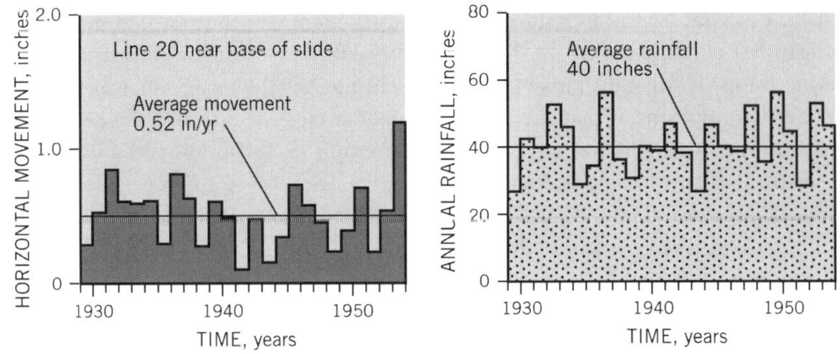

Figure 8.32 Washington Park Reservoirs Slide:
(a) average annual downslope movement for points on survey line 20 from 1930 to 1954
(b) annual rainfall over the same period

the same period. It can be seen that the movements vary around the average of 0.52 inch per year, showing no trend to increase or decrease with time.

Velocity of Shear Movements Decelerating

Shear movements in clay soils that are slowing down with time have improving stability. A long-term slowing velocity indicates that either the driving forces are being reduced or that resistance to sliding is building up.

Two case histories will suffice. Pelton Upper Slide, more fully described in Case History 6, is a double wedge-type failure, which started rapidly (as a delayed failure) and gradually slowed to zero over an 18-year period. The reason for the decreasing velocity was that the upper wedge, which is the driving force, moved downward to form a graben as the lower wedge moved out horizontally. The drop in the level of the upper wedge gradually decreased the "push" of the landslide until equilibrium was restored and movements stopped. In this case, the driving force decreased with time.

The second example of decreasing rates of movement is the clay shale in the foundation of Oahe Dam on the Missouri River. This huge earth dam (900 million cu. yd.) was built in the period 1955 to 1959. The right side of the dam is founded on a clay alluvium underlain by Pierre Shale, known as a "clay shale." The average properties of the bentonite clay shale are LL 141, PL 34 with natural water contents near the plastic limit, i.e. a highly overconsolidated clay. Long after construction, horizontal shear displacements continued at depths of 30 to 50 feet below the top of the clay shale in an area about 1000 feet wide on the right flank of the dam.

A section through the center of the movement area is shown on Figure 8.33, including total horizontal shear movements at the slip zone (1971–1999) and graphs showing the rates of horizontal movement from 1978–1999. In the 1990s, the Corps of Engineers Omaha District performed two in-depth studies of the problem area and then convened a Board

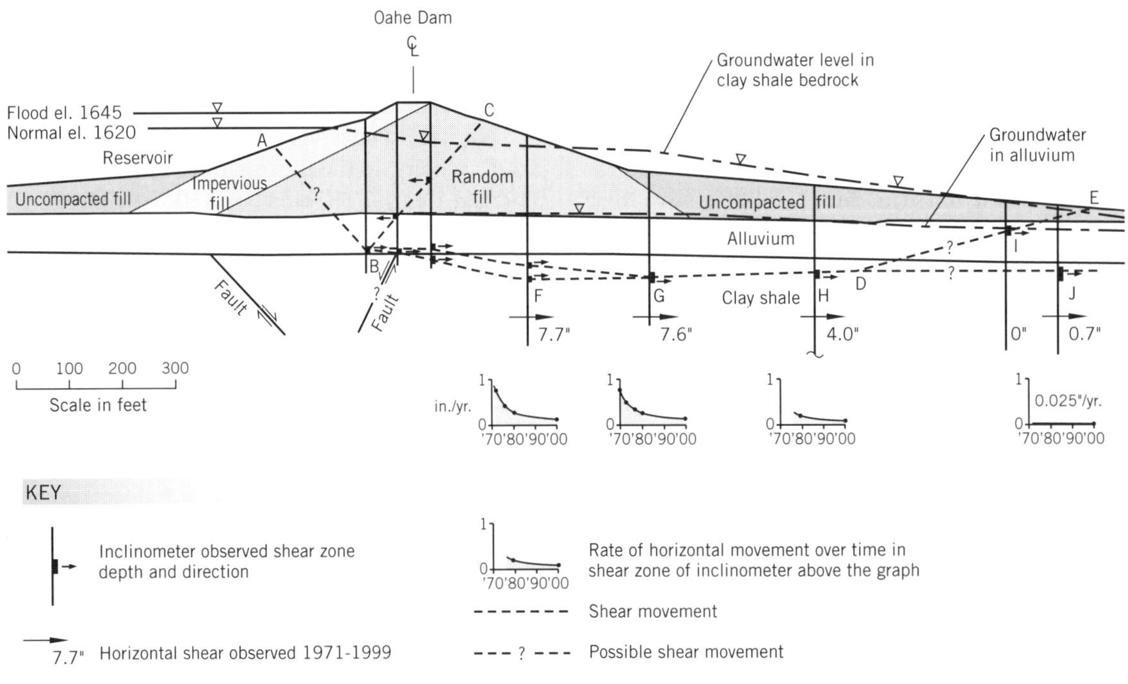

Figure 8.33 Oahe Dam, South Dakota: Cross-section through the right abutment showing the shear movements and rates of movement within the clay shale bedrock after construction.

of Consultants (the author and Professors Duncan and Morgenstern) to help in interpreting the data.

The piezometer data showed that there were elevated pore pressures within the clay shale, up to 100 feet higher than in the alluvium overlying the shale (Figure 8.33). The high pore pressures are slightly below reservoir pool levels, and it is possible that a slow leak exists between the pool and the foundation shale in this part of the dam.

The stability analyses for the dam were based on a triple-wedge model comprising an upper wedge (ABC), shear movement through the clay shale (BFGHD), and a toe passive breakout (DE). The reverse scarp (BC) was consistent with upslope movements in inclinometers near the dam crest, but there has been no confirmation of movement within the headscarp (AB). This may reflect a more distributed strain within the embankment dam. Another possible explanation for the absence of a headscarp observation is an ancient landslide feature in the clay shale (shown by Fault on the section, Figure 8.33) that may have been reactivated by the weight of the embankment dam. At the downstream end of the movement, shear was observed in the inclinometer at I (possibly the breakout wedge) and minor shear movement at J (possibly the horizontal shear strain petering out).

The main focus of attention was the central block of embankment experiencing shear movements at inclinometers F, G, H. They were installed around 1971, 12 years after the embankment was completed, and both F and G moved about 7.6 inches between 1971 and 1999. The inclinometer at H moved about 4.0 inches during the same period. The rates of movement have declined from around 0.8 inch per year in 1971 to 0.16 inch per year in 2000 at F and G, with proportionately decreasing rates at H.

Two additional borings were drilled at G and H specifically to examine the cores in the shear zone. They showed that the clay shale was generally massive, with very few fissures, and bentonite layers within it did not appear to be sheared or slickensided. Some slickensides were noted elsewhere in the cores. There was no discrete shear zone and shear was occurring over a depth of about 8 feet according to the inclinometer observations. The clay was visually anisotropic, indicating less resistance in the horizontal plane due to fissility. It was concluded that the overall shear movement was being accommodated by slip along several parallel horizontal plane surfaces.

Allowing for shear movements that occurred before 1971, the Board of Consultants decided that the upstream part of the central block possibly could have reached residual strength but the downstream part of the block probably was at a higher strength comparable to a fully softened condition. Any toe breakout wedge has to shear across the bedding planes, which would provide higher strengths due to anisotropy.

The conclusion of this study was that the continuing shear movements were mobilizing the passive resistance downstream and that very adequate reserves of strength were still available. This accounts for the decreasing rate of shear movements in the clay shale foundation. After making some estimates of clay properties, largely based on other sites with similar clay shales, a stability analysis indicated that about 60 percent of the overall resistance to sliding is provided by passive resistance.

This second example demonstrates that deep-seated passive resistance requires large shear displacements to fully mobilize the resistance. The decreasing rates of shear movement observed by the inclinometers reflect this build-up of passive resistance.

To summarize this section, decreasing shear velocity shows favorable changes in the driving or resisting forces, i.e., decreasing driving force or increasing resistance force. Remedial treatment generally will not be required (assuming estimated future movements can be tolerated).

CHAPTER

9

Slope Stability Analyses

This chapter discusses stability analyses for landslides. The three most commonly used methods in practice are the circular arc, double wedge, and triple wedge. The margin of stability is measured by the factor of safety F, which can be defined by:

$$F = \frac{\text{total available shear (or moment) resistance}}{\text{shear force (or moment) needed for static equilibrium}} \quad \text{Eq. (1)}$$

Calculations of slope stability are based on force and/or moment equilibrium. Both conditions should be satisfied and some procedures (e.g., Spencer, Janbu's Rigorous) achieve this objective. A slope is on the point of failing when F = 1.00 because the resistance is in exact balance with the destabilizing force.

Stability analyses involve the application of mathematics to conditions of nature. A high degree of precision is impossible because the calculations are a compromise between mathematical formulae and the real nature and variability of soil properties at the site under study. Nevertheless, the extent to which landslides can be modeled by geotechnical engineering is impressive, and is based on more than 70 years of scientific research.

Today stability analyses are done by computer, and many younger engineers have never performed hand calculations of stability. However, hand calculations can help engineers to better understand the mechanics of failure. On a landslide, it is sometimes preferable to perform the calculations by hand because the geometry of the failure is known and does not require numerous potential rupture surfaces to be examined, unlike a slope stability design. Complex details can be easily included in a hand calculation.

Stability calculations are usually two-dimensional based on a cross-section through the slope. Therefore, loads, stresses, and weights are calculated for unit width (e.g., 1 foot or 1 meter) normal to the cross-section.

9.1 MEASUREMENT OF SOIL DENSITY

Density affects both driving forces and resisting forces (in effective stress analyses) so any error tends to be offset on both numerator and divisor sides of the factor of safety equation. Furthermore, the range of densities is fairly small compared to strength or groundwater level variations, so a significant error in the stability analysis is not likely to occur due to error in the assumed or measured densities of the soils.

Fine-Grained Soils

Clays, silts, and fine sands (the latter having 50 percent or more by weight of particles passing the No. 60 sieve) taken below the groundwater table can have the saturated density estimated by taking a water content test. In most site investigations water content specimens are taken from SPT samples or the base of thin-wall tube samples.

The saturated density is determined by measuring or assuming a value for the specific gravity G of the soil grains and then calculating the voids ratio e for saturated soil. A value of G = 2.68 is about average for soils. A graph relating wet density γ to water content w for G = 2.68 is shown on Figure 9.1. This graph also includes densities for G = 2.60 and G = 2.75, the approximate range for soils.

The relevant formulas are:

$$e = wG \quad \text{Eq. (2)}$$

$$\gamma = \frac{(G + e)\gamma_w}{(1 + e)} \quad \text{Eq. (3)}$$

where γ = wet density; G = specific gravity of solids; e = voids ratio; w = water content; γ_w = density of water.

Coarse-Grained Soils

Density estimates are more difficult in coarse-grained soils. One approach is to take some known ranges of density for

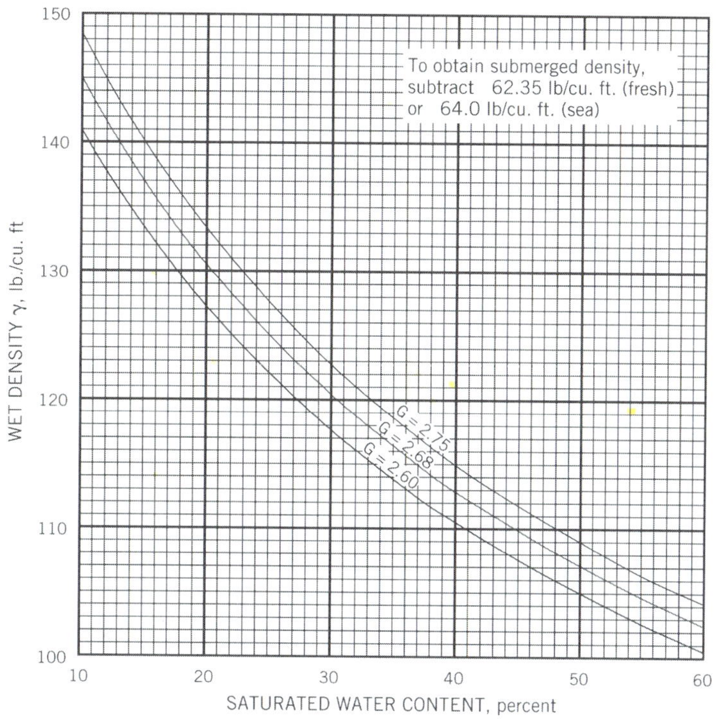

Figure 9.1 Determination of soil density from saturated water content.

certain soils and compare them to the soils under study, making adjustments for relative density, particle angularity, and gradation. For these variables:

- Wet density increases with increasing relative density.
- Wet density increases for well-graded soils compared to uniformly graded soils.
- Wet density increases for well-rounded stones compared to angular stones.

The density ranges for Brasted Sand, an angular, uniform, fine sand with about 10% of nonplastic fines passing the No. 200 sieve, is given in Table 9.1. The density range for a well-graded, well-rounded sandy gravel is given on Table 9.2. Other density ranges for sand can be estimated from Table 7.5 and Figure 7.18 of Chapter 7.

One of the more difficult densities to estimate is an angular shot rock. It cannot be easily measured in the field but some commonly assumed swell factors are 35–45% for hard shot rock and 25–35% for soft rock. For example, a swell factor of 40% for rock with a specific gravity G = 2.68 gives a compacted density of 120 lb./cu. ft. Rockfill may be damp, if water is used to aid compaction, or it may be dry. It is rarely saturated unless it is below the water table. Many geotechnical engineers tend to overestimate the density of hard shot rock. The angular stones and relatively uniform gradation of hard rock fragments produce a compacted fill with above-average void spaces.

Table 9.1 Densities of Brasted Sand at Various Relative Densities

Relative Dry Density (RDD) %	Saturated Density γ lb./cu. ft.	Water Content w %	Dry Density γ_d lb./cu. ft.	Porosity n %	Voids Ratio e
0	120.8	29.5	93.2	44.2	0.792
20	123.4	26.7	97.3	41.8	0.717
40	125.9	24.2	101.3	39.4	0.649
60	128.4	21.9	105.3	37.0	0.587
80	130.9	19.7	109.4	34.6	0.528
100	133.4	17.7	113.3	32.2	0.475

Note: Specific gravity of grains G = 2.68

Table 9.2 Densities of a Well-Graded Rounded Sandy Gravel at Various Relative Densities*

Relative Dry Density (RDD) %	Saturated Density γ lb./cu. ft.	Water Content w %	Dry Density γ_d lb./cu. ft.	Porosity n %	Voids Ratio e
0	128.5	21.0	106.2	35.7	0.556
20	131.3	18.6	110.7	33.0	0.493
40	134.2	16.3	115.4	30.2	0.432
60	137.0	14.3	119.8	27.5	0.379
80	139.8	12.4	124.4	24.7	0.328
100	142.7	10.6	129.0	22.0	0.281

*Specific gravity G = 2.65 *Source:* Mangla Dam, Pakistan filter soil

9.2 TOTAL STRESS AND EFFECTIVE STRESS ANALYSES

Stability of clay slopes is analyzed by one of two approaches: total stress analysis or effective stress analysis.

Total Stress Analysis

In total stress analysis, the immediately available shear strength of the soil is measured. This analysis assumes that all factors affecting the undrained strength of the soil are accounted for in the measured strength.

For normally consolidated silts and clays, a total stress analysis is an appropriate design technique. The immediately available shear strength can be measured by in-situ field vane shear tests in several borings to determine a profile of shear strength vs. depth. Alternatively, the shear strength can be measured by taking thin-wall tube samples and testing the soil undrained in the laboratory under the total stresses of the site.

The shear strength of clay is represented by cohesion c in the total stress analysis. Cohesion c is one-half of the unconfined compressive strength in an undrained triaxial test. It should not be confused with the cohesion intercept c′ obtained from effective stress shear tests. The two parameters are unrelated. The term s_u or c_u are sometimes used by others to indicate the undrained shear strength of clay.

Effective Stress Analysis

In an effective stress analysis, the computations account for water pressures within the pores of the soil. The *principle of effective stress* states that strength and deformation are controlled by the effective stresses acting on the soil. Effective stress is defined by:

$$\sigma' = \sigma - u \qquad \text{Eq. (4)}$$

where σ' = effective stress
σ = total stress
u = pore water pressure

The main benefit of an effective analysis is that *changes* in the landslide geometry and groundwater conditions can be studied in the analysis. Many geotechnical practitioners think that all stability analyses should be conducted in effective stresses. However, there are situations where it is difficult to perform:

- Transient pore pressure changes, such as cuts made into stiff overconsolidated clays
- Small landslides in which the effective stress stability is very sensitive to the position of groundwater
- Slopes with negative pore pressures near the outer surface; negative pore pressures are difficult to measure

Acceptable procedures for stability analyses on clay slopes are:

Short-Term Stability

For *normally consolidated* clays and silts, use total stress analysis. Effective stress can be used but is more difficult because it needs estimates or direct measurements of the pore pressures within the soils.

For *overconsolidated* clays and silts, no good method is available because the soil strength undergoes fairly rapid changes with time. Empirical rules are sometimes available to practitioners for a particular clay stratum in their locality based on past experience.

Long-Term Stability

For all clay types, use effective stress analysis.

Critical Stability

It is preferable for the geotechnical practitioner to think in terms of the *critical* conditions of stability rather than "short-term" or "long-term." Many situations occur in practice where such conditions may be different to both the short-term or long-term treatment of the landslide. An example is temporary works that may be needed as part of remediation; stability temporarily would be more critical than either the short- or long-term. Another example is a site where both soft and stiff clays are present.

In general, fine-grained soils that are soft and are likely to develop positive pore pressures under increased shear stresses will have a more critical stability in the short term. When pore pressures dissipate, the soil becomes stronger. Conversely, fine-grained soils that develop negative pore pressures under increased shear stresses will lose strength over time, and their lowest stability will occur when groundwater equilibrium conditions are reached. However, the *amount of shear displacement* is important for these soils and needs to be taken into account. Large shear displacements will cause a clay to reach residual strength (i.e., the lowest possible shear resistance.) A minor shear displacement will result in a mobilized shear strength between peak and residual strength, as previously discussed in Chapter 8.

9.3 LANDSLIDE SHEAR SURFACES

Landslide Geometry

The *shape* of the landslide shear surfaces have to be measured for modeling in a stability analysis. For this reason, it is advisable to position most borings down the line of maximum slope at or near the mid-width of the landslide to obtain a good model of the landslide geometry. Too often practitioners will distribute holes over the whole area of a landslide, which may prevent a reliable cross-section being available for stability analysis. Therefore, a line of holes should be the first consideration, with supplementary holes as needed to check other parts of the landslide, depending on the shape of the slide and budgetary constraints.

The more common shapes of the shear surfaces are: single plane (or slightly curved) on steep cuts; circular arc in uniform strata such as fills, soft clay, and smaller landslides; double and triple wedges on translational landslides, usually in the medium to large size categories. These shapes can be analyzed by numerical or graphical procedures, and computer software programs are available to perform the calculations. Analyses of the more common shear surfaces are presented later in this chapter. On larger slides, the shape of the shear surface may be more complex, due to varying soil or bedrock conditions over the length and width of the landslide.

Discrete Shear Zone

The shearing surface within a landslide usually occurs as simple shear over a narrow zone, often termed a *discrete shear zone.* Typically, the zone ranges from a few inches thick to about 4 feet thick, although in isolated cases, shear can be spread over multiple slip zones totaling 8 feet in thickness or more. However, the discrete shear zone in stiff clays generally is too narrow for a sample to be obtained for laboratory triaxial tests. To test for residual strength in the ring shear apparatus, only a small disturbed sample at natural water content is needed. In the field, it may be possible to perform a direct shear test on the slip plane by excavating an adit or test pit to reach the shear zone.

In stiff clays, the discrete shear zone is softer than the clay above and below it. The water content of the shear zone is higher due to dilatancy during shear. Therefore, samples taken from the clay overburden above a discrete shear zone will provide undrained shear strengths that are too high. If this is not recognized by the geotechnical practitioner, an erroneously high factor of safety will be obtained from stability analyses, possibly causing the slope to be deemed stable when it is either marginally stable or unstable. This is a common error in studying preexisting landslide conditions such as ancient landslide terrain.

9.4 BACK ANALYSES

Back analysis is probably the most valuable tool available for landslide studies. It provides confidence in ensuring the reliability of remedial work and allows the practitioner to use less-conservative factors of safety for landslides than for slope stability calculations where no failure has occurred.

General Procedure

Back analysis involves an analysis of the stability at the onset of slope instability, knowing that the static factor of safety F then is at 1.00. This removes one of the unknowns in the stability calculations. If the soil density and shape of the shear failure surfaces can be measured with good accuracy, it leaves groundwater levels and shear strength on the slip surfaces as the principal uncertainties. The shear strength presents the greatest difficulty because of its variability in terms of stresses, strains, and drainage characteristics. Earlier chapters have attempted to provide enlightenment on this subject. However, even if sampling procedures and laboratory testing were completely reliable, the measurements still would be made on an extremely small part of the entire slip surface of the landslide. By contrast, a back analysis can be used to determine the average shear strength parameters acting on the entire slip surface.

In practice, the shape of a landslide is obtained from borings instrumented with inclinometer casings, and groundwater levels at the slip surface are measured by piezometer tips at or slightly above the shear zone (see Chapter 4). Using this field data, the stability analysis back-calculates the average shear strength properties acting on the shearing surface at the time of the landslide activity. For effective stress analyses, it is common to assume $c' = 0$ but analyses can be made by putting $c' = 200$ psf or 400 psf for first-time slides in overconsolidated clays where appropriate.

Analyses of proposed remedial treatments are comparative studies with the back analysis. Unless a very significant error has been made in back analysis, such as water levels that are clearly wrong, the proposed improvement to the landslide is almost certain to be a true improvement and should result in permanent stability for the site. Parametric analyses can be made by varying the strengths and groundwater levels to check that the calculated factor of safety is sufficient to account for uncertainties. A reliability analysis also can be performed (see Chapter 10, Section 10.4).

If the landslide is a construction-related reactivation of an ancient or preexisting landslide, the shear surfaces may comprise (i) first-time sliding from the headscarp to the old slip surface, (ii) residual strength sliding along the old slip surface (usually the longest segment within a translational landslide), and (iii) first-time sliding from the old slip surface back to the ground at the bottom of the landslide. Thus, there are two or possibly three different shear strengths acting on the slip surfaces of the entire landslide. The suggested procedure for this situation is to calculate the first-time slide strengths based on laboratory shear strength measurements and determine the average residual strength along the old slip surface by a back analysis.

Effect of Velocity on Back Analysis

A factor of safety F of 1.00 implies that the resisting forces (or moments) of the landslide exactly balance with the driving forces (or moments). Back analyses are routinely performed on the assumption that the static factor of safety is 1.00 in the landslide. However, the assumption is only correct at the *onset* of movement when the velocity of the landslide is zero. If the relationship between resisting and driving forces continues to deteriorate, the landslide movements increase in velocity and a factor of safety calculation based on *static* forces will fall below 1.00.

Consider the following two examples:

1. When an intermittently active landslide begins to move at the start of a wet season, the balanced condition is met and F = 1.00. However, if groundwater levels continue to rise as the wet season progresses, the resistance decreases with essentially no change in the destabilizing forces (this assumes that the soils above the groundwater level are already fully saturated, which is the usual situation). The landslide velocity increases as the groundwater levels rise above the

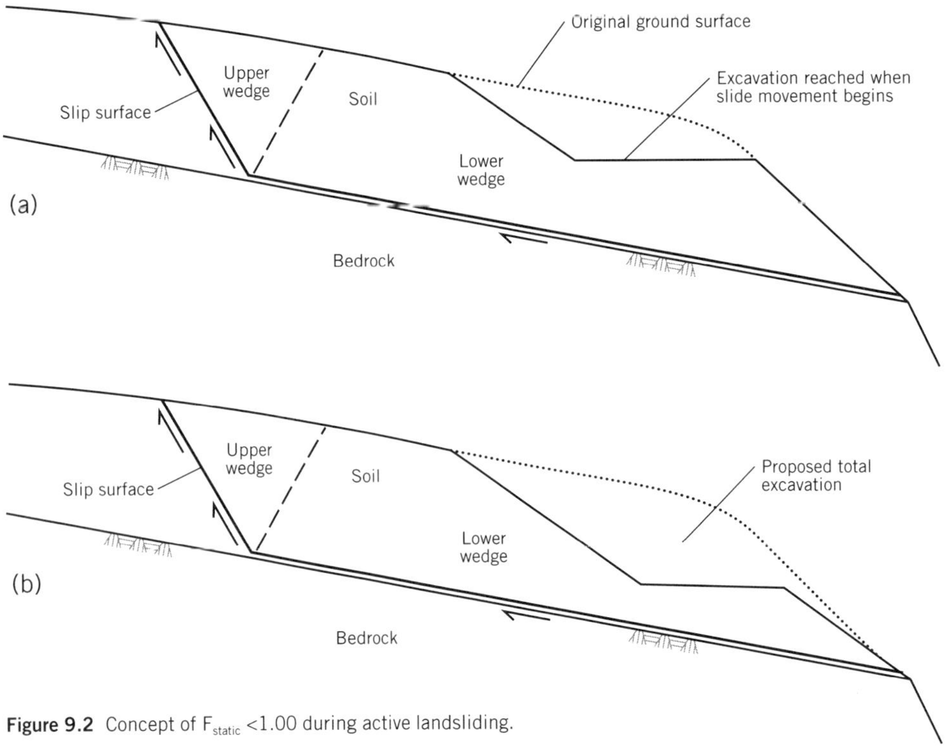

Figure 9.2 Concept of F_{static} <1.00 during active landsliding.

threshold for initiation of movement. Thus, the loss of *static* resistance from rising pore pressures is compensated by some form of *dynamic* resistance. The net effect is that the factor of safety F, in terms of a *static* mathematical analysis, falls below the threshold level of 1.00.

2. A cut for a highway is made into a natural soil slope, as shown in Figure 9.2(a). As the cut proceeds, a double wedge failure develops. As described in Section 9.7, a double wedge failure comprises an upper wedge that "pushes" the landslide and a lower wedge that provides a net resistance to instability. Thus, in this example, excavation for the road prism has reduced the resistance while leaving the principal driving force of the upper wedge unchanged. At the onset of the failure, the static factor of safety is 1.00. However, if the road cut excavation is continued, as in Figure 9.2(b), the resistance to sliding further decreases. Therefore, in terms of a *static* stability analysis, the factor of safety must fall below 1.00. However, many engineers perform a back analysis of an actively moving landslide assuming that the factor of safety in a static stability analysis is still 1.00.

A fortuitous set of events in 1990 allowed the *static* factor of safety F to be measured (with good reliability) for different velocities at the Pelton Upper Slide in central Oregon. This project is described in some detail in Case History 6 and in two published papers (Cornforth and Vessely, 1992a, 1992b).

The landslide is on a bench about 175 feet above a recreation park near Pelton Dam. A railroad fill was built across the site in about 1910 but was immediately abandoned. In 1975, 65 years after construction, a rapid double wedge failure occurred, with the railroad fill being on the upper (driving) wedge of the landslide. There was a horizontal slip surface below the lower wedge (Figure 9.3a). Later site investigations showed that the horizontal slip surface passes through a weathered siltstone-sandstone stratum. Since the site has never been used by the owner and is distant from all human encroachments, the cause of the landslide is attributed to a delayed failure within the siltstone-sandstone resulting from the earlier railroad fill construction. Average soil properties on the lower slip plane were LL 82, PL 57, PI 25, w 54.

Over time, the velocity of the landslide (lower wedge) declined from an initial (highest) rate of 3.6 inches per week to zero after 13 years (1988). During this period, 27 survey stakes on the landslide were surveyed at frequent intervals (see Case History 6) to record the movements. As the lower wedge moved out horizontally (eventually to more than 17 feet), the upper wedge settled into the void left behind by the horizontal movement; thus, the settling upper wedge gradually reduced the "push" on the lower wedge until movements ceased.

When movements stopped, the *static* factor of safety was exactly 1.00. Therefore, by knowing the earlier ground profiles, the *static* factors of safety at any other time can be calculated and plotted against the landslide velocity. (Note: At the onset of the landslide, the shear strength along the horizontal shear surface would rapidly reach residual strength and would thereafter remain at residual strength.)

In 1990, the owner (Portland General Electric, a public utility) decided to improve the stability of Upper Pelton Slide to make the recreation park below it safer for public use.

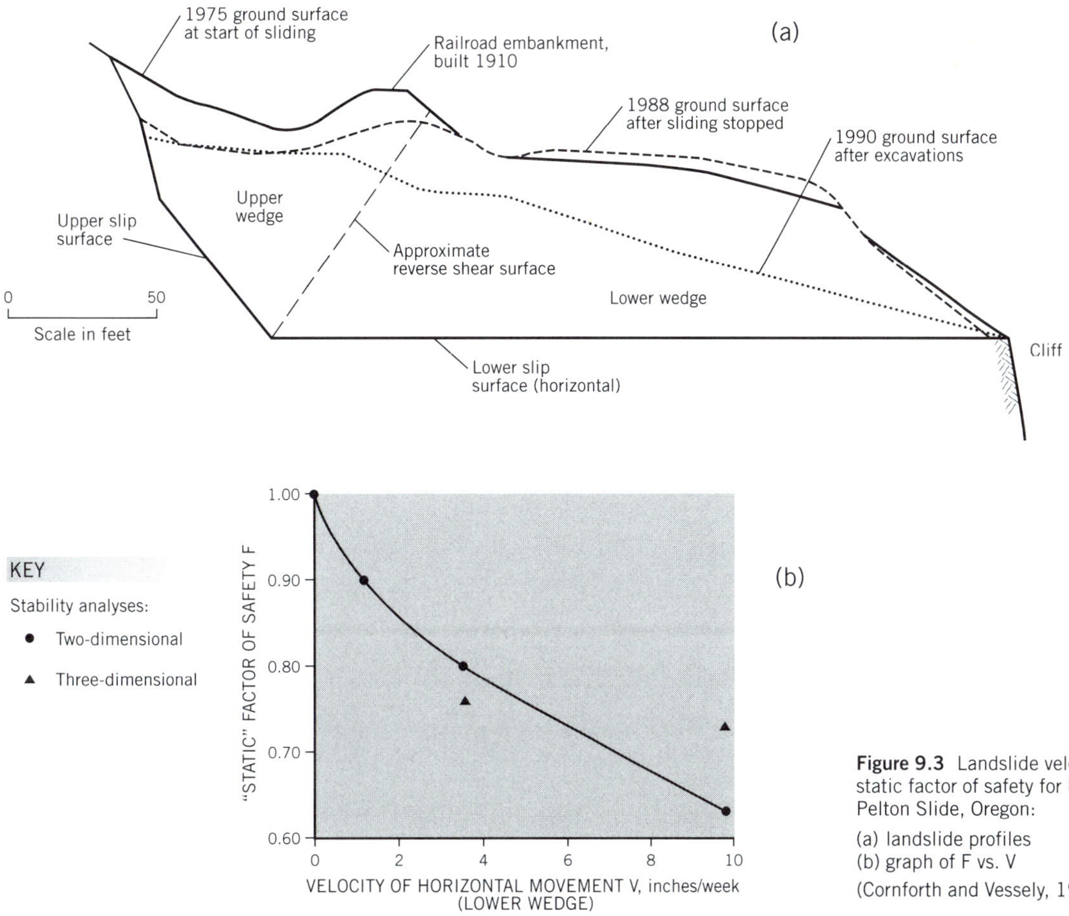

Figure 9.3 Landslide velocity vs. static factor of safety for Upper Pelton Slide, Oregon:
(a) landslide profiles
(b) graph of F vs. V
(Cornforth and Vessely, 1992b)

Unfortunately, their geotechnical consultant (not this author!) advised them to unload the slide by excavating part of it. Soils were removed from what was the lower wedge, thereby reducing the resistance to sliding. The landslide immediately reactivated, and the velocity increased as the excavation proceeded. Additional survey data were collected during this period, allowing the *static* factors of safety to be calculated for the worsening condition. Eventually, the landslide was stabilized by the cut-and-fill technique described in Case History 6.

The graph relating the horizontal velocity of the lower wedge to the calculated *static* factor of safety F is shown on Figure 9.3(b). The two-dimensional calculations are considered to be a reliable measure of the change in *static* factor of safety at this site. Comparable results also were obtained in three-dimensional stability analyses.

The graph on Figure 9.3(b) shows that the calculated *static* factor of safety falls significantly below 1.00 as the velocity increases. For example, when the velocity of the slide was 1 inch per week (0.14 inch per day) the *static* F decreases to 0.90. Therefore, the shear resistance would have to be increased by 11.1% just to bring the factor of safety back to static equilibrium at 1.00. At this example velocity, the value of F used in a back analysis should be 0.90 and not 1.00.

With the lack of other case histories, it is impossible to know whether the graph of static F vs. velocity (Figure 9.3b) can be reliably applied to other sites. However, until more research becomes available, it is suggested that the graph be used as a first estimate of the reduction of static F due to velocity.

In summary, the designer faced with remediation of an actively moving landslide has the choice of (a) using the *existing* conditions and allowing for a static factor of safety of less than 1.00 in the back analysis or (b) knowing or estimating the site conditions at the *onset* of the movement and using a static factor of safety of 1.00 in the back analysis. These approaches should ensure that an adequate factor of safety for stability is being provided in the remedial design.

9.5 VERTICAL CUT IN CLAY

Maximum Height

A vertical cut in clay can be analyzed by a single wedge, Figure 9.4(a). If the potential rupture surface BC is inclined at angle α to the horizontal:

$$\text{Along BC, driving force} = W \sin \alpha$$
$$BC = H \operatorname{cosec} \alpha$$
$$AC = H \cot \alpha$$
$$\therefore W = \tfrac{1}{2}\gamma H^2 \cot \alpha$$
$$\text{Resisting force } S = c\,H \operatorname{cosec} \alpha$$

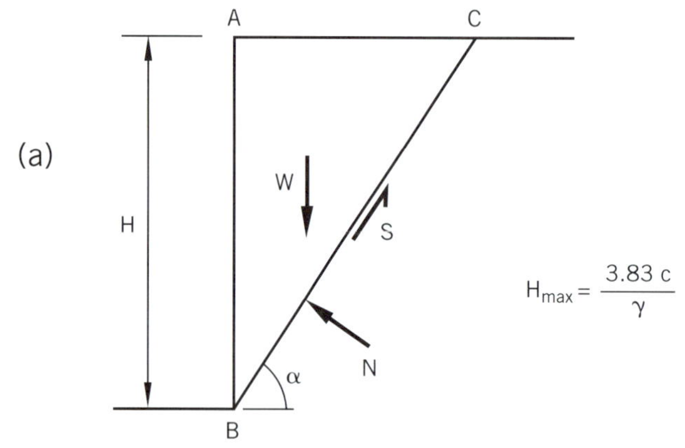

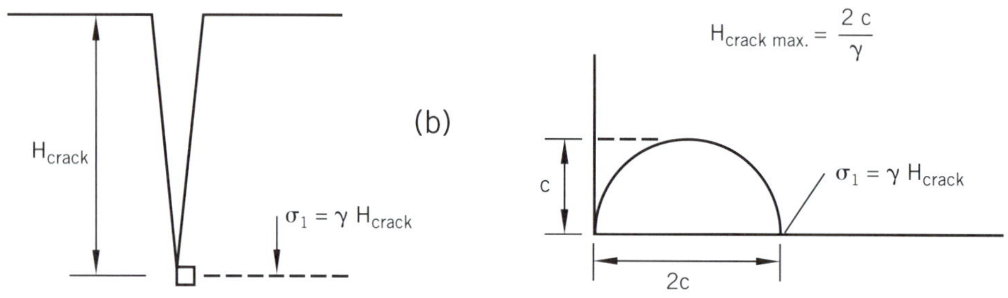

Figure 9.4 Maximum heights:
(a) vertical cut slope
(b) tension crack in clay

To determine the maximum height to which the slope can be cut for a given shear strength c:

$$F = \frac{cH \operatorname{cosec} \alpha}{\frac{1}{2}\gamma H^2 \cot \alpha \sin \alpha} = 1$$

$$\text{Simplifying, } c = \tfrac{1}{4}\gamma H \sin 2\alpha$$

$$H \text{ is maximum when } \frac{dc}{d\alpha} = 0$$

$$= \tfrac{1}{2}\gamma H \cos 2\alpha$$

$$= 0 \text{ when } \cos 2\alpha = 0;\ \alpha = 45°$$

$$c = \tfrac{1}{4}\gamma H_{max}$$

$$\text{or } H_{max} = \frac{4c}{\gamma}$$

If a tension crack of depth h_t occurs at the top of the wedge, point C, it can be shown that

$$H_{max} = \frac{4c}{\gamma} - h_t$$

More rigorous analyses indicate that the numerator is 3.83 rather than 4.

$$\therefore H_{max} = \frac{3.83c}{\gamma} \qquad \text{Eq. (5)}$$

Knowledge of the maximum height of a vertical face of clay is more useful to stability studies than to landslides. However, it has some relevance to the stability of a headscarp and maximum vertical cuts for drains or cutoffs.

Determinations of H_{max} need to be applied with considerable caution. It is an appropriate expression for normally consolidated clay or structurally compacted clay fill when applied with a factor of safety of at least 2. It should not be applied to uncompacted clay fill or clays with low plasticity. In the case of overconsolidated clay, especially fissured clay, it will provide non-conservative information. Note that the critical angle α is 45°; temporary cuts normally should be laid back to 1:1 (horizontal:vertical) or flatter unless they are braced or supported against cave-ins.

Tension Crack

Consider the tension crack, maximum height H_{crack} on Figure 9.4(b). A crack cannot exist if the minor principal stress σ_3 is equal to or greater than zero. It equals zero when the major principal stress σ_1 reaches $\gamma\, H_{crack}$. On the Mohr diagram:

$$\sigma_1 = \gamma\, H_{crack} \text{ when } \sigma_3 = 0$$

$$\sigma_1 - \sigma_3 = 2c$$

$$\therefore \gamma\, H_{crack} = 2c$$

$$H_{crack} = \frac{2c}{\gamma} \qquad \text{Eq. (6)}$$

In stiff clays, Eq. (6) indicates that very deep tension cracks can develop in a landslide. However, at the initiation of sliding, such a deep crack is unlikely because a landslide event usually coincides with heavy rainfall and percolating water will quickly soften the clay on the sides of the crack, thus lowering shear strength c locally.

The depth of tension cracks to be included in analysis is controversial. A suggested limit by Terzaghi (1943) is $\frac{H}{3}$ where H is the vertical height of the slope, or H_{crack} as calculated in Eq. (6), whichever is smaller. In most practical cases, it can be assumed that the tension crack extends down to the water table (usually high) and is water-filled at the onset of the failure.

Active and Passive Failures

The concept of active and passive slip planes is attributed to Rankine and is useful to studies of translational landslides. On Figure 9.5(a), a cohesionless soil is shown in a semi-infinite mass with a horizontal plane surface, and an element of soil with vertical sides is immediately below the surface. The vertical stress at the base of the element is σ_v.

If the horizontal stress is decreased while the vertical stress remains constant, the full shear strength of the soil will be mobilized. At this stress condition, the *active* Rankine state has been reached. If the horizontal stress is increased while holding the vertical stress constant, the horizontal stress becomes the major principal stress while the vertical stress is the minor principal stress; this is the *passive* Rankine state.

These two extreme conditions are shown on the Mohr diagram, Figure 9.5(c). The orientation of the slip planes are shown on Figure 9.5(b). For a horizontal ground surface, the active planes are at $(45° + ½\phi)$ to the horizontal; for the passive planes, the angles are at $(45° - ½\phi)$ to the horizontal.

Measurements by the author on numerous headscarps have consistently shown headscarps inclined at 60° to 70° to the horizontal. If it is assumed that the soil has a strength of $\phi' = 40°$ against first-time sliding, the Rankine slip line for the active case would be at 65° to the horizontal. A similar rupture surface can be observed on the reverse scarp of a graben at the head of a translational landslide.

The breakout slope at the toe of a translational landslide usually erupts to the surface on a relatively flat inclination to the horizontal. Although eruption angles have only occasionally been measured, the Rankine slope (25° to the horizontal for $\phi = 40°$) appears to be reasonable for observed conditions. The low angle of eruption typically produces an elongated hump at the toe.

Terzaghi (1943) showed that the theoretical inclinations change slightly for a sloping ground surface (Figure 9.5d,e).

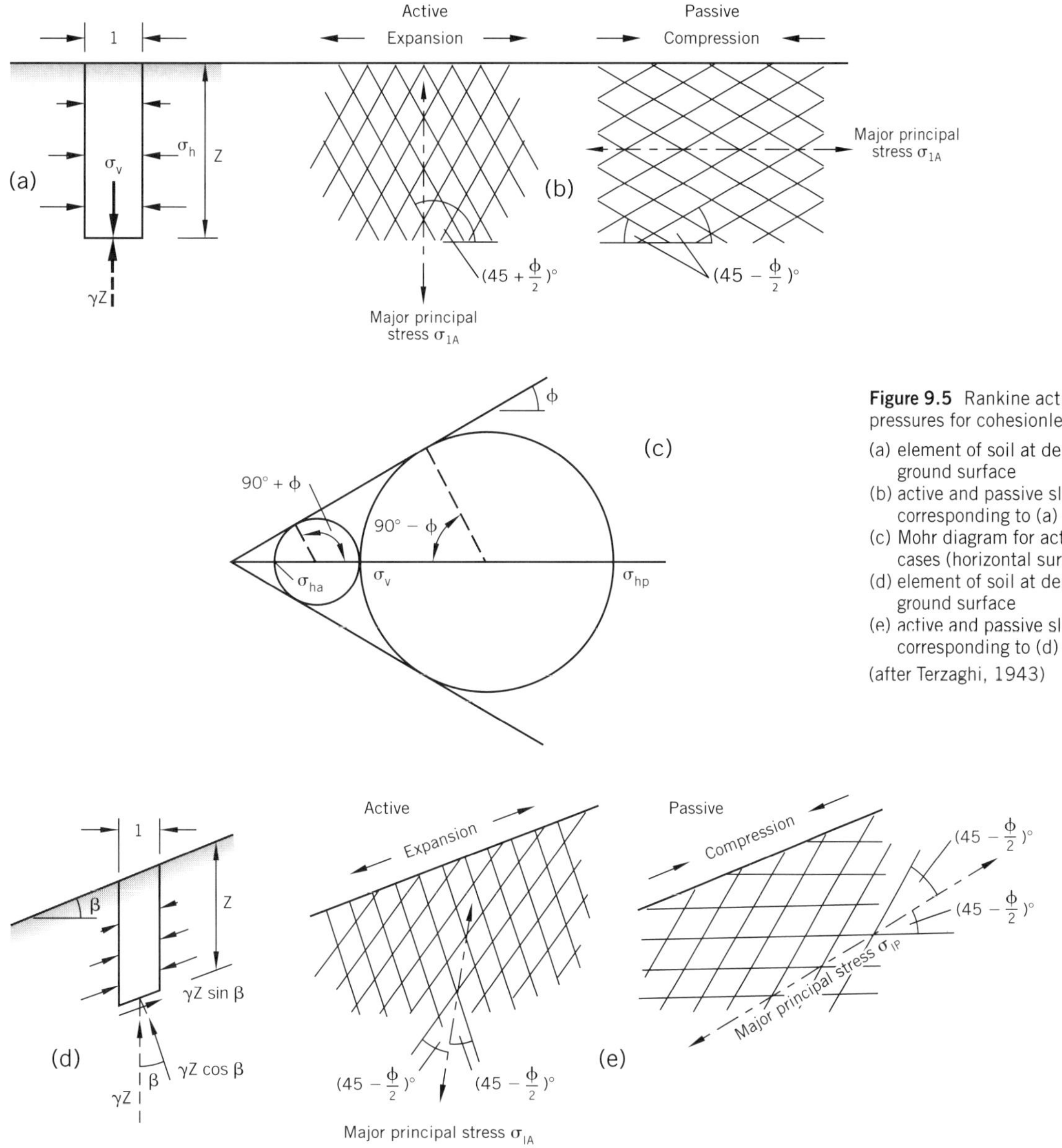

Figure 9.5 Rankine active and passive pressures for cohesionless soil:
(a) element of soil at depth Z below horizontal ground surface
(b) active and passive slip planes corresponding to (a)
(c) Mohr diagram for active and passive cases (horizontal surface)
(d) element of soil at depth Z below sloping ground surface
(e) active and passive slip planes corresponding to (d)
(after Terzaghi, 1943)

9.6 INFINITE SLOPE ANALYSIS

In many large landslides, the major portion of the slide mass is moving approximately parallel to the ground surface. The nature of the slide is controlled by a geological feature such as bedrock, a weak interbed, clay seam, etc. If the landslide is very long relative to its depth, the soil resistance contributed by the entry and exit slip surfaces may be minor.

Stability Equation

On Figure 9.6(a), the slip surface DC, the ground surface AB, and groundwater (height h above the slip surface) are parallel to each other and inclined at angle β to the horizontal. For an infinite slope, the interslice forces P_L and P_R are equal and cancel out.

Weight of soil slice $W = \gamma z b$

Pore pressure u acting at center of slice base:

$$u = \gamma_w h \cos^2 \beta$$

On Figure 9.6(a), groundwater flow is parallel to the slip surface and all flow lines are at angle β to the horizontal. Therefore, line EP (length x) from the center of the slice to the groundwater is an equipotential line. The observed vertical height of water in a piezometer with its tip at E is h_s.

$$x = h \cos \beta$$
$$h_s = h \cos^2 \beta$$

Referring to Figure 9.6:

$$\text{Water force } U = u \frac{b}{\cos \beta} = \gamma_w b h \cos \beta$$

$$N' = \gamma b z \cos \beta - \gamma_w b h \cos \beta$$
$$= b \cos \beta (\gamma z - \gamma_w h)$$

$$S = \frac{c' b}{\cos \beta} + b \cos \beta (\gamma z - \gamma_w h) \tan \phi'$$

$$F = \frac{c' \sec \beta + \cos \beta (\gamma z - \gamma_w h) \tan \phi'}{\gamma z \sin \beta}$$

$$F = \frac{c' \sec^2 \beta + (\gamma z - \gamma_w h) \tan \phi'}{\gamma z \tan \beta} \quad \text{Eq. (7)}$$

Eq. (7) can also be written as:

$$F = \frac{c' + (\gamma z - \gamma_w h) \cos^2 \beta \tan \phi'}{\gamma z \sin \beta \cos \beta} \quad \text{Eq. (8)}$$

At failure, F = 1.00 If c′ = 0, the value of ϕ′ can be determined from Eq. (7) as follows:

$$\tan \phi' = \frac{\gamma z \tan \beta}{(\gamma z - \gamma_w h)} \quad \text{Eq. (9)}$$

Influence of Groundwater on Slope Stability

The general expression for the infinite slope analysis is given by Eq. (7). The effect of varying groundwater levels provides insights into slope stability.

(1) Water head is zero

If h = 0 and c′ = 0

$$F = \frac{\tan \phi'}{\tan \beta} \quad \text{Eq.(10)}$$

The slope fails if β = ϕ′. This is the situation for loose dry sand where the angle of repose β is equal to the shear strength of the sand ϕ′.

(2) Water head at the ground surface

If h = z and c′ = 0

$$F = \frac{\gamma' \tan \phi'}{\gamma \tan \beta} \quad \text{Eq. (11)}$$

For limiting equilibrium, F = 1.00 and the expression becomes:

$$\tan \phi' = \frac{\gamma}{\gamma'} \tan \beta$$

Example: γ = 120 lb./cu. ft. and slope angle β is 18° to the horizontal:

$$\tan \phi = \frac{120}{(57.6)} \tan 18°$$
$$= 0.677$$
$$\phi' = 34°$$

Conclusion: If groundwater is at or near the ground surface, the shear strength ϕ′ on the slip surface is approximately twice the slope angle β.

In summary:

(i) If groundwater is deep and at or below the slip surface, the shear strength ϕ′ will be approximately equal to the average slope angle β.

(ii) If groundwater is close to the ground surface, the shear strength ϕ′ will be approximately twice the average slope angle β.

(3) Ratio of water head h to depth of landslide z

It can be seen from Eq. (7) that, for a given shear strength ϕ′ and slope angle β, the factor of safety F goes down as the water head h rises toward the surface, i.e., as ratio h/z increases towards unity. Example:

β = 20° to the horizontal
c′ = 0, ϕ′ = 35°
z = 40 feet, γ = 120 pcf

Equation (7) becomes:

$$F = 1.92 - (0.025)h$$

If h = 20 feet, F = 1.42
If h = 30 feet, F = 1.17

This result reflects the well-known fact that rising groundwater levels decrease stability. Conversely, any decrease in groundwater level (or methods to prevent it rising during wet seasons) is a sound remedial technique for long translational landslides.

Stability Design Charts

Charts have been prepared to solve the infinite slope analysis quickly based on Eq. (9). The only simplifying assumptions are

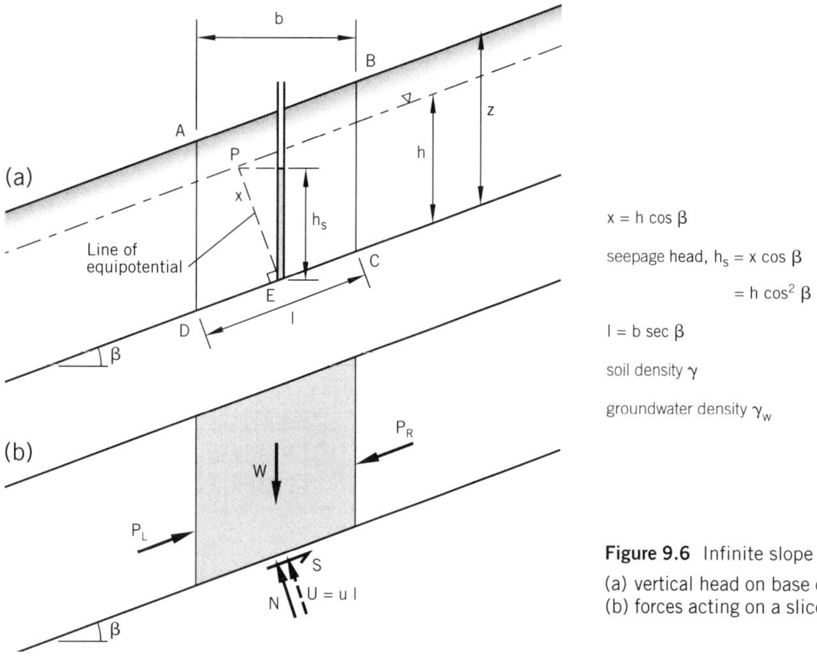

Figure 9.6 Infinite slope analysis:
(a) vertical head on base of slice
(b) forces acting on a slice

that the wet density of the landslide is 124.7 lb./cu. ft., twice the density of water, and that cohesion intercept $c' = 0$. The density is similar to that of many fine-grained soils in landslides and corresponds to a saturated water content of around 26 percent.

For back analysis, the factor of safety is F = 1.00 and Eq. (9) becomes:

$$\tan \phi' = \frac{\gamma z \tan \beta}{(\gamma z - \gamma_w h)}$$

Substitute $\gamma = 2\gamma_w$ and $h = rz$ gives

$$\tan \phi' = \frac{2 \tan \beta}{(2 - r)} \qquad \text{Eq. (12)}$$

This relationship is shown on the chart, Figure 9.7, and is used in the following examples.

Example 1

Depth of sliding mass z = 20 feet
Height of groundwater measured during movement h = 14 feet
Ratio h:z = r = 0.7
Angle of inclination β is 11° to the horizontal

$$\tan \phi' = \frac{2(0.1944)}{1.3}$$
$$= 0.2991$$
$$\phi' = 16.7°$$

On Figure 9.7, the broken line shows how the same result is obtained from the chart.

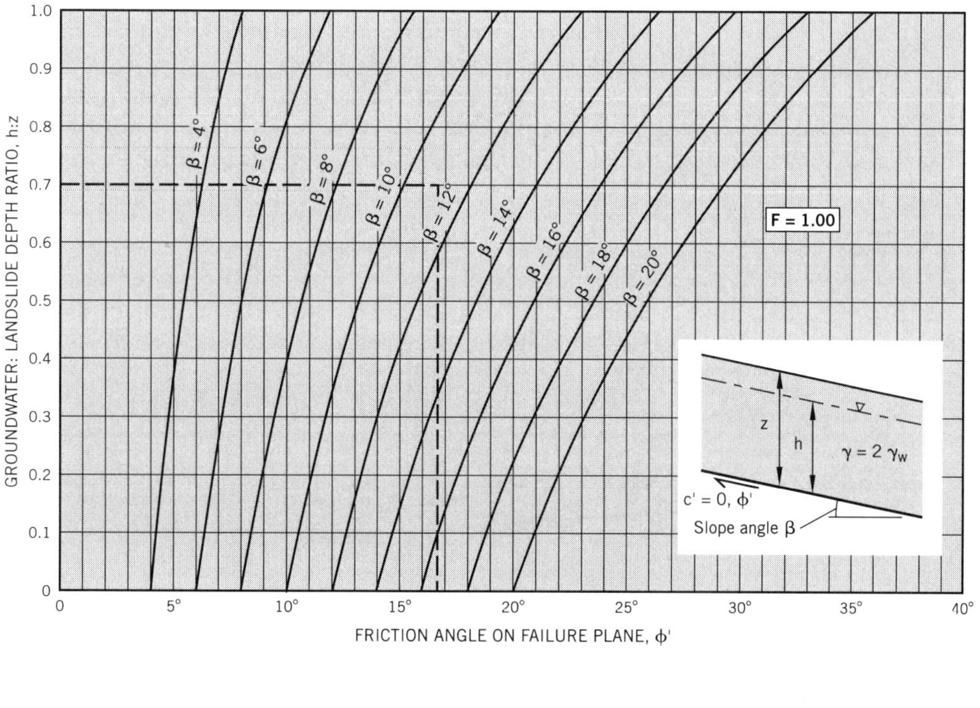

Figure 9.7 Infinite slope analysis. Determination of ϕ' on failure plane based on slope angle β and groundwater/landslide depth ratio.

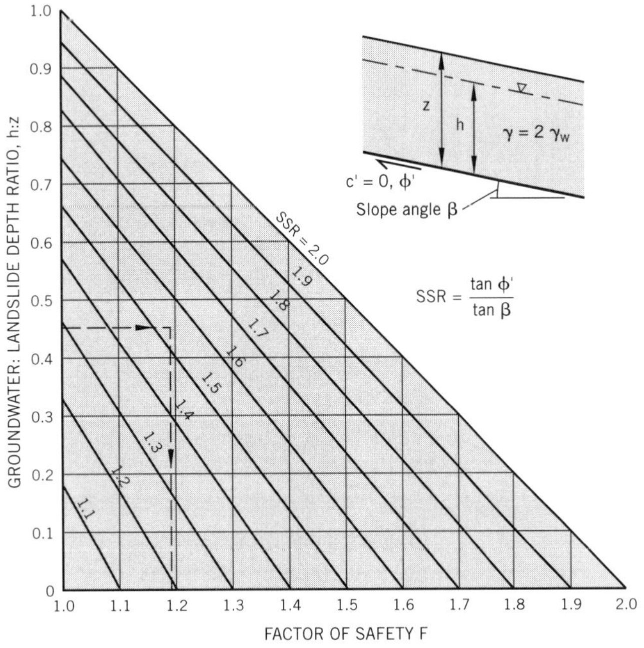

Figure 9.8 Infinite slope analysis. Determination of factor of safety F for known groundwater: landslide depth ratio (h:z) and known shear strength:slope inclination ratio (tan ϕ'/tan β).

Having obtained the shear strength on the deep failure plane corresponding to failure, a second chart Figure 9.8 can be used to either (i) determine the factor of safety produced by remedial drawdown of groundwater, or (ii) determine the amount of drawdown needed to achieve a pre-selected factor of safety.

Using equation (7) and substituting $\gamma = 2\gamma_w$;

$$F = \left(1 - \frac{h}{2z}\right) \frac{\tan \phi'}{\tan \beta} \qquad \text{Eq. (13)}$$

To use the chart, first calculate the strength-to-slope ratio SSR = tan ϕ'/tan β

Example 2

Trench drains are used to draw down groundwater levels in the landslide described under Example 1. If the average groundwater level is dropped from the existing 0.7z to 0.45z, what is the resulting factor of safety?

By calculation, SSR = tan ϕ'/tan β = 0.2991/0.1994 = 1.54
F = (1 − 0.225)(1.54)
= 1.19

On Figure 9.8, the broken lines show how the result is obtained directly from the graphs.

Example 3

Using the data from Example 2, how much would the average drawdown have to be to achieve a factor of safety of 1.30?

Ratio SSR = 1.54
From the chart, Figure 9.8: For F = 1.30, h/z = 0.31.
Depth of groundwater to achieve F = 1.30 = (0.31)20
= 6.2 feet
Original groundwater depth = 14 feet
Therefore, required average drawdown = 14 − 6.2
= 7.8 feet

By calculation:

$$\frac{h}{2z} = 1 - \frac{F}{(\tan \phi' / \tan \beta)}$$

$$= 1 - \frac{1.30}{1.54}$$

$$= 0.156$$

$\therefore$ h/z = 0.312 (Remainder of the calculation is the same as given above)

9.7 DOUBLE WEDGE ANALYSIS

Double Wedge Failure Characteristics

Landslides that can be modeled as a double wedge are fairly common. The upper wedge is usually a steeply inclined headscarp that pushes against a lower wedge inclined on a flatter shear plane. The lower wedge slip surface is controlled by a geological feature such as bedrock surface. Figure 9.9 shows a typical cross-section. The toe may move over a cliff or river bank, or simply exit into a cut slope.

A double wedge configuration should be thought of as comprising a driving (upper) wedge and a resisting (lower) wedge. On Figure 9.9, the upper wedge ABD shears along

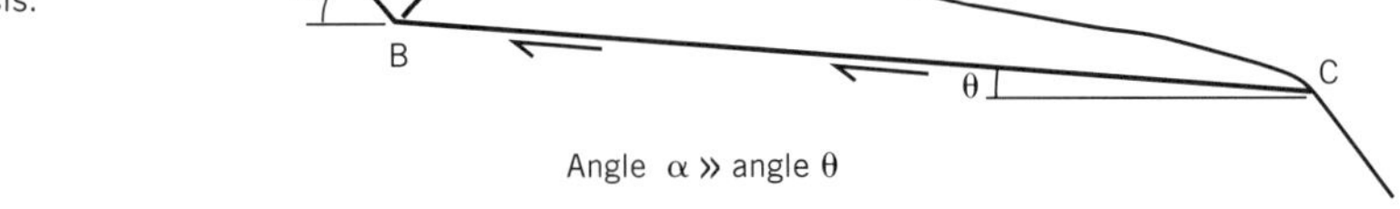

Figure 9.9 Double wedge analysis. Typical section.

plane AB and the lower wedge BDC shears along plane BC. Because the two slip surfaces usually have significantly different angles of inclination α and θ, a double wedge failure develops a graben AD at the head of the landslide. The lower wedge moves outward and the upper wedge drops during the slide movement, causing a reverse (antithetic) scarp to develop at point D. Note that surface BD is a shear plane with similar characteristics to the headscarp AB.

As the upper wedge sinks to accommodate the downslope movement of the lower wedge, there is an area of intense shear and fracture occurring near point B. Projects built directly above the *upper* wedge are likely to experience distress during active movements. The lower wedge may stay relatively intact and thus may not damage structures on its surface, except near the edges of the landslide.

Estimate of Landslide Depth

The graben is a distinguishing feature of a translational landslide such as a double wedge failure. At a preliminary reconnaissance level, the maximum depth near the top of the landslide can be approximately evaluated by measuring the width of the graben. Figure 9.10 shows a graben feature. Assuming that the reverse scarp develops at the same angle α to the horizontal as the headscarp, the vertical depth of the graben can be estimated by recreating the conditions at the start of the landslide. Triangle A′B′D′ on Figure 9.10(a) shows the recreated initial condition, and Figure 9.10(b) is the relevant geometry.

It can be shown that:

$$E'B' = A'D' \frac{\sin(\alpha - \beta)\sin(\alpha + \beta)}{\sin 2\alpha \cos\beta} \qquad \text{Eq. (14)}$$

Distance A′D′ can be approximately obtained by measuring the distance AD along the slope of the depressed graben. Headscarps typically occur at 60° to 70° to the horizontal, probably reflecting the Rankine wedge of Section 9.5. As an example, assume an outer slope angle β of 20° to the horizontal and angle α of 65° to the horizontal then, from Eq. (14), $E'B' = 0.98\,A'D'$.

As an estimate, therefore, the maximum vertical depth E′B′ to the base of the upper wedge is approximately equal to the length of the graben measured along the slope of the graben. This relationship is useful for obtaining a preliminary understanding of the landslide geometry during initial studies, and for planning site investigations.

Secondary Cracks and Scarps

Cracks and scarps within the mass of a long translational landslide are common. They are referred to as secondary features and should be interpreted with caution.

Secondary scarps (Figure 9.11) are sometimes interpreted to be evidence of retrogressive landsliding; i.e., the slide first

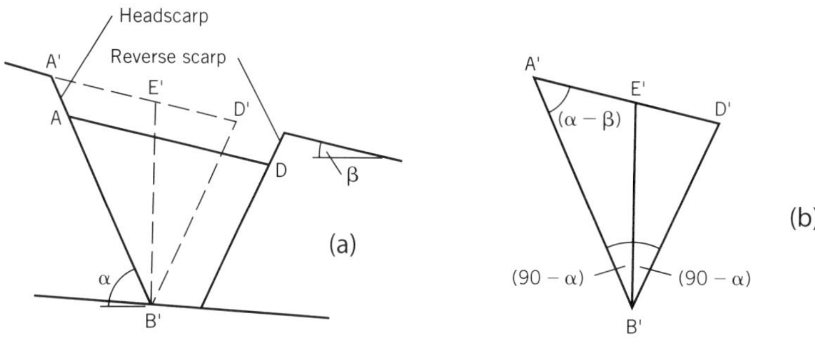

Figure 9.10 Estimation of maximum depth to the slip surface below a graben:
(a) recreating the initial conditions at onset of sliding
(b) triangle geometry

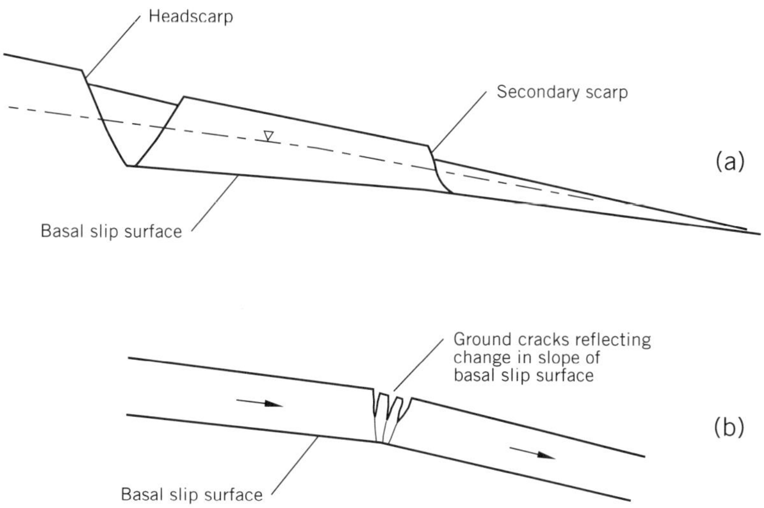

Figure 9.11 Interpretation of secondary cracks and scarps on translational landslides:
(a) secondary scarp resulting from faster velocity lower down the slope or increased inclination of the basal slip surface
(b) concentrated secondary cracks reflecting change of inclination of the basal slip surface

occurs near the base of the slope and its movement undermines the slope above, thereby triggering movement in the upper slope. Another interpretation of a secondary scarp feature is that the lower slope is moving at a faster velocity than the upper slope during the landslide due to local conditions (e.g., locally higher groundwater levels, steeper basal slip surface, erosion at the bottom of the slope, etc.). Unless there is evidence that the landslide has regressed upslope during an observable time interval between movements, it is more likely that a landslide with secondary scarps occurred as a single event; i.e., the upper movement did not occur as a result of failure beginning at the bottom of the slope.

Secondary cracks are common, especially on shallow-depth landslides. On long translational landslides, concentrated cracking running across the slope probably indicates an increasing gradient of the basal rupture surface on the downhill side, shown in Figure 9.11(b).

Examples of Double Wedge Landslides

In Part C, the following case histories involve double wedge failures:

No. 1—Washington Park Reservoirs Slide
No. 5—Pelton Lower Slide
(*) No. 6—Pelton Upper Slide
No. 10—Hagg Lake Perimeter Road, Slide 4
(*) especially recommended reading

Double Wedge Stability Computations

As shown on Figure 9.12(a), the double wedge analysis first subdivides the landslide by an imaginary vertical line BE that passes through the intersection point B of the two planar slip surfaces. The two wedges, AEB and ECB, are analyzed as free-body diagrams with a thrust force P acting on the vertical plane BE of each body (Figure 9.12b).

The normal forces N_1 and N_2 are dependent on the unknown P values. This can be overcome by drawing polygons of forces for each wedge (Figure 9.13). Forces W and U are plotted and the mobilized shear resistance is determined by dividing the maximum resistance by the factor of safety F. The value of P is obtained for the polygons of the two wedges. By varying F in each polygon, graphs are drawn for values of P vs. F (Figure 9.13c). The intersection of the graphs provides the common P to satisfy equilibrium, and thus obtain F.

To make the computations, the angle δ for the P force has to be estimated. Since the vertical line BE is near the center of the graben, there is unlikely to be much shear stress developing on the vertical plane at this location (unlike a vertical wall which is rigid compared to the abutting soil). Therefore, angle δ should be small and is arbitrarily assigned a value of around 10° to the horizontal. However, a concern in the double wedge analysis is that the computed F is affected by the assumed value of angle δ.

As alternatives to the suggested δ value given previously (and in the following section on triple wedges), the designer could:

- Take δ as the average slope of the ground surface and slip surface (Lowe and Karafiath, 1960); in this case, the slip surface slope would be the average of the slopes below adjoining wedges.
- Use a parametric approach for values of δ between 0° and ϕ' and select $F_{minimum}$.

Most calculations show that the factor of safety F increases as angle δ increases. Therefore, a horizontal (δ = 0) or near-horizontal (δ = 8°–12°) interslice force should be reasonably conservative. The results can be compared with a computer-generated analysis using Spencer's Method or Janbu's Rigorous Method—see Section 9.10.

For landslide analysis, F = 1.00. In many landslides, the lower wedge is shearing along a preexisting landslide plane with an unknown residual shear strength ϕ'_r. By assigning a

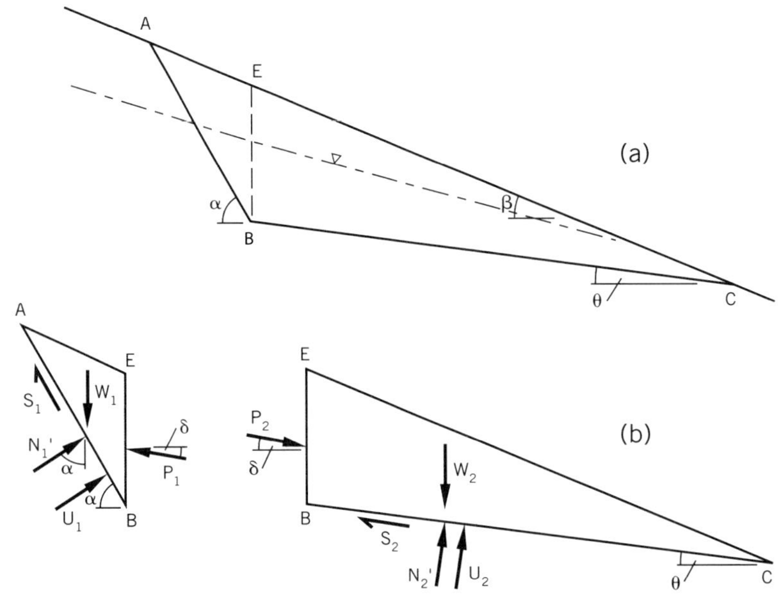

Figure 9.12 Double wedge analysis:
(a) separation into upper and lower wedges by vertical line BE
(b) forces acting on the wedges

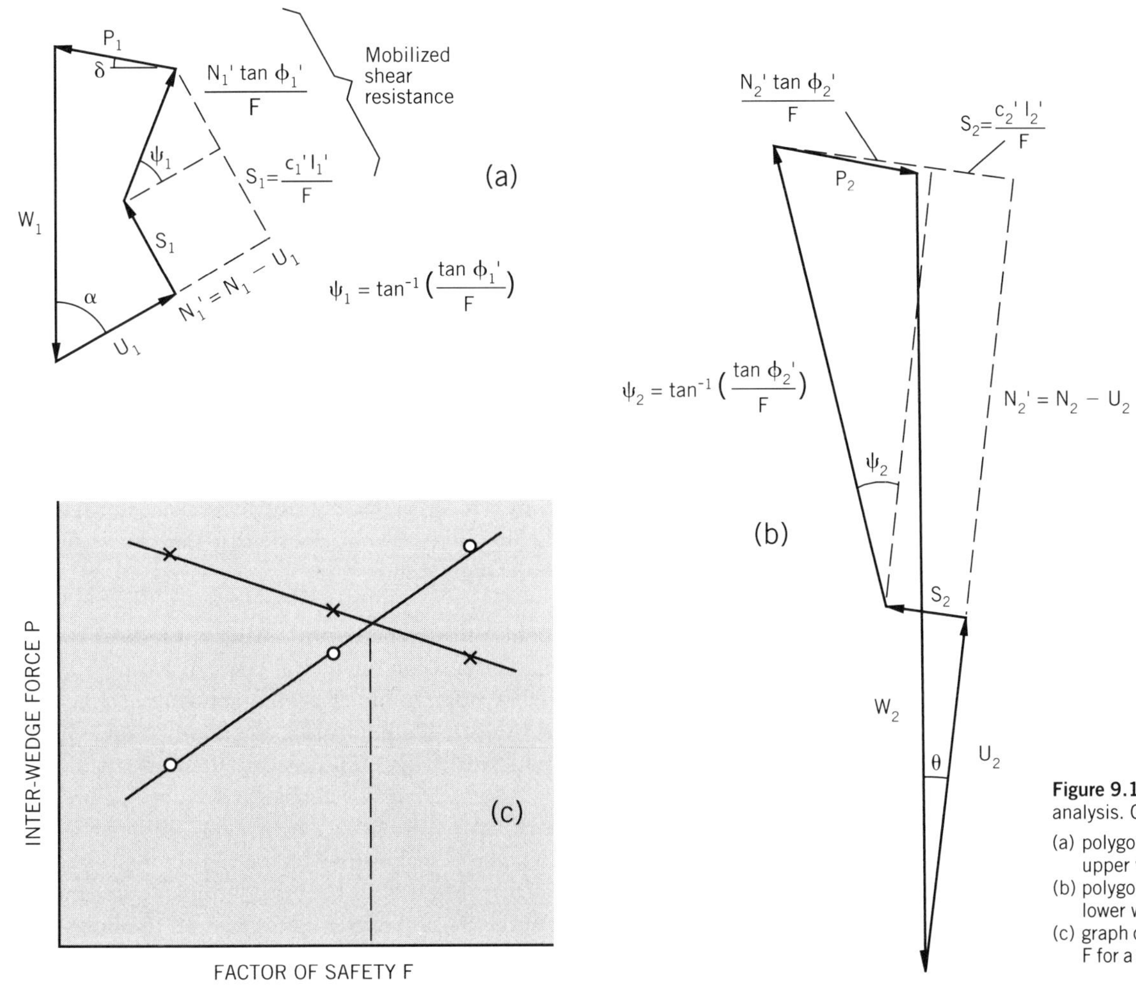

Figure 9.13 Double wedge analysis. Graphical solution:
(a) polygon of forces for the upper wedge
(b) polygon of forces for the lower wedge
(c) graph of P vs. F to determine F for a common P

"first-time" shear strength to slide plane AB, the double wedge analysis can be used to determine ϕ'_r by back analysis. There are three ways to make this calculation: Method A and Method B are graphical, and Method C is analytical.

Example of Double Wedge Computation

For the double wedge shown on Figure 9.12, use the following data: γ = 120 lb./cu. ft.; ϕ' on plane AB = 35° (c' = 0); α = 60°; β = 22°; θ = 7½°; δ =10°; length BC = 280 feet; groundwater profile as shown on figure.

Upper wedge	W_1 = 268 kips/ft.
	U_1 = 41 kips/ft.
Lower wedge	W_2 = 1284 kips/ft.
	U_2 = 433 kips/ft.

Determine the residual strength ϕ'_r (assuming c'_r = 0) along the ancient slide plane BC.

Graphical Solution for Double Wedge, Method A

For the upper wedge, graphically determine the interwedge force P_1 for various factors of safety F (Figure 9.14 and Table 9.3). The engineer then selects an initial ϕ'_r for the lower wedge and completes the polygon of forces for the same range

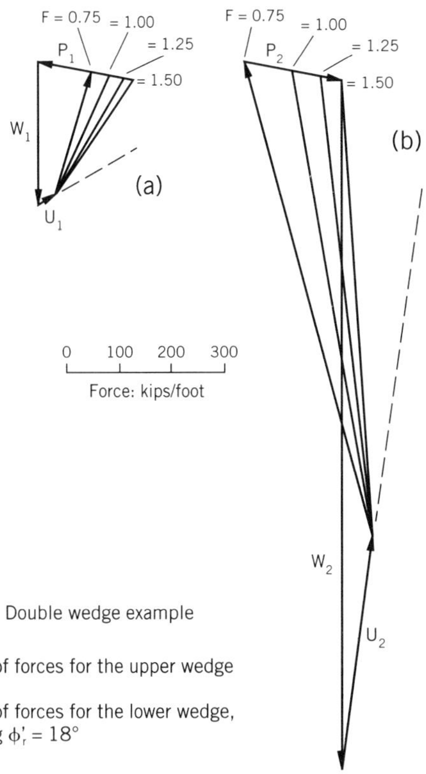

Figure 9.14 Double wedge example (Method A):
(a) polygon of forces for the upper wedge ϕ'= 35°
(b) polygon of forces for the lower wedge, assuming ϕ'_r = 18°

Table 9.3 Double Wedge Analysis: Calculations for Graphical Method A

	Assumed Factor of Safety F			
	0.75	1.00	1.25	1.50
UPPER WEDGE				
$\phi_1' = 35°$				
$\tan \phi_1'/F$	0.934	0.700	0.560	0.467
ψ_1	43°	35°	29°	25°
P_1 (from polygon)	104	140	167	186
LOWER WEDGE				
(1) $\phi_r' = 15°$				
$\tan \phi_r'/F$	0.357	0.268	0.214	0.179
ψ_2	20°	15°	12°	10°
P_2 (from polygon)	132	48	5	(negative)
From graph, Figure 9.15(a), F = 0.80				
(2) $\phi_r' = 18°$				
$\tan \phi_r'/F$	0.433	0.325	0.260	0.217
ψ_2	23½°	18°	14½°	12°
P_2 (from polygon)	192	98	42	5
From graph, Figure 9.15(a), F = 0.905				
(3) $\phi_r' = 21°$				
$\tan \phi_r'/F$	0.512	0.384	0.307	0.256
ψ_2	27°	21°	17°	14½°
P_2 (from polygon)	260	148	80	42
From graph, Figure 9.15(a), F = 1.02				

of F used for the upper wedge. In the first trial for an initial ϕ_r' of 15°, the P_1 and P_2 forces are equal when F = 0.80 (Figure 9.15a). To determine the ϕ_r' that corresponds to F = 1.00, the value of ϕ_r' used in the analysis has to be raised. The second trial uses ϕ_r' = 18°, which is still too low because it gives F = 0.905 in the analysis. A third trial is made assuming ϕ_r' = 21°, giving a computed F = 1.02.

The computed ϕ_r' values are plotted against F on Figure 9.15(b). This graph shows that the ϕ_r' corresponding to F = 1.00 is 20½°.

It can be seen from the worked example (Figure 9.14) that the lower wedge is larger in weight than the upper wedge. To obtain good accuracy in the computation, it is important to use a large protractor (8-inch diameter or larger), large drawings (minimum 8½ inch × 11 inch or A4), and fine pen or pencil (preferably not more than 0.2 mm line thickness). It is also recommended that the two wedges be drawn to the same scale.

Graphical Solution for Double Wedge, Method B

For a landslide, a second graphical procedure can be used to more quickly determine ϕ_r'. The polygon of forces are rearranged so that the interwedge force can be joined at the upper and lower wedges. The general arrangement is shown on Figure 9.16 based on the double wedge slope failure given on Figure 9.12.

The graphical analysis of Method B is more awkward to construct than Method A because the pore-water pressure load U_2 of the lower wedge is not drawn off the intersection with weight W_2 in the lower wedge. Instead, its position

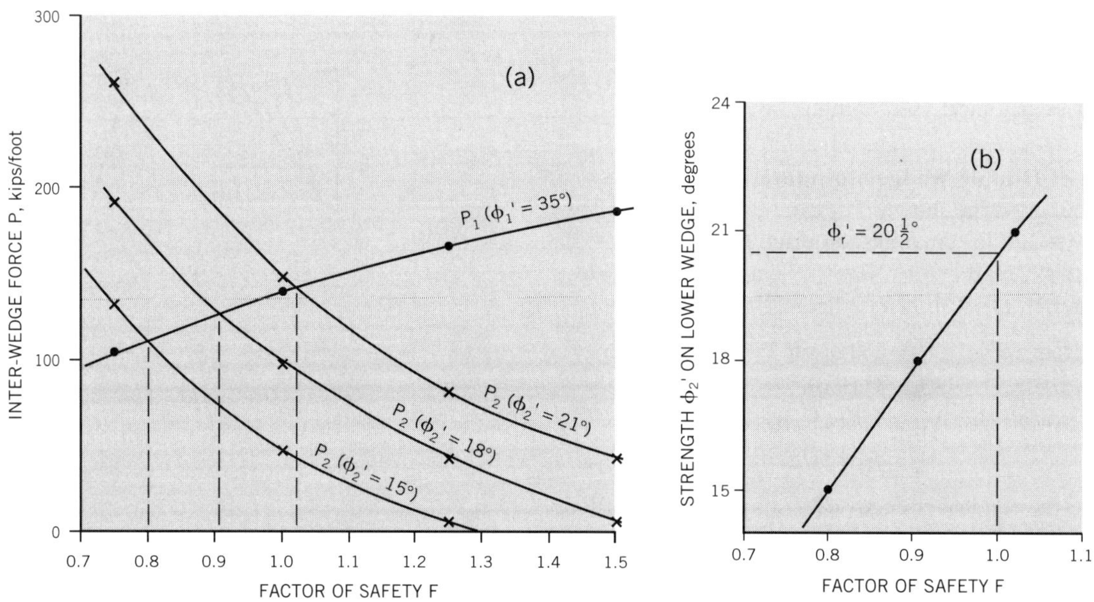

Figure 9.15 Double wedge example (Method A):
(a) P vs. F for various assumed values of ϕ_2' on the lower wedge
(b) determination of ϕ_r' on the lower wedge corresponding to F = 1.00

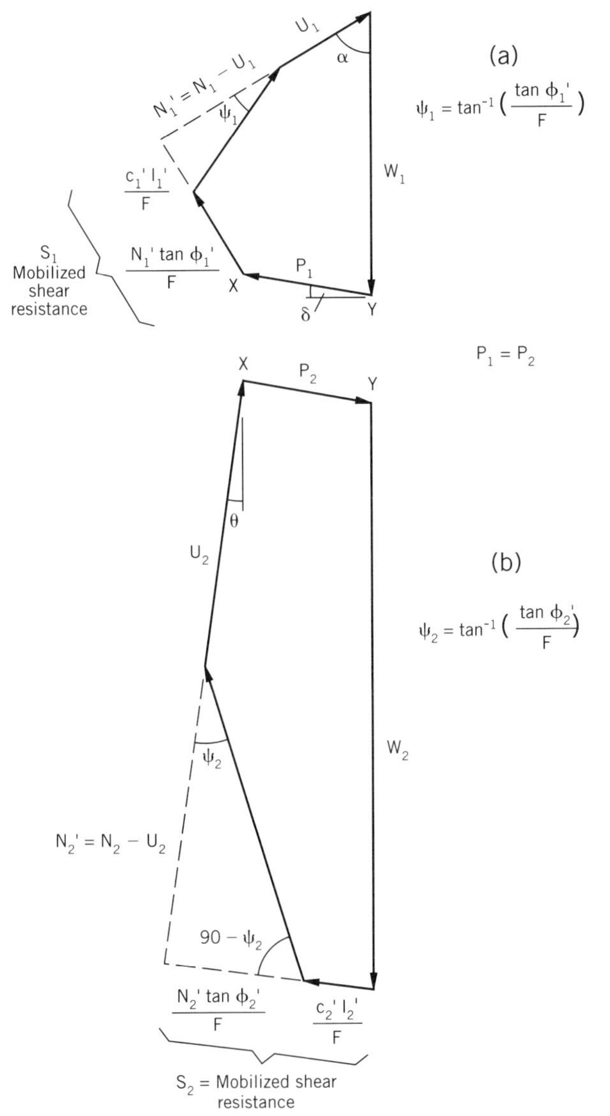

Figure 9.16 Double wedge graphical solution (Method B). Polygon of forces for:
(a) upper wedge
(b) lower wedge

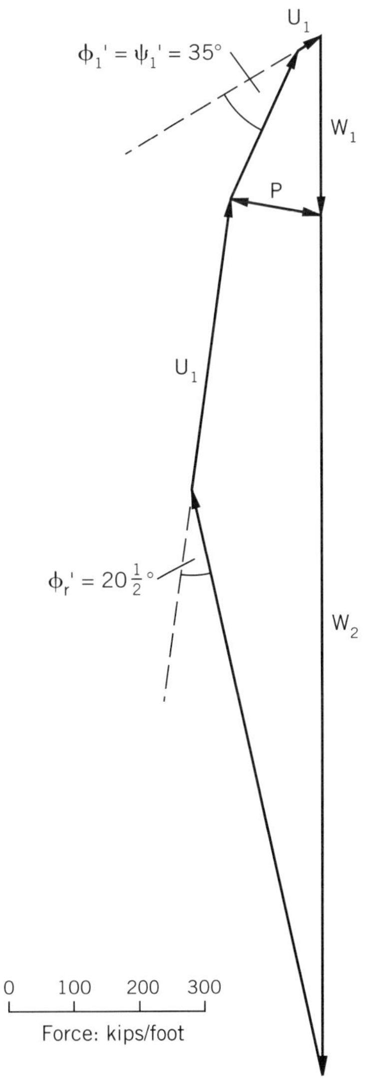

Figure 9.17 Double wedge analysis. Graphical Method B.

depends on the unknowns P and angle ψ_2. However, when it is used to determine the residual strength ϕ'_r (for $c'_r = 0$) on the lower wedge, it is extremely fast. For this analysis, the shear strength ϕ' along the failure plane of the upper wedge has be measured by laboratory tests or assumed.

For the worked example used earlier, the shear strength of the upper wedge $\phi'_1 = 35°$ and the cohesion is zero for both wedges. First the upper wedge is drawn. For $F = 1.00$, $\psi_1 = \phi'_1 = 35°$; $\alpha = 60°$; $\delta = 10°$ (assumed); $\theta = 7\frac{1}{2}°$; $W_1 = 268$ kips/ft.; $W_2 = 1{,}284$ kips/ft.; $U_1 = 41$ kips/ft.; and $U_2 = 443$ kips/ft.

Next, draw the lower wedge using the common line XY of Figure 9.16. (Note: lines XY are not identical if Method A is being used; the objective in Method A is to find equality in the lines XY).

Draw U_2 from point X; then draw a straight line from the lower end of line U_2 to the bottom of W_2. The angle ψ_2 is then ϕ'_r. The polygon of forces for this example is on Figure 9.17. The measured ϕ'_r is $20\frac{1}{2}°$, in agreement with the value calculated from Method A.

Analytical Solution for Double Wedge, Method C

The back-analysis computation shown in the worked example of graphical Methods A and B can be solved analytically for $c' = 0$ on all wedges. For $F = 1.00$, $\psi = \phi'$.

Referring to the force diagrams shown on Figure 9.13: $P_1 = P_2 = P$. Problem: Determine ϕ'_r for the lower wedge.

Upper Wedge

$$W_1 = U_1 \cos\alpha + \frac{P\cos\delta - U_1 \sin\alpha}{\tan(\alpha - \phi'_1)} + P\sin\delta$$

$$P = \frac{U_1 \sin\alpha + (W_1 - U_1\cos\alpha)\tan(\alpha - \phi'_1)}{\cos\delta + \sin\delta\tan(\alpha - \phi'_1)} \qquad \text{Eq. (15)}$$

$P = 144$ kips/ft.

Lower Wedge

$$W_2 = U_2\cos\theta - P\sin\delta + \frac{U_2\sin\theta + P\cos\delta}{\tan(\phi'_r - \theta)}$$

$$\tan(\phi'_r - \theta) = \frac{U_2\sin\theta + P\cos\delta}{W_2 - U_2\cos\theta + P\sin\delta} \qquad \text{Eq. (16)}$$

Substitute $P = 144$ kips/ft from the upper wedge calculation.

$$\phi'_r - \theta = 12.7^\circ$$
$$\phi'_r = 20.2^\circ$$

This result agrees with the graphical results of Methods A and B.

Remedial Analysis of a Double Wedge Failure

Remedial treatments of double wedge and triple wedge analyses are similar. They are presented at the end of the triple wedge section, which follows this section. Groundwater lowering and buttress construction have been given as examples. Most other remedial treatments can be analyzed in a similar fashion.

9.8 TRIPLE WEDGE ANALYSIS

Triple Wedge Failure Characteristics

The triple wedge is another common slip surface for larger landslides. Like the double wedge, it is usually created by a geological feature (such as bedrock, a weak ancient slip plane, or a soft sediment) that underlies the main translational segment of the landslide. The strength of this midsection is often difficult to determine from laboratory shear tests, especially if failure is occurring through a thin shear zone.

Figure 9.18 shows a typical triple wedge configuration. The upper wedge ABD is the driving wedge (i.e., more driving force than resistance to sliding) and the other two wedges BDHC, HCG generally have net resistance to landsliding. The upper and lower wedges are active and passive failures, respectively.

After significant movement occurs (Figure 9.18b), the upper wedge develops a graben below the headscarp as the wedge ADB drops into the space created by the downslope movement of the middle wedge BDHC. Point D moves to D′ and point B moves to B′. The characteristics of the upper wedge have been described in the preceding section.

At the lower end of the landslide, the middle wedge thrusts into the lower wedge, causing the toe to heave up as the slide moves up the stable inclined plane CG. Point C moves to C′, and point H moves to H′. A second passive-type failure develops along CH. This heave mechanism suggests that the ground directly above point C is rising nearly vertically, and there should be relatively minor shear occurring on this vertical plane. For this reason, it is recommended that the interslice force acting between the middle and lower wedges be assumed horizontal on low to moderate slopes. On steeper slopes, a downslope inclination of 10° to the horizontal is suggested.

Where a triple wedge failure is encountered, the site investigation should attempt to measure the ingress (headscarp) and egress (toe) angles of inclination. There is a surprising lack of data available in case histories on the inclination of the egress plane because it is not normally exposed when the landslide overrides the ground downslope. However, it can be examined by test pits. Toe heave is more apparent when the depth to the bedrock or weak layer is close to the surface.

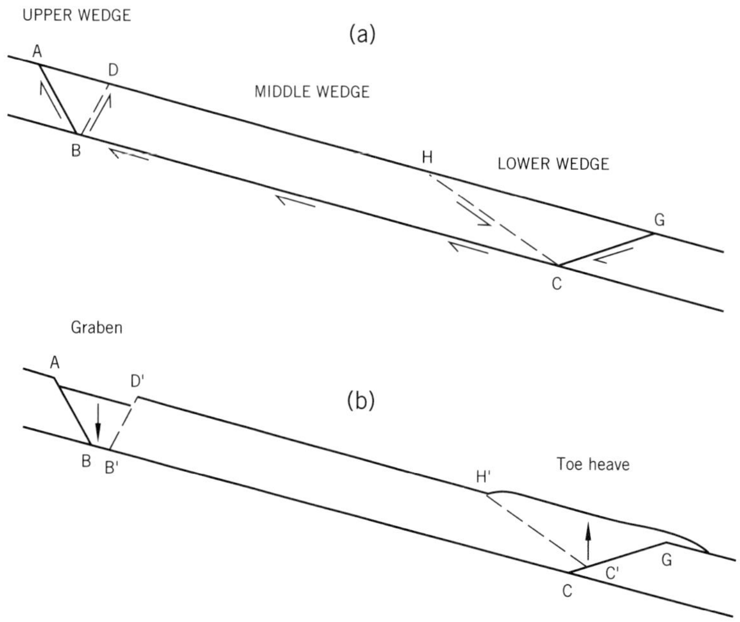

Figure 9.18 Typical triple wedge failure:
(a) at start of sliding
(b) after significant downslope movement

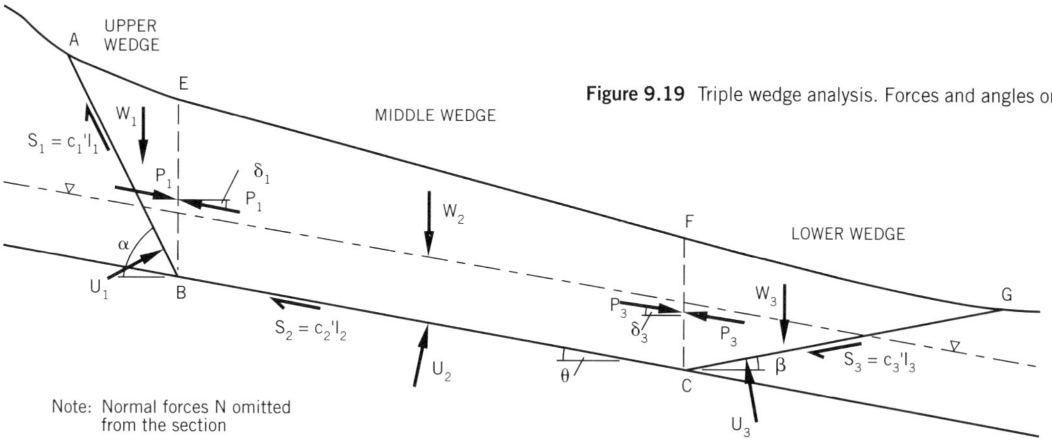

Figure 9.19 Triple wedge analysis. Forces and angles on each wedge.

Triple Wedge Computations

The triple wedge computation is similar to the double wedge analysis, except that there are two unknown interwedge forces rather than one. The general force diagram for a triple wedge is shown on Figure 9.19. The force polygons of Figure 9.20 show how the polygons are constructed for each wedge.

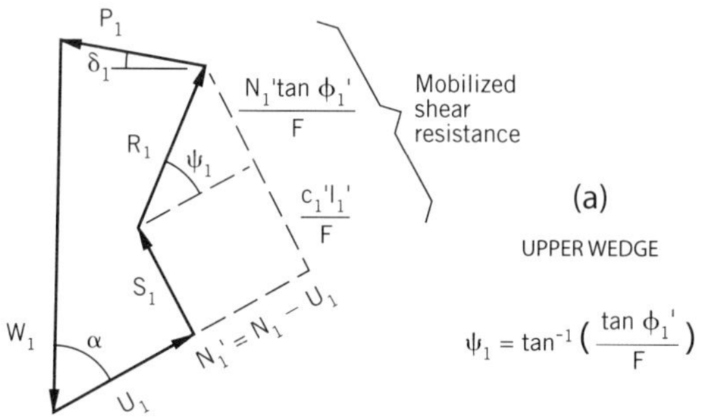

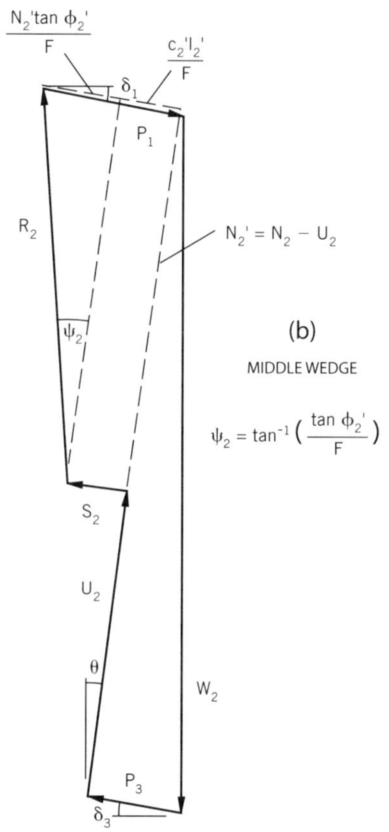

Figure 9.20 Triple wedge analysis. Polygons of forces for:

(a) upper wedge
(b) middle wedge
(c) lower wedge for the typical section shown on Figure 9.19

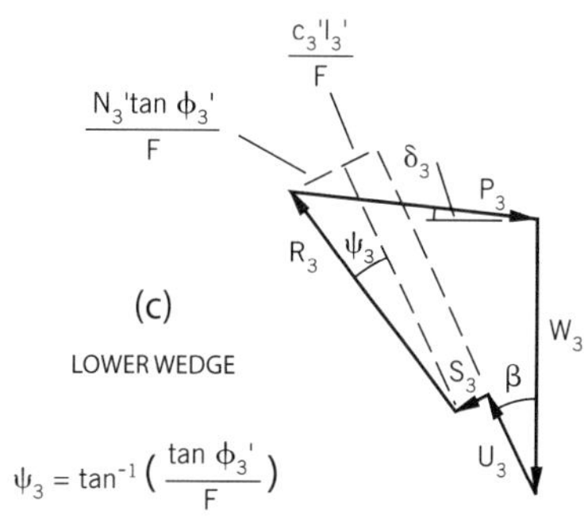

For a landslide, the factor of safety is 1.00 and angle ψ is replaced by the fully mobilized shearing resistance, ϕ' or ϕ'_r. The force polygons for a landslide can be redrawn as shown on Figure 9.21 to align the interwedge P forces for each wedge. The inclusion of cohesion intercept c′ is shown on Figure 9.21(b). However, in most effective stress analyses, the cohesion intercept c′ is zero; in this case, the polygon shown on Figure 9.21(a) is used, as discussed in the graphical and analytical methods presented below.

In many landslide investigations, the key issue is to determine the shear strength below the middle wedge. This strength is usually very difficult and often impossible to deter-

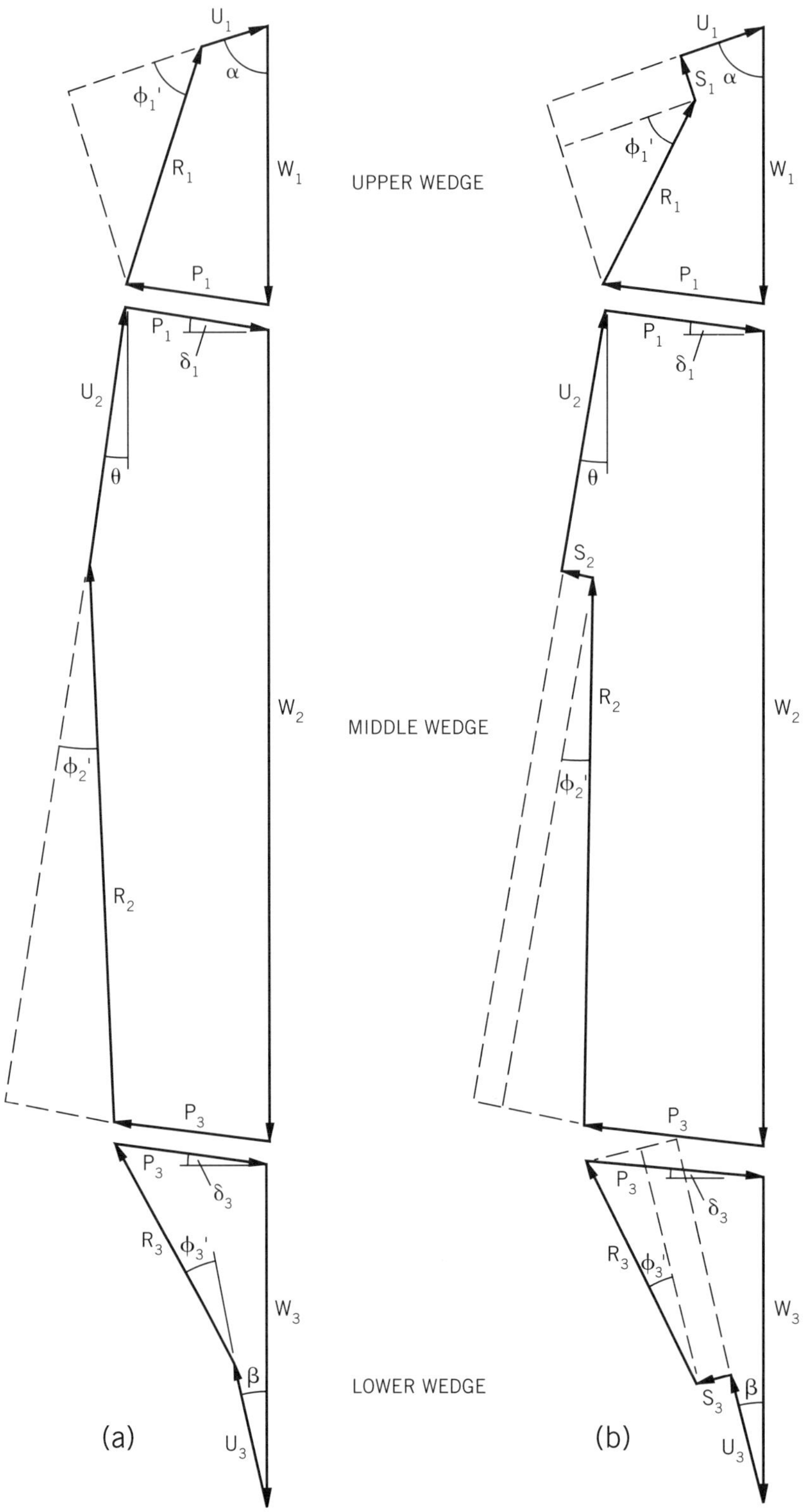

Figure 9.21 Triple wedge analysis for a landslide (F=1.00), aligning the interwedge force P:
(a) with friction only
(b) with cohesion and friction components

mine from a site investigation because of the shallow depth of the discrete shear zone and the distinctly different shear strength properties on it compared to the surrounding ground. This parameter is the focus of the analyses given below, although the analyses are applicable to other situations.

The strength of the overburden soils can be determined from laboratory shear tests on undisturbed samples. If such samples are unavailable, the strengths of sands and gravels can be estimated by the methods described in Chapter 7.

Graphical Solution for Triple Wedge, Method A

Starting at the top, the polygon of forces for the upper wedge is drawn, using the measured or estimated values of overburden shear strength (Figure 9.20). Next, draw the force diagram for the middle wedge with a first trial value for the residual shear strength. Complete the force diagram for the triple wedge by drawing the force polygon for the lower wedge. At the end of the first trial, it is likely that the total polygon of forces does not close at the bottom of the lower polygon because the first trial value of residual shear strength is incorrect. This lack of closure requires error force E to be added to P_3 to achieve closure. However, by examining the polygon, it should be clear whether the initial trial strength was too high or too low.

Two further trials should be undertaken so that error E is both positive and negative. Plot E vs. assumed strength ϕ'_2 below the middle wedge to determine by interpolation the shear strength ϕ'_2 at which the polygons close, corresponding to F = 1.00.

As recommended earlier, the inclination of angle δ to the horizontal is assumed to be 10° for the upper/middle wedge and 0° for the middle/lower wedge. The analysis can be repeated using other assumptions for the inclination of the interwedge forces to determine the sensitivity of the results to these assumptions.

Rapid Graphical Solution for Triple Wedge, Method B

Draw in weights $W_1 + W_2 + W_3$ as a vertical line, Figure 9.21(a). Complete the upper wedge and lower wedge polygon of forces to obtain P_1 and P_3. Put U_2 on the middle wedge. Finally, draw a line R_2 from the base of the U_2 force to force P_3. The angle between this line and the extension of force U_2 is ϕ'_2, the required answer. If c'_2 is not zero, the cohesion force must be added before drawing the line for R_2 (Figure 9.21b).

Analytical Solution for the Triple Wedge, Method C

The triple wedge analysis for a landslide can be solved analytically. The terms are defined on the section, Figure 9.19, and the force polygons of Figure 9.21(b).

Upper Wedge:

Resolve forces vertically:

$$W_1 = U_1 \cos\alpha + P_1 \sin\delta_1 + c'_1 1_1 \sin\alpha + \frac{P_1 \cos\delta_1 + c'_1 1_1 \cos\alpha - U_1 \sin\alpha}{\tan(\alpha - \phi'_1)}$$

Rearranging terms gives:

$$P_1 = \frac{U_1 \sin\alpha - c'_1 1_1 \cos\alpha + (W_1 - U_1 \cos\alpha - c'_1 1_1 \sin\alpha)\tan(\alpha - \phi'_1)}{\cos\delta_1 + \sin\delta_1 \tan(\alpha - \phi'_1)} \quad \text{Eq. (17)}$$

Lower Wedge, treated in a similar manner:

$$W_3 = U_3 \cos\beta - c'_3 1_3 \sin\beta + \frac{P_3 \cos\delta_3 - U_3 \sin\beta - c'_3 1_3 \cos\beta}{\tan(\beta + \phi'_3)} - P_3 \sin\delta_3$$

which can be expressed:

$$P_3 = \frac{U_3 \sin\beta + c'_3 1_3 \cos\beta + (W_3 - U_3 \cos\beta + c'_3 1_3 \sin\beta)\tan(\beta + \phi'_3)}{\cos\delta_3 - \sin\delta_3 \tan(\beta + \phi'_3)} \quad \text{Eq. (18)}$$

Middle Wedge:

$$W_2 = U_2 \cos\theta + c'_2 1_2 \sin\theta + P_3 \sin\delta_3 - P_1 \sin\delta_1 + \frac{P_1 \cos\delta_1 + U_2 \sin\theta - c'_2 1_2 \cos\theta - P_3 \cos\delta_3}{\tan(\phi'_2 - \theta)}$$

which rearranges to:

$$\tan(\phi'_2 - \theta) = \frac{P_1 \cos\delta_1 + U_2 \sin\theta - c'_2 1_2 \cos\theta - P_3 \cos\delta_3}{W_2 - U_2 \cos\theta - c_2 1_2 \sin\theta + P_1 \sin\delta_1 - P_3 \sin\delta_3} \quad \text{Eq. (19)}$$

The analytical procedure is:

1. Calculate the interwedge force P_1 of the upper wedge using Eq. (17).
2. Calculate the interwedge force P_3 of the lower wedge using Eq. (18).
3. Substitute P_1 and P_3, calculated as above, in Eq. (19) to determine $(\phi'_2 - \theta)$ and hence ϕ'_2.

Triple Wedge Failure: Example Stability Computation

The triple wedge analysis section shown on Figure 9.22 is based on a case history in Landslide Technology files of a coastal landslide. Sliding beneath the middle wedge was occurring along an ancient landslide surface of sedimentary claystone. For analysis purposes, this lower stratum can be assumed to have negligible thickness. The overburden was sand, and sliding had been initiated by sea erosion.

The data are as follows: Sand overburden: γ = 125 pcf; $c' = 0$, $\phi' = 34\frac{1}{2}°$ ($= \phi'_1 = \phi'_3$); $\alpha = 60°$; $\theta = 4°$; $\beta = 14°$; groundwater as shown. Draw vertical lines BE and CF to separate the landslide into three wedges:

W_1 = 431 kips/ft. U_1 = 33 kips/ft.
W_2 = 2,233 kips/ft. U_2 = 557 kips/ft.
W_3 = 110 kips/ft. U_3 = 41 kips/ft.

Assume δ_1 = 10° to the horizontal; $\delta_3 = 0$ and $c'_1 = c'_r = c'_3 = 0$. Determine the residual strength ϕ'_r of the ancient slide plane beneath the middle wedge by rapid graphical method (B) and the analytical computation method (C).

Graphical Method B: $W_1 + W_2 + W_3$ = 2,774 kips/ft. The polygon of forces is shown on Figure 9.22(b). The main advantage of the graphical procedure over an analytical calculation is that it demonstrates the relative importance of the three wedges on the outcome, especially the dominance of the middle wedge on this problem and most other triple wedge failures. It can also be seen that the angle of inclination δ of the interwedge forces is not too critical; its influence has often been exaggerated in technical literature. Finally, the lower wedge, although very small in this example,

Figure 9.22 Triple wedge analysis. Example:
(a) section
(b) graphical procedure, Method B

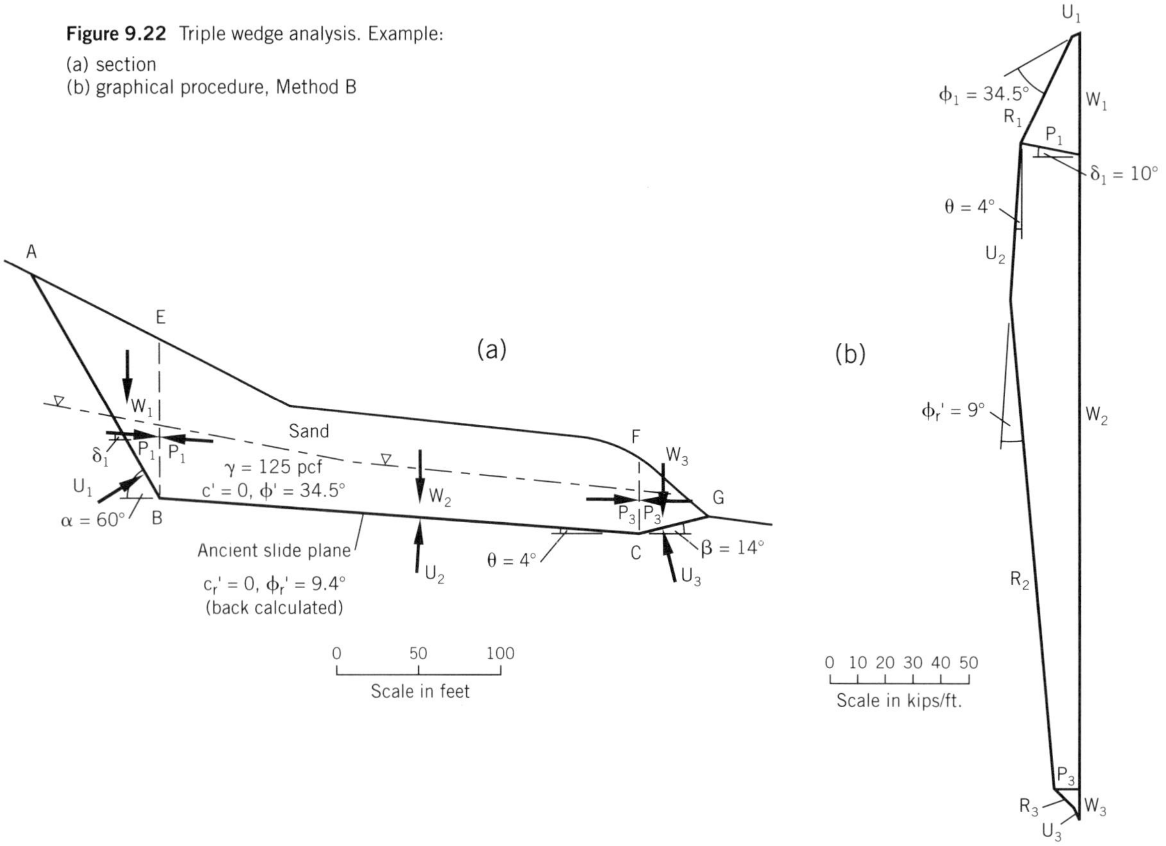

significantly affects the calculation of residual strength on the ancient slide plane. Therefore, the depth of the ancient slide plane near the toe of the landslide should be established by a site investigation. In many landslides, it is not measured because the toe has a steep slope or is heavily cracked, making access difficult.

The principal disadvantage of the graphical method is that the force polygons for the upper and lower wedges are relatively small and require a large scale and precision to achieve acceptable results.

Graphical procedure result: $\phi'_r = 9°$
Analytical Computations: $(\alpha - \phi'_1) = 25\frac{1}{2}°$; $\beta + \phi'_3 = 48\frac{1}{2}°$

Following the procedure described earlier:

$$P_1 = 214 \text{ kips/ft.}$$
$$P_3 = 89 \text{ kips/ft.}$$
$$\tan(\phi'_r - \theta) = 0.0938$$
$$(\phi'_r - \theta) = 5.4°$$
$$\phi'_r = 9.4°$$

Triple Wedge Failure: Remedial Analysis

For remedial analysis of a triple wedge failure, the factor of safety will be increased above 1.00. In this case, the stability analysis must use the general polygon of forces, Figure 9.20, where the mobilized shearing resistance is reduced by the factor of safety F.

Thus $$\psi = \tan^{-1}\left(\frac{\tan\phi'}{F}\right) \qquad \text{Eq. (20)}$$

and $$S = \frac{cl}{F} \qquad \text{Eq. (21)}$$

The strengths along all three shear planes have been determined as described in the previous paragraphs. In most cases, back analysis will have been used to determine ϕ'_2 (or ϕ'_r) of the middle wedge.

Drainage Remediation

For remediation by drainage, the pore-water load U decreases. By inserting the reduced values of U_1, U_2 and U_3, the factor of safety can be obtained graphically for trial values of F to calculate error E (drawn horizontally from the bottom of the lower polygon). By plotting F vs. E, the value of F is obtained corresponding to E = 0.

Alternatively, the result can be obtained analytically by substituting Eq. (20) and Eq. (21) above in the formulas of Method C described earlier. As an example, suppose the groundwater levels shown on Figure 9.22(a) are uniformly reduced by 10 feet. What is the factor of safety, as calculated by the analytical technique, Method C?

The general equations for the three wedges, when $F > 1$, become:

Upper Wedge

$$P_1 = \frac{U_1 \sin\alpha - \frac{c'_1 l_1}{F}\cos\alpha + \left(W_1 - U_1\cos\alpha - \frac{c'_1 l_1}{F}\sin\alpha\right)\tan(\alpha - \psi'_1)}{\cos\delta_1 + \sin\delta_1 \tan(\alpha - \psi'_1)} \qquad \text{Eq. (22)}$$

Middle Wedge

$$P_{3M} = \frac{P_1\cos\delta_1 + U_2\sin\theta - \frac{c'_2 l_2}{F}\cos\theta - \left(W_2 - U_2\cos\theta + P_1\sin\delta_1 - \frac{c'_2 l_2}{F}\sin\theta\right)\tan(\psi'_2 - \theta)}{\cos\delta_3 - \sin\delta_3\tan(\psi'_2 - \theta)} \qquad \text{Eq. (23)}$$

Lower Wedge

$$P_{3L} = \frac{U_3\sin\beta + \frac{c'_3 l_3}{F}\cos\beta + \left(W_3 - U_3\cos\beta + \frac{c'_3 l_3}{F}\sin\beta\right)\tan(\beta + \psi'_3)}{\cos\delta_3 - \sin\delta_3\tan(\beta + \psi'_3)} \qquad \text{Eq. (24)}$$

To perform the analysis:

1. Simplify equations (22), (23), (24) because:
   $c'_1 = c'_2 = c'_3 = 0$
   $\delta_3 = 0; \sin\delta_3 = 0; \cos\delta_3 = 1$
   Restate as equations (25), (26), (27)
2. Calculate interwedge force P_1 for several trial factors of safety F.
3. Calculate P_3 for the middle and lower wedge separately (P_{3M}, P_{3L}) for the same range of trial factors of safety F.
4. If force equilibrium is satisfied, $P_{3M} = P_{3L}$. For each trial F, calculate the error $E = P_{3M} - P_{3L}$.
5. Plot E vs. F to determine the F corresponding to $E = 0$.

$$c'_1 = c'_2 = c'_3 = 0 \qquad \delta_3 = 0;\ \cos\delta_3 = 1;\ \sin\delta_3 = 0$$

$$P_1 = \frac{U_1\sin\alpha + (W_1 - U_1\cos\alpha)\tan(\alpha - \psi'_1)}{\cos\delta_1 + \sin\delta_1\tan(\alpha - \psi'_1)} \qquad \text{Eq. (25)}$$

$$P_{3M} = P_1\cos\delta_1 + U_2\sin\theta - (W_2 - U_2\cos\theta + P_1\sin\delta_1)\tan(\psi'_2 - \theta) \qquad \text{Eq. (26)}$$

$$P_{3L} = U_3\sin\beta + (W_3 - U_3\cos\beta)\tan(\beta + \psi'_3) \qquad \text{Eq. (27)}$$

$$E = P_{3M} - P_{3L} \qquad \text{Eq. (28)}$$

$\phi'_1 = \phi'_3 = 34\frac{1}{2}°; \phi_r = 9.4°$	$\delta_1 = 10°$	$\sin\delta_1 = 0.174$	$\cos\delta_1 = 0.985$
$U_1 = 20$ kips/ft.; $W_1 = 431$ kips/ft.	$\alpha = 60°$	$\sin\alpha = 0.866$	$\cos\alpha = 0.5$
$U_2 = 375$ kips/ft.; $W_2 = 2{,}233$ kips/ft.	$\theta = 4°$	$\sin\theta = 0.070$	$\cos\theta = 0.998$
$U_3 = 17$ kips/ft.; $W_3 = 110$ kips/ft.	$\beta = 14°$	$\sin\beta = 0.242$	$\cos\beta = 0.970$

The calculations are tabulated on Table 9.4, and the graph of E vs. F for the groundwater lowering design is shown on Figure 9.23. For the groundwater lowering of 10 feet in the example, the factor of safety increases from 1.00 to 1.12.

Table 9.4 Triple Wedge Analysis: Error E for Various F in the Groundwater Lowering Example

	Trial Factors of Safety			
	F = 1.00	F = 1.10	F = 1.20	F = 1.30
$\tan\phi'_1/F$	0.687	0.625	0.573	0.529
ψ_1	34.5°	32.0°	29.8°	27.9°
$(\alpha - \psi'_1)$	25.5°	28.0°	30.2°	32.1°
$\tan(\alpha - \psi'_1)$	0.477	0.532	0.582	0.627
P_1	204	224	241	257
$\tan\phi'_r/F$	0.166	0.150	0.138	0.127
ψ'_2	9.4°	8.6°	7.9°	7.3°
$(\psi'_2 - \theta)$	5.4°	4.6°	3.9°	3.3°
$\tan(\psi'_2 - \theta)$	0.095	0.080	0.068	0.058
P_{3M}	47.3	95.1	134.5	169.0
$\tan\phi'_3/F$	0.687	0.625	0.573	0.529
ψ'_3	34.5°	32.0°	29.8°	27.9°
$(\beta + \psi'_3)$	48.5°	46.0°	43.8°	41.9°
$\tan(\beta + \psi'_3)$	1.130	1.036	0.959	0.897
P_{3L}	109.8	101.0	93.8	88.0
$E = P_{3M} - P_{3L}$	−62.5	−5.9	40.7	81.0

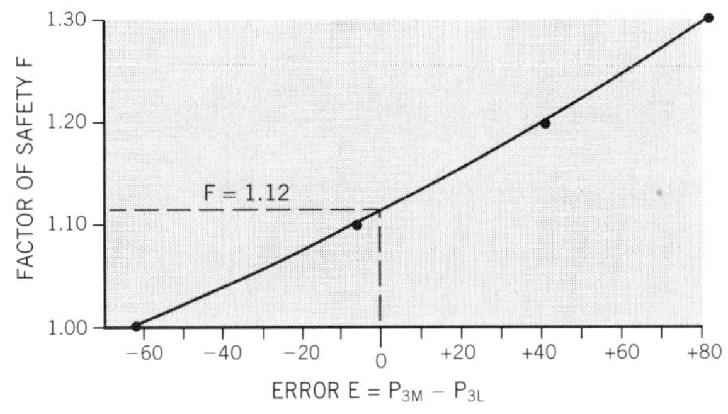

Figure 9.23 Triple wedge analysis. Result of groundwater-lowering remediation stability analysis.

Buttress Remediation

A buttress adds resistance to the lower wedge. Since there is no change in inclination of the underlying slide plane, it can be added to the lower wedge, allowing for a change in the

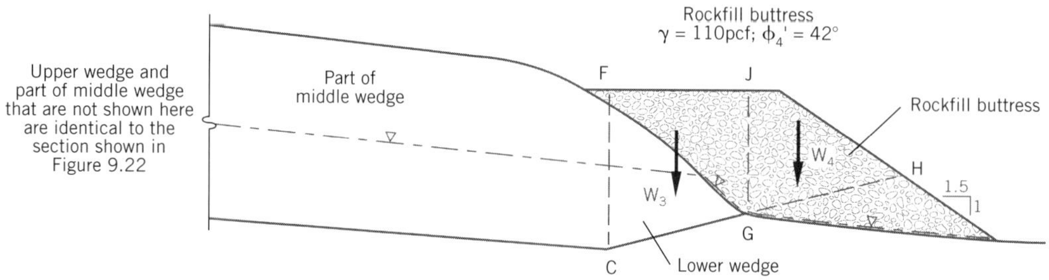

Figure 9.24 Triple wedge analysis. Rockfill buttress placed at toe of section shown on Figure Figure 9.22.

strength properties. Figure 9.24 shows a buttress placed at the toe of the triple wedge landslide of Figure 9.22, and will be used for an example set of buttress remediation calculations.

All dimensions and angles for the triple wedge landslide are as before. The slide plane through the rockfill buttress GH is an extension of the egress plane CG of the landslide. Note that almost all the benefits of a buttress design apply to this example: (i) segment FJG applies weight to plane CG which is a reasonably strong sand overburden; (ii) segment JHG provides rockfill strength; (iii) slope GH is a counterweight to the sliding direction of the landslide; and (iv) the rockfill is fully drained, deriving maximum benefit from cohesionless rockfill.

For this example, determine the factor of safety provided by the toe buttress.

The analysis is performed as a triple wedge, but the weight of the lower wedge and buttress is split up at vertical line GJ into W_3 and W_4, where W_3 is the weight acting on the sand slide plane CG and W4 is the weight acting on the rockfill plane GH.

$$c_1' = c_r' = c_3' = c_4' = 0 \qquad \text{Assume } \delta_3 = 0$$

Ignore the small weight of buttress on the middle wedge near point F.

Equations (25) and (26) for P_1 and P_{3M} remain unchanged. The equation for interwedge force P_{3L} becomes:

$$P_{3L} = U_3 \sin\beta + (W_3 - U_3 \cos\beta)\tan(\beta + \psi_3') + W_4 \tan(\beta + \psi_4') \qquad \text{Eq. (29)}$$

With the original water table

U_1 = 33 kips/ft.;	W_1 = 431 kips/ft.
U_2 = 557 kips/ft.;	W_2 = 2,233 kips/ft.
U_3 = 17 kips/ft.;	W_3 = 203 kips/ft.
U_4 = 0;	W_4 = 138 kips/ft.; ϕ_4' = 42°

The calculations are summarized on Table 9.5, and the graph of E vs. F for the buttress design is shown on Figure 9.25. The calculated factor of safety F for the buttress is 1.59. This is probably too high for the project requirements, and the

Table 9.5 Triple Wedge Analysis: Error E for Various F in the Buttress Example

	Trial Factors of Safety			
	F = 1.20	F = 1.35	F = 1.50	F = 1.65
$\tan\phi_1'/F$	0.573	0.509	0.458	0.417
ψ_1'	29.8°	27.0°	24.6°	22.6°
$(\alpha - \psi_1')$	30.2°	33.0°	35.4°	37.4°
$\tan(\alpha - \psi_1')$	0.582	0.649	0.711	0.765
P_1	248	271	292	309
$\tan\phi_r'/F$	0.138	0.123	0.110	0.100
ψ_2'	7.9°	7.0°	6.3°	5.7°
$(\psi_2' - \theta)$	3.9°	3.0°	2.3°	1.7°
$\tan(\psi_2' - \theta)$	0.068	0.052	0.040	0.030
P_{3M}	166	216	257	291
$\tan\phi_3'/F$	0.573	0.509	0.458	0.417
ψ_3'	29.8°	27.0°	24.6°	22.6°
$(\beta + \psi_3')$	43.8°	41.0°	38.6°	36.6°
$\tan(\beta + \psi_3')$	0.959	0.869	0.798	0.743
$\tan\phi_4'/F$	0.750	0.667	0.600	0.546
ψ_4'	36.9°	33.7°	31.0°	28.6°
$(\beta + \psi_4')$	50.9°	47.7°	45.0°	42.6°
$\tan(\beta + \psi_4')$	1.230	1.099	1.000	0.920
P_{3L}	353	318	291	270
$E = P_{3M} - P_{3L}$	−187	−102	−34	21

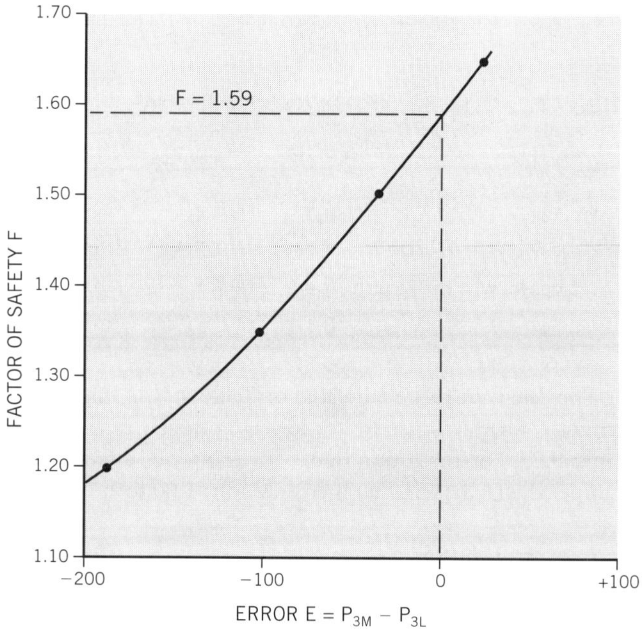

Figure 9.25 Triple wedge analysis. Result of buttress remediation stability analysis.

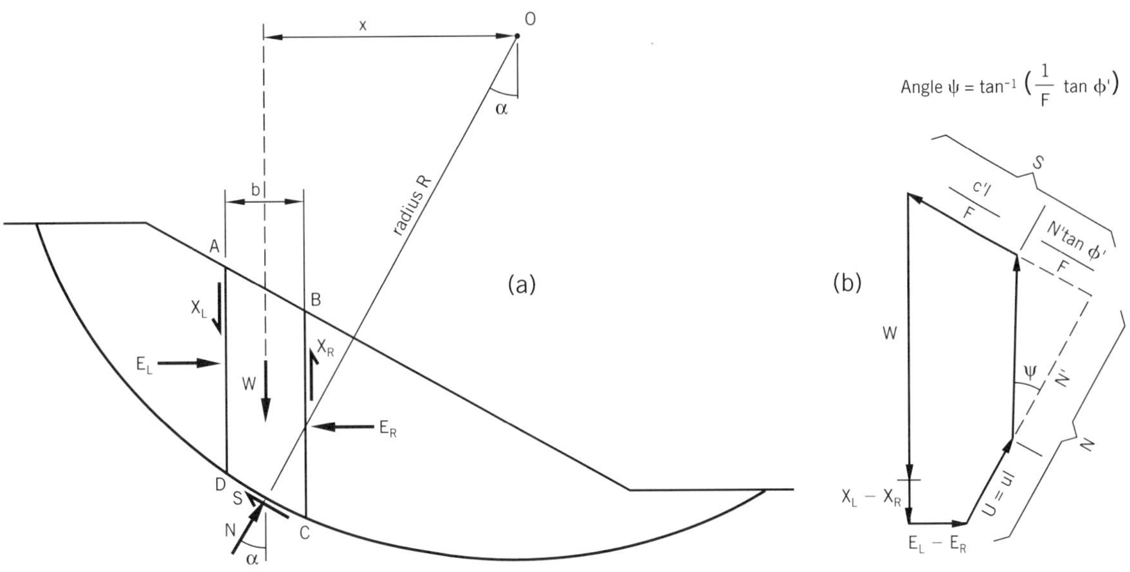

Figure 9.26 Circular arc analysis:
(a) forces on a slice
(b) polygon of forces

size of the buttress could be reduced to provide a factor of safety around 1.30.

It should be noted that the buttress is more effective than the horizontal drains for this landslide. The buttress provides a shear strength of 42° to resist sliding along a slip plane with a residual strength of 9.4°. It requires only a small buttress. By contrast, horizontal drains only marginally improve the effective normal stress and shear strength on the weak residual plane of the landslide and, by comparison, provide limited effectiveness.

9.9 CIRCULAR ARC ANALYSIS

Method of Slices

Slope failure along an approximately circular arc is a well-understood geotechnical phenomenon. It can be observed in the bowl-shaped scarp left behind after a flow slide, and generally occurs in shear slides within soils of relatively uniform properties.

The circular arc is analyzed in frictional soils by subdividing the arc into vertical slices, Figure 9.26. This allows the normal stress and frictional strength to be computed at the base of each slice. Ideally, the width of the individual slices should be very small but in practice the arc is usually subdivided into 6 to 12 slices. For ease of calculation, the position of the vertical lines can be chosen to pass through selected points, such as (i) the top edge of an embankment, (ii) where the slope reaches a river, or (iii) where there are changes in strata.

The slice ABCD on Figure 9.26 has a weight W and moment arm x about the center of rotation O. The radius R from the center of rotation is at angle α to the vertical and is positive towards the uphill side.

$$\text{Distance } x = R \sin \alpha$$

$$\text{Length of arc DC, } l = b \sec \alpha$$

$$\text{Pore-water load acting on base of slice } U = ul$$

Mobilized shear resistance is obtained by dividing the components of shear strength by the factor of safety F. The interslice forces are shear forces X_L, X_R and normal forces E_L, E_R.

$$\text{Shear force } S = \frac{1}{F}\{c'l + (N - ul)\tan\phi'\}$$

Equating moments about center of arc O for all the slices in the section:

$$\Sigma Wx = \Sigma S R$$

Substitute for S

$$F = \frac{R}{\Sigma Wx}\Sigma[c'l + (N - ul)\tan\phi']$$

Substituting $x = R \sin \alpha$, the equation becomes:

$$F = \frac{\Sigma[c'l + (N - ul)\tan\phi']}{\Sigma W \sin\alpha} \qquad \text{Eq. (30)}$$

Equation (30) is the original method for performing a circular arc analysis using slices. It ignores the static equilibrium which has to be satisfied for each individual slice where the interslice forces E and X are present. Today, this type of analysis is known as the Ordinary Method of Slices and provides conservative values of the safety factor F, especially for deep circles.

Bishop Routine Method

Bishop (1955) used the same moment equilibrium approach but takes into account the interslice forces. On Figure 9.26:

Resolving forces normal to the slip surface:

$$N = (W + X_L - X_R)\cos\alpha - (E_L - E_R)\sin\alpha$$

Equation (30) becomes:

$$F = \frac{\Sigma[c'l + \tan\phi'(N - ul) + \tan\phi'\{(X_L - X_R)\cos\alpha - (E_L - E_R)\sin\alpha\}]}{\Sigma W \sin\alpha} \qquad \text{Eq. (31)}$$

The ordinary method of slices ignores the term $\tan\phi'\{(X_L - X_R)\cos\alpha - (E_L - E_R)\sin\alpha\}$. This term is only zero if

ϕ' and α are constant along the surface, i.e., on a plane surface. In all other cases, omission of these terms is an approximation in which accuracy decreases as the central angle of the arc increases.

By resolving the forces vertically and rearranging terms, the Bishop solution becomes:

$$F = \frac{\Sigma[c'b + \{W - ub) + (X_L - X_R)\} \tan\phi'] \dfrac{\sec\alpha}{1 + \frac{1}{F}(\tan\phi' \tan\alpha)}}{\Sigma W \sin\alpha} \quad \text{Eq. (32)}$$

For static equilibrium $\Sigma \ (X_L - X_R) = 0$ Eq. (33)

and $\Sigma \ (E_L - E_R) = 0$ Eq. (34)

It can be shown (Bishop, 1955) that

$$\Sigma(E_L - E_R) = \Sigma \frac{m}{F} \sec\alpha - (W + X_L - X_R)\tan\alpha \quad \text{Eq. (35)}$$

where m is the numerator in equation (32). Therefore, values of $(X_L - X_R)$ must be chosen to satisfy equation (33) and make equation (35) equal to zero.

In practice, Bishop's Routine Method is to put $(X_L - X_R) = 0$ on all slices and compute F from equation (32). He noted there were many variations of $(X_L - X_R)$ which satisfy equation (35), but they give variations of F which are insignificant. The major part of the gain in accuracy is obtained by taking $(X_L - X_R) = 0$. In a more rigorous analysis, the X forces have to be addressed.

The equation used for Bishop's Routine Method of analysis is, therefore:

$$F = \frac{\Sigma[c'b + (W - ub)\tan\phi'] \dfrac{\sec\alpha}{1 + \frac{1}{F}(\tan\phi' \tan\alpha)}}{\Sigma W \sin\alpha} \quad \text{Eq. (36)}$$

Equation (36) has factor of safety F on both sides of the equation. The solution for F is reached by an iterative process.

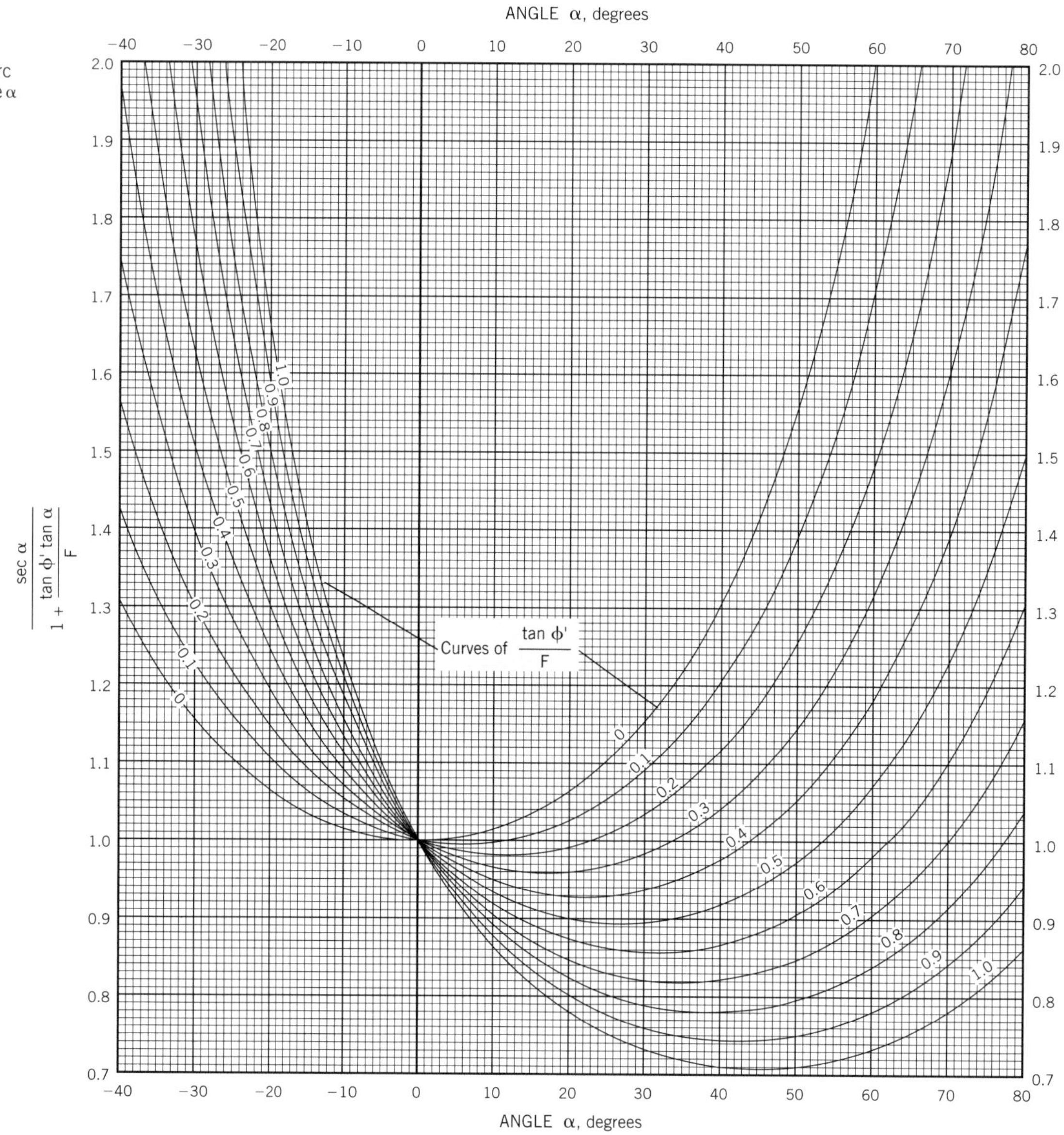

Figure 9.27 Circular arc analysis. Graph of angle α vs. $\dfrac{\sec\alpha}{1 + \dfrac{\tan\phi' \tan\alpha}{F}}$

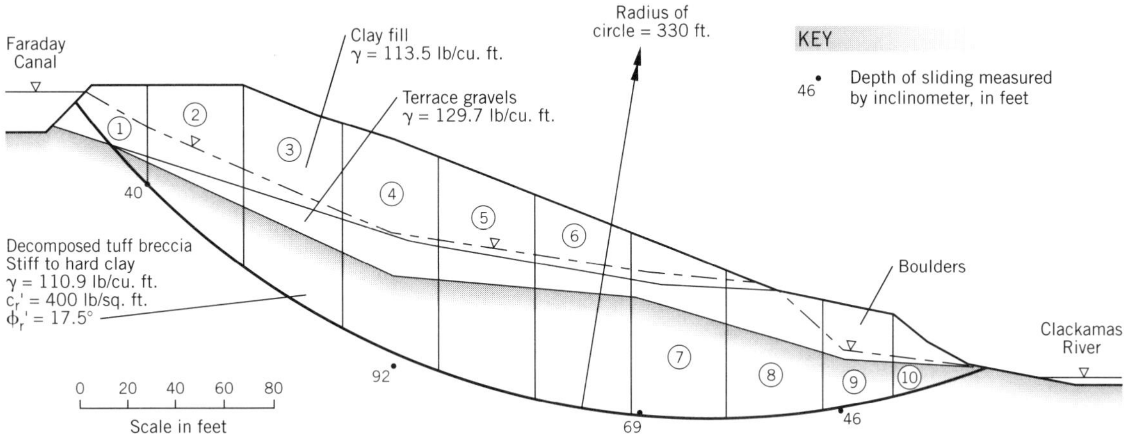

Figure 9.28 Circular arc analysis. Section of Faraday Slide (Case History 8).

Fortunately, convergence is rapid, and it normally requires only two iterations to reach the value of F for the computation. The analysis is also termed *Bishop's Simplified Method.*

The term $\dfrac{\sec\alpha}{1+\frac{1}{F}\tan\phi'\tan\alpha}$ has been changed to $\frac{1}{m_\alpha}$ in some textbooks; m_α can also be expressed as $m_\alpha = \cos\alpha + \sin\alpha\,\frac{\tan\phi'}{F}$. A graph of $\frac{1}{m_\alpha}$ vs. α is shown on Figure 9.27 for values of $\frac{1}{F}\tan\phi'$.

Bishop's Routine Method: Example Analysis

The example of a circular arc analysis is taken from the Faraday Slide, near Estacada, Oregon (Case History 8). The soil properties, depth of sliding, and the circular arc used in the analysis are shown on the section, Figure 9.28.

The stability calculations for the landslide are shown on Figure 9.29 which includes Table 9.6. These calculations are based on equation (36). Note that column (15) of the table is $1/m_\alpha$.

The landslide is subdivided into 10 slices. For each slice, the horizontal width b is measured, column 2. The average height of each stratum is put in column 3, using triangles, rectangles or trapezoids to obtain the answers. The width is multiplied by the average height to obtain the area increment for each stratum (column 4) and this is converted to weight, using the density of the stratum (column 5). The individual stratum weights for each slice are summed in column 6. Angle α is determined for each slice (column 7). The slice can be split into trapezoids and triangles to determine the point at which the weight W of the slice cuts the circular arc. With practice, this result can be achieved quickly. Columns 8 and 9, sin α and W sin α, are computed to obtain ΣW sin α at the bottom of column 9.

For U.S. units of measure, it is recommended that weights be recorded in kips (i.e., 1,000 lb. units). A weight of 77,600 lb. would be entered in column 6 as 77.6 kips. Angle α should be measured to the nearest 0.1° using a large protractor; record sin α to three decimal places.

The end slices can be more difficult to analyze in tabular form. When the width of the triangle or trapezoid is less than width b of the slice, the actual width can be entered separately on the table. An example is 15½ feet for two of the strata in Slice 1 of Table 9.6. The area calculations for Slice 1 are given on Figure 9.29(b).

The second half of the table deals with shearing resistance. The height of the pore-water is measured off the cross-section and averaged for the slice, column 11. Columns 12 and 13 calculate the friction developed at the base of the slice. Add columns 10 and 13 to obtain the total shear resistance in column 14.

When the cohesion intercept $c' = 0$, column 10 is left blank. In such a case, column 14 is identical to column 13.

It is advisable to sum up column 14 and divide by ΣW sin α from column 9 to get F approx. The first trial calculation is made with F_1 being slightly higher than F approx. Using the $\frac{1}{F}\tan\phi'$ for the base of each slice, obtain the multiplier $\dfrac{\sec\alpha}{1+\tan\alpha\,\frac{\tan\phi'}{F_1}}$ from the graph, Figure 9.27, for the corresponding value of α for the slice. Enter the result in column 15, Trial 1.

In the worked example, $F_1 = 1.00$ was selected for the first trial. After completing column 16, summing the calculations and dividing by ΣW sin α, the result was $F = 0.97$. Therefore, $F_2 = 0.97$ was selected for the second trial, and the recalculations in columns 15 and 16 confirmed that $F = 0.97$ for this example.

The tabular format of Table 9.6 also can be used to calculate the average undrained shear strength c along the circular arc of a landslide. This is a total stress analysis, and is useful for comparing the undrained strength of the landslide with the undrained strength of the soil based on soil classifications. In stiff, overconsolidated clay, the soil on the slip surface is weaker than the main body of the stratum due to strain-softening.

For calculation of average undrained shear strength c, columns (1) through (9) of Table 9.6 (the driving moment)

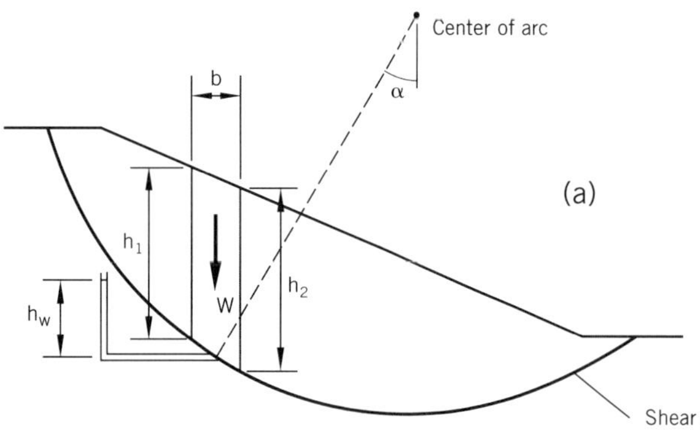

$$u = \gamma_w h_w$$

$$\bar{h} = \tfrac{1}{2}(h_1 + h_2)$$

$$F = \frac{1}{\Sigma\, W \sin\alpha}\, \Sigma \left[\left\{ c'b + \tan\phi'\,(W - ub) \right\} \frac{\sec\alpha}{1 + \dfrac{\tan\phi' \tan\alpha}{F}} \right]$$

Table 9.6 Stability Calculations for Faraday Slide

(1) Slice	(2) Width b feet	(3) Average stratum height $\bar{h}$ feet	(4) Stratum area A sq. feet	(5) Stratum weight W $A\gamma$ kips	(6) Slice Weight W $= \Sigma A\gamma$ kips	(7) Angle α degrees	(8) sin α	(9) W sin α (6) x (8) kips	(10) Cohesion c' b kips	(11) Pore-water load u b = $\gamma_w \bar{h}_w b$ kips	(12) Effective load (W - u b) (6) - (11) kips	(13) Friction (12) x tan ϕ' kips	(14) Cohesion + Friction (10) + (13) kips	(15) $\sec\alpha / \left(1 + \frac{\tan\phi' \tan\alpha}{F}\right)$ Trial 1 $F_1 = 1.00$	(15) Trial 2 $F_2 = 0.97$	(16) Arc resistance (14) x (15) Trial 1 kips	(16) Arc resistance (14) x (15) Trial 2 kips
1	29	See (b) 15½ (b) 15½ below	593 19 70	67.3 2.5 7.8	77.6	48.0	0.743	57.6	12.6	25.3	52.2	16.5	29.1	$\frac{\tan\phi'}{F} = 0.315$ 1.105	$\frac{\tan\phi'}{F} = 0.325$ 1.095	32.2	31.9
2	40	35 4.5 16.75	1400 180 670	158.9 23.4 74.3	256.6	40.3	0.647	165.9	17.4	77.4	179.2	56.6	74.0	1.03	1.025	76.2	75.9
3	40	39.25 10.25 28	1570 410 1120	178.2 53.2 124.2	355.6	32.2	0.533	189.5	17.4	107.3	248.3	78.4	95.8	0.985	0.98	94.4	93.9
4	40	37.75 14.25 32.5	1510 570 1300	171.4 73.9 144.2	389.5	24.4	0.413	160.9	17.4	121.1	268.4	84.8	102.2	0.96	0.96	98.1	98.1
5	40	31.75 11.75 40.5	1270 470 1620	144.1 61.0 179.7	384.8	17.0	0.292	112.5	17.4	137.3	247.5	78.2	95.6	0.955	0.955	91.3	91.3
6	40	23.25 68 46.5	930 320 1860	105.5 41.5 206.3	353.3	9.9	0.172	60.8	17.4	146.0	207.3	65.5	82.9	0.965	0.965	80.0	80.0
7	40	13 9.75 44.25	520 390 1770	59.0 50.6 196.3	305.9	2.8	0.049	14.9	17.4	144.8	161.1	50.9	68.3	0.995	0.995	68.0	68.0
8	40	24 3 18 31.5	72 720 1260	8.2 93.4 139.7	241.3	−4.2	−0.073	−17.6	17.4	118.6	122.7	38.8	56.2	1.025	1.025	57.6	57.6
9	30	20.5 18	615 540	79.8 59.9	139.7	−10.4	−0.181	−25.2	13.1	42.1	97.5	30.8	43.9	1.08	1.08	47.4	47.4
10	39	30 9.5 7.5	285 293	37.0 32.4	69.4	−15.4	−0.266	−18.4	17.0	20.7	48.7	15.4	32.4	1.14	1.14	36.9	36.9

Notes: 1 kip = 1000 lbs. All units are per depth foot depth (17) Σ W sin α = 700.9 Σ = 680.4 (18) Σ = 682.1 = 681.0

$F = \frac{(18)}{(17)}$ = 0.97 = 0.97

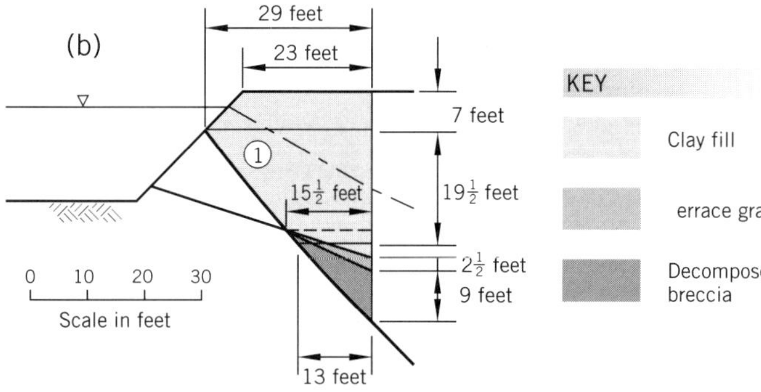

Area calculations:

$\frac{1}{2}(23 + 29)(7) + \frac{1}{2}(29 + 13)(19\frac{1}{2}) = 593$ sq. ft.

$\frac{1}{2}(2\frac{1}{2})(15\frac{1}{2}) = 19$ sq. ft.

$\frac{1}{2}(9)(15\frac{1}{2}) = 70$ sq. ft.

Figure 9.29 Circular arc analysis. Terms and formulas.

remain unchanged. To obtain the circular arc resistance ($\phi = 0$ analysis) only columns (10) (15) and (16) are needed. Column (15) is simply sec α which converts horizontal width b into length l along the arc of the slip surface.

For F = 1.00, the average undrained strength along the circular arc is 1,660 lb./sq.ft. For comparison, very stiff clays have undrained strengths in the range of 2,000 to 4,000 lb./sq.ft., and hard clay is above 4,000 lb./sq.ft. (Chapter 8, Table 8.4). Therefore, the clay on the slip surface is weaker than the surrounding very stiff to hard clay, which is consistent with the sharply lower SPT blow counts measured near the slip surface.

Spencer (1967) made the assumption that the interslice forces are parallel; i.e., angle θ is constant on all slices (Figure 9.30). This procedure allows both moment and force equilibrium to be satisfied, and Spencer's method (available on most computer software programs for stability analyses) is the only circular arc analysis that achieves this desirable objective.

One set of calculations by Spencer (1967) is illustrated on Figure 9.31. The section (Figure 9.31a) represents a typical 1 vertical:2 horizontal slope. Several values of interslice angle θ were chosen and the values of F were calculated to satisfy the force (F_f) and moment (F_m) equilibrium equations. F_{mo} is the factor of safety corresponding to Bishop's Routine Method ($\theta = 0$). The correct factor of safety is where the two graphs intersect on Figure 9.31(b); F = 1.070 in this example. However, it can be noted that the Bishop's Routine Method of $\theta = 0$ gives a value of F = 1.039 which is close to Spencer's solution. The graphs on Figure 9.31(b) also demonstrate that the calculated F values by moment equilibrium are not particularly sensitive to angle θ, but calculations of force equilibrium are much affected by the assigned θ.

Spencer observed that the line passing through the point of action of the interslice forces is close to the lower third point at each interslice boundary. It is shown by the broken line on Figure 9.31(a) and implies an approximately triangular pressure distribution on the vertical slice boundary.

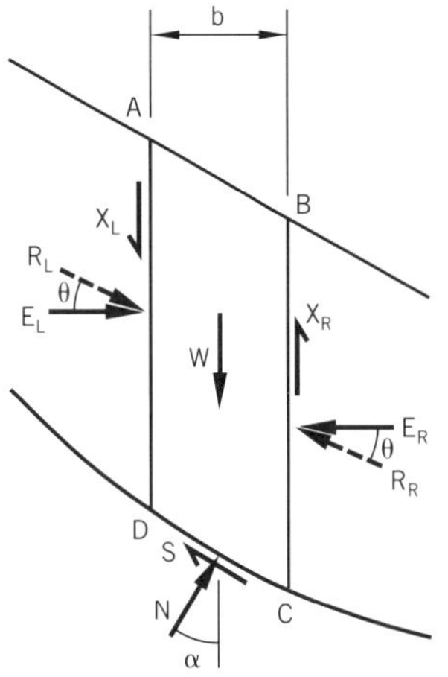

Figure 9.30 Circular arc analysis. Spencer's Method.

Figure 9.31 Spencer's Method:
(a) example slope
(b) variation of F_m and F_f with θ
(adapted from Spencer, 1967)

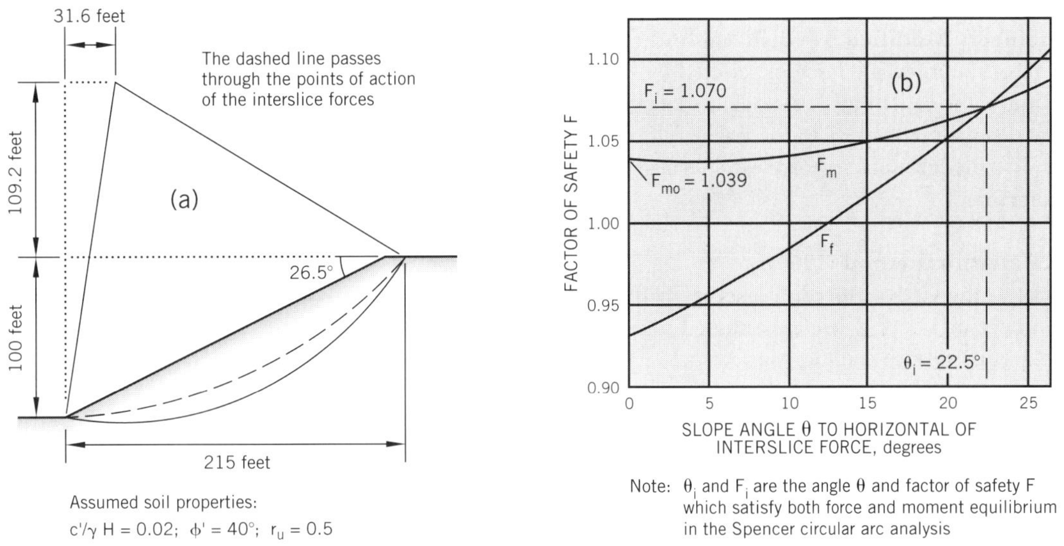

9.10. OTHER CIRCULAR AND NON-CIRCULAR STABILITY ANALYSES

Nash (1987) provides an excellent summary of the numerous slope stability analyses available today. Bromhead (1986) also examines the subject in detail.

Other methods of analysis consider the circular arc or a more generalized method of slices on a non-circular slip surface. They examine equilibrium of forces, equilibrium of moments, or both. The main issue in all methods concerns the assumed position and direction of the interslice forces, i.e., the relationship between interslice shear force X and interslice normal force E. For the Bishop Routine Method, $X_L - X_R = 0$ is assumed. For the Spencer circular arc method, the assumption is made that ratio $X/E = \tan\theta =$ constant throughout the slope.

Non-circular analyses of slope stability can be solved by moving the center of rotation of a circular arc to accommodate the actual slip surface of the landslide. This introduces a moment caused by the normal force N at the slip surface rotating around the assumed center of rotation O (in the circular arc analysis, Figure 9.26, the force N passes through O and there is no moment).

A brief synopsis of some better-known methods is given below.

Spencer Method (1967)

Already described in the previous section, the Spencer method can be modified to calculate stability of non-circular slip surfaces.

Morgenstern and Price Method (1965)

Circular arc or non-circular failure path. This method allows the forces to vary across the landslide and formulates equations of equilibrium by resolving parallel to and normal to the base of the slice. The assumed relationship of the interslice forces is given by:

$$\frac{X}{E} = \lambda \, f(x)$$

where f(x) is a specified function (e.g., constant or trapezoidal) relating angle θ across the slide from left to right. λ is a scaling factor.

Janbu's Rigorous Method (1957)
Non-circular failure surface. Assumes a line of thrust for the interslice force, then calculates force and moment equilibrium to determine F. The line of thrust passes from head to toe of the landslide at about one-third the height between the failure surface and the ground surface.

Janbu's Simplified Method (1957)
Non-circular failure surface. This is a force equilibrium method. It assumes interslice forces are zero but provides a correction factor, dependent on the depth/length ratio of the landslide, to account for them.

Corps of Engineers Modified Swedish Method (1970)
The inclination of the interslice forces is selected by the analyst, and the same value is used in all slices. The Corps recommends that the inclination be equal to the average slope of the embankment. The method satisfies force equilibrium but not moment equilibrium.

Lowe and Karafiath's Method (1960)
This is identical to the Corps procedure except for the inclination of the interslice forces. These are assumed to be at the average of the ground surface and shear surface below the vertical boundary of the slice. Therefore, the inclination of the interslice forces varies from boundary to boundary of each slice.

Comparison of Methods
A conclusion reached in comparing methods is that the moment equilibrium approach gives factors of safety that are less sensitive to angle θ (the angle of thrust to horizontal of the interslice force) than force equilibrium solutions. Since an estimate of thrust of the interslice forces has to be assumed in all analyses, it seems that a method satisfying moment equilibrium (such as the Bishop Routine Method) should be used for preference. A second outcome of comparisons between methods is that there are only relatively small differences in the calculated factor of safety F. The results obtained by Fredlund and Krahn (1977) are given in Table 9.7. The lower factors of safety computed by the Ordinary Method of Slices (which excludes m_α and is not recommended) is shown in the far right column of the table for comparison.

Computer Analyses
Computers are used routinely to perform stability analyses of landslides. A list of slope stability software providers has been compiled in Table 9.8; other suppliers may be available.

9.11 SPECIAL CASES: (a) PARTLY SUBMERGED SLOPE

A body of water, on the outside of a slope, improves stability. Many slopes receive such support from a river, sea, lake, or reservoir. If the outside water level fluctuates, the factor of safety is lowest when the water level is at its lowest. In addition to losing support, a river in flood or a sea at high tide may raise the groundwater level within the slope. When the outside water level drops rapidly, groundwater within the slope may be slow to drain; in this case, the higher-than-normal groundwater levels within the slope also decrease slope stability.

The suggested method of analysis for a partly submerged slope is to use *the buoyant weight for all soil below the outside*

Table 9.7 Comparison of Results for Different Stability Analysis Methods*

	Computed Factor of Safety F				
	Bishop Routine	Spencer	Janbu Simplified	Morgenstern–Price f(x) = constant	Ordinary Method of Slices
(1) 2:1 slope 40 feet high c′ = 600 psf, ϕ′ = 20°	2.08	2.07	2.04	2.08	1.93
(2) As for (1) above with thin weak layer c′ = 0, ϕ′ = 10°	1.38	1.37	1.45	1.38	1.29
(3) Same as (1) except with piezometric line	1.83	1.83	1.83	1.83	1.69
(4) Same as (2) but with piezometric line in both layers	1.25	1.25	1.33	1.25	1.17

*abstracted from Fredlund and Krahn, 1977

Table 9.8 Slope Stability Software Providers

Software Provider	List Price*	Methods of Analysis
1. XSTABL—Version 5 (1996) Interactive Software Designs, Inc. 953 N. Cleveland Street Moscow, ID 83843 www.xstabl.com e-mail: ssharma@uidaho.edu Author: Sunil Sharma, Ph.D.	$450	Bishop's Simplified (Routine) Spencer Janbu General Limit Equilibrium Force Equilibrium Morgenstern–Price
2. UTEXAS4 Shinoak Software 3406 Shinoak Drive Austin, TX 78731 www.shinoak.com e-mail: sales@shinoak.com Author: Stephen Wright, Ph.D.	$3,000 (prepaid) A simplified (and less expensive) version is also available	Bishop's Simplified (Routine) Spencer Corps of Engineers Lowe and Karafiath
3. GSLOPE (1990) Mitre Software Corporation Suite 200, 9636—51st Avenue Edmonton, Alberta Canada T6E 6A5 e-mail: info@mitresoftware.com	$895	Bishop's Simplified (Routine) Janbu Simplified Non-Circular Analysis using Fredlund & Krahn
4. SLOPE/W Geo-Slope International #1400, 633–6th Avenue SW Calgary, Alberta Canada T2P 2Y5 e-mail: info@geo-slope.com www.geo-slope.com	(Version 2004) $3,995 Also available @ $3,995 each: Seep/W for seepage Sigma/W for stress distribution $7995 for purchasing all three	Ordinary Method of Slices Bishop's Simplified (Routine) Janbu Simplified Spencer Morgenstern–Price Corps of Engineers Lowe & Karafiath General Limit Equilibrium Finite Element Stress Probabilistic Analyses

*prices shown are for 2004. Contact the vendors to obtain specifications, current prices, and other conditions for software use.

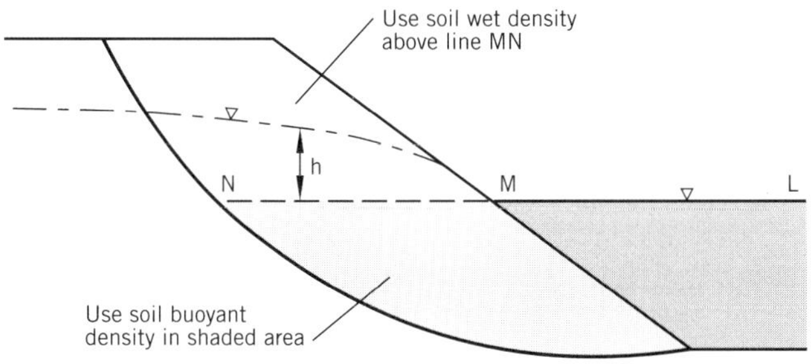

Figure 9.32 Partly submerged slope stability analysis.

water level. On Figure 9.32, the outside water surface LM is projected into the slope as MN. For the shaded area below MN, the buoyant density is used in the driving moment calculations. For shearing resistance at the base of each slice, use the excess head h above line MN for pore water pressure calculations.

9.12 SPECIAL CASES: (b) PARTLY CONSOLIDATED SOILS

When construction fills are built above soft slow-draining soils in the foundation, the fill load is initially carried by the pore-water within the soil, creating excess (positive) pore water pressures. As the pore pressures dissipate, the soil consolidates and strengthens. If outside conditions do not change, the most critical time for stability is when the fill has just been completed and the highest pore pressures exist in the foundation. This is often termed the "end of construction" stability.

For stiff overconsolidated clays in the foundations, the pore pressures are depressed by the induced shear stresses of loading. The soil is at its strongest condition when first loaded but weakens with time. Slope failure can occur at any time during the weakening process. The limit is when the pore pressures rise back to their long-term equilibrium levels. This is termed "long-term" stability.

There are situations where instability can occur in partly consolidated soils. Two examples are given below:

Clay Fill Construction: A compacted clay fill develops excess pore pressures. If the rate of fill construction is much faster than the rate of pore pressure dissipation in the previously placed fill, the high pore pressures may cause failure of the clay fill. This problem can develop during the construction of dam embankments and often requires the pore pressures to be monitored throughout construction. If the pore pressures become dangerously high, fill placement has to be stopped until the pore pressures dissipate to a safer level.

Tidal Fluctuations: Silt that is actively sedimenting in a marine environment may set up on slopes which are marginally stable at low tide. If a fill is built over the marine silt, it may be capable of remaining stable for a short time after construction but can fail shortly afterward at an extreme low tide. This type of failure was reported by Bjerrum (1971) for marine silts in the fjords of Norway. Seed (1983) also describes flow slides occurring at extreme low tide due to fill construction.

Example Calculations of Excess Pore Pressures in a Partly Consolidated Soil Stratum

A fill was built above deep sediments of marine silt at Skagway, Alaska, in October 1994 (more fully described in Case History 7). The fill was built over a 19-day period and was in place for a further 19 days before slope failure occurred at extreme low tide. During this short time, the marine silt close to the drainage boundaries had time to partly consolidate and strengthen prior to the failure.

For this type of situation, a *total stress analysis* can be performed on the silt after failure. At Skagway, the undrained strengths were measured shortly after failure by field vane tests, which is suitable for soft, saturated, normally consolidated clays and silts. The measured strengths reflect any strengthening achieved from partial consolidation of the soils prior to failure.

For an *effective stress analysis*, an estimate has to be made of the excess pore pressures remaining in the silt at the time of failure. This is a fairly difficult assignment but is useful in quantifying the effect of different variables on slope stability. For example, it is possible to examine the contributions of tide levels and, at Skagway, artesian pressures on the calculated factor of safety in addition to partial consolidation of the foundation silts.

The procedure adopted for the analysis of partial consolidation at Skagway is described below. The general four-step approach should be equally applicable to other sites where the foundation soils have partly consolidated at failure.

Step 1. *Determine the vertical stress distribution within the marine silt due to the fill load*

The riprap stockpile at Skagway had an unusual shape, consisting of a windrow gradually rising to a maximum height of 25 feet near the mid-length. Knowing that the fill stresses would become distributed at depth and to avoid analyzing the stability section with the largest loading, the stockpile shape was transformed to an equivalent uniform loading over its full length of 300 feet. For 6,000 cu. yd. of riprap, this gives an equivalent maximum height of 10½ feet for a base width of 70 feet and 1½:1 (H:V) side slopes. Below the oddly shaped riprap stockpile, other fill placed mostly below water has a uniform cross-section across the entire width of the stockpile. It had an estimated volume of 6,700 cu. yds., approximately one-half of the total fill load of 12,700 cu. yds.

The bulk density of the riprap was assumed to be 120 lb./cu. ft. The foundation pad was assigned a bulk density of 125 lb./cu. ft. and a submerged density of 61 lb./cu. ft. below mean sea level.

The vertical stress changes with depth (Figure 9.33) were based on the Boussinesq stress distribution, and were obtained from the computer program Sigma/W (Geo-Slope). The talus and bedrock below the marine silt were arbitrarily assigned a modulus of elasticity 100 times greater than the silt.

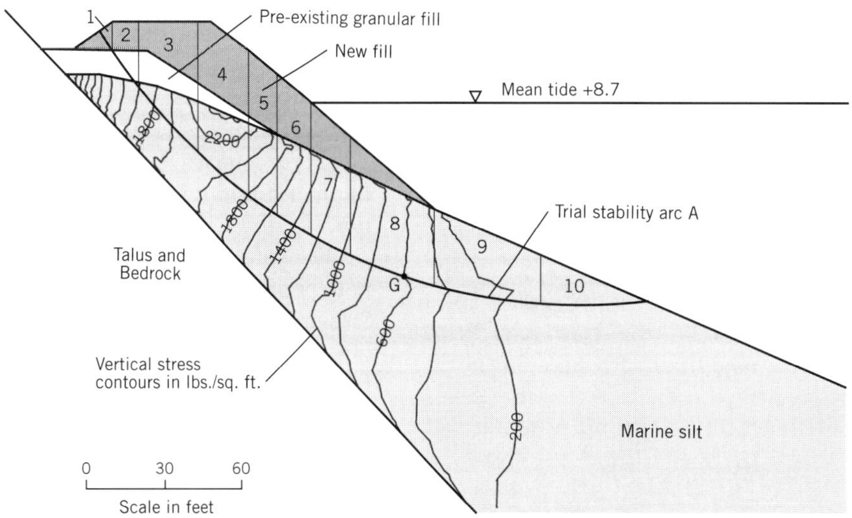

Figure 9.33 Boussinesq vertical stress distribution within the marine silt at Skagway, Alaska, due to new fill loading.

It can be assumed that the load of the fill is initially carried by the pore water rather than the soil structure of the marine silt. Thus, the initial pore pressures within the marine silt will be equal to the distributed vertical stresses of the fill. This is the principle of Terzaghi's one dimensional consolidation theory, and such a pore pressure response can be expected in a normally consolidated silt with water contents above the liquid limit.

Step 2. *Determine a "composite" coefficient of consolidation for the loaded soil stratum*
The second step is to find out how much the soil has consolidated under the fill load prior to failure. This allows the remaining excess pore pressures to be calculated.

The partial dissipation of the excess pore pressures between the time of fill placement and slope failure depends on the coefficient of consolidation c_v of the marine silt. The soils on the slope have variations in gradation, reflecting factors such as the variable sediment load of the Skagway River flowing into the harbor, turbidity in the sea around the slope, and other seasonal factors. It is reasonable to expect that layers will be laid down parallel to the outer slope face. Since the outer slope is a drainage boundary, the flow of water during pore pressure dissipation will be normal to the outer slope. As shown on Figure 8.13 (Chapter 8), the "composite" permeability for a multilayered stratum with flow at right angles to the bedding planes is given by:

$$\frac{n}{k} = \frac{1}{k_1} + \frac{1}{k_2} + \frac{1}{k_3} + \ldots \frac{1}{k_n}$$

where k = coefficient of permeability
n = number of layers

This principle has been used to calculate a "composite" value of the coefficient of consolidation c_v at Skagway because parameter c_v is predominantly controlled by permeability k. Thus:

$$\frac{n}{(c_v)_c} = \frac{1}{(c_v)_1} + \frac{1}{(c_v)_2} + \frac{1}{(c_v)_3} + \ldots \frac{1}{(c_v)_n} \qquad \text{Eq. (37)}$$

Five specimens, taken by thin-wall sampler tubes, were tested by laboratory triaxial dissipation tests to measure c_v. All specimens were presaturated by back pressures before being consolidated. The results were c_v values of 890; 29,700; 4,130; 27,800; and 790 sq. ft./year. Assuming each result represents a layer of equal thickness, the analysis gives a "composite" c_v of 1,850 sq. ft./year. It should be noted that the composite figure of 1,850 sq. ft./year is much closer to the lower values measured in the individual tests than the higher values. This is because it takes longer for water to pass through the less permeable strata during drainage, and such layers control the overall rate of drainage.

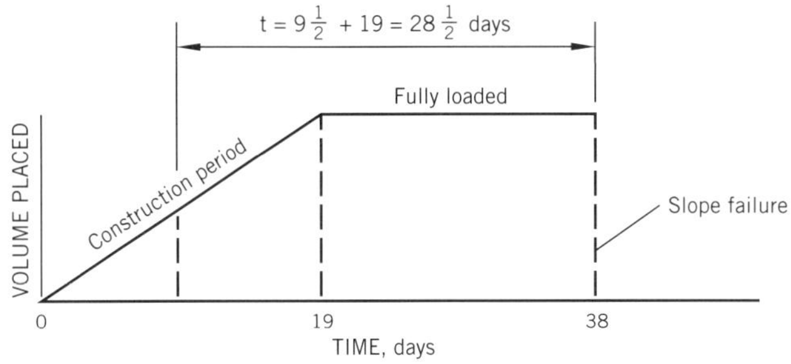

Figure 9.34 Assumed rate of fill loading at Skagway, Alaska.

Step 3. *Determine the excess pore-water pressures in the slope at the time of failure*
Using the composite c_v determined in Step 2, the excess pore pressures within the marine silt at failure can be estimated. First, the time of loading is assumed to be one-half of the fill construction period plus the time under full loading (Figure 9.34). This period is approximately 29 days. Next, the time factor equation is rearranged to:

$$H = \sqrt{\frac{c_v t}{T}} \qquad \text{Eq. (38)}$$

where H = drainage distance from a point within the marine silt to the drainage boundary in feet
T = time factor from the theory of consolidation
t = elapsed time before failure = 29 days = 0.0795 year
c_v = "composite" c_v of 1,850 sq. ft./year

$$H = \sqrt{\frac{147.0}{T}} \text{ at the Skagway site}$$

For 50% dissipation, time factor T = 0.38 (see Chapter 8, Figure 8.8). Therefore, H = 19.7 feet for 50% dissipation after loading for 29 days.

The data are presented in Table 9.9. It shows that the excess pore pressures set up in the marine silt by the fill are 80

Table 9.9 Distance H from the Drainage Boundary for Various Dissipation Levels
c_v = 1,850 sq. ft./year; t = 29 days after loading

H in feet	Pore Pressure in Marine Silt		H in feet	Pore Pressure in Marine Silt	
	Percent Dissipation	Percent Nondissipation		Percent Dissipation	Percent Nondissipation
54	0.3	99.7	19.7	50	50
38.3	5	95	17.6	60	40
33.6	10	90	15.7	70	30
30.3	15	85	14.0	80	20
28.0	20	80	11.9	90	10
24.4	30	70	8.6	99	1
21.9	40	60			

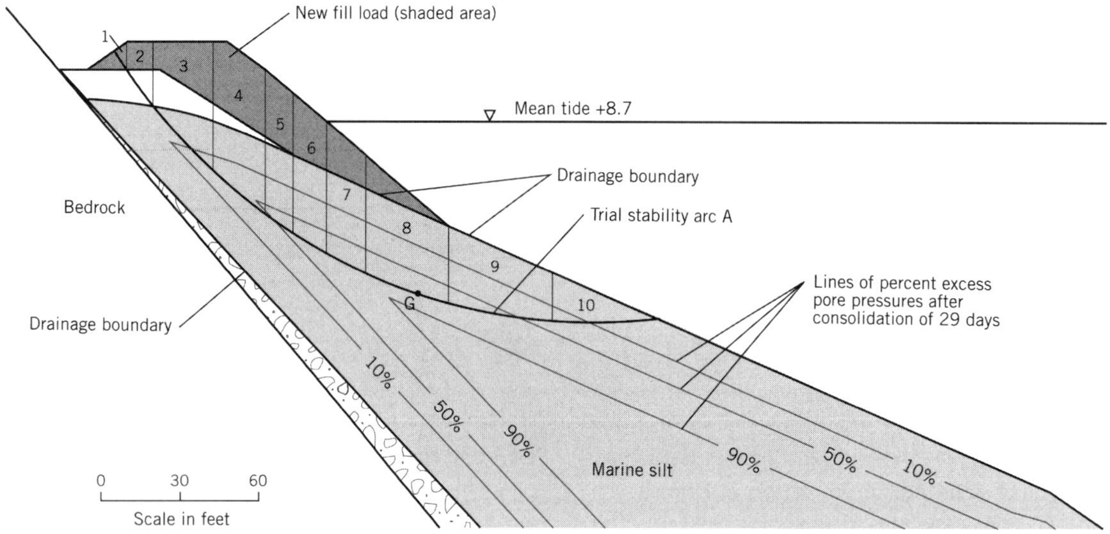

Figure 9.35 Section showing percent of excess pore pressures remaining in the marine silt at Skagway, Alaska, at the time of failure.

percent dissipated at a distance of 14 feet from drainage, but only 20 percent dissipated at a distance of 28 feet. Since the stability analysis requires the *non-dissipated* pore pressures, the data from the final column of Table 9.9 are used in the stability analyses.

Selected lines of percent non-dissipated pore pressures are plotted on Figure 9.35 using the stability analysis section from Skagway. These percent non-dissipated values are multiplied by the vertical stress increments given on Figure 9.33 to obtain the calculated excess pore pressure at that location. For example, at point G of the trial arc shown on Figure 9.33, the vertical stress increase caused by the fill is 600 psf. Assuming that the initial pore pressure on first loading is 100 percent of the applied vertical stress, the initial pore pressure also would be 600 psf. Based on Figure 9.35, the undissipated pore pressure at point G after 29 days of loading is about 70% of the initial pore pressure; i.e., 420 psf.

For most projects, these calculations would complete the determination of excess pore pressures in a soft, normally consolidated soil during loading. It is based not only on a 100 percent response of pore pressure on first loading by the fill, but also on the assumption that pore pressure coefficient A at failure is 1.00. These are normal expectations for such soils (e.g., Bishop and Henkel, 1957). At Skagway, however, the consolidated-undrained triaxial tests with pore pressure measurements had a small amount of dilatancy during shear. The excess pore pressure at failure averaged about 50 percent of the applied deviator stress of the test, i.e., $A_f = 0.5$.

Table 9.10 Calculation of Excess Pore Pressures (Head) Acting on Base of Slices for Trial Arc A due to 29 Days of Loading by the Fill at Skagway, Alaska

(1)	(2)	(3)	(4) = (2) × (3)	(5) = 0.5 (4)	(6) = (5)/64
	Pore pressures created by fill				
Slice	Increase in Vertical Stress (psf)	Percent Undissipated at Base of Slice	Undissipated Pore Pressure (psf)	50% Undissip. Pore Pressure (psf)	Undissipated Head (feet)
3	1925	10%	193	96	1.5
4	2000	35%	700	350	5.5
5	1675	52%	871	435	6.8
6	1400	63%	882	441	6.9
7	1100	72%	792	396	6.2
8	700	75%	525	263	4.1
9	275	55%	151	75	1.2
10	100	10%	10	5	0.1

Explanations:

Col. (2): Increase in vertical stress obtained for recently placed fill using Boussinesq distribution with depth.

Col. (3): Percent undissipated pore pressure at base of slice assumes initial $\Delta u = \Delta\sigma_v$ and dissipation based on c_v = 1,850 sq. ft./year.

Col. (5): 50% dissipation based on A_f = 0.5 (Skempton Pore Pressure Coefficient A at failure, based on testing of the marine silts at Skagway).

Accordingly, the calculated excess pore pressures calculated for the Skagway slope were reduced to 50 percent to reflect the likely pore pressure conditions at the time of the slope failure. Therefore, for the example of point G on Figure 9.35, the excess pore pressure at failure is estimated to be 0.5 (420) = 210 psf.

Table 9.10 shows the calculated values of excess head above sea level in the marine silt on trial arc A at Skagway after 29 days of loading by the new fill.

Step 4. *Use the calculated excess pore pressures in an effective stress stability analysis*

In a circular arc analysis, the average excess pore pressure acting on the base of a slice can be calculated from Steps 1 to 3. This should be added to any pore pressure preexisting at this depth before the fill load was placed on the stratum.

9.13 SPECIAL CASES: (c) ARTESIAN PRESSURES

Artesian conditions occur where groundwater within a more permeable layer or stratum is impeded from draining by a less permeable layer or stratum. The artesian layer can be identified and the pressure head measured during a site investigation.

Example Calculations of Artesian Pressures

The landslide at Skagway, Alaska, provides an example of how artesian pressures can be treated in a stability analysis. A schematic drawing of the stability analysis section is shown on Figure 9.36. Four deep borings passed through the marine silt of low permeability into the very permeable talus beneath and recorded artesian fresh water flows of up to 50 gpm. The average static head was elevation +18.0 feet. Thus, the groundwater within the talus is trapped by the relatively impermeable marine silt. If the slope drains towards a lake, a flow net can be obtained for the steady-state crossflow from the artesian interface to the lower level of the lake. Such a flow net can be drawn by hand or can be determined by a computer analysis, such as SEEP/W (Geo-Slope, Calgary, Canada).

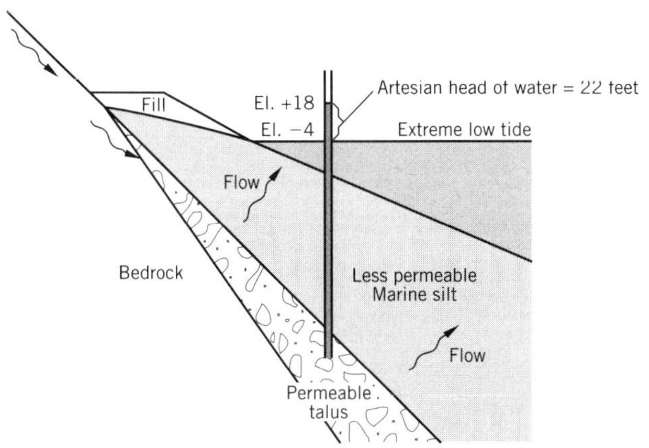

Figure 9.36 Schematic drawing of the artesian head condition within the talus/bedrock at Skagway, Alaska, at extreme low tide.

The flow net can be refined to account for anisotropic behavior. In most sedimentary deposits, the permeability in the bedding plane (usually termed the x-axis) is higher than the permeability in the direction of deposition (y-axis). Using a transformed section (details in textbooks on groundwater flow and seepage) a flow net can be constructed to account for anisotropy.

At Skagway, Alaska, the artesian pressure dissipates through a thick marine silt stratum to a fluctuating sea level. The twice-daily tidal range varies from about 14 feet to 26 feet.

The composite value of the coefficient of consolidation c_v of the marine silt, based on laboratory triaxial dissipation tests, was calculated to be 1,850 sq. ft. per year. Using this value, it can be shown from the time factor equation that only the outer 1 foot of the silt will consolidate and rebound within a 6-hour time period, and silt more distant than 3 feet from the drainage boundary is essentially unaffected by the tide.

Long term, the artesian pressure causes a crossflow from the talus to the *mean* sea level. Many thousands of tides have occurred within the existing framework of geological conditions, and the equilibrium (steady state) flow will mathematically approach this mid-tide level. As already mentioned, the daily tide fluctuations, affecting only the thin outer mantle, have no practical effect on this long-term crossflow. Low tide reduces slope stability due loss of lateral support on the outer slope but does not affect the internal pore pressures within the slope. Thus, the soil within the slope shears undrained ($\phi = 0$) and is essentially a rapid drawdown from mean tide to low tide.

The suggested method of analyzing a Skagway-type artesian condition where the artesian pressure crossflows to a fluctuating tide is a three-step procedure as follows:

Step 1. *Obtain the flow net for the steady state crossflow from the artesian source to the mean tide level (or average water level in a lake or river).*

The equipotential lines for the flow net at Skagway are shown on Figure 9.37. The flow lines of the net are not shown but would be at right angles to the equipotential lines. To allow for anisotropy due to sedimentation, the permeability in the bedding planes parallel to the outer slope was assumed to be 10 times higher than in the direction of flow at right angles to the bedding planes; i.e., $k_x = 10\ k_y$, as shown on Figure 9.37. On this drawing, BC is the artesian pressure boundary at a constant head of elev. +18.0 feet and DE is the drainage surface of the sea at a constant head of elev. +8.7 feet mean sea level, a difference of 9.3 feet. All elevations are based on mean low low water (MLLW), but this term will be omitted when referring to elevations elsewhere in this section.

Step 2. *Perform a stability analysis on the trial arc for the ground conditions at mean tide level.*

Draw in a trial arc, such as arc A on Figure 9.37, and take off the average head acting on the base of each slice within the marine silt stratum. This is shown on Table 9.11 and, by subtracting the mean sea level of elev. +8.7 feet from each measurement, the excess head of water above sea level is obtained.

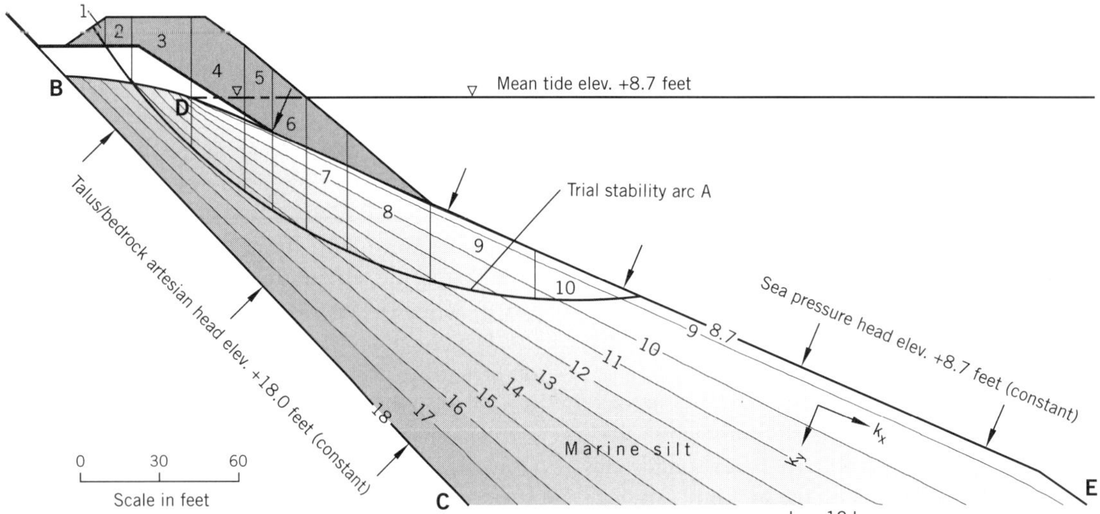

Figure 9.37 Equipotential lines for steady state crossflow from the artesian pressure in the talus/bedrock to mean sea level at Skagway, Alaska.

Table 9.11 Calculation of Excess Head Acting on Base of Slices for Trial Arc A due to Artesian Pressure at Skagway Alaska

Slice	Artesian head: Steady crossflow elev.	Artesian head: Head above sea level feet
3	14.4	5.7
4	14.7	6.0
5	14.4	5.7
6	14.2	5.5
7	13.75	5.05
8	12.8	4.1
9	11.3	2.6
10	9.7	1.0

Assumptions:

1. Steady state crossflow from elev. +18.0 feet (artesian head) to elev. +8.7 feet (mean sea level)

2. $k_v = 10\ k_h$ (anisotropic)

A stability analysis is performed for this trial arc and sea level. The total shear resistance of the marine silt along the arc can be obtained from the stability analysis.

Step 3. *Perform another stability analysis on the trial arc for ground conditions at low tide, where stability is most critical. The drop in sea level reduces the outside support on the slope.*

It can be assumed that the shear strength within the marine silt remains unchanged as the tide fluctuates for the reasons given earlier (i.e., $\phi = 0$ analysis in the marine silt). Therefore, take the total shear resistance calculated in Step 2 and use it unchanged in the stability calculations for low tide (or any other tidal level). In a tabulated set of calculations (see Table 9.6 on Figure 9.29), the shear resistance calculated in column (16), the final column of the table, is summed for all slices that have their bases passing through the marine silt. Any slices that shear through free-draining (ϕ-dependent) granular soils, such as slices 1 and 2 of Figure 9.37, have to be adjusted in the normal procedure for a circular arc analysis; i.e., unlike the silt stratum, they change strength if the groundwater level changes from mean tide to low tide. At Skagway, however, there is no change in groundwater in slices 1 and 2, so no adjustment was necessary except for minor changes in strength due to the change in F.

Excluding any change in shear resistance in the granular soils, the only change in the stability analysis is in the driving moment, W sin α. As the sea level drops, this term increases. At high sea levels, the term decreases due to increased buoyancy.

To sum up this technical approach to a fluctuating sea level, groundwater pressures in the relatively impermeable soil stratum are governed by the crossflow from the artesian source to the mean sea level. These groundwater levels are unaffected by short-term tidal variations. They also control the immediate shear strength c within the impermeable stratum. When the tidal level drops to more critical levels of stability, the shear strength within the stratum remains unchanged ($\phi = 0$). This shear strength along the arc can be obtained from a preliminary stability analysis at mean sea level.

Combined Artesian Pressure and Undissipated Pore Pressure Calculations

When artesian pressures and undissipated pore pressures from a fill loading are both present in a stability analysis, the two pressure heads can be combined to give the excess head above the water level outside the slope. For the Skagway shallow arc A, which has been used as an example in Section 9.12(b) and in the current section, the results of the combined analysis are shown on Table 7.2 of Case History 7. For that arc, which passes through the middle of the consolidating marine silt, the dissipation occurs toward the sea.

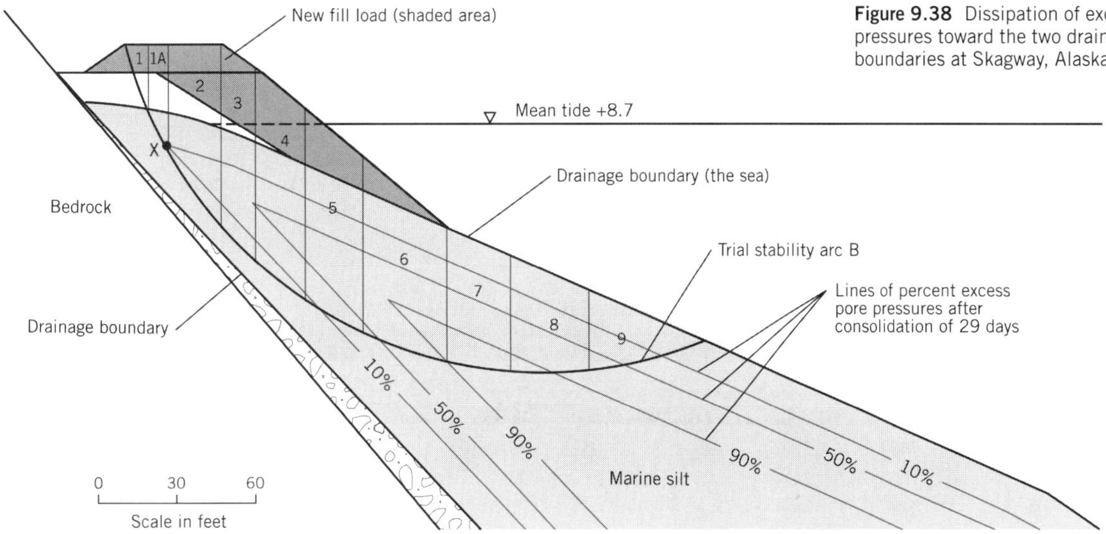

Figure 9.38 Dissipation of excess pore pressures toward the two drainage boundaries at Skagway, Alaska.

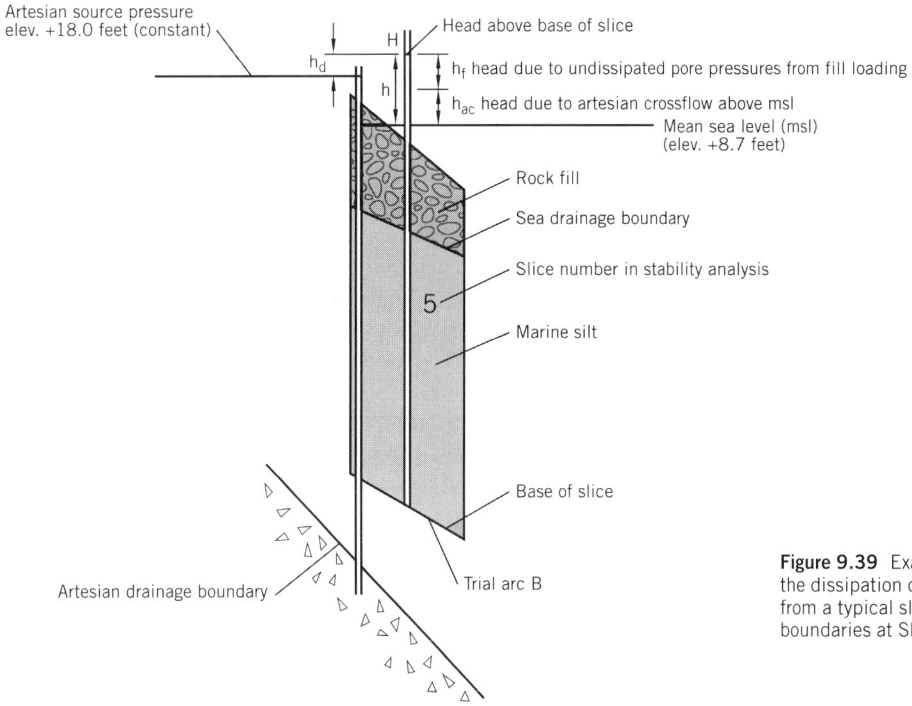

Figure 9.39 Example slice to show the dissipation of excess pressure head from a typical slice to the two drainage boundaries at Skagway, Alaska.

A more general case is illustrated on Figures 9.38 and 9.39 (deeper trial arc B from the Skagway slide). The upper level of the trial arc passes close to the artesian drainage boundary and the lower end of the arc reaches the sea (Figure 9.38). To simplify the analysis, it is assumed that drainage towards one boundary does not affect drainage to the other boundary. For example, at point X of slice 1A, the undissipated pore pressure contour is 10% for both the artesian boundary and the sea boundary; no allowance is made for interference between the two drainages, even though such a correction would reduce the combined effect to slightly less than 10%.

The suggested three-step procedure to determine the excess head h for each slice is summarized in Table 9.12. As shown on Figure 9.39 for slice 5 of arc B, the head h rises above sea level due to the artesian crossflow h_{ac} and the undissipated excess pore pressures of the fill loading h_f. If point H is above the artesian pressure source elevation, some dissipation will occur from H toward the artesian boundary, a difference of h_d. However, if point H is below the artesian source elevation, dissipation of excess head h can only occur toward the sea.

Step 1. *Calculate the excess head in the partly consolidated soil due to the fill.* This procedure is described in Section 9.12(b). For the Skagway example, the vertical stress is obtained from Figure 9.33 (for arc B, which is shown on Figure 9.38 but not on Figure 9.33), and a reduction of 50% in pore pressure is

Table 9.12 Calculation of Excess Head Acting on Base of Slices for Skagway Slide, Trial Arc B, for Drainage to Two Boundaries

(1)	(2)	(3)	(4)	(5)	(6)	(7)	(8)	(9)	(10)	(11)	(12)	(13)	(14)
	Artesian Head		Fill Loading			Dissipation to Artesian Source				Dissipation to Sea			
Slice	Steady Cross Flow	Head Above Mean Sea Level h_{ac}	Increase in Vertical Stress $\Delta\sigma_v$	Initial Increase in Pore Pressure $\Delta u_i = \Delta\sigma_v$ (head)	Pore Pressure Head after Strain $\Delta u_f = \frac{1}{2}\Delta u_i$	Total Head on Base of Slice before Dissip.	Difference in Head from Slice to Source h_d		Undissipated Head on 11/3/94	Excess Head due to Fill $h_{f'}$		Undissipated Head on 11/3/94	Total Excess Head h
	(feet elev.)	(feet)	(psf)	(feet)	(feet)	(feet elev.)	(feet)	%	(feet)	(feet)	%	(feet)	(feet)
1A	15.0	6.3	1700	26.56	13.28	28.28	10.28	5	**0.51**	13.28	5	0.66	6.81
2	16.5	7.8	2000	31.25	15.63	32.13	14.13	0	**0**	15.63	70	10.94	7.8
3	16.8	8.1	1800	28.13	14.06	30.86	12.86	0	**0**	14.06	93	13.08	8.1
4	16.9	8.2	1420	22.19	11.10	28.00	10.00	0	**0**	11.10	98	10.88	8.2
5	16.4	7.7	1000	15.63	7.81	24.21	6.21	20	**1.24**	7.81	98	7.65	8.94
6	15.3	6.6	600	9.38	4.69	19.99	1.99	65	**1.29**	4.69	98	4.60	7.89
7	13.9	5.2	300	4.69	2.34	16.24	<0			2.34	97	**2.27**	7.47
8	12.3	3.6	170	2.66	1.33	13.63	<0			1.33	92	**1.22**	4.82
9	10.4	1.7	80	1.25	0.63	11.03	<0			0.63	60	**0.38**	2.08

Abbreviations: elev. – elevation, dissip. – dissipation of excess pore pressures

(5) = (4)/64; (6) = ½ (5); (7) = (6) + (2); (8) = (7) – 18.00; (10) = (8)x (9); (11) = (6); (13) = (11) x (12); (14) = (3) + bold value from either (10) or (13)

allowed for dilatancy of the Skagway silt after strain. The result is given in column (6) of Table 9.12.

Step 2. *Calculate the undissipated head of the partly consolidated soil.* Treat dissipation to the artesian source and dissipation to the sea separately. For dissipation to the artesian source, subtract the artesian source elevation from the *total* head on the base of the slice. This is the head difference h_d that can dissipate, column (8) on Table 9.12.

Step 3. *Calculate the total excess head h for use in the stability analysis.* This result is obtained by adding the undissipated head due to the fill (i.e., the bold values in column (10) and (13) of Table 9.12) to the artesian steady state crossflow head above sea level, column (3).

9.14 SPECIAL CASES: (d) PILE RESISTANCE

Piles can be used to stabilize landslides by construction of a shear pile wall. A row of closely spaced piles constitute the "wall." To stabilize the entire landslide, the pile wall is usually constructed at or near the base of the slide, similar to a buttress support. In other cases, it may support a specific facility, such as a road or building, allowing the landslide downslope to continue moving.

Design methods for shear pile walls are still evolving. Most landslides develop a discrete shear zone at the slip surface where the soil is weaker than the ground above and below. The pile design is usually governed by the bending stresses developed across the discrete shear zone. This topic is covered in Chapter 22, Section 22.1.

Piles embedded in soft, normally consolidated soils are not subject to "strain-softening" of overconsolidated soils. Such piles can be designed to resist a shear force where the slip surface intersects the pile. This type of analysis is described below.

An example of a pile shear failure calculation is the Skagway flow slide that has been cited earlier. At this site (Figure 9.40), the piles of an existing wooden dock were intercepted by stability arc A, believed to be the approximate location of the initial shear failure that triggered a much larger flow slide. The timber piles of the dock at this location were 80–110 feet long (typically 90 feet), with 8 piles per bent and bents spaced at 8-foot intervals. Thus, for a two-dimensional stability analysis (1 foot deep), there is the equivalent of one pile in the stability cross-section. The question was: How much resistance did the timber piles provide?

The structural analysis was performed by Dr. Lee Schroeder using the computer program LPILE. A pile near the

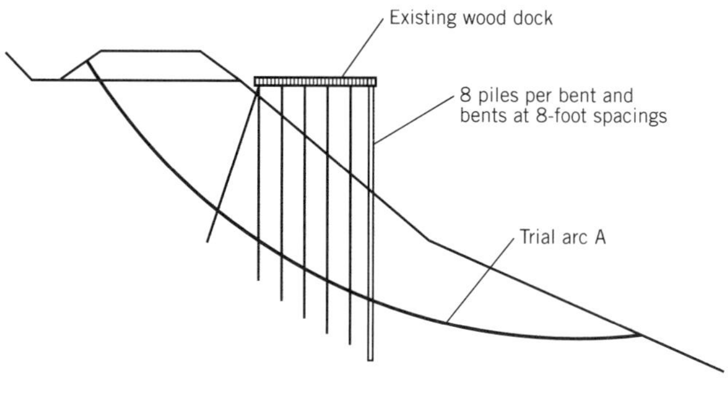

Figure 9.40 Wood piles of a dock penetrating into normally consolidated soils (Trial Arc A at Skagway, Alaska).

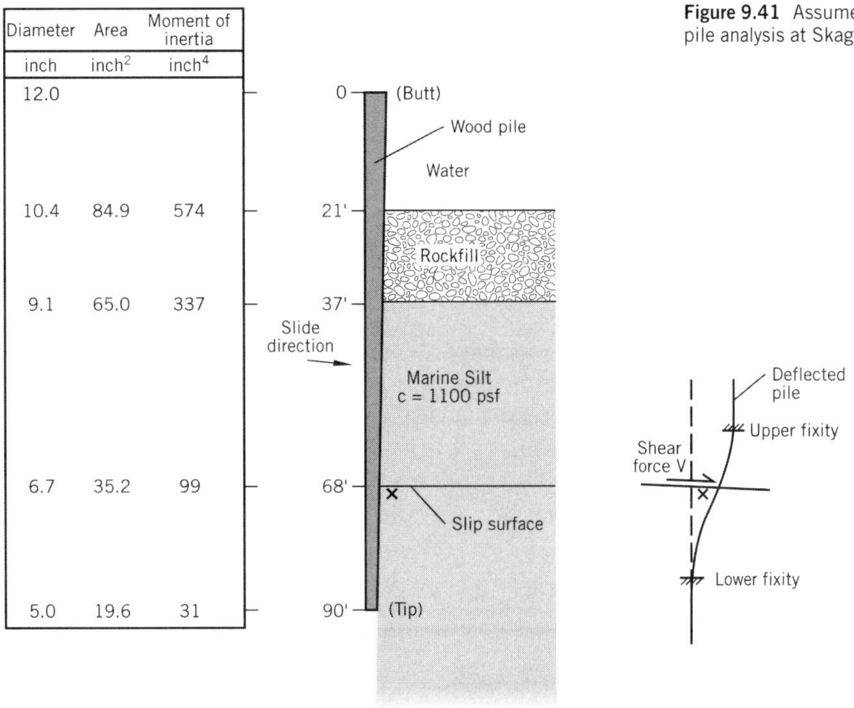

Figure 9.41 Assumed conditions for the shear pile analysis at Skagway, Alaska.

center of the dock width was selected for the analyses, and the profile and pile properties are shown on Figure 9.41. There is sufficient pile embedment length above and below the slip surface to develop fixity.

The limiting strengths according to the Western Wood Products Association are 6,220 psi in bending and 330 psi in shear. Modulus E = 1,500,000 psi.

Maximum shear stress = 1.33 x average shear stress V
(well-known assumption in strength of materials)

$$= \frac{1.33V}{35.2}$$

For a limiting shear stress of 330 psi, V = 8,712 lb. Values of V above and below 8712 lb. were analyzed by LPILE and gave the results shown on Table 9.13.

Checking the ultimate lateral bearing capacity for the piles:

$$p_u = 9\,c\,b \qquad \text{Eq. (39)}$$

$$(p_u)_x = 9\,(1100)(6.7)/144 = 461 \text{ lb./inch at slip surface X}$$

$$(p_u)_t = 9\,(1100)(5.0)/144 = 344 \text{ lb./inch at tip of pile}$$

For a shear force of 8,712 lb. (which is the limit for shear stress as previously calculated) the bending stress of 6,328 psi slightly exceeds the limit of 6,220 psi. Therefore, the *maximum* shear force that the timber pile could provide at depth X is around 8,600 lb. The soil lateral bearing values are adequate to resist soil movement without failure in all cases.

The same analysis for the pile section above the failure surface would show essentially the same results. Therefore, relative to its initial position, the pile section at the failure surface would be displaced twice the values shown. For the limiting shear force of 8,600 lb., the displacement should be approximately 2 (3.25) = 6.5 inches. Therefore, the full resistance of around 8,600 lb. would only develop after 6½ inches of shear displacement has occurred along the soil slip surface.

TABLE 9.13 Maximum Response Values for Applied Shear Force, V

Applied Shear Force (lb.)	Deflection (inch)	Bending Moment (in.-lb.)	Bending Stress (psi)	Shear Stress (psi)	Soil Reaction (lb./in.)
9757	4.45	225,000	7613*	370*	266
8712	3.27	187,000	6328*	330*	247
7667	2.36	152,000	5143	290	228
6534	1.65	119,000	4027	248	198

*equals or exceeds strength values from Western Wood Products Association. Soil reactions for all cases are less than the limiting value of 461 lb./in. at the slope failure surface.

At Skagway, the initial shear failure rapidly transformed into a massive flow slide. It is possible that the shear displacement was less than 6½ inches when this occurred, in which case the pile resistance would have been lower than 8,600 lb. However, a pile resistance of 8,600 lb. was used in stability analyses. Since it represents less than 3 percent of the total resistance along the slip arc, it has only a marginal effect on the calculated factor of safety.

9.15 SPECIAL CASES: (e) RAPID DRAWDOWN ANALYSIS

Rapid drawdown occurs when a river, reservoir, or tide level drops so quickly that the groundwater in the adjoining slopes has insufficient time to adjust to the steady state condition as

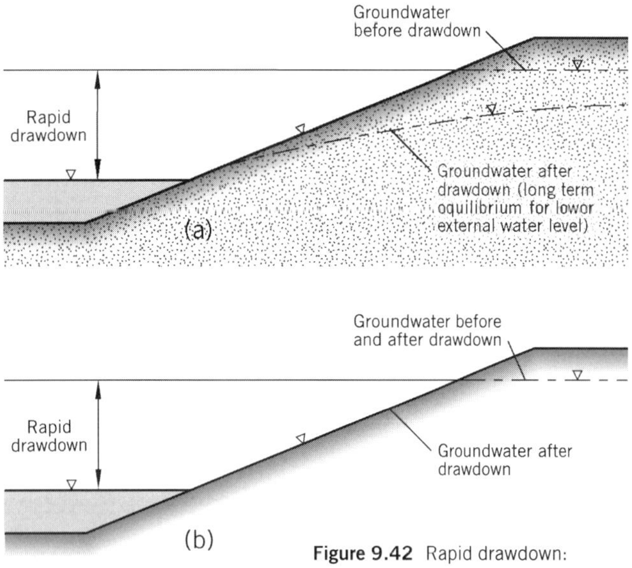

Figure 9.42 Rapid drawdown:
(a) drained slope
(b) undrained slope

the water recedes. Therefore, relatively high groundwater levels remain in the slope and decrease slope stability, sometimes causing slope failures. Examples of slopes subject to rapid drawdown include: (i) pumped storage projects where the reservoir is filled and emptied frequently; (ii) reservoirs drawn down for irrigation or water supply; (iii) streams that fall rapidly after a flood; and (iv) coastal sites with large tidal fluctuations.

The issue of rapid drawdown has interested earth dam design engineers for decades, and there are theoretical ways to estimate transient flow nets. Bromhead (1986) gives a good summary of these techniques and reference sources.

In practical landslide work, the transient analytical procedures are rarely employed. It is generally assumed that slopes will either drain fully or not at all. If there is doubt about the ability of the soil to drain rapidly during drawdown, it should be assumed that the soil will behave undrained.

The groundwater profiles for the drained and undrained conditions within a slope are illustrated on Figure 9.42. If the time factor analysis (discussed in the next few paragraphs) indicates that the soil will fully drain during drawdown, a steady-state flow net can be drawn for the groundwater surface shown on Figure 9.42(a) and the stability analysis can be performed with pore pressures taken from the equipotential lines. As a precaution, some residual excess pore pressures can be added to represent a less-than-complete drawdown within the slope.

Will the Slope Drain?

The first issue to be determined is whether the slope drainage will keep pace with the fall in water level outside the slope. This depends on the rate of drawdown and the properties of the soil in the slope, especially the coefficient of consolidation c_v. Another factor is the presence of more permeable layers within the slope which may speed up the response to drawdown.

The rate of drainage can be assessed through the dimensionless time factor T given by:

$$T = \frac{c_v t}{H^2}$$

where c_v = coefficient of consolidation; t = time for drawdown
H = drainage path (see also Chapter 8, Section 8.4).

As recommended by Duncan, Wright, and Wong (1990), the soil can be treated as fully drained if the time factor T is 3 or greater, for which the dissipation of excess pore pressure at distance H is more than 99%. For soils in which T is less than 3, they recommend that the soil be conservatively treated as undrained during drawdown, except that drained strengths should be used wherever they are smaller than the undrained strengths (as discussed later).

Using a few simplifying assumptions, some general guidelines can be provided to roughly identify the transition between soils that drain and those that do not drain during practical drawdown situations. The U.S. Corps of Engineers typically limits the rate of drawdown in reservoirs behind earth dams to about 3 feet per day. Similarly, rivers in flood can typically recede at rates of around 3 feet per day. Therefore, as a simplifying assumption, time t will be based on a drawdown rate of 3 feet per day.

Failures due to drawdown generally are fairly shallow landslides. The two examples cited in Duncan et al. (1990) have drainage paths H which are about 45 percent of the total reservoir drawdown. For simplification, assume H = 0.5 R where R is the external water level drawdown (Figure 9.43).

Rearranging the time factor equation

$$c_v = \frac{TH^2}{t} \qquad \text{Eq. (40)}$$

For the soil to be drained, T > 3

$$c_v = \frac{3H^2}{t}$$

Substituting H = 0.5 R and t (in days) = 0.33 R

$c_v > 2\frac{1}{4}$ R sq. ft. per day (i.e., 820 R sq. ft. per year)

Figure 9.43 Slope failure and drawdown assumptions made to estimate the minimum c_v required for soils to be fully drained during rapid drawdown.

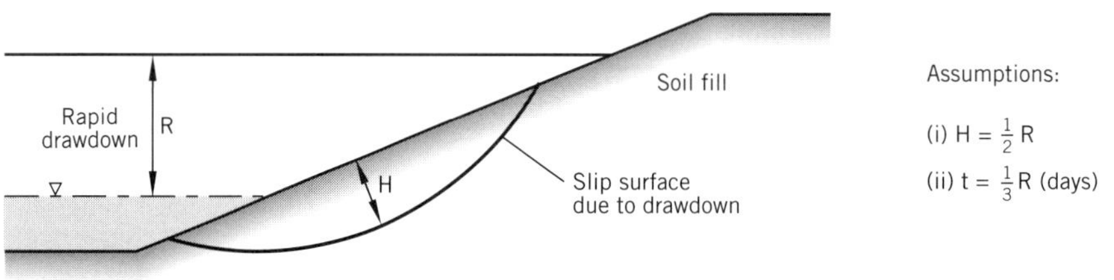

For a rapid drawdown R of 40 feet, the c_v of the soil in the slope would need to be greater than 32,800 sq. ft. per year. Referring to the chart, Figure 8.11 in Chapter 8, the soil would have to be a noncohesive soil (i.e., Cohesive Index CI = 0) to meet the requirement for drainage during rapid drawdown. For other rates of drawdown and drainage path H, the required c_v to achieve drainage can be calculated.

To measure parameter c_v on native soils or compacted fill, a triaxial test has to be performed on a specimen without side filter drains and presaturated by the application of back pressures. The specimen should be at least 4 inches high and 3 to 4 inches in diameter. If the rate of pore pressure dissipation at the non-draining end of the specimen is slow, the test can be ended prematurely in the knowledge that c_v is too low for drainage during rapid drawdown.

Rapid Drawdown Calculations

Two methods of calculating rapid drawdown stability will be presented. Both use a two-step approach. First, the shear strengths along a potential failure surface (such as a circular arc) are calculated for a slope with steady-state groundwater equilibrium *before* drawdown occurs. Second, drawdown of the external water occurs, which increases the driving moment in the factor of safety calculation. Since the soil is fully saturated and is incapable of draining during drawdown, the shear resistance determined in the first step remains unchanged. The soil shears undrained.

The difference between the two methods is in the calculation of shear resistance in the slope. Method A, as proposed by the author, uses an effective stress stability calculation to determine the shear resistance along the slip surface under the equilibrium groundwater conditions before rapid drawdown commences. This analysis includes allowance for interslice forces acting on the sides of each slice (e.g., Bishop's Routine Method). The strengths are based on the soil's effective stress parameters obtained from consolidated-undrained triaxial tests with pore pressure measurements or consolidated-drained tests. Once the soil resistance has been determined in this manner, the shear resistance along the slip surface cannot change ($\phi = 0$ analysis) during drawdown.

Method B, which has been developed by Duncan, Wright, and Wong (1990) as an outgrowth of earlier procedures, delves into soil behavior to specify the undrained soil strengths that will develop during rapid drawdown. Compared to Method A, the approach is much more complicated and time-consuming, but at least one computer program (UTEXAS) has the software to perform the calculations.

As an example of a slope failure in rapid drawdown, the failure at Skagway, Alaska, is used to demonstrate the calculations by Methods A and B. It should be pointed out that this is only one case history, and comparisons may be different at other failure sites.

At Skagway, daily tidal changes are relatively large. There is also an artesian crossflow passing between bedrock and the sea, the artesian pressure gradient increasing from high to low tide. However, as discussed in Section 9.13, the soil does not have time to react to tidal fluctuations except in the outer 3 feet of the slope (which can be neglected for practical purposes). Therefore, the steady-state artesian crossflow is from full artesian pressure at the talus/bedrock to mean sea level. For example calculations, the circular trial arc A will be used. This trial arc and the equipotential lines for the artesian condition are shown on Figure 9.37.

The steady state equilibrium conditions for the Skagway slope occur at mean sea level of +8.7 feet. Rapid drawdown occurs when the sea level falls below mean sea level to low tide. As described more fully in Case History 7, the top of the slope was loaded by a very large fill in mid-October, 1994, and failed catastrophically three weeks later during an extreme low tide of elev. –4.0 feet. Thus, the rapid drawdown was 12.7 feet from mean to extreme low tide.

Method A

The analysis is illustrated on Figure 9.44. The two-step procedure for each trial arc is as follows:

Step 1. *High water before drawdown. Determine the available shear resistance of the clay stratum.* It is assumed that the slope has been subjected to the high external water level for sufficient time for the clay strength to have reached equilibrium with the internal pore pressures and external water. The total shear resistance along the arc, $\bar{c}\ R\ \psi$ (Figure 9.44a), can be obtained from a circular arc effective stress stability analysis using either a computer program or the final column of a tabular hand calculation such as that shown on Figure 9.29. The effective stress shear strength parameters are obtained from consolidated-drained or consolidated-undrained triaxial tests with pore pressure measurements. For soft normally consolidated clay, another option is to measure the immediate (undrained) shear strength of the clay from laboratory or field vane tests.

Step 2. *Low water after drawdown. Calculate stability during rapid drawdown, assuming no change in the clay's shear resistance.* For the after-drawdown condition, calculate the new driving moment $W_2 R \sin\theta_2$ (the R subsequently cancels out in calculating F). The change is caused by the loss of buoyant weight in the slope resulting from the drawdown, as shown shaded on Figure 9.44(b). For the section shown, it can be noted that the center of gravity of the soil mass moves upslope. Thus, both W_2 and θ_2 are larger than the comparable values of W_1 and θ_1 in the pre-drawdown condition of Step 1. The increase in driving moment is entirely responsible for the drop in factor of safety F. The clay resistance remains unchanged ($\phi = 0$ analysis).

Note: If a free-draining layer is above or below the non-draining layer, it should be treated as a ϕ-dependent strength in the usual way.

Figure 9.44 Rapid Drawdown Analysis Method A: unchanged shear resistance during drawdown.

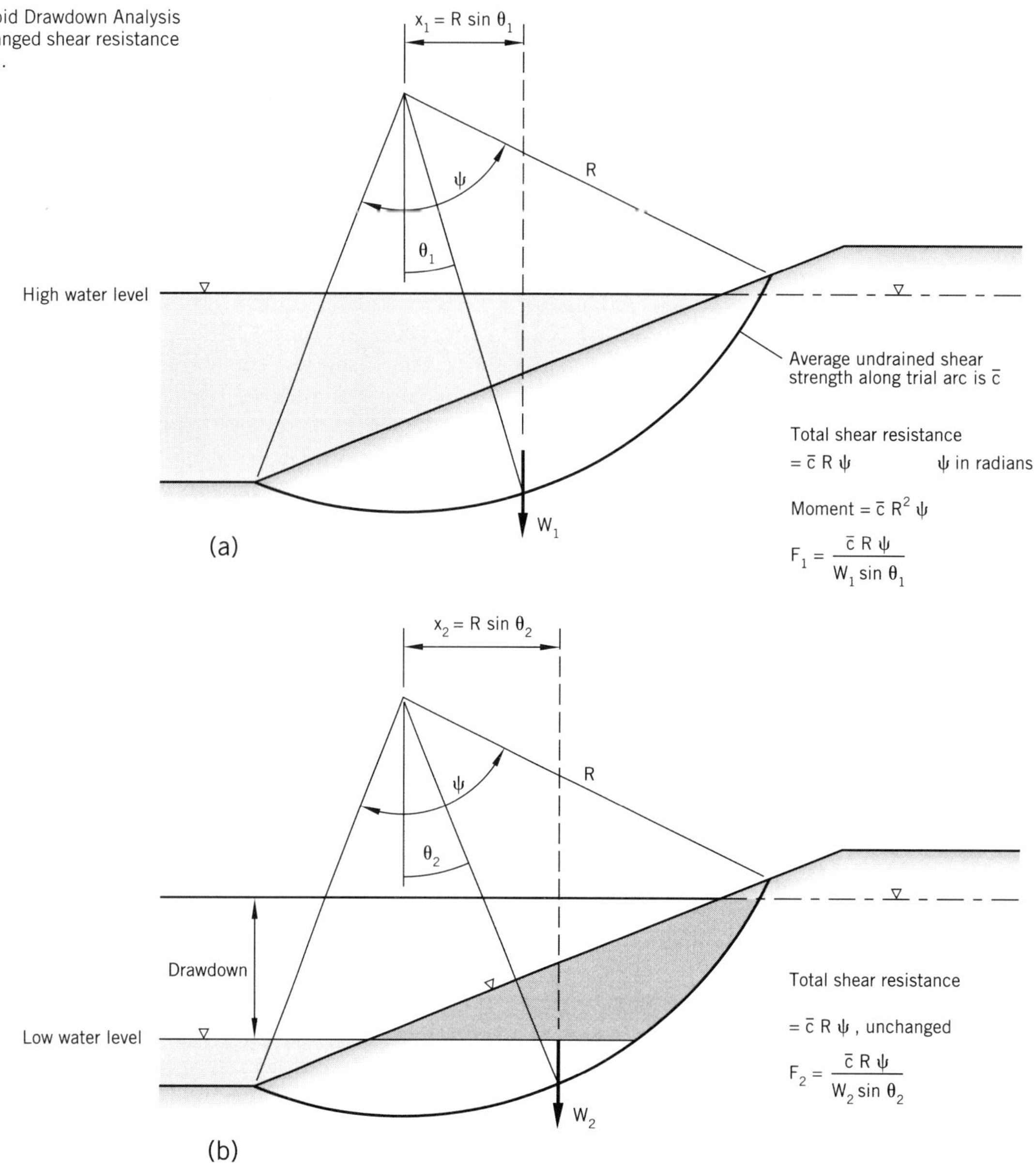

Skagway Example. The stability of arc A was performed using Bishop's Routine Method.

The results of the calculations by Method A were:

Mean tide level, before rapid drawdown $F = 1.09$

Extreme low tide level, after rapid drawdown $F = 0.96$

Slope failure occurred at extreme low tide.

Method B

Method Development

This method of analysis was proposed by Duncan, Wright, and Wong (1990) as a further development of earlier design procedures. The method has been widely adopted in the United States. A brief overview of these earlier procedures is helpful in understanding Method B.

The Corps of Engineers (1970) used a two-stage analysis based on the consolidated-drained and consolidated-undrained shear strength envelopes (Figure 9.45). The first stage, using the pre-drawdown conditions, determines the effective normal stress on the slip surface. These are used to calculate the undrained strengths in the second stage drawdown analysis.

The consolidated-undrained Mohr envelope uses the minor principal stress σ'_{3c} at the end of the consolidation stage of the test and the maximum deviator stress at failure $(\sigma'_1 - \sigma'_3)_f$. These two parameters are considered incompatible because the effective minor principal stress after consolidation σ'_{3c} is not the effective minor principal stress when the maximum deviator stress is reached.

Lowe and Karafiath (1960) introduced a diagram of undrained strengths related to the effective normal stress σ'_n

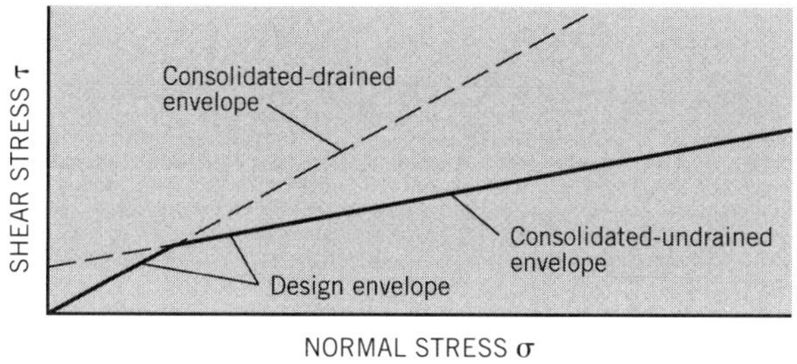

Figure 9.45 Corps of Engineers shear strength envelope for slow-draining soils during rapid drawdown (after U.S. Corps of Engineers, 1970).

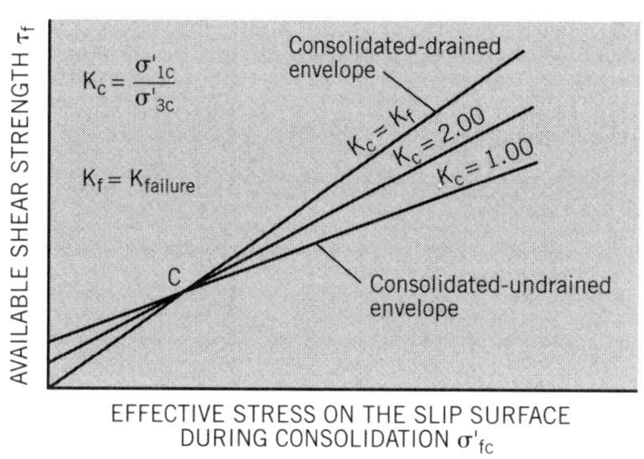

Figure 9.46 Strength envelopes developed by Lowe and Karafiath (1960) for rapid drawdown that depend on the principal stress ratio K_c acting on the slip surface during consolidation (modified from the original—see text).

on the potential slip surface. It takes into account that the undrained strength of soils varies with the effective principal stress ratio $K_c = \sigma'_1/\sigma'_3$ during consolidation. The value of K_c varies from 1.0 for isotropic consolidation (the normal procedure in a triaxial test) to the principal stress ratio K_f corresponding to failure (Figure 9.46). K_c values between 1.0 and failure represent anisotropic consolidation stress paths. Thus, a soil consolidated to a K_c value of 2.0 develops a higher undrained strength than a soil with K_c of 1.0 if it is to the right of the crossover point C on Figure 9.46. In practice, almost all the soil along the slip surface of a trial arc will have K_c values higher than 1.0 so the ratio at the site has to be determined as part of the analysis.

To directly obtain a Lowe and Karafiath (LK) envelope for a particular soil, it would be necessary to consolidate soil specimens isotropically and anisotropically in the laboratory and then fail them under undrained conditions. The upper limit of K_c is at failure and would require consolidating the soil at effective stresses corresponding to failure (impossible to do in practice but this result can be approached by using slightly lower deviator stresses than failure during consolidation).

The Duncan, Wright, and Wong (1990) rapid drawdown method recommends using the LK strength envelope and deriving it entirely from isotropically consolidated-undrained triaxial tests. The result of a typical test from the Skagway Alaska slide is shown on Figure 9.47 to demonstrate the development of LK envelopes for $K_c = 1.00$ and $K_c = K_f$. The effective stress circle provides the shear stress τ_f and effective normal stress $(\sigma'_n)_f$ on the failure plane, and thus a point on the $K_c = K_f$ envelope. The same value of τ_f applies to the $K_c = 1.00$ envelope except that the effective normal stress during consolidation is given by σ'_{3c}.

Assuming $c' = 0$, the effective normal stress and shear stress on the failure plane of the test specimen can be obtained from a line tangent to the effective stress circle at failure and passing through the origin. Since it is difficult to read the exact point where the tangent touches the circle, it

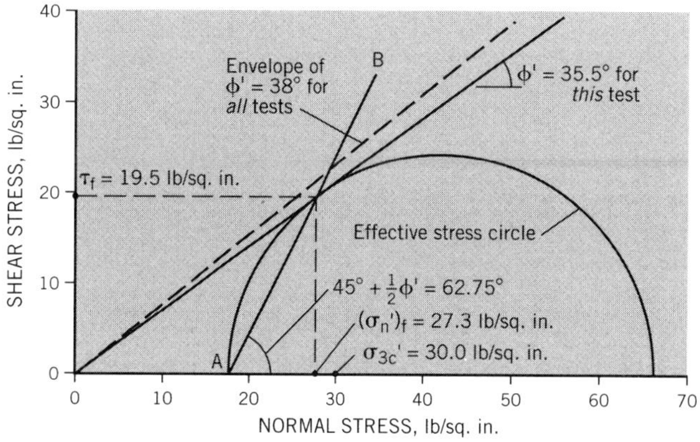

Figure 9.47 Determination of the shear stress at failure (τ_f) and effective normal stress (σ'_n) from a consolidated-undrained triaxial test with pore pressure measurements.

is best to construct line AB on Figure 9.47 to determine the tangent point.

It can be noted that the tangent to the circle on Figure 9.47 is at 35½° but the tangent for all the effective stress consolidated-undrained tests from Skagway is 38° (Figure 9.48b). The envelope of $\phi' = 38°$ does not touch the effective stress circle of Figure 9.47.

The suggested procedure for developing the LK envelopes is to draw two Mohr envelopes for the consolidated-undrained tests: a total stress envelope (Figure 9.48a) and an effective stress envelope (Figure 9.48b). For the Skagway data, it can be observed that the total stress envelope has much greater scatter in the individual results.

The *total stress* envelope then is converted to a LK envelope (Wright, 1991) by the formulas:

$$c_{LK} = c_R\left(\frac{\cos\phi_R \cos\phi'}{1-\sin\phi_R}\right) \qquad \text{Eq. (41)}$$

$$\tan\phi_{LK} = \left(\frac{\sin\phi_R \cos\phi'}{1-\sin\phi_R}\right) \qquad \text{Eq. (42)}$$

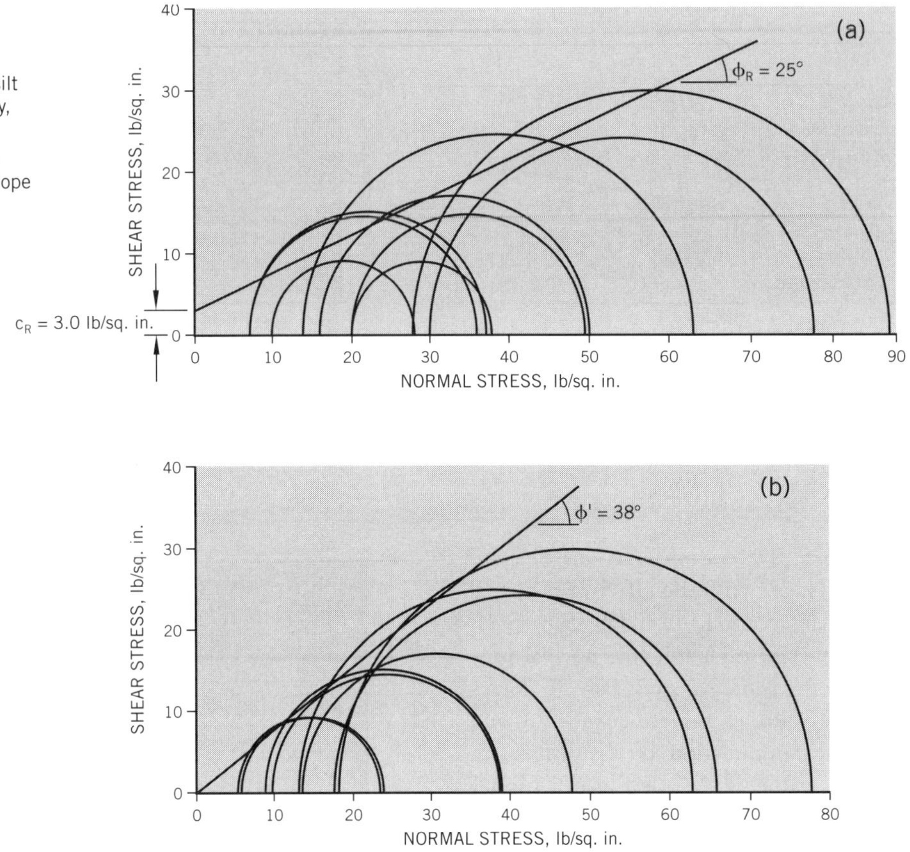

Figure 9.48 Mohr envelopes for nlne consolidated-undrained triaxial tests on marine silt specimens from Skagway, Alaska:

(a) total stress envelope
(b) effective stress envelope

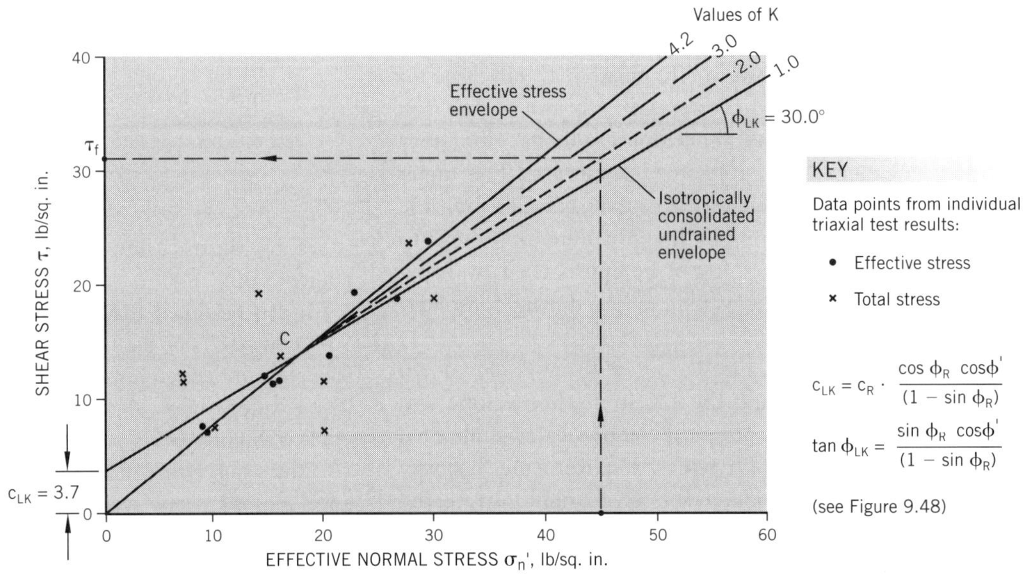

Figure 9.49 Lowe and Karafiath (LK) strength envelopes based on isotropically consolidated-undrained triaxial tests on marine silt specimens from Skagway, Alaska.

Using the Skagway results of Figure 9.48, c_{LK} = 3.7 lb./sq. in., ϕ_{LK} = 30.0°. The effective stress envelope at failure of c' = 0, ϕ' = 38° is the line for $K_c = K_f$ = 4.20 on the LK graph and requires no conversion. The two LK envelopes for the Skagway marine silt are shown on Figure 9.49. After determining the two limits (K_c = 1.00 and $K_c = K_f$) as described above, intermediate values of K_c are calculated by linear interpolation (Figure 9.49).

Method B Calculations

The stability calculation is a two-or three-step procedure:

Step 1. *High water before drawdown. Determine the effective normal stress σ'_n and shear stress τ at the base of each slice in a trial arc. Calculate principal stress ratio $K_c = \sigma'_1/\sigma'_3$ for these conditions. Determine the undrained strength of the soil from Lowe and Karafiath (LK) envelopes.*

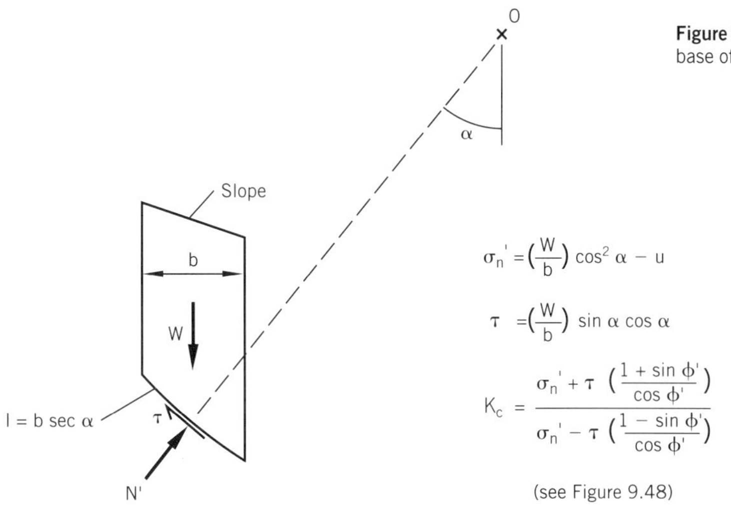

Figure 9.50 Calculation of σ_n' and τ on the base of a circular arc slice.

The effective normal stress σ'_n and shear stress τ acting at the middle of the base for each slice in a trial arc can be calculated by hand or be obtained from a computer printout. The equations are:

$$\sigma'_n = \left(\frac{W}{b}\right)\cos^2\alpha - u \qquad \text{Eq. (43)}$$

$$\tau = \left(\frac{W}{b}\right)\sin\alpha\cos\alpha \qquad \text{Eq. (44)}$$

where α is the angle between the center of the arc and the vertical, b is the horizontal width of the slice, and W is the weight of the slice including buoyant weight below the outside water level (Figure 9.50).

Pore pressure u can be obtained from the equipotential lines of a steady state flow net. More commonly, it is assumed to be $\gamma_w h_w$ where h_w is the excess head above the outside water level as could be measured by a standpipe piezometer at the middle of the slice base.

The principal stress ratio K_c at the base of the slice is calculated (Wright, 1991) from:

$$K_c = \frac{\sigma'_n + \tau\left(\frac{1+\sin\phi'}{\cos\phi'}\right)}{\sigma'_n - \tau\left(\frac{1-\sin\phi'}{\cos\phi'}\right)} \qquad \text{Eq. (45)}$$

This calculation assumes that the orientations of the principal stresses after consolidation (i.e., before drawdown) are the same as they would be at failure. The values of K_c just calculated may exceed K_f, especially for steeply inclined slip surfaces. In such cases, the strength at the base of the slice is governed by the effective stress strength ($K_c = K_f$).

Using the effective normal stresses σ'_n determined in Step 1, obtain the undrained shear strength available from the LK envelope corresponding to the K_c value at the base of each slice.

Step 2. *Low water after drawdown. Using the undrained shear strengths calculated in Step 1, determine the slope stability after drawdown.*

Multiply the shear strength on each slice by the length of arc (b sec α) and sum the results to obtain the total shear resistance along the trial arc. Determine the factor of safety for the slope with the external water at the drawdown level.

Duncan et al. (1990) recommend using drained strengths wherever the drained strength after drawdown is lower than the calculated undrained strength obtained in Step 2.

The reasoning behind this conservative assumption is that dilatant soils may cavitate or drain when negative pore pressures develop and therefore may not attain an undrained strength that is higher than the drained strength at such locations. This correction lowers the calculated factor of safety. Since this correction is based on conditions *after* drawdown, a third stability computation (Step 3) is often required for each trial arc. A check is made of each slice, and the drained strengths are substituted for the undrained strengths where they are lower. To make this third analysis, pore pressures within the slope after drainage have to be specified.

Skagway Example. The stability analysis of trial arc A was performed according to the procedures given for Steps 1 and 2. The shear resistance included the undrained strengths without reducing them to the lower drained strengths at low effective stresses. The results for Method B are:

Mean tide level, before rapid drawdown	F = 1.11
Extreme low tide level, after rapid drawdown	F = 0.98

As stated previously, slope failure occurred at extreme low tide. Two points can be made: (i) the results are almost identical to those obtained by Method A (which is quicker and simpler), and (ii) the calculated result for extreme low tide is in good agreement with the slope failure occurrence.

All the slices in this stability analysis had effective normal stresses that were less than 14 lb./sq. in. (i.e., were to the left of point C on Figure 9.49.) Duncan et al. (1990) have recommended that the lower drained strengths be used when the drained envelope falls below the undrained envelope on the

LK plot. Following this Step 3 directive, the result of Method B becomes:

Extreme low tide level, after rapid drawdown F = 0.87

Further Comment on Method B

Since most rapid drawdown failures are shallow in depth, there is a high probability that the LK drained envelope will be below the LK undrained envelope in the relevant stress range of the slope stability analysis. The present Design Method B requires the undrained shear strengths to be discarded in favor of lower and more conservative drained strengths.

9.16 SPECIAL CASES: (f) THREE-DIMENSIONAL ANALYSIS

Almost all landslide stability analyses are based on a two-dimensional cross-section of the slide, generally through or near the mid-width. In practice, this section is usually the deepest and most critical section. Based on various theoretical studies (e.g., Azzouz and Baligh, 1978; Cavounidis, 1987) it is widely understood that a three-dimensional study will produce a slightly higher factor of safety than a two-dimensional analysis through the center. Theoretical solutions have been developed to analyze slopes in three dimensions where the slip surface can be approximated to certain shapes (Azzouz and Baligh, 1978; Chen and Chameau, 1983).

Skempton (1985) recommends a correction in the stability analysis to allow for a three-dimensional slide. It is to reduce the shear stress by a factor:

$$\frac{1}{1 + KD/B} \qquad \text{Eq. (46)}$$

where D = average depth of landslide
B = average width of landslide
K = earth pressure coefficient (typically 0.5 for at rest)

The correction is typically about a 5% increase in factor of safety F.

The subject of three-dimensional stability is inadequately supported by actual field studies. Many landslides grow wider in the downslope direction, and it is reasonable to expect that, once movement has been initiated, there is partial or even total loss of contact along the sides, i.e., little or no contribution to stability. Vertical cracks, several inches wide, can be observed on the flanks of many landslides and suggest that most side support is lost after significant downslope movements have occurred. However, it is also possible to observe pressure ridges at the edges of landslides; these suggest that a local passive failure developed during the slide movements, which would increase sliding resistance.

In general, the author does not recommend three-dimensional stability studies of landslides. To investigate two or three sections adds significantly to the cost of site explorations and, for a given budget, is likely to substitute quantity for quality.

As stated previously in several parts of this book, landslide analysis and remediation is a comparative process and does not depend on perfection of the landslide analysis. Any errors (variations from reality) are carried forward into the remediation analysis of the same section and generally do not affect the outcome and decisionmaking of the study. Therefore, a better use of the available funds for site explorations is to ensure that a high quality single cross-section has been obtained through the center of the landslide for analysis.

There are some landslide situations where more than a single cross-section should be considered. These include:

- Where the depth of slippage changes significantly between the center and edges of the slide. This possibility can often be detected during the geological reconnaissance of the site and can be confirmed by additional borings.
- Where the width of the slide exceeds the length by a factor of two or more.
- Where a change in the surface geology or groundwater conditions across the width of the slide is evident.

For these situations, a common technique in landslide practice is to study two or three cross-sections. These are analyzed independently and are then combined into a single analysis. The sections should cover equal areas of the slip surface; as examples, sections made at the quarter-point widths of the slope (Figure 9.51) for a two-section analysis, or sections taken at the one-sixth, centerline, one-sixth widths for a three-section analysis.

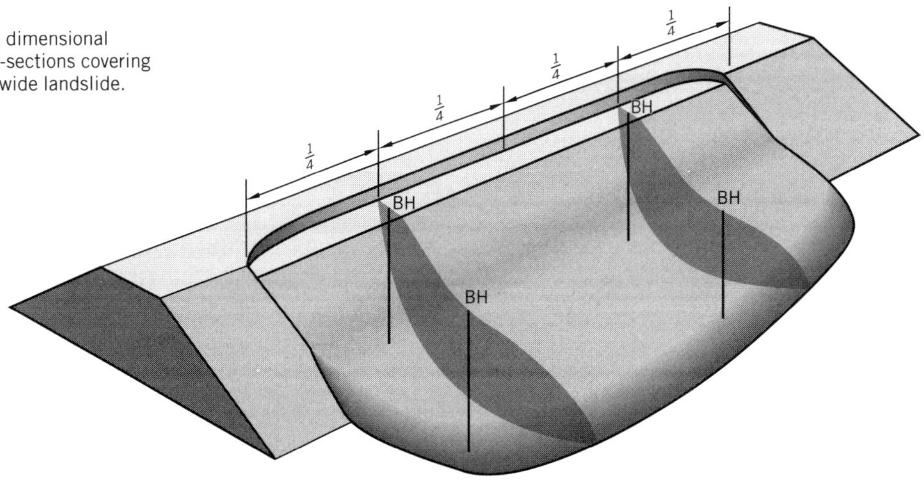

Figure 9.51 Three dimensional analysis. Two cross-sections covering "equal areas" of a wide landslide.

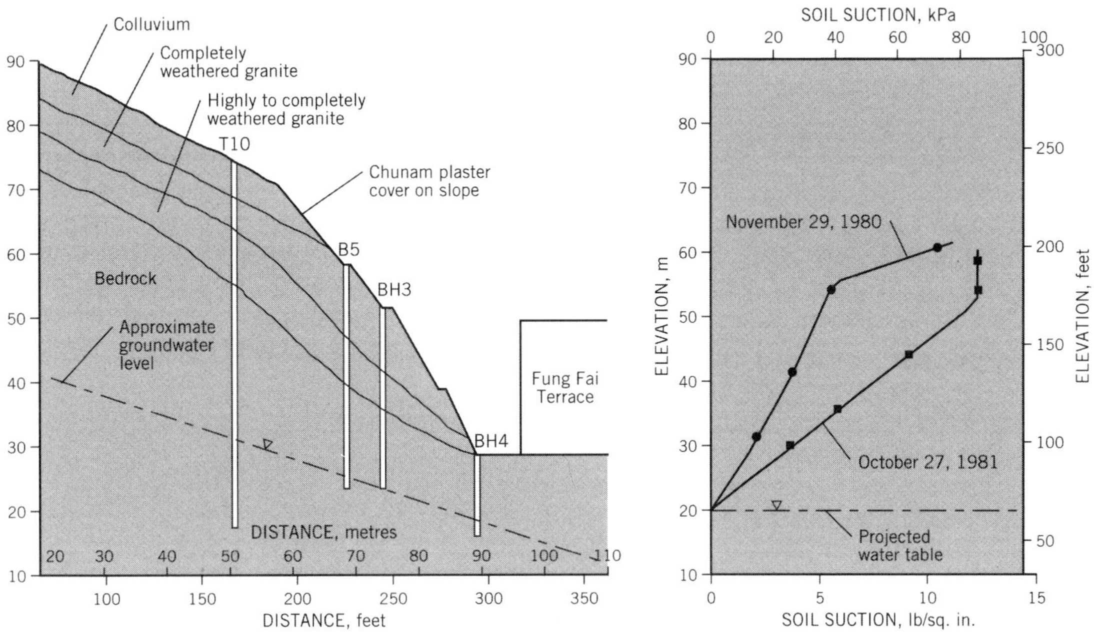

Figure 9.52 Soil suctions measured at Fung Fai Terrace, Hong Kong (after Fredlund, 1987).

9.17 SPECIAL CASES: (g) UNSATURATED SOILS

Partly saturated clays, silts, and fine sands above groundwater level generally have negative porewater pressures. They account for near-surface stability at sites where a conventional effective stress stability analysis, without allowance for soil suction, would show a factor of safety below 1.00. For most slope stability analyses, where only a very minor part of the slip surface passes through unsaturated soils near the surface, the extra strength attributable to soil suction is ignored.

In areas of the world where steep cuts are made into partly saturated residual soil slopes to accommodate structures, there is a need for establishing rational shear strength parameters in partly saturated fine-grained soils. A brief overview will be presented below.

Measurement of Soil Suction

Soil suction pressures can be measured in the field by a tensiometer. The sensor consists of a high air entry ceramic porous tip with the tip cavity and leads being filled with de-aired water. The leads are small bore capillary tubes (inside a rod) that are connected to a vacuum dial gage at the ground surface. The system can measure suctions of up to about 12–13 psi. At greater suctions, air that diffuses through the porous ceramic tip comes out of solution and cavitates. A tensiometer can be installed in a small diameter pre-bored hole or inserted at the bottom of a conventional site investigation borehole. The borehole is backfilled with impermeable grout or bentonite plugs.

Clays can develop negative pore pressures that exceed the stress range of a tensiometer. Greater suctions can be measured indirectly by thermal conductivity sensors which, for practical purposes, can approximately double the range of measurable suction to about 30 psi. A problem with this technique is the time required for the sensor to provide a stable reading.

An example of tensiometer readings taken at a site in Hong Kong (Ching et al., 1984) are shown on Figure 9.52. The probe was inserted through small openings in the face of a steep slope and the variations in readings at the same elevation have been attributed to the close proximity of the slope face. Suction forces at the upper part of the slope could not be measured because the tensiometer readings had reached their limit. At this site, the steep cutting had an average inclination of 60° to the horizontal and was protected by a thin layer of chunam plaster.

Failure Criterion for Unsaturated Soils

Fredlund et al. (1978) proposed an extended Mohr-Coulomb failure criterion for unsaturated soils:

$$\tau = c' + (\sigma_n - u_a) \tan \phi' + (u_a - u_w) \tan \phi^b \qquad \text{Eq. (47)}$$

where σ_n = total normal stress
u_a = pore air pressure
u_w = pore water pressure
ϕ^b = friction angle equal to the slope of matric suction $(u_a - u_w)$ vs. shear strength τ graph when $(\sigma_n - u_a)$ is held constant. The effective angle of friction ϕ' remains the same for all suction values (Figure 9.53).

Unlike the failure criterion for saturated soils, which has one stress state variable $(\sigma_n - u_w)$, the failure criterion for unsaturated soil has *two* stress state variables $(\sigma_n - u_a)$ and $(u_a - u_w)$. One of the main advantages of Eq. (47) is that when the soil reaches saturation $(u_a = u_w)$, the equation is identical to the Mohr–Coulomb criterion for saturated soils. Thus,

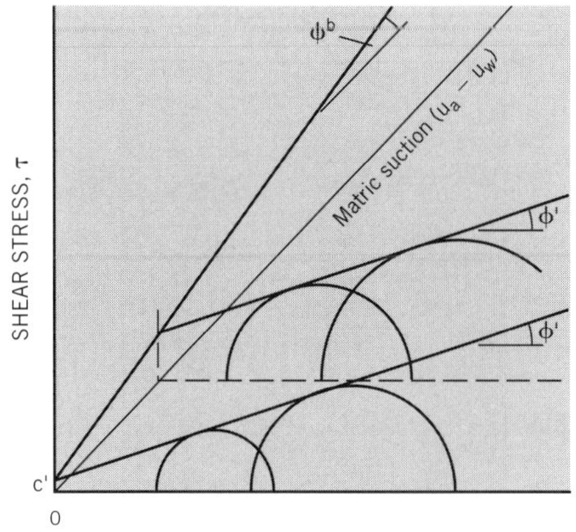

Figure 9.53 Mohr circles at failure for an unsaturated soil (after Fredlund, 1987).

there is a "seamless" transition from the unsaturated to the saturated soil profile. There are computer software programs available (e.g., Geo-Slope) that can handle the transition.

The stress circles at failure can be plotted on a diagram (Figure 9.53) with the two stress state variables ($\sigma_n - u_a$, and $u_a - u_w$) plotted on the horizontal axis and shear strength τ as the ordinate. The term ($u_a - u_w$) is the *matric suction* and, in most practical situations, the pore air pressure can be assumed to be zero (i.e., atmospheric).

Equation (47) can be visualized as a two-dimensional graph of τ vs. ($\sigma_n - u_a$) with the third variable of matric suction ($u_a - u_w$) added to it (Figure 9.54). The intercept on the ordinate of this graph (corresponding to ($\sigma_n - u_a$) = 0) increases as the matric suction ($u_a - u_w$) increases. This ordinate (termed cohesion c by Fredlund (1987), can be plotted against matric suction to give the friction angle ϕ^b (Figure 9.55).

The angle ϕ^b can be considered as a friction angle or as a component of the cohesion intercept. In the latter case, the combined intercept c is given by:

$$c = c' + (u_a - u_w) \tan \phi^b \qquad \text{Eq. (48)}$$

[*Important note:* Professor Fredlund's parameter c is not the same as the immediate shear strength c used throughout this book.]

By using Eq. (48), an unsaturated soil can be treated in an identical manner to a saturated soil in stability calculations. Eq. (47) becomes:

$$\tau = \underbrace{c' + (u_a - u_w) \tan \phi^b}_{= c} + (\sigma_n - u_a) \tan \phi' \qquad \text{Eq. (49)}$$

On saturation $u_a = u_w = u$ which provides the conventional shear strength equation for saturated soils:

$$\tau = c' + (\sigma_n - u) \tan \phi'$$

Since $u_a = 0$ and soil suction u_w is always negative (relative to atmospheric), it follows from Eq. (49) that the value of parameter c is always greater than c′ obtained on saturated soils. Also, experience shows that parameter ϕ^b is always less than ϕ'.

Unsaturated Shear Strength Measurements

Laboratory tests to measure the strength of partly saturated soils can be performed in a modified triaxial apparatus. On the lower pedestal of the triaxial cell, a high air entry ceramic porous stone has to be epoxy-glued into the metal base to measure negative pore water pressures during a shear test. A coarse porous stone is placed on top of the test specimen to measure the pore air pressures.

Specimen set-up and testing procedures for a three-stage triaxial shear test are described in Fredlund (1987). Only the test itself will be briefly described here. The test begins by adding water to the test specimen to reduce the initial suction. In the first stages, a low cell pressure is applied and the specimen is consolidated. In the shear phase of the test, the stresses σ_3, u_w and u_a are held constant as the deviator stress is applied. The rate of strain is typically slow and is set by the permeability of the high air entry stone or the soil properties. Since u_a and u_w are held constant throughout the shear test, parameters ($u_a - u_w$) and ($\sigma_3 - u_a$) are constant. The

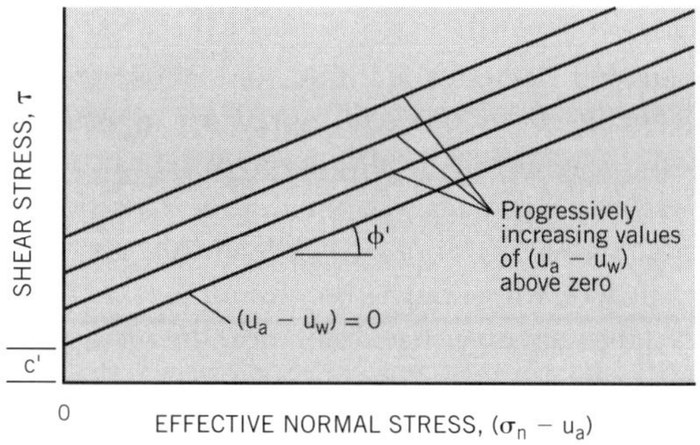

Figure 9.54 Failure surfaces for an unsaturated soil viewed parallel to the ($u_a - u_w$) axis (after Fredlund, 1987).

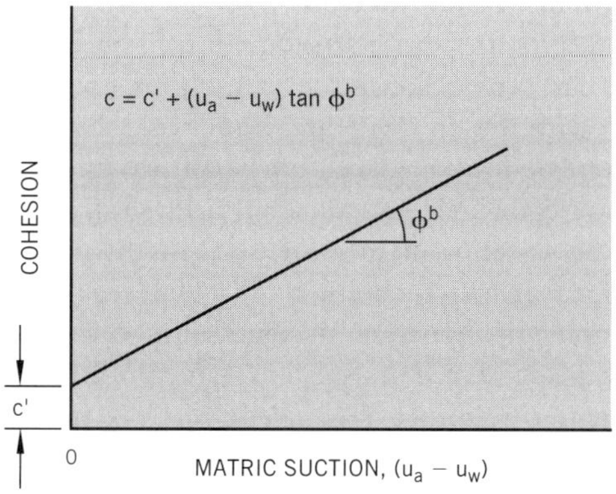

Figure 9.55 Cohesion ordinate c vs. matric suction ($u_a - u_w$) corresponding to ($\sigma - u_a$) = 0 (after Fredlund, 1987).

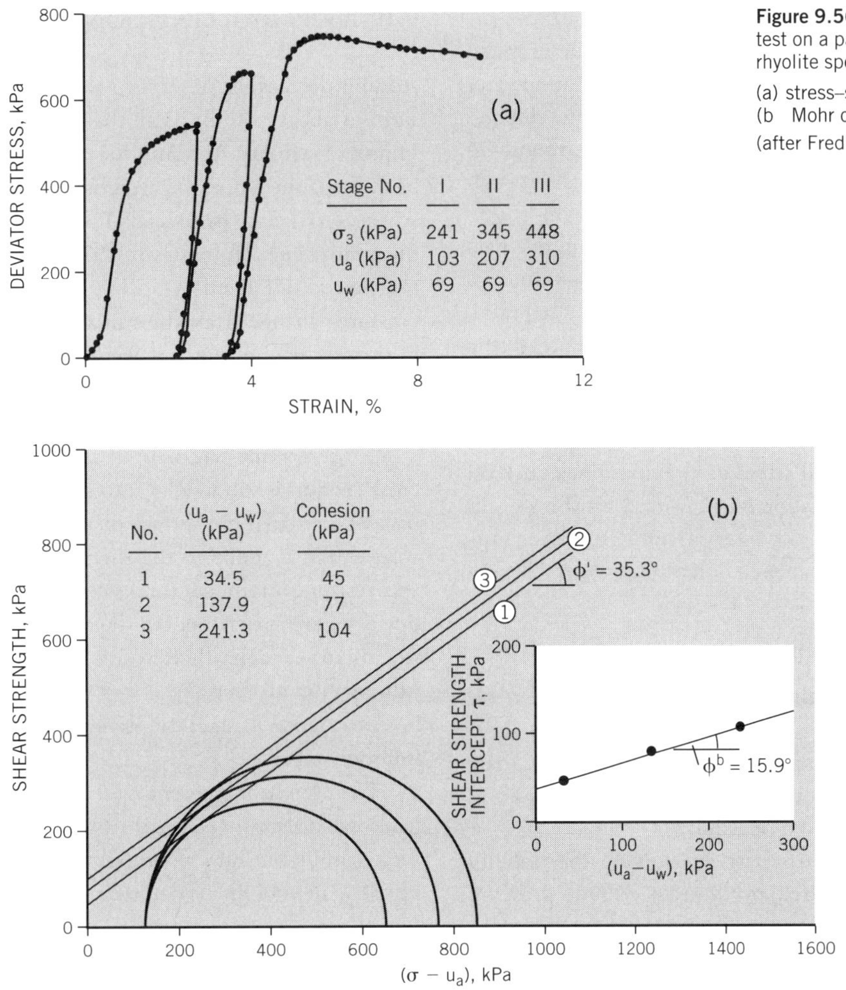

Figure 9.56 Three-stage triaxial shear test on a partly saturated decomposed rhyolite specimen from Hong Kong:
(a) stress–strain curves
(b Mohr circles and ϕ^b determination
(after Fredlund, 1987)

stress–strain curve is obtained up to the peak value, after which the specimen is unloaded to zero. For stages II and III, progressively higher cell pressures σ_3 are applied and, after consolidation against the same suction u_w of stage I, the shear phase is repeated.

The results of a three-stage shear test on a sample of decomposed rhyolite from Hong Kong are shown on Figure 9.56 (Fredlund, 1987). The metric units have not been converted to U.S. units, but it can be noted that the cell pressures are 35, 50, and 65 psi in the three stages. It is seen that the deviator stress increases as $(u_a - u_w)$ increases in the three stages of the test, while $(\sigma_3 - u_a)$ is held constant. The presentation of two graphs on Figure 9.56(b) is an interesting feature of the shear tests on partly saturated soils. Unless additional tests are run, the constant ϕ' value has to be assumed to obtain the cohesion c in the multistage test.

Slope Stability Analyses

Stability analyses on partly saturated soils can be performed by increasing the cohesion intercept c' to the value of Fredlund's cohesion c using Eq. (48). This requires knowledge of ϕ^b and the matric suction profile of the slope under study. Ho and Fredlund (1982a) recommend the three-stage triaxial test procedure, described above. The data given in Fredlund (1987) indicates that ϕ^b ranges from about 12° to 22°.

Assuming that a full-scale investigation of partly saturated shear strength and full suction profile is not feasible, what would be a reasonable way to account for the higher strength available in a partly saturated soil above the water table, compared to a fully saturated soil? One possible approach would be to take some measurements of negative pore pressures in the slope and extend the data linearly from the water table; next, assume a conservative lower bound ϕ^b of about 12°. Using the value of Fredlund's c parameter obtained in this way and ϕ' obtained from conventional effective stress tests on saturated soils, a stability analysis for the unsaturated soil can be made.

The suggested approach given in the previous paragraph is intended as a way of possibly simplifying what is currently in the realm of applied research. Clearly, there is a need for a way to take account of the extra "cohesion" provided by partly saturated soils for many slope stability or back-calculated landslide analyses where the present knowledge can be applied.

Summary

In summary, knowledge of the shear strength of unsaturated soils is desirable in (i) arid climates where the groundwater levels are deep below the surface and (ii) temperate climates where shallow instabilities occur in soils above the groundwater due to differential weathering or other geological processes that have created a thin mantle of weak soils (e.g., colluvium). However, the necessary field and laboratory data needed for the analysis are difficult to obtain at a commercial level due to the sophisticated measuring techniques required. The need to predict the past, current, or future changes in the soil suction profile is another obstacle. Given this situation, a simple technical approach at the present time may be to estimate the soil suction profile from limited field measurements and adjust the data, if needed, based on previous research-level projects. In the absence of laboratory test data for the soils in the slope, a lower bound (i.e., conservative) ϕ^b of 12° is suggested for calculating the increased cohesion intercept due to soil suction.

9.18 STABILITY CHARTS

Stability charts provide a rapid method of determining the factor of safety for simple slopes. The shear strength properties of the soil have to be measured or estimated.

A summary of numerous charts for solving the stability of slopes has been published by Duncan (1996) drawing information from publications by Janbu (1968), Duncan et al. (1987) and Hunter and Schuster (1968). The earliest stability charts were published by Taylor (1948). Duncan's summary focuses primarily on the total stress analysis ($\phi = 0$) of clay slopes, including determination of the coordinates of the center for the most critical (least stable) stability arc, and correction factors for surcharge, submergence, seepage, and tension cracks (with and without water filling the crack). Charts are also provided for clays in which the undrained strength of the clay increases with depth, infinite slope analysis, and soils with $\phi > 0$. Duncan suggests ways to convert real slopes into an equivalent simple and homogeneous slope for such analyses. Other sources for stability charts include Morgenstern and Price (1965), Janbu (1954, 1996), and Michalowski (2002), among many available offerings on this subject. Nash (1987) lists 11 publications of slope stability charts.

Stability charts are not used in the study of most landslides because the shape of the slip surface and other parameters are specific to a slide. Therefore, simplification of the section or a search for the most critical slip surface are not needed. On small landslides, a stability chart can be used to estimate the undrained shear strength mobilized in a failed clay slope. In this case, the cost of a site investigation is high relative to the technical effort needed to provide remedial recommendations.

9.19 NEUTRAL LINE CONCEPT

Adding the weight of a fill to the top of a landslide decreases slope stability; adding fill to the bottom of a landslide improves stability. At some location in between the top and bottom of the slope the weight of a fill changes from being adverse to being beneficial. This location has been termed the *neutral line* (Hutchinson, 1978). The position of the neutral line is not fixed in a clay slope but moves as the pore pressures change from the initial loading to long-term equilibrium. This concept has value in showing where fill loads will be effective in landslide remediation by buttresses, slope flattening, etc.

To determine the neutral line position, a slope stability cross-section is subdivided into slices as, for example, the portion of a circular arc shown on Figure 9.57. An increment of weight δW is added to the slice, and its effect on the driving and resisting moments (or forces) are calculated. The neutral line is where the changes in the driving and resisting moments exactly offset each other; i.e., there is no change in the preexisting factor of safety F_0. *Above* the neutral line, fill loads will be detrimental to stability; *below* the neutral line, fill loads will improve stability.

The weight increment δW can be included in any of the hand calculations described in this book. Alternatively, a line force can be put into a computer stability analysis and moved until the neutral line is determined.

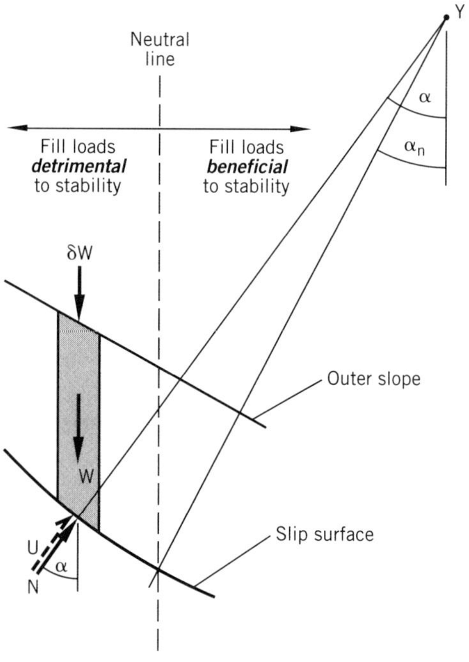

Figure 9.57 Hutchinson's neutral line concept.

For the special case of a circular arc analysis, it can be shown (e.g., Bromhead, 1986) that the neutral point on a slip surface is given by:

$$\tan \alpha_n = (1 - \bar{B} \sec^2 \alpha_n) \frac{\tan \phi'}{F_0} \qquad \text{Eq. (50)}$$

where α_n = angle between the center of the slice base and the center of rotation Y (Figure 9.57) where the neutral line passes through

ϕ' = friction angle of the soil at the base of the slice

$\bar{B} = \Delta u / \Delta \sigma_1$

F_0 = factor of safety before adding the load increment

Therefore, for a circular arc analysis of a normally consolidated clay slope, the neutral line changes as follows:

- *Short term.* A saturated clay will not change in strength immediately, so there will be no increase in shear resistance along the slip surface on first loading. The neutral line will be directly below the center of rotation, i.e., $\sigma_n = 0$.
- *Long term.* The clay consolidates, and eventually $\bar{B} = 0$. The initial factor of safety F_0 increases to F_1. According to equation (50):

$$\alpha_n = \tan^{-1}\left(\frac{\tan \phi'}{F_1}\right)$$

The neutral line has moved upslope. Therefore, if a fill load is placed over the slip surface between $\alpha_n = 0$ and $\alpha_n = \tan^{-1}\left(\frac{\tan \phi'}{F_1}\right)$, it will initially be detrimental to stability but will ultimately provide improved stability.

This conclusion should not deter a geotechnical practitioner from placing part of a buttress or slope-regrading fill on the "wrong" side of the neutral line. Such construction is often necessary in order to build the buttress at the base of a slope. However, the neutral line concept is helpful in visualizing where weighting is more efficient in providing improved slope stability.

CHAPTER

10

Stability Margin

This chapter presents several approaches to the issue of how much safety should be incorporated into landslide remediation. The subject has received surprisingly little attention in the technical literature, yet the decision can affect the cost of remediation by millions of dollars.

The first responsibility is to safeguard the users of the property and the general public from death or injuries from a landslide event. In this regard, a fast-moving catastrophic landslide is much more dangerous than the more common, slow-moving type and requires more concern for future safety. In general, however, an ultra-conservative approach to remediation is a disservice to the owner and may be so expensive that it precludes any attempt to stabilize the slide. The geotechnical practitioner has to select a middle ground in which a reasonable level of study, good scientific understanding, and experience are combined to provide an optimum treatment that is both cost-effective and achieves stability.

The standard of care is difficult to define because larger landslides are unique in terms of their geology and causation. Therefore, the legal definition of what a like professional operating in the same region would do in the circumstances is hindered by the fact that other professionals may adopt an entirely different solution, though not necessarily a better one.

The author suggests the following: *Landslide stabilization measures shall provide permanent stability under current and reasonably foreseeable future conditions.*

Under this definition, the designer should be expected to anticipate future storms (with consequent groundwater buildup) or future erosion from an existing source. What cannot be anticipated are unknown future developments such as land use changes or a road-widening project.

In this chapter, the margin of stability is first discussed in terms of the safety factor, which is the more common method of assessing stability of slopes. Then two other approaches are presented: original profile analysis and the observational method, both of which can be applied in appropriate circumstances. The chapter concludes with a section on reliability analysis, which is becoming increasingly popular. It uses probability techniques to supplement deterministic (single answer) calculations of safety factor.

10.1 FACTOR OF SAFETY

Setting an Appropriate Factor of Safety

The factor of safety (F) of a landslide can be defined in terms of either forces or moments. The definition is:

$$F = \frac{\text{Resisting forces (or moments)}}{\text{Driving forces (or moments)}}$$

where resisting forces are the shear strength of the soils and rocks along the slip surface, plus any introduced resistance, such as tieback anchors or a gravity wall. The driving forces are the gravity loads of the landslide mass, plus any loading from structures, live loads (e.g., traffic), or earthquakes.

The designer performs calculations of the resisting forces (or moments) and the driving forces (or moments) of the landslide to determine the factor of safety F. The result is the *calculated* static factor of safety.

It is important to recognize major differences between a *conventional slope stability* study and the *stability analysis of a landslide.* A conventional slope stability analysis is frequently required for civil engineering projects. The main concern of the designer is to prevent a slope failure caused, for example, by a fill being built above original ground. Given the uncertainties of the soil strength, the shape of the potential slip surface, and future groundwater conditions, it is customary to require a calculated static factor of safety of at least 1.50 for these circumstances. Similarly, for a high hazard earth structure, such as an earth dam retaining a reservoir, the design factor of safety for static loading is usually set at 1.50. It is not surprising, therefore, that some engineering consultants have a mindset that landslide remediation should also require a calculated factor of safety of 1.50. However, an F of 1.50 is inappropriate for most landslides.

The major difference between a landslide and general slope stability is that for a landslide, the *actual* factor of safety F is 1.00 at the start of landslide movements. Furthermore, the movement defines the shear zone so that: (i) the geometry of the landslide can be determined, and (ii) the pore pressure acting on the slip surface can be directly measured. Because the shear strength on the slip surface can be difficult to measure, a knowledge of the other factors involved in the landslide stability calculations allows the geotechnical practitioner to calculate the strength being mobilized along the slip surface. This calculation is termed *back analysis* and is a huge benefit compared to the lesser information available to an engineer for a general slope stability study.

The same back-calculated strength and groundwater parameters are carried forward into the landslide remediation analysis. Therefore, even if there are errors in the measured pore pressures and back-calculated strengths, the same errors occur in the remedial calculations. Since the original and remedial landslide calculations are comparative "before" and "after" studies, it is almost certain that slope stability will be improved. The only question is whether the proposed improvement is sufficient to allow for errors in modeling the landslide and foreseeable future conditions affecting the site.

The calculated factor of safety has to be set at a high enough level to overcome analysis errors. Errors most commonly arise when: (i) actual field conditions differ from the measured/assumed conditions, and (ii) mathematical analyses differ from actual soil behavior. Two significant factors affecting the likely magnitude of errors are the size of the landslide and the amount of data collected about the landslide.

The size of the landslide is a factor because a larger landslide can be more easily modeled than a smaller one. In very large landslides, the overall shear resistance can be determined with a high level of confidence from back analysis, and the geometry of the landslide is usually well defined. Therefore, a relatively modest improvement of safety factor should provide assurance of future stability. At the opposite extreme of very small landslides, effective stress analysis is difficult to perform due to (i) negative pore pressures near the ground surface (insignificant on most larger landslides), and (ii) greater sensitivity to pore pressure fluctuations. Therefore, the calculated factor of safety needs to be higher for smaller landslides than for larger landslides.

The scope of the landslide investigations also affects the reliability of the stability analysis. The amounts of field geology, subsurface explorations, and field monitoring of subsurface movements and groundwater levels have to be considered. It is self-evident that a more thorough field and laboratory study allows the designer to use lower factors of safety in remediation because the degree of uncertainty has been reduced compared to a lesser investigative effort. Sometimes, a more limited scope is necessary due to time and/or budget constraints. In such cases, the geotechnical consultant should determine an acceptable minimum effort and use a higher design factor of safety.

Table 10.1 Suggested Guidelines for Factor of Safety in Landslide Studies According to Level of Information and Landslide Size

	Minimum Study*		Normal Study*	
Landslide Size	Borings	Calculated Minimum F	Borings	Calculated Minimum F
Very small	1**	1.50	1	1.50
Small	1	1.50	2	1.35
Medium	2	1.40	4	1.25
Large	3	1.30	6	1.20
Very large	4	1.20	8	1.15

*see text
**occasionally zero, depending on circumstances

Table 10.1 provides suggested guidelines for setting minimum factors of safety according to the size of landslide and level of studies. These are subject to the judgment of the practitioner, who may have special knowledge of the region, and are not intended to represent the standard of practice in the profession. Landslide sizes are given in Chapter 1.

Table 10.1 lists a *Normal Study* and a *Minimum Study*. The Normal Study is the number of borings and minimum static factors of safety that are suggested for typical landslides of the given sizes. The Minimum Study represents the minimum field exploration effort that, in the author's opinion, needs to be undertaken and, since less information is obtained, requires higher calculated factors of safety than a Normal Study landslide of the same size. Table 10.1 guidelines are for a General Site Investigation as described in Chapter 3, Section 3.1. Additional borings and studies may be needed for more specific remedial analyses. Exceptions to the suggested guidelines of Table 10.1 may be justified where the analysis is being based on Original Profile Analysis or the Observational Method, as described later in this chapter.

In Table 10.1, it is assumed that: (i) the landslide has been field-examined by professional staff, the borings are sampled and soils/rocks are described by a well-qualified geotechnical engineer or engineering geologist; (ii) all borings are within the active slide area; (iii) borings are instrumented with inclinometers in all but very small landslides; (iv) piezometers are installed; (v) the shear strength has been obtained from back-analysis computations or from laboratory/field measurements that confirm F = 1.00 at the onset of landslide movement; (vi) relevant stability analyses are used; and (vii) the study is deterministic. Where inclinometers and piezometers are installed in the same hole or close together (see Chapter 4), it should be considered as one boring for purposes of implementing Table 10.1.

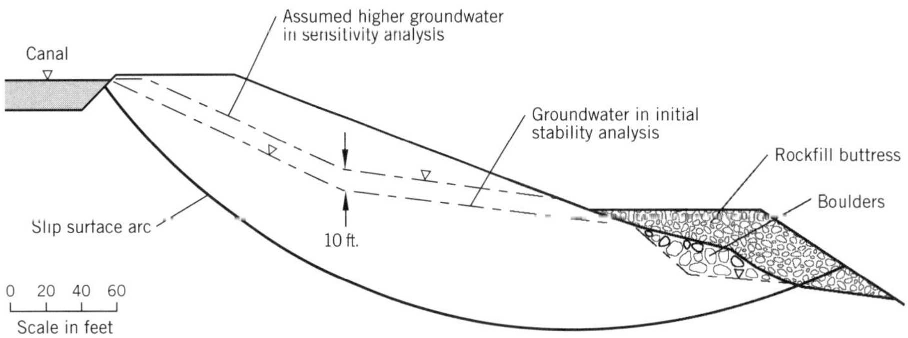

Figure 10.1 Sensitivity analysis for Faraday Slide.

A second way in which the designer can understand the effect of errors is to make a sensitivity analysis or reliability analysis as part of the study. In a sensitivity analysis, the various parameters in the stability analysis are raised or lowered to determine their effect on safety factor F. These variations are set by the designer based on judgment and experience. For example, the groundwater levels can be raised by 6 feet above the highest levels observed during the wet season. All other parameters are held constant. By calculating the change in F, and the likelihood that such a change in groundwater level could conceivably occur, the designer develops knowledge about the factor of safety that is appropriate for the site.

A reliability analysis (see Section 10.4) is similar to a sensitivity analysis but uses probability theory to calculate the likelihood of a future failure. This technique is gaining acceptance by many consulting engineers, clients, and regulators.

Suppose the factor of safety cannot be raised to the suggested levels cost effectively? This is a common problem on larger landslides. There are two other stability improvement options: (i) discuss with the owner a *marginal stabilization* approach in which it is agreed that certain remedial work will be performed with the understanding that it may be insufficient to achieve complete stability under all future conditions (see Chapter 14, Section 14.6). Obviously, this approach is not an option where there is a risk of a rapid or catastrophic failure because it would pose a high threat to life; (ii) perform *selective stabilization* in which only part of the total landslide is stabilized and the remainder is allowed to continue moving (see Chapter 14, Section 14.5). The latter solution may apply, for example, to a road or railway crossing the top of a landslide or to protect a building immediately above a headscarp from future retrogressive movements.

Other alternatives to full remediation are discussed in Chapter 14.

Example Application

The stability analysis of Faraday Slide will be discussed. It is described in Case History 8 and the stability calculations are presented in Chapter 9, Section 9.9.

The slide has a surface area in plan of about 160,000 sq. ft., putting it in the medium size category (Chapter 1, Table 1.1). It was investigated by 5 borings, each with an inclinometer. Piezometers were installed in separate holes alongside the inclinometers. The observations showed that slippage was occurring along a circular arc, almost entirely within decomposed tuff breccia (stiff clay). The peak and residual strengths of the breccia were measured by stress-reversal direct shear tests.

Long-term surface observations of this landslide showed that it was developing into a delayed failure. The shear strength along the slip surface must have been close to residual strength. The strength measured in the laboratory $c'_r = 435$ psf, $\phi'_r = 17.5°$ gave a factor of safety F of 0.98 using winter groundwater levels. This appears to be in excellent agreement with the landslide conditions because the actual strength was probably slightly above residual.

Analysis of the proposed rockfill buttress remediation gave a factor of safety F of 1.26 immediately after construction. This analysis conservatively does not allow for any consolidation and strengthening of the clay foundation during construction.

The close agreement between the measured laboratory strengths and the known F of 1.00 made it unnecessary to perform sensitivity analyses on the shear strength. The presence of the buttress should prevent any further loss of strength with time along the slip surface because there would be no further shear strain after the slope is stabilized. However, there was a concern with the possibility that the canal above the slope could raise groundwater levels within the slide area in the future, or that an intense storm could raise groundwater levels above those assumed in the analysis. Accordingly, a sensitivity analysis (Figure 10.1) was performed by raising groundwater levels by 10 feet within the clay (but not above the ground surface, and no rise was assumed in the highly pervious rocks at the base of the slope). This analysis showed that F would reduce by 0.08; i.e., change F from 1.26 to 1.18. This leaves an acceptable margin of safety should such an unlikely event occur. Over the longer term, the stabilized slope will have an increased factor of safety. The effective stress analysis for the slide and buttress combination, after excess pore pressures in the buttress foundation have dissipated, is F = 1.40.

In summary, the Faraday Lower Slide qualifies as a Normal Study of a medium landslide (suggested minimum F = 1.25). It had a calculated F of 1.26 immediately after construction. A sensitivity analysis of a worst case groundwater rise scenario reduced F to 1.18, which is still an acceptable margin against a highly unlikely event. Faraday Slide has remained stable since the buttress was built in 1986.

10.2 ORIGINAL PROFILE ANALYSIS

Principle of Original Profile Analysis

This concept was developed by the author (Cornforth, 1995) as a method of determining the amount of resistance required to reestablish equilibrium in a slope that has been destabilized by construction activities. The landslide caused by the construction is back-analyzed to obtain the shear strength along the slip surface. The original profile of the ground prior to construction is then drawn on the section and the factor of safety is calculated using the strength obtained from the back calculations. It is higher than 1 because the ground has been destabilized by construction. The analysis thus gives the minimum factor of safety needed to reestablish equilibrium in the slope. This is a practical substitute for the designer arbitrarily selecting a design factor of safety.

In practice, the geotechnical consultant calculates the net loss of resistance (in the case of a cut) or the net increase in driving force (in the case of a fill). As a minimum requirement, the remediation must provide the same force to offset the destabilizing force resulting from construction.

The principle of Original Profile Analysis is illustrated on Figure 10.2. A cut AC is made into a hillside for a road and a double-wedge failure ACDE occurs behind the cut slope. A site investigation determines groundwater levels and the location of the slip surface. A back analysis calculates the shear strength parameters on the slip surface. In some cases, two or more strengths may be needed (e.g., AE and DE on Figure 10.2(a)).

The second step is to reinsert the original profile by putting triangle ABC back into the section. Since the lower wedge of a double wedge provides net resistance, the excavation of triangle ABC represents a net loss of resistance and thus causes the landslide to occur. The net loss is the shear resistance along AB minus the driving force of the soil in ABC. Therefore, applying a force P to the face of slope AC equal to the net loss of resistance should return the slope to its original condition of stability.

The required resisting force P can be achieved by a retaining wall, shear piles, or buttress. The extra resistance can also come from groundwater lowering or other forms of remediation. The benefits of Original Profile Analysis is that it provides the designer with an answer to the question: How much force is enough to reestablish stability? If, for example, the Original Profile Analysis indicates that the calculated factor of safety before construction was 1.14, the designer knows that a remedial factor of safety of 1.14 or higher is sufficient and does not need to provide values of 1.25 or some other arbitrary number. However, a designer normally would increase the calculated factor of safety F or force P determined by Original Profile Analysis to allow for uncertainties of design.

The principle of Original Profile Analysis can be applied to other landslide situations. For example, if the slope failure of Figure 10.2 had been caused by a fill being placed near D, a similar analysis would apply. First, there would be a back analysis to provide shear strength parameters, followed by analysis of the original profile before the fill was built. Force P would be calculated for the current conditions of the landslide.

Some words of caution are needed. The Original Profile Analysis assumes that the shear strengths of the soils have not been changed by the landslide. In the case of soils with a preexisting slip condition, such as colluvium or ancient landslide terrain, this assumption is valid. However, if the construction activity causes a first-time slide, part or all of the slip surface may have changed from the higher strength of a first-time slide to a lower residual strength or, for cohesionless soils, from a higher strength to the ultimate strength. In these cases, the soil strengths at the initiation of the sliding need to be used in the Original Profile Analysis. This would result in the need for a larger force P or factor of safety F to reestablish the original slope stability.

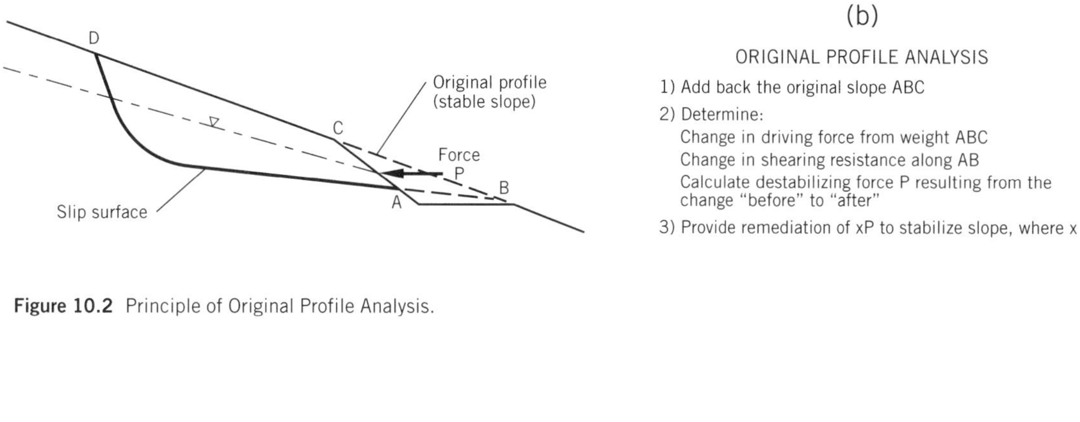

Figure 10.2 Principle of Original Profile Analysis.

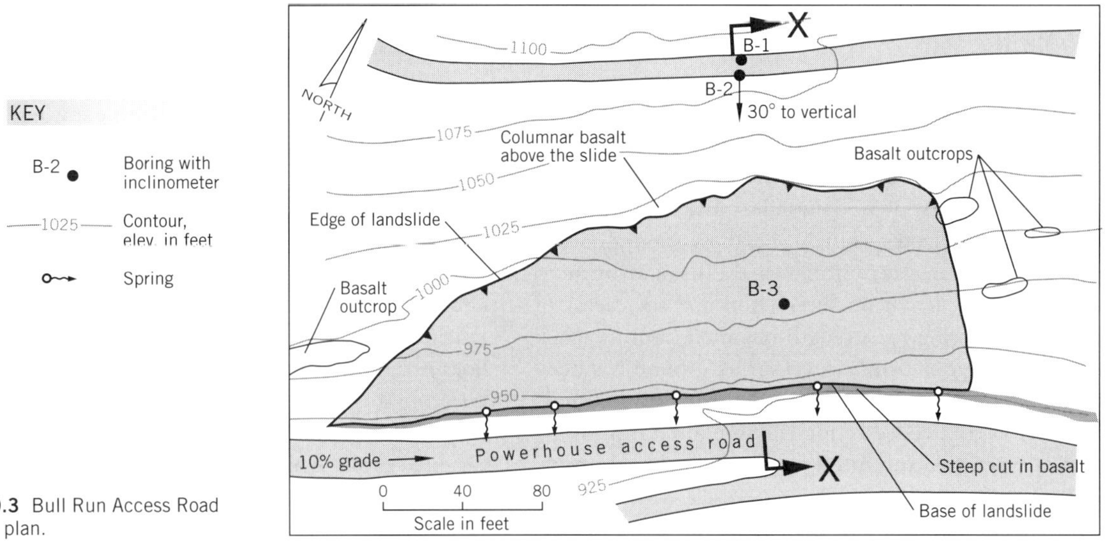

Figure 10.3 Bull Run Access Road Slide site plan.

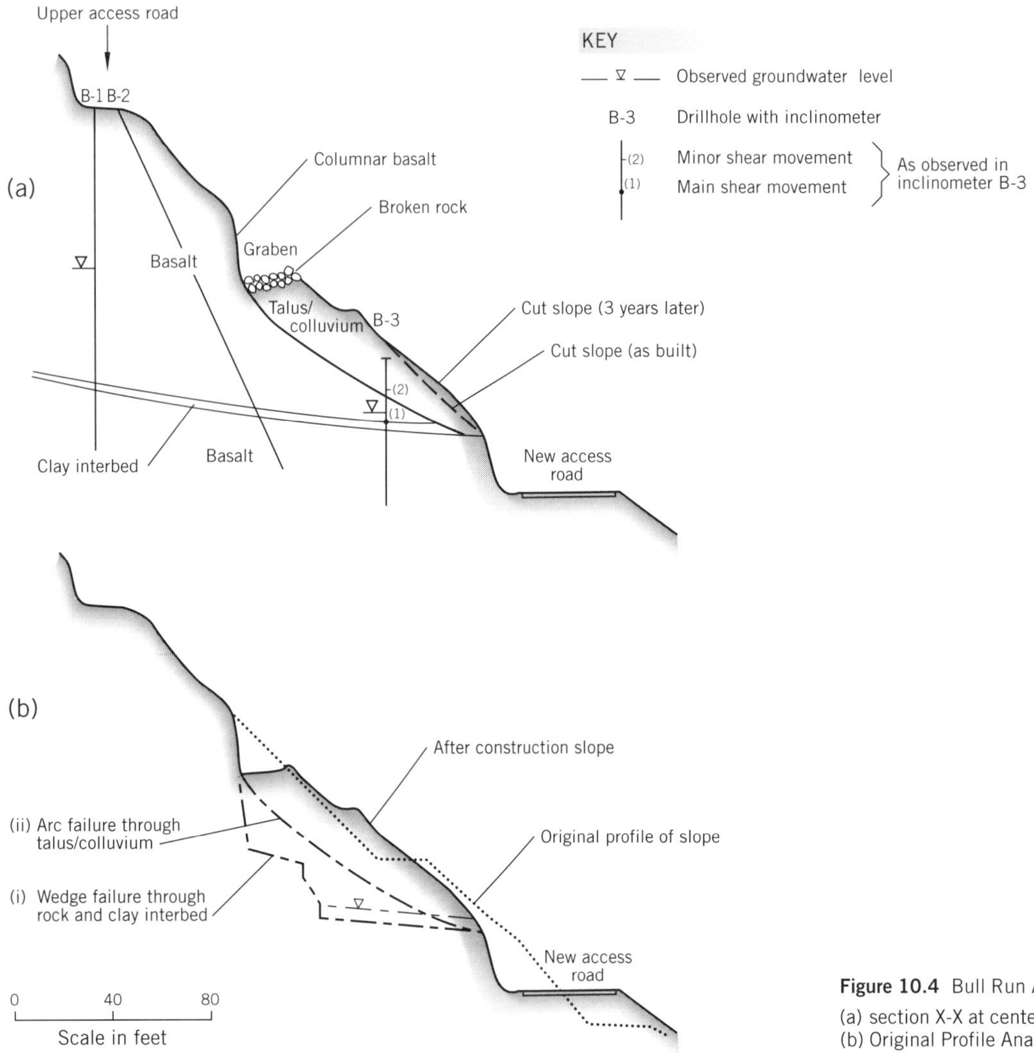

Figure 10.4 Bull Run Access Road Slide:
(a) section X-X at center of slide
(b) Original Profile Analysis

Figure 10.5 Bull Run Access Road Slide showing drill rig on the steep slope at boring B-3. The slope is temporarily covered by wire netting. (Photo: 1983).

Example Application

When a powerhouse was built at Bull Run Dam, Oregon, in 1979, an access road was cut into the steep valley slope downstream of the dam. The excavation caused a medium-size landslide over a width of about 320 feet (Figure 10.3). Between October 1979 and February 1983, ground surveys showed movements on the talus slope of up to 13.8 feet horizontal and 6.3 feet vertical. Rocks rolled off the talus slope onto the road, causing a danger to powerhouse workers.

Three drillholes were put down to investigate the subsurface conditions at the slide: two drillholes from the road above (one hole at a 30° angle to the vertical) and a third hole from the steep talus slope at the middle of the slide (Figure 10.4a). The drill rig had to be lifted onto the slope by a helicopter (Figure 10.5). Inclinometer casings were installed in all three holes.

The site explorations showed that the rock slope contained a clay interbed 1½ to 3½ feet thick between two basalt flows. On the outer part of the slide, a coarse talus overlies colluvium. The inclinometer in drillhole B-3 (at the middle of the slide) showed that slippage was occurring within the clay interbed at a depth of 25 feet, and a less active slippage was occurring within the talus at a depth of 11 feet. The interbed was very stiff, black, silty clay, intensely slickensided.

The slide models were taken to be: (i) a deep-seated detachment of rock with the slip plane passing through the clay interbed, and (ii) slippage occurring through the talus/colluvium.

The wedge failure for the clay interbed is shown on Figure 10.4(b). The bedrock block failure is sliding along the clay interbed, probably having detached along joints within the overlying basalt bedrock. The back analysis gave residual shear parameters of $c_r' = 0$, $\phi_r' = 9½°$. Reinstating the original stable slope into the analysis (Original Profile Analysis), the calculated factor of safety was 1.05.

The relatively shallow slide in the talus/colluvium was assumed to be an arc failure surface (Figure 10.4b). A shear strength of 40° in the talus/colluvium and a groundwater level averaging 5 feet above the interbed gives a factor of safety of 1.00 in the back analysis. Replacing the original slope in the Original Profile Analysis, the factor of safety F is 1.14. Therefore, F of 1.15 was selected for design. The analysis showed that a force P of 14,500 lb./lin. ft. was needed to reestablish a factor of safety of 1.15 and this was rounded out to 15,000 lb./lin. ft. for design purposes.

The slope was stabilized by a concrete rib retaining wall that was anchored to the steep rock face of the road cut and provided cantilever support to the talus slope above the rock cut (Figure 10.6). The space behind the wall and above the ground anchor was backfilled with a 6-inch minus well-graded granular backfill having less than 5 percent (of the 1-inch minus fraction) passing the No. 200 sieve in a wet sieve analy-

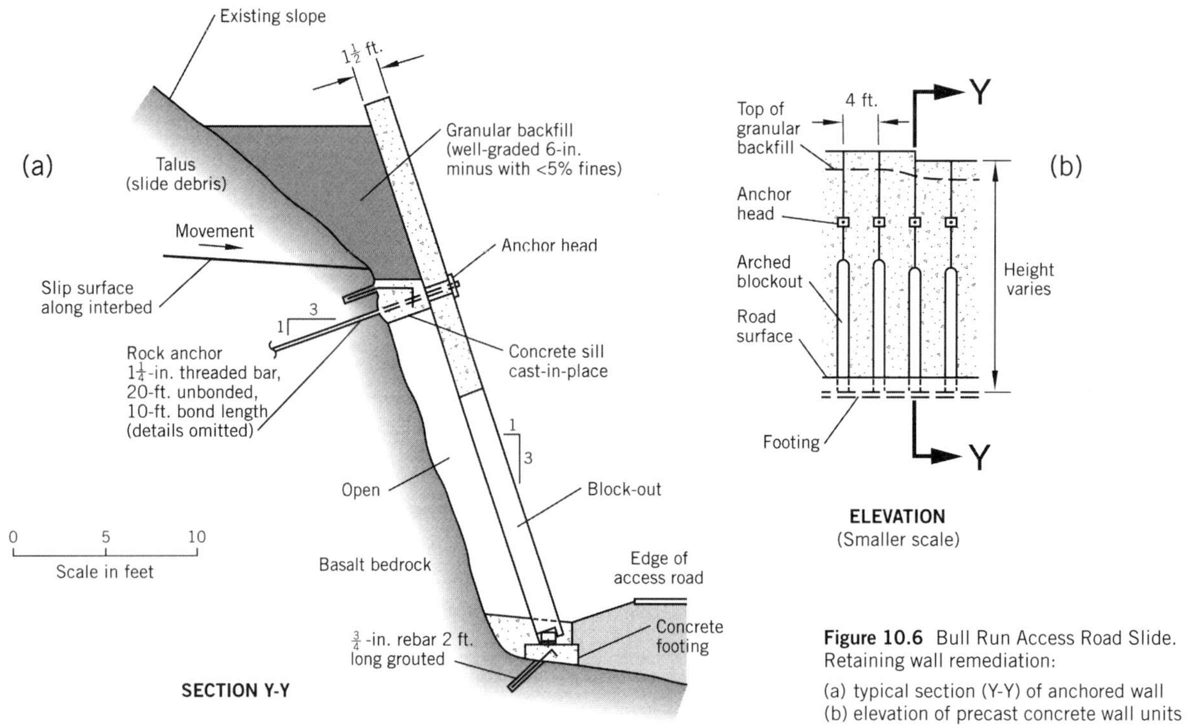

Figure 10.6 Bull Run Access Road Slide. Retaining wall remediation:
(a) typical section (Y-Y) of anchored wall
(b) elevation of precast concrete wall units

sis. A concrete sill was cast-in-place to support the granular fill. The rock anchors were 30 feet long and were bonded over 10 feet. The concrete retaining wall is 304 feet long and the average rib length is 27 feet. The wall was constructed in 1983 at a cost of $530,000. This price included (i) $22,000 for the installation of 1225 lin. ft. of rock anchors in the columnar basalt at the top of the slide to prevent further toppling failures and (ii) $60,000 to clear numerous large basalt blocks from the surface of the slide. (The ENRCCI (Chapter 13) for September 1983 is 4140.)

The Original Profile Analysis gave the soil design load for the wall. The wall design included a structural factor of safety for this soil pressure. The wall, completed in 1983, has remained stable.

This example is the most common way of using Original Profile Analysis. The procedure calculates the amount of force needed to reestablish stability without the need for the designer to arbitrarily select a factor of safety for remediation.

10.3 OBSERVATIONAL METHOD

The third approach to designing stable slopes is to use observational methods to substitute for, or supplement, conventional calculations. It can circumvent some of the uncertainties or conservatism of mathematical analysis and often provides direct feedback on soil behavior. However, on major projects, the flexibility needed to successfully implement the strategy may create contractual problems, as discussed later.

A simple example of the direct application of the observational method is to determine the stable cut slopes of an area by surveying existing cut slopes. This approach is acceptable, provided: (i) the stratum is relatively uniform; (ii) the survey is conducted in the same area as the proposed cuts; and (iii) variable groundwater is not a factor in stability.

A hypothetical example, using a residential development on a hillside, illustrates how this observational approach can be put into practice. Typically, residential developments require modest sidehill cuts and the cost of a detailed site investigation and stability analyses often cannot be justified. If a geological reconnaissance (or geology map) reveals that the ground conditions are uniform in the area, a survey of existing cut slopes can be made. The height and slope angles can be determined by the methods described in Chapter 3, Section 3.2. The age of the cut can be estimated from the vegetation and general surroundings as either recent (less than 5 years) or old (more than 5 years). Older cut slopes give a more reliable indication of long-term stability, especially in stiff clay slopes that might be subject to delayed failure (see Chapter 2, Section 2.15).

The collected data can be assembled into a table (e.g., Table 10.2) and plotted on a graph, Figure 10.7. A line is drawn separating stable from unstable slopes on the graph. The geotechnical consultant then sets the criteria for the proposed development. In this example, the selected design criteria are that cuts of up to 5 feet high can be made at 1:1 and higher cuts should be limited to 1½ horizontal : 1 vertical.

The procedure is simple and inexpensive. It uses existing field data to draw reasonable conclusions without the need for a detailed study.

An example of misuse of available field experience can be briefly presented here. A speaker at a geotechnical conference described a project for widening an existing road. The precise

Table 10.2 Summary of Cut Slope Condition Survey (Hypothetical)

Height	Measured Angle	Estimated Age	Observed Stability		Notes
4	34	0–5	S		
5	55	5+		U	Local "pop out" failure
5	45	0–5	S		
8	45	5+		U	Local "pop out" failure
9	47	0–5		U	Slope cracks
11	36	0–5	S		
11	35	0–5	S		
14	35	0–5	S		
17	38	0–5		U	Eroded soil at base
20	34	5+	S		
22	44	5+		U	Slump along most of cut
30	26	5+	S		

S – stable U – unstable

details are unavailable, but the gist of the case history was that the original road had been built about 70 years earlier and had cut through relatively competent soils to depths of about 50 feet. The side slopes were about 1¾ horizontal : 1 vertical and had remained stable. For a proposed road widening project, site investigation borings revealed that the soils behind the proposed road cut were essentially the same as those observed at the existing cut slope earlier. The presenter went on to describe the sampling, laboratory tests, and slope stability analyses. Based on this study, the design slope for the new cut was set at 2½ horizontal : 1 vertical, much flatter than the existing stable slope, due to a conservative interpretation of the shear strength data.

This example demonstrates the absurdity of relying on calculations to design a stable slope when the field evidence clearly indicated that a slope of 1¾ horizontal : 1 vertical is stable at the site. The observational method would simply ascertain that the conditions were essentially the same from a technical standpoint and then use the experience of the 70 year old cut slope as the design for the new cut slope. There would have been significant savings in construction cost through excavating less volume and a lower right-of-way purchase.

An example of the observational method being used to check slope stability is the railroad crossing of the Great Salt Lake in Utah. The foundation conditions were relatively consistent but there was some uncertainty about the ability of the foundation materials to support the railroad fill. The designers decided to build test fills and observe any failures that may occur. Such failures would be studied to refine the shear strength properties of the foundation. However, no failures occurred under the weight of the fill using the standard side slopes of the railroad's design practice. Therefore, no modification was needed and the fill was completed successfully. Had a detailed geotechnical study indicated a precautionary need for stage construction or flatter side slopes, the costs and time for completion would have been adversely impacted. The observational method demonstrated that stability was adequate.

Some potential benefits of an Observational Method are: (i) to save construction time; (ii) to reduce construction cost; (iii) to observe and control during construction some uncertainty that arose during design; or (iv) to take advantage of existing performance information available at the site. One key requirement, as described by Peck (1969), is to have a contingency plan in place before any site work commences. Should the assumptions or ground behavior be less successful than anticipated, the backup plan would be implemented.

Many examples of the observational method require installation and monitoring of field instruments such as inclinometers and piezometers. This requires advance planning and cooperation with the contractor during the actual construction phase. Furthermore, the method usually requires the option of being able to alter the design or make other contractual changes during construction such as a temporary halt in the work. Such extra costs may exceed any savings that would have resulted from the successful use of the observational method. Therefore, the technique has to be selected carefully with well-written specifications and full understanding by the owner of the potential financial consequences of any contract changes resulting from it.

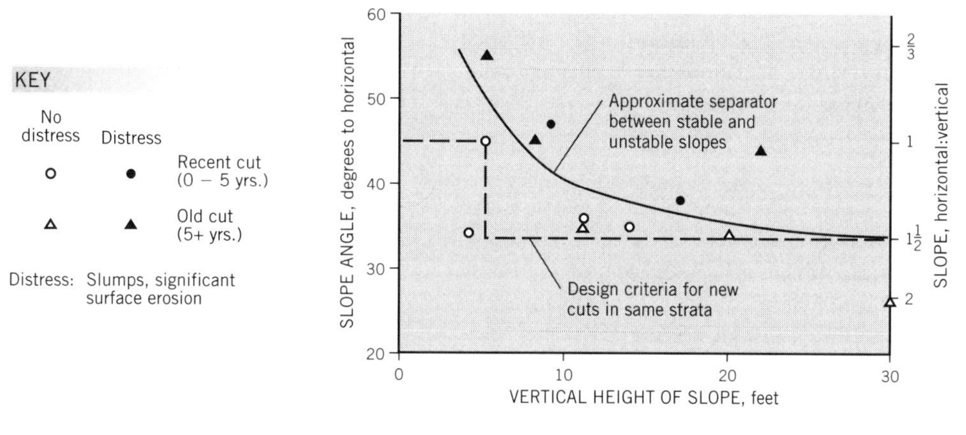

Figure 10.7 Example of the observational method for determining cut slope design.

For stability work, the main use of the observational method has been to *prevent* instability when building over weak foundations. The Salt Lake railroad crossing has already been described. Other similar examples include:

1. The Republic Steel ore yard in Cleveland, Ohio (described in Bjerrum et al., 1960) in which ore heaps were being built over soft ground. An elaborate observational program was set up that included contingency plans to hold loads constant or decrease them quickly if needed to prevent a foundation failure. This observational program was conceived and implemented under the direction of Karl Terzaghi. Its success allowed the maximum utilization of the ore yard without causing a foundation failure and consequent economic losses to the steel company.
2. Transportation of the Saturn rockets over a four-mile causeway from the assembly building to the launching pads at Cape Kennedy (Peck, 1969). The huge rockets are transported vertically by track-type crawlers. On one causeway, the penetration resistance (300 lb. weight, 18 in. drop) of the sand foundation was less than 2 blows per foot at depths of 20 to 40 feet below the surface. There was concern that the speed and weight of the rocket transporter could temporarily liquefy the very loose sand.

   The adopted observational approach was to pass the empty transporter over the weak foundation, allowing the sand to develop (and quickly dissipate) the excess pore pressures resulting from the load. Based on these observations, the transporter was incrementally loaded and passed over the foundation until it was able to support the full weight of the transporter and rocket. With each passage, densification occurred and, in general, the pore pressures measured were smaller than in the prior loading. The program proved to be very successful.

10.4 RELIABILITY ANALYSIS (TAYLOR SERIES METHOD)

Reliability analysis is a method, using a statistical technique, of determining the degree of uncertainty in a calculation such as the factor of safety of a slope. The application of the method to geotechnical engineering calculations has been presented by Duncan (2000), drawing from various resources.

Calculation Method

For a slope stability calculation, the steps are as follows:

1. Calculate the deterministic value of the factor of safety F using the best estimate values for each of the variables in the factor of safety equation. This is termed the *most likely value* of the factor of safety: F_{MLV}.
2. Estimate the standard deviation from the mean σ for each of the variables in the stability equation, i.e., soil density, pore water pressure, and shear strength parameters. This set of calculations will be discussed later.
3. Use the Taylor series technique (Wolff, 1994; U.S. Corps of Engineers, 1998) to estimate the standard deviation of the factor of safety (σ_F) and the coefficient of variation (V_F) of the factor of safety.

$$\sigma_F = \sqrt{\left(\frac{\Delta F_1}{2}\right)^2 + \left(\frac{\Delta F_2}{2}\right)^2 + \ldots\ldots + \left(\frac{\Delta F_N}{2}\right)^2} \qquad \text{Eq. (1)}$$

$$V_F = \frac{\sigma_F}{F_{MLV}} \qquad \text{Eq. (2)}$$

$$\Delta F_1 = \left(F_1^+ - F_1^-\right), \Delta F_2 = \left(F_2^+ - F_2^-\right) \text{ etc.}$$

   where F_1^+ is the factor of safety calculated with the value of the first variable parameter *increased* by one standard deviation from the best estimate value of that parameter (all other variables being kept at their original most likely values in the factor of safety equation).

   F_1^- is the factor of safety calculated in the same manner, except that the first variable parameter is *decreased* by one standard deviation from the best estimate value.

   The values of the other variables in the factor of safety equation are dealt with similarly to obtain ΔF values for each variable. The ΔF values are substituted in Eq. (1) to obtain σ_F, the standard deviation of the factor of safety F_{MLV} and the coefficient of variation V_F is obtained from Eq. (2).
4. With F_{MLV} and V_F determined, the probability of failure P_f can be obtained from Figure 10.8, which has been drawn from tabulated data given in Duncan (2000). Duncan obtained the probability of failure by first calculating the lognormal reliability index β_{LN} using the formula:

$$\beta_{LN} = \frac{\ln\left(\dfrac{F_{MLV}}{\sqrt{1+V_F^2}}\right)}{\sqrt{\ln\left(1+V_F^2\right)}} \qquad \text{Eq. (3)}$$

   The standard cumulative normal distribution function (reliability) corresponding to the β_{LN} value is obtained from a textbook on probability and reliability (e.g., Dai and Wang, 1992). The probability of failure is one minus the reliability; for example, if the reliability index is 0.98, the probability of failure is 1 − 0.98 = 0.02 or 2 percent.

Methods of Estimating Standard Deviation σ

The key part of a reliability analysis is the estimates of standard deviations for each of the geotechnical parameters. Duncan (2000) suggests four methods:

1. ***Use the formula definition***

$$\sigma = \sqrt{\frac{\Sigma\left[(x-\bar{x})^2\right]}{N-1}} \qquad \text{Eq. (4)}$$

where σ = standard deviation

x = value of the parameter

$\bar{x}$ = average value of the parameter

N = number of values of the parameter

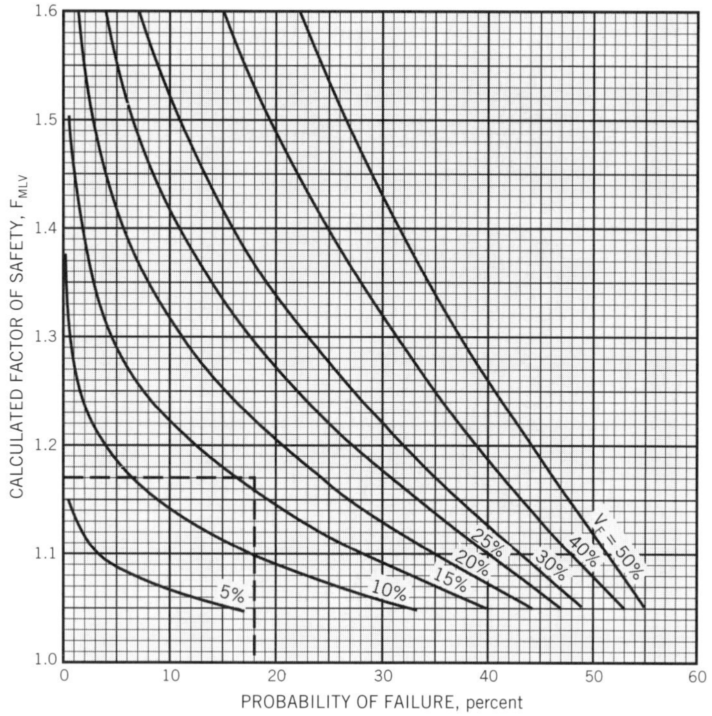

– – – – Example (see text)

Figure 10.8
Determination of the probability of failure P_f from calculated values of F_{MLV} and V_F.

This is the correct method of obtaining σ but in many geotechnical analyses there are insufficient numbers of data points to obtain σ for a normal distribution curve.

2. ***Published values***

$$\sigma = V\bar{x} \qquad \text{Eq. (5)}$$

where V = coefficient of variation
$\bar{x}$ = average value of the variable parameter

Duncan (2000) includes a table of V values for many geotechnical properties measured in the laboratory or in situ. This table is not reproduced here because it has been compiled primarily from four sources; the coefficients of variation are dependent on field and laboratory techniques that cannot be judged. Therefore, published values appear to be the least satisfactory method of estimating σ and are not recommended by the present author.

3. ***Three-Sigma Rule.*** This method of estimating σ is based on the fact that 99.73% of all values in a normal distribution fall within three standard deviations of the mean. Therefore, each standard deviation σ is obtained by estimating the *highest and lowest conceivable values* (HCV, LCV) of each parameter and dividing the difference by 6.

$$\sigma = \frac{(HCV - LCV)}{6} \qquad \text{Eq. (6)}$$

The highest HCV and LCV numbers are based on judgment. According to Duncan (2000), there is a tendency for geotechnical engineers to understate these estimates due, in his opinion, to their lack of experience in making such estimates. He recommends that they make a conscious effort to make the HCV to LCV range as wide as possible to overcome the tendency to make the range too small. Since the three-sigma rule is based largely on judgment, it can be applied when only limited or no data are available.

4. ***Graphical Three-Sigma Rule.*** The Three-Sigma Rule can be extended to linear and nonlinear relationships such as a Mohr strength envelope or undrained shear strength c vs. depth graph. Two examples from Duncan (2000) are shown on Figure 10.9(a)(b).

The procedure is:

(i) Draw a straight line or curve through the average variation of the parameter; in linear relationships, this could be the statistical line of best fit or a line selected by the consultant to represent the best average based on judgment.

(ii) Draw lines or curves that represent the highest and lowest conceivable bounds for the data. Points that are believed to be erroneous can be disregarded.

(iii) Draw lines or curves that are one-third of the distance from the average line towards the highest and lowest conceivable bounds. These lines or curves represent the average, plus and minus one standard deviation. They can be used in the Taylor series method in the same manner as single value parameters.

The average + sigma envelope is used to calculate F^+ and the average − sigma envelope is used to calculate F^-. Duncan recommends that the envelope be used in preference to using separate standard deviations for the strength parameters c and ϕ. The graphical three-sigma rule provides a simple means to characterize the uncertainty in shear strength.

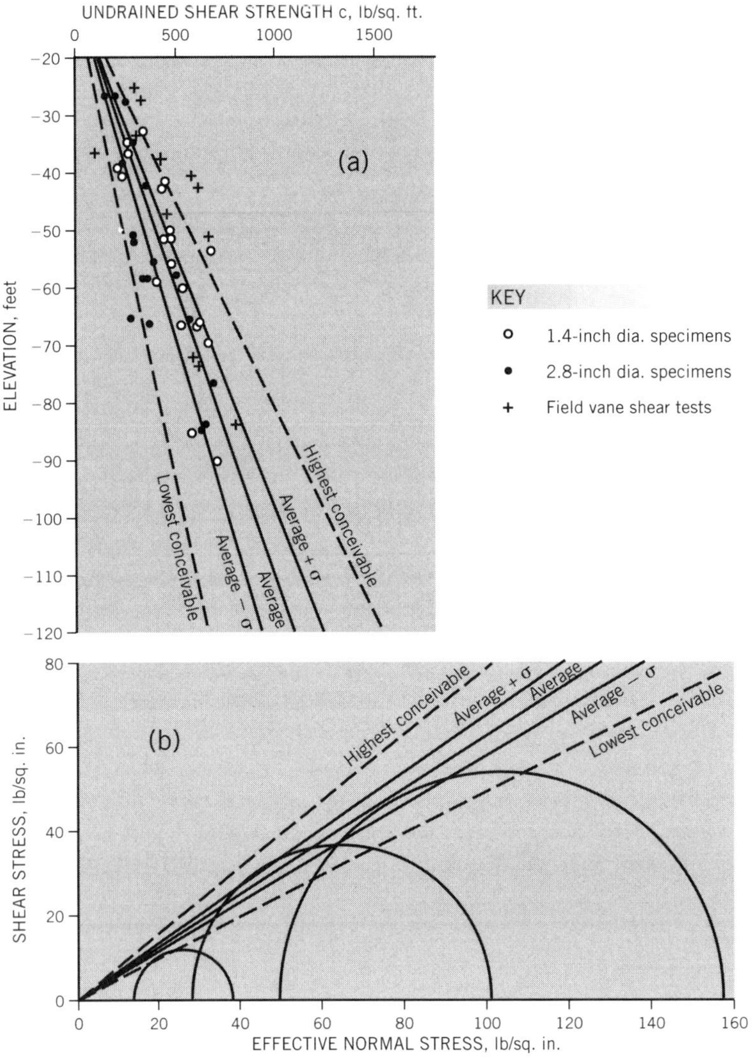

Figure 10.9 Three-sigma rule method of estimating the standard deviation of a graph:
(a) strength vs. depth for San Francisco Bay mud at the LASH terminal (after Duncan and Buchignani, 1973)
(b) effective stress envelopes for isotropically consolidated-undrained triaxial tests on claystone test fill, Los Vaqueros Dam (after Duncan, 2000)

Example Calculation

Duncan (2000) includes an illustrative example of a reliability calculation for a failed slope that has been published previously (Duncan and Buchignani, 1973). It involved the construction of a ship terminal at the Port of San Francisco where a long deep trench was to be excavated through San Francisco Bay Mud (Bay Mud) and then backfilled with sand to provide a stability buttress/shear key (Figure 10.10). The Bay Mud is a normally consolidated slightly organic silt or silty clay of marine origin (typically LL 50, PL 30). The undrained shear strengths were measured by unconsolidated-undrained triaxial tests and in situ vane shear tests, as shown on Figure 10.9(a).

A more comprehensive description of the project is given in the previously cited reference, but a brief summary of the relevant facts are: The proposed trench was 2,000 feet long. After about 500 feet had been completed, a slope failure 250 feet long occurred, as shown on the cross-section of Figure 10.10. Later, a second failure occurred over an additional 200 feet, but the remainder of the trench remained stable for four months, during which the trench was backfilled with sand.

The original stability analysis was based on the average shear strength properties shown on the graph, Figure 10.9(a). It showed that a temporary excavation slope had a calculated factor of safety of 1.17 for side slopes of 0.875 horiz : 1 vert. The reliability analysis is summarized in the table of Figure 10.10. With a standard deviation σ_F of 0.18 and a coefficient of variation V_F of 16% relating to the factor of safety F_{MLV}, the broken lines on Figure 10.8 show that the probability of failure is 18%. It is likely that such a calculation, if made in 1970 at the time of construction, may have caused the designers to conclude that the design slope was too risky. The failure showed that the real margin of safety was zero in some parts of the slope. The perception during design was that the probability of failure was much less than 18%.

Comments on Risk-Based Stability Analyses

Setting a rational design value on the factor of safety in stability calculations requires skill and experience. It depends on many factors such as the variability of the geotechnical parameters, location, consequence of failure, temporary or permanent construction, past experience with the soil type, etc. All the factors pertinent to a particular site have to be considered by the geotechnical consultant and come together in making a professional judgment as to the appropriate value of factor of

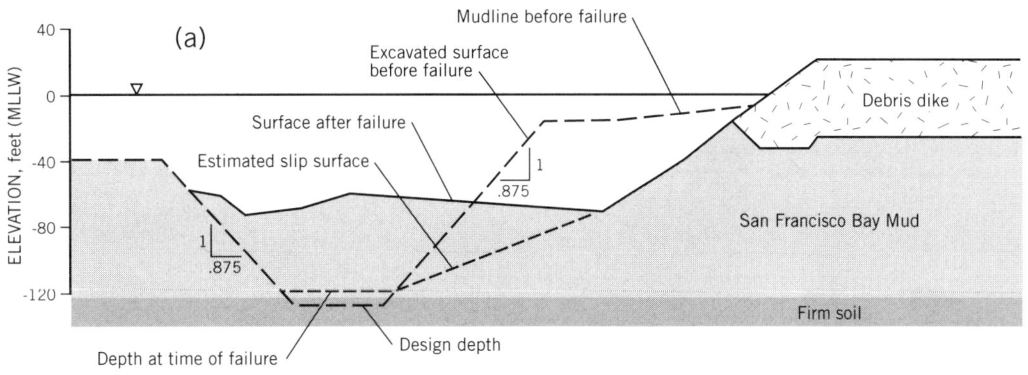

(b)

Variable (1)	Values (2)	Factors of safety (3)	ΔF (4)
Bay mud undrained strength			
Most likely value plus σ	See Figure 10.9(a)	$F^+ = 1.33$	
Most likely value minus σ	See Figure 10.9(a)	$F^- = 1.02$	0.31
Bay mud buoyant unit weight			
Most likely value plus σ	41.3 pcf	$F^+ = 1.08$	
Most likely value minus σ	34.7 pcf	$F^- = 1.28$	0.20

$F_{MLV} = 1.17$

Standard deviation of factor of safety

$\sigma_F = \sqrt{(0.31/2)^2 + (0.20/2)^2} = 0.18;$

Coefficient of variation of factor of safety $V_F = 0.18/1.17 = 16\%$

Probability of failure = 18% (see Figure 10.8)

Figure 10.10 Reliability analysis example calculation, LASH Terminal, Port of San Francisco:
(a) cross-section of trench before and after slope failure
(b) Taylor series reliability analysis calculation table
(after Duncan, 2000)

safety for the circumstances at that particular site. Most engineers develop this knowledge during their university training followed by many years of tutelage under experienced competent engineers.

These thoughts about training and the development of professional judgment are pertinent in the light of the trend towards risk-based procedures in geotechnical engineering. The laudable aim is to provide another tool for the professional engineer to check the reliability of assumptions and calculations.

The author's concern is that risk-based analyses will become a substitute for professional judgment. Many clients of consulting engineers have staff who are attracted to the idea of a mathematical answer to the issue of risk rather than a judgment that incorporates the other factors listed in the first paragraph of these comments. It is not difficult to imagine that, in the future, decisionmaking on the basis of reliability calculations may become required by owners and clients.

There are aspects of risk-based stability analyses that are somewhat disturbing. In terms of statistical probability, there is no absolute zero. Thus, as engineers, it may be acceptable to dismiss a small statistical number (for example, 1%) as being negligible, but it may not seem unimportant to others. A plaintiff lawyer may regard a 1 in 100 chance of a landslide as being a totally unacceptable risk to society or a particular client. Thus, the tool of a reliability index may be used against a very competent practitioner by a third party who doesn't understand (or want to understand) the process of engineering design. As an example, consider the circular arc landslide shown in Figure 10.1 and more fully described in Case History 8: Faraday Slide. The slide is well modeled and the shear strength parameters have been back analyzed for a factor of safety of 1.00. In constructing a buttress fill at the toe of the slope, the continuation of the discrete shear zone has to pass through it. Therefore, there is *no possibility* of a future shear failure even though a statistical risk analysis probably would indicate a small risk due to the mathematics of the theory.

The second concern about statistical risk-based analysis is the assumption that the variable parameters have a normal distribution (or at least, a non-skewed distribution) that allows a standard deviation to be calculated. This may be incorrect because either the parameter itself does not have a normal distribution within a stratum (such as permeability or coefficient of consolidation) or the measured value is incorrectly interpreted by an engineer or geologist with insufficient knowledge and experience of soil properties. In the latter case, an analysis by probability theories of a stiff clay slope might conclude that a slope is more stable than it actually is as a result of accepting mathematical calculations over geological evidence and knowledge of stiff clay behavior.

CHAPTER

11

Erosion Control

Erosion is responsible for initiating many of the landslide occurrences described in Chapter 2. Erosion control can prevent or remediate landslides. This chapter describes filter design, riprap slope protection, and geotextile fabrics. These techniques protect soils from (i) migrating into or through a coarser soil, or (ii) being directly eroded by water.

11.1. FILTER DESIGN

Filter Criteria

The *base soil* is the soil that is to be protected from erosion.

The *filter* is the sand, gravel, or rockfill that is designed to prevent movement of soil from the base into the filter by flowing water or gravity.

A *graded filter* is a set of filter layers, each layer being successively coarser and capable of preventing migration of soil from the finer layer in contact with it.

Filters must provide two essential functions:

1. Have void spaces small enough to block the passage of soil grains from the protected soil; this prevents piping (loss of fine soil into or through a coarser soil)
2. Be significantly more permeable than the protected soil so that water passing through the filter system flows freely out of the filter layer

These two functions can be met if the two soils in contact with each other meet *filter requirements.* The original filter requirements were published by Terzaghi, but later research has created modifications.

The most widely accepted filter requirements in current use are:

Piping Criterion: D_{15} filter should be *less* than 5 d_{85} base
Permeability Criterion: D_{15} filter should be *more* than 5 d_{15} base

where D_{15} is taken from the gradation curve of the coarser soil (filter) as the grain size corresponding to 15% finer by weight; d_{15} and d_{85} are taken from the gradation of the finer soil (base) as the grain sizes corresponding to the 15% and 85% finer by weight

The piping criterion is the more critical requirement. A summary of research data on filter experiments (Terzaghi, Peck, and Mesri, 1996) is shown on Figure 11.1. It can be seen that the piping ratio D_{15}/d_{85} can be as high as 8 without causing failure. However, the widely accepted ratio of 5 is recommended for design purposes, because some segregation is likely to occur during placement. The permeability criterion is believed to provide the filter with 20 to 25 times the permeability of the base.

The Soil Conservation Service (1986) recommends that filters have a maximum particle size of 3 inches and a maximum of 5 percent passing the No. 200 sieve; the fines should be nonplastic.

The soils to be protected have a *range* of gradations. These should be bracketed to show the coarsest range and finest range. Likewise, the filter will have a specified range of gradation. In the piping criterion, the d_{85} of the *finest* soil to be protected should be compared with the *coarsest* D_{15} of the filter. For the permeability criterion, it is d_{15} of the *coarsest* base soil that is compared to the D_{15} of the *finest* gradation of filter soil.

Illustrative Example of Filter Design

The design of a filter is demonstrated on Figure 11.2. Material A is a supplier's specification for a 1-inch minus rounded gravel backfill. Curves B and C are materials that would be in contact with Material A. Do they meet filter requirements?

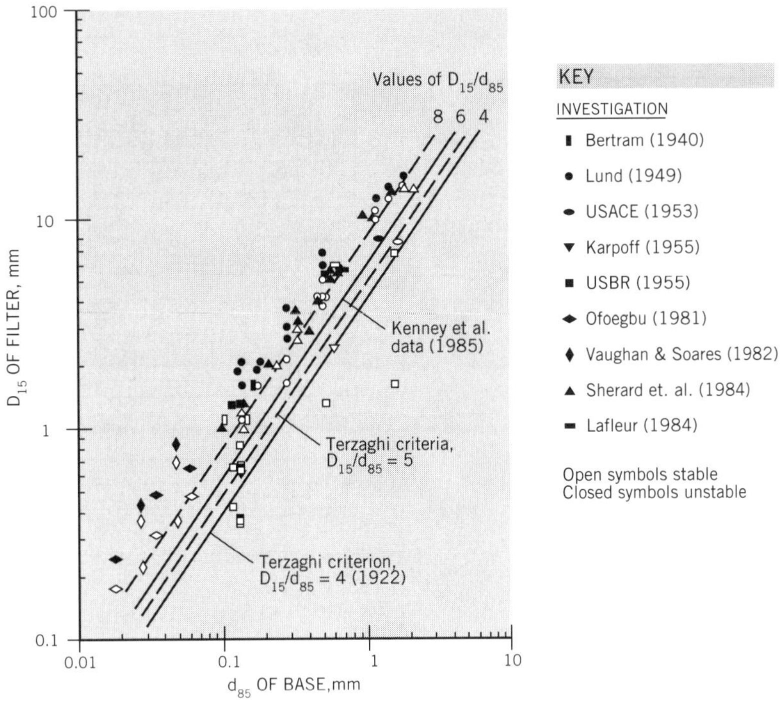

Figure 11.1 Research on piping ratio D_{15}/d_{85} (after Terzaghi, Peck, and Mesri, 1996).

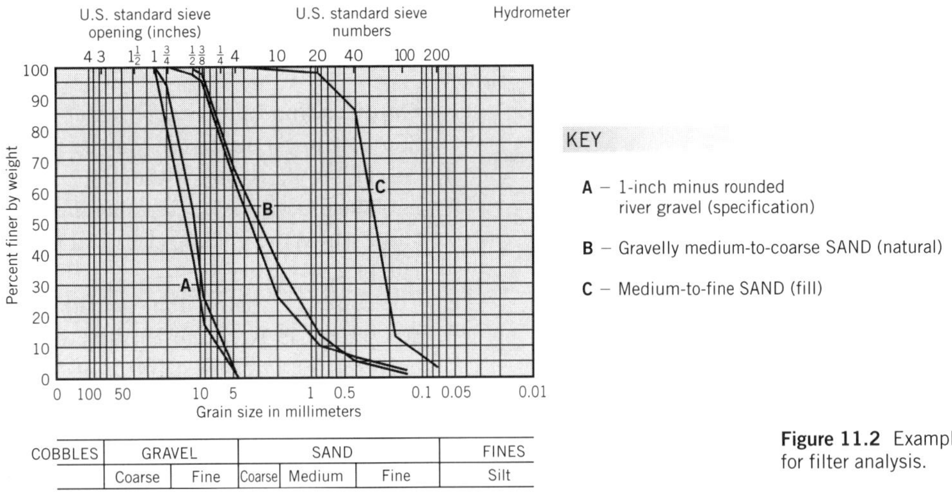

Figure 11.2 Example gradations for filter analysis.

Materials A and B

Piping criterion: D_{15} filter material A = 8.9 mm
d_{85} base material B = 7.2 mm

D_{15} of filter Material A is less than five times the d_{85} of the base, material B. Acceptable for the piping criterion. Note that the D_{15} is the *coarser* gradation range for filter material A and d_{85} is the *finer* of the two gradations for base material B.

Permeability criterion: D_{15} filter material A = 7.0 mm (the finer side of the range)
d_{15} base material B = 1.1 mm (coarser side of the two gradations)

$$D_{15} > 5\ d_{15}$$

Conclusion: Material A meets the filter requirements with respect to material B.

Materials A and C

Piping criterion: D_{15} filter material A = 8.9 mm
d_{85} base material C = 0.4 mm

D_{15} filter is more than 20 times d_{85}. Therefore, material A does *not* meet the filter requirements with respect to material C.

Segregation

All broadly graded filters are prone to segregation during placement. Larger stones tend to be carried to the end of the spread or roll around the edges of a dozer blade. This causes lenses or pockets of coarse material that are not representative of the filter design, and will allow the base material to move through the voids of the coarser soil.

For design purposes, the maximum stone size of the filter should be limited to 2 inches or 3 inches. It is also advisable to

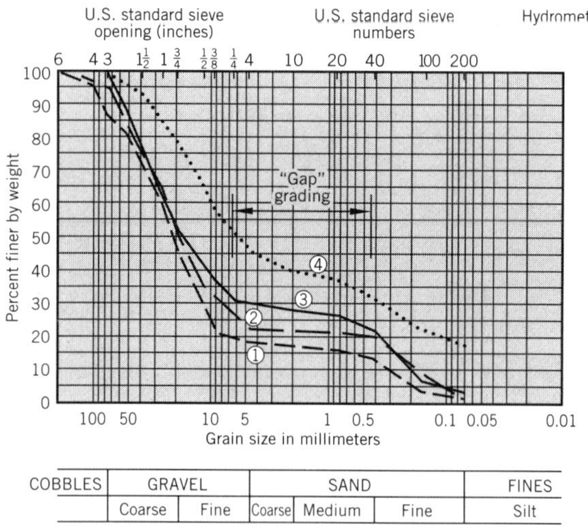

Figure 11.3 Gap graded gravel samples (Pasco Gravel) from Richland, Washington.

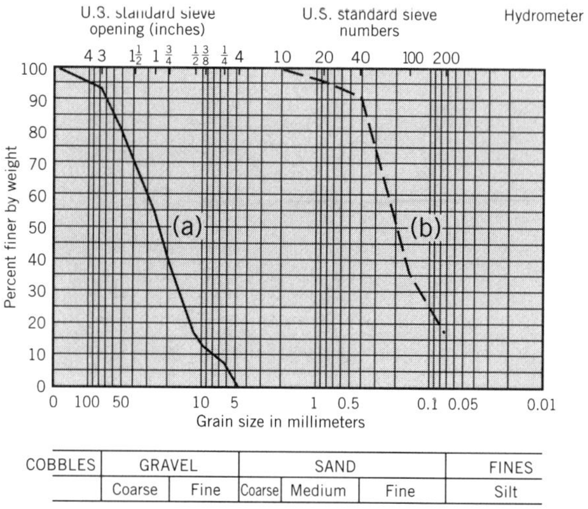

Figure 11.4 Separated gradation curves for gap graded gravel Sample 2 (Figure 11.3):

(a) coarse fraction
(b) fine fraction

Table 11.1 Calculation of Internal Stability to Seepage Flow

For internal stability $D_{15} \leq 5\ d_{85}$

Gradation: gravel No. 2 on Figure 11.3

		Separated Gradations		
Sieve Size	X Percent Passing	D Coarser Gradation	d Finer Gradation	Internal Stability Calculation
6"	100	100		
3"	95	93		D_{15} = 11 mm
2"	85	81		
1"	65	55		d_{85} = 0.37 mm
¾"	52	38		$D_{15} = 30d_{85}$
½"	35	17		
⅜"	32	13		(unstable)
¼"	28	8		
No. 4	22	0		
No. 10	22	0	100	
No. 20	21		96	
No. 40	20		91	
No. 100	8		36	
No. 200	4		18	

X_c = Cutoff point of gap-graded soil = 22%

Coarse Gradation $D = (X - X_c)\dfrac{100}{(100 - X_c)}$

Fine Gradation $d = (100)\left(\dfrac{X}{X_c}\right)$

specify a gradation that is more uniform than well-graded. Sherard et al. (1984a) recommends limiting the coefficient of uniformity (D_{60}/D_{10} ratio) to a maximum of 20 to control the broadness of the filter gradation.

During construction, spreading can be kept to a practical minimum by: (i) using relatively thick lifts of around 12 inches; (ii) dumping directly against the soil to be protected; (iii) spread using a front end loader or backhoe working from a stockpile rather than dump trucks with dozer. In critical erosion control situations, such as shoreline protection, the filter layer needs to be sufficiently thick to overcome any local segregation (see Case History 2).

Gap Grading of Protected Soil

Some naturally occurring soils are gap-graded and need to be checked to determine if they are internally stable against piping. The procedure is to separate the coarser and finer fractions at the gap in the grading chart and replot the data as two separate gradation curves.

Figure 11.3 shows four gradation curves for Pasco Gravel at a project in Richland, Washington. An example of the calculation for internal stability of Sample 2 (Figure 11.3) is shown on Table 11.1, and the resulting two gradation curves for the separated fractions are shown on Figure 11.4. The ratio D_{15}/d_{85} is 30 for this sample, indicating that it is internally unstable and the finer fraction of fine sand can easily pass through the voids of the fine-to-coarse gravel fraction. The four samples shown on Figure 11.3 all fail the filter test, having piping ratios ranging from 12 to 30. The upper limit for self-filtering is generally agreed to be at a piping ratio of 8 to 9 (Sherard et al., 1984a).

The Pasco Gravel in its natural state is very compact, and fine sand infills the void space of the fine to coarse pebbles. For the construction of a deep storm sewer at Richland, Washington, the Pasco Gravel formation was first dewatered by deep wells, and then was excavated and stockpiled for use as backfill (Figure 11.5a). A storm sewer pipe was laid in the bottom of the trench and surrounded by 1 inch to ⅜ inch pea gravel to a height of 1 foot above the crown. The stockpiled gravel was backfilled above it, and compacted in thin lifts to the surface. Despite good inspection of the backfilling operation, substantial settlement occurred after the dewatering was discontinued, and settlements continued for several years after

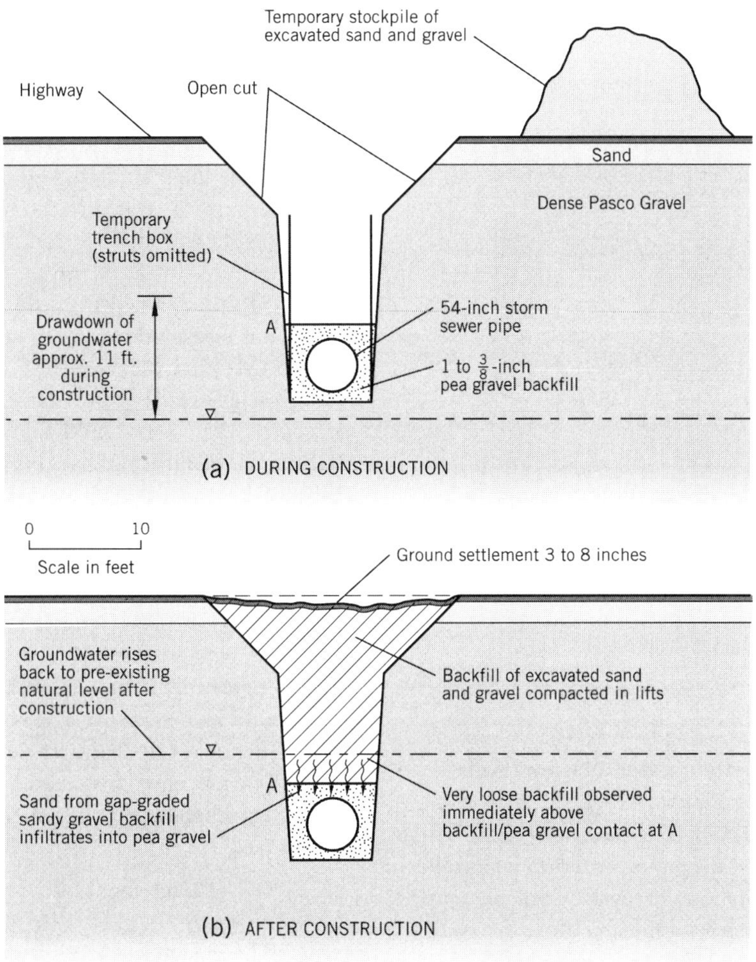

Figure 11.5 Trench excavation and backfill at Richland, Washington.

construction. Later field investigations showed that the SPT dropped to very low blow counts (zero in some boreholes) near the contact between the gravel backfill and the pea gravel pipe surround (Figure 7.2b).

What appears to have occurred at the Richland project is that, on saturating from rising groundwater, the sand within the Pasco Gravel backfill easily migrated by gravity into the openwork pea gravel. These two materials, shown in Figures 11.2 (sample A) and 11.4(b), do not meet filter requirements. Once the sand from the compacted sandy gravel backfill starts moving into the pea gravel, the Pasco Gravel backfill itself becomes internally unstable due to the sand loss; it gets loose near the contact between the backfill and the pea gravel.

Laboratory experiments confirmed that the fine sand within the Pasco Gravel moved down into the voids of the openwork pea gravel when water passed upward through the contact between the two materials. In the laboratory, the downward movement of sand through the gap-graded gravel mostly occurred through local "channels" rather than uniformly across the contact.

Filter Protection of Silts and Clays

Filtration experiments by Sherard et al. (1984b) have shown that filters for *silts and clays* can have piping ratios of more than 9 and still perform satisfactorily; in some cases, the base soil was retained at piping ratios as high as 50. The issue of providing reliable filters for silt and clay is more critical for cores of earthfill and rockfill dams than for landslide remediation.

It is generally agreed that sand, and gravelly sand filters containing fine sand sizes, are suitable filters for even the finest clays. The D_{15} size of the filter should be approximately 0.5 mm (No. 40 sieve) or less. For sandy silts and sandy clays, the conventional filter criterion $D_{15}/d_{85} < 5$ is always conservative and reasonable.

Sherard et al. (1984b) recommend the sand and gravelly sand filters shown on Figure 11.6 for clays. The two types of filter are considered about equal in their filter properties. The Soil Conservation Service (1986) design guidelines for filters are in general agreement with Figure 11.6; they limit the D_{15} size to around 1.0 mm for the filters of sands and clays.

Ripley (1986) recommends that the maximum particle size in the filter should be ¾ inch (to prevent segregation on placement) and that at least 60% by weight of the particles should be finer than 0.25 inch. A maximum limit of 2% should be set for material passing the No. 200 U.S. sieve. It seems to be generally agreed by research workers that filters need to be wet during placement to reduce segregation.

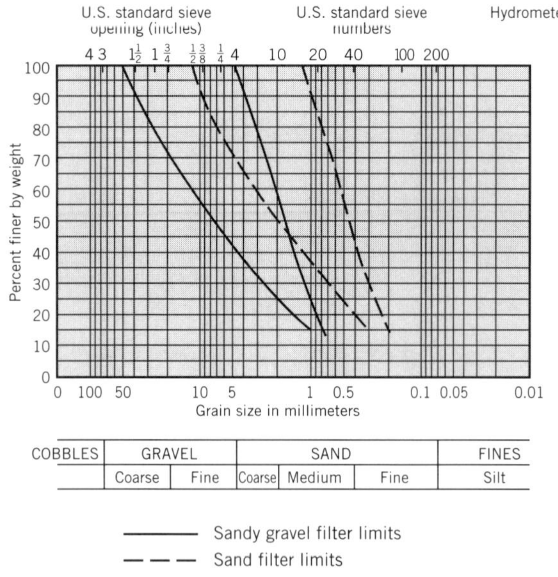

Figure 11.6 Sand and sandy gravel filters for use in protecting clays (after Sherard, Dunnigan and Talbot, 1984b).

Filter Protection of Broadly Graded Natural Soils

Many natural soils contain a broad range of particle sizes. Some examples are residual soils, colluvium, and glacial till. In such soils, the d_{85} is large and the piping criterion would allow a relatively coarse filter to be used. However, broadly graded soils have the potential for the soil fines to enter the voids of the filter even if the coarser particles cannot penetrate the voids. For this reason, the design of a filter for a broadly graded soil (for purposes of facilitating drainage) should exclude the coarser fraction.

Sherard et al. (1963) recommends that the filter be designed for the portion of the gradation curve that is finer than the 1-inch sieve. The Federal Highway Administration (1980) follows the same guideline. The Soil Conservation Service (1986) recommends excluding all gravel sizes (i.e., coarser than No. 4 sieve) from the base material gradation. In each case, the gradation of the fines fraction is increased by a multiplier similar to the procedure shown on Table 11.1 for a gap-graded base material.

Drainage Pipes Encased in Filters

Drainage pipes are often placed within a filter layer to increase flow rate of collected water. Pipes are available in a wide range of diameters and with various hole diameters or slots.

The Corps of Engineers (1986) recommends that perforated pipe be designed as follows:

$$\text{Hole diameter} \leq D_{50} \text{ of the filter}$$
$$\text{Slot width} \leq 0.83\, D_{50} \text{ of the filter}$$

The pipe diameter should be at least 4 inches in most landslide applications, or larger if needed to handle anticipated high water flows at critical times. The gradient along the pipe should be at least 2 percent. Flexible corrugated plastic pipe of 4 to 12 inches diameter is commonly used in trench drains and buttresses. Typical product data are shown on Table 11.2.

Additional Discussion of Filter Requirements

It is generally agreed that permeability of soils is mostly governed by the fines. Similarly, the pore channels through which the water flows is related to the size of the smaller particles of the filter (Sherard et al., 1984a). This source indicates that an approximate estimate of the permeability of dense sand and gravel can be obtained from the expression $k = 0.35\ (D_{15})^2$ where k is measured in cm/sec and D is in mm. It goes on to demonstrate that water under a given gradient travels at the same average velocity through the pore channels of dense sand or gravel filter as it would through a pipe with a constant diameter of about $0.16\ D_{15}$. This is referred to as the *equivalent hydraulic diameter.*

When water flows through a pipe of irregular cross-sectional area, such as the pores of a sand or gravel filter, the head loss is largely controlled by the smallest dimensions of the flow path (where head loss is greatest due to friction). Sherard and colleagues suggest that the equivalent hydraulic diameter of $0.16\ D_{15}$ probably is close to the diameter of the smaller pores in the sand and gravel. It agrees well with the experimental results (Figure 11.1) that particles with diameter d_{85} of $0.12\ D_{15}$ will be prevented from entering by the pores of the filter.

Three other filter requirements that are widely accepted in practice:

1. D_{50} of filter should be equal to or less than 25 d_{50} of the base.
2. The gradation curves of the filter and base soils should be approximately parallel to each other.
3. The filter should be compacted to at least 80 percent relative density.

Table 11.2 Typical Pipe Perforations for 3-Inch to 24-Inch Single Wall Pipe

Inside Diameter		Perforation Type	Slot Length or Diameter		Slot Width	
inches	(mm)		inches	(mm)	inches	(mm)
3	(75)	Slot	0.875	(22.2)	0.125	(3.18)
4	(100)	Slot	0.875	(22.2)	0.125	(3.18)
5	(125)	Slot	0.875	(22.2)	0.125	(3.18)
6	(150)	Slot	0.875	(22.2)	0.125	(3.18)
8	(200)	Slot	1.250	(31.8)	0.125	(3.18)
10	(250)	Slot	1.250	(31.8)	0.125	(3.18)
12	(300)	Slot	2.50	(63.5)	0.125	(3.18)
12	(300)	Circular	0.375	(9.52)	—	
15	(375)	Circular	0.375	(9.52)	—	
18	(450)	Circular	0.375	(9.52)	—	
24	(600)	Circular	0.375	(9.52)	—	

1. Data supplied by Advanced Drainage Systems, Inc. (ADS) Columbus, Ohio (1996).
2. ADS pipe is perforated for water entry with slots or circular perforations. The perforations are uniformly spaced along the length and circumference of the pipe.
3. ADS pipe is manufactured to comply with the perforation requirements specified in the following industry standards: ASTM F405, ASTM F667, AASHTO M252, AASHTO M294, and SCS Code 606.

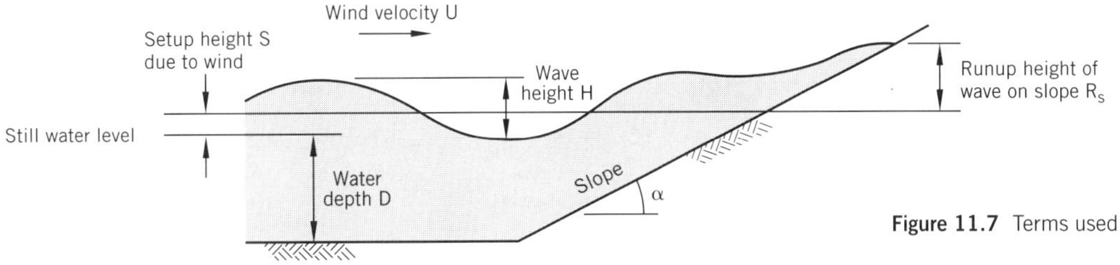

Figure 11.7 Terms used in riprap design.

Sherard's research takes issue with items (1) and (2), claiming that they are irrelevant. He also concludes that crushed rock filters perform in a similar manner to filters comprised of rounded stones. Milligan (2003) disagrees with item (3), arguing that too much compaction makes the filter susceptible to cracking, especially in filters composed of hard crushed rock. He recommends that customary placement methods are usually sufficient to produce an acceptable relative density without any additional compactive effort. Milligan also recommends using natural materials (rounded river soils) over crushed materials wherever possible. The reason is that crushed materials have micro cracks in the stones that cause them to break down further when they are placed and loaded.

11.2 RIPRAP DESIGN

Riprap is a layer of heavy rock pieces that protects a soil slope from erosion due to waves or rapidly flowing current. Riprap protection is required for two different situations:

- Direct impact of **waves** onto a slope. The design is based on the wave height generated by winds.
- **Current** flow during flood that erodes streambanks.

(A) Riprap Design for Waves

The following procedures are based on Corps of Engineers and Soil Conservation Service designs for embankment dams but should be generally applicable to lakes, reservoirs, and coastal slopes that are impacted by wind-generated waves. The two design methods generally agree but there are minor differences. The terms used in the design are shown on Figure 11.7.

The step-by-step procedure for the design of riprap protection in the United States will be demonstrated by a hypothetical lake near Sacramento, California (Figure 11.8; Table 11.3). A bridge approach fill is to be protected from instability by riprap at Point A. The approach fill is a cohesionless compacted sandy gravel with side slopes of 2 horizontal : 1 vertical. This will be referred to as the *example site in Sacramento.*

A "shortcut" design is given at the end of this section. It avoids the necessity of making most of the calculations by using some simplifying assumptions. In most cases, this shortcut procedure is sufficient for landslide applications.

The 8-step approach for designing riprap to resist wave action is: (1) develop a wind-duration curve for the site; (2) determine the effective fetch of the wind; (3) determine the wind velocity—duration curve for the water body; (4) determine the deep-water wave height and wave period; (5) calcu-

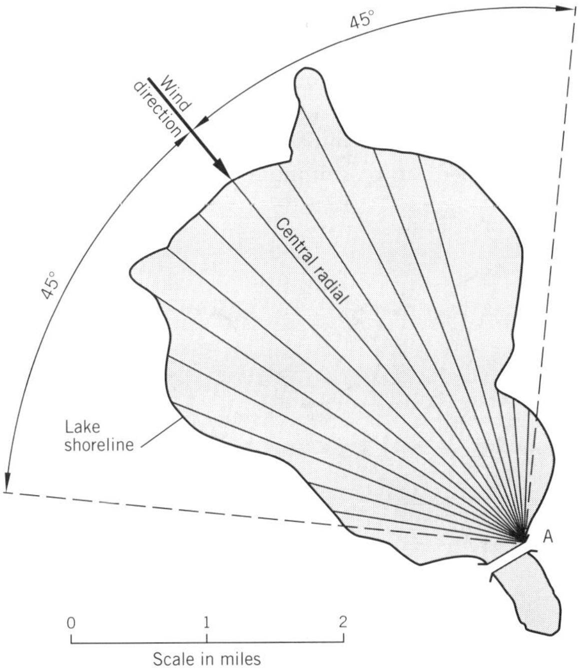

Table 11.3 Effective Fetch F_e Computations

F	α	cosα	cos²α	F cos²α
1.42	42	0.743	0.552	0.784
1.63	36	0.809	0.654	1.066
2.75	30	0.866	0.750	2.062
2.92	24	0.914	0.835	2.438
3.26	18	0.951	0.904	2.947
3.35	12	0.978	0.956	3.203
3.35	6	0.995	0.990	3.317
* 3.36	0	1	1	3.36
3.23	6	0.995	0.990	3.198
3.53	12	0.978	0.956	3.375
3.01	18	0.951	0.904	2.721
2.88	24	0.914	0.835	2.405
1.05	30	0.866	0.750	0.788
1.00	36	0.809	0.654	0.654
0.93	42	0.743	0.552	0.513
		13.512		32.831

* central radial

$$F_e = \frac{\Sigma F \cos^2\alpha}{\Sigma \cos\alpha}$$

$$= \frac{32.831}{13.512} = 2.43 \text{ miles}$$

F – radial fetch distance in miles

α – angle (in degrees) from central radial

F_e – effective fetch for site A

Figure 11.8 Computation of effective fetch F_e for a hypothetical site near Sacramento, California.

Table 11.4 Determination of Design Wind Direction

Month	Wind Velocity	Azimuth	(A) × (B)	Month	Wind Velocity	Azimuth	(A) × (B)
	(A)	(B)			(A)	(B)	
January	32	360	11,520	July	50	225	11,250
February	40	360	14,400	August	45	90	4,050
March	40	315	12,600	September	30	315	9,450
April	36	315	11,340	October	55	360	19,800
May	60	315	18,900	November	60	360	21,600
June	61	315	19,215	December	63	315	19,845

Design wind direction $= \frac{\Sigma\,(A)\,(B)}{\Sigma\,(A)} = \frac{173{,}970}{572} = 304°$

late the wave runup and wind setup heights; (6) calculate the weight of riprap needed; (7) determine the gradation and thickness of riprap; and (8) determine the height of riprap protection needed on the slope.

(1) Wind Velocity Duration Curve for the Area

The first step of the analysis is to determine the maximum winds that are likely to occur at the site based on past records. The U.S. Weather Service publishes annual climatological data summaries that include wind data for local areas. This is available from the National Climatic Center, Asheville, North Carolina, part of the National Oceanic and Atmospheric Administration (NOAA). Another source is "Climatic Atlas of the United States," published by U.S. Department of Commerce. Other sources of wind data for particular areas in the United States include: the Corps of Engineers; U.S. Forest Service (or state); military bases; Federal Aviation Agency; Environmental Protection Agency; State Bridge Authority; State Department of the Environment; etc.

The wind data depend on the topography around the site, vegetative cover, and other factors related to the site environment. Wind velocity is measured at a standard height of 25 feet above ground.

Method 1. Wind velocity-duration curves present the highest average wind speeds that can be expected from a specific direction during a specified time period. Typically, they are drawn with a family of direction curves (such as the eight points of the compass) with wind speed on the ordinate of the graph and length of time on the absissa. If sufficient data have been collected, they can be analyzed statistically to provide recurrence data. The Corps of Engineers (1978) recommends that only wind records that lie in a 90-degree quadrant coincident with the effective fetch calculations (e.g., Figure 11.8) should be used to calculate the site winds.

Method 2. If using the U.S. Weather Service tables "normals, means, and extremes" for the local area of concern, obtain the speed and direction for the fastest mile each month. Multiply the azimuth of the wind by the corresponding velocity for each month. Divide these sums by the sum of the wind velocities to obtain the design wind direction. An example is given in Table 11.4.

Method 3. If the computed wind direction in Method 2 points away from the riprap design area, an alternative procedure is to use regional wind speeds. The maximum wind speeds for various cities around the United States are reproduced on Figures 11.9 (winds of one minute duration) and 11.10 (winds of one hour duration). The data are abstracted from Corps of Engineers' manual ETL 1110 – 2-221 (1976).

In the example, Figure 11.9 shows that the maximum speed of one minute duration at the example site in

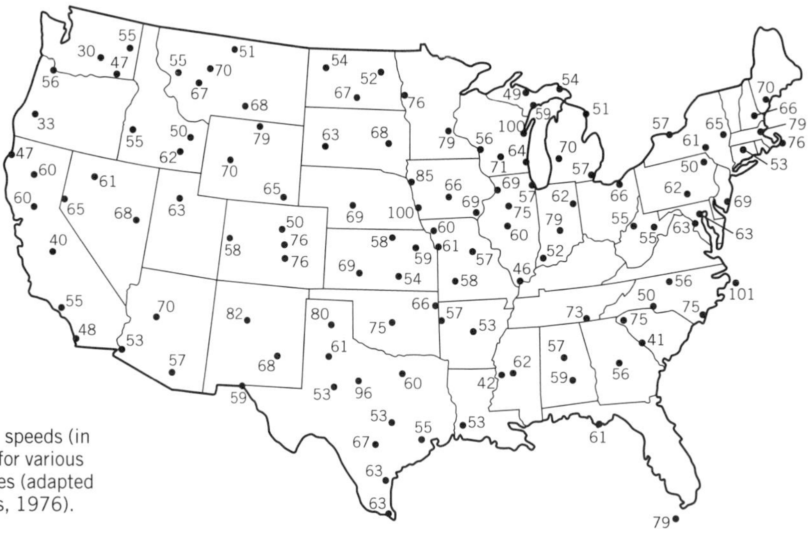

Figure 11.9 Maximum wind speeds (in m.p.h) of 1 *minute* duration for various cities within the United States (adapted from U.S. Corps of Engineers, 1976).

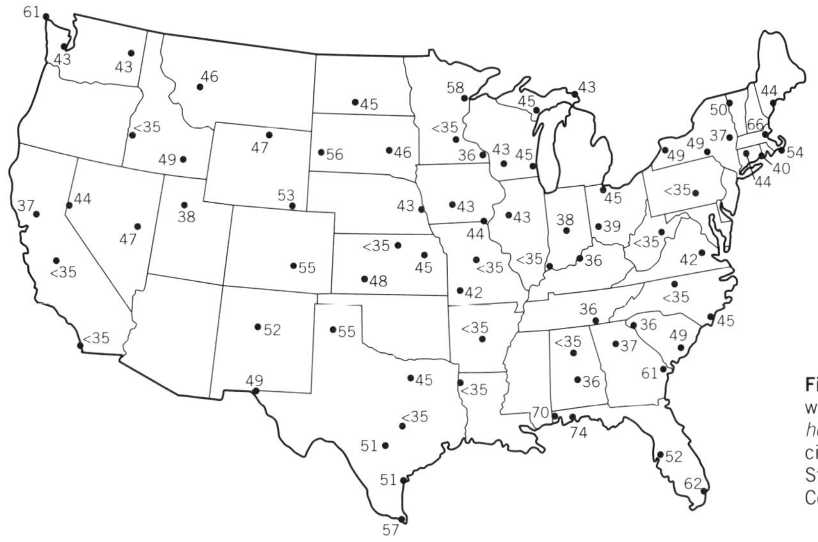

Figure 11.10 Maximum wind speeds (in m.p.h.) of 1 *hour* duration for various cities within the United States (adapted from U.S. Corps of Engineers, 1976).

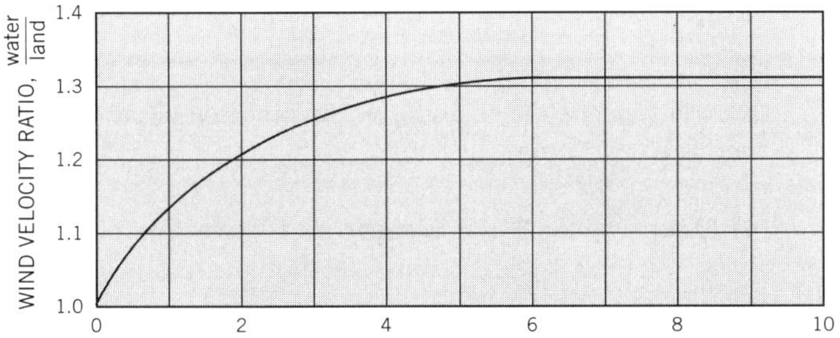

Figure 11.11 Ratio of water-to-land wind velocities as a function of effective fetch F_e (after U.S. Corps of Engineers, 1973).

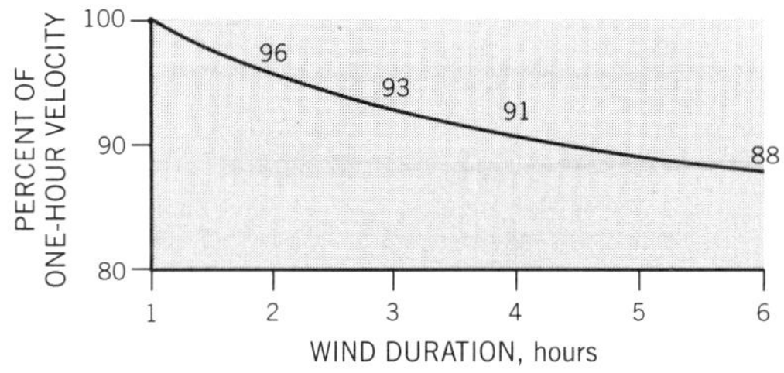

Figure 11.12 Extension of maximum wind speeds beyond one hour duration (after U.S. Corps of Engineers, 1960).

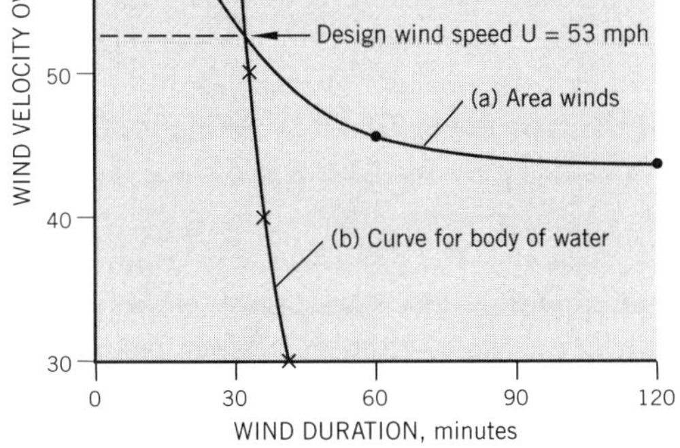

Figure 11.13 Determination of the design wind speed for a hypothetical site near Sacramento, California.

Table 11.5 Wind Velocity—Duration Data for Hypothetical Sacramento, California Site

Duration of Wind Speed	Wind Speed (mph)	
	Land	Water
1 minute	60	73.8
1 hour	37	45.5
2 hours		43.7

Sacramento, California, is 60 mph. The maximum wind speed sustainable for 1 hour is 37 mph (Figure 11.10).

Wind speeds over water are faster than over land because a water surface is relatively smooth and uniform. Research and field studies (U.S. Corps of Engineers, 1962) have produced the graph, Figure 11.11, which is suitable for use in lakes or reservoirs that are surrounded by terrain of moderate irregularities and surface roughness. Since the calculated effective fetch (described later) for the Sacramento lake example is 2.43 miles, the wind ratio for water to land is 1.23.

Wind velocities for durations of longer than 1 hour can be obtained from Figure 11.12 (U.S. Corps of Engineers, 1960). It is now possible to calculate the wind velocities–duration curve for the example Sacramento site, as shown on Table 11.5. These data provide the wind velocity–duration curve for the winds at this site (curve a on Figure 11.13).

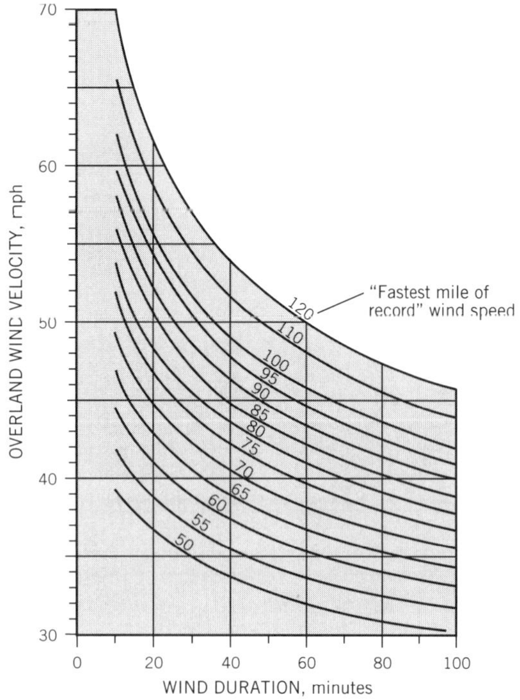

Note: Data may need to be converted to overwater wind speeds (Figure 11.11)

Figure 11.14 Maximum wind velocity–duration curves for use with the *fastest mile of record* data (after U.S. Corps of Engineers, 1960).

Method 4. Obtain the fastest mile of record under the "normal, means, and extremes" of the U.S. Weather Service local data bulletin. Using the relationships on Figure 11.14, obtain the overland wind velocity–duration curve for graph (a) on Figure 11.13. Adjust the data, using Figure 11.11 to obtain the wind speeds over water. This is an alternative to Method 3 and has not been calculated for the Sacramento example.

(2) Effective Fetch F_e

Waves are generated by winds passing over the surface of the water in a particular direction. Because lakes, rivers, and enclosed bodies of water are limited in length over which the wind can act, the effective fetch has to be calculated. The Corps of Engineers method of calculating effective fetch is to draw radials over a 90° arc from the point of interest (i.e., where riprap is to be placed). In the current example, Figure 11.8, the radials are 6° apart with the central radial aligned in the most unfavorable direction for wind and wave attack. The length of each radial pointing in a different direction to the design wind is multiplied by the square of cosine of the included angle (from the central radial). The effective fetch F_e is defined by:

$$F_e = \frac{\Sigma F \cos^2 \alpha}{\Sigma \cos \alpha} \qquad \text{Eq. (1)}$$

where F = individual fetch length of each radial
α = angle between the individual radials and the central radial

The calculated effective fetch F_e for the hypothetical Sacramento site is 2.43 miles (Table 11.3, Figure 11.8).

(3) Wind Velocity–Duration Curve for the Body of Water

The effective fetch is entered on Figure 11.15 to obtain a range of wind velocities and durations for the body of water. For the example of F_e = 2.43 miles, shown by a broken line on Figure 11.15, the relationships are:

WIND SPEED (MPH)	DURATION (MINUTES)
80	27
60	30
50	33
40	36
30	41

This graph is drawn on Figure 11.13(b). Where the two wind velocity–duration curves intersect is the design wind U that will generate the largest waves. In our example, the design wind is U = 53 mph.

(4) Deep-Water Wave Height and Wave Period

"Deep" water is when the water depth in the reservoir or lake is more than 50% of the wave length. The deep-water wave length is:

$$L_O = 5.12\,T^2 \qquad \text{Eq. (2)}$$

where L_O = wave length, feet
T = wave period, seconds

The deep-water wave height can be forecast from Figure 11.16. This is the *significant wave height* H_s defined as the average height of the one-third highest waves of a wave group. The height is measured as the vertical distance between the crest and the preceding trough. For the Sacramento site example, the significant wave height H_s is 3.5 feet.

The *wave period* T is the time required for a wave crest to traverse one wave length and can be determined from Figure 11.17. For the Sacramento site example, T =3.6 seconds and, from Eq. (2), the wave length L_O is 66 feet. Therefore, if the body of water is more than 33 feet deep (½ L_O), it is deep water and the wave is unaffected by the floor of the reservoir or lake. Most reservoirs and lakes fall into the deep water category.

Where shallow water waves develop, estimates based on deep water conditions give conservative wave height predictions. Therefore, shallow water projects can be designed using the deep water formulas.

(5) Wave Runup R_s and Wind Setup, S

The wave runup on the slope is needed to determine the height of riprap H_R needed for a project. The height H_R above the maximum still water reservoir pool or lake level is a combination of the wave runup and wind setup (see Figure 11.7).

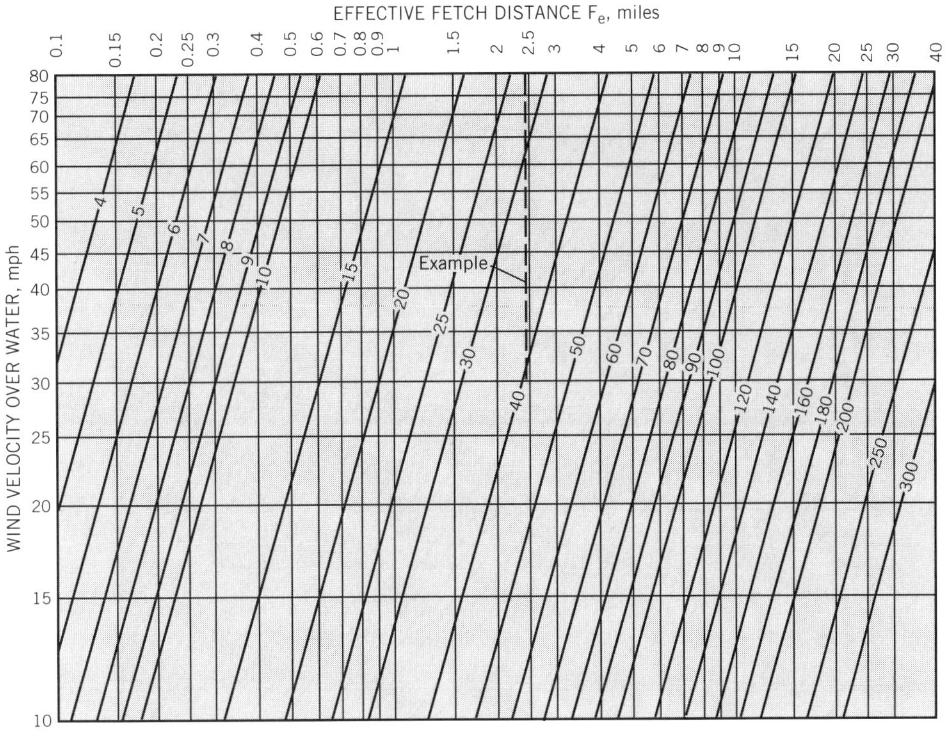

NOTES: (1) Angled lines represent duration of wind in minutes

(2) — — — Example (see text)

Figure 11.15 Wind velocity–duration relationships for lakes (after U.S. Corps of Engineers, 1976).

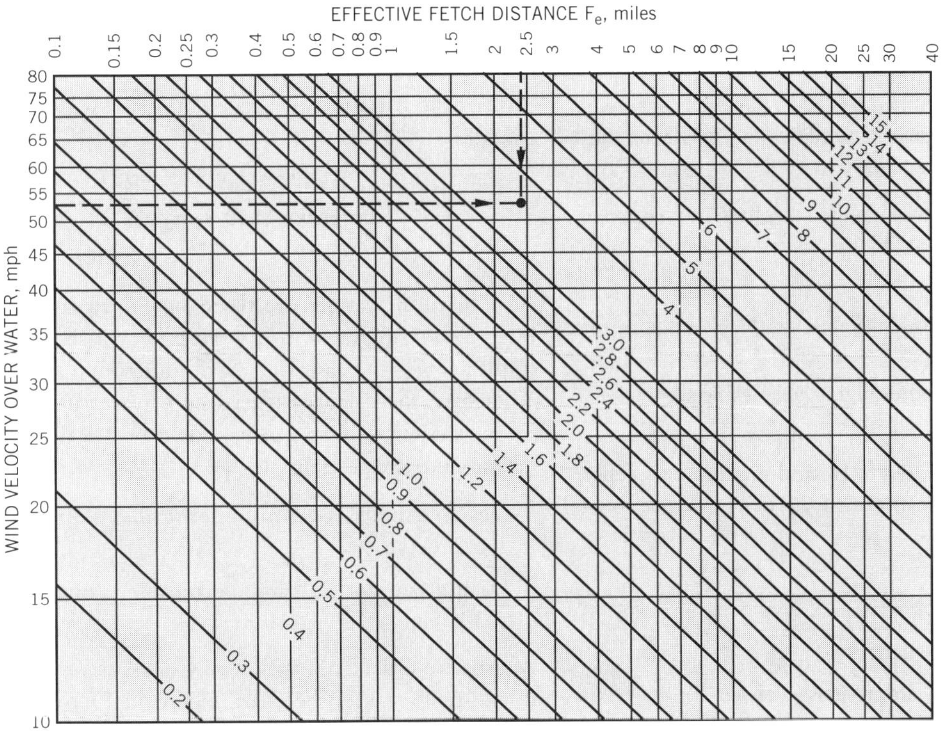

NOTES: (1) Angled lines represent significant wave height, H_s feet

(2) — — — Example (see text)

Figure 11.16 Significant wave height H_s for deep-water lakes (after U.S. Corps of Engineers, 1976).

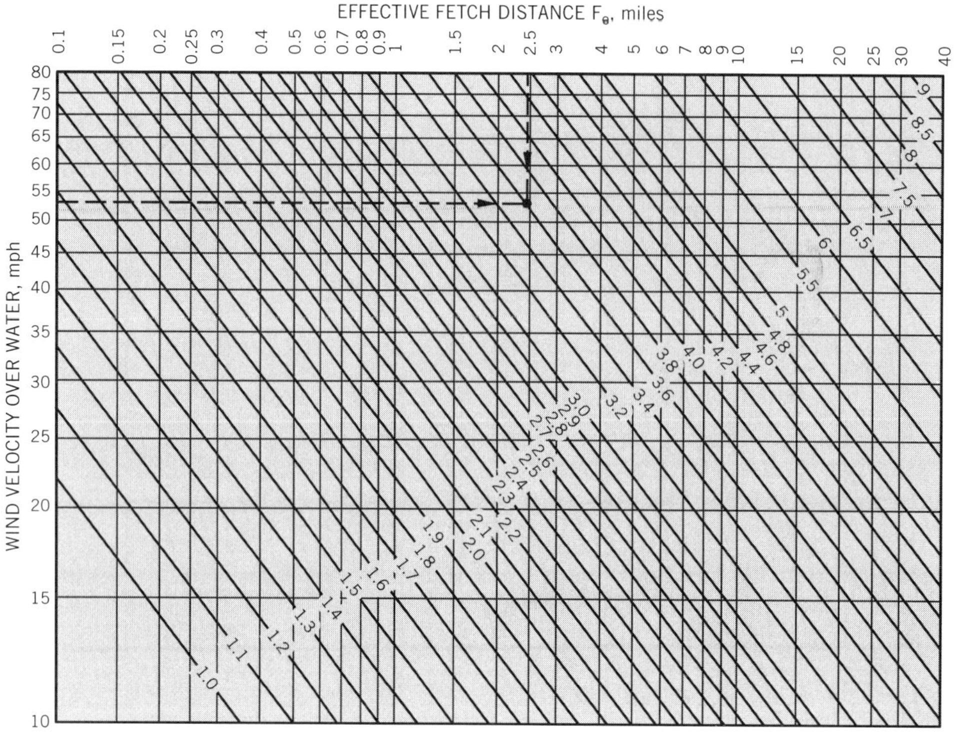

NOTES: (1) Angled lines represent wave period, seconds
(2) — — — Example (see text)

Figure 11.17 Wave period T for deep-water lakes (after U.S. Corps of Engineers, 1976).

$$H_R = R_S + S \qquad \text{Eq. (3)}$$

where R_S = wave runup, caused by the significant wave
S = wind setup, caused by wind stresses on the surface of the water

The *wave runup* R_S depends on the steepness and nature of the slope that is being attacked by the waves. Slopes are subdivided into:

Impermeable—earth embankment with a riprap outer protection

Permeable—rockfill or other highly permeable slope

When the water depth at the toe of the slope is more than 3 times the significant wave height H_S and the slope is steeper than 5 : 1 (horizontal : vertical), the runup height can be forecast from:

I. Impermeable slope

$$R_S = \frac{H_S}{0.4 + \cot \beta \sqrt{H_S/L_O}} \qquad \text{Eq. (4)}$$

where R_S, H_S, L_O are previously defined
β = slope angle to the horizontal

In the Sacramento example, assuming deep water conditions and a slope angle β of 26.6° (2 horiz : 1 vert.).

L_O = 66 feet; H_S = 3.5 feet.

Therefore, R_S = 4.1 feet

II. Permeable slope (rubble mound)

For the hypothetical Sacramento lake with slopes at 2 : 1 (H :V).

$$H_S/T^2 = 3.5/(3.6)^2 = 0.27$$

From Figure 11.18, $R_S/H_S = 0.75$

R_S = 2.6 feet

These calculations are based on the significant wave H_S but 13% of all waves in a wave train will be higher. These higher waves increase the maximum runup by about 50%. Therefore, the calculated maximum wave runup R_m caused by the design wave spectrum should be 1.5 times R_S. In the Sacramento design example,

Design runup R_m for impermeable slope = (1.5)(4.1) = 6.2 feet

Design runup R_m for permeable slope = (1.5)(2.6) = 3.9 feet

Alternative Calculation: The Soil Conservation Service (1978) design method uses the chart, Figure 11.19, to estimate wave runup for a single layer of riprap (impermeable slope case). These graphs have been derived from multiple sources. For the Sacramento lake example, H_S/L_O = 0.053. For a 2:1 (H:V) slope, ratio R_m/H_S = 1.70; R_m = (1.70)(3.5) = 6.0 feet, which is about the same as the value of 6.2 feet described above from Eq. (4) with a 50% add-on for the highest waves.

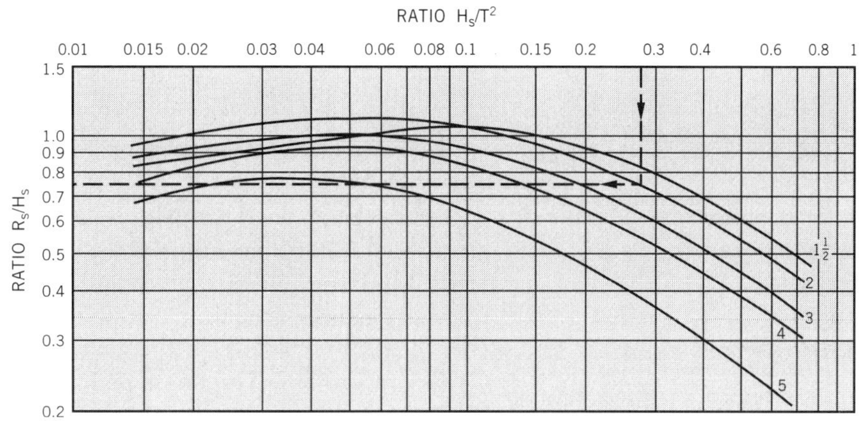

Numbers on curves denote slope, e.g. 3 is 3 horizontal : 1 vertical

– – – – Example (see text)

Figure 11.18 Estimation of wave runup on permeable (rock rubble) slopes (after U.S. Corps of Engineers, 1973).

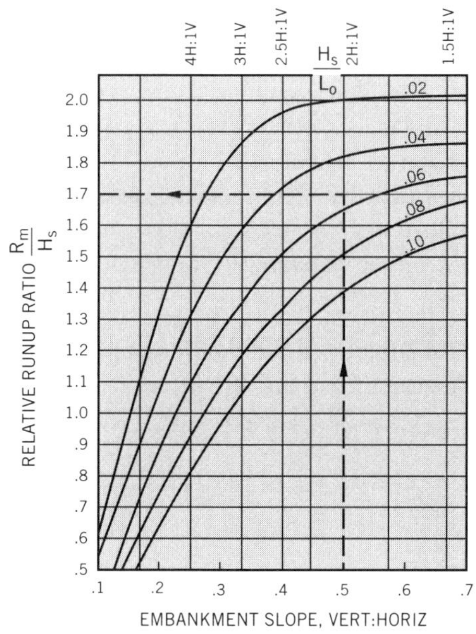

Note: Multiply $\frac{R_m}{H_s}$ by 1.2 for blocky-shaped rock or flagstone

– – – Example (see text)

Figure 11.19 Estimation of wave runup on a single layer of uniform angular riprap on an impermeable base (after U.S. Soil Conservation Service, 1978).

The *wind setup* S is caused by wind piling up water against a slope or structure. It is the still water level on which the wave is superimposed. The wind setup level S (U.S. Corps of Engineers, 1962) can be computed from:

$$S = \frac{U^2 F}{1400\ D}$$

where U = design wind speed

F = wind fetch, which can be considered equal to 2 F_e used for wave generation calculations

D = average water depth (feet) along the fetch line

This can be rewritten

$$S = \frac{U^2 F_e}{700\ D} \qquad \text{Eq. (5)}$$

For the Sacramento example, assuming D = 50 feet

$$S = \frac{(53)^2(2.43)}{(700)(50)} = 0.20 \text{ feet}$$

It can be noted that the wind setup is fairly insignificant compared to wave runup. Therefore, the average water depth D can be roughly estimated where bathymetry is unavailable. The example calculations are summarized on Table 11.6.

Table 11.6 Summary of Calculations for Hypothetical Sacramento, California Site

Design wind		U = 53 mph	
Effective fetch		F_e = 2.43 miles	
Significant wave height		H_S = 3.5 feet	(deep water)
Wave period		T = 3.6 seconds	(deep water)
Wave length		L_O = 66 feet	(deep water)
Wave runup:	(1) impermeable slope	R_S = 4.1 feet	(deep water)
	(2) permeable slope	R_S = 2.6 feet	(deep water)
Maximum wave runup:	(1) impermeable slope	R_m = 6.2 feet	(deep water)
	(2) permeable slope	R_m = 3.9 feet	(deep water)
Wind setup		S = 0.2 feet	
Deep water		> 33 feet	

Weight of Riprap

The principal resistance to wave attack is provided by the size of rock used for riprap. The rock pieces must be placed and keyed together so that the riprap slope will not be damaged by waves or allow the underlying soils to be eroded. In addition to weight, other factors affecting the design of the riprap are: (i) gradation, (ii) surface roughness, (iii) slope of the face, and (iv) filter conditions below the riprap.

The wave forces acting on a riprap slope are a combination of drag and inertia components. The force exerted by a wave can be written as:

$$F_q = \frac{C_q l^2 \gamma_w H}{2\pi\delta} \qquad \text{Eq. (6)}$$

where C_q = a coefficient combining drag and inertia
l = linear dimension of a rock
γ_w = density of water
H = wave height
δ = H/L, the wave steepness

For dumped riprap, the principal resistance to the wave force is the buoyant weight of the rock; the friction between stones is neglected. By equating the wave force of Eq. (6) to the buoyant weight of the rock, which represents incipient failure, an empirical equation was developed by Hudson (1961) of the form:

$$W_r = \frac{\gamma H_S^3}{K_\Delta (G-1)^3 \cot\beta} \qquad \text{Eq. (7)}$$

where W_r = weight of stone required (= W_{50})
γ = density of rock
H_S = significant wave height
G = specific gravity of rock
β = slope angle to the horizontal
K_Δ = 3.2 for dumped riprap
(*Note:* the Corps of Engineers uses K_Δ = 4.37)

Equation (7) can be rewritten

$$W_{50} = \frac{19.5\, G\, H_S^3}{(G-1)^3 \cot\beta} \qquad \text{Eq. (8)}$$

The nomograph for this Soil Conservation Service (1978) equation is given on Figure 11.20.

For the Sacramento example, using an assumed specific gravity G = 2.70 for the rock, the W_{50} size is 230 lb. for the Soil Conservation Service procedure and 168 lb. for the Corps of Engineers formula.

(7) Gradation and Thickness of Riprap

There are two main schools of thought concerning riprap gradation:

- *Uniform size riprap,* normally placed by a large backhoe to provide a well-keyed outer surface that resists plucking. Uniform riprap needs careful attention to the backing layers to ensure that filter requirements are met for the relatively large void spaces of the riprap. It is always angular (freshly blasted) rock fragments.
- *Well-graded riprap* in which larger rocks have their void spaces partly infilled with smaller rock. Well-graded riprap is dumped and segregation has to be avoided during final placement.

Sherard et al. (1967) reported that well-graded riprap has performed better than uniform size riprap on dams in the

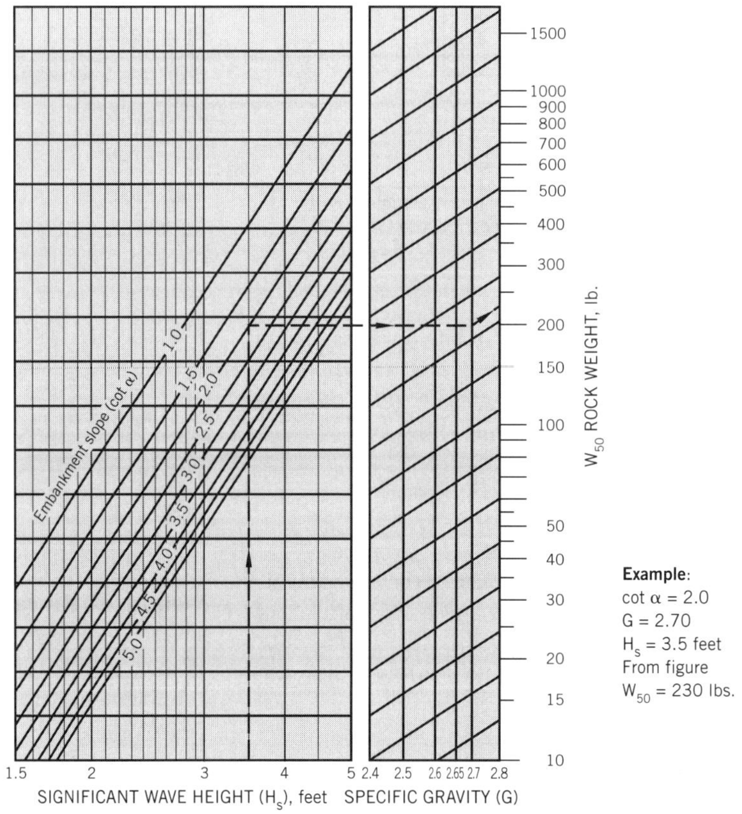

Figure 11.20 Riprap sizing chart (after U.S. Soil Conservation Service, 1978).

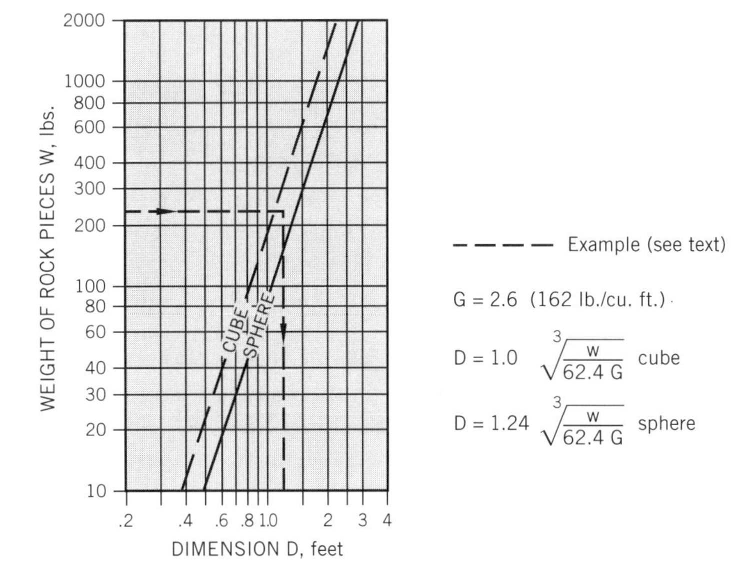

Figure 11.21 Conversion of rock weight to linear dimension (after U.S. Soil Conservation Service, 1978).

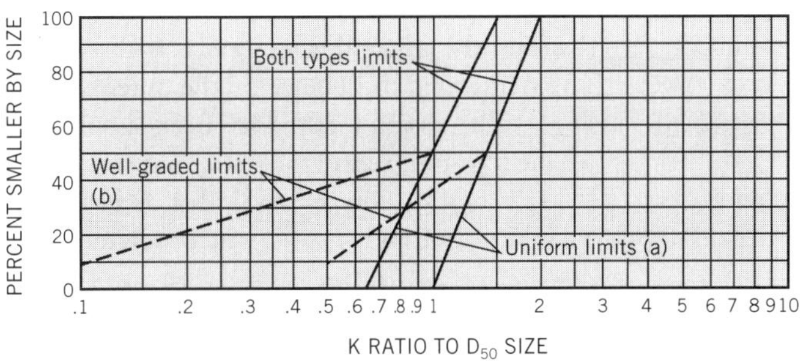

(a)
Example:
Uniform
D_{50} = 15 inches

Percent smaller	Sizes inches
100	22-30
85	20-27
50	15-21
15	11-16
0	10

(b)
Example:
Well-graded
D_{50} = 15 inches

Percent smaller	Sizes inches
100	22-30
50	15-21
10	$1\frac{1}{2}$-$7\frac{1}{2}$

Figure 11.22 Gradation templates for:
(a) uniform riprap
(b) well-graded riprap
(after U.S. Soil Conservation Service, 1978)

United States. The problem with uniform size riprap is that if one rock becomes displaced, the underlying filter layer is exposed and the erosion protection has been lost. Floating logs and ice have been implicated in these failures. In a well-graded filter, the movement of individual stones does not destroy the integrity of the riprap protection. The layer tends to remain intact and may actually become more stable.

Soil Conservation Service Design The W_{50} weight of the rock pieces can be converted to dimension D using the shape relationships shown on Figure 11.21 (Soil Conservation Service, 1978). For the example of the site at Sacramento, California, and assuming that the rock pieces are midway between a cube and sphere in shape, the D_{50} size is about 15 inches. This information is converted to gradation sizes using the templates of Figure 11.22.

The minimum thickness of a riprap layer should be twice D_{50} measured normal to the outer slope. In the example calculations, the minimum riprap thickness would be 30 inches.

Corps of Engineers Design The Corps of Engineers recommends the following gradation:

$$W_{max} = 4\,W_{50}; W_{min} = (0.125)\,W_{50}$$
$$T = 20\,(W_{50}/\gamma)^{1/3}$$

where T = thickness of riprap layer in inches
γ = density of stone material in lb./cu. ft.
W_{50} = median weight stone in lbs.

For the Sacramento site, the Corps design would be

W_{50} = 168 lb.
W_{max} = 672 lb.
W_{min} = 21 lb.
Thickness = 20 inches

The Corps also recommends the following: (i) minimum thickness should not be less than 12 inches, or W_{100}, or 1.5 W_{50} sizes; (ii) increase the calculated thickness by 50% when placing riprap under water to provide for the poorer control inherent in underwater construction; and (iii) increase the thickness by 6 to 12 inches if the riprap slope is likely to be hit by floating debris or large waves from boat wakes.

(8) Height of Riprap Protection

Practices vary on the amount of riprap that needs to be provided at a site, depending on the importance of the structure being protected, public safety, and other factors.

- *Maximum level.* This should be at least the maximum height of the lake or reservoir plus wind setup and maximum wave height. For the example used throughout this section, Table 11.6 gives the wind setup plus maximum wave runup for an impermeable slope to be 6.4 feet. The maximum lake level could be set at the 100-year flood level. In the case of embankment dams, the maximum reservoir level could be conservatively set for the probable maximum flood (PMF). This usually means several feet above spillway level. Allowing for wind setup and maximum wave runup, it is usually prudent to riprap the upstream slope of a dam up to the crest.
- *Minimum level.* This should be set at 1.5 H_S below the low water level of the body of water. The base of the riprap should be placed on a bench; this facilitates construction of the riprap layer and keys the base of the riprap layer.

Simplified Riprap Design for Waves (Shortcut Method)

The Soil Conservation Service (1978) has suggested a shortcut method of riprap design that does not require a study of design winds.

The shortcut design is:

$$D_{50} = 0.85\sqrt{F_e} \qquad \text{Eq. (9)}$$

where F_e = effective fetch, in miles
D_{50} = stone size in which 50% passing by weight through screen, in feet

The simplification is based on the following assumptions: (i) design wind 50 mph; (ii) G = 2.60; (iii) slope 2½ : 1 (H :V); and (iv) rock shape between cube and sphere (v) a few minor factors in the full procedure formulas.

Equation (9) is a condensation of the theoretical and empirical relationships that constitute the framework of the full procedures. It can be used to obtain quick, approximate, and generally conservative estimates of required rock size. It may be all that is needed for certain riprap designs.

Using the Sacramento lake as an example:

$$F_e = 2.43 \text{ miles}, D_{50} = 0.85\sqrt{2.43} = 1.33 \text{ feet (16 inches)}$$

This is slightly more conservative than the D_{50} of 15 inches calculated by the full procedure.

(B) Riprap Design for Stream Flow

The design of riprap for channels and stream flow depends on the channel alignment, cross-section, gradient and velocity distribution, in addition to the same requirements for rock quality as riprap designed to resist wave action. Riprap protection should be designed so that any flood that could be expected to occur during the service life of the project would not cause damage exceeding normal maintenance.

The design method presented here is based primarily on the Federal Highway Administration (FHWA) "Design of Riprap Revetment" 1989 publication with additional input from the U.S. Corps of Engineers (COE) manual EM 1110-2-1601 (1994).

Riprap Size

Riprap in channels has to resist hydrodynamic drag and lift forces created by the water velocities adjacent to the stone. The forces resisting motion are the submerged weight of the stone and any force components caused by contact with other stones in the revetment. Riprap at bends in the stream are designed conservatively for the point having the maximum force or velocity. Since local boundary shear is difficult to estimate, the design uses local velocity and local flow depth to quantify the stream forces. It assumes that the riprap thickness is at least D_{100} (i.e., the maximum size of stone within the riprap gradation).

The FHWA method uses a stability factor SF that not only provides a margin of safety of the resisting force over the destabilizing force, but also compensates for uncertainties in the design parameters and items such as ice or debris impacts against the riprap during service. It also corrects for the higher velocities and turbulence on the outer bank of river bends.

The design method has been simplified to an equation for the D_{50} stone size of the riprap and corrections for specific gravity G and stability factor SF that differ from the assumed values of G = 2.65 and SF = 1.20 incorporated into the basic formula.

$$D_{50} = \frac{V_a^3\, 10^{-3}}{\sqrt{d_{avg}}\,\sqrt{K_1^3}} \qquad \text{Eq. (10)}$$

where D_{50} = median riprap stone size, feet
V_a = average velocity in the main channel, feet/sec
d_{avg} = average flow depth in the main channel, feet
K_1 = factor related to design shear on the slope

The main channel is the area through which the river is flowing, excluding flood plains.

$$K_1 = \sqrt{1 - \left(\frac{\sin^2 \beta}{\sin^2 \phi}\right)} \qquad \text{Eq. (11)}$$

where β = slope of the riverbank to the horizontal
ϕ = angle of repose of the riprap (usually assumed $\phi = 40°$)

Equation (10) can be solved by the nomograph of Figure 11.23. Corrections for variations from the assumptions used to create Equation (10) are given below:

$$C_1 = \frac{2.12}{\sqrt{(G-1)^3}} \qquad \text{Eq. (12)}$$

where C_1 = correction for different specific gravity
G = specific gravity of the stone (different from G = 2.65)

$$C_2 = \sqrt{\left(\frac{SF}{1.2}\right)^3} \qquad \text{Eq. (13)}$$

where C_2 = correction for different stability factor
SF = stability factor selected (different from SF = 1.20)

Correction factors C_1, C_2, if required, can be multiplied together to arrive at a single factor $C = C_1\, C_2$ which is then multiplied by the riprap size computed from Equation (10) or Figure 11.23.

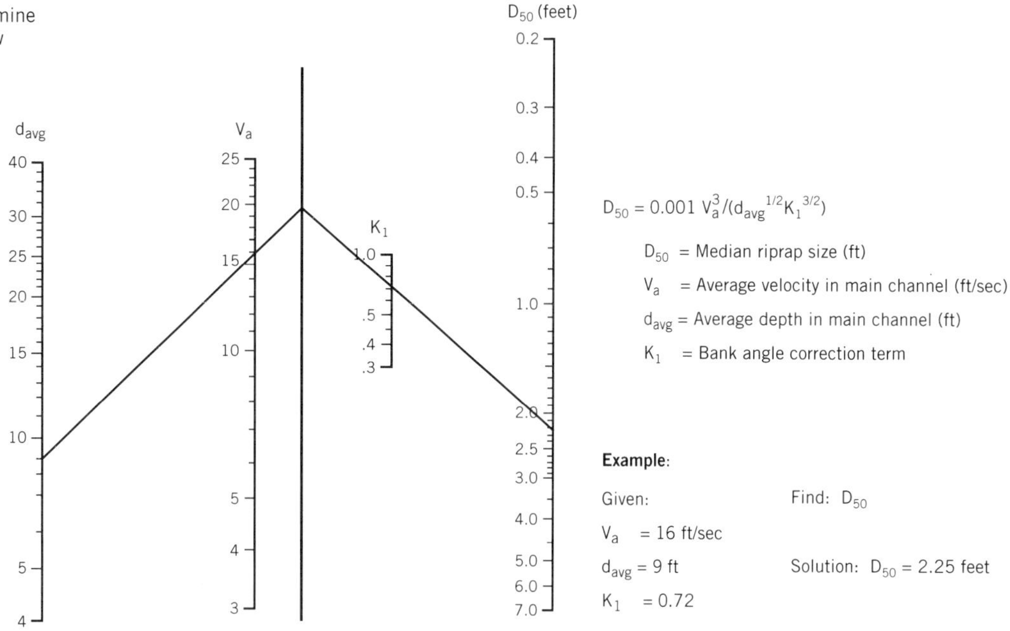

Figure 11.23 Nomograph to determine the D_{50} riprap size from stream flow information (after Federal Highway Administration, 1989).

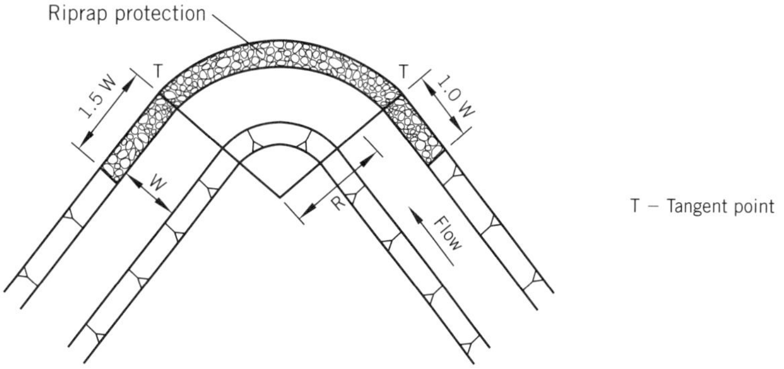

Figure 11.24 Extent of revetment protection (after Federal Highway Administration, 1989).

The selection of the stability factor is up to the judgment of the design engineer. The FHWA guidelines are given in Table 11.7.

The R/W ratio is a measure of stream curvature. R is the radius of the curve at the centerline of the stream and W is the width of water surface. The designer selects these values based on a plan, using the maximum curvature to design riprap for the outer bend where erosion protection is most needed. Figure 11.24 shows the recommended length of the protection. It is important to extend the riprap past the two ends of the bend, especially at the downstream end. Many site-specific factors have a bearing on the actual length of bank that should be protected. Figure 11.24 is a starting point and the actual treatment should take account of local factors, such as the erosion problem area, intrusions into the stream (such as bridge abutments) etc.

Equation (10) is intended for direct use on straight or mild curves in channels of uniform cross-section. The FHWA guidelines indicate that the equation can be used for other common situations by adjusting the velocity and/or stability factor as described below.

River Bends. See Table 11.7.

Table 11.7 Guidelines for the Selection of Stability Factors

Condition	Stability Factor Range
Uniform flow. Straight or mildly curving reach (curve radius/channel width > 30); Impact from wave action and floating debris is minimal; Little or no uncertainty in design parameters.	1.0–1.2
Gradually varying flow. Moderate bend curvature (30 > curve radius/channel width > 10); Impact from waves or floating debris moderate.	1.3–1.6
Approaching rapidly varying flow. Sharp bend curvature (10 > curve radius/channel width); Significant impact potential from floating debris and/or ice; Significant wind and/or boat generated waves (1–2 ft); High flow turbulence; Turbulently mixing flow at bridge abutments; Significant uncertainty in design parameters.	1.6–2.0

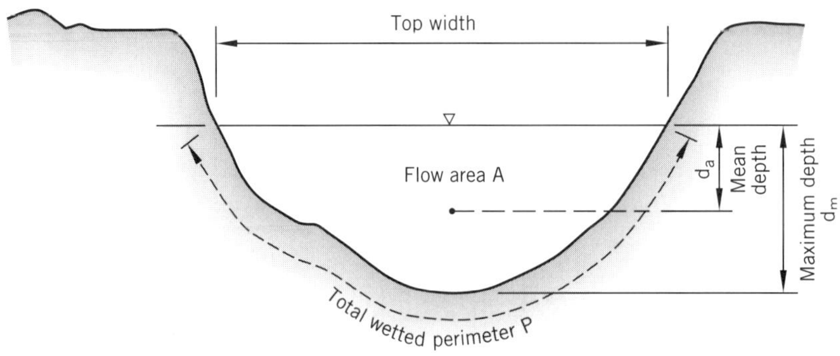

Figure 11.25 Terms used in river channel hydraulics.

Steep Channel Slopes. Flow through steep gradients are characterized by high velocities and turbulence. Both the velocity and stability factor (Table 11.7) have to be carefully evaluated.

It is recommended that the value of Manning's n coefficient (Arcement and Schneider, 1984) be calculated from the formula:

$$n = 0.39\, S_f^{0.38}\, R^{-0.16} \qquad \text{Eq. (14)}$$

where S_f = friction slope
R = hydraulic radius; i.e., flow area A divided by wetted perimeter P (Figure 11.25)

Ice Damage. Where ice flows have historically caused problems, a stability factor of 1.2 to 1.5 should be used to increase the design rock size. However, local experience is probably the best guide. In most instances, the design rock size is sufficient to withstand ice impact and crushing forces.

Wave Damage. See earlier design analysis.

Abutments, Bridge Piers. Not covered here because it rarely involves landslide stabilization.

Gradation and Thickness of Riprap

The recommended gradation for river protection is shown on Table 11.8. It is very similar to the Corps of Engineers gradation, requiring a fairly well-graded riprap. It can be specified by stone size or weight.

Table 11.8 Gradation for Riverbank Protection*

Percent Smaller Than	Stone Size (feet)	Stone Weight (lb)
100	1.5 to 1.7 D_{50}	3.0 to 5.0 W_{50}
85	1.2 to 1.4 D_{50}	2.0 to 2.75 W_{50}
50	1.0 to 1.15 D_{50}	1.0 to 1.5 W_{50}
15	0.4 to 0.6 D_{50}	0.1 to 0.2 W_{50}

D_{50}—stone size corresponding to 50 percentile of material by weight
W_{50}—stone weight corresponding to 50 percentile of material gradation

*FHWA, 1989

The thickness criteria are the same as previously described for riprap protecting against wave action.

Base of Riprap Slope

The toe of the riprap is subject to greater erosive forces than other areas of the slope. In most situations, the riprap should extend to the river bottom plus an allowance for future scour. The bottom of the river should be assumed to be the deepest point of the cross-section, irrespective of whether it occurs at the base of the slope to be protected. (It could move there in the future.)

In the past, riprap has been extended 5 feet below the deepest point of a river to allow for scour. Research by Blodgett (1986) at 21 sites on nine different rivers found that the maximum depth of scour ranged from 9 feet in gravel beds to 13 feet in fine sand beds. The maximum depth can be estimated from:

$$d_S = \frac{6.5}{D_{50}^{0.115}} \text{ for } D_{50} > 0.005 \text{ feet (1.5 mm)} \qquad \text{Eq. (15)}$$

where d_S = estimated maximum depth of scour in feet
D_{50} = median grain size of the streambed material in feet

The value of d_S has to be added to the deepest point elevation of the river to obtain the total depth to the base of the riprap slope.

In practice, these scour depths are extreme, and some discretion to set the base of the riprap between 5 feet and the depth calculated from Equation (15) should be used, depending on the site circumstances.

Available Design Procedures

In the United States, rock riprap design procedures have been produced by:

U.S. Corps of Engineers	EM-1601
U.S. Bureau of Reclamation	USBR-EM-25
Federal Highway Administration	HEC-11, HEC-15
California Dept. of Transportation	Cal-B and SP (bank and shore protection)
American Society of Civil Engineers	Manual 54

11.3 FABRICS

Geotextile fabrics are frequently used as an alternative to a soil fine filter in filter systems. Some of the advantages and disadvantages are listed below:

Advantages

1. Easy to install on prepared flat surfaces.
2. Less expensive than a soil filter, especially when the filter has to be specially manufactured and hauled from a different source than other earthwork materials.
3. Material quality of fabric can be specified and reliably obtained.
4. Fabric has tensile strength (soils do not).

Disadvantages

1. Difficult to lay fabric on uneven surfaces or below water.
2. Lingering concerns about long-term life, especially in situations where it cannot easily be repaired or replaced.
3. Other concerns: (i) bacterial activity could affect the permeability, (ii) exposure to sunlight on site can damage the fabric, and (iii) cohesionless soils may move downslope beneath the fabric from wave action.
4. May provide a weak layer behind riprap on steeper slopes that could cause slope instability.
5. Safety concerns for personnel in placing fabric at landslide sites.

The first and last two items on the disadvantages list are of concern when repairing erosion bowls left behind by flow slides. The headscarps typically are vertical at the top and very steep immediately below the top. The fabric usually has to be staked at the top, and there is some risk to personnel in this situation, particularly for cohesionless soils. However, a backhoe can shape the upper slope to a flatter angle before fabric is laid.

Fabrics, like the fine filter layer they replace, have a dual function: drainage and filtration. They must allow water to pass through but retain soil. Therefore, the cloth must prevent piping and clogging, and have sufficient tensile strength to

Table 11.9 Geotextile Properties*

Geotextile Property	Test Method	Minimum Values					
		Drainage Geotextile		**Riprap Geotextile**		**Subgrade Geotextile**	Embankment Geotextile
		Type 1	Type 2	Type 1	Type 2		
Grab tensile strength minimum in each principal direction	ASTM D 4632	80 lb. (355 N)	180 lb. (800 N)	200 lb. (890 N)	**260 lb.** (1155 N)	**180 lb.** (800 N)	230 lb. (1025 N)
Grab elongation	ASTM D 4632	15%			**15%**	—	
Burst strength, diaphragm method	ASTM D 3786	130 psi (900 kPa)	290 psi (2000 kPa)	320 psi (2200 kPa)	**435 psi** (3000 kPa)	**290 psi** (2000 kPa)	435 psi (3000 kPa)
Puncture strength	ASTM-D 4833 or ASTM D 3787	35 lb. (155 N)	80 lb. (355N)	80 lb. (355N)	**110 lb.** (490 N)	**80 lb.** (355 N)	110 lb. (490 N)
Apparent opening size (AOS), U.S. standard sieve	ASTM D 4751	**212 μm or smaller opening**				**600 μm sieve or smaller opening**	
Permittivity	ASTM D 4491	**0.5 s^{-1}**				**0.01 s^{-1}**	
Ultraviolet stability	ASTM D 4355 at 500 hours	—		**70% strength retained**		—	—
Manufacturers: examples of many available**	Contech SI Geosolutions Æ TNS Adv. Tech Æ	C-40 NW Geotex 311 E 040	C-70 NW Geotex 311 E 070	C-80 NW Geotex 801 E 070	**C-120 NW** **Geotex 1201** **E 100**	**C-200** **Geotex 200** **STW-200**	C-120 NW Geotex 1201 E 100

Most common geotextile specifications used in landslide work are shown in **bold.**

**Source:* Standard Specifications for Highway Construction 1996 Oregon Dept. of Transportation

***Source:* Geotechnical Fabrics Report "Specifier's Guide 2001"

avoid being torn by construction materials placed on it and stretching from settlement in service. Quality control of placement is essential.

Fabric is manufactured in widths of 12½ to 15 feet. It is laid from top to bottom of the slope. The edge overlaps range from 1 to 3 feet (2 feet is a good compromise for most projects). Securing stakes are placed at 2 to 5 foot intervals at the mid-width of the overlap. *Materials placed on top of the fabric should be placed from bottom to top of the slope* and some slack should be provided to help the fabric to extend a little in the event of ground settlement below. Dropping materials onto the fabric should be avoided.

A lighter fabric is used for fine-grained soils, sand, and gravel; a heavier fabric is needed for rockfill and riprap applications. Table 11.9 lists the specification requirements in bold lettering for the two types of fabric. There are many manufacturers capable of providing fabrics that meet these specifications.

CHAPTER

12

Earthquake-Induced Landslides

Earthquake engineering is a specialist subcategory of civil and geotechnical engineering. It has experienced much research over the past 30 years, and standards of practice are still evolving. The intent of this chapter is to present a general understanding of liquefaction analysis and a method for estimating the probable surface displacements of preexisting landslides during a strong earthquake. Landslides caused by earthquakes are discussed in Chapter 2, Section 2.16.

12.1 LIQUEFACTION ANALYSIS

Basic Concepts

If loose, saturated cohesionless soils are subjected to earthquake vibrations, the soil contracts and develops positive pore-water pressures unless drainage occurs rapidly. Should the pore pressure rise to the level of the overburden stress, the vertical effective stress becomes zero. The soil loses most of its static strength and significant deformations occur. When such deformations are large, the soil is considered *liquefied*. For more limited deformations, the behavior is often termed *cyclic mobility*. When the cyclic motion of the earthquake stops, there will be an excess pore pressure remaining in the soil that can spread laterally as it dissipates and cause detrimental consequences to adjoining layers. Much of the damage caused by liquefaction occurs shortly after the earthquake shaking has finished. The near-disaster of the Lower San Fernando Dam in 1971 (Seed and Harder, 1990) is a good example of this effect.

Large earthquakes produce numerous strong cyclic stresses in the soil. A liquefaction design analysis compares the resistance of the soil to cyclic stresses to the input cyclic stresses of the design earthquake. For this purpose, earthquake engineers use a "normalized" cyclic loading termed the *cyclic stress ratio* CSR defined by:

$$CSR = \frac{\tau_o}{\sigma_o'} \qquad \text{Eq. (1)}$$

where τ_o = shear stress at the depth in the ground
σ'_o = effective vertical overburden pressure at the same depth

Liquefaction occurs when the pore pressure u equals the vertical soil pressure σ_v. If these parameters are combined as a ratio r_u

$$r_u = \frac{u}{\sigma_v} = \left[\frac{\Delta u}{\sigma_v'}\right] \qquad \text{Eq. (2)}$$

and r_u = 100% at initial liquefaction

The factor of safety F against liquefaction can be defined:

$$F = \frac{\text{CSR causing } r_u = 100\%}{\text{CSR induced by the earthquake}} = \frac{(CSR)_l}{(CSR)_e} \qquad \text{Eq. (3)}$$

The cyclic stress ratio for liquefaction is also termed the *cyclic resistance ratio* (CRR).

Soil Resistance to Liquefaction

The most widely used method for evaluating the liquefaction resistance of cohesionless soils under level ground conditions is on the basis of field experience with soils of similar composition and penetration resistance. The original relationship (Seed et al., 1985) has undergone slight modifications from time to time, most recently by Youd and Idriss (2001), as shown on Figure 12.1. To the left of the curves are points representing liquefaction or marginal liquefaction. To the right, the points represent no liquefaction. The curves represent the boundary between soils that liquefy from those that do not liquefy.

The fines content is an important property. Soils with higher amounts of fines (passing No. 200 U.S. sieve) are less prone to liquefaction than cleaner soils.

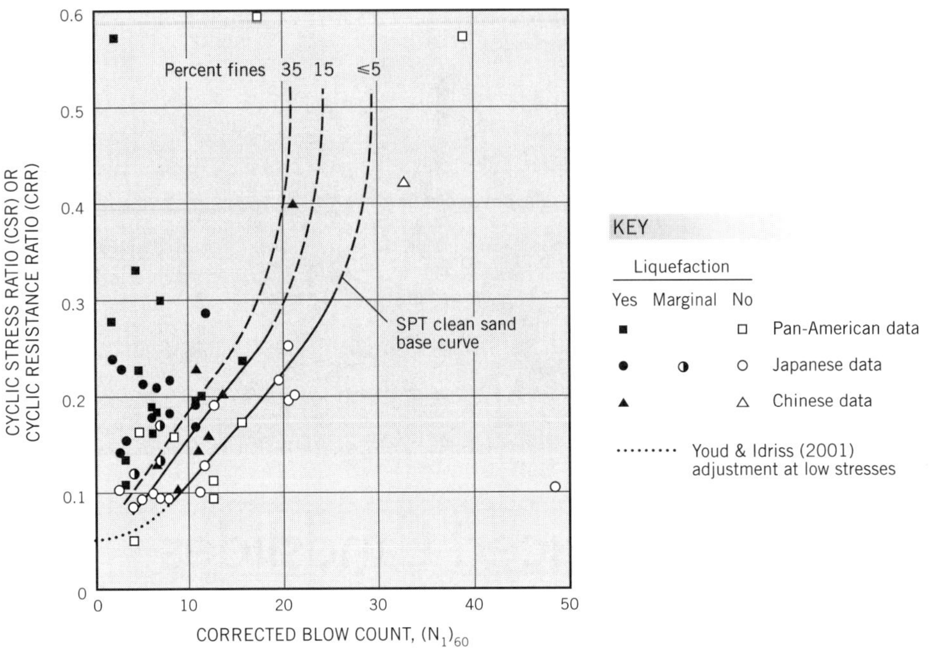

Figure 12.1 Relationship between cyclic stress ratio and corrected blow count $(N_1)_{60}$ for a magnitude M7.5 earthquake (after Seed et al., 1985).

The abscissa of the chart is the corrected SPT blow count $(N_1)_{60}$. These corrections are described in Chapter 3, Section 3.8.

The relationships shown on Figure 12.1 are applicable to soil on level ground (slope angle $\alpha = 0$) for earthquakes with magnitudes of approximately 7.5 and an effective overburden pressure of 1 ton/sq. ft. For other site conditions, the average cyclic shear stress required to cause liquefaction ($r_u = 100\%$) in the soil has to be adjusted as follows:

$$\tau_{av} = \left[\frac{\tau_{av}}{\sigma_0'}\right]_{\alpha=0} K_\alpha \cdot K_\sigma \cdot \sigma_0' \cdot C_M \qquad \text{Eq. (4)}$$

where $\left[\frac{\tau_{av}}{\sigma_0'}\right]_{\alpha=0}$ is the CSR obtained from Figure 12.1. The other terms in Eq. (4) are explained below:

K_α

This correction is based on the understanding that a preexisting static shear stress on a horizontal plane is beneficial to liquefaction resistance in a soil. The term α is the ratio of static driving shear stress on a horizontal plane to the initial effective overburden stress. Parameter α can be obtained by use of the computer program FEADAM (Duncan et al., 1984) for slopes and dams. However, current published relationships between K_α and α have shown a lack of similarity between different research studies and there is a need for additional research on this topic. The National Center for Earthquake Engineering Research (Youd and Idriss, 2001) has recommended that the published relationships not be used in routine engineering practice or by nonspecialists in geotechnical earthquake engineering. Therefore, a value of $K_\alpha = 1$ should be used until better guidance is provided by earthquake engineering specialists.

K_σ

Cyclically loaded laboratory data indicate that liquefaction resistance increases with increasing confining stress. The most recent values of this correction (Hynes and Olsen, 1999) are given on Figure 12.2. These values are considered minimal (conservative) for use on both clean and silty sands, and for gravels.

C_M

This factor is based on the knowledge that larger magnitude earthquakes tend to produce a longer duration of shaking and thus more cycles of loading. Seed et al., (1975, 1982) provide procedures for converting an irregular cyclic load history to an "equivalent" number of uniform loading cycles with an amplitude equal to 65% of the maximum amplitude of the irregular load history. After processing many recorded earthquake time histories using these techniques, the typical numbers of equivalent uniform loading cycles for earthquakes of different magnitudes are listed on Table 12.1.

A recent change in the Magnitude Correction Factor C_M has been agreed upon by a group of research workers (Youd and Idriss, 2001). These are shown in Column 3 of Table 12.1.

An important factor in the resistance of sand to liquefaction is the age of the deposit. Seed (1979) observed increases of up to 25% in the laboratory cyclic resistance ratio of 100-

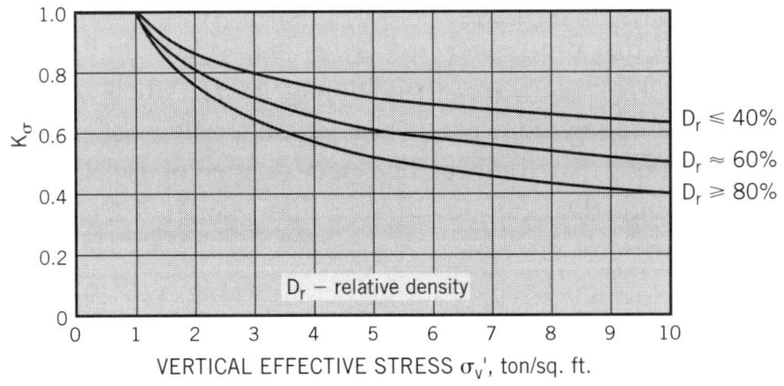

Figure 12.2 Correction factor K_σ (after Hynes and Olsen, 1999).

Table 12.1 Relationship of Resistance Factor C_M, Number of Representative Cycles, and Earthquake Magnitude

Earthquake Magnitude M	Number of Representative Cycles at 0.65 a_{max}*	Magnitude or Duration Correction Factor C_M**
5.5		2.2–2.8
6.0	5–6	1.76–2.1
6.5		1.44–1.6
7.0		1.19–1.25
7.5	15	1.00
8.0		0.84
8.5	26	0.72

*Seed et al., 1975, 1982
**design engineer to choose C_M from within given range (see Youd and Idriss, 2001)

day-old reconstituted sand specimens compared to freshly prepared specimens. According to the observations of Youd and Idriss (2001): "Sediments deposited within the past few thousand years are generally much more susceptible to liquefaction than older Holocene sediments; Pleistocene sediments are even more resistant; and pre-Pleistocene sediments are generally immune to liquefaction."

No corrections have been developed for "aged" sand and the phenomenon is still poorly understood within the profession. For recent fills, aging effects are minimal, and no allowance should be made for age in calculating liquefaction resistance. For older deposits, engineering judgment must be used.

Cyclic Stress Induced by Earthquakes

The cyclic stresses produced by an earthquake at a site require an evaluation of the design earthquake and the distance of the epicenter from the site. These are usually provided by specialists in seismology. The results provide the peak ground accelerations on bedrock. From bedrock, these accelerations may increase, especially in thick deposits of soft soils (Idriss, 1991). However, for routine studies in stiff soils, it can be a fairly insignificant adjustment. A ground response analysis, such as the computer programs SHAKE (Schnabel et al., 1972; Idriss and Sun, 1992), FLAC, (Itasca Consulting Group), or DESRA (Lee and Finn, 1978), can be used to obtain the peak ground acceleration a_{max} in the sediment under study.

The estimation of the cyclic shear stresses can be made with a reasonable degree of accuracy by using the simplified procedure recommended by Seed and Idriss (1971). The maximum shear stress is given by:

$$\tau_{max} = \frac{\gamma h}{g} \cdot a_{max} \cdot r_d$$

where γ = soil density
h = depth of soil element
a_{max} = peak ground surface acceleration
r_d = factor varying from 1.0 at the ground surface to 0.9 at a depth of 30 feet (use a linear change with depth)

As previously stated, experience has shown that the average uniform cyclic shear stress τ_{av} is about 65% of the maximum, τ_{max}.

Therefore, the average cyclic shear stress is given by $\tau_{av} = 0.65\,\frac{\gamma h}{g} \bullet a_{max} \bullet r_d$

Factor of Safety against Liquefaction

The factor of safety against liquefaction can be calculated for any point in the vulnerable stratum by dividing the soil resistance by the average cyclic stress induced by the earthquake. To allow for the uncertainties in the Seed and Idriss simplified procedure the results can be interpreted as follows:

FACTOR OF SAFETY AGAINST LIQUEFACTION	INDUCED PORE-WATER PRESSURE RATIO r_u %
<1.1	100
>1.3	≈0

(under more precise evaluations, F of 1.00 would correspond to r_u = 100%)

The data can be reviewed to note any areas with F close to 1.00, and the spatial relationships around such points. The geotechnical consultant then can decide on the need for preventative compaction or other ground improvement at the site. Areas with low F values may be surrounded (confined) by denser soils in which case a liquefaction-caused landslide is unlikely to develop.

Residual Strength after Liquefaction

In slopes that have failed during earthquakes, back analysis studies have shown that a liquefied soil retains a low shear strength, currently (and unfortunately) termed the residual strength. Obviously, it should not be confused with the residual strength of clays.

Seed (1986) developed a technique for estimating the residual undrained strengths for post-liquefaction stability calculations. It uses another "corrected" penetration resistance $(N_1)_{60\text{-}cs}$, which provides an equivalent clean sand blow count.

$$(N_1)_{60\text{-}cs} = (N_1)_{60} + N_{corr}$$

where N_{corr} = 1 for 10% fines; 2 for 25% fines; 4 for 50% fines; 5 for 75% fines

An updated relationship (Seed and Harder, 1990) is shown on Figure 12.3. Seed and Harder suggest that the lower bound or near lower bound relationships be used for residual undrained strength analyses due to the limited number of case studies and other uncertainties.

Baziar and Dobry (1995) reexamined 20 case histories (originally compiled by Stark and Mesri, 1992) that had experienced large permanent horizontal surface ground movements (D_H) from earthquake-induced lateral spreading. The earthquakes had moment magnitudes of 8.0 or less. All silty sands with fine contents of 10% or more were included except for three sites (two of which involved gravelly sands).

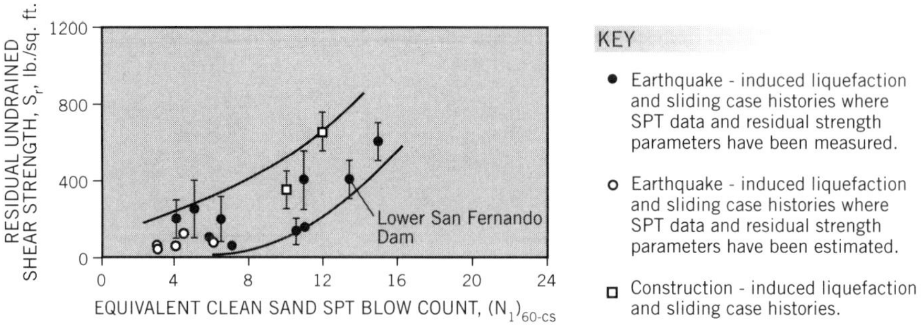

Figure 12.3 Undrained residual strength S_r based on case histories (after Seed and Harder, 1990).

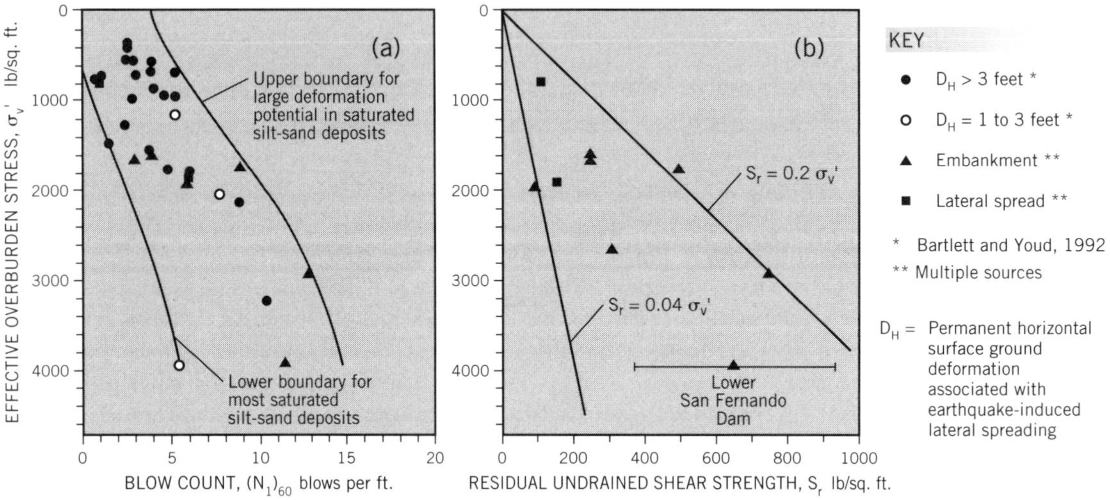

Figure 12.4 Saturated silt-sand deposits that have undergone large deformations:
(a) $(N_1)_{60}$ blow counts
(b) residual shear strength S_r
(after Baziar and Dobry, 1995)

The first graph from this data, Figure 12.4(a) suggests that there is an upper boundary limiting the likelihood of large deformations in these soils, depending on the $(N_1)_{60}$ blow count and the effective overburden stress. The second graph, Figure 12.4(b), indicates that the residual strength s_r is a function of the effective overburden stress σ_v' and the s_r/σ_v' ratio ranges from 0.04 to 0.2.

12.2 PSEUDO-STATIC ANALYSIS

For soils that do not lose significant strength under seismic loading, instability can occur due to an inertia force from the earthquake. It can be modeled by a horizontal force kW where k is termed the *seismic coefficient* (Figure 12.5a). This allows a dynamic shaking event to be analyzed by a *pseudo-static* stability analysis.

Seed (1979) has recommended the following values of k for embankments constructed of soils that do not build up large pore pressures nor show more than 15% strength loss due to earthquake shaking (usually clays, silty clays, sandy clays, or very dense cohesionless soils):

EARTHQUAKE MAGNITUDE	SEISMIC COEFFICIENT k	MINIMUM F
6½	0.10	1.15
8¼	0.15	1.15

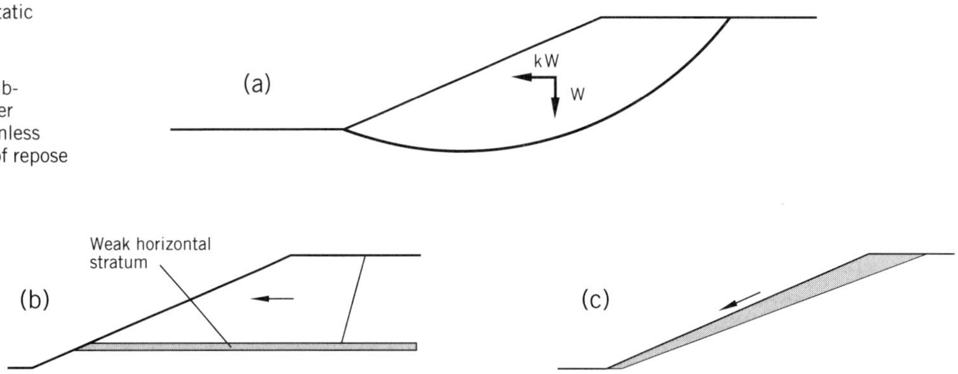

Figure 12.5 Pseudo-static stability analysis:
(a) coefficient k
(b) failure through a sub-horizontal weak layer
(c) failure of a cohesionless slope at the angle of repose

Although the seismic coefficient is an acceleration and is frequently so presented (e.g., 0.1g), it should not be confused with the particle acceleration a propagated through the ground by an earthquake. Many instances have occurred in practice in which engineers mistakenly refer to coefficient k as a substitute for a. In pseudo-static stability analyses, coefficient k is an artificial device to aid analysis of nonliquefying soils. It is usually one-half or less than the maximum particle acceleration a_{max}, which is a property of the wave form emanating from an actual or idealized earthquake.

Landslides that occur in nonliquefying soils are commonly of the types shown on Figure 12.5(b) and (c). The slip surface passes along a horizontal or near-horizontal weak layer, often encountered in sedimentary soils or rocks (e.g., flood plain soils). The horizontal seismic force increases the horizontal shear stresses to well above the levels experienced under normal static loading conditions. There is some indication that the distance from the slope face to the headscarp of the instability is a function of the seismic wave length, but no serious study has been made of this mechanism.

The second type of failure, Figure 12.5(c), involves steep slopes of granular soils at or near the angle of repose. Since the soil is at the margin of stability on the outer slope, severe seismic shaking causes shallow raveling over the outer surface.

12.3 DISPLACEMENT OF MARGINALLY STABLE SLOPES

The downslope movement to be expected on marginally stable slopes during a strong earthquake is an issue that is being studied with increasing frequency in earthquake-prone areas. The observational method—how much landslides have moved during past earthquakes—is probably the most reliable guide at present. Techniques are being developed for a rational design approach (one method is described in this chapter), and further developments and refinements can be expected in the future.

Observational Data

The Magnitude 7.1 Loma Prieta earthquake struck the San Francisco Bay area on October 17, 1989. Shortly after the earthquake, engineers and earth scientists performed a reconnaissance of landslide damage caused by the earthquake. These observations were summarized by the California Division of Mines and Geology (Manson, et al., 1992). Approximately 50 landslides were documented, and seismic displacements were estimated based on ground fractures at the headscarp for each slide; of these, 12 slides were active prior to the earthquake.

The observed movements are plotted against the distance from the epicenter on Figure 12.6. The slides have not been differentiated according to soil type, size, or slide mechanism and there is significant scatter. However, the expected trend toward less displacement at increasing distance is apparent; more importantly, the data provide good data on the likely *range* of movements at a particular distance from the epicenter.

On the extreme right of the graph is the Penetencia Water Treatment Plant (PWTP), 24 miles (39 km) from the epicenter. Inclinometers at this site recorded movements of 0.2 to 0.75 inch during the Loma Prieta earthquake. This slide has been subjected to two other earthquakes: the 1984 Morgan Hill (M6.2), and the 1986 Mount Lewis (M5.7), with movements ranging from 0.5 to 0.75 inch during each earthquake (Salah-Mars, et al., 1995). The latter two earthquakes had epicenters about 11 miles (18 km) from the treatment plant.

Other evidence of relatively small movements of landslides during strong earthquakes include data from the 1949 M7.1

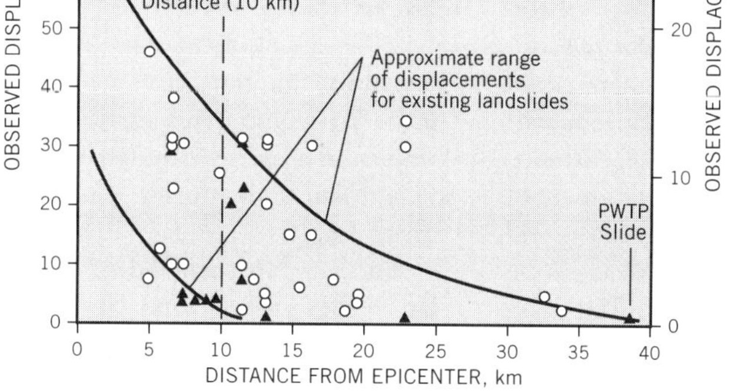

Figure 12.6 Observed ground displacements due to the magnitude M7.1 Loma Prieta earthquake.

KEY

○ First-time landslide

▲ Existing landslide

PWTP Penetencia Water Treatment Plant (see text)

Note: Slides have not been differentiated according to soil type, size or slide mechanism

Figure 12.7 Newmark analysis:
(a) sliding block analogy
(b) acceleration–velocity–displacement curves

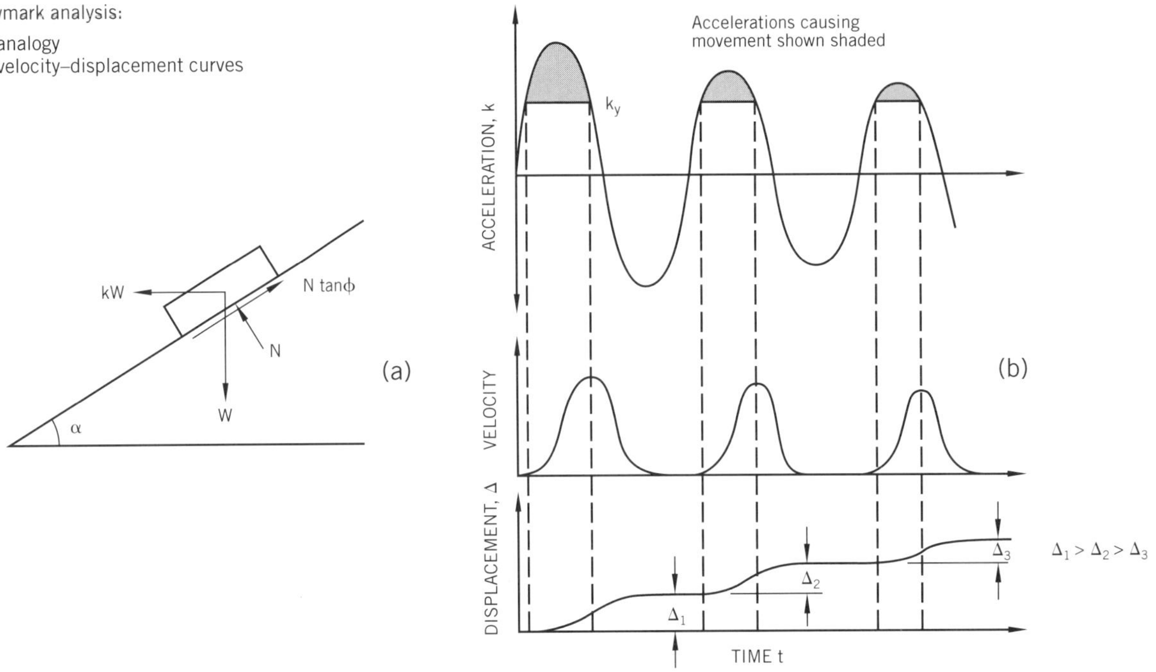

Olympia, and 1965 M6.5 Seattle earthquakes in Washington State (Chleborad, 1994), the 1976 M7.5 earthquake in Guatemala (Harp et al., 1981), and the 1991 M7.0 Racha earthquake in the Republic of Georgia (Jibson et al., 1994).

Analysis

Newmark (1965) proposed a method for calculating the displacement of earth dams and slopes in response to earthquake shaking. It is based on the concept of a block on an inclined plane that is subjected to a sinusoidal wave and calculates the acceleration k_y needed to exceed static equilibrium (Figure 12.7). When the ground acceleration of the seismic wave exceeds k_y, the ground moves. At other times within the cycle, including the opposite direction of the wave, the ground remains stationary. By examining or re-creating the significant representative cycles of ground motion (see Table 12.1) the total displacement during an earthquake can be calculated by integration. It does not apply to soils that liquefy or lose a significant portion of their static strength under seismic loading.

This technique has been applied to the slopes of earth dams to estimate their possible movement during an earthquake. In recent years, the Newmark approach has been extended to include natural slopes and landslides (Jibson, 1993). However, calculations for many slopes that have experienced a major earthquake show that the calculated movements are much larger than has been observed. For example, the PWTP slide (Figure 12.6) has a calculated displacement of 27 inches when using a Newmark analysis with a low yield acceleration k_y of 0.001 and a scaled Loma Prieta acceleration time history. This can be compared to the actual displacement, measured by inclinometers, of 0.5 to 0.75 inch. Similarly, the broad band of data for the Loma Prieta landslide (Figure 12.6) indicates that movements in the range of 1.5 to 12 inches occurred at an epicentral distance of 6.3 miles (10 km). A Newmark analysis, assuming $k_y = 0.001$, gives an estimated movement of 90 inches, well above the observed range.

An uncertainty is the actual factor of safety of these marginally stable slopes at the time of the earthquake. If the yield accelerations k_y are greater than 0.001, the Newmark analysis will provide lower estimated movements.

A probable explanation of the apparently poor correlation between observed and calculated ground movements is that the Newmark analysis assumes that the residual strength along the slip surface is the same as that measured in the laboratory by shear box or ring shear tests. Research data indicate that the residual strength of clay increases at the rate of about 10% per log cycle of shearing velocity (Kulhawy and Mayne, 1990). The maximum velocity in an earthquake is about 7 to 8½ orders of magnitude higher than in the laboratory direct shear tests. Therefore, assuming a 10% per cycle of velocity increase, the actual strength would be 70% to 85% higher. This increases k_y applicable to an earthquake, and provides an estimate of landslide movement that fits more closely with observed values.

Example Calculation

At the Washington Park Station Slide (Case History 4), block samples of clay were obtained from the shear zone of an ancient landslide during a shaft excavation (Vessely and Cornforth, 1998).

The samples consisted of stiff to very stiff, mottled silty clay with scattered sand-sized nodules. The samples were highly slickensided.

Grain size distribution for one of the specimens gave 70 percent passing the No. 200 sieve, and 30 percent passing 0.002 mm in the hydrometer test. The soil properties were LL 85, PL 59, w 45%.

To prepare ring shear test specimens, a steel "cookie cutter" template was made to the shape of one quarter of a ring shear specimen. Four clay pieces were cut and combined in the apparatus to form the annular specimen. After consolidating the specimen to the estimated effective stress in the field, shearing was commenced at a rate of 0.005 inch/min to measure the residual strength at a "slow" rate of shear. Once residual strength was obtained, the shearing was stopped, the torque was removed, and the specimen was allowed to rest for a period ranging from one hour to several days. Shearing was resumed at a higher rate until the strength became relatively constant. Following each stage of shearing at progressively higher rates, the specimen was returned to the "slow" shear rate.

A cumulative shear stress-horizontal displacement plot for one specimen is shown on Figure 12.8. The selected residual stress of each shearing rate is shown by broken lines. At a shear rate of 1.78 inch/min, for example, the shear stress drops to 0.40 ton/sq. ft. on three occasions. These low points were taken as the residual stress for this strain rate. The graph of residual shear stress vs. shear rate (Figure 12.9) shows that the rate of increase in strength averages 24% per cycle of shear rate on a logarithmic scale, which is well above the 10% rate of increase measured by others.

A back analysis of the slide showed that the average residual strength on the ancient landslide terrain is around 7 degrees. Inclinometers measured intermittent (winter) movements averaging 0.08 inch per year. If this movement occurs over a 3-day period, due to a high intensity winter storm, the velocity during shear is 3×10^{-7} inch/second. The velocity during a M6.5 earthquake with an epicentral distance of 6 miles (10 km) probably averages about 8 inch/sec. Therefore, the earthquake velocity is 7.3 orders of magnitude greater than the landslide velocity. Accepting the 10% per log cycle increase in strength rather than the higher 24% per log cycle measured in the laboratory ring shear tests, the residual strength ϕ'_r increases from 7° to 12°. Therefore, the yield acceleration k_y is 0.12 instead of 0.001, and the estimated slope movement for a postulated M6.9 earthquake located 1 km from the site is 20 inches. This estimate is in reasonable agreement with the observed data at the M7.1 Loma Prieta earthquake (Figure 12.6).

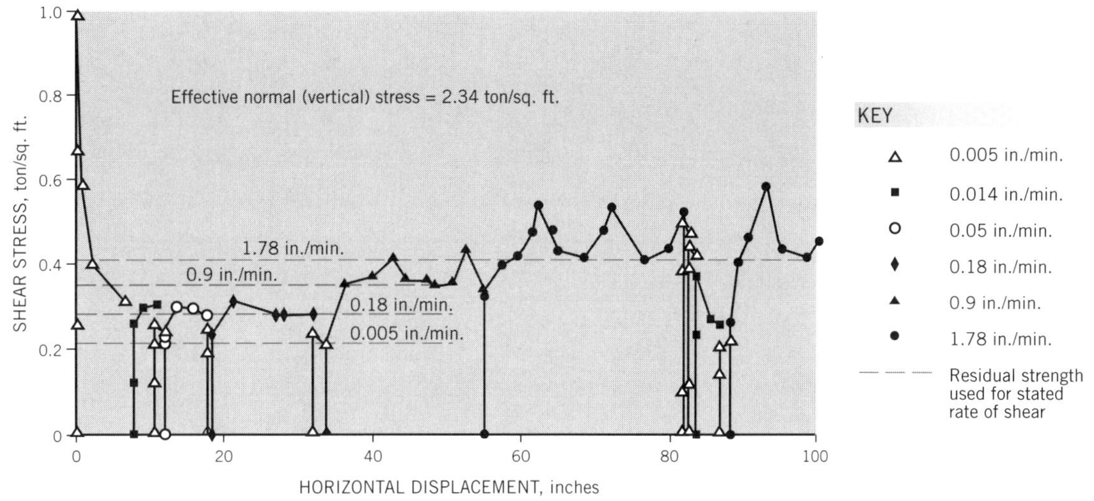

Figure 12.8 Residual strength tests at various shear velocities: test specimen from Washington Park Station Slide, Portland, Oregon (after Vessely and Cornforth, 1998).

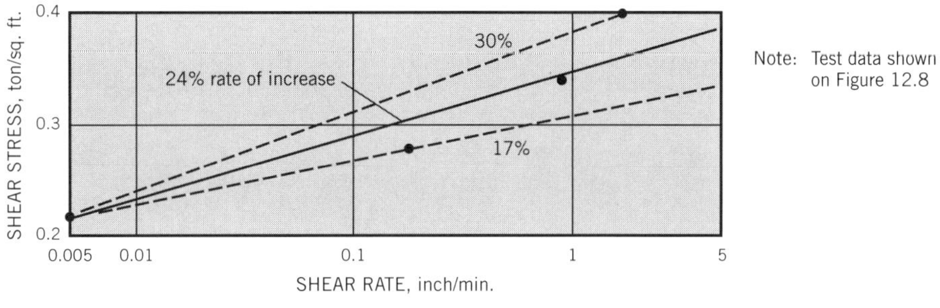

Figure 12.9 Increase in residual strength with increasing shear velocity: test specimen from Washington Park Station Slide, Portland, Oregon (after Vessely and Cornforth, 1998).

Pelton Park Slide

Although a modified Newmark analysis seems to be a reasonable way of estimating the likely movements of dormant landslides during a major earthquake, the observed behavior of Pelton Park Slide (Case History 6) provides a significant anomaly. This large landslide, adjoining the reservoirs behind Pelton Dam in arid central Oregon, has a static factor of safety that is very close to 1.00 under normal circumstance (see case history for more details). When a M5.6 crustal earthquake occurred 72 miles away in May, 1993, Pelton Park Slide began to move again at the rate of 1 inch per week and *continued moving for more than 5 months*. The slide is a double wedge failure with an approximately horizontal slip plane below the lower wedge. The clay on the slip plane has a residual strength ϕ'_r of around 8°. Although earthquake motions are especially likely to create instability along a horizontal slip surface, it is very clear that: (i) the measured movements (up to 21 inches), and (ii) duration of movements, are incompatible with the Newmark analysis previously described. Other mechanisms besides the residual strength aspect of stability must have played a role at this site.

PART B Remedial and Preventative Options

This section of the book presents numerous ways to stabilize or prevent landslides. Some methods apply only to smaller landslides and cut slopes, while others are intended for larger landslides.

The intent is to inform the reader of the available options, where they can be successfully applied, the design approach, construction methods, and relative costs. The scope of these descriptions falls well short of making the reader an "expert" in the subjects, but provides enough information for a reader to become knowledgeable enough for selection and decisionmaking. In this respect, the presentation exceeds the brief treatment provided in most landslide textbooks. It has been the author's experience that clients are primarily concerned with solutions and expect the geotechnical consultant to provide the most cost-effective remediation (or prevention) for the particular site circumstances. This need has been the guiding principle in choosing what to include and exclude from the summaries. For most subjects, reference is made to a more exacting treatise on the subject, such as the handbooks on specific treatments prepared by the U.S. Federal Highway Administration.

Well-established procedures, such as gravity retaining walls, are presented briefly. Emerging technologies, such as deep soil mixing and reinforced earth, are treated in more detail so that their potential use (and avoidance at unsuitable sites) can be better understood within the profession. However, these newer methods are likely to undergo significant changes in both design and construction in the years ahead. In general, the practices described in Part B are believed to be current at the year 2000. This should be sufficient to provide the reader with enough information for comparing alternatives at a conceptual level.

The author has made considerable effort to check that the information provided herein is valid and accurate. However, it should be understood that the author disclaims any legal responsibility for the consequences of any inadvertent errors that may occur within the text.

CHAPTER

13

Common Issues in Remediation

This chapter covers some topics that are common to several remedial and preventative techniques. They avoid repetition and duplication in later chapters.

13.1 WHAT IS SUFFICIENT REMEDIATION?

The purpose of landslide remediation (see Chapter 10) is to restore permanent stability to a landslide against existing and reasonably foreseeable future conditions. This broad statement covers potentially contentious issues. What is "permanent" stability and "reasonably foreseeable" future conditions? Lawyers in their advocacy roles may put significantly different interpretations of what a geotechnical consultant should provide in remediation to protect the owner and third parties (usually the public) from future landslide hazards. This book is primarily concerned with *actual* landslides and not *potential* landslides (as, for example, in highway design slope stability). Therefore, the following remarks address only landslides.

The geotechnical consultant investigating a landslide event has one considerable advantage over a slope stability analysis where failure has not occurred. During a landslide, the factor of safety, defined as the ratio of resisting forces to destabilizing forces, has the known value of 1.00. Therefore, if the landslide can be modeled sufficiently by surveys, drillholes, inclinometers, and piezometers, the principal uncertainties in the stability analyses are the shear strengths and piezometric levels on the slip surfaces at the time of failure. Techniques for measuring and interpreting such data have already been presented.

Since *any* improvement (reduction in destabilizing forces and/or increase in resisting forces) may achieve stability, the main issue is how much is sufficient, given the lack of precision inherent in modeling landslide masses by a few drillholes with instruments. One approach, generally favored by regulatory authorities, is to set a minimum value of the factor of safety F, such as 1.30. The author generally opposes this approach because it ignores the effect of different variables on the reliability of stability analyses. For example, it is much more difficult to accurately model a small landslide than a large landslide; therefore, the design factor of safety for remediation of a small landslide should be higher than for a large landslide. Furthermore, the cost of providing a relatively high factor of safety for a large landslide can be prohibitive, which sometimes leads to the incorrect conclusion that a large landslide cannot be stabilized. It would be more accurate to state that a large landslide cannot be stabilized to the high factor of safety that has been (arbitrarily) set. This topic is discussed more fully in Chapter 10.

The author's opinion is that the factor of safety needed to provide permanent stability should be set by the professional judgment of the geotechnical engineer. It should take into account: (i) the type of landslide movement, (ii) the amount of study performed in characterizing and modeling the landslide, (iii) the size of the landslide, (iv) the potential consequences to life and property of continuing landslide activity, and (v) the experience of the geotechnical engineer. These variables can be put into a matrix (Table 13.1) in helping to set an appropriate factor of safety for remediation. Assuming that a back analysis (Chapter 9) has been performed and the site has been explored for geology, subsurface conditions, and soil properties, the factor of safety can range from about 1.15 to 1.50 for landslide remediations taking the above variables into account (see Table 10.1 in Chapter 10).

There are situations in landslide practice where any improvement in stability is preferable to the status quo. The owner and geotechnical practitioner can agree to a marginal stabilization, recognizing that future movements may occur but at lesser velocities (see Chapter 14: "Alternatives to Full Remediation of a Landslide").

Table 13.1 Calculated Factor of Safety Selected for Remediation

Variable	Factor of Safety Should be Relatively Higher	⟷	Lower
Type of landslide movement	very fast	⟷	very slow
Level of study performed	minimal	⟷	sophisticated
Size of landslide	small	⟷	large
Potential consequences of future instability to life and property	very significant	⟷	insignificant
Experience of geotechnical consultant	limited	⟷	very experienced

Referring back to the purpose of landslide remediation at the start of this section, the term *permanent* stability refers to the cessation of movement or reduction to a low rate of creep after remediation has been applied. The remediation site could be the back of a wall, the base of a shear key, or a dewatered area. In many cases, the area upslope of the remediation may have been cracked or loosened by earlier landslide movements. This condition requires time for the ground to heal, and some additional movements are likely to occur in these areas over the ensuing first years after remediation. It is important that affected property owners be warned of this probability, otherwise rumors may circulate that the landslide is still unstable when, in fact, it has been stabilized.

The term *reasonably foreseeable* future conditions should include normal groundwater fluctuations, including storm and snowmelt extremes. If the landslide occurred during an extreme rainfall event, the possibility has to some extent been covered by the back analysis. However, assumption of an even higher spike in groundwater levels should be checked in remedial calculations. If the groundwater is normally very close to the surface, the assumption of it reaching the surface is a reasonable "worst condition." By contrast, a deep groundwater table generally will never rise to the ground surface. The extent to which groundwater could rise depends on the hydrologic and hydrogeologic conditions. The author's past experience of several sites, where extensive records have been kept, indicates that the peak rise *above normal high winter levels in clay soils of moderate slopes* is probably no more than 6 feet. However, the assumptions used in a back analysis or parametric study should be based on local experience, if available, and judgment of the behavior of the subsurface conditions. The figure of 6 feet is given here only as a guide where no other data are available. Normal high winter levels are described in Chapter 5, Section 5.3.

Conditions that are *not reasonably foreseeable* should be self-evident. Examples would be developments that introduce new sources of water into the landslide area (or at a faster rate) after the landslide has occurred. Other examples are changes in land use, cuts below the landslide, fills above the landslide, etc.

Another common feature of clay slopes is surficial creep caused by seasonal softening of the clay in winter and desiccation cracking in summer. These surface movements can cause separation between concrete slabs and soil, minor distortion of wood structures, walls, etc. Surficial creep is not corrected by deep-seated landslide remediation.

13.2 GROUNDWATER LOWERING

Permanent Dewatering

There are many ways to lower the groundwater levels within a landslide, thereby increasing slope stability; the options are presented in Chapters 17 and 18. The theoretical solutions to these drainage options are based on assumptions of (i) homogeneous soils and (ii) either isotropic properties or anisotropic properties of fixed proportions (such as permeability coefficient in the horizontal direction being a multiple of the permeability coefficient in the vertical direction). In practice, many landslides that have been heavily disturbed by recent or historic movements meet neither of these criteria. They can be a jumbled mass with pockets or zones of more permeable soils within a generally impermeable mass. The permeable soils may be connected or isolated, and may occur at different levels. Permeable zones create high groundwater levels within a landslide that can be drained by horizontal drains, trench drains, and similar approaches to tap into and relieve the permeable zones. However, it is very difficult to assign a "composite" permeability that is representative of the landslide mass as a whole.

Most theoretical solutions to slope drainage are based on the long-term equilibrium conditions obtained after the drains have achieved drawdown. In the short term, however, the time required for the landslide to reach the steady-state long-term condition is of concern and often overlooked in the analysis. During this interim period, the drawdown of dewatering wells is less than it will eventually become. The engineering geologist has to allow for this transition by using experience or trials (such as pump tests) to predict the duration of the transition period from the existing groundwater condition to the long-term steady state.

Estimates of groundwater lowering can be based on theoretical work backed by rational judgment of other factors affecting the outcome. It is important to recognize that long-term equilibrium may take months or years to be achieved and that an unconservative bias (more limited drawdown) is occurring immediately after drainage measures have been installed. Therefore, groundwater lowering may not immediately stabilize a landslide. Future groundwater monitoring can track the progress of groundwater lowering and measure its ability to permanently lower the groundwater table and suppress seasonal fluctuations. Such monitoring should be included as part of the remediation process.

Temporary Dewatering

Dewatering is often necessary *during construction* of remedial measures for landslides to prevent loss of ground, protect the safety of workers and the public, and preserve the original integrity of the ground (i.e., avoid softening or loosening of strata). In the case of temporary works, it may be necessary to specify the type and extent of dewatering to prevent problems occurring at the construction site that are foreseeable by the geotechnical practitioner. This opinion is somewhat controversial because the traditional practice in construction work is to defer the methods of construction to the contractor. The supposed benefit is that one of the bidding contractors may have a brilliant idea that will expedite construction and save the owner and contractor money. The reality is that contractors have very little time to prepare their bids and, in the absence of a specified method, will opt for the least expensive dewatering technique, usually sump pumps. To do otherwise would increase their bid price relative to other bidders who also price for sump pumps, and likely would lose them the contract.

The successful bidder begins the construction work, gets into trouble, and makes claims for the extra cost of providing a more sophisticated dewatering system on the basis of "defective plans and specifications" or "changed conditions." Delays, arguments, and lawyers then increase the construction costs; worse yet, any injuries/deaths, or damages/delays to third parties can change the situation to multi-year litigation. The cost of such claims is almost always much higher than the "extra" cost of specifying a more appropriate and costly dewatering system in the original contract. Furthermore, the geotechnical engineers involved in the design team will be brought into the controversy because, amongst the various parties, the geotechnical consultant will be judged to have the most knowledge of the preexisting ground conditions. Thus, it is strongly recommended that the plans and specifications for a project requiring temporary dewatering should spell out the requirements and perhaps include a separate item in the bill of quantities to draw attention to its importance to the success of the project. In particular, an acceptable method (or alternative methods) should be written into the specifications with an explanation of the technical reasons for dewatering. An alternative approach is to require that the contractor employ the services of an experienced groundwater specialist to design and implement the site dewatering.

Case History 3 describes a major flow slide at Bonners Ferry, Idaho. It could have been prevented if dewatering prior to excavation had been regarded as a *design* requirement rather than as a temporary construction requirement.

13.3 FILTER AND DRAINAGE LAYERS

Filter Materials

Filter materials are used in a wide variety of applications in landslide remediation work, including:

- Separation layer between a steep headscarp and rockfill in an erosion bowl
- The first layer covering an active spring or seepage area
- Separation between rockfill and an overlying cover of soil fill
- Layer separating soil below groundwater with a rockfill above
- Surrounding soil for drainage pipes

All these prevent migration of fines from one soil to another by either water, gravity, or both. A geotextile fabric is often used for the same purpose, but a soil filter layer provides better shear strength than fabric, especially for steep slopes. Filter layers need to be free-draining, so the amount passing the No. 200 sieve should be set at a practical minimum (generally 3 to 5% by weight, based on wet sieving) and be nonplastic. It is often the intermediate layer between a crushed rockfill and native soils and must satisfy filter requirements with both materials.

A good general gradation specification for a free-draining filter material is given on Table 13.2. A soil filter layer, such as clean sandy gravel, should be at least 12 to 15 inches thick measured normal to the slope. A thicker layer is frequently used to simplify placement for the contractor, e.g., a 6-foot wide (horizontal) strip placed by a small dozer or front end loader.

A clean, free-draining filter layer can be compacted by one of several methods, depending on the amount of material and space available for the equipment. These methods include vibrating rollers, vibrating plates, smooth wheel rollers, etc. In general, a high level of compaction is not required and may even be detrimental (e.g., behind retaining walls, by increasing lateral pressures). As a separator between cut slopes in native soils and a rockfill, filter materials often are placed in small heaps or windrows, and pushed up the slope just ahead of the rockfill placement.

Table 13.2 Typical Gradation Specification for Filter Material

Sieve Size	% Passing by Weight
2"	100
1"	55–75
1/4"	30–45
No. 16	7–20
No. 200	<3

Fines should be nonplastic. Fines content can be raised to 5% by weight in noncritical applications.

A soil to be avoided for drainage purposes is a gap-graded gravel. These soils are likely to segregate during handling, causing local pockets of sand or open-graded gravel. During use, the sand is likely to migrate through the gravel and reduce the permeability at places of high hydraulic gradient such as drain outlets or drainholes in pipes. For additional comments on gap-graded gravel, see Chapter 11, Section 11.1.

Geotextiles (Filter Fabric)

Geotextiles are frequently used as a separator between soils of dissimilar gradation. The most common application is between rockfill and native soil.

Many geotechnical practitioners prefer to specify filter fabric rather than graded filter soils when placing rockfill against native soils. This avoids the need to specify and import a material that may be difficult to manufacture or be available only from a remote source. Geotextiles generally have a lower shear resistance than soil filter layers and could provide a surface of weakness. However, geotextile manufacturers are continually improving the adhesion (shear resistance) of fabrics.

Geotextiles need to have strong resistance to tearing and puncture. Two classes of filter geotextiles specified for landslides are described in Chapter 11, Section 11.3. The heavy duty geotextile is used for contact with rockfills.

The ground surface on which the geotextile is to be placed should be as smooth as practicable. Geotextile should be placed with a minimum overlap of 12 inches. On slopes, the geotextile is placed with the overlaps along the line of maximum slope. Fill should be placed over the geotextile starting at the base of the slope and spreading upwards. The geotextile is usually staked to prevent it from shifting. The upper end is often embedded in a shallow, backfilled trench. At least 1 foot of soil cover should be placed over the geotextile before equipment is allowed to pass over it. On slopes, compaction of fill above the geotextile can be achieved by winching a roller up and down the slope from the top.

Drainrock

Where it is essential to move water out of the soil quickly, a more open-graded material can be used. The ideal drainage soil has a uniform particle size with no fine sand, silt, or clays present. Pea gravel, which is rounded alluvial stones washed from the sea or river beds, meets this criterion. It is usually available in sizes from about 1 inch down to 3/16 inch in the United States. It has a narrow range of densities and is easily placed and compacted (if needed). However, pea gravel has significantly lower strengths than crushed hard quarry rock (sometimes important) and must be separated from fine-grained soils by a geotextile filter fabric.

Crushed hard quarry rock is another good drainage material. The angular stones create large void spaces. Although crushed rock is more compressible under overburden loads due to fracturing of the angular point-to-point contacts, this does not usually cause a technical problem in most landslide applications.

13.4 HARD, CRUSHED ROCKFILL PROPERTIES AND CONSTRUCTION

The best quality of rockfill is produced by a hard rock quarry where the rock is blasted into fragments (shot rock) and subsequently processed to different gradations. Some typical gradation requirements are listed on Table 13.3.

The following guidelines are suggested for the design properties of rockfill incorporated into landslides:

Fully saturated density	126–131 lb./cu. ft.
Moist density (fully drained)	102–110 lb./cu. ft.
Strength at effective overburden stresses of *less* than 1 ton/sq. ft.	$\phi' = 42°$
Strength at effective overburden stresses of *more* than 1 ton/sq. ft.	$\phi' = 40°$

The lower strength ϕ' at higher overburden stresses allows for the crushing of the angular edges of the rock fragments and a curved Mohr envelope.

Outer slopes of rockfill usually are designed for 1½ : 1 (horizontal : vertical). In some situations, where open ground is limited or there is a need to maximize the overburden stress applied by a rockfill (for example, above a shear key) the outer slope may be designed for 1¼ : 1. These design slopes assume

Table 13.3 Typical Gradation Specifications for Crushed Rockfill

18-inch minus		8-inch minus	
Rock or Sieve Size	% Passing by Weight	Rock or Sieve Size	% Passing by Weight
18"	100		
12"	60–100	8"	100
6"	30–80	2"	20–50
½"	0–10	½"	0–10
No. 200	<3	No. 200	<3

Material passing the No. 200 sieve should be nonplastic. Fines content can be raised to 5% by weight in noncritical applications.

that the rock is a hard, crushed, quarry rock, and will be fully drained.

Rockfill should be spread in 12- to 24-inch lifts on most landslide work, with each lift compacted by at least four passes of the tracks from a heavy (Caterpillar D-8 or larger) bulldozer. For higher densities, a 10-ton vibrating smooth drum roller can be used.

Oversize rocks brought to the fill can be pushed to the outer edges. However, care should be taken to ensure that rocks do not roll down steep outer slopes to endanger adjoining property, and large rocks should not be left "hanging" on the slope edge. Many lawsuits in the United States have resulted from rocks falling or bouncing down slopes (during rainstorms or snowmelt), damaging property, or injuring people below the slope.

Rockfill should be free-draining. Fines passing the No. 200 sieve should be nonplastic, and limited to about 3 percent by weight. This specification requirement is difficult to enforce because of the need to obtain a large representative sample for a test.

For many rock types, a limit of 3% fines cannot be obtained without washing the rockfill during processing. In less critical applications, the engineer may raise the fines limit to 5%. This latter figure is fairly common in contract specifications, and generally reduces cost compared to a tighter limit on fines.

The rock material by weight passing the No. 4 sieve should be limited to about 30% (or even 25%). This criterion is needed to ensure point-to-point contact between the coarser rock fragments. It should be mentioned that 30% by weight has the appearance of a higher percent when seen on a site. In a rockfill stockpile with 30% by weight of gravel-size (or larger) pieces, the sand fraction will appear to cover about one-half of the surface of the stockpile.

Many contractors offer to show the owner's representative the quarry that is being proposed as the rockfill source. It is a good idea to take this preview of the proposed source because it can prevent problems of unsuitable materials being brought to a site. A prior review of the quarry does not relieve the contractor of meeting the contract specifications.

In specifying and selecting a good rockfill source, the following issues should be considered:

- Shotrock should not be platelike in shape. The long dimension should not be more than three times the shortest dimension. Platelike rock pieces tend to compact with the long dimension horizontal and provide a preferred orientation of slippage.
- Any overburden soil, highly weathered rock, or decomposed rock at the quarry should be stripped off prior to blasting so that unsuitable materials do not contaminate the shot rock.
- Crushed rounded river gravel is not an acceptable substitute for shot rock. The strength of crushed river gravel will be substantially lower than shot hard rock from a quarry. Therefore, river rock should only be used where its use has been anticipated and accounted for in the design analysis.

Rockfill is the material of choice for most landslide remediation projects involving earthworks or drainage. There are two main disadvantages to its use, namely:

1. *Cost.* A hard rock quarry may be many miles from the landslide site. Royalties, blasting, and processing add to these costs. The availability of more than one quarry source is helpful in keeping prices competitive.
2. *Need for Filter Fabric or Filter Layer.* The fine-grained soils in many landslides are incompatible with rockfill unless a separation layer is placed between them. Without the separator, seepage will likely carry fine-grained soils into the open voids of the rockfill, causing internal erosion behind or above the rockfill. The drainage capabilities of the rockfill may be adversely affected.

13.5 TEMPORARY EXCAVATIONS AND CLOSELY-SEQUENCED CONSTRUCTION

Many types of remedial work require a contractor to temporarily excavate into a landslide slope. Some examples:

- Excavation for a shear key
- Removal of landslide debris prior to constructing a replacement buttress
- Excavation for a trench drain

These and similar situations involve temporary removal of support to a landslide and have the potential to reactivate movements upslope of the excavation.

The possibility of further activating landslide movements by temporary cut slopes may seem self-evident. However, geotechnical engineers are primarily focused on either the landslide itself or stability after remedial works are in place. Stability during the intermediate event of a temporary excavation into the slope may be overlooked. Unfortunately, upslope movements caused by temporary excavations do occur and often have damaging consequences. It takes a sense of awareness and carefully controlled construction methods to avoid or minimize movements towards a temporary excavation.

In terms of awareness, the engineer needs to carefully assess the landslide slope and proximity of roads, buildings, buried utilities, etc. to the proposed excavation. If the risk of damage is high, the proposed landslide remedial technique may not be acceptable.

The second strategy is to schedule the work at the time of year when groundwater levels are seasonally near their lowest levels. In the Pacific Northwest, for example, this occurs in late August to early October. The schedule requirement should be

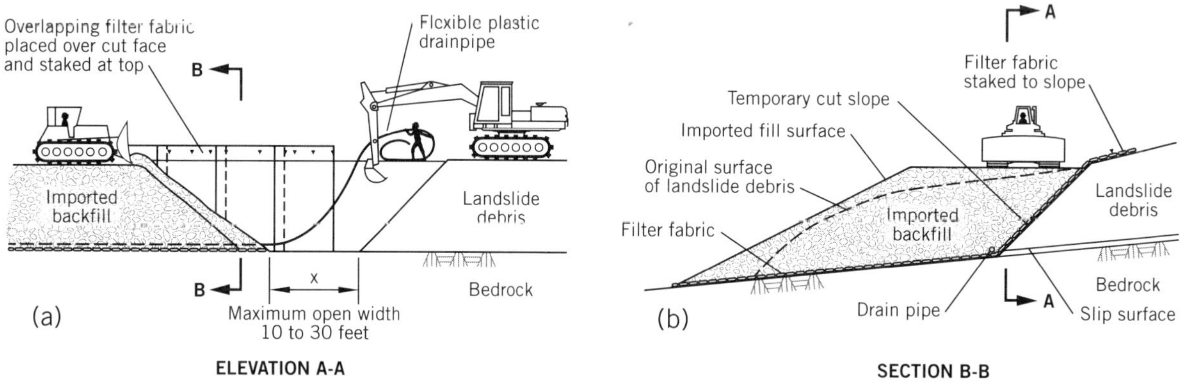

Figure 13.1 Closely-sequenced construction method.

considered part of the design, and its importance should be stressed with the owner. A low groundwater level provides a moderate safety margin for a landslide that occurred under winter (seasonally high) groundwater conditions, and partially offsets the destabilizing effect of the proposed temporary excavation.

The third and most important strategy is to require and strictly enforce *closely-sequenced construction* methods. The concept is illustrated on Figure 13.1, and a field project is shown on Figure 15.10. In this technique, excavation and backfilling at any part of the slope occur in a close sequence such that the length of open excavation is kept to a practical minimum at all times during the work day. When site work is suspended overnight or at weekends and holidays, the excavation is temporarily backfilled with previously excavated soils. These soils can be quickly excavated again at the start of the next shift. The need to totally backfill the open excavation during non-working hours is discretionary, depending on the site conditions, public safety, etc. On many sites, backfilling to zero base distance across the temporary excavation (distance x on Figure 13.1a) is sufficient.

The excavator (usually a large backhoe) moves across the landslide to make the required excavation. The slope temporarily cut by the backhoe is kept as steep as practicable for safety, typically about 1¼ horizontal to 1 vertical. Meanwhile, a bulldozer continually pushes imported backfill (usually rockfill) into the excavation from the other side or, in the case of a trench drain, the backfill is placed into the excavation by a front end loader or backhoe. The backfill slope should approximate the angle of repose. The maximum open distance separating the base of the cut slope in the landslide debris from the base of the replacement backfill should be restricted by the contract specifications to 10 to 30 feet, depending on circumstances. In this way, most of the landslide remains fully supported at all times, and any local slope failure along the temporarily exposed face should be minor. The gap at the excavation base allows sufficient room for (i) a width of filter fabric to be placed across the backslope and base, (ii) a flexible plastic drain pipe to be installed at the base of the cut slope during construction, and (iii) minor delays that may occur when the rate of excavation is faster than the arrival on site of imported backfill.

Closely-sequenced construction is often performed as a two-stage excavation. The first stage is to make about one-half of the required excavation cut, usually involving soils that are mostly above groundwater level. The side slopes, normally 1:1 (horizontal:vertical), will usually remain stable in stiff soils. Any concern about temporary slope stability in this first cut should be addressed by a stability analysis during the design phase. The second stage is to perform the closely-sequenced excavation and backfill as quickly as possible. Depending on the ground conditions, dewatering of the slope by deep wells before construction may be required.

Some cracks may develop upslope of the excavation during construction and the site should be monitored for safety and possible distress on adjoining properties. It is advisable to perform structural surveys (visual inspection, with photographs) on houses and other structures on the upslope side of the excavation. Wherever needed, inclinometers and survey stakes should be installed and measured before, during, and after construction. Monitoring not only measures movements but can also show proof that zero movement (excluding healing of cracks within the slide mass) has occurred after remediation. Claims associated with landslide reconstruction activities can be genuine, fraudulent, or perceived. Using factual data is the best way to equitably handle the issue.

Field inspection during closely-sequenced construction by a member of the design engineer's staff should be mandatory. This should ensure that the contractor adheres to the technique and that the engineer's staff is available to provide advice if local movements occur or unexpected obstructions are encountered in the excavation. These observational field costs are inconsequential compared to the costs that are likely to be experienced if a significant problem develops on a site without observation.

It is advisable to discuss the purpose of closely-sequenced construction at both pre-bid and pre-construction meetings. At the latter, the design engineer should encourage the contractor to schedule the earliest excavations at less critical sections of the landslide remediation.

The *speed* of construction should be emphasized in closely-sequenced construction. In most landslides, some seepage can be expected near the bottom of the excavation. The longer the excavation is open, the more likely it is that some sloughing will occur and it may become difficult to lay fabric against the landslide face. On most projects, contractors make slow progress in the early days and speed up later.

The benefit of closely-sequenced construction for temporary excavation and backfill in landslide remediation cannot be overemphasized. In many landslide situations, the method should be specified and strictly adhered to.

13.6 CONCEPTUAL CONSTRUCTION CONTRACT COSTS

When comparing landslide remediation options, relative cost is one of the more important factors affecting choice. This cost includes the site investigation studies, preparation of plans and specifications, construction contract, and the extent of site attendance by design staff and resident engineer staff during construction. The most significant expenditure is the construction contract, and geotechnical engineers usually provide their clients with a conceptual level estimate of the contract price when comparing remedial alternatives.

Construction costs for landslide repairs depend on many factors: availability of material sources, availability of mechanized earthmoving equipment, site access, spoil and water disposal, extent of environmental restrictions, and the regional availability of specialized ground improvement equipment (if needed). Thus, relative costs will vary in different parts of the world, and some remedial options may be irrelevant in less-developed countries. Nevertheless, the various techniques for remediation and prevention discussed in the later chapters include information on construction costs. They are based on the files of Landslide Technology, Portland, Oregon, and other published data on past construction costs in the United States. It is likely that the relative costs of the different options will be similar elsewhere in developed nations. For this reason, the construction costs have been included together with the date of construction and the Engineering News Record Construction Cost Index (ENRCCI) at that date. To obtain the approximate current cost of construction in the United States, find the current ENRCCI from the Engineering News Record magazine, or www.enr.com, and determine the ratio between the current and past value of the index, as given on Figure 13.2. This ratio is multiplied by the actual costs given in the text to provide an approximate current construction cost.

Example:

Cost range of horizontal drains (1995–1999) = \$12.20 to \$20.95/ft.

ENR Construction Cost Index (mid-1997) = 5863

Assume that the ENRCCI becomes 7830 in the year 2015

$$\text{Ratio} \frac{\text{ENRCCI (2015)}}{\text{ENRCCI (1997)}} = 1.335$$

Approximate construction costs range for horizontal drains in the year 2015 = <u>\$16.29 to \$27.97 per lin. ft. of drain</u>

The construction cost data provided in this book are believed to be accurate for the projects for which costs have

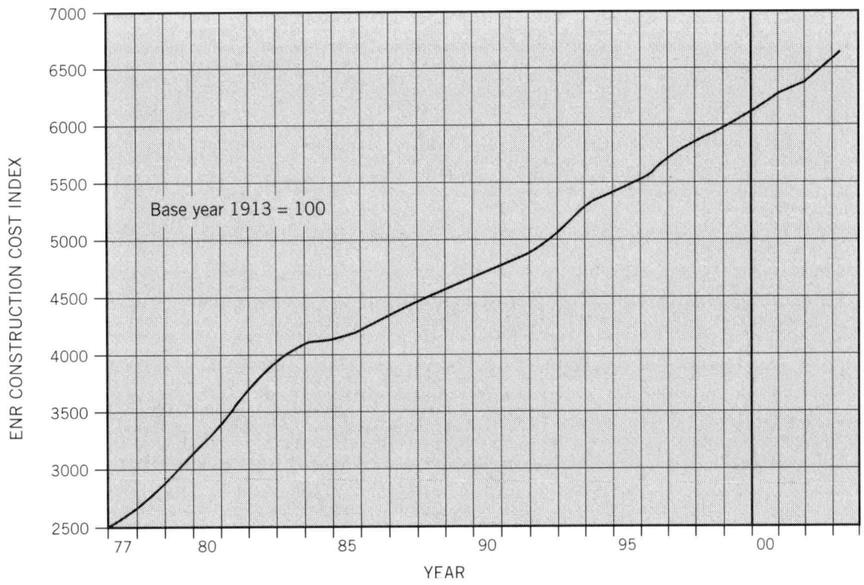

Figure 13.2 Engineering News-Record Construction Cost Index (ENRCCI).

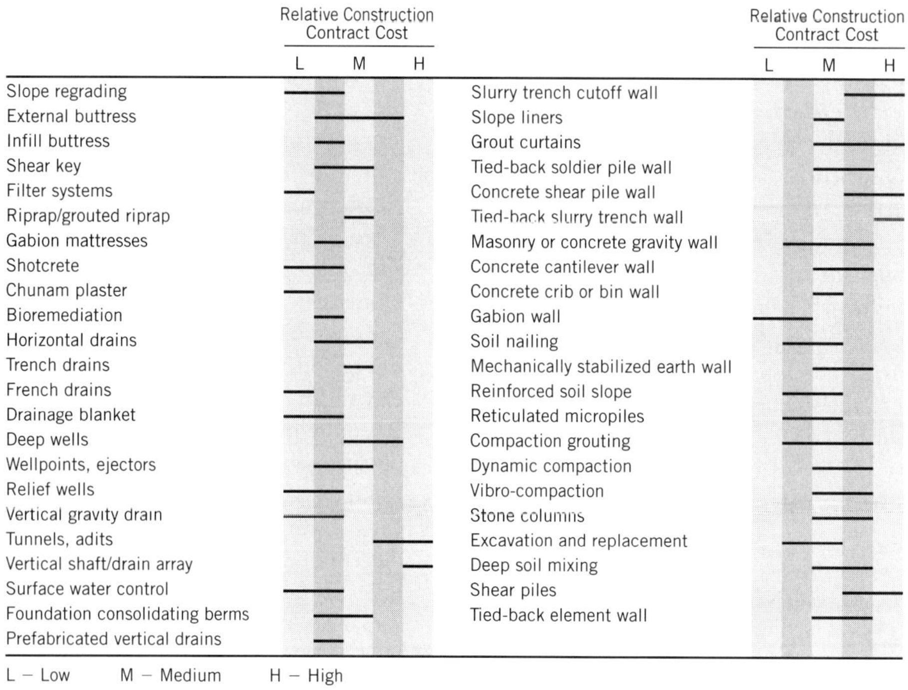

Figure 13.3 Relative construction costs for various remedial or preventative landslide options.

been stated. However, the author accepts no responsibility for future interpretation of the data by others.

Actual bids for contractor construction costs depend on additional variables to those listed earlier, including: equipment available to a contractor, level of work activity in the area, backlog of individual contractor, cost of borrowing, perceptions of risk involved with the project, past experiences with the contracting agency, time of year, availability and haul distance for imported materials and waste disposal, and other factors related to the specific project. Landslide remediation projects commonly have a spread of at least 30 percent between low and high bids.

In the next chapters, 45 different techniques for prevention or remediation of landsides are presented and discussed. Figure 13.3 lists the approximate relative construction costs for these options.

CHAPTER

14

Alternatives to Full Remediation of a Landslide

Under normal circumstances a landslide is stabilized with a margin of safety that meets the standard of practice. However, there are many situations where full stabilization is impractical due to factors such as the large size of the landslide, environmental and ownership restrictions, excessive cost, etc. In these cases, it is often concluded that nothing can be done, but some action is often preferable to no action and can achieve worthwhile objectives. This chapter discusses some alternative options to full remediation.

14.1 NO ACTION

The option of "do nothing" is, of course, plausible in many situations. It is sometimes presented in a geotechnical report as if the writer has just discovered Archimedes Principle! When it is included as a technical recommendation, there is a need to discuss the technical issues and likely consequences that may not be readily apparent to the owner.

An example of taking no action might be an owner who has monitored a landslide over a number of years. A review of the data may show a slowing, constant, or accelerating trend. The landslide consultant can offer to review the data and discuss the likelihood of damaging or dangerous consequences in the future. If the rate of movement is slowing or constant, and the technical reasons are understood, no action may need to be taken (see Case History 1). By contrast, the owner may not be aware that an accelerating rate of movement in a stiff clay may be the prelude to more catastrophic (delayed) failure later (Case Histories 6 and 8).

14.2 MAINTENANCE

Railroad owners and highway agencies have maintenance costs built into their annual budgets. Since the staff already is included in the budget, additional maintenance is usually a marginal extra cost, i.e., for materials or possibly emergency/overtime situations. It can be scheduled for low-demand periods. For a landslide that is moving slowly, or moving periodically during wet weather, maintenance can be the best option. It may be the *only* option on very large landslides where remedial costs are extremely high.

When compared to the capital cost of landslide remediation (often exceeding $1 million), the cost of maintenance is usually an economical alternative. However, there is the considerable nuisance effect of annual road closures on the public, especially in a high-traffic area. Of more concern is the risk that a major movement may occur that could endanger public safety and property. Therefore, as a minimum, a landslide that is moving sufficiently quickly to require continual road resurfacing or shoulder reconstruction should be examined in some depth to determine the causative factor involved. If causation is likely to become more dangerous in the future (such as river erosion, sea erosion, springs, etc.), then maintenance may not be a suitable option because of the increasing risk with time. Similarly, if the amount and frequency of repairs are increasing for no apparent reason, it may be indicative of delayed failure developing in a stiff clay, and some action will need to be taken.

A good candidate for maintenance is a slow-moving landslide in which the rate of movement does not change significantly with time. An example might be a colluvial slope moving along an ancient slip surface at residual strength.

Before selecting maintenance as the option of choice, the following issues need to be considered:

- Is there a moderate to high risk of significant and abrupt loss of ground that could endanger the public? An example might be a landslide caused by erosion at the outside bend of a river. During flood, more rapid erosion could undermine the road (or railroad) and quickly cause significant loss of ground at a time when remediation would be very difficult.
- Is continued movement likely to regress and endanger upslope structures or facilities?
- Will continued movement break culverts, water, stormwater, or sewer pipes within or close to the landslide? Such pipes can often tolerate small movements, but major breaks may cause flow slides during storms.
- Are there other sensitive utilities within or close to the slide area? Breaks in underground cables, high voltage lines, pressure gas pipelines, etc. can be dangerous to the public and very damaging to the local economy should failure occur.

The decision to favor maintenance over remediation is usually dictated by cost. However, the economic and potentially life-threatening risks listed above need to be considered because the potential for litigation from unsafe conditions and economic losses to third parties are relevant to engineering cost comparisons.

14.3 OBSERVATIONS

Monitoring Landslides

Observations in the form of inclinometer measurements or surface survey points are commonly employed in landslide analysis. Owners may elect to continue taking readings rather than to undertake remediation. This is a cost-effective approach because controlled monitoring may be all that is needed for a slow-moving landslide. Once the pattern of movement has been established by monthly or quarterly readings, for example, the frequency of readings can be progressively reduced to yearly or longer intervals.

There needs to be a sufficient number of control points such that a loss of one or more points will not materially affect the landslide control measure. For example, on medium to large landslides, three or more lines of survey hubs could be installed, each line having 4 to 10 hubs. The reference point itself *must* be stable, which is self-evident but a critical requirement in landslide areas. An engineering geologist should check that the reference point is on stable ground, such as an in-place rock outcrop or stable ground at some distance from the landslide. Where inclinometers are being used, the stable reference point is the casing segment anchored below the landslide slip surface.

Readings taken on inclinometers, piezometers, survey points and any other on-site devices should have the data calculated, plotted, and interpreted within 1 to 7 days of the field measurements. The elapsed time between observation and interpretation will depend on the quantity of data and need for the results, but it is important that the process takes place as soon as practicable. There are some owner-clients who want their own staff to take the readings for various reasons: cost-saving, availability of staff at the site, training, etc. With a few exceptions, this is not a good practice and should be discouraged. Too often, the assigned personnel are disinterested in the tasks and make careless errors. Worse, the raw data are often put in a desk drawer and forgotten about for long periods; it is then too late to correct for missing or questionable readings. Therefore, the consultant responsible for the geotechnical work should also be responsible for collecting and analyzing the field observations.

Observations of movements or potential movements of hillsides are sometimes required in construction contracts and are included in the bid price of each contractor. In the United States (and elsewhere), a contractor may collect bid prices from several engineering consultants and include the lowest price (or an even lower price of his own) in the bid document. The successful bidder may then re-open the "bidding" for the instrumentation subcontract to obtain a lower price. From the perspective of past experience, the author can relate that the results have varied from generally unsatisfactory to disastrous. Many of the more reputable geotechnical specialists simply refuse to participate in a bidding format. The best way to handle these requirements is to remove this requirement from the contract and treat field instrumentation as a necessary part of the project's design. In this way, the tasks of reading and interpreting are left with the design team. However, contractual language and insurance require careful attention to avoid potential problems.

Monitoring is especially useful for landslides subject to third-party oversight, such as regulatory authorities or adjoining landowners, to show that the rates of movement are non-threatening. It is also a good practice to take additional readings on inclinometers *after* a landslide has been stabilized to demonstrate that remediation has been successful. Such an offer can sometimes facilitate a speedier resolution of a lawsuit involving a landslide, especially to parties skeptical of the remediation outcome.

Examples of Observation Usage

1. At the *Washington Park Reservoirs Slide,* Portland (Case History 1), the hillside has been monitored for more than 100 years, possibly a record for any landslide in the United States. Some survey hubs have been damaged or destroyed over this period, but most have survived. The ancient slide moves intermittently in response to winter rainfall. However, when the data were subdivided into 22-year intervals, the rate of movement over these periods was approximately uniform, indicating that the underlying clay is at residual strength and is unlikely to change in the future unless there are manmade changes. Although there has been extensive damage to a pump house at the base of the slope, the landslide movements

of around 0.3 inch per year are considered to be minor and no remediation has been undertaken. In addition to surface survey hubs, there are eight inclinometers that are read quarterly.

2. The *Washington Park Station Slide* (Case History 4) is on a hillside close to the Washington Park Reservoirs Slide, and is built over similar ancient landslide terrain. High-speed elevator shafts pass through the landslide into the underlying bedrock; the two shafts have very limited tolerances for lateral movements at the landslide slip plane. Although marginal stabilization (see later) has been provided, the owner is relying on an in-place inclinometer and other conventional inclinometer installations to monitor the rates of movement. If the rate of movement exceeds ¼ inch in 10 years, additional remedial treatment will be implemented. However, there is a good chance that this threshold will not be reached, in which case there will have been a substantial cost saving.
3. The *Leaburg Canal Hydroelectric project* at Leaburg, Oregon, covers a hillside of weathered andesite on which six landslides, ranging from medium to large, were mapped. One of these landslides, Percy Slide, continued to move at an increasing velocity and has been stabilized. The other five landslides have been monitored since 1987 using inclinometers. All five were read at 3-month intervals initially. After three years, two slides with intermittent movements were reduced to annual readings; three landslides showing no movement were reduced to 3-year intervals, and later to 5-year intervals.

   Observations are a cost-effective way of managing this landslide complex because it keeps the owner informed of the slope stability status and satisfies federal regulators that the project is not endangered. To stabilize all the landslides would have rendered the project uneconomic, and it probably would have been closed down.

14.4 AVOIDANCE

Avoiding a landslide can be less expensive than remediation. Five possible ways of achieving this objective for landslides affecting roads are:

- Moving the road into the stable hillside
- Realigning the road onto the valley floor
- Relocating the facility elsewhere
- Bridging over the landslide
- Tunneling through the stable hillside behind the slide

Cutting into the stable hillside is often performed on a sidehill cut-and-fill road where the outer edge of fill becomes unstable. The road alignment is moved into the hillside to put the road section onto cut. Possible disadvantages include: (i) unacceptable horizontal curves, (ii) right-of-way purchase, (iii) relocation of utilities, and (iv) destabilizing the uphill side of the cut.

Relocating the alignment into the slope is often accompanied by preventative measures such as a wall to support the inner slope. Another consideration is that the landslide will regress back into the realigned road which, with the curves introduced by the original realignment, would increase the risk to road users.

Realigning the road onto the valley floor takes the road off the hillside to flat ground. This alternative has become less attractive in recent years because the flat ground is often floodplain or wetlands. Permits to build over wetlands can take years to obtain or may be refused.

When Interstate 5 was built through Kelso, Washington, in the mid-1960s, a 1-mile stretch of the proposed freeway alignment passed along the base of the hillside identified as ancient landslide terrain. Four separate active landslides were identified as impacting the alignment. The state decided to relocate the freeway onto the floodplain of the Coweeman River below the slope rather than to stabilize the four slides.

Relocating the facility is a possible alternative in some cases. However, many projects are built on or adjacent to active landslides or ancient landslide terrain through necessity.

Natural gas pipelines, water transmission pipelines, and high voltage electrical transmission lines can be routed to avoid landslide areas. A thorough geological reconnaissance, including ground traverses, topographic and terrain map reviews, aerial photograph reviews, etc., can avoid these problem areas during the planning stage. Established pipelines in rural areas can often be relocated at a reasonable cost because pipelines are usually buried at a shallow depth.

Bridging over a landslide, when only the headscarp affects the road, seems attractive to many owners because it does not involve adjoining land and there are few environmental constraints. In most cases, however, it is a risky option because the untreated landslide may widen in the future, jeopardizing the abutments of the bridge. If selected as a viable option, the bridge generally will need to have pile-supported abutments and no intermediate supports within the landslide. As the cost of a bridge structure increases rapidly with increasing span length, the cost of a bridge can be uneconomic except for relatively short lengths (typically less than 150 feet).

Tunneling through the stable hillside behind the slide. This expensive option is usually a "last resort" choice for roads and requires unusual circumstances. The huge Thistle Slide in Utah (April 1983) required construction of a tunnel to reestablish the Denver & Rio Grande Western Railroad tracks.

14.5 SELECTIVE STABILIZATION

Technique

The upper reaches of a landslide can be selectively stabilized to protect a particular facility while allowing the remainder of the landslide to be untreated. This option can be chosen when the cost of stabilizing the entire landslide is prohibitive and selective stabilization can be done economically. It also becomes a viable option when downslope landowners are unwilling to cooperate in stabilization of the entire slide.

The main benefits of selective stabilization to the owner are that the treatment usually involves no purchase of adjoining property and requires less-extensive remediation than would be needed to provide a sufficient degree of safety to the entire landslide. It can also avoid environmental issues that cause delays to remediation. Using selective stabilization, the cost of providing remediation to a specific facility is significantly lower.

Selective stabilization is often chosen for roads and railways where the top of a landslide is at or near the right-of-way. A potential disadvantage of the selective stabilization option is that the slope below may continue to move and reduce any support it provides. Thus, the design of a retaining wall has to assume that some or all of the downslope ground will move away from the wall, leaving a vertical crack at the front of the wall. The depth of such a crack depends on the soil type, its strength, and the depth to the slip surface. One possible remedy to this effect is to perform a staged partial excavation in front of the wall and backfill the excavation with rounded gravel. If a crack develops in the ground, the rounded gravel will fall into the crack and provide continued support on the downslope side of the wall (Figure 14.1).

The recommended design procedure is to assume that some support will be lost on the downslope side of the remediation. In the worst case, all the support from the landslide soils is lost, and this can happen for shallow landslides or stiff clays. In many other situations, total loss of support is highly unlikely to occur. It is suggested that the calculated factor of safety be dropped to 1.1 or 1.15 for this extreme case and the main design calculation be based on some judgment of the probable loss of support over the design life of the remedial structure.

Another option is to require field inspections at prescribed time intervals after construction and to take appropriate action to reinstate support if needed. However, future inspections are not generally recommended because they can become overlooked as personnel change jobs and memories fade. Unless failure inspections are mandated by a regulatory agency, reliance upon future inspections can lead to problems from neglect.

A separate stability calculation of the lower *unremediated* landslide may show that the ground will remain stable. For example, a wall supporting the upper portion of a circular arc slope failure may remove most of the original driving moment of the landslide. The curved slip surface at the base of the slide may have sufficient shear resistance to remain stable indefinitely once the upper part of the slide has been remediated.

Examples of Selective Stabilization

1. A riverbank landslide at *Mohler, Oregon,* reached the shoulder of the county road, the headscarp being 5 feet high (Figure 14.2a). Full remediation would have required some type of support at the river's edge over a width of 180 feet. At the highway however, the affected width was less than 90 feet. In addition, to stabilize the ground below the highway would have required

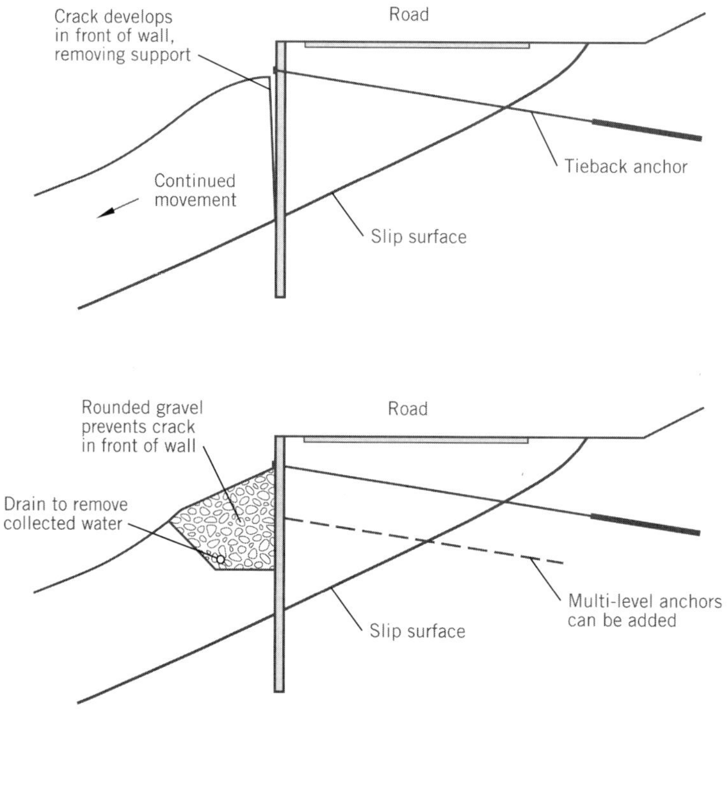

Figure 14.1 Use of cohesionless gravel to provide some continuing lateral support below a selectively stabilized landslide.

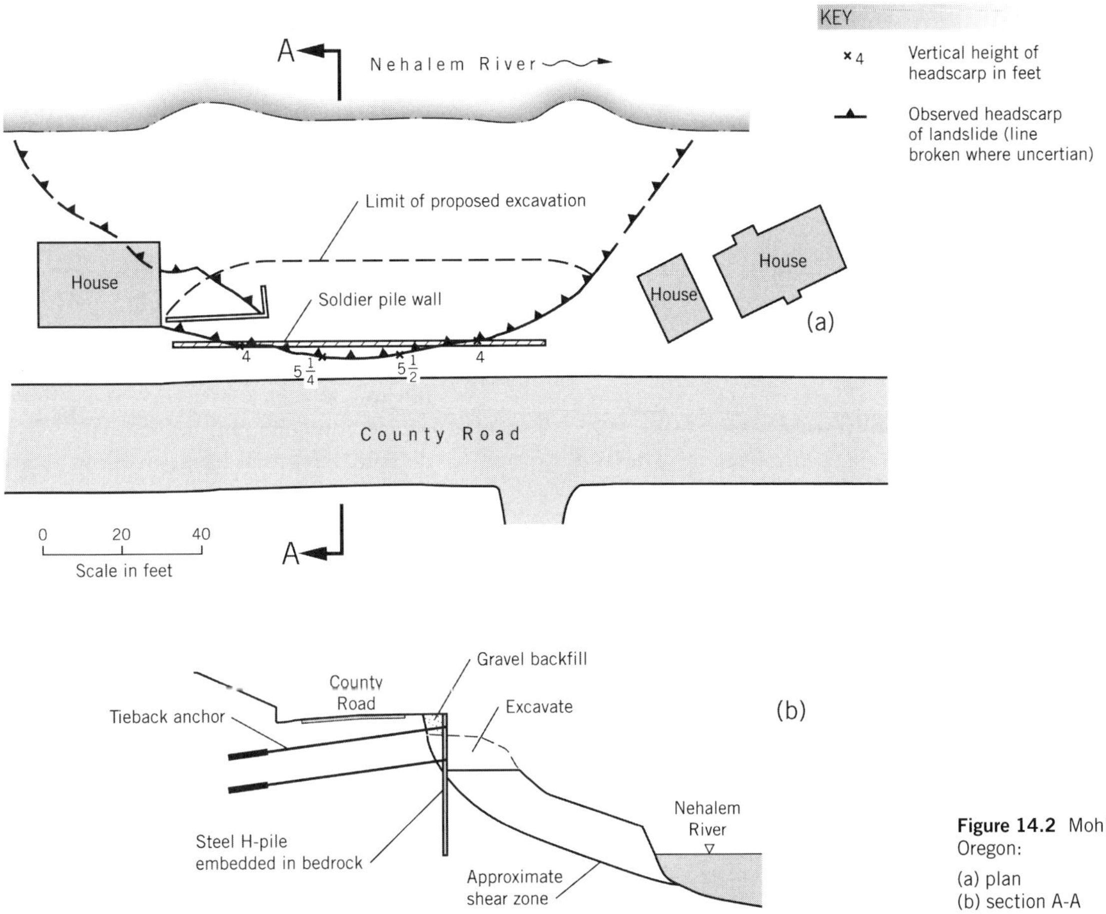

Figure 14.2 Mohler Slide, Oregon:
(a) plan
(b) section A-A

negotiations with three property owners and difficult environmental negotiations for construction work in the river. Instead, the county chose selective stabilization of the road. A tied-back soldier pile was built to stabilize only the county road.

2. *Hagg Lake Perimeter Road Slide 4* (Case History 10). Here, the road is near the head of a long landslide extending down to the lake. Instead of attempting to stabilize the entire slide, a keytrench was excavated through about 20 feet of slide debris to bedrock, backfilled with shot rock, and the road was reconstructed on top of the rockfilled keytrench. The ground downslope of the road has continued to move slowly towards the lake, but the road above the keytrench (and ground further upslope) has remained stable.
3. *Goat Lick Slide* in Glacier National Park, Montana (Case History 9). The road is at the top of a deep-seated landslide extending down to the Flathead River. The headscarp of the slide crossed both lanes of the road over a width of more than 200 feet. Stabilization of the road was achieved by constructing a tied-back shear pile retaining wall near the outer shoulder.

14.6 MARGINAL STABILIZATION

Technique

Marginal stabilization is the implementation of stabilization measures in the knowledge that stability will be improved but at a level that is lower than the desirable factors of safety discussed in Chapter 10.

The goal of marginal stabilization is to slow down or stop landslide movements, thereby improving safety to the facility and its users. In most cases, marginal stabilization is a decision to do *some* improvement over "do nothing." It is accepted that further ground movements may occur, possibly requiring additional treatment. If the facility is left open to the public, the risk to personal safety has to be minimal.

Projects for marginal stabilization have to be selected with care and with the full consent of the owner (and regulatory agency, if applicable). The proposed treatment and its limitations need to be fully explained and documented so that there can be no possible confusion later that it was the result of corner cutting or ignorance. Typically, marginal remediation projects require continued monitoring to check on the adequacy and effectiveness of the remedial measures.

Lest the subject of low factors of safety appear onerous, it can be pointed out that there are many thousands of slopes with marginal slope stability that provide little risk to the public. These include natural slopes of talus, colluvium, and ancient landslide terrain, plus manmade slopes such as roads built from spread rockfill and gravel where the outer slopes stand at the angle of repose. These slopes are stable or moving so slowly that they do not constitute a safety threat.

The author's experiences with marginal stabilization have been excellent. Where applied carefully and appropriately, many worthwhile improvements have been achieved at a reasonable cost. Some examples are given below:

Examples of Marginal Stabilization

1. *Ditch Camp Slide* near Sandy, Oregon. The water pipelines serving Portland, Oregon, cross a hillside of ancient landslide terrain. The old slide debris is a heterogeneous mixture of stiff clay and waterbearing gravels. In 1965, the local county highway department straightened out the road at the base of the slide area by excavating a large "hump" of ground. In the following winter, the slope above the excavated area moved downslope by several feet during a prolonged period of wet weather, requiring emergency measures to preserve the alignment (on trestles) of the pipelines above the road.

   Stabilization was achieved by rebuilding the "hump" with a free-draining rockfill that replaced the original weight and provided improved drainage. The calculated factor of safety improvement was inadequate but achieved the objective. This "temporary" stabilization lasted for more than 30 years, with minor creep movements, before an extensive program of groundwater lowering was implemented to further improve long-term stability. Ditch Camp Slide is an example (and first use of) Original Profile Analysis (Chapter 10, Section 10.2).
2. *Pelton Park Slide,* near Madras, Oregon (Case History 5). This very large landslide is a public recreation park adjoining the reservoir behind Pelton Dam in central Oregon. The deep-seated slip plane is horizontal and the toe is well below the reservoir. It is a double wedge configuration, and the upper driving wedge is defined by ground cracks. Slow movements were occurring, and there was no realistic way for providing a normal design factor of safety at a reasonable cost.

   Marginal stabilization was undertaken by excavating a trench at the head of the landslide to reduce the lateral loading provided by the upper (driving) wedge. The calculated improvement in the factor of safety was 2%. Although insufficient by conventional standards, it was enough to stop the ongoing movements, and only minor creep has occurred subsequently. This has allowed the park to remain open to the public. However, movements were subsequently reactivated by an earthquake. Further excavations at the head of the slide stopped the movements.

CHAPTER

15

Earthworks

15.1 EARTHWORKS OVERVIEW

Earthworks are a relatively simple form of remediation for unstable slopes. They can be analyzed with a high degree of confidence that stability will be achieved provided: (i) the landslide slip surface has been modeled with sufficient accuracy, and (ii) the shear strengths are measured or back-analyzed correctly.

Cost for Materials

The cost of remediation depends mainly on the availability of the materials needed for reconstruction. For soils available on site, the remediation is low cost because the soils have only to be excavated and moved from one part of the site to another part.

The cost for importing soil or rockfill to a site is made up of several components:

- Royalty fee for excavating materials from the quarry or pit
- Blasting for hard rock, including incidental overburden removal
- Excavating and loading
- Hauling on public highways with legal trucks
- Dumping, spreading, and compacting

Since the total cost of many earthworks remediation projects is heavily dependent on the cost of imported materials, the quality, available quantity, and the haul distance are very important considerations in making a conceptual cost estimate during the design stage.

For rockfill or filter rock, a gradation specified by a geotechnical specialist may be difficult to obtain from a quarry or pit without the need for additional screens and possibly much wastage. This extra cost of production can only be justified on very large projects. It is usually preferable to select the gradation of a rockfill around the local or state highway department standard gradations. Quarries are set up to meet these standards, although the standards can be rather broader than is desirable for landslide remediation. In such cases, samples can be taken from several local quarries, and a suitable gradation can be chosen.

Bidding Earthwork Contracts

One of the main benefits of an earthworks remediation is that the equipment needed for the project is usually readily available and there are plenty of qualified contractors. Bid prices for a contract are likely to have a narrow spread. Any special requirements need to be emphasized in the contract documents so that they cannot be overlooked, which could cause a possible underbid by the low bidder. Underbidding should be avoided because it will likely lead to friction during construction between the contractor, owner, and the owner's geotechnical engineers, with claims after the contract is completed. A pre-bid meeting is a useful way to present the needs of the project and provide a "level playing field" for all contractors preparing a bid.

All parties to an earthworks contract should work toward obtaining a fair bid price with an appropriate level of profit for a contractor that performs a well-executed project. Any changes due to unforeseeable site conditions, or as needed by the owner, can be handled by change orders negotiated during the course of the work. On most smaller projects, change orders usually are not required or are minor. Unresolved claims during construction should be avoided because they are very time-consuming to resolve later and usually involve lawyers, thereby escalating the costs to all concerned.

Private owners may choose to pre-select three or four earthworks contractors and limit the bidding to the selected few. This procurement method has several advantages to both the owner and contractor. The owner can pick contractors that have a good track record of quality work, on-time completion, and infrequent claims. Since it is an earthworks contract for which they are all competent, the spread of bid prices should be relatively narrow—within 5–10% for the two lower bidders is a reasonable expectation, and is often closer. Because the competition is limited, the contractor's interest level will be higher, and they are likely to make a serious effort in their bidding. Finally, the risks of having to deal with a serious underbid (due to a contractor's mistake in bidding) or an incompetent contractor, are much reduced. It is important to first check that the pre-selected contractors are available and interested in the project. If they are very busy or do not have the appropriate pieces of equipment available at the time, it is possible that they will still bid (so that they will not be taken off the select list of that client or engineer), but their bid will be high. This decision reduces the number of serious bids and is not in the owner's interests. Enquiries prior to selecting the bidders can avoid this situation.

Bid Documents

The geotechnical engineer responsible for the landslide mitigation design has to prepare the plans and specifications for the proposed contract. These items are outside the scope of this book because specification writing varies immensely. The only advice being proffered here is that the specifications should avoid the extremes of: (i) being vague, brief, ambiguous, and non-quantified, and (ii) being long and repetitious.

The vague use of terms (such as "rock," "granular fill," or "well graded") are likely to produce materials or methods that significantly differ from what the engineer had in mind and may require change orders or compromises in the finished project. By contrast, a highly detailed set of specifications, although seeming to cover all the bases, usually constitutes a thick volume. It is unreasonable to expect a contractor to pick out the finer nuances of the current project from this volume during a two- or three-week bid period. Also, thick specifications are normally recycled from job to job and often have discrepancies where a change has been made that has not been cross-checked with related statements elsewhere in the document.

In summary, specifications should be written in a concise and readable narrative that clearly explains the objectives of the project and the owner's requirements for the finished product. Unusual or special issues need to be emphasized within the specifications and on the plan drawings, as well as at the project pre-bid meeting.

Some suggestions for specifications of earthwork materials are presented in Section 15.7. These generic statements are provided for guidance, and are not intended for verbatim use by readers on specific projects.

15.2 SLOPE REGRADING

Technique

Removing weight from the head of a landslide reduces the driving force. Adding weight to the toe of a landslide provides increased resistance. Regrading a slope uses these principles to reinstate stability to a landslide (Figure 15.1).

Appropriate Uses of Regrading

This low-cost earthmoving technique is especially suitable for small landslides. Minor regrading can be accomplished by a backhoe, and occasionally by a small dozer or front end loader.

On larger landslides, such as a double wedge or triple wedge failure surface, excavation of soils from the upper wedge reduces the driving force. In long translational landslides, the upper wedge is usually a small part of the total volume of unstable soils and partial excavation can dramatically improve stability. At Upper Pelton Slide, Case History 6, soil excavated from the upper wedge was placed over the lower wedge and stabilized the slide (Figure 15.2). At Pelton Park Slide, Case History 5, where similar ground conditions were encountered but on a larger scale, partial excavation of the upper wedge provided marginal stabilization. Regrading can be combined with dewatering to provide greater stability than

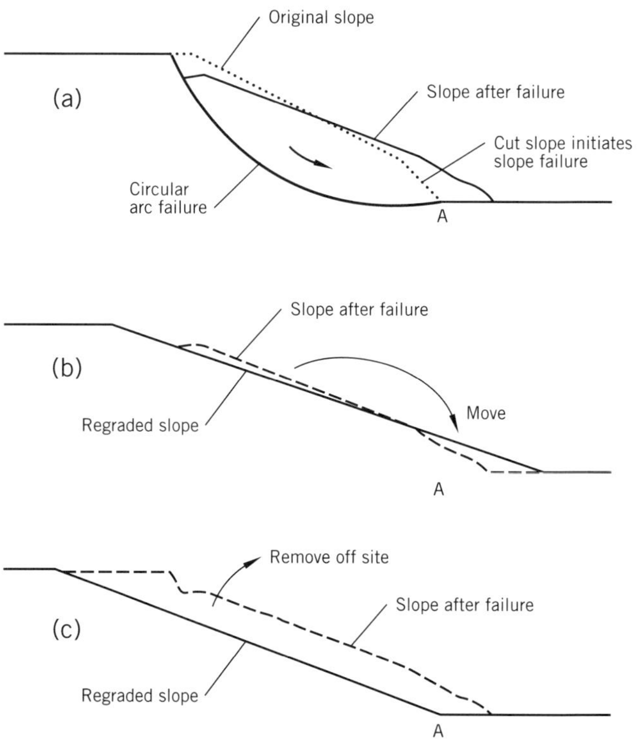

Figure 15.1 Slope flattening to stabilize a landslide:
(a) failure due to a cut slope
(b) slope regrading using available onsite soils
(c) slope regrading by removing upslope soil to maintain the original cut slope base

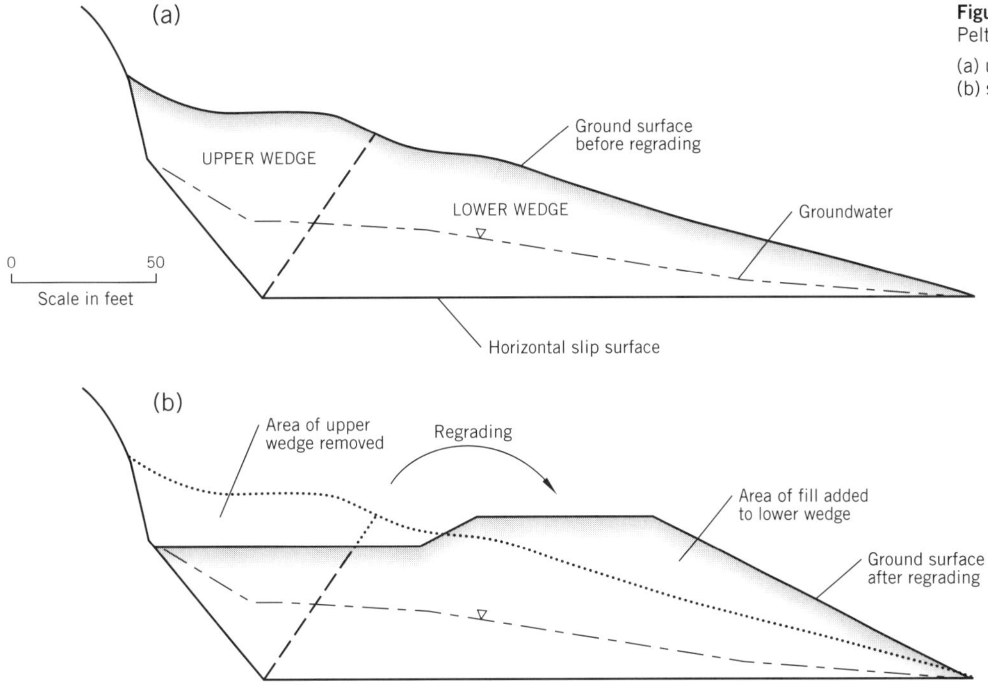

Figure 15.2 Slope regrading. Pelton Upper Slide:

(a) unstable, before regrading
(b) stable, after regrading

can be achieved from regrading or dewatering alone. Other remedial/preventative treatments also can be combined with regrading.

Limitations to Regrading

Slope regrading is not a viable option for most medium (or larger) landslides where land has to be acquired above or below the slide area to accommodate the regrading. The required land area is often owned by several parties. There are usually other restrictions such as roads, houses, or long uphill slopes. Other issues include the quantities to be hauled offsite, logging of trees, and potential surface erosion of the regraded slope surface.

On long translational landslides, most of the slope represents the lower, or middle and lower, wedges that provide net resistance in the stability analysis. If soil is excavated from the middle or lower wedge in the mistaken belief that it is "unloading" the landslide, the ground movements are likely to continue and probably accelerate. As previously explained in Chapter 9, Section 9.6, slope stability in the infinite slope analysis *decreases* when surface soils are removed because the ratio of groundwater height:landslide height (h:z) ratio increases (Figure 15.3).

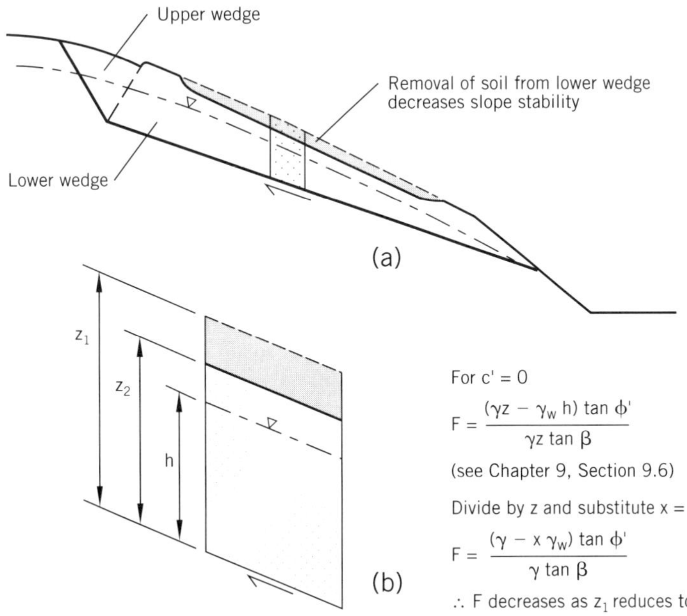

For c' = 0

$$F = \frac{(\gamma z - \gamma_w h) \tan \phi'}{\gamma z \tan \beta}$$

(see Chapter 9, Section 9.6)

Divide by z and substitute $x = \frac{h}{z}$

$$F = \frac{(\gamma - x\,\gamma_w) \tan \phi'}{\gamma \tan \beta}$$

$\therefore$ F decreases as z_1 reduces to z_2 (i.e. ratio x increases)

Figure 15.3 Slope regrading:

(a) excavation on the lower wedge of a double wedge decreases stability
(b) infinite slope analysis

Design

Two possible treatments for a slump block landslide are shown on Figure 15.1. The original base of an oversteep cut slope is at A, and the circular arc slope failure overrides it (Figure 15.1a). One possible option is to regrade the slope as a balanced cut-and-fill construction using all the available soil (Figure 15.1b). This saves the cost of hauling soil to waste (*Note:* If the landslide debris is waterlogged or badly broken up, it usually has to be hauled offsite and be replaced by a rockfill or cover of filter materials). If the original point A has to be re-established, a flatter stable cut slope can be constructed by excavating a significant part of the slide mass (Figure 15.1c) and taking ground from the upper slope.

Stability of the regraded slope can be calculated by first back-analyzing the slumped ground before remediation. Using the average shear strength along the circular arc, various regrading options can be checked for their stability assuming that any later slope failures would develop the same strength. In stiff dilatant clays, a long-term (c' = 0) stability analysis may be required for the project.

Excavation of the upper wedge of a double or triple wedge failure requires the ground above the headscarp to be left in a stable condition. This may need additional remediation to support the headscarp, especially if the slope continues uphill or there are structures or roads close to the headscarp. In the field, the upper wedge can be identified as the area between the headscarp and the reverse scarp below it.

Construction

No additional comments needed.

Examples

Case Histories 5 and 6, previously cited.

15.3 EXTERNAL BUTTRESS

Technique

An external buttress supports the base of a landslide, thereby increasing the resistance to sliding. An example of a buttress supporting a circular arc landslide is shown on Figure 15.4. The arc AB rotates around point 0. The external buttress BEDF is assumed to be built of a free-draining, hard, angular, rockfill. In this example, the buttress improves stability in three ways:

1. It applies a counter-rotation moment to the circular arc where the slip surface is rising, MC.
2. It improves the total shear resistance by making the shear surface pass through the lower part of the rockfill, BC.
3. It increases the effective stress and shear resistance of the lower part of the landslide where the buttress overlies the natural ground, GB.

Appropriate Use of an External Buttress

An external buttress at the toe of a landslide is very easy to build, using widely available earthmoving equipment.

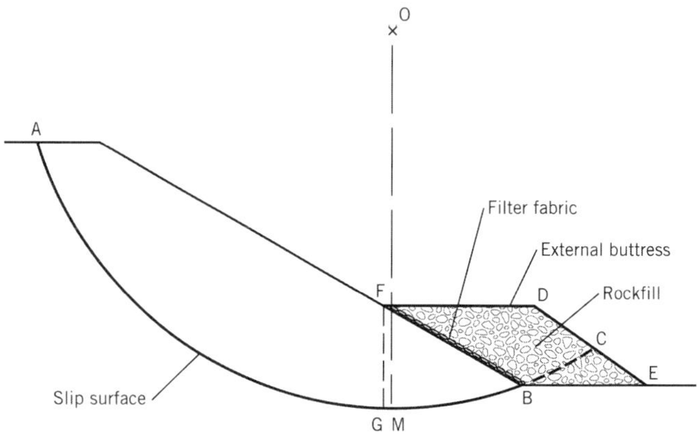

Figure 15.4 External buttress supporting the base of a circular arc landslide.

Buttresses are effective on circular arc and translational slides provided there is available ground below the slide for the buttress foundation.

Limitations to Use of an External Buttress

- The toe of the landslide should be located at the base of the slope. If the slip surface passes deep beneath the toe, most of the buttress benefits are lost (Figure 15.5). There is likely to be no counterbalancing effect, the slip surface does not pass through the rockfill of the buttress, and increased shear resistance due to the buttress weight may not develop quickly enough to produce stability. To the contrary, the buttress may create a new slide downslope.
- The base of the hillside is often a flood plain area where environmental restrictions apply. Permission to build a buttress may not be forthcoming, and permits may require further studies of the affected habitat, causing unacceptable delays in construction. Environmental restrictions also apply to the construction of a buttress into lake, river, or wetland areas.
- The land needed for a buttress may belong to others outside the right-of-way or landowner's control. Therefore, the land may not be available or, if the right of eminent domain is invoked, there may be a long delay before construction can begin.
- There may be insufficient area at the base of the landslide to build an external buttress. In this case, a replacement buttress (Section 15.5) may be feasible. For circumstances in which there are no good alternatives, a marginal stabilization approach (Chapter 14, Section 14.6) could be adopted.

Design

The landslide is back-analyzed to determine the shear strength properties on the slip surface. The external buttress is put into the landslide cross-section and stability is recalculated until a buttress large enough to provide the design factor of safety (Chapter 10) has been achieved.

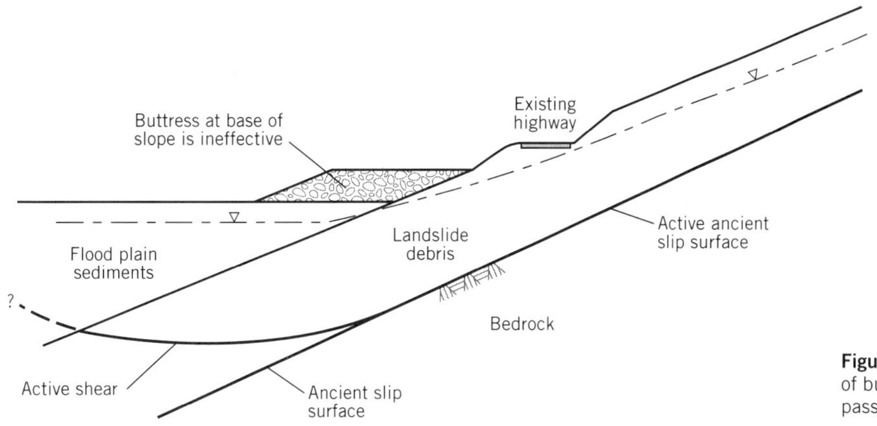

Figure 15.5 Ineffective location of buttress where the slip surface passes deep below the slope base.

The counter-rotation effect shown on Figure 15.4 increases as the weight and center of gravity of the buttress moves away from the vertical OM. Modeling the landslide slip surface by inclinometers placed between A and B will define the bottom of the arc M. If the buttress is built too high, point F will move to the left of the vertical OM and the buttress will become progressively less efficient for counter-rotation as the buttress height increases.

In designing the buttress (Figure 15.4), point D should be as far as possible downslope of point B. To make the best use of rockfill, slope DE should be as steep as practicable, such as 1¼:1 or 1½:1 (horizontal:vertical), depending on the conditions immediately downslope of the buttress. The design properties and placement of rockfill are presented in Chapter 13, Section 13.4. At many sites the distance BE is limited by either the site conditions (e.g., near a river) or by owner restrictions (e.g., need for building space). In extreme situations, the steep outer slope of the buttress can be increased by mechanically stabilized earth techniques (Chapter 20) or roller-compacted concrete.

The extra weight provided by a buttress above arc GB (Figure 15.4) increases the shear resistance along this part of the arc after the foundation soils have dissipated excess pore pressures set up by the buttress weight. If the foundation soils are slow-draining soils, it should be remembered that increased soil strength will not develop immediately but may require years. This fact is often overlooked in effective stress stability analyses of a landslide with a buttress in place.

For clay soils, the average undrained shear strength $\overline{c}$ can be calculated from the back analysis of the landslide. This value should be used for the end-of-construction analysis of stability. If this factor of safety is satisfactory, the long-term stability will improve in softer cohesive soils that develop positive pore pressures on first loading.

Variations in Buttress Design

Some alternative buttress designs include:

1. *Replace the upper layers with common fill.* As shown on Figure 15.6(a), the lower part of the buttress is built of rockfill to provide high strength where the continuation of the circular arc failure passes through the buttress; the rockfill also acts as a drain at the base of the slope. Common fill, usually locally available and less expensive than rockfill, is placed above the rockfill to add counterbalance weight. Soil fill normally has a higher wet density than rockfill because it holds more water in the void spaces (i.e., 120–125 lb./cu. ft. versus about 102–110 lb./cu. ft. for moist rockfill). Another benefit of using soil is that the outer slope can be landscaped, making the buttress more visually attractive. The upper outer edge (point D, Figure 15.4) can be rounded for aesthetic appeal (see Chapter 16, Section 16.8).

Disadvantages of using a compacted fill are: (i) the soil has to be placed in thinner lifts and requires moisture control for compaction; and (ii) the outer slopes probably will be flatter, requiring more land and lessening the benefit of weight over the rockfill shear surface.

Some engineers are concerned that thinning the depth of rockfill may cause the slip surface to move higher and bypass the rockfill altogether. This should be checked during design by using "first time" failure strengths in the landslide mass where the trial slip surface *deviates from the existing slip surface.* In general, a slip surface passing through a dilatant soil develops a discrete shear zone that is appreciably weaker than the soils above and below this zone. Therefore, it is unlikely that a new failure surface will develop, assuming that a sufficient depth of rockfill is built above the toe of the landslide. For guidance purposes, the rockfill should be at least 12 feet above the level where the slip surface meets the inner face of the rockfill buttress.

2. *Provide riprap on the face of the outer slope* to protect the buttress from river floods or high tides (Figure 15.6b). The top of the riprap has to be above the flood and wave runup height as presented in Chapter 11, Section 11.2. The rockfill usually will provide adequate backing for the riprap. In some cases a second layer, provided by quarry spalls, may be needed.

Construction of a buttress into a river channel constricts the channel width. This may increase the height of floods both above and below the point where the buttress is located. The

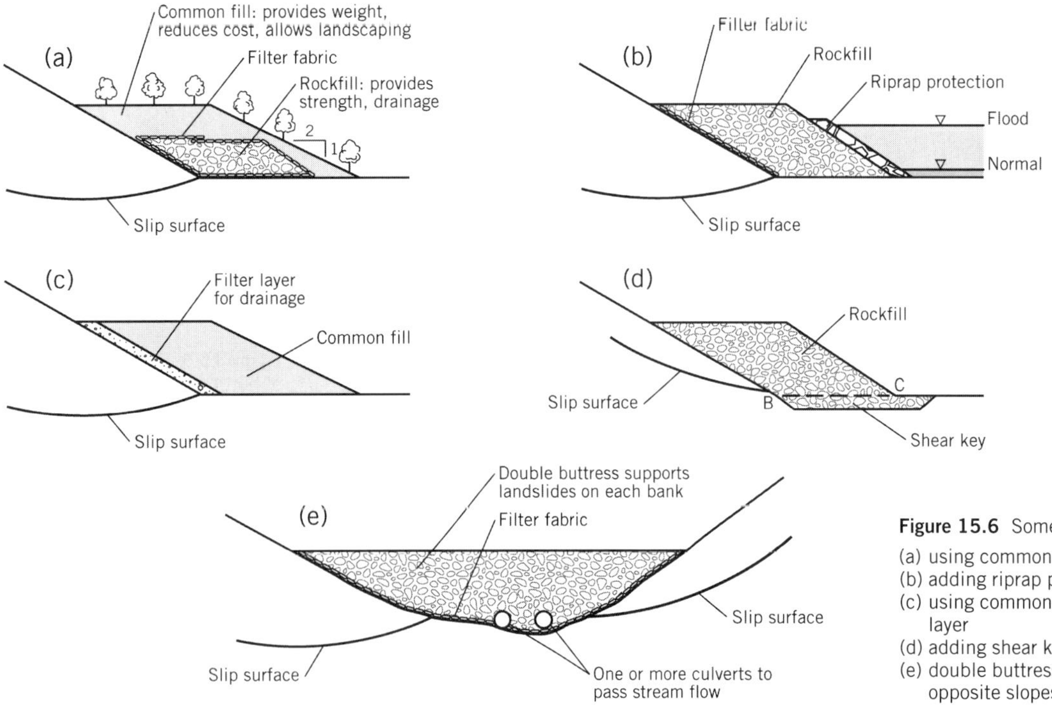

Figure 15.6 Some variations in buttress design:
(a) using common fill to partly replace rockfill
(b) adding riprap protection adjacent to a river
(c) using common fill with an internal drainage layer
(d) adding shear key at base of slope
(e) double buttress to support landslides on opposite slopes

hydraulics of the river channel needs to be checked in most cases. Also, attention has to be paid to the upstream and downstream edges of the buttress to reduce turbulence caused by a an abrupt decrease in the channel cross-section. Case History 8 briefly addresses these issues.

3. *Build the buttress of common fill with an internal drainage layer* (Figure 15.6c). Most landslides have high groundwater conditions at the base of the slope. Where good quality rockfill is unavailable or limited in quantity, an alternative to rockfill is to build a drainage blanket against the slope and use common fill for the bulk of the buttress. Pore pressures should not develop in the buttress during wet weather. A drainage pipe should installed at the base of the drainage layer to carry off the collected water. The drainage blanket must satisfy filter requirements for both the common fill and landslide debris.

4. *Provide a shear key* to ensure that the slip surface passes through the rockfill of the buttress (Figure 15.6d). Where the slip surface at the toe of the landslide is near horizontal or dipping towards the valley, there is a risk that the continuation BC of the slip surface will pass below the buttress. Although the ground below the toe is often bedrock or very hard stable soil, a shear key can be excavated to deepen the buttress. The depth of the shear key can vary from a minimum of about 2 feet into rock (unless it is too hard to rip) to perhaps 6 feet or more in soils, depending on the outcome of a stability analysis of the section. Case History 10, Slide 3, encountered this situation.

5. *Construct a double buttress* to stabilize landslides on both sides of a valley (Figure 15.6e). Streams that are rapidly downcutting may cause active instabilities on both slopes. A buttress built across the streambed can buttress each slope. This solution is only feasible for small side drainages and is comparable with a highway fill crossing a swale. One or more culverts are needed to pass stream flow. There are likely to be environmental issues concerning stream encroachment.

This treatment can be a very attractive option. In addition to stabilizing the two slopes, it may be possible to provide landscaping over the treated area or reduce a sharp bend in a highway that has to cross the stream and pass along both banks.

Several hydraulic issues must be addressed. Inlet grates are essential to prevent the entrances to the culverts being blocked by logs, branches, and other debris. If the culverts are subjected to an extraordinary storm, there is a possibility that their capacities will be exceeded, creating a temporary pond on the upstream side. Another concern is that culverts speed up the flow of water compared to a natural stream. Higher velocities will increase erosion below the culvert outlets, especially since the outflow is concentrated by the culvert opening. If the buttress is long, it may be advisable to build drop catch basins at intervals along the culvert to reduce water velocities. A civil engineer with surface water management experience generally is needed to assist the geotechnical consultant in providing a satisfactory design. Finally, frequent inspection is needed to check that the system is working properly, especially after storms.

6. *Construction from common fill* is an economical procedure for small landslides. If the ground has slumped but not broken up, a small buttress can be built to provide extra stability to the

slump block. This is similar to the slope regrading illustrated on Figure 15.1(b), except that additional fill is brought to the slope to build a toe buttress.

7. *Use of the buttress as a highway or railroad alignment.* In some situations a buttress not only provides an excellent bed for roads and rail tracks but may improve (reduce) the preexisting horizontal curves resulting from past movements of the hillside.

8. A buttress combined with a protective facing (wall or riprap) may stabilize an unstable cliff face and *provide a level surface for waterfront facilities,* such as a dock or promenade.

Construction

Construction of an external buttress is usually straightforward. Fill is brought to the site in trucks, tipped onto the surface of the previously placed materials, and spread and compacted. For a rockfill, spreading by the dozer is all that is usually required (see Chapter 13, Section 13.4). Soil fills require moisture control, thin lifts (8- to 9-inch loose measure) and compaction equipment appropriate for the soil type. Relative compaction to 95 percent of standard Proctor maximum dry density (roughly 92 percent of modified Proctor) should be sufficient. Higher lift thicknesses are used for rockfill (Chapter 13).

The base of the buttress usually requires minor excavation, including clearing of vegetation, and compaction. A permanent drainage outlet usually has to be built first. Ramps or access roads have to be provided for the trucks hauling materials to the site. Silt control fences usually are required.

Examples

Case Histories 8 and 10 include external buttress remediations.

15.4 INFILL BUTTRESS

Technique

An infill buttress (Figure 15.7) reinstates the original slope profile after a flow slide using free-draining rockfill or gravel. Since the eroded soils were mostly saturated at the time of failure and the replacement materials will not develop pore pressures, the slope is more stable. There is no need to perform a stability analysis.

Development of Erosion Bowls

The most common cause of an erosional failure in soils is a *concentrated water source,* i.e., a large water flow impacting a relatively small area of erodible soils. A spring in a hillside is a natural cause. There are numerous manmade causes. A few examples are: (i) a culvert crossing beneath a road fill that separates at a joint; (ii) surface runoff crossing a road after the inner ditch overflows (Figure 15.8); (iii) broken water or stormwater pipes; (iv) a culvert plugged with debris at the upper end that develops a pond behind it during a storm; subsequently, the debris dam breaks, flushing water rapidly through the culvert; (v) trash, tree limbs, and other obstructions put into a drainage channel that block the channel, ponding water during a storm and later unleashing a torrent of water. Most of these causes are avoidable with good maintenance practices.

An erosion bowl commonly develops from an initial small failure that undermines the slope around and above it. If the soil is being saturated by a heavy supply of water, further collapses occur until a much larger mass of loose saturated soil is involved. Eventually the flow slide begins, either as one event or as a series of retrogressive slides moving further back and laterally from the original source. Although sand and loose cohesionless fill are particularly susceptible to this type of fail-

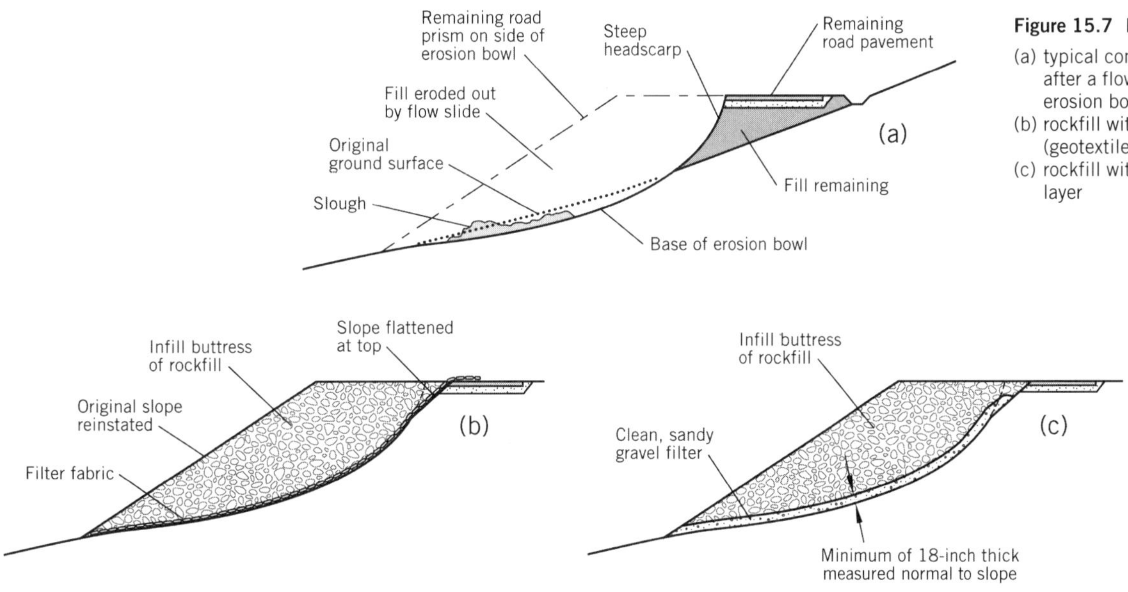

Figure 15.7 Infill buttress:
(a) typical conditions encountered after a flow slide (center of erosion bowl)
(b) rockfill with filter fabric (geotextile) separator
(c) rockfill with sandy gravel filter layer

Figure 15.8 Flow slide caused by water overflowing from a ditch into a road fill (Gerber-McKee Slide, Amboy, Washington).

ure, gravels, and slightly cohesive silts can also develop into flow slides.

A characteristic of most flow slides emanating from concentrated water sources is that almost all of the landslide debris goes downslope, leaving behind a bowl-shaped cavity with near-vertical upper slopes. Any debris within the bowl usually has fallen in after the flow slide, often associated with continuing seepage. Although the empty bowls appear to be dangerous to walk into, they are surprisingly stable once the drying process has begun. If there is no seepage from the sides, they can often be left open for months without any significant deterioration of the walls. Erosion bowls typically range from 20 to 100 feet wide.

Appropriate Uses of an Infill Buttress

This type of buttress applies only to erosional failures. It can be built under emergency conditions as a temporary or permanent remediation for lost ground. If rockfill is dumped into the cavity without a fabric or filter layer (during emergencies) there is a risk that internal erosion will develop in the future as fine-grained soils in the back slope pass into the voids of the rockfill.

For permanent reconstruction, it is advisable to wait until the storm has passed and drier weather returns. This decision has to be made on a case-by-case basis. The sides of an erosion bowl are usually very steep near the original ground surface. The sides may continue to slough into the erosion bowl or become a safety hazard.

Design and Construction

Remediating an erosion bowl is technically simple. The bowl should be backfilled with a free-draining shot rockfill or a free-draining well-graded sandy gravel. Rockfill is the preferred option. The upper edge of a steep-faced erosion bowl may be flattened to 1:1 (horizontal: vertical) as a safety precaution (Figure 15.7b). A filter fabric (geotextile) is laid over the steep surface, which is usually smooth unless there has been sloughing of the sides. For a relatively shallow bowl, a fabric layer may create a weak zone in the reconstructed cross-section. Instead, the design engineer can substitute a thin soil filter layer of sandy gravel satisfying filter requirements (Figure 15.7c). Another rock product that is widely available for base course construction of roads—1-inch minus crushed aggregate—can be used as a filter. The author usually prefers a natural soil filter rather than a filter fabric for infill buttress construction.

Rockfill is placed into the erosion bowl and compacted in lifts in a large bowl; usually, there is no compaction in a small bowl. Compaction is achieved by the tracks of a dozer, front end loader, or backhoe that is used to place the rockfill in 12- to 24-inch lifts. Vibratory compaction is not recommended because the steep side slopes of the bowl could become unstable from shaking.

In a small erosion bowl, the rockfill is often tipped or chuted into the bowl, and the outer slope is dressed as needed. Alternatively, a small bowl can be backfilled by a large backhoe standing outside the bowl.

Other design options that may be appropriate at some sites include: (i) reinstating the slope with compacted native fill (this assumes that the cause of the flow slide has been corrected and will not recur); (ii) providing an inner filter layer against the slope and common fill on the outside, as shown on Figure 15.6(c); and (iii) adding a reverse filter and thin layer of topsoil on the outside of the slope for landscaping. However, the width of a slope failure is usually too narrow for spreading, watering, and compacting native soils without difficulty, and it is much simpler to use rockfill for the entire remediation. For infill buttresses built entirely of rockfill, there is usually no need to include a drain in the base of the backfill.

Difficulties in Constructing an Infill Buttress

The two main problems encountered in remediating an erosion bowl are soft ground on the floor of the bowl (slough) and access:

1. *Slough.* Soft, saturated, landslide debris (slough) can be excavated by a backhoe working from the top or bottom of the slide area, depending on access. Although it is preferable to clean out all slough, minor amounts can be left in place. If needed, some lime or cement can be intermixed with the wet soil to improve the consistency and strength.
2. *Access.* Innovative approaches are needed at sites with very difficult access. At a site near Sandy, Oregon, a flow slide (caused by spring erosion) took out about 100 feet of a trestle-supported flume built against a steep slope. The top of the erosion bowl was 80 feet below the top of the slope. A dozer was lowered into the bowl by a crane, and bags of cement were chuted down to the bowl area. Sloughed soils in the floor of the erosion bowl were intermixed with cement and pushed to the interior (back) of the bowl. Subsequently, this mixture was buried by rockfill that also was chuted down to the bowl from the top of the slope.

At another site, downstream of Bull Run Dam No. 2, Oregon, an erosion bowl occurred with its base near the mid-height of a steep slope. A maneuverable conveyor belt was set up at the top of the slope with its outer end cantilevered over the slope. Fine rockfill (3-inch minus) was dropped from a height of almost 100 feet into the erosion bowl. This procedure pushed out most of the slough as a mud wave, sending it over the edge and down the slope below. Once this fine rockfill had covered the exposed sides of the bowl, a coarser rockfill (90 percent passing 8-inch size) was used to infill the remainder of the bowl. The coarse rock was dozed over the top edge of the slope into the bowl. Below the erosion bowl, loose, waterlogged soils had to be excavated and replaced with rockfill as part of the contract. A backhoe was able to work uphill from the base by building the outer skin of rockfill and then using it as a pad to slowly climb the hillside. Once it reached the erosion bowl, the backhoe completed the backfilling of the bowl and shaped the outer surface. The backhoe came downhill again with the aid of cables anchored to a dozer at the top of the hill.

Examples

This type of remediation is relatively simple. No additional examples will be presented.

15.5 REPLACEMENT BUTTRESS

Technique

A replacement buttress is constructed by first excavating loose soils from the bottom of a landslide and then replacing them with stronger, free-draining rockfill (Figures 15.9, 15.11, 15.12). The slip surface of the landslide passes through the rockfill, which has a higher strength than the soil it replaces. The rockfill in the replacement buttress is drained, which improves stability both within the buttress and immediately upslope.

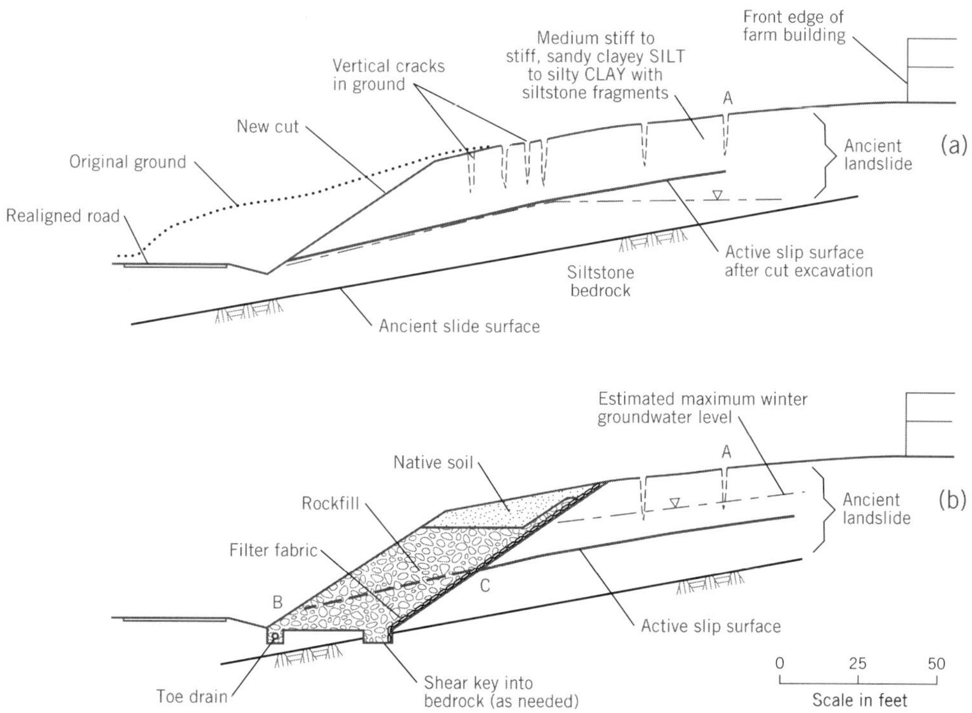

Figure 15.9 Example of a replacement buttress:
(a) landslide conditions
(b) replacement buttress
(Lorane Road Slide, Oregon)

Appropriate Uses of a Replacement Buttress

A replacement buttress is more expensive to construct than a comparable-size external buttress but has distinct advantages in certain situations. They include the following:

- On highway alignments, a replacement buttress allows the stabilized slope face to remain in line with adjoining cut slopes.
- A replacement buttress can avoid the necessity of purchasing more land below the slide; this benefit is especially useful where the base of the landslide is at the edge of a flood plain or other environmentally-sensitive site.
- A replacement buttress can be built into a hillside without any risk of destabilizing the slope below because the original profile of the slope has been maintained and drainage has been improved.

Concerns in the Use of a Replacement Buttress

These include:

- Excavation at the toe of the landslide temporarily undermines the landslide and may cause movements of the ground above the temporary cut slope toward the excavation; closely sequenced construction of the buttress is essential.
- Construction of the replacement buttress generally has to be scheduled for the dry season when the groundwater levels are lower than they were at the time of active landsliding; alternatively, temporary groundwater lowering by deep wells may be needed prior to, and during, reconstruction work.
- It is more difficult to predict final construction costs compared to an above-ground buttress.

Design

The landslide is back-analyzed to determine the shear strength properties of the slip surface. It is commonly assumed that cohesion intercept $c' = 0$. If more than one soil type is involved in the failure surface, it may be necessary to assign soil properties to the soils having short lengths along the shear surface in order to back-calculate the average strength of soils along the major length.

The buttress prism is inserted into the stability section and adjusted as needed to provide the design factor of safety. Suggested properties for hard angular rockfill are presented in Section 13.4. Other soils may be used for backfill, but their effectiveness depends on the strength of the backfill compared to the back-calculated strength of the landslide slip surface.

The back face of the excavation is usually designed for a cut slope of 1:1 (horizontal:vertical). Cuts are made at each end to transition into the full width of the landslide to be remediated.

The base of the buttress should extend at least 5 feet *below* the slip surface within a similar soil formation or at least 2 feet (or refusal) into a hard bedrock. The rockfill should also extend at least 12 feet *above* the slip surface. It reduces the likelihood of the slip surface bypassing the rockfill after remediation; this is highly unlikely in stiff dilatant soils in which the slip surface occurs along a weakened shear zone. As part of final design, it may be necessary to install and monitor additional inclinometers within the proposed buttress footprint to determine the precise location of the slip surface.

A perforated pipe drain is usually installed at the back of the buttress (i.e., toe of the temporary excavation cut). It prevents any significant buildup of groundwater within the buttress and allows water to be taken out of the slope and disposed of by a storm sewer or stream. The disposal of groundwater should be a design item and not left to the whims of the contractor.

The outer surface of the buttress can be designed with a layer of compacted native soils to provide a natural appearance or landscaping of the finished slope. However, such treatments add little to the shear resistance, and requires a deeper initial cut into the landslide to provide the length of rockfill needed for stability of the section. If soil is used on the outer slope, a covering of filter fabric over the rockfill will be needed, and the outer slope will have to be made flatter than that of a rockfill.

Construction

The access to the main treatment area is excavated first; it is usually constructed within stable ground at one end of the landslide. The drain outlet typically is constructed at this time.

The excavation into the base of the landslide temporarily undermines the slope above it. It is often necessary to make the cut in two vertical stages. The first stage typically removes about one-third to one-half of the required volume of excavation, often being above the groundwater table. This first cut will usually remain stable without support. It provides a level bench for the backhoe to move across in the more critical second stage when the full depth of excavation is undertaken.

The second stage of excavation and backfill must adhere to *closely-sequenced construction* as described in Chapter 13, Section 13.5. The excavation proceeds across the face of the landslide from one side to the other (Figure 15.10). As the landslide debris soils are excavated, the backfill is placed as rapidly as practicable to minimize the amount of unsupported face. Sloughing of soils below the water table can be a problem but usually can be overcome by a good backhoe operator and rapid construction. This may require minor deviation from perfection to achieve the required results. Dewatering wells can be installed in advance of construction if excessive sloughing is anticipated due to permeable water-bearing soils at the site.

During closely-sequenced construction the design engineer's site representative has to continually check that: (i) the excavation is completed to the design depths and slopes; (ii) the filter fabric is installed satisfactorily; and (iii) the slope is surveyed before backfilling to determine pay quantities. Cross-sections of the slope typically are taken at about 25-foot intervals along the trench.

If the base of the excavation has a slope along the length of the trench, excavation normally begins at the lowest point

Figure 15.10 Closely-sequenced construction. Excavation is proceeding from left to right across the landslide face. On right, a backhoe loads a dump truck. On left, rockfill is being dumped to build the replacement buttress. Filter fabric (dark gray) is on the slope of native soils behind the rockfill. (Lorane Road Slide, Oregon).

and works toward the higher point. It is usually possible for a flexible plastic drainpipe to be fed into the base of the excavation from the level bench (see Figure 13.1). However, workers have to spend some time at the base of the excavation to place the filter fabric at the bottom of the cut slope. The time spent on this operation should be kept to a practical minimum.

The rockfill is usually built up to the same level as the bench on which the backhoe sits. After the second stage (closely-sequenced) construction has been finished, a third stage occurs when the upper section of rockfill or compacted native soils is added to complete the outer slope profile.

The need for closely-sequenced construction has to be emphasized in the pre-bid meeting of potential bidders and in the bid documents. For this type of work, a list of pre-qualified contractors is preferable to an open bid, but this is not generally permitted on public works construction. The main difficulties experienced on these projects are quality of rockfill, access to the site, good scheduling to avoid delays during the closely- sequenced construction phase, dewatering needs, and occasional weather delays.

Cost overruns can occur due to claims (for the above reasons) or quantities being increased as the excavation is opened up. The owner should be advised of this possibility in advance to ensure that funds are available for a reasonable contingency. The circumstances at each site can be quite variable, and final construction costs may be 25–30% above the original bid price. Assuming that the cost with contingency is still competitive with alternatives, the likelihood of a higher-than-bid final cost should not deter implementation of a very effective landslide remediation technique.

Monitoring

Inclinometers will have been installed as part of the landslide investigation. It is good practice to take readings of the inclinometers before, during, and after the construction of a replacement buttress to provide factual data on the ground behavior.

A structural survey of properties close to the landslide, both upslope and downslope, should be made before any reconstruction begins. It should be remembered that past movements may have opened up ground cracks that may take years to heal after remediation is completed. Inclinometer data can show that stabilization has been achieved, but little can be done to prevent post-stabilization movements due to healing of cracks upslope from the buttress. If needed, an inclinometer can be installed within the rockfill buttress during or immediately after construction to confirm that the ground is stable at the buttress location.

Examples

Landslide Technology has stabilized numerous landslides using the replacement buttress technique. Three examples are briefly described below.

Lorane Road Slide, Oregon

A road-improvement project involved cutting 47 feet horizontally into the hillside and lowering the grade level by 2 feet (Figure 15.9a). Despite dry summer weather, a landslide developed in ancient landslide terrain with many near-vertical ground cracks extending 90 feet upslope from the 30-foot high cut. Inclinometers (not shown on the section) showed that slippage was occurring 16 to 20 feet below the surface with a second (inactive) ancient slip surface at a depth of about 40 feet. Groundwater was at or below the active slip surface. As the slide continued to regress upslope, as evidenced by additional cracking, a farmhouse and outbuildings were being threatened by the reach of the slide.

A replacement buttress was selected to stabilize the landslide and maintain the alignment of the road-improvement project. To estimate the maximum level of groundwater in winter, an Original Profile Analysis (see Chapter 10, Section 10.2) was performed, based on the assumption that the original ground was just stable at the maximum winter groundwater level. (There was no prior history of landslides on this road at this location.) This analysis showed that the water could rise to within 10 feet of the ground surface, as shown on Figure 15.9(b).

Back analysis of the landslide at the time of sliding in August 1994 gave a residual strength of $c'_r = 0, \phi'_r = 12.5°$ on the slip surface.

Design calculations using this residual strength and the estimated winter groundwater levels gave a required horizontal length of 35 feet for the rockfill buttress to provide a factor of safety of 1.30 against failure of either the actual slip surface of August or the deeper slip surface of the ancient landslide. The deeper surface, assuming the same residual strength, was slightly more critical and governed the design.

Filter fabric was placed on the cut slope as a separator between the rockfill and the native clays and silts of the landslide. The upper part of the buttress was completed with compacted native fill. Continuation of the active slide surface BC (Figure 15.9b) passes through the rockfill towards the base of the buttress. A shear key was excavated at the heel of the buttress to intercept the ancient slip surface.

Two stages of vertical cut excavation were needed. For the deep second stage, a closely-sequenced construction procedure was followed, the maximum width between the bases of the excavated soils and the rockfill being limited to 20 feet. A photograph of closely-sequenced construction is shown on Figure 15.10.

The rockfill buttress is 250 feet long and has 50-foot long transition zones at each end. The volume of the buttress is approximately 12,000 cu. yds. and was built in 1994 at a cost of about \$200,000. The Engineering News-Record Construction Cost Index for September 1994 is 5437.

Skyline Drive Slide, West Linn, Oregon

This short (160 feet) but relatively wide (325 feet) slide affected two houses below Skyline Drive. These high-value houses have expansive views of the river, valley, and mountains. The distress, observed at the headscarp and houses, had progressively increased over the previous two to three years.

Inclinometers showed that slippage was occurring within a clayey silt seam that is at the top of the weathered zone above basalt bedrock (Figure 15.11a). The shear zone is overlain by old terrace gravel. Fill had been placed on the slope to bury the footings of pillars supporting the downslope side of one house. Back analysis of sections through the landslide gave residual shear strengths of $c'_r = 0, \phi'_r = 19°$. Groundwater was low within the gravel stratum but emerged as a springline where the eroded edge of the gravel meets the basalt outcrop (point A, Figure 15.11a).

The remediation buttress was a hybrid: part replacement buttress and part external buttress (Figure 15.11b). It takes

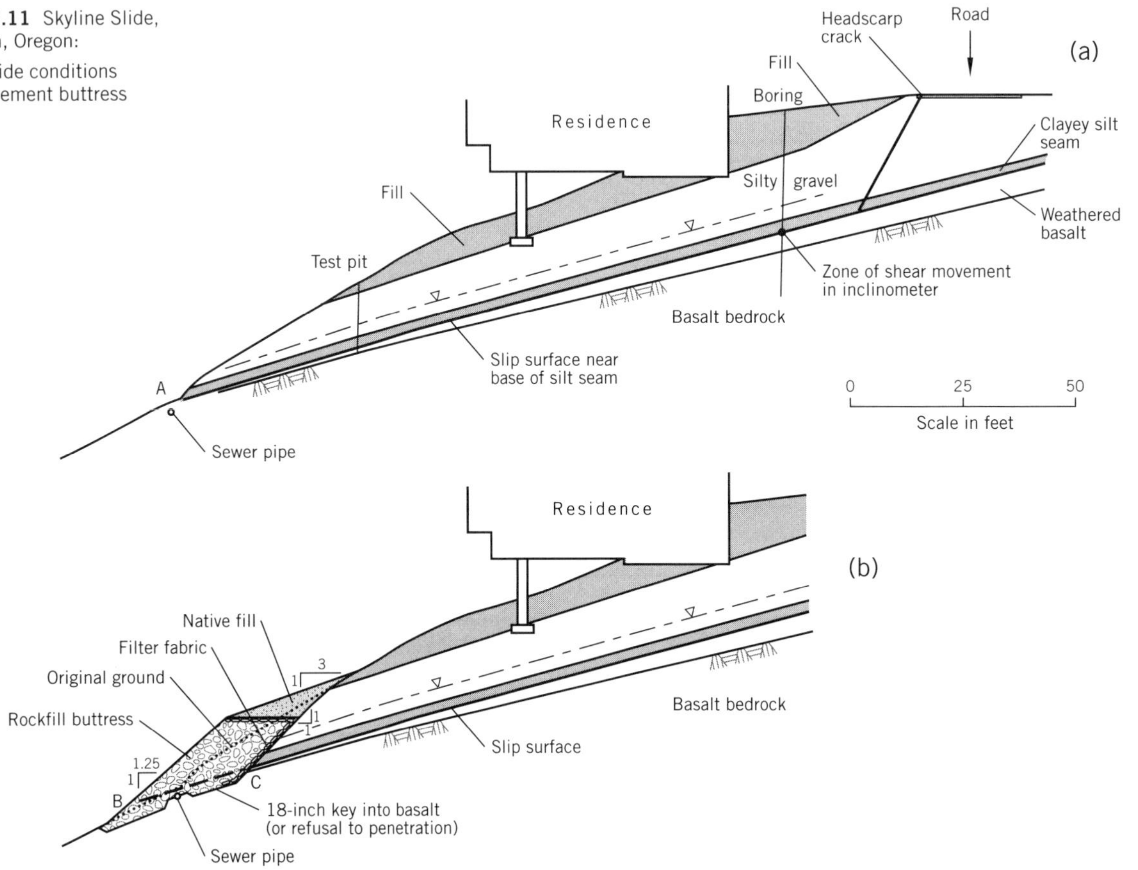

Figure 15.11 Skyline Slide, West Linn, Oregon:
(a) landslide conditions
(b) replacement buttress

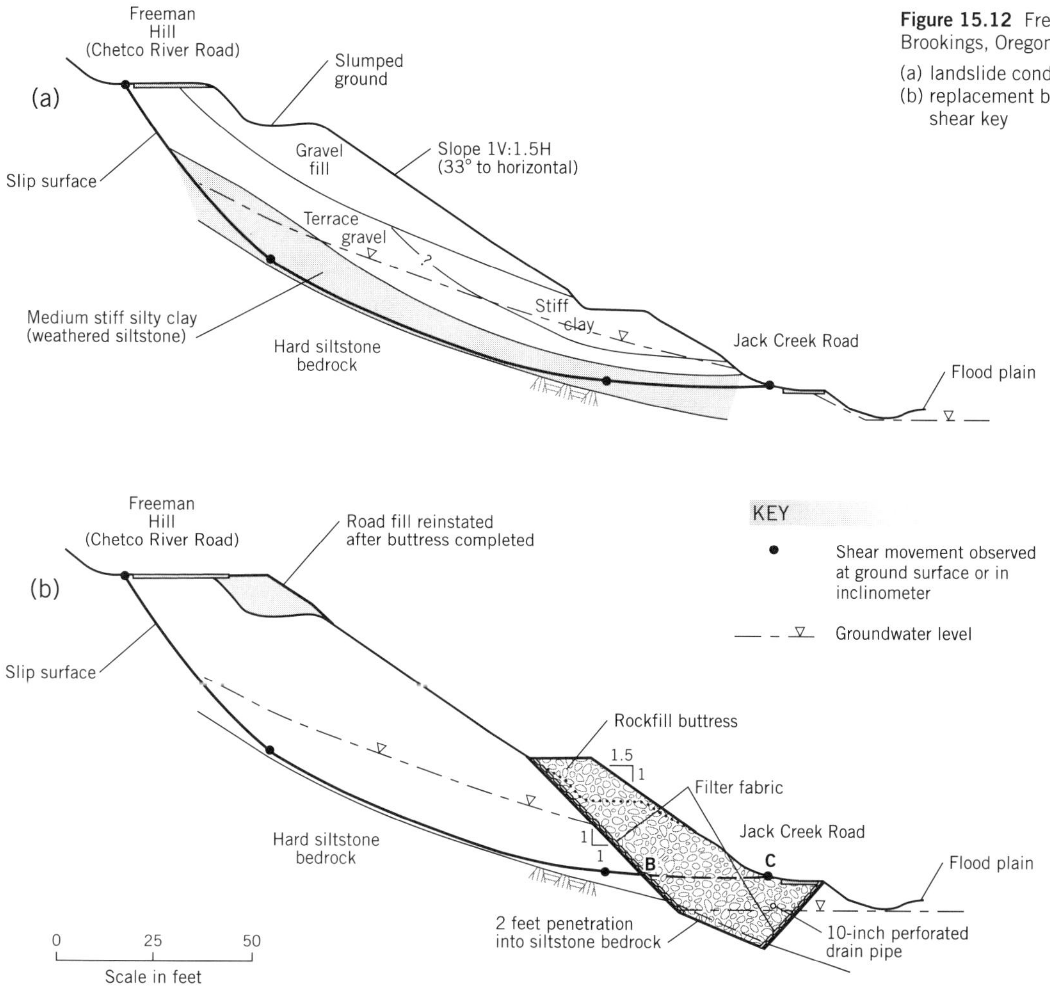

Figure 15.12 Freeman Hill Slide, Brookings, Oregon:
(a) landslide conditions
(b) replacement buttress and shear key

advantage of the eroded face of the gravel to excavate and key into the underlying hard basalt to provide the foundation for the buttress. The temporary excavation for the buttress undermines the slope, with the houses being close to the cut face.

A single stage closely-sequenced construction was used to rapidly excavate the base of the landslide and build the rockfill buttress. Access was difficult, and construction began from the end that provided about 50 feet of completed buttress to support the landslide before reaching the ground below the first house. The work was performed in late summer when groundwater levels are seasonally at their lowest levels. Rockfill lifts (15 inches thick) were compacted by four passes of the dozer tracks. Some native fill, temporarily stockpiled, was used to construct a soil surface above the rockfill. The calculated factor of safety was 1.20.

Built: Fall 1993. Cost: about \$220,000. Excavation: 2130 cu. yd. Rockfill: 4510 cu. yd. Soil fill: 400 cu. yd. Engineering News–Record Construction Cost Index: 5255.

Freeman Hill Slide, Brookings, Oregon.

This slide occurred on a main road in Brookings and required constant maintenance each summer. The cross-section, Figure 15.12(a), shows gravel fill overlying terrace gravels and the weathered zone of the underlying siltstone bedrock. The slip surface passes through the medium stiff silty clay of the weathered siltstone. Groundwater is relatively low but rises during the winter to trigger the annual movements. The main road is about 80 feet above the flood plain at the base of the slide.

Permission to build a conventional external buttress on the edge of the flood plain to support the landslide would have taken years of negotiation and may not have been granted. Instead, a replacement buttress and shear key was constructed into the base of the slope. One lane of traffic had to be kept open throughout the construction period. The width of the buttress was approximately 190 feet.

Back analysis of the slope gave an average shear strength (for $c' = 0$) of $\phi' = 28\frac{1}{2}°$. The calculated factor of safety after remediation is 1.20 for $\phi' = 42°$ in the rockfill.

The remedial replacement buttress/shear key (Figure 15.12b) was built in the early fall of 1996. Construction was difficult because of the need to cut a temporary 1:1 excavation directly below an existing slope of 33° to the horizontal.

The contractor excavated the buttress area in two vertical stages. The first stage removed the upper wedge of soil to create a bench for the backhoe during the second stage. The second stage followed a closely-sequenced construction procedure to build the shear key and main section of the rockfill buttress. Some minor problems were experienced from side slope sloughing and seepage. These were exacerbated by some delays in the backfilling operation. As added assurance for stability, the county decided to raise Jack Creek road, at the base of the slope, by a few feet.

Built: October, 1996. Cost: $221,000 bid, excluding extra work. Engineering News–Record Construction Cost Index: 5719. Excavation: 5250 cu. yd. Rockfill for buttress: 10,350 tons. Reinstatement of road prism at top of slope: 1,430 tons of rockfill and base rock.

15.6 SHEAR KEY

Technique

A shear key (Figure 15.13) is a backfilled trench constructed across the width of a landslide to intercept a slip surface. Stabilization is achieved through the higher shear resistance of the backfill compared with the original ground, and the benefits of drainage. It is a very effective remedial treatment, especially where a hard angular rockfill replaces clay at residual strength. Shear keys are often used in association with a buttress; the buttress provides overburden weight to increase the shear strength across the base of the shear key.

Appropriate Use for a Shear Key

Key trenches are most commonly used where the slip zone is 10 to 30 feet below the ground surface at the proposed stabilization part of the slope. If the slip zone is less than 10 feet deep, the rockfill strength may be only marginally higher than the in-place landslide strength and there is also a higher risk that the slip zone will subsequently break out to the surface upslope from the remediation. If the slip zone is more than about 30 feet deep, the rapidly increasing volume with increasing depth makes the technique less cost-effective compared to alternative treatments. Deep excavations also increase the risk of upslope instability when the trench is open.

Key trenches are especially suitable for clay and clayey silt overburden in landslides with low residual strength. The rockfill strength of around $\phi' = 42°$ is much higher than the residual strength of clay, and provides a significant increase in factor of safety for a relatively small excavation volume.

The backfill can be any soil stronger than the soil on the failure surface, including sands and gravels. If the landslide comprises sands and silts below the water table, dewatering before construction is needed.

Shear keys are well suited to remediate shallow translational landslides such as colluvium. These landslides frequently occur above siltstone and hard clays, or hard rock with a weathered surface.

For small landslides, a conventional strutted excavation (similar to a sewer trench) backfilled with rockfill may be sufficient to stabilize the slope. Drainage through a pipe at the bottom of the backfill should be provided to drain the rockfill and increase the shear resistance. This technique is useful where space at the base of a slope is limited.

Limitations of Shear Key Construction

Limitations include:

- Excavation temporarily undermines the landslide and may cause movements of the ground above the temporary cut slope towards the excavation; closely-sequenced construction (Chapter 13, Section 13.5) is essential.
- Construction of the shear key generally has to be scheduled for the dry season when the groundwater levels are lower than they were at the time of active landsliding (except when groundwater lowering is performed ahead of construction).
- The required top width of the trench can thwart the use of a shear key at some sites due to property ownership, adjacent facilities, or natural features.
- Deep shear keys may be unable to drain by gravity if they adjoin a reservoir, flood plain, or similar features. Rockfill below the groundwater level has a buoyant weight, reducing the shear resistance of the shear key compared to one that is fully drained.

Design

Additional site investigation borings are usually needed to determine the position of the slip surface across the area selected for the shear key. This entails a "Specific" site investigation phase, as described in Chapter 3, Section 3.1. At many sites the slip surface is approximately planar. In such cases, contours representing the slip surface can be plotted on a topographic map of the site to show the likely depth of exca-

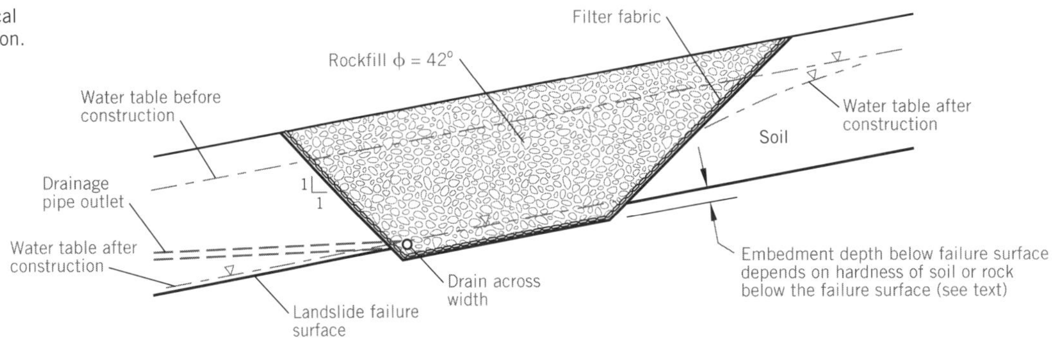

Figure 15.13 Typical shear key remediation.

vation at all parts of the shear key. This technique can also be used to find the most favorable location for the shear key, i.e., where the overburden is thinner than elsewhere.

A back analysis of the landslide is performed to determine the shear strength properties along the slip zone. The width and depth of a shear key to achieve the selected factor of safety is obtained by additional stability calculations.

A typical shear key section is shown on Figure 15.13. The base of a shear key is typically 15 to 30 feet wide and has temporary excavation slopes of 1 : 1 (horizontal : vertical) on the sides.

The preferred backfill option for a shear key is angular shot rockfill obtained from a hard rock quarry. The shear strength properties and construction methods for hard rockfill are given in Chapter 13, Section 13.4.

Rockfill requires a geotextile (filter fabric) at the contact with native soil to prevent fines from entering the large voids of the rockfill. This requirement is especially important on the uphill side of the trench where groundwater enters the rockfill, and near the drainage pipe where hydraulic gradients may become high during storms. The geotextile fabric on the downhill side of the trench is sometimes omitted if the water table has been permanently lowered by the key trench. However, geotextile fabric is a minor cost item, and it is recommended that all rockfill/soil interfaces be provided with a geotextile fabric separator (Figure 15.13).

Where slippage is occurring at a discrete shear plane, rockfill should extend at least 12 feet *above the slip surface*. This should prevent the landslide from breaking out above the rockfill.

The base of the shear key is excavated below the landslide slip surface. If hard bedrock underlies the slip surface, it may be feasible to key only a few inches (say ½ foot) into the bedrock surface. In soils, embedment probably should be at least 3 to 5 feet below the slip surface to avoid any risk of the slip surface re-routing below the shear key. Generally, landslides have a discrete shear zone, the soil within the shear failure being appreciably weaker than the soils above and below. One benefit of a shear key is that the geotechnical engineer usually can observe the exact position of the slip surface during construction and adjust the base depth if needed. If the ground immediately below the slip surface shows evidence of shear disturbance, the engineer should evaluate the situation and possibly recommend deepening the shear key to extend it past the entire disturbed zone.

A drainage pipe is usually placed at the lowest elevation of the shear key to collect groundwater and conduct it away from the landslide. The drain is sized according to the potential inflow from groundwater and surface runoff; typically, the drain is in the 6-inch to 12-inch diameter range and is made of flexible plastic for ease of construction. If significant amounts of water are expected to drain from the shear key in winter, the drain outlets should be routed downslope to an existing ravine or stream.

Shear keys built alongside a river, reservoir, or flood plain may be unable to completely drain. In these cases, the buoyant density has to be used for rockfill below the water table, which provides a lower shear strength than a fully drained rockfill. This design issue also applies to deep shear keys where it may not be practicable to provide a gravity drain outlet.

Due to the shape of a key trench section, the cross-sectional area increases rapidly with increasing depth below the surface. Formulas for calculating the cross-sectional area of a key trench are given on Figures 15.14 and 15.15. On Figure 15.14, the base length L is parallel to the ground surface at depth D (infinite slope of a translational slide). On Figure 15.15, the base length L is horizontal with depth D at the centerline of the trench. In both formulas the side slopes are 1:1.

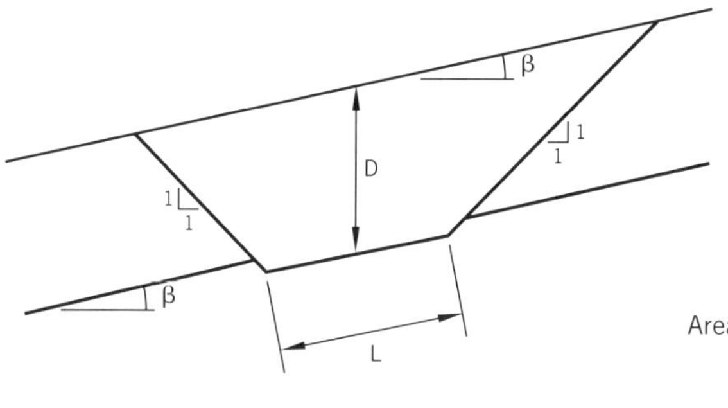

$$\text{Area} = D\left\{L\cos\beta\,(1-\tan^2\beta) + D\right\}$$

Figure 15.14 Section for calculation of shear key area for constant depth D and 1:1 side slopes.

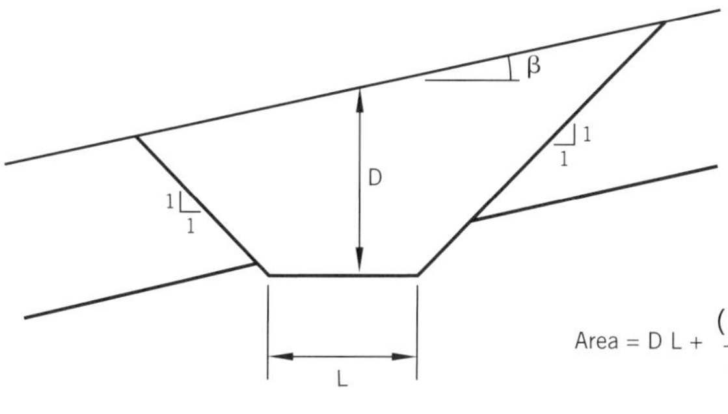

$$\text{Area} = DL + \frac{\left(D - \frac{1}{2}L\tan\beta\right)^2}{2(1+\tan\beta)} + \frac{\left(D + \frac{1}{2}L\tan\beta\right)^2}{2(1-\tan\beta)}$$

Figure 15.15 Section for calculation of shear key area for horizontal base and 1:1 side slopes.

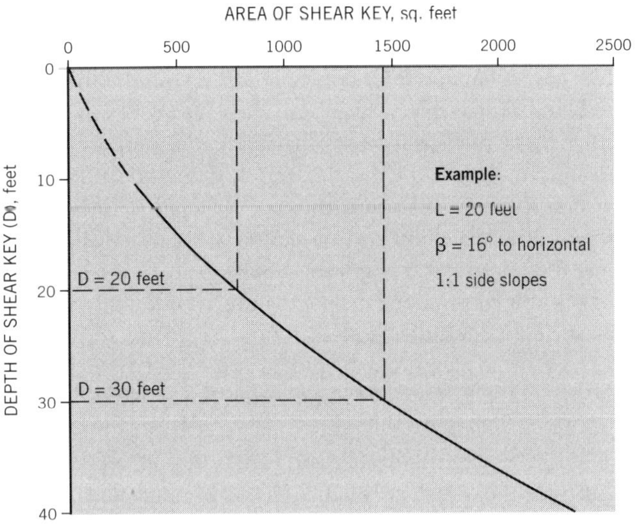

Figure 15.16 Example calculation to demonstrate the rapid increase in area (and volume) with increasing depth of a shear key.

The rapid increase in area with depth is illustrated by an example calculation on Figure 15.16 for a sloping base. For this example, increasing the vertical depth of cut by 50% from 20 to 30 feet almost doubles the area (and volume) of the shear key. The practical limit for a shear key construction is about 30 feet below the surface at the center of the trench. At greater depths, groundwater inflow and uphill stability issues become increasingly difficult to overcome, and volumes escalate. In general, shear keys are a medium cost remediation alternative.

Variations in Shear Key Design

Two common variations in shear key designs include:

1. *Build a buttress above the shear key.* As shown on Figure 15.17, the buttress adds overburden pressure to the rockfill and increase the shear resistance. The outer edge of the buttress, point X on Figure 15.17, should extend to, or slightly past, the vertical line above point C (where the slip surface BC passes out of the shear

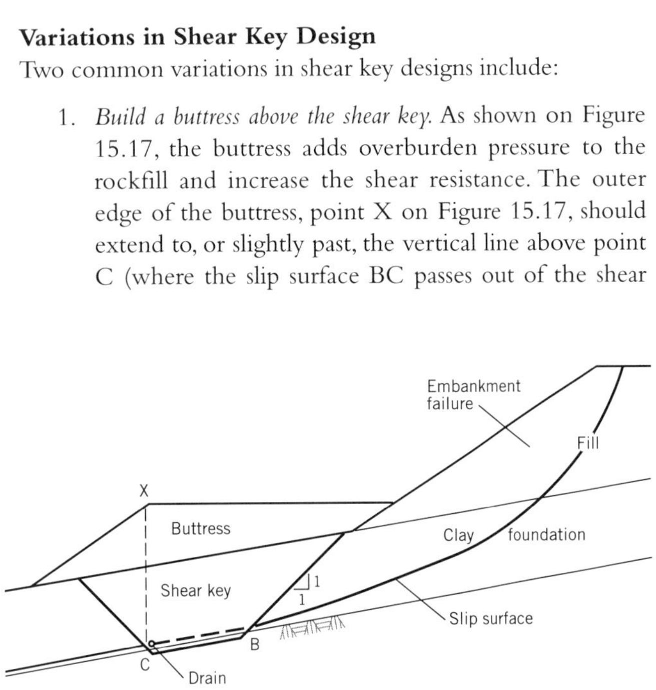

Figure 15.17 Shear key combined with a buttress.

key) to obtain the full benefit of the buttress weight. The buttress can be constructed of common fill. Combinations of shear keys with buttresses are sometimes built as a preventative measure in new construction of fills over a weak foundation.
2. *Replace the upper part of the key trench with native soils (Figure 15.18).* The soil can be temporarily stockpiled near the site. The soil provides weight to increase the normal stress (and, therefore, the shear strength) acting on the rockfill at the slip zone. However, there should be sufficient rockfill cover above the slip zone to prevent a shallower breakout of the landslide through the soil backfill, as described earlier.

   Soil fill at the ground surface can be landscaped. Geotextile fabric should be placed above the rockfill before placing soil fill above it (see Case History 8).

Construction

Excavation for a shear key cuts a slot across the width of the landslide and temporarily undermines the slope. To prevent reactivation, the remedial work should be scheduled near the end of the dry season when groundwater levels are at or near their lows. Construction of the shear key should adhere to closely-sequenced construction practice, as described in Chapter 13, Section 13.5. An excavated trench should *never* be cut across the width of a landslide without concurrent backfilling or full structural trench support.

Shear keys penetrating into waterbearing gravels, sands, and silts should be dewatered before the trench is excavated. In silty soils it may require several weeks of dewatering to lower the groundwater to the desired levels. When close to buildings, dewatering may cause settlements or other undesirable consequences; these possibilities may preclude the use of a shear key at such locations.

A typical construction sequence for a shear key is as follows:

The contractor builds an access road to the site, usually at both ends of the trench. Dewatering wells, if specified, are installed to lower the groundwater levels. The site is cleared of trees and topsoil. A first-stage excavation and creation of a working bench is achieved by excavating the upper soil to about 12 feet above the slip zone. Temporary cuts are 1:1 on each side of the trench.

Beginning at one end of the trench (the downhill side, if there is a longitudinal slope to the trench), an outlet drain is built downslope from the shear key to daylight on the natural slope. A backhoe then digs out the first section of the trench for a length of about 30 feet. Fabric is laid across the trench from uphill to downhill and is staked at the uphill/downhill edges. Lengths of fabric are added, with 2 feet overlap at the edges, to cover the exposed earth on the sides and base of the trench.

Rockfill is then dozed into the hole to form a ramp entering the excavation from the end of the trench. This first lift has minimal compaction. Thereafter, the backhoe excavates

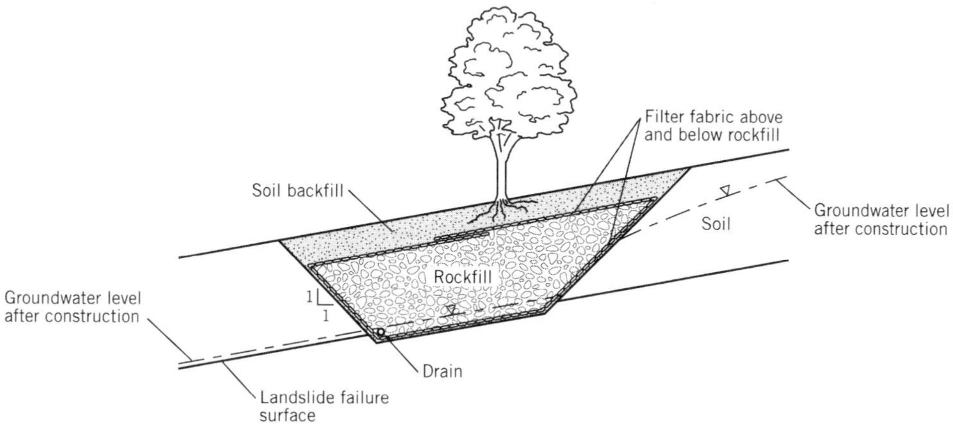

Figure 15.18 Shear key comprised of a rockfill overlain by soil backfill.

another length along the trench, fabric is placed, and rockfill is dozed into the excavation on sloping lifts, 12 inches thick, to infill the shear key. Trucks carry off the slide debris from the backhoe's end of the excavation. In this way, excavating and backfilling proceed across the hillside until the shear key is completed. When the rockfill is completed, fabric can be placed across the top and a soil layer may be added to allow landscaping of the surface.

Pay quantities can be determined from an initial site survey followed by cross-sections taken at 25- or 50-foot intervals as the excavation progresses to determine in-place excavation and backfill quantities. An alternative is to base the contractor's compensation on truck loads, weighed or by volume. Agreement on quantities between the contractor and the owner's representative needs to be made daily.

The construction of a shear key remediation should be performed under the full-time, on-site observations of a geotechnical engineer from the designer's office. The geotechnical engineer's role is to ensure that the project is constructed to meet the intent of the design, as spelled out in the contract plans and specifications. The engineer may also assist with quantity surveys and make other technical decisions to facilitate the work. However, the engineer should not accept responsibility for safety issues on the construction site that are customarily the responsibility of the contractor.

Monitoring

It is strongly advised that inclinometers installed on or around the landslide be read during and after construction of the shear key, especially where structures, buried utilities, and other facilities sensitive to ground movements are close to the slide area. Should a problem or claim for damage arise, the inclinometers will provide factual data on the ground behavior.

Structures close to the landslide, including those to the sides, probably should have structural surveys of existing distress made before and after remediation. A meeting with property owners in the vicinity of the slide may help to reduce anxiety and create understanding of the stabilization process.

After Remediaton

Ground above a shear key is often badly cracked by slide movements. As cracks heal over time, ground displacements may occur upslope of a shear key for months or years after construction. Usually, nothing can be done about ground cracks but, if severe enough, sand or grout infilling of cracks may help to reduce post-stabilization movements.

Examples of Shear Key Projects

West Percy Slide, Leaburg, Oregon

Percy Slide is a fairly complex landslide in which an ancient landslide of very stiff, sandy, clayey silt is sliding over weathered andesite bedrock. The more active portion of the slide had been stabilized earlier by an external buttress, but the untreated west side continued to creep downslope and was dragging on the west end of the buttress.

Back analysis of a section through the middle of the slide area, Figure 15.19(a), gave a shear strength ϕ'_r of 18° for $c'_r = 0$ using assumed winter groundwater levels 10 feet above the measured levels. For the design shear key, Figure 15.19(b), the calculated factor of safety was 1.26.

The construction required a deep cut (AB) into the sloping hillside at a slope of 1:1. Mature trees had to be taken out and the owner wished to replant the slope with trees after the shear key was built. Accordingly, the rockfill was 18 feet deep and the ground above it was backfilled with the native soils excavated from the trench. The key trench was 325 feet wide and had a longitudinal slope of 6½ percent. A drain (pipe) in the base of the trench had an outlet cut through the downslope side CD.

Shear key construction was scheduled for the dry season. Trench excavation was carried out in two stages. The first stage took the ground surface down to 18 feet above the design base BC (Figure 15.19b). The second stage was closely-sequenced construction: the backhoe excavated the landslide soils, and rockfill was infilled behind it as the construction moved across the width of the landslide from the lower toward the upper level of the longitudinal slope of the trench base.

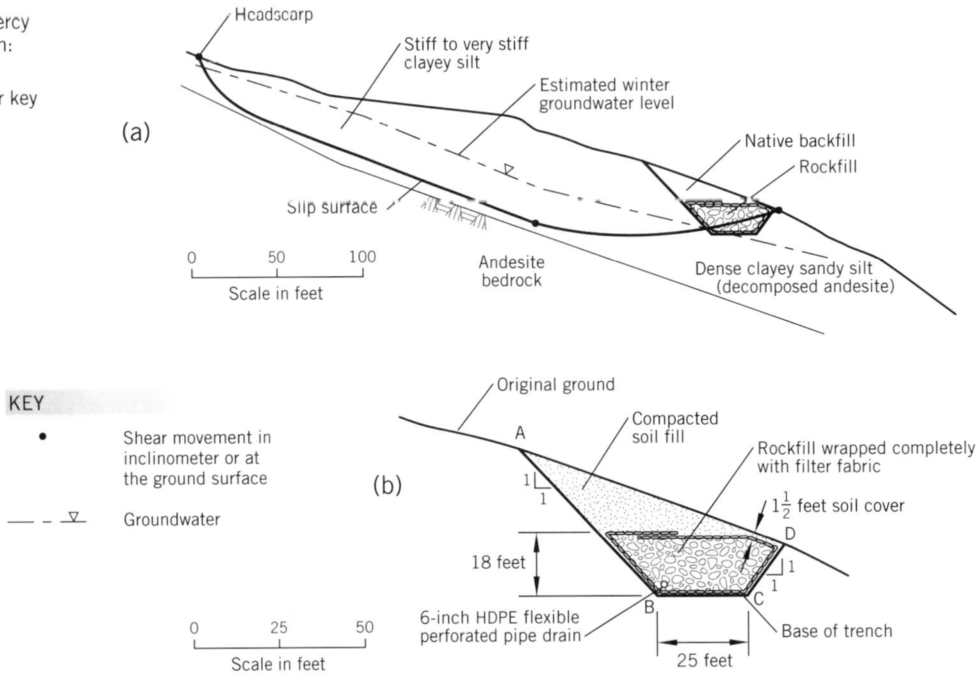

Figure 15.19 West Percy Slide, Leaburg, Oregon:
(a) stability section
(b) details of the shear key

The maximum length permitted between the bottom of the excavation slope and the bottom of the replacement rockfill was 25 feet.

During remediation, the slip surface of the landslide could be seen on the slope faces being excavated by the backhoe, including the uphill slope AB of the trench. The disturbed zone was about 1 foot thick; within it a 1-inch thick seam of soft clay was observed.

Construction: 1997. Volume of excavation: 21,000 cu. yd. Volume of rockfill: 9,800 cu. yd. Volume of compacted soils: 10,850 cu. yd.; Bid price: $406,700. Engineering News-Record Construction Cost Index: 5851

Hagg Lake Perimeter Road Slide 4, Gaston, Oregon

This translational landslide is described in Case History 10. A shear key was constructed near the top of the landslide to support the road and stabilize the upper part of the slide.

Freeman Hill Slide, Brookings, Oregon

This slide was remediated by a combination of a shear key with a replacement buttress (see Figure 15.12 and earlier description of this slide remediation).

15.7 EARTHWORK SPECIFICATIONS FOR COMPACTED FILL

This section provides comments on writing technical specifications for compacted fill. It is intended as a guide, and is not intended to be model specifications or all-inclusive. Practices vary by region, and the recommendations are based on Landslide Technology's practices in the northwest United States. The appropriateness of the techniques for a particular site has to be evaluated by the design geotechnical engineer.

Properties and specification requirements for filter soils and hard, crushed rockfill (shot rock) have been previously given in Chapter 13, Sections 13.3 and 13.4. Riprap design and specifications are presented in Chapter 11, Section 11.2.

Subgrade Preparation for Fill

Trees, stumps, roots and other shrubs should be removed from the fill area (termed "clearing and grubbing"). Topsoil is often left in place, but can be specified for removal. If the ground surface is disturbed, it should be scarified and rolled. The purpose of rolling is to detect soft areas which, if found, should be overexcavated and replaced with fill. A smooth-drum roller is the best equipment for this job but an alternative is to allow the contractor to drive a loaded dump truck or scraper over the subgrade to detect soft spots.

Springs or lines of seepage are often found on landslides. Before placing fills, the springs should be cleaned of loose debris and covered by at least 12 inches of drainrock (meeting filter requirement, Chapter 11, Section 11.1). If there are significant flows of water (> 2 gpm), it is advisable to collect the springwater into plastic or concrete pipes and drain downslope off the site. In spring areas, perforated pipe is used to pick up water, and solid pipe is used to move the water downslope. A 6-inch diameter HDPE pipe is usually sufficient for small amounts of drainage, and should be laid at a slope of at least 2 percent. It may be advisable to put the pipe in a shallow trench, embedded in drainrock, to avoid damage from heavy construction equipment.

Large boulders on the surface of a landslide can be broken up by a steel ball dropped by a crane or with a hydraulic hammer. If there are numerous large boulders, they can be left in place and infilled with a 12-inch minus crushed rock that is carefully placed to prevent the formation of large voids. The infill materials can be tamped by the back of a backhoe bucket or vibrated by a vibrating hammer or tamper.

Compacted Fills

Compacted fills are commonly referred to as *engineered fill* or *structural fill*. Organics must be excluded from such fill. Fill is brought to the site, spread in horizontal lifts, and compacted. For clays, silt, and sand fills, the uncompacted lift thickness is set at a maximum of 8 to 9 inches.

On sloping ground steeper than 4:1 (horizontal: vertical) the fill is keyed into the hillside by excavating shallow steps (benches) into the native soil. This procedure avoids an unbroken sloping contact between the fill and subgrade that could become a plane of weakness.

Clay fills can be constructed from heavily overconsolidated clay in which the natural water content is close to the plastic limit. A criterion that the water content should be less than 1.2 times the plastic limit has worked well on many of the fat clays encountered in southern England; this provides an undrained shear strength of at least 1000 lb./sq. ft. in the compacted clay. However, a suitable criterion has to be developed for clays of the specific locality. Clay fill placed at the optimum moisture content for compaction is stiff, and loaded scrapers and other construction vehicles should cause only minor ruts on the fill surface. As a general rule, it is not feasible to dry out clays that are too wet for compaction. This is sometimes required by contract specifications, and invariably leads to costly claims. The best option is to waste clay that is above optimum moisture content and find an alternative fill source. Clay fills cannot be constructed under wet winter conditions.

Clay fills are compacted by a kneading action. A sheepsfoot or clubfoot roller provides excellent compaction. A rubber-tired roller is also very good; on small projects, a loaded scraper or dump truck may be sufficient. A minimum of three passes is recommended to ensure that all areas of the fill surfaces receive sufficient compaction.

Quality control of clay fills is governed by one of two standardized compaction tests: the standard Proctor (ASTM Designation D 698) or the greater compactive effort of the modified Proctor (ASTM Designation D 1557). Since these standard tests are occasionally revised, the controlling revision should be stated (e.g., ASTM D-1557-91) in case disputes arise. In the United States, the modified Proctor has almost superseded the standard Proctor because the results of the test are in closer conformance with the compaction achieved by modern construction equipment.

The compaction required in clay fills for landslide remediation depends on the purpose of the fill. If it merely supplies weight (e.g., on top of a buttress) or is being used for landscaping purposes, a minimum compaction level of 85 to 90% of modified Proctor is sufficient. For clay fill that is to provide strength for stability, the specified minimum level of compaction may be set as high as 95% of modified Proctor. In practice, the average compaction achieved is likely to be 1 to 3 percentage points above the minimums. However, any laboratory triaxial shear tests on compacted specimens for design purposes should be tested at the minimum specified relative compaction.

Density of the compacted fill is usually measured by a nuclear densometer, which obtains readings that can be converted to wet density and moisture content. The nuclear densometer should be periodically checked for accuracy.

There are potential problems with compaction control of clay fills. One problem is that some clays change properties when they are dried out and rewetted. Clays containing the clay mineral halloysite are especially affected. It is preferable to air dry clays slowly on the laboratory floor (with frequent remixing to keep the moisture evenly distributed) rather than to oven dry samples before the test. Compaction points are obtained at and around optimum by adding water or air drying. At least five moisture content points are needed to properly define a compaction curve.

A second problem with compaction control is the presence of oversize stones. The ASTM–D 1557-91 test specifies a maximum particle size of ¾ inch in the larger 6-inch diameter mold. Oversize stones have to be removed (difficult to do in a stiff/hard clay) and then corrected for in the maximum dry density calculation. There are ASTM procedures to make these corrections, but they become significant changes in density as the oversize fraction increases. An associated concern is that a reading of the nuclear probe could be taken very close to a stone buried in the fill, possibly giving a high density reading that is unrepresentative of the general fill. When the percentage of oversize stones exceeds about 10 to 15 percent by volume of the sample, which is not uncommon in colluvium and other landslide soils, it is preferable to switch from a field density criterion to a "methods" criterion for compaction control. In a "methods" specification, the control is exercised through specifying the compaction equipment and number of passes for each lift of fill. In the case of clay fills, a moisture content criterion for the fines can be added.

Silt fills are similar in most respects to clay fills. However, weakly cohesive and non-cohesive silts require special attention to filter requirements when placed in contact with other soils and rocks.

Sand fills have different characteristics to silt and clay fills because of their cohesionless nature. *Fine* sand responds well to compaction by the wheels of a rubber-tired roller, scraper or vehicle. Compaction of fine sand is only slightly improved by vibratory compaction equipment.

Medium to coarse sands require vibratory compaction such as four passes by a 10-ton vibratory roller. Sands require both weight and vibration to become denser, and the upper surface often rebounds from the energy of the vibrating roller. Compaction tests have to be taken a few inches below this surface layer to measure the degree of compaction achieved. To obtain high relative compaction, a good supply of water needs to be available to continually spray the sand as it is being compacted.

Sand can be difficult to compact in the laboratory compaction test because the drop of the compaction test hammer causes the adjacent sand in the mold to jump and loosen. If this happens (generally occurring in very clean sands), the test does not achieve a dense state. An alternative is to base the field compaction on relative density (see Chapter 7) rather than on relative compaction. The maximum density achieved in the relative density test will be higher than in the modified Proctor compaction test. The relative density result is based on the difference between maximum and minimum densities. In specifying relative density, a criterion for acceptable compaction in the field could be set at 70% relative density. A second alternative is to revert to a "methods" specification, in which the design engineer specifies the type and size of equipment, minimum number of passes, lift thickness, and moisture requirements.

Gravel fills and sandy gravel (without cohesive fines) generally are compacted by vibratory methods. If a clayey gravel is to be used, it should be treated as a clay fill (with gravel) when specifying compaction requirements.

Clean gravels and sandy gravels should be monitored by a method specification. A minimum of four passes of a vibrating roller is desirable for each lift. The uncompacted lift thickness is usually 12 inches.

A well-graded sandy gravel is easily segregated into component sizes by spreading equipment. The larger stones tend to move around the end of a dozer blade or be dropped out at the end of each spread. This segregation can be partly alleviated by using scrapers or by dumping from a truck on the move to avoid large heaps. Segregation can affect the ability for adjoining soils to act as filters.

Clean sandy gravel, gravel, and rockfills are known as "all weather" fills. They can be placed and compacted during inclement weather because rainfall passes through the fill without ponding on the surfaces or softening the soil. "All weather" fills are especially valuable for emergency repairs at landslides.

CHAPTER

16

Erosion Control Measures

Erosion is one of the principal causes of landslides. Ground is lost, leading to collapse of the undermined and steepened slopes, followed by a shear failure or a flow slide. Some erosive agents have been described in Chapter 2, and include:

- Sea attack on sands, gravels, and weak rocks
- Waves generated on rivers, lakes, reservoirs
- Strong currents in rivers and ocean shores
- Seepage and springs
- Rainfall runoff affecting silts and sands
- Weathering of soft rocks from freeze/thaw, rainfall, etc.
- Development of sinkholes through soils of widely different gradations
- Concentrated water flows, such as overflowing ditches, blocked conduits, broken pipelines and culverts

Cohesionless soils are especially vulnerable to erosion because their strength is derived from the weight of overburden acting on them. At the ground surface, there is essentially no resistance to erosion unless vegetation is present. Gravel, sand, silt, and slightly cohesive silt can be eroded. Clays with liquid limits above 40 and plasticity index higher than 15 are resistant to erosion in most circumstances.

The remediation of erosion-caused landslides includes replacement of lost ground combined with slope armoring, walls, or other erosion-resistant facings to prevent future erosion. Most erosion failures can be prevented, and this chapter provides techniques to control erosion before a major slide occurs or to reinstate an eroded slope.

16.1 FILTER SYSTEMS

Technique

Filter systems serve two principal functions in preventing or remediating landslides. They (i) allow seepage pressures within the slope to be relieved without loss of ground, and (ii) protect a slope of fine-grained soils from erosion by rainfall and other external water sources.

Filter systems can be graded filters or a single filter backed by filter cloth (geotextile), as shown schematically on Figure 16.1.

Uses and Limitations of Filter Systems

Filter systems have several uses:

- To provide a filter blanket that covers the surface of a slope to protect exposed fine-grained soils from erosion.
- To act as a separator between fine-grained soils and rockfill to prevent internal erosion.
- To prevent surface erosion at springs or lines of seepage on a slope.

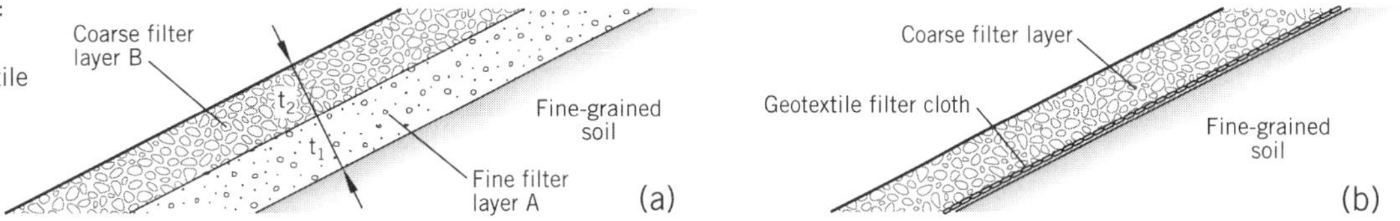

Figure 16.1 Filter systems:
(a) graded filters
(b) single filter with geotextile filter cloth

- To protect drainage pipes from clogging with fine-grained sediment.
- To protect slopes from wave action and stream currents; this normally requires an outer layer of riprap, excluded from this section but presented in Section 16.3.

The main difficulties in using filter systems are (i) availability of suitable soils for filter systems, (ii) the need for additional washing or processing to meet tight gradation specifications, and (iii) avoidance of segregation during placement.

Graded Filters: Design

The fine-grained soil to be protected from erosion is covered by successive layers of coarser soils. Each layer has to meet the filter requirements. On Figure 16.1(a), filter layer A has to meet the filter requirements for both the fine-grained soil below and the coarse filter layer B above. The design of the filters is described in Chapter 11, Section 11.1.

A graded filter system has to release the pore pressures within the slope so that there is no pore pressure gradient within the outer layer B arising from groundwater levels within the protected fine-grained soil. This result is achieved by strict adherence to meeting the permeability criterion in filter design (Chapter 11, Section 11.1).

The thicknesses t_1, t_2, measured normal to the slope (Figure 16.1a) need to be sufficient to ensure that the intent of the design is met and there is a representative amount of each gradation in place (see Table 16.1). The two concerns are: (i) segregation of the soils during spreading; and (ii) thin zones created by uneven surfaces below. Segregation can occur when construction spreading pushes the coarser stones to the ends and edges of the spread. Short spread distances minimize this effect.

The hardness and durability of the stones are important requirements because the filters should not fracture or crush during placement or deteriorate during service. These standards are set by national specifications, such as ASTM and AASHTO in the United States.

Filter layers usually have to be imported and the required specifications may require extra processing at the rock quarry crushing plant. Wherever possible, gradations should try to fit with a commonly-available standard, such as state highway department specifications, to avoid the cost of extra processing. Samples should be collected from the quarries for analysis. It is important that the filter materials are not gap-graded and that the percentage of fines passing the No. 200 sieve are small (less than 5% by weight on wet sieving) and nonplastic. In critical applications, the fines passing the No. 200 sieve may be restricted to only 3% but this can lead to extra construction costs to comply. It should also be noted that *wet* sieving (the recommended procedure) generally gives a higher minus 200 percent than *dry* sieving.

Table 16.1 Recommended Thicknesses of Filter Materials in a Two-Layered Graded Filter

	Thickness t Measured *Normal* to the slope		
Type of Filter	Minimum	Normal	Critical
Fine filter A	6 inches	12 inches	up to 3 feet
Coarse filter B	12 inches	18 inches	up to 4 feet

Minimum: low slopes, small jobs; Normal: most basic projects; Critical: higher slopes, risk of failure unacceptable.

Note: For a typical 2 horizontal : 1 vertical slope, the horizontal width is 2.22 times the thickness normal to the slope.

Graded Filters: Construction

Filter layers should be placed from the bottom towards the top of a slope. This technique provides better control of the layer. Spreading from the top causes stones to roll down the entire slope at many sites. Also, it is advisable to dump filter soils in short heaps, rather than a single large heap, to minimize segregation during dumping and spreading. Another option is to use controlled dumping from a moving truck so that the filter soils form a windrow at the base of the slope to be treated. Since filter soils are always cohesionless, tracked lightweight spreading equipment should be used to reduce ruts and other unevenness of the surface.

Compaction of the filter layer can be achieved by winching a smooth-drum roller up and down the treated slope. For small quantities of filter, hand-operated power tampers can be used. However, for many landslide applications the filter layers are placed against a relatively steep slope face (typically 2:1 or 1:1, horiz:vert.) and compaction may cause excessive disturbance and be counterproductive. No compaction of thin filter layers is often the best choice.

At sites of heavy seepage flows, such as a spring actively eroding the ground above and beside it, the fine filter and coarse filter may be added sequentially such that the finer of the two filters is rapidly overlain by the coarser filter before the spring can wash out the fines fraction. In such cases, the coarse filter may be riprap, or riprap becomes a final (third) layer on the outside of the remediation.

Field observations during placement by the geotechnical engineer or owner's representative is essential.

Graded Filters: Hypothetical Example

A hypothetical example of a landslide stabilized by a graded filter system is demonstrated on Figure 16.2. The slope has water-bearing sandy colluvium overlying stiff clay. Groundwater level is near the mid-height of the colluvium.

The owner makes a 1:1 cut into the slope to build a water storage tank. It exposes the sand/clay contact (Figure 16.2a) and erosion of the sand begins at the line of seepage. The eroded ground regresses upslope as the sand face caves off and washes away. Eventually, a larger landslide occurs due to the loss of support at the steep face (Figure 16.2b). The slide covers much of the proposed structural pad and spurs the owner into action to stabilize the landslide area.

Two possible remedial schemes involving two-layer filter systems are shown on Figure 16.2(c) and (d). In Figure

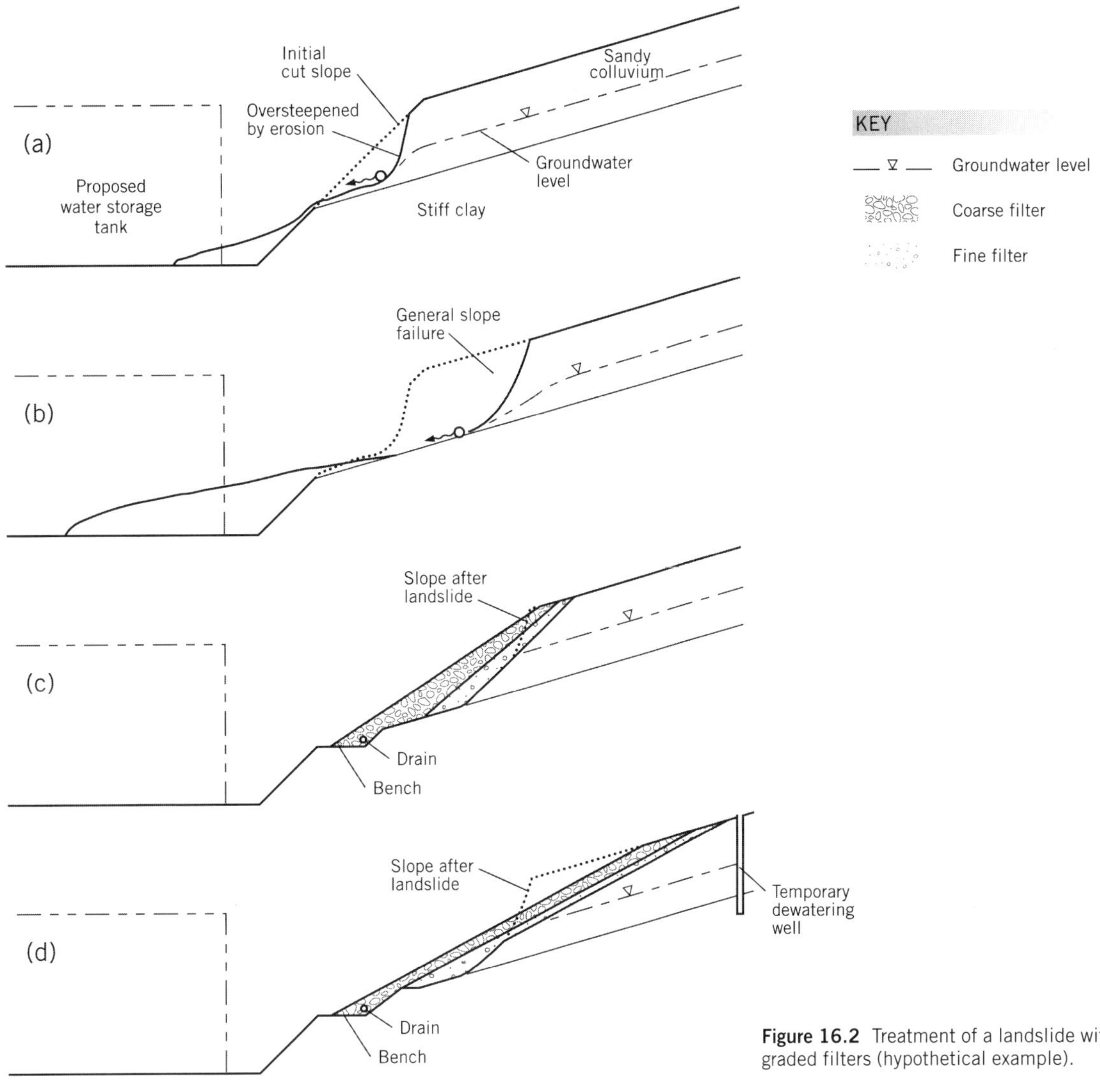

Figure 16.2 Treatment of a landslide with graded filters (hypothetical example).

16.2(c), the contractor first cuts back the steep headscarp to a 1:1 temporary slope. Assuming the soil is excavated from the top (requiring an access road) a thick layer of fine filter could be placed by a backhoe, possibly with an outer slope of about 40° (angular crushed rock).

The contractor then could work from the bottom of the slope by first cutting a bench in the clay to key the toe of the outer rockfill. A drain would be installed on the inside of the bench to collect seepage water and carry it away from the structural pad to a storm drain system. The outer rockfill could be placed to a slope of about 32°. Groundwater seepage would enter the fine filter, pass down the clay surface, and be collected by the drain. Stability analyses are needed to check that the design is satisfactory.

Dewatering of the site during construction may be needed. Dewatering wells could be installed above the work area to dry up the groundwater flow before remediation is begun.

The second option, Figure 16.2(d), involves more excavation and would take longer to accomplish. It would likely require dewatering as previously discussed. The native sandy soil would be cut back to a stable slope (shown as 2:1, horizontal to vertical). The two-layer filter system is thinner and essentially: (i) prevents erosion below the water table; and (ii) drops the groundwater levels at the outer face of the slope. Stability calculations would be needed to check that the filter layers are thick enough to provide long-term stability. A benefit of this second option is that a flatter final slope of 2:1 (horizontal:vertical) would be less intimidating to people working at the water tank site below the slope.

These two options are intended to demonstrate principles and not actual designs. No stability calculations have been made and each of the options may require thicker filter layers than shown to provide an adequate factor of safety. In addition to stability concerns, the whole issue of constructibility has to be examined for a particular site, and the input of earthwork contractors should be sought in many cases. There are also issues of land ownership, threat to adjacent property, and time of reconstruction that affect the decision-making process.

Single Filter Backed by Geotextile Filter Cloth: Design

A geotextile filter prevents the fines of the fine-grained soil from passing through the geotextile, yet has a relatively high permeability. The selection of geotextiles for particular applications is discussed in Chapter 11, Section 11.3.

A major advantage of geotextile filter is that it is relatively inexpensive, easy to install on most slopes, and readily available. Fine filter soil, by contrast, can often be difficult to obtain in some areas and generally is expensive because it usually has to be manufactured to specification and hauled from a hard rock quarry.

There are some situations where a geotextile filter has limitations for landslide work. First, the soil surface beneath the geotextile needs to be smooth; however, in many landslide situations, such as the head of an erosion bowl, it would be dangerous to try to smooth out overhangs, roots projecting into the bowl, and other surface irregularities, before anchoring the geotextile. (*Note:* High strength or thick, nonwoven geotextiles are sometimes specified to overcome this particular problem). Second, many erosion sites have springs or running water present. Geotextile is easier to place on a damp or dry surface. Third, the use of geotextile on a steep headscarp may introduce an interface of weakness between the native soil and the coarse filter (usually rockfill). The significance of this weaker zone depends on the individual circumstances, but the past failure of a landfill slope (Seed et al., 1990) at the soil/geotextile contact has highlighted this weakness. In each of the three situations just described, a filter layer generally is preferable to a geotextile.

Single Filter Backed by Geotextile Filter Cloth: Construction

Geotextile filter cloth is laid from top to bottom of a slope in one continuous piece. It is temporarily held in place by stakes, but may require anchoring by burial in shallow trenches at one or both ends of the treated area. Edge overlap is usually 2 to 3 feet. The roll widths range from 12 to 15 feet. If two strips are used to form one length up the slope, the adjoining edges can be sewn or overlapped in a shingle style. A single continuous strip is preferable.

16.2 REVERSE FILTERS

Technique

Reverse filters are simply a variation of the graded filter system in which the objective is to *protect the coarse rock layer* rather than an erodible fine-grained layer. In a reverse filter layout, the coarse filter is on the inside (being protected) and the fine filter is on the outside. The integrity of the rockfill for use as a drain is the key objective of the filter system.

Uses and Limitations of Reverse Filters

There are several situations associated with landslide remediation and prevention in which an open rockfill has to be protected from infiltration by fines, mostly to ensure good drainage. Some examples are:

- *Interior drain* (Figure 16.3). A graded filter and reverse filter are built together with fine-grained soils on each side. This creates a highly permeable drain that improves slope stability. An interior drain can be built into buttresses (Figure 16.3) and shear keys to improve drainage, or can be built on the outside of a rockfill to allow planting or improved appearance.
- *Sinkhole repair* (Figure 16.4). This is perhaps controversial but reverse filters have shown effectiveness in remediating sinkholes; the longer-term effectiveness for this use is unproven at this time.

The limitations in use are the same as those listed for filters in Section 16.1.

Design of Reverse Filters

For filter design, see Chapter 11, Section 11.1.

Interior Drain

Use the thicknesses given for a two-layered graded filter in Table 16.1 except that the coarse filter can be doubled in width. As shown on the example application of Figure 16.3, a toe drain pipe should be included along the base of the landslide. At the top of the interior drain, the drain can be sealed to exclude surface runoff, or the surface runoff can be allowed to feed into the interior drain. Most engineers seem to favor sealing off the top to prevent the interior drain from becom-

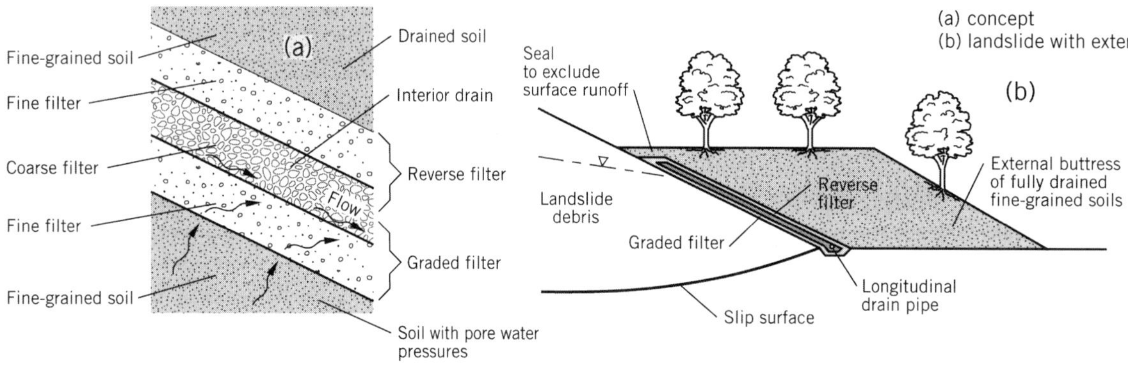

Figure 16.3 Interior drain produced by reverse filters:
(a) concept
(b) landslide with external buttress

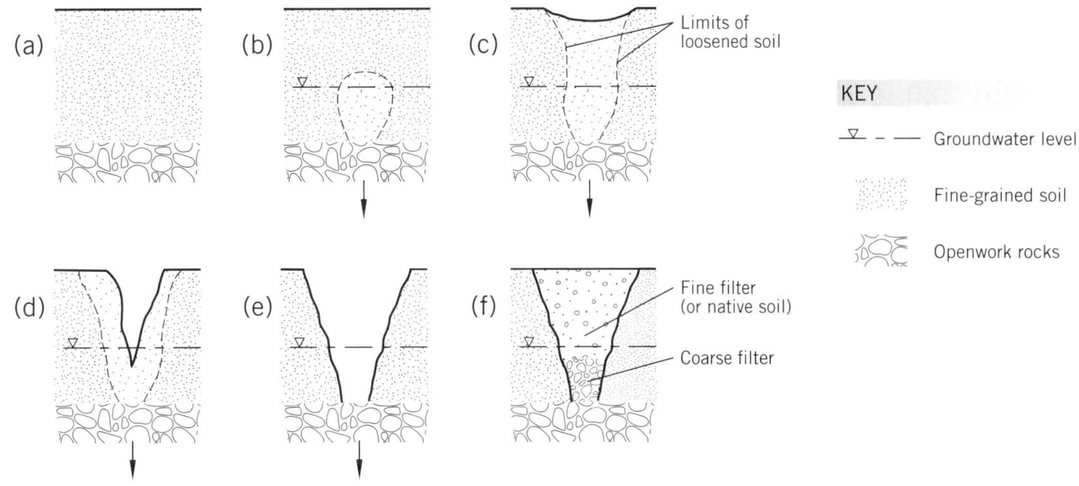

Figure 16.4 Development of a sinkhole and treatment with reverse filters.

ing contaminated with debris or being overloaded with water during a storm. A separate drainage system can be provided for surface runoff.

Sinkhole Repair

Sinkholes occur in many landslides as a result of either: (i) a heterogeneous mixture of soils and rocks that create internal erosion, or (ii) soils falling into broken pipes or other subterranean cavities. Nevertheless, they can cause excessive leakage and increase groundwater levels at some distance from the sinkhole. In most cases, the sinkhole is simply the surface manifestation of internal erosion deeper in the ground. Water pressures from a reservoir or canal may provide the groundwater flow that triggers the internal erosion.

The direct cause of the sinkhole (excluding broken buried pipes) is due to finer-grained materials being washed into larger void spaces by water flow and/or gravity. An example would be a cluster of boulders on an old riverbed that has been covered by fill or bank failure debris.

Large landslides often have clusters of broken rock brought down from higher ground and are engulfed by finer-grained soils. The seasonal rise of groundwater or slow infiltration of rainfall runoff over many years can cause the fines to infiltrate the cluster of broken rock and eventually lead to a sinkhole at the ground surface.

Figure 16.4(a) shows a cluster or layer of rocks overlain by a finer-grained soil; the two materials do not meet filter requirements with respect to each other, but are sufficiently compact and damp that there is little infiltration of the fines into the large voids of the broken rock. At some later date, a project causes groundwater levels to rise above the contact between the two materials. Fine-grained materials start to migrate down into the rock voids, initially producing a loose zone, shown enclosed by broken lines on Figure 16.4(b). With passage of time, the zone of ground being fed into the voids reaches the surface, and a ground depression develops, Figure 16.4(c). As internal erosion continues, a sinkhole is seen at the ground surface and the passageway may reach the layer of broken rocks, Figure 16.4(d)(e).

The reverse filter approach to repair a sinkhole is to infill the lower part of the sinkhole with a fairly coarse graded filter that is intermediate in gradation between the clustered rock and the finer-grained overburden above it. The specified filter should be designed to satisfy the filter requirements with each of the other two materials. In practice, the size of the rock fragments may not be known and it may have to be guessed based on local geological conditions. A better option is to dig an exploratory test pit or put down a boring to check the nature of the rocky deposit.

The upper part of the sinkhole is backfilled with a fine filter or, if appropriate, the native soil at the site. If the sinkhole has reached the coarse layer, Figure 16.4(e), the coarse filter will cover the open rock fragments and prevent further loss of finer-grained soils into it, Figure 16.4(f). If the sinkhole is less developed and loose sand still covers the open rock fragments, Figure 16.4(d), ground settlement could occur as the lowermost loose sand continues to erode into and through the open rock. However, eventually the reverse filter will reach the contact and perform the task of preventing further losses. Thus, the sinkhole may have to be periodically "topped up" with the fine filter or native soil until settlements stop.

Assuming that the sinkhole has a conical shape and is wider at the ground surface than below, a benefit of the reverse filter approach is that continued erosion below or bypassing the coarse filter simply causes the coarse filter to sink and continue to wedge into the disturbed ground. Once the openwork gravel or boulders has been reached, the coarse filter overlain by fine filter should effectively stop further erosion.

The above procedure, based on a limited number of cases, does seem to work short term, with the longer-term permanency still unproven. Sinkholes are extremely difficult to analyze. Owners may try grout, tamped clay, pouring concrete in the hole, etc. in attempts to correct the ground loss or leakage. Most of these efforts fail because the groundwater erodes around the plug of concrete or clay. The reverse filter approach allows the leakage source to be eventually sealed as continued erosion puts the coarse filter backfill into direct contact with the underlying gravels or rocks.

Other methods of treating sinkholes are available. Compaction grouting seals the voids of the coarse rock to prevent infiltration. In other cases, excavation is made to reach the coarse rock layer and remove or seal it.

Construction of Reverse Filters

See "Graded Filters: Construction" in Section 16.1.

A key requirement in constructing multiple filter layers is to avoid cross-contamination between different filters within a limited construction area. Good control by the contractor, and continuous observation by the engineer's onsite staff, is needed.

16.3 RIPRAP SLOPE ARMOR

Technique

Riprap is large, angular shot rock that is heavy enough to resist displacement and uplift from strong river currents or waves. In most cases, it is part of a two- or three-layer graded filter system.

Uses and Limitations of Riprap Slope Armor

As a preventative measure against erosion, riprap is used for jetties, marinas, oceanfront structures, earth dams, highway and railroad embankments adjoining or crossing water, outer bends of eroding rivers, and other locations vulnerable to waves or stream action. In landslide remediation, riprap is needed to repair slope failure into a river or sea and for buttresses built into river channels.

The main limitation in using riprap is the difficulty of obtaining the larger sizes of rock needed for riprap. As a substitute, specially manufactured concrete shapes are available to form an interlocking outer surface. Some public entities dislike riprap, and may not permit it to be used for aesthetic or environmental reasons.

Design of Riprap

Details of riprap design are given in Chapter 11, Section 11.2. Different design approaches are needed for: (i) the direct attack on shorelines from waves, and (ii) the velocity damage caused by rivers in flood.

Riprap has to be carefully specified, and is difficult to manufacture at a hard rock quarry. There is a significant amount of wastage, especially if a single size riprap is specified, but the spalls can usually be reprocessed or used as a backing filter layer.

Some common specification requirements for riprap are that the stone used should be angular, hard, dense, durable and resistant to weathering and water action. It should contain no overburden soils or organics. The longest dimension of each stone should not exceed three times its shortest dimension. Other test requirements (density, durability, and resistance to weathering) are listed in Table 16.2.

A fairly common procedure by public agencies in the United States is to separate riprap specifications into classes. A typical set of specifications is: Class 1: maximum weight 33 lbs.; Class 6: maximum weight 3,500 lbs.; with the other classes intermediate between these limits. An "equivalent" cubic dimension can be specified and provides a field inspector with a visual means of assessing compliance with the specification. For a 33 lb. stone, the equivalent dimension is 7 inches; for a 3,500 lb. stone, the dimension is 34 inches.

Table 16.2 Laboratory Test Requirements for Riprap Stones

Specific gravity (bulk-saturated, surface dry)	greater than 2.57 (density 160 pcf)
Absorption 24-hour	maximum 5%
Los Angeles abrasion test (ASTM C-131, AASHTO T96)	maximum loss 20% at 500 revolutions
Sulphate soundness test* (AASHTO T104)	maximum loss 10% after 5 cycles
Freeze/thaw test**	maximum loss 10% after 10 cycles

Purposes: *exposure to salt water **resistance to weathering

The design of the edges of riprap need special attention to protect them against undercutting by eddies. For the same reason, any transitions in bed thickness also need attention in design. The thickness of riprap should be doubled at the edges, and the outer edge may need rounding at the upstream end. The toe detail should extend below the anticipated scour depth of the river in flood. The outer slope can be designed to a maximum of 1½ horizontal : 1 vertical for hand-placed or machine-placed riprap, although some procedure manuals limit the maximum slope to 3 horizontal : 1 vertical. A flatter slope is required for *dumped* riprap.

Construction of Riprap Slopes

The quality of riprap depends on the blasting operation at the hard rock quarry. The spacing of the blast holes, rock joint characteristics, and size of the explosive charge determine the sizes and gradations of rock. It often requires experimentation to optimize the final product. Sorting can be achieved by screens. Larger pieces can be broken by a wrecking ball.

Quality control of the actual gradation has to be assessed visually at the site. It is sometimes helpful to obtain a sample of 5 to 10 tons of riprap that meets the specification and dump it at the construction site so that the field inspector can use it as a reference for materials being brought in by the contractor.

Dumped riprap is tipped onto the filter layers directly from trucks. It is not spread, which would cause some segregation of sizes. Similarly, chutes should not be used. Dumped riprap requires some handwork, usually by men with steel bars or by the bucket of a backhoe or dragline. In general, there should be minimal reworking of dumped riprap.

An interlocked surface of large, hand-placed riprap is built from bottom to top of the slope using a clamshell, skip, or other device. Every effort must be made to avoid segregation, and the riprap cannot be built in layers. Rearranging of stones may be needed to produce riprap that conforms to the specified gradation. Construction tolerances are ±6 inches from the drawings below water and ±4 inches above water. The final product should be a dense uniform mass of rocks with no apparent voids or pockets.

Examples

Case History 2 (Beaver Slope Erosion) is an example of riprap used to repair a riverbank failure caused by erosion. Case History 8 (Faraday Slide) describes riprap used to protect a buttress from river currents.

16.4 GROUTED RIPRAP

Technique

Grouted (mortared) riprap is achieved by filling the large void spaces with concrete or a cement mortar. The grouted riprap provides a sloping rock wall or hard shell that does not allow water to pass through. It creates a heavy rock mass that cannot be lifted by waves or turbulent flow.

Uses and Limitations of Grouted Riprap

Grouted riprap can be used at sites with high exposure to erosion from the sea or actively-eroding river bends. A slope of grouted riprap below the 100-foot high spillway at Bull Run Dam No. 2, Oregon, is shown on Figure 16.5. The slope protection was built around 1962 and is still effective (year 2004) in protecting the soil slope behind it from the very turbulent water of major floods passing over the spillway chute.

The more common difficulties in using grouted riprap protection are: (i) obtaining permits to construct in rivers; (ii) gaining access to the site; and (iii) placing the grout in the dry.

Design of Grouted Riprap

The grout must totally infill the spaces between the large rock fragments. Normally, the riprap and grout are placed concurrently. A backing layer of a graded filter is usually built behind the grouted riprap. The filter acts as a drain for any seepage coming out of the slope behind the structure. Drainage pipes should be placed below the filter soils so that seepage water is conducted away from the site laterally (i.e., parallel to the width of the slope). Weepholes can be built through grouted riprap, but it is possible that turbulent water outside the slope will pull filter materials through the weepholes and cause internal erosion and possibly failure of the slope protection.

The base of the slope protection should extend below the river bed or beach level. The depth of burial should exceed any future scour depths when the river is in flood.

The design should be the same as for hand-placed riprap. The grout simply provides additional protection. Should small amounts of grout be eroded at the surface, the riprap should remain stable. However, periodic inspections should be made to check the facing of grouted riprap and any flaws should be repaired quickly.

Permission to build a grouted riprap slope protection within a river channel may require time-consuming negotiations and strict guidelines set by regulators. It may be possible to build it during the dry season by a similar procedure to a replacement buttress described in Chapter 15, Section 15.5, provided that slope stability can be achieved.

In another design choice, the Federal Highway Administration has used steel dowels, grouted into predrilled holes, to provide extra resistance to erosion (Figure 16.6). The dowels tie the grouted riprap to the bedrock of the slope base.

Construction of Grouted Riprap

Construction of the grouted riprap face has to be done under dry conditions and this is a major factor in design and construction. Construction should be scheduled for low river

Figure 16.5 Grouted riprap (on right of photo) remediating an erosional landslide scar adjacent to the stilling basin of the 100-foot-high Bull Run Dam No. 2 spillway, Oregon.

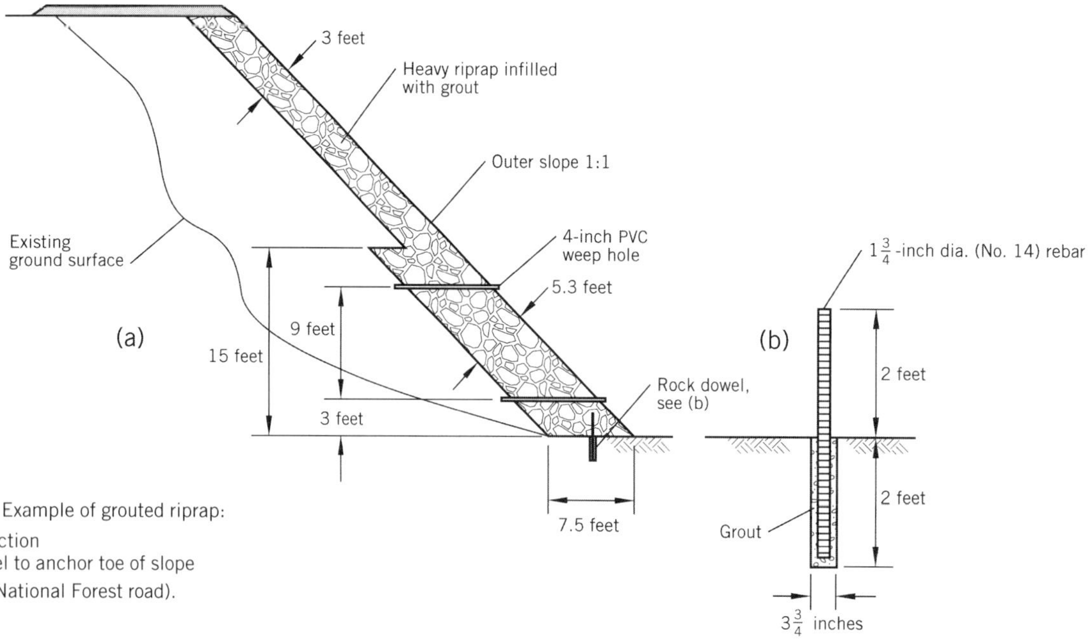

Figure 16.6 Example of grouted riprap:
(a) typical section
(b) rock dowel to anchor toe of slope
(Wenatchee National Forest road).

flow conditions. In addition, a temporary diversion and cofferdams may be needed to allow the site to be dewatered. Sufficient room has to be provided to allow the base of the grouted riprap to be keyed into either bedrock or the estimated maximum depth of scour.

The grout must fill the void spaces within the pre-washed riprap without displacing the individual stones. The riprap can be raised in 3- to 4-foot lifts before grouting each completed lift. The grout should contain sufficient aggregate (e.g., $\frac{3}{8}$-inch minus) to provide an erosion-resistant bond.

Example

A grouted riprap design was used for a Washington State forest road to build an embankment fill into an outer river bend (see typical section, Figure 16.6a). In this case, a very steep 1:1 outer slope was used and the toe of the riprap protection was anchored to hard rock at the base of the slope by rock dowels (Figure 16.6b). The dowels were spaced at 4-foot intervals horizontally. The dowels were grouted up from the base of the drillhole and allowed to cure for 7 days before placing riprap above them. Weepholes were provided at 9-foot vertical intervals and 20-foot horizontal intervals within the grouted riprap near the bottom of the embankment.

16.5 GABION MATTRESSES

Technique

Gabions (see also Chapter 19, Section 19.9) are rectangular baskets made of wire mesh that are filled with small rocks. They are built by hand, first filling the wire mesh basket and then wiring the lid (made of the same mesh) into place.

A gabion *mattress* has a shallow depth compared to the width and length. Several mattresses can be joined together (Figure 16.7) to provide erosion protection to a slope.

A gabion mattress is a possible substitute for riprap. It functions as a massive block to resist the uplift forces of currents or waves.

Uses and Limitations of Gabion Mattresses

A gabion mattress is economical where larger riprap stones are unavailable or the quantities needed are small. Therefore, the principal advantage of a gabion mattress is that smaller stones can be effectively used to provide erosion protection to a slope.

There are some disadvantages for use in rivers:

- The wire basket is susceptible to corrosion.
- The gabion mattress is susceptible to damage from debris abrasion; plastic covered wire is easily stripped
- High labor costs are required to fabricate and fit into place.
- Compared to a riprap slope, they are less flexible and are more difficult and expensive to repair.

Experience has shown that once the wire has been broken the gravel is soon washed away. Also, mattresses laid on steep slopes tend to distort, thereby exposing the filter layer behind the mattress to river erosion. There has also been concern about vegetation growing through the mattresses. This does not occur where the rocks are frequently submerged, and is not considered to be a major problem.

Design of Gabion Mattresses

The mattresses are laid end to end and side to side on the prepared slope. They need a fairly solid foundation because their ability to flex is limited. The slope has to be well prepared for the mattresses, and a filter layer is placed beneath it. The toe design and edge designs require careful attention. The mattress or a conventional gabion protects the base of the slope from

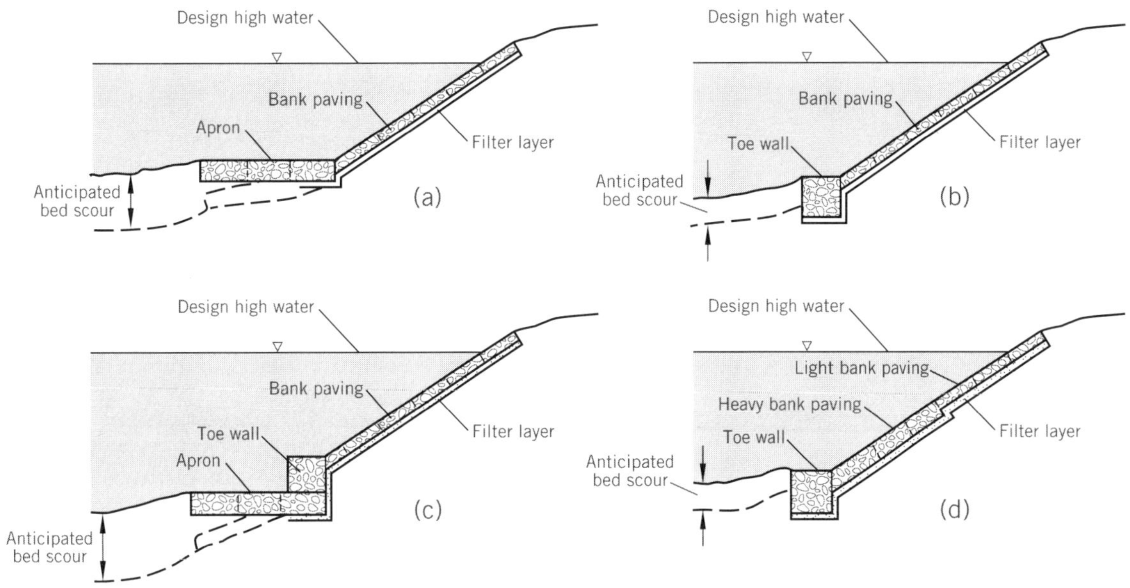

Figure 16.7 Gabion mattress designs:
(a gabion mattresses on slope and apron
(b) gabion mattresses on slope, gabion baskets at toe
(c) gabion mattresses on slope, gabion baskets at toe with apron mattress
(d) as for (b), but with variable thickness of gabion mattresses
(after FHWA manual HEC-11, 1989).

Table 16.3 Commercially Available Gabion Sizes

Thickness (feet)	Width (feet)	Length (feet)	Wire-mesh Opening Size (in. x in.)
0.75	6	9	2.5 x 3.25
0.75	6	12	2.5 x 3.25
1	3	6	3.25 x 4.5
1	3	9	3.25 x 4.5
1	3	12	3.25 x 4.5
1.5	3	6	3.25 x 4.5
1.5	3	9	3.25 x 4.5
1.5	3	12	3.25 x 4.5
3	3	6	3.25 x 4.5
3	3	9	3.25 x 4.5
3	3	12	3.25 x 4.5

Table 16.4 Design Criteria for Gabion Mattresses*

Bank Soil Type	Maximum Velocity (ft./sec.)	Bank Slope (horiz:vert)	Minimum Required Mattress Thickness (inches)
Clays, heavy cohesive soils	10	flatter than 3:1	9
	13–16	flatter than 2:1	12
	any	steeper than 2:1	18
Silts, fine sands	10	flatter than 2:1	12
Shingle with gravel	16	flatter than 3:1	9
	20	flatter than 2:1	12
	any	steeper than 2:1	18

*U.S. Federal Highway Administration, 1989

erosion. The toe is typically 12 to 20 inches deep. Examples are given on Figure 16.7. At the edges (upstream, downstream) the mattress should be thicker than elsewhere (usually double).

Table 16.3 lists the commercially-available basket sizes and the wire-mesh opening sizes. In mattress baskets there are internal diaphragms to maintain the shape of the units. The criteria used to design gabion mattresses for particular soil types are given on Table 16.4 (FHWA, 1989).

The median stone size for mattresses up to 1 foot thick are in the range of 3 to 6 inches. The stones should be well-graded with at least 70 percent by weight larger than the wire-mesh openings.

In mattresses of 1½- to 3-foot thickness, the median size of stones can be increased to 12 inches but no stone should be larger than the mattress thickness. The stone quality should be the same as for riprap (see Section 11.2).

The zinc-coated wire used in gabions is 13½ gage and somewhat heavier at edges and corners. For highly corrosive conditions (salt water, industrial areas, peaty environment) a PVC coating should cover the galvanized wire.

The filter layer should be at least 4 to 6 inches thick. A filter fabric (geotextile) can be substituted for the filter layer.

Construction of Gabion Mattresses

Gabion mattresses are normally fabricated on site. The units are laid on the surface, filled with rocks and the top laced on, then laced to the adjoining unit. For shallow water installations, the mattress units can be placed by a crane using a lifting frame, or the units are filled individually on the shoreline and dragged down the slope into the water (Figure 16.8). In deep water, a

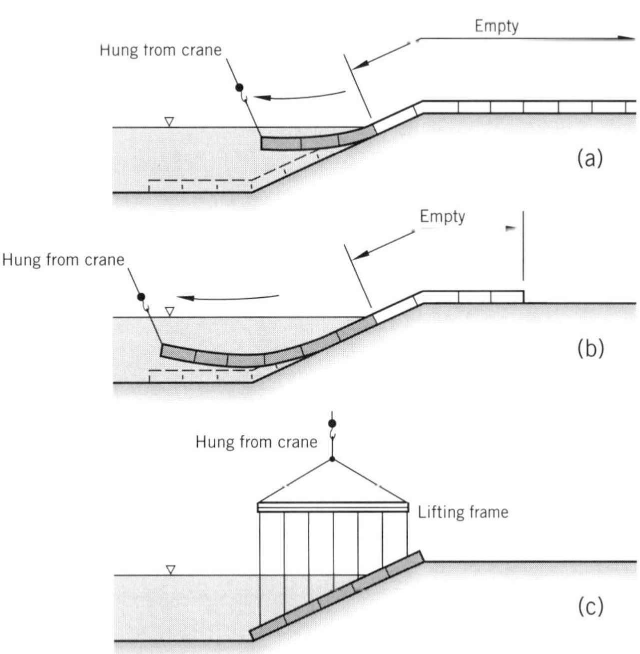

Figure 16.8 Underwater placement of a gabion mattress in shallow water:
(a) (b) filling units and pulling downslope
(c) lifting into place with a frame
(after FHWA manual HEC-11, 1989).

Figure 16.9 Underwater placement of a gabion mattress in deep water from a pontoon or barge (after FHWA manual HEC-11, 1989).

barge or pontoon is used to fabricate the mattress, after which it is launched into the water near the shoreline (Figure 16.9).

16.6 SHOTCRETE

Technique

Shotcrete, a soil-cement mixture that is sprayed (Figure 16.10) onto a soil or weak rock face, provides a resistant covering to prevent: (i) erosion from weathering, or (ii) infiltration of rainfall. The latter use maintains suction (negative pore pressures) in partly saturated soils and prevents the buildup of positive pore pressures.

Uses and Limitations of Shotcrete for Landslides

The treatment is usually applied to stiff or dense soils and to weak and fractured rocks. Some applications of shotcrete to maintain slope stability include the following:

- Thin interbeds between more competent rock strata.
- Temporary facings in soil cuts of dense sand, colluvium, or weathered rock.
- Facing on steep soil slopes, such as a soil-nailed slope, mechanically stabilized earth wall, or an anchor block wall.

Shotcrete can substitute for conventional reinforced concrete at remote sites, sites with difficult access, or sites where concrete formwork is expensive to install.

There are three major limitations on the use of shotcrete for civil engineering projects:

1. *Permanence.* shotcrete is subject to deterioration over time, including damage from weathering.
2. *Groundwater.* Seepage within the treated slope is confined and may build up pressure sufficient to break through the relatively thin layer of shotcrete; shotcrete is unsuitable for slopes with free water running over the surface, but minor seepage can be handled.
3. *Aesthetic.* The light, bright color of shotcrete is considered unattractive by most people; rust streaks from steel or wire reinforcement can also disfigure the appearance. However, this problem can be overcome by a final thin cover layer referred to as a "flush coat." The flush coat is unreinforced and can be mixed with colored dye. Shotcrete also can be sculpted to resemble a natural rock structure.

Shotcrete Process

Shotcrete is a mixture of sand, fine gravel, cement, reinforcement fibers, and occasionally other admixtures (such as accel-

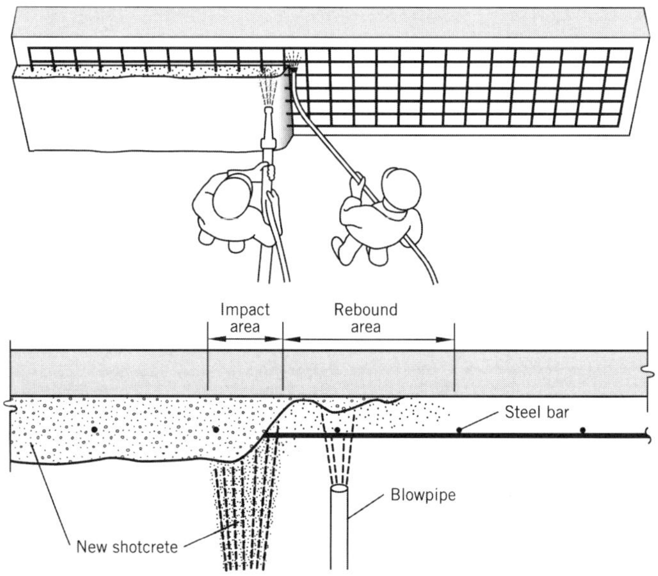

Figure 16.10 Shotcrete application: nozzle operator on left guns the shotcrete; blowpipe operator on right cleans off rebound.

erators or air entrainment), applied pneumatically to a slope at high velocity through a hosepipe and nozzle. It builds up a layer, 2 to 4 inches thick per application, which is strong, somewhat ductile, and relatively impermeable, thus providing erosion resistance to the outer face of a slope.

There are two methods of manufacturing shotcrete: dry mix and wet mix. Each method has advantages and disadvantages, as listed on Table 16.5. The dry-mix procedure pneumatically conveys a slightly moistened mixture to the nozzle where most of the water is added under the control of the nozzleman. In the wet-mix procedure, all the ingredients including the required water are mixed before entering the delivery hose. It is delivered to the nozzle either pneumatically or by a displacement pump.

Table 16.5 Comparison of the Dry-Mix and Wet-Mix Processes for Shotcrete*

Dry-Mix Process	Wet-Mix Process
Mixing water is instantaneously controlled at the nozzle by operator to meet variable field conditions.	Mixing water is controlled at plant and measured at time of batching.
Longer hose lengths are possible.	Normal pumping distances.
Limited to accelerators as the only practical admixture.	Compatible with all ordinary admixtures. Special dispensers for addition of accelerators are necessary.
Use of air-entraining admixture is not beneficial. Resistance to freezing and thawing is poor.	Air entrainment is possible. Acceptable resistance to freezing and thawing.
Intermittent use is easily accommodated within prescribed time limits.	Best suited for continuous application of shotcrete.
Exceptional strength performance possible.	Lower strengths, similar to conventional concrete.
Lower production rates.	Higher production rates.
Higher rebound.	Lower rebound.
Equipment maintenance costs tend to be lower.	Equipment maintenance costs tend to be higher.
Higher bond strengths.	Lower bond strengths, yet often higher than conventional concrete.

*U.S. Corps of Engineers, 1993

The *dry-mix* procedure is illustrated on Figure 16.11. The dry aggregate is usually passed through a premoisturizer to add about 3–6% moisture content before entering the "gun" (Figure 16.11). Compressed air at around 40 psi pushes the mixture through the conveyor to the nozzle where water at about 60 psi is applied through a water ring to uniformly inject water into the mixture. The nozzleman is able to control the workability of the shotcrete as it is directed onto the slope. A typical plant layout for delivery of dry-mix shotcrete is shown on Figure 16.12.

The *wet-mix* procedure (not illustrated) is similar to conventional concrete. The shotcrete ingredients are mixed in a truck or at a batching plant (which can be offsite). Air-entrainment admixtures can be included in the mix to improve freeze-thaw resistance. Wet-mix shotcrete can be applied at a faster rate of production than a dry mix.

When shotcrete is sprayed onto a face, some of the mix bounces off the face. This *rebound* ranges from 5–25% of a dry mix and 5–10% of a wet mix. The rebound, mostly composed of coated aggregate, cannot be reused or incorporated into the final shotcrete facing. It is removed from the surface by a shotcreting assistant using a blowpipe (Figure 16.10).

Shotcrete is reinforced, primarily to reduce shrinkage cracks during curing, but also to provide some ductility and toughness. *Welded wire fabric* is a steel mesh, typically with openings of 4 inches, that is attached to the slope face before shotcreting begins (see Figure 16.10). Steel mesh reinforcement is going out of favor because of: (i) the time needed to set the mesh at the correct distance off the face so that it is properly encased; (ii) the difficulty of preventing voids forming behind the mesh; and (iii) the relatively high rebound caused by the vibration of the mesh when the shotcrete hits it. Welded wire reinforcement will not be discussed further in this section.

Fiber reinforcement consists of short fibers of steel or polypropylene that are added to the mix. Other fibers (glass, and various other synthetics) are also available. Fiber lengths

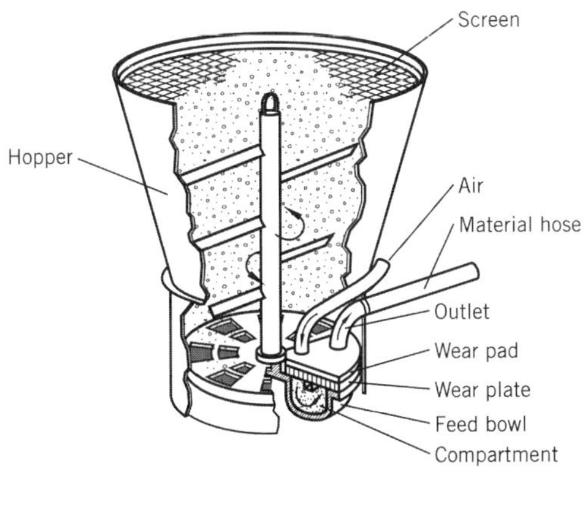

Figure 16.11 Typical dry-mix gun used for shotcrete (Mahar, Parker & Wuellner, 1975).

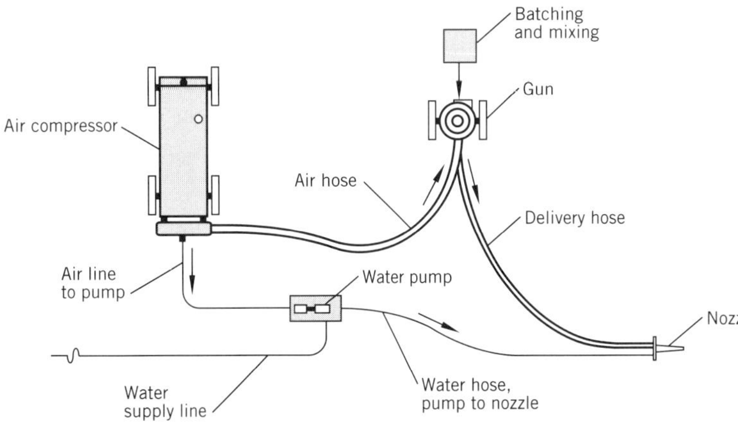

Figure 16.12 Typical plant layout for dry-mix shotcreting (Crom, 1966).

range from ½ inch to 3 inches long, but are typically 1 inch long and have a length : diameter ratio of about 50.

Steel fibers are of many shapes: round, flat, or irregular, with additional deformations along the length or at the ends to improve anchorage, including hooks, bulbs, flattened surfaces, etc. These shape variations produce the desired holding properties with less material. Steel fiber reinforcement is typically 1–2% by volume of the shotcrete (roughly 100 lb./cu. yd.).

Collated fibrillated-polypropylene (CFP) fibers are applied at the rate of 1–2 lb. per cu. yd. An advantage of CFP is that it is less of a hazard to workers from rebound and does less damage to the delivery hoses. Polymer filament fibers are inert and noncorrosive; therefore, they do not rust if exposed at the outer face of the shotcrete.

Another product used in a shotcrete mix is *silica fume*, which is not technically a reinforcement but a substitute for part of the cement. Silica fume, primarily silica dioxide, is a byproduct of silicon alloy production and has pozzolanic properties that improve strength and durability. The powder is much finer than cement and fills the microscopic voids between cement particles. This increases density, reduces permeability, and increases adhesion and cohesion during shotcrete placement. This cohesiveness reduces the rebound by about one-half of other mixtures. Silica fume is typically 14% by mass of the cement used in a mix. A major advantage of silica fume is that it can be sprayed onto the slope in a thicker layer that can eliminate two-layer construction at many sites. It is generally mixed by the dry-mix procedure because it does not perform well in a truck mixer.

Shotcrete Mix Design

The soil-cement mix for shotcrete uses a well-graded gravelly sand with a maximum particle size ranging from ⅜ inch to ¾ inch and 2–10% by weight passing the No. 100 sieve. This gradation minimizes segregation during pumping. The finer aggregate mix has more shrinkage on drying. The coarser aggregate mix produces more rebound and increases the risk of creating pockets of aggregate or voids behind a mesh reinforcement.

Cement content in shotcrete is rich: typically 500–700 lb./cu. yd. This helps adherence to steep surfaces. Water : cement ratio is typically 0.30 to 0.40 (3 inch minimum slump).

Accelerators are sometimes required for rapid strength development (e.g., seawalls, canal linings) to prevent fresh shotcrete from being washed away. However, they reduce the 28-day strength by 25–40%.

Shotcrete Slope Protection Contracts

The shotcrete mix design is often delegated to the contractor. A typical project requirement is that shotcrete cores should attain a minimum compressive strength of 1,800 psi at 3 days and 2,500 psi at 7 days as determined by AASHTO-T22. For design acceptance and production testing, the shotcrete contractor typically prepares test panels in molds 18 inches square and 4 inches deep. Cores of 3-inch diameter are taken from the panels.

The total design thickness of a shotcrete application is usually 3 to 6 inches. If thicker than 3 inches, it is applied in two coats, allowing about 24 hours between successive applications. After the shotcrete is placed, a coat of curing compound is sprayed over the surface to retain moisture during the curing process. If desired, it can be removed by sandblasting 7 days later.

The outer surface of a shotcrete surface has numerous fiber ends sticking out; if made of steel, these ends will rust and discolor the surface. One way to avoid this effect is to add a thin (about 1 inch thick) outer finish coat of unreinforced shotcrete that can be tinted to a selected color. This thin layer, often referred to as a "flash coat," provides a more aesthetically-pleasing finish at highly visible sites.

Weepholes should be provided through the shotcrete surface to relieve minor water seepage that could accumulate behind the shotcrete treatment. If reinforcement of the holes is needed, PVC pipes can be used. The pipes should be installed before shotcreting. Temporary protection (tape) is used to cover the exposed ends of the pipe during gunning.

Shotcrete is usually paid for by the measured area of the treated surface. The price is inclusive of all incidental costs.

The quality of work on a shotcrete project is highly dependent on the skill of the nozzle operator. Therefore, a requirement that the nozzle operator have at least 2 years of full-time experience should be required together with a list of prior projects and owner references. In a bid job, the right of the owner to reject or change the nozzle operator should be included in the job specification.

Shotcrete Construction

The surface to be shotcreted needs cleaning and other preparatory work. Any vegetation, spongy areas, or loose materials have to be removed.

Noncorrosive pins or other depth guides are installed in the slope. These gaging pins are laid out on a 5-foot square grid. The surface to be treated is thoroughly dampened so that the slope materials do not suck water out of the shotcrete.

The nozzleman (Figure 16.10) holds the nozzle perpendicular to the surface and about 3 feet away from it. This produces a satisfactory shotcrete density and the least amount of rebound. A steady circular or elliptical motion is used during spraying. Concurrently, a helper with an air blowpipe removes any rebound adhering to the surface, especially surfaces that are about to receive shotcrete.

In the dry-mix procedure, the nozzleman has to achieve the optimum water addition, which is evident by a slight gloss to the surface. If the surface is too dry, rebound increases; if the surface is too wet, the shotcrete will slough or sag. To avoid rebound adhering to untreated surfaces, the nozzleman works from bottom to top. The shotcreted area is tapered at the edges. Shotcrete cannot be applied on windy days, during rainfall, or when temperatures fall below 40°F.

16.7 CHUNAM PLASTER

Technique

Chunam plaster (Figure 16.13) is used in some countries of southeast Asia for slope protection against erosion and water infiltration. It covers the surface of a slope with a hard waterproof shell. This protects the slope from erosion and preserves negative pore pressure conditions at the face to enhance stability.

Use and Limitations

Chunam plaster is a labor-intensive alternative to shotcrete. In Hong Kong, it has been used extensively on steep existing slopes and on temporary cut slopes. However, since the late 1980s, it has been phased out in Hong Kong and replaced by shotcrete or vegetation as a means of preventing water infiltration (private correspondence, Hong Kong Geotechnical Engineering Office).

The main concerns in using chunam plaster are: (i) its effectiveness in preventing infiltration decreases as it ages, and (ii) it reflects sunlight excessively. For these reasons, it is declining in popularity.

Design of Chunam Plaster

The recommended mix (by weight) for chunam plaster is:

- 1 part cement
- 3 parts hydrated lime
- 20 parts clayey weathered granite or volcanic soil, free from organics

The cement and lime are mixed dry before adding soil. The minimum amount of water to provide the required workability is then added to the mix. The plaster is put onto a prepared surface in two layers.

The thickness of the plaster for permanent facing is 1.6 to 2 inches, but this can be reduced for temporary protection of slopes. To reduce glare in sunlight, colorants such as manganese dioxide or ferrous oxide powder (at 3% by weight of the cement content) are added to the outer layer.

A design practice in Hong Kong is to extend the chunam above the slope crest to connect to a drain system further upslope. As shown on Figure 16.13, both a surface ditch and a subsurface drain (running parallel to the slope crest) tie into the chunam plaster.

Construction of Chunam Plaster

The slope to be treated is stripped of vegetation and graded to a smooth surface. Bamboo dowels, 1-inch diameter and 1 foot long, are embedded in the slope at 5-foot centers (on a staggered pitch) to leave 1 inch protruding from the slope; this holds the chunam in place during the plastering.

The base layer is applied, not less than 0.8 inch thick, and the surface is scored with a trowel to provide a key for the second layer. A delay of 24 hours is sufficient between layers. The chunam is placed on the slope surface as compactly as possible so that the treatment will be free of cracks and impermeable. No construction joints are provided.

16.8 BIOREMEDIATION

Technique

Living plants can be used to prevent surface erosion and remediate very shallow landslides. These techniques will be described in this section under the broad term *bioremediation*. Advantages and disadvantages of various plant types are listed in Table 16.6.

Bioremediation can be considered in the following circumstances:

- For surficial erosion sites where plants can resist erosion from runoff or flowing streams; the plants sometimes need extra support from erosion mats or nets.
- For very shallow landslides where tree roots provide tensile resistance to sliding.
- To beautify a landslide remediation project built of stones or concrete.

A major drawback of all bioremediation projects is that plants need time to grow and effectively produce ground covers or deep roots. As interest in these methods increase in the engineering community, several aids to fostering stability during this interim growth period have been developed. They include meshes and nets made of geosynthetics or natural fibers that will hold the soil in place until the plants are established.

The main focus of this section will be on the prevention or remediation of slopes prone to erosion. First, however, there will be a few comments on the other two benefits of bioremediation listed above.

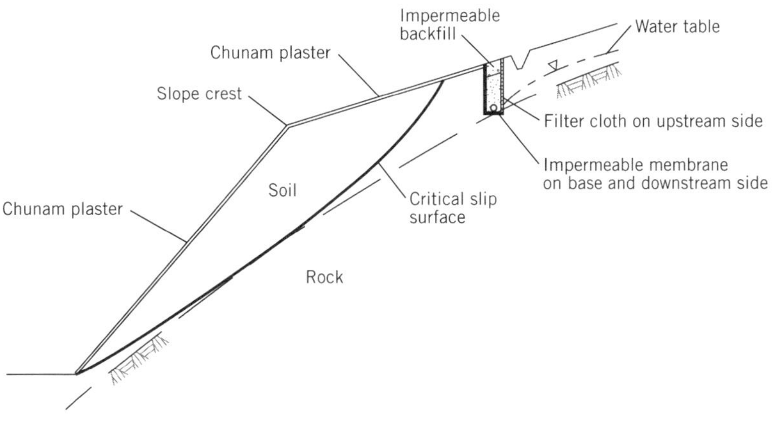

Figure 16.13 Typical use of chunam plaster to protect a soil slope in Hong Kong (after Geotechnical Control Office, 1984).

Table 16.6 Suitability of Plant Types for Geotechnical Use*

Type	Advantages	Disadvantages
1. Grasses	Versatile and cheap; wide range of tolerances; quick to establish; good dense surface cover	Shallow rooting; regular maintenance required
Reeds and sedges	Establish well on riverbanks, etc.; quick growing	Hand-planting expensive; difficult to obtain
2. Herbs	Deeper rooting; attractive in grass sward	Seed expensive; sometimes difficult to establish; many species die back in winter
Legumes	Cheap to establish; fix nitrogen; mix well with grass	Not tolerant of difficult sites
3. Shrubs	Robust and fairly cheap; many species can be seeded; substantial ground cover; deeper rooting; low maintenance; many evergreen species	More expensive to plant; sometimes difficult to establish
4. Trees	Substantial rooting; some can be seeded; no maintenance once established	Long time to establish; slow growing; expensive
Willows and poplars	Root easily from cuttings; versatile, many planting techniques; quick to establish	Care required in selecting correc-tive type; cannot be grown from seed

*after Coppin & Richards, 1990

Benefits of Trees to Slope Stability

Tree roots can anchor a soil mantle to fractured rock, partly disintegrated rock, or a competent soil (e.g., glacial till). As shown on Figure 16.14, the tap root (main vertical root directly below the bole of the tree) is the key. Research studies have measured the root tensile strength of different tree species at different ages. Sinker roots also provide some resistance to shear.

Tree roots are only significant contributors to slope stability over a relatively short depth of up to about 5 feet beneath the surface. Much of the interest in their contribution is related to the controversial issue of clearcutting of forested slopes. As described in Chapter 2, Section 2.13, the incidence of debris flow slides appears to increase (above natural frequency of occurrence) a few years after clearcutting, and this has been attributed to decay of tree roots and consequent loss of resistance at the soil overburden/bedrock contact. This issue generally concerns the harvesting of mature trees on steep slopes where a thin mantle of soil overlies bedrock.

For general remediation of landslides, trees have both beneficial and adverse effects and these are summarized on Figure 2.20 of Chapter 2. As a practical matter, trees cannot be used to remediate landslides except as part of a long-term solution.

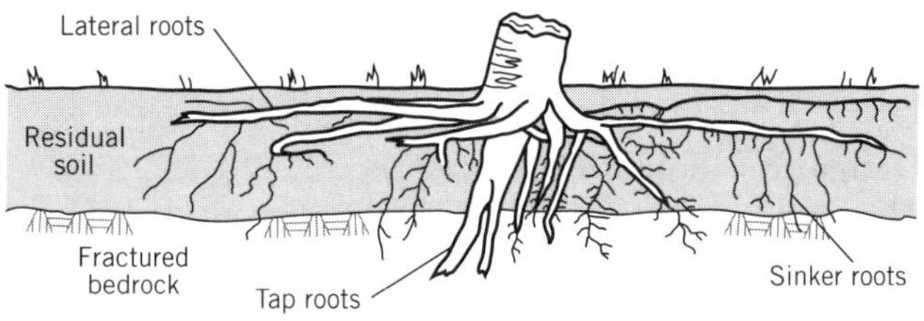

Figure 16.14 Lateral, tap, and sinker roots of a woody root system.

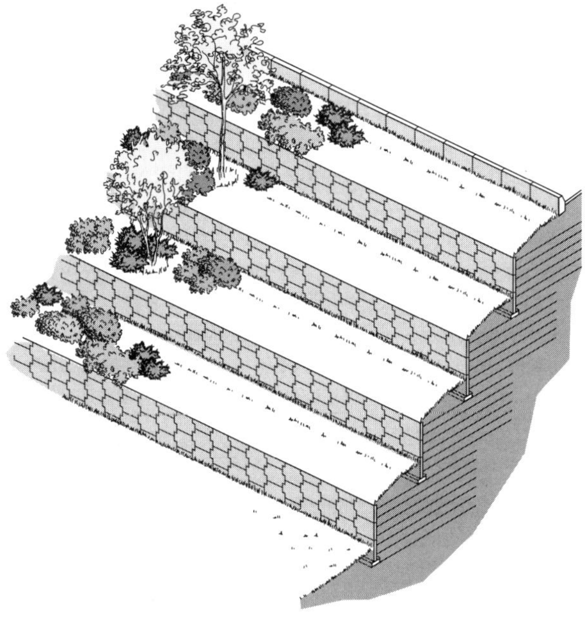

Figure 16.15 Tiered retaining structure of mechanically stabilized earth with landscaped benches (after Walkinshaw, 1975).

Beautification

The aesthetic benefits of plants are receiving increasing attention for many civil engineering projects, especially in highway construction and retaining walls. Although proponents of bioremediation list various technical benefits from plantings, a more realistic appraisal is that such benefits are insignificant except for some benefit against surface erosion. However, plants can be installed wherever there are openings in a structure, and often these can be created to accommodate plants. Plantings enhance the overall appearance or hide more ugly features. Some examples of structures that can benefit from the marginal extra cost of planting include gabion walls, bin walls, crib walls, masonry walls, cut slopes, buttresses with a soil cover, mechanically stabilized earth walls, etc.

These improvements in appearance can be helped by more imaginative design on the part of the engineer. Instead of using linear slopes, various alternatives can be considered. *Tiers* or *terraces* can be created, as shown by the example on Figure 16.15, where plants are installed on the level tiers. Tiers

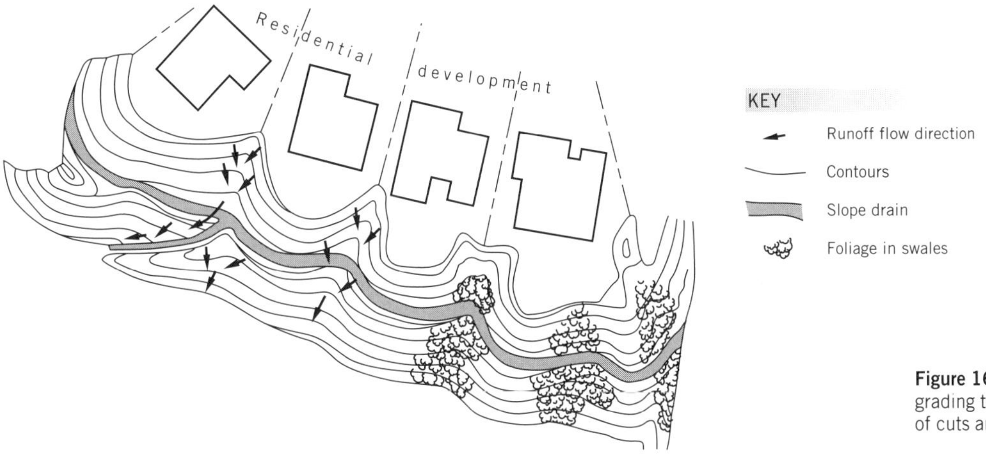

Figure 16.16 Representation of "landform" grading to create a more natural appearance of cuts and fills (after Schor, 1992).

are helpful to construction of mechanically stabilized earth walls, soil nail walls, and anchor block walls.

Another beneficial landscaping technique is *landform grading* of cut and fill slopes (Figure 16.16). In the past, highway departments have used "contour" grading in which cut and fill slopes are given deliberate curvatures, especially at the ends, top, and base of manmade slopes, to soften the appearance of linear slope gradients and angular intersections. Landform grading, pioneered by Schor (1992) in California, has evolved from contour grading to emulate natural landforms and planting. Mounds and valleys are built into the earthwork design. Trees and shrubs are planted in valleys, which receive the runoff from the slopes above, and ground cover is planted on the drier convex slopes (Figure 16.16). Planting follows the patterns that occur in nature rather than to lay out plants in a random or uniform coverage. Technical benefits to this approach include reduction of irrigation water needs and easier collection of runoff water. This type of design is especially suitable for planned hillside communities, but the concept can be adopted to some degree on most major cut-and-fill earthworks where there is a need to improve aesthetics.

Erosion Control Measures

Erosion control can be as simple as hydroseeding a bare slope to more complex designs involving geosynthetic boxes staked to the slope, infilled with soil, with plants selected for the particular microclimate, azimuth, slope gradient, underlying soil properties, and groundwater conditions of the site. A horticultural specialist should be part of the design team.

At sites where erosion is already occurring (or is expected to occur on newly created slopes), some artificial help may be needed to establish the plants. If the plants will eventually take over the role of controlling erosion, such aids may be nets with a very limited design life. In more severe cases, grids are used as a permanent part of the erosion protection measures and provide some mechanical resistance to surface stability.

Formulas are available to estimate the level of erosion risk (e.g., Gray and Sotir, 1996). At a qualitative level, Table 16.7 lists various types of rolled erosion control products and their applications. Table 16.8 lists several such available products and their manufacturers/suppliers.

Table 16.7 Performance Classification of Rolled Erosion Control Products

Required Performance Characteristics	Suitable Type of Rolled Erosion Control Product	Properties and Typical Applications
Low-velocity degradable	Single net, organic fiber erosion control blankets Biodegradable natural fiber and photodegradable geosynthetic nets	One-to two-season longevity Limited capacity in resisting damage and resisting erosion under severe conditions Slopes of moderate grade, length, and runoff; channels where potential for damage during installation is minimal
High-velocity degradable	Double net erosion control blankets or high-strength nettings and/or increased quantities of organic fibers Dense, open-weave geotextiles or meshes (e.g., coir, jute, polypropylene)	Similar to low-velocity degradables in installation and function, but designed for more severe site conditions Heightened durability and longevity (1–5 years) Steeper slopes and high-velocity channel linings where natural, unreinforced vegetation is expected to provide permanent stabilization
Long-term nondegradable	High-strength geosynthetic mattings and cellular containment, e.g., erosion control revegetation mats, turf reinforcement mats, geocellular containment systems	Provide immediate, high-performance erosion protection followed by permanent reinforcement of established vegetation Steep slopes with very erodible soils; channel linings subjected to relatively high, long-duration flows

Source: Adapted by Gray & Sotir (1996) from Austin and Driver (1995).

Table 16.8 Examples of Erosion Control Products

Trademark or Registered Name	Manufacturer/Supplier	Brief Description
	Fabrics and Meshes (Two-Dimensional)	
POLYJUTE®	Synthetic Industries	Woven polypropylene mesh
Terram 42A	ICI Fibres Geotex GP	Woven polyester mesh
Geojute	Belton Industries	Woven jute mesh
DeKoWe 700 & 900	Belton Industries	Woven coir mesh
S75, SC 150, C125	North American Green Inc.	Polymeric netting sewn together with straw and coconut fibers
	Mats and Cellular Webs (Three-Dimensional)	
Enkamat®	Akzo Industrial Systems	Mat of heavy nylon monofilaments
Tensar Mat®	Netlon Ltd.	Multilayered polyethylene mat
C350	North American Green Inc.	Coconut matting with permanent three-dimensional reinforcement net structure
LANDLOK®	Synthetic Industries	Three-dimensional mat of polyolefin fibers and nets
PYRAMAT™	Synthetic Industries	Three-dimensional woven geotextile
Armater®	Akzo Industsrial Systems	Polyester expanding cellular grid
Geoweb®	Presto Products	Polymeric expanding cellular grid

Source: Gray & Sotir (1996)

Hydroseeding

This procedure is widely used to revegetate cut slopes alongside roads, canals, and reservoirs to reduce erosion on bare soil. Hydroseeding has to be performed under the following conditions: air temperatures between 32°F and 90°F; soil temperature above 55°F, dry weather, and winds of less than 10 mph.

The construction procedure is to lightly scarify any compacted or firm soil surfaces to a depth of about 4 inches. Select topsoil, loose and damp, is spread over the surface to a depth of 12 inches. Fertilizer is uniformly spread at a typical rate of about 5 lb./1000 sq. ft. The fertilizer can be ammonium nitrate or a commercially-available mix (e.g., 16–20–10 formula). It is sometimes included in the hydroseed mix.

The grass seed depends on the particular site and is often matched to the native species. It is usually mixed with wood-cellulose and an adhesive (tackifier) to bind the seed/fiber together and bond it to the soil. The hydroseeding equipment should have a built-in agitation system. A typical mix per 1,000 sq. ft. of ground coverage is: seed 2 lbs; mulch 5 lbs; tackifier 1 lb; fertilizer 4-5 lbs. Hydroseeding should produce a 90 percent vegetated cover within two months. Additional fertilizer (e.g., 3 lb./1000 sq. ft.) can be added at that time.

Cellular Grids

Cellular grids allow vegetation to become established on steeper slopes. An example of a grid made from wood boards is shown on Figure 16.17. The grid structure is fastened to stakes driven into the slope. Topsoil and seed infill the openings of the latticework, after which the surface is covered with straw and secured by a jute or coir (coconut fiber) or polypropylene netting.

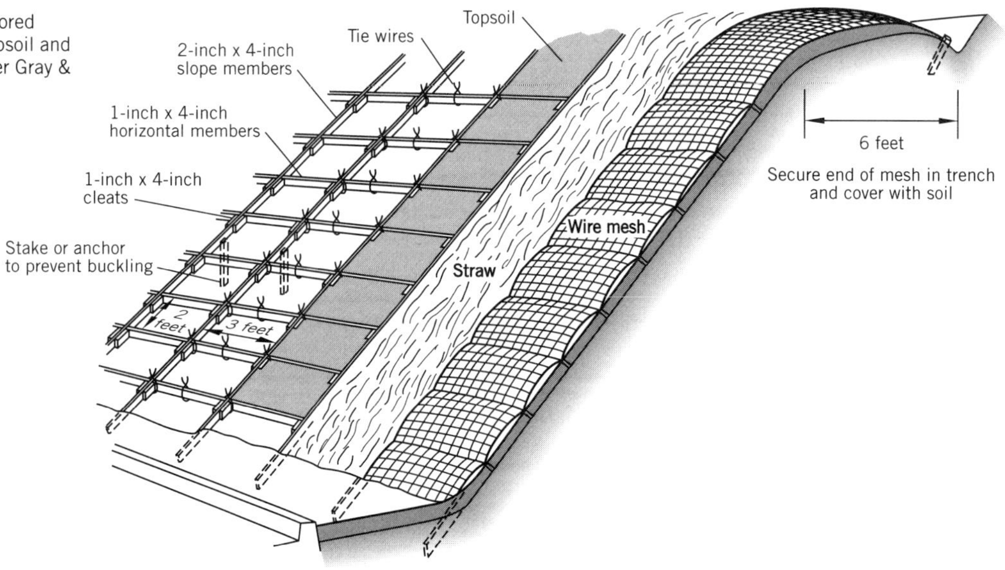

Figure 16.17 Anchored grid used to hold topsoil and slope plantings (after Gray & Sotir, 1996).

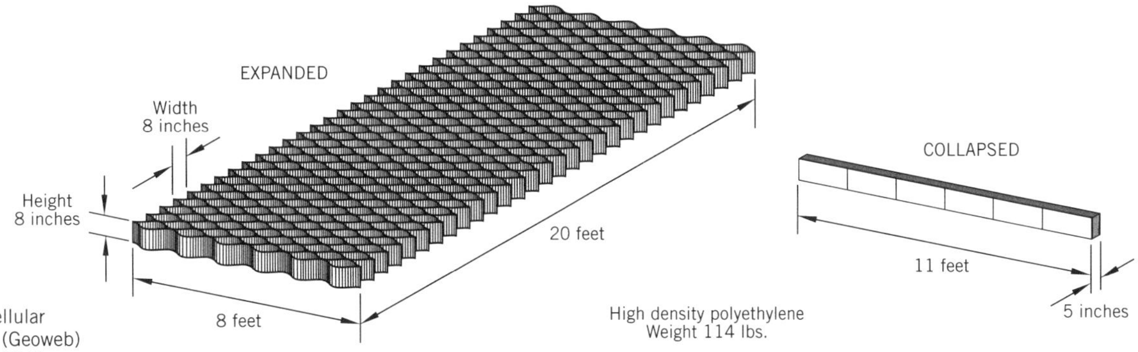

Figure 16.18 Geocellular containment system (Geoweb) (after Gray & Sotir, 1996).

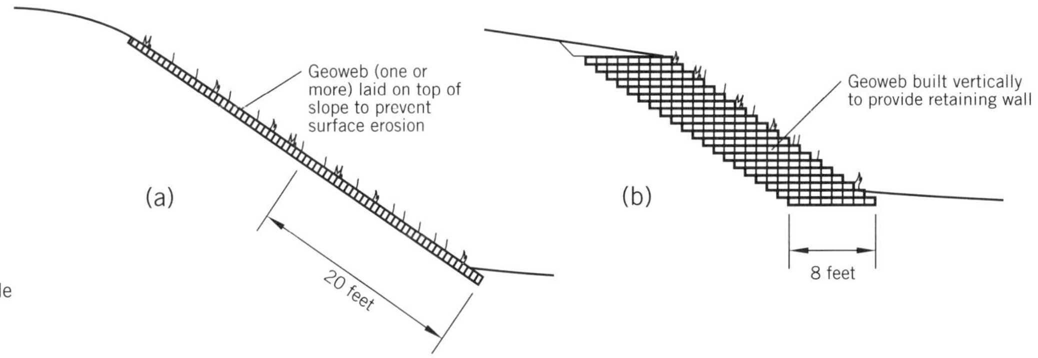

Figure 16.19 Geowebs:
(a) laid on a slope
(b) laid horizontally to provide deeper protection
(after Gray & Sotir, 1996)

Another type of cellular grid is a mat of honeycombed cells that is delivered to the site in a collapsed form and is opened up for use (Figure 16.18). The Geoweb system shown on this drawing can be 4 inches or 8 inches high. The cells are filled with soil, then seeded, fertilized, and covered with a temporary mulch. This type of mattress grid is usually placed on top of a carefully graded slope (Figure 16.19a). However, the grid can be stacked vertically to create a wider structural member capable of resisting shallow sliding (Figure 16.19b).

Geosynthetic Rolled Mats

These mats are manufactured from synthetic fibers or filaments (UV-stabilized) to produce a three-dimensional mattress or blanket that is rolled out at the site. An example is shown on Figure 16.20. They are designed for permanent use at sites where high water velocities exceed the limit of erosion resistance that can be expected from mature natural vegetation. Many types are available. The reinforcement net shown on Figure 16.20 is an entanglement of fibers that provides sufficient depth to allow soil and plant roots to develop while the mat is anchored to the ground surface by large staples or pegs.

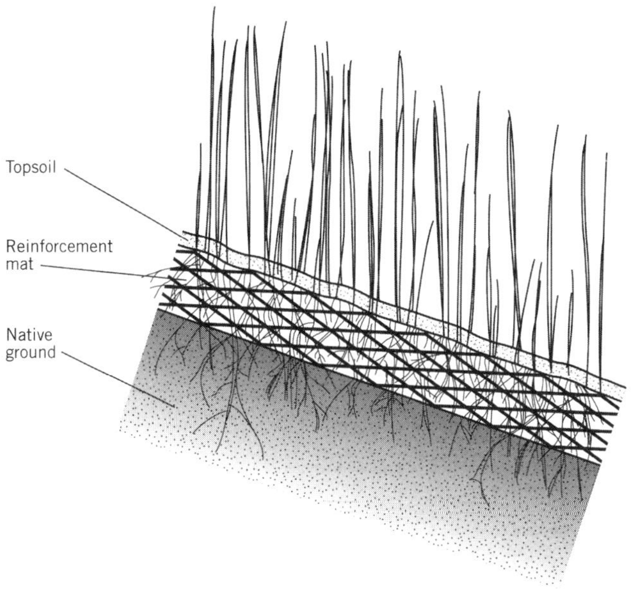

Figure 16.20 Turf reinforcement mat used to foster development of plant roots (after Gray & Sotir, 1996).

Example Application of Erosion Control

The closed St. John's Landfill in Portland, Oregon, has a surrounding earth dike that separates the landfill from sloughs of the Columbia River (Figure 16.21a). During periods of high water, the outer slopes of the dike were being eroded at three locations. There were concerns that continued erosion could undermine the landfill above the dike and release refuse and leachate into the slough.

The solution selected by Landslide Technology was to first restore the eroded slope with improved armor and later construct a cement-bentonite slurry trench cutoff wall to provide a barrier to leachate leakage. The design section is shown on

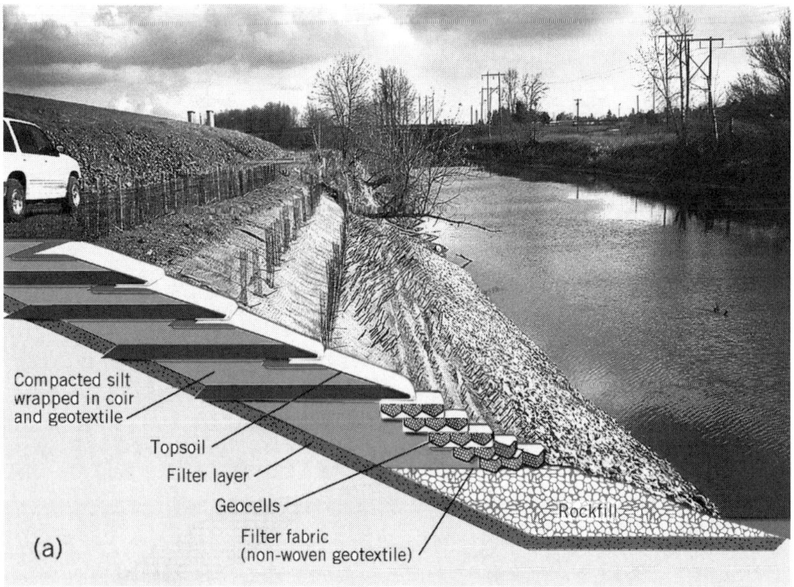

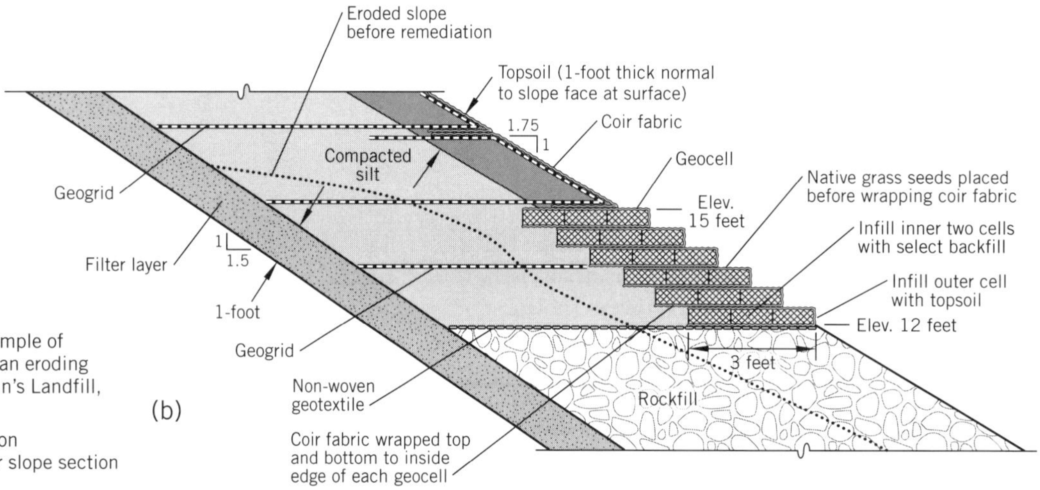

Figure 16.21 Example of bioremediation of an eroding slope face, St. John's Landfill, Portland, Oregon:
(a) photo-simulation
(b) details of lower slope section

Figure 16.21(b). The slurry trench wall (not shown) is further inland below the access road (below vehicle on extreme left).

To repair the eroded slopes, a backhoe excavated a cut slope at 1½ horizontal : 1 vertical) to provide a bench 10 feet wide at around elevation 0. This work was completed in late summer 2000 during the seasonally low river level (typically elev. +7 feet). A filter layer, 1 foot thick, was placed on the excavated bench and cut slope, and the lower part of the slope was backfilled with well-graded 12-inch minus rockfill to elev. +12 feet, the average tide level below which plants do not grow. During this phase, the contractor was required to use closely-sequenced construction procedures (Chapter 13, Section 13.5) to minimize the risk of slope instability of the cut slope.

Geocells (Presto Perforated Geoweb) were built into the outer slope immediately above the rockfill. At this level, close to mean tide, the slope is frequently inundated and the stacked geocells provide extra erosion resistance at this more vulnerable location. The cells were filled with silt and wrapped in coconut fiber (coir) mats.

The upper slope was backfilled with silty soils reinforced at intervals with a geosynthetic fabric to provide a reinforced soil slope (see Chapter 20, Section 20.3). The geogrids were Tensar BX 1120 made from polypropylene, and have biaxial strength (high in both directions). They also have good protection against ultraviolet light damage. The outer 1 foot was built with organic soils and the fabric was wrapped around it (after seeding with native grasses) to retain the earth. A short length of coir was wrapped around the outer face to promote vegetation growth.

In summary, the rockfill at the toe of the slope provides scour protection and helps to maintain slope stability. The reinforced silt in the upper slope provides a suitable growing medium for plants. As the native plants mature in the upper slope, they add to the scour protection and eventually blend into the natural slough environment. Other benefits are that fish and other wildlife can use the vegetation, and the overall aesthetics are improved.

Costs. Length of treated shoreline 1150 feet
Contractor's construction cost $540,000 ($470/lin. ft.)
Engineering News–Record Construction Cost Index: 6233 (August 2000)

16.9 CONCRETE BLOCK SYSTEMS

Technique

Precast concrete elements frequently are used to protect slopes from erosion. These systems generally are proprietary (patented).

Concrete block systems are prefabricated alternatives to natural stone slope protection and usually require the same design and construction procedures as the stone system that they imitate. These systems have been available since about 1980 and appear to be effective. New systems are likely to be developed in the future, and only a brief overview will be presented here. Manufacturers can be contacted via an internet search. Mention of specific products for illustration purposes should not be construed as an endorsement by the author. The products fall into three broad categories: articulated concrete block mattresses, unconnected concrete blocks, and interlocking angular blocks.

Articulated Concrete Block Mattresses

The blocks are of various shapes (depending on the manufacturer's design) and are connected by steel wire cables, synthetic fiber ropes, or a geotextile backing sheet (Figure 16.22). The concrete elements (cells) that make up a mattress section are usually available as either "open" or "closed" design. The "open" design has a preformed opening in the middle that can be used for planting. If needed for aesthetics, the concrete can be colored.

The advantage of articulated concrete block mattresses include: (i) flexibility, (ii) ease of assembly, (iii) they can be placed underwater, and (iv) they allow vegetation to grow through the gaps in the mattress sections placed above normal water level.

In design, special attention needs to be given to edge fixity where stream or wave action can undermine the completed mattress. Gravel- or rock-filled termination trenches can be used for this purpose. An earth anchor can be installed near the top of the slope to secure the mat against slippage. The toe of the mattress can be buried in a gravel-filled toe trench or the mattress can be extended onto the stream floor to allow for possible future scour.

The bank should be graded to a smooth surface, including infilling of holes, removal of shrubs, surface boulders, other hard projecting rock, etc. A soil filter blanket or geotextile filter fabric should be placed above the native soil to prevent erodible soils from being pulled through the spacings of the concrete block revetment by waves or river currents. Vulnerability to ultra-violet rays of the underlying fabric needs to be considered where the concrete mattresses are supported directly on a geotextile fabric.

Concrete block mattresses are pre-assembled in sections, typically 8 feet wide and up to 16 feet long, that are trucked or barged to the site. The mattress sections are lifted into place on the prepared (smooth) surface by a crane or backhoe using a spreader bar. After being laid side by side, the sections are connected to cover the slope treatment area.

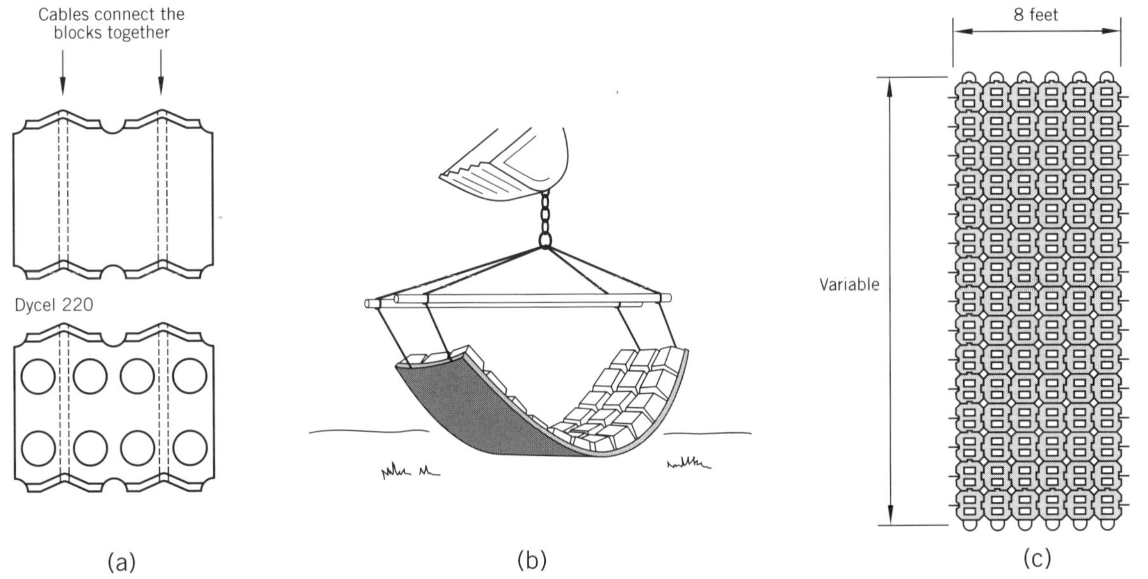

Figure 16.22 Articulated concrete block mattresses:
(a) Dycel blocks
(b) Cable Concrete mattress being lowered using spreader bar
(c) Petraflex mattress

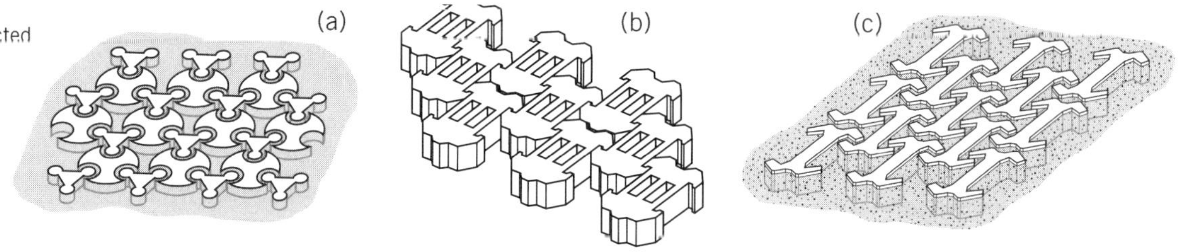

Figure 16.23 Unconnected concrete block systems:
(a) Tri-Lock
(b) Geolink
(c) Ankalok

Representative products for block mattress construction include:

- Dycel (RPC Products, U.K.). Panel typical size: 8 feet by 16 feet. Individual blocks are approx. 19 inches x 16 inches, 4 inches to 16 inches thick.
- Cable Concrete (Bethlehem Precast, U.S.A.). Truncated, pyramid-shaped blocks bonded to an underlying geotextile fabric.
- Petraflex (Petratech, U.S.A.). Concrete blocks, bi-directionally connected by cables. Size approx. 16 inches x 16 inches, 4 to 9 inches thick. The panel size is 8 feet wide, variable length.
- Armorflex (Armortec, U.S.A.). Concrete blocks are connected by cables passing through preformed ducts. The sizes range from 13.0 x 11.6 inches to 17.4 inches in plan; thicknesses are 4.75 to 9.0 inches.

Unconnected Concrete Blocks

These blocks are hand placed and thus labor intensive. However, the systems require no cables or ropes that could break or corrode over time.

The unconnected blocks rely on weight and/or interlock (in the plane of the revetment surface) to provide stability from waves or currents. Sand/gravel/topsoil can infill the open spaces between the blocks, and such areas can be planted within the slope above normal water level. These designs are usually restricted to slopes flatter than 3:1 (horizontal: vertical). A filter blanket or geotextile filter fabric is needed as backing for the systems. A toe berm may be constructed to provide support at the base of the treated slope. Site preparation is the same as for articulated concrete block mattresses. Representative products (Figure 16.23) in this group include:

- Tri-lock (CCI Industries Ltd.). There are two block designs: a lock block and a key block. Each block is keyed to two other blocks in a pattern similar to a jigsaw puzzle. They are available in a variety of sizes and weights.
- Ankalok (Ruthin Precast Concrete Ltd., U.K.). A double anchor block design that interlocks with six surrounding units.
- Geolink (Petratech, U.S.A.). Similar to the Ankalok.

Interlocking Angular Blocks

These are comparable in behavior to riprap. Precast concrete blocks are used where severe wave action occurs, such as coastal protection or sea jetties. This is a specialized area of geotechnical work and is outside the scope of this book.

The shape of the blocks provides interlocking with adjoining blocks. They also obtain stability benefit from weight, which can exceed 30 tons per element. Product names include Core-Loc, Accropode, A-Jack, Tetrapod, and Dolos.

16.10 TRENCHFILL REVETMENT

An innovative technique to construct a protective revetment in a floodplain or deltaic environment has been developed by the U.S. Corps of Engineers (New Orleans District). It is used to realign a channel by making the river do most of the construction work.

A trench is dug along the proposed future bank alignment of the river and is backfilled with gravel or stone. The river in flood erodes the silt and sand between the river and the gravel-filled trench. Once the trench has been reached by river erosion, the gravel wall collapses and forms a gravel slope, thereby protecting the sand behind it from further erosion and realigning the river bank to the designer's preferred location. The technique has been successfully used by the Corps on the Mississippi River delta.

CHAPTER

17

Dewatering Systems

17.1 COMMON DEWATERING ISSUES

Since the strength of soils depends on the effective stresses acting on the soil, lowering the pore pressures on the slip surface of a landslide will increase the stability of the slope. Dewatering a slope (depressurizing is a better term) is a common method of stabilizing landslides. It is most effective where the geological conditions are conducive to manmade interference with the natural groundwater regime. Examples would be the relief of artesian groundwater pressures (or ground that would become artesian after a proposed excavation is made) or at sites where permeable ground feeds water into an adjacent slope.

Landslide Dewatering Concerns

Although drainage and groundwater pressure relief measures are commonly selected for landslide stabilization, the techniques have had mixed success in some situations. The principal weakness is the nature of many landslides where a lack of lateral continuity can make it difficult to contact permeable strata or there is insufficient permeable ground for the proposed drainage measure to be effective. For example, it is a common experience in horizontal drains to have widely different flows from a set of drains installed in an array. Some drains have zero flow. If the entire array has no flow, the horizontal drain system becomes a severe embarrassment to the designer. Another example would be relief wells to lower artesian pressure. If the pressure drop, as monitored by nearby piezometers, is confined to the area immediately around the relief well, the benefit to a large landslide would be minimal. Thus, the *element of uncertainty* can be fairly high in dewatering projects caused primarily by uncertain ground conditions. It is important for the geotechnical practitioner to conduct a thorough site investigation, possibly including in-situ packer tests and a pumping test, before recommending slope dewatering for major landslides. It is also important to fully inform the owner of the uncertainties of the outcome and reach a mutual decision to move ahead with a dewatering option. A backup plan, which could be a contingency allowance to provide more dewatering points (relatively common) or a different remediation technique, needs to be in place in case the original dewatering plan is insufficient to provide an adequate margin of safety for stability. Thus, for dewatering projects, the geotechnical consultant and owner need to have very good communications and a mutual understanding of the risks to success. The potential benefit is that a successful dewatering project may be significantly less expensive than the alternatives.

In some emergency situations, drains are installed quickly in an attempt to alleviate water pressures in a slope. The risk that ground movements will not be arrested should not deter engineers from taking such actions.

A second concern is the *limited benefit* of groundwater lowering on the factor of safety. The effect can be demonstrated by consideration of the infinite slope analysis, previously presented in Chapter 9, and shown again on Figure 17.1(a). For discussion purposes, the shear strength parameters are $c' = 0$, $\phi' = 35°$, the slope angle β is 19.3° and the soil density is 124.8 lb./cu. ft. These values were chosen as being typical values and to simplify the mathematics. There is a linear relationship between factor of safety F and the ratio of groundwater level h to depth to failure z (h/z). For the above assumptions, F = 1.00 when h/z = 1 (i.e., groundwater at the ground surface) and increases to a maximum of F = 2.00 when h/z = 0 (i.e., groundwater is at the slip surface), as shown on Figure. 17.1(b). For this example, it requires a 20% reduction in groundwater level to raise the factor of safety from 1.00 to 1.20. However, in practice, it is difficult to lower the groundwater level by an average of 20%, especially in fine-grained soils.

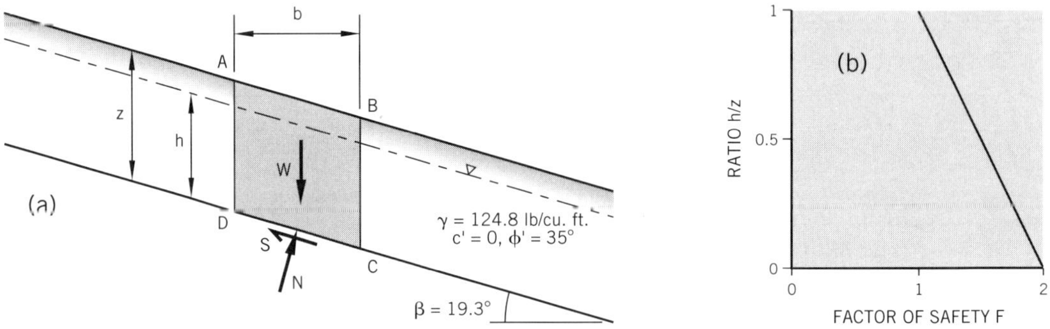

Figure 17.1 Infinite slope analysis:
(a) example
(b) effect of groundwater level on factor of safety

A second conclusion of this illustrative example is that groundwater lowering should not be used as a remedial treatment of landslides where the groundwater is close to the slip surface. In such cases, it is almost impossible to improve the factor of safety to an acceptable level for permanent stability. Although this would seem to be self-evident, the author has witnessed a situation where an engineer had recommended drainage when groundwater was within two feet of the slip surface. This approach appears to stem from a widely-held belief that any drainage will be beneficial to landslide stability.

The third concern of groundwater lowering is the *potential for affecting adjacent properties*. If a large drawdown is achieved, the area of influence is likely to reach nearby properties. One possible effect is to cause ground settlement where relatively weak compressible soils are near the ground surface. A more likely scenario is that there will be a *perception* that dewatering has caused ground settlement by neighboring landowners who are observing the dewatering project. Any cracks in a house or pathway will be attributed to dewatering even though, on many stiff clay foundations, cracks routinely appear and disappear seasonally due to surface dessication and rewetting. If such claims are likely (especially when there is opposition to a project), structural surveys should be conducted before and after the project dewatering. Ground survey reference points can also be established and monitored. These pre- and post-monitoring procedures should be able to separate legitimate from bogus claims for damages. However, the potential risk for collateral damage should be recognized by the owner and geotechnical engineer before the dewatering technique for landslide stabilization is implemented, and may affect the decision to proceed.

Area-wide groundwater lowering also could affect domestic water wells in close vicinity to a dewatering project. Other possible adverse effects may be present at specific sites. It is simply a consideration that should not be overlooked during the planning for temporary or permanent dewatering.

17.2 HORIZONTAL DRAINS

Technique

Horizontal drains are installed in a slope to either: (i) lower groundwater levels generally within the slope, or (ii) tap into and relieve groundwater aquifers feeding water into the slope (Figure 17.2). After installation, horizontal drains reduce the level of groundwater buildup produced by heavy precipitation or snowmelt. The drains are not horizontal but are inclined (typically 5° to horizontal) so that water is removed by gravity to the outer surface where it is usually collected and conducted to a drain or watercourse.

Horizontal drains can be expected to have highly variable flows in certain geological conditions, especially landslides in which soils have been intermixed and cracked. To be effective, relatively large numbers of drains need to be installed; to remain effective, a program of inspection, flow monitoring, and periodic cleaning is strongly advised. It is a medium cost remedial treatment for landslides and has minor environmental impacts (except during installation, when settling ponds and other devices are needed to control the muddy water).

Horizontal drains may be ineffective in clays and other fine-grained soils. The possibility of poor performance should be considered when assessing the relative merits of horizontal drains to other remedial options.

Horizontal Drains and Their Use for Stabilizing Landslides

Horizontal drains are small-diameter slotted plastic pipes that are installed in holes drilled into soils or rocks, usually at angles above horizontal. Some of the more common situations that can be stabilized by horizontal drains are illustrated on Figure 17.2. The first example, Figure 17.2(a), shows a landslide where large blocks of rock have dislodged a fine-grained soil at the top of the slope. The graben in the finer soils provides a bowl that ponds water. The pond feeds water into the slope below, especially during periods when groundwater seasonally rises (heavy storms, snowmelt from higher ground). Horizontal drains can tap into the pond and remove the stored water.

In the second illustrative example, shown on Figure 17.2(b), an existing ancient landslide moves intermittently during seasonal high groundwater levels. Here, a series of horizontal drains can lower the phreatic surface provided there are sufficient pervious soils, ground cracks, rock clusters, etc. to be effectively drained. In the example shown, the slide debris is a mix of materials from the gravel stratum at the head of the landslide and the underlying claystone. There are local pockets of gravel in the debris where groundwater can be drained but, since the clay is dominant, many horizontal drains will be

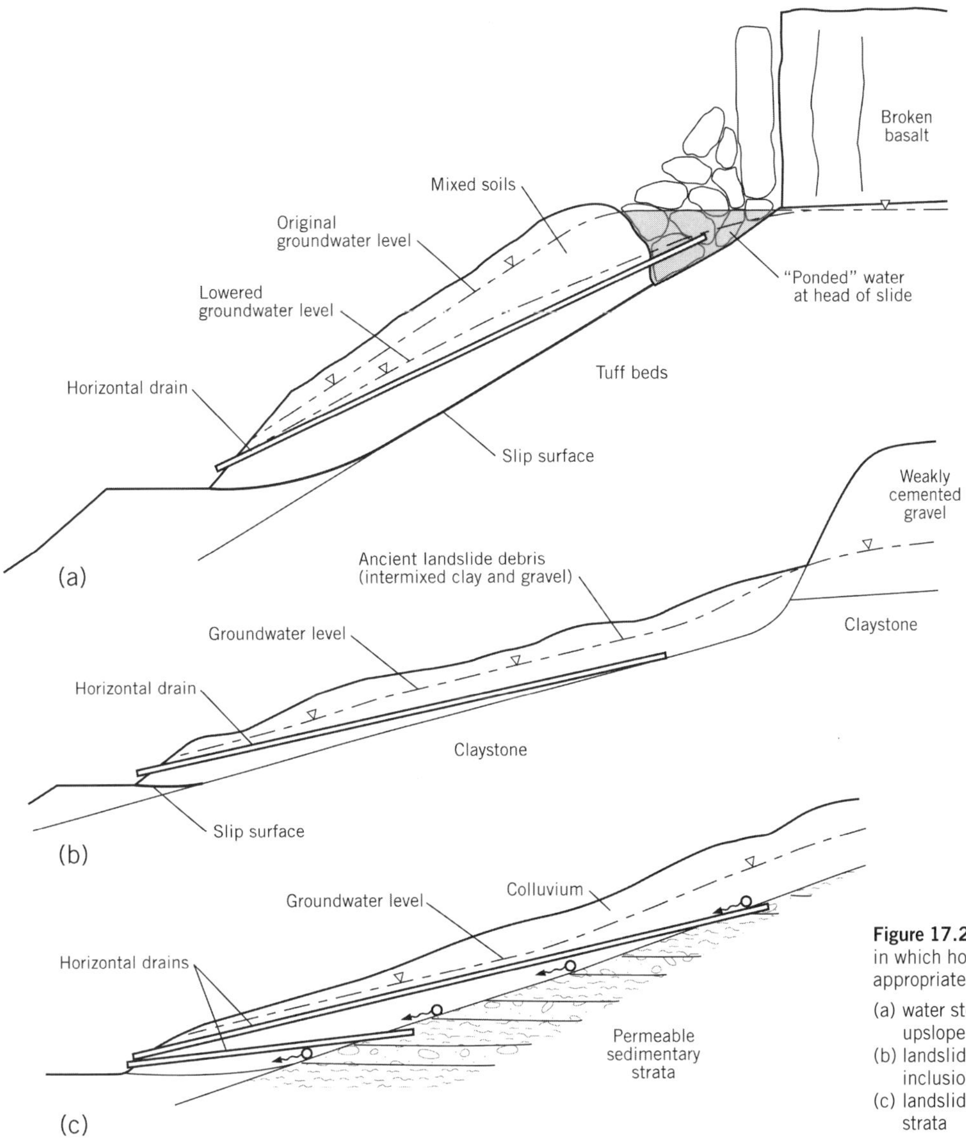

Figure 17.2 Examples of geological conditions in which horizontal drains may be an appropriate remedial option:
(a) water stored in highly permeable ground upslope
(b) landslide of mixed soils with waterbearing inclusions
(c) landslide covering deep-seated permeable strata

required with no assurance in advance that significant groundwater lowering can be accomplished.

The third example is a similar landslide in which colluvium covers in-place sedimentary soils or rocks (Figure 17.2c). The more permeable beds of the in-place materials feed water into the colluvium and maintain the high groundwater levels in the landslide. Horizontal drains at different inclinations to the horizontal can target the permeable beds or generally lower pressures along the slip surface. Another example of this situation could be a highly permeable layer oriented approximately parallel to the outer slope, creating an artesian condition near the base of the slope (especially after a cut is made into the slope).

The literature contains many examples of using horizontal drains to promote slope stabilization. A paper by Barrett (1979) relating to the construction of Interstate 70 in Colorado is of interest. Three examples are summarized below. Since they are presented for illustrative purposes only, many details of the case histories have been omitted, and drawings have been modified to cover only the essential points.

The project shown on Figure 17.3 used horizontal drains to relieve the head of water within a buried river channel higher up the slope. There, waterbearing gravels collected water from the spring snowmelt and fed water into the fine-grained soils, creating an artesian flow at the base of the cut. The horizontal drains were able to tap into the gravel and reduce groundwater levels in the slope. Although this case history is similar to Figure 17.2(a) described earlier, the difference is that there was no surface indication of the buried river channel. It was only discovered by drillholes made higher up the slope, thus showing the value of exploring for the source of unusually high groundwater levels.

The second example from Barrett (1979) is the use of horizontal drains to temporarily lower the water table so that a permanent structure (wall, buttress, etc.) can be safely built (Figure 17.4). In the example, horizontal drains were drilled into a recurrently-active ancient landslide 150 feet deep. This

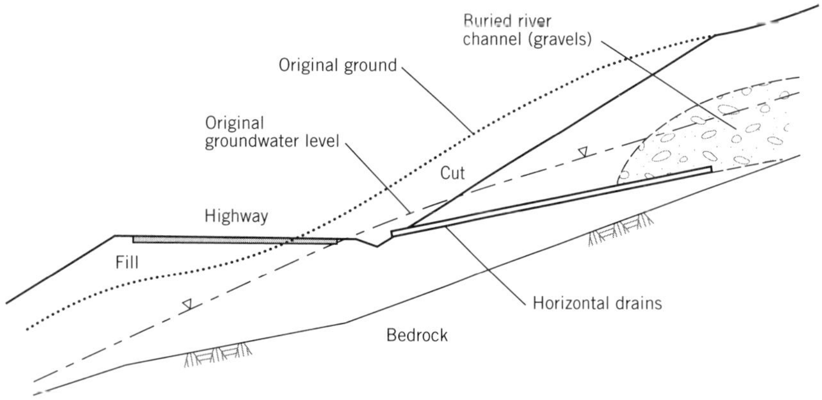

Figure 17.3 Use of horizontal drains to relieve water pressures from an upslope buried river channel. Interstate 70 near Dowd Junction, Colorado (modified from Barrett, 1979).

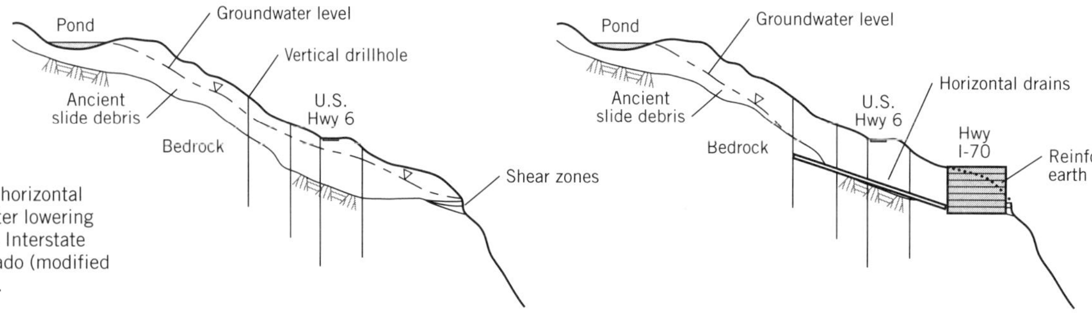

Figure 17.4 Use of horizontal drains for groundwater lowering during construction. Interstate 70, Vail Pass, Colorado (modified from Barrett, 1979).

silt overlies broken-up bedrock. The latter included very large blocks of rock that caused great bursts of water when the drillholes penetrated through them. Sufficient drains were installed to lower the water levels by 50 feet on average within about 300 feet of the project. By using sequenced construction and building under winter (frozen) conditions, they were able to build a vertical reinforced earth wall, founded on bedrock, at the toe of the slope. This example shows how horizontal drains can provide both dewatering for temporary works and an improved factor of safety after construction.

The third and final example taken from Barrett (1979) is the use of horizontal drains to provide a road underdrain for conveying modest quantities of water. At this site, shown by the section on Figure 17.5, the base of a cut slope was becoming progressively softer due to seepage from thin beds of very permeable rock behind the slope. Because groundwater levels were near the base of the cut, an interceptor drain was installed at the ditchline. Five horizontal drains were drilled underneath one lane of the highway to convey water to the median strip and then into the inlet of the culvert beneath the other (outer) lane. This technique avoided the need to dig a 20-foot deep trench across the inner lane, which would have disrupted traffic and would have probably produced a bump in the road after backfilling the trench.

Another appropriate use for horizontal drains is in certain emergency situations where local instabilities occur on highways during severe storms. In many cases, the cut or fill slope will slump down and groundwater will flow from cracks in the exposed scarp. After the storm has abated, and there is no longer a high risk of further movements immediately, horizontal drains can be installed rapidly to drain water from inside the slope. Typically, there is insufficient time to perform site explorations, but the observation of groundwater issuing from the slope face is sufficient justification that horizontal drains will succeed in picking up groundwater and improving stability. Horizontal drains can be rapidly deployed and effective in such situations.

This section will be concluded by a brief discussion of *inappropriate* uses of horizontal drains. Perhaps the most obvious case is where groundwater is at or only slightly above the slip surface of the landslide. In such cases, the potential benefit of groundwater lowering is almost nonexistent and should be

Figure 17.5 Use of horizontal drains as collectors for an interceptor drain at base of cut slope. Interstate 70 near Edwards, Colorado (modified from Barrett, 1979).

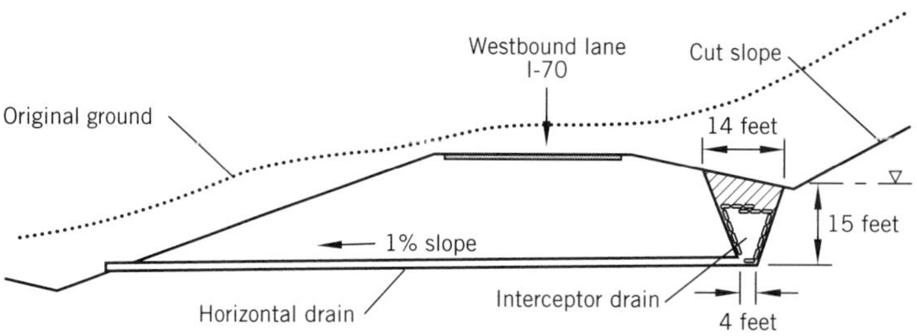

avoided. However, the piezometers indicating low groundwater levels should be carefully checked (by adding water or pumping out in the case of standpipes) to confirm that they are reading correctly. Another possibility is that the piezometer tips may have penetrated through the landslide into a more permeable rock or soil stratum that is locally draining water from the tip area, thereby giving misleading information. Similarly, vibrating wire piezometers have occasional malfunctions, and a very low reading should be confirmed by other piezometers in the vicinity.

The second circumstance in which horizontal drains generally should be precluded from use is where the geology and subsurface explorations show that the slope is composed largely of intact clay. Horizontal drains placed into clay will have a very limited zone of drawdown influence and will have no significant effect on the landslide stability. There has been a long and ignoble history of horizontal drains being installed in clay formations without benefit. Usually, it is tried because a maintenance person recalls that horizontal drains "worked" at an unrelated site.

There is an occasional jackpot event where the "try it" philosophy overcomes conventional judgment. The author was very surprised at a site where the owner and another consulting engineer pursued a horizontal drain strategy despite evidence that the landslide was composed of clay. The first very long drain reached some blocky materials just outside the more active part of the landslide and produced a heavy flow of water. In this local part of the site, groundwater levels fell by as much as 30 feet. Although this provided some local stability improvement to a large landslide, other horizontal drains in the landslide produced zero flow. In this case, the horizontal drains achieved some success despite the seemingly inappropriate ground conditions of clay.

The third scenario in which horizontal drains may be insufficiently effective is in deep-seated arc failures. Horizontal drains need a downhill gradient. On low to moderate slope angle landslides with a deep shear surface, the drains only reach the upper part of the slip surface. Therefore, it is possible that much of the lower slip surface is unaffected and receives less benefit than the higher ground. In such cases, additional slope stability remediation measures may be needed to achieve the desirable factor of safety for the landslide. For example, Section 17.12 describes how horizontal drains can be installed in a vertical shaft to reach deep-seated slip surfaces.

Horizontal Drain Design

In the United States, horizontal drains are made from polyvinyl chloride (PVC) pipe Schedule 80 conforming to ASTM D 1785. The pipe is 1.5-inch internal diameter, 1.9-inch outside diameter, and is sold in 10–20 foot lengths that are solvent-welded on site to achieve the desired length. Drainage occurs through two longitudinal rows of slots, the rows being 120° apart on the pipe circumference. Flow rates can reach up to 45 gpm.

Plastic pipe is manufactured in slot widths of 0.01, 0.02, and 0.05 inch. The widest slots are more permeable but allow fine sand and silt to pass through the opening. The finest slot width of 0.01 inch is more easily plugged and requires more frequent cleaning. The middle width of 0.02 inch (23 slots per linear foot) is used for most applications. Using filter requirements, the Corps of Engineers recommend:

$$d_{85} \text{ soil} > 1.2 \text{ slot width}$$

where d_{85} is the particle size in which 85 percent of the soil by weight is finer. If the soil has significant amounts of fine sand, non-cohesive silt, or dispersive fines, it is advisable to use 0.01 inch slots.

The entire length of the horizontal casing is slotted except for the lower 5 to 20 feet at the outlet end where solid (unslotted) pipe is used. The solid pipe prevents tree roots from invading the pipe in this area, and prevents soil erosion at the outlet. Most specifications require the slots to be oriented upwards in the hole. Except for the possibility that soil caving off the drillhole onto the drain will be looser (and slightly more permeable) than the native undisturbed soil supporting the drain below, there does not seem to be any compelling reason to require this slot orientation. An argument can be made that placing the slots facing downward (as for perforated drain pipes) would cause less fine soil to pass through the slots into the drain. The author believes that specifying the orientation of the slots is of little importance in practice because the twist in the drain can cause significant deviation from the intended orientation when hundreds of feet of drain are being installed.

Horizontal drains can be of any length from about 15 feet to 900 feet. Common lengths for landslide work range from 75 to 300 feet. The inclination to the horizontal can be as low as 2° and is typically 2° to 10°. However steeper inclinations of 25° or more can be accomplished. Drillers prefer to start the hole from a steep face. This helps to hold the designated vertical alignment. A steep face can be accommodated on relatively flat landslide slopes by digging an access trench for the tracked drill rig.

The drains generally are installed either: (i) parallel to each other, or (ii) in a fan-shaped array from a common exit location. The parallel installation method is often used when access follows a linear alignment, such as a road cut, railroad, or canal. The drains typically are spaced 3 to 10 feet apart and are drilled normal to the slope and in the direction of slide movements. In the case of canals and rivers, the outlet ends of the drains may be allowed to flow freely into the water below.

Arrays of drains are more frequently used on landslides than parallel drains. Fan arrays make it easier to collect the water, minimizes the number of drill rig moves, and lessens disturbance to the ground. Individual drain outlets are about 3 feet apart. Locations are often dictated in part by landslide features.

The drill rig normally requires a horizontal pad of at least 30 feet in the direction of slope, but cribbing can be used to minimize the required slope cut. Current practice in the United States requires a sediment pond and other erosion control measures to be provided by the driller.

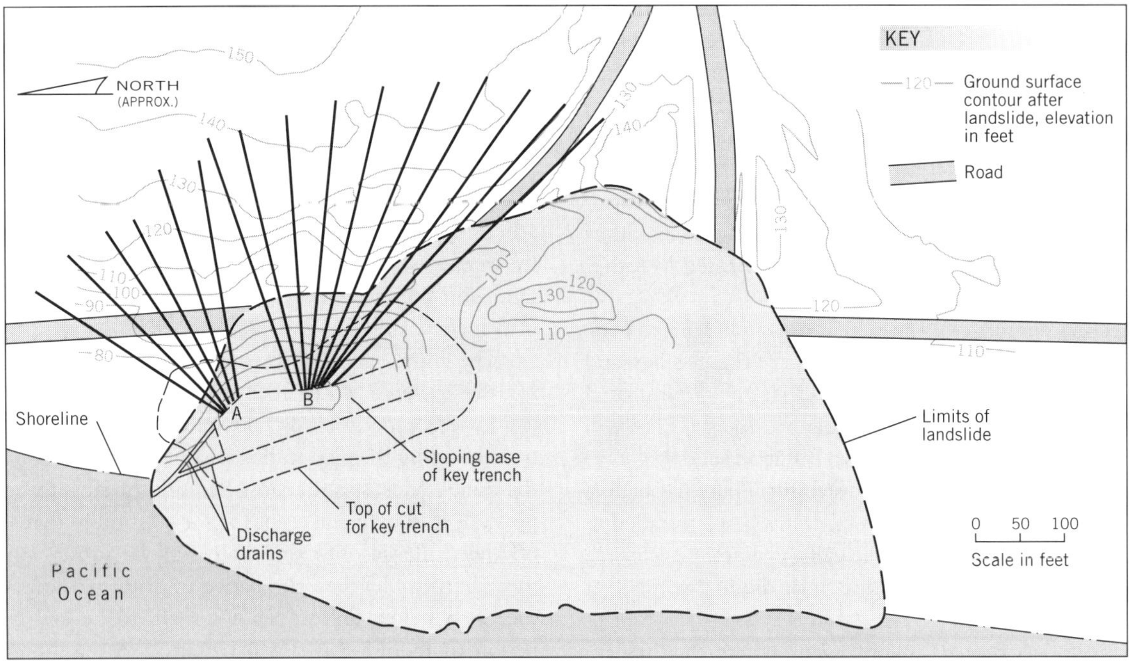

Figure 17.6 Two arrays of horizontal drains used to lower groundwater levels at Sandlake Road Slide, Oregon.

An example from an Oregon coast project, shown on Figure 17.6 has two arrays of drains. An access road was put into a graben feature. As a first stage of remediation, horizontal drains were extended into the upper hillside to lower groundwater feeding into the landslide. Subsequently, the landslide was stabilized by a combination of measures including a shear key (shown adjoining the horizontal drains on Figure 17.6), a trench drain (not shown), and some slope regrading. The horizontal drains were drilled at 4° to 15° to the horizontal through broken siltstone and clayey silt slide debris.

Attempts have been made to provide a design method to optimize the number and spacing of horizontal drains (Kenney et al., 1977; Prellwitz, 1978; Long, 1986). All methods are based on groundwater flow principles. The major difficulty with theoretical design methods is that the permeability is assumed to be constant throughout the ground. In most applications of horizontal drains to stabilize landslides the ground is broken up (such as at Sandlake Road Slide, cited earlier) or the slide debris has pockets and layers of permeable ground intermixed with impermeable soils.

The decision to use horizontal drains should be based on a geological and engineering appraisal of the causes of the landslide, its current condition, and the likelihood that significant amounts of water can be drained from the slope with a consequent drop in groundwater levels. Ground conditions can be highly variable with the occurrences of layers or pockets of permeable ground occurring in random fashion in many landslides. In rocks, significant sources of groundwater usually flow through open joints, faults, and other discontinuities rather than through intact rock. The number, direction, length, and point of origin of horizontal drains are largely judgmental decisions on the part of the geotechnical practitioner and are often influenced by site access and environmental constraints in addition to geology.

A practical method is to use the *observational approach* to design. A first layout of drains is made on the basis of the known site geology. This could comprise one or several arrays of drains at different locations on the landslide and represent part (say one-third to one-half) of the project budget. Depending on the outcome of this first phase, the technique can be abandoned due to unsatisfactory results or the concentration of drains can be increased where the best results have been obtained.

Since horizontal drain remediation is often a lower cost alternative, it can be a very attractive option for many landslides. However, because clays and other fine-grained soils are normally present in high groundwater level situations, there is always a risk that horizontal drains will be less successful than expected or even totally unsuccessful. It is up to the various parties—owner, geologist, engineer—to assess the risk and reach a decision about using horizontal drains at a particular site.

Experience has shown that a relatively large number of drains are usually required. Some drains may produce large flows; others may have no flow.

Whenever possible, piezometers should be installed in advance of construction to provide base readings of the existing groundwater levels. The number of piezometers should be at least four, but six to ten is preferable for measuring an average drawdown and providing input to stability analyses.

Calculations of the factor of safety help to set the initial goals for drawdown in the slope. Since drawdown is unlikely to be uniform and there are a limited number of places on the slope where piezometers are installed, the factor of safety should be used for guidance rather than being the determining factor.

The objective of all remedial measures for landslides is stability under existing and anticipated future conditions. To this end, inclinometers and/or survey points should continue to be monitored in the months and first years after remediation to confirm that movements have stopped or creep movements are at an acceptable level.

The designer needs to be aware of the possible effects of horizontal drains on adjacent properties. These effects include settlement due to drawdown on soft surface soils, drawdown of water-producing aquifers, and the possibility that a wayward drain will reach the ground surface or penetrate into an underground structure. Such occurrences are rare, but adjacent property owners are usually very sensitive to construction work in their vicinity and may claim for real or perceived damage from the horizontal drain work. Therefore, the designer may need to make structural surveys and set ground settlement stakes in more critical areas outside the project before construction begins.

The outlet end of the drain is designed to ensure that the collected water only passes through the horizontal drain and not around it. A typical schematic is shown on Figure 17.7. The PVC drain is solid (unslotted) over the outer 10 to 20 feet. It can be protected by a sleeve of galvanized pipe that keeps out roots, avoids erosion at the outlet, and prevents breakage of the PVC drain near the outlet. The annular space between the PVC pipe and the sleeve is grouted over the outer 3 to 5 feet (or as much as 20 feet) by a cement-sand-bentonite (or cement-bentonite) seal (Figure 17.7b).

Another important design feature is the removal of drain water off site. In a drain array, the outlet end of each drain is fitted with a plug that can be removed to clean the drain (Figure 17.7a). Above the plug is a T-joint that diverts the water through a 1½-inch diameter flexible pipe into a collector pipe (see Figure 17.8). All the drain outlets in the array

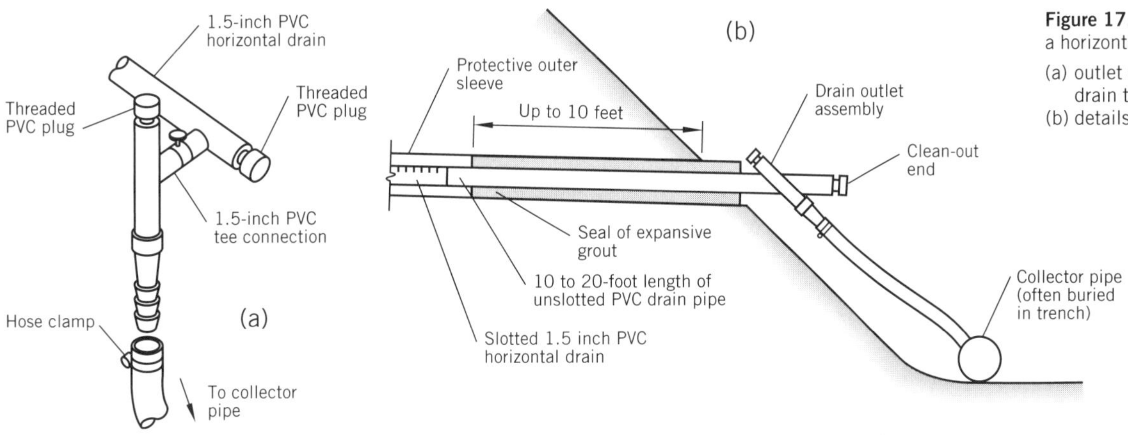

Figure 17.7 Discharge end of a horizontal drain:
(a) outlet assembly to allow the drain to be jet cleaned
(b) details of seal at outlet end

Figure 17.8 Horizontal drains connected to a corrugated plastic collector drain (Sandlake Road Slide, Oregon).

feed into the collector pipe. This pipe is installed at the start of construction. Water is taken by the collector pipe to a designated discharge point off the landslide.

The main collector pipe for a fan array of drains is typically a PVC or corrugated metal pipe (6-inch to 8-inch diameter). It can be staked to the ground or buried in a shallow trench, the latter generally being preferable to prevent vandalism or disturbance. The downhill slope of the collector drain should be at least 2 percent.

In some circumstances, horizontal drains can be allowed to spill directly onto the ground below the outlet end. Open-ended discharge can occur into a stream, canal, or a lined ditch. Another arrangement is to pass the drains through a vertical concrete wall with a concrete apron or riprap below to dissipate the water energy and prevent erosion.

Horizontal Drain Construction

The drilling equipment is usually mounted on a crawler tractor and is hydraulically powered. In a typical model, the rig can provide about 2,200 ft.-lb. of torque and a thrust of 9,000 lb. on the drill bit.

The construction of a horizontal drain through soils is illustrated on Figure 17.9. The hole is usually drilled by a tricone (rock roller bit) attached to NQ drill rods. The hole is approximately 4½ inches in diameter. Casing is usually needed to support the hole. Another option is to use a 4-inch diameter "casing" drill rod with a "knock-off" bit (see below) to both drill and case the hole simultaneously. Drilling is relatively fast and no samples are taken. Normal water flush, in which the drill water passes through the drill rods and returns along the outside, is the most common technique. Circulation has to be maintained to prevent trapped cuttings from increasing the torque needed. There are various additives to the drill water that can lower the resistance to cutting (e.g., liquid soap) or help to carry out the cuttings. Water is fed at rates of 5 to 40 gpm to cool the drill bit. Air flush is occasionally used for passing through erosive soils or rock drilling. A drag bit may be preferred in soft ground to make faster progress. If boulders or other hard obstructions are encountered, the driller may switch to a down-the-hole hammer or rotary coring methods to advance the hole.

The tri-cone bit is disposable. It is attached to the drill rods (or casing) by an adapter with a J-shaped notch at the end. When the hole has reached the required length, the direction of rotation is reversed to disconnect the bit from the adapter, leaving the bit at the end of the hole (Figure 17.9b).

The horizontal drain, with the upper end capped, is inserted through the drill rods. The drill rods are withdrawn (Figure 17.9c). The annular space between the drain and the walls of the drillhole is left open, but in many landslides the soils collapse loosely around the drain during and after the casing withdrawal. The drain can be partially withdrawn to check that the slotted pipe is working properly (Figure 17.10).

The final step of construction is to install and grout a protective sleeve at the outlet of the drain (Figure 17.9d and Figure 17.7b). Details were given in the prior section. The initial flow from the drain is measured. In general, flow rates decline appreciably within a few days of installation but are likely to temporarily rise during storms. After the initial flow rate reading, the drain is connected to the collector pipe (Figure 17.8). A permanent individual identification marker is attached to each drain.

The overall production rate to drill and install a horizontal drain is about 200 feet per day. This excludes construction of access roads.

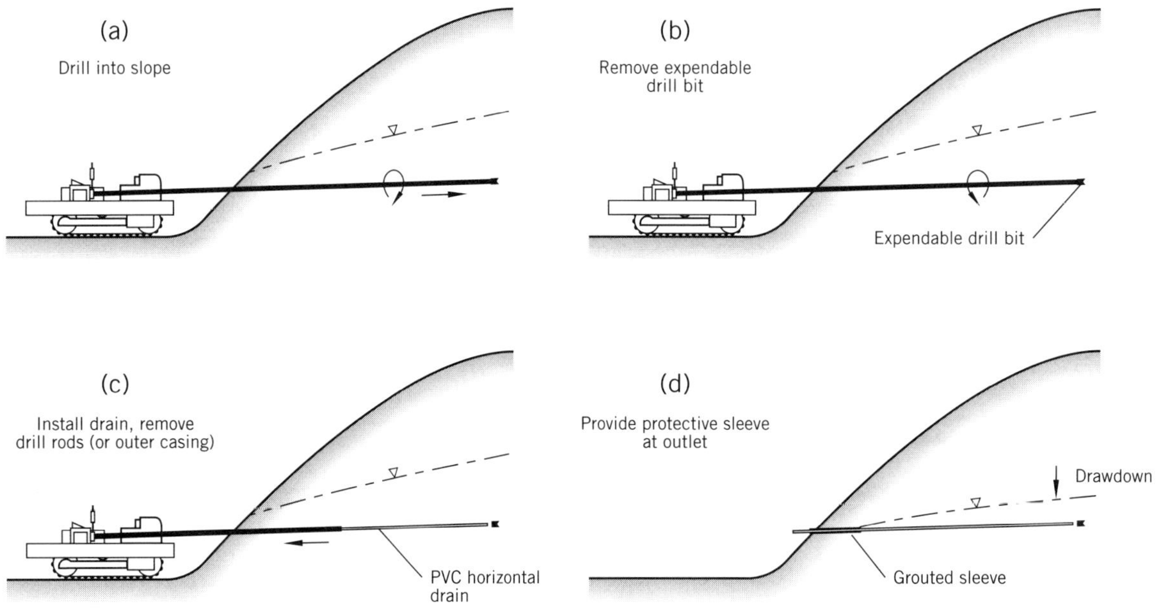

Figure 17.9 Construction of a horizontal drain (modified from Royster, 1980).

Figure 17.10 Horizontal drain partially withdrawn to show water coming out of drain slots (Sandlake Road Slide, Oregon).

Control of Horizontal Drain Alignment

Horizontal drains can be installed to lengths of up to 1,000 feet, but difficulties in the control of the slope and azimuth increase as the length of hole increases. The drilling bit bends downward under the cantilever of the drill rods. This change of alignment can also be affected by layers of varying stiffness, boulders, and other obstructions. Therefore, the initial angle of inclination to the horizontal has to be set according to the proposed length of the drain and the intended "target" within the slope. On long drains, the slope may become negative at the upper end. This does not prevent the drain from lowering groundwater levels provided that there is a positive water head above the "crest" of the longitudinal drain profile.

The elevation and direction of a horizontal drain can be checked at selected points during installation. For the *pressure head* test, a solid PVC open-ended pipe is inserted up the drill stem; the other end of the pipe is attached to a flexible standpipe tube. By passing water up the solid pipe until it overflows at the tip, the elevation of the tip can be measured by the water level in the standpipe (Figure 17.11). Since the length of the pipe is known, the average slope (grade) from the tip to the outlet can be calculated. A common procedure is to measure the tip elevation in increments of 50 feet or 100 feet along the drain.

A major drawback of the pressure head test is that it cannot measure negative slope after a horizontal drain changes slope from uphill to downhill. If this occurs, additional pres-

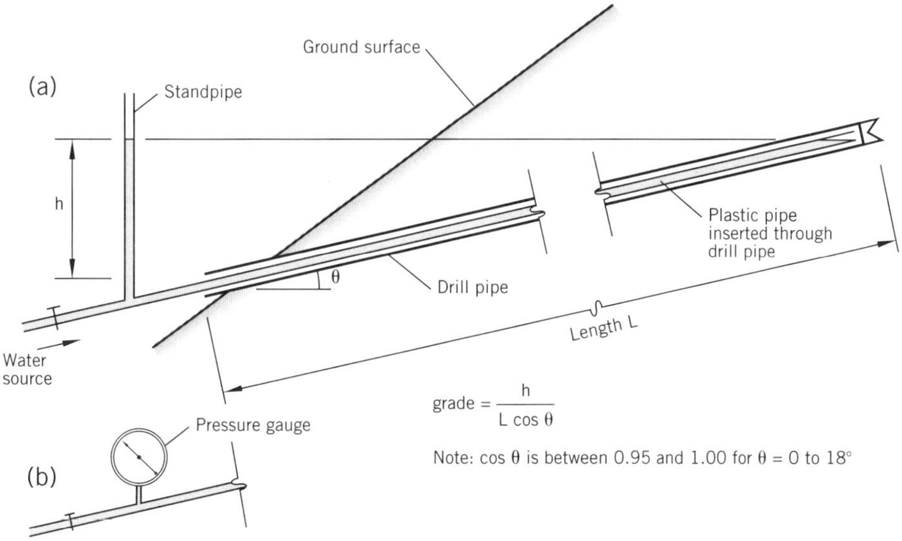

Figure 17.11 Measurement of the slopes and elevations within a horizontal drain:
(a) standpipe
(b) pressure gauge

Table 17.1 Horizontal Drain Inspection (Example)

Horizontal Drain No. HD-9 (Lower Fan)
Date: Saturday, September 19, 1998
Inspector: B. Black Site: Cornelius Pass Road Slide, Portland, Oregon
Initial inclination of drain = 12° to horizontal = 21.2 percent slope
Initial azimuth 279°

A Test Length into Slope feet	B Outlet Elev. feet	C Pressure Gauge Reading* psi	D Water Head ** feet	E Elev. at Test (B + D) feet	F Avg. Slope of Hole (D/A) percent
0	+ 324	0	0	+ 324	21.2
50	+ 324	6.25	14.4 – 1.8 = 12.6	+ 337	25.2
100	+ 324	10.6	24.5 – 1.8 = 22.7	+ 347	22.7
150	+ 324	13.1	30.3 – 1.8 = 28.5	+ 353	19.0
200	+ 324	16.0	37.0 – 1.8 = 35.2	+ 359	17.6
end hole 270	+ 324	19.8	45.7 – 1.8 = 43.9	+ 368	16.3

*Note: 1 psi = 2.31 feet of head
**Correction for height difference from lower end of drillhole to gauge = –1.8 feet

Field Observations: Boulder at 7–11 feet
Some return passing through adjacent drain HD-8
Good water return throughout
Easy drilling in clayey silt (decomposed basalt)

Flow rate upon completion of installation: <1 gpm Date: 9/23/98

Pajari Test Results:	Test Length, feet	Inclination	Azimuth
Date: 9/21/98	100	8°	272°
	200	7°	280°

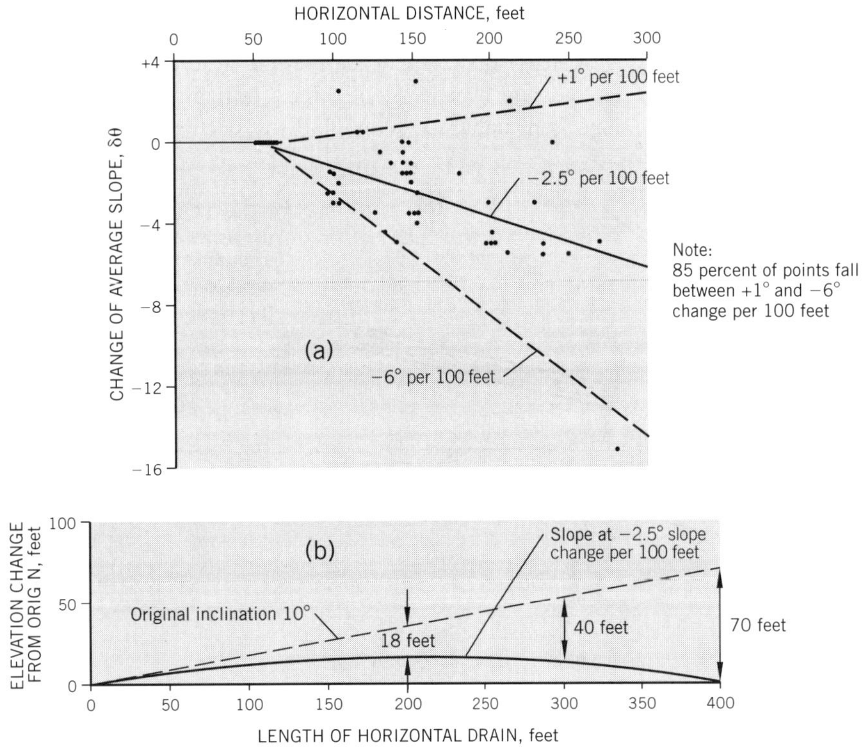

Figure 17.12 Results of measured horizontal drain gradients, Cornelius Pass Road, near Portland, Oregon (Landslide Technology data):
(a) changes of average slope vs. horizontal distance
(b) profile of a hypothetical drain with grade changes of −2.5° per 100 feet of length

sure head tests further up the drain will measure the same elevation; i.e., they measure the "crest" level of the curved drain.

When the pressure head becomes too high for a standpipe, a pressure gauge can be used to measure head. It is advisable to cross-check the readings of the standpipe and the pressure gauge at one or two levels in the measuring program.

An inspector's log for recording the relevant data during the installation of a horizontal drain is shown on Table 17.1 and includes an example of such a log. The calculations of average slopes should be made in the field to ensure that the answers from successive measurements are reasonably consistent. It is helpful to plot the drain profile on the geological cross-section to observe the position of the drain relative to the subsurface and groundwater conditions. If the drains are deviating from the intended line (or final target), it may be possible in the field to make adjustments of initial inclination for subsequent drains.

The results of numerous grade readings taken by the pressure head test at a site near Portland, Oregon, are plotted on Figure 17.12. Allowing for possible higher inaccuracies of slope measurement at the outlet (where the drain has more flex), the reference point has been established at the first measuring point at 50–65 feet into the drillhole. The change of average slope inclination with distance is shown on Figure 17.12(a). Although there is significant scatter, an approximate "line of best fit" indicates that the inclination drops about 2.5° per 100 feet on average. Thus, a "line of best fit" horizontal drain, with an initial inclination of 10° to the horizontal, becomes horizontal at 200 feet and drops to the same elevation as the outlet end at 400 feet (Figure 17.12b). At 200 feet, the upper end of the drain is 18 feet below where it would have been at a constant 10° slope inclination. Zaruba and Mencl (1969) reported drops of 7 to 10 feet at a distance of 200 feet in their measurements.

To measure the inclination and azimuth of a horizontal drain at a *point* along the length, a more sophisticated device is required. Such instruments have been developed for the petroleum and mining industries. An example is the *Pajari directional borehole surveying instrument* (Figure 17.13) marketed by Pothier Enterprises Ltd., Delta, B.C., Canada. It is a combination of a tiltmeter and compass, both of which are initially in the "locked" position. When a timing mechanism is activated by the user, the tiltmeter and compass are free to move. In this condition, the instrument is inserted at the end of a PVC rod to the specified depth in the drillhole and set in position before the final 10 minutes of the timer. The two components then "lock", the instrument is withdrawn, and the measurements of inclination and azimuth are read off the locked scales.

The Pajari can measure a negative slope, which sometimes occurs in very long holes. However, since it only measures slope and azimuth at one point, it does not measure the shape of the drillhole over its length. An approximate method of determining this profile is to plot the measured Pajari slope at each point and extrapolate in each direction to make the extrapolated lines intersect the lines from the next Pajari measurement at the halfway point between measurements. An example is given on Figure 17.14, which shows the profile of a horizontal drain calculated from Pajari slope readings at two points and average slope readings measured by water head

Figure 17.13 Pajari directional borehole surveying instrument.

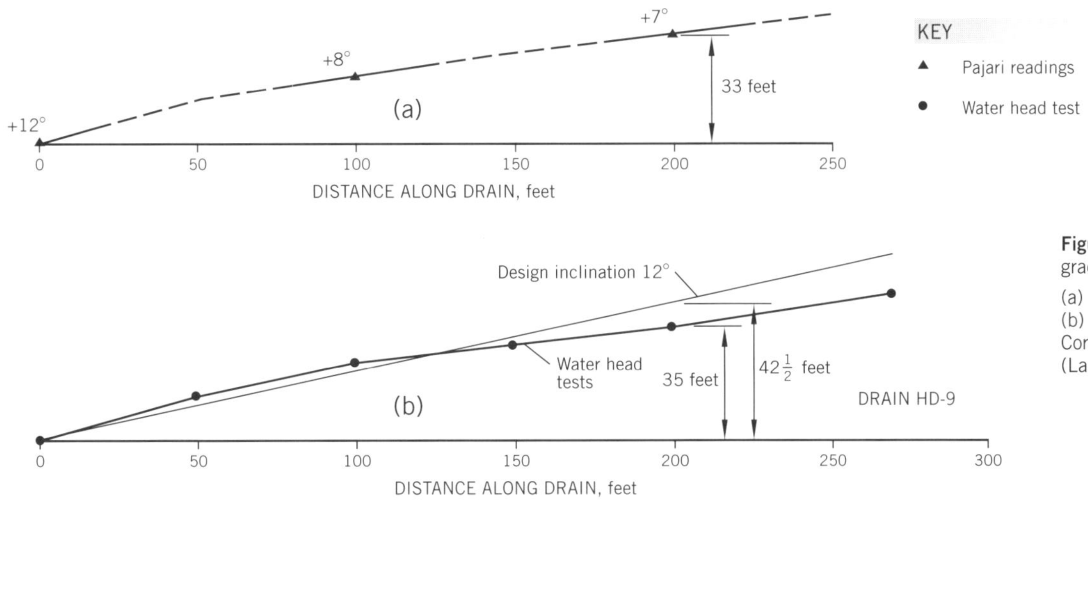

Figure 17.14 Example of grade measurements:
(a) Pajari tests
(b) water head tests.
Cornelius Pass Slide, Oregon (Landslide Technology data)

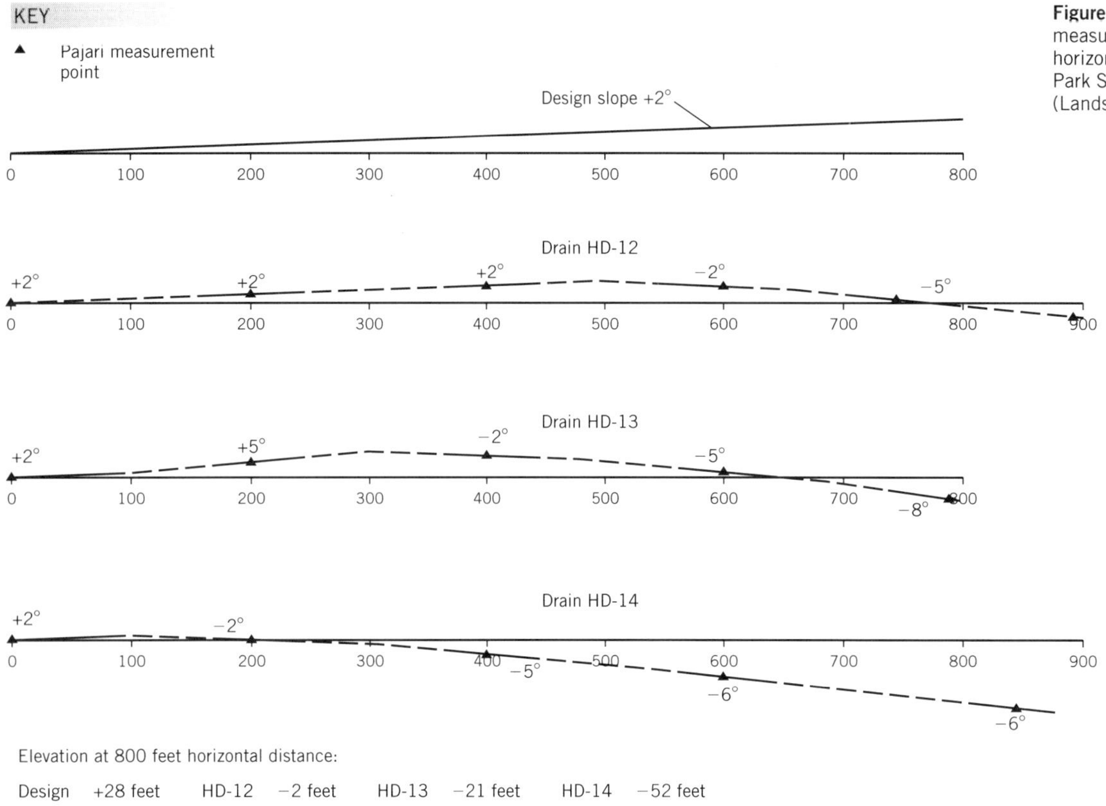

Figure 17.15 Pajari measurements on very long horizontal drains. Washington Park Slide, Portland, Oregon (Landslide Technology data).

tests. The agreement is very good in this case. At a length of 200 feet up the horizontal drain, the elevation rise is measured at 35 feet by the pressure head tests and at 33 feet by the Pajari test extrapolation technique. At a constant inclined slope of 12°, the drain at 200 feet should be 42½ feet higher.

An example of the bending of very long horizontal drains is shown on Figure 17.15. The three drains (HD-12, HD-13, HD-14) were in the same array and essentially passed through the same soils. The initial inclination of 2° to the horizontal was the same for each drain and, had they stayed on this slope, the elevation at a distance of 800 feet would have been 28 feet *above* the outlet. Instead, the Pajari readings indicate that the drains developed negative slopes and were 2 to 53 feet *below* the outlet level at 800 feet.

Measurements taken by the Pajari instrument are time-consuming and, with the drill crew on standby, add significantly to the field costs. They can only be taken after the drilling has been completed because steel drilling rods affect the compass readings.

Maintenance of Horizontal Drains

Owners should be made aware that horizontal drains will require periodic cleaning and monitoring to maintain their effectiveness. The usual method for cleaning horizontal drains is by high pressure water jets. Jet cleaning is fast, easy to perform, and effectively scours the slots. A jet nozzle has a brass head with small diameter holes passing out of the sides of the head. An example is shown on Figure 17.16. The nozzle is attached to a hosepipe and is fed up the horizontal drains. The

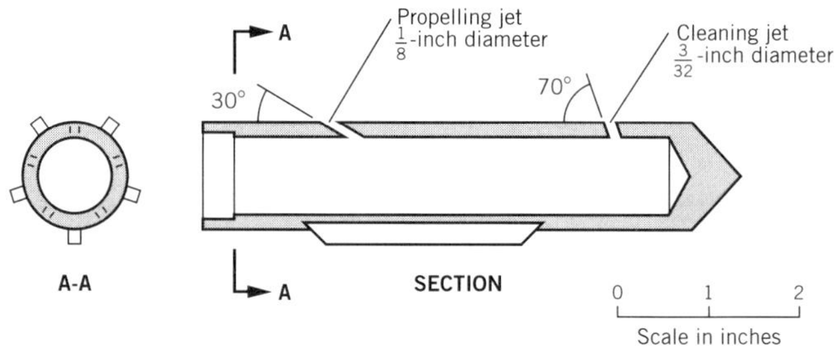

Figure 17.16 Nozzle for high-pressure jet cleaning of horizontal drains (after Ford, 1974).

nozzle holes that face backward propel the hosepipe up the drain. The nozzle holes that direct the water jet laterally clean out the drain slots of clogging materials such as fine-grained soils or mineral deposits. The latter may be difficult to dislodge and may require aggressive chemical treatment.

The pump pressures are around 300 to 400 psi with a flow of at least 30 gpm. The nozzle moves along the drain at about 30 feet per minute. Excessive pressures or too many passes can damage the drains. If an obstruction is found, it can be removed either by using a nozzle with a forward-pointing jet or by using a cutter head. Tree roots (especially from willows), silt and rust (in steel pipes) are the principal materials to be removed. Tree roots and precipitates are more likely to be found near the outlet end than elsewhere in the drain. A galvanized metal sleeve pipe can provide good protection against roots (see earlier). Significant quantities of silt probably indicate a break or crack in the drains.

Table 17.2 Cost of Horizontal Drains

Site (date of construction)	Number of Drains	Length of Drains (feet)	Total Length (feet)	Contract Price	Construction Cost per foot of drain
Sandlake Road Slide (1999)	26	100–500	8,790	$183,522	$20.95/ft.
Cornelius Pass Slide (1998)	24	115–280	4,090	$64,418	$15.75/ft.*
Ditch Camp Slide (1997–98)	92	80–650	29,160	$373,140	$12.20/ft.
North Fork Santiam Slide (1997)	56	135–325	11,496	$199,915	$17.39/ft.
Washington Park Slide (1995)	2	885–925	1,810	$37,580	$20.36/ft.

*no outlet casing or collector drain needed

Forrester (2001) has suggested that jet cleaning should be scheduled at three months, one year, four years, and every four years thereafter, from the time of installation. If calcium carbonate clogging occurs, the four-year intervals should be reduced to two years. There is general agreement in the literature that the routine interval should be between two and eight years, depending on the subsurface conditions, judgment, and experience in the locality. However, the owner usually makes the decision in setting the frequency of visits and, as a result, maintenance is often neglected. Poor maintenance practices need to be taken into account when recommending horizontal drains for long-term stability.

Records should be kept for each maintenance visit. This should include: flow from each drain before and after cleaning, type and amount (subjective) of materials flushed out during cleaning, obstructions (if any) during cleaning and their location within the drain, repairs made or needed, etc. A preprinted form can summarize the information, making the inspection relatively simple and consistent between different personnel.

The effectiveness of horizontal drains with time has been difficult to quantify because the flow rates generally are affected by storm events and the level of moisture in the ground. These vary from year to year. Smith (1980) surmised from his study of California highway experiences over 40 years that there is a slight decrease in efficiency over a period of years. However, he concluded "...if proper cleaning and maintenance is performed on a regular basis, a drain installation will function effectively for at least as long as the expected life of the highway."

Cost of Horizontal Drain Installation

Examples of the construction cost of horizontal drains taken from the files of Landslide Technology are given on Table 17.2. These are projects for the Pacific Northwest and include all costs associated with mobilization of equipment, construction of drilling pads, drilling and installation of horizontal drains, measurements during drilling of elevation by pressure heads, subsequent orientation/slope measurements by Pajari instrument, control of site water, construction of effluent discharge, cleanup, etc. The costs do not include engineering design, field observations, or follow-up inspections/flow measurements.

The costs are provided as a guide to illustrate the approximate range of construction cost for horizontal drains in the Pacific Northwest region of the United States, mainly for comparison with other types of landslide remediation. Adjustments will be needed for other regions and abroad. The construction costs can be updated via the ENR Construction Cost Index, as described in Chapter 13, Section 13.6.

17.3 TRENCH DRAINS

Technique

Trench drains provide very good remedial treatment at a relatively modest cost in landslides with high groundwater levels. The drains generally are built of free-draining rockfill separated from native fine-grained soils by a geotextile filter fabric (Figure 17.17). The principal concern during construction is local failure of the trench sides; this can be partly overcome by rapid construction techniques. In unstable ground, full sup-

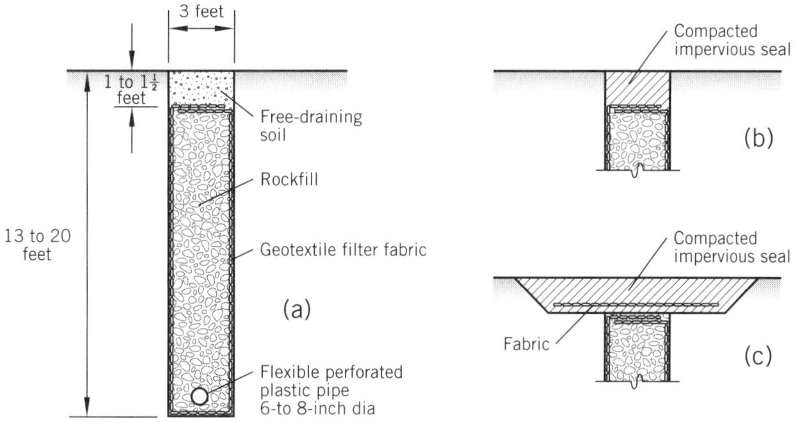

Figure 17.17 Typical trench drain details:
(a) free-draining soil at surface to capture surface runoff
(b) (c) impervious seal to exclude surface water (two variations)

port of the trench may be required, possibly accompanied by temporary dewatering by wells.

Trench drains are most effective in shallow landslides (up to 20 feet deep) in which the drains fully penetrate through the landslide debris into the underlying stable ground. Although trench drains can be placed in any direction on the slope, they are usually differentiated as either: (i) *counterfort* drains placed parallel to each other *in the line of* maximum slope to depress groundwater levels and provide additional strength from the granular backfill, or (ii) *interceptor* drains placed *across* the slope, usually at or near the head of the landslide, to collect and remove groundwater before it reaches the main body of the landslide. The top of a trench drain can be made impermeable (to exclude surface runoff) or permeable (to collect both runoff and groundwater). In clay soils, it may take several weeks after construction for long-term drawdown conditions to be reached.

Trench drains that partially penetrate through a landslide also depress groundwater levels but do not benefit from the strength of the drain backfill on the slip surface. Other options include: (i) construction of very deep trench drains using more sophisticated construction methods involving higher costs, and (ii) extending drainage from the base of a trench to a deeper permeable stratum or artesian zone using sand drains or wick drains. Trench drains also can be built into new cut slopes of clay or silt as a preventative measure against future instability.

One of the keys to success in trench drains is the rapid removal of water collected by the drains; this requires attention to backfill gradation and the trench gradient in design and construction. A flexible perforated pipe in the base is recommended.

Trench Drains and Their Use for Stabilizing Landslides

Trench drains are trenches of moderate depth (less than 20 feet) that are backfilled with a free-draining soil and usually include a drainage pipe near the trench base. As a landslide remedial measure, they are most effective where groundwater is close to the ground surface, a condition that commonly occurs in flat-lying landslides within depressed ground. The soils are usually slow-draining silts, clayey silts, and clays. When a network of trench drains are built into a landslide mass, there is drawdown of both wet and dry season groundwater levels. Therefore, trench drains can reduce both the seasonal buildup of groundwater and the spikes of groundwater caused by storms or snowmelt.

Shallow Landslides

Landslides of shallow depth, such as colluvium or solifluction sheets above claystones and siltstones, allow the trench drains to penetrate through the landslide into the bedrock or harder layer. For trench excavation by a backhoe, the depth is limited to about 20 feet below the surface. Deeper trenches can be made but require full shoring and possibly temporary dewatering; the cost of construction may become prohibitive.

Trench drains constructed *parallel to each other in the direction of slope* are illustrated on Figure 17.18. The base of the trench drain is dug a short distance into the stable stratum underlying the landslide. This type of drain (termed a *counterfort drain* in Britain) provides two benefits to slope stability: (i) it reduces the groundwater levels in the fine-grained soils of the landslide mass between the drains and (ii) the base of the drain itself becomes part of the landslide slip surface; with very

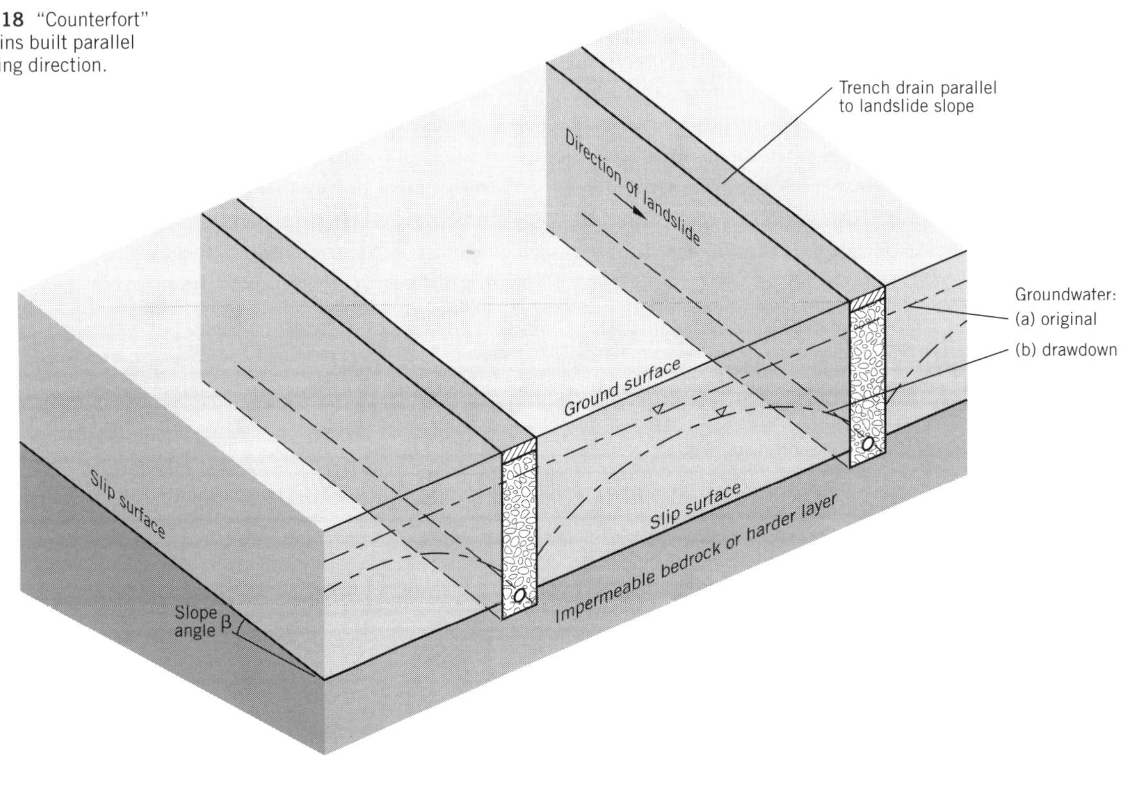

Figure 17.18 "Counterfort" trench drains built parallel to the sliding direction.

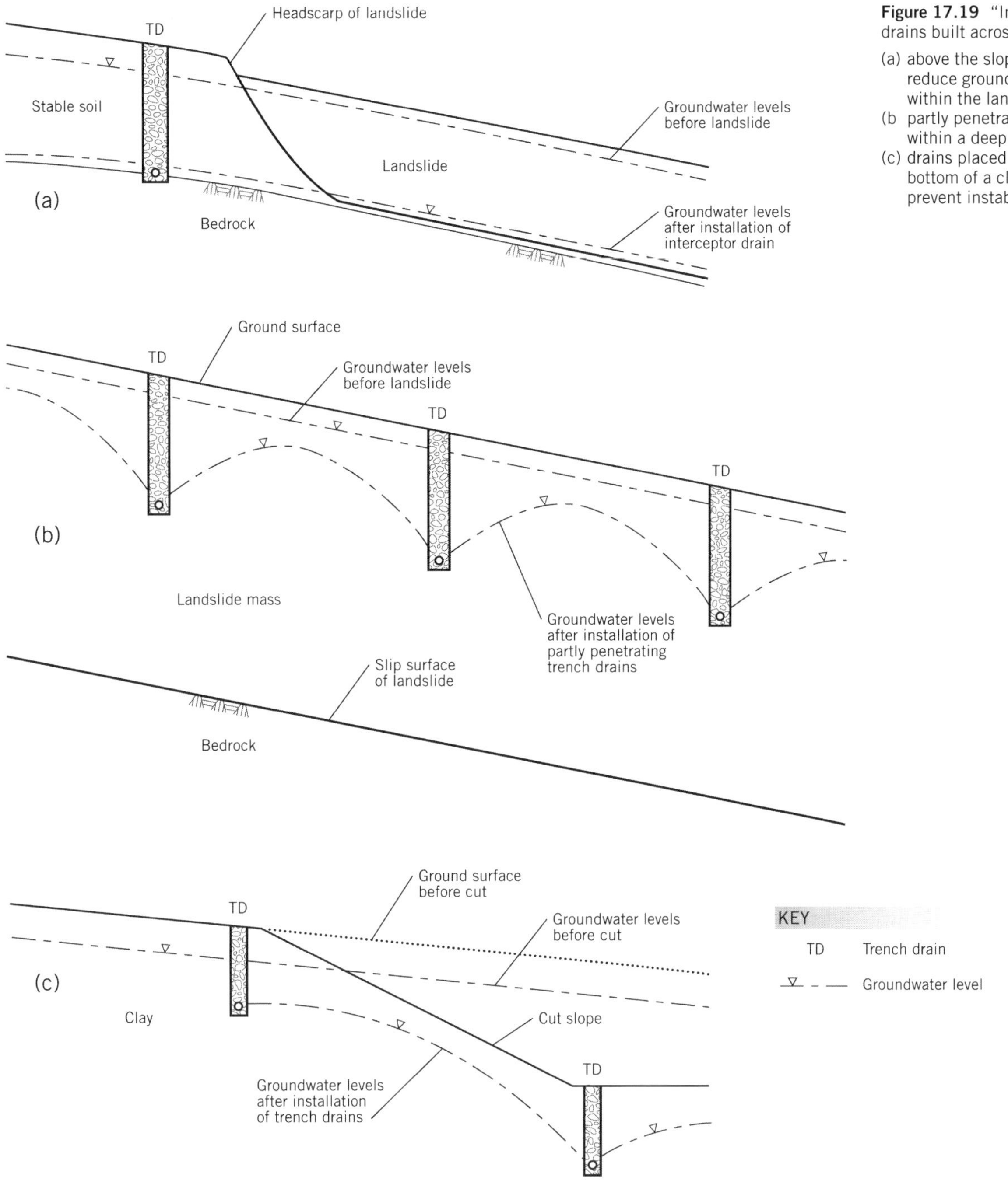

Figure 17.19 "Interceptor" trench drains built across the slope:
(a) above the slope failure to reduce groundwater levels within the landslide
(b partly penetrating drains within a deep landslide
(c) drains placed at the top and bottom of a clay cut slope to prevent instability

low groundwater above the base and the gravel backfill having a much higher shear strength than the native soils, the drain makes a significant contribution to the shear resistance of the landslide. Thus, a set of counterfort drains built into a shallow landslide with a high preexisting groundwater table is a very effective remediation.

Trench drains constructed *transverse to the direction of slope* are shown conceptually on Figure 17.19. For a shallow landslide, the base of the trench is excavated a short vertical distance into the top of the underlying stable ground. These drains are termed *interceptor drains* because they intercept the flow of groundwater passing into the landslide mass. After collecting the water, it is conveyed from the landslide by an outlet drain. One interceptor drain at the top of a landslide may be sufficient to drain the entire landslide mass if there is no recharge further downslope from the bedrock or base layer (Figure 17.19a).

A major drawback of an interceptor drain is that the excavation for the drain may reactivate the area immediately upslope. An excavation made into a preexisting landslide cuts a "slot" normal to the direction of sliding. Since almost all landslides (including translational slides) have a net balance of slightly more driving force at their upper end and thus rely on net resistance at the lower end, the cutting of a trench through

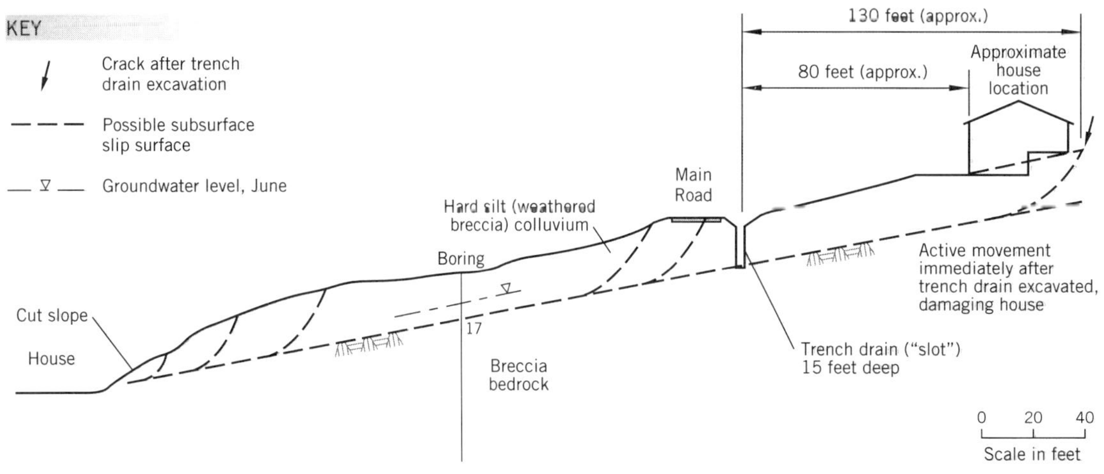

Figure 17.20 Case history of a trench drain excavation ("slot") causing upslope movements.

the landslide reduces support to the ground above the trench. If deep enough, the trench excavation may renew landslide activity even when groundwater levels are seasonally low.

A cautionary example of this effect will be briefly presented. At a site in Washington state, an excavation for a house site at the base of a slope had caused landslide movements further upslope. Cracks were mapped within the slope and up to the main road above. A section through the middle of the landslide area is shown on Figure 17.20. A geotechnical consultant recommended that the county construct a 450-foot long interceptor trench drain parallel to the road below the inside ditch, and borings confirmed that hard bedrock was 12 to 15 feet below the ditch. The interceptor drain was designed to reduce groundwater levels in the unstable slope below. The overburden soil was a colluvium of hard silt mixed with fragments of the parent breccia rock. In late spring, the contractor started trenching for the drain using a trench box to protect the workers. Almost immediately, there were reports that the uphill side of the trench was periodically collapsing, and a house uphill from the trench experienced severe cracking. Observable cracks at this house were 130 feet upslope from the trench, but there is anecdotal evidence that cracking extended several hundred feet further upslope.

Subsequently, the trench drain was completed by closely-sequenced construction (see Chapter 13, Section 13.5), but the cost of repairing the upslope damage (more than one-half of the house value) and lawsuits far exceeded the construction contract. Like many lawsuits involving landslides, the owner and consultant each made errors of judgment, but the key point is that an excavation made *across* a landslide involves a significant risk, and the risk needs to be carefully evaluated before construction. In this example, closely-sequenced construction or a fully braced excavation, allowing for landslide loading, could have prevented both the damages and lawsuits.

Some of the judgmental errors included: (i) failure to make inquiries of the upslope resident—he had been experiencing periodic ground movements during his seven years in residence; this would have warned the consultant of the existing marginal stability of the colluvium; (ii) failure to require closely-sequenced construction or a fully braced excavation when making a cut across a landslide; and (iii) refusal by the county to pay for the geotechnical consultant to have an engineer on site during construction of the trench drain.

Deep Landslides

Trench drains that only partially penetrate into the landslide can be built in the direction of slope or transverse to the slope. In either case, the drains simply lower groundwater levels and help to reduce any spikes of groundwater during storms or snowmelt. Unlike the fully penetrating trench drain, the gravel backfill does not provide additional shearing resistance.

Parallel trench drains constructed transverse to the slope can provide progressive drawdown (Figure 17.19b). It is a similar effect to a wellpoint or suction well system. However, as already pointed out, trenching carries a fairly high risk of reactivating the uphill segment of the landslide.

A deep landslide for trench drain remediation is one that cannot be trenched in one pass of a backhoe using closely-sequenced construction techniques. Depending on the availability of excavating equipment, this limit ranges from 13 to 20 feet below the original ground. Slightly deeper trenches can be constructed in open areas by first digging an access path down to the water table, which is usually several feet below the ground surface in summer. However, the risk of instability increases as the depth increases, and it is generally necessary to fully support the excavation on deeper trenches when trenching *in the direction of* the landslide. For deep trenching *across* the landslide, fully shored trench excavation is mandatory and may also require dewatering prior to trenching. The excavation can be made between the shoring by clamshell or by hand excavation (power shovels) and bucket lift.

If the ground below a deep landslide is permeable, vertical sand drains, stone columns, or wick drains can be installed through the base of the trench drain to allow vertical drainage between the trench drain and the permeable stratum. The vertical sand drains can be installed from the ground surface before constructing the trench drain. Similarly, artesian pressures at depth can be relieved into the base of a trench drain (Figure 17.21).

Preventative Use

Interceptor drains are sometimes built into the top and bottom of clay cut slopes for major roads, such as freeways (Figure 17.19c). The upper interceptor drain is constructed before the cut is made; the lower drain is put in at the ditchline immediately after the cut slope has been completed.

This design technique can prevent landslides from occurring during construction and longer term. As examples: (i) permeable layers in a clay stratum that are exposed by a cut slope can cause instability due to local erosion; the trench drain dries up the permeable layers; and (ii) in stiff clays, an initially depressed groundwater profile caused by a cut will eventually recover and could cause a delayed failure; interceptor drains keep the groundwater levels low during this return to equilibrium groundwater status. Because the trenches are cut cross-slope, closely-sequenced construction should be followed.

Design of Trench Drains

Trench Drain Materials

A trench drain is typically 3 feet wide and 13 to 20 feet deep (Figure 17.17). A geotextile filter fabric separates the drain rock from the native soils. A non-woven fabric of subgrade geotextile, listed on Table 11.9 of Chapter 11, is suitable for a trench drain.

The geotextile needs to be stored off the ground and kept dry. It also needs to be protected from ultraviolet rays (sunlight) and soil contaminants. During construction the strips of geotextile, spanning from one side of the trench to the other, have to be staked or pinned at each end. The geotextile is supplied in rolls 12½ to 15 feet wide and the overlap between successive strips should be 1 to 2 feet (use 2 feet for deeper trenches) to avoid the possibility of gaps developing when the trench is backfilled.

After infilling the trench with rockfill to within 12 to 18 inches of the surface, the fabric is wrapped over the top (Figure 17.17). The top of the drain can be completed in one of two ways. The more common method is to compact impervious fill to keep out surface water (Figure 17.17b). The drain then functions as a subsurface drain only. A wide strip of compacted impervious fill, with fabric at the bottom, can prevent cracking and loss of seal at the top edges of the trench drain (Figure 17.17c). The property owner or environmental considerations may dictate this option. The second method is to add free-draining soils at the surface so that the drain can capture and dispose of both surface runoff and groundwater. Since the top of the drain can become contaminated by fines, the fabric is overlapped at 12 to 18 inches below the surface to preserve the integrity of the subsurface drain. Some geotechnical consultants are concerned that permeable soils at the top of the drain add runoff water into the landslide. However, with provision of a rapid method of taking the collected water off the site, the combined surface and subsurface water quantities should not create a problem.

The trench backfill should be uniform size gravel. The most common sizes are: 4 inch to 2 inch; 2 inch to 1 inch; 1½ inch to ¾ inch. Crushed hard quarry rock with angular edges is the best backfill. The uniform size and angular stones provide a high permeability. As the stone size increases, there is an increasing risk that the geotextile will be punctured, possibly causing the rockfill to become contaminated and plugged. This is a small risk, in the author's opinion, and the coarser gradation range is preferred because it will allow the collected water to be rapidly taken out of the landslide. This requirement is especially important if both groundwater and surface water are being collected by the drains. The critical times for the system are during storms or snowmelt.

The second important requirement is to specify and provide site control against contamination of the rockfill from quarry fines. Soil permeability is governed by the fines content. In the case of quarried rock, the crushing process can produce much fine dust that can intermix with or cling to individual

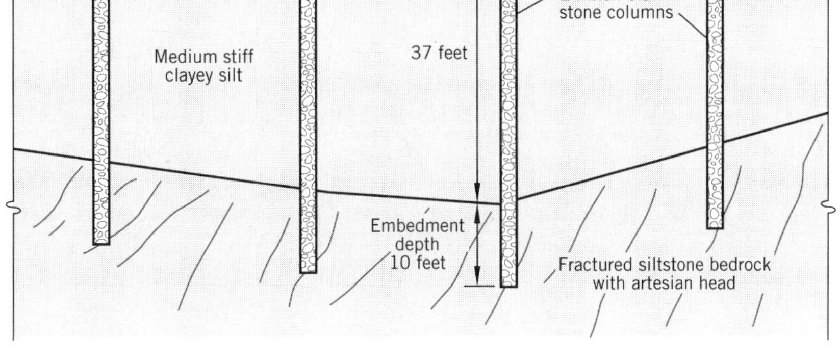

Figure 17.21 Vertical stone columns below trench drains to relieve artesian pressure in bedrock, Meadow Lake Road Slide, Oregon (Landslide Technology data).

gravel stones. In a drain, the fines can be washed by flow or gravity to the bottom of the trench and reduce the permeability in this part of the drain. Fines can also clog the openings of a perforated pipe where high hydraulic gradients occur.

The author recommends that fines passing the No. 200 U.S. sieve should be limited to 3 percent by weight. It should be remembered that 3 percent by weight is probably equivalent to about 4–5 percent by volume, which is what the site inspector observes. The fines should be nonplastic. Many quarries have difficulty in meeting this specification and samples should be required from the contractor's proposed source, before construction begins, to ensure compliance. In the event that no local quarry is able to meet the specification fines limit, the geotechnical consultant may have to increase the limit to 4 or 5 percent by weight. However, the tighter specification limit is preferable. One option is to require the crushed rock to be washed at the quarry.

Compaction requirements for quarried rock should be minimal, partly due to the difficulty of getting compaction equipment into the trench and partly because of the time required for compaction, which increases the risk of trench face instability. The backhoe can ensure that no large voids occur from arching by tapping the layer of gravel with the back of the bucket.

An alternative to hard crushed quarry rock is to use washed rounded rock, often termed *pea gravel* or *drain rock* by suppliers. This material is available in sizes ranging from 1½ inch to ¼ inch. It is easily placed in a trench and requires no compaction because the relative density range is very limited. The main disadvantage of drain rock is a lower strength compared to crushed quarry rock. The angle of repose is around 35°, which produces a wider unsupported area of trench than angular rockfill when using closely-sequenced construction. It also ravels easily to the base of the slope and may interfere with the fabric placement.

Clean medium-to-coarse sand has been used as a backfill in trench drains. It is acceptable if the trench drain is being used to capture only groundwater in slow-draining soils. In such cases, the water quantities are small. A perforated pipe at the bottom of the trench should be wrapped with filter fabric to prevent the sand from entering the drain.

In all types of trench drains, it is advisable to include a flexible plastic perforated pipe at the bottom of the trench (Figure 17.17). The size of the pipe can be determined from hydrology and hydraulic calculations. Typically the pipes are 6-inch to 10-inch diameter. Although a drain is a nuisance during closely-sequenced construction, it can usually be manipulated from the edge of the trench as the work proceeds (see Chapter 13, Figure 13.1). If a choice has to be made between a drainpipe and a need for a fast rate of construction, it is preferable to sacrifice the drainpipe. However, in this event, a coarser gradation of rockfill should be chosen.

The base of the trench should be sloped so that water does not pond in the trench. A minimum slope is 2 percent, but steeper slopes are better. The drain outlet has to be designed to prevent local erosion during outflow.

Trench Drain Support

When a trench drain is cut *in the direction of the slope,* cracks may develop for a fairly short distance from the sides of the trench and local sloughing can occur. However, in the case of trench drains cut *across* the slope, cracks and slide movements can extend upslope for distances of many times the depth of the trench. In addition to the example already described, the author has seen similar examples where relatively shallow cuts became unstable for surprisingly long distances upslope in translational infinite slope-type configurations. Thus, there is a need for very careful evaluation of the uphill instability risk where trench drains are cut across the slope below roads, railways, canals, structures, buried utilities, pipelines, etc. In such areas, top-down braced trench support probably will be needed rather than closely-sequenced construction. "Top-down" refers to providing support as excavation proceeds and not first digging the trench followed by later support.

There is no established procedure to calculate the earth pressures to be resisted by a braced support. In addition to the pressure diagrams readily available in soil mechanics textbooks (e.g., Terzaghi, Peck, and Mesri, 1996), it is suggested that the lateral trench force can be estimated by: (i) back-calculating the residual strength ϕ'_r of the entire landslide, then (ii) include the trench, perform stability analyses for various double wedge or circular arcs for the slope above the trench, allowing for vertical or near-vertical ground cracks (no strength) filled with water to the level of existing groundwater or higher. Use the back-calculated ϕ'_r over all potential slip surfaces. Select the most conservative lateral trench force needed for stability of the braced excavation and upper slope.

Estimation of Groundwater Drawdown between Parallel Trench Drains

When trench drains are completed, there is immediate drawdown within the drain itself to near the base of the trench (usually to the invert level of the drainpipe). In the landslide mass between adjacent trench drains, the groundwater table undergoes a transition from the preexisting levels to a steady-state condition. Since landslides with high groundwater levels are predominantly fine-grained clays, silts, and fine sands, the transition may take from weeks to years for equilibrium to be reached.

The actual groundwater drawdown curve will be influenced by heterogeneous conditions within a landslide, which are impossible to predict. There will also be seasonal rise and fall of groundwater levels, with upward spikes due to storms or snowmelt.

For design purposes, theoretical solutions to the long-term steady-state groundwater drawdown levels between adjacent parallel trench drains have been made by Hutchinson (1977), Stanic (1984) and Bromhead (1986). All are based on trench drains constructed in the direction of the landslide's maximum slope (Figure 17.18). Other assumptions are: (i) the infinite slope model applies; i.e., the ground surface, groundwater levels before drainage, and slip surfaces are planar and parallel to each other and there are no end contributions to

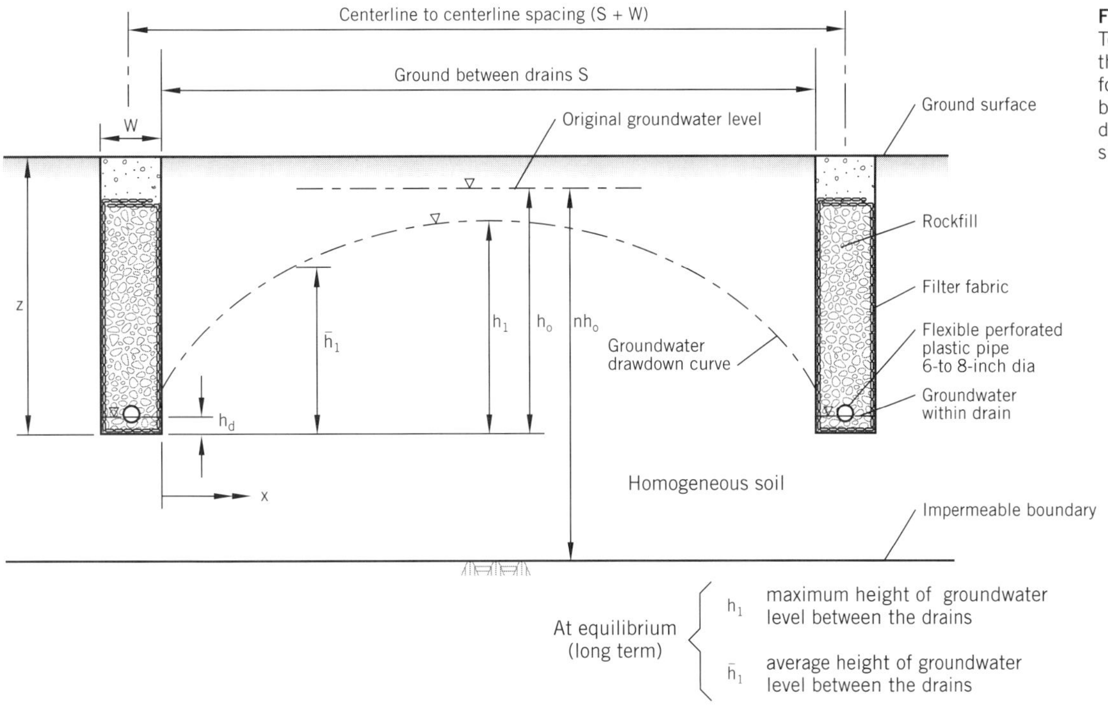

Figure 17.22 Terminology used in theoretical calculations for estimating drawdown between parallel trench drains oriented in the sliding direction.

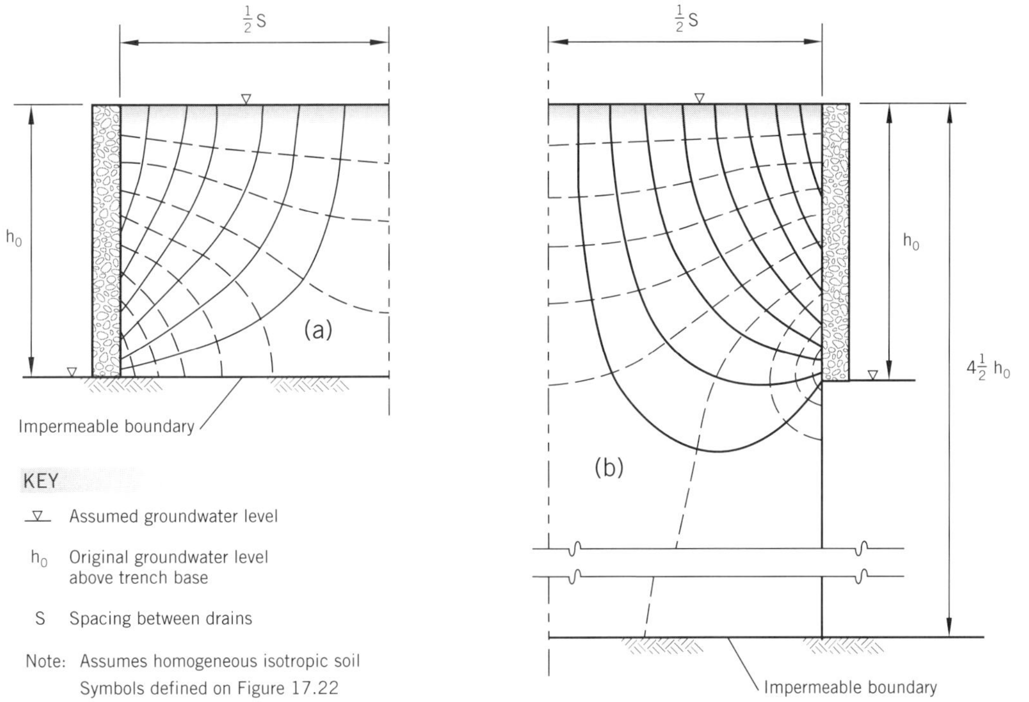

Figure 17.23 Groundwater flow net to a trench drain:
(a) impermeable boundary at the base of the trench drain
(b) impermeable boundary at depth (n = 4½)
(after Hutchinson, 1977)

stability upslope or downslope; and (ii) the landslide mass is homogeneous and isotropic (Bromhead's method allows for some anisotropy in permeability).

The objective of all the theoretical solutions is to predict the drawdown of groundwater between the drains as it relates to the original height of groundwater and the spacing between the drains. Armed with this information, the geotechnical designer can determine the spacing of drains needed to achieve a defined long-term factor of safety (such as a 25 percent improvement) or examine the relationship between the cost of drains and the factor of safety benefit.

Since each method uses slightly different terminology, and to remain consistent with terms used throughout this book, the section and definitions shown on Figure 17.22 will be used. The slope angle is β to the horizontal.

Hutchinson (1977) used finite-element techniques to study the flow patterns toward trench drains. The flow nets on Figure 17.23 are (a) for drains that fully penetrate the landslide

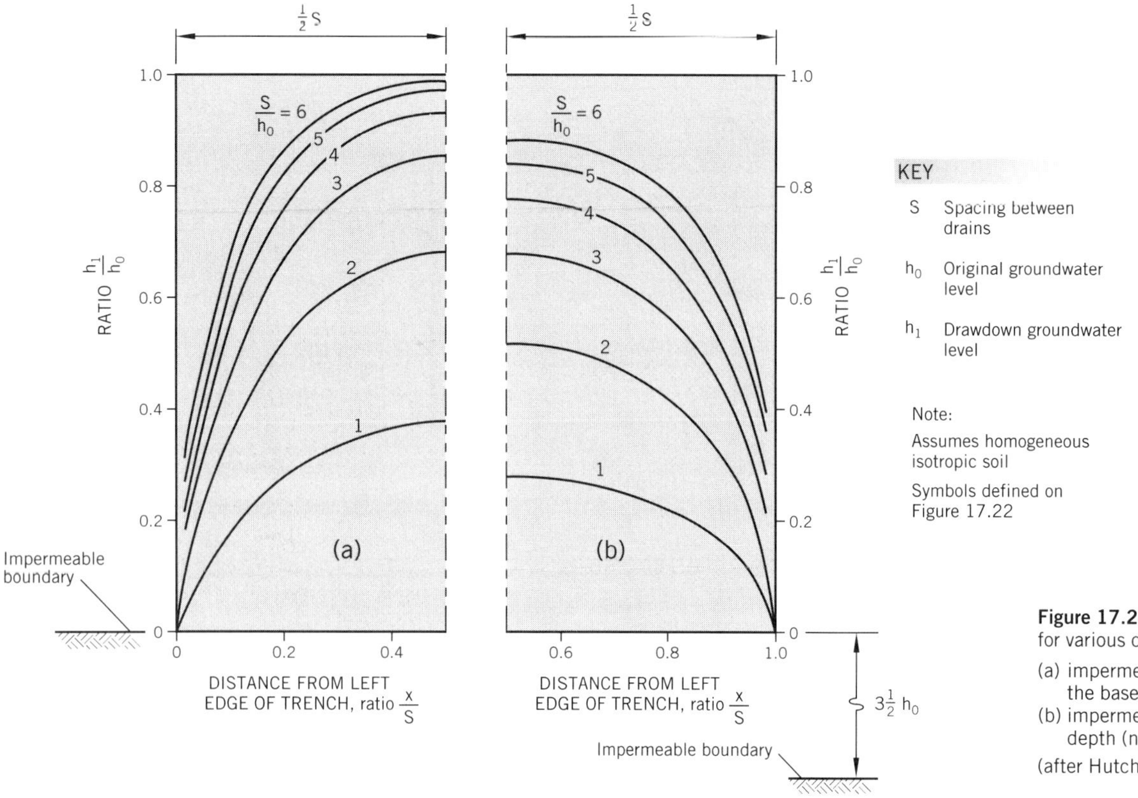

Figure 17.24 Drawdown curves for various drain spacings:
(a) impermeable boundary at the base of the trench drain
(b) impermeable boundary at depth (n = 4½)
(after Hutchinson, 1977)

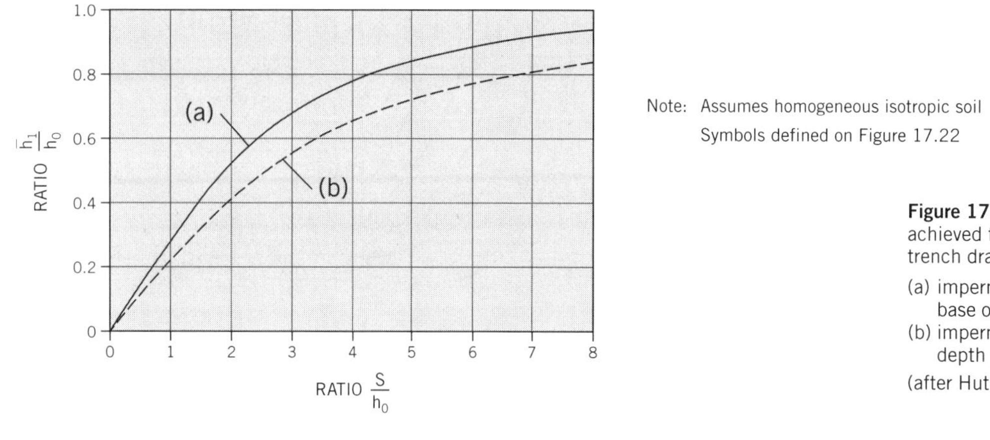

Figure 17.25 Average drawdown achieved for various spacings of trench drains:
(a) impermeable boundary at the base of the trench drain
(b) impermeable boundary at depth (n = 4½)
(after Hutchinson, 1977)

to a hard impermeable stratum and (b) for a partly penetrating drain in which the underlying impermeable stratum is at a considerable depth below the base of the trench drain. For a similar set of conditions, Figure 17.24 shows the groundwater drawdown profiles for various ratios of trench spacing: original head at the base of the trench drain. Finally, Figure 17.25 provides the average head (as a fraction of the original head h_o) for various S/h_o ratios for the two positions of the impermeable boundary. Hutchinson noted that the observed drawdown at three sites gave reasonably good agreement but were generally flatter curves than is predicted from these charts.

Stanic (1984) modeled the domain boundaries and reached a computer-based solution of Poisson's elliptic partial differential equation. His solution is used to calculate the average and maximum groundwater levels for different slope angles as a ratio of the original (pre-drains) groundwater level (Figure 17.26). It can be noted that slope angle has only a minor effect on the groundwater regime on slopes flatter than 4 horizontal : 1 vertical. Stanic recommends that the spacing between drains, ratio $S : h_o$, should not be more than 4.0 because experience has shown that sliding can continue to occur in the landslide mass *between* the drains when the $S : h_o$ ratio exceeds 4.

The graph of Figure 17.26(a) gives the average piezometric head of the drawdown for different $S : h_o$ ratios. The data can be entered into a stability analysis to calculate the estimated factor of safety achieved by the trench drains.

Bromhead (1986) extended the work of Hutchinson to provide curves of average drawdown head for several values of n, the depth to the underlying stratum as a fraction of the original height of groundwater above the drainage trench base

h_o. In addition, he accounts for anisotropy in which the horizontal coefficient of permeability is higher than the vertical permeability, a condition that is likely in layered strata. Colluvium is sometimes layered, whereas jumbled landslide debris is not layered.

If $R = S/h_o$, Bromhead introduces a transforming parameter $R_s = R\sqrt{(k_v / k_h)}$ where k_v, k_h are the coefficients of permeability in the vertical and horizontal directions, respectively. Thus if $k_v = k_h$, $R_s = R$. When $k_v < k_h$, $R_s < R$.

As an example, if $k_h = 4k_v$, $R_s = \frac{1}{2}R$ and the trench drains will have the effect of lowering groundwater more effectively than in isotropic soil in which the permeability is the same horizontally and vertically.

The graph on Figure 17.27 gives the average drawdown head at the level of an impermeable boundary. In many practical situations, the shear surface of the landslide is at or near the top of an impermeable hard layer.

Bromhead (1986) also provides similar charts for the average head vs. R_s for different reference levels from top to bottom of the trench drain (e.g., one-quarter of the depth of the trench drain), which could be useful if a landslide shear surface occurred at these higher levels (unlikely). Another chart covers a *permeable* base stratum. In this case, groundwater is draining vertically toward the underlying stratum as well as laterally toward the trench drains. In practice, it is likely that sand drains or wick drains would be put into the base of a trench drain rather than to leave the trench drain in a partly penetrating condition. Neither of these charts are reproduced here but are available in the cited reference for such unusual landslide conditions.

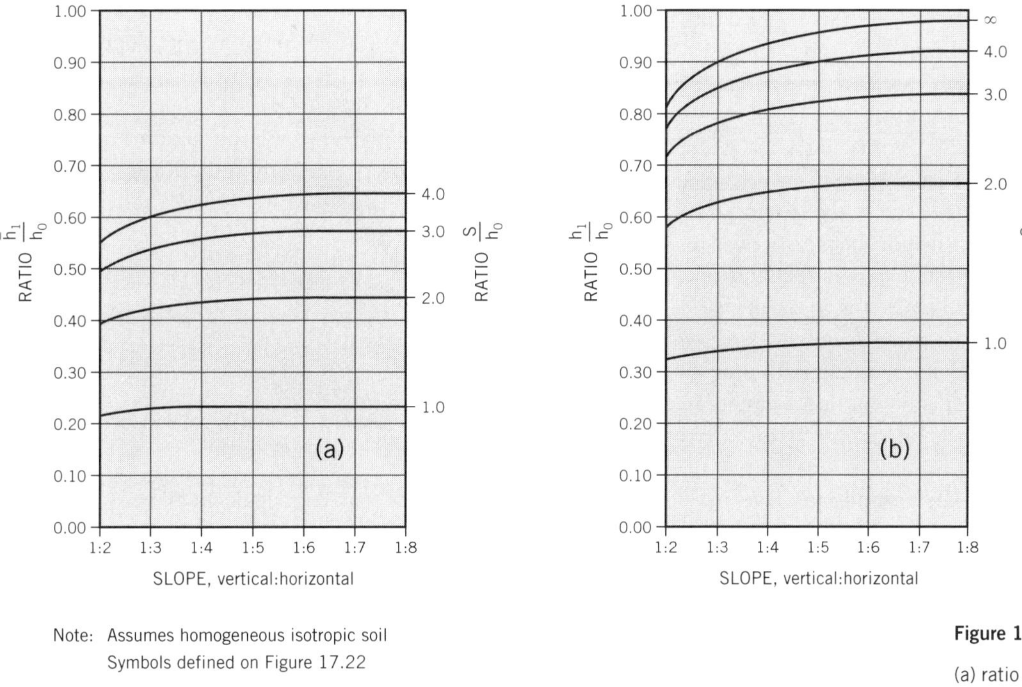

Figure 17.26 Stanic (1984) method:
(a) ratio $\frac{\bar{h}_1}{h_o}$ vs. slope for various $\frac{S}{h_o}$ ratios
(b) ratio $\frac{h_1}{h_o}$ vs. slope for various $\frac{S}{h_o}$ ratios

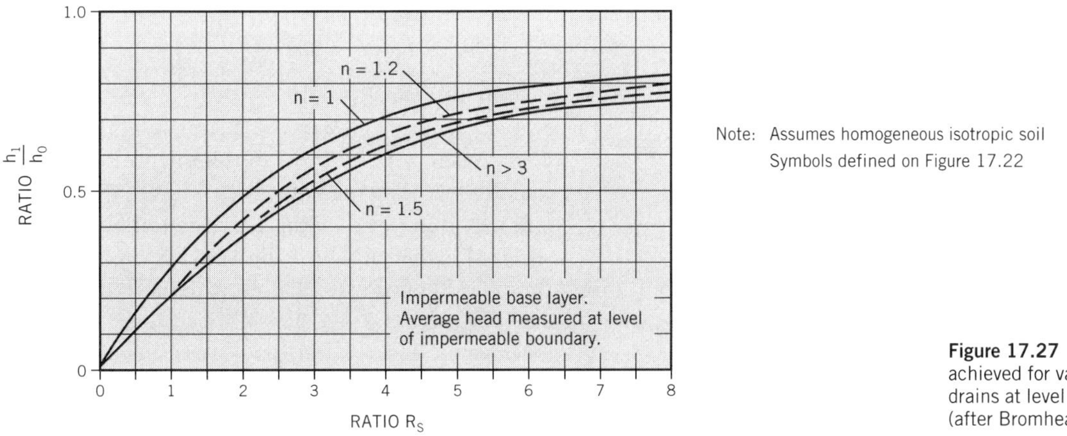

Figure 17.27 Average drawdown achieved for various spacings of trench drains at level of impermeable boundary (after Bromhead, 1986).

Comparison of Groundwater Drawdown Estimates

Ground slope	tan $\beta = 0.2$ (11.3° to horizontal)
Original groundwater level	$h_o = 13$ feet
Spacing between trenches	$S = 47$ feet

Assume isotropic ground conditions and impermeable boundary at invert level (base) of trench drain. Calculate the long-term (steady state) average groundwater level above base of trench drain.

Ratio $S/h_0 = 3.6$

(i) Hutchinson, 1977	$\bar{h}_1/h_0 = 0.75$	(Figure 17.25)
	$\bar{h}_1 = 9.7$ feet	
(ii) Stanic, 1984	$\bar{h}_1/h_0 = 0.62$	(Figure 17.26)
	$\bar{h}_1 = 8.1$ feet	
(iii) Bromhead, 1976	$\bar{h}_1/h_0 = 0.67$	(Figure 17.27 for n = 1)
	$\bar{h}_1 = 8.7$ feet	

The estimated average drawdown $h_0 - \bar{h}_1$ for this example ranges from 3.3 feet (Hutchinson) to 4.9 feet (Stanic method). The variation is sufficient to show that "designing" the spacing of trench drains does not call for extensive fine-tuning of theoretical solutions to reach a conclusion.

There are other considerations. If the spacing is too wide, there is a risk that the ground between the drains will continue to move, as noted by Stanic. However, as the spacing narrows, the cost of trench drain remediation begins to escalate. Based on these considerations and simply by inspection of a slope section (for example, Figure 17.22) it is apparent that *a spacing of about 3 to 4 times the depth of the drains makes a good choice for fully penetrating drains.* In addition, the strength provided by the drain itself is a significant contributor to stability in counterfort (fully penetrating) trench drains, as discussed below.

It should be remembered that the groundwater level *within* the drains will be at approximately pipe invert level. This drawdown should be computed separately and added to the average drawdown determined from the theoretical methods. Referring to Figure 17.22:

$$\bar{h} = \frac{h_d W + \bar{h}_1 S}{(W + S)} \qquad \text{Eq. (1)}$$

Effect of Backfill on Stability of Fully Penetrating Trench Drains

The preceding section provided methods of estimating the steady-state drawdown in fine-grained soils of a landslide with high groundwater levels. If the trench drain fully penetrates through the landslide debris and extends a further 6 to 12 inches into the underlying bedrock or hard impermeable stratum, the drawdown at the pipe invert of the drain should be close to the base of the landslide. The gravel backfill of the trench drain becomes part of the landslide shear surface. For a drain constructed *across* the slope as an interceptor, the effect of the backfill on stability is directly analogous to a shear key. For a drain built *in the direction of* sliding as a counterfort, the extra resistance comes from either shear across the width of the drain or as skin friction down the side of the drain (in the latter case, the drain itself remains stable). Both scenarios of failure can be checked out for specific landslides. For purposes of this discussion, it will be assumed that potential failure will occur across the trench drain rather than in skin friction along the sides.

When the trench drain is built, consolidation of the granular backfill should be immediate. The fine-grained soils of the landslide will consolidate more slowly as drawdown develops. The net effect should be that the consolidating clays and silts will cause downdrag on the sides of the trench drain. Therefore, it should be satisfactory to assume that the vertical effective stress at the base of the trench drain will be at least equal to the density times the height of the drain, and possibly higher due to the downdrag effect.

The strength available along the base of the trench drain can be calculated, assuming groundwater at pipe invert. The landslide mass and trench drain contribute proportionately to their respective widths across the slope. The next section illustrates a typical set of calculations for an infinite slope analysis.

Example Calculations of Trench Drain Design

Refer to section and elevation shown on Figure 17.28. It will be assumed that groundwater can temporarily reach the ground surface during strong winter storms and this is the time when instability occurs.

1. *Back analysis of residual strength.* Refer back to Chapter 9, Section 9.6.

   Ratio h : z = 1.0
   $\beta = 11.3°$
   Strength of landslide slope at failure (Figure 9.7) for $c_r' = 0$
   $\phi_r' = 21.8°$

   Note: This value can be calculated directly from the infinite slope formula, Eq. (9) in Chapter 9.

2. *Summer stability before trench drain construction.* On Figure 17.28, groundwater is 3 feet below the surface in summer. The stability before construction can be calculated from the infinite slope equation (Eq. (7) in Chapter 9) or from the chart, Figure 9.8.

   Using the chart:

   SSR = tan (21.8°)/tan (11.3°)
   = 2.00
   Ratio h : z = 13/16
   = 0.81

   From the chart F = 1.19

3. *Stability developed by long-term drawdown in the landslide debris.* Assume, for purposes of this calculation, that the trench drain exists as shown on Figure 17.28 but is infinitely thin (zero width). The identical problem was analyzed earlier in which the Hutchinson, Stanic and Bromhead drawdown methods gave estimated average drawdowns of 3.3, 4.9, and 4.3 feet, respectively. The mean result is a 4.2 feet drawdown, and the average height of groundwater reduces from 13.0 to 8.8 feet above drain invert level.

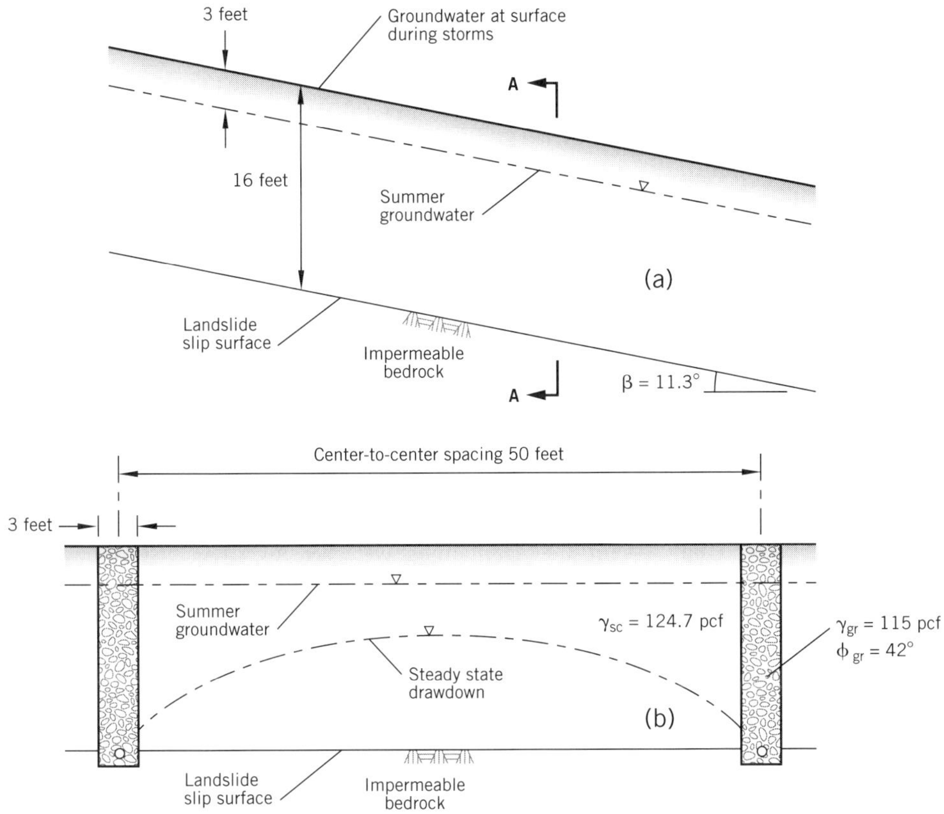

Figure 17.28 Example for calculating the effect of drawdown and trench backfill on the stability of a full depth penetration, counterfort trench drain:

(a) section
(b) elevation A-A

$$\text{Ratio } h:z = 8.8/16 = 0.55$$

From Figure 9.8, for SSR = 2.0

$$F = 1.45$$

Should groundwater level rise by 3 feet above the steady-state level during severe storms, the ratio h : z rises to 0.74 and F drops to 1.26; this is slightly higher than the summer stability prior to trench drain construction.

4. *Stability developed by a combination of long-term drawdown in the landslide debris and the backfill strength in fully penetrating trench drains.* Using the infinite slope formula:

$$F = \frac{(\gamma z - \gamma_w h)\tan\phi'}{\gamma z \tan\beta} \qquad \text{Eq. (2)}$$

$$= \frac{\{(124.7)(16) - (62.35)(8.9)\}(47)(\tan 21.8°) + (115)(16)(3)(\tan 42°)}{\{(124.7)(16)(47) + (115)(16)(3)\}(\tan 11.3°)}$$

$$F = 1.62$$

For this example, the shear strength of the drains raises the factor of safety for long-term drawdown to 1.62 from 1.45. The improvement would be still higher if the residual strength of the landslide mass was lower than assumed in this example.

5. *Stability under extreme storm event assuming no drawdown has occurred in the landslide mass, but fully penetrating trench drains installed.* Assume groundwater in the landslide mass at the ground surface. Using the infinite slope formula:

$$F = \frac{(62.35)(16)(47)(\tan 21.8°) + (115)(16)(3)(\tan 42°)}{\{(124.7)(16)(47) + (115)(16)(3)\}(\tan 11.3°)}$$

$$F = 1.20$$

This calculation suggests that fully penetrating counterfort drains alone, without *any* drawdown in the landslide mass between the drains, may provide sufficient stability against an extreme storm event. Such a conclusion would have to be evaluated for the specific conditions at other landslides.

Construction of Trench Drains

Clays and Silts

Construction of an open cut (unsupported) trench drain through clays and silts is described in the next few paragraphs. It normally begins at the drain outlet and proceeds upslope. There are two benefits: (i) by having a drainage outlet in place below, seepage from an open trench excavation is continuously removed from the bottom of the work area without sump pumps, and (ii) the initial trench cuts are shallow, allowing the contractor and consultant to observe the stability of the vertical side slopes of the trench before deeper cuts are reached, thus providing time to adjust future excavation procedures, if needed.

The benefits of a speedy and efficient production rate cannot be overemphasized in trench drain construction. Rapid excavation and backfill lessens the sloughing of the trench sides and thus the quantity of backfill in the drain. Closely-sequenced construction (Chapter 13) should be mandatory, written into the job specifications, and enforced by inspection at the site.

Excavation is usually performed by a large-tracked backhoe moving upslope from the trench. After a length of trench has been exposed, a width of filter fabric is laid across the trench (see Figure 13.2), overlapping the upslope edge of the already-built section of the drain. The ends of the fabric on opposite sides of the trench are staked to the ground. Granular backfill is placed in the drain using any method that provides reasonable care to avoid tearing or disrupting the fabric and sides of the trench. Contractors may use a second backhoe, front end loader, small crane with clam, bottom-opening bucket, etc. Backfilling should be done from the downstream end of the drain rather than from the side, where the weight of the equipment might destabilize the trench. Any fabric damaged by backfilling or side sloughing should be replaced immediately.

It is common practice to include a flexible perforated drainpipe in the base of the trench drain. The pipe can be manipulated from the top and does not require precise placement. It is sufficient to provide a few inches of backfill beneath the pipe invert. The trench has to be cleaned out when caving occurs, and the pipe may be damaged or broken by the backhoe bucket. Two corrective options are: (i) use a trench box to allow a worker to join a new length of pipe to the broken end; or (ii) wrap a short length of fabric to the end of the new length of pipe such that it can be pushed over the end of the broken end of the pipe in the trench. The broken end and the new length of pipe can closely abut each other, and the fabric facilitates the joint (like a bell and spigot connection) to prevent rockfill entering the pipe.

The backfill should be tamped by the back of the backhoe bucket to ensure that no large voids have been left in the trench due to arching. A high level of compaction is not needed, and vibratory densification would likely increase sloughing of unsupported trench walls.

When the backfill has reached to within 12 to 18 inches of the trench top, the geotextile fabric is wrapped over the backfill with an overlap. Completion of the drain by impervious or pervious fill to the ground surface should be done shortly thereafter.

Trenching is always a potentially dangerous form of construction, and safety of site personnel has to be a primary concern. Nobody should be allowed to enter an unsupported trench deeper than 4 feet below the surface. The installation of geotextile fabric in the trench should be performed by two workers, each no closer than about ½ H to the edge of the trench, where H is the trench depth. Control of the fabric placement can be provided by the backhoe operator, who can look down into the trench. No stockpiles of waste or backfill should be allowed near the sides of the trench. The trench should be temporarily backfilled with spoil overnight and on weekends/holidays. Fabric or plywood sheets can temporarily separate the rockfill and temporary backfill. The public should be excluded from the work area.

There should be close observation of the extent of sloughing, which depends on the soil types and soil layering encountered in the excavation. Some contractors prefer to use a trench box to temporarily support the sides of the trench and/or keep the base free of sloughed materials. However, a trench box is a mixed benefit because it hinders placement of the fabric liner and slows down the speed of construction.

Since it is inevitable that some sloughing will occur during construction, it is important to keep good site records to achieve an equitable cost settlement with the contractor (see next section). Photographs, sketches, and measurements taken across the top of the trench by survey rod can be used to record the as-constructed shape of the trench. Truck loads should also be counted (or weighed) to determine the quantities of aggregate placed in the drains.

Waterbearing Strata

Landslides containing waterbearing strata may be encountered. "Waterbearing" implies that the soil will seep when exposed in a trench excavation. In the presence of such waterbearing soils, careful evaluation is needed before proceeding to recommend and install trench drains.

Dewatering of the site ahead of construction is usually required to control seepage, improve stability, and make the site safer. Without dewatering, ground seepage is likely to cause the trench wall to collapse or liquefy, and the trench bottom would fill with water. The types of appropriate dewatering methods should be specified in the contract, rather than being left to the whims of the contractor. This approach not only prevents potentially serious consequences resulting from unacceptably risky dewatering methods, such as sump pumping, but also ensures that all contractors are aware of the requirements when preparing their bid price. Temporary dewatering is discussed in more detail in Chapter 13, Section 13.2.

A fully-supported trench also may be needed under some ground conditions to prevent excessive side collapse, and as a further safety precaution. This requires some type of trench shoring and bracing, and the services of a contractor with sufficient past experience in designing and installing such temporary trench support. It is strongly recommended that the need for properly-designed support be explicitly specified in the contract documents if site investigations show that ground support is necessary.

A braced excavation makes it difficult to install geotextile fabric around the exposed faces and bottom of the trench prior to backfilling. It may require a short length of the trench to be temporarily unsupported while a fabric width is placed in the trench. In soil with a high probability of collapse if not fully supported, this problem may preclude use of trench drains. In this circumstance, and provided the required depth of drain is shallow, an alternative technique may be to flatten the side slopes of the trench, similar to shear key construction.

Trench support and/or temporary dewatering will add significantly to the cost of constructing trench drains through waterbearing strata. Typical construction costs for trench drains, given on Table 17.3, do not include costs of sophisticated temporary dewatering or a fully-supported trench during

Table 17.3 Cost of Trench Drains

Site (date of construction)	Nominal Trench Width	Average Total Depth	Total Length*	Cost/lin. ft.of Drain**	
				Actual	Year 2000
(1) Hagg Lake, Slide 6, Oregon (8/1980)	3 feet	14.8 feet		$ 74.60	$140.56
(2) Zimmerly Slide, Washington (7/1997)	4 feet	c. 14 feet		$ 76.92	$ 81.67
(3) North Fork Slide, Oregon (7/1997)	3 feet	c. 16½ feet	1,804 feet	$112.28	$119.21
(4) Sandlake Slide, Oregon (8/1999)	3 feet	c. 11 feet	1,024 feet	$ 81.84	$ 83.64

*total includes allowing ½ length of outlet drain

**corrected from the date of construction to year 2000 using the ENR Construction Cost Index (see Chapter 13, Section 13.6)

Factors affecting construction cost:

Notes: (1) Long haul distance, had to cross paved road, outlet riprap
(2) Short haul distance, off-road construction, used trench box, relatively easy excavation
(3) Two-thirds alongside highway requiring traffic control, long haul, environmental requirements significant
(4) Access easy, long haul distance
All the above sites had fine-grained (clay, silt, or fine sand) soils

construction. However, there is some offsetting cost savings, when compared with open cut construction, resulting from reduced overbreak.

Maintenance of Trench Drains

Maintenance of trench drains should be minimal. Periodic measurements of flow can be made at the outlet over the first two years, with the frequency of visits decreasing with time. The flow rate should decrease after construction during dry weather and there should be no evidence of sediment in the water (which may indicate a tear or gap in the filter fabric). Any erosion at the drain outlet should be repaired.

Example of Trench Drain Stabilization

See Case History 11: Hagg Lake, Slide 6.

Cost of Trench Drains

It is difficult to specify and provide for an equitable payment for unsupported trench drains (in silts) because of the uncertainties in the quantities of excavation and backfill that will be required to construct them. The irony is that a badly organized contractor will exacerbate the extent of side collapse of the trench sides and is likely to financially benefit from poor workmanship. However, the loss of ground from sloughing varies from site to site and cannot be predicted with reliability by the consulting engineer. A test trench may be of assistance in making a preliminary assessment before writing the specifications.

The payment can be set up in several ways, including the following:

- *Base quantities of excavation and backfill on the neat lines.* Keep good site records (photos, measurements, etc.) and negotiate the extra costs after the contract is completed.
- *Use previous method, but include a bid price per cu. yd. of additional excavation and backfill.* This procedure is not recommended because savvy contractors may "unbalance" their bid by setting an unrealistic price on the extra costs, knowing that the unit price will have very little effect on the overall contract price but that relatively large quantities of trench overbreak will yield significant extra payments at highly inflated unit prices.
- *Estimate the total quantities, including overbreak, and pay for the measured quantities.* This provides a basis for comparing bid prices, and also obtains a realistic cost estmate. This is probably the fairest method of setting up a contract where bidding is required.
- *Negotiate a contract with a selected contractor, including unit costs for overbreak.* This allows the owner to choose a contractor with the most experience, and usually makes it easier to further negotiate the cost of special measures (such as dewatering or bracing) should they become necessary in the course of the work. Unfortunately, this payment option is not open to public agencies.

Regardless of the payment method chosen for the contract, the site engineer and the contractor need to keep very good records of the quantities, and agree upon the figures on a daily basis. It can be expected that the overbreak quantities will be significant. Disputes can arise if the engineer and contractor cannot agree as to whether the overbreak is due to the subsurface conditions or are being exacerbated by poor workmanship. It is almost inevitable that the contractor's progress will be relatively slow initially and will speed up as the construction method develops into a routine. Some difficulties and tension may occur. The best way to defuse this possibility is to require mandatory attendance of all bidders at a pre-bid meeting at which the design engineer explains the objectives, likely behavior of the ground, and needed safety/construction precautions. The contract documents need to be carefully worded so that the designer's intent is fully understood by bidders.

Trench drains are a very cost-effective remediation technique when used intelligently. The potential for disputes during and after construction can be largely avoided by advance awareness of the likely dispute issues. The owner should also be made aware of the uncertainties involved with the final quantities, and an appropriate contingency should be made available in case of need. On-site inspection services by the engineer's representative is essential.

Table 17.3 provides data on the construction cost of some trench drain projects in the northwest United States to show the approximate range of costs for this work. Note that the costs include allowance for mobilization, clearing and grubbing, and other incidentals associated with field work. See Chapter 13, Section 13.6, for additional comments on costs.

Slurry Trench Drains

A recent development is to use the slurry trench techniques of Chapter 18, Section 18.1 to construct drainage trenches. At first glance, the term "slurry drain" appears to be an oxymoron, but the slurry is a biodegradable product that breaks down within a few weeks of use and allows slurry-laden gravels or rockfills to become highly permeable again. Slurry trench drains have been built at several environmentally-sensitive sites. However, there is no reason to prevent the method being adopted for use on trench drain construction for landslides and other civil engineering projects.

Compared to the trench drains described earlier in this section, a slurry drain has the following advantages:

- The trench can be much deeper—up to 65 feet deep using a backhoe with an extended reach.
- Smaller excavation quantities and greater safety due to less side sloughing.
- Relatively fast construction, with no need for dewatering, sheeting or temporary shoring.

The disadvantages are:

- The slurry has to be circulated through the trench to "develop" the drain, and the remaining slurry has to be disposed of to a wastewater facility.
- There is difficulty in using this technique on sloping ground.

Cost comparisons are not available due to the limited experience with slurry trench drains.

Technique

Day and Ryan (1992) describe the use of a guar gum-based slurry that remains effective for about 1 day unless treated with additives. The slurry has to meet viscosity and filtrate tests (e.g., filtrate loss <25 ml) and pH requirements. The head of slurry should be kept at least 3 feet above the adjacent groundwater table.

The slurry maintains the sides of the open trench during excavation. After the design depth has been reached, the trench is backfilled with a free-draining gravel or shot rock. At completion, the slurry is treated to initiate degradation and the trench is flushed by pumping and recirculating a minimum of three pore volumes of the trench using degraded slurry. The remaining degraded slurry has to be evaporated, solidified, or disposed of to a wastewater facility.

The Biological Oxygen Demand (BOD) and Chemical Oxygen Demand (COD) of the degraded slurry are similar and are in the range of 3,000 to 6,000 mg/l. Normally, the BOD decreases to about 1,000 mg/l within one week and eventually (about 6 months) reverts back to background levels.

The pea gravel or crushed rock backfill is placed by a tremie pipe through the slurry or by pushing into the trench from one end. A woven geotextile fabric can be placed around the permeable backfill. It has to be pushed through the slurry using a weighted frame or with concrete pieces.

A drainage pipe can be installed at the bottom of the trench using special pipelaying equipment similar to that used to install cable. The perforated pipe itself (typically 6-inch dia.) has to be flexible and corrugated. The pipelaying machine travels over the slurry-filled trench behind the excavator to lay the pipe. Concurrently, bedding and backfill are placed around the drain through a tremie pipe.

Usually, trench drains are designed to outflow by gravity. However, deep wells with submersible pumps can be installed.

17.4 FRENCH DRAINS

Technique

French drains are shallow, rockfilled trenches cut into the surface of a slope to channel surface runoff. Their purpose is to reduce surface erosion and prevent minor slumps. It is a low-cost construction expedient used primarily for preventative purposes in hillside cuts, but occasionally to control surface water in landslides. French drains have no significant benefit to deep-seated instability.

Appropriate Applications for French Drains

The principal application is to prevent surface erosion in slopes of sands and silts. They can collect water from small springs or light seepage.

Design of French Drains

Layout of french drains is usually based on judgment. A common design consists of trenches cut down the line of steepest slope and spaced 50 to 100 feet apart. These main drains are fed by "chevron" drains that are angled across the slope face in a "herring-bone" pattern at about 45° to the slope (Figure 17.29). The chevron drains intercept the surface water runoff.

Calculations of potential surface runoff can be based on conventional hydrology using historical records of rainfall intensity or snowmelt. These computations should help the designer to decide whether pipes are needed within drains.

The drains are typically 2- to 3-feet square in section and are backfilled with open-graded, crushed, hard, angular rockfill of 6-inch to 2-inch size (Figure 17.30) or rounded drainrock. A geotextile filter fabric is used to line the trench and prevent infiltration of fines into the rockfill; fabric can be omitted on clay slopes where erosion is low. The top of the french drain is left open (i.e., not covered by filter fabric). A perforated pipe in the base of the trench would be placed holes facing down; most french drains omit pipes. The end of the main drains may require appropriate erosion protection at the ditchline.

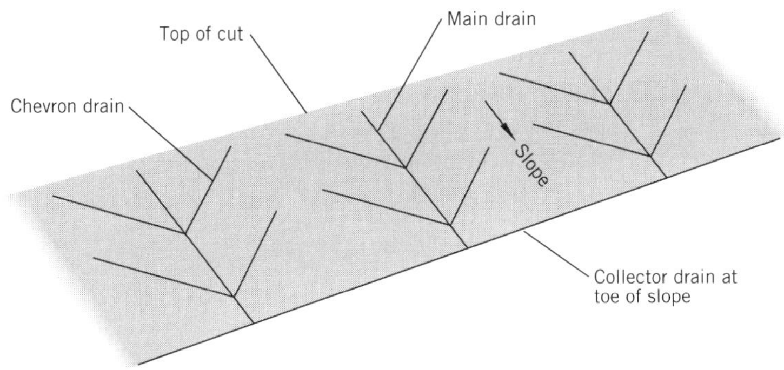

Figure 17.29 Typical chevron layout of french drains on a cut slope.

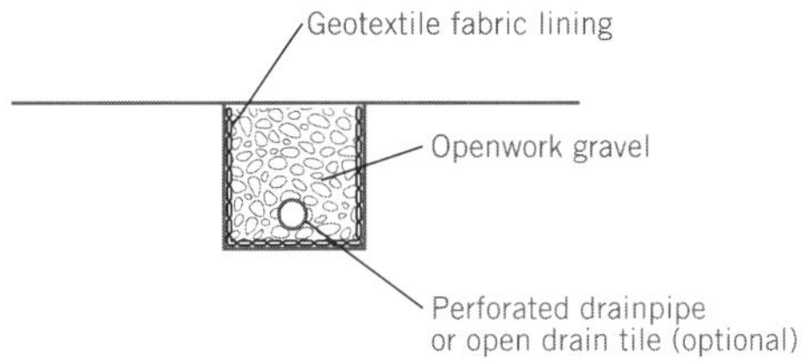

Figure 17.30 Section of a french drain.

Maintenance or Improvement of French Drains

Two potential problems with french drains are: (i) heavy slope erosion causing fine-grained silts and sands to choke the drains and destroy their effectiveness; and (ii) insufficient flow capacity during heavy storms, causing collected water to overflow the drains.

If the drains are being choked by sediment, corrective measures are: (i) replace the french drains with a drainage blanket over the entire slope; or (ii) if there is no filter fabric, reconstruct the drains using a fabric liner.

If heavy rainfall or snowmelt exceeds the capacity of the drains to carry the collected water away, some corrective options are: (i) reconstruct the french drains, making them deeper; (ii) provide a pipe in the base of the drains; and (iii) construct an interceptor drain at the top of the cut slope or landslide.

17.5 DRAINAGE BLANKET

A drainage blanket covers the outer slope of a landslide with coarser soils to allow seepage forces to dissipate before reaching the surface. A common example of treating an unstable road cut is shown on Figure 17.31. In this example, the slope is unstable because the cut intercepts the groundwater, creating a seepage line and erosion. After cleaning out the sloughed soils from the ditch, a two-layer filter system is constructed to blanket the slope. The progressively coarser soils cause the groundwater within the blanket to drop, thereby eliminating erosion and reestablishing slope stability.

The design of filter systems is described in Chapter 11, Section 11.1. Depending on the gradation of the unstable soil in the original slope, a single filter layer may be sufficient. This type of remediation can be undertaken during wet season (high groundwater) conditions. However, in many cases, the sloughed material can be continually removed (to keep the ditch open) and repairs can be put off until drier conditions occur in the following summer. If seepage is low at the time of repair, a filter fabric can be substituted for the fine filter.

The thickness of the drainage blanket can be determined by a simple slope stability calculation. For small cut slopes, 12- to 18-inch thick layers, measured normal to the slope surface, are often sufficient, and the blanket can be laid directly on the existing slope (Figure 17.31). For deeper cuts, it may be necessary to excavate into the slope to create room for the filter blanket. In such cases, temporary dewatering or other measures to protect upslope facilities from being undermined may be needed. If the seepage line is part way up the slope, a "notch" may be cut at the permeable/impermeable contact and only the upper part of the slope would be blanketed.

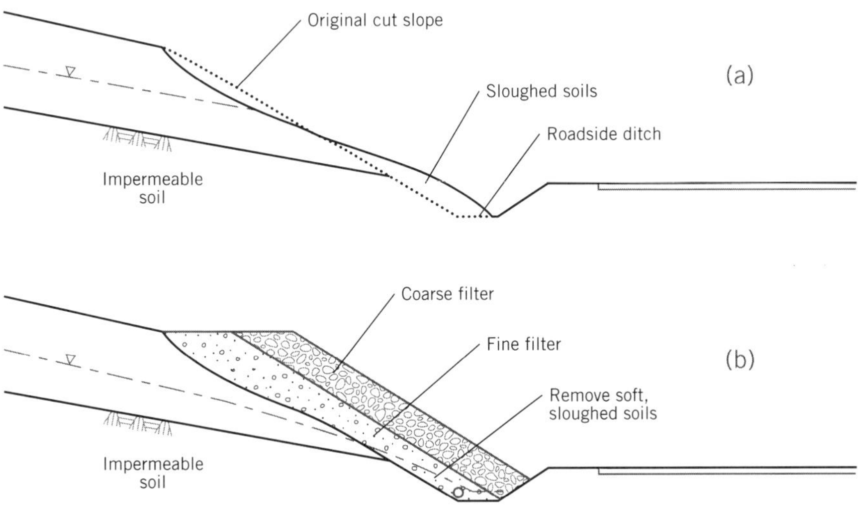

Figure 17.31 Drainage blanket remediation:
(a) failed slope
(b) filter layers placed on the outside of the slope

Drainage blankets are a relatively inexpensive remedial treatment and use readily-available construction equipment and materials. However, the gradations of the filter layers should be tightly specified (especially fines content), and observation during construction by the engineer's representative is strongly advised.

17.6 DEEP WELLS

Technique

Deep wells can be used to provide temporary, and occasionally permanent, stability to landslides. Temporary drawdown of the groundwater is often required to construct retaining walls, drainage trenches, etc. Long-term pumping is rarely undertaken because of the cost of power and maintenance, and the possibility that an untimely interruption of power (for example, during a storm) could occur at a critical time for stability.

Construction dewatering by wells is a mature technology in the United States and numerous specialist contractors are equipped to provide design and installation services. Except for large landslides in a complex geohydrologic environment, the design of dewatering schemes for landslides is often by rule-of-thumb rather than through detailed calculations. For most landslides, the length of time for temporary dewatering does not call for sophisticated analysis. Two excellent textbooks that can be recommended for more information on the subject are Driscoll (1986) and Powers (1992). A fairly brief summary of the groundwater technology and well construction techniques will be described in this section.

There are three main advantages to using a deep well system for dewatering:

1. The wells can extract water from below the levels that can be reached by gravity methods.
2. The capacity to drain can be changed at any time by adding or removing wells.
3. The wells are adjustable by changing the pumping rate or through switches.

For landslide use, the disadvantages include:

1. There is a need for continuous power supply.
2. Continual maintenance is needed.
3. Remote sites must be be made vandal-proof.
4. Water containing sand cannot be pumped.
5. The wells are susceptible to disruptions during electrical storms.

Radial Drainage to a Deep Well

A compact submersible pump with an attached electrical motor is lowered down a drillhole to a specified depth. The pump lowers the water level in the well, which creates radial drainage toward the well from the surrounding ground (Figure 17.32). The resulting cone of groundwater depression brings the groundwater level below the depth of a proposed excavation, thereby providing slope stability. Groundwater lowering also avoids steep hydraulic gradients near the base of the excavation that could cause ground disturbance or quicksand conditions. Multiple wells at close spacing create "interference" between adjacent drawdown curves so that the groundwater levels between wells are reduced.

The theories of groundwater flow to a pumped well depend on numerous assumptions about homogeneity, isotropy, compressibility, etc. which are at odds with the true nature of geological strata. Nevertheless, these theories have proved to be very useful in providing approximate numerical solutions for design purposes.

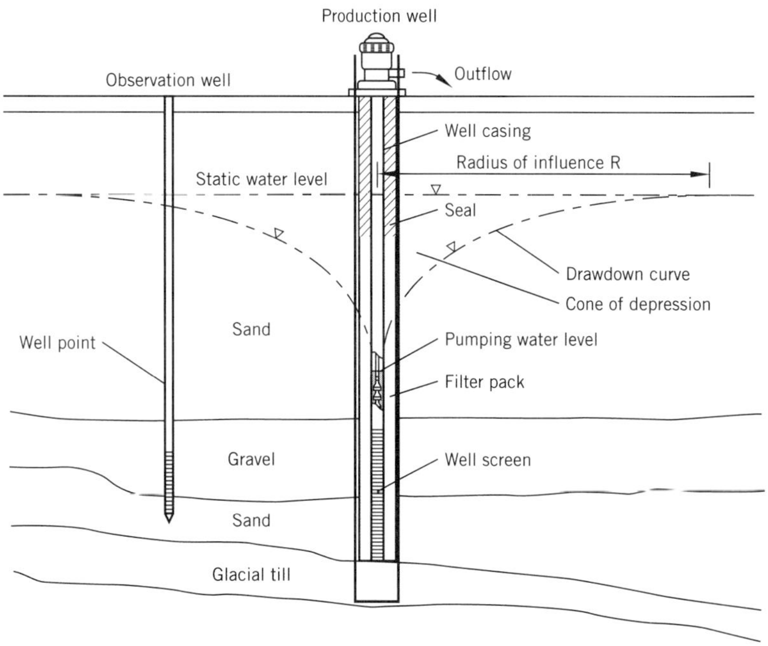

Figure 17.32 Drawdown of groundwater to a pumped well under unconfined conditions (after Driscoll, 1986).

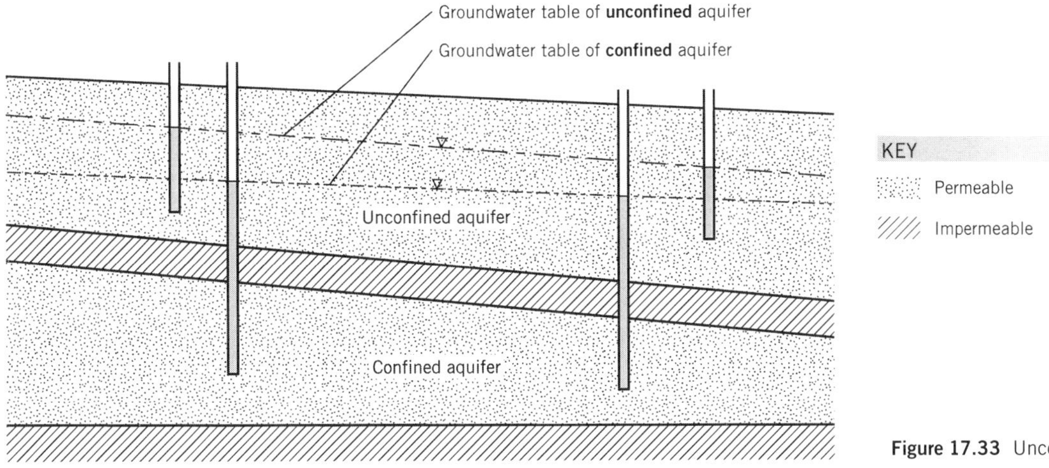

Figure 17.33 Unconfined and confined aquifers.

A key factor is whether the flow to the well is unconfined or confined. For most landslide situations, dewatering involves the near-surface soils and mostly occurs **unconfined,** as shown on Figure 17.32. The cone of depression is open to atmospheric pressure. The radius of influence R is the horizontal distance from the center of the well to the limit of the cone of depression. This radius increases with increasing soil permeability. Wells installed in fine-grained soils of low to moderate permeability have to be spaced more closely than coarse-grained soils of higher permeability to achieve significant drawdown away from the well. Therefore, deep wells are generally ineffective in clays. Conversely, gravels with a large radius of influence R require large pumps to handle the flow quantities and may depress groundwater levels for some distance outside the project boundaries (and possibly cause ground settlement).

In practice, deep wells are spaced 20 to 400 feet apart, depending on the ground characteristics, to interact with adjacent wells and achieve the drawdown objectives. For example, in a slightly clayey, sandy silt, deep wells spaced about 30 feet apart can lower the groundwater sufficiently for landslide remediation. However, it usually takes several days or weeks to draw the water table down in fine-grained soils. The quantity of water extracted in such soils is very low. For gravels, the quantity of water to be handled by the pump and disposal area is often the principal concern. The volume of voids within the cone of depression is a measure of the water to be extracted by the well before equilibrium drawdown conditions are attained. Once steady state has been reached, the permeability of the ground is the principal parameter affecting the rate of flow to an established well.

The *coefficient of storage* S (dimensionless) is the volume of water released from storage per unit of aquifer storage area per unit of head. It is also termed the *specific yield* of the aquifer. Another term, the *coefficient of transmissivity* T is the rate at which water flows through a vertical strip of aquifer 1 foot wide under a hydraulic gradient of 1. This latter term is similar to the coefficient of permeability k of soil mechanics but is multiplied by the depth b of the aquifer.

The terms S and T can be obtained from a constant-rate *pump test* performed on site. In a typical test, observation wells are installed at varying distances from the pumped well (for example: 20 feet, 50 feet, 200 feet) in a line to measure the drawdown curve (Figure 17.32). The well is pumped at a constant rate for 24 or 72 hours and readings of groundwater levels in the pumped well and observation wells are recorded at regular intervals. The details of performing and interpreting pumping tests are covered in numerous textbooks on groundwater hydrology (e.g., Driscoll, 1986) and will not be covered here because pumping tests are relatively infrequent on landslide studies. However, based on the values of parameters S and T obtained from the test, a hydrogeologist can estimate:

- Drawdown in the aquifer at various distances from a pumped well.
- Drawdown in the well any time after pumping starts.
- How multiple wells can affect each other (interference effects).
- Drawdown in the aquifer at various pumping rates.

The number, spacing, and depth of the wells can be estimated for the dewatering project based on the test. On a major dewatering site or where subsurface conditions are highly variable, two or more tests are needed.

A **confined** aquifer (Figure 17.33) is overlain by an impermeable barrier of soil or rock and behaves differently to an unconfined aquifer. An artesian layer is an example of a confined aquifer that may be encountered in a landslide. In addition to "confined" and "unconfined" aquifers, the variability of soils within an aquifer and "leaky" confinement are just two of the many other variables that can affect the interpretation of a pump test and design of a dewatering system. The advice of an experienced geohydrologist is recommended for any major dewatering project involving pumped wells.

Deep Well Construction

A deep well can be installed by many different drilling techniques. The choice is usually dictated by local practices, size of hole, and the subsurface conditions.

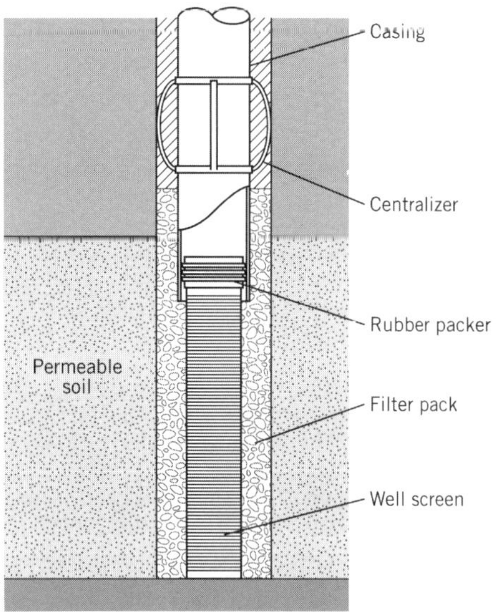

Figure 17.34 Connection of the well screen to the casing (after Driscoll, 1986).

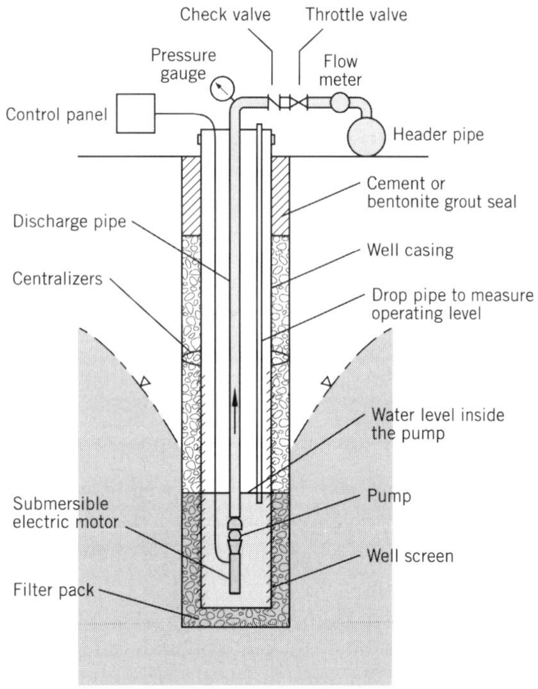

Figure 17.35 Typical details of a deep well installation (modified from Powers, 1992).

A common method for protecting the well screen during installation is to advance the casing to the full depth of the well with the screen above the casing bit. The casing is then partially withdrawn to expose the required length of well screen. A rubber packer seals the gap between the casing bottom and the well screen (Figure 17.34).

A typical deep well installation is shown on Figure 17.35. At the ground surface, the riser pipe of the well is connected to a header pipe that removes the groundwater to waste, and a flow meter and pressure gauge normally monitor the groundwater removal. There are also a throttle valve to control surge in the line and a check valve to allow the pump to be taken out of service temporarily (if needed) for repair and maintenance. The control panel, usually housed in a vandalproof enclosure on unsecured sites, supplies power to the electric motor of the pump.

The well itself is centered in the hole using centralizers; this ensures that a filter pack provides an annulus of relatively uniform thickness between the screen and borehole wall. The level of the groundwater within the well is measured by a drop pipe. The filter material may extend up the full length of the hole but is usually sealed in the upper few feet to exclude surface water. Some deep wells do not have a filter pack around the screen (see later).

After the casing, well screen, pump, and filter material backfill are in place, the well is "developed" before being put into service. Well development involves surging water around the well screen, typically by intermittent pumping. This function can also be performed before the filter material is added. Development removes drilling debris, mud cake, and fines within the soil close to the screen. Besides pumping, other methods of well development include flow reversal (first moving water into the well, then back out into the filter and soil) or by airlift. A well screen with a developed natural gravel surround is shown on Figure 17.36.

Design Details of a Deep Well

The key components of a deep well are described below.

Well Screen

The screen is a filtering device that permits water to pass through, supports the outside ground, but prevents sediment from reaching the pump. Although many types of screen are available, the best screen is the *continuous-slot* screen fabricated from wire and providing slot sizes ranging from 0.003 inch to 0.25 inch. The wire is wound around a circular array of longitudinal rods and is welded to the rods to produce a rigid cage.

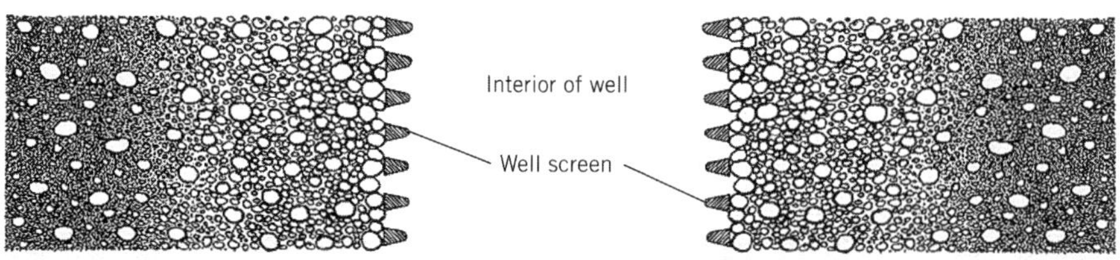

Figure 17.36 Well development. Surging of water around a screen removes the soil fines, leaving coarser aggregate against the screen openings. Example above is for a screen installed in a natural gravel stratum.

A characteristic of the wire is the triangular section (Figure 17.36) that creates an opening that is narrowest at the outer face and widens toward the inside. The design prevents clogging of the wire mesh as sand particles pass through it during the development of the well and in its subsequent use as a well filter. The screen cages are manufactured in diameters ranging from 2 inches to 36 inches.

The screen length depends on the aquifer thickness but is typically one-third to one-half of the thickness for homogeneous unconfined flow. In a confined aquifer, about 80-90% of the aquifer should be screened. All screens should extend to the bottom of the aquifer.

The slot size is selected so that most of the finer material in contact with it will pass through the screen and be pumped from the well during the well development. This provides a gradation of soil on the outer 1 to 2 feet away from the screen that is coarser than the soil further out (Figure 17.36). The design usually requires 60% of the soil to pass through the slot and 40% to be retained by the slot size. For example, a fine sand may require a No. 10 slot (0.01 inch) and a coarse sand and gravel may need a No. 50 slot (0.05 inch). This type of well, in contact with the native soils of the aquifer and termed a *naturally developed* well, must be designed for the grain size of the finest layer in contact with the screen. The screen can be custom-designed to be suitable for different layers of soils over the screen length.

Filter Pack

The alternative design is to use a filter pack around the outside of the screen (Figure 17.35). It requires a larger diameter borehole but this can sometimes be obtained by using an underreaming tool when drilling the hole. The filter pack generally has to be a clean, well-rounded, mainly quartz, sand or gravel of uniform size (uniformity coefficient of 2 to 3 – see Chapter 7, Section 7.1). The well screen opening should be designed to retain 90% of the filter pack (finest gradation in the range) after development. Filter packs are mostly used for fine-grained sediments such as alluvial sands, laminated sands, etc. The filter pack has to completely surround the screen, the annulus being at least 3 inches but not more than 8 inches wide. The use of a filter pack speeds up the time taken to develop the well; it also increases the effective hydraulic diameter of the well.

Pump

The submersible turbine pump (Figure 17.37) operates at a speed of about 3,500 rpm. It is most efficient (minimum operating cost) when pumping at about 50% of the maximum capacity. The pump is not designed to pump sand (hence the need to carefully develop the well first). The pump bowl size depends on the well yield (Table 17.4).

Riser Pipe

The well casing can be made of plastic (PVC, ABS), galvanized steel, or stainless steel (for a harsh environment). Steel casing is generally arc-welded or threaded; plastic casing can be threaded or solvent-welded.

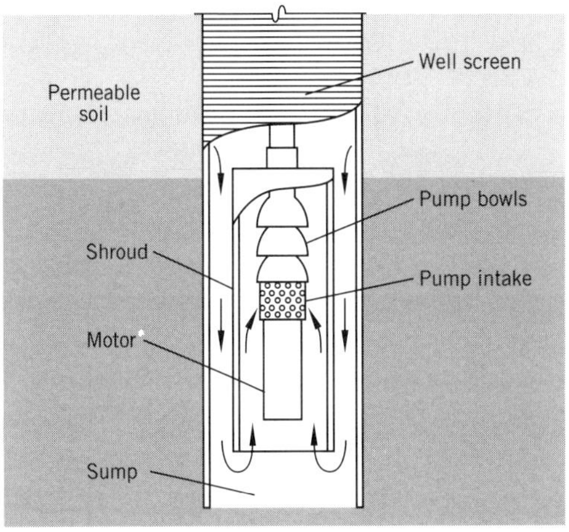

Figure 17.37 Submersible pump inside a well screen (after Driscoll, 1986).

Table 17.4 Recommended Well Diameters for Various Pumping Rates*

Well Yield (gallons/minute)	Nominal Size of Pump Bowl or Motor (inches)	Optimum Size of Well Casing ** (inches)
< 100	4	6
75–175	5	8
150–350	6	10
300–700	8	12
500–1,000	10	14
800–1,800	12	16

*after Driscoll, 1986 **excluding any filter pack

17.7 WELLPOINT AND EJECTOR SYSTEMS

Wellpoints

The technique of groundwater lowering by wellpoints is used extensively for temporary excavations through waterbearing sandy gravels, sands, and cohesionless silts. Several shallow wells are installed below the groundwater level; each well is connected by a flexible hose to a common header pipe that is laid horizontally at the ground surface (Figure 17.38). A pump lifts the water from the wells by creating a partial vacuum in the header and riser pipes.

The maximum drawdown obtainable from this system is the difference between the suction head provided by the pump and the height from the static water level to the pump level (Figure 17.39). In theory, the system is capable of providing 28.5 feet of drawdown at sea level. However, vacuum losses in the header pipes and valves, pump inefficiency, and hydraulic friction losses within the pipes combine to reduce the actual head drop. Most wellpoint installations are designed to lower groundwater by about 15 feet.

Wellpoint systems are relatively easy to install, and the number of wells can be increased or decreased during construction according to the encountered site conditions. The principal drawback is their limited drawdown capability of only 15 feet.

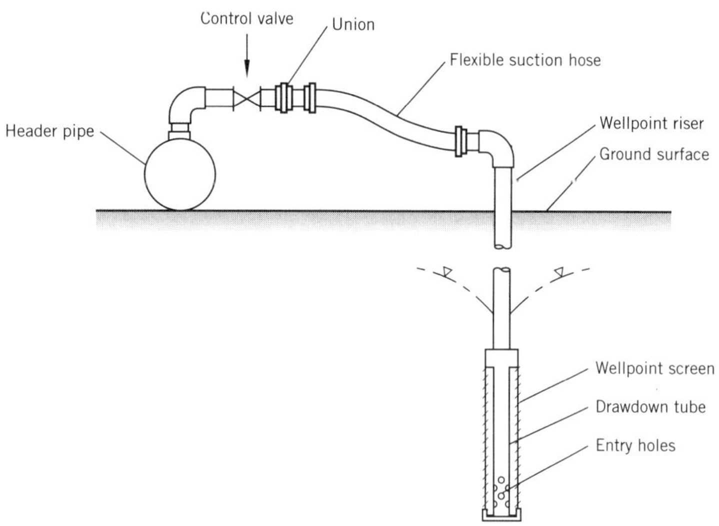

Figure 17.38 Wellpoint system (after Powers, 1992).

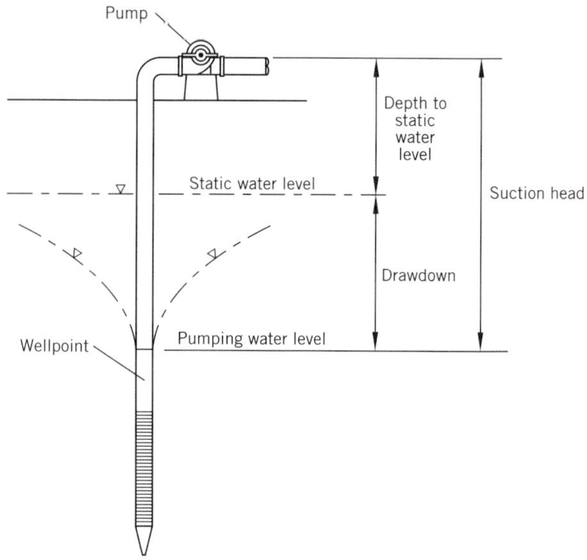

Figure 17.39 Drawdown obtainable from a wellpoint system (after Driscoll, 1986).

Wellpoint Equipment and Well Installation

The well consists of an intake screen made of heavy wire mesh, slotted plastic, or perforated metal, which is rugged enough to withstand frequent installation and removal. A drawdown tube is behind the intake screen and allows groundwater to be taken from the well at the lower end of the tube. Above the intake screen is a vertical riser pipe (Figure 17.38) that extends to the ground surface.

A header pipe collects the groundwater from the individual wells and connects to the pump. Since both water and air are collected by the system, the wellpoint pump is equipped with an automatic float chamber to separate air and water. The air is handled by a continuous prime vacuum pump, and the water flows are removed by a centrifugal water pump. There are various valves to isolate parts of the system so that repairs and maintenance can be performed during the dewatering period. Standby pumps have to be installed and be ready to operate at all times; a standby diesel power generator is also essential.

There are three general methods of installing the wellpoint tips:

1. *Self-jetting.* This is the most common method for penetrating into uniform sands. The tip has an automatic ball valve that opens when the water jet is applied through the riser pipe (Figure 17.40a). When the jetting is stopped, the groundwater pressure pushes the ball up and closes the valve.
2. *Driving.* The well is fitted with a driving point at the lower end (Figure 17.40b) and the well is driven into the ground to the required depth.

Figure 17.40 Wellpoints for:
(a) self-jetting into the ground, and
(b) driving into the ground
(after Driscoll, 1986)

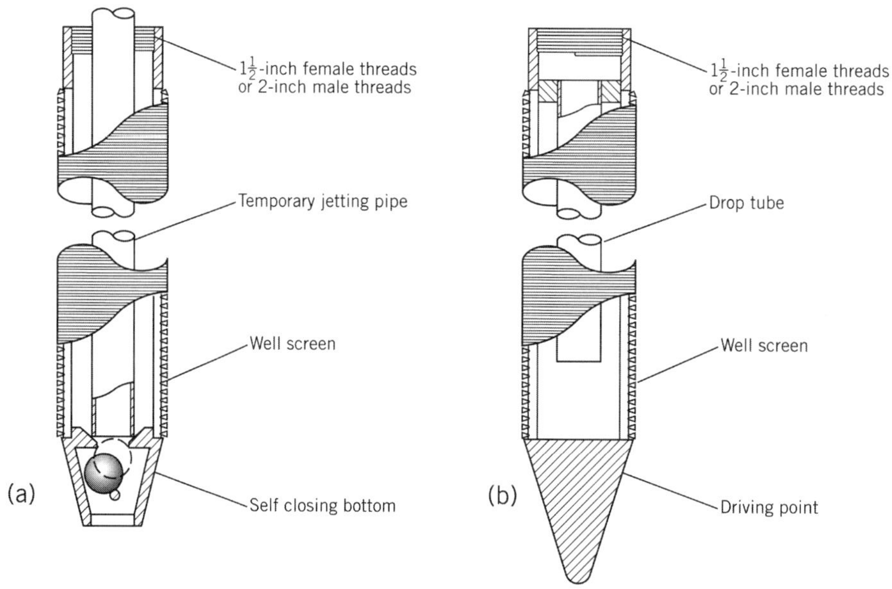

3. *Bored hole.* A borehole is made to a larger diameter than the well using hollow stem auger, flight auger, or rotary drilling equipment. The wellpoint is placed in the hole and the screen is surrounded by a filter sand. This procedure is advisable in layered soils where the sand pack provides vertical connection between permeable layers and also increases the effective diameter of the wellpoint.

Contractors specializing in wellpoint installations often have their own installation methods. For example, a chain attached to the outside of a well can create a larger diameter hole during self-jetting; this allows sand to be added around the well and helps to develop the well before use.

The well screen should extend at least 3 to 5 feet below the base of the excavation. Larger screens significantly increase the efficiency and yield of the well. However, if a clay or impermeable bedrock is at the base, the screen should be installed so that the top of the screen only extends about 6 inches above the impermeable interface to reduce the amount of air drawn into the system. This reduces the capacity of the well but pulls down the groundwater level closer to the interface.

The wells can be developed by surging water through the screen to remove silt and fine sand and increase yield. Surging is achieved with the aid of a plunger or pump. Any silt or sand in the bottom of the screen can be removed by an air-lift pump.

After the wells have been installed, the system is "tuned." This involves throttling (via the control valves) wellpoints that are drawing excessive air. The dewatering system efficiency improves as a result of tuning.

The general purpose wellpoint comprises 1.5-inch diameter wells, each capable of withdrawing 10–15 gpm through suction. Larger diameter wellpoints are available to discharge larger quantities of water. Very large wellpoints are usually termed *suction wells*.

Wellpoints are usually spaced 3 to 12 feet apart. In most waterbearing soils, a larger wellpoint is used if the required spacing falls below 7 feet.

The design and layout of the wellpoints are normally performed by a specialist dewatering contractor. It involves calculating the estimated flow quantities from groundwater equations (Powers, 1992, is a very good reference source) and dividing the total quantity by the length of the header pipe. The size and spacing of the wells is then determined, allowing for friction losses through the pipes and fittings. Heterogeneous soils require closer spacings to improve the likelihood of the wells intercepting significant pockets of waterbearing soils. One of the major benefits of a wellpoint system is that wellpoints can easily be added or removed at any time after the initial installation.

Uses of Wellpoint Systems

Wellpoints can be utilized for shallow temporary excavations into waterbearing sands and gravels that would otherwise be unstable. For landslides, such projects include shear keys, replacement buttress, temporary removal of landslide debris, trench drains, etc. A common procedure is to lower the groundwater levels by a multiple-stage excavation as shown on Figure 17.41. Additional wellpoints and header pipes are installed as the excavation deepens, and the depressed groundwater levels are at a short depth below the excavated surface. Each "step" between benches is about 12 feet high.

Wellpoint systems are especially useful for the following ground conditions:

1. Finely layered strata in which waterbearing sands or gravels are interlayered with more impermeable layers of silt and clay. If lateral continuity of the strata is poor,

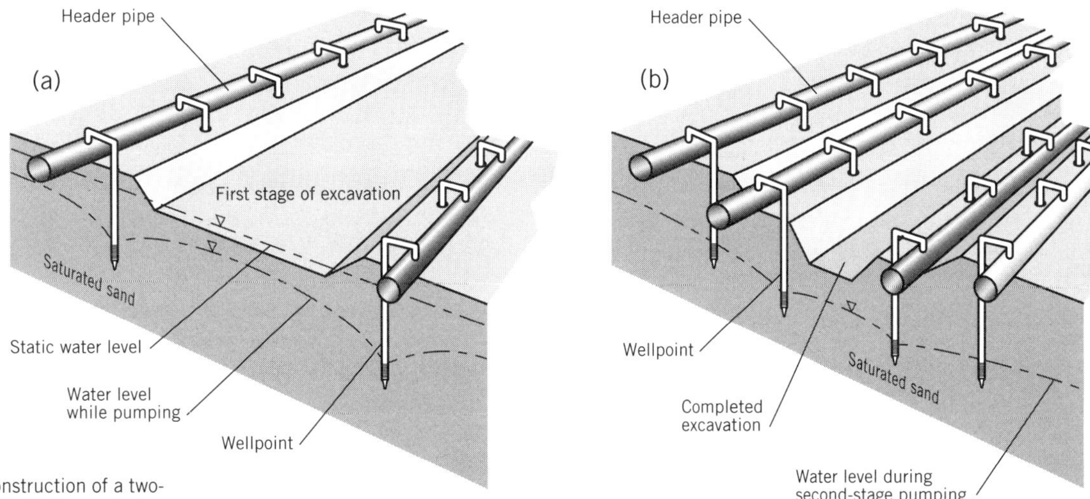

Figure 17.41 Construction of a two-stage wellpoint dewatering project:
(a) first stage excavation
(b) completed excavation
(after Driscoll, 1986)

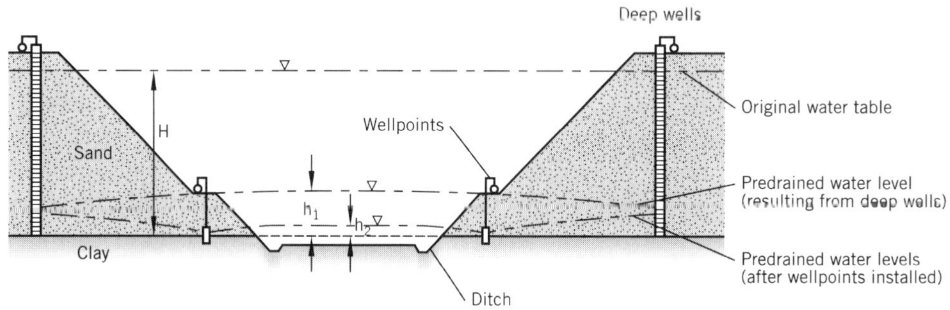

Figure 17.42 Use of wellpoints to supplement deep wells in a temporary excavation (after Powers, 1992).

such as in river alluvium or landslides with broken up ground, a dewatering system of closely spaced (3 to 6 feet apart) wellpoints can intercept local pockets of free-draining soils.

2. Excavations through waterbearing sands and gravels that are designed to reach or pass below an impervious stratum, such as clay or rock. In this situation, a ring of deep wells around the excavation can remove the major portion of the enclosed groundwater (Figure 17.42). This allows most of the excavation to be made "in the dry" but a low dome of groundwater, height h_1 at the center of the excavation, will remain. Wellpoints, installed after the main excavation has reached the low dome, can remove almost all of the remaining water above the impermeable floor.

The small head of water left at the base of the excavation may be capable of causing local toe instability. Sandbags or a perimeter gravel drain (Figure 17.43) can complete the dewatering of the excavation. These are laid against the toe of the excavation slope to allow the groundwater to seep into the excavation interior without eroding and undermining the slope. The water is removed from the excavation area by sump pumps.

Fly ash, varved clays, and other noncohesive silts can be stabilized by closely spaced, sealed, vacuum wellpoints (Figure 17.44). The wellpoint is surrounded by filter sand and a vacuum pump is applied to the wellpoint. Water is squeezed toward the filter sand from the external atmospheric pressure of about 1 ton/sq. ft. This increases the effective stress on the soils. Silts and silty sands with permeabilities in the range of

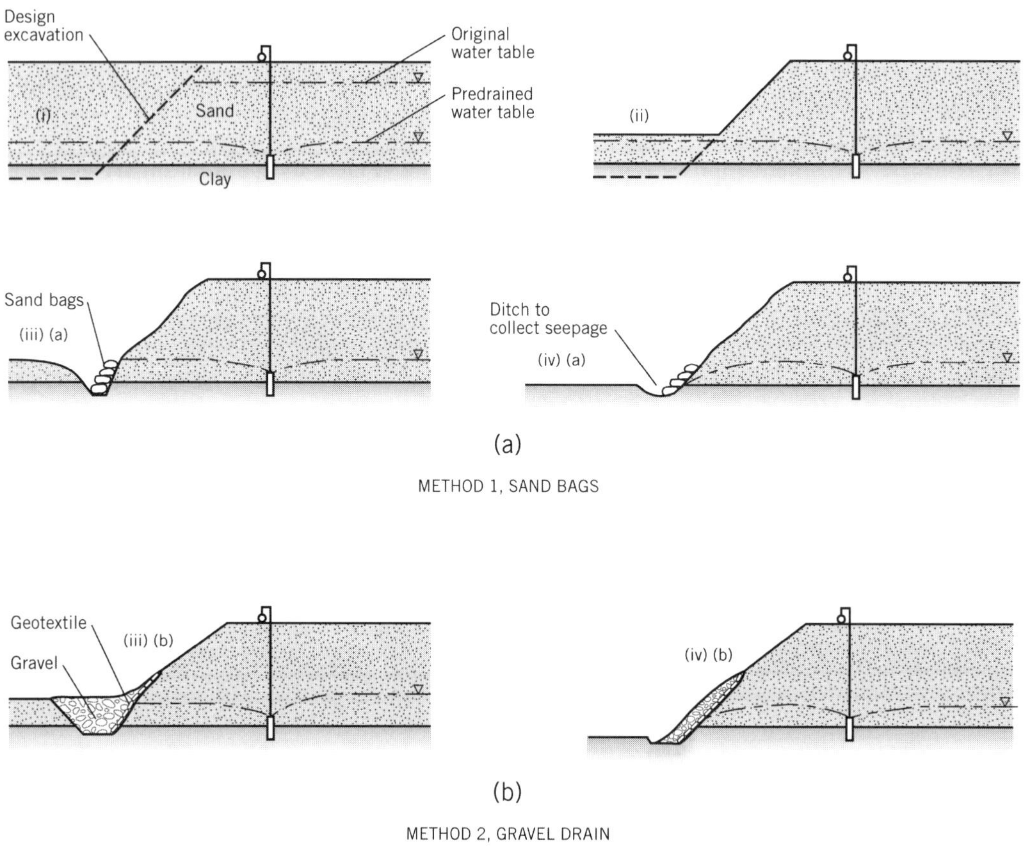

Figure 17.43 Methods of stabilizing slope excavation toes from remaining seepage after drawdown by pumps (after Powers, 1992).

10^{-3} to 10^{-5} cm/sec can be effectively treated by this technique. The quantity of water removed by the pump is very low, and the improvement in soil stability may require several weeks of pumping. The soil being treated in this manner must be saturated up to the ground surface. Applicable soil profiles include that shown on Figure 17.44 and saturated silts overlain by saturated clay.

Ejector Systems

Ejector systems (also known as jet-eductor systems) are similar in many ways to wellpoints but are capable of much greater drawdowns. The cost of installation and operation is generally less than for deep pumped wells so they can be used economically at close spacings. The main disadvantage is their poor efficiency which, according to Powers (1992), is usually around 10%. By contrast, wellpoints operate at about 40% efficiency. Ejector systems are particularly useful for deep drainage of fine-grained soils where the quantities of water to be removed (and thus the power usage) are low but deep drainage is required. For this application, wells typically are 5 to 10 feet apart.

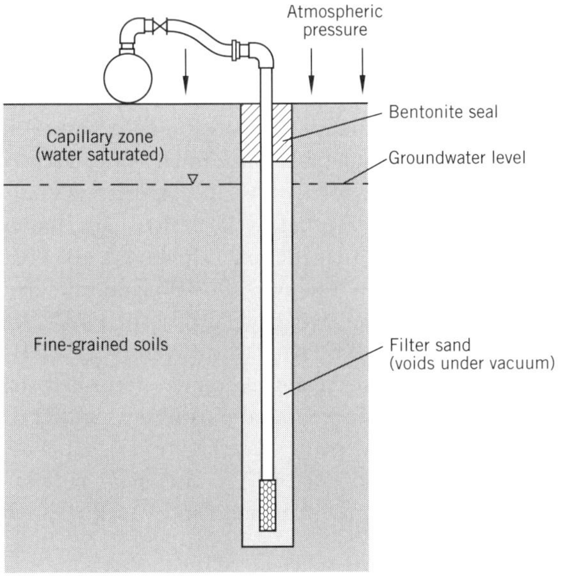

Figure 17.44 Sealed vacuum wellpoint (after Powers, 1992).

There are two types of ejector installations:

1. The *single pipe ejector,* which can be installed in a small diameter hole similar to a wellpoint, and
2. The *two pipe ejector,* which requires a larger outer casing to accommodate two separate pipes (for supply and return).

Only the single pipe ejector will be discussed below.

Water is pumped via a supply header down the outside of a double tube riser pipe and returns to the surface through an inner riser pipe (Figure 17.45a). The returning water is a mix of the supply water and groundwater picked up deep in the ground by the ejector. The combined water passes through the return header to a water tank where surplus water is discharged to waste.

Fresh water or recirculated water can be used. If the available water has iron precipitates present, it is preferable to use fresh water. The nozzle of the ejector is a small opening (Figure 17.45b) and is easily clogged. For the same reason, a strainer is needed in the supply line to filter out sand or other foreign materials. The return water typically contains air, and air vents are needed in the return header.

The ejector body puts the supply water through a fine nozzle followed by a venturi section (Figure 17.45b). This cre-

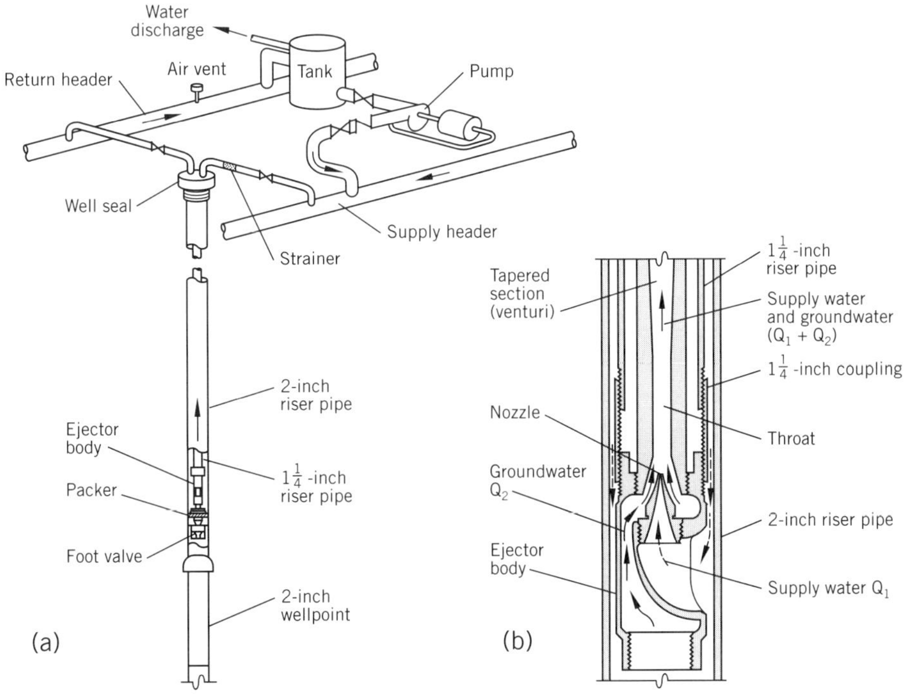

Figure 17.45 Single pipe ejector system:
(a) system requirements
(b) detail of ejector
(after Powers, 1992)

ates the vacuum that draws groundwater into the inner riser pipe from the tip. The ejector requires adequate submergence to prevent cavitation. According to Forrester (2001), the maximum economical lift of an ejector system is about 130 feet, which is sufficient for almost all landslide applications.

17.8 RELIEF WELLS

Vertical Relief Wells

Relief wells are installed to lower groundwater pressures under artesian conditions. As explained in more detail in Chapter 2, artesian heads develop when a soil stratum is overlain by more impermeable soils that effectively trap the pressure head in the lower stratum. These higher pressures at depth are at risk of developing sand boils at the ground surface (and, if untreated, create erosion and undermining of the slope above) or a general instability from high pore pressures. In ground that is stable under natural conditions, an excavation cut toward a stratum with artesian pressure can trigger the failure. Failure can be prevented by installing pumped wells or simply by reducing the artesian head through vertical relief wells.

The other common use of relief wells is at the downstream toe of dams or levees where permeable soils are overlain by more impervious soils. In many cases the high (artesian) underseepage threatens the stability of the dam or levee during floods or high reservoir conditions. A line of pressure relief wells can reduce the underseepage pressures to acceptable levels. Levees on the Mississippi River have been treated in this manner for about 50 years.

Geologists distinguish between: (i) an *artesian* condition where the groundwater head within the more permeable stratum (aquifer) is above the surface of the aquifer (but can be below the ground surface), and (ii) a *flowing artesian* condition where the groundwater head of the more permeable stratum is above the ground surface. Pressure relief wells are normally provided for flowing artesian conditions, and will be simply termed "artesian head" in the remainder of this section. In landslides, artesian heads can often rise 10 to 20 feet above the ground surface.

The design and construction of relief wells has considerable variation. Since the permeabilities of the artesian zone and overlying stratum vary widely, the quantity of water flowing from a well reflects the permeability and thickness of the more permeable stratum. It can range from a "gusher" to less than 1 gpm. The spacing of wells has to be adjusted accordingly, with closer spacings for a less permeable artesian stratum.

Examples of two relief well designs are shown on Figure 17.46. The first design, attributed to Forrester (2001), was used on the Mt. Ousley landslide in New South Wales, Australia, where a coal seam under artesian pressure was about 33 feet (10 m) below ground level. Large (26-inch) diameter auger holes were drilled through the artesian zone and were lined on the sides and base with cylindrical "socks" of geotextile (Figure 17.46a). For each relief well, a 3-inch diameter PVC riser pipe, perforated at the lower end, was placed in the center of the lined hole

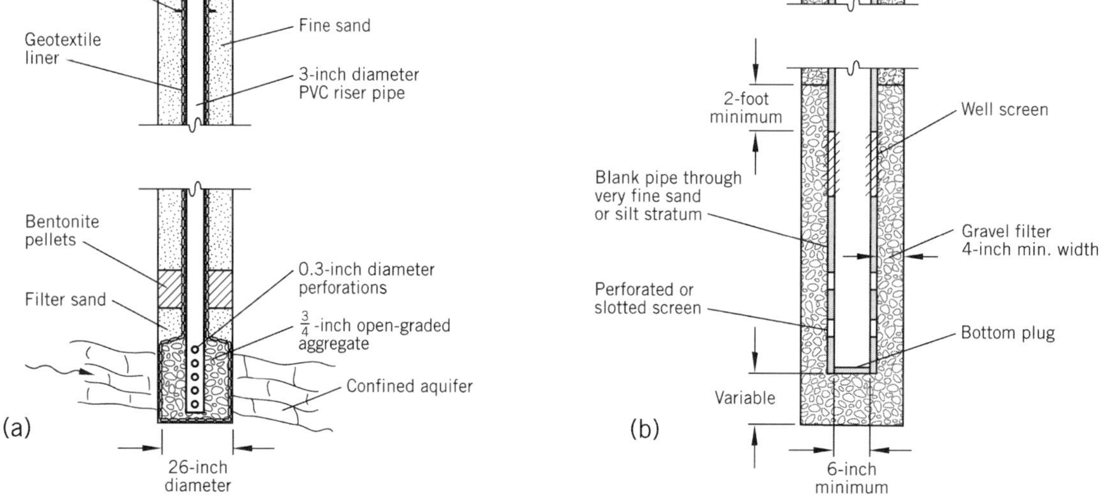

Figure 17.46 Relief well installations:
(a) Forrester (2001)
(b) U.S. Corps of Engineers (1992)

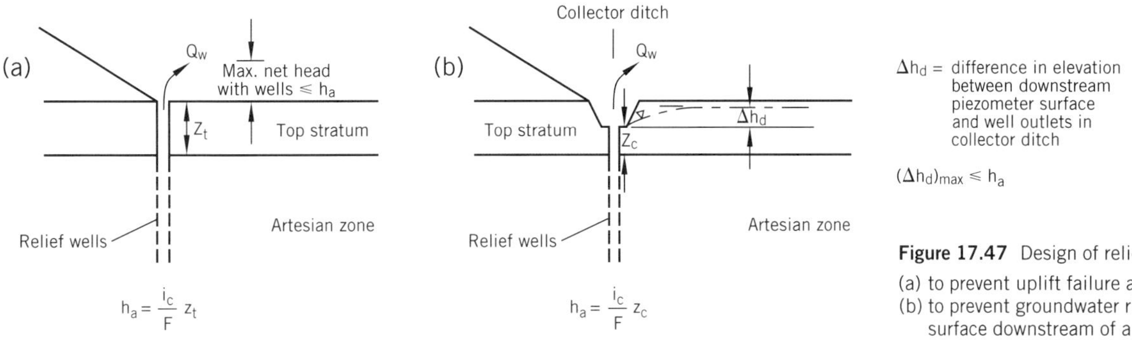

Figure 17.47 Design of relief wells:
(a) to prevent uplift failure at the ground surface
(b) to prevent groundwater rising to the ground surface downstream of a collector ditch
(after U.S. Corps of Engineers, 1992)

and open-graded coarse gravel (¾ inch) was infilled to the top of the artesian zone. This was covered by a sand layer, a bentonite seal, and further sand to the top of the hole. These materials were placed down the outside of the geotextile sock, as shown on Figure 17.46(a), which was tied to the riser pipe near the top. A T-section and PVC outlet pipe were used to drain the collected water to a stormwater sewer.

The second pressure relief well installation, Figure 17.46(b), is typical of the design used for levees by the U.S. Corps of Engineers (1992). It has a perforated or slotted well screen and gravel filter surround through the artesian zone, and a sand backfill through the more impermeable upper stratum. This design includes a V-notch weir at the top of the well to monitor the flow rate coming out of the well. The screen and riser pipes range from 6- to 18 inch-diameter.

The artesian pressure in a well can be measured by incorporating valves in the riser and outlet pipes so that the flow can be shut down. A pressure gauge then can be put on the top of the vertical riser pipe via a "quick-connect" coupling to measure the water pressure.

One of the main difficulties in constructing a relief well is in placing backfill through rising artesian flow conditions. One technique is to add casing to the top of the hole until the groundwater flow stops and backfill in the same manner as for a piezometer.

Design

The design of a relief well system has two objectives:

1. To lower groundwater pressure within the artesian zone to either restore or prevent landsliding of the slope.
2. To prevent piping and heave (e.g., sand boils) at the ground surface above the artesian conditions that could disturb and erode the surface soils, possibly leading to undermining of the slope and/or a flow slide.

The Corps of Engineers (1992) has developed a relatively simple procedure to design against piping and ground heave (Figure 17.47). It assumes horizontal flow towards the relief wells in the waterbearing artesian zone, and vertical flow through the top stratum of more impermeable soils. As shown on Figure 17.47(a) the artesian head h_a remaining after the wells are in place is given by:

$$h_a \leq \frac{i_c}{F} z_t \qquad \text{Eq. (3)}$$

where h_a = maximum net head with wells installed
i_c = critical upward hydraulic gradient
z_t = thickness of the top stratum, transformed (see later)
F = factor of safety = 1.5 or higher

The critical hydraulic gradient is given by:

$$i_c = \frac{\gamma'}{\gamma_w} \qquad \text{Eq. (4)}$$

where γ' = submerged density of the upper stratum
γ_w = density of water

In most soils i_c is approximately 1.00 but can be less in soils of high moisture content (organic soils, clays, etc.). Therefore, it can be seen that any residual artesian head needs to be substantially less than the thickness of the relatively impermeable top stratum. This need increases further if the criterion of acceptance is that the head downstream of the wells is not permitted to reach the ground surface (Figure 17.47b). The formula for piping is the same but the depth z_c has been reduced by a collector ditch.

The concept of a "transformed" thickness z_t for the top stratum is an attempt to allow for variable permeabilities of a layered stratum. Knowing the vertical permeability of each layer, the stratum is transformed to a single layer thickness equivalent to the least permeable layer. For example, Table 17.5 is based on an upper stratum of three layers, the lowest permeability being 1 x 10^{-4} cm/sec in the clay layer. The sandy silt layer (k = 2 x 10^{-4} cm/sec) is twice as permeable as the clay and is equivalent to one-half of the clay thickness. Thus, the transformed thickness is one-half of its actual thickness and is equivalent to 4.0 feet of clay. The more permeable silty sand is reduced to one-tenth of its actual thickness. The transformed thickness is that the 18-foot thick top stratum is equivalent to a single layer of clay 9.5 feet thick with a permeability of 1 x 10^{-4} cm/sec.

Table 17.5 Example Calculation of Transformed Thickness*

Layers in Top Stratum	Actual Thickness (feet)	Actual Permeability (cm/sec.)	F_1**	Transformed Thickness (feet)
Clay	5	1×10^{-4}	1	5.0
Sandy Silt	8	2×10^{-4}	0.5	4.0
Silty Sand	5	10×10^{-4}	0.1	0.5
	z = 18			$z_b = 9.5$

*after U.S. Corps of Engineers, 1992

$**F_1 = \frac{k_b}{k_n}$ for $k_b = 10^{-4}$ cm/sec.

z = actual thickness of top stratum

z_b = "transformed" thickness of top stratum

This analysis has been developed for the ground conditions encountered in a particular region. It is not difficult to imagine that, had a thin layer with a very low permeability been measured, the entire transformed thickness would become extremely thin. A more logical analysis, for example, might be to set the permeability of the upper zone to 0.01 times the permeability within the artesian zone when calculating the transformed thickness of the upper layer. Clearly, judgment needs to be made when computing the transformed thickness.

In practice, heaving and sand boils occur at localized areas due to variable subsurface conditions. The factor of safety F in Eq. (3) reduces the likelihood of sand boils occurring because a weak spot was missed by site borings. Infilled river meanders, borrow pits, burrowing animals, and other discontinuities within an impervious cover can also create hazardous spots on the downstream side of a levee on a flood plain (Mansur et al., 2000).

Calculations of flow towards a line of relief wells is outside the scope of this book, but is discussed in detail by the Corps of Engineers (1992). It requires field pumping tests and relatively sophisticated groundwater analyses. For many landslide projects, artesian pressures are handled by arbitrarily selecting well spacings and monitoring their effectiveness with piezometers. If more drawdown is needed, additional wells are installed.

Maintenance

Relief well screens are subject to deterioration from bacteria and dissolved metals, predominantly iron. Calcium carbonate deposits can encrust the gravel pack. Therefore, the discharge capabilities of wells decline with time and the wells must be restored or replaced if their functionality is to be retained. Table 17.6 provides factors that are conducive to corrosion and encrustation in a filter and gravel pack. A groundwater chemical analysis should be carried out to check for the presence of these indicators.

The efficiency of a relief well system can be monitored with: (i) piezometers placed between wells, and (ii) flow measuring devices at the discharge point. A loss of relief well discharge capacity can be mostly restored by high pressure jet cleaning and/or treatment with chemicals (Forrester, 2001). The frequency of maintenance depends on the rate at which clogging occurs at a particular site.

Mansur et al. (2000) analyzed the performance of hundreds of relief wells installed behind levees on the Mississippi River. After 40 years of service, they noted that wells with little sediment had lost only about 10% of their specific capacity during this period, whereas other wells subjected to floodwaters had average reductions ranging from 30% to 60%. They recommended redevelopment of relief wells by surging, pump tests, and pressure cleaning (with or without chemicals) wherever the specific capacity has fallen to less than 80% of the original specific capacity.

The Corps of Engineers well design criteria proved to be very satisfactory in preventing ground heave and sand boils during a major flood in 1993. Relief wells were installed at locations where the uplift gradient i was greater than 0.67; the measured uplift gradients during the 1993 flood at these sites ranged from 0.47 to 0.64 (average 0.54). Elsewhere, of seven sites where significant sand boils were observed, the uplift gradients i ranged from 0.58 to 0.84 (average 0.70).

Inclined Relief Wells

Relief wells have always been constructed as *vertical* wells for releasing or reducing an artesian head at depth. However, consideration should be given to installing them as *inclined* wells to increase the drawdown effect.

The principle of inclined relief wells is illustrated on Figure 17.48 for a permeable stratum with artesian flowing head. For a piezometer installed vertically within the permeable soil at point A, the pressure head rises to a height h_1 above the ground surface at B.

If a piezometer is drilled normal to the slope from C to the same point A, the head will be at the same elevation. However, a relief well constructed along AC will not only be shorter and less costly than a vertical well AB but, more importantly, the head drop h_2 will be greater.

Table 17.6 Indicators of Corrosive and Encrusting Groundwater*

Indicators of Corrosive Water	Indicators of Encrusting Water
1. A pH less than 7	1. A pH greater than 7
2. Dissolved oxygen in excess of 2 ppm	2. Total iron (Fe) in excess of 2 ppm
3. Hydrogen sulfide (H_2S) in excess of 1 ppm detected by a rotten egg odor	3. Total manganese (MN) in excess of 1 ppm in conjunction with a high pH and the presence of oxygen
4. Total dissolved solids in excess of 1,000 ppm indicates an ability to conduct electric current great enough to cause serious electrolytic corrosion	4. Total carbonate hardness in excess of 300 ppm
5. Carbon dioxide (CO_2) in excess of 50 ppm	
6. Chlorides (CL) in excess of 500 ppm	

Note: ppm = parts per million *after U.S. Corps of Engineers, 1992

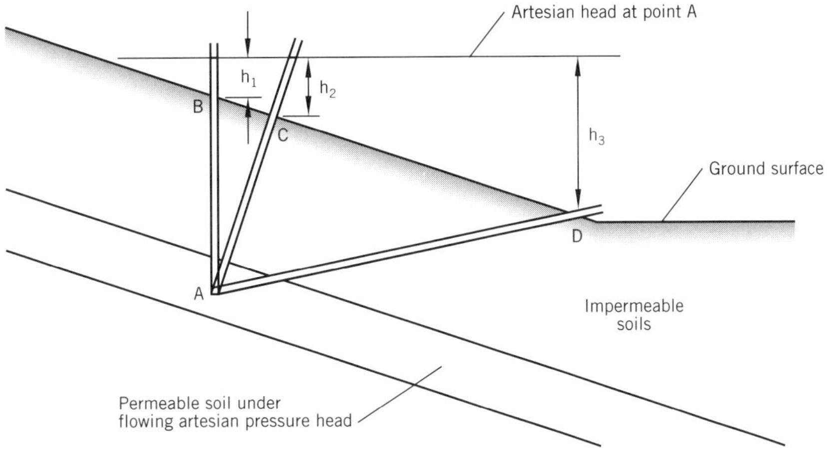

Figure 17.48 Reduction in artesian head using inclined relief wells.

A third location for the relief well is at the bottom of the slope, point D, where the maximum head drop h_3 can be obtained. As shown on Figure 17.48, a well AD is essentially a horizontal drain with a negative slope into the hillside. It would require casing into the permeable soil to prevent a soil collapse, and the hole diameter may be restricted by the available equipment for drilling near-horizontal holes. However, holes up to 8 inches diameter are routinely available for horizontal drains and ground anchors. Surprisingly, the concept of inclined relief wells appears to have been rarely used in the treatment of artesian pressures.

17.9 VERTICAL GRAVITY DRAINS

One of the simplest dewatering techniques, if the subsurface conditions are favorable, is to drain the landslide to a lower aquifer using vertical gravity drains. This remediation technique was considered as an option at the Upper Pelton landslide, described in Case History 6 and illustrated on Figure 17.49. At Upper Pelton, the groundwater was perched by an impervious stratum of sandstone-siltstone beds through which the basal failure plane passed. At a depth of about 15 feet below the failure plane was the top of a thick basalt bedrock, which in turn was underlain by relatively pervious silts, sands and gravels (slightly cemented). Groundwater in this lower bed corresponded approximately to reservoir level, the flat gradient indicating a relatively free-draining stratum.

A vertical gravity drain, approximately 250 feet deep, could have transferred the perched groundwater of the landslide through the basalt to the underlying pervious stratum. A line of drains located near the head of the Upper Pelton slide should easily drain the loose mix of ancient landslide debris that is above the slippage plane. The Upper Pelton slide is a double wedge failure (the two wedges being separated at the dotted line shown on Figure 17.49). Drainage would have

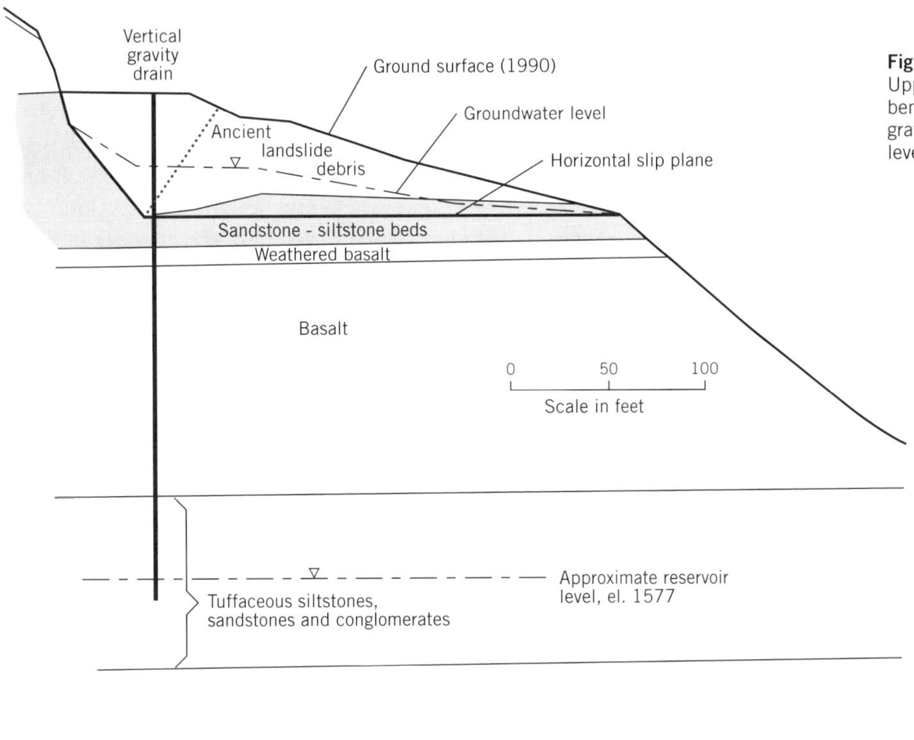

Figure 17.49 Section through the Upper Pelton Slide showing the benefits of installing a vertical gravity drain to reduce groundwater levels in the landslide.

reduced the driving force on the upper wedge and increased the resistance along the horizontal slip plane of the lower wedge, thereby stabilizing the landslide.

Vertical gravity drains at this site could have been installed by a down-the-hole hammer, Becker Drill, or by rotary coring. This remediation option was not selected because there was a simpler regrading remediation available. There was also concern that state regulators might object to the transfer of one groundwater regime to another (plus the inevitable delays of getting such a permit). In other states or countries, such a technical solution to relieving perched groundwater within a landslide may not be a problem.

17.10 TUNNELS AND DRAINAGE ADITS

Deep drainage of landslides can be provided by tunnels or adits. An **adit** is a short length of tunnel that is constructed into the slope, usually at right angles to the face. A **tunnel** normally implies a longer conduit. To simplify the narrative, both will be referred to as tunnels hereafter.

For landslide remediation, tunnels usually serve one of the following objectives:

- To provide general groundwater lowering within the landslide mass
- To reach a specific area of high permeability or recharge and drain it so that groundwater levels further downslope are reduced

It is desirable that a tunnel drain by gravity so that long-term pumping is not required.

Tunnel Construction by Hand Methods

Drainage tunnels for landslide remediation are usually excavated by hand methods, beginning at the downslope entrance. The walls and roof are supported by heavy treated timbers to form ribs at short spacings. Lagging is needed in sands or waterbearing strata. Typically, the tunnel is only made large enough to accommodate movement of the workers. Muck is removed by small wheeled carriages that travel over a simple rail track. Sections are typically 4 feet wide by 6 feet high (tight) or 5 feet wide by 7 feet high.

A thorough site investigation is needed to understand both the ground and groundwater conditions before any tunnel construction begins. It is very dangerous work due to the potential for ground collapse or a breakthrough into more waterbearing soils. The excavated soil has to be removed from the tunnel continuously and often away from the site altogether. Seepage water within the tunnel can make life miserable for the miners. Dewatering should be provided if there is significant seepage; this improves safety and makes excavation easier and quicker. Electric power and fresh air have to be brought into the tunnel on most projects.

When tunnel excavations are completed, a perforated pipe (of sufficient size to carry the maximum expected flow) can be laid in the bottom. The temporary wood or steel supports can be removed as the tunnel is backfilled with free-draining gravel; the gravel should meet the filter requirements for contact with the surrounding soils. One option is to blow in fine pea gravel (¼ inch to ½ inch sizes) although it is sometimes too uniform in gradation to act as a filter.

In some cases, tunneling contractors build adits or tunnels with untreated timber and leave the timber in place at the end of the contract. The tunnel itself may not be backfilled. The wood supports deteriorate, the soil collapses into the tunnel, and the drainage benefit is lost. Even if pressure-treated wood is used for the tunnel construction it will eventually rot. Once a ground collapse occurs, the ground above the tunnel is loosened and sinkholes may develop, especially near the tunnel entrance. Therefore, the landslide specialist should ensure that a tunnel is not only constructed for the intended purpose but also will not deteriorate and cause problems in the future.

Hand excavation of a drainage tunnel has the advantage that the tunnel can change horizontal alignment or branch off to the sides without the need for a vertical shaft at the change point. Another potential benefit is that the miners are able to see the ground conditions as they progress. This may allow them to excavate the slip zone as they move upslope. If the tunnel is subsequently backfilled with a crushed rock, a combination of improved drainage and higher shear strength on the slip zone can be achieved.

Tunnel Jacking

Tunnel jacking can be used to construct a tunnel that is fully supported at all times. At the downslope entrance to the tunnel a large block of concrete is cast-in-place to provide the resistance to thrust. In most cases, this requires the excavation of a starter pit. Once the resistance block is completed (and hardened), a length of casing is lowered into place and is jacked into the soil using two or more hydraulic jacks. When the initial length of casing has penetrated to the full depth, the soil is mucked out and a second length is butt-welded to the lower end of the first casing. The combined length of casing then is jacked into the soil and the sequence continues until the pipe has reached the required length. Except at the leading edge of the pipe, the pipe walls fully support the soil and there should be no loss of ground above the pipe. A heavy-duty cutting shoe is used for the leading pipe section.

The technique is simple and effective in the right circumstances. The pipe has to be large enough for workers to enter and dig out the soil from the cutting end. Therefore, a minimum inside diameter is about 3 feet, and this is sufficient for most landslide drainage projects. The main drawbacks are the limit on length (due to thrust required), control of direction, and problems with obstructions. Therefore, pipe jacking is usually limited to lengths of about 100 feet, although much longer drives have been achieved.

Microtunnel Construction

A fairly recent development in "trenchless" tunnel construction is to use a remote-controlled microtunneling boring machine (MTBM). Dobbels and Lyman (2002) describe

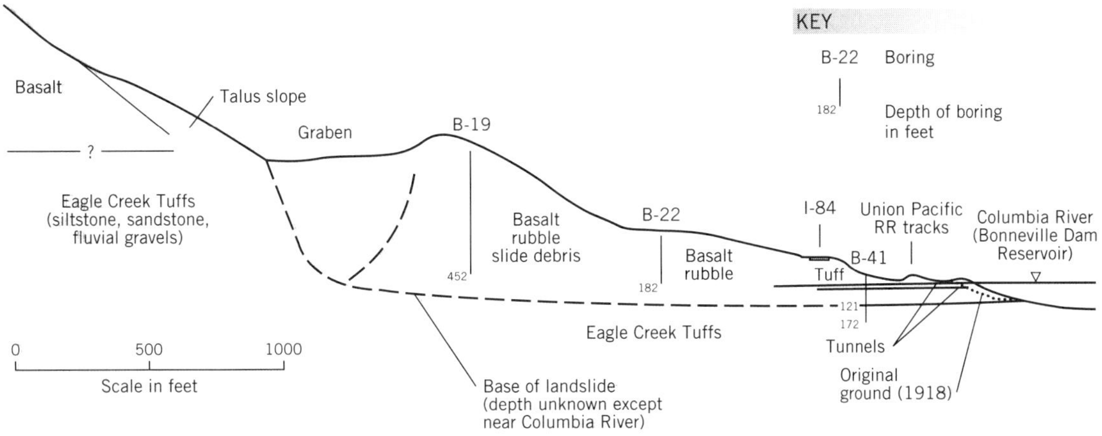

Figure 17.50 Ruckel Slide cross-section.

microtunneling as "a remotely controlled pipe jacking process that utilizes a pressure-balance tunneling machine."

The MTBM is a sophisticated piece of equipment that is similar to the large tunnel boring machines used to build large diameter tunnels for roads and railways. The cutting head generally mixes the excavated soil with bentonite and water in a chamber behind the head. The pressurized slurry is equal to the groundwater pressure to prevent caving at the face. The soil-bentonite slurry is pumped to the ground surface where the soil is separated out and clean bentonite slurry is returned to the cutting head. In clay soils, water alone can be sufficient to retain the face. The MTBM also has a laser guidance system that projects a beam onto a target to hold the excavation to line and grade.

The MTBM requires vertical shafts at the beginning and end of each run and can only follow a straight line (machines currently are being developed to follow curves). The shafts allow the machine to be put underground and taken out afterward. The entry shaft is also used to jack the steel tubes of the tunnel. The thrust wall is built and has to be capable of withstanding loads in excess of 1,000 tons on longer drives (Dobbels and Lyman, 2002). An entry ring is also needed to mine through silt and sand below the water table; the ring provides a watertight seal as the pipe is pushed into the ground outside the shaft. Thus, the entry shaft is larger than the removal shaft.

The main advantages of tunneling with a MTBM is that the alignment can be closely controlled, good progress is made under uniform soil conditions, dewatering is not needed, and it is not disruptive to the activities on the ground above the tunnel. Microtunneling drive lengths for pipelines smaller than 3 feet diameter are commonly 400 to 750 feet.

There are several areas of concern: (i) hard obstructions (e.g., boulders or old piles) encountered along the alignment can create major problems and may require a rescue shaft to be dug to remove the obstruction in front of the cutter head; (ii) the cutter head cannot be changed during a drive, so a change in soil conditions can seriously hamper progress; a thorough site investigation should prevent this potential problem; (iii) a major cost item is the need to build shafts; the shafts have to be large enough for the machine to be lifted in or out, and the location of shafts can present difficulties, especially in an urban environment; (iv) it requires a very experienced worker to operate the machine, and (v) very loose sands and gravel or very soft silt and clay are not ideal for microtunneling, nor are the variable soil conditions often encountered in landslides.

Microtunneling construction is continuing to develop newer techniques to overcome some of these disadvantages and improve efficiency/productivity. These should be very beneficial for future dewatering applications in landslide remediation.

Examples of Drainage by Tunnels

1. Washington Park Reservoir Slide, described in Case History 1, is an example of using a network of tunnels to lower the groundwater levels of a landslide.
2. The Ruckel Slide on the Oregon shore of the Columbia River about 45 miles east of Portland, Oregon (Figure 17.50) is an example of tunnels being constructed to tap into an aquifer at the upper reaches of a landslide. The slide occurred where soft tuffaceous sedimentary rocks—sandstone, siltstone, mudflows, and fluvial gravels—are overlain by much stronger basalts. The age of the landslide is unknown but is likely to have occurred during the downcutting phase of the Columbia River about 1 million years ago; it also would have been affected by the late glacial floods that came down the river about 12,000 years ago.

Ruckel Slide is a very large landslide—8,000 feet wide at the river edge and extending 2,300 feet upslope to 500 feet above river level (Figure 17.50). The undercutting of the soft sedimentary rocks by the river apparently undermined the hard basalt above, causing a massive slide. Borings show that the upper part of the slide is mostly broken-up blocky basalt; the lower part close to the river mainly comprises tuffaceous fine-grained sediments. During the original failure, the blocky basalt pushed out the weaker sediments and infilled the space left behind as the silts and sands moved toward the river. These landslide conditions may be typical of situations where weaker rocks underlie harder rocks and are subsequently undermined by erosion from a river or sea.

The slide has been very active. After the Union Pacific Railroad tracks were built over the slide toe, continual maintenance was required. In the period 1913-18, movements of up to 40 feet were recorded and one report described it as "the worst piece of track on the Union Pacific...." Another comment at that time was "...the whole country seems to be moving steadily toward the river like a glacier. It is all broken and full of fissures...."

The apparent cause of the continual landslide activity is that the upslope blocky basalt acted as a reservoir to continuously feed water into the tuffaceous soil below; as the landslide moved outward, the river eroded away the toe.

The Union Pacific Railroad built eight horizontal tunnels into the landslide from river level between 1918 and 1924 (Figure 17.50). A typical tunnel was 4 feet by 6 feet in section and 600 feet long. The tunnels were advanced with much difficulty and risk to the miners. It was reported that flows of water would issue from the tunnels in such volume that work had to be suspended for up to several days to allow the flows to subside. In January 1923, the combined discharge from the eight tunnels was 350 cu. ft./sec. The drainage tunnels were very successful in removing the water stored within the blocky basalt debris. Landslide movements ceased in 1921.

Bonneville Dam was built in the 1930s and the reservoir flooded the drainage tunnels. The Corps of Engineers built a new set of tunnels above the earlier ones (Figure 17.50). In 1981, an exploratory tunnel was drilled 2,300 feet through the slide to terminate in solid rock (Hannan and Devine, 1998). It encountered very little groundwater, a testimony to the effectiveness of the earlier drainage measures.

17.11 VERTICAL SHAFT WITH DRAINAGE ARRAY

Technique

This system of drainage consists of a vertical shaft with arrays of horizontal drains drilled around the circumference. Although it is expensive to construct, the system allows deep drainage by horizontal drains where the failure surface is deep-seated.

The principle of a vertical shaft with drainage array remediation is illustrated on Figure 17.51, which shows a circular arc failure. If conventional horizontal drains are installed into the slope, the maximum length of the failure arc that can be intercepted by the drains is length AB. This is only about one-third of the total arc length AC. By installing a vertical shaft near the lowest elevation of the failure arc, an array of horizontal drains radiating from the shaft can intercept groundwater over the majority length of the arc BC.

The groundwater is collected into the shaft. The main difficulty is groundwater disposal. The water can be pumped from the shaft, but this is usually a short-term expedient for temporary works because of the power costs of long-term pumping. On major projects, such costs may be acceptable when compared to alternative treatments. The second option is to provide gravity (passive) drainage to open ground at a lower elevation than the shaft base. A large drain or tunnel may be feasible from the shaft to an outlet further downstream.

Example: Arizona Inn Landslide Remediation

Landslide Study

The Arizona Inn landslide, a large slide on the southern Oregon coast, was stabilized by a vertical shaft and horizontal drain design (Peterson et al., 1998). A plan and section of the slide are shown on Figure 17.52.

The landslide has a long history of intermittent large movements toward the Pacific Ocean. In the past, unloading and regrading of the upper portion of the slide plus surface drains and short horizontal drains had all been ineffective. The landslide covers 32 acres (700 feet wide at beach level and 1,900 feet long) and consists of about 4 million cu. yd. of slide debris. The vertical relief from head to toe is about 700 feet. The most recent major slide (1993) required traffic on the main coastal road, Highway 101, to make a 280-mile detour.

The slide debris is highly sheared sandstone and mudstone, which form a fine-grained melange. Slickensided surfaces are observed throughout the landslide and underlying rock mass.

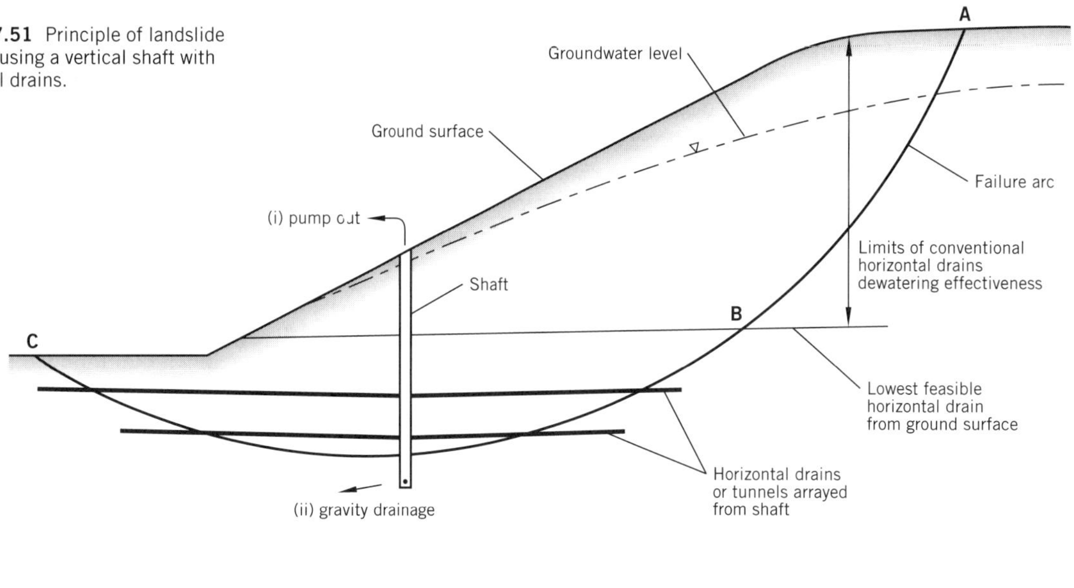

Figure 17.51 Principle of landslide drainage using a vertical shaft with horizontal drains.

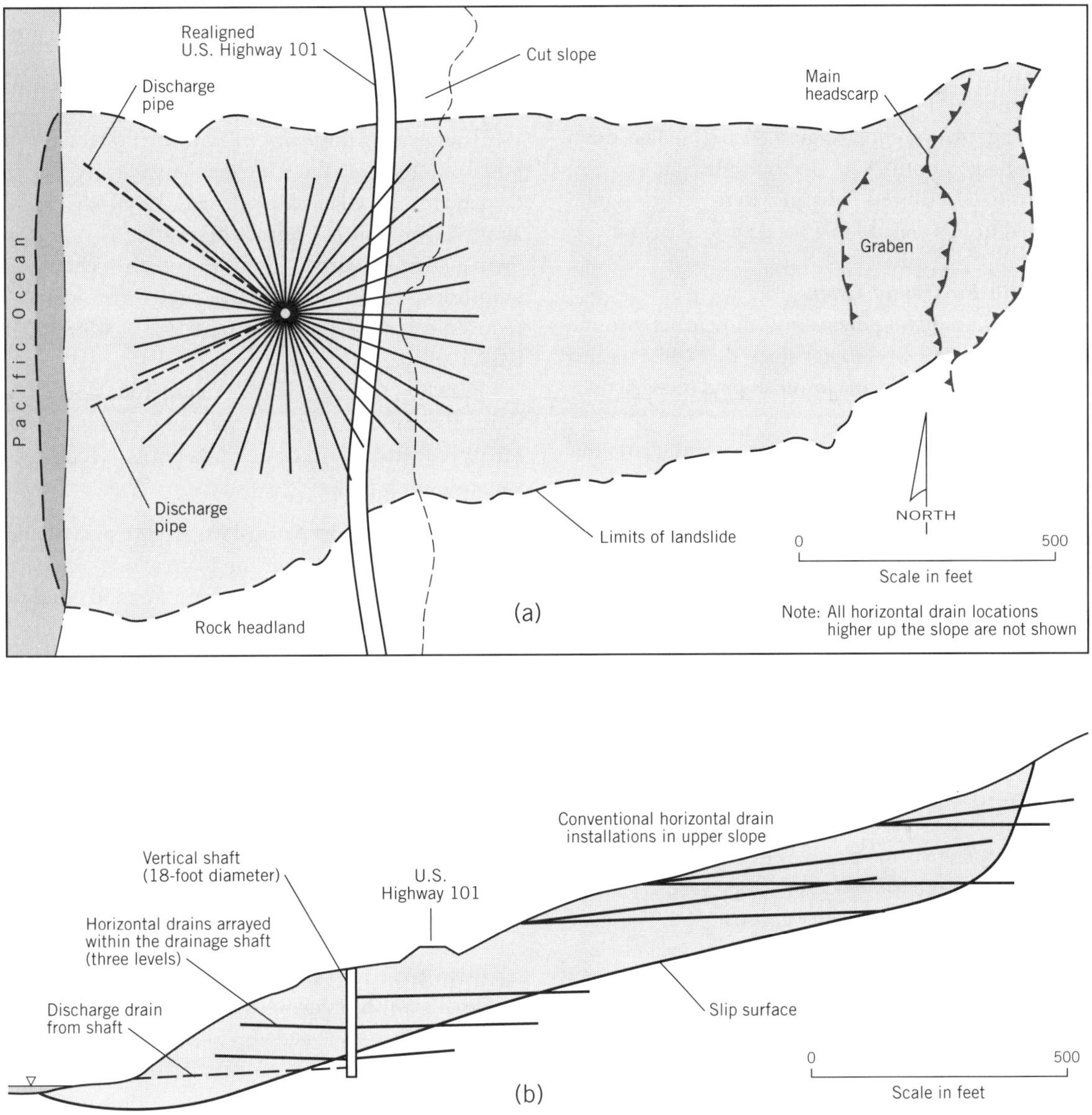

Figure 17.52 Arizona Inn Slide, near Brookings, Oregon:
(a) plan
(b) section through center of landslide
(after Peterson et al., 1998)

The upper part of the slide has a failure plane at about 12° to the horizontal and typical depth of 150 to 160 feet below the surface (Figure 17.52b). The lower slide has a steeper slip surface of up to 19° to the horizontal and multiple shear zones. Shearing extends to at least 44 feet below sea level and the toe is at an unknown location offshore.

Two pumping tests, together with numerous piezometers installed throughout the site, showed that the slide mass was very heterogeneous in hydrologic characteristics. Most of the soils had permeabilities of less than 10^{-4} feet per minute. A few zones, located generally near the failure zone in the upper part of the slide, had somewhat higher permeabilities on the order of 10^{-3} feet per minute. Artesian groundwater (flowing above ground) was encountered at depth. Groundwater recharge is attributed to two sources: surface infiltration from rainfall, primarily into the more permeable upper parts of the slide, and deep pressurized groundwater from more distant sources. Annual rainfall is around 80 inches.

The objectives of the landslide remediation were: (i) to reduce the peak piezometric heads by 100 feet in the upper part of the slide and facilitate rapid removal of groundwater during high rainfall periods, and (ii) to reduce the peak piezometric heads by 50 feet in the lower part of the slide area.

The total program of dewatering comprised:

- *Upper part of slide.* Three tiers of horizontal drains from 11 drill pad locations. At each pad, 10 drains (2-inch diameter slotted PVC) were arrayed into the slope with 9° azimuth separation between drains. The 110 drains were

200- to 700-feet long (average 500 feet). Additional horizontal drains were placed in the road cut.

This horizontal drain program has been omitted from the site plan, Figure 17.52(a), but some drains are shown on the cross-section, Figure 17.52(b). The drain inclinations ranged from 2° to 7° above horizontal.

- *Lower part of slide.* Three radiating tiers of horizontal drains drilled from a vertical shaft, as described below.

Vertical Shaft with Horizontal Drains

A vertical shaft with radiating drains was chosen because the base of the landslide has a steep outer slope that makes access to the slope base difficult. Also, the lower slope is being actively eroded by the sea.

The vertical shaft (Figure 17.52b) is 18 feet diameter, which provided sufficient room for the horizontal drain installations. The 205-foot deep shaft penetrated through the slip surface by 20 feet. The shaft walls were supported by lattice girders and a shotcrete lining.

The shaft was excavated over a 9-week period. Overbreak was minimal and no blasting was required. A small tracked excavator worked within the shaft. Before excavation began, dewatering wells were installed to control groundwater infiltration.

At the base of the shaft, two outlet drains were constructed between the shaft and the beach. A third drain was built higher up the shaft as a precaution should the lower two drains become inoperative for any reason. The inlets of the pipes are protected by special structures to prevent discharge of debris to the beach during and after construction. The shaft has a removable cover to allow future inspection, monitoring, and maintenance.

The arrays of horizontal drains were installed at three different levels in the shaft (Figure 17.52b). Each drain had 10° of separation in azimuth and had an initial slope of 2° above horizontal. The drains were 200 to 500 feet long. No drains were oriented toward the sea at the upper level because of the short distance between the shaft and the outer slope. The total length of the 91 drains was 29,350 feet (design-actual footage unknown). All the horizontal drains were drilled from a moveable platform suspended from the shaft collar by cables, beginning with the lower tier.

The project appears to have been very successful in attaining its objectives. Groundwater levels reportedly have met the drawdown objectives. Only minor creep movements were measured in the five years following construction.

17.12 CONTROL OF SURFACE WATER AND WATER-CARRYING PIPES

Numerous very small-to-medium size landslides occur during heavy-intensity storms. Most of these slides are preventable through awareness of vulnerable areas, hydrologic and hydraulic design, and good maintenance practices.

To put the issue in perspective, the February 6–9, 1996, rainstorm in western Oregon caused 7 to 10 inches of rainfall in Portland and is estimated to be a 1 in 100-year return interval event. A team of graduate geology students at Portland State University examined most of the more than 400 landslides recorded in Portland during the storm. The damage from the landslides (Burns, 1998) was estimated to be about $4 million (17 homes destroyed, 64 others extensively damaged). The survey also showed that about 60% of the landslides would not have occurred, or would have had reduced damages, if better surface water control measures had been in place.

Based on the experience of numerous storm-related landslides, the author offers the following thoughts and guidance for preventative actions involving surface water or water/sewer pipes:

- **Homes and other buildings:** (i) repair any downspouts emptying onto the ground surface; (ii) check that sprinkler systems are turned off and in good condition during the wet season; (iii) repair any broken or cracked water supply, stormwater and sewerage disposal pipes; and (iv) remove any obstructions in gutters, edge of driveways, etc.
- **Canals:** inspect weekly for changes, such as seepage, sloughing, and obstructions to flow
- **Roads:** (i) *roadside ditches* should be well maintained and be deep and wide enough to carry the flow of a 50-year or 100-year storm; if heavy storm flows cannot be computed, make reasonable estimates to ensure that ditches do not overflow at critical locations, such as steep cross-slopes; (ii) *culverts* are especially vulnerable locations; at least monthly during the wet season, the upstream side should be inspected for buildup of debris and cleaned out as needed. At least annually, culverts crossing moderate to steep slopes should be monitored to check for cracks or separations in the pipe, or earth entering through joints. If there is evidence that the culvert is undersized, it should be replaced as soon as practicable; (iii) *road cuts and fills* should be checked for excessive seepage and preventative action taken, if needed; (iv) *water-carrying pipes* should not be placed on the outside shoulder of a road fill; (v) *hard rock slopes* adjoining roads, railroads, etc., should be provided with fences, netting, fallout ditches, etc. as needed to prevent loose rocks washing onto the road during storms; slopes with a history of frequent rockfalls during storms should be scaled at least once every 10 years. At critical locations the slope should be prioritized (see Chapter 23, Section 23.2) for more effective preventative action to reduce or halt future rockfalls; and (vi) *surface water* from large paved areas, such as shopping malls or new housing estates should not be discharged into nearby ravines without an appropriate engineering study of the likely effect on erosion and/or slope stability.

This list is not intended to be all-inclusive, but to help the reader to be aware of the circumstances that commonly cause landslides (and consequent economic losses) in hilly terrain.

17.13 DEWATERING THROUGH CONSOLIDATION

Soft ground can be consolidated by a preload fill to prevent instability when a fill is built over it. The technique is well known and has been undertaken on many projects, often to reduce post-construction settlement rather than stability. For slope stability purposes, the technique is preventative and is used in new construction.

A hypothetical example is shown on Figure 17.53. An embankment can be built over a soft foundation by "stage construction." The first stage fill, Figure 17.53(a), is placed over the foundation area. Stability analyses will determine the maximum height of preload fill that can be safely placed without causing base failure or lateral spread in soft clay layers (with, for example, a factor of safety F of at least 2.0). The preload fill is left in place for several weeks or months to allow the soft foundation to consolidate and strengthen. Vertical drains, such as prefabricated vertical drains or sand drains, can be installed in the foundation soils to speed up the rate of consolidation.

Settlement plates (Figure 17.53a) are placed at selected locations of the preload area to measure total and rates of settlement. A settlement plate is a 12- or 18-inch square steel plate with a steel riser pipe attached to the center. Flags on top of the riser pipes reduce the risk of construction equipment hitting the pipes during fill placement. Two sets of initial readings should be taken to get a reliable zero reading. Subsequent readings are taken as the fill builds up and thereafter at about weekly intervals during the consolidation period. Preloading is stopped when primary consolidation has reached 85–90%. To obtain the records, including the rate of fill placement, it is recommended that an independent surveyor working with the design geotechnical engineer should take these measurements rather than the contractor's staff. Assuming that the fill is brought to the site at a uniform daily rate (excluding weekends), the time t = 0, equivalent to a single instantaneous load, is the midpoint of the fill construction period. It is important to ensure that the reference benchmark is distant from the preload fill and is not being affected by the settlements. Compliance with this requirement can be difficult on flood plains or landslides.

The settlement data can be plotted on a root time graph and should exhibit a good linear relationship during the initial primary consolidation phase. Chapter 8 discusses the rate of settlement of clays.

Piezometers can be installed as an additional control method. Piezometer tips can be located at one-quarter, one-half, three-quarters, and full depth of the compressible layer. It is preferable to select vibrating wire (VW) piezometers rather than standpipes because the VW piezometer leads can be placed in a shallow backfilled trench and taken to a terminal outside the fill area before the earthworks contractor arrives on site. If vertical drains have been installed in the foundation soils, the piezometers need to be put midway between drains or omitted altogether due to the difficulty of reliably knowing the sensor location relative to the drains.

Once the foundation settlement has attained a high percentage of the primary consolidation, the highway or railroad fill can be completed in a second stage construction (Figure 17.53b). Although one option is to use the outer preload fill to construct the upper embankment, it is usually better to leave the outer preload fill in place to support the higher fill, thereby acting as an external buttress ("counterberm"). Since the preload fill is to become part of the final embankment, it must be placed and compacted as a structural fill.

The advantage of preloading is that it is simple to construct by readily available equipment. Preloading sometimes can take place while design plans and specifications are being finalized. The main drawback is that the time needed to reach 85–90% primary consolidation is uncertain and, if longer than

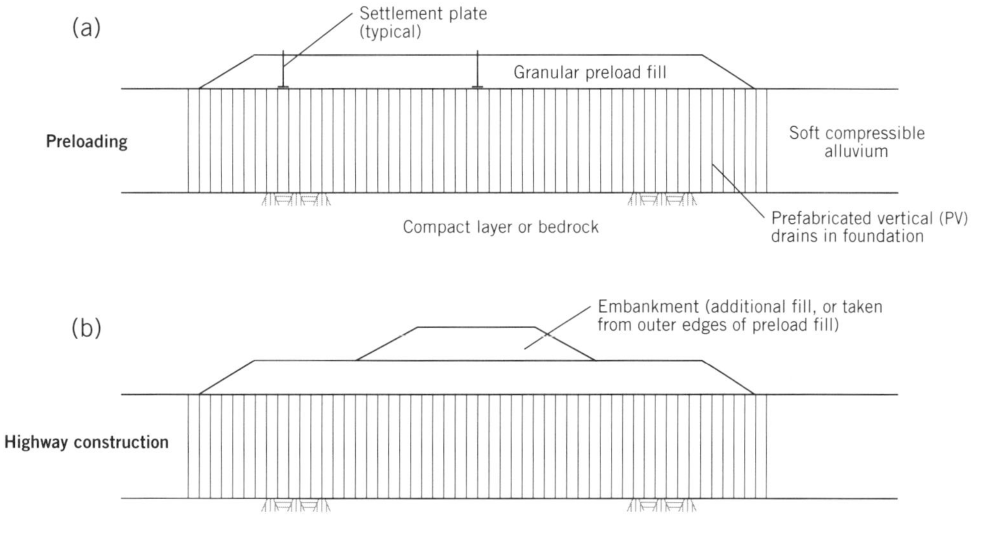

Figure 17.53 Dewatering (consolidating) a soft compressible alluvial foundation by preloading:
(a) preload fill in place
(b) highway embankment built above the preload fill (staged construction)

anticipated, is likely to interfere with the project completion. The use of vertical drains may be able to offset this uncertainty. Another potential problem is that the preload plate riser pipes may be disturbed by vandals on open sites or by careless equipment operators during construction. It is advisable to install a few extra settlement plates as a contingency against such losses during the observation period.

17.14 PREFABRICATED VERTICAL DRAINS

Prefabricated vertical drains (hereafter referred to as PV drains) can be used in landslide work to relieve artesian or other uplift pressures or to increase the rate of consolidation and thereby increase shear resistance in soft soils. A third use is to reduce the potential for liquefaction in silts and fine sands (see Case History 3). *Strip drains* are PV drains laid horizontally as a substitute for a soil or fabric drainage layer.

PV drains are now used almost exclusively over their predecessors—sand drains and fabric-encased sand drains—because they are quicker, easier, and cheaper to install. These alternative technologies will not be described in this chapter, but the general principles are identical.

PV Drain Description and Installation

A PV drain consists of a plastic core surrounded by a geotextile filter fabric. The drain has a shallow rectangular cross-section (referred to as "band shaped," similar to an elastic band) in which the aspect ratio of width a : thickness b is typically 25-30 (Figure 17.54). There are at least 50 types of PV drain being manufactured worldwide. They were formerly known as "wick" drains.

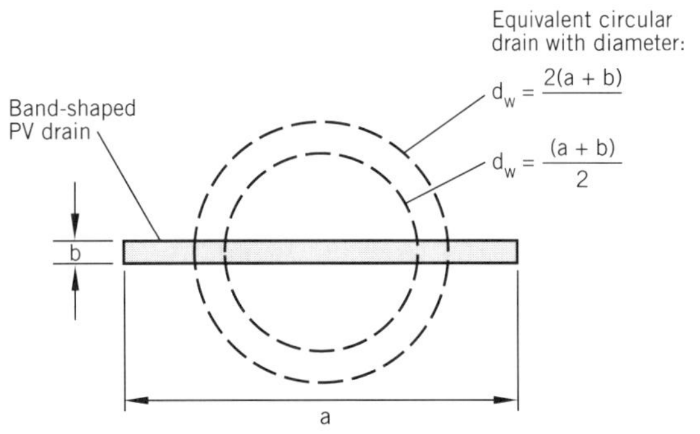

Figure 17.54 "Equivalent" circular drain of a band-shaped PV drain (after Rixner et al., 1986).

The *plastic core* can be a grooved channel, mesh, or protruding studs. Its function is to provide a flow path for water, resist buckling, and support the surrounding filter fabric. The jacket material or *filter fabric* separates the plastic core from the surrounding soil and acts as a filter to prevent soil entering the core while allowing water to pass through. Both components of the PV drain are equally important to success as a drain. A typical drain is 4 inches wide and 0.16 inch thick, although drains of up to 12 inches wide are available.

Drains typically are installed at 3- to 10-foot spacings in a square or triangular pattern (Figure 17.55). Typical lengths are 30 to 60 feet, although drains reportedly have been installed to depths of 200 feet. Lengths above 70 feet require a crane, and lengths exceeding 120 feet need both a large crane and a specialized installation mast.

There are many different ways of installing PV drains, including jetting, vibratory hammer, augering, rotary drilling, and combinations of such methods. However, almost all methods use a steel mandrel to protect the drain as it is being installed. An example is shown on Figure 17.56. The PV drain is supplied in rolls and is threaded through the mandrel. Typically, a crane is used to hold the drain on a reel and lift the mandrel into position (Figure 17.56a). The mandrel is inserted into the ground at an approximately constant rate of advancement, typically 0.5 to 2 feet per second. The mandrel has to be capable of protecting the PV drain and penetrating vertically through the ground. In addition to these requirements, the cross-sectional area of the mandrel should be kept to a practical minimum to keep the soil disturbance low. Typically, the mandrel area is around 10 sq. in.

All PV drain types require an anchoring system at the bottom to hold the drain in place while the mandrel is withdrawn. Anchorage is usually provided by a length of pipe or rebar attached to the bottom of the drain. Obstructions above the compressible layer, such as dense sand, stiff clay, or boulders may require pre-drilling. Once the PV drain has reached the final depth, the top of the drain is cut to leave sufficient length

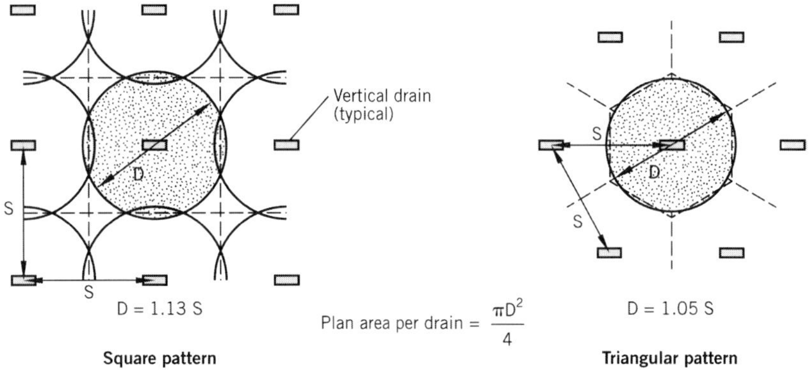

Figure 17.55 Layout of vertical drains in square and triangular patterns (after Rixner et al., 1986).

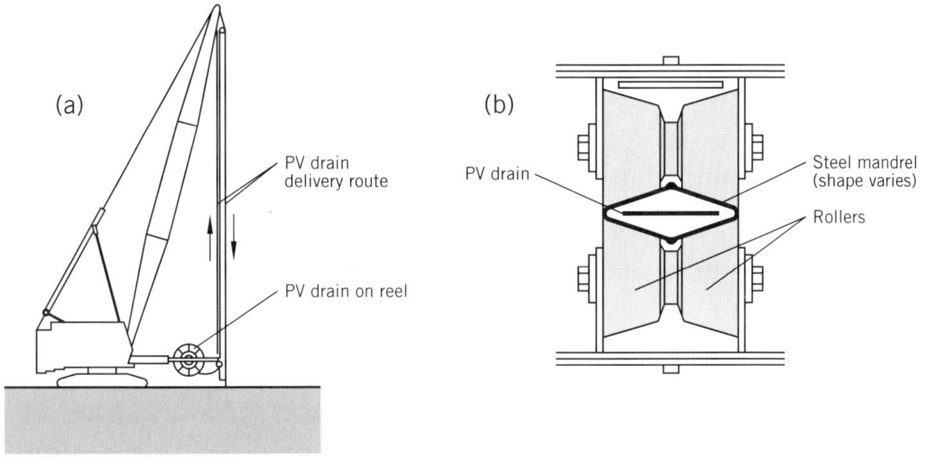

Figure 17.56 Installation of PV drains:
(a) mandrel lifted by crane
(b) cross-section of mandrel and drain
(after Rixner et al., 1986)

above ground for embedment in a drainage blanket or other pervious stratum.

Slopes steeper than about 5 horizontal : 1 vertical are an obstacle to PV drain installation. They require benches at least 20 feet wide to provide access for the equipment. The cost of constructing access roads may make the drainage option uneconomical compared to alternatives.

PV drains are usually installed by a specialty contractor. The cost of the PV drain materials is about 40–50% of the installed cost. The installed cost, excluding any surface drainage blanket, predrilling, or mobilization cost, was around \$0.75 to \$1.00 per lin. ft. in the United States in 1986 (Haley and Aldrich, 1986). Current costs can be obtained directly from installers or can be estimated from the Engineering News-Record Construction Cost Index, as explained in Chapter 13, Section 13.6.

Advantages and Limitations of PV Drains

Some of the advantages cited for PV drains over sand drains are: (i) lower cost; (ii) generally less disturbance to the compressible soil when displacement methods of installation are compared; (iii) can be installed in a nonvertical orientation; (iv) eliminates the need for sand and disposal of spoils; (v) can withstand significant lateral displacement or buckling without losing function; (vi) can be placed very close together if needed for more rapid consolidation; and (vii) can be installed below water.

The main limitations of the technique are: (i) displacement installation may destroy thin pervious lenses around the drain, thereby reducing drainage towards the PV drain; (ii) installation will disturb sensitive soils and thus may locally reduce the available shear strength prior to consolidation; (iii) organic soils may be less responsive to vertical drains, and (iv) concerns that the plastic core may be crimped or buckled by large ground settlements below a buttress.

Design

The design of the drain spacing is based on the earlier work of Barron (1948) for radial drainage to sand drains. The band shape of the PV drain is converted to an "equivalent" circular drain (Figure 17.54). This is given by the formula:

$$d_w = \frac{2\,(a + b)}{\pi} \qquad \text{Eq. (5)}$$

where a, b are defined on Figure 17.54

Haley and Aldrich (1986), based on finite element studies, suggest that Eq. (5) be modified to:

$$d_w = \frac{(a + b)}{2} \qquad \text{Eq. (6)}$$

The "ideal" design ignores the effects of soil disturbance and drain hydraulic resistance to give the formula:

$$t = \frac{D^2}{8c_h}\, F_n \log_e \frac{1}{(1 - \bar{U}_h)} \qquad \text{Eq. (7)}$$

where t = time required to reach the average degree of consolidation, $\bar{U}_h$
($\bar{U}_h$ is commonly set at 90% in design)
D = drain influence zone (see Figure 17.55)
c_h = coefficient of consolidation in the horizontal plane (for a vertical drain)
F_n = drain spacing factor
$= \log_e\left(\frac{D}{d_w}\right) - \frac{3}{4}$ (see Figure 17.57) Eq. (8)
d_w = "equivalent" diameter of a circular drain (typically 2 to 3 inches)

Various attempts have been made to calculate the effect of soil disturbance, notably by Hansbo (1979). These adjustments increase the time to reach a specified average degree of consolidation. However, given the variable nature of soil deposits, which makes it difficult to measure a single coefficient of consolidation for a stratum, and the many different installation procedures, it is advisable to avoid too much reliance on theoretical calculations for drain spacings and use calculations only as a guide. For example, a designer can use the "ideal" calculation, ignoring soil disturbance and drain hydraulic resistance (i.e., Eq. (7)), and then reduce the spacing by an arbitrary amount to allow for the disturbance/hydraulic resistance factors. Another option is to base the time for consolidation on

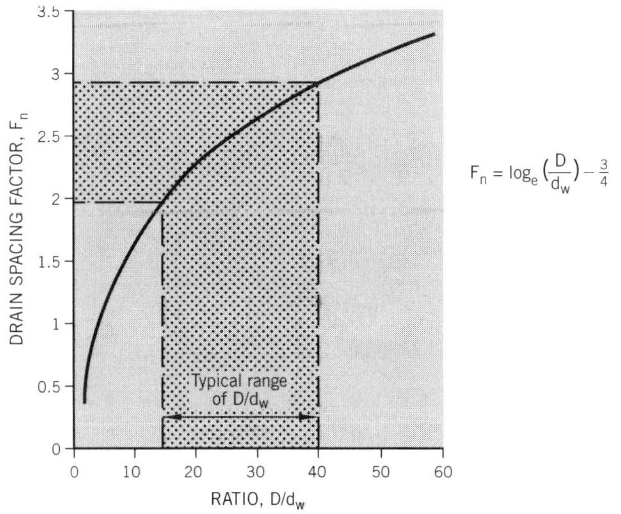

Figure 17.57 Determination of drain spacing factor ("ideal" case) (after Rixner et al., 1986).

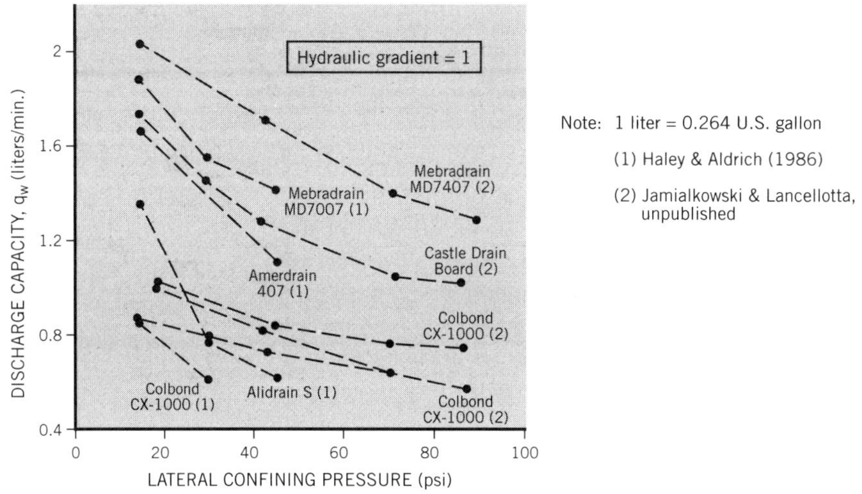

Figure 17.58 Typical values of vertical discharge capacity for various PV drains (from Haley and Aldrich, 1986).

the coefficient of consolidation c_v rather than on c_h in Eq. (7). For sedimentary soils, c_v is generally lower than c_h and thus increases the estimated time of consolidation. It is also much easier to measure c_v than c_h in the soils laboratory. An alternative is to use the soil disturbance procedures outlined in Hansbo (1979) or Haley and Aldrich (1986).

The vertical discharge capacity of the individual drains is not usually a problem for most fine-grained soils. Some experimental data for selected PV drains are shown on Figure 17.58.

Field Performance

The effectiveness of PV drains in accelerating the time for primary consolidation can be monitored by field measurements. Settlement plates can be installed and periodically surveyed, and piezometers can be placed at different levels within the compressible stratum to measure pore pressure dissipation. However, vertical alignments of the PV drains and piezometers have to be carefully considered, especially if the drains are closely spaced, when interpreting the field data.

CHAPTER

18

Seepage Barriers

Seepage barriers are an alternative to dewatering of landslides. A typical scenario consists of a permeable stratum underlain by an impermeable stratum. If groundwater levels in the permeable soils are raised by the construction of a canal, reservoir, or urban runoff (as examples) the increased lateral flow can destabilize slopes downstream of the source (Figure 18.1). A seepage barrier can be constructed as either a preventative or remedial treatment.

Groundwater barriers have been used extensively in recent years to contain contaminated groundwater emanating from landfills and industrial sites. This has lead to increasing innovation by specialist contractors who build the barriers. Such contractors have experienced design and construction staff, and their store of knowledge should be used.

As part of site investigations, in-situ tests should be carried out to measure permeabilities in the subsurface soils. These data are used to determine the length and depth of the seepage barrier, and which technique is likely to be more cost-effective and constructible at the site.

It is important to install piezometers on the downstream side of a seepage barrier so that the effectiveness of a barrier can be measured. Baseline readings should be obtained prior to construction of the barrier.

18.1 SLURRY TRENCH CUTOFF WALLS

Technique

A slurry trench cutoff wall probably is the most effective method of eliminating lateral seepage. It is also expensive because specialized equipment and labor are needed to construct the wall.

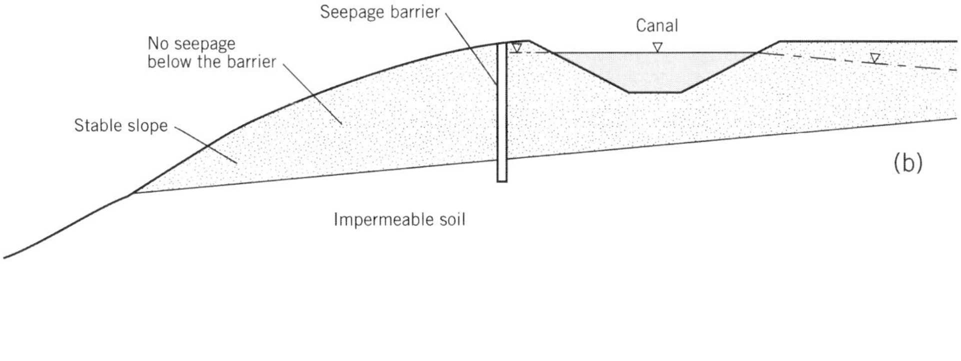

Figure 18.1 Example of seepage barrier construction to stabilize a landslide:
(a) canal feeding water into a slope
(b) seepage barrier in place

The technology was developed in the 1960s by European firms such as Soletanche and Bachy. It is an outgrowth of drilling practice in which a slurry of bentonite and water is used to prevent unsupported cohesionless soils from collapsing into a drillhole. The slurry is believed to form a thin plaster ("cake") on the walls of the hole. The slurry, which is slightly denser than groundwater, must always be at a higher level than the outside groundwater level.

The slurry trench method is constantly evolving as specialist contractors develop new techniques to speed up the process, deal with difficult ground conditions, or improve the watertightness of the joints at the ends of the panels. The basic procedure for constructing a wall will be described, followed by some of the more common variations that are available.

Slurry Trench Wall Construction

The earliest slurry trench walls were a method for constructing a *structural concrete* or *diaphragm* wall. They had a variety of applications, including underground retaining walls, seepage cutoff walls below dams, etc. This is the basic technique described below. However, alternative methods have evolved that are more flexible and/or less expensive; these variations are described at the end of this section.

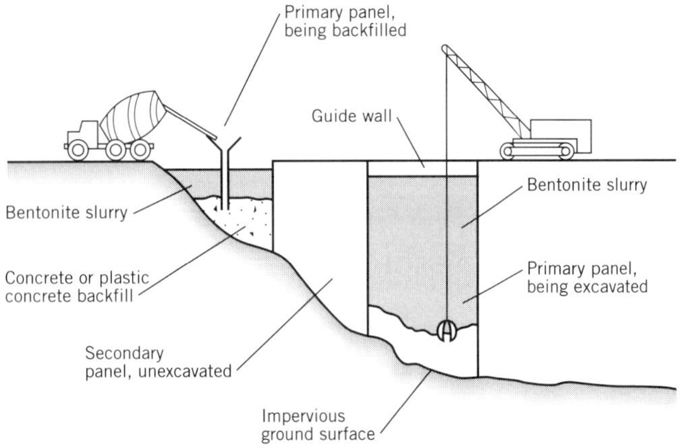

Figure 18.2 Construction of a concrete cutoff wall by alternate panels (after Millet et al., 1992).

A slurry trench wall is built as a series of connected "panels" with vertical ends (Figure 18.2). After first constructing a concrete guide wall at the ground surface, the soil is excavated by a grab using bentonite slurry to support the sides of the trench. When excavation is completed, a vertical end stop is installed and tremie concrete is placed from the bottom upward. The bentonite slurry is continually recirculated to remove suspended soil and control the quantity of slurry within the trench. Additional concrete panels are constructed as needed for the length of the cutoff wall.

The most common procedure is to construct *alternate* panels using cylindrical (pipe) end stops (Figure 18.3). The primary panels have a vertical pipe at each end of the panel. After the concrete has set and hardened, the pipes are pulled out and the intermediate secondary panel is filled with concrete. The joint between the primary and secondary panels is half round which generally provides a tight fit.

An alternative procedure is to build *successive* panels, as shown on Figure 18.3(b). In this technique, the end stop pipe of the previously placed panel is withdrawn and the adjoining panel is excavated. When excavation of the panel is completed, another end stop is installed and the panel is concreted.

The concrete panels typically are 2 to 3 feet wide and 6 to 30 feet long. The vertical height is almost unlimited but the difficulties of keeping the walls vertical and properly interlocked increases with depth. Depths of up to 100 feet are routine for specialized contractors. The wall base penetrates 2 to 4 feet into the underlying impermeable stratum; in hard rocks, embedment may be limited to just a few inches.

Construction Details

Guide Wall

The first step in constructing a slurry trench wall is to provide a guide wall at the ground surface. A typical guide wall design is shown on Figure 18.4. It consists of two parallel concrete beams constructed along the line of the slurry trench wall with

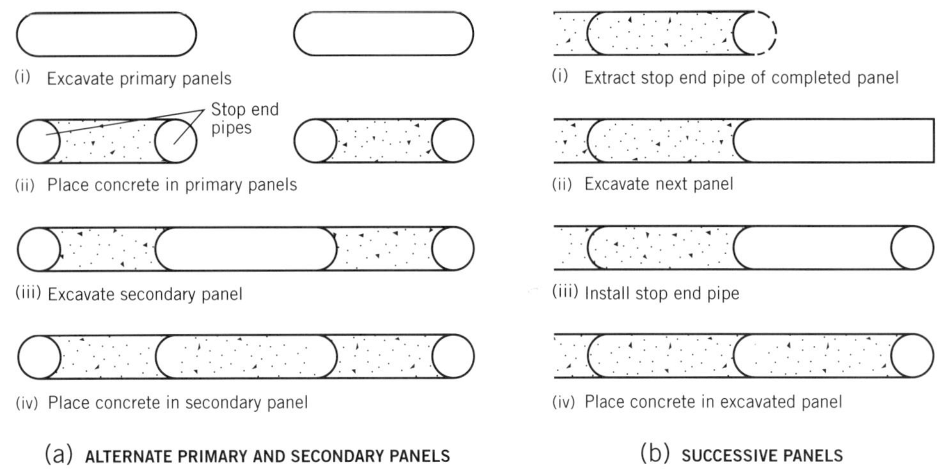

Figure 18.3 Construction sequences used to build slurry trench walls with stop end pipes.

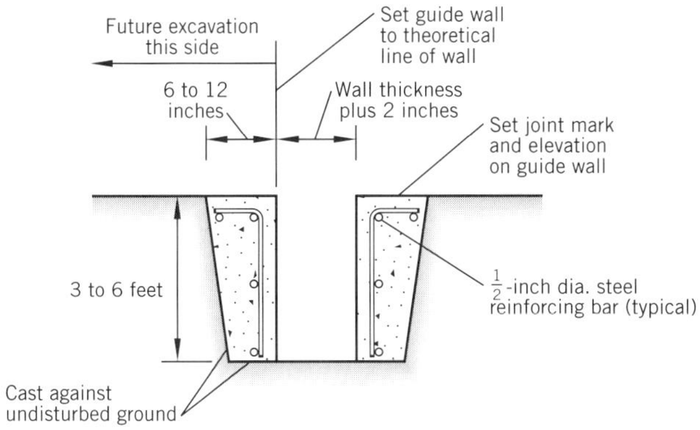

Figure 18.4 Typical guide wall details for a slurry trench wall (after Tamaro and Poletto).

the space between them being slightly greater than the wall width. Only the inside faces of the guide walls are formed, and nominal reinforcement is used. The concrete is cast against the earth below and behind the reinforcement. The guide wall is critical to providing good quality control and adds significantly to the overall cost of the slurry trench wall. It can be left in place or removed at the end of construction.

Excavating Equipment

The most widely used excavating machine is a mechanically operated (cables) clamshell (Figure 18.2). The digging head is specially shaped to excavate a bite equal to the width of the trench. Hydraulically operated clamshells are also used but are more prone to breakdowns. Clamshells can be attached to a kelly bar to provide more downward thrust. Clamshell buckets typically weight 9 to 13 tons.

Verticality of the excavation is one of the key elements in constructing deep slurry trench walls. It is needed to maintain the minimum section of the wall and to achieve continuity between adjoining panels. Most projects specify verticality within the range of 1:80 to 1:400, although the latter is difficult to achieve and likely will increase bid prices. Several field techniques are available to check the verticality of the trench, and the number, depth, and frequency of these measurements (plus corrective procedures, if needed) should be specified in advance of construction. Walls that exceed the verticality tolerance typically are backfilled with lean concrete and re-excavated.

Boulders, hard strata, buried objects and caving ground are the principal impediments to rapid construction. The site investigation before construction needs to be thorough to forewarn designers and contractors of any adverse subsurface conditions. Heavy chisels are used to break up obstructions and hard strata. The length of the panel is often shortened when caving ground is encountered. Other major concerns to contractors include loss of slurry into highly permeable ground, and lateral blowouts where construction is adjacent to a steep slope.

Conventional backhoes can be used to construct relatively shallow slurry trench walls up to about 60 feet deep (using extender arms). A preliminary dozer pit can extend the depth which can be reached by the backhoe. Typically, the wall is thicker (up to about 5 feet) when a backhoe is used as the excavator.

Another type of excavating machine is the hydromill which consists of a large frame structure with counterrotating cutter wheels at the bottom (Figure 18.5). Hydraulic motors drive the rotary drum cutters. The excavated materials and slurry are recovered by reverse circulation to a desanding plant

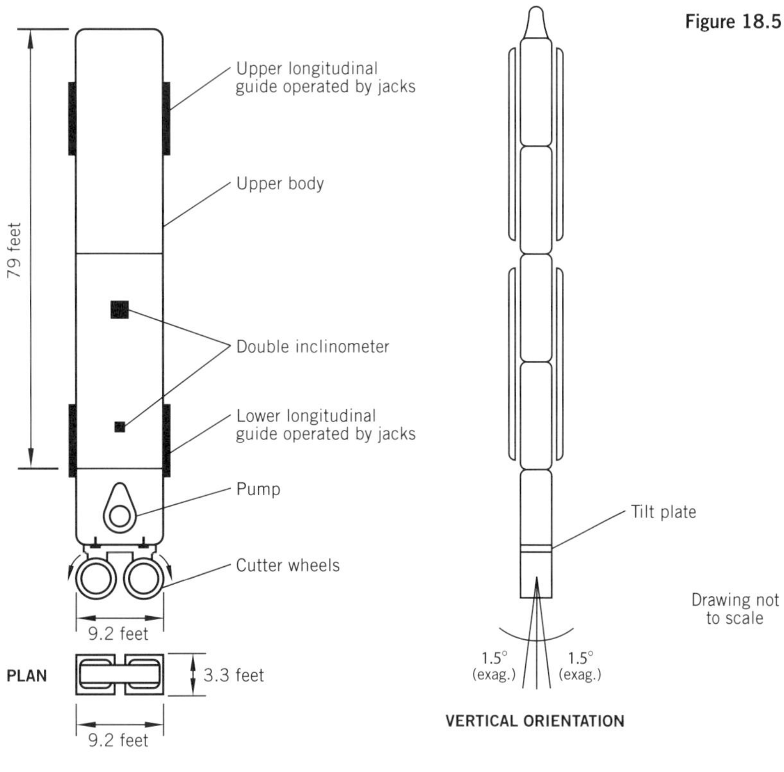

Figure 18.5 Hydrofraise 12,000 machine.

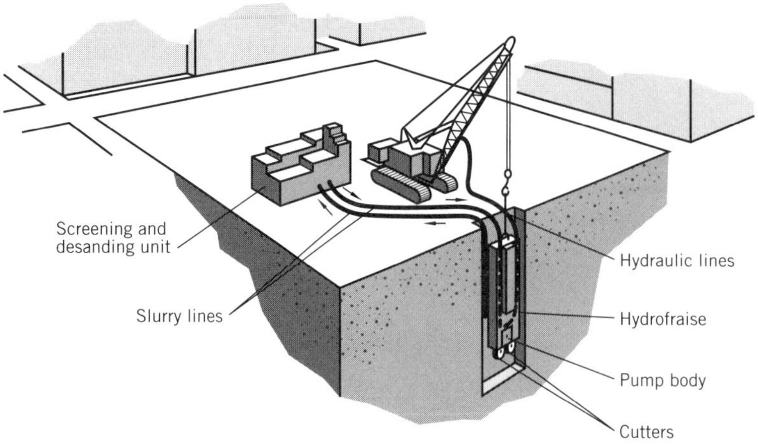

Figure 18.6 Hydrofraise excavating a slurry trench wall.

(Figure 18.6). Cleaned or fresh slurry is pumped back to the trench. The hydrofraisse (Figure 18.5) is fitted with two inclinometers to precisely control verticality. These large machines are up to 80 feet high.

Hydromill machines have difficulty in penetrating through cobbles and boulders because the cutters jam. However, they penetrate through other hard strata with relative ease and are extremely fast under favorable soil conditions. They are equipped with special (pic) cutter teeth to provide a serrated vertical surface directly into the concrete of the adjoining panel for good joint characteristics.

Slurry

Sodium bentonite is used in the United States; most parts of Europe use calcium bentonite. The water has to be free of dissolved salts. The slurry should be mixed and stored on site for 24 hours to allow the bentonite to fully hydrate before use.

Bentonite is not always compatible with mix water or contaminated groundwater. It is incompatible if seawater, chlorides, or chlorinated organic contaminants are present. Where these conditions exist, attapulgite can be used as a substitute because it has a different mineral structure.

The bentonite content is typically 5%. A typical set of requirements are: Density 65–75 lb/cu. ft; viscosity of 32 seconds using the Marsh Cone Funnel; maximum filtrate loss of 15–30 cc using a standard filter press; pH between 6.5 and 10.5. The filtrate requirement measures the ability of the slurry to maintain a stable trench. If the pH exceeds 10.5, flocculation of the bentonite may occur. Before placing tremie concrete, the sand content in the slurry should not be more than 5%, as measured 5 feet above the bottom of the trench. On some sites, the sand-laden slurry in the trench is replaced with fresh slurry after the panel excavation is completed.

The slurry should fill the trench to within 1 to 2 feet of the top and should be 1 to 2 feet above the maximum groundwater level outside the trench. Loose sand and other debris on the bottom of the trench is cleaned out with an airlift (Figure 18.7).

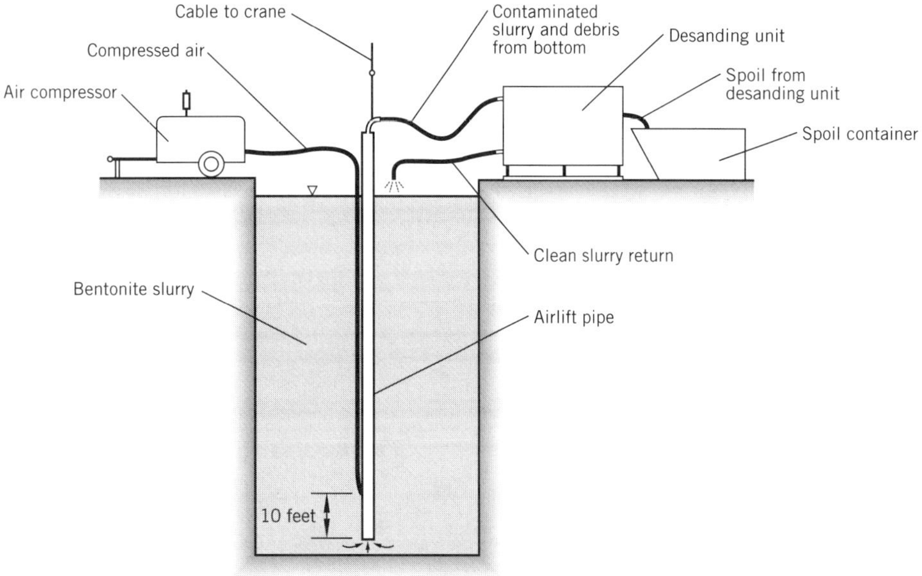

Figure 18.7 Loose sediment removal by an air lift (slightly modified from Schoenwolf and Dobbels, 1992).

Synthetic biodegradable polymer-based slurries are replacing bentonite slurries in some applications (Tallard, 1992). Environmental regulations are making it difficult to dispose of bentonite slurry spoil. By contrast, a polymer slurry can be reused many times until the end of the project. After separation of the solids and neutralization of the liquid pH, the slurry can be released without environmental harm. Handling, storing and processing the polymers is much easier and quicker than for bentonite. Since the sand fraction quickly settles out of a polymer slurry, there is less wear on the excavating equipment. There are several other operational benefits to using polymer slurry rather than bentonite, and it is likely that polymers will have increased future use in slurry wall construction.

Concrete

The concrete has to have good workability, and a slump of 7–9 inches is specified. The maximum particle size is around ¾ inch, and 28-day strengths are usually in the range of 3,000 to 5,000 lb./sq. in. Various other additives, such as plasticizers, air entrainment, fly ash, etc. are specified as needed for the site and working conditions.

The bottom of the tremie pipe should be 6 to 15 feet below the top of the fresh concrete (except at the bottom of the trench). Construction joints should be cleaned before fresh concrete is placed against previously placed concrete.

Normal practice is to withdraw the end pipes during the later stages of the concrete pour. They have to be lifted slowly and continuously, preferably just after the initial set. The placing of concrete is the most critical element of the entire operation. Enough concrete has be available to complete the entire panel as a continuous pour; horizontal cold joints are unacceptable for a concrete barrier.

Watertight Joints

The half round joint provided by an end stop pipe is intended to reduce leakage and achieve continuity of the wall. If carefully installed, it should be satisfactory for most landslide projects.

Specialized contractors have devised various end forms to allow waterstops to be incorporated into the joints. A good example is the CWS joint developed by Enterprise Bachy of France (Vanel, 1992) and illustrated on Figure 18.8 for a diaphragm wall construction. A cutoff wall uses the same technique. On the left, Figure 18.8(a), is the conventional way of withdrawing a pipe end stop. The pipe is pulled out lengthwise. Selecting the time to withdraw the pipe is difficult: if it is too early, the concrete will collapse; if it is too late, the adhesion may prevent it from being extracted.

The CWS end form, Figure 18.8(b), is extracted laterally without sliding. A special rail is temporarily sealed in the concrete of the previous panel; the rail acts as a guide for the excavating tool, either a grab or hydromill. After the excavation is finished, the excavating tool latches onto the rail and pulls it sideways to detach it from the earlier concrete panel.

The CWS device allows one or more waterstops to be installed between two adjoining panels (Figure 18.9). Two waterstops (as illustrated) is the most common arrangement. The plastic or rubber waterstops are wedged into the longitudinal grooves of an attachment to the end form. One half of the waterstop is concreted into the primary panel. The lateral extraction of the end form then uncovers the other half of the waterstop, allowing it to be concreted into the later (secondary) panel. If needed, a grout pipe can be included in the system to allow post-construction grouting (Figure 18.9b).

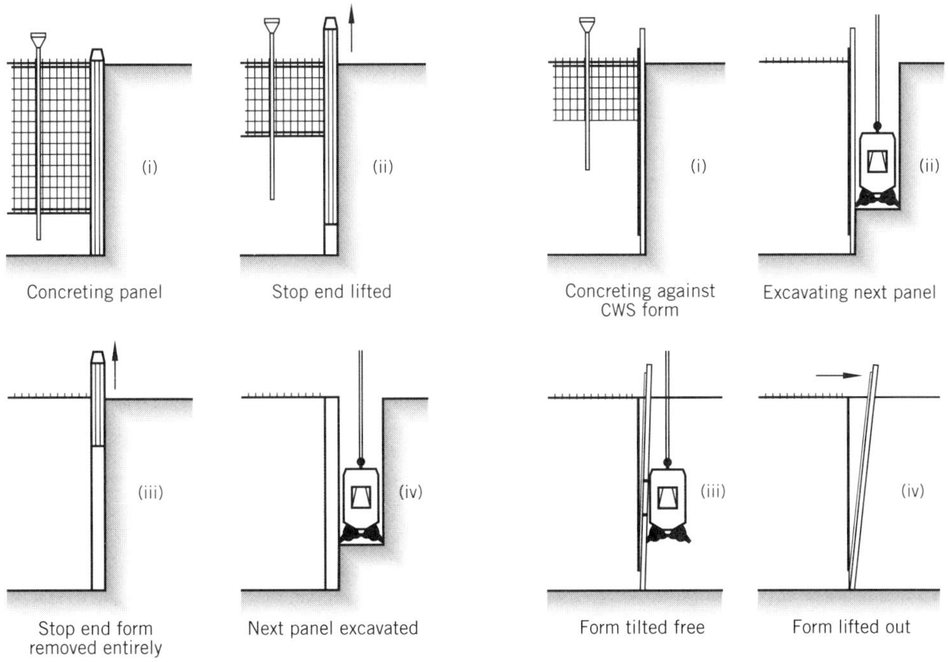

Figure 18.8 End construction:
(a) stop end pipe construction
(b) CWS end form to provide watertight joint between panels in a diaphragm wall
(after Vanel, 1992).

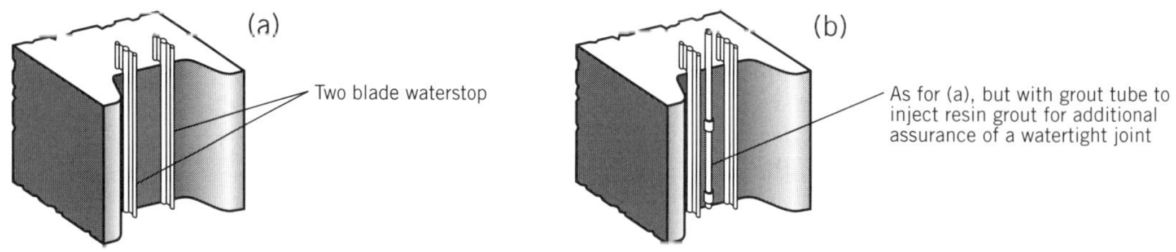

Figure 18.9 Rubber or plastic waterstops used to achieve a watertight joint between concrete panels (after Vanel, 1992).

Lined Walls

A seepage barrier's effectiveness can be enhanced by incorporating an *HDPE liner* in the wall. This has been accomplished in a variety of ways. A common method is to attach the liner to a rigid rectangular frame and lower the counterweighted frame into the center of a cement-bentonite mix inside the trench. This is difficult to accomplish in windy weather. The HDPE liner, with a permeability of about 10^{-12} cm/sec, typically is 100 mil thick and is encased by the concrete. As in the case of a conventional concrete wall, the joints between wall panels are a potential source of leakage to the barrier and need special attention in design and construction.

Alternative Slurry Trench Cutoff Walls

Three other types of cutoff walls are used and each provides more ductility and lower strength than a structural concrete wall (Millett et al., 1992). These alternative wall types are shown schematically on Figure 18.10. The first method, which produces the most ductile backfill, is a **soil-bentonite** mix (Figure 18.10a). It is usually excavated by a backhoe in a trench 5 to 7 feet wide. The soil preferably has a wide range of particle sizes including about 10–20% by weight of fines passing the No. 200 sieve. The bentonite content is 2–4% by weight. The soil-bentonite mix has a density 5 lb./cu. ft. higher than the density of the bentonite slurry and has a slump of 4–6 inches. The mix is initially tremied into the trench to produce a slope from the base up to one end of the trench. Then additional soil-bentonite mix is pushed into the trench by a dozer and slides down slopes of 5:1 to 10:1 (horizontal:vertical) under its own weight; free fall through the bentonite slurry should not be permitted. The coefficient of permeability achieved by the procedure has been measured to be on the order of 10^{-7} cm/sec (Millett et al., 1992).

The second method is to mix a **cement-bentonite** slurry and use it to support the sides of the trench during excavation (Figure 18.10b). Later, the cement sets and hardens to the consistency of a stiff clay. In this technique, the cement is added to the hydrated bentonite-water slurry. Permeabilities of less than 10^{-6} cm/sec are achieved.

A variation of these first two wall types is to mix bentonite, cement, and soil. In the United States, the Corps of Engineers (Sacramento, California District) has installed many miles of slurry walls by this technique to stabilize levees. The soil is usually the cohesionless materials excavated from the trench. The trench width is 2.5 to 3 feet, and end stops are not installed. Instead, when contractors excavate alternate panels, they dig into the ends of the primary panels and replace them with a fresh slurry mix.

The third method, **plastic concrete,** (Figure 18.10c), is a mixture of cement, bentonite, and aggregate placed by the tremie method through a bentonite slurry. The addition of bentonite to the concrete mix adds some ductility to what is essentially a concrete wall. The bentonite content is 4–12% by weight. A typical project (Erwin and Glenn, 1991) specified an 8-inch slump, compressive strength of 500-1,000 lb./sq. in. at 28 days, and a minimum permeability of 10^{-7} cm/sec. After performing laboratory mix design studies, this project used 55 percent coarse aggregate, 45 percent sand, and a cementi-

Figure 18.10 Four types of slurry trench cutoff walls.

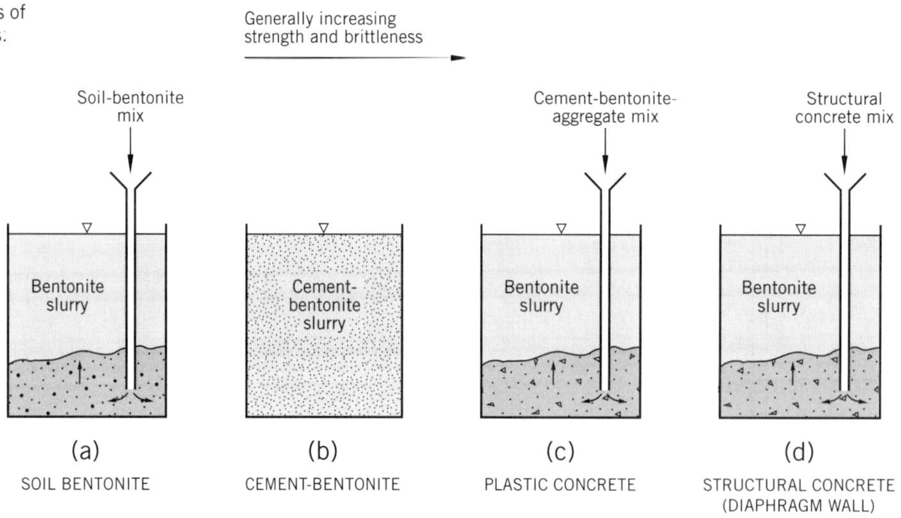

tious mix of 350 lb./cu. yd. The mix was composed of 70 percent cement, 20 percent flyash and 10 percent bentonite to obtain a compressive strength of about 800 lb./sq. in. at 28 days. Plastic concrete projects normally require laboratory mix tests to obtain the desired properties from available aggregate sources.

18.2 SLOPE LINERS

General Comments

Liners can be constructed on the slopes of containment structures (such as dams, levees, ponds, etc.) to reduce or eliminate seepage into the adjacent ground. Because this method creates an impervious barrier at the water source, no purchase of additional land is needed.

The liner materials are impervious but leakage from the *constructed project* is the main concern. A good analogy is to consider a plastic bag. If the plastic bag is filled with water, it can retain the water within it almost indefinitely. However, if a small hole pierces the bag, the water will be lost quickly even though the hole occupies only a very small percentage of the bag's surface area. Similarly, a liner treatment relies on the continuity of the entire impervious lining for effectiveness in preventing leakage.

Many types of liners are used on slopes and ditches including: (i) precast concrete slabs; (ii) cast in-place concrete panels; (iii) shotcrete; (iv) plastic (geosynthetic) sheets; (v) asphalt; (vi) half round plastic, metal, or concrete pipes; (vii) soil-bentonite layers; (viii) compacted clay layers; and (ix) combinations of the above.

The principal causes of leaks are:

- Cracks within the joints of the liner system
- Tree and shrub roots passing through the liner
- Burrowing animals piercing through the liner
- Accumulations of trapped gas (e.g., marsh gas) breaking the liner seals
- Buried structures or utilities passing through the liner and causing leaks at the connections between the structure and liner
- Loss of support from soil erosion or human interference
- Damage from floating debris

Unfortunately, it is almost impossible to make a large area of liner "leakproof." Modern landfill construction requires very stringent construction procedures for geosynthetic liners, and typically uses a double liner design to catch and remove the small amounts of leachate that passes through the upper (primary) liner.

Many liner types (panels, slabs, half pipes) require a flexible sealant to prevent seepage through the joints. The sealing materials: (i) may change properties over time (such as becoming more rigid); (ii) may separate at the liner/sealant under loading/unloading by water; (iii) may experience cracks developing from thermal contraction/expansion, freeze/thaw damage, etc. These fairly common problems during service require continual maintenance to keep leakage under control.

Other liner materials made with fewer or no joints also can be troublesome to maintain. Concrete or shotcrete may crack and deteriorate with time. Soil-bentonite liners require near-perfect mixing of the materials and very stringent placement; otherwise, "windows" of untreated (or undertreated) soil creates the hole-in-the-plastic-bag situation described earlier.

A buried sheet of very impermeable, flexible plastic provides the most reliable seal for a near-surface seepage barrier. The construction technique is described later in this section.

In summary, slope liners require good design and substantial quality control/quality assurance on site to obtain good results. A perfect seal from a single layer liner is nearly impossible to achieve, but a lesser result can be acceptable at many landslide sites if it achieves the objective of reducing the seepage to levels that do not cause slope instability. To confirm that the liner is performing satisfactorily, downslope monitoring with piezometers is essential at critical sites. A leak test, if appropriate for the situation, is also desirable. The extra costs of these observational checks may be substantially less than alternative forms of seepage barriers. Such pre- and post-construction monitoring should be considered part of the project cost.

Flexible Membrane Liners

A flexible membrane liner can be built from several types of material. Common materials in use include high density polyethylene (HDPE), polyvinyl chloride (PVC) and polypropylene (PP). Of these, HDPE has a low interface friction that limits its use on steeper slopes. PVC has good stretch capabilities but is not certified in the United States for potable water applications. Polypropylene fulfills both interface friction and potability functions but has a higher cost. In recent years, smooth-textured liners are being replaced with textured membranes to overcome low interface friction. Friction Seal and Gundline are two trade names for textured membranes. Some values of friction, attributed to Field and Stone (1995), are given on Table 18.1.

A liner design in common use is to combine a textured flexible membrane liner with a nonwoven geotextile. This combination sticks to each other (akin to a velcro joint) and

Table 18.1 Interfacial Friction Angles of Composite Liners*

Interface	Typical friction-angle ranges (degrees)
Nonwoven geotextile/smooth geomembrane	10–13
Geonet/smooth geomembrane	10–12
Compacted sand/smooth geomembrane	15–25
Compacted clay/smooth geomembrane	5–20
Nonwoven geotextile/textured geomembrane	25–35
Geonet/textured geomembrane	16–23
Compacted sand/textured geomembrane	20–35
Compacted clay/textured geomembrane	15–32

*Field and Stone, 1995

installers have to use a slip sheet (typically a woven geotextile) to align the two materials at the site. Once the panels have been aligned correctly, the slip sheet is pulled out and the components bond together. The materials can be obtained in very large rolls (e.g., 15 feet x 300 feet), thereby keeping the length of joints to a practicable minimum. The nonwoven geotextile can be placed on one or both sides of the membrane liner and generally acts as a protective cushion. It can also behave as a filter layer between the membrane and gravel backfill.

Liner Protection

Two examples of flexible membrane liners are shown on Figure 18.11. The liners need a cover to protect the plastic from ultraviolet light damage and damage from other sources. A clay cover, Figure 18.11(a), provides an extra layer of impermeable material and can be used where clay is available. However, a sandy clay that does not shrink on drying is preferable to a fat clay. In general, clay has good resistance to erosion from flowing water and is unlikely to crack or change properties if it remains under water. It is susceptible to instability on steeper slopes, especially from rapid drawdown.

Another choice is to use graded rockfill as shown on Figure 18.11(b). Here, the fine rockfill cover immediately above the composite liner is a well-graded 1-inch minus crushed rock obtained from a hard rock quarry and having less than 3 percent by weight of nonplastic fines passing the No. 200 sieve (see also Chapter 13, Section 13.4). The outer cover of rockfill is much coarser: a well-graded 14- to 2-inch size rockfill with not more than 5 percent passing the 2-inch sieve. The section shown on Figure 18.11(b) also has a thin leveling course of 1 inch minus crushed hard rock underneath the composite liner. This allows the liner to be placed on a smooth compacted surface and reduces the chance of tearing the fabric. However, it also puts a premium on the workmanship of the joint seals and other seals at the ends of the liner because any water passing through or around the liner is immediately picked up by this permeable layer and spreads to all parts of the underlying slope being protected. The alternative is to produce a very smooth subgrade, free of projecting stones and roots, by scarifying, raking, and compacting the native soils or by importing a clay fill for this purpose. This foundation soil can be placed by conventional earthwork procedures, including replacement of any soft spots in the foundation with firmly compacted soil.

Rockfill or clay placed *above* the membrane requires very careful field construction. Fill must be spread from *bottom to top* of the slope (Figure 18.12) to minimize the risk of tearing the liner. These cover soils should be spread by a wide-track dozer with a maximum ground pressure of 4.5 psi (for example, a Caterpillar D5 LGP dozer). The dozer must work up

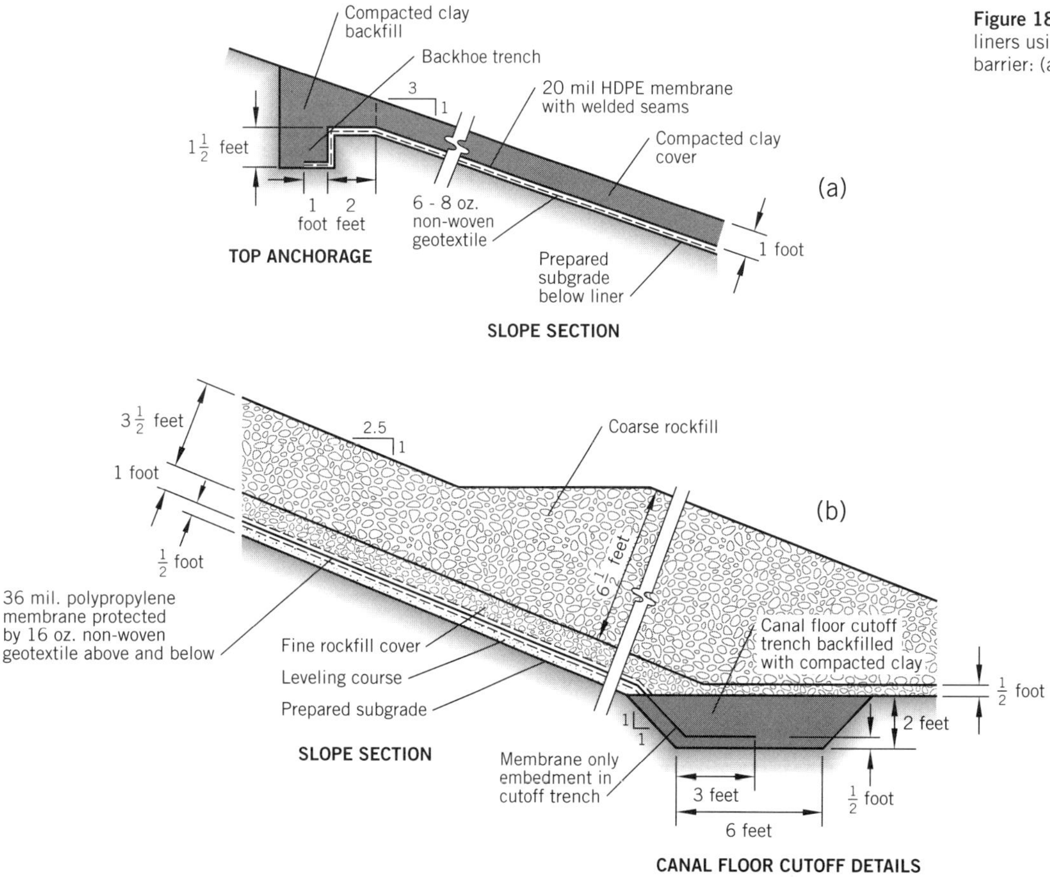

Figure 18.11 Typical designs for slope liners using a geomembrane as a seepage barrier: (a) clay cover (b) rockfill cover.

Figure 18.12 Slope liners. Fine rockfill being placed over a composite polypropylene / non-woven geotextile on a 3:1 canal slope. The small dozer spreads the fine rockfill from bottom to top of the slope with no turns permitted. Section shown on Figure 18.11(b).

and down the maximum gradient of the slope with no turns. If needed, it can be winched up steeper slopes by an anchor dozer positioned at the top of the slope. Lift thicknesses should be limited to around 24 inches for rockfill, 9–12 inches for 2 inch minus or finer soils. The thicknesses should be controlled by grade stakes and a spotter. In the case of clay fills, the moisture content and compacted density also have to be monitored by site inspectors. Rockfill needs minimal control—two passes of the dozer treads should be sufficient to provide nominal compaction.

Finally, a slope stability analysis is needed to check that the cover will not slip off the slope either along the soil/geocomposite interface or within the soil itself. If appropriate to the site, a rapid drawdown analysis is likely to be the most critical stability case (see Chapter 9, Section 9.15). The friction angle of the materials can be measured in a shear box, or a conservative value can be chosen from the published literature, such as Table 18.1. The stability analysis probably will control the design thickness of rockfill cover needed for the project. The rockfill gradation should be sized to resist the flow of water in canals or rivers (see Chapter 11, Section 11.2).

Sealing of Geocomposite Liner

Sealing of the joints is a key element of successfully constructing a water barrier using a plastic membrane. The manufacturer can perform shop welding to reduce the number of field welds. There are ASTM specifications covering membrane manufacture, properties, and field testing of seams that will not be presented here. Typically, the supplier provides experienced personnel on-site to weld seams, test for watertightness, and provide patches if needed to correct any observed deficiencies.

The fabric is placed as a continuous strip from top-to-bottom of the slope with no intermediate joints. Adjacent lengths of fabric are overlapped by 12 inches. During installation, sandbags or other heavy objects are used to hold the fabric in place prior to welding the seams using single hot edge welding procedure. The seams typically have a nominal 4-inch overlap and 2-inch seam. The edges have to be clean and dry before the work is begun.

Quality control and quality assurance tests are performed on all field seams. Nondestructive air lance tests (ASTM D 4437) are routine. In addition, qualification welds are prepared every four hours and are destructively tested for heat bonded seam strength and heat bonded adhesion strength by the contractor. The minimum requirements for these parameters are set by the appropriate ASTM standards. Seams that fail these tests are repaired and retested.

Limitations of Geocomposite Liners

Plastic liners can be used on low to moderate slopes using a textured flexible membrane. The liners cannot be placed during wet or windy conditions, so the technology is unsuitable for a typical landslide emergency situation. Although extensive field testing is usually performed to check the watertightness of seams, it is difficult to check the actual performance in most landslide prevention/remediation sites. However, it should be remembered that small amounts of leakage may be acceptable at some sites if the membrane serves to substantially reduce leakage into the adjoining ground. This desired result can be checked by a leak test and/or piezometer measurements in the ground downslope of the membrane liner.

Costs

The installed costs in the United States for various geomembranes (year 1998) are given below:

	COST PER SQ. FT.
High density polyethylene (HDPE)—60 mil thick	$0.50 to 0.75
Polyvinyl (PVC)—40 mil thick	$0.32 to 0.42
Polypropylene (PP)—36 mil thick	$0.65 to 0.95

These prices do not include buffering materials (e.g., nonwoven geotextiles) above and/or below the membrane or cover and supporting soil layers.

Bureau of Reclamation Study

The U.S. Bureau of Reclamation conducted a 10-year study in which 34 canal-lining test sections were constructed and evaluated over time (Swihart and Haynes, 2002). At the end of the 10-year period, 7 of the test sections had failed while the remaining test sections were described as being in fair to excellent condition. Each test section covered about 30,000 sq. feet. The test sections were grouped into the following subgroups:

- *Fluid-applied membranes.* Many test sections failed due to poor quality control and adverse weather conditions during late fall and early spring construction period (i.e., outside the irrigation season).
- *Concrete.* Excellent durability but random cracking lowered the seepage reduction to only 70 percent
- *Exposed membrane.* Excellent effectiveness (90 percent seepage reduction), but the exposed geomembranes showed signs of loss of physical properties in the liners, thereby reducing the estimated service life to around 10–20 years. Also vulnerable to surface damage.
- *Geomembrane with concrete cover.* This combination gave the best performance because the geomembrane provides the water barrier and the concrete protects the geomembrane from weathering and damage (animals, vandalism, etc.). Seepage reduction was around 95 percent. Construction cost was about $2.50 per sq. ft. The concrete cover is easily maintained and the geomembrane requires no maintenance. Service life is estimated to be 40–60 years.

18.3 GROUT CURTAINS

General

Techniques of permeation grouting can be broadly subdivided into chemical grouts and cement grouts (the latter used in preference to the term "cementitious"). Many mixes are available in each category and only a limited coverage will be provided here. In general, chemical grouting is more expensive than cement grouting and may have undesirable reactions with the soil and groundwater. However, chemicals are solutions that, with low viscosity, can penetrate into a finer gradation of soils than cement.

A grout curtain can be constructed by three different grouting techniques: slurry grouting, chemical grouting, and jet grouting. In *slurry grouting,* a cement-water mix (usually with additives such as fly ash, bentonite, or sand fillers) is injected under pressure into the voids of a gravel or sandy gravel. The particulates of the slurry infiltrate the void spaces and, on hardening, produces a conglomerate or sandstone which is relatively impervious. For larger voids, such as a boulder cluster or eroded pockets of soil, the grouting contractor can add fillers to create more bulk in the grout and lower the overall cost. For sands, microfine cement can be used to penetrate into smaller void spaces.

Chemical grouts permeate into the voids of sands and replace the pore water. The liquid chemicals combine to form an impermeable and stable gel.

In *jet grouting,* a probe equipped with high-pressure jets erodes into the native soils and mixes them with the cement slurry (where close to the jets) or permeates into the ground voids (where distant from the jets). This newer technique has gained rapid acceptance because it can penetrate into all soil strata and can handle layered strata better than the other two grout techniques. The final product of a jet grouting wall is similar to a concrete slurry trench wall.

Slurry, chemical, and jet grouting are performed by specialist contractors that have extensive knowledge of their use and limitations. Geotechnical consultants should discuss their proposed projects with such contractors before making a final decision to use these grouting techniques. Although cement and chemical grouting have been used for more than 50 years, the methods continue to evolve as improved drilling and injection equipment become available. There are also numerous textbooks and technical papers dedicated to research on the performance of grouting.

On major projects, it is advisable to test the proposed method with a pilot program. Although this cannot be expected to check the effectiveness of the barrier itself, it can help to finalize the proposed drilling and grouting techniques based on the actual subsurface conditions of the site and may avoid significant technical and contractual problems during the actual work.

One caveat that seems to apply to almost all grouting projects: the quantity of grout injected into the ground generally exceeds the engineer's estimate. Therefore, it is advisable to include a fairly generous contingency for this outcome when estimating project costs or when comparing grouting costs with alternatives at a conceptual level.

Construction of a Grout Curtain by Permeation Grouting

Both slurry and chemical grouts require injection of fluids into the ground from boreholes. The grout curtain is formed by drilling a line of holes at close spacing so that the grout from one hole extends to the grout of the adjacent hole, thus creating the curtain or wall of grouted ground. For a high level of effectiveness, the grout curtain needs to be complete, allowing no paths for seepage to pass through or below the

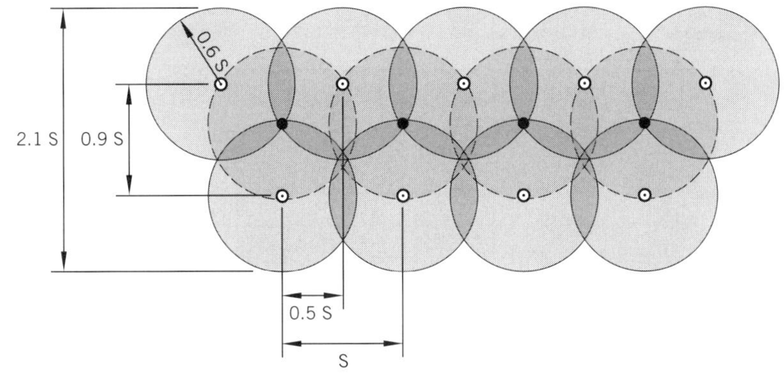

Figure 18.13 Overlapping grout holes to create an impervious grout curtain (example).

grouted area. To essentially eliminate leakage, a three-line system of grouting, as shown on Figure 18.13, is desirable. The two outer lines of grout holes are installed first (primary holes). After the grout has set and hardened, a line of secondary grout holes are put down between the original two lines. The secondary grouting infills any "windows" left by the primary grouting sequence. Should the grout "take" in the secondary phase be similar to the primary holes, it probably will require additional (tertiary) grouting to achieve a curtain seal.

Large quantities of grout injected into the ground usually means that grout is being lost to areas beyond the target area. It indicates that the ground is highly permeable or there is a significant crossflow of water, and grout is escaping before it has had sufficient time to set. Experienced grouting contractors can add fillers or decrease the set times to deal effectively with highly variable or highly permeable ground. However, the results generally are less reliable when these procedures have to be followed and the frequent delays in progress and additional materials will increase costs. A level of uncertainty is always present on grouting projects and needs to be communicated in advance by the consulting engineer to the owner. Grouting contracts frequently exceed the initial cost estimate and an appropriate contingency needs to be anticipated, depending on the perceived degree of difficulty.

Although the three-line grout curtain provides the likelihood of a complete water barrier, a single or double line of grout holes may be an acceptable and less costly alternative. These lesser options may be appropriate where: (i) the grout curtain is of shallow depth, such that any out-of-plumb holes are unlikely to affect the completeness of the grout curtain; (ii) the subsurface conditions are relatively uniform; or (iii) partial effectiveness of the grout curtain is an acceptable option. The benefit of specifying a single or double line of grouting is that the cost is lower than the triple line, and it is usually possible to return to the site and add another line of grout holes later should the need arise.

Seepage Barriers for Landslides

Grouting to create a seepage barrier for landslide stabilization (or prevention) can be considered in the following situations:

- Openwork gravel or highly broken hard rock
- Alluvial gravels and medium to coarse sands
- Landslide debris containing pockets and lenses of waterbearing sands and gravels.

Openwork gravel or *highly broken rock* can have large volumes of water and possibly crossflow. The challenge for grouting is to fill and seal the void spaces. The grout has to be injected at a low pressure and must set or gel rapidly so that it is not diluted or washed away within the waterbearing stratum. An example would be to use a cement-based grout with fillers and rapid set time.

Selection of an experienced grouting contractor is needed. Consideration should be given to limiting competition to one or two contractors that can demonstrate past success in similar conditions. An inexperienced contractor may lose valuable time and is likely to submit claims for extra compensation despite forewarning of difficult ground conditions.

Alluvial gravels and medium to coarse sands can be grouted using cement or chemical grouts. The choice of grouts will depend on the soil gradations.

Landslide debris containing *pockets and lenses of waterbearing sands and gravels* are similar to gravels and sands except that the ground may contain stiff clays or hard rocks intermixed with permeable soils. The grouting equipment must be able to penetrate through cohesive materials and/or rock fragments and achieve an acceptable rate of progress.

The permeability of the grouted mass controls the leakage through the seepage barrier. An old "rule of thumb" for grouting is that soils with a permeability of less than 10^{-4} cm/sec cannot be grouted by cement or silicate grouts. Although chemical grouts can penetrate into finer-grained soils, a permeability of 10^{-4} cm/sec or less should be sufficient for landslide seepage barriers. A review of three well-documented case histories (Terzaghi, Peck, and Mesri, 1996) led to the conclusion that the coefficient of permeability achieved in field-grouted sediments ranged from 10^{-4} to 10^{-5} cm/sec irrespective of the coefficient of permeability of the untreated sediments.

Cement-Based Grouting

Cement-based grouting is a long-established technique and is suitable for clean gravels and coarse sands. Type II cement, which is moderately sulphate resistant, is the most commonly specified cement for grouting (Henn, 1996). There are several

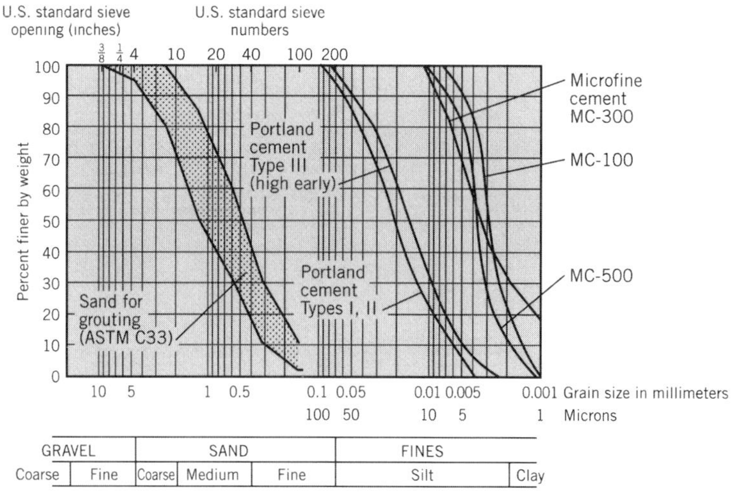

Figure 18.14 Grain size distribution curves for portland cement Types I, II, III, microfine cements MC -100, -300, -500 and sand for grout filler.

other choices of cement: Type I, Type III (high early strength) Portland cement, and microfine cements. Ordinary Portland cement will be termed "cement" hereafter. The microfine cements, defined as having 100 percent of the particles less than 15 microns (0.015 mm) are capable of permeating into soils of finer gradation and compete with chemical grouts. Microfine cements are sometimes combined with silicates to provide a quick-setting grout that can penetrate into smaller void spaces than conventional cements. However, microfine cement is much more expensive than Types I–III cements. As an alternative, the CE MILL process (Bruce et al., 1993) uses a colloidal wet refiner to reduce the particle size of conventional cement.

The gradations of various cements are shown on Figure 18.14. In cement grouting, the objective is to make cement particles enter the voids of the soil. This depends on the gradation of the grout and soil. Groutability ratio (GR) is defined as:

$$GR = \frac{D_{15}\ \text{Soil}}{D_{95}\ \text{Grout}} \qquad \text{(Mitchell, 1970)} \qquad \text{Eq. (1)}$$

This can be interpreted (Weaver, 1991) as GR > 24 : grouting consistently possible; GR < 11 : grouting impossible. Values of GR between 11 and 24 indicate possible but uncertain success in grouting.

A typical cement-based grout mix consists of cement, bentonite, and water. For coarser gravels, fillers such as sand or crushed rock may be added to reduce costs. Fly ash is also frequently added to cement-based grouts. Sand filler reduces the shrinkage of the grout after setting. The main drawback is that sand is very abrasive and increases wear on the pumps and other parts of the distribution equipment.

Other additives put into cement-based grouts (Henn, 1996) include:

- *Accelerators.* These shorten the set time. Calcium chloride in liquid form (1 to 6% by weight of cement) is added to the mixing water. There are many other accelerators. Sodium hydroxide can provide a rapid set; for example, a mixture of a 50% solution of sodium hydroxide (2% by weight of cement) can set the cement within one hour (Weaver, 1991).
- *Dispersants.* These prevent the cement particles from flocculating and lower the viscosity of the grout, allowing it to penetrate into gravel more easily.
- *Gas-producing agents.* These offset shrinkage in the grout as it sets and hardens. Finely divided metals (e.g., zinc, aluminum, and magnesium) react with the alkalis in cement to produce hydrogen. Special products on the market, such as Interplast–N, serve a multifunction purpose as expanders, fluidizers, and water-reducers. This product is added at 1% by weight of cementitious materials (cement plus fly ash, if used).

Cement-based grouting is often considered to be part science and part art. Certainly there is a need for some flexibility on site to change techniques according to the ground conditions actually encountered, and these conditions cannot usually be fully anticipated from a set of borings in a conventional site investigation. In some cases, owners and their engineers prefer to use a performance specification rather than a methods specification. Another option, especially recommended for larger projects, is to begin with a pilot program to determine the more effective grouting techniques for the site.

A typical grouting system is shown on Figure 18.15. It is advisable to circulate the grout rather than use one-way injection. Water : cement ratios typically begin at 3 or 4 : 1 but are usually reduced to around 2 : 1 or 1 : 1 in gravels. The grouting pressure should not be excessively high (generally less than the overburden pressure in soils); high pressures may fracture the ground rather than forcing the cement into the gravel pores. To avoid this situation, uplift should be continually checked, especially at shallow depths. Quantities of grout and pressure in the lines should be monitored and recorded. Data acquisition systems are available to measure these items automatically. An inspector with prior grouting experience should be on site full-time.

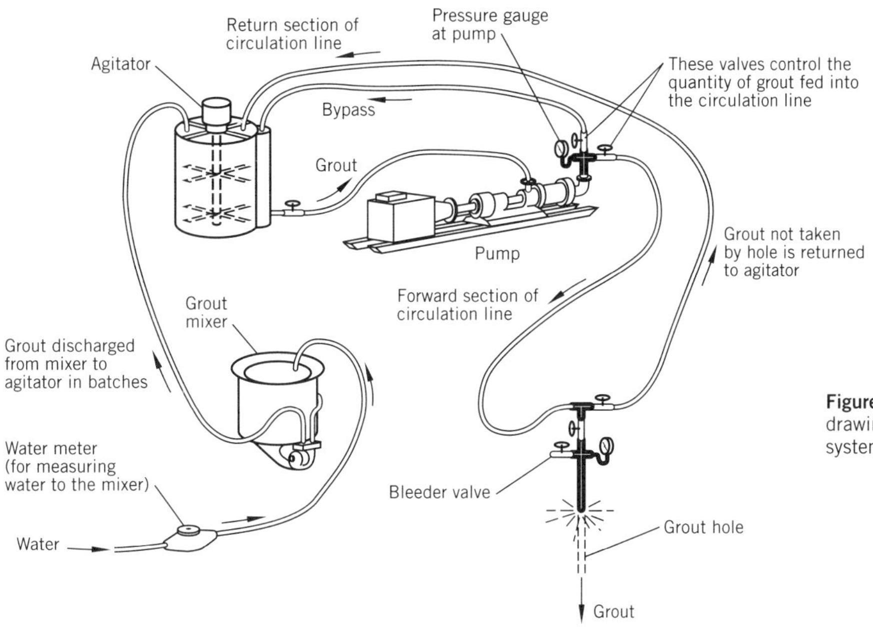

Figure 18.15 Schematic drawing of a field grouting system (after Houlsby, 1990).

The sleeve-port grouting system can be used with cement-based gravels. This technique allows isolated sections of the hole to be selectively grouted and also permits later grouting at higher pressures, if appropriate. The sleeve-port grouting system begins by drilling a slurry-supported hole of about 6 inches diameter. The PVC grout casing is 1 to 3 inches outside diameter and typically has one sleeve port per 2 to 3 feet of length. A weak cement–bentonite grout replaces the drilling mud. Using a double packer device (Figure 18.16), a high pressure of about 300 lb./sq. in. is used to open up the rubber sleeve over the port and create cracks through the annulus of cement grout. Once grout begins to move into the soil the grout pressure is reduced to 30 to 75 lb./sq. in. for penetration of grout into the gravel voids.

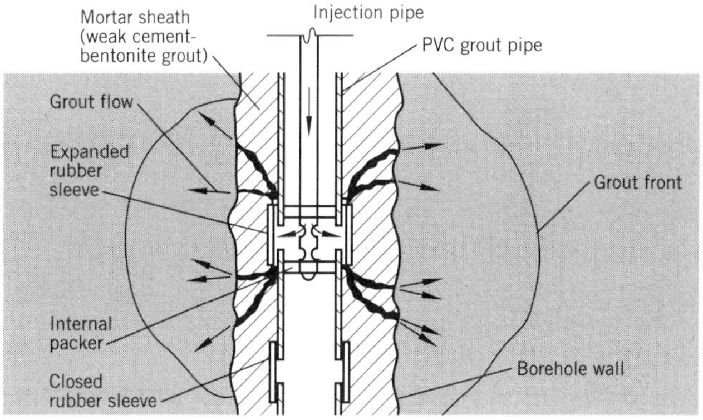

Figure 18.16 Schematic drawing of sleeve-port grouting.

Chemical Grouting

Chemical grouts are a mix of reactive solutions that form an impermeable gel within the void spaces. They include sodium silicates, acrylates, lignins, urethanes, and resins. The most common chemical grout is gel derived from sodium silicate (Na_2SiO_3) and will be the only chemical grout discussed below. It is considered safe, compatible with environmental concerns, and provides good results in sands and other cohesionless soils. It is also less expensive than other chemical grouts for most situations.

The soil gradations that can be grouted by chemical grouts are shown on Figure 18.17. The sensitivity of chemical grouts to the silt fraction should be noted.

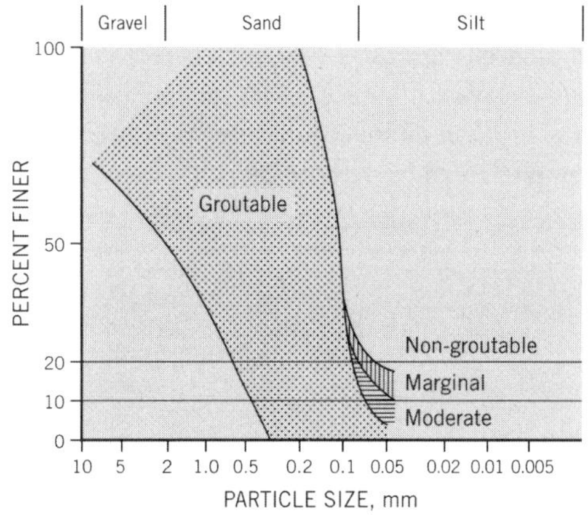

Figure 18.17 Soil gradations for grouting with chemicals (after Karol, 1983).

Grout Design

A sodium silicate grout usually has three components:

- *Sodium silicate.* The base material—typically 20–50% by volume.

- *Reactant*. Grout component that reacts chemically with the base material to form a gel—most commonly ethylene carbonate formamide (often simply referred to as "amide") makes up 2–10% by volume.
- *Accelerator*. Sodium aluminate, sodium bicarbonate, or calcium chloride. One of these chemicals is used in a mix to control the gel time. The sodium aluminate is better for shutting off seepage (U.S. Corps of Engineers, 1997) and can be used in acidic soils.

If the grout has to be able to resist wet/dry or freeze/thaw cycles, the concentration of sodium silicate needs to be at least 35% by volume.

The *gel time* is the time that elapses between mixing the grout fluid and the formation of a gel. Rapid gel times are needed: (i) in highly pervious soil where there is groundwater flow or (ii) for filling voids. The gel time can be shortened in several ways: decreasing silicate concentration, increasing reactant concentration, increasing accelerator concentration, adding cement as a filler (for high permeability soils or large voids), increasing temperature, and/or the presence of soluble salts in the groundwater (chlorides, phosphates, sulphates, etc.). Gel times can be varied from a few minutes to several hours.

In difficult ground, such as a crossflow or large voids, the grout should ideally gel at the moment when all the grout has been pumped through the system. In practice, some contingency time has to be allowed so that the gelling does not clog the pumps and hose; otherwise, the grout would have to pass through prematurely gelled soil. Attempting to grout highly permeable ground can be a difficult and frustrating experience.

The *viscosity* of the grout is a property measuring its resistance to flow. The more viscosity increases, the harder it becomes for the grout to penetrate into the soil. Thus, for a given gel time, the distance traveled by the grout from the injection point is reduced. Viscosity of grout fluid is measured in centipoise (cP), with one centipoise being the viscosity of water. For example, grout with a viscosity of 2cP will penetrate at half the rate of water at the same injection pressure (i.e., such a grout requires double the pressure to achieve a penetration rate equal to that of water).

Table 18.2 Groutability of Soil with Chemicals*

Coefficient of permeability of soil k (cm/sec.)	Groutability
$> 10^{-1}$	Use chemical grout with fillers (or cement slurry grout)
10^{-1} to 10^{-3}	Suitable for all chemical grouts
10^{-3} to 10^{-5}	Difficult to grout with grout viscosity above 10cP

*Karol, 1983

In sodium silicate grouts, the viscosity depends on the silicate concentration in the grout mix. At a concentration of 40% by volume (a fairly typical grout mix), the viscosity is 4 to 6cP and increases rapidly at higher concentrations (U.S. Corps of Engineers, 1997).

Finally, another important property of chemical grouts is *syneresis*, which is the contraction of the gel over time due to loss of liquid. Chemical grouts require year-round saturation to prevent shrinkage. For durability, the silicate should be at least 35% of the liquid volume.

Some general guidelines for the soils that can be successfully grouted with chemicals are given in Table 18.2.

Field Procedures

The sodium silicate and reactant/accelerator are usually mixed in separate tanks and use separate pumps for delivery (shown diagrammatically on Figure 18.18). This provides better control of the gel time and prevents the grout from gelling prematurely in the tanks, pumps, or delivery hoses. Meters are used to control the mix. The pumps are either progressive helical cavity (Moyno) pumps or piston pumps. The positive-displacement pump moves the grout at a constant rate of delivery and pressure, and has less pulsation than piston pumps. Reversible air motors can be used to unclog plugged lines.

Grouting is usually performed using packers to seal off parts of the hole. The simplest method is to use a single packer, which is inflated by air or mechanical means, and inject grout under pressure into the hole below the packer. The packer level can be raised successively and grout injected from bottom to top of the hole. A second option is to use two packers and inject grout between the packers.

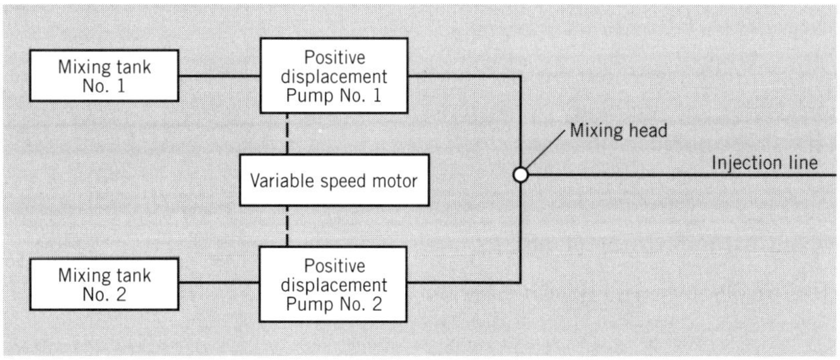

Figure 18.18 System for preparing chemical grout for injection (after U.S. Corps of Engineers, 1997).

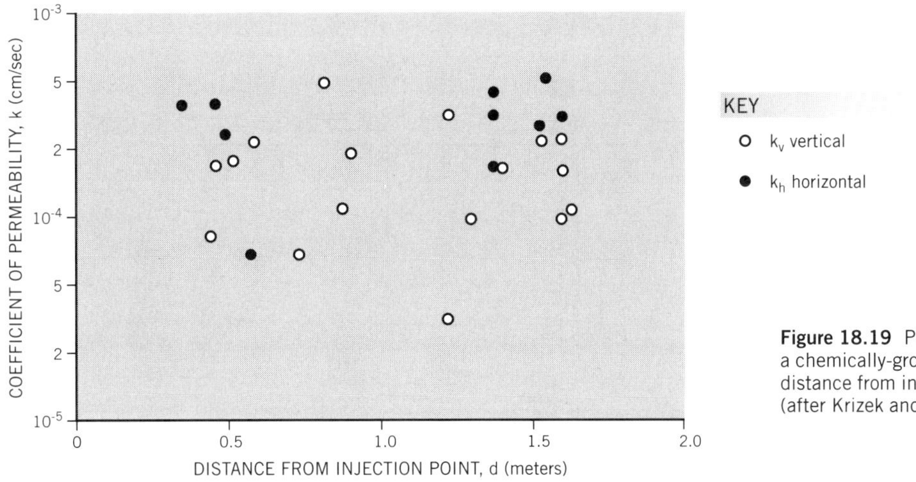

Figure 18.19 Permeabilities of a chemically-grouted sand vs. distance from injection point (after Krizek and Spino, 2000).

To grout in collapsible soils, the hole can first be plugged with a fast-setting grout. Next, the grout plug is drilled to accommodate the packers and the packer is installed to allow the chemical grout to be injected.

Another common grout technique is to use a sleeve-port pipe (termed *tubes à manchette* in Europe). The perforated PVC pipe has rubber sleeves around the perforations that act as one-way valves during grouting (Figure 18.16). The PVC pipe, with pre-set sleeved intervals, is first grouted into the hole with a weak cement or cement-bentonite grout. The packer and grout pipe device is inserted into the PVC pipe and two packer seals straddle one of the sleeved openings. A high pressure (e.g., 300 psi) is applied which cracks open the cement grout, after which the pressure is lowered and chemical grout is injected into the surrounding soil through the cracks. The grout pipe and packers can then be moved to another level in the PVC pipe and the process is repeated. The sleeve-port pipe is a flexible arrangement that can allow, if desired, multiple injections (at higher pressures or less viscosity) through the same port.

The spacing of grout holes ranges from 1½ to 8 feet. In general, faster gel times require closer spacing of holes. The spacing of injection ports in the grout pipes range from 1 to 3 feet. Baker (1982) states that the volume of chemical grout Vg can be calculated from the formula:

$$V_g = V_z\,(nF)\,(1 + L) \qquad \text{Eq. (2)}$$

where V_z = total volume of design grout zone
n = soil porosity
F = void filling factor (typical range 0.85 to 1.0)
L = grout loss outside the design limits (typical range 0.05 to 0.15)

This assumes that the soil within the treated zone can be permeated by the chemical grout. Silt and clay strata should be excluded from the calculation. However, in highly variable strata, grout volumes can deviate from the preliminary estimate by a wide margin. Owners frequently complain that engineering consultants underestimate the grout quantities and there is a corresponding cost overrun. The best advice is to make the volume calculations and then add an appropriate contingency (which may be as high as 50% of the target volume) to allow for unanticipated ground conditions and/or grout take.

Achievable Field Permeability Example

Krizek and Spino (2000) studied the properties of a field-injected sand from Odenton, Maryland, which was excavated and cut into chunks, then cured for 17 to 24 months before testing in the laboratory. The sand was a relatively homogeneous stratum with an effective grain size (d_{10}) of 0.25 mm, coefficient of uniformity from 1.6 to 2.1, water content of 11 percent, and 7 percent by weight passing the No. 200 U.S. sieve; the average void ratio of the adjacent untreated sand was 0.68. The permeability of the in-situ deposit was not measured but, from various empirical relationships, it was estimated to be on the order of 10^{-2} cm/sec. The injected grout consisted of 50% sodium silicate, 5% ethyl acetate and 5% formamide by volume.

The permeability of the grouted sand was measured in the laboratory on 27 specimens (Figure 18.19) and ranged from 0.4 x to 5.0 x 10^{-4} cm/sec. Thus, the permeabilities of the grouted sand specimens were almost two orders of magnitude lower than the empirically inferred permeability of the ungrouted sand. There was no apparent change in permeability at increasing distances from the injection point for a distance of 5.3 feet (1.62 m). After these tests, the grout was dissolved in sodium hydroxide and it was found that the grout content was nearly constant throughout the grouted mass, although more scatter was observed beyond 4 feet from the injection point. The studies showed that the chemical grout occupied 65% ± 9% of the available voids within the sand.

Jet Grouting

Techniques

A high-energy water jet erodes the soil and mixes it with the cement slurry, leaving a hard, impermeable soil-cement (soil-

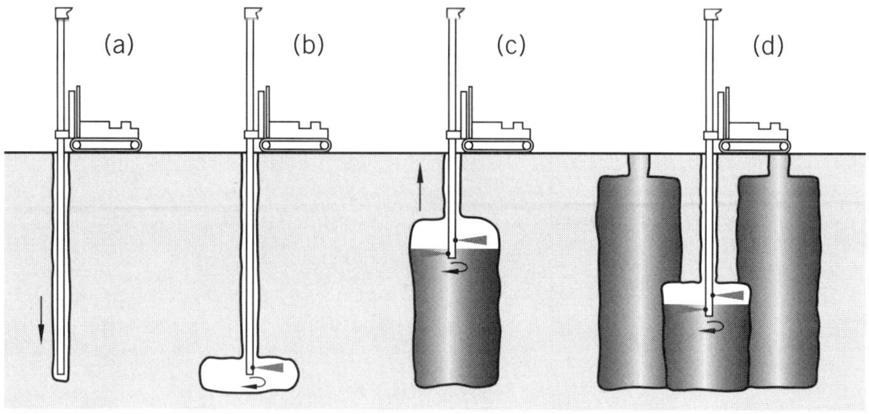

Figure 18.20 Jet grouting process:
(a) drilling to base of treatment zone
(b) start of erosion with water jet
(c) constructing soilcrete column
(d) alternate treatment panels to construct curtain wall or foundation of treated ground
(courtesy of Hayward Baker, Inc.)

crete) column to replace the original soil. Grouting takes place from bottom to top of the hole.

The procedure is shown schematically on Figure 18.20. A drillhole, typically 6 inches diameter, is made into the cohesionless soils using drilling mud to stabilize the hole. After reaching the full depth of treatment, grouting occurs through two diametrically opposite holes near the base of the grout pipe.

The spray comes out of the nozzles as a planar sheet of grout. Normally, the grout pipe is rotated as it is raised towards the ground surface and makes a cylinder of soilcrete. To construct a seepage barrier wall, the jet pipe is lifted without rotation and sprays within the plane of the desired wall. For this application, the jet holes are typically 4 to 5 feet on centers. Alternate holes along the wall are drilled and grouted first; the intermediate holes are grouted later. Using this technique (Figure 18.20d), there is usually no doubt that a solid wall has been created. The typical thickness of a wall is 1 foot.

Grouting pressures typically range from 6,000 to 7,000 lb./sq. in. The grout pipe is 4½ inches diameter and up to 120 feet long. It is usually impractical to change the pipe lengths during high-pressure grouting, so an appropriate size of crane is needed to lift the grout pipe. During grouting, the grout pipe is lifted in 6- to 12-inch increments and is held for 10 to 20 seconds at each level.

There are three jet grouting systems in use: single rod, double rod, and triple rod grouting (Figure 18.21). In *single rod* grouting, grout is pumped through the hollow rod and exits horizontally at a speed of about 650 ft/sec (445 mph). This erodes gravels for a radial distance of 1 to 2 feet and mixes the grout with the soil. The *double rod* system has separate supplies of grout and air down concentric nozzles. The air shrouds the grout jet and increases erosion. The radial distance of erosion is about 1½ feet in medium to dense soils and more than 1½ feet in loose soils. Double rod is the system most commonly used. The *triple rod* system sends a combination of air and water to erode the soil and grouts through a separate and lower nozzle at a lower velocity (Figures 18.20, 18.21). This provides a higher quality soilcrete. The radial distance of treatment for this system ranges from 1½ to more than 2 feet. Greater widths of soilcrete have been achieved by the SuperJet process using cement slurry and air.

In permeable soils, jet grouting has two effects. Close to the pipe, the high pressure and velocity breaks up the ground, mixing the sand and gravel with the cement grout. Further

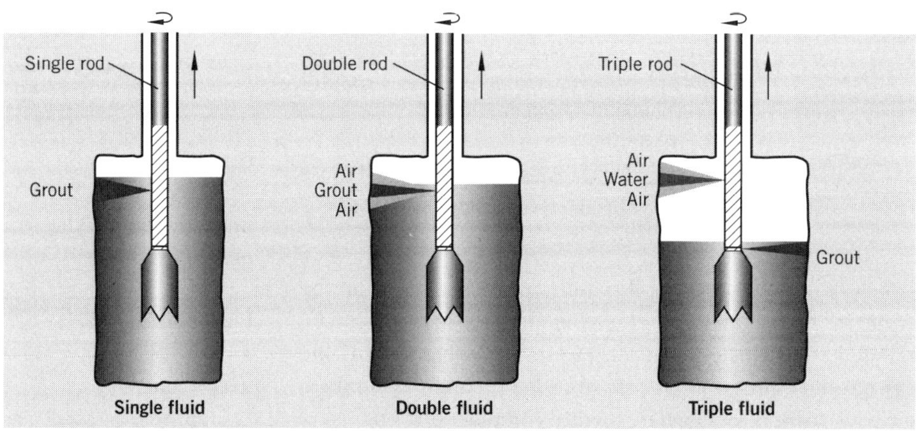

Figure 18.21 Three grouting methods used in jet grouting (courtesy of Hayward Baker, Inc.).

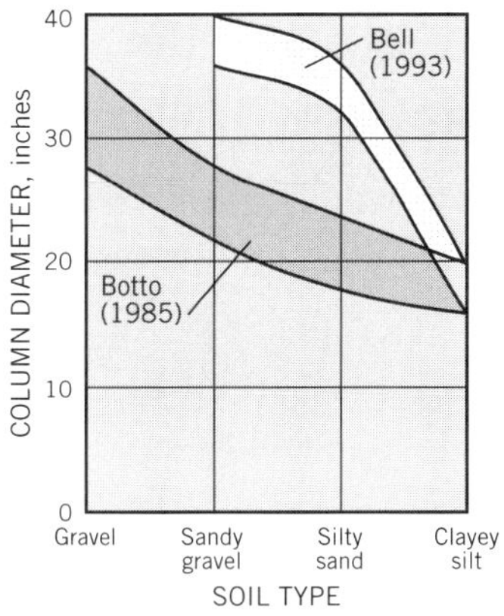

Figure 18.22 Soilcrete column sizes in different soil types.

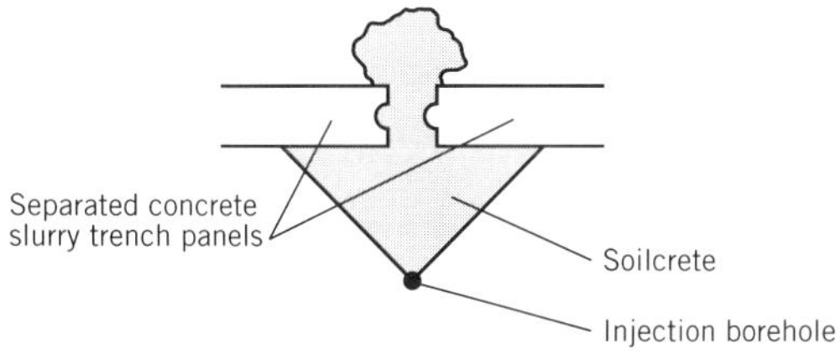

Figure 18.23 Jet grouting used to seal a separated joint in a concrete slurry trench wall.

away, the grout permeates into the pores of the gravel without destroying the original gravel structure.

The diameter of the soilcrete column depends on the soil type, being larger for gravel than clay. Some reported data is shown on Figure 18.22 (Croce and Flora, 2000).

The design strength of soilcrete is usually an unconfined compressive strength of 3,000 lb./sq. in. after 30 days. Rock core samples are taken to verify that the strength has been attained.

Jet grouting can also be used to seal leaks on existing underground structures. For examples, it can seal sheetpile walls or separated joints in a concrete slurry trench wall (Figure 18.23).

The principal advantages of jet grouting are: (i) it is effective in almost all soils except openwork gravels; (ii) it can cut through layered strata; (iii) it is relatively fast; and (iv) it requires no further maintenance. Drawbacks of the system are: (i) it generates large amounts of spoil, typically 80–100% of the actual grout treated, which has to be removed from the site; and (ii) it requires large quantities of cement.

Approximate Construction Costs

Typical costs for single fluid jet grouting in the western United States (year 2000) are:

Mobilization costs	\$15,000 to \$25,000
Cost of treated ground	\$150 to \$250 per cu. yd.

18.4 SOIL MIX WALLS

Technique

A cutoff wall is achieved by augering into the native soils and intermixing a bentonite cement grout (and other additives) with the soils to form soil-cement columns. Typically, two, three, or four columns of treated soil are formed simultaneously, and overlapping columns create a continuous wall. The technique competes with slurry trench walls but requires less excavation. The augered holes are supported at all times by the soil-cement admixture.

Deep soil mixing has several applications in geotechnical engineering. They include seepage barriers, retaining walls, foundation treatment, liquefaction mitigation, in-situ reinforcement piles, and environmental remediation. In appropriate circumstances, these applications provide competitive alternatives to concrete piles, conventional grouted soils, gravity walls, etc. Most of this section describes the soil mix technology in general terms that are applicable to all uses. At the end, an example is given of the design and construction of a seepage barrier built by deep soil mixing. In Chapter 23, there is a brief discussion of deep soil mixing for liquefaction mitigation.

Development of Deep Soil Mixing

Soil-cement walls are one application of a rapidly evolving technology termed deep soil mixing. Although the concept has been attributed to earlier work originating in the United States, impetus came from Japanese contractors beginning around 1985 (excluding a few earlier projects). U.S. specialist contractors began their own developments of the technique around 1992. These methods now are used throughout the world, but especially in southeast Asia and Scandinavia.

All the systems are proprietary. To a high capital-intensive development cost must be added continuing maintenance and upgrade costs. Therefore, site mobilization costs for deep soil mixing are high, and a project has to be sufficiently large for the technology to be competitive with alternative methods. There are other limitations, as discussed later. However, the large number of deep soil mixing machines—there were at least 24 different methods available in the year 2000—make it very difficult to compare the various offerings available from different contractors. Each system is sufficiently different that it may have an advantage at a particular site. Therefore, a performance-based specification is preferable to a methods specification for this type of work.

Deep Soil Mixing Technology

The soil mixing equipment typically comprises 1 to 4 hollow stem augers that drill into the ground simultaneously to create a panel of overlapping soil-cement columns (Figures 18.24, 18.25). The auger flights are short and discontinuous; in between the flights are paddles for mixing the reagents into the churned up soil. In some cases, the reagents are

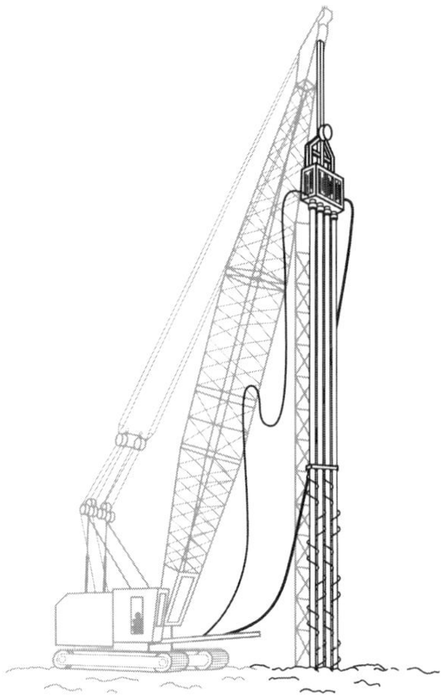

Figure 18.24 Sketch of a deep soil mixing machine with four augers mounted in a line.

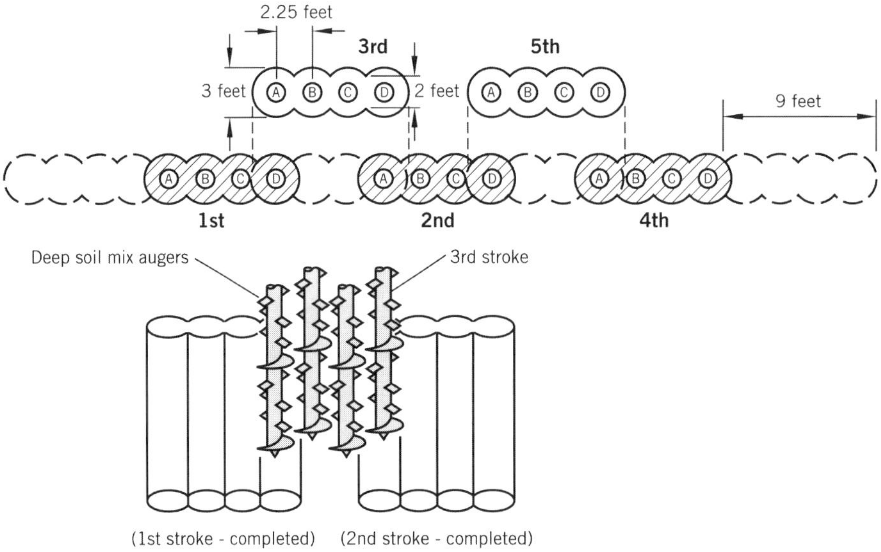

Figure 18.25 Installation technique for constructing a soil mix wall (after Bahner and Naguib, 1998).

injected through the walls of the auger where a double-wall system is used. However, in most systems, the cementitious reagents are injected into the soil at the bit during drilling and/or withdrawal of the auger. The auger flights break up the soil and the paddles blend the soil and reagent. As the augers advance, the mixing paddles continue the mixing process. The lower 10 feet of each stroke is double mixed (repeated) to improve homogeneity of the mixture. In a group of two or more augers, adjacent augers rotate in opposite directions. The directions of rotation are reversed when the augers are withdrawn.

The group of augers are held and guided by a large crawler crane (Figure 18.24). Vertical alignment of the augers can be controlled by accelerometers attached to the leads; these provide continuous feedback to the crane operator. Horizontal alignment can be obtained by a laser beam and target arrangement. The auger group can be set in any suitable alignment, such as perpendicular or parallel to the crane tracks. The parallel alignment allows a wall to be constructed along a narrow access road, such as a berm, dike, or top of a dam. In such cases, the augers can be aligned within or outside the tracks of the crane.

Deep soil mixing machines require high torque capabilities. The maximum depth of penetration depends on the soil types, number of augers in the group, auger diameter, and available power. Under favorable circumstances, depths of 140 feet have been reached. A more common range is 80–100 feet.

Augers typically are 2 to 3 feet diameter but can be as large as 5 feet diameter. In a multiple-auger arrangement, alternate panels are constructed. The gap is closed by a panel that redrills the outer columns of the previously constructed adjoining panels (Figure 18.25). This ensures continuity of the wall and requires retarders in the mix to prevent the cement from setting up too quickly.

The augers are rotated at about 20–30 rpm during drilling. The penetration rate is typically 2 ft./min. during penetration and twice as fast on withdrawal. However, such rates can vary significantly from these typical values, depending on the technology and ground conditions.

Portland cement is the most common reagent used in deep soil mixing. The resulting soil-cement column is sometimes referred to as *soilcrete*. Cement-bentonite slurry is used in walls designed for seepage cutoff. The usual practice is to perform pre-construction test mixes in the laboratory to obtain a design mix for the particular soil conditions at a site. An advantage of the deep soil mixing process is that it causes vertical intermixing of the native soils and thus breaks up horizontal layering that may be present in-situ. In addition to cement and bentonite, fly ash, sand, furnace slag, dispersants, retarders, etc. may be included in the mix. A schematic layout of a batch plant for deep soil mixing is shown on Figure 18.26.

Grout is provided to the bit by progressive cavity (moyno) or positive displacement piston pumps. Flowmeters measure the quantities being used.

Deep soil mixing uses the native soils as part of the ground improvement and thus generates much less excavated waste than other techniques, such as slurry trench walls, that totally excavate the native soils. The swell factor ranges from about 10% for sandy soils to as much as 40% in clays for which the proportion of reagent in the soil is higher. The excess soil, a soil-cement slurry mixture, has to be hauled off site. A common procedure to avoid this requirement is to dig a shallow containment trench along the top of the cutoff wall alignment. The trench holds the surplus soilcrete. Excavated soil from the trench usually can be used for other site construction purposes.

Deep Mixing Techniques

As previously stated, there are numerous proprietary installation techniques in use. Bruce (2001) provides a good summary. These techniques can be broadly differentiated as:

- *Wet methods.* The reagent is supplied to the soil as a slurry or grout. This is the most common procedure in the United States and has been already described.
- *Dry methods.* The reagent is air-fed to the bit as a dry powder and allows pore water within the soil to hydrate the reagent. Commonly used in Scandinavia to treat soft clays with lime to form soil-lime columns.
- *Jet mixing.* The **wet** jet mixing procedure is similar to high pressure jet grouting methods; several variations are available. One of the more popular techniques is termed SWING (Spread Wing) and was developed in Japan. With the blade retracted, the 2-foot dia. pilot hole is drilled to the base of the ground to be treated. The blade is expanded and either rotary mixing (up to 6-feet diameter) or air jetting (up to 12 feet diameter) is performed. The benefit of the retractable blade is that selected depths can be treated to produce a large diameter column. The column size is controlled by the applied pressure.

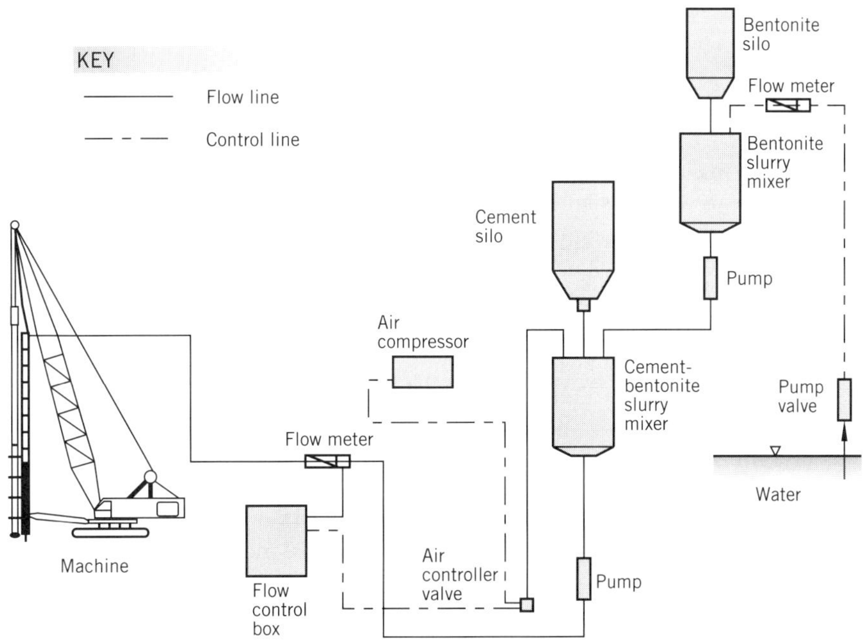

Figure 18.26 Schematic layout of a deep soil mixing batch plant (after Bahner and Naguib, 1998).

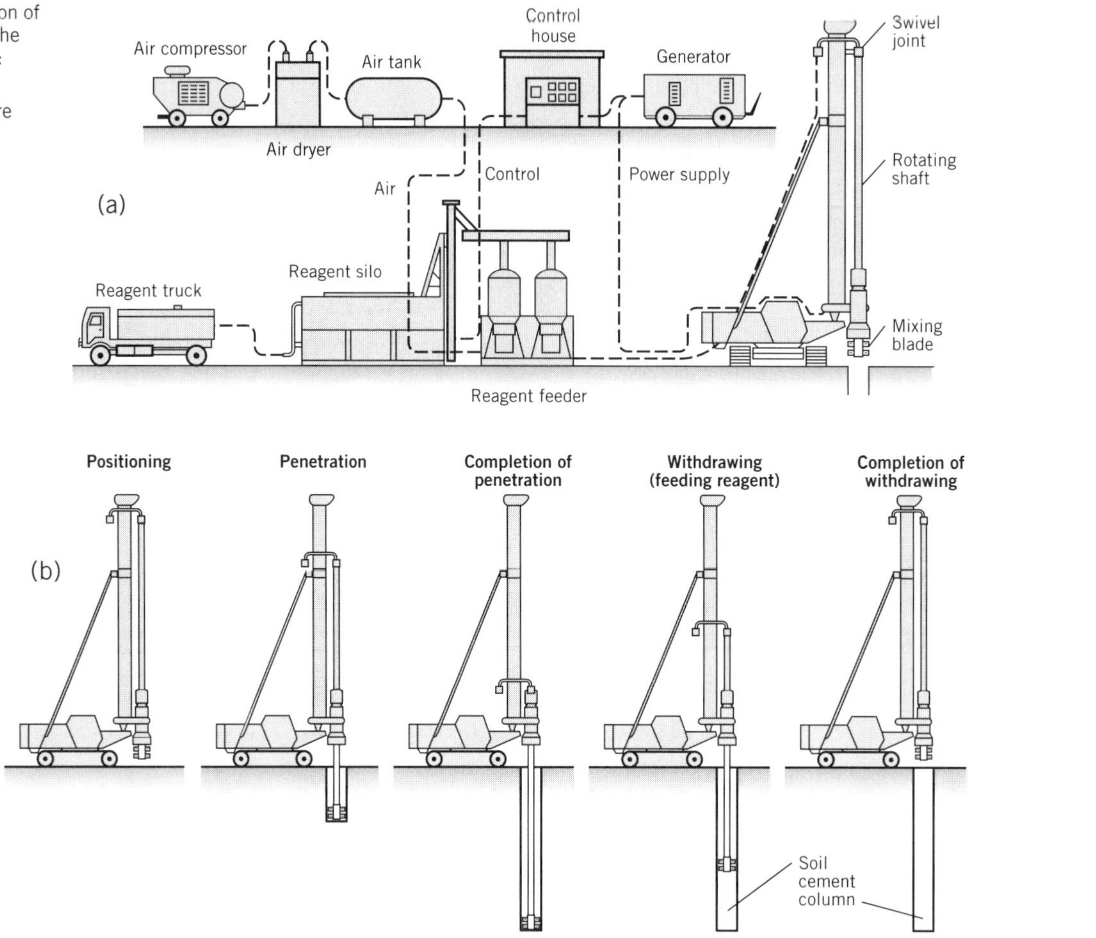

Figure 18.27 Construction of soil-cement columns by the Dry Jet Mixing procedure:
(a) equipment
(b) construction procedure
(after DJM Association of Japan, 1993)

The **dry** jet mixing procedure is illustrated on Figure 18.27. The hole is drilled using compressed air to avoid clogging. During withdrawal, dry reagents are injected into the soil during reverse rotation of the mixing blade. The system normally uses cement, but lime can be used in clays. The machines have one or two shafts and produces small amounts of spoils.

Suitable Soil Conditions for Deep Mixing

The ideal soils for deep mixing applications are *loose to medium dense silts, sands and gravels.* These are typical ground conditions requiring treatment to either reduce seepage or prevent liquefaction, the two uses described in this book. Cohesionless soils generally require significantly less reagent volumes than clays.

Boulders present some difficulties and, if abundantly present, may render treatment impossible as a practical matter. Occasional boulders have been treated in several ways. The most common procedure is to predrill the wall alignment at close intervals (e.g., 3-foot dia. holes at 4-foot centers) and break down the boulders. Another option is to excavate a test pit, remove the boulder, and backfill with sand. A third option is to construct a bypass panel that connects back into the main wall and does not jeopardize continuity of the barrier.

A similar problem may be encountered with *rubble fill,* which is often near the surface in urban environments. Large concrete fragments may be both difficult to remove and highly permeable.

Stiff and highly plastic clays require considerable torque for an auger to penetrate into the soil. Plastic clays are difficult to break up and blend for uniform mixing, and generally produce larger quantities of spoils than cohesionless soils. Pre-augering can help to churn up the clay before treatment is undertaken. Pre-augering also can be applied to very dense sands and gravels.

Soils with high acidity ($pH<4.5$) and high organic content ($> 4\%$) are also unsuitable for mixing with cement.

Despite concerns for drilling through boulders, obstructions, and stiff clays, deep mixing specialist contractors have been able to overcome difficult situations. Yang and Takeshima (1994) describe case histories involving construction of deep mix cutoff walls through: (i) dense to very dense sand and clayey silt underlain by a cemented zone of glacial till with cobbles and boulders; (ii) coral reef with cavities underlain by sand, gravel, cobbles, and boulders; and (iii) miscellaneous fill (gravel, cinders, brick, concrete, wood) underlain by organic silt, Boston Blue Clay, and hard silt. All three projects were successfully completed.

Advantages and Limitations of Deep Soil Mixing

Advantages:

- Can be applied to most waterbearing soils
- No construction joints in the wall
- No possibility of drilling wall collapse during construction
- Quantity of reagent per cu. yd. of treated soil can be tightly controlled
- Lesser volume of spoils than for full excavation techniques
- Relatively vibration-free
- Uniform treatment of layered soils
- No effect on soils outside the treatment area

Disadvantages:

- Not competitive for smaller projects due to relatively high mobilization costs
- Uses large specialized equipment
- Restricted competition in many locations
- Needs significant headroom
- Has difficulties with boulders, rubble fill, and other obstructions (e.g., utilities)
- Generally unsuitable for fat clays (mixing difficulties)
- A sealed connection to an existing concrete structure can be difficult

Design

Design begins with a thorough site investigation to observe the soil strata, compactness, groundwater levels, etc. A common procedure for a cutoff wall design is to develop three or more mixes of cement, bentonite, and any other additives for the soils encountered at the site. Samples are prepared and are tested for permeability and unconfined compressive strength (UCS) after moist curing for 7 to 28 days. A typical criterion is that the coefficient of permeability k should be less than 1×10^{-6} cm/sec. Strength generally is of less importance for a seepage barrier, but a UCS of around 100 psi may be specified.

A test panel is desirable before construction begins to provide confidence that the design parameters are suitable. As an example, the test panel could be 12 feet long and 12 feet deep and could be excavated on one side about one week after construction so that the integrity of the soil-cement barrier can be evaluated. If needed, adjustments in the mix can be made.

Bruce (2001) provides data on mix design specifications for most of the deep mixing procedures available at that time. There is considerable variation, but some general figures are:

Cement content: 100–400 kg./cu. m. of soil (i.e., 170–670 lb. per cu. yd. of soil—about 5 to 20% by weight)

Water: cement ratio : 0.8 to 1.5, but mostly around 1.0

Volume of grout: volume of soil: typically 20–50%

Unconfined compressive strength (after 28 days): 30–200 psi (clay); 200–700 psi (sand)

In general, sands require a lower percentage of cement than clays. These figures are for guidance only; specialist contractors should be able to provide more reliable ranges for their particular method.

Quality Control/Quality Assurance

Field records should include the mixing duration, grout mix, grout injection rates and pressures, total grout injected, mixing tool rpm during penetration and withdrawal, and any problems encountered due to subsurface conditions and equipment malfunction. Samples are taken from the augers to measure permeability and unconfined compressive strength. Bentonite slurry is tested for density and viscosity (Marsh cone).

Field testing of soil-cement columns include wet sampling during construction and large diameter rock coring of the columns 7, 14, or 28 days after construction. Many indirect tests are used, including CPT, pressuremeter, etc.

Examples of Deep Mix Cutoff Wall

Case History 1

Walker (1994) describes the design and construction of a seepage cutoff wall at Lockington Dam, near Piqua, Ohio. The wall, installed to a maximum depth of 21 feet, involved raising a pre-existing dam core to protect it from a Probable Maximum Flood event.

The base machine was a 175-ton capacity Manitowoc 4100 crane. The crane leads guided four hydraulically driven overlapping auger flights equipped with mixing paddles. The flight augers were 3 feet diameter, and the cement-based slurry was injected through the bits during penetration.

The contract specifications required a wall permeability of 1×10^{-6} cm/sec or less and a 6% by weight cement content in the soilcrete. A mix design demonstrating these properties had to be submitted prior to construction. In total, 12 different grout-soil mixtures were tested in the pre-construction laboratory program. Permeability was tested by the ASTM D-5084 (flexible wall triaxial) procedure. Samples were moist–cured for durations of 14 to 21 days prior to testing.

Although permeability is the key consideration for a cutoff barrier, the cement content was considered to be equally important at this project because construction was to be undertaken through partly saturated fill soils. It was reasoned that a grout with significant cement content would be able to withstand erosion should the dam ever be overtopped. Because increased cement content could affect permeability, desiccation tests were performed (including oven drying); these showed no deterioration or cracks. The selected grout mix had a cement : water ratio of 33% by weight, bentonite : water ratio of 4% by weight, and grout : soil ratio of 32% by volume. The cement content : soilcrete ratio was 5.9% by weight, the measured (lab) permeability was 10^{-7} cm/sec, and 7-day unconfined compressive strength was 52 psi.

The contractor first built a 3 feet wide by 3 feet deep trench along the wall centerline to contain the materials (about 20% of the wall volume) displaced during mixing. The dam crest was 24 feet wide and the leads were oriented so that the panels of treated soil were parallel to the crane tracks. The soils were mostly loose to medium dense gravelly sand and silty sand/sandy silt. Falling head permeability tests in the more pervious upper strata gave measured permeabilities as high as 4×10^{-2} cm/sec. Grout was provided by four Moyno L-10 grout supply pumps, one for each auger, and had to be pumped about 3,000 feet. The amount of grout injected was controlled by a technician using automatic flowmeters. Slurry densities and bentonite viscosities were measured by mud balance and marsh cone, respectively. Samples of the mix were taken at prescribed intervals within a wet soil-cement column by a specially designed tube; it was closed after sampling by a pneumatic device. As anticipated, a few of the permeability tests exceeded the 10^{-6} cm/sec target; the maximum value from 23 tests was 3.5×10^{-6} cm/sec, and about one-third were between 1.0×10^{-6} cm/sec and the maximum. This deviation from specification was deemed acceptable. Peak production was 7,000 sq. ft. of completed wall per shift. There were no significant unexpected field problems.

An interesting feature of this case study is that the cutoff wall had to be tied into the concrete walls of the spillway structure near the mid-length of the dam. The required water-tight seal at the soil-structure contact was achieved by using a jet grout to connect the soilcrete wall and concrete structure. The jetting was performed before the soilcrete in the adjoining panel had set. Using a pressure of 3,000 psi, the concrete face was scoured and, simultaneously, the closest soil-mix column was cut into by the jet. The same grout mix used for the deep soil mixing infilled the gap.

Case History 2

Day and Ryan (1995) describe a seepage barrier for an environmental site near Houston, Texas. The wall was 2,300 feet long and had an average depth of 52 feet. Subsurface conditions were 10 feet of sand fill over interbedded sand and silty clay layers. The design requirement was a final permeability of 1×10^{-7} cm/sec. Since strength was not required, only bentonite was used as an additive. After performing laboratory tests of various mixes, the results showed limited variation in properties because vertical blending of the soils produced an ideal base mix. The selected design was a 6% bentonite slurry with a grout : soil ratio by volume of 35%.

During construction, samples of the mixed soil were collected at depths of 20 feet and 40 feet of the fluid soil grout mix using a specially designed sampler. The samples were reconsolidated in the laboratory at pressures corresponding to the stresses in the ground. All permeability tests met the criterion of being less than 1×10^{-7} cm/sec.

CHAPTER

19

Retaining Walls

19.1 RETAINING WALLS OVERVIEW

Retaining walls can be constructed by either "top down" or "base up" methods. Both techniques are presented in this chapter, but emphasis is given to the "top down" construction methods for stabilizing landslides. In this technique, the wall is built starting from the ground surface of the landslide. The disturbed ground downslope of the wall is excavated after the wall support is in place. Therefore, the landslide equilibrium is not changed because the wall support replaces the soil support. "Top down" walls can provide some assurance that the landslide will remain stable during remediation work, which is beneficial to safety and adjoining property owners.

"Base up" retaining walls are conventional walls founded on firm ground and provide resistance through gravity or cantilever action. In landslide situations, there is usually a need to excavate some of the landslide debris at the bottom of the slide before the retaining wall can be constructed. As this undermines the landslide, it may require dry weather conditions or dewatering to lower groundwater levels within the landslide before construction of the retaining wall can begin.

Retaining Wall Systems

Retaining walls can be broadly separated into four categories:

1. *Gravity walls* in which the foundation of the wall provides the resistance to sliding and overturning. This group includes proprietary interlocking systems, masonry walls, gabion walls, and mass concrete walls.
2. *Cantilever walls* in which the support is provided from a vertical or inclined cantilever, usually of relatively modest height. This group includes cantilever concrete retaining walls, sheetpile walls, etc.
3. *Tied-back walls* in which the vertical or inclined wall face is restrained by ground anchors to limit outward deflections. This group includes tied-back soldier pile walls, ground-anchored systems, and tied-back slurry trench walls.
4. *Reinforced soil walls* in which soil is reinforced by metal strips, plastic strips, grids, soil nails, or fabric reinforcements to allow the outer face to stand at relatively steep slopes and provide internal stability.

In recent years, the Federal Highway Administration has published several manuals on these topics. They cover all aspects of design, construction, contract specifications, costs, and performance, and are readily available to the public. The primer by Sabatini et al. (1997) is especially recommended for its summary of wall systems.

FHWA manuals generally provide more detailed coverage than is possible here; the objective in this textbook is to present the remedial alternatives with enough information to allow a geotechnical specialist to study the options, make choices, and, if needed, perform a preliminary design.

Earth Pressures

The concepts of active and passive earth slope failures are presented in Chapter 9, Section 9.5. Active shear surfaces typically occur at the headscarp of landslides. Failures at the toe often have a shallow angle to the horizontal, similar to a passive wedge failure.

For gravity retaining walls, the active and passive wedges are shown on Figure 19.1. It takes very little outward movement to reach the active (minimum) lateral pressures, but much larger movements are needed to develop passive (maximum) lateral pressures. Most design procedures for retaining walls consider the active and passive pressures as the starting point and modify them in a semi-empirical manner to correlate with observed loadings in the field. Theory can also account for such variables as: (i) the friction δ developed at the back of the wall, which cannot exceed angle ϕ in cohesionless soils; (ii) the backslope angle β of the ground behind the wall (important in landslide studies); and (iii) groundwater pressures.

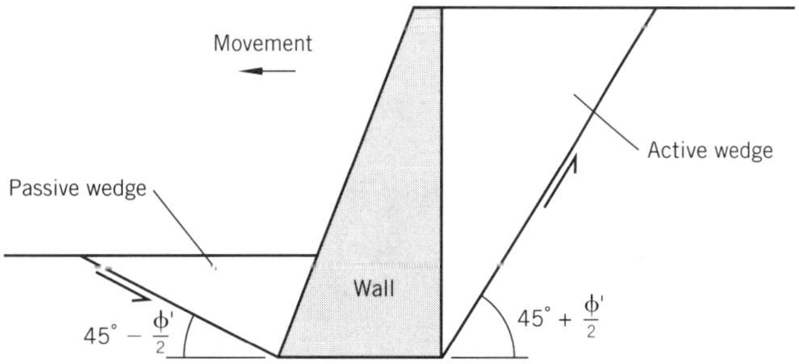

Figure 19.1 Development of active and passive wedge failures due to outward movement of a smooth retaining wall.

To briefly recap the Rankine theories for the wall shown on Figure 19.1, the formulas are:

Cohesionless soils (ϕ, c' = 0)

$$K_A = \frac{1 - \sin\phi}{1 + \sin\phi} = \tan^2(45 - \tfrac{1}{2}\phi) \qquad \text{Eq. (1)}$$

$$K_P = \frac{1 + \sin\phi}{1 + \sin\phi} = \tan^2(45° + \tfrac{1}{2}\phi) \qquad \text{Eq. (2)}$$

Cohesive soils (cohesion c, ϕ = 0)

$$K_A = 1 - \frac{2c}{\gamma H};\ P_A = \gamma H - 2c \qquad \text{Eq. (3)}$$

$$K_A = 1 - \frac{2c}{\gamma H};\ P_P = \gamma H + 2c \qquad \text{Eq. (4)}$$

These results lead to some interesting conclusions about net lateral resistance to movement in clays, assuming that the undrained shear strength c remains unchanged with depth. As shown on Figure 19.2, Case I, a shear pile wall develops an ultimate resistance of 4c, which is constant below a depth of

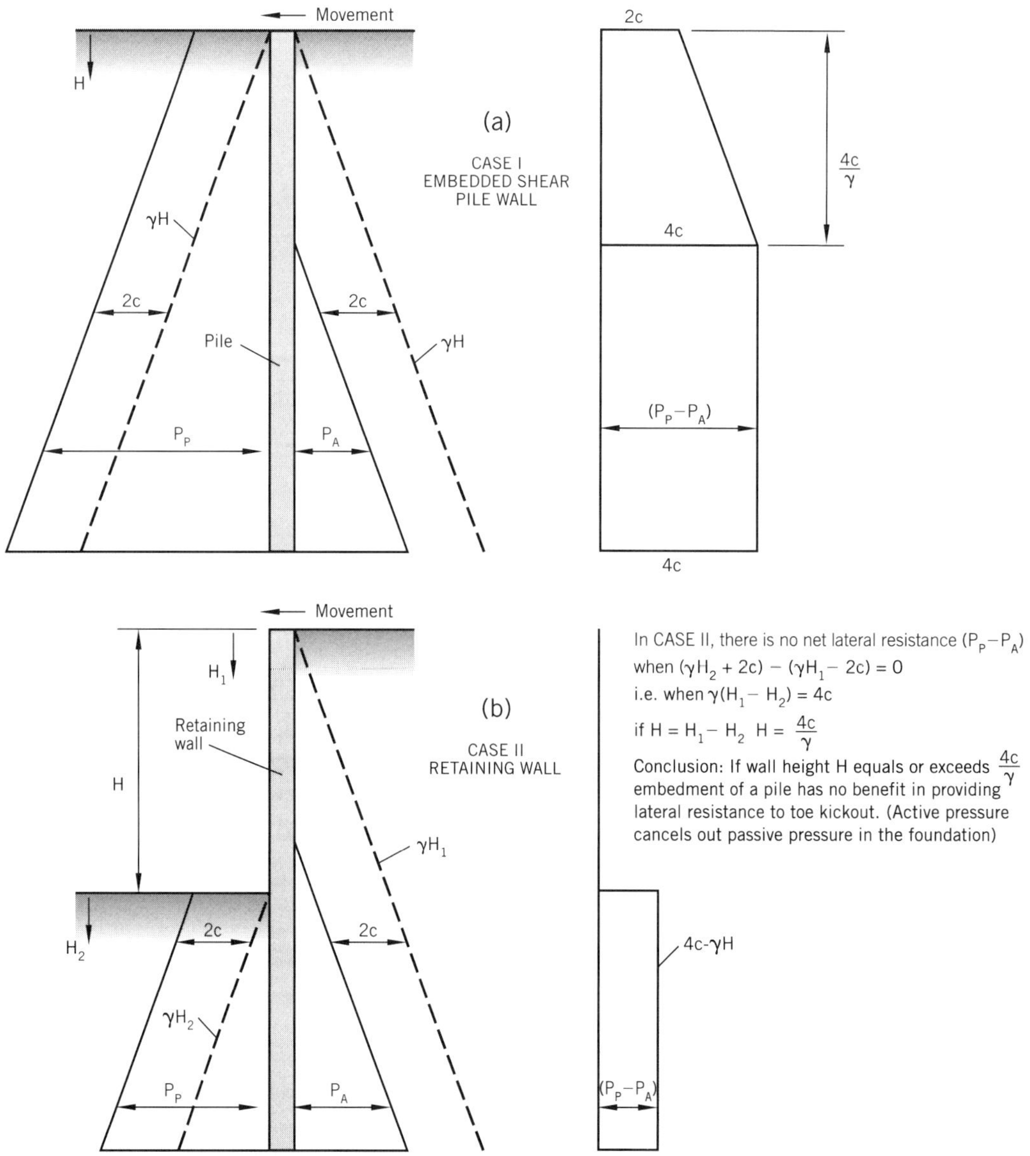

Figure 19.2 Effect of active and passive earth pressures on resistance to lateral movements in clays.

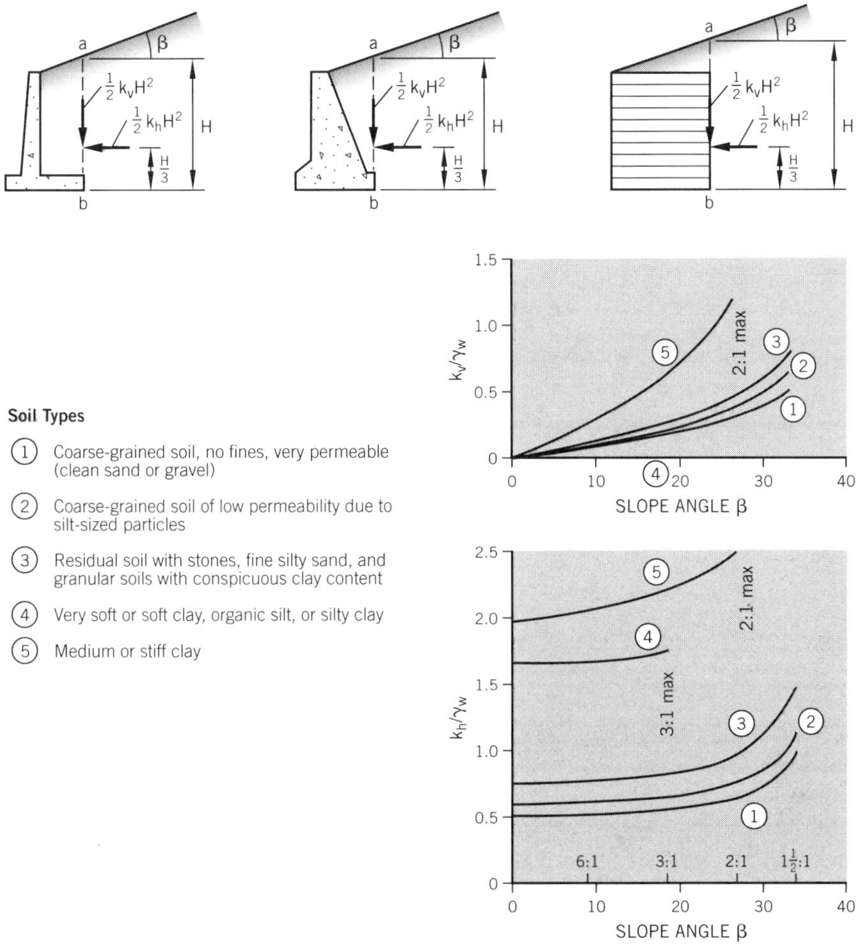

Figure 19.3 Charts for estimating backfill pressures against retaining walls for sloping backfill surface (after Terzaghi, Peck, and Mesri, 1996).

$4c/\gamma$. However, for a wall in which the soil is excavated on one side (Figure 19.2, Case II), ultimate resistance below the base of the wall decreases with increasing wall height. It becomes zero if $(H_1-H_2) = H = 4c/\gamma$; i.e., embedding a pile will provide no benefit in terms of kickout resistance. For example, if γ = 120 pcf and c = 800 psf, the wall height at which there is no benefit is 26.7 feet.

In landslide practice, long piles are usually embedded in a hard stratum where the undrained shear strength c is much higher than the corresponding strength in the landslide. Also, such piles usually have ground anchors, which alter the soil-structure interaction. However, the principle of net lateral resistance discussed above applies to gravity retaining walls in clay where deep embedment of the base may not significantly benefit stability.

For a sloping backfill, as might occur on a landslide, Terzaghi, Peck and Mesri (1996) provide the charts shown on Figure 19.3. These are semiempirical design charts based on small retaining walls that have generally behaved satisfactorily in the past. The figure shows three types of retaining walls, including a reinforced soil wall. The height H is obtained by drawing a vertical line ab through the heel of the retaining wall or back of a reinforced soil wall. The total earth pressures against vertical section ab can be obtained from the two charts for five categories of soils. For type 5 soil, height H can be reduced by 3 feet from the actual value. The pressure distribution is assumed to increase linearly with depth and the point of application is at height H/3. For additional or unusual backfill slopes, or for surcharge loading, textbooks on earth pressure theories should be consulted.

Many walls are backfilled with select, cohesionless, granular fill. The active earth pressure coefficients K_A for such soils are shown on Figure 19.4 as a function of β/ϕ where β is the backslope and ϕ is the friction angle of the soil fill. It can be noted that K_A rises rapidly as angle β approaches the same value as angle ϕ.

More recent interest in earth pressures has been focused on lateral pressures caused by compaction behind retaining walls (see below) or on non-yielding face elements placed on walls as they are built up from the ground (Chapter 20). These design methods often include the coefficient of earth pressure at rest K_0. Since K_0 is not commonly measured in soils laboratories, it usually has to be estimated. Two relationships are being used:

1. $K_0 \cong 1 - \sin\phi'$, a theoretical relationship attributed to Jaky (1948)
2. $K_0 \cong 1.5K_A$, an empirical relationship

Both provide reasonable answers, although, as shown for Brasted Sand in Chapter 7, Section 7.9, the values of K_0 given by these relationships were somewhat lower than the K_0 values measured experimentally.

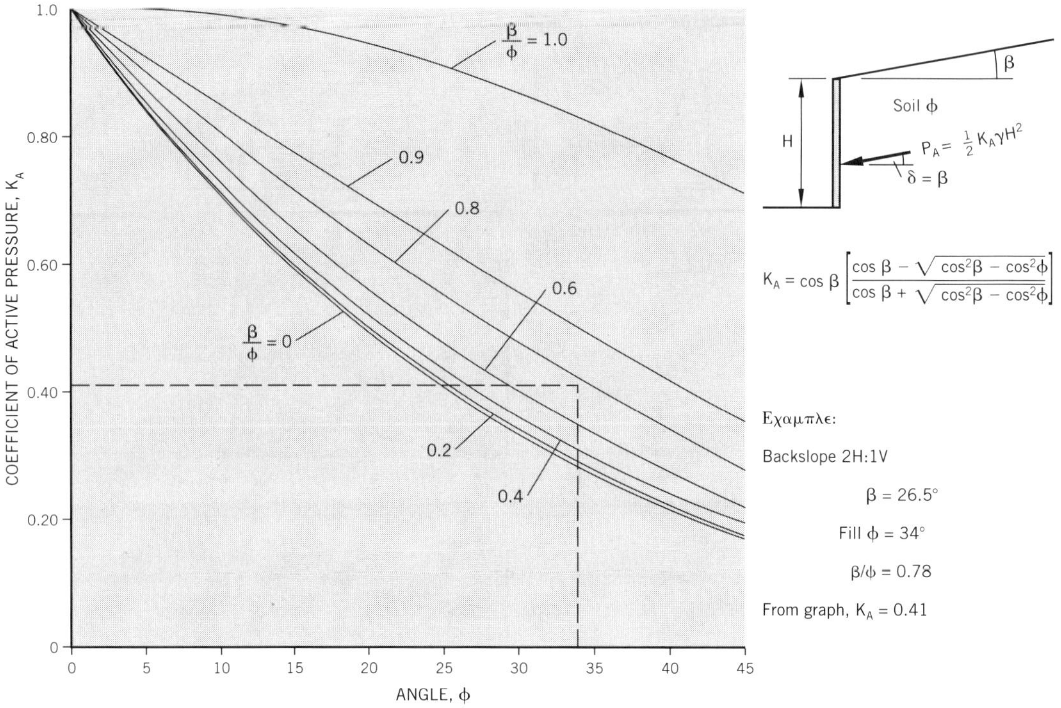

Figure 19.4 Effect of backfill slope angle β on the coefficient of active earth pressure K_A.

Actual Lateral Pressures on Retaining Walls

Current design methods assume that gravity retaining walls move and that such movements will develop active earth pressures. However, the actual lateral pressures behind a wall are complex, depending on several factors, including:

- The structural rigidity of the wall
- The points of application of external loading, such as water, struts, anchors, or thermal expansion (e.g., bridge deck)
- The method of placing the backfill, especially compacted fill immediately behind the wall
- The response of the foundation to loading

The case of lateral loading caused by compaction against *rigid* vertical retaining walls has been studied by Duncan and Seed (1986) and Duncan, Williams, et al. (1991, 1993). Such a situation would require a massive wall on a solid foundation such as bedrock. However, the analysis is useful in providing an approximate upper limit for lateral pressures induced by compaction.

The estimate of postcompaction lateral pressures imposed by a compacted granular backfill against vertical non-deflecting structures is a four-step procedure to obtain a graph similar to that shown on Figure 19.5. The rationale is:

- ab is the lateral passive pressure.
- bc is the at-rest pressure of the backfill plus the residual compaction-induced horizontal stress.
- cd is a line representing the envelope of residual lateral pressures resulting from placing and compacting successive lifts of fill.

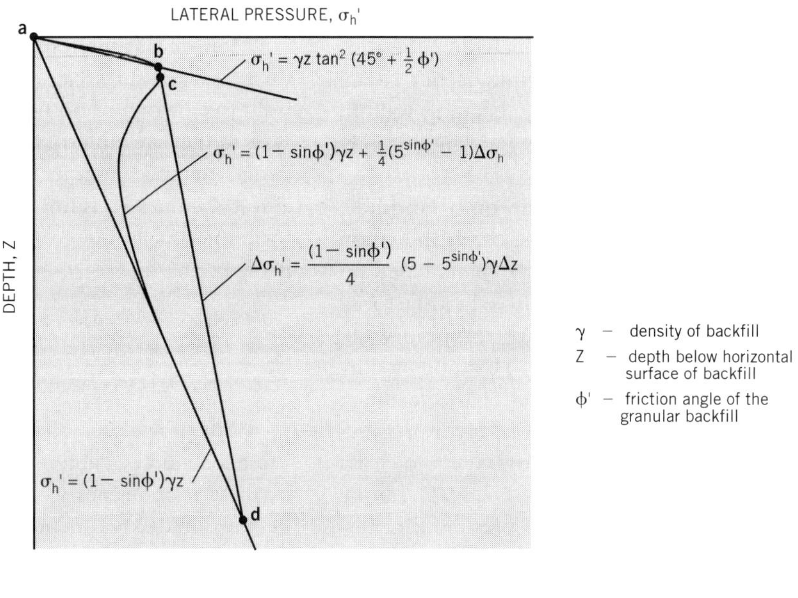

Figure 19.5 Diagram for calculating the lateral pressure of a compacted granular soil against a non-yielding vertical retaining wall (after Duncan et al., 1991).

- Below d the overburden pressure exceeds the increase in stress due to compaction. The overburden pressure line ad is $K_0\gamma z$. Using the expression $K_0 = 1 - \sin \phi'$ (Jaky, 1948), it becomes $\sigma'_h = (1 - \sin \phi')\gamma z$.

The calculation of $\Delta\sigma_h$ due to the compaction rollers is complicated. Duncan, Williams et al. (1991, 1993) have produced charts that allow for different lift thicknesses, distances from the wall, compactor type and dimensions, and friction angle of the backfill. In practice, the non-yielding wall is uncommon and actual measured horizontal stresses in walls are variable, depending on the factors mentioned earlier.

One apparent anomaly is that high concrete gravity (rigid) walls on bedrock have performed satisfactorily despite the fact that they were designed to support active (K_A) earth pressures rather than at-rest (K_0) earth pressures. According to Duncan et al. (1990), this good performance can be explained as follows: As the backfill settles due to successive lifts of fill, it exerts a downward shear force on the back of the wall. Thus, the overturning moment due to at-rest (K_0) pressures is partly offset by the restoring moment due to these shear forces.

Backfill and Drainage

Backfill generally is a free-draining granular soil. This eliminates groundwater pressures acting on the wall; it also is easier to place and compact free-draining cohesionless soils. If rockfill is used, a geotextile filter cloth is required as a separator between native fine-grained soil and rockfill. Pea gravel backfill requires the same treatment as rockfill. Although seepage quantities may be low, a drainage pipe should be placed at the base of the backfill to remove any groundwater collecting in the backfill.

Where a wall is constructed tightly against the retained soil, such as a tied-back soldier pile wall, commercially-available thin drains can be installed at the back of the wall. If the wall itself is free-draining (e.g., wood lagging), the drains can be attached to the outside face.

Some walls built by "top down" construction techniques do not have a drainage system. They are designed to resist groundwater pressures. Examples are concrete slurry trench walls and shear pile walls.

External Stability of Gravity Retaining Walls

Gravity-type retaining walls have common requirements for external stability. These are: (i) resistance to sliding along the base; (ii) resistance to overturning; (iii) allowable bearing pressure; and (iv) resistance to slope failure behind and below the wall. Item (iv) is covered in Chapter 9. In seismic regions, seismic stability can be added to this list. The minimum factors of safety customarily assigned to retaining wall design are listed on Table 20.1 of Chapter 20.

Resistance to Base Sliding

Horizontal resistance along the base of a retaining wall is provided by friction between the base and the underlying foundation soil (Figure 19.6a). The friction between a concrete base and soil is likely to be less than the shear strength of cohesionless soil in the foundation. Terzaghi, Peck, and Mesri (1996) suggest the following values for base/soil friction angle ϕ':

- $\phi' = 30°$ for coarse-grained soil containing no silt and clay
- $\phi' = 24°$ for coarse-grained soil containing silt
- $\phi' = 20°$ for thin layer of granular fill overlying cohesive soils

The resultant force R of wall weight W and backfill load P must remain within the middle third of the wall base to maintain a positive pressure at the uphill edge of the base at U. If R is exactly at the downslope third point of the base (Figure 19.6b), a triangular distribution of normal effective stress occurs across the base of the wall. The maximum stress, σ'_v (max), below the downhill toe at D, should not exceed the bearing capacity of the foundation soil. The resistance to sliding at any point across the base width is calculated as $\sigma'_v \tan \phi'$.

For retaining walls built over silt and clay, a thin layer of crushed sand or sand/gravel should be placed between the native soils and the wall base. If the clay foundation has undrained shear strength c, the resistance to sliding will be based on the diagram of Figure 19.6(c) where the back of the wall develops resistance in the granular fill immediately below

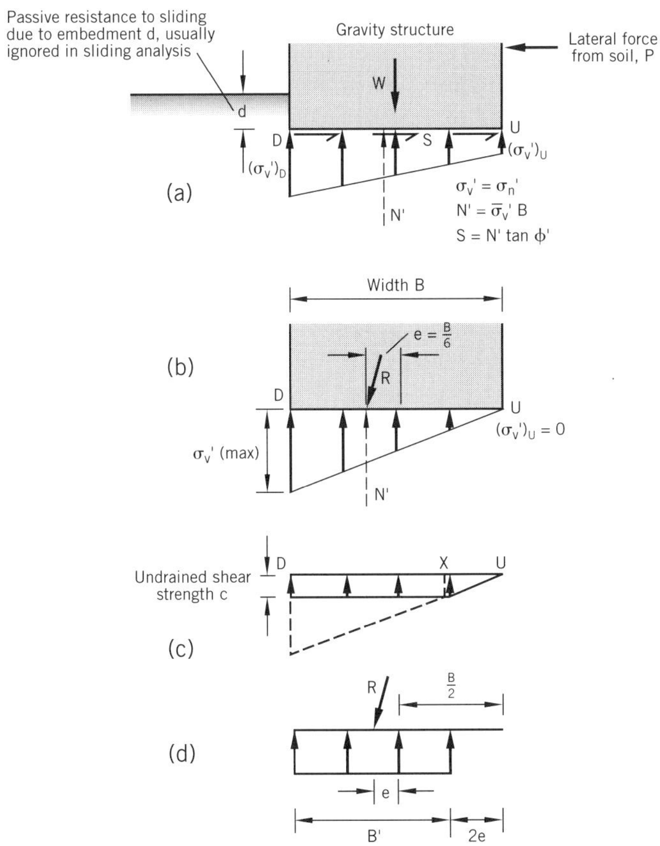

Figure 19.6 Pressure diagrams for footings of gravity retaining walls:
(a) stresses on base
(b) stresses with resultant force R at edge of middle third
(c) resistance to sliding if $c < \sigma_v' \tan \phi'$
(d) approximate equivalent uniform soil pressure for wall foundation subject to eccentric loading
(after Terzaghi, Peck, and Mesri, 1996)

the base and the remaining width DX has resistance of clay strength c multiplied by DX. Resistance developed over width XU will be based on $\phi'=20°$.

Positive resistance to sliding is also obtained from embedment depth d at the downhill side of the wall. However, depth d is usually shallow and the ground in front of a wall is often disturbed by construction or subject to seasonal softening. In most cases, passive resistance due to wall embedment is ignored in calculating sliding resistance.

If the wall base is too narrow to provide adequate factors of safety against sliding and/or overturning, the wall can be founded on piles. Batter piles can be used to resist horizontal forces. According to Peck et al. (1948), if the weight of the backfill above a clay subsoil exceeds about one-half of the bearing capacity, progressive horizontal movements of the retaining wall are likely to be excessive, even with batter piles. In such cases, an alternative to conventional backfill needs to be found (e.g., additional superstructure, lightweight fill, etc.).

Bearing Pressure

In practice, the triangular distribution of stress shown on Figure 19.6(b) is unlikely to develop due to yielding of the soil beneath the toe. Instead, Meyerhof (1953) has recommended using the assumption that the contact pressure is uniformly distributed over part of the base, as shown on Figure 19.6(d). This approximation can be used to calculate the bearing capacity of a foundation subjected to eccentric loading.

19.2 GROUND ANCHORS (TIEBACKS)

Technique

Ground anchors, and the technology of grouting a steel tendon into soil or rock, is being increasingly used to provide slope stability. These include tied-back concrete retaining walls, and anchor block walls. This section describes the design and construction of anchors.

Anchors apply tensile loads to the face of a wall or slope. As a structural member, they prevent outward movements of the face. On cantilever walls, an anchor (or series of anchors) reduces the maximum bending moments and allows a wall or trench to retain an almost unlimited height of soil behind it. Anchors constructed on open face slopes restrain the outward movement of soil at the face, helping to provide stability against shallow slope failures. Finally, anchors passing through the slip surface of a landslide can create an uphill pull to resist the gravity forces and also increase the shear strength at the slip surface.

An anchor is drilled through overburden soils to an anchorage zone of bedrock or firm soils. For simple cut slopes, the anchorage is a short distance behind the active wedge or potential slip surface. In landslide analyses, the anchor has to be constructed within the stable ground behind the slip surface (Figure 19.7). An additional contingency margin of at least 10 feet should be provided between the slip surface and front edge of the anchor in soil landslides, but less when the anchor is within bedrock. The geotechnical engineer can select a contingency margin such that a front edge bearing failure would not occur within the landslide debris.

An anchor is constructed by first drilling a deep angled hole into the ground. Next, a steel bar or stranded wire is inserted to the full depth of the hole and is grouted by a cement grout (normally neat cement, although sand aggregate is sometimes included for larger diameter holes). There are four methods in use: gravity grouted, pressure grouted, post grouted, or underreamed. Only the first two methods will be described here.

Gravity grouted anchors allow the grout to be installed by tremie methods and require a downsloping drillhole. The technique is commonly used for rock or very stiff/hard clays. Casing is not usually needed but can be provided.

Pressure grouted anchors are employed for cohesionless soils or weak fractured rock. The hole is cased. As the casing is withdrawn, the grout is injected under pressures higher than 50 psi. The pressure may enlarge the hole diameter and produce higher normal stresses on the hole wall, both effects contributing to higher pullout resistance.

Tieback anchors have been in use since the 1960s and the technology is mature. Early concerns about the longevity of anchor loads are resolved. Numerous specialist foundation contractors in the United States have the equipment and trained personnel to design, install, and test ground anchors. Geotechnical and structural engineers can provide the appropriate loading diagram, oversee the field installations and lock-off loading, and conduct any follow-up studies after construction to check the performance of the loaded anchors. The structural engineer or geotechnical engineer may specify the technical requirements in the contract specifications. The specialty contractor then provides the working drawings and design calculations for their anchor system and is responsible for the anchors achieving the specified design loads.

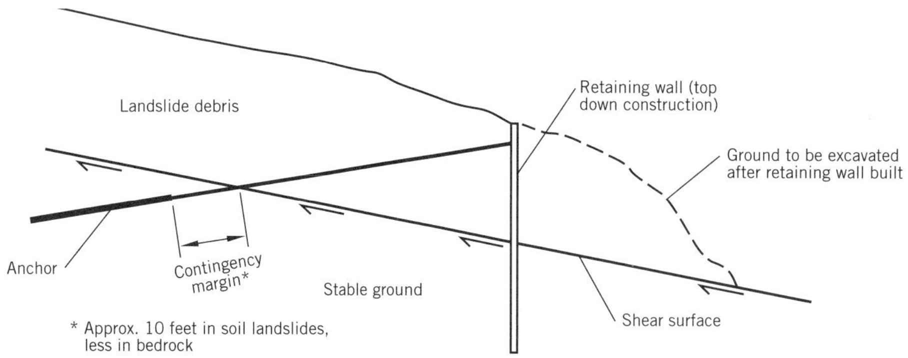

Figure 19.7 Ground anchor installed through a landslide.

Design

A ground anchor comprises a tendon, which provides resistance to pullout, and an anchorage on the face of the wall. The tendon has a *bonded length,* which has to provide the loadcarrying capacity and an *unbonded length* that passes through unstable or potentially unstable ground between the wall and the deep ground anchor. The tendon usually dips at about 15° to the horizontal to allow tremie grouting of the deep anchor. Inclinations vary from about 10° to 30° in practice. The inclination puts a pulldown load on the face of the wall so the dip should not be excessive. However, the bonded length (deep ground anchor) should be at least 15 feet vertically below the ground surface to provide an adequate pullout resistance in cohesionless soils.

The vertical pulldown force on the wall face is T sin α where T is the tensile load of the ground anchor and α is the angle of inclination to the horizontal. Therefore, a vertical soldier pile design has to include an analysis of the pile end bearing capacity.

The anchor drillholes are usually 3 to 8 inches diameter and design loads generally are in the range of 40 to 150 tons per anchor, and occasionally up to 300 tons per anchor. The anchor bond length typically ranges from 15 to 40 feet in soil and 10 to 30 feet in rock; very little additional resistance is achieved by using longer bond lengths. A factor of safety is used to reduce the ultimate stress to an allowable stress.

The tendon is constructed from either a single steel bar or stranded wires. In stranded wire tendons, a cluster of several wires (0.5 inch or 0.6 inch dia.) forms a strand (Figure 19.8a). The wires are made of high-tensile steel conforming to ASTM A-416. The number of strands in a tendon are deter-

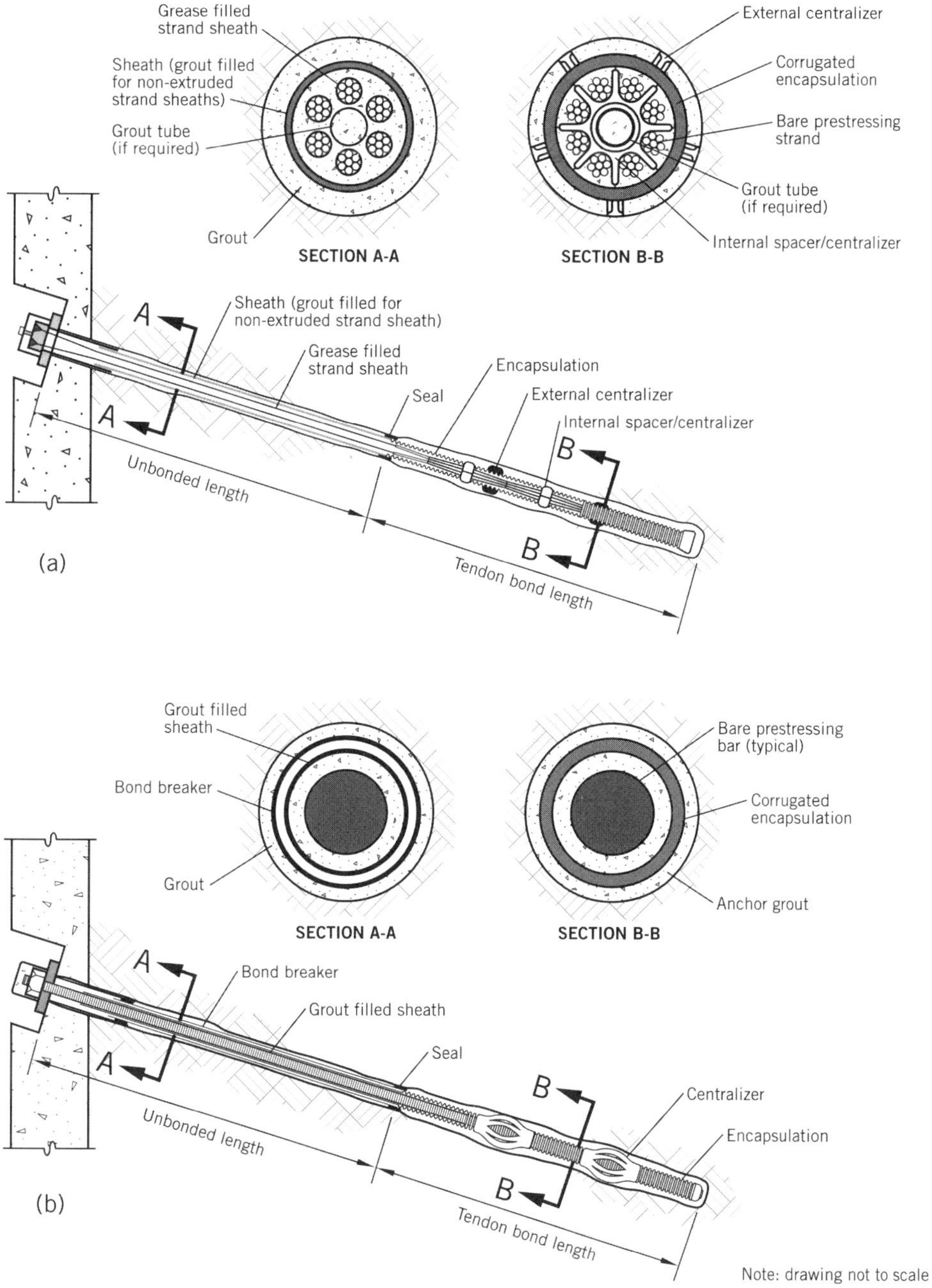

Figure 19.8 Example of Class 1 corrosion protection:
(a) stranded wire tendon
(b) bar tendon
(after Sabatini et al., 1999)

mined by the design load. For example, a single strand of 0.6 inch dia. wire can carry a design load of 17.6 tons. Therefore, it requires 6 strands to achieve a design anchor load of 100 tons. Stranded tendons are more commonly used in landslide stabilizations than bar tendons. A benefit of a stranded tendon is that it can be manufactured to any specified length and does not require field couplings.

Bar tendons (ASTM A-722) are available in sizes ranging from 1 inch to 2.5 inch in diameter and in lengths of up to 60 feet (sufficient for most anchors, but can be insufficient for some landslides). Bars are easier to stress than strand tendons, and their load can be adjusted after lockoff if needed. A bar tendon of 1⅜ inch diameter has a maximum design load of 70 tons.

Class 1 corrosion protection must be provided for permanent ground anchors installed in landslides. Some examples of anchor designs are shown on Figures 19.8, 19.9, and 19.10. The anchor used at Goat Lick Slide, Montana, Case History 9,

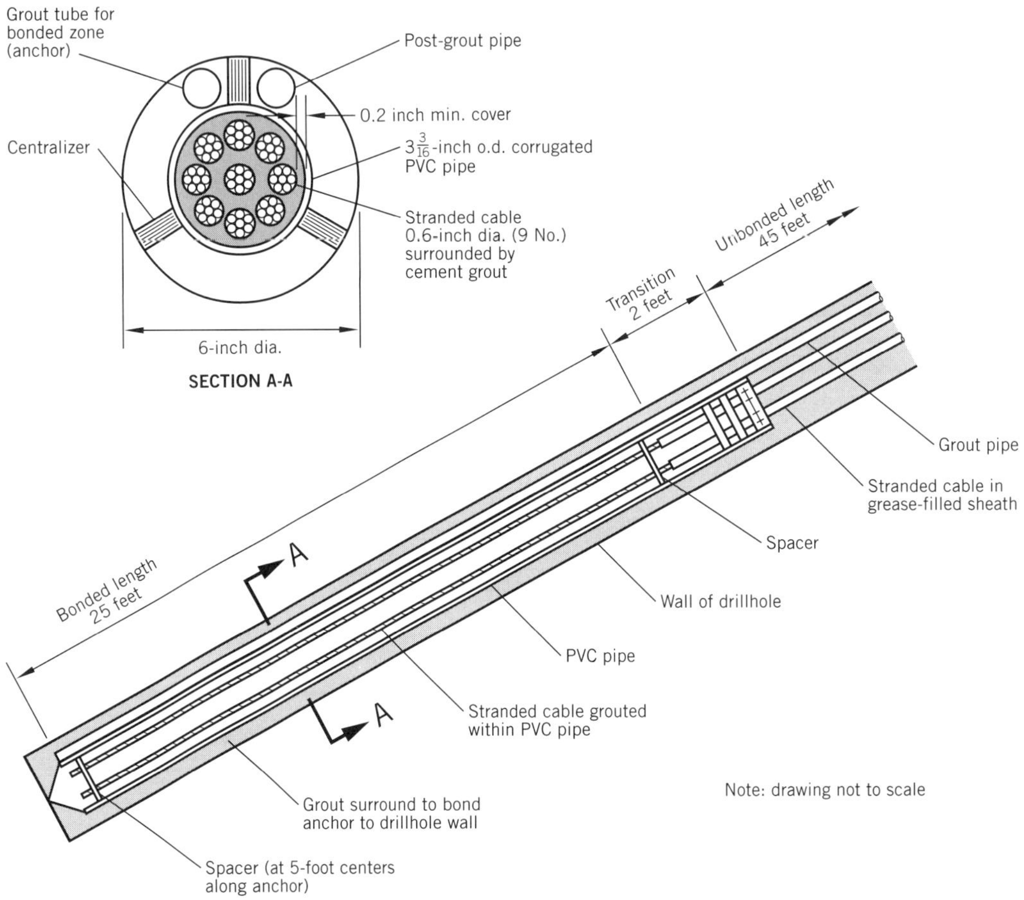

Figure 19.9 Example of tieback anchor in bedrock. Dwyidag System International, U.S.A., Inc. anchor design used at Goat Lick Slide (Case History 9).

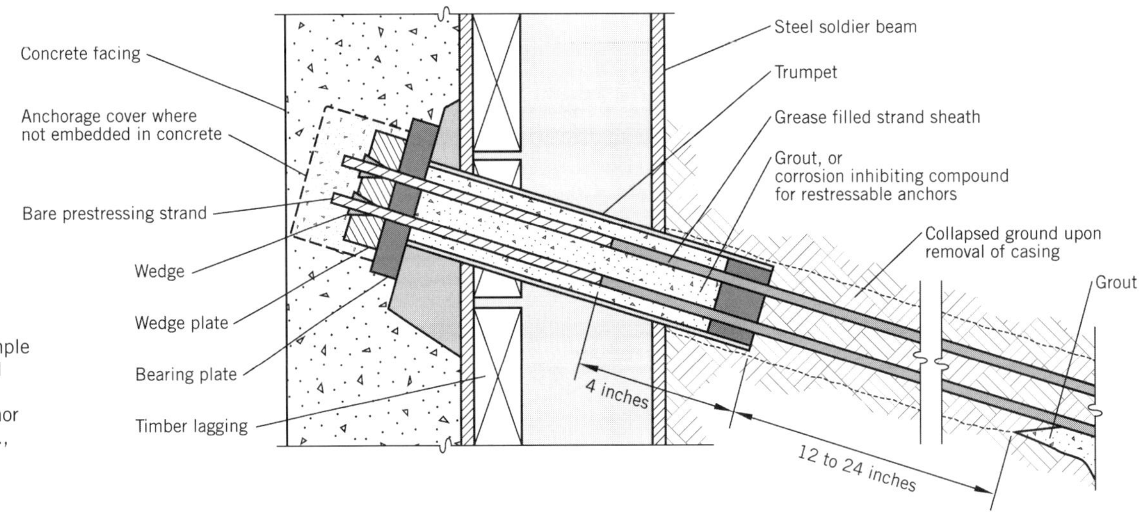

Figure 19.10 Example of anchorage to wall detail for a strand tendon ground anchor (after Sabatini et al., 1999).

is shown on Figure 19.9. The other drawings are examples provided by Sabatini et al. (1999).

Corrosion protection is a critical requirement for permanent ground anchors. The *unbonded* or free length of the tendon is encapsulated by a seamless polypropylene sheath with the space between the steel strands and the sheath being filled with a grease that lubricates and provides anti-corrosion properties (e.g., Chevron Polyuria EP Grease, No. 2 grade or similar). The *bonded* length (anchor) is corrosion-protected by encapsulating the tendon in a grout-filled corrugated plastic or deformed steel tube. This grouting can occur before placing the tendon into the drillhole. In stranded tendons, spacers are used to ensure that each steel cable is separated by 0.25 inch to 0.5 inch to securely bond the cable to the encapsulating grout. The grout-filled tube is placed in the drillhole to the required anchorage depth (using centralizers to provide at least 0.5 inch annular space between the tube and drillhole wall) and is grouted in place starting at the lower end. Normally, both the bonded and unbonded length of the tendon are grouted at the same time, and this is a strongly recommended procedure over the alternative of grouting the bonded length first, testing, and grouting the unbonded length later. It ensures a continuous outer protective layer of cement grout.

At the wall, a "trumpet" (Figure 19.10) of steel or PVC is used as a transition from the anchorage face to the unbonded length corrosion protection. One end of the trumpet is fastened and sealed to the bearing plate and the other is fitted with a watertight seal, such as an O ring, which fits around the protective tube but allows the tendon to move within the trumpet. The annular space between the trumpet and tendon is filled with anti-corrosion grease. The exposed outer end of the anchor head is covered by at least 3 inches of concrete (often it is buried in the structure) or a corrosion inhibitor-filled cap.

Grouting and other corrosion protection measures in this area have to be carefully performed. Experience has shown that ground anchors are more likely to fail in service within 6 feet of the wall than at the deeply embedded anchor zone.

The grout typically uses Type I cement and a water : cement ratio of 0.4 to 0.55 by weight (Sabatini et al., 1999). This mix can achieve a minimum compressive strength of 3000 psi by the time of anchor stressing. Additives are occasionally included in the mix, depending on the circumstances.

The tendon is test-loaded to ensure that it can carry the design load, after which it is "locked off" at some pre-set load (typically 60–100% of the design load) by driving steel wedges into the trumpet to hold the load and anchor the tendon to an anchor plate. The void space between the trumpet and the stressing wires is filled with grout (or grease). A center hole load cell can be temporarily or permanently attached to the anchorage at the retaining wall.

Tendons are designed for a design load that does not exceed 50–60% of the minimum guaranteed ultimate tensile strength (GUTS) of the steel in the tendon. The GUTS of steel is 150,000 psi for bar and 270,000 psi for strand. Tendon design capacity t is computed as :

$$t = AR\,\sigma_u \qquad \text{Eq. (5)}$$

where A = cross-sectional area of steel
R = percent of allowable stress (50 or 60%)
σ_u = GUTS of steel

Some representative *ultimate* loads are:

Standard tendon	diameter 0.5 inch	20.6 tons per strand
	diameter 0.6inch	29.3 tons per strand
Bar tendon	diameter 1.0 inch	63.7 tons
	diameter 1.375 inch	118.5 tons

The grouted anchor acts in skin friction. The pullout is:

$$P_u = \pi D L c_a \qquad \text{Eq. (6)}$$

where P_u = ultimate anchor pullout resistance
D = diameter of drillhole
L = bonded length
c_a = adhesion between grouted anchor and soil/rock

Some representative values of average *ultimate* bond stress for various materials (after PTI, 1996) are listed on Table 19.1. Allowable ground anchor loads are estimated by dividing the ultimate load with a factor of safety of 2.

Ideally, the values of t and P_u should be similar to avoid overdesign. Values of c_a can also be obtained from soil mechanics literature for friction piles. However, where grout is injected under pressure to form an anchor, field measurements show that the capacity per unit length can be substantially higher than calculations based on conventional skin friction. Specialist contractors usually have compiled substantial amounts of data on the capacities achieved by their grouting methods in different soils and rocks, and base their anchor designs on such data. The grout usually includes chemical additives to control bleed or retard set. The strength of the grout is specified (e.g., 3,500 psi at 7 days). There is usually only a limited time available between the installation of the anchor and the need to stress to its working load.

Installation

The installation of ground anchors requires an experienced contractor. It is common practice for the project specifications to require minimum years of relevant experience and numbers of completed projects for both the engineer/site superintendent and drill operators assigned to the project. Example: 3 years, 5 projects for supervising engineer; 1 year for operators, with descriptions of past work and contact persons for verification, all subject to prior approval by the owner. There are many other submittals required for ground anchor projects, including proposed methods, GUTS, load cell calibration, grout mixes, etc.

The drillhole for the anchor can be advanced by any drilling technique that will drill at an economical speed through overburden soils or rock, such as rotary drilling, percussion drilling, or core drilling. A continuous flight auger is one of the more popular rotary drilling options. Temporary

Table 19.1 Presumptive Average Ultimate Bond Stress for Ground/Grout Interface along Anchor Bond Zone*

Rock		Cohesive Soil		Cohesionless Soil	
Rock Type	Average *Ultimate* Bond Stress (psi)	Anchor Type	Average *Ultimate* Bond Stress (psi)	Anchor Type	Average *Ultimate* Bond Stress (psi)
Granite and basalt	250–450	Gravity-grouted anchors (straight shaft)	4.4–10	Gravity-grouted anchors (straight shaft)	10–20
Dolomitic limestone	200–300	Pressure-grouted anchors (straight shaft)		Pressure-grouted anchors (straight shaft)	
Soft limestone	150–200	• Soft silty clay	4.4–10	• Fine-med. sand, med. dense – dense	12–55
Slates and hard shales	120–200	• Silty clay	4.4–10	• Med.–coarse sand (w/gravel), med. dense	15–95
Soft shales	30–120	• Stiff clay, med. to high plasticity	4.4–15	• Med.–coarse sand (w/gravel), dense–very dense	36–140
Sandstones	120–250	• Very stiff clay, med. to high plasticity	10–25		
Weathered sandstones	100–120	• Stiff clay, med. plasticity	15–36	• Silty sands	25–60
Chalk	30–160	• Very stiff clay, med. plasticity	20–50	• Dense glacial till	45–75
Weathered marl	22–36	• Very stiff sandy silt, med. plasticity	40–55	• Sandy gravel, med. dense–dense	30–200
Concrete	200–400			• Sandy gravel, dense–very dense	40–200

Note: Actual values for pressure-grouted anchors depend on the ability to develop pressures in each soil type. *after Sabatini et al., 1999

casing may be needed for caving soils. In a marine environment, such as a quay wall, the leads can be cantilevered over the water to allow a hole to be drilled behind the bulkhead. A down-the-hole hammer can be used to drill past boulders and other obstructions, or to drill into bedrock. Tolerance is $\pm 3°$ from the specified inclination.

Tendons have to be very carefully handled and stored at the construction site to prevent damage and corrosion. After the hole has been drilled, the tendon has to be inserted without being driven or forced into the ground; if necessary, the hole may have to be cleaned out or redrilled to accept the anchor.

The grout is of neat cement (usually with chemical additives) and should be pumped in one continuous operation using a positive displacement pump, such as a piston pump or progressive cavity (Moyno type) pump. A stroke counter and pressure gauge are used to control and monitor the quantities and pressures used in grouting. The grout is injected from the lowest point of the drillhole. After grouting, the tendon is not loaded until the grout has attained sufficient strength to carry the test load (usually 3 to 5 days). The bearing plate needs to be placed perpendicular to the axis of the tendon.

Testing and Stressing Anchors

Each anchor is tested to check that it can hold the design load without excessive movement. The normal test is termed a *proof* test (line AB on Figure 19.11) and a cyclic loading sequence, usually performed on the first anchors installed, is termed a *performance* test (a typical sequence is shown on Figure 19.11). The maximum load (point B) is held for 10 minutes to obtain a set of creep rate measurements. All other load/unload points are held just long enough to measure the movement. The anchor is deemed acceptable if, at the maximum test load, the following conditions have been satisfied: (i) the total movement *is less than* 0.04 inch after 10 minutes, and (ii) the total movement *exceeds* 80% of the theoretical elastic elongation of the *unbonded* length of the tendon. The second of these two criteria is to ensure that the unbonded length is not providing pullout resistance.

The maximum test load, in both performance and proof tests, ranges from 133% to 200% of the design load but is more commonly closer to the low end of this range. The maximum load on Figure 19.11 is at 150% of the design load.

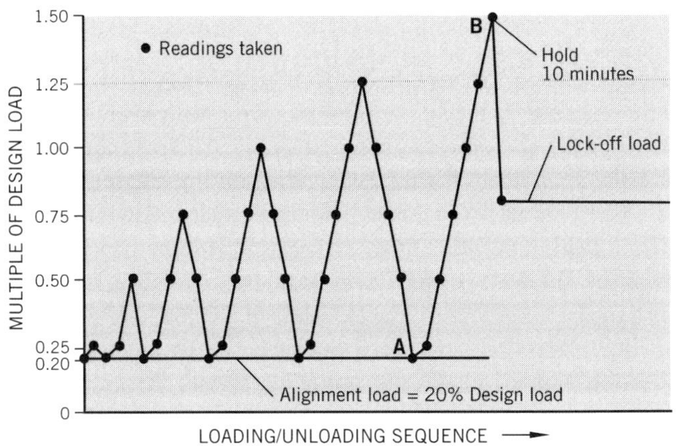

Figure 19.11 Performance Test load/unload sequence. Proof Test—leg AB only.

The theoretical movement of the unbonded tendon length is calculated from:

$$\delta \geq 0.8\ PL/AE \qquad \text{Eq. (7)}$$

where δ = total movement under load
P = test load
L = free (unbonded) length of tendon
A = area of steel tendon
E = Youngs modulus for steel used in the tendon

If the anchor fails the proof test, the options are: (i) hold the test load for an additional 60 minutes; the anchor becomes acceptable if the creep rate is less than 0.08 inch per log cycle of time and the rate of creep is linear or decreasing; (ii) allow the anchor to stabilize for 10 minutes and permit the anchor to be used in the finished work at one-half of the stabilized load; or (iii) install a replacement anchor. If several anchors fail, the construction procedures probably should be modified or the anchor type should be changed.

There is a possibility that the uppermost anchor of a soldier pile could experience a passive failure during proof or performance testing. If it is overloaded in this way, the inward movement will be excessive and the lagging may be damaged. To check the passive resistance, use the formula (Weatherby 1998):

$$F_p = 1.125\ K_p\ \gamma\ H_1^2\ S \qquad \text{Eq. (8)}$$

where F_p = passive resistance
K_p = passive earth pressure coefficient
H_1 = vertical depth to the upper anchor
S = horizontal spacing between adjacent anchors
γ = soil density

Apply a factor of safety F = 1.5. The result needs to be higher than the ground anchor test load.

The final lock-off load is a fairly arbitrary decision. It is typically set at 60–100% of the design load.

Cost of Anchors

Weatherby (1998) has compiled some approximate 1997 costs for installing ground anchors as part of a soldier pile wall. For landslides, the anchor costs range from \$15 to \$60 per sq. ft. of wall, with an average cost of \$20. In landslides, the length of the anchor can be highly variable. The anchors ranged from \$1,000 to \$6,000 each, including the soldier pile connection and testing. Each ground anchor was assumed to support 100 sq. ft. of wall. The Engineering News-Record Construction Cost Index (ENRCCI) for the year 1997 is 5800.

Cheney (1984) provides a more detailed breakdown of costs for 1982 as follows:

Anchors. First 50 feet—\$30 per lin. ft. Add \$2 per lin. ft. for lengths from 50 feet to 120 feet. For double corrosion protection, required in permanent anchors for landslides, add \$4 per lin. ft.

Soldier pile. \$10 per lin. ft. plus \$0.20 per lb. of steel

Permanent wall facing. \$10 per sq. ft. of exposed face. Exposed aggregate facing—add another \$2 per sq. ft.

Mobilization. \$10,000

The ENR Construction Cost Index for 1982 is 3800.

For specific sites, specialty contractors should be contacted for approximate costs in preparing a conceptual construction cost estimate, but the above figures can provide a preliminary estimate when first considering alternatives. The use of the ENRCCI data for updating construction costs is explained in Chapter 13, Section 13.6.

19.3 ANCHOR BLOCK AND ELEMENT WALLS

Technique

Multiple isolated anchors tied to concrete bearing pads at the ground surface are referred to as *element walls* (Figure 19.12). On many projects of this type the anchor blocks are almost touching each other or are connected together horizontally by a waler. These are termed *anchor block walls.* This latter term will be used throughout the remainder of this section.

The unbonded length of the steel anchor passes through the landslide mass and applies the full anchor load T at the slip surface (Figure 19.13a). Because the anchor usually crosses the slip surface at an oblique angle, the anchor provides: (i) a pullback component load P parallel to the slippage, and (ii) an increased normal load perpendicular to the slip surface, thereby increasing the shear resistance by an increment δS (Figure

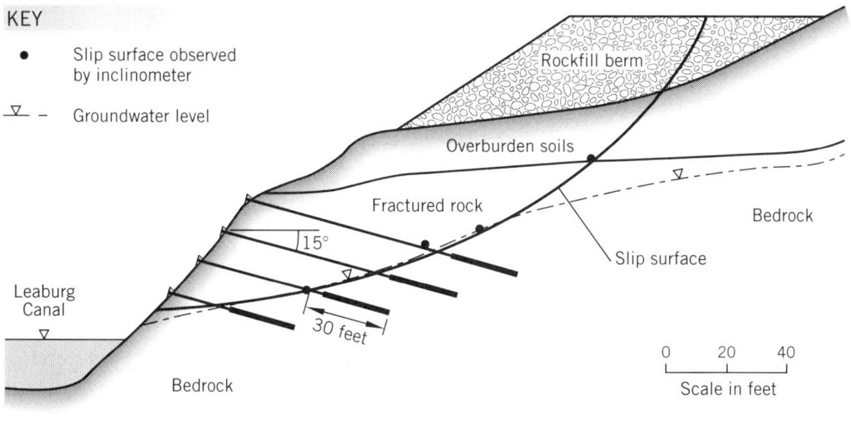

Figure 19.12 Conceptual tied-back elements for Percy Slide remediation, Leaburg, Oregon.

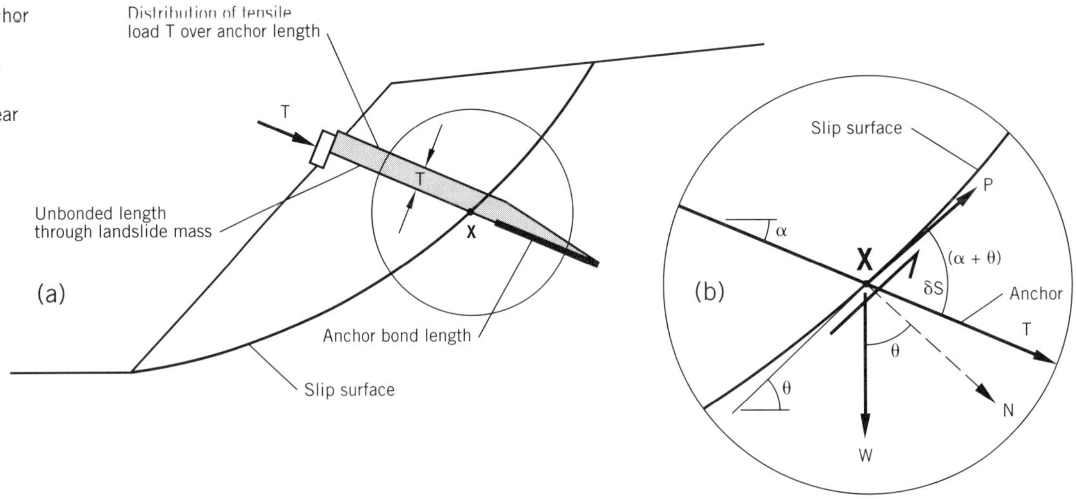

Figure 19.13 Effect of anchor tensile load T on stability:
(a) full load T acting at slip surface X
(b) pullback load P and shear strength δS benefits

19.13b). This dual function can produce a maximum total benefit to stability when the anchor is inclined at an optimum inclination to the horizontal (as discussed later).

Anchor block walls are a versatile remediation technique that can provide innovative solutions to difficult landslide conditions. They can be installed on uneven terrain relatively quickly with minimal disruption to the surrounding area (such as roads, railroads). Potential difficulties include the required "standup" time of the slope during construction and the unsightly appearance of the blocks on the slope surface. "Standup" time is the time that elapses between excavating a steep bearing face and completing the anchor installation. These two concerns can be overcome in most situations.

Advantages and Limitations of Anchor Block Walls

Advantages of anchor block walls relative to alternative systems include:

- They can be installed on uneven surfaces if needed; therefore site preparation costs can be minimal.
- Construction is rapid and provides only minor disruption of the surrounding area; the main requirements are access roads and level areas for anchor drilling and for a crane to lift concrete blocks into position.
- The bearing pads can be made of precast concrete brought to the site; this requires no concrete mixing plant at or close to the site.

The principal *limitations* of the anchor block technique are:

- It needs a relatively competent bearing surface; if settlement occurs after stressing the anchor, tension will be reduced or lost.
- The "standup" time of the ground can be a concern. During this time the slope is unsupported and may collapse.
- Anchor block walls can be visually unattractive at sites accessible to the public; additional treatment may be needed to improve their appearance. Common treatments are to attach precast concrete panels to the face (Figure 19.14) or use colored/sculpted shotcrete to produce a more attractive wall. Another common procedure is to simply bury the anchor blocks and leave a natural soil slope outside the wall (Figure 19.15).

Design

The design of ground anchors is described in Section 19.2. For conceptual level studies, designers often assume that a design load of about 125 U.S. tons (250,000 lb.) is feasible at most sites for anchors.

The anchor block design, including any waler beams and structural facing elements, is mostly a structural design and falls outside the scope of this book. Both the anchor head-to-block and block-to-ground bearing strengths have to be satisfactory. From a geotechnical standpoint, the allowable bearing pressure at the ground surface has to be satisfactory. Because the loads are concentrated, a relatively competent foundation is needed to control settlement during stressing and subsequent service. A very dense, granular soil, such as glacial till or weathered bedrock, is suitable. If post-construction settlement occurs, it will de-stress the tendons.

It is customary practice to space the anchors evenly over the face of the anchor block wall and apply equal loads to each anchor. This practice should produce relatively uniform settlements of each anchor block and prevent local overstressing.

There are two other issues that need to be addressed during the design stage of an anchor block wall: continuing movement of the landslide during construction, and site access. If a landslide is actively moving, anchors that are installed and stressed early in the program may become overstressed and fail. Therefore, using information provided by inclinometers (almost always installed on such projects during the site investigation phase), the pattern of movements should become known. Whenever possible, anchors should be installed during a dormant period. However, if active landslide movements are occurring, anchors can be installed and proof tested in the normal way. The anchor

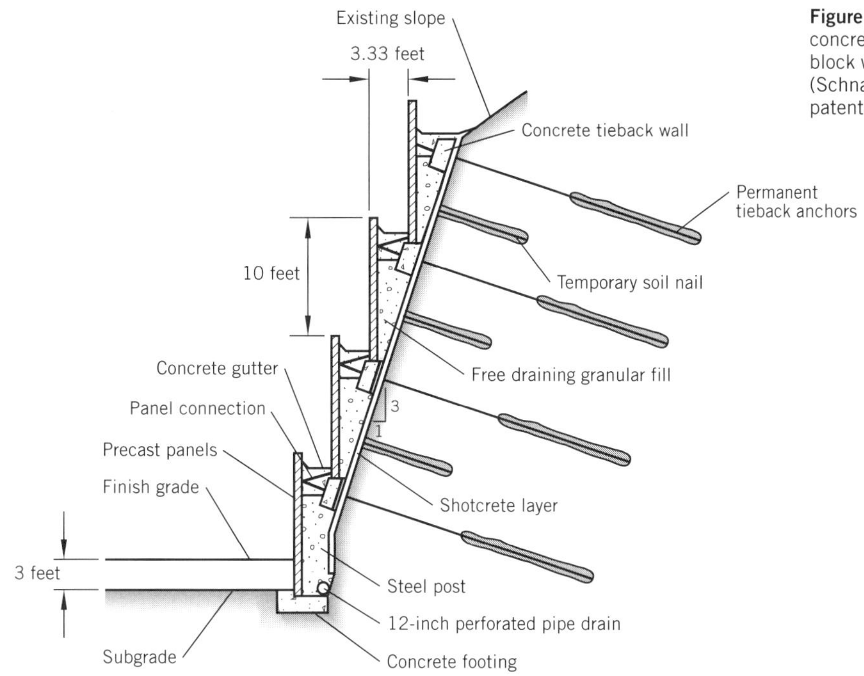

Figure 19.14 Example of precast concrete panels attached to an anchor block wall to improve aesthetics (Schnabel Foundation Company patented system).

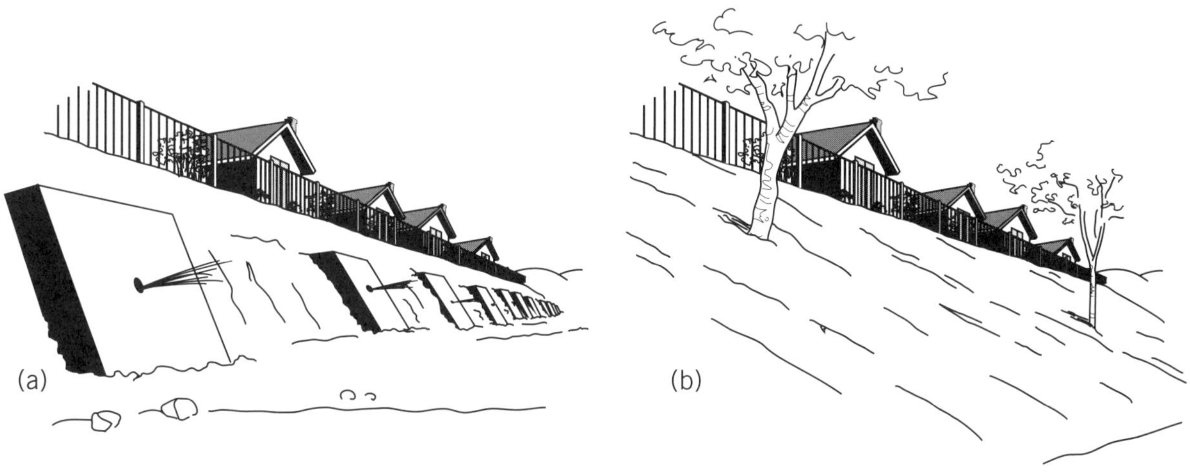

Figure 19.15 Tied-back element wall:
(a) anchors installed and ready for tensioning
(b) final landscaped slope (anchor blocks buried)

group then should be permanently stressed (lockoff) at a later time when the entire group can be stressed over a period of one or two days.

Many anchor block walls are constructed on relatively steep slopes. Access is needed for drill rigs and erection equipment. The geotechnical engineer needs to check that the proposed construction is feasible at the site. It is advisable to discuss the project with one or more specialty contractors to avoid overlooking a flaw that could add significant unexpected costs or compromise safety. Oblique aerial photographs can be very helpful in viewing potential construction problems at sites with difficult access.

Factor of Safety

Factor or safety F is commonly defined, depending on the analysis, as:

$$F = \frac{\text{Resisting force (or moment)}}{\text{Driving force (or moment)}} \qquad \text{Eq. (9)}$$

Should the anchor force T be considered an external load reducing the driving force (Figure 19.13a) or should it be considered as part of the resisting force at the slip surface (Figure 19.13b)? The choice affects the calculated value of F.

The author recognizes the anchor force T as only affecting the *resistance* to sliding. Thus, the dual benefits of the anchor (pullback load + incremental shear strength gain) are both applied to the numerator side of the stability equation for F.

$$F = \frac{W \cos\theta \tan\phi'_r + T\{\cos(\alpha + \theta) + \sin(\alpha + \theta)\tan\phi'_r\}}{W \sin\theta} \qquad \text{Eq. (10)}$$

where W = weight of landslide soil block, ϕ'_r = residual strength along the slip surface, and all other terms are defined on Figure 19.13(b).

Optimum Anchor Inclination

An inclined anchor provides a dual benefit of: (i) pullback resistance, and (ii) incremental shear strength improvement to slope stability. The relevant force diagram is shown on Figure 19.13(b). The question arises: Is there an "optimum" anchor inclination α at which there is a maximum benefit to the stability factor of safety?

To investigate this issue, the Richardson Highway Slide, near Fairbanks, Alaska (Figure 19.16) provides an example. The slip surface is essentially a plane inclined at approximately 20° to the horizontal. For this long, translational landslide, the end effects (top and bottom of the slide) and groundwater have a minor influence on stability. Therefore, the stability analysis is comparable to a soil block on a sliding plane, making the mathematics relatively simple.

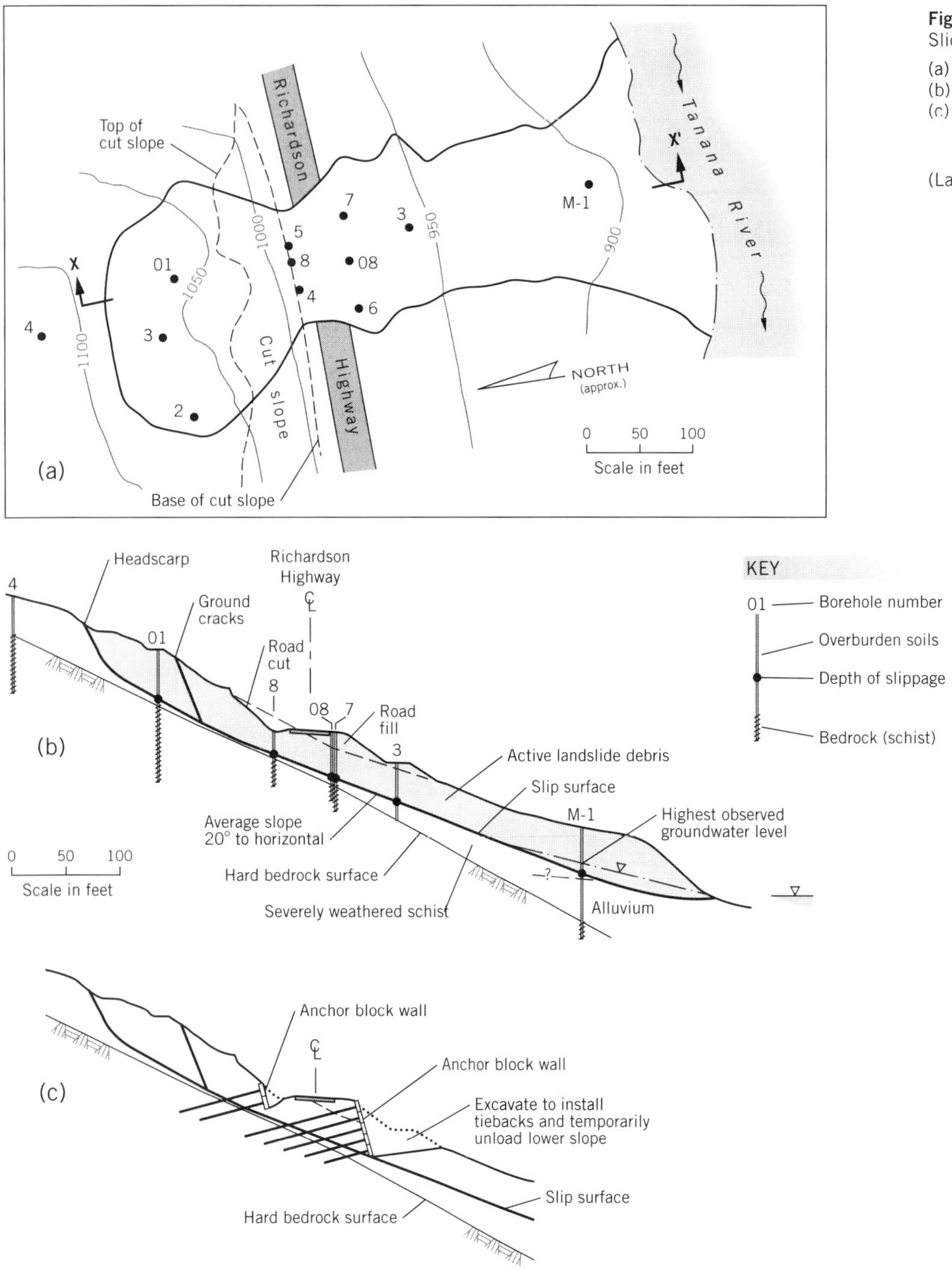

Figure 19.16 Richardson Highway Slide near Fairbanks, Alaska:
(a) site plan
(b) cross-section X-X'
(c) conceptual layout of anchor block walls above and below the highway
(Landslide Technology)

Considering the factor of safety equation (Eq. (10), F is maximum when $dF/d\alpha = 0$

$$\frac{dF}{d\alpha} = \frac{T}{W \sin\theta}\{-\sin(\alpha+\theta) + \cos(\alpha+\theta)\tan\phi_r'\} \quad \text{Eq. (11)}$$

$$= 0 \text{ when } \tan(\alpha+\theta) = \tan\phi_r'$$

$$\text{i.e., } \alpha = \phi_r' - \theta$$

For the Richardson Highway slide, where there is essentially no groundwater within the slide mass, slip surface angle ϑ = residual angle ϕ_r' and $\alpha = 0$ (horizontal). This result is shown graphically on Figure 19.17 where the resistances due to pull-back force and incremental shear resistance are plotted separately and as a combined graph. It can be seen that the maximum combined resistance to sliding (R_{max}) is 1.06 T, where T is the tensile load of the anchor. Thus, the "optimum" angle α for achieving maximum resistance is a horizontal angle in this example. (This subject has not been studied in detail, but it can be noted that groundwater level at the ground surface would reach the same conclusion.)

As the angle α approaches zero, the length of the anchor increases (Figure 19.18). There is clearly a trade-off between the increased resistance to sliding and the extra construction cost of longer anchors. For discussion purposes on this issue, let it be assumed that the cost of constructing an anchor is directly proportional to its length. The shortest anchor length (L_{min}) occurs where the anchor enters bedrock at a right angle. At the Richardson Highway slide, this shortest depth is 40 feet and an additional length of 15 feet can be included for the anchor load zone making a total minimum length of 55 feet. Because the slip surface is inclined at 20° to the horizontal, the anchor inclination for the shortest anchor length is α = 70° (Figure 19.18). All other anchor lengths can be calculated as the length to intercept the slip surface plus 15 feet for anchorage. Thus, a ratio L/L_{min} can be obtained for any angle α. By dividing the total anchor resistance R (from Figure 19.17) by the relative length of the anchor (L/L_{min}) a resistance "normalized" to the length of an anchor is obtained (Figure 19.19).

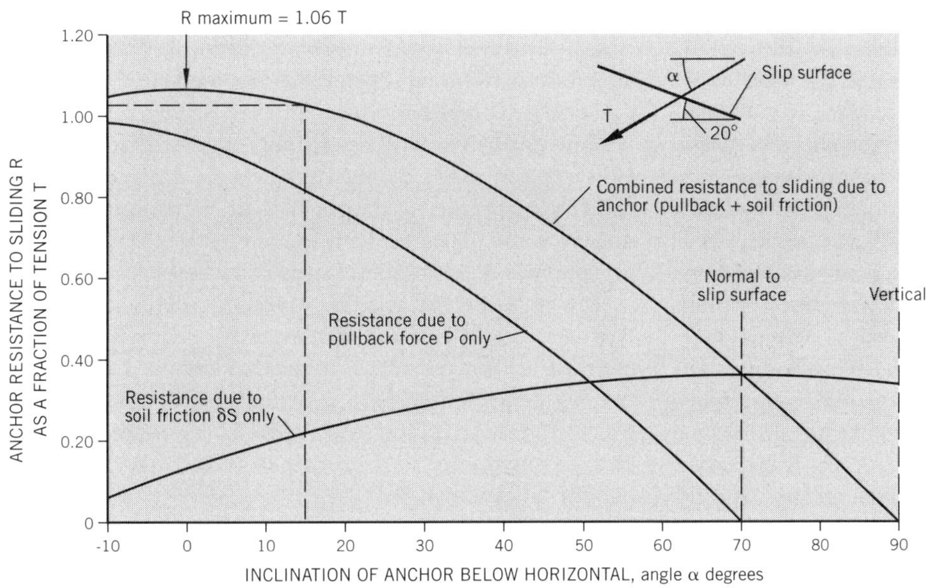

Figure 19.17 Resistance to sliding developed by an anchor as a function of inclination α, Richardson Highway Slide, Alaska.

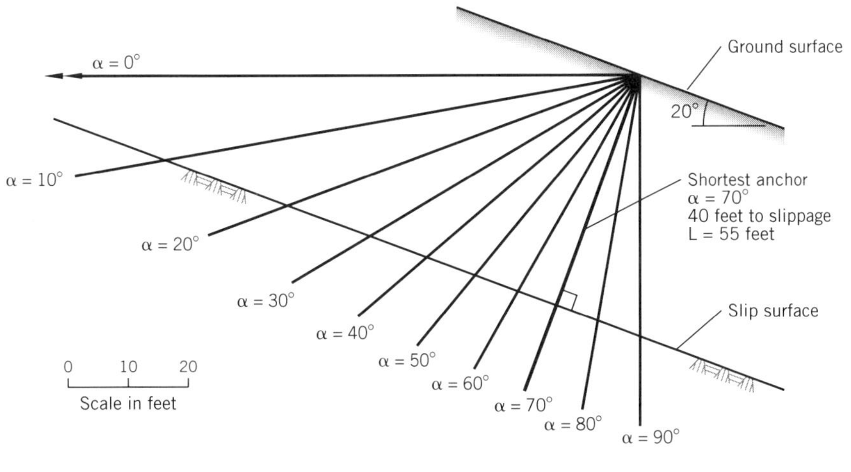

Figure 19.18 Effect of inclination angle α on the anchor lengths at Richardson Highway Slide, Alaska.

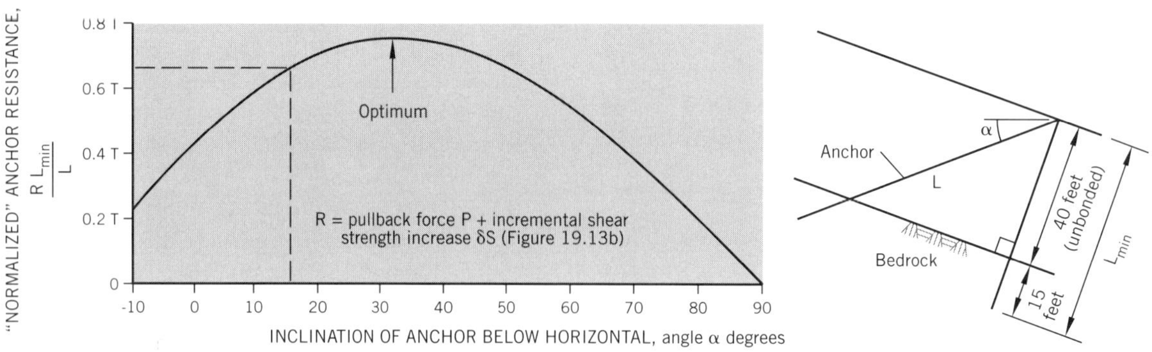

Figure 19.19 Determination of the approximate "optimum" inclination of ground anchors.

As an example: For α = 15° to the horizontal, L/L_{min} = 1.54 and the total resistance to sliding (pullback + incremental shear strength) is R = 1.028T (Figure 19.17). The "normalized" resistance is 1.028T/1.540 = 0.668T (Figure 19.19).

This calculation exercise shows that the maximum anchor resistance per unit length of anchor occurs where the angle of inclination is about 30° to the horizontal. However, the construction cost of anchors is not directly related to the anchor length, as assumed in this exercise. Therefore, the true "optimum" anchor inclination probably lies between α = 0° (for total resistance without regard for anchor length) to α = 30° (where construction cost linearly relates to anchor length).

As a practical matter, anchors are usually installed at angles of 15° to 30° to the horizontal. This allows grout to be placed by gravity methods. Also, since the anchor head block needs to be perpendicular to the anchor length, the temporary bearing face stability benefits from being non-vertical. At a slope of 2 vertical : 1 horizontal (26.5° to the vertical) the bearing face is 63.5° to the horizontal, which approximately corresponds to the backslope of the Rankine wedge for many soils. At this slope, the cut face has a good probability of remaining stable in the short term, thereby helping "standup time" during construction.

To sum up this topic for the assumed site conditions: Anchors produce the maximum resistance to sliding when installed horizontally, but total anchor lengths increase rapidly as angle α approaches zero. If the anchor resistance is "normalized" by taking into account anchor length, the optimum angle of inclination is 30° to the horizontal. Since construction cost for anchors is not directly related to length, the optimum angle with regard to construction cost is likely to fall between these 0° and 30° limits. These calculations support the current industry practice of installing anchors at 15° to 30° to the horizontal, which is close to the optimum for achieving value for money.

Construction

Anchor block walls are constructed directly on the ground surface (with minor modifications to prepare the bearing pad) or are "top down" construction in cut slopes. In the latter case, "standup time" may be a factor to avoid local collapse of the exposed face. For widely spaced bearing pad elements, the slope faces between the pads may require support against shallow sloughing that is independent of a deeper landslide condition.

Both temporary and permanent near-surface slope support can be provided in several ways. These include reinforced shotcrete, short grouted soil nails perpendicular to the soil face, grouted reinforcement bars installed from the top of the slope into position just behind the proposed cut face (similar to reticulated piles) and various forms of lagging.

Anchor blocks can be formed and cast at the site. However, cast-in-place concrete blocks need significant time to set up forms and for the concrete to cure, and may not be suitable for an active landslide. An alternative technique is to precast the concrete blocks offsite and lift the blocks into place using a crane. Precast concrete is manufactured under closely controlled and supervised conditions but requires more site preparation than cast-in-place blocks. Controlled density fill (weakly cemented sand) is commonly used to seat precast blocks. On sites with difficult access or remote from a concrete batching plant, another alternative is to spray shotcrete into the block forms.

Staged construction may be necessary. The example given below describes a project where anchor blocks were installed in two stages.

Anchor Block Wall Example

The Richardson Highway slide is about 65 miles southeast of Fairbanks, Alaska. It is close to the Arctic Circle in an arid area of extreme temperature variations throughout the year. The slide is located on an outside bend of the Tanana River.

The landslide is about 550 feet long and crosses 140 feet of the Richardson Highway. The road is of cut-and-fill construction (Figure 19.16a,b). The width of the slide is controlled by structural features of the schist bedrock. There is abundant evidence of past landslide activity, including ancient slump features upslope of the currently active slide and river alluvium buried below the landslide toe.

Several inclinometers were installed in the landslide. They showed (Figure 19.16b) that slippage was occurring at a fairly consistent depth of about 40 feet below the ground surface.

The near-planar slip surface had an average slope of about 20° to the horizontal. Several borings had discontinuous permafrost conditions immediately below the slip surface.

The subsurface conditions consist of severely weathered schist overlying highly fractured hard micaceous quartzitic schist. The colluvium comprises silt, sand, rock fragments, and blocks of schist. The colluvium also includes zones with organics and tree limbs. The Tanana River is described as "very active" and carries surges of glacial runoff and eroded sediments.

Piezometers installed at several parts of the landslide encountered no groundwater levels except near the slope toe. It is thought that perched water may develop during the spring runoff and trigger landslide activity. However, movements (about 3 inches per month) appear to occur continually from late spring to fall, according to state maintenance personnel. It seems likely that the highly broken schist combined with the arid climate severely limit groundwater buildup in the main body of the landslide. River erosion plus some groundwater at the toe apparently are responsible for activating the landslide.

Anchor block walls were designed and built both above and below the road for landslide remediation. This treatment maintained the existing road alignment, accommodated traffic during construction, and was designed to resist earthquake forces. A conceptual-level design (Figure 19.16c) called for two anchor rows above the road in the cut slope and five anchor rows within a deep temporary cut below the road. The proposed construction procedure was to install and stress the anchors on the uphill side of the road first. With these anchors supporting the landslide, the anchors on the downhill side of the road could be installed by a "top down" construction procedure.

Earthquake load stability was based on a peak ground acceleration of 0.22 g and a horizontal seismic coefficient k_h of 0.11. A pseudo-static slope stability analysis gave a factor of safety F of 1.03 for temporary earthquake loading. The minimum long-term static factor of safety was set at 1.25 with short-term (during construction) minimum factors of safety of around 1.10.

The detailed anchor block wall designs were performed by Dr. John Byrne (Ground Support PLLC, Redmond, Washington) on behalf of the successful specialty ground anchor contractor, Malcolm Drilling Co., Kent, Washington. The number of rows in the lower wall was reduced from five to four (Figure 19.20).

Some details of the lower anchor block wall are shown on Figure 19.21. These include treated timber lagging wedged behind the anchor blocks, runoff drains, insulation (for the severely cold climate), and a shotcrete protective layer. The permanent stranded-wire anchor is shown schematically on Figure 19.22. The anchor design loads are given on Table 19.2.

The upper wall, constructed first, required only minor amounts of excavation within the existing road cut. The lower row of anchor blocks in the upper wall were placed and stressed as a group. The ditch in front of them was temporarily backfilled to provide a berm for construction of the upper row. All anchors in the upper wall were in place and stressed during the dormant season (before mid-May).

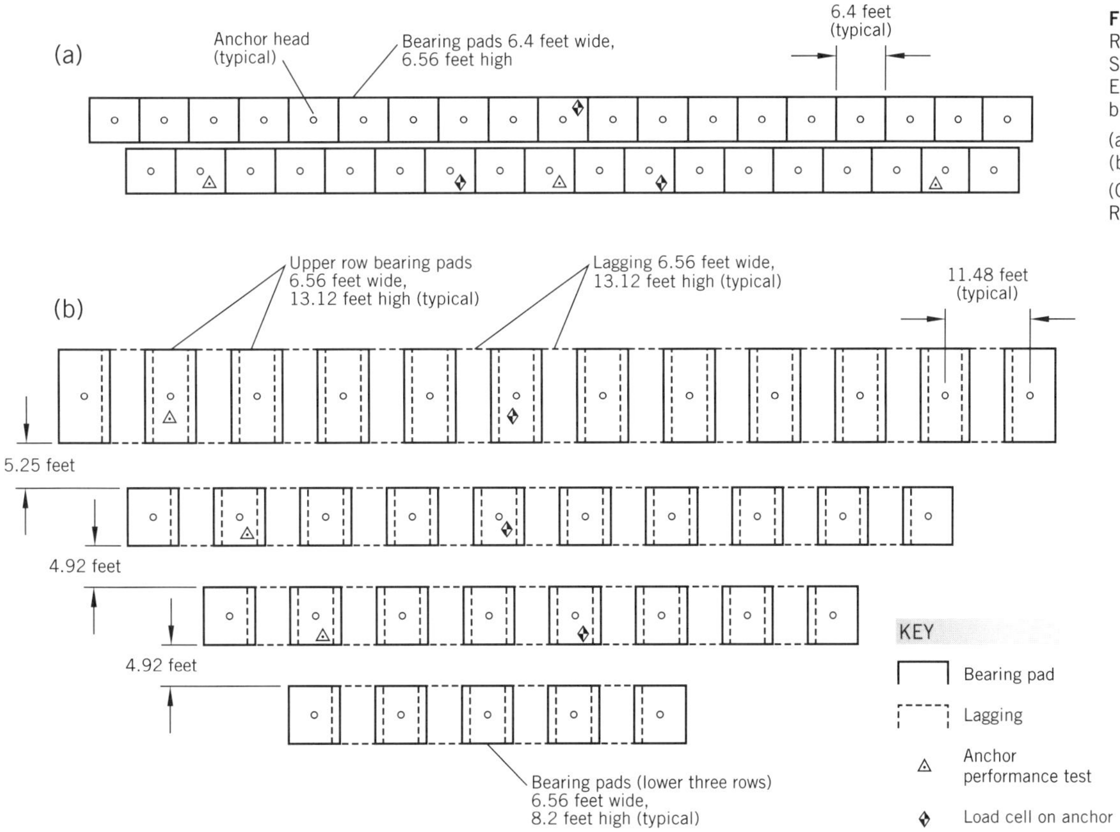

Figure 19.20 Richardson Highway Slide remediation. Elevation of anchor block walls:
(a) upper wall
(b) lower wall
(Ground Support, Redmond, Washington)

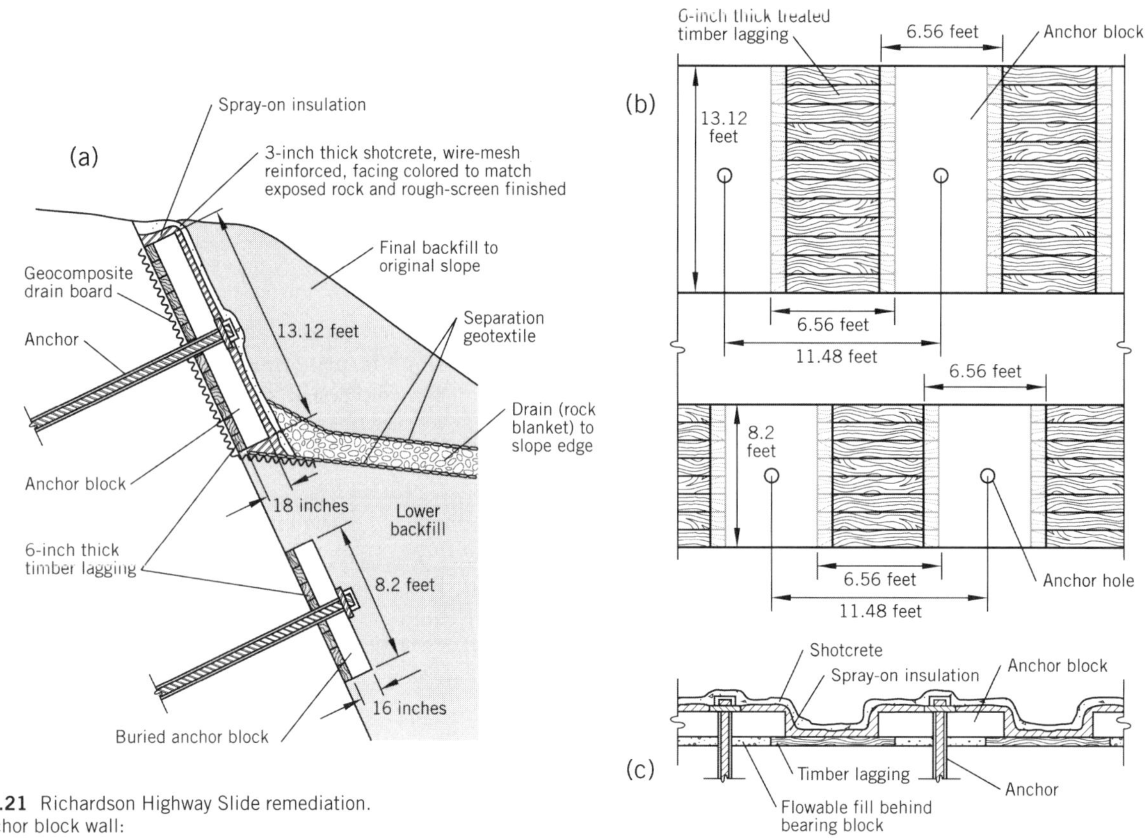

Figure 19.21 Richardson Highway Slide remediation. Lower anchor block wall:
(a) section
(b) elevation
(c) plan showing outer protection
(Ground Support, Redmond, Washington)

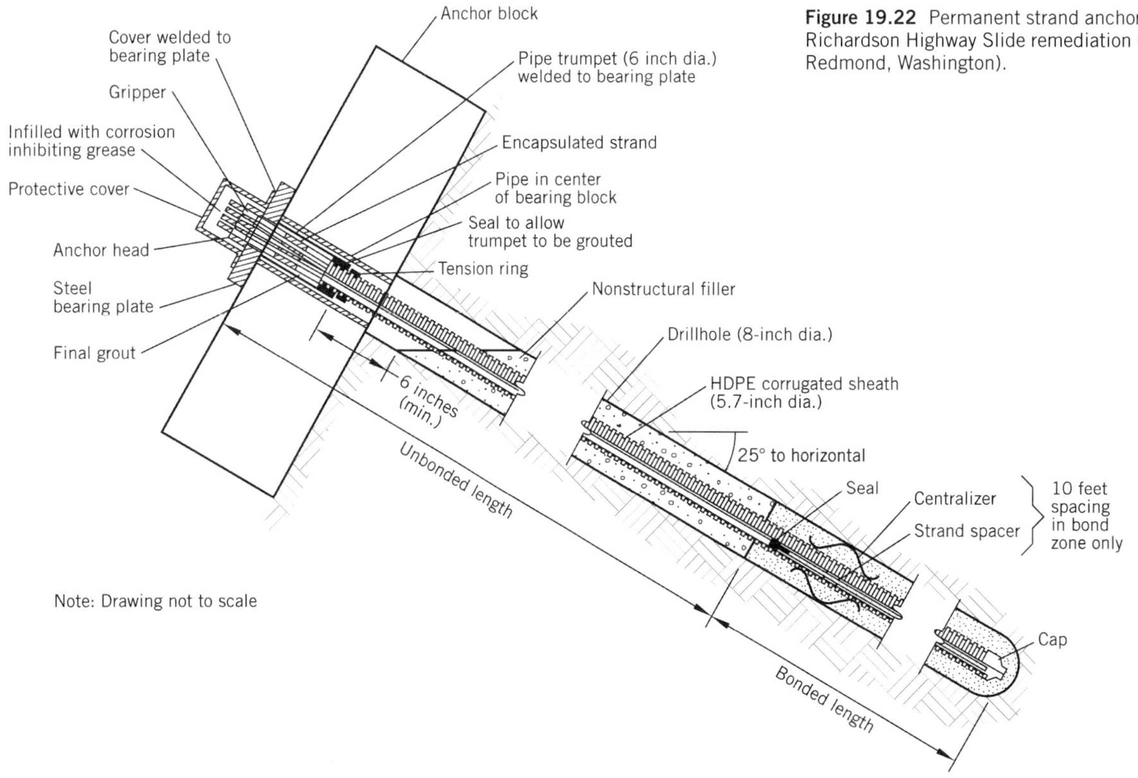

Figure 19.22 Permanent strand anchor design used at Richardson Highway Slide remediation (Ground Support, Redmond, Washington).

Table 19.2 Anchor Information, Richardson Highway Slide, Alaska

Location	Number of Anchors	Load per Anchor (kips)	Bond Length (feet)	Strands per Anchor
Upper wall	Total 37 (2 rows)	520	35	15
Lower wall	Total 35 (4 rows)	175	15	5

1 kip = 1,000 lbs.

After the upper part of the landslide had been stabilized by the upper wall, the lower wall was built "top down." Each anchor row was installed and locked off before moving onto the next lower row.

For the upper row of the lower wall (Figure 19.21a), a full-face coverage of drain mat was attached to the face and was held in place by the timber lagging boards. The concrete block holds the lagging in position (Figure 19.21 a, b, c) and the gap behind the block was filled with "flowable" fill. The anchor then was installed and locked off.

When all 12 anchors in the upper row had been installed, the row was completed by applying insulation and protective shotcrete over the entire face of anchor blocks and lagging. The same construction process was repeated for the lower three rows except that the drain mat, insulation layer, and shotcrete cover were omitted. The insulation and shotcrete treatment of the upper anchor row was provided as a precaution against freeze-jacking and overstressing of the anchors, should the anchors become exposed by future landslide movement below the treated ground.

After completion of all anchors in the lower wall, the slope was reinstated to its original profile in two stages. The lower part was backfilled to the base of the upper row and a drain was installed to remove any water collecting behind the upper anchors. The final backfill was added above the drainage blanket and against the upper row of anchor blocks.

19.4 TIED-BACK SOLDIER PILE WALLS

Technique

Tied-back soldier pile (also termed soldier *beam*) retaining walls are designed for both temporary and permanent support of cut slopes in soils. In landslide remediation, they can be placed at the base of an unstable slope to stabilize the entire landslide or can be placed part way up the slide to provide selective stabilization of a particular facility, such as a road or canal. It is "top down" construction that allows the slope to be supported at all times during remediation. Therefore, the wall can be built at any time of the year, including shortly after slope failure.

Another benefit of soldier pile walls is that the geotechnical designer has a broad choice in terms of the margin of safety built into the wall. It is simply a matter of increasing the number of ground anchors and/or raising the pullout resistance of anchors to increase the design factor of safety.

The earth pressures are semi-empirical, combining conventional active and passive formulas with historical field observations of loads on strutted excavations. The result is an *apparent earth pressure diagram* that does not attempt to predict the actual earth pressures, which depend on many factors occurring during the construction process, but provides a relatively conservative envelope that should exceed the actual pressures and should, therefore, lead to a safe design. This section will first present the usual design approach for soldier pile walls built to support cut slopes made into stable ground. It will next discuss the extension of this approach to stabilization of landslides.

Construction of the wall requires access for heavy equipment and sufficient room for maneuverability of the equipment at the site. Specialty contractors often provide the design as well as installing the wall. For landslide stabilization, the geotechnical designer should provide the pressures that the wall has to be capable of resisting. In practice, it is unlikely that the wall will ever be subjected to these pressures. The factor of safety provided in landslide stabilization requires resistance to be substantially higher than is likely to occur in service.

Ground anchor design and construction are presented in Section 19.2. This topic will not be further discussed in this section.

A tied-back soldier pile retaining wall is a relatively expensive form of remediation, but is usually less costly than a conventional retaining wall for deep cuts. The probability of success is very high. For a more detailed analysis, Weatherby (1998) provides excellent coverage.

Typical Applications of Soldier Pile Walls

Soldier pile walls are built by "top down" construction methods. They are very adaptable to many landslide conditions where the ground is marginally stable, such as shortly after a landslide has occurred. If removal of the landslide debris below the wall is essential, the "top down" construction procedure ensures that the slope behind the wall will remain stable throughout the construction process. This protects the public, construction personnel and equipment, and any "at risk" facilities further downslope. In the United States, there are many specialty contractors available to design and install soldier pile walls.

There are numerous potential applications for the use of soldier pile walls in landslide stabilization or prevention. Some applications are provided below, but there are undoubtedly other situations where a soldier pile wall will be cost-effective compared to alternative treatments.

Temporary Support

Although rarely done in practice due to the extra costs involved, a soldier pile wall can be built to temporarily support a landslide so that another form of remediation can be installed. The principal advantage is that the vertical face provides room for construction on the downhill side. Two exam-

Figure 19.23 Use of temporary soldier pile walls to stabilize a landslide:

(a) support to allow construction of replacement buttress
(b) support to allow construction of a gravity retaining wall

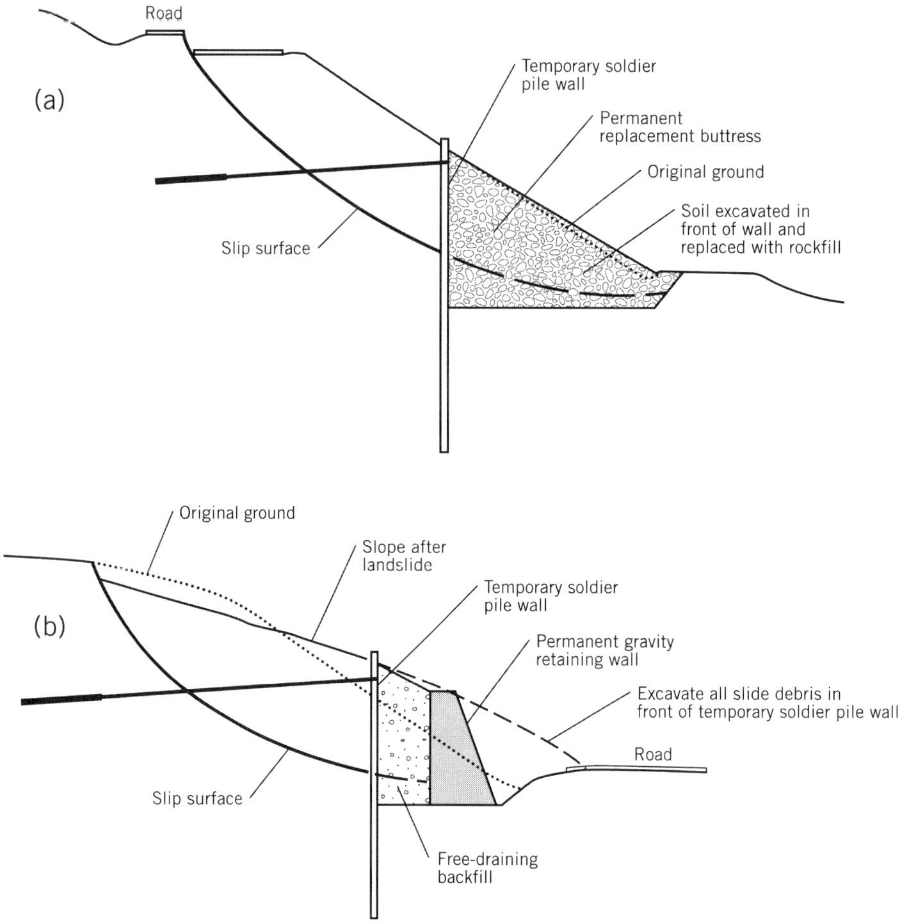

ples are illustrated on Figure 19.23. In example (a), the soldier pile wall creates space for a replacement buttress of rockfill at the bottom of the slope. The rockfill has a higher strength than the soil on the slip surface of the landslide. If a 1:1 temporary cut slope had been used, it would have extended a long way upslope, necessitating more excavation and backfill, summer construction, and temporary closure of the entire width of road near the top of the slope. In example (b), the temporary soldier pile wall provides space for the construction of a gravity wall (many types available) with free-draining backfill behind the wall. For the designers, the gravity wall may be preferable to a permanent tiedback soldier pile wall due to corrosive soil, aesthetics, drainage, concern over freezing conditions, etc. The soldier piles can be recovered or left in place at the end of construction.

Permanent Support

The situation illustrated on Figure 19.24 is a fairly common landslide occurrence. A cut is made into a sloping hillside to provide level ground for a structure, such as a building or water tank (Figure 19.24a). The cut passes through waterbearing strata (often colluvium or residual soils) and into an impermeable bedrock or hard clay. Construction occurs in the drier summer months; a small spring or line of seepage develops at the interface between the two strata. When winter arrives, the seepage becomes vigorous, sending sloughed soils to the bottom of the cut slope and slowly undermining the slope above. A storm then causes a general slope failure, often extending a considerable distance upslope from the cut (Figure 19.24b). Water-softened slide debris piles up against the structure and may cause extensive damage and loss of use.

One solution to this landslide is to construct a soldier pile wall at the base of the original cut slope (Figure 19.24c). A temporary access road is created to provide a pad for the crane or piledriver. The soldier piles are installed and the tieback anchors are placed as the slide debris is progressively excavated. Finally, the damaged structure is repaired. The soldier pile wall allows the repairs to be made shortly after the failure without having to wait for lower groundwater conditions in the following summer.

Another application of soldier pile walls is to repair slopes alongside rivers that have been damaged by floods or surface runoff. An example is shown on Figure 19.25, based on an actual landslide that was stabilized by other means but a soldier pile wall was considered as a viable option. The failure was a flow slide caused by flood conditions in the river and surface runoff overflowing from a ditch. The river bank comprised fine sandy soil eroded from sandstone cliffs further up-

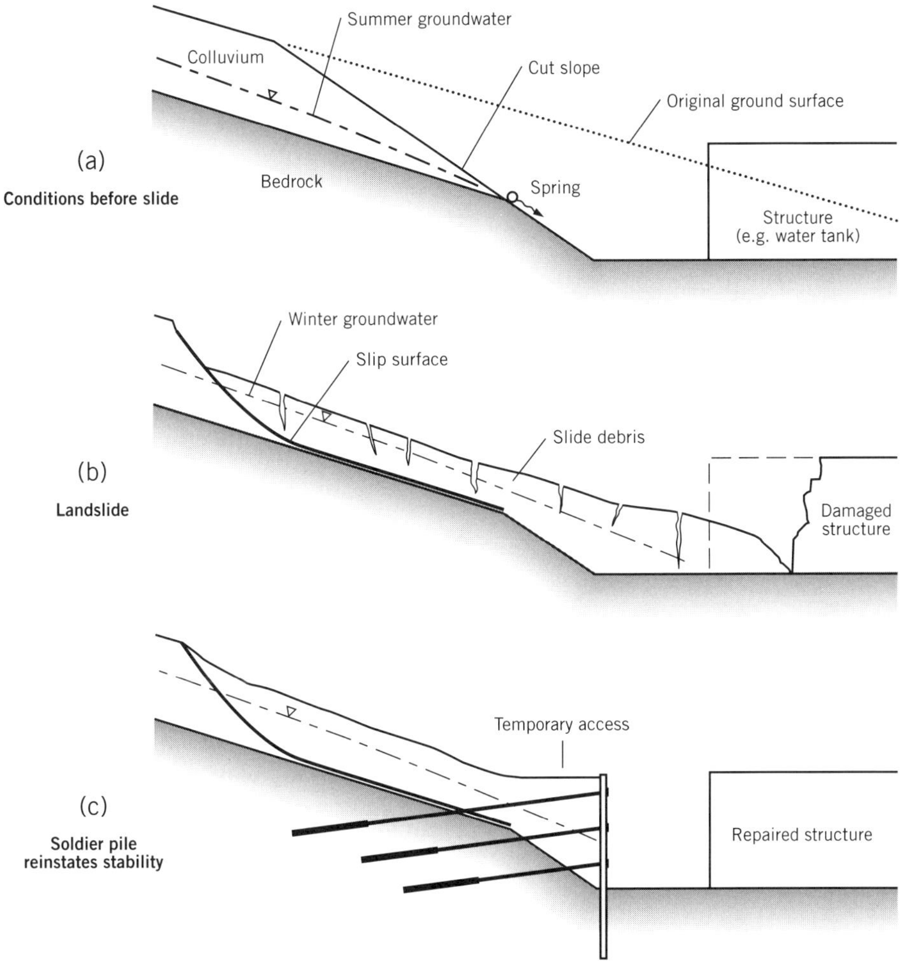

Figure 19.24 Applications for a soldier pile wall; restoration of a failed cut slope.

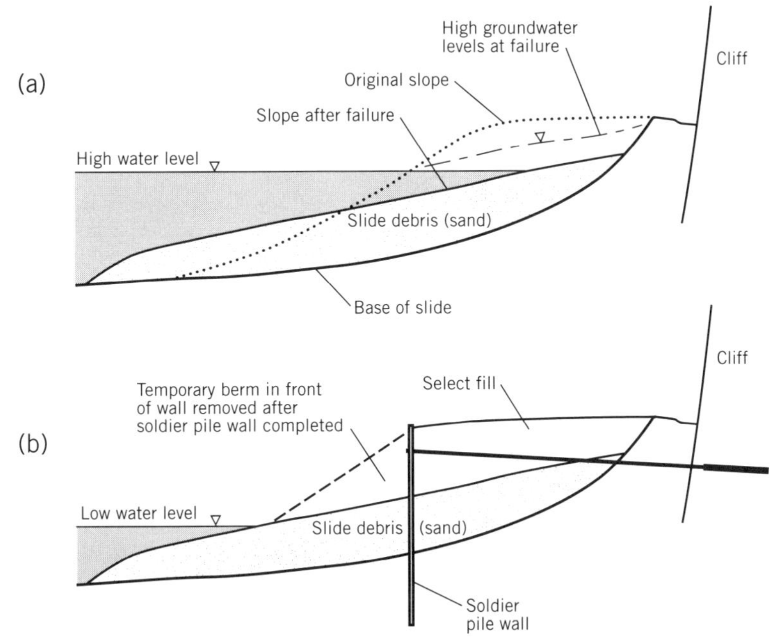

Figure 19.25 Applications for a soldier pile wall; restoration of a failed river bank:
(a) slope failure
(b) possible repair by use of a soldier pile wall

slope. The stabilization procedure would have been to first construct a berm of select free-draining fill to replace the lost ground (Figure 19.25b). The soldier pile would have been built with ground anchors extending into the sandstone cliff. Fill in front of the wall would have been excavated to approximately restore the river bed to its former section.

Another example, in this case selective stabilization, is the landslide slump affecting a road at Mohler, Oregon, when the Nehalem River was in flood (Figure 19.26). The shoulder of the road slumped, leaving a dangerous headscarp 5½ feet high next to the road. The strip of land alongside the river was privately owned and may have required significant time to be acquired by the county. Instead, the county elected to selectively stabilize the road rather than the entire landslide. Conceptually, the procedure was as follows (Figure 19.26b,c):

1. After the high river level and groundwater levels subside, add granular fill over the slump to temporarily restore the original section for construction access.
2. Drive soldier piles with sufficient anchorage D below the landslide slip surface.
3. Install lagging and tieback anchors while progressively excavating the soil prism ABC from the front of the wall. Soil prism ABC could be replaced after completing the wall to reinstate the original conditions downslope.

It should be mentioned that specialty contractors are able to install tieback anchors from the riverward side of a wall. The drill rig is either mounted on temporary scaffolding on the river bank or can be put on a platform suspended by a crane.

Another concern in many landslides is the possibility of future retrogressive (uphill) movements behind the headscarp (Figure 19.27). A soldier pile wall built at the headscarp can safeguard a structure further upslope.

Inappropriate Uses

Soldier pile walls should not be considered for the following situations:

- A landslide that is actively moving at the time of construction, such that the ground anchors would be damaged before stability is restored. (*Note:* Many landslides are stable during the drier months of the year, allowing construction to take place.)
- A landslide with no access for heavy construction equipment.
- A landslide with insufficient working space at the proposed wall location.
- A landslide where the slip surface is at a considerable depth below the lowest ground anchor on the wall (i.e., the base of the excavation is too high above the slip surface).

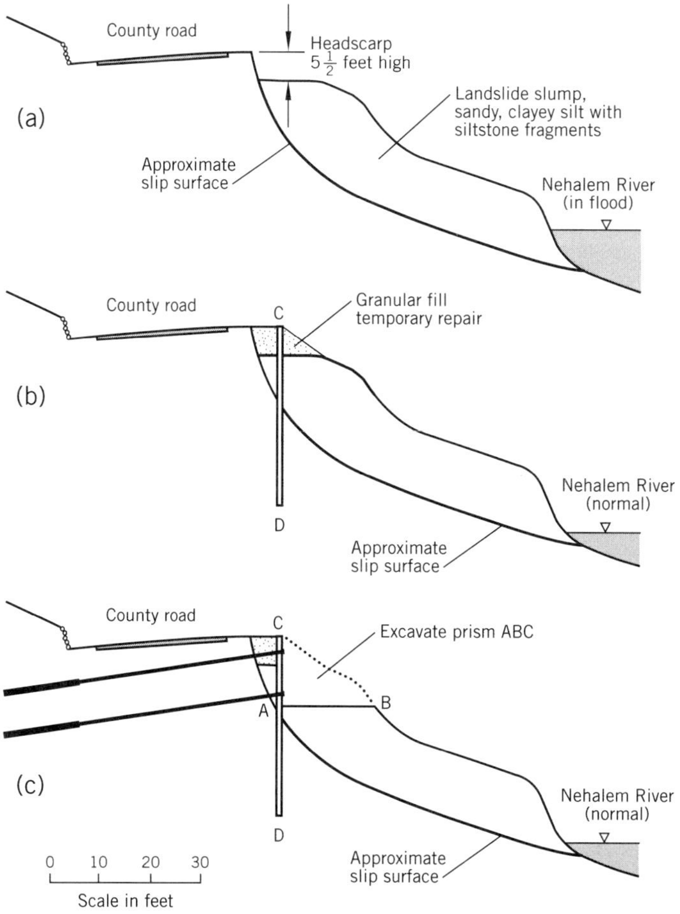

Figure 19.26 Applications for a soldier pile wall. Selective stabilization of a road at Mohler, Oregon:
(a) slope failure during river flood
(b) soldier pile installation
(c) final appearance of wall section

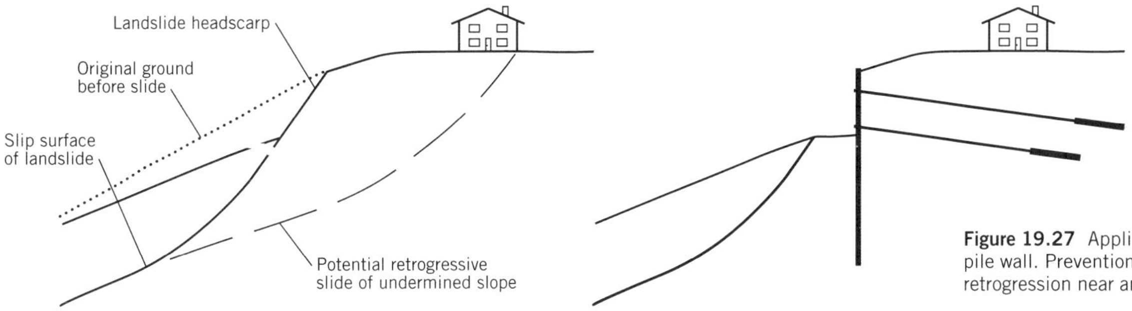

Figure 19.27 Applications for a soldier pile wall. Prevention of headscarp retrogression near an upslope structure.

Construction of Soldier Pile Walls

A soldier pile usually consists of either a wide flange steel H section or a steel double channel section. In constructing a soldier pile wall, the piles are installed first. They can be driven into the ground, using armored tips if needed to penetrate into a harder layer or to pass through boulders within the landslide debris. Alternatively, a hole can be drilled and the H section placed in it and backfilled with concrete. In this case, structural or lean-mix concrete is used from the base of the pile to the excavation level, and a very weak concrete is placed from the excavation level to the ground surface. The weak concrete is chipped off during excavation to expose the pile for installing anchors and lagging.

Advantages of a predrilled hole are: (i) the concrete provides long-term corrosion protection to the pile surfaces in contact with the soil, and (ii) the vertical alignment of the pile can be checked before the concrete is placed. Predrilling is, therefore, preferable to driving soldier piles in most cases.

Installing the soldier pile is the first step (Figure 19.28). Excavation on the downslope side of the wall is taken to about 2 feet below the level of the first tieback anchor (Figure 19.28, step 2). During excavation, lagging boards are inserted between adjacent soldier piles to retain the earth. The lagging spans the distance between adjacent soldier piles, and is held in place with wedges or shims. A small gap between boards allows drainage. The lagging has to be a close fit to the vertical face of soil behind it so that the soil face does not move significantly when the tieback anchors are stressed later.

The tieback anchor is installed and, within a few days, it is prestressed to the selected "lockoff" load. All anchors are test-loaded to at least 125 percent of the design load (see Section 19.2) before being accepted. In temporary applications, the tieback load is transferred to the adjacent soldier piles by a horizontal wale structural member that is usually attached between the flanges of the H section. In some installations, especially permanent walls, the tieback is fixed directly to the H pile; i.e., as concentric to the web as possible.

The construction sequence continues to the full depth of excavation in front of the wall (Figure 19.28, step 3). Lagging is installed after each 4- to 5-foot depth of excavation. This depth interval has to be shortened if the ground starts to cave. The final steps are to add drainage measures to the outside of the lagging and then construct a permanent concrete facing (typically 8 to 12 inches thick) to the wall (Figure 19.28, step 4).

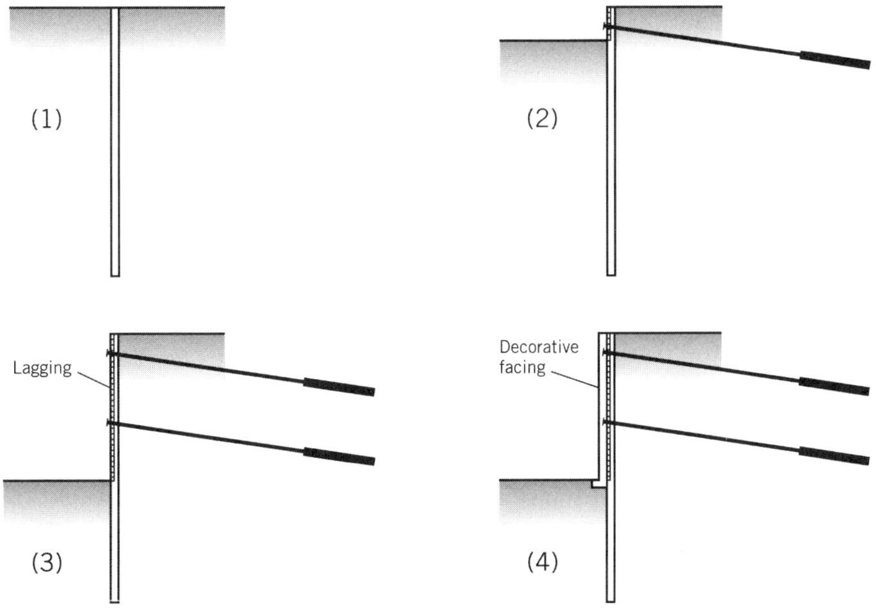

Figure 19.28 Construction sequence for a soldier pile retaining wall (see text).

Soldier piles typically are spaced 6–10 feet apart horizontally (4 feet minimum to avoid overlapping stress patterns). Anchor spacing is 8–12 feet apart vertically. The lagging is pressure-treated timber boards or reinforced shotcrete sprayed on the open face. Some soils, such as loose sands, gravels, or soft clay, have a very short "standup" time and need to be supported quickly. The permanent facing is either cast-in-place concrete or precast concrete panels attached to the wall.

On larger walls, the horizontal walers are often installed on the outer face of the soldier pile. However, this thickens the section and may expose the pile to corrosion. It is also aesthetically unattractive on permanent walls and, if a cast-in-place concrete facing is used, will require a thicker concrete facing.

Design of Soldier Pile Walls

Pressure Diagram

The design lateral earth pressures for soldier pile walls are based on the classical pressure diagrams described in Terzaghi and Peck (1948). These are semi-empirical pressure distributions originally intended for the design of struts in braced excavations such as trenches (Figure 19.29). The pressure diagrams are not intended to represent the actual pressures behind the supported excavation, which are complex and dependent on the contractor's method of installation. Rather, they approximate the shape and magnitude of the earth pressures after the wall has been completed with some built-in conservatism; i.e., the pressure diagrams should envelop (slightly exceed) the actual pressures.

The Terzaghi and Peck pressure diagrams have served the profession well and are generally applicable to ground anchors. Some modifications are gaining acceptance as reasonable adjustments to the diagrams (Figures 19.30. 19.31). Henkel (1971) indicated that the active earth pressure increases if the soil in the foundation is a soft clay because the clay yields and moves toward the excavation (Figure 19.30). This condition occurs when the base stability factor N exceeds 5.14. The increment ΔK has to be added to the normal active

Figure 19.29 Apparent earth pressure diagrams proposed by Terzaghi and Peck (1967).

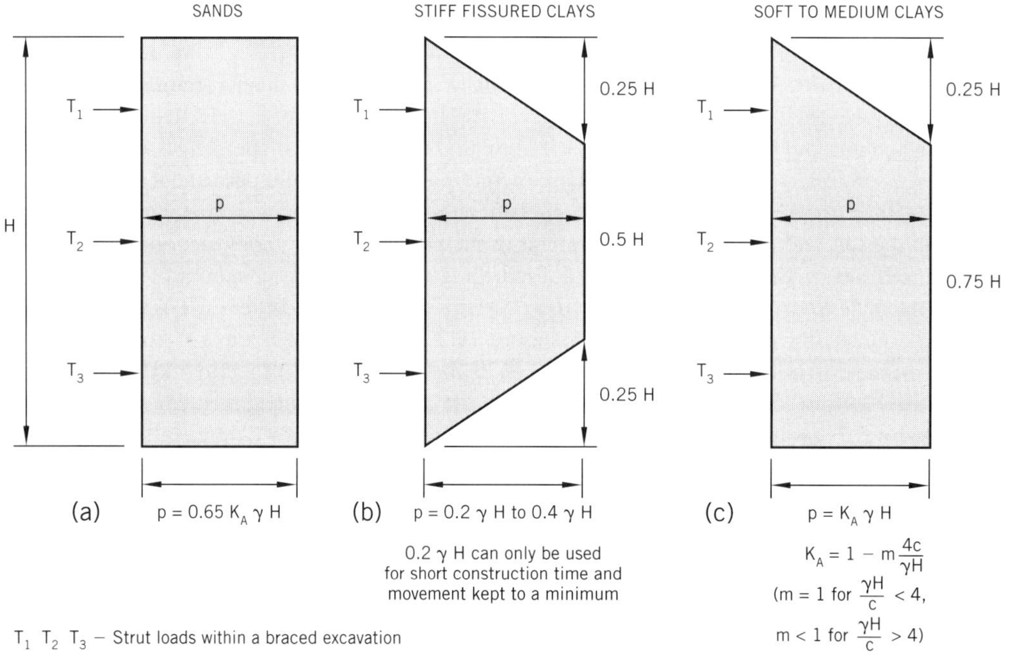

Figure 19.30 Correction to the active pressure on a soldier pile wall resulting from soft clay in the foundation (after Henkel, 1971).

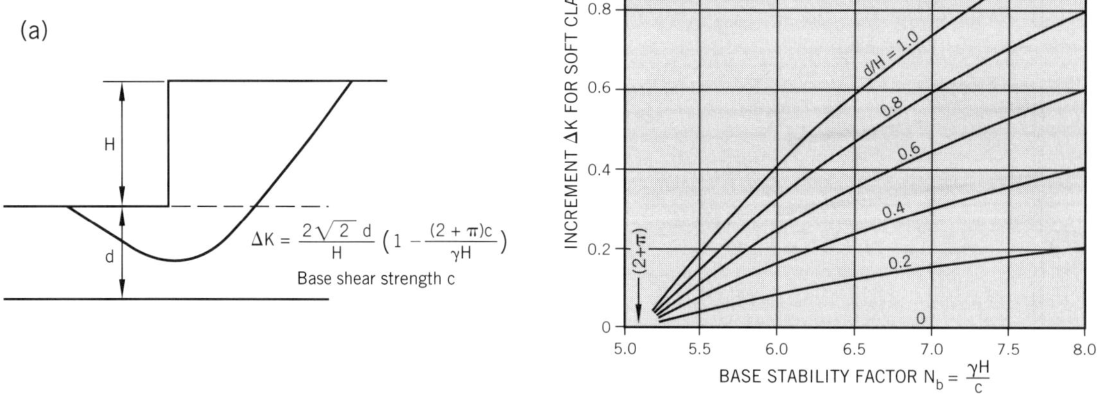

earth pressure for softer clays. In practice, permanent soldier pile walls (or other walls relying on ground anchors) are not built where the bottom of the wall is underlain by weak clay. However, temporary walls occasionally are built in these materials.

A chart (Long et al., 1998) to quickly determine the lateral load for soft to medium stiff clay is given on Figure 19.32. The average or selected undrained shear strength $\bar{c}$ of the clay is determined, then divided by a factor of safety of 1.3. Using the reduced shear strength, draw a vertical line to the wall height H and read off the earth pressure factor X. The total lateral load is XH^2. Next, distribute the load on the wall according to the apparent earth pressure diagram of Figure 19.31 for soft to medium clay. *Note:* If the wall height is *to the left* of the vertical line representing the shear strength on Figure 19.32, treat the clay as a stiff clay for design purposes.

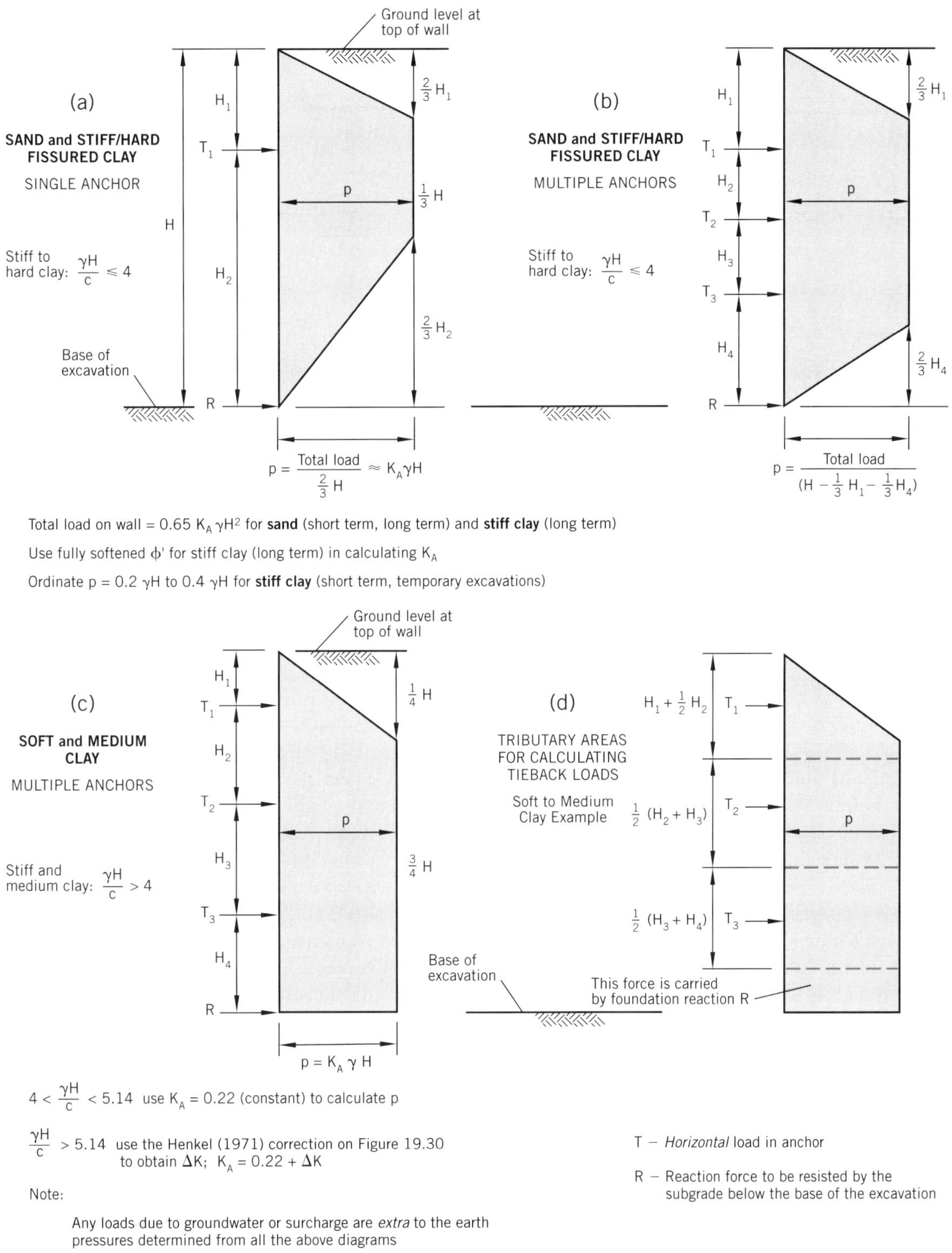

Figure 19.31 Apparent earth pressure diagrams behind an anchored wall (after Sabatini et al., 1999).

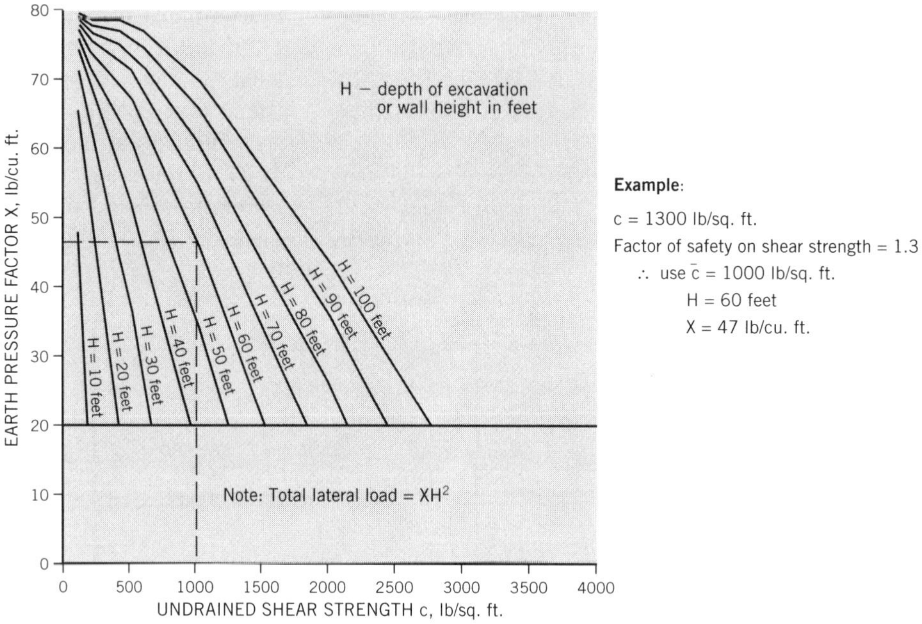

Figure 19.32 Earth pressure factors for soft to medium stiff clay, no base failure (after Long et al., 1998).

More recently, the pressure diagram for sands has been modified by Sabatini, Pass and Bachus (1999), as shown on Figure 19.31(a)(b). The justification for the changes are: (i) there can be no lateral pressure acting at the top of the wall in cohesionless soils; (ii) stresses are highest around the anchor points so the maximum stress p has to be reached above the first anchor and below the lowest ground anchor (the set distances of the Terzaghi and Peck (1967) diagrams do not take account of the actual strut levels); and (iii) earth pressures reduce below the lowest anchor in medium dense to very dense sands due to the passive resistance that is developed below the base of the excavation.

For stiff to hard fissured clays, the lateral pressures are harder to quantify because such clays generally lose strength with time after a cut has been made. A lower strength applies more lateral pressure. Sabatini et al. (1999) have reviewed several recommended pressure diagrams from different authors. They show that the maximum pressure ordinate p ranges from 0.15 γH to 0.4 γH, similar to the Terzaghi and Peck (1967) range. These values are subject to modification by the geotechnical designer based on local experience and practice. Observed lateral stresses measured in the field fall within this range. In the absence of local experience, the more conservative value of 0.4 γH probably should be used in temporary wall design. This provides a total load of around 0.33 γH^2, the exact load being dependent on the anchor locations (Figure 19.31).

For long-term conditions, Sabatini et al. (1999) recommend that the lateral pressures for stiff to hard clays be based on the *drained, fully softened* strength of the clay, i.e., $c' = 0$. The total load on the wall is the same formula as for sand, 0.65 $K_A \gamma H^2$ where K_A is defined as $(1 - \sin \phi') / 1 + \sin \phi')$. After comparing the short-term temporary loading with the drained loading, choose the higher of the two values for design purposes. The long-term earth pressures apply to most permanent landslide applications of a soldier pile wall.

For soft to medium stiff clays, Sabatini et al. (1999) recommend the loadings shown on Figure 19.31(c). These are significantly different from the Terzaghi and Peck diagram of Figure 19.29(c) in terms of the maximum pressure p, and take account of the Henkel (1971) correction given on Figure 19.30.

The lateral pressure on the back of a soldier pile wall is increased by a slope behind the wall, which commonly occurs in landslide work. The active pressures behind a wall with sloping ground at angle β to the horizontal are shown on Figure 19.4. (*Note:* negative backslope angle, non-vertical wall faces, passive pressures, and skin friction $\delta < \phi'$ are not shown but are available from other textbooks on soil mechanics). Groundwater and surcharge loading also have to be included (e.g., 250 psf uniform pressure for traffic loading, if applicable).

In landslides where the soldier pile wall is close to the headscarp, pressure diagrams based on the earth pressure theories described above may be sufficient. On landslides where the wall is much further downslope, the lateral pressures to be resisted by the wall will depend on limit equilibrium calculations (Figure 19.33). The lateral force required to establish equilibrium in the landslide can be calculated by an original profile analysis (Chapter 10, Section 10.2). The design lateral force can be determined as some multiple of this force (e.g., a factor of 3). Thus, if the net lateral force P_1 needed to re-establish static equilibrium at the soldier pile wall site (due to the loss of support from ground ABC) is 20 tons per foot width, the design lateral force P would equal $3P_1$, i.e., 60 tons per foot width (Figure 19.33c). For soldier piles at 7-foot centers, the total lateral force to be resisted by each soldier pile is 420 tons.

Another design approach is to calculate the lateral load P_1 needed at the soldier pile wall site by the original profile analysis and then add the lateral load that achieves a selected

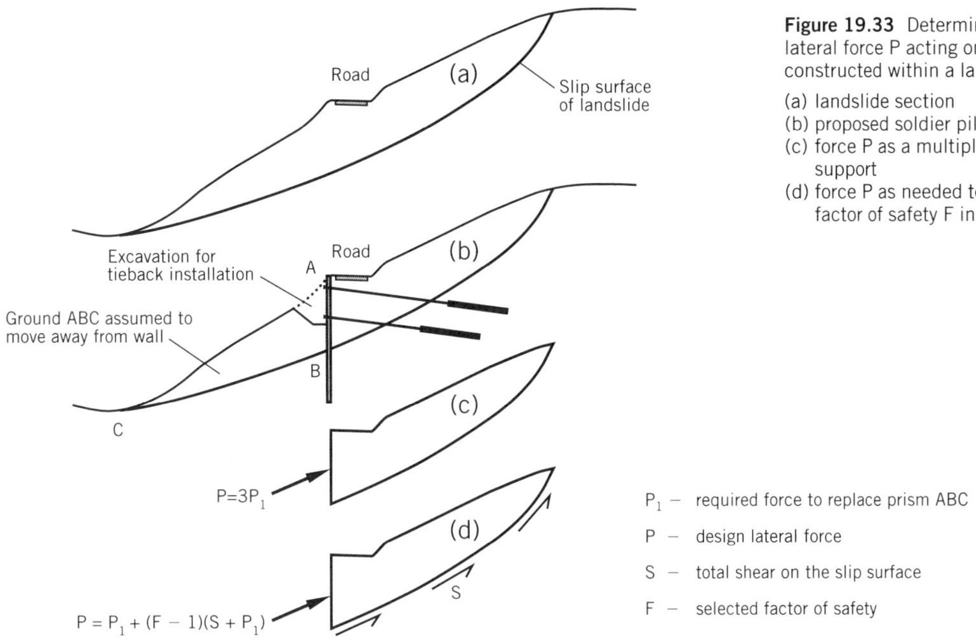

Figure 19.33 Determination of the design lateral force P acting on a soldier pile wall constructed within a landslide:
(a) landslide section
(b) proposed soldier pile wall
(c) force P as a multiple of the net loss of support
(d) force P as needed to achieve a selected factor of safety F in the supported landslide

static factor of safety along the slip surface of the landslide behind the wall (Figure 19.33d). For example, if the original profile analysis indicates that a lateral force P_1 of 20 tons per foot width is needed to replace the ground downhill of the wall and 45 tons per foot width is needed to increase the factor of safety of the landslide above the wall from 1.00 to 1.25, the total resistance needed from the wall is 65 tons per foot width. Performing both these design approaches should help the geotechnical designer to determine an appropriate lateral force for design of the wall. Once the force has been established, the lateral pressure diagram can be drawn based on the shapes given on Figure 19.31.

The suggested design procedure to analyze the lateral wall forces for a landslide is:

Step 1. Calculate the lateral pressure diagram based on the procedures for a cut into a stable slope (Sabatini et al., 1999). Allow for any slope above the wall, groundwater, and surcharges.

Step 2. Calculate the lateral pressure using the principle of original profile analysis (Chapter 10, Section 10.2) to determine the net loss of support (gravity force minus shear resistance) of the ground downhill from the wall. This value can be obtained from a computer stability analysis for the force on the slice corresponding to the wall location. Multiply this last force by a factor of safety of 3 to 4.

Alternatively (or as an additional calculation), determine the lateral force on the wall needed to increase the factor of safety of the landslide behind the wall to some selected value.

Determine the design lateral force based on one of these sets of calculations.

Step 3. Select the larger of the lateral forces calculated by Step 1 or Step 2.

Step 4. Determine the apparent earth pressure diagram (Figure 19.31) for the soil type of the landslide debris behind the wall.

The *tieback anchor loads* T_1, T_2, T_3, etc. can be calculated by the tributary area method, as shown on Figure 19.31(d). If the anchor is inclined at angle α to the horizontal, the inclined tensile force $T_i = T_h \sec \alpha$ where T_h is the horizontal component calculated from the apparent pressure diagram. Since this value is per foot width, the actual tendon load is $T = S_h T_h \sec \alpha$ where S_h is the horizontal spacing between ground anchors.

The lateral loads below the base of the wall can be analyzed by the relationship of Broms (1965) shown on Figure 19.34. For driven soldier piles, width of pile b is the flange width; for drilled-in soldier piles backfilled with concrete, use the diameter of the concreted hole for b. Note that there is no resistance for a depth of 1.5 b in cohesive soils to allow for ground disturbance during construction. Usually, the soldier pile is driven a sufficient distance into the ground to develop fixity at the lower end. A pressure diagram similar to that shown on Figure 19.35 is obtained.

An alternative set of passive calculations use the Wang and Reese (1986) approach, as summarized by Weatherby (1998). Computer programs are available to model the wall-soil interaction.

The wall needs to develop enough passive resistance to carry the reaction force R (Figure 19.31) at the base of the wall with a factor of safety of at least 1.5. The active pressure of the embedded portion of the pile has to be added to force R. A reversal of curvature in the deflected shape of soldier pile wall below ground indicates that fixity is reached.

The analyses of bending moments, walers, and lagging design, etc. is not described here. These are structural design procedures (refer to Weatherby (1998) and other textbooks directed at wall design).

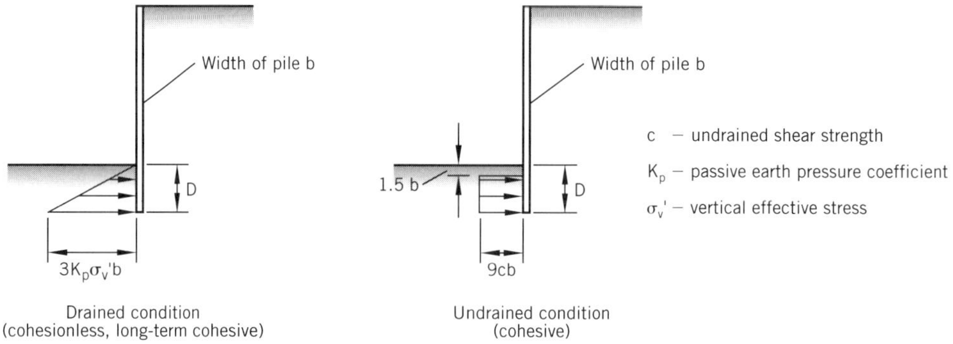

Figure 19.34 Ultimate passive resistance diagrams at the base of a soldier pile wall (after Broms, 1965).

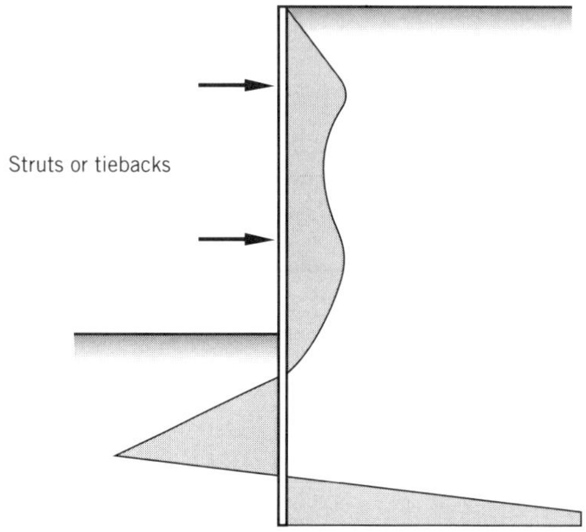

Figure 19.35 Shape of active and passive pressures on an anchored wall with lower end fixity.

Analysis of Vertical Cut

The calculation is shown on Figure 19.36 (Weatherby, 1998). The anchor load T contributes to wall stability as well as the wall embedment. The solution is found iteratively by varying angle α and wall depth of embedment d to obtain the maximum P_{REQ}. This load is redistributed as an apparent pressure envelope to calculate the ground anchor loads and bending moments in the wall.

Outer Facings

The outer facing of a wall normally uses cast-in-place or precast concrete for a permanent installation. For a tall wall, a battered outer face is less threatening to people below the base of the wall. Curved walls, battered walls, or walls of variable height generally have a cast-in-place face. Precast concrete panels offer a variety of architectural finishes. The precast panels can be made from reinforced or prestressed concrete. Handling and lifting stresses have to be considered in their design. The facings are designed to resist apparent earth pressures rather than landslide pressures.

Groundwater Control

Soldier pile walls are free-draining, but a shotcrete or cast-in-place concrete facing can impede groundwater flow. Prefabricated drains, typically 16 inches wide with a geotextile fabric on one side, are installed against the lagging at the mid-span between soldier piles and at construction and expansion joints. The drains are extended to the base of the excavation and have vertical overlap joints of at least 16 inches. A footing drain below finished grade, or weepholes near the base of the wall, collect and discharge any groundwater.

For precast concrete panels, any spaces behind the wall can be infilled with free-draining aggregate to provide the dual function of drain and backfill. Water is collected by weepholes or a footing drain, as already described. For a permanent shotcrete facing, the geotextile drain is placed with the fabric side against the soil.

Cost of Soldier Pile Walls

The approximate cost (1997) of installing permanent soldier pile walls in the United States has been collected by Weatherby (1998) using the following assumptions: Driven soldier piles at 8-foot centers. Drilled soldier piles at 10-foot centers. Ground anchors, including connections and testing: \$1,000 to \$6,000 each. Each anchor supports 100 sq. ft. of wall.

Table 19.3 summarizes the costs in terms of dollars per sq. ft. of exposed wall face. Specialty contractors often quote approximate costs in these terms.

In making cost comparisons between soldier pile walls and other remedial treatments, a geotechnical engineer should also consider savings from faster construction and reduced inconvenience to the owner and public. Design costs may be part of a design-construct package offered by a specialty geotechnical contractor.

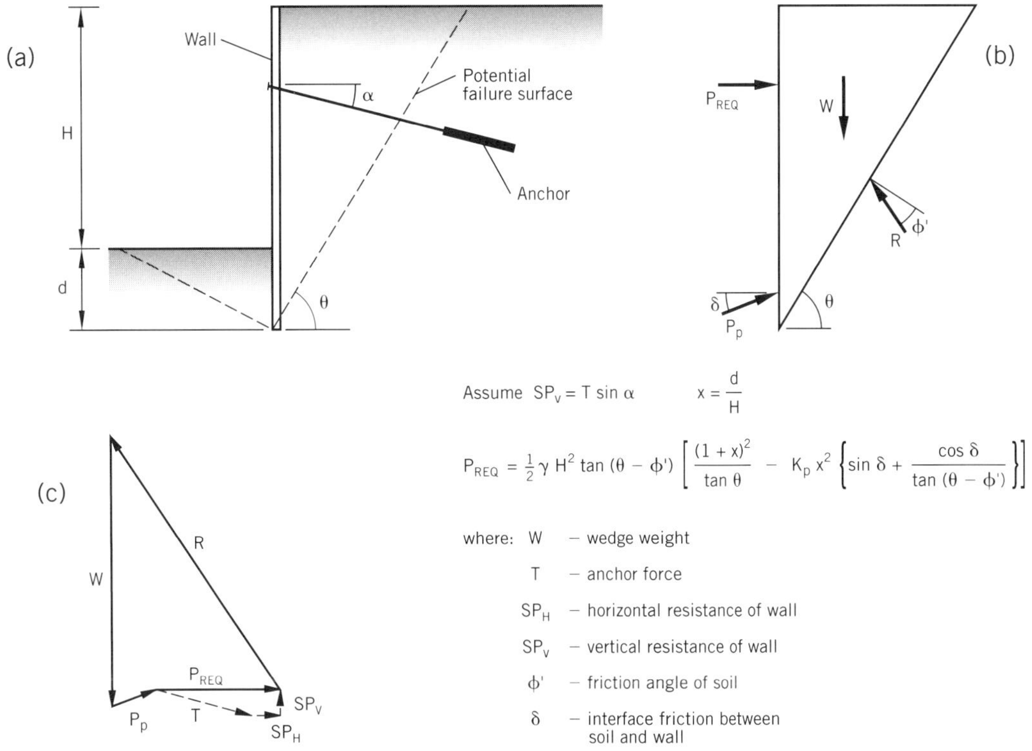

Figure 19.36 Force equilibrium method of determining the force P_{req} to be resisted by an anchored vertical wall (after Weatherby, 1998).

Table 19.3 Approximate Cost of Constructing Permanent Soldier Pile Walls per Square Foot of Wall Face*

	Driven Soldier Piles			Drilled Soldier Piles		
Cost Elements	Min. $	Avg. $	Max. $	Min. $	Avg. $	Max. $
Soldier piles	3	4	6	10	15	30
3-inch temp. lagging and drainage	5	8	12	5	8	12
Anchors	15	20	60	15	20	60
Cast-in-place facing	11	17	25	11	17	25
Total costs / sq. ft.	34	49	103	41	60	127

Engineering News-Record Construction Cost Index (1997): 5860. *Weatherby, 1998

19.5 CONCRETE SHEAR PILE WALLS

Technique

Heavily reinforced concrete piles can be designed to stabilize landslides. Such shear piles are relatively easy to construct and do not require specialty subcontractors. The pile wall also may be buried within the landslide, making remediation less obtrusive than other techniques.

The reinforced concrete piles pass through the landslide slip surface and are anchored at their lower end in the stable ground below, which is often bedrock. The pile length within the landslide mass can be free-standing (i.e., cantilevered above the stable ground) or tied-back by ground anchors (as discussed in Section 19.2). Actual examples are illustrated on Figure 19.37.

The cast-in-place piles are "top-down" construction, which is especially beneficial to landslide remediation. Construction starts at the ground surface and soil on the downhill side is not excavated, if required, until the pile wall is completed. Shear piles are a relatively expensive remedial treatment, but are very effective for partial stabilization in which a road, railroad, or canal has to be protected and access to the landslide slope is difficult.

Pile Wall Configurations and Construction

Reinforced concrete cylinder piles placed in a row to form a wall (Figure 19.38) have various configurations, namely:

1. A *tangent* pile wall, where the piles are side-by-side in a straight line.
2. A *secant* pile wall, where the piles are side-by-side but intersect the adjoining piles on each side.
3. A *staggered* pile wall, where the piles are side-by-side in a staggered alignment.
4. A *spaced* pile wall, where soil or weak unreinforced concrete fills the gap between adjacent piles.

All of these configurations can be used to stabilize landslides. However, it should be recognized that the force that has to be arrested is typically much larger than the Coulomb or Rankine pressures experienced by a wall built into originally stable ground; therefore, the piles have to be closely spaced and heavily reinforced.

Reinforcement for the piles is either by a steel H-beam, with the web oriented in the direction of sliding, or by a cage

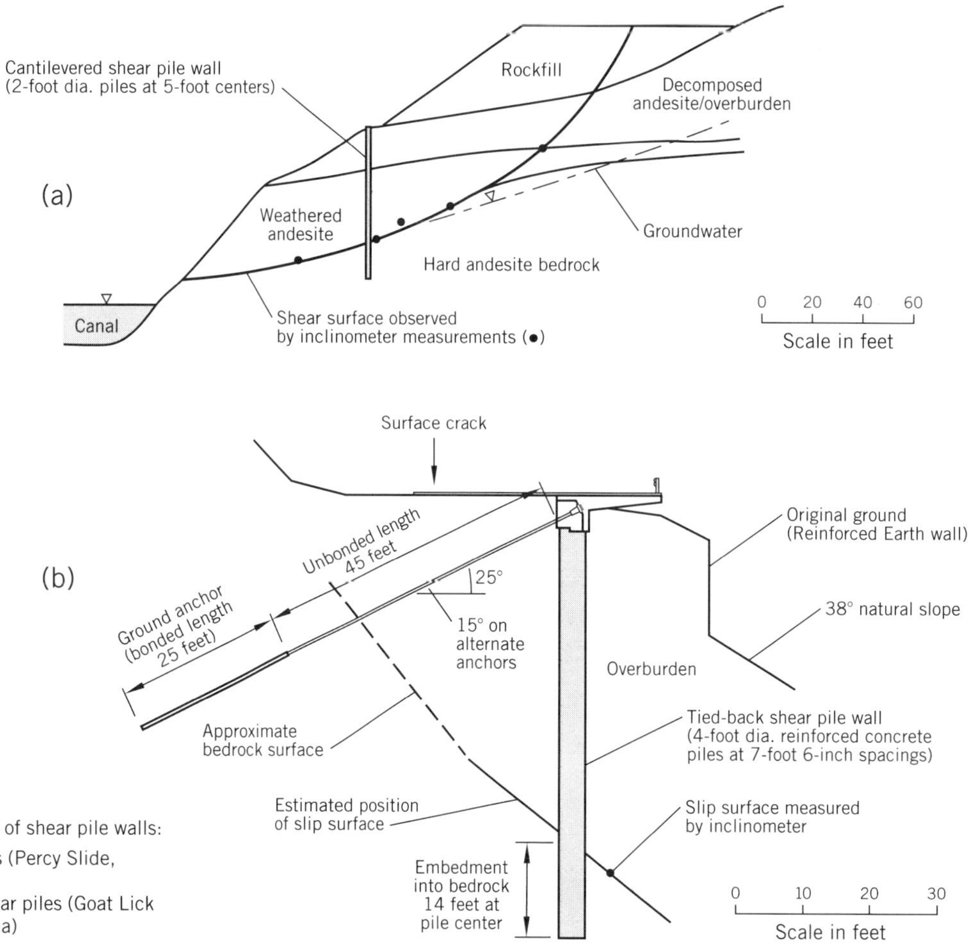

Figure 19.37 Examples of shear pile walls:
(a) cantilever shear piles (Percy Slide, Leaburg, Orgeon)
(b) tied-back anchor shear piles (Goat Lick Slide, Essex, Montana)

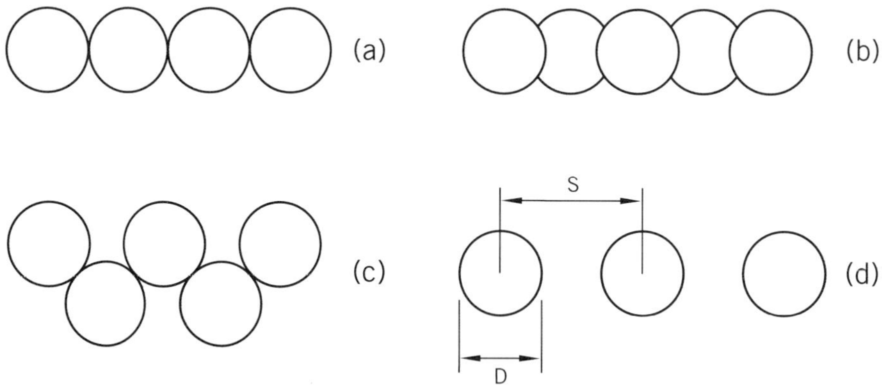

Figure 19.38 Reinforced concrete pile arrangements (in plan):
(a) tangent pile wall
(b) secant pile wall
(c) staggered pile wall
(d) spaced pile wall

of reinforcing bars arranged near the perimeter of the cylindrical pile section (Figure 19.39). Design of the reinforcement is structural engineering, but it is understood that reinforcing bars are more efficient than a H-section. Reinforcement bars can be varied along the length of the pile according to the bending stresses (Figure 19.39c).

Shear piles are usually cast-in-place. The hole is drilled by a helical auger attached to a Kelly bar or by a bucket auger. A chisel or downhole hammer is used to break up boulders and other obstructions. The same equipment can usually penetrate a short distance into harder bedrock or underlying stable ground. In very hard rock, it may be necessary to use a large coring barrel.

Temporary casing will be needed to support the sides of the hole when passing through caving soils or cohesionless soils below the water table. The casing can be installed by either reaming out the hole with cutting tools attached to the outside of the bucket auger or by using a larger auger bit. The

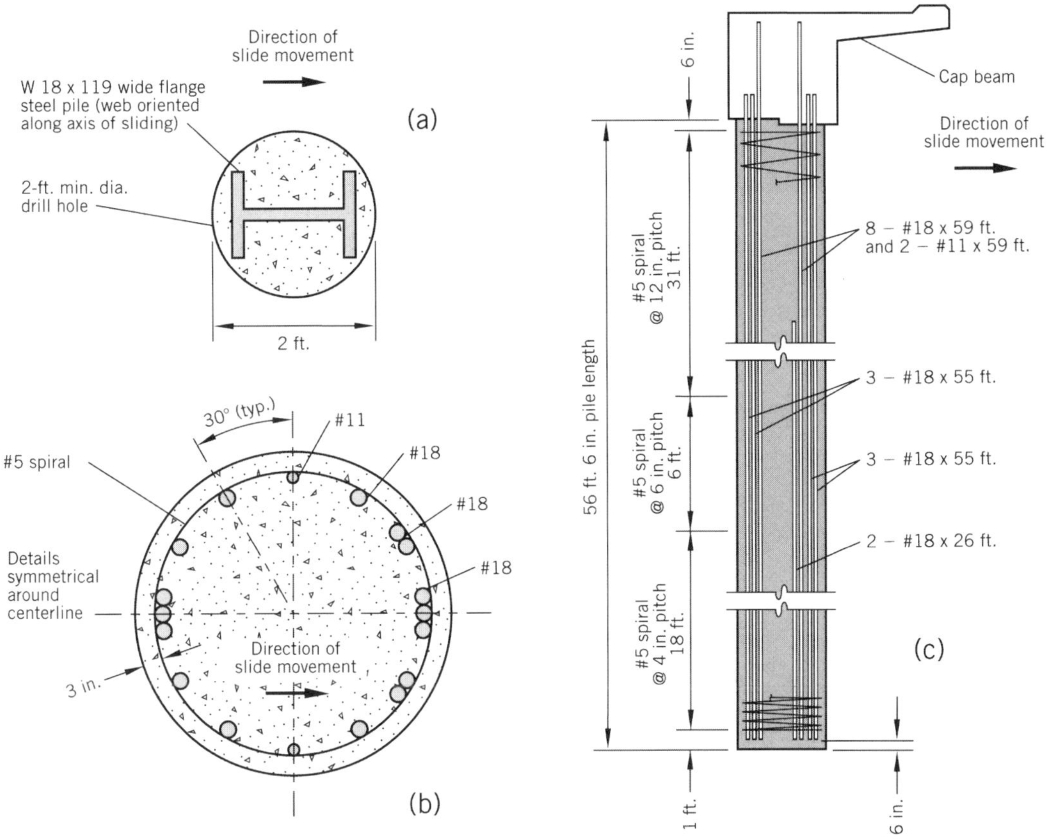

Figure 19.39 Examples of steel reinforcement in shear piles:
(a) plan view; wide flange H pile (Percy Slide, Leaburg, Oregon)
(b) plan view; rebar reinforcement (Goat Lick Slide, Essex, Montana)
(c) elevation (Goat Lick Slide)

casing can be hammered into the ground, but this may damage other piles recently constructed in the vicinity. A third alternative is to use a relatively thin-wall casing and leave it permanently in the ground as part of the pile. In this case, the casing is not considered as part of the resistance to sliding but, in reality, makes some contribution.

Another option is to use precast concrete pipe segments to form the piles. The hole can be drilled by a large diameter auger and the pipe segments lowered into the hole to provide a casing. Alternatively, soil can be excavated from inside the pipe segment, and the pipe is hammered into the ground to case the hole. In the latter technique, a cutting shoe is required for the lead (lowest) section of the pipe. Precast concrete pipe construction method is primarily used for large-diameter piles. Once the pipe has been installed to the full depth, reinforcement and mass concrete are installed to form the structural pile.

Geotechnical Design

Reinforced concrete piles provide additional shearing resistance to the landslide. Landslides develop a discrete shear zone and the piles placed through it are subject to simple shear. The soil immediately above the shear zone is severely strained and fails in bearing for some distance up the pile, this distance being longer for weaker soils. The pile design generally is governed by the bending moment that develops. The lateral behavior of piles is analyzed as a beam on an elastic foundation. The theoretical approach is more fully discussed in Chapter 22, Section 22.1.

In the late 1980s, when Landslide Technology first became interested in using reinforced concrete piles to stabilize landslides, the available literature on the subject was very limited and confusing to understand by a practitioner. Most technical papers on lateral loading of piles were directed at calculating deflections rather than resisting shear forces at depth. However, Reese and Wang (1989) published the computer program LPILE that numerically solves the beam on elastic foundation equation of Hetenyi (1946) using finite-difference techniques. This program is of enormous help in calculating the bending moments, shear forces, deflections, etc. of shear piles once the subsurface soil profile and design loads have been determined.

Landslide Technology, with assistance from Dr. Lee Schroeder, has developed design procedures for shear pile walls constructed in landslides (Cornforth, Schroeder, Hill and Vessely, 1996). Later, the LPILE procedure was modified by Kenji Yamasaki to allow for a full-depth crack in front of the wall (see Chapter 22, Section 22.1).

Reese, Wang, and Fouse (1992) also describe a procedure for stabilizing slopes with shear piles. However, the two procedures differ in several respects, mainly because the Reese method is based on improving stability in ground that is susceptible to landsliding rather than to an actual landslide condition. Thus, it is more appropriate for slope stability studies rather than landslides.

A typical application of a shear pile wall design could be a road, affected by a landslide, as shown by a section through the center of the slide, Figure 19.40(a). The road and upper part of the landslide is to be supported by a shear pile wall with tieback anchors at the top of the piles to prevent lateral movements at the road, Figure 19.40(b). This is an example of selective stabilization (see Chapter 14, Section 14.5) because only the upper part of the slope is being remediated. The lower part of the slide may continue to move downslope in the future.

The design procedure is as follows:

Step 1. Perform a back analysis of the landslide. Determine the interslice force P_1 due to the lower part of the landslide that has to be replaced by the shear pile wall.

Step 2. Select the factor of safety F for the stabilized slope. Assuming that the factor of safety of the landslide is originally at about 1.00, the shear pile wall must provide an additional design force P_2 at the discrete shear zone of the slip surface to raise the factor of safety to the selected design value.

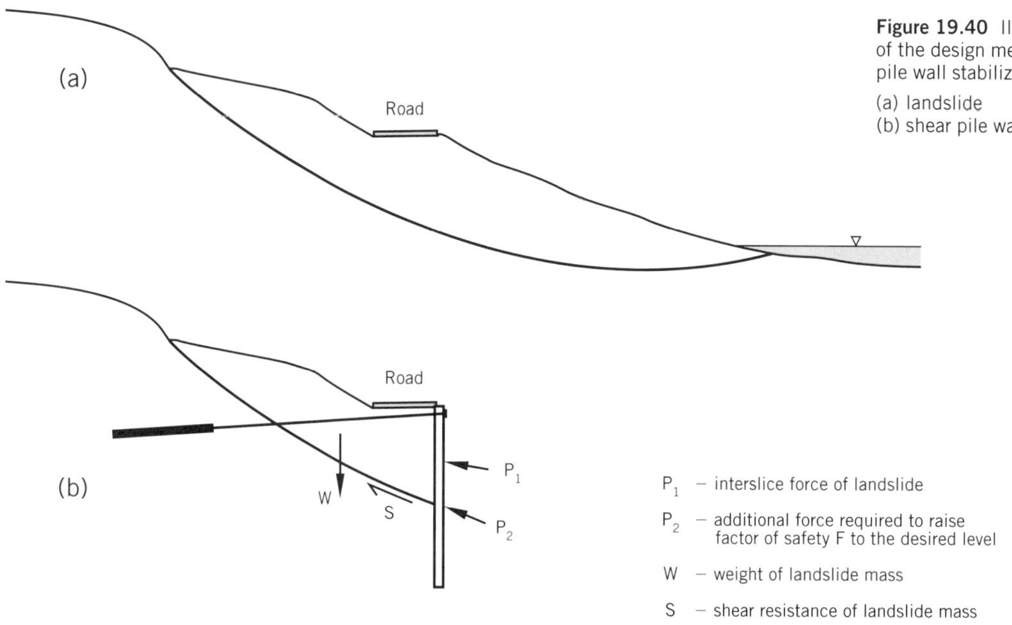

Figure 19.40 Illustrative example of the design method for a shear pile wall stabilization:
(a) landslide
(b) shear pile wall supporting road

Figure 19.41 Yamasaki design procedure for shear piles where downslope ground is unremediated.

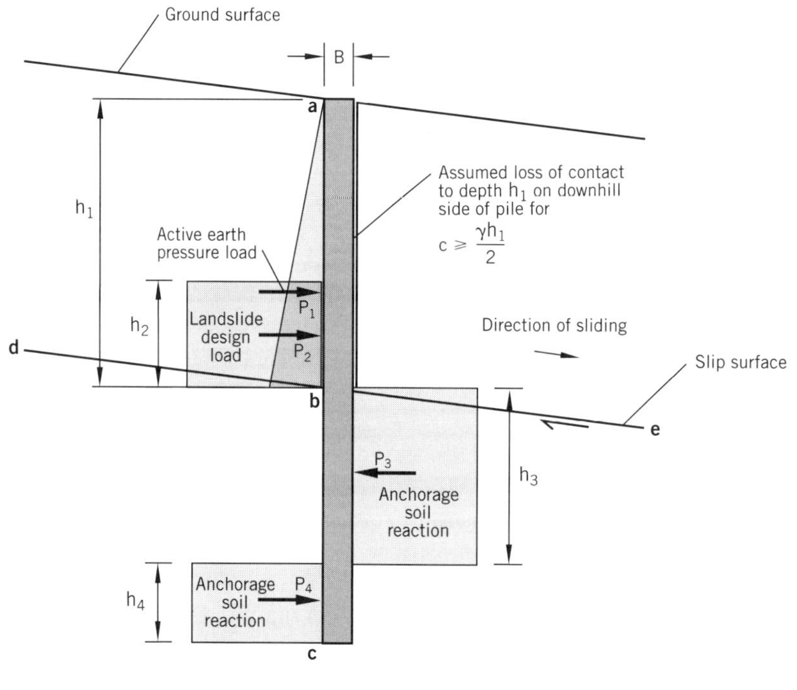

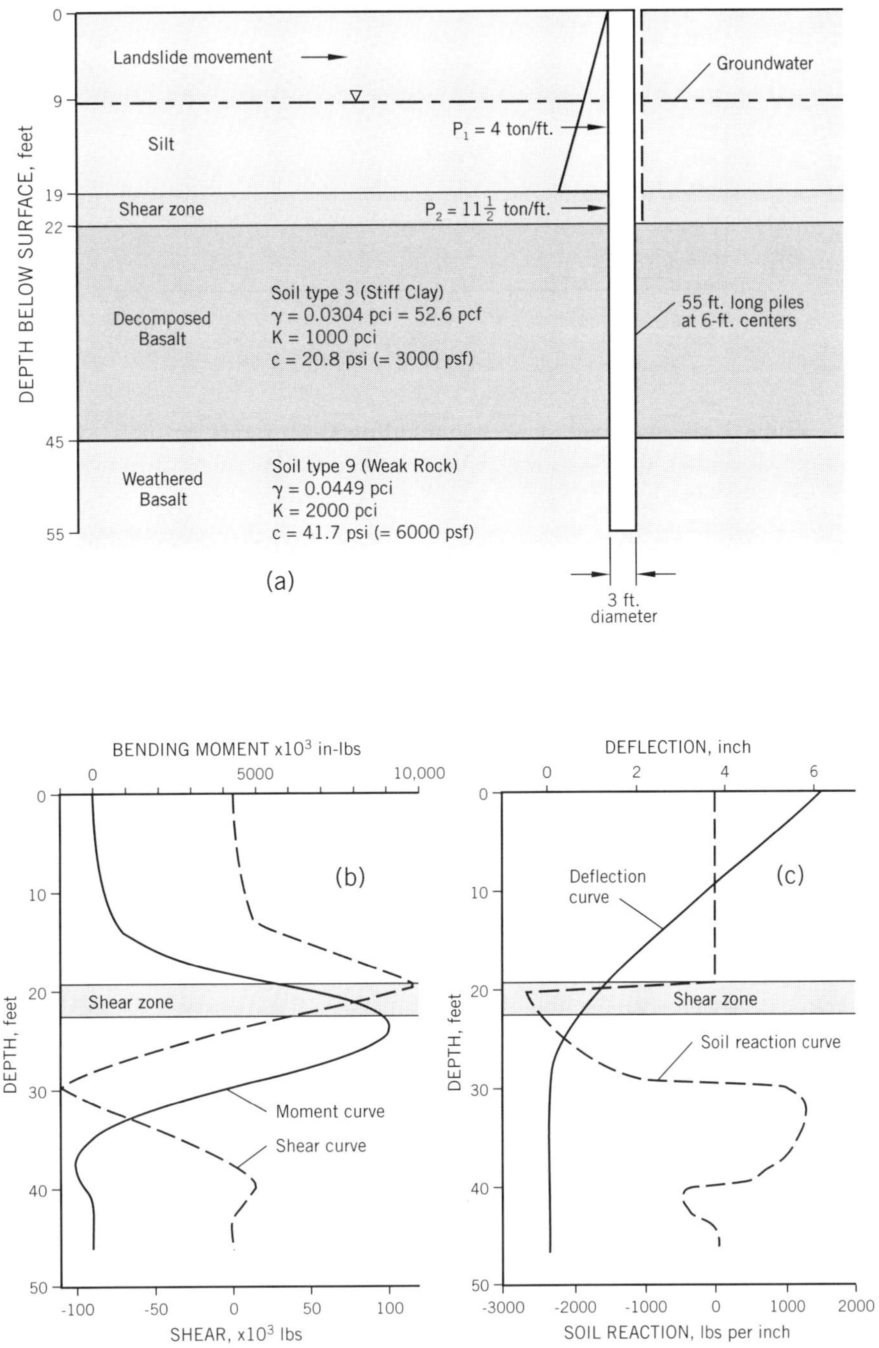

Figure 19.42 Example of shear pile wall design:
(a) soil conditions
(b) moment and shear curves
(c) reaction and deflection curves
(Cornelius Pass Slide, near Portland, Oregon)

Step 3. Follow the pile design procedure described in Chapter 22, Section 22.1. Use the computer program LPILE to calculate the maximum bending moment.

Step 4. Use conventional reinforced concrete design practice to determine the steel reinforcement needed for the piles.

Example Calculations

Refer to the basic pile design forces diagram, Figure 19.41. Force P_1 is the interslice force and P_2 is the landslide design load.

The pile wall design (Figure 19.42) has no tieback anchor. Each 3-foot diameter shear pile supports a 6-foot width of soil loading.

From a back analysis of the existing marginally stable landslide (F = 1.00), the interslice force is calculated to be 8,000 lb/ft at the pile wall location (Figure 19.43). This is assumed to be a triangular pressure diagram of force P_1 (Figures 19.41, 19.42). To provide a factor of safety F of 1.25 for the landslide, a further stability analysis shows that the pile wall must provide a design load resistance of 23,000 lb/ft. This resistance has to be provided by the pile

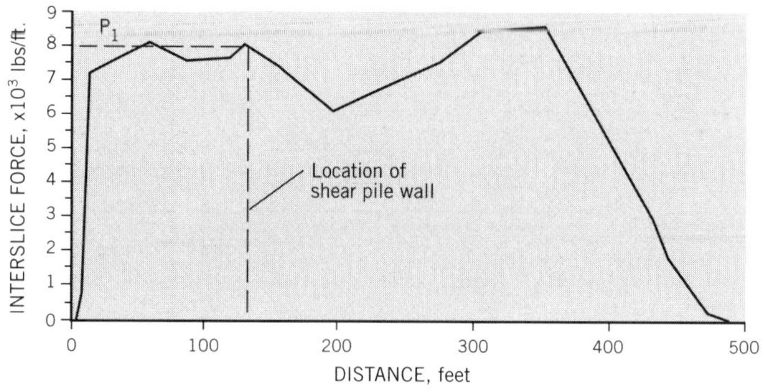

Figure 19.43 Determination of interslice force P_1 from back analysis of landslide (XSTABL computer program, Spencer's Method) (Cornelius Pass Slide, near Portland, Oregon).

at the discrete shear zone. It is assumed that the landslide above the shear zone acts as a large block with shear strains confined to the discrete zone below the block. However, since the discrete shear zone has essentially no extra shear strength capacity, the pile achieves resistance from the soil immediately above the discrete shear zone (load P_2 acting over height h_2 in Figure 19.41). Thus load $P_1 = 4$ tons per foot width and $P_2 = 11\frac{1}{2}$ tons per foot width. Since the pile-to-pile spacing is 6 feet in this example, each shear pile has calculated design loads of $P_1 = 24$ tons and $P_2 = 69$ tons.

It is assumed that the soil downslope from the pile wall loses contact with the pile wall; i.e., a full depth crack develops to a depth of 22 feet below the surface. The assumed soil and weathered rock properties are shown on Figure 19.42(a).

The results of the computations by the computer program LPILE show that the maximum bending moment is 9,000,000 lb-in. (750,000 lb-ft.) at a depth of 24 feet, which is 2 feet below the base of the shear zone (Figure 19.42b). The maximum deflection of the pile would be 6.1 inches under the design load (which probably would never be reached)—see Figure 19.42(c).

The calculations also show that anchorage can be achieved at a depth of about 45 feet. This is 10 feet shorter than the 55 feet assumed for the trial analysis. Thus, the pile tips would reach the surface of the weathered basalt but would not have to penetrate into it. The reader is referred to Chapter 23, Section 23.1, for a more detailed explanation of shear pile design.

This design was not implemented because a less-expensive alternative was selected. The trial design described above could have been adjusted to evaluate the effect of larger pile diameters and/or a continuous wall of piles instead of spaced piles.

A very effective design of a shear pile wall is described in Case History 9: Goat Lick Slide, Montana. It is a spaced pile wall with the space between structural piles being infilled by shallow *unreinforced* concrete piles. These infill piles act as lagging to support the ground when excavation is made in front of the wall later. It also has the advantage that it allows groundwater to pass underneath the shallow piles and thus not impede groundwater flow going down the slope; i.e., it prevents buildup of groundwater pressures behind the shear pile wall. A three-dimensional drawing of such a wall is given in Case History 9, Figure CH-9.5.

Concluding Comments

Design procedures are still evolving and numerical solutions of soil-pile interaction should continue to improve and more accurately represent the forces and moments on a shear pile. Shear piles are a high cost remediation that are best suited to narrow landslides or selective remediation (to protect a particular structure or facility).

Shear pile walls are often designed with tieback anchors to control movement at the top of the pile and reduce bending moments. Using "top down" construction and with many contractors available to perform the work (in industrialized countries), shear piles are particularly useful at very difficult landslide sites with limited options for remediation.

19.6 TIED-BACK SLURRY TRENCH CONCRETE WALLS

A tied-back slurry trench wall is an example of "top down" construction in which the slope is supported at all times. It is relatively expensive compared to most other stabilization options, but is suitable for the repair of medium landslides or for prevention of landslides/lateral movements of slopes in urban environments.

The techniques for constructing a slurry trench wall are described in Chapter 18, Section 18.1, and will not be repeated here. The main differences between a cutoff wall and a structural retaining wall are: (i) the need for a reinforcing cage within a structural wall; (ii) the need to provide cutouts to allow ground anchors to be installed through a structural wall; and (iii) the need to later excavate the ground on one side of the structural wall. These walls are commonly referred to as *diaphragm* walls.

Construction Details

Reinforcing Cage

Typical details are illustrated on Figure 19.44. The design is kept relatively simple so that it can be altered in the field if necessary. Other structural attachments to the cage include: (i) steel for lifting and handling the cage into position; (ii) spacers to provide a minimum (typically 3 inches) of concrete cover-

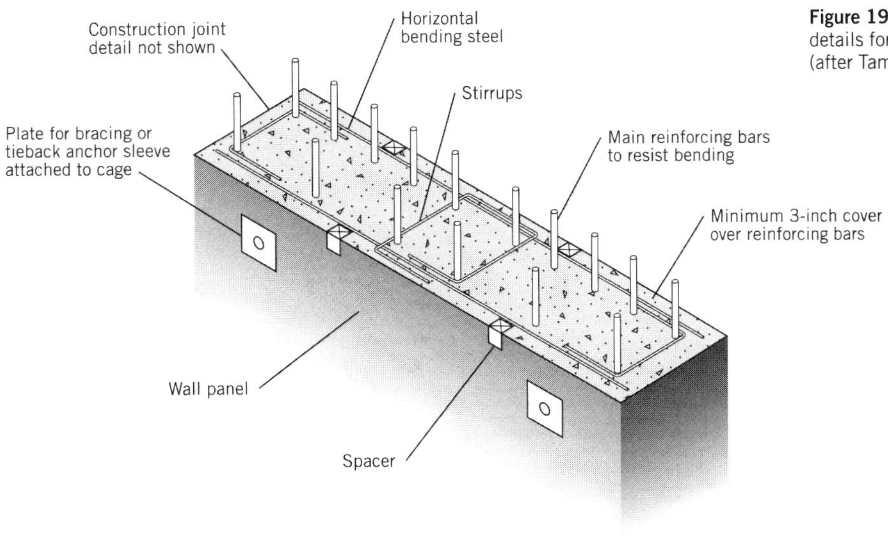

Figure 19.44 Typical reinforcement cage details for a slurry trench diaphragm wall (after Tamaro and Poletto, 1992).

age over the steel bars; (iii) sleeves and trumpets to allow tieback anchors to pass through the wall; and (iv) bearing plates for struts, if needed. One concern about this construction technique is that slurry may adhere to the reinforcing bars and affect the concrete-to-steel bond. Some designers use bond reduction factors of 0.6 to 0.8 to allow for this possibility.

Joints

The vertical joint between adjacent concrete panels can be formed by using steel wide flange beams (Figure 19.45a) or a keyed joint (Figure 19.45b). The wide flange beam provides structural capacity and can be connected to the tieback anchors. However, complicated joint designs are expensive, difficult to install, and may interfere with the installation of the reinforcing cage.

External Finish

After the outer face of the wall has been exposed, any defective joints are chipped out, cleaned, and packed with rapid-setting grout mixes. Occasionally repairs extend into the soil behind the wall. The exterior face can be shotcreted or sandblasted to obtain a smooth finish, or other facing can be added (e.g., precast panels).

Quality Control

Full-time inspection by a representative of the design engineer is required to check that the wall is being built correctly. A simple chart (Figure 19.46) can be produced to compare the design quantities of concrete with the actual amounts placed. Significant variations from the design quantities may require further investigation.

Tieback Anchors

See Section 19.2.

Performance

Typical performance data are shown on Figure 19.47. In this example, the maximum lateral movement of a 40-foot deep excavation was only 0.4 inch.

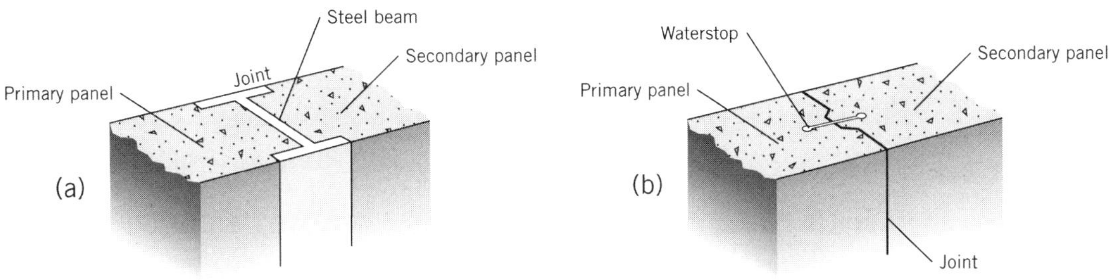

Figure 19.45 Alternative vertical joints for a slurry trench diaphragm wall:
(a) steel H-beam
(b) keyed joint with waterstop
(after Tamaro and Poletto, 1992)

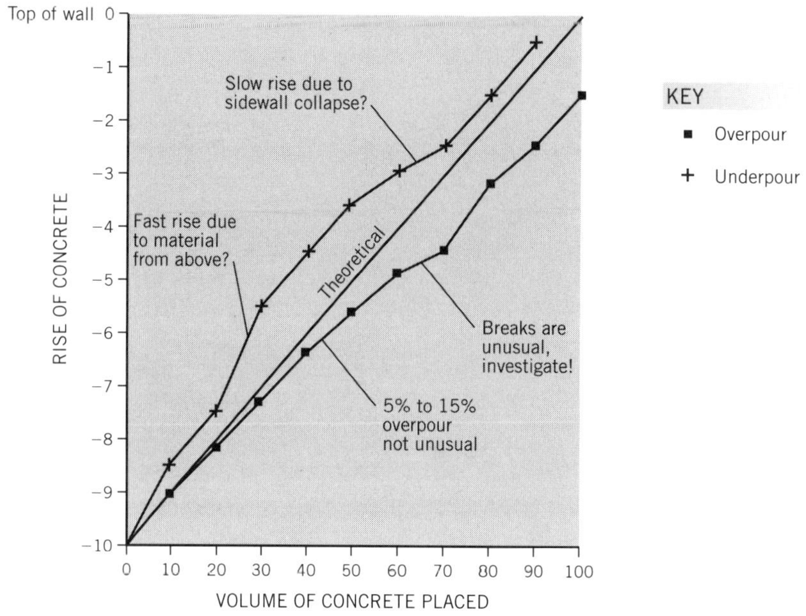

Figure 19.46 Hypothetical field record to illustrate quality assurance and detect possible problems (after Tamaro and Poletto, 1992).

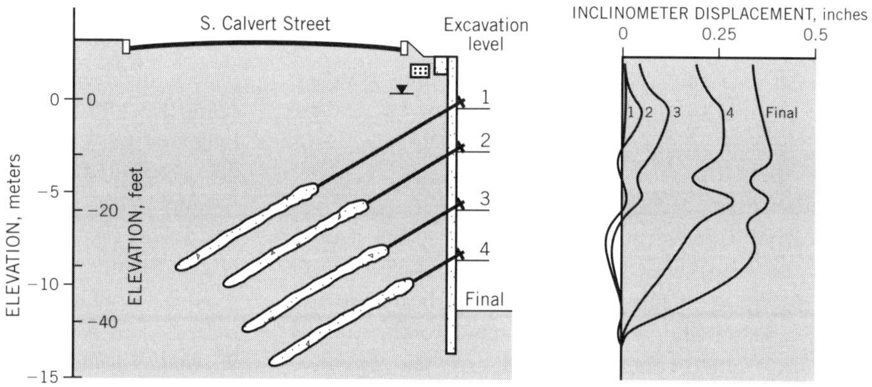

Figure 19.47 Typical slurry wall lateral displacements as excavation progresses to lower tieback levels (after Gifford and Wheeler, 1992).

19.7 MASONRY AND CONCRETE GRAVITY WALLS

This type of wall (Figure 19.48) is well-known and requires minimal description. It is trapezoidal in shape and requires a granular backfill. Stability is achieved by resistance to base sliding and overturning moment. Since the wall section increases as the square of the height, it is usually economical to heights of about 12 feet. Base width is 50–70% of height.

Concrete gravity walls require a firm base but can be built on an uneven surface. The base width increases directly with height. They are rarely used for landslide stabilization because gabion walls are usually a more cost-effective option for a low-height retaining wall.

Cost: Mostly dependent on volume of concrete; for masonry walls, labor of construction is a key factor.

19.8 CONCRETE CANTILEVER WALLS

A concrete cantilever wall is a relatively thin, heavily reinforced wall section that is cantilevered off a horizontal reinforced concrete base (Figure 19.49). A shear key at the back of the base is often provided to improve base shear resistance. The wall is built from the base upward and requires formwork and other time-consuming procedures. Counterfort walls can be built into the back of the cantilever section to stiffen taller walls and reduce deflections.

Concrete cantilever walls are designed by a structural engineer. A geotechnical engineer can provide technical input on lateral pressures, base resistance, and potential erosion in front of the wall. The design and construction of cantilever walls are well-known. Only a brief description is provided here.

Cantilever retaining walls typically can be built to about 30 feet high, and up to 50–60 feet high with counterforts. The

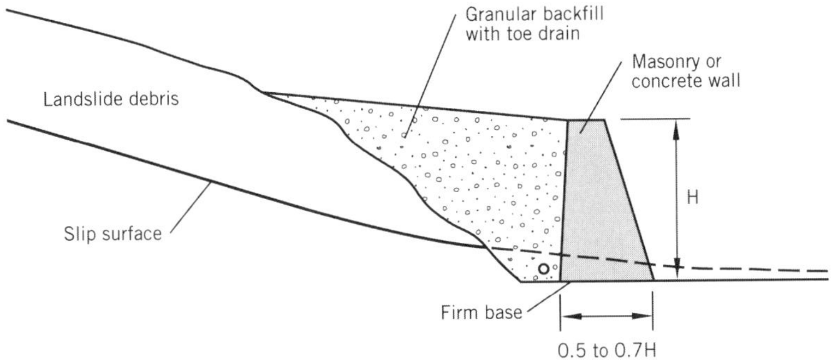

Figure 19.48 Concrete or masonry gravity retaining wall.

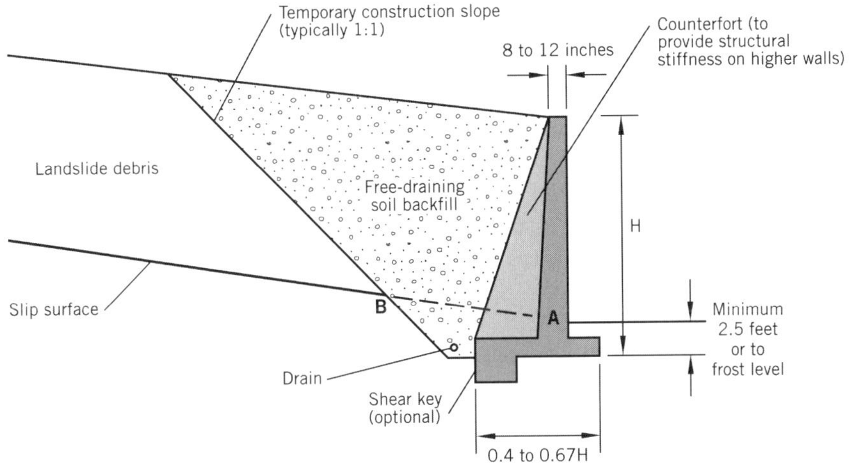

Figure 19.49 Concrete cantilever retaining wall.

base width ranges from 40% to 70% of the height. Construction costs (1997) range from about \$25 to \$60 per sq. ft. of exposed face (Sabatini et al., 1997). As with all "base up" construction, good drainage should be provided behind the wall to reduce lateral loads on the wall.

In general, cantilever retaining walls are economical to greater heights than a conventional gravity wall (Section 19.7). They require a very firm base and are a good choice when founded on bedrock or very dense strata. They are more commonly used to retain slopes in an urban environment. In general, these walls are too time-consuming to build for emergency landslide repairs; in such situations, there may be a relatively high risk to construction workers and the public from continued landslide movements.

For landslide remediation, the wall should intercept the extension of the slip surface (AB on Figure 19.49). It must be capable of resisting the landslide force with an appropriate safety factor.

One potential weakness in a long concrete wall is the vertical joint between concrete pours. It can allow relative deflection between adjoining wall sections. This is an important detail for sea defense walls.

19.9 CONCRETE CRIB WALLS

A crib wall is a form of gravity wall in which the wall mass (of granular fill) is retained by a "crib" of concrete or treated timber elements. They are commonly used to provide a steep outer slope for roads in steep terrain, especially where small slope instabilities have occurred.

A typical section through a crib wall is shown on Figure 19.50 for repair of a hypothetical slump failure. The back of the crib wall must intersect the slip surface of the landslide to be effective in stabilizing the slope. The landslide force can be back-calculated by limit equilibrium methods, and the design load is based on the higher of the landslide force (with factor of safety added) or the lateral force in conventional slope stability. The landslide force usually is more critical.

Each crib unit is made from beam elements that are laid horizontally across the slope face and at an inclined low angle into the slope (Figure 19.50). The lattice of longitudinal and transverse elements form each crib. The crib wall is backfilled as it is built from the base upward. The crib infill should always be slightly ahead of the backfill behind the wall. Suppliers of crib components in the United States include Criblock (concrete) and Permacrib (treated timber).

Crib walls can be built to heights of up to 30–35 feet; the base width is 50% to 70% of the height. Since crib walls can become badly disturbed and structurally damaged by differential settlements, the base must be on firm ground. Free-draining granular backfill, spread and compacted in approximately horizontal lifts (e.g., 12-inch lifts for coarser granular fills), should be placed within and behind the wall. Pad type vibrators are commonly used for compaction. Open-faced crib

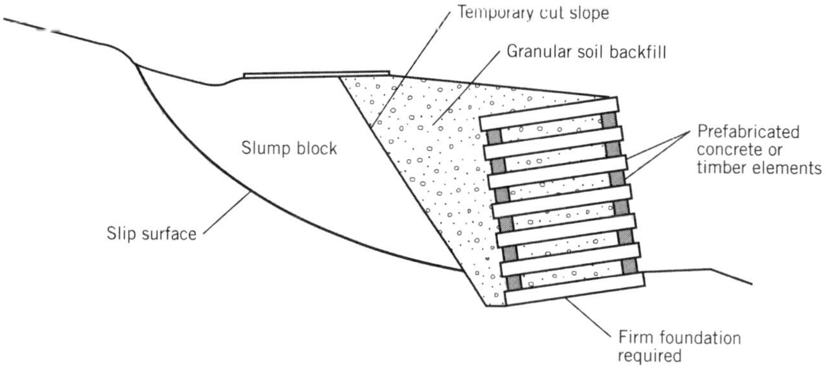

Figure 19.50 Crib retaining wall.

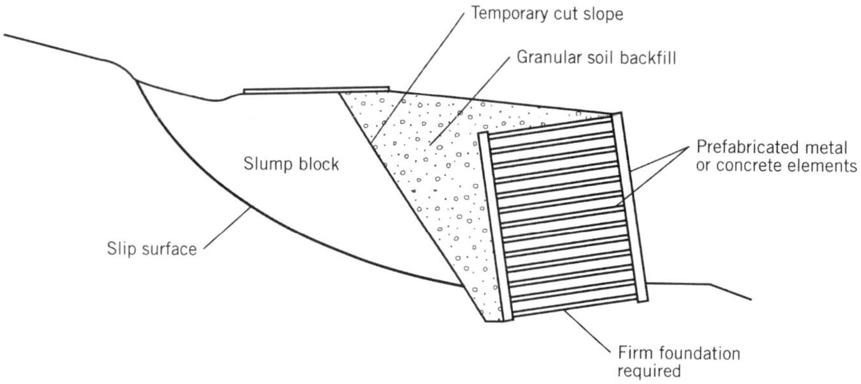

Figure 19.51 Bin retaining wall.

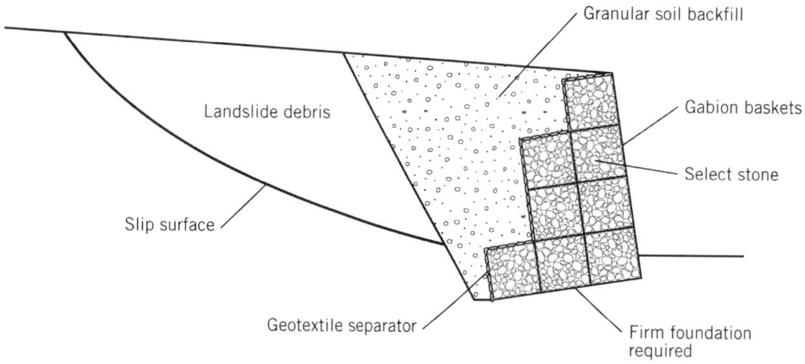

Figure 19.52 Gabion retaining wall.

walls may need a coarse rockfill or screen to prevent loss of fill through the openings between structural members.

Crib walls are relatively easy to build and do not require special equipment or expertise. The costs are moderate—around \$25 to \$35 per sq. ft. of wall face in the 1990s (Sabatini et al. 1997).

19.10 BIN WALLS

A bin wall is a form of gravity wall in which the wall mass (of granular fill) is retained by a steel or concrete "bin". Except for the bin feature, they are essentially the same in construction and cost as a crib wall (Section 19.9) and serve the same purposes. Proprietary systems in the United States include Stowal, Stresswall, Doublewal, and Evergreen (all concrete bins) and Contech, Syra Steel (both steel bins). Bin walls probably are less flexible than crib walls and thus need a very firm foundation.

A typical section through a bin wall is shown on Figure 19.51. The similarities with a crib wall are obvious. Refer to the previous section (19.9) for relevant comments. This includes the typical heights, base width : height ratio, construction methods, field compaction, and construction costs.

Bin walls have different appearances, which may affect the choice. The steel bins are bolted together on site. The steel members can be galvanized for corrosion protection. Concrete bin walls are built from interlocking precast reinforced concrete modules that are assembled like building blocks at the site. The bins are infilled with compacted granular soils.

19.11 GABION WALLS

Gabions are rectangular baskets made of wire mesh that are filled with small rocks (typically 4-inch to 8-inch sizes). A gravity-type retaining wall can be constructed by stacking and tying together numerous gabions in a stepped arrangement (Figure 19.52). An alternative is to step the outer face to create a battered slope. The principle is that the gabion creates a large gravity-type block from an infilling of relatively small stones. The principal advantages of gabion walls over most other gravity structures are their flexibility, perviousness, and simplicity of construction. The principal drawback is their labor-intensive construction, especially for higher walls.

Gabion walls provide a simple solution to many small landslide projects. Because they are assembled by hand at the site, they can be constructed in places with limited access or headroom. They can also be built to moderate heights at more open sites, including along the edge of small streams, but become uneconomic with other types of wall as the required height and length of wall increases. The walls depend on the basket for structural integrity, and corrosion of the wire mesh is a concern. Vandalism is a potential problem at some sites.

A wall made from gabions has to be constructed on a moderately firm, nearly level foundation. A gabion wall can accept more differential movements than other gravity retaining structures built from concrete elements. The gravity structure must satisfy the usual structural requirements of stability against overturning and base sliding. The backfill behind the wall should be designed as a free-draining filter between the native soil and the coarse rock fragments or cobbles within the gabion. The backfill is added as the gabion wall height increases; typically, the wall is backfilled after every three levels of gabions have been built.

The dimensions of the gabion baskets, rock requirements (size, gradation) and the quality of the stone have been presented earlier (Section 16.3, Riprap Slope Armor, and 16.5 Gabion Mattresses) and will not be repeated here. The most common size of gabion basket for block wall construction has a width and depth of 3 feet.

Commercially–available gabions in the United States include: Hilfiker, Maccaferri. The maximum wall height is around 25 feet but they normally are used in smaller applications of less than 12 feet. The base width is about 50% to 70% of wall height. Total construction costs range from about \$25 to \$50 per sq. ft. of wall face (1995). ENR Construction Cost Index for 1995 is 5484.

CHAPTER

20

Earth Reinforcement Systems

Earth reinforcement involves the insertion of various tensile-resisting materials into the soil to improve stability. The materials include steel rods, steel angles, metal strips, geosynthetics of cloth, and grids. Rapid growth in use of these methods has been experienced because they are flexible, easy to construct, and inexpensive relative to conventional gravity retaining walls or soldier pile walls. Some systems are patented, which may limit their general availability and use.

All methods of soil reinforcement in this chapter—soil nailing, reticulated micropiles, and mechanically stabilized earth—change the soil near the outer face of a slope into a unit that serves as a gravity retaining structure. The reinforcing elements pass through unstable (or potentially unstable) soils and provide tensile pullout resistance. Some issues of soil-structure interaction remain subject to debate.

Soil-structure interaction of embedded materials has generated much interest from researchers. It has produced many alternative methods of analysis. For example, the U.S. Federal Highway Administration (FHWA) manual on soil nailing (Byrne et al., 1996) lists 10 different methods of analysis. This has created a lot of confusion for practitioners who have insufficient time to fully investigate the different methods and rationally select from their attributes/shortcomings. Field measurements of loads and deflections are shedding light on these issues. However, methods of construction and differing soil conditions produce variable loading of soil, and there is a need for broadly acceptable design standards to emerge similar to those developed for lateral loads on strutted excavations.

For mechanically stabilized earth and soil nailing, the available methods of analysis can be placed into three groups:

- *Limit equilibrium analyses.* These are conventional slope stability calculations with potential slip surfaces modeled as circular arc, linear, bi-linear, log spiral, etc. The critical slip surface without reinforcement approximately represents the surface of maximum tensile load (see Figure 20.1); the analysis examines the same slip surface and others to find the lowest factor of safety after reinforcement.

  Most of these methods assume that the reinforcement within the resistant zone is only in tension, ignoring any contribution from shear or bending to the total resistance. This is conservative but seems advisable because such contributions only develop as the ground approaches failure after significant movements have occurred. In practice, significant movements are avoided by using a pullout factor of safety of around 2.0.

  A weakness of limit equilibrium methods is that they assume the same proportional contribution from each anchor in the resistant zone. Actual observations of failure (for example, the Clouterre full-scale soil nail experiments in France) have indicated that failure occurs at the bottom of the wall first and progresses upslope.
- *Finite element analyses.* These can more closely model the actual behavior of the wall. However, the technique is expensive and complicated to apply in practice. It is likely to remain a research tool that may lead to better analyses by simpler methods.
- *Empirical analyses.* These methods typically convert field observations into a rational load diagram that encompasses observed data and thus provides conservative results for design. This approach was the outcome for the design of strutted excavations (Terzaghi and Peck, 1948), which has similar uncertainties about loads induced by construction. Empirical relationships have been suggested, but currently are based on limited field data. This appears to be the best method of analysis of horizontal and near-horizontal pressures but may require further refinement to account for all soil types.

Reticulated piles present a special technical problem because the piles enter the stable ground at different inclinations. At the time of writing, there is no consensus for design.

20.1 SOIL NAILING

Technique

Soil nailing is a reinforcement technique in which closely-spaced parallel steel bars are installed into the face of a slope or vertical cut to improve stability. The technique has evolved rapidly since the early 1980s.

The soil nails provide pullout resistance and are in tension over their entire length (Figure 20.1). For a vertical cut slope with a horizontal backslope, the nail lengths typically are 60–100% of the wall height, depending on the soil conditions, and spaced at 6-foot centers vertically and 5- to 8-foot centers horizontally. Soil nails installed for permanent slopes require corrosion-resistant treatment similar to soil anchors.

The most common installation method is to insert the steel bars into a drillhole and grout from the bottom up by gravity or low pressure. This procedure is described in some detail in this section. Other construction methods include jet grouting or driving nails into the slope. Most of these other methods are proprietary (patented) applications. For driven nails, only launched nails are described herein.

Soil nailing is a "top down" construction procedure. It is suitable for temporary and permanent cut slopes and for shallow depth landslide remediation. Excavation to expose the steep cut face proceeds with a series of steps 3 to 6 feet high. After each step has been opened up, soil nails are installed in a row and the face is shotcreted to prevent collapse or erosion of the exposed face. The end of the nail projecting out of the slope face is then attached to the shotcrete facing by a metal plate and bolt. Except for a seating load, no prestress is applied (unlike soil anchors). After the full depth of excavation has been reached, a second layer of shotcrete (or cast-in-place concrete) is added to strengthen the wall and protect the nail anchorages.

A key requirement of soil nail walls is the *standup time* of the soil face during excavation of the steps. It restricts soil nailing to relatively competent soils above the groundwater table. However, where appropriate, soil nailing generally is less expensive than alternatives. Drainage measures, during and after construction, are also important.

Advantages and Limitations of Soil Nail Walls

Advantages

Soil nail walls share the same advantages as soldier pile walls. These include safety against slope failure during construction and less interference to construction activities below. Specifically, the main benefits are:

- Construction of the wall is a "top down" procedure that supports a steep slope (usually vertical) as excavation proceeds.
- No additional depth of excavation is required behind the final wall face for formwork or base width; as a consequence, there is no need for backfill behind the wall.
- There are no temporary supports in front of the wall.

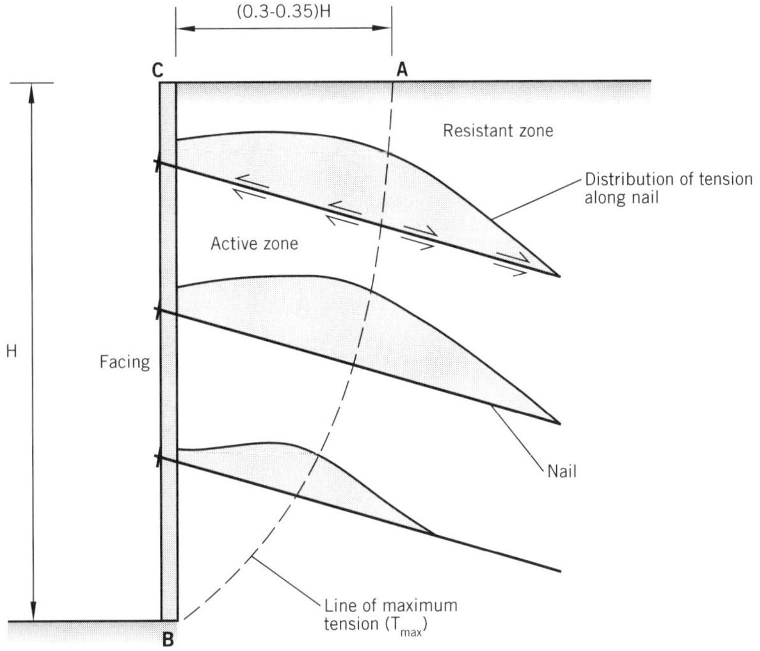

Figure 20.1 Soil nail wall tensile loads.

Compared to soldier pile wall construction, the relative advantages of soil nail walls are:

- Construction cost is about 10 to 30 percent lower in ground suitable for soil nails.
- Equipment is smaller and more mobile, making access easier to difficult sites.
- The completed structure can accommodate significant total and differential settlements.
- Soil nail walls can be used on sites with unusual or changing grades (attractive for landslide work).
- Design can be modified or optimized during construction to avoid boulders, buried pipes, or other unexpected obstructions.
- Soil nail walls require less embedment below the final cut grade than soldier piles; this makes them a more attractive option for sites where bedrock is at a shallow depth below the base of the cut.

Limitations

The principal limitations of soil nailed walls are that they require fairly competent soils with real or "apparent" (negative pore pressure) cohesion and the absence of free groundwater within the soil face, as described below:

- *Unsuitable soils.* These include soft clays (shear strengths of less than 1,000 lb./sq. ft.), loose sands and gravels (SPT blow count less than 10), waterbearing sands and gravels, organic soils, loose fills, rubble fills, and frost-susceptible silts.

    Clays need very careful evaluation during design. Creep movements and sensitivity to disturbance are generally unacceptable. There is also the possibility of stiff expandable clays developing reduced pullout resistance over time due to changes in moisture content.
- *Standup time.* The soil slope must be capable of remaining stable in a vertical cut 3 to 6 feet high before the soil nails are installed. Problems with standup time are most common near the original ground surface where there may be loose fill, buried utilities, or locally softened ground.

    Cotton (1992) discusses multiple options for handling face instability during construction. These techniques include: (i) reducing the open cut height and installing additional nails; (ii) battering the slope face, where feasible; or (iii) placing a 1- to 2-inch thick "flash coat" of shotcrete over the surface to bind the soil together at the outer slope face before installing the anchor.
- *Groundwater.* Soil nailing has to occur above groundwater level. Although drains are routinely placed on the slope face during construction, any seepage is likely to be troublesome during construction and should be avoided. Surface runoff water should also be intercepted before reaching a soil nailed wall, and any collected water should be disposed of rapidly. Frozen water can damage the shotcrete facing and place additional loads on the anchor plates due to soil expansion.

Other concerns for soil nailing construction, especially in urban areas, include:

- For a permanent wall, underground easements may be required if the wall depends on the nails for long-term stability.
- Soil nails may interfere with utilities.
- Unacceptable horizontal movements may affect adjacent structures.
- Shotcrete facing may deteriorate with time.
- Usually requires a specialty contractor.

Tensile Loading of Soil Nails

Soil nails are considered to be passive elements within the soil. They rely on the bond between nail and soil to restrain the soil slope from becoming unstable. No prestress is needed except for a minor loading to seat the anchor plate to the shotcrete.

Theoretical analyses and field measurements have shown that a soil nail penetrating into a slope face develops a tensile force that increases to a maximum value and then decreases at further distances from the face. As shown on Figure 20.1, the locus of maximum tension defines a curved surface that extends from the ground surface behind the wall at A to the toe of the slope at B. For a vertical cut slope of height H and horizontal backslope, the curved surface AB reaches the ground at a distance AC behind the wall of about 0.3H to 0.35H. This curved surface separates the *active zone* from the *resistant zone* behind it.

Nail tensions develop gradually as the excavation proceeds. As observed in the Clouterre research project (Clouterre, 1991), most of the nail tensile loading occurs during the first three excavation steps following a nail installation. Thus, the nails near the bottom of the wall carry less tensile load than elsewhere and, as shown on Figure 20.1, the point of maximum tensile load moves closer to the face BC. This behavior is attributed to the restraining effect of the excavation base, resulting in less outward deformation and less development of tensile loading. Usually, the bottom row of soil nails carries no load after installation but may later develop tension due to long-term deformations of the ground.

Soil Nail Wall Design

Soil nails tie the active zone to the resistant zone by providing the necessary tensile strength. The tensile strength of the steel bar must be sufficient to carry the maximum load. The nails must be embedded far enough into the resistant zone to resist pullout. Within the active zone, the combination of nail head strength (i.e., connection system) and pullout resistance of the nail length EX (Figure 20.2) must be sufficient to provide the required nail tension at the potential slip surface X. These three conditions have to be checked, and the least stable of the three controls the contribution to stability at any part of the design slip surface.

Soil nail walls have four design components:

1. The steel bar must be capable of carrying the maximum tension without fracturing.

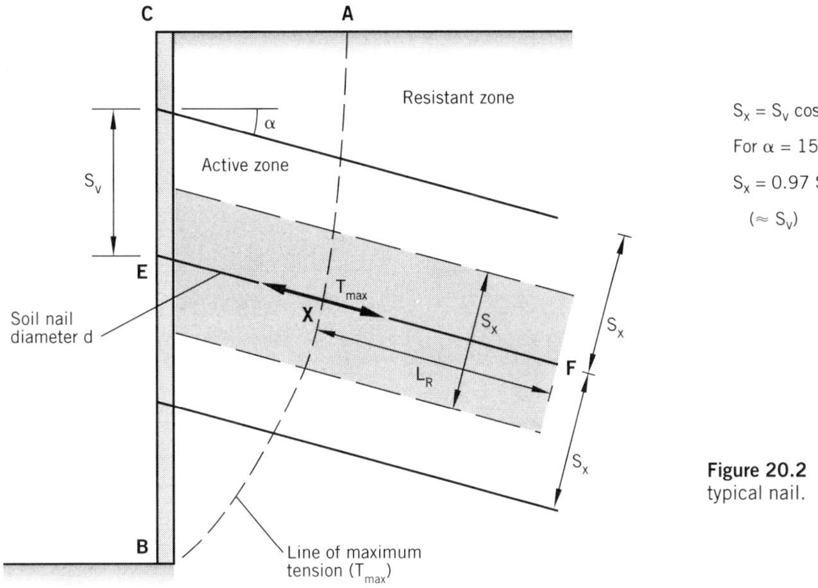

Figure 20.2 Design analysis of a typical nail.

2. The length of soil nail behind the potential or actual slip surface must be capable of providing adequate pullout resistance
3. The anchor at the wall surface has to carry the shear load without undergoing punching shear into the concrete wall face and soil
4. The wall facing between nails must be reinforced to carry wall loadings

Components (3) and (4) require structural design that are covered in Byrne et al. (1996). Soil loadings are described in this section.

In common with other gravity retaining structures, soil nail wall design requires that stability be checked for: (i) sliding along the base; (ii) overturning failure; (iii) bearing capacity failure; (iv) presence of unstable weak layers at shallow depth below foundation level; and (v) overall slope stability outside the reinforced soil.

Maximum Tensile Load (T_{max})

Each soil nail is providing tensile force to a volume of soil equal to the product of the vertical spacing S_v between rows, horizontal spacing S_h between adjacent nails in a row, and the length of the nail. As shown on Figure 20.2, for a sloping nail the shortest distance between rows S_x is, for practical purposes, the same as S_v.

Collected data on the maximum tensile loads of soil nails from eleven sites are plotted on Figure 20.3 (Byrne et al., 1996). The nail loads were calculated from strain gauge monitoring of test sections and attempted to include the portion of

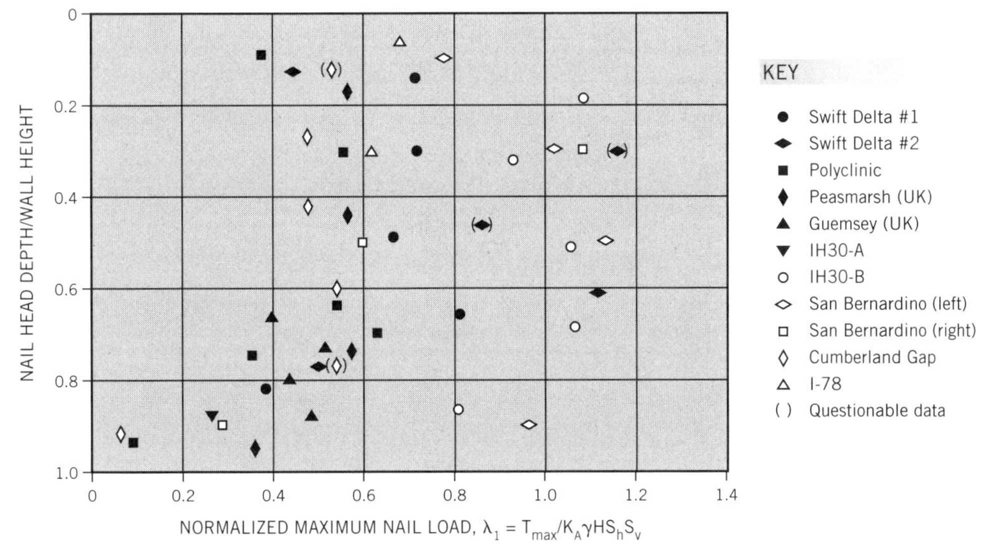

Figure 20.3 Normalized maximum nail load measured in test sections (after Byrne et al., 1996).

TABLE 20.1 Ultimate Bond Stress q_u for Soil Nails in Soils and Rocks*

Soil Type	Ultimate Bond Stress, q_u (lb./sq. in.)	Rock Type	Ultimate Bond Stress, q_u (lb./sq. in.)
Silt, nonplastic	3.0–4.5	Marl/Limestone	43.5–58.0
Med. dense sand/silty sand	7.0–11.0	Phyllite	14.5–43.5
Dense silty sand & gravel	11.5–14.5	Chalk	72.0–86.5
Very dense silty sand & gravel	17.5–34.5	Soft dolomite	58.0–86.5
Loess	3.5–11.0	Fissured dolomite	86.5–144.5
Stiff clay	6.0–8.5	Weathered sandstone	29.0–43.5
Stiff clayey silt	6.0–14.5	Weathered shale	14.5–21.5
Stiff sandy clay	16.5–29.0	Weathered schist	14.5–25.5
		Basalt	72.0–86.5

*Byrne et al., 1996

the nail load attributed to the grouted annulus. The data are presented in a dimensionless format where:

$$T_{max} = \lambda_1 K_A \gamma H S_h S_v \qquad \text{Eq. (1)}$$

where T_{max} = maximum tensile load in soil nail
K_A = active earth pressure coefficient
γ = density of soil
S_v = vertical spacing of soil nails
S_h = horizontal spacing of soil nails
H = total vertical height of wall
λ_1 = dimensionless factor

The dimensionless factor (termed here λ_1 for convenience) is obtained by dividing T_{max} by $K_A \gamma H S_h S_v$ and ranges from about 0.4 to 1.1 within the upper two-thirds to three-quarters of the wall height, the mean value being 0.75 (Figure 20.3). Below these levels, the maximum nail loads decrease significantly and approach zero at the base of the wall.

Tensile Load (T_F) at the Wall Connection

A similar graph to Figure 20.3 was obtained from the eleven test sites for the tensile loads measured at the nail head. This graph is not reproduced here, but it showed a fairly uniform load in the upper two-thirds of the wall (albeit with significant scatter), which decreases in the lower one-third of the wall. The dimensionless factor, which can be termed λ_2, typically ranges from 0.2 to 0.6, the mean value being around 0.42. This is 56% of the mean of the maximum nail loads.

All of these data are subject to the uncertainty of soil properties, natural variability, etc. However, it is in general agreement with results obtained elsewhere. For example, the Clouterre (1991) project reported nail head loads in the range of 0.4 to 0.5 of the maximum nail load for the most heavily loaded nails in the upper portion of walls. Byrne et al. (1996) recommends $\lambda_2 = 0.5$ and a nail head service load T_F:

$$T_F = 0.5 K_A \gamma H S_h S_v \qquad \text{Eq. (2)}$$

All terms have been previously defined.

It is customary practice to specify uniform lengths and diameter for the steel tendons over the entire height of the wall. This makes field construction simpler. However, shorter lengths of nail can be designed for the lower parts of the wall (Byrne et al., 1996).

Pullout Resistance

Based on the Clouterre (1991) research project, the ultimate total pullout resistance Q is relatively independent of the depth below the ground surface. This has been attributed to decreasing soil dilation with depth that offsets the increasing overburden pressure.

For analysis of the pullout resistance behind the curved surface of maximum tensile load (Figure 20.2):

$$Q = q_u \pi d L_R \qquad \text{Eq. (3)}$$

where Q = total ultimate pullout resistance
q_u = ultimate pullout resistance (bond stress) per unit length of nail
d = diameter of soil nail
L_R = embedded length in the resistant zone

A list of ultimate bond stress values for various soil and rock units is given in Table 20.1. On soil nail projects, field pullout tests are performed to check the values of q_u for the particular site conditions.

Factors of Safety

For yielding of the steel tie bar:

$$F = \frac{f_s \pi d^2}{4T_{max}} \qquad \text{Eq. (4)}$$

where f_s = yield strength of the tie bar
d = diameter of soil nail

For pullout resistance

$$F = \frac{\pi d L_R q_u}{T_{max}} \qquad \text{Eq. (5)}$$

Suggested minimum factors of safety are listed in Table 20.2.

TABLE 20.2 Minimum Factors of Safety for Reinforced Structures*

Stability	Design Consideration	Minimum Factor of Safety F		
		SNW**	MSEW***	RSS***
External	Sliding along the base	1.50	1.50	1.30
	Deep-seated stability	1.30/1.50	1.30	1.30
	Overturning stability		1.30	1.30
	Bearing capacity failure (toe)	2.50	2.50	2.50
	Seismic stability (pseudo-static)		75% of static F	
Internal	Pullout resistance of reinforcing elements	2.00	1.50	1.50
	Allowable tensile strength:			
	(a) steel strips		0.55 F_y	0.55 F_y
	(b) steel grid†		0.48 F_y	0.48 F_y
	(c) geosynthetic		T_a	T_a
	(d) steel rods	1.67		

* Based on the cited references
** Byrne et al. (1996)
*** Elias et al. (2001)
† (when connected to concrete panels or blocks)
SNW Soil nail wall
MSEW Mechanically stabilized earth wall
RSS Reinforced soil slope
F_y = Yield stress of reinforcement steel
T_a = Allowable tensile strength

Design Procedure

Soil nail walls are designed by conventional slope stability analyses that include the tensile resistance of the soil nail lengths outside the trial arc. A moment equilibrium or force diagram approach (e.g., Terzaghi, Peck, and Mesri, 1996) can be used.

Design practices vary between practitioners in different countries and there is no set standard. In 2000, two soil nail design computer programs were being used in the United States:

1. GoldNail, developed by Golder Associates, Redmond, Washington
2. SNAIL, developed by the California Department of Transportation, Division of New Technology, Sacramento, California

The two U.S. design methods take account of only the tensile resistance in the soil nails. Since a soil nail is relatively thin and flexible, the contribution of shear and bending moment is considered negligible (calculations have indicated additional resistance of the order of 2%).

The stability analyses generally check many potential failure surfaces. These include potential slip surfaces within, outside, and partially within and outside the reinforced soil mass (Figure 20.4).

Construction of Grouted Soil Nail Walls

A nail installation sequence is shown on Figure 20.5. The six-step procedure to build a wall is:

Step 1. A vertical cut, 3 to 6 feet high, is made into the ground. The height of the cut is determined by the design spacing of the nails and the standup capability of the soil. As a precaution against sloughing, a low berm can be left to reduce the unsupported vertical height of cut.

Step 2. A drillhole, 4 to 12 inches diameter, is made into the slope at a dip angle of 10° to 20° to the horizontal. The angled hole facilitates grouting by gravity or low pressure. Many types of drilling equipment can be used, including open hole in stable soils.

Step 3. A steel bar, 0.75 to 1.38 inch diameter, is placed in the hole. Centralizers maintain an annular gap of at least 1.5 inches between the bar and sides of the hole. The steel bar can be made from mild- or high-tensile steel. A cement-based grout encases the bar. Grouting proceeds from the bottom up. The grout should be designed to attain a minimum compressive strength of 3,000 lb./sq. in. within 7 days. It is typically a neat cement grout with a water : cement ratio of 0.4 to 0.5 (Byrne et al., 1996), or sand–cement grout in larger diameter holes.

Vertical and horizontal spacings of soil nails range from 3 to 8 feet, and often are the same spacing in each direction. However, vertical spacing between nails is commonly set at

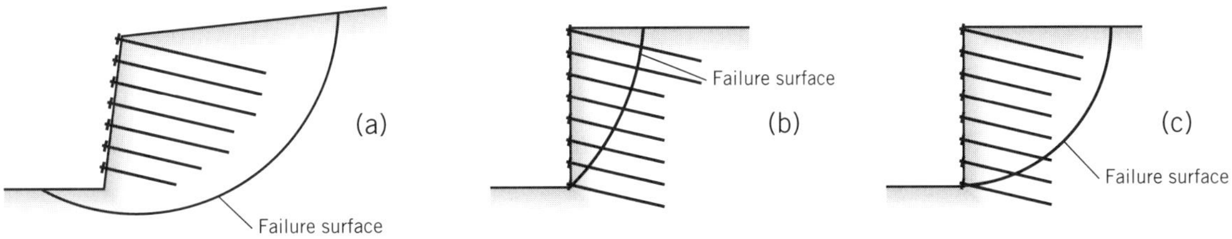

Figure 20.4 Potential failure surfaces that need to be studied in soil nail design:
(a) external failure
(b internal failure
(c) mixed failure

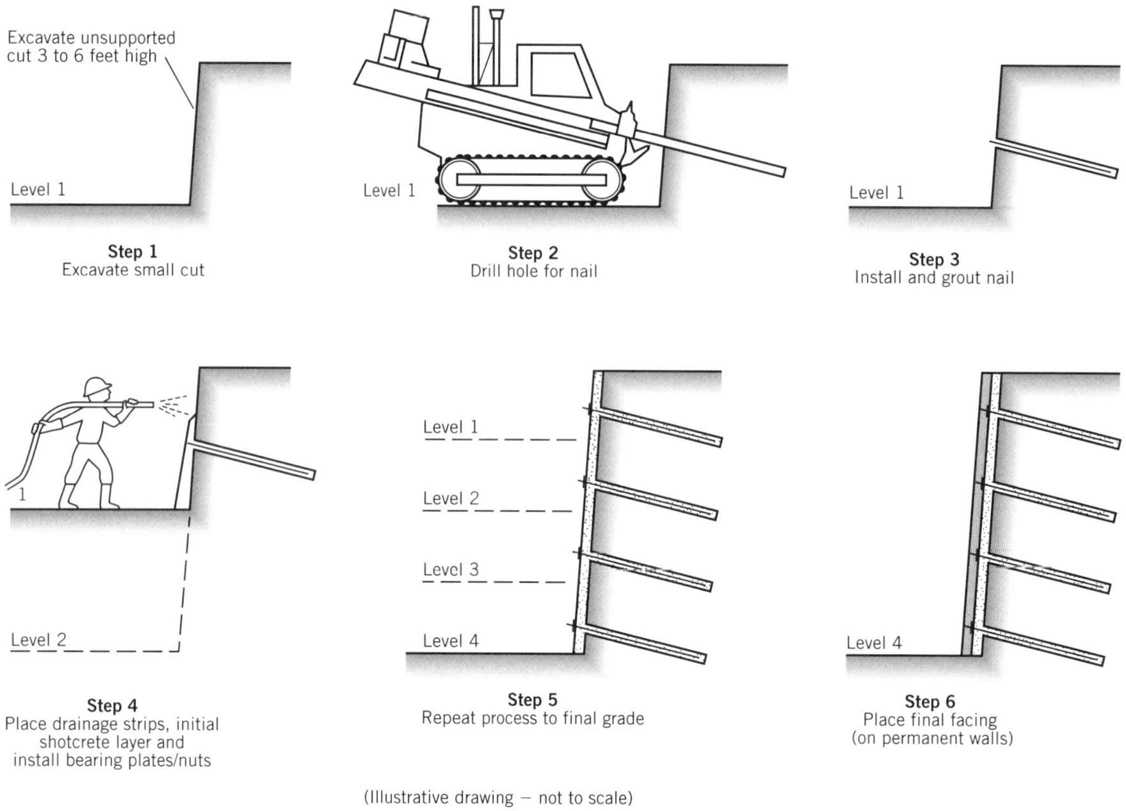

Figure 20.5 Typical soil nail wall construction sequence (after Byrne et al., 1996).

6 feet, which is the width of welded wire mesh reinforcement. In this case, the horizontal spacing is adjusted to provide an efficient design, typically with horizontal spacings from 5 to 8 feet.

The nail pattern is normally a square grid. This facilitates the installation of strip drains midway between the vertical lines of nails. An alternative pattern preferred by some engineers is to use an offset grid. For example, if the soil nail rows are numbered from top to bottom, the odd-numbered rows have their nails vertically aligned; the even-numbered rows are also vertically aligned but are offset horizontally to positions midway between the off-numbered rows. This provides a more uniform coverage pattern on the wall face.

Step 4. Prefabricated vertical (PV) drains are laid against the exposed soil face to collect minor seepage. The 12-inch or 16-inch wide vertical strips are placed midway between nail lines.

An initial layer of shotcrete, 4 to 6 inches thick, is sprayed onto the vertical surface, encasing a welded wire reinforcement mesh (see Figures 20.6 to 20.8). A typical installation also includes continuous horizontal waler reinforcement bars at each horizontal row and local vertical bearing bars at the anchor plate. Shotcrete usually is applied on the same day as the excavation.

The outer end of each nail is bolted to a steel anchor plate, 6- to 8-inch square and typically 0.5 to 0.75 inch thick. The plate connects the soil nail to the shotcrete facing. A light load, typically about 10% of the design working load, is applied to the anchor plate by a torque wrench to seat the plate into the unset plastic shotcrete.

Step 5. The excavation level is lowered by another 4 to 6 feet and steps 1 to 4 are repeated to install another row of nails. Construction in this manner is continued to the base of the wall. A minimum of 3 to 4 days is allowed for the grout to strengthen before proceeding to a lower level.

Step 6. For permanent structures, a facing of cast-in-place concrete is applied to structurally stiffen the wall and permanently encase the nail anchor plates. For temporary structures, the outer face (if needed) is constructed from a second layer of reinforced shotcrete. The total thickness of a completed permanent wall is usually about 10 inches. Occasionally precast concrete panels are attached to the outside of the wall.

Additional Comments on Soil Nail Walls

Lateral Movements

Soil nails should be installed as rapidly as possible to limit cave-ins during construction and restrict lateral movement at the soil face. A small amount of movement is needed to develop pullout resistance. Numerous pullout tests have established that relative displacements on the order of 1 to 5 mm (0.04 to

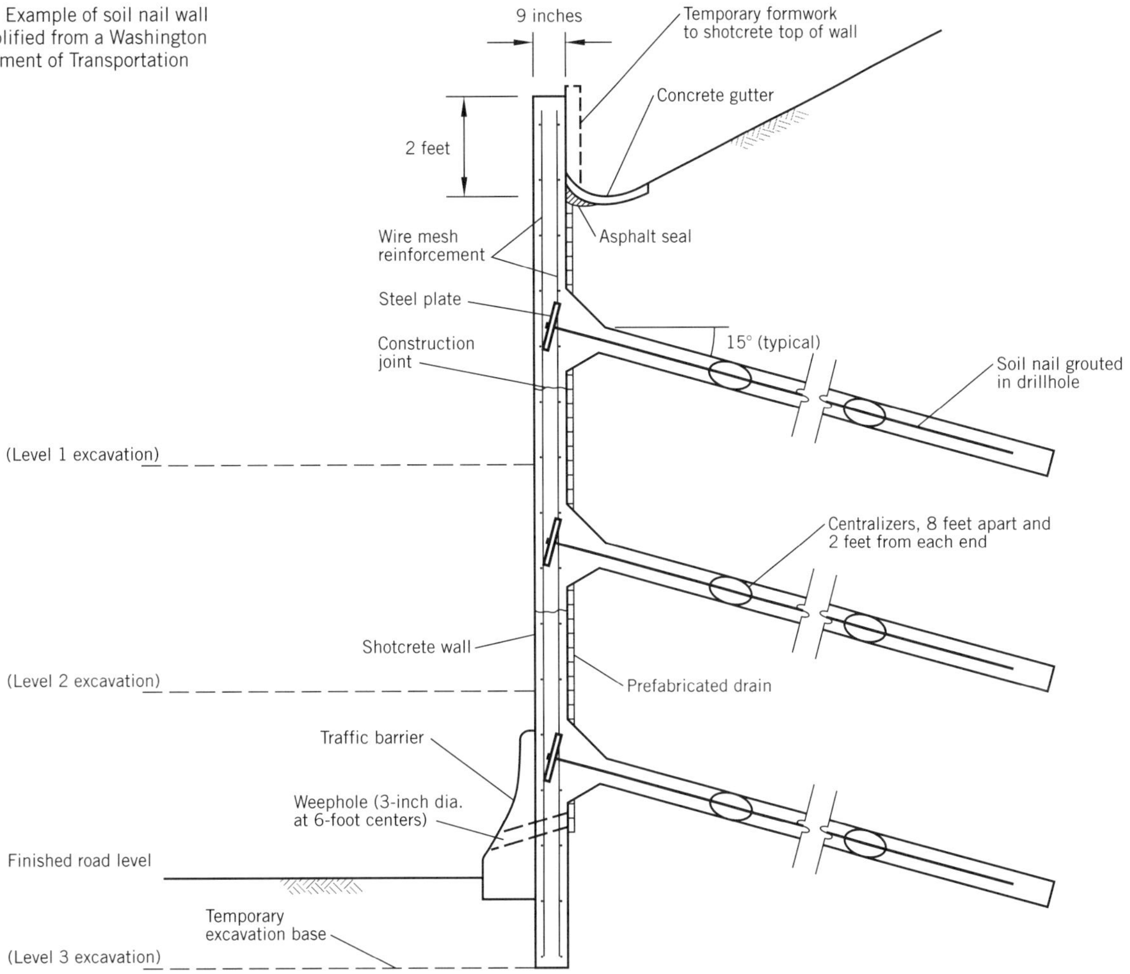

Figure 20.6 Example of soil nail wall details (simplified from a Washington State Department of Transportation project).

0.2 inch) are sufficient to mobilize the full pullout resistance. (Byrne et al., 1996).

Field observations have shown that the maximum outward movement of a soil nail wall occurs at the top of the wall face. For walls designed to a reasonable factor of safety (F = 1.5 against instability) the Clouterre (1991) studies showed that horizontal and vertical movements at the wall top were about equal and ranged from about 0.1% of the wall height H in stiff soils and weathered rocks to 0.3%H for clayey soils. Data collected by Clough and O'Rourke (1990) showed that horizontal movements at the wall top averaged about 0.2%H, thus confirming the Clouterre range.

Start of Construction

The upper row of nails is considered critical to successful construction. For this reason, many specialist contractors limit the first row to no deeper than 2.5 feet below the top of the ground. The upper part of a high soil nailed wall is often stepped.

Drainage

Drainage strips pick up minor seepage at the slope face. They should extend to the base of the wall and be connected to a footing drain or weepholes (Figure 20.6). If heavier seepage is encountered during construction, short horizontal drains can be inserted to intercept groundwater before it reaches the outer face.

Surface water has to be tightly controlled to prevent erosion of the soil in the vertical face during construction, to prevent softening, and to prevent freezing after construction. A collector ditch should be built directly above and behind the wall (Figure 20.6). A gravel-filled drainage ditch is often built at the base of a wall.

Shotcrete Facing

A "wet mix" shotcrete should be specified. It provides good quality control and can be air-entrained to improve freeze-thaw durability. The exterior face can be hand textured to provide a pleasant appearance.

The shotcrete layer has several important functions: (i) it protects the exposed soil face from erosion by rainfall or freezing; (ii) it helps to limit the lateral expansion of the soil towards the excavation; (iii) it provides structural continuity between the wall and the nail; and (iv) it prevents changes in water content and density of the soil near the face.

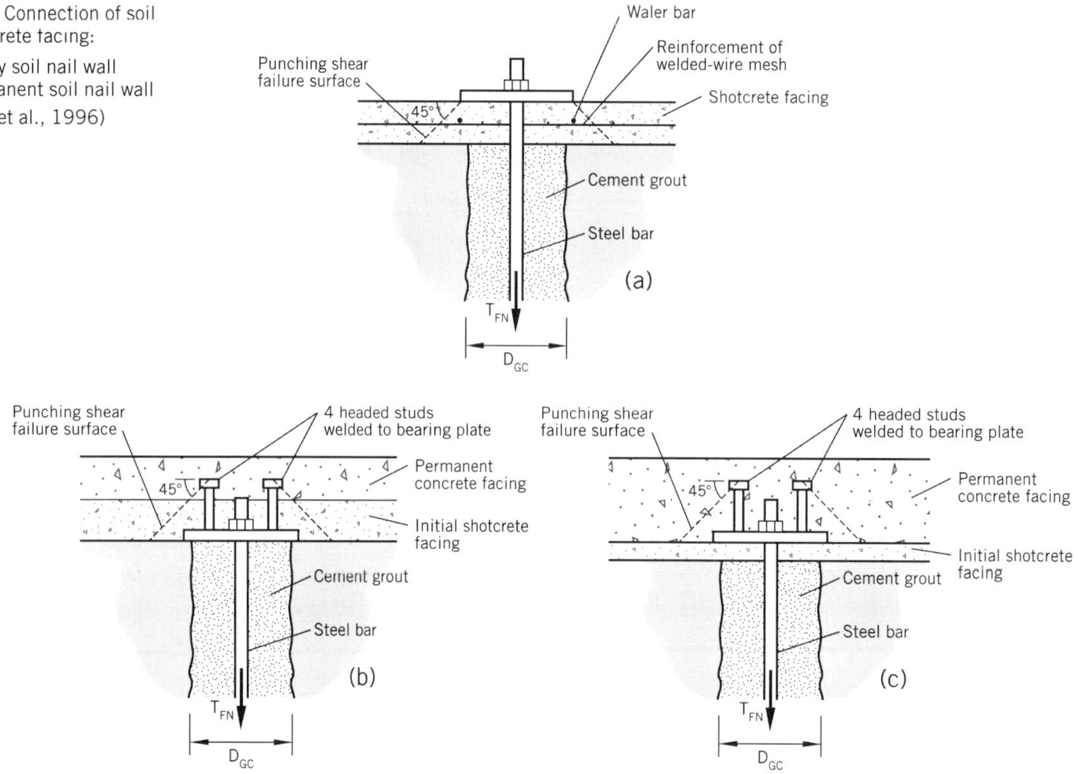

Figure 20.7 Connection of soil nails to concrete facing:
(a) temporary soil nail wall
(b) (c) permanent soil nail wall
(after Byrne et al., 1996)

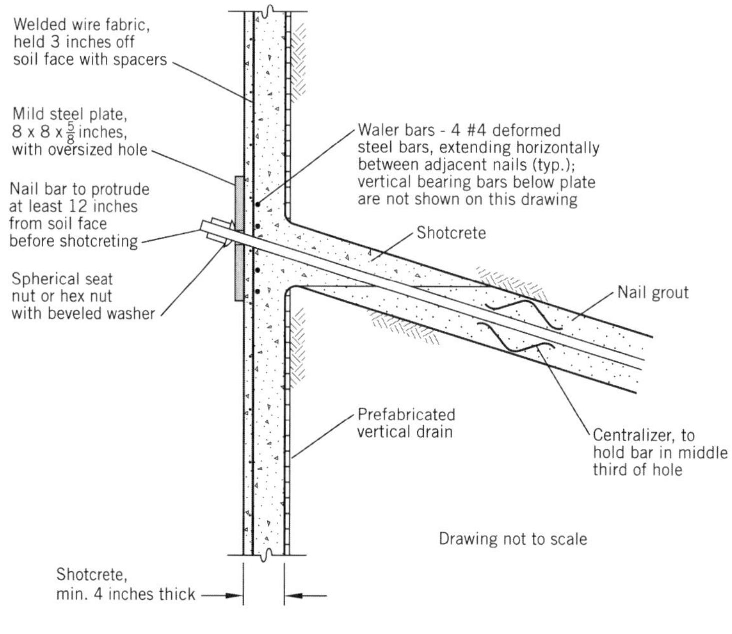

Figure 20.8 Example of soil nail/wall face connection details (after Cotton, 1992).

Soil Nail-to-Wall Connection

In U.S. practice, there are two methods to make the connection (Figure 20.7). For temporary walls, the steel anchor plate bears on the outside of the shotcrete facing using an external nut and washer (Figure 20.7a). For permanent walls, the plate is embedded within the wall. Resistance to punching shear is obtained from a headed-stud connection. On Figure 20.7(c) a thin initial shotcrete facing is constructed, after which the bearing plate is bolted to it. The headed studs develop punching shear resistance within the permanent cast-in-place facing.

On temporary shotcrete facings, it is common practice to install deformed horizontal waler bars continuously along each nail row; these are located behind the face bearing plate (Figure 20.8). Another common practice is to provide short vertical bars behind each bearing plate (e.g., No. 4 or No. 5 bars, 2.5 to 3 feet long).

Corrosion Protection of Steel Nails

Corrosion protection is needed for *permanent* soil nail walls. The requirements are the same as for tieback anchors and should include:

- Minimum grout cover of 1.5 inches, using centralizers.
- In non-aggressive ground, a resin-bonded epoxy coating on the nail surfaces, i.e., minimum coating of 0.012 inch (0.3 mm) using an electrostatic process.
- In "aggressive" ground or critical applications, the nails should be double-encapsulated, as described for ground anchors in Chapter 19, Section 19.2. "Aggressive" soil conditions occur where: (i) pH of soil is below 5; (ii) resistivity is below 2,000 ohm-cm; (iii) sulphates are above 200 ppm; or (iv) chlorides are above 100 ppm.

Seismic Stability

Soil nail walls appear to be suitable for seismic regions due to their massive and flexible design. After the Loma Prieta (California) magnitude 7.1 earthquake in 1989, eight soil nail walls in the area reportedly had no damage to the wall facings and nine nails, tested to 150% of their design load, showed no loss of carrying capacity (Felio et al., 1990).

Field Testing

Since there are numerous nails in a typical soil nail wall, it is unnecessary to field test each nail to verify the carrying capacity. The common practice is to proof test about 5% of the nails to 1.5 times the design load. A few selected nails are subjected to 2.0 times the design load to verify the ultimate load-carrying capacity. The test procedure is essentially the same as that described for soil anchor testing (see Chapter 19, Section 19.1). For test purposes, the selected soil nails are grouted only within the anchor zone (resistant zone, Figure 20.1) before the test. Later, the active zone is also grouted so that the nail can be used as an integral part of the wall. An alternative procedure is to provide a bond breaker (PVC pipe) and grout the hole to the full depth before performing the test.

Performance Monitoring

Inclinometers and surface settlement points are routinely installed on soil nail projects to monitor vertical and lateral movements during and after wall construction. Strain gauges can be put onto the tendons to allow tensile loads to be calculated. Load cells can be attached to selected soil nail heads.

Grouted Soil Nail Wall Costs

The cost of building a soil nailed wall depends on such factors as: (i) site access; (ii) height and length of the wall; (iii) corrosion protection used; (iv) whether the wall is temporary or permanent; (v) availability of experienced contractors in the site vicinity; and (vi) type of facing. Of these factors, the major one is the selected facing. According to Byrne et al. (1996), the addition of a cast-in-place or precast concrete facing over a 4-inch thick shotcrete facing may be 40 to 50 percent of the total wall cost.

For sites amenable to soil nailing and with normal working conditions, the typical construction cost range for soil nail walls in 1992 (Chassie 1992) were:

Temporary walls	\$20–30 per sq. ft. of wall face area
Permanent walls (roadway cut slopes)	\$30–40 per sq. ft. of wall face area

Higher costs can be expected for small projects, very difficult ground conditions, difficult access, remote sites, highly congested urban sites, etc.

Other Types of Soil Nail Walls

Two other installation methods for constructing soil nail walls are: *jet grouted nails* and *launched (driven) nails*. Both are proprietary (patented) systems and may not be available in many parts of the world. Each has certain advantages for use, as discussed next. Other types of driven soil nails are not presented.

Jet Grouted Nails

Louis (1986) developed a technique that installs a soil nail while enlarging the hole by jet grouting (Figure 20.9). The

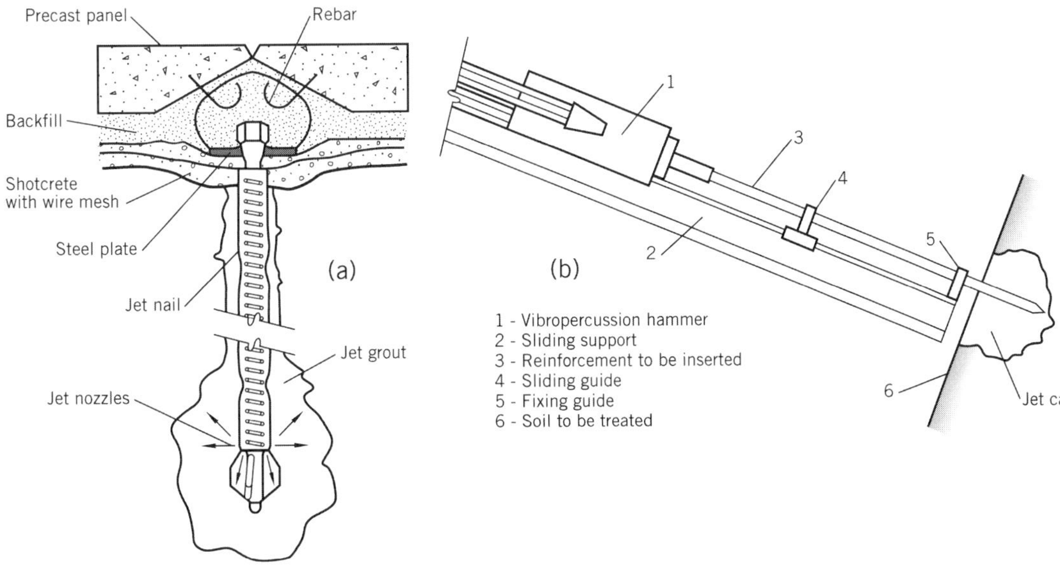

Figure 20.9 Jet grouted soil nails: (a) completed nail (b) nail driver (after Louis, 1986)

TABLE 20.3 Jet Nailing: Typical Grout Bulb Diameters and Ultimate Pullout Resistance for Various Soil Types*

	Gravel	Sand	Silt	Clay
Grout bulb diameter (inches)	24	16	12	8
Ultimate pullout resistance (kips/ft)	85	37	14	5

Note: 1 kip = 1,000 lbs.

*after Louis, 1986

equipment consists of a vibropercussion hammer, operating at 70 Hz, which injects cement grout at pressures of more than 200 psi. The jet grouting produces a large-diameter "bulb" that results in a high pullout resistance for the nail. The design relies mostly on empirical formulas for calculating resistance; these are based on field experience. Table 20.3 shows typical values. The central steel rod ranges from 1.2 to 1.6 inches in diameter.

Launched Nails

Soil Nailing Ltd. of Cardiff, Wales (U.S. office: Easley, South Carolina) has developed equipment for launching nails into

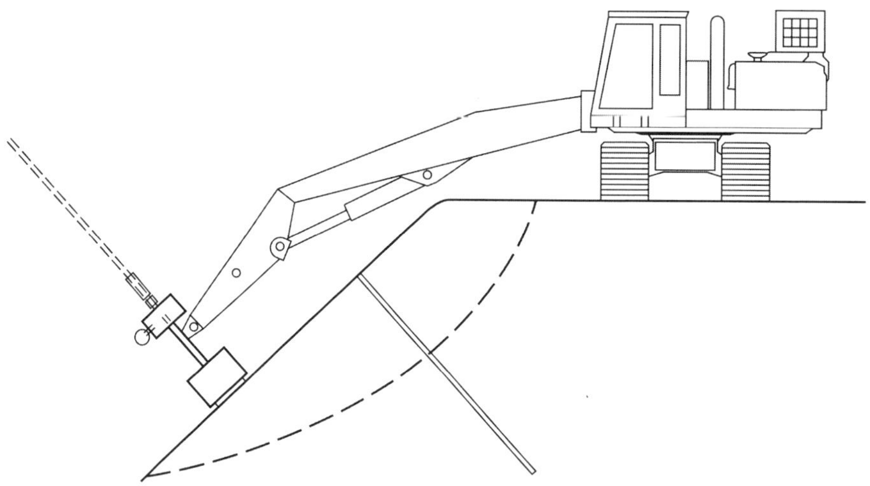

Figure 20.10 Soil nail launcher mounted on the boom of a hydraulic excavator (after U.S. Forest Service, 1994).

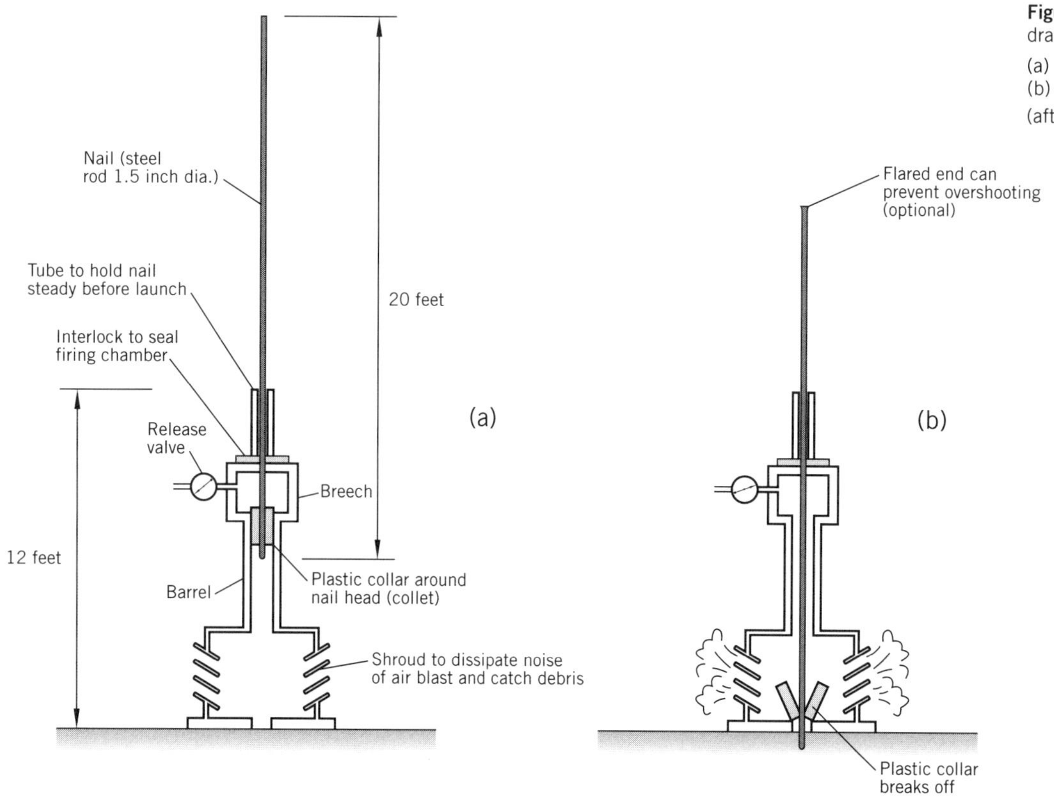

Figure 20.11 Schematic drawings of soil nail launching:
(a) nail set to fire
(b) nail being fired
(after U.S. Forest Service, 1994)

slopes as projectiles. The nails are made of plain or galvanized steel, 1½ inches diameter and 20 feet long. The nail launcher can be attached to the boom of a crawler-mounted excavator (Figure 20.10) and drives the nails into the soil approximately normal to the slope face using an articulated knuckle joint at the end of the boom.

The installation procedure is shown schematically on Figure 20.11. The steel nail has a plastic collar (collet) fitted just behind the nail tip. It is then placed into the breech chamber and compressed air (at up to 2,500 psi) is released against the collet to fire the nail down the barrel at an initial speed of around 200 mph. At the launch, the nail is held at the tip; this puts the nail in tension and prevents buckling. Before entering the ground, the high-velocity nail passes through a shroud where the plastic collar is broken into pieces to separate it from the nail. The nail penetrates into the ground to depths that depend on the initial compressed air pressure in the breech chamber and the resistance of the ground.

The nail can pass into many ground conditions up to the full length. However, the nail has difficulty in penetrating through cobbles, boulders, and stiff clays, and it is not suitable for these soil types. Any nail length left sticking out of the ground is cut off at ground level. In softer ground, a flared end can be added to the tail of the nail to prevent it from penetrating past the ground surface.

The launching technology, also referred to as *ballistic* or *rocket* nailing, is a 1989 development. For guidance on design, the manufacturer or field agents should be contacted. However, spacings of 3 feet to 4 feet are commonly used in soils, and typical minimum pullout resistances are around 2,000–3,000 lb. per nail. Estimated 1992 construction costs, including mobilization, ranged from $80 to $135 per nail. For typical small landslides, using 15 to 50 nails, the total cost range in the U.S. is about $2,000 to $6,000 per site for 1992 installation prices (1992 ENR Construction Cost Index is 5000).

Launched nail applications are limited to shallow depth (less than about 14 feet below the surface) landslides or potential slip surfaces. A typical application might be a rural road where the outer fill was built by side-casting, which often causes long-term maintenance. Many other sites can benefit from launched soil nails.

The main advantages are: (i) the speed of installation (15 nails per hour, but more commonly 4 to 6 nails per hour on average); (ii) flexibility to nail both cut or fill slopes; (iii) minimal disturbance to the ground; and (iv) they are economical. There are, however, several disadvantages including: (i) uncertainty about the depth of penetration into the ground before and during construction; (ii) concern about corrosion resistance in long-term applications; (iii) a maximum boom length of 35 feet above or below the excavator's platform; (iv) noise—a very loud bang (114 dB) at launching; and (v) availability of the equipment where needed. It is likely that future developments will address these shortcomings. The technology seems to be promising, especially for rapidly dealing with small, shallow landslides on roads and embankments.

20.2 MICROPILES

Technique

Micropiles are essentially an outgrowth of the technology used in the construction of ground anchors. They are small-diameter bored piles in which steel reinforcement is grouted into a borehole to form the pile. The loads are carried by skin friction between the soil and grout, but load-bearing is provided by the steel; they can carry loads in either compression or tension. End bearing capabilities are negligible and usually ignored.

The piles can be designed for two broad uses:

- As a conventional pile of small diameter
- As a composite soil/pile mass in which the piles reinforce the soil mass three-dimensionally into a gravity-type structure

Micropiles are typically 4 to 10 inches in diameter, 70 to 100 feet long, and carry loads of 35 to more than 100 tons in tension or compression. Their main attribute is flexibility of use. The piles can be drilled into almost all ground conditions at any preselected direction. The construction equipment requires little headroom and can gain access to restricted spaces. They have been widely used for residential underpinning, seismic retrofit, and many other difficult circumstances.

Micropiles for landslide stabilization typically are installed at various angles to the vertical (Figure 20.12) to knit soils together into a composite mass of soil and piles. The individual piles are not directly loaded but provide a framework for achieving a coherent mass of soil. Such piles are frequently referred to as *reticulated* piles or *root* piles to provide an analogy to the roots of a tree. Other common terms in English include minipiles, pin piles, and needle piles.

Micropiles typically are designed and constructed by specialty contractors using a performance specification. In developing a project use for micropiles, geotechnical consultants

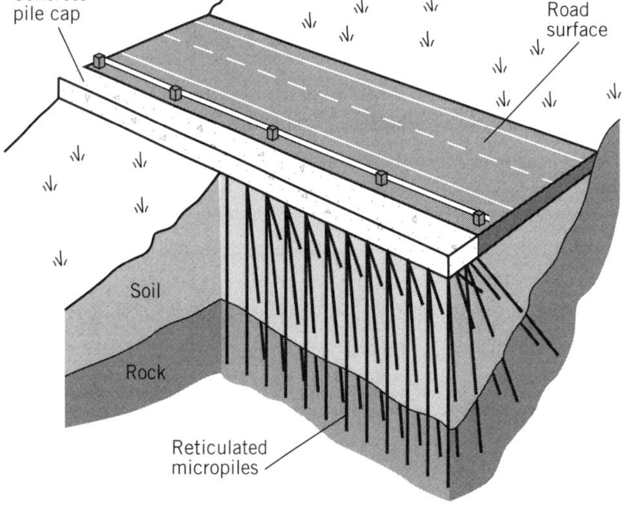

Figure 20.12 Reticulated micropiles.

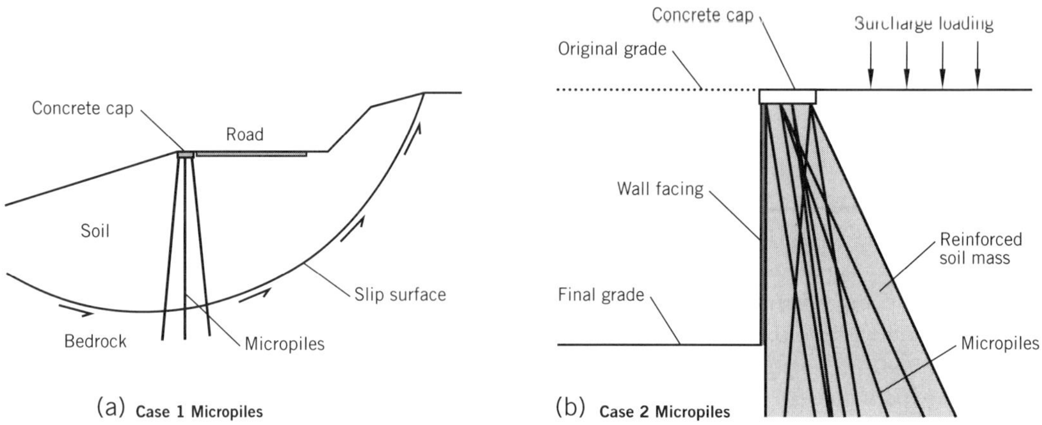

Figure 20.13 Slope stabilization with micropiles:
(a) Case 1 slip surface reinforcement
(b) Case 2 reticulated pile soil mass reinforcement
(after Armour et al., 2000)

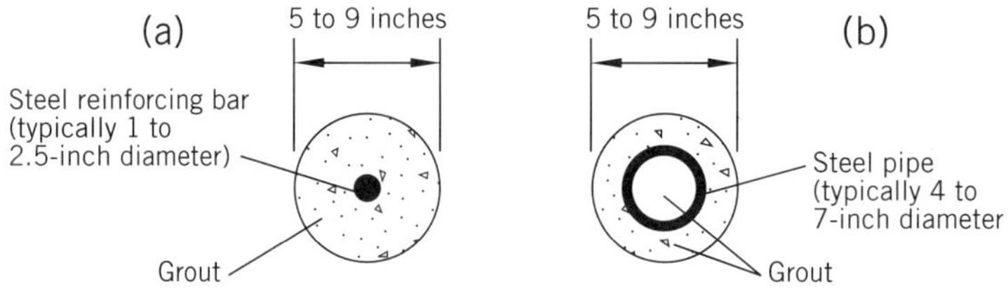

Figure 20.14 Typical sections for micropiles:
(a) grouted steel bar
(b) grouted steel pipe

should seek the advice of specialty contractors in determining feasibility for the project and writing appropriate specifications.

An excellent reference book on United States practice for construction of micropiles is the Federal Highway Administration manual *Micropile Design and Construction Guidelines Implementation Manual* by Armour et al. (2000).

Classification of Micropiles

The Federal Highway Administration (FHWA) has adopted a classification system for micropiles based on how they transmit the loads to the ground and the method of grouting (Armour et al., 2000). Case 1 micropiles are directly loaded and the micropile resists the applied loads, vertically and/or laterally (Figure 20.13a). Case 2 micropiles are an interlocking, three-dimensional network of reticulated piles. These piles are not as heavily reinforced as Case 1 micropiles because they are not directly loaded but knit together a larger block of soil (Figure 20.13b).

The grouting classification is:

Type A—gravity grouting by tremie method.

Type B—grout is injected under pressure as the temporary casing is withdrawn.

Type C—similar to Type A, but before the grout has hardened, additional grout is injected.

Type D—similar to Type A, but the first filling of grout is allowed to harden before injecting the second grout.

The micropile itself is constructed from a steel bar or pipe embedded in grout (Figure 20.14). Mild- or high-tensile strength steel is used, depending on the load requirements.

Advantages and Limitations of Micropiles for Slope Stabilization

Micropiles have advantages that they can be constructed under the following site conditions:

- Low headroom
- Restricted access
- Almost all soil conditions, including boulders and voids
- Sensitive sites requiring minimal disturbance to the adjoining ground or structures during construction
- High groundwater levels, although unsuitable for flowing artesian conditions

Case 1 micropiles (Figure 20.13a) are designed to carry loads in tension or compression. In landslide applications, such piles are too slender to provide much resistance at the slip surface, especially resistance to bending. In this regard, they are comparable with ground anchors and soil nails, and will resist instability primarily through pullout resistance. However, the vertical or near-vertical angle of installation is not efficient for this purpose. In general, Case 1 micropiles are only suitable where the landslide: (i) is relatively small (i.e., small to medium size); (ii) is experiencing creep rates of movement and thus requires only a marginal improvement in stability; and (iii) meets one or more of the advantages listed above that provides specific benefit over alternative treatments. The relatively poor bending stress resistance of Case 1 micropiles can be overcome to some extent by: (i) using large numbers of piles, and (ii) selecting a high-tensile steel pipe section (Figure 20.14b) over a reinforcement bar as the reinforcement.

Case 2 reticulated micropiles (Figure 20.13b) can provide temporary or permanent stabilization to near-surface landslides and cut slopes. The numerous angled micropiles transform a soil mass into a gravity-type retaining wall. In many cases, pressure grouting is not required; thus, the construction equipment is relatively simple to set up and move around a site.

Design

Case 1 (load-bearing) micropiles are conventional friction piles in which the skin friction depends on the type of grouting. Table 20.4 provides nominal grout-to-ground bond strengths for Type A through Type D micropiles.

For landslide projects, Case 1 micropiles have to be able to provide either: (i) *bending* resistance at the discrete shear zone of the slip surface, or (ii) *pullout* resistance from the stable ground behind the slip surface (this becomes the bonded length).

For micropiles consisting of a single grouted tendon **bar** it is recommended that stability be checked by the design principles for soil nails (this chapter, Section 20.1 preceding) and ground anchors (Chapter 19, Section 19.2). The soil/pile interaction should be similar to these other tensile loading methods. If the micropile is a steel **pipe** embedded in grout, it could be conservatively treated in the same manner as a shear pile (Chapter 22, Section 22.1).

Structural design is outside the scope of this book. Armour et al. (2000) provides a step-by-step design procedure.

Case 2 (reticulated) micropiles have no accepted design procedure at the time of writing. The block of soil reinforced by reticulated micropiles needs to extend deep enough into the slope to include the unstable soil or potential slip surface. For a vertical face (Figure 20.13b) the reinforced block should resemble a gravity retaining wall. Assuming that the piles provide internal stability, the block can be analyzed for the customary external stability considerations of overturning, sliding at the base, general slope stability, and possible instability along a deeper weak layer (if present). The number of piles required to tie together the soil mass can be a matter of judgment taking account of past precedent from published literature. The block of treated soil probably should extend at least 0.25 H below the base of the cut, where H is the height of the vertical cut.

Inclined outer slopes, or an inclined backslope above a vertical cut, would require a slightly different approach. However, it is evident that reinforcement by micropiles should extend through and behind the active wedge on a trial basis, after which stability analyses (for external stability of the mass) can be made until a satisfactory design is reached. The concept of an "imaginary" gravity retaining wall (as depicted on Figure 20.13b) can help to estimate a reasonable volume of soil to be treated.

Examples

Micropiles were used for a demonstration project to stabilize Forest Highway 7 in Mendocino National Forest, California (Palmerton, 1984). The plan and section (Figure 20.15) shows that 2.3 piles per lin. ft. of pile cap were installed. The landslide crossed a two-lane road and was stabilized by the micropiles. Although this case history has been cited as an example of Case 2 micropiles, in which the piles perform the dual function of reinforcing the soil mass and providing additional shear capacity, it seems likely that additional shear resistance at the slip surface is the principal benefit.

Two examples of using Case 1 micropiles to provide slope stability are shown on Figure 20.16. In the first example, Figure 20.16(a), a 600-foot long wall had vertical cut heights ranging from 13 to 31 feet (Ueblacker, 1996). Here the micropiles provide vertical compression and inclined tension pile support for the retaining wall. The second example, Figure 20.16(b), involves four rows of type 1A micropiles that passed through the slip surface of the landslide to anchorage in competent bedrock below (Bruce, 1988).

Table 20.4 Typical Grout-to-Ground Bond Strengths for Micropile Design*

Soil/Rock Description	Typical Range of Grout-to-Ground Bond Nominal Strengths (psi)			
	Type A	Type B	Type C	Type D
Silt & Clay (some sand) (soft, medium plastic)	5–10	5–14	7–17	7–20
Silt & Clay (some sand) (stiff, dense to very dense)	7–17	10–28	14–28	14–28
Sand (some silt) (fine, loose to medium dense)	10–20	10–28	14–28	14–35
Sand (some silt, gravel) (fine–coarse, med.–very dense)	14–31	17–50	21–52	21–55
Gravel (some sand) (medium–very dense)	14–38	17–50	21–52	21–55
Glacial Till (silt, sand, gravel) (medium–very dense, cemented)	14–28	14–45	17–45	17–50
Soft Shales (fresh–moderate fracturing, little to no weathering)	30–80	N/A	N/A	N/A
Slates and Hard Shales (fresh–moderate fracturing, little to no weathering)	75–200	N/A	N/A	N/A
Limestone (fresh–moderate fracturing, little to no weathering)	150–300	N/A	N/A	N/A
Sandstone (fresh–moderate fracturing, little to no weathering)	75–250	N/A	N/A	N/A
Granite and Basalt (fresh–moderate fracturing, little to no weathering)	200–600	N/A	N/A	N/A

Type A—Gravity grout only

Type B—Pressure grouted through the casing during casing withdrawal

Type C—Primary grout placed under gravity head, then one phase of secondary "global" pressure grouting

Type D—Primary grout placed under gravity head, then one or more phases of secondary "global" pressure grouting

*after Armour et al., 2000

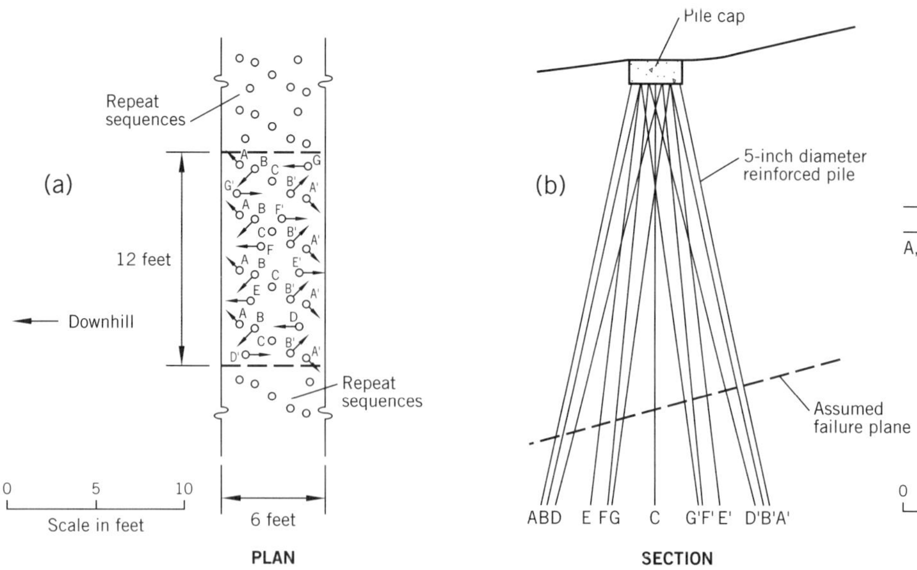

Pile	Batter V:H
A, A', B, B'	10:3.4
C	Vert.
D, D'	10:2.6
E, E'	10:1.3
F, F'	10:1
G, G'	10:1.4

0 10 20
Scale in feet

Figure 20.15 Use of reticulated micropiles to stabilize a landslide at Mendocino National Forest, California:
(a) plan
(b) section
(after Palmerton, 1984)

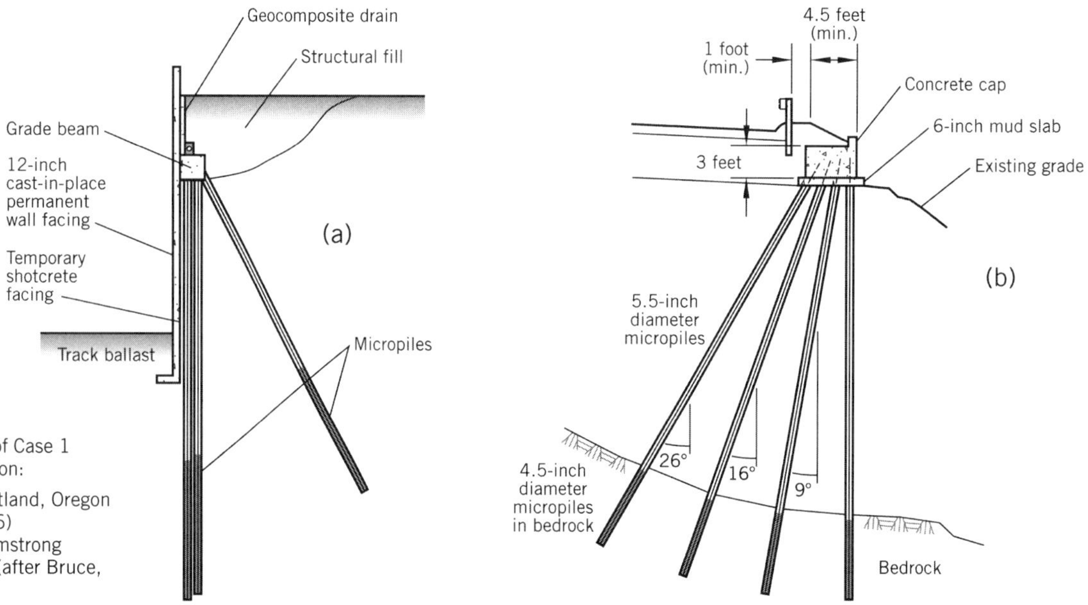

Figure 20.16 Examples of Case 1 micropile slope stabilization:
(a) cut slope support, Portland, Oregon (after Ueblacker, 1996)
(b) slope stabilization, Armstrong County, Pennsylvania (after Bruce, 1988)

Construction

Construction of a micropile requires three components: drilling the hole, installing the steel reinforcement, and grouting. It is usually accomplished as a continuous operation on the same day. The procedure is shown schematically on Figure 20.17. Common practice in the United States is to leave the drill casing in place from the ground surface down to the bond length of the pile.

Drilling

Numerous drilling techniques are used to construct micropiles, but the most common is by power rotary drill rigs using coring. Large drill rigs can be brought to open sites and smaller rigs can operate where there is restricted overhead or more difficult access. In extreme cases, frame-mounted drill rigs can install piles at sites with as little as 10 feet of headroom.

Open hole drilling is possible within clays. For collapsible soils, temporary casing or drilling fluid is used. Bentonite drilling fluid is usually avoided because the cake on the borehole wall is likely to affect the soil/grout bond. Bentonite slurry is difficult to dispose off site; polymer drilling muds, which break down quickly after use, can be substituted. Bentonite drilling muds are occasionally used with high pressure grouting.

Water is normally the flushing medium. Air is also used,

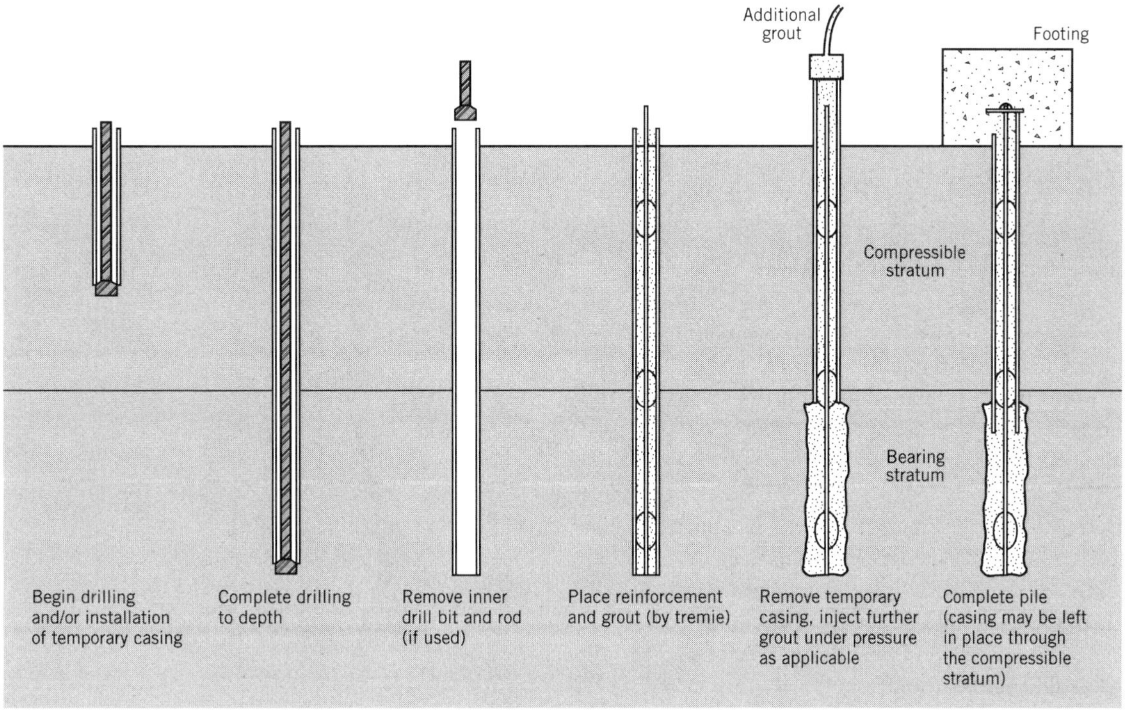

Figure 20.17 Typical micropile construction sequence (after Armour et al., 2000).

but high air pressures can fracture the adjacent ground and cause ground heave.

Grouting

The grouting operation is probably the key component of successful micropile construction because it significantly affects the soil/pile bond. In the United States, the most common grout is made from neat cement and water. A high strength and stability is required and the grout must be pumpable, sometimes over significant distances. Typically, the design water : cement ratio is in the range of 0.45 to 0.50 by weight. This can provide unconfined compressive strengths (UCS) of 4,000 to 5,000 lb./sq. in. (Figure 20.18) after 28 days. Additives are rarely necessary, but plasticizers are needed for long pumping distances or very hot weather.

Outside the United States, and especially in Europe (e.g., Italy, Britain) sand is often added to the grout mix with a sand : cement ratio of 1:1 or 2:1. However, this practice is becoming less common because of the trend towards higher grouting pressures in which sand is less desirable as an ingredient for pumping. The water : cement ratio can sometimes be as high as 0.60 in sanded mixes.

It is essential to the integrity of the pile that there be no significant loss of grout from any part of the pile. The grout is needed for both load transfer and protection of the steel reinforcement from corrosion. Current practice is to grout to refusal. If grout is lost into a permeable stratum or crack, it requires redrilling through the previous grout and regrouting, possibly with a sand filler.

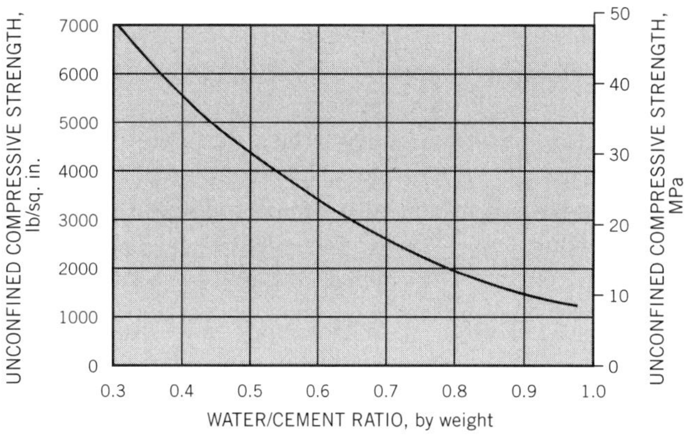

Figure 20.18 Unconfined compressive strength of neat cement grout after 28 days (after Barley and Woodward, 1992).

The grout is placed by tremie methods beginning from the lowest point of the drillhole. The tremie tube initially is tied to the reinforcing steel with duct tape and raised as the grout is fed into the hole. Any temporary casing is also withdrawn in stages over the bond length and left in place over any unbonded length above.

The grouting equipment consists of mixers, pumps, hoses, and measuring devices. The best mixer is a high speed, high shear, colloidal mixer. The usual practice is to put the water in first, followed by cement, aggregate, and any other admixtures. The materials are mixed for about 2 minutes and then kept under slow agitation in a holding tank and used within one hour. The quantity of water in the mix is controlled by a flowmeter and all other ingredients are measured by weight.

As previously described, micropiles are classified by Types A, B, C, D, depending on the grouting procedures. Type A requires no pressure. For low-pressure grouting (Type B), a constant pressure rotary-screw type pump (Moyno) is commonly used. At higher pressures (Types C, D) a fluctuating pressure piston or ram pump supplies the grout to the hole.

Specific techniques for the different micropiles are:

Type A Micropiles. These are constructed by gravity filling. The drillhole is filled with grout tremied in from the bottom upward, and the reinforcement is installed through the grout. This type of grouting is referred to as *primary treatment*. It is commonly applied in stiff/hard clays and sound rock.

Type B Micropiles. After the primary treatment has been completed, additional grout is injected as the temporary casing is withdrawn. The objective is to enhance the grout/soil bond, especially in cohesionless soils. The higher pressure may be restricted to the load transfer (bond) length only or it may be extended over the full length of the pile.

The pressure cap is attached to the top of the drill casing. U.S. practice is to inject the secondary grout at pressures of up to 150 psi to achieve permeation in coarse-grained granular soils or fractured rock. Another benefit, termed *pressure filtration,* is believed to result from pressure grouting of cohesionless soils. Some of the grout water penetrates into the surrounding soil, lowering the water/cement ratio of the grout and giving it higher strength. It also produces a cakelike paste at the grout/soil interface that improves bond. Another possibility, as in compaction grouting, is that the pressure may densify the surrounding soil; this would be a further benefit for Case 2 (reticulated) piles but is an unproven concept. The applied pressures typically are 1 psi per lin. ft. of depth in loose soils and 2 psi per lin. ft. of depth in dense soils.

Types C and D Micropiles. These use postgrouting to achieve even higher grout pressures than Type B piles. In Type B, the pressures are limited by leakage around the casing or by hydrofracture of the adjoining ground.

Postgrouting techniques always use neat cement-water mixes (for ease of pumping) and usually have slightly higher water : cement ratios of 0.50 to 0.75 by weight. Some of this excess water is expelled into the soil by pressure filtration (previously described). Both Type C and Type D micropile construction begin with gravity (primary) grouting; the difference is in the high-pressure injection process. In Type C, the high-pressure grout is administered once through a preplaced grout pipe (or hollow reinforcement tube) at pressures of at least 150 psi. The secondary grout in Type D construction is not injected until after the primary treatment has set. It passes through a sleeved grout pipe (Figure 20.19). The injection site is isolated by double packers and grout is supplied at pressures of 300 to 1,200 psi to rupture (crack open) the primary grout and inject additional grout behind it. The benefit of this system (known as *tube à manchette* after the French developers) is that several phases of injection can take place at the same location using progressively higher pressures and thus develop a larger bulb of grouted soil. This increases the resistance of the pile to penetration or pullout. The final volume of grout may be as much as 250% of the primary volume.

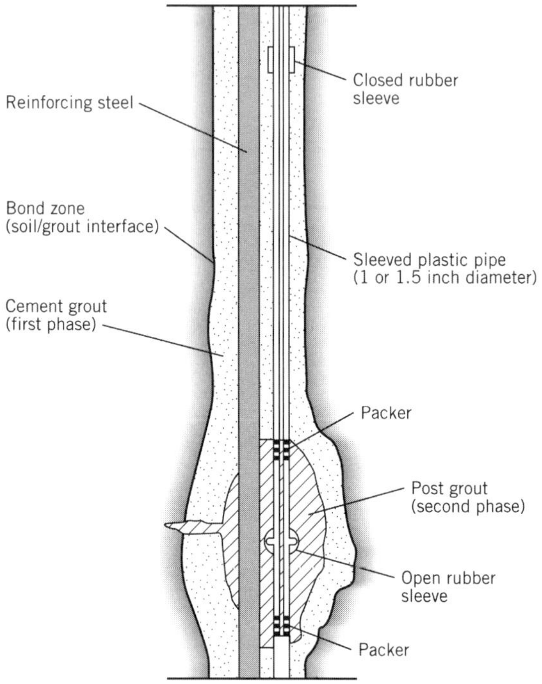

Figure 20.19 Postgrouting injection using a sleeved tube (after Armour et al., 2000).

Another high-pressure grouting technique is the *circulating loop* arrangement. Again using a sleeved tube, the grout pressure is gradually increased until the grout breaks out through one or more of the sleeves. This procedure is repeated by further pressure raises.

Reinforcing Steel

Several options are available, including:

Reinforcing Bars

These are usually single bars of 1 inch to 2.5 inches diameter, but groups of bars are sometimes used. Yield strengths: 60,000 to 75,000 psi. The bars are supplied in lengths of up to 20 feet. Centralizers, usually plastic, keep the reinforcing bar in position in the center of the drillhole.

Continuous Thread Bars

These are used throughout the world for micropile reinforcement. Examples: Dywidag Threadbar (GEWI pile) and Williams All-Thread Bar. The steel bars have a coarse pitch, continuous ribbed thread on the outside of the bar (Figure 20.20) that has several benefits. It improves the steel/grout

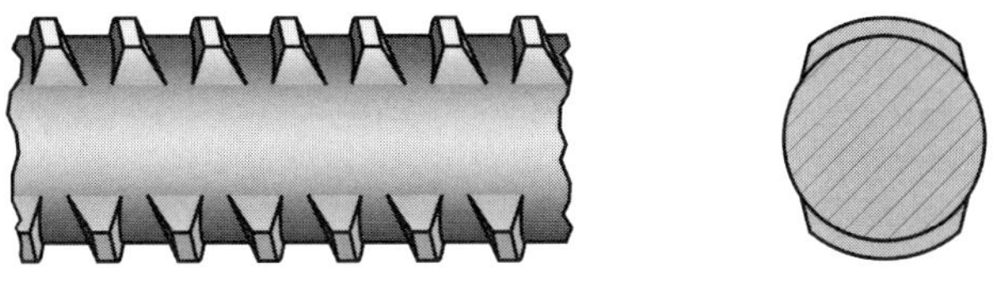

Figure 20.20 Dwyidag continuous thread steel bar (Dwyidag Systems International).

bonding; more importantly, it allows the bar to be joined with a coupler to retain full tension/compression capacity of the bar. Finally, it allows the bar to be connected to a surface anchor plate using a hex nut.

Continuous Thread Hollow Core Bars

The hollow core allows the bar to be used for drilling the hole for the pile. A drill bit is mounted on the tip. The bar is drilled into the ground using grout flush passing through the hollow core. Manufacturers include Dywidag, Ischebeck, and Chance.

Steel Pipe Casing

The casing provides higher resistance to lateral loading than central bars and can be considered as a possible technique for improving resistance at the slip surface of landslides. The steel pipe can serve as the drill casing and is left permanently in the hole as reinforcement. Another option is to install a permanent smaller diameter pipe inside the temporary borehole casing and grout the annular space as the temporary casing is withdrawn. API (N-80) casing, with a yield strength of 80,000 psi, is commonly used by specialist contractors in the United States. It is available in sizes ranging from 5.50 to 9.62 inches o.d. A combined casing/triple bar arrangement is shown on Figure 20.21.

Quality Control/Assurance

The field inspector should keep a log of each micropile installation including details of the pile drilling and grouting. Armour et al. (2000) provide a suitable log and example.

Grout is a vital component of a micropile. A set of three grout cubes (2 inch side length) should be taken directly from

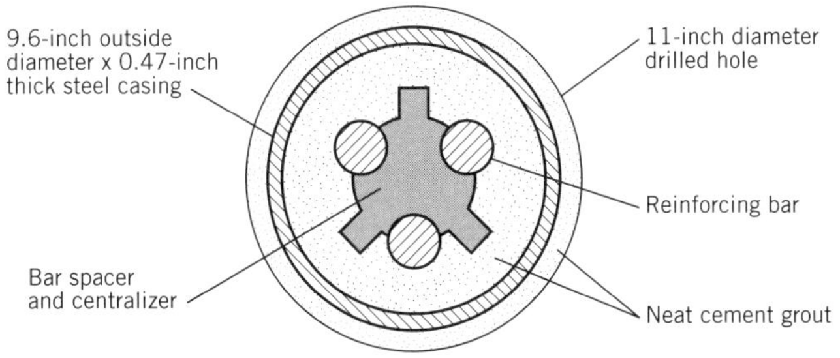

Figure 20.21 Example of a heavily reinforced micropile using steel casing and reinforcement bars (after Armour et al., 2000).

the grout plant and, after curing, a cube from the batch should be tested in *unconfined compression* at 3, 7, and 28 days (ASTM test C-109). One set of cubes should be obtained for each day that grouting operations are undertaken, or for every 10 piles, whichever occurs more frequently.

The water content of the grout mix is another key control on grout quality. It is commonly checked by a *specific gravity* test; for neat cement grout density, the Baroid Mud Balance test (ASTM C 188) is used. The test should be made at least once per pile—a density in the range of 112 to 119 lb./cu. ft. (1.8 to 1.9 Mg/m^3) corresponds to a water : cement ratio of around 0.45, which is often specified for micropile grout.

Load-bearing micropiles can be test loaded in compression or tension. A tension (pullout) test is less expensive and is usually more relevant to landslide work. A verification test is usually performed at the start of a project on a sacrificial pile. At a load of 1.33 times the design load (DL), the load is held constant for 60 minutes as a creep test. It is then loaded to a maximum of 2.50 times the design load; this load is maintained for 10 minutes. Loads are added in increments of 0.25 DL.

Proof tests are carried out on about 5 percent of production piles. The loads are applied in increments to a maximum of 1.67 DL. If a pile fails the verification test, the pile design may need to be modified or the construction procedures may need change. For a production pile, the options for inadequate pile capacity include: (i) construct a replacement pile; (ii) allow the pile to be used at 50% of the maximum load reached in the test; (iii) increase the bond length; or (iv) postgrouting. In a creep test, the creep should be linear or decreasing with time under a constant load, and the creep rate should be less than 0.04 inch (1 mm) from 1 to 10 minutes and less than 0.08 inch (2 mm) from 6 to 60 minutes.

Reticulated micropiles may require minimal load testing or none at all on small projects.

Construction Costs

Construction costs for installing micropiles depend on numerous variables including location, access to site, subsurface conditions, pile design, field testing requirements, and contractor availability. As a general guide for conceptual cost estimating, Armour et al. (2000) give a range of $45–90 per lin. ft. of pile in the United States. This reference includes guidelines for refining the estimate to take account of pricing variables, some of which have been listed above. The price range assumes: (i) no physical, environmental or access restrictions; (ii) no unusual subsurface conditions; (iii) average pile load (110 tons) and lengths (50 feet); (iv) one verification pile load test and proof testing 5% of production piles; (v) one mobilization/demobilization; (vi) continuous pile drilling operations; and (vii) prevailing labor rates and contractor overhead/margin percentages. The rates are based on 1996 costs. ENR Construction Cost Index for mid-1996: 5600.

A typical contract document pays for micropile installation on a per pile basis. Mobilization and pile loading tests are paid as separate items.

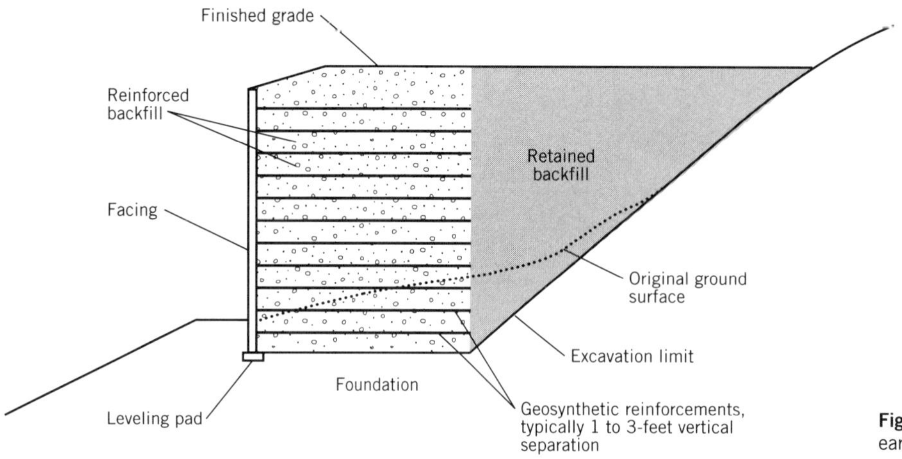

Figure 20.22 Mechanically stabilized earth wall: terms.

20.3 MECHANICALLY STABILIZED EARTH WALLS

Technique

Mechanically stabilized earth (MSE) walls use horizontal reinforcement to provide tensile resistance within a soil mass (Figure 20.22). The reinforcing elements can be made of galvanized steel (sometimes epoxy coated) or fabric—high density polyethylene (HDPE), polyester (PET) or polypropylene (PP). Reinforcement lengths typically are 70 to 90 percent of the wall height. Vertical separation ranges from about 1 to 3 feet.

Metal reinforcements are more rigid and their inclusion within the soil restricts the lateral deformation of the soil. They can be used in temporary or permanent walls to heights of up to 100 feet.

Fabric reinforcement undergoes more deformation than metal inclusions. It is also more susceptible to ongoing degradation with time. The concern for long-term deterioration of the fabric has historically restricted the use of fabric walls to short-term temporary construction (up to about 50 feet high) or permanent walls (up to 15 feet high) in non-critical environments.

The different deformability properties of steel and geotextiles produce somewhat different tensile loadings behind the wall. Both types of wall need an outer facing to protect the soil from erosion and vandalism, and, for geotextiles, to reduce degradation of the reinforcement itself.

MSE walls are "bottom up" construction built with fill. In general, they are significantly less expensive to build than concrete retaining walls for heights above 10 feet provided a high quality aggregate backfill is readily available.

For landslide applications, MSE walls can be built to support shallow slides. They can also be incorporated in a buttress remediation where available land is restricted (e.g., edge of a flood plain, urban area, river edge, etc.). Also, the full height of the buttress (due to the vertical face) is a valuable stability benefit where the slip surface passes through the buttress. MSE walls are beneficial for road widening projects and developing level building sites on slopes.

A *reinforced soil slope* is the term used for mechanically stabilized earth where the outer slope is less than 70° to the horizontal. The design is similar in principle to that for MSE walls and Elias, et al. (2001) provides detailed procedures. Some possible applications are shown on Figure 20.23. However, since rockfill slopes can be built *without* reinforcement to slopes of 1:1 (45° to the horizontal) there is a limited need for reinforced soil slopes in landslide practice. Some additional comments are given at the end of this section.

Advantages and Limitations of MSE Walls

MSE walls generally compete with concrete gravity retaining walls and are more cost-effective at a majority of sites where both types of wall could be built. The principal *advantages* of MSE walls are:

- The walls are more flexible than concrete walls and can tolerate moderate foundation settlements without distress. Wrap-around walls (geosynthetics, wire mesh) are more tolerant of settlements because the outer facing can be added at the end of construction after most of the settlement has occurred.
- The walls are built by readily available construction equipment and do not require special labor skills.
- Many alternative MSE wall systems are available, providing healthy price competition.
- The walls are built more rapidly than concrete walls.

For landslide remediation and prevention, the following benefits can be mentioned:

- Free-draining granular backfill improves slope drainage.
- In some cases, landslide debris may be acceptable for use as fill in MSE construction.
- The walls require less land area than an unreinforced soil slope.

The *limitations* of MSE walls include:

- The walls require significant space behind the wall to accommodate the reinforcement.

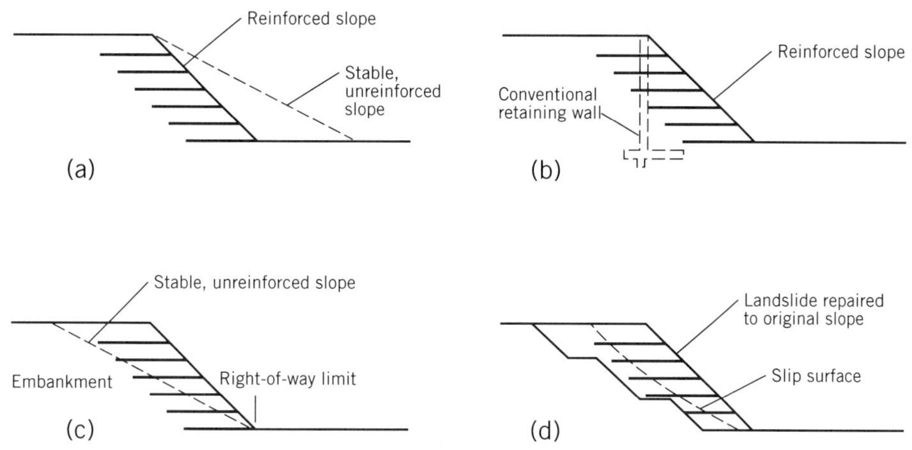

Figure 20.23 Soil reinforcement applications on slopes:
(a) steeper outer fill slope
(b) alternative to retaining wall
(c) embankment widening with limited right-of-way (sliver fills)
(d) shallow slide repair
(after Elias et al., 2001).

- Good-quality granular fill needs to be available.
- Aggressive soil conditions can cause deterioration of the reinforcements.
- Design practices are not standardized, requiring more vigilance from geotechnical practitioners responsible for a structurally safe design.
- They cannot be used at sites where utilities have to be buried within the reinforced backfill.

Concern about geosynthetic reinforcement deterioration has historically restricted the use of fabric walls to short-term temporary construction or permanent walls of up to about 15 feet high in noncritical environments. *Non-critical* walls are defined (Allen and Holtz, 1991) as sites where, should a failure of the wall occur, there would be: (i) a low risk of life loss, and (ii) minimal disruption of services and/or economic losses. In addition, such a wall should not be supporting another structure, and the soils in contact with the wall should not be chemically aggressive to the geosynthetic.

When fine-grained soils are used as backfill in the reinforced wall, the issue of long-term creep needs careful study. This is not covered herein, and the reader should review available publications and the results of continuing research on this topic. The best advice is not to use fine-grained soils at all and specify granular backfill.

Reinforcement Types

Reinforcement systems are subdivided into those that are *extensible* and those that are *inextensible*. An extensible reinforcement deforms without rupture to deformations that are greater than occur in the soil. This group includes *geotextiles* and *geogrids*. Inextensible reinforcements include *steel strips* and *bar mats*. The differences in reinforcement deformability properties affect the location of the maximum tensile force behind the wall. Another significant behavioral difference is that geotextile fabrics provide their resistance to pullout solely by friction between the soil and fabric. Steel reinforcements rely on a combination of friction and passive resistance (from ribs, transverse bars) for pullout capacity.

Available types of reinforcement are listed on Figure 20.24. Key properties (typical ranges) are given on Table 20.5. These products are subject to change in specifications, and this table should be used only as a general guide. Current specifications can be obtained directly from suppliers.

Galvanized steel is used for strip reinforcements. The zinc coating provides a sacrificial layer of protection. In the design calculations of tensile resistance the deterioration of the metal by corrosion is estimated for the life expectancy of the wall.

Geosynthetic reinforcements are in the form of geogrids or geotextiles. Geogrids are made from either high density polyethylene (HDPE) or PVC-coated polyester (PET). Geotextiles are either polyester (PET) or polypropylene (PP). Geotextiles can be woven or non-woven.

In the United States, reinforcement manufacturers include: The Reinforced Earth Co., McLean, Virginia (galvanized ribbed steel strips); Hilfiker Retaining Walls, Eureka, California (welded wire mesh); Tensar Earth Technologies, Atlanta, Georgia (HDPE geogrids); Foster Geotechnical, San Diego, California (metal bar grid); Contech Construction Products, Middletown, Ohio (geosynthetic fabrics). Many other suppliers are available (see Elias et al., 2001) and can be contacted via an Internet search. Several suppliers offer more than one reinforcement type.

Facings

Numerous types of facings are available for MSE walls. They are needed to retain the soil at the outer edge of the wall. The three main types are:

1. Precast concrete panels that are attached to the reinforcing elements. The panels have decorative shapes, including rectangle, cruciform, hexagonal, diamond, etc.
2. Modular block wall units that are specially designed and manufactured for retaining wall facings. These units are usually constructed without mortar and are connected vertically by shear pins, lips, or keys. They are termed *segmental retaining walls* and provide a pleasing appearance as well as being functional (Figure 20.25a).

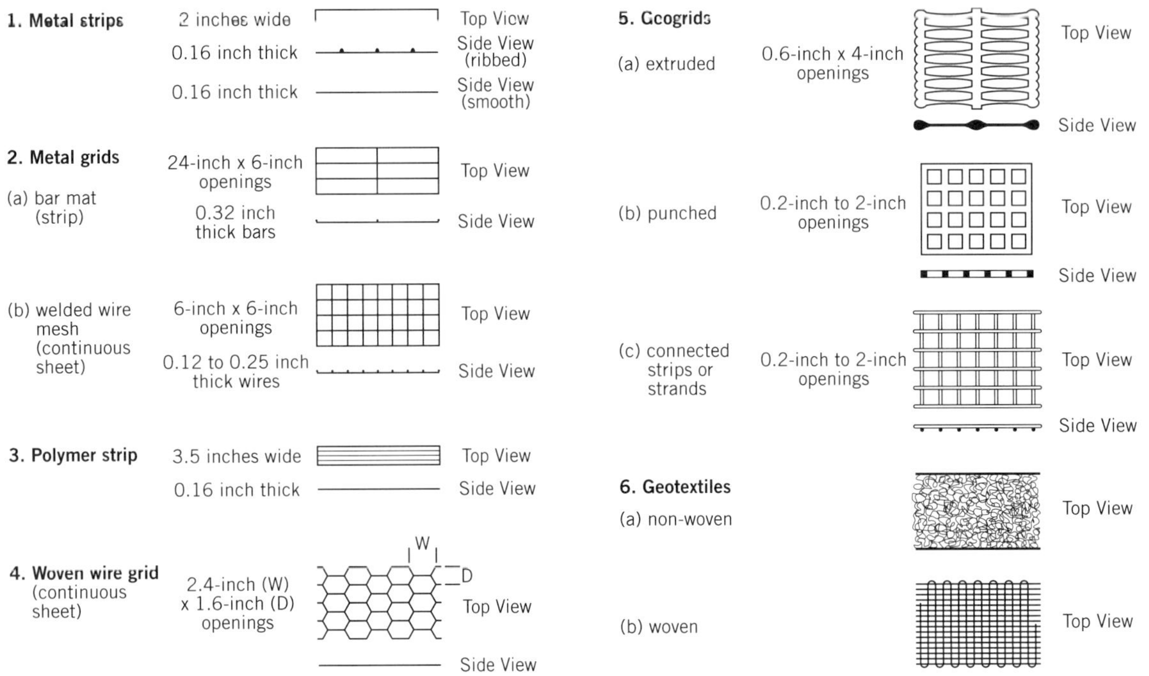

Figure 20.24 Various types of soil reinforcement (after Mitchell and Christopher, 1990 and Schlosser and Delage, 1987).

Table 20.5 Representative Mechanical Properties for Various Types of Soil Reinforcement*

Type	Subgroup	J Modulus (lb./inch)	Tensile Capacity (lb./inch)	Stiffness Ratio (lb./in./in.)
1. Metal Strips	Ribbed/smooth	4.75 x 10^6	3,100	13,000
2. Metal Grids	(a) Bar mat (strip)	460,000–515,000	340	5,800–8,700
	(b) Welded wire mesh (continuous sheet)	57,000–230,000	55–170	2,500–10,000
3. Polymer Strips		110,000	1,700	440
4. Woven Wire Grid	Woven Wire Mesh (continuous sheet)	11,400–57,000	110–170	440–2,300**
5. Geogrids	(a) Extruded (b) Punched (c) Connected strips or strands	430–11,400	30–300	18–440
6. Geotextiles	(a) Non-Woven	12–4,600**	12–150	0.44–190
	(b) Woven	430–57,000	30–240	18–2,500

*after Mitchell & Christopher, 1990, and Schlosser & Delage, 1987 **confined

Notes:

1. J Modulus is given in force per unit width of reinforcement

   $J = E (A_c/b)$ where A_c = total cross section of reinforcement material

   b = width of reinforcement

   E = modulus of the material

   Tensile capacity is allowable value *without* reduction for durability considerations

   Stiffness ratio assumes vertical spacing of 2 feet between reinforcements

2. The listed data should be used only as a general guide. Manufacturer's current specifications should be obtained prior to design.

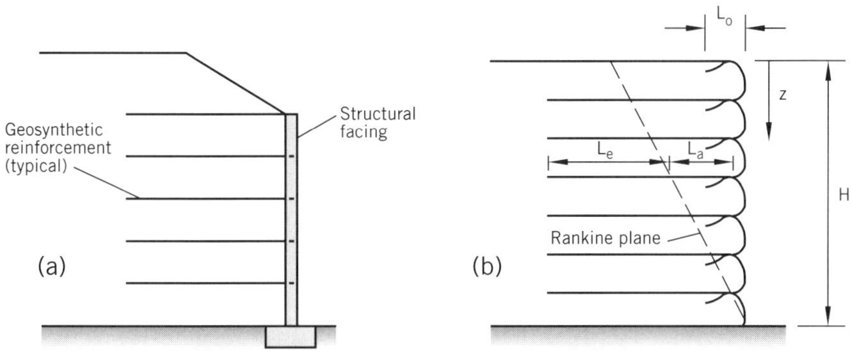

Figure 20.25 Geosynthetic walls:
(a) with structural facing
(b) with wrap-around geosynthetic facing
(after Allen and Holtz, 1991)

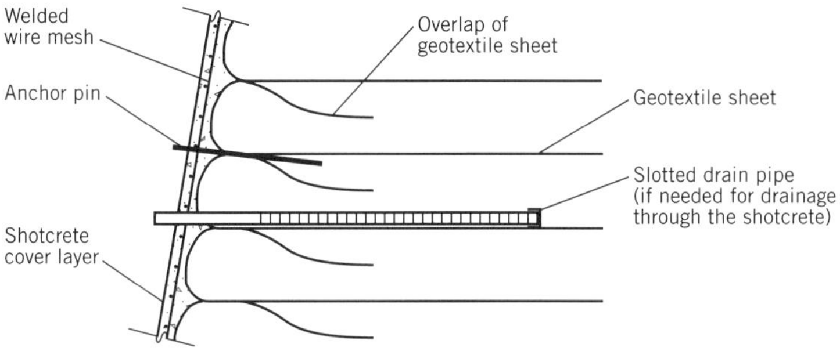

Figure 20.26 Shotcrete finish for wrap-around fabric wall.

3. Wrap-around facings, where the soil reinforcement material (geosynthetic, wire mesh) is bent through a right angle at the wall face to retain the soil. The facing is bent again, after backfilling, to anchor the reinforcement into the layer above (Figure 20.25b). This creates an outer face that is unattractive, and the face is usually finished with shotcrete, precast concrete panels, or cast-in-place concrete. An example of a finish detail is shown on Figure 20.26. Gabions, treated timbers, and other inexpensive facings have been used for temporary wall protection.

Backfill Soil

Most MSE walls are built using select, cohesionless, backfill soils. A common specification is to require a well-graded, free-draining, granular soil with the following gradation limits:

U.S. SIEVE SIZE	PERCENT PASSING
4 inch	100
No. 40	0–60
No. 200	0–15

Coefficient of uniformity should be greater than or equal to 4, and the plasticity index should be less than 6. Organic content should not exceed 1%. For epoxy-coated steel reinforcement, and also for geosynthetics, the maximum particle size should be reduced to ¾ inch to prevent reinforcement damage.

The soil particles should be hard and durable. Using the magnesium sulphate soundness test (AASHTO T-104), the loss should be less than 30% after 4 cycles.

For landslide remediation and prevention, good drainage is highly desirable at most sites, and a lower fines content in the backfill is needed. The author recommends restricting the backfill to nonplastic soils with less than 3–5% by weight passing the No. 200 sieve.

Other specification limits on soils used for backfill are given on Tables 20.6 and 20.7. Material suppliers of proprietary MSE systems usually have their own criteria for backfill.

There is interest in using cohesive soils as a backfill for MSE walls. However, their low strength, poor drainage, and creep properties make clays undesirable for this work.

Table 20.6 Recommended Limits of Electrochemical Properties for Backfills When Using Steel Reinforcements

Property	Criteria	Test Method
Resistivity	>3000 ohm-cm	AASHTO T-288-91
pH	Between 5 and 10	AASHTO T-289-91
Chlorides	<100 ppm	AASHTO T-291-91
Sulfates	<200 ppm	AASHTO T-290-91

Table 20.7 Recommended Limits of Electrochemical Properties for Backfills When Using Geosynthetic Reinforcements

Base Polymer	Property	Criteria	Test Method
Polyester (PET)	pH	3 to 9	AASHTO T-289-91
Polyolefin (PP & HDPE)	pH	more than 3	AASHTO T-289-91

Design

MSE walls have to be designed for both *external* and *internal* stability. The external stability is essentially the same as that for any other gravity-type retaining structure. Internal stability refers to the ability of the soil reinforcement to maintain stability within the reinforced soil mass. The internal stability is the main focus of this summary.

External Stability

The earth pressures acting externally on a MSE wall with horizontal and sloping backslopes are shown on Figure 20.27. The example of a level backslope surface (Figure 20.27a) also includes a surcharge load, but live loads (such as traffic) are usually ignored. The "equivalent" uniform loading near the toe of the wall (the Meyerhof correction for bearing capacity calculations) is shown on Figure 20.27(b). The wall facing load is included where it is significant (e.g., gabions or segmental block facings).

The design calculations require calculations of: (i) overturning; i.e., eccentricity e must stay within the middle third ($F \geq 2.0$); (ii) sliding along the base ($F \geq 1.5$); (iii) bearing capacity failure ($F \geq 2.0$); (iv) overall stability analysis; i.e., circular arc, wedge, or failure along a weak zone within the foundation ($F \geq 1.3$ temporary or ≥ 1.5 permanent); (v) seismic stability in seismic regions ($F \geq 1.1$); and (vi) settlement. Most of these calculations are routine foundation engineering. Elias et al. (2001) give a good set of design procedures for MSE walls including seismic pseudo-static stability. If the calculations indicate that the minimum factors of safety are not being achieved, the general remedy is to lengthen the soil reinforcement.

Highway design standards generally require the minimum reinforcement length L to be the larger of 8 feet or 0.7 H, where H is the vertical height of a MSE wall with a horizontal backslope. A wall with a sloping backslope requires longer reinforcement, generally in the range of 0.8 to 1.1 H, where H is the vertical height of the wall at the face. Most MSE walls are built with uniform lengths of reinforcement over the full depth of the wall.

Internal Stability

MSE walls have horizontal passive reinforcements that pass through the active zone of the soil mass and into the resistant zone behind it (Figure 20.28). The role of the reinforcement is to tie the active zone to the resistant zone by providing the necessary tensile (pullout) strength. The maximum tensile force (T_{max}) approximately coincides with the boundary between the active and resistant zones. Thus, the total length of a reinforcement element L can be subdivided into the length within the active zone L_a and the remaining length L_e within the resistant zone behind the potential failure surface. For sloping backslopes (not shown) the maximum tension lines of Figure 20.28(a)(b) are extended linearly to the sloping ground surface.

For each reinforcement level in the wall, internal stability calculations are needed to determine:

- The maximum tensile load (T_{max})
- The tensile load at the connection to the facing
- The pullout resistance (P_r) of the reinforcement within the resistant zone
- The allowable tensile load within the reinforcing element to prevent fracture

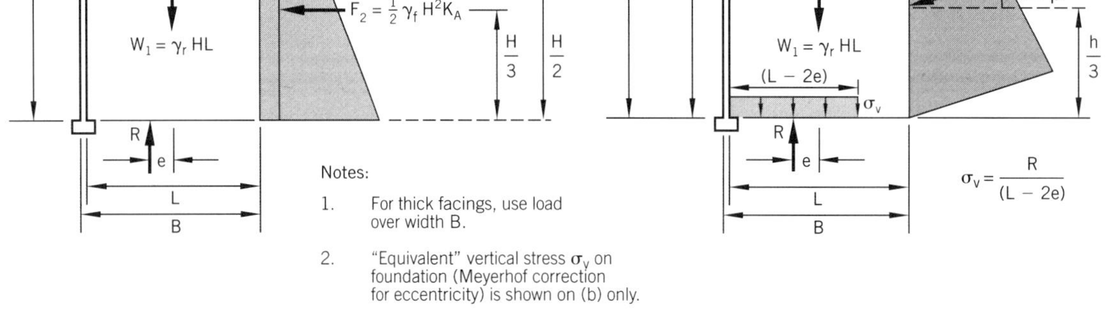

Figure 20.27 External lateral pressures acting on an MSE wall:
(a) horizontal backslope
(b) sloping backslope

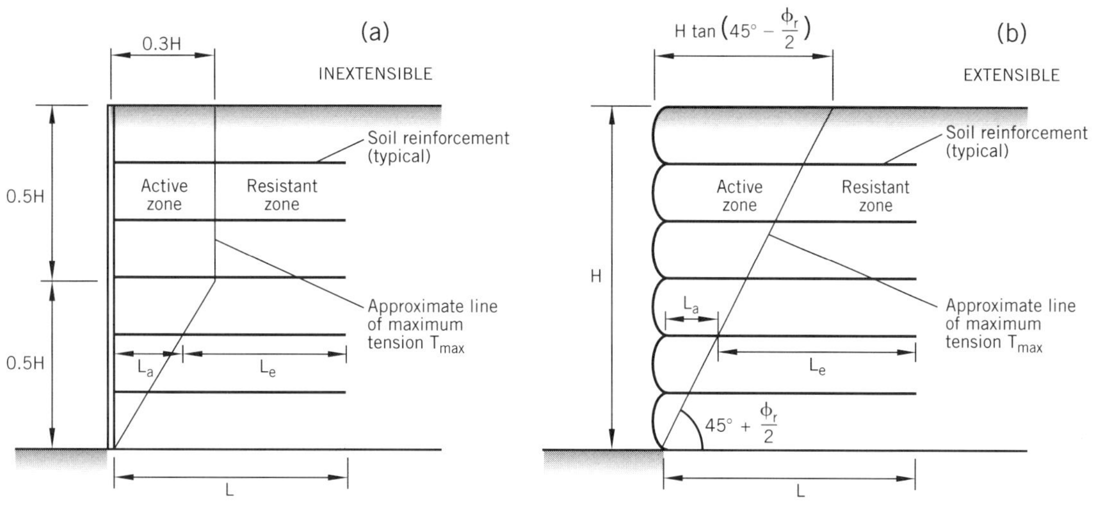

Figure 20.28 Assumed line of maximum tensile load for MSE walls with horizontal backslope:
(a) inextensible reinforcement
(b) extensible reinforcement

At each level of reinforcement, the pullout resistance P_r must exceed the pullout force T_{max} by a specified factor of safety, usually 1.50 for permanent structures.

Much of the knowledge on these subjects has been obtained from small-scale laboratory and full-scale field experiments combined with theoretical analyses such as finite element studies. Several design approaches have been put forward, and it is likely that further refinements will occur in this evolving technology. The method described by Elias et al. (2001) in a Federal Highway Administration publication is presented below.

Maximum Tensile Loads

Experimental data have shown that horizontal metal reinforcements within a soil mass produce a different failure surface than the classical Mohr-Coulomb wedge. For such inextensible reinforcements (steel strips, welded bar grids, etc.) the approximate location of the maximum tension line, separating the active from the resistant zones, is shown on Figure 20.28(a). It is approximately bi-linear and passes through the toe of the slope.

The maximum tensile loads developed within *inextensible* reinforcements of a MSE wall are higher than active pressures. It is understood to be a function of reinforcement shape and modulus, and the density of such reinforcements within the soil. The effect of such higher-than-active lateral pressures decreases with depth below the wall top, allegedly due to an offsetting relationship between soil dilation and overburden stress.

Geotextiles and other *extensible* reinforcements deform in a similar way to unreinforced soil behavior. The Mohr-Coulomb wedge, Figure 20.28(b), approximately defines the line of maximum tension, T_{max} for these reinforcements.

The diagram shown on Figure 20.29 represents practice in the United States for calculating the maximum tensile forces occurring in soil reinforcements. It was prepared from observed field data "normalized" as a function of the active earth pressure coefficient, K_A. It has been termed the *simplified coherent gravity* method (Elias et al., 2001). It is based on the assumption that the vertical stress acting on the reinforcement is equal to the stress imposed by the overburden weight γ H directly above it. For a vertical wall face with a horizontal backslope, $K_A = \tan^2 (45° - \frac{1}{2} \phi_r)$ where ϕ_r is the friction angle of the reinforced soil fill. Active earth pressure coefficients for other geometries and conditions (e.g., sloping backslope, inclined wall face, friction between wall face and soil, etc.) can be found in Elias et al. (2001) and many other widely available sources on earth pressure calculations.

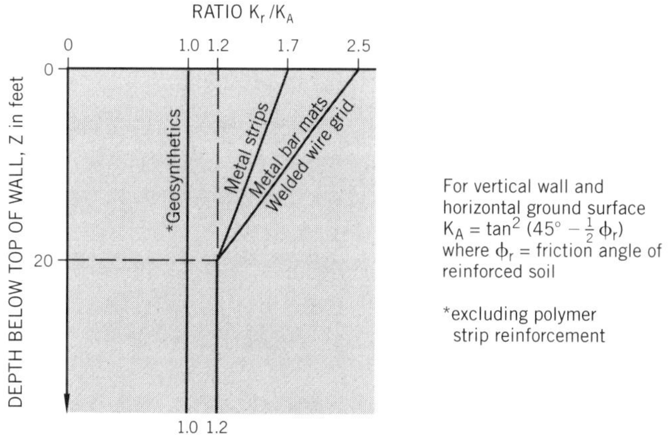

Figure 20.29 Lateral earth pressure ratios K_r/K_A for various reinforcement types.

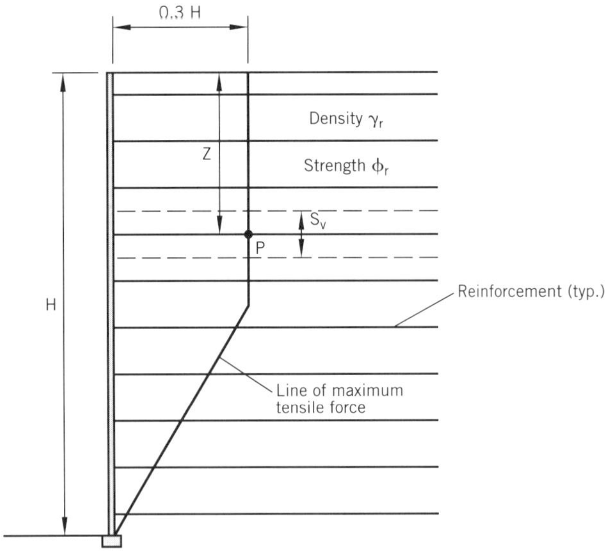

Figure 20.30 Calculation of maximum tensile force at point P on a reinforcement.

For a vertical MSE wall with a horizontal backslope and no surcharge loading, Figure 20.30, the calculations are:

$$\sigma_v = \gamma_r Z \quad \text{Eq. (6)}$$

where σ_v = vertical stress at the reinforcement level, point P
Z = depth of reinforcement below top of wall
γ_r = density of fill within the reinforced soil wall

$$\sigma_h = K_r \sigma_v \quad \text{Eq. (7)}$$

where σ_h = horizontal stress acting on the reinforcement at the potential failure line, point P
K_r = earth pressure coefficient at P obtained from Figure 20.29

For sheet reinforcement,

$$T_{max} = \sigma_h S_v (= K_r \sigma_v S_v) \quad \text{Eq. (8)}$$

where T_{max} = maximum tensile force per unit width of wall
S_v = vertical spacing between reinforcements

For discrete reinforcements (metal strips, bar mats, geogrids, etc.)

$$T_{max} = \sigma_h S_v / R_c = K_r \sigma_v S_v S_h / b \quad \text{Eq. (9)}$$

where R_c = coverage ratio b/S_h
b = gross width of reinforcing strip, sheet, or grid (Figure 20.31)
S_h = center-to-center horizontal spacing between strip, sheet, or grid
[*Note:* R_c = 1.0 for continuous reinforcement where each layer covers the entire horizontal surface of the reinforced soil mass]

Tensile Load at Wall Face

For design of the maximum tensile load at the connection of the reinforcement with the facing, conservatively use T_{max}.

Pullout Resistance

The pullout resistances of reinforcements are fairly complex. Geotextiles have a planar surface, and resistance develops as adhesion between the soil and geotextile. Experimental data show that frictional interaction is fully mobilized at shear displacements of only 0.04 to 0.2 inch (1 to 5 mm).

In the case of ribbed metal strips and metal bar grids, pullout resistance is a combination of friction along the longitudinal bars (in the direction of pulling toward the wall face) and passive resistance in front of the transverse bars or raised surfaces of the ribs. Passive resistance requires at least 1 inch (25 mm) of shear movement to be fully mobilized, much more than is needed for full frictional resistance.

Another factor in the soil/reinforcement behavior is the dilatancy of compacted granular fill. According to Schlosser (1990), dilatancy (soil volume increase during shear) is restrained by the surrounding soil. It develops a three-dimensional friction mechanism and induces an increased normal force on the reinforcement, thus increasing the pullout resist-

Figure 20.31 Dimensions and terms for:
(a) strips
(b) grid reinforcements
(isometric views)

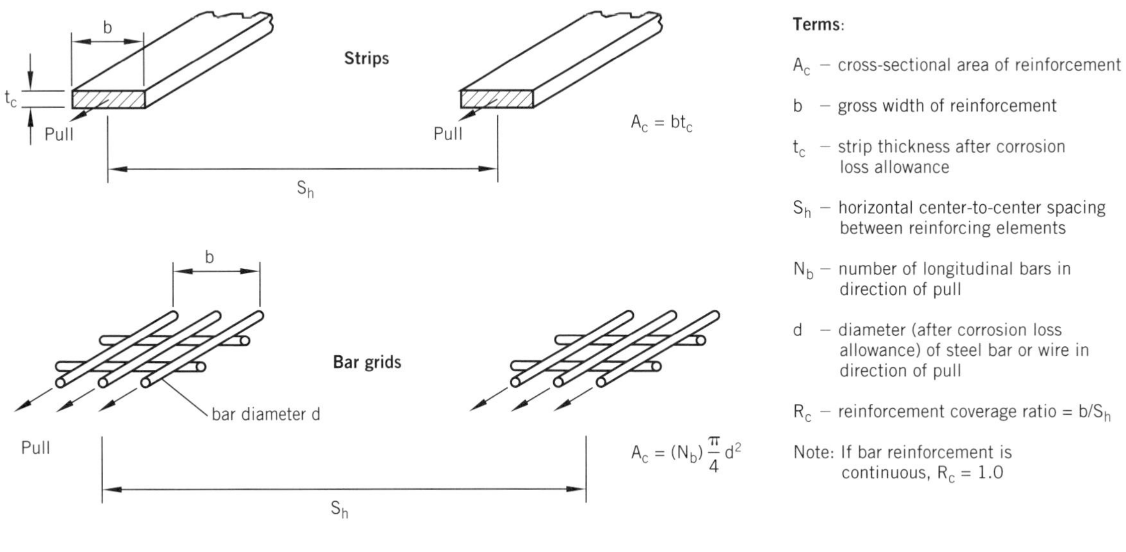

ance. To account for this behavior, which he terms the *restrained dilatancy* effect, Schlosser developed the concept of an "apparent" coefficient of friction $\mu^\star$ as a function of overburden pressure γ H, i.e., $\mu^\star = \tau / \gamma H$. The value of $\mu^\star$ decreases with initial normal stress in an approximately linear rate to a depth of 20 feet after which it remains constant. Parameter τ is the pullout friction strength.

In summary, pullout resistance is made up from three possible components when discrete elements are used for soil reinforcement, namely: friction, passive resistance, and a three-dimensional soil-reinforcement interaction component. The following paragraphs briefly describe U.S. practice (Elias et al., 2001).

The pullout resistance (P_r) required per unit width of reinforcement is given by:

$$P_r = F^\star \alpha \sigma'_v L_e C \qquad \text{Eq. (10)}$$

where $F^\star$ = pullout resistance factor
α = scale effect factor
σ'_v = effective vertical stress at the soil-reinforcement interface
L_e = embedment or adherence length of reinforcement within the resistant zone behind the potential failure surface (see Figure 20.25)
C = unit perimeter for the reinforcement ($C = 2$ for strips, grids, sheets; $C = \pi$ for bars)

Pullout resistance factor $F^\star$ is determined by friction for geosynthetic sheets and by a combination of friction and passive resistance for other inclusions. Most specialty system suppliers can provide recommended pullout parameters for their products when used with select backfill, and there are test procedures to measure the parameters in the laboratory. Semi-empirical relationships for use with the standard specifications for granular backfill (given earlier) are summarized in Table 20.8.

Scale effect factor α accounts for non-linear stress reduction over the embedded length of highly-extensible reinforcements. It can be obtained from pullout tests on reinforcements with different lengths. In the absence of test data, use $\alpha = 0.6$ for geosynthetic sheets and $\alpha = 0.8$ for geogrids. For inextensible reinforcements use $\alpha = 1.0$.

For geosynthetic sheets (i.e., continuous reinforcement over the full width) and specified granular fill (assume $\phi_r = 34°$), the pullout resistance is:

$$P_r = (0.45)(0.6)\,\sigma'_v L_e (2)$$
$$= 0.54\, L_e \sigma'_v$$

The pullout force is given by:

$$T = K_A \sigma'_v S_v = 0.28\,\sigma'_v S_v \text{ for } \phi_r = 34° \text{ and level backslope}$$

Therefore, the factor of safety F is:

$$F = \frac{0.54\, L_e}{0.28\, S_v}$$

$$\text{i.e., } L_e \geq 0.52\, F\, S_v \qquad \text{Eq. (11)}$$

Example $S_v = 2$ feet, factor of safety $F = 1.50$
$L_e \geq 1.56$ feet

Table 20.8 Semi-Empirical Relationships to Estimate the Pullout Resistance Factor F* for Soil Reinforcements

Reinforcement Type	Semi-Empirical Relationships for F*	
Ribbed steel strips	Top of structure	$F^\star = 1.2 + \log C_u = 2.0$ max.
	Below 20 feet	$F^\star = \tan \phi_r$
Steel grids, transverse spacing $S_t \geq 6$ inches	Top of structure	$F^\star = 20\, t/S_t$
	Below 20 feet	$F^\star = 10\, t/S_t$
Geosynthetic geotextile or geogrid sheets	At all levels	$F^\star = 0.67 \tan \phi_r$ $= 0.45$ (for assumed $\phi_r = 34°$)

$C_u = D_{60}/D_{10}$ = uniformity coefficient of backfill (if unknown, use $C_u = 4$)
t = thickness of transverse bar
S_t = horizontal spacing between transverse bars
ϕ_r = friction of fill within the reinforced soil mass

In practice, length L_e should be designed to extend at least 3 feet behind the Mohr-Coulomb potential failure surface. Therefore, length L (Figure 20.28) in feet is given by:

$$L = H \tan (45° - \tfrac{1}{2}\phi_r) + 3 \qquad \text{Eq. (12)}$$

For H = 15 feet, L = 11 feet (i.e., 0.73 H for this wall height)

Note: For geotextile reinforcements in general, it is usually assumed that the friction angle between the soil and geotextile ϕ_{sg} can be calculated from:

$$\phi_{sg} = 0.67\, \phi_r \qquad \text{Eq. (13)}$$

where ϕ_r = friction angle of the backfill material

Eq. (13) is considered to be conservative. Actual values can be measured in the laboratory for major projects and for the following situations (Allen and Holtz, 1991):

- Widely spaced reinforcement layers
- Use of geosynthetic strips instead of continuous sheets
- Fine-grained soil backfill that might creep under shear load

Fracture Resistance of Reinforcing Elements

Steel. The allowable tensile force (T_a) per unit width of reinforcement is determined from:

$$T_a = \frac{F F_y A_c}{b} \qquad \text{Eq. (14)}$$

where F = factor of safety = 0.55 (strips) or 0.48 (grids)
b = width of the reinforcing element (Figure 20.31)
F_y = yield stress of steel
A_c = design cross-sectional area of steel (actual area less the anticipated corrosion losses during the design life)

The loss of steel cross-sectional area due to corrosion depends on the degree of aggressiveness of chemicals within the backfill soils. See Elias et al. (2001) for more details.

Geosynthetics. The tensile capacity of geosynthetic reinforcement is affected by many variables, including creep, temperature, construction damage, and age (including attack from microorganisms, chemicals, hydrolysis, etc.). For design purposes, reduction factors (RFs) are applied to the ultimate tensile strength (ASTM test D 4595) to arrive at the allowable tensile strength.

$$T_a = \frac{T_{ult}}{(RF)F} \qquad \text{Eq. (15)}$$

where T_{ult} = ultimate or yield tensile strength of the reinforcement
RF = reduction factor
F = overall factor of safety = 1.5 (permanent walls) = 1.2 (temporary walls)

$$RF = (RF_{CR})(RF_D)(RF_{ID}) \qquad \text{Eq. (16)}$$

The *creep reduction factor* RF_{CR} is the ratio of the ultimate load to the maximum sustainable load in the design life. Typical RF_{CR} values: polyester 1.6 to 2.5; polypropylene 4.0 to 5.0; high density polyethylene 2.6 to 5.0. The *durability reduction factor* RF_D covers several ageing factors and ranges from 1.1 to 2.0; use 2.0 in the absence of specific information. The *installation damage reduction factor* RF_{ID} ranges from 1.1 to 3.0, depending on backfill gradation and geosynthetic material; for geogrids, a default value of 1.8 can be used and for geotextiles, a default value of 3.0 is recommended for this factor.

The value of T_a is usually obtained from the manufacturer and includes reduction factors. However, it does not allow for the overall factor of safety F. More details on reduction factors, applicable tests, etc. for geosynthetic walls is provided in Elias et al. (2001).

Base Embedment

The toe of the MSE wall, point D, needs to be embedded below the adjacent ground in front of the wall. The minimum recommended depths of embedment are shown on Figure 20.32. For walls adjoining streams, the wall base should extend at least 2 feet below the depth of maximum potential scour.

Vertical and Lateral Movements

The facing of a MSE wall can be damaged by differential settlements and the permissible limits are listed in Table 20.9. Rigid concrete panels are more susceptible to damage than flexible welded wire. Total settlements have to be considered where a wall is close to a structure.

Lateral displacements of a wall are more difficult to estimate and are higher for extensible reinforcements than for inextensible reinforcements. Increasing the length : height (L:H) ratio decreases lateral movement of the wall. In general, vertical and lateral movements of a MSE wall are not a concern in landslide remediation.

Seismic Design

MSE walls have performed very satisfactorily to date (2000) during major earthquakes, and there are no known failures of internal stability. However, failure of a modular block facing occurred during the Chi Chi earthquake of September, 1999 in Taiwan (Leshchinsky, 2000). The usual design is by pseudo-static procedures. Good coverage of design procedures is provided by Elias et al. (2001). The minimum factor of safety is 1.1.

Construction of MSE Walls

Rigid Facings (Facing Panels)

The subgrade of the wall is prepared and an unreinforced concrete leveling pad, typically 12 inches wide and 6 inches deep, is constructed along the facing line (Figure 20.22). The pad is a guide for the erection of panels and is not a structural foundation. For a modular block wall facing, it is common practice to use a compacted gravel pad in lieu of concrete.

After the concrete pad has set and hardened (one day), the first row of facing panels is set up on the leveling pad using shoring to maintain the correct alignment and provide stability. The first layer of granular backfill is placed behind the facing. The total depth of a layer between reinforcements typically is 1 to 3 feet thick. However, individual lift thicknesses during backfilling should not exceed 12 inches. Backfill is compacted to a minimum density of 95% of the maximum dry density measured in the standard Proctor compaction test (AASHTO designation T-99).

Backfill is dumped in the middle and back of the reinforced wall section, and is bladed toward the wall face as needed. The retained backfill (Figure 20.22) behind the reinforced fill section is placed concurrently.

After reaching the lowest connection level, the first row of reinforcing elements is laid out and attached to the back of the

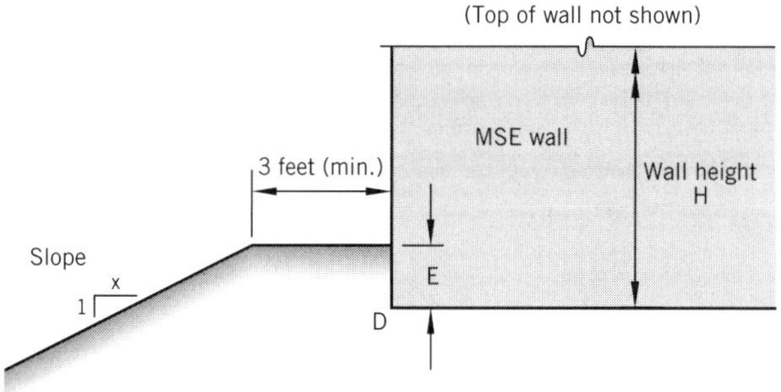

Slope in front of wall	Embedment E	Minimum E
Horizontal	H/20	1 foot
x = 3	H/12	1.5 feet
x = 2	H/7	2 feet
x = 1.5	H/5	3 feet

Figure 20.32 Recommended minimum embedment at the front of an MSE wall (Reinforced Earth Company, 1995).

Table 20.9 Recommended Limiting Differential Settlements for MSE Walls

Wall Facing	Differential Settlement Limit	Wall Facing	Differential Settlement Limit
Precast concrete facings:		Full height concrete facing:	1/500
joint width ¾ inch	1/100	Modular block units:	1/200
joint width ½ inch	1/200	Welded wire	1/50
joint width ¼ inch	1/300	Geosynthetics	1/50

facing panel. The strips or grids are placed perpendicular to the back of the panel. Construction of additional face panels, reinforcement, and fill continues until the wall is completed.

Erection of the wall facing is a critical part of the construction process. Because the backfill will deflect the wall panels outward during construction, the panels are initially set with a slight backward batter of about ¼ inch per foot of panel height. Full-height panels need special care to maintain specified tolerances. The first row of panels needs to be continuously braced until several layers of reinforcement and backfill have been placed.

Each system has its own method of attaching the grids or strips to the back of the precast concrete panels. In the Reinforced Earth Company system, for example, a tie-strip is cast into the back face of the panel and a single ½-inch bolt acting in double shear attaches the ribbed strip to the panel. A rubber pad separates the top of one panel from the bottom of the panel above. At the inside (back face) of the panels, a 1½-foot wide strip of glued nonwoven filter cloth spans the gap between joints (horizontal and vertical). The filter cloth allows groundwater to seep out but prevents soil from passing through the cracks.

Field control of moisture content and density is essential. For soils containing 30% or more retained on the ¾ inch sieve, a methods specification usually replaces the compaction test specification. Typically, a methods specification requires three to five passes with a vibratory roller compactor. For moisture control, Elias et al. (2001) suggest that the fill be placed and compacted at about 2 percent *dry* of optimum moisture content. A moisture content wet of optimum often makes it difficult to maintain an acceptable facing alignment, particularly with soils having a high fines content.

Large, vibrating rollers are used to compact soils over most of the wall footprint. However, heavy compaction equipment near the wall face can cause excessive panel movements. Instead, walk-behind vibratory rollers or vibratory plate compactors (three passes) are recommended for compaction within 3 feet of the face. If fine-to-medium sand is being placed as backfill, a smooth-drum static weight roller can be used; sheepsfoot rollers, which can damage the reinforcement, should be avoided.

Flexible Facings

Flexible-faced MSE walls use the reinforcing material to provide the outer facing of the wall. These materials include geotextiles, geogrids, and welded wire mesh. In this construction procedure, a concrete leveling pad is not needed, and construction begins from a level grade.

Geosynthetics often have anisotropic strength characteristics, so the reinforcement has to be laid out with the higher strength direction perpendicular to the face of the structure. Adjacent sheets of fabric are overlapped by a minimum of 6 inches. Geogrid and wire mesh can be tied together by a butt connection. Flexible reinforcements require pretensioning to remove slack and then are either staked or covered by fill to maintain the taut conditions. These procedures reduce the future horizontal movements as the wall is built.

A wrap-around facing construction technique is shown on Figure 20.33. Temporary framework provides the correct alignment for the fabric. After about one-half of the total layer has been placed and compacted, the geosynthetic is wrapped over an outer "windrow" of fill. The remainder of the layer then is placed and compacted such that the outermost tail of fabric is "tucked in" and buried to hold it in place. This fabric tail is about 3 feet long.

The wall is built to the full height by placing and compacting granular fill as previously described for a rigid facing wall. At the filling completion, almost all the settlement and lateral deformation of the wall has occurred. A facing layer is added to protect the exposed flexible fabric from weather and vandalism.

Tolerances

Several alignment tolerances need to be constantly checked and evaluated during construction. These include:

- Adjacent facing panel joint gaps: ¾ inch ± ¼ inch
- Wrapped face walls (e.g., welded wire, geosynthetic facings): 1 inch per 5 feet horizontally, ½ inch per 5 feet vertically; bulging limited to 1 to 2 inches
- Reinforcements: elevation within 1 inch of the connection elevation

Facing elements that are out of alignment require removal of the fill and reinforcement followed by resetting the panels. This is costly and time-consuming, and may require rapid decisionmaking by the geotechnical consultant. Therefore, it is important that a representative of the designer be on site throughout construction to closely monitor the quality of

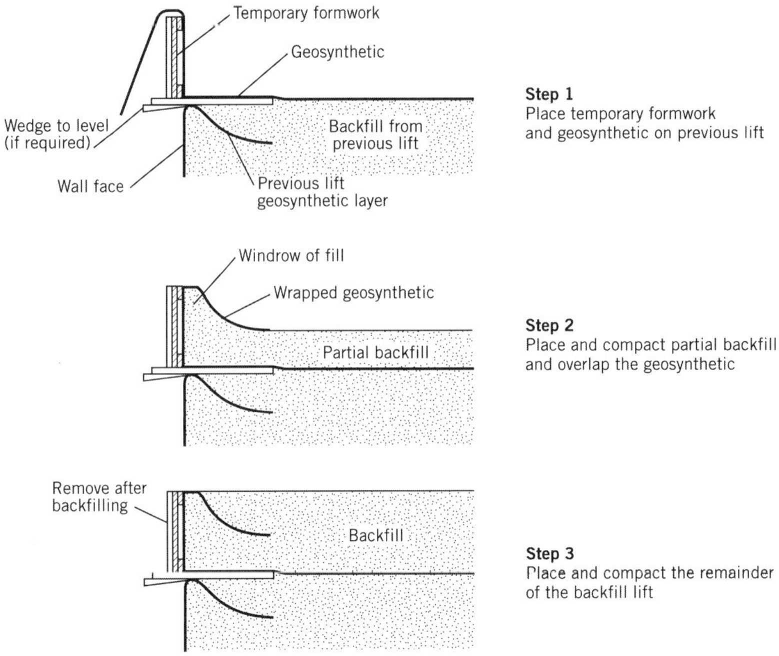

Figure 20.33 Construction of a fabric-wrapped outer face.

workmanship. It is also advisable, where possible, to use a contractor with past experience of reinforced soil construction. Elias et al. (2001) provides a checklist of possible causes for out-of-alignment situations based on observable site conditions.

Quality Control/Assurance

The key requirement is for the owner to have a geotechnical field inspector on site full-time during construction to check that vertical and horizontal alignment tolerances are not exceeded and to ensure that other good construction practices are being followed. Daily records need to be kept, and a good photographic record should be obtained in case problems develop during or after construction. The customary field density and gradation tests should be performed. Chemical tests may be needed on the backfill soils.

MSE Wall Costs

Many variables affect the cost of building an MSE wall; these include the need for cut and fill into native soils, availability and haul distance for imported backfill, the wall facing selected, etc. Due to the ability of MSE walls to accept differential settlements, the lower cost of constructing MSE walls over concrete walls becomes very significant on less competent foundation soils, where concrete walls require a deep or specially prepared foundation.

According to Elias et al. (2001), typical construction costs for MSE walls in the United States range from \$19 to \$37 per sq. ft. of face. Costs rise with increasing height of wall, as shown on Figure 20.34 (Koerner et al., 1998). This chart is based on several retaining wall systems built in the United States for state and federal transportation projects in the mid-

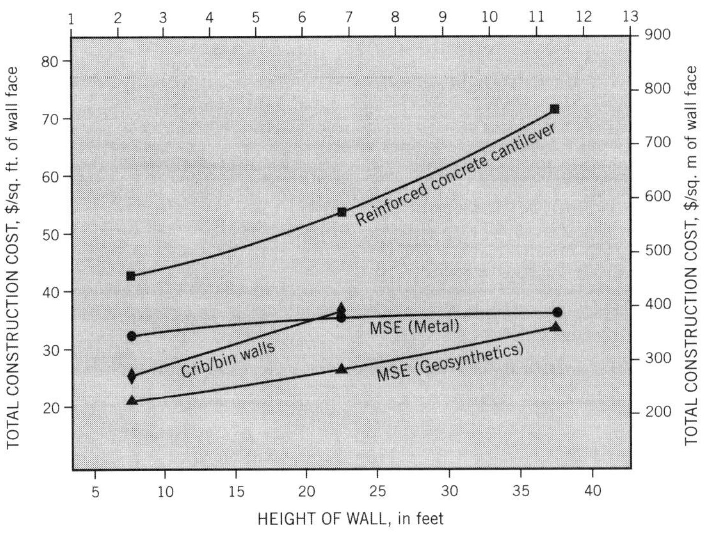

Figure 20.34 Approximate construction contract costs for various types of retaining walls (after Koerner et al., 1998).

1990s. The Engineering News Record Construction Cost Index for 1995 is 5450.

Reinforced Soil Slopes

Several computer software programs have been developed to analyze reinforced soil slopes (RSS). As previously stated, RSS relates to slopes of less than 70° to the horizontal. As noted by Elias et al. (2001), one such computer program (RSS) may be downloaded free of charge from the FHWA Geotechnical Information Center at www.fhwa.dot.gov/bridge or a disk copy may be purchased from the Center for Microcomputers in Transportation (McTrans) at www.mctrans.ce.ufl.edu. The program is supported by FHWA for all state and federal agencies. For private sector users and others, a supported licensed version is available from the developer GEOCOMP through its web page at www.geocomp.com/software.htm.

An example application of a RSS to provide erosion protection in a riverbank is described in Chapter 16, Section 16.8. The wrapped fabric face of the slope allowed native grasses and other vegetation to establish and provide a natural appearance.

CHAPTER

21

Liquefaction Mitigation Techniques

Soils that are subject to liquefaction during a major earthquake range in gradation from slightly clayey silts to sandy gravels. The most vulnerable soil is clean sand. When sand liquefies, and loses most of its strength, landslides can occur. Where a potential slope failure exists, steps can be taken to densify sand to resist major earthquake shaking. Since these procedures are expensive, treatment is usually restricted to key facilities.

This chapter describes several options for densifying sands and gravels below the groundwater table that are preventative measures against instability during a major earthquake. Most require specialized ground improvement equipment and resources. These procedures continue to be improved and/or upgraded, and close involvement with specialty contractors during the design phase is strongly recommended. Each method has benefits and drawbacks. Many owners and engineers choose to use a performance-based specification rather than a methods specification. This allows the specialty contractor to select the equipment and procedure based on their knowledge and past experience within a continually changing technology.

The chapter has been kept brief, with the focus on techniques and examples of use. Some construction costs have been provided, but it is advisable to obtain such estimates directly from specialty contractors.

Verification tests involve either cone penetration tests (CPT), standard penetration tests (SPT), or Becker penetration tests (BPT, for gravels) before and after treatment. The after-treatment data are usually collected within days or weeks of completing the ground improvement to fit in with a project timetable. However, there have been numerous studies of the "aging" of cohesionless strata that show that sand builds up increasing resistance for years after the sand has been treated. An example is given on Figure 21.1. Therefore, the improvement typically shown by before and after treatment testing is probably conservative and the actual benefits one year or later will be better than is indicated by verification testing performed shortly after treatment.

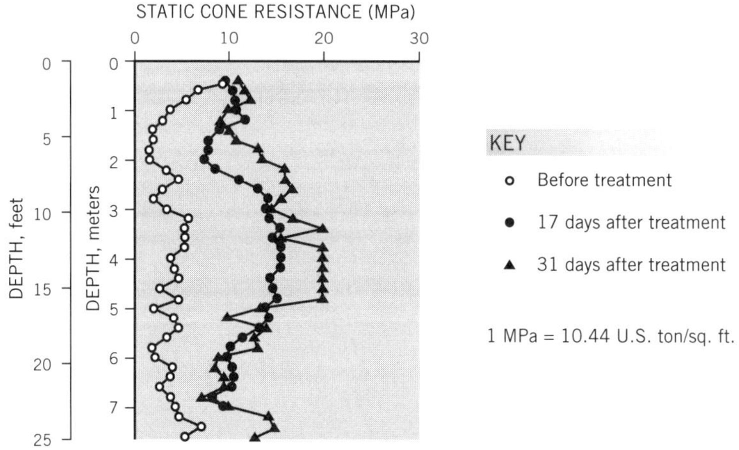

Figure 21.1 Cone penetration resistance increases for sand subjected to dynamic consolidation (abstracted from Mitchell and Solymar, 1984).

21.1 COMPACTION GROUTING

Technique

Compaction grouting creates a cylindrical or teardrop grout bulb around the injection point to displace and densify the surrounding soil (Figure 21.2). The technique uses a stiff (limited mobility displacement or LMD) grout subjected to a high pressure. The intent is not to permeate into the pores of the soil but consolidate the soil to a higher density. Loose to medium dense sands develop excess pore pressures that dissipate relatively quickly. However, time has to be allowed for consolidation to occur.

The injection grout pipe is usually 2 to 4 inches diameter and is drilled or driven into the soil. Hole spacings range from 5 to 12 feet, depending on the required result. Secondary grouting is often used to supplement a grid of primary holes.

Compaction grouting requires special equipment that can operate at high pressures to initiate flow into the soil or rock and keep the grout mobile in the pumps and lines. The equipment must be durable to withstand the abrasiveness of the grout. Essentially, the grout has to be pumpable yet relatively immobile once it leaves the end of the grout pipe.

The soil-cement grout normally includes silty sand (typically 25–30% by weight) and water to provide a low slump of 1 to 2 inches. The cement content is 4–10% by weight. Design mixes often include other materials such as gravel, fly ash, and bentonite.

Grouting is performed in stages, usually 2 to 3 feet vertically, starting from the bottom of the hole (Figure 21.3). The grout pressures range from 150 to 1,000 lb./sq. in., but are typically 650 to 750 lb./sq. in. According to Baker (1985), the influence of the compaction grouting extends to about 20 times the volume of the injected grout.

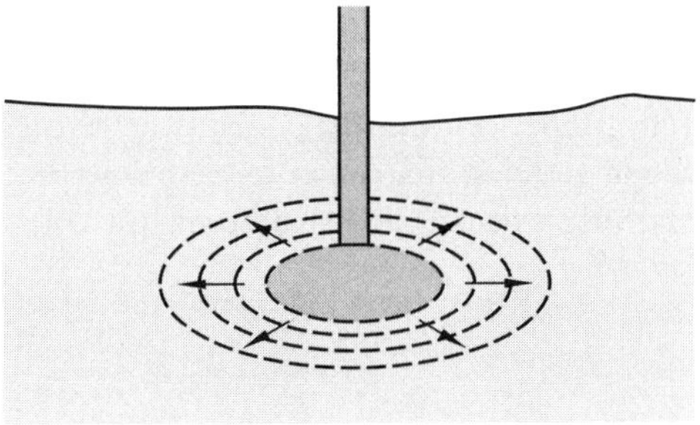

Figure 21.2 Schematic representation of compaction grouting (after Essler et al., 2000).

Records are kept of the grout pressures, grout take (often by counting the strokes of the grout pump) and other site observations, especially ground heave. Because the ground is being displaced around the injection point, measurements of movements need to be taken on any structures close to the grouting operation.

Grout refusal is the point reached where additional grout injection does not advance the purpose of the grouting. Some criteria include:

- Target injection volume or maximum volume limit per injection
- Maximum safe injection pressure or set pressure limit per injection
- Observed damage, undesired movement, or maximum allowable uplift

A typical set of limits (e.g., Scherer and Gay, 2000) are: 15 cu. ft. of injected grout per foot of lift; max pressure of 650 psi at point of injection; ½ inch of surface heave.

Benefits and Effectiveness

Some advantages of compaction grouting relative to alternative treatments include: (i) speed; (ii) can be performed in tight conditions; (iii) can reach considerable depths, and (iv) no waste disposal. The effectiveness of compaction grouting in raising the relative density of cohesionless soils is measured by the *replacement ratio,* RR. This is defined by:

$$RR = \frac{\text{Compaction grout volume}}{\text{Treatment volume}}$$

The densification procedure typically gives RR values of 5 to 15%. An alternative term is *displacement ratio,* and the amount required to achieve the design relative density can be calculated in advance of grouting.

The technique has limited effectiveness near the ground surface where lower overburden pressures and less lateral confinement cause the grout to fracture and heave the soil instead of densifying it. Typically, the upper 20 feet is affected in this

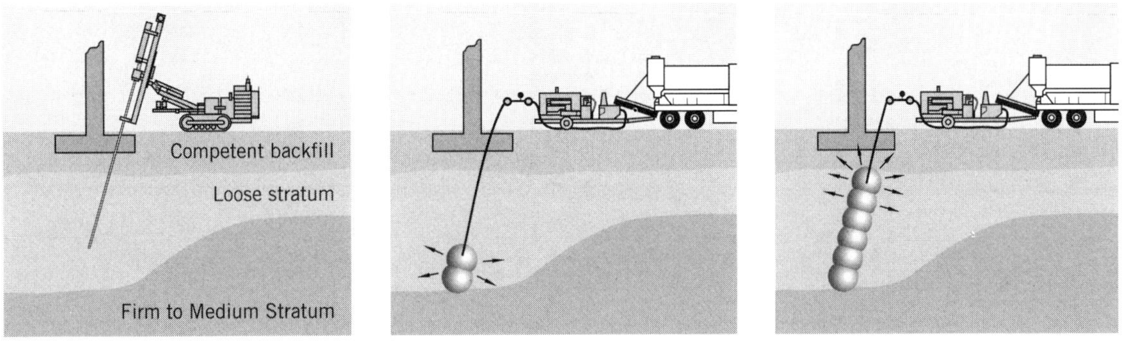

Figure 21.3 Compaction grouting of a loose stratum (after Hayward Baker, Inc.).

way. Another potential drawback is that pressure grouting cannot be reinitiated from the same location once the grout bulb has set and hardened. Occasionally, compaction grouting has been satisfactorily accomplished without using any cement in the mix (e.g., Shuttle and Jefferies, 2000).

Example

A well-documented case history by Ivanetich et al. (2000) demonstrates the technology and practical issues of sand densification by compaction grouting. Although the example refers to a bridge seismic retrofit, these issues would be similar for ground that would be vulnerable to landsliding from liquefaction during a major earthquake. The details are abbreviated from the original technical paper.

The site was the Laurel Street Bridge in Santa Cruz, California, where two bridge piers were underlain from the ground surface to a depth of 90 feet by interbedded layers of gravelly sand, sand, and silty sand. Although the piers were pile-supported, this upper layer of alluvium posed a problem of potential liquefaction. During the Loma Prieta earthquake (M7.1) of October 1989 substantial ground distress was noted around the bridge piers, including sand boils and ground cracking. Pier 2 had the weakest subsurface soils and will be discussed below.

Compaction grouting was selected to increase the relative density of the sand and gravel. The pattern of grout holes (Figure 21.4a) consisted of primary holes on a 11-foot square grid with secondary holes at the center of each primary square.

Rotary percussive track drills with air flush were used to advance the drill bit. The 3-inch dia. casing, lagging slightly behind the bit, was easily tapped into the ground and provided a good seal for the grout.

The grout mix consisted of sand, cement, and water. The sand was a well-graded silty sand with 23% by weight of non-plastic fines passing the No. 200 sieve. The cement was Type II, averaging 5% by weight. The target strength was an unconfined compressive strength of 50 psi but the actual strengths were 200 to 250 psi. The grout strength has no influence on the compaction grouting densification goals but the grout columns, 1 to 3 feet diameter, provide some additional reinforcement to the soil mass.

The stage grouting was bottom up. The casing was lifted in 2-foot increments, and grout pressures of 700 psi were applied at each level. Due to the limited headroom, hydraulic rams and wedges were used to extract the casing during grouting. The grout injection was monitored by pressure gauges and pump stroke counters at the header. The pumping rate averaged 2 cu. ft./minute.

The predetermined criteria to stop pumping at each stage included the following:

- grout flow ceases under 700 psi pressure
- sudden drop of grout pressure (possibly indicating soil shear)
- surface heave greater than 0.125 inch
- maximum grout volume of 25 cu. ft. per stage
- signs that the casing was becoming "locked" into the ground

Pier 2 foundation accepted a total grout volume of 74,000 cu. ft. for a replacement ratio of 9.5%. The estimated grout take during design was 5 to 7% for this pier.

Verification testing included both SPT and CPT measurements. The SPT measurements, before and after compaction grouting, are shown on Figure 21.4(b). These are average $(N_1)_{60}$ blow counts at 10-foot intervals without regard for soil characterization. The estimated blow counts needed to prevent liquefaction ranged from 30 (clean sand) to 20 (>25 percent fines). It can be seen that these goals were exceeded by the compaction grouting program at depths below 10 feet.

The following problems occurred during the compaction grouting program:

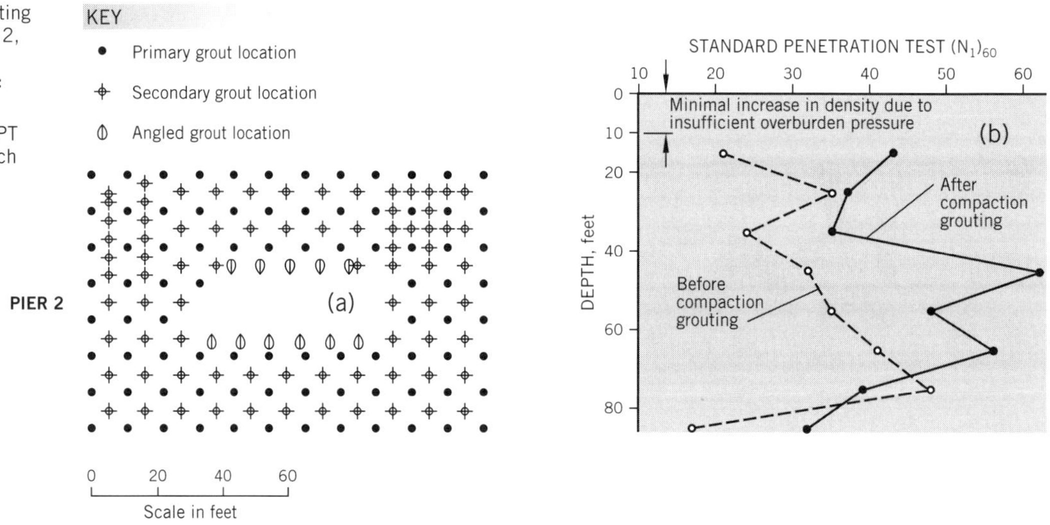

Figure 21.4 Compaction grouting layout and test results for Pier 2, Laurel Street Bridge seismic retrofit, Santa Cruz, California:
(a) foundation grouting plan
(b) pre-grout and post-grout SPT blow counts (average at each depth)
(after Ivanetich et al, 2000)

Upper 10 feet. The sands had minimal increases in density because there was insufficient overburden pressure and confinement. Ground heave usually occurred when grouting reached within 25 feet of the surface.

Casing locking. As grouting proceeded, the casing would become locked into the ground. This required continual checks and, in some cases, required the grouting to be stopped and the casing raised to the next stage before the design grouting pressures had been applied. At other times the casing was subjected to a slow continuous pull as grouting continued.

21.2 DYNAMIC COMPACTION

Technique

This procedure for densifying soils is extremely simple. A large weight, 10 to 30 tons, is lifted by a crane to heights of 50 to 110 feet above the ground. The weight is allowed to free-fall and makes a crater in the ground surface, thereby increasing the density of the soil within the zone of influence of the dynamic loading (Figure 21.5).

The depth of influence D, in feet, is given by the formula:

$$D = n\sqrt{WH}$$

where W = weight, in tons
H = height of drop, in feet
n = a factor between 0.3 and 0.6, depending on the soil, but typically around 0.5

Example: W = 20 tons; H = 100 feet;
depth of influence D = 22 feet for n = 0.5

The technique is very effective in cohesionless soils that rapidly dissipate the pore pressure set up by dynamic energy. As a preventative measure for sands and gravels subject to liquefaction and slope instability, the U.S. Bureau of Reclamation has used dynamic compaction at three of their major dams (Dise et al., 1994). They have used the Lampson thumper LDC-350 lifting device, which can lift 20 to 30 ton tampers to heights of 110 feet. Their studies concluded that the greatest improvement in reducing liquefaction potential occurred in the upper 28 feet of the ground.

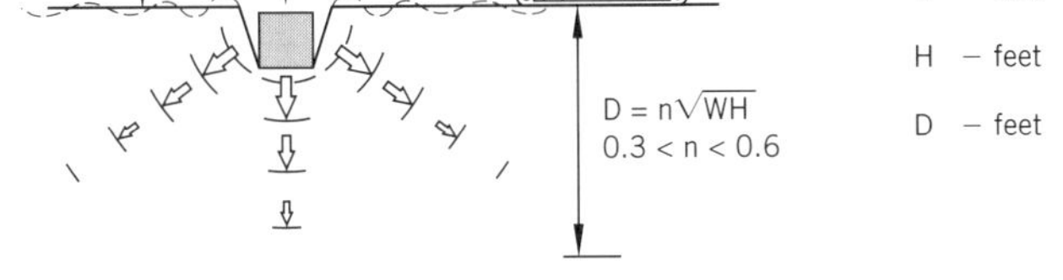

Figure 21.5 Dynamic compaction using a falling weight.

Conventional cranes cannot exceed 20 tons for single line pull capacity. The usual limit set for cranes is 16 to 18 tons.

Most tampers are made of steel, but some use concrete. Various shapes of tampers have been tried; there is no strong evidence that any shape is better than others.

The impact grids typically are in squares ranging from 7 x 7 feet to 20 x 20 feet. Contractors generally start the work at wide intervals and then decrease the gap as the work progresses. One theory is to use high energy impacts to cause deep compaction and then use less energy (i.e., smaller weights with less drop) to densify soils closer to the surface.

According to Welsh (1986) the maximum improvement of the soil occurs over a depth of one-third to one-half of the effective depth (defined earlier). Lesser improvements occur: (i) near the surface due to disturbance from the impact, and (ii) at greater depth due to reduced energy. To improve the sands and gravels near the surface, an "ironing" pass can be made using less energy. The tamper is raised only 15 to 20 feet and is dropped on an overlapping grid. Alternatively, the surface soils can be densified by conventional earthworks means such as vibratory rollers.

At each tamping location, there are preset criteria to limit the number of drops. As an example, typical criteria are to stop tamping when: (i) the crater depth exceeds 1.5 times the tamper height; (ii) groundwater appears at the crater base; or (iii) 10 to 15 tamps have been made. The craters then are infilled with either native soils or imported granular fill. After backfilling, the tamping may be resumed if the preset maximum numbers of tamps had not been reached.

The ground depression caused by dynamic compaction depends on preexisting conditions. As a guide, the range is typically 5–15% of the effective depth of densification. Verification testing of the treatment effectiveness generally involves before and after determinations of relative density using the standard penetration test, cone penetration test or, in the case of gravels, Becker Penetration tests. An example of improved penetration resistance for a sand subjected to dynamic compaction is shown on Figure 21.1.

Dynamic compaction is relatively inexpensive, and effective in sands and gravels to depths of about 30 feet. Some further improvement of penetration resistance with time is likely to occur due to aging.

There are, however, some major drawbacks to using dynamic compaction, especially in urban areas. According to one source, it "rattles the neighborhood." The low frequency vibrations of around 4 to 10 cycles per second can produce peak particle velocities of 0.5 inch per second at distances of 100 to 150 feet from the falling weight. This has caused numerous complaints and litigation in the past. It is essential that a good public relations program, vibration monitoring, and structural inspections (before and after) be undertaken in urban areas close to structures.

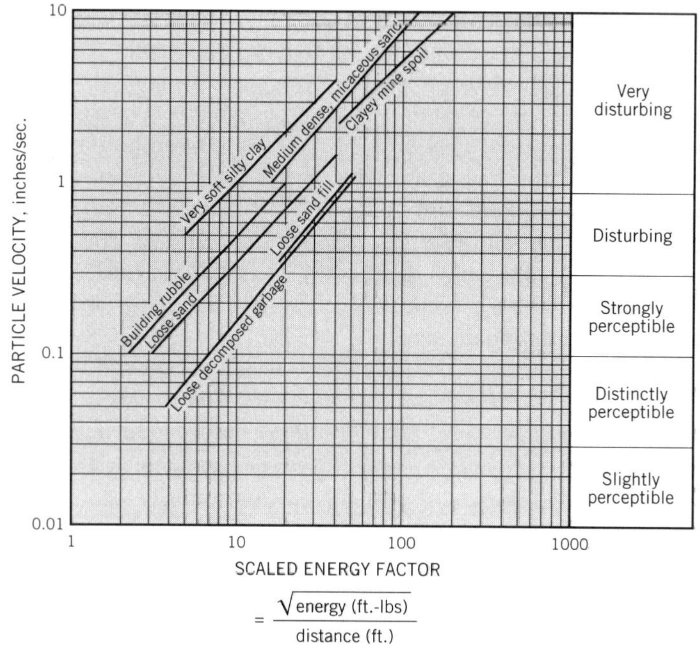

Figure 21.6 Experiences of Lukas (1986) in measuring the particle velocities at the ground surface due to dynamic compaction of different soil types.

A widely used reference on dynamic compaction is the report prepared by Lukas (1986) for the U.S. Federal Highway Administration. His experiences of the vibrations set up by dynamic compaction on various soils can be used as a general guide in planning a project (Figure 21.6). Particle velocities of more than 0.5 inch/second are likely to damage older buildings. However, as with all vibration projects, there are three levels of concern: (i) structural damage, which can affect the safety of a building; (ii) "architectural" damage, which causes some damage to facades but does not compromise structural integrity; and (iii) perceptions of vibrations in which some people are very sensitive to even small levels of vibration. Further discussion on this subject is outside the scope of this book. However, the effect of vibrations on the surrounding area is a major issue for many dynamic compaction projects.

Significant other concerns are: (i) site safety to protect the crane and its operator; (ii) groundwater needs to be at least 5 to 6 feet below the surface at the start of compaction; and (iii) limitations on the depth that can be compacted by this technique. For deeper compaction needs, other options include vibro-compaction, vibro-replacement, and compaction grouting.

Cost

The cost of dynamic compaction in the United States in 1987 (ENR Construction Cost Index 4400) generally ranged from \$1.00 to \$2.00 per sq. ft. of surface area. The cost of providing more sand fill to the site is extra. In 1987, mobilization costs started at around \$15,000.

21.3 VIBRO-COMPACTION

Technique

Vibro-compaction, formerly known as vibroflotation, is probably the most effective method for achieving high relative densities at depth in clean sands and gravels (Figure 21.7). The effectiveness is reduced, however, if more than about 5–10% by weight of the soil passes the No. 200 sieve.

The vibro-compaction probe has a hollow steel tube containing an eccentric weight mounted on a vertical shaft (Figure 21.8). The shaft is driven by an electrical motor placed

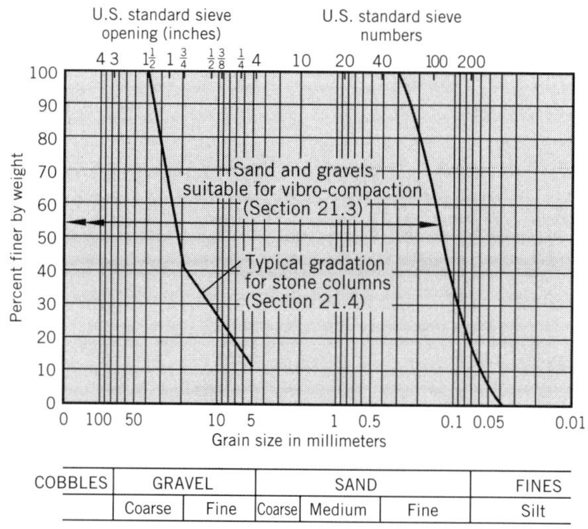

Figure 21.7 Gradation of soils suitable for vibro-compaction and stone column construction.

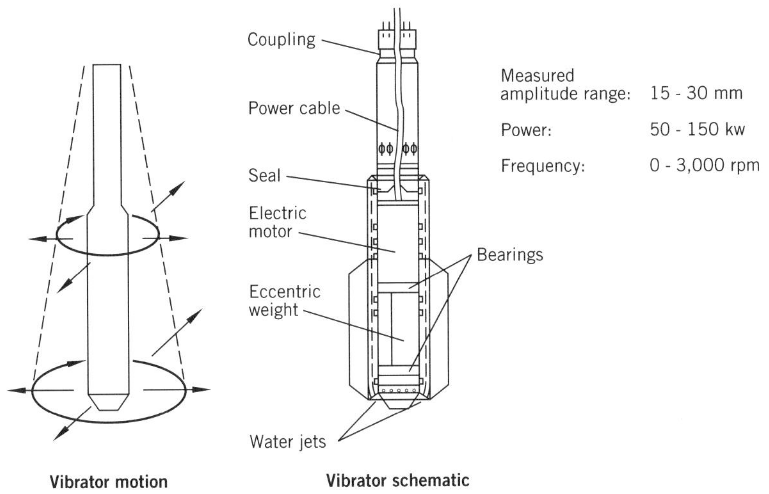

Figure 21.8 Vibro-compaction.

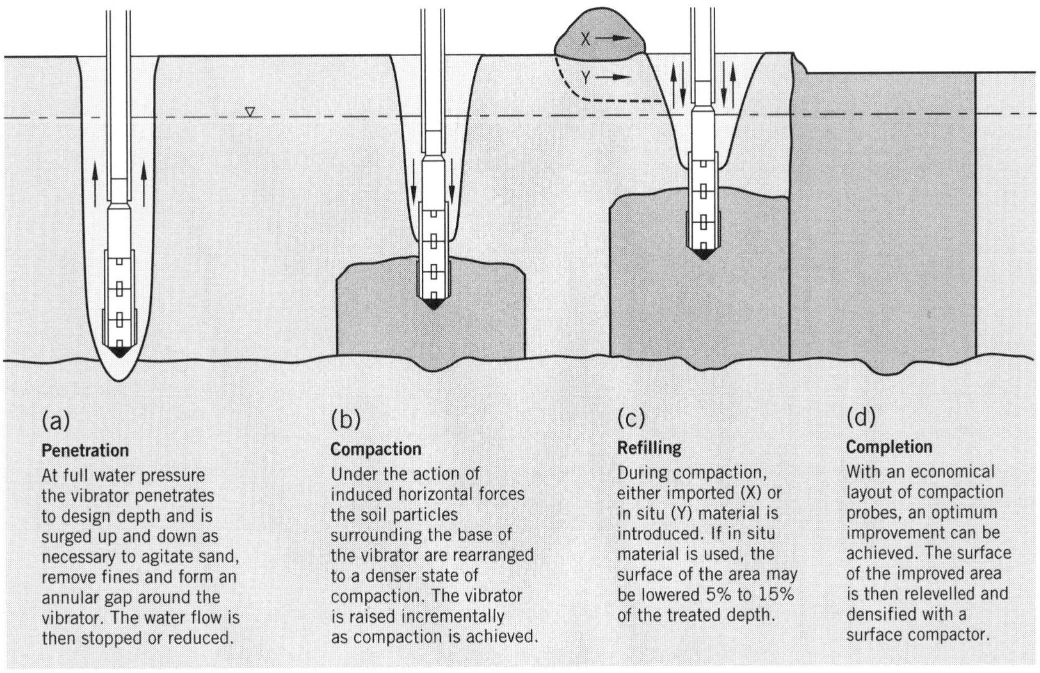

Figure 21.9 Densification of clean sands and gravels by vibro-compaction (Hayward Baker, Inc.).

directly above it within the probe. Attached to the probe by couplings are follower tubes that extend from the probe to the ground surface. Vibrations emanate from the probe radially in a horizontal plane (Figure 21.8).

A typical probe is the Vibroflot V23, which operates at 2,800 rpm and has a vibration amplitude of 0.75 inch. This unit is 11.5 feet long and 1.15 feet diameter, has a power consumption of 130 kW, and weighs 2.1 tons. It applies a dynamic force of up to 30 tons.

The principal change to this technology in recent years has been toward larger machines with greater dynamic force. Another improvement has been the introduction of a variable frequency control to obtain an optimal relationship with the site soils.

The vibroflot is suspended from a 60 to 100 ton crawler crane. It is lowered into the ground at a rate of about 3 to 6 feet per minute using a water pressure of around 115 lb./sq. in. and flow rate of 800 gpm to facilitate penetration (Figure 21.9). The probe is surged up and down as necessary to agitate the sand, remove fines, and form an annular gap around the vibrator. After reaching the full depth of the ground to be treated, the water flow is reduced or stopped. The vibrating probe then is raised to the surface in vertical increments of 1 to 3 feet, each level receiving vibratory compaction for about 20 to 30 seconds. Water flow should be kept to a practical minimum during withdrawal to achieve the best results.

The construction process is monitored through the rate of incremental withdrawal of the vibro-probe and the quantity of

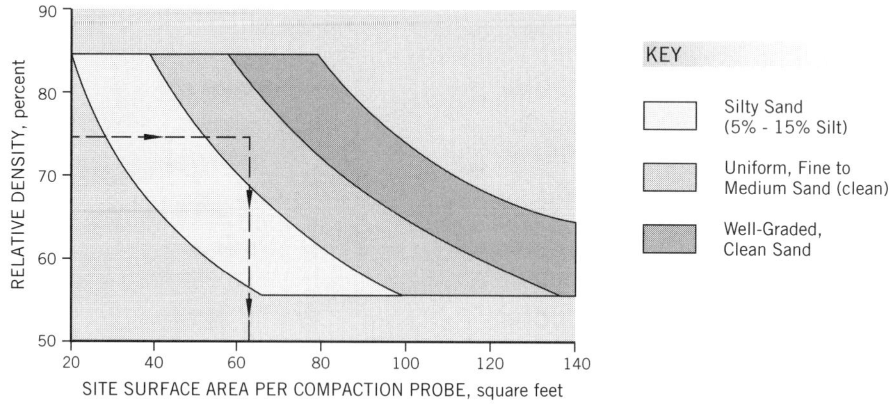

Figure 21.10 Approximate method of determining the spacings of compaction probes in vibro-compaction.

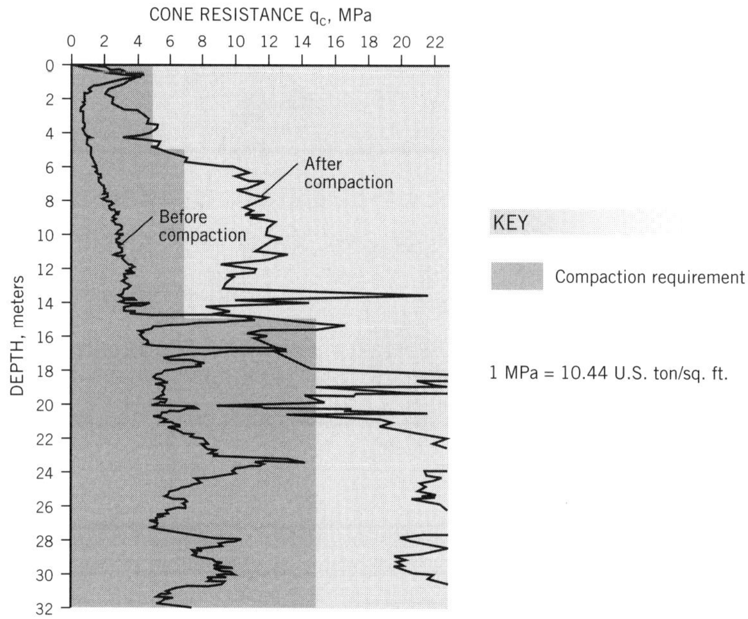

Figure 21.11 Depth-dependent acceptance criteria for vibro-compaction at a site near Dresden, Germany (after Degen, 1997).

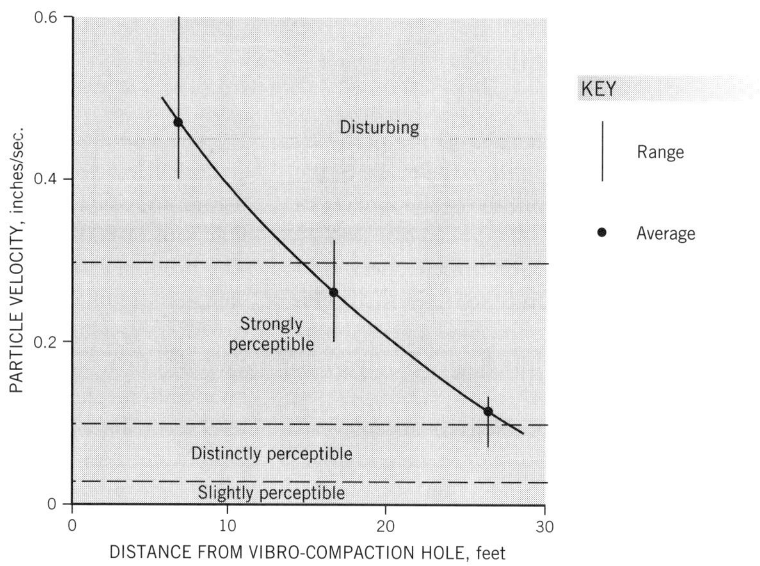

Figure 21.12 Average peak particle velocities measured near a vibroflot on a flat estuarine plain (after O'Brien, 1997).

replacement fill needed. The amperage buildup of the vibrating unit should be closely watched because it indicates the resistance of the soil as the backfill is supplied. A relatively high amperage implies that a high relative density is being achieved.

Ground cracks and subsidence occur over a 7- to 10-foot radius from the hole. During the compaction process, sand (on-site or imported) is pushed into the cone of subsidence (Figure 21.9c) to make up the ground.

The spacing of holes ranges from 5 to 14 feet. They are arranged with the centers on a triangular or square pattern. The zone of improved soil reaches 5 to 13 feet away from the vibrator, depending on the soil type and vibroflot power. An approximate guide to design spacing (Hayward Baker, Inc.) is shown on Figure 21.10. For example, if a relative density of at least 75% is specified for a stratum of clean, uniform, fine to medium sand, the graph indicates that a surface area per compaction probe of about 64 sq. ft. is needed; i.e., for a square grid, the probe spacings should be at 8-foot centers.

A typical treatment depth for vibro-compaction is 15 to 50 feet, but depths of over 100 feet are common. Degen (1997) reports successful densification of reclaimed mining fill near Dresden, Germany, to depths of up to 184 feet (56 m). The main difficulty at this site was lifting out the vibro unit, which had a dead weight of 19 tons plus skin friction. The cone penetration test, using a depth-dependent criteria, was used for quality control. A partial chart, to a depth of 105 feet (32 m), is shown on Figure 21.11.

Vibro-compaction raises concerns about structural damage and the annoyance factor of vibrations when it is performed close to structures. Data collected by O'Brien (1997) for a vibroflot producing 50 to 80 kW of power is shown on Figure 21.12. The site was an estuarine plain of slightly silty fine sand in north Wales.

Vibratory Hammer Probes

This section would be incomplete without mentioning vibratory hammer probes. These devices have vibratory hammers at the *top* of a 30-inch diameter steel pipe. The vibrations are vertical (unlike a vibroflot) at operating speeds of 900 to 1,100 cycles per minute. Examples are the Terra-Probe and Toyomenka SVS. The advantage of the top-drive hammers is a shorter cycle time, but allegedly they need a much closer pattern spacing and achieve a lesser degree of densification than a vibroflot.

21.4 STONE COLUMNS (VIBRO-REPLACEMENT)

Technique

Stone column technology is an outgrowth of vibro-compaction and often replaces it. It extends vibratory construction methods to cohesive soils, fills, and mixed strata. For liquefaction avoidance purposes, stone columns can be easily installed in silts as well as sands and gravels.

In loose sands and gravels, stone column construction has several benefits in reducing the risk of liquefaction strength losses:

- Vibrations densify the surrounding soil similar to vibro-compaction.
- The stone column acts as a drain with a short drainage path in the event of a major earthquake.
- The stone column functions as a reinforcing element, making the foundation stronger.

Stone column construction is schematically illustrated on Figure 21.13. Using water jetting, the vibroflot penetrates to the design depth. The water jets are adjusted to maintain an

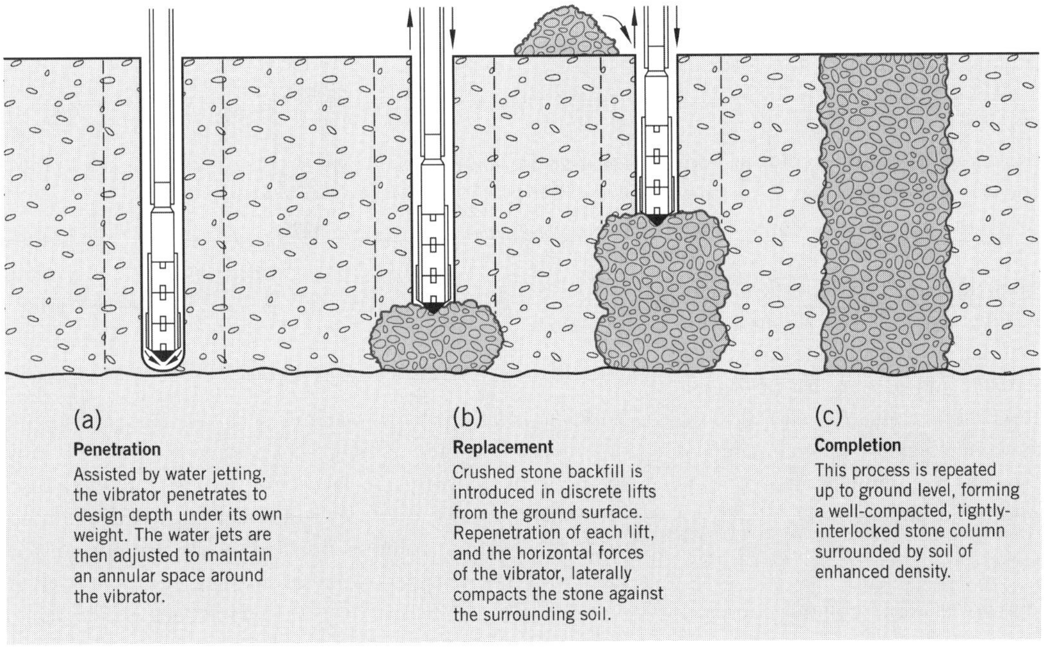

Figure 21.13 Stone column construction by the top feed (wet) method (after Hayward Baker, Inc.).

annular gap around the vibrator unit. Crushed stone backfill is fed into the hole from the ground surface and falls through the annular gap. The vibrator compacts the rockfill and forces it against the surrounding soil. Incremental lifts of stone are built until the stone column reaches the ground surface.

The above procedure, known as the *wet top feed,* is the most common and cost-effective method. The water jet is used to remove the soft material during the initial penetration. It generates spoil that usually has to be hauled offsite.

An alternative procedure is known as the *dry bottom feed* and is shown schematically on Figure 21.14. The probe is advanced to the maximum depth of treatment using compressed air to displace the soil. Stones are fed to the tip of the vibrator by a chute attached to the follower tubes (Figure 21.15). Backfill is continually provided to the developing column from a hopper attached to the top of the vibrating unit. Except for the supply tube and hopper, the equipment is the same as that used for vibro-compaction (Section 21.3 previous) and will not be described again here. The vibrating unit is in the ground throughout the stone column construction. The procedure is entirely dry and, since no flushing water is used, there is no waste water or other spoil to be disposed offsite. (However, water flush can be used, if needed.) This lack of waste products has advantages at many sites.

Treatment by stone columns is possible to depths of up to about 90 feet. Columns typically are spaced at 6- to 12-foot centers, and the diameter of a stone column in cohesionless soils is around 3 feet. A typical gradation for crushed rock backfill is:

SIEVE SIZE	PERCENT PASSING
1½ inch	100
¾ inch	40
No. 4	10
No. 200	0-5

The fines should be nonplastic. See Figure 21.7. Many contractors use zero sand content in the crushed stone backfill.

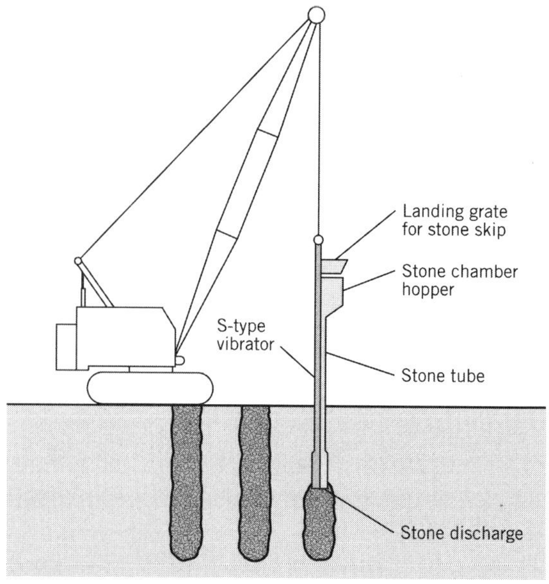

Figure 21.14 Stone column construction by the bottom feed (dry) method (after Hayward Baker, Inc.).

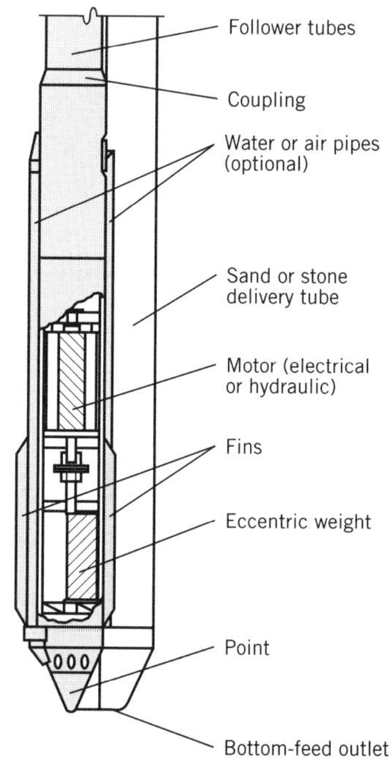

Figure 21.15 Vibrating probe for stone column construction by the bottom feed method (after Slocombe, Bell and Baez, 2000).

Cost

The construction costs associated with stone column construction depend heavily on the availability of a hard stone source and the haul distance to the site. For a 42-inch diameter column in soft ground and a relatively short haul to a hard rock quarry, a typical construction cost (2001) in the Pacific Northwest is around $16–26 per lin. ft. (ENR Construction Cost Index in 2001 is 6400). It does not include any testing (SPT, CPT) of the ground before and after treatment or contractor mobilization/demobilization costs. As with all construction cost estimates, these numbers are approximate and should be adjusted for site specific requirements after contacting contractors specializing in this treatment.

Example

A well-documented case history of stone column construction through saturated silty fine sand at Albany Airport, New York, has been given by Soydemir et al. (1997). The main area of concern for potential liquefaction was a layer of very loose to medium dense glacio-lacustrine deposits 12 feet thick immediately below a surface crust 5 feet thick. A decision was made to use stone columns installed by the dry bottom feed method. Three test areas were studied and Test Area A at the proposed Concourse Building is described below. The data of the technical paper have been simplified for this summary.

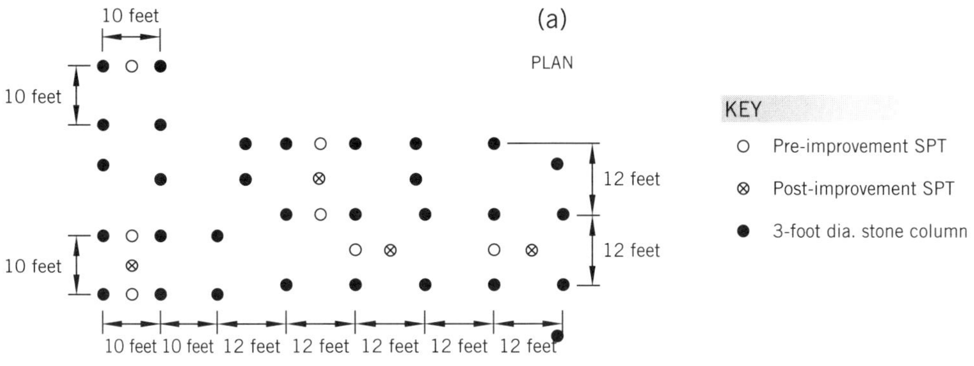

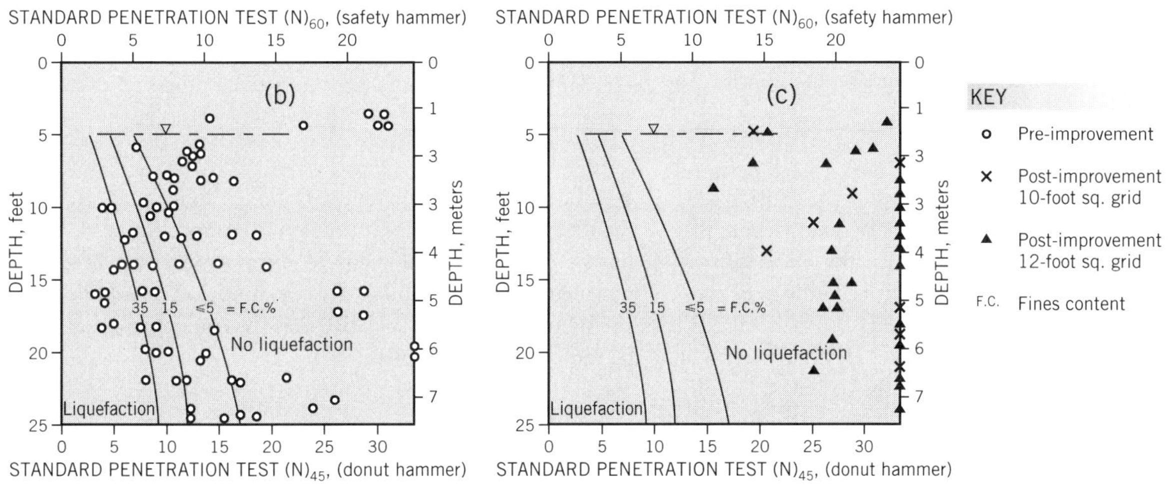

Figure 21.16 Stone columns example. Test area A results at Albany Airport, New York:
(a) plan of test area
(b) pre-improvement SPT data
(c) post-improvement SPT data (modified from Soydemir et al., 1997).

Test Area A has 31 stone columns, typically on square grids of either 10 feet or 12 feet (Figure 21.16a). The 3-foot diameter piles were backfilled with crushed rock, the approximate gradation being shown on Figure 21.7. Based on the design earthquake analyses, the curves separating liquefaction from non-liquefaction are plotted on Figure 21.16(b) and (c) for various fine contents from ≤ 5% to 35%. The sand gradations at this site covered this entire range. However, a glance at Figure 21.16(b) shows that many of the SPT blow counts for the existing ground fall within the probable liquefaction areas of the graph. (At this time, drillers in the area were using both the donut hammer and safety hammer with different energy levels at impact—the authors used the appropriate scales according to the hammer being used for the SPT.)

The post-improvement data, Figure 21.16(c), shows that the relative densities (as defined by SPT) were acceptable in all the tests performed. Three days were allowed between stone column installation and conducting a post-improvement test boring in the immediate vicinity. Based on these results, the initial design 10-foot square grid was increased to a 12-foot square grid for the production stone columns in this area, with further confirmation being obtained from additional compliance test borings during the course of the work.

21.5 EXCAVATION AND REPLACEMENT

Technique

A relatively simple preventative treatment for saturated liquefiable sand is to excavate and replace it: (i) with the same sand recompacted to a higher (non-liquefiable) relative density, or (ii) with imported sand fills or rockfill. The cost of this procedure is similar to a shear key, discussed in Section 15.6, and for practical purposes may be limited to depths of up to 30 feet below ground. It could be combined with deep compaction grouting or vibro-compaction, both of which are less effective at densifying soils near the ground surface.

Examples

Loose to medium dense saturated sands commonly occur on flood plains or hydraulic fills. An illustrative example is shown on Figure 21.17 in which loose sand below the water table is underlain by clay. An embankment fill built over such flood plain sediments would lose most of the foundation support during a major earthquake and could fail. The excavation probably would extend outside the footprint of the embankment to include the extension of the fill side slopes down to the clay stratum (Figure 12.17b). The excavation would require extensive dewatering from wellpoints (or similar).

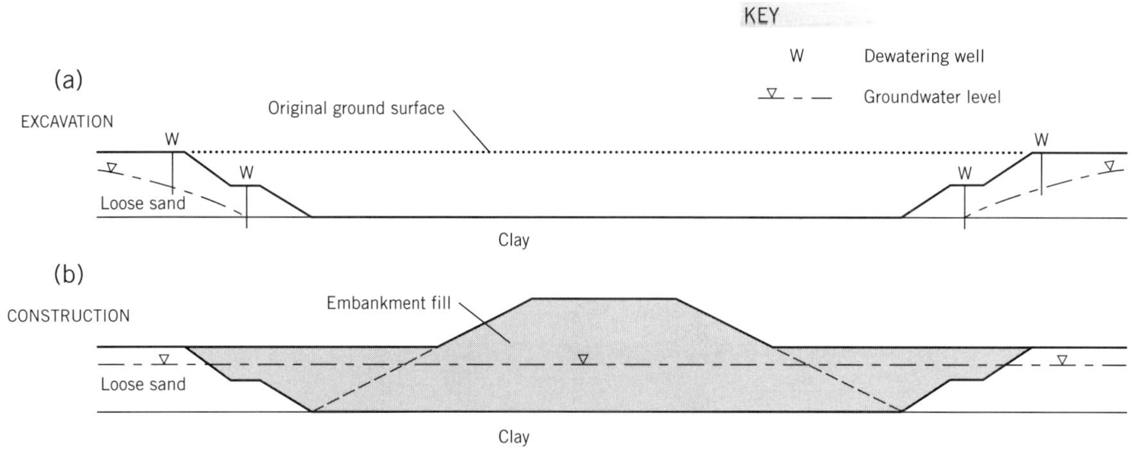

Figure 21.17 Excavation and replacement. Example of an embankment on a flood plain.

Deeper wells may be needed both outside and inside the excavation to prevent uplift through the clay base, depending on the actual subsurface conditions. Once the excavation is completed, the structural fill can be built according to conventional earthwork procedures.

A second illustrative example (Figure 21.18) concerns a preexisting landslide that is supported at the base by a liquefiable sand. Should the sand lose most of its strength during a major earthquake, the landslide could be reactivated. As shown on the figure, the old landslide has interfingered with the sand alluvium. One possible preventative treatment for this situation is to temporarily support the landslide with a vertical tieback wall so that all the loose sand can be excavated (Figure 21.18b). The tieback wall would be designed to carry the landslide forces (as determined from a stability analysis) rather than the lateral forces of a simple excavation. Again, dewatering of the loose sand prior to excavation is essential. The replacement granular fill (native or imported) would be compacted in lifts to a high relative density. The tieback wall could be removed or left in place.

21.6 DEEP SOIL MIXING

Technique

This technique has been described in Chapter 18, Section 18.4, to which the reader is referred for more details. Deep soil mixing is a very good technology for in situ treatment of

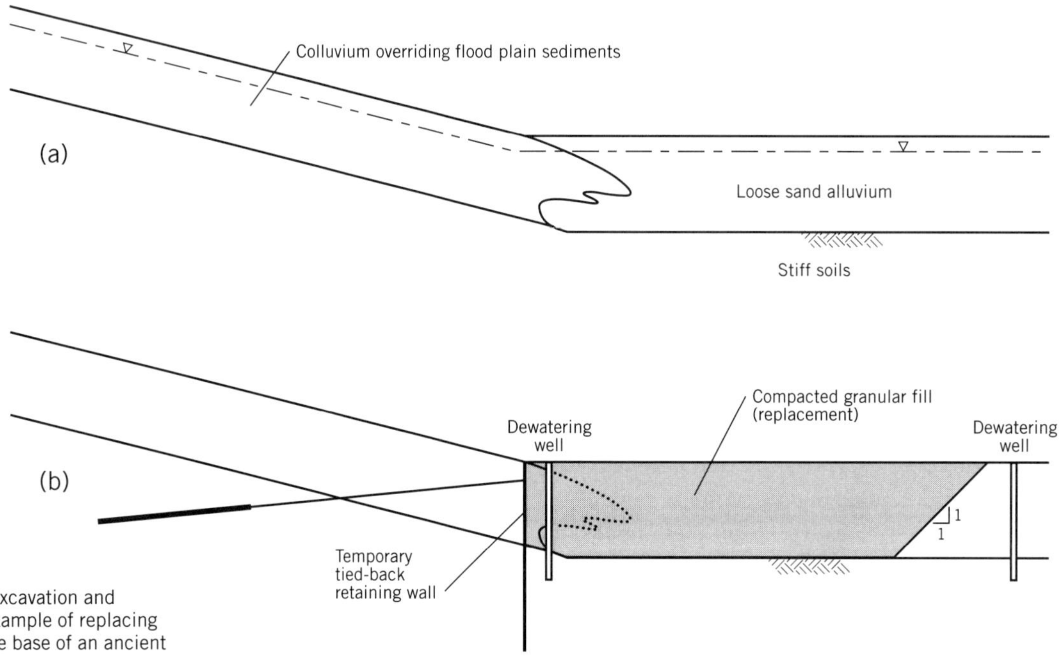

Figure 21.18 Excavation and replacement. Example of replacing loose sand at the base of an ancient landslide.

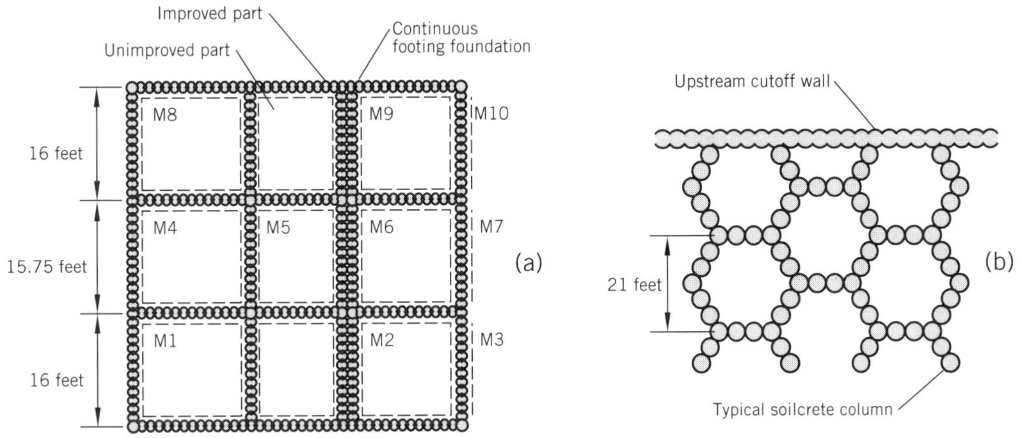

Figure 21.19 Examples of cell construction using deep soil mixing to isolate liquefiable soils:
(a) building in Kagoshima, Japan (after Babasaki, 1991)
(b) Jackson Lake Dam, Wyoming (after Pujol-Rius et al., 1989)

loose to medium dense sands and other cohesionless soils. The hole is fully supported at all times as the cement-based grout or slurry (reagent) is injected into the soils. The required quantity of reagent per cu. yd. of treated soil is lower for sandy soils than for clays.

Examples

The liquefiable soils are isolated into cells (Figure 21.19). This limits the development of cyclic shear strain and prevents liquefaction. For a building foundation (Figure 21.19a), it provides a solid underpinning down to a firm nonliquefiable base stratum.

Liquefaction mitigation of Jackson Lake Dam project in Wyoming is illustrated by the plan on Figure 21.19(b). In this case, a cutoff wall was constructed across the full width of the dam at the upstream side and a honeycomb of cells was built over the entire dam foundation footprint. The cell walls were constructed from panels of deep soil mixing soil-cement columns.

A third example is the reconstruction of Torishima dike with a grid constructed of deep soil mix columns (two overlapping columns per stroke) following the Kobe, Japan, earthquake (Figure 21.20). The treated zone extended through the liquefiable loose sand and underlying soft silt to a nonliquefiable base stratum.

Shallow soil mixing is also available. Typically, it allows a single shaft large diameter auger head (6 to 12 feet diameter) to cover a large area with each stroke. It commonly uses two

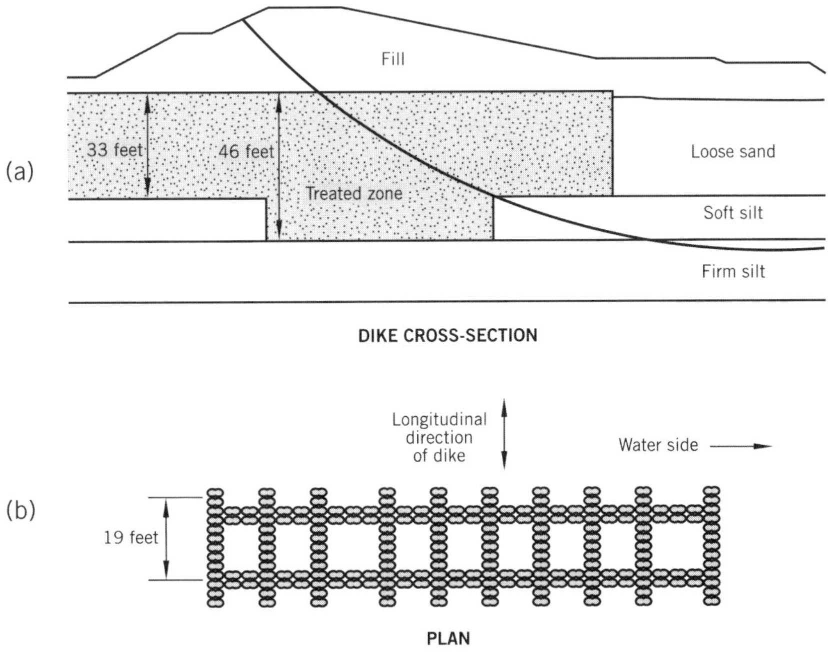

Figure 21.20 Reconstruction of Torishima Dike with a modular grid of deep soil mixing columns (after Yang et al., 1998).

advance/withdrawal strokes of the auger to provide good mixing of the soil and reagent. The grout is injected into the soil at the bit.

There are other soil foundation improvement treatments besides the lattice type shown on Figure 21.19. These include blocks (entire foundation area infill), walls and piles.

Design

Laboratory tests are performed to determine the cement volume needed to prevent liquefaction. According to Zen et al. (1987), a cement content of 5% by volume should render soils nonliquefiable. Assuming that a large foundation block has been treated, a stability analysis can be performed on the block based on the assumption that the surrounding soil liquefies during an earthquake. This requires high lateral (fluid) pressures on one side of the block and residual strength of sand providing resistance on the other. An inertia force is applied by the treated ground. More analysis details are provided in Zen et al. (1987).

CHAPTER

22

Slip Surface Strengthening

The slip surface of landslides in stiff clays or highly weathered rocks is substantially weaker (softer) than the materials above or below it. This zone has been referred to as the *discrete shear zone* in many parts of this book. For example, the general mass of a landslide in stiff clay may have average undrained shear strengths of 3,000 lb./sq. ft., but a back-calculated analysis of the slide may show that the slip surface has an average shear strength of only 600 lb./sq. ft. This information can be easily obtained for such landslides.

This chapter describes two ways to improve the resistance at the weak shear zone. The first way is to provide very strong reinforcement piles (concrete shear piles, stone columns) passing through the clay, and the second way is to replace a length of the discrete shear zone with a stronger material (deep soil mixing, rubblizing the zone, fully penetrating trench drains, and jet grouting). All these treatments have received limited application in landslides, but offer good potential for future development.

22.1 ISOLATED SHEAR PILES (DOWEL PILES)

Technique

Shear piles are reinforced concrete cylindrical piles that pass through the landslide and are anchored at their lower end in stable soils or bedrock. The pile anchorage provides lateral bearing resistance near the base of the moving ground (Figure 22.1). Design of the reinforcement steel is controlled by the maximum bending moment developed in the pile.

Shear piles can be placed in a line to provide a wall supporting a key facility such as a road or railroad within a landslide. This provides selective stabilization (see Chapter 14, Section 14.5). Shear pile walls are presented in Chapter 19, Section 19.5, and Case History 9 is an example.

The present section discusses *isolated* shear piles installed across a broad area of a landslide to increase the shear resistance at the slip surface. An illustrative example is shown on Figure 22.2. Ideally, the piles are spread out to cover approxi-

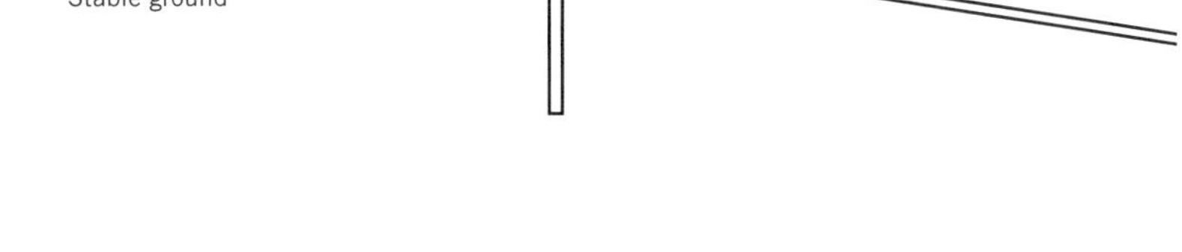

Figure 22.1 Concept of shear pile stabilizing a landslide.

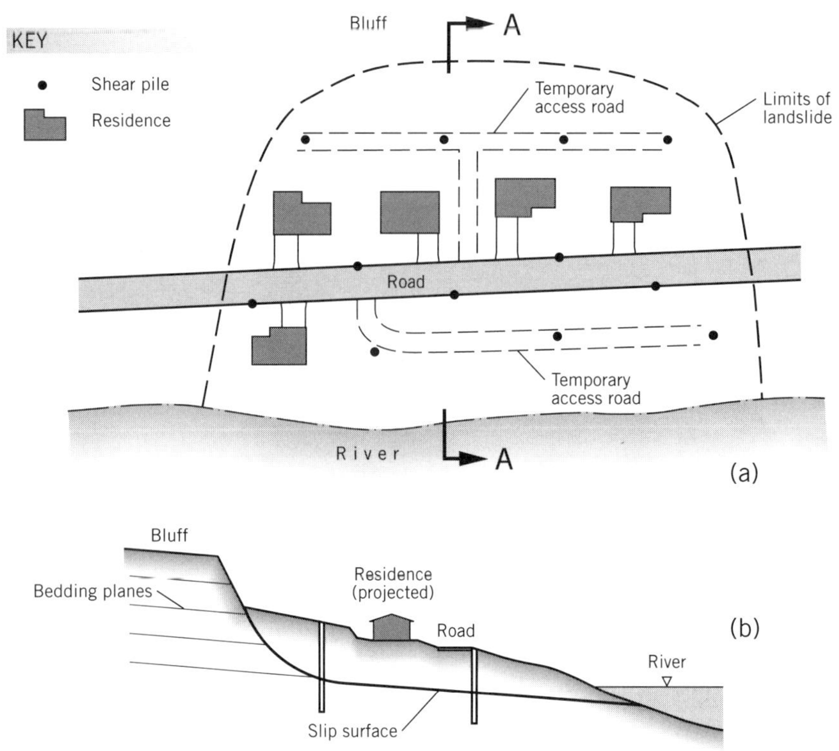

Figure 22.2 Illustrative example of using isolated shear piles (dowels) to stabilize a landslide:
(a) plan
(b) section A-A

mately equal areas of the landslide slip surface. In practice, the pile locations have to take account of access difficulties, property ownership, existing structures, etc. These isolated piles are sometimes referred to as *dowel* piles.

In deep-seated landslides, the shear piles need only be reinforced above and below the shear zone to the points of fixity (Figure 22.3a). These distances depend on the stiffness of the materials in which the piles are embedded. However, in many practical situations, the height to the fixity point is less than the height to the ground surface. Allowing for a margin of error in the structural calculations, the remaining height of the pile above fixity (plus error margin) can be backfilled with unreinforced lean concrete, gravel, or excavation spoils. This procedure has the advantages of: (i) lower construction cost, and (ii) being able to leave the site in a "natural" condition, which is environmentally beneficial at sensitive sites such as ocean or river frontage. In cases where the distance between the upper point of fixity A and the ground surface is short (Figure 22.3b), the pile probably should be reinforced over its full length because of the relative ease of construction using a full-height reinforcement cage.

Concrete shear piles are infrequently used in the United States. A major reason has been the lack of a suitable design procedure (see later). For landslides that move seasonally, shear piles are usually installed during a dormant period. This allows the piles to be installed, and the concrete to harden, before landslide loads are applied to the piles.

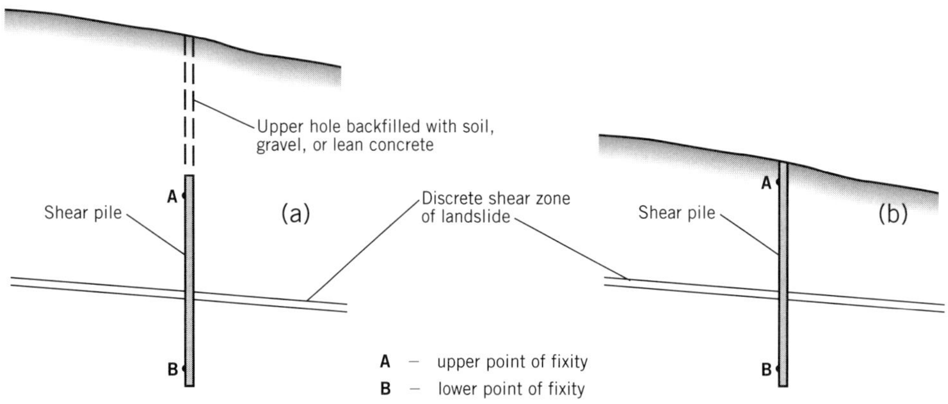

Figure 22.3 Types of isolated shear piles:
(a) buried shear pile
(b) full depth shear pile

Benefits and Disadvantages

The *benefits* of isolated shear piles are:

- Lateral deflection is usually not a concern, so the piles do not have to be anchored near the top.
- The top of the pile can be buried below the ground surface so that it is invisible after construction or can be covered by a structure, playing surface, road, etc.
- Individual shear piles provide more passive resistance per pile than a group of piles (see later).
- There is flexibility in selecting installation locations.

Some *disadvantages* of isolated shear piles are:

- They are relatively expensive.
- They cannot be installed in moving landslides. The piles are cast-in-place, and there is a high risk of the piles becoming damaged from movement before sufficient piles have been installed to provide stabilization.

Pile Loading Calculation

The first step in shear pile design is to determine the additional resistance required from the piles. After a site investigation and field instruments have modeled the landslide conditions, a back analysis will provide the existing shear resistance along the slip surface corresponding to a static factor of safety F of 1.00. This assumes marginal stability; for active landsliding, the static F will be less than 1.00 (see Chapter 9, Section 9.4).

The extra shear resistance of the piles provides the improvement in stability. A relatively small contribution should stop movements in seasonally slow-moving landslides, but the design factor of safety needs to be set high enough to provide stability against reasonably foreseeable future conditions. Such considerations depend on the circumstances at specific sites, but may include: (i) exceptionally high groundwater levels; (ii) seismic loading; and (iii) permanent decrease in shear strength due to strain softening (e.g., for stiff clays, from first-time sliding to residual strength). In addition, some contingency in the factor of safety should be included for uncertainty and lack of precision in the back analysis.

Based on the selected factor of safety, calculate the shear resistance required from the piles. For example, if the selected F = 1.25, the total resistance of the group of piles should be designed to provide an additional 25% of the existing total soil resistance on the slip surface of the landslide as determined by back analysis. *Note:* This is a different concept to the usual stability analysis where only a two-dimensional cross-section is analyzed. For example, given the situation depicted on Figure 22.3, it may be necessary to consider two or three cross-sections to model the lateral forces acting on the shear piles.

Design Principles

The forces and moments acting on a shear pile are shown on Figure 22.4 in which the pile has been separated at the top and bottom of the shear zone to show free body diagrams. In the upper sliding ground, the bearing pressure on the pile (P_u) cannot exceed the ultimate value of the bearing strength of the soil in contact with it. This can be determined from conventional bearing capacity theories for deep foundations. In weak soils, the pile length subjected to the limiting pressure is greater than in stiff soils, thus increasing the bending moment in the pile. The permissible maximum bending moment cannot exceed the allowable stresses for reinforced concrete design.

Referring to the simplified concept of Figure 22.4, there are equal and opposite forces P_R at the top and bottom of the shear zone, assuming that there is no resistance mobilized in the shear zone itself. Moment equilibrium of the pile across the shear zone requires that a bending moment M develops, as shown on Figure 22.4. If the upper sliding ground and the lower stable ground were of equal soil strength, it would be reasonable to presume that an equal distribution of moment would develop in the pile at the top and bottom of the shear zone. For illustration purposes, it is assumed here that the two moments are equal. Therefore, the boundary condition consists of a force P_R and moment $aP_R/2$.

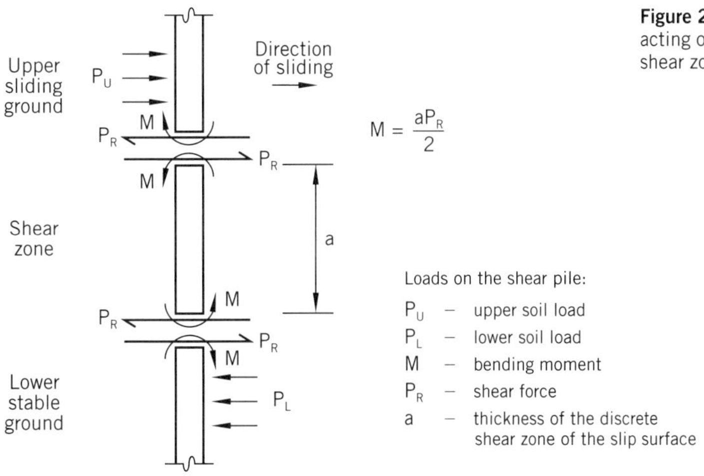

Figure 22.4 Forces and moments acting on a pile at the discrete shear zone.

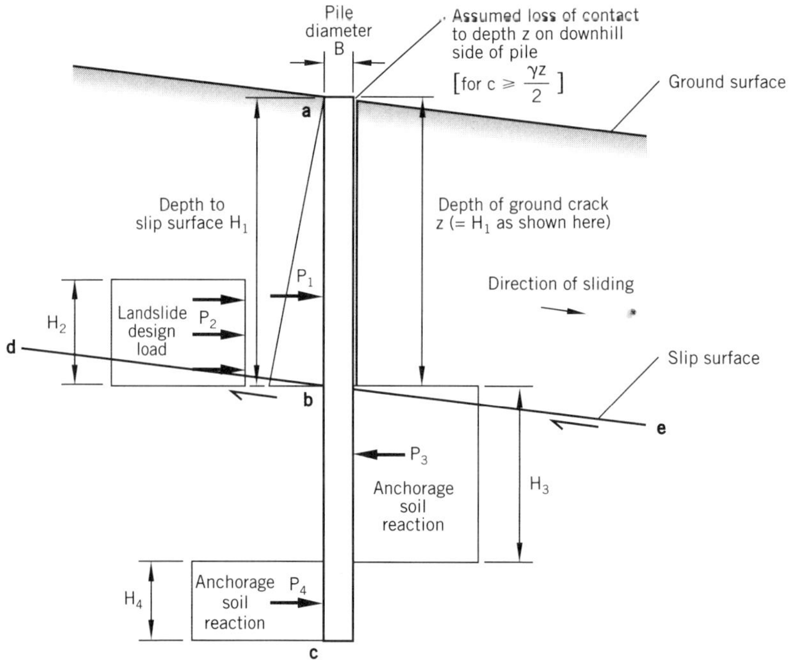

Figure 22.5 Pile design forces.

A design based solely on the bending moment in simple shear across the discrete shear zone would only occur in the exceptional situation where the pile is anchored in hard, unyielding rock above and below the shear zone. This could conceivably occur if shearing was occurring along a thin interbed sandwiched between two hard rock strata.

The more general case is that the pile has to obtain resistance to shear from the bearing capacity of the ground in contact with it (Figure 22.5). This is a mathematical problem of soil-structure interaction, and is treated as a beam resting on an elastic foundation, solved by Hetenyi (1946):

$$EI\frac{d^4y}{dx^4} + Q\frac{d^2y}{dx^2} - p + W = 0 \qquad \text{Eq. (1)}$$

where

Q = axial load on the pile (if any)
x = length coordinate
y = lateral deflection of the pile at a point x
p = lateral soil reaction per unit length
EI = flexural rigidity of the pile
W = distributed load along the pile

This differential equation is solved using load-deflection (p-y) curves for the soils, a subject that has been extensively researched over the past 30 years. At any depth x of the buried pile, the relation between the lateral soil reaction p and lateral deflection y is postulated for the soil at that depth. A schematic drawing of this concept is shown on Figure 22.6. Once the p-y curves have been established, the above equation can be solved by difference-equation techniques using iteration to converge to the required points on nonlinear p-y curves. The solution must satisfy conditions of equilibrium and compatibility. In 1989, Reese and Wang published the computer program LPILE to solve the differential equation.

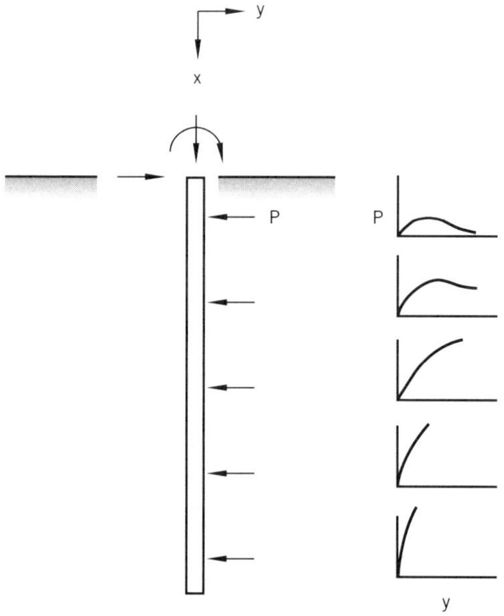

Figure 22.6 Family of p vs. y curves with depth for a laterally loaded buried pile (after Reese et al., 1975).

P-y curves vary with the soil properties, depth below ground, and loading conditions (e.g., static, cyclic, temporary or sustained, etc.) (Figure 22.6). Thus, it is difficult for most geotechnical engineers to gain the necessary experience to select appropriate curves for soils and rocks. Computer program LPILE (Reese and Wang, 1997) allows the geotechnical designer to either: (i) input soil properties from which the computer will generate p-y curves; (ii) input actual p-y curves obtained from sources such as field experiments on test piles; or (iii) select from several soil/rock types descriptions for which the computer has sets of p-y curves. This is an excellent program and likely will be improved further in the future to more closely match p-y curves to subsurface conditions.

The computer program can also accommodate tied-back anchors. These are commonly used for pile walls (Chapter 19, Section 19.4 and Case History 9), but are not likely to be installed where isolated (dowel) piles are used for general landslide stabilization purposes.

Pile Design Procedure

Landslide Technology, with assistance from Dr. Lee Schroeder, has been designing and installing reinforced concrete piles to stabilize medium to large landslides since about 1990. Kenji Yamasaki also contributed to the current procedure.

Referring to Figure 22.5, the length of pile ab is embedded within the landslide, and length bc is anchored into the underlying stable ground. The discrete shear zone (given no finite thickness on this figure) is represented by the line de that crosses the cylindrical pile at b, depth H_1 below the ground. Width B is the diameter of the cylindrical pile.

The landslide design force (for example, 25% of the soil resistance along the slip surface to provide F = 1.25) is P_2 which, for design purposes, acts immediately above the discrete shear zone. In resisting the landslide, the cylindrical shear pile applies a passive force to the uphill ground and this causes yielding of the soil over a depth H_2. The bearing pressure against the pile P_2 cannot exceed the ultimate passive resistance of the soil nor can the bending moment induced in the pile exceed the allowable for reinforced concrete design.

The lateral bearing capacity in the landslide can be obtained from bearing capacity formula for deep foundations available in many textbooks on soil mechanics. For clays at large depth ($H_1 \geq 10$ B) the bearing capacity factor N_c is 9; at the ground surface it is 6, and at intermediate depths the value can be linearly interpolated between these values. Thus, at depth, the lateral bearing capacity q_u = 9c where c is the undrained shear strength of the clay in the landslide. For soft to medium stiff clays, the laboratory measured shear strength can be used in the analyses. In stiff clays, a more conservative interpretation should be selected to allow for possible loss of undrained strength over time for loaded stiff clay.

The very low shear strength within the discrete shear zone can be: (i) treated as a separate resistance zone using the back-calculated residual strength; (ii) integrated into the P_2 calculation (which allows for its depth in the moment); or (iii) ignored, but the depth is included in the moment calculation. Determination of the thickness of the discrete shear zone is made by inclinometer measurements, as described later.

The resistance of the stable anchorage below the slip surface is high in most landslides. In hard clays, weathered rock, or sound rock, the properties are usually estimated. Such data are readily available in textbooks.

The shear force at the upper boundary of the shear zone, force P_R on Figure 22.4, is transformed to a uniform distributed load P_2 on Figure 22.5, where H_2 is determined from the maximum bearing capacity of the soil in the landslide (i.e., 9c for isolated single piles in clay if $H_1 > 10$ B).

The depth H_2 in cohesionless sands and gravels (a less likely scenario in landslides) can be calculated by bearing capacity formulas available in many basic textbooks on soil mechanics of deep foundations. It is suggested that the value of friction angle ϕ be taken as the ultimate strength (angle of repose). It reflects the significant movement that occurs before the full mobilization of passive resistance.

Depending on the location of the shear piles, it may be prudent to assume that the landslide downslope of the piles has not been stabilized and will pull away from the piles. Therefore, a full depth crack ($z = H_1$) could develop at the front (downslope) side of the isolated pile or wall, extending to the slip surface. The full-depth tension crack on the downslope side of piles only applies if shear strength c exceeds $0.5\gamma H_1$. Similarly, loose sands and gravels may remain in contact with the pile after small ground movements occur in the landslide downslope of the pile.

The assumption of a deep tension crack on the downhill side is conservative for isolated shear piles because the ground adjoining such piles acts as a unit and does not separate into stable and potentially unstable ground on the upslope and downslope sides, (as is the case for a shear pile retaining wall). Therefore, a geotechnical designer may wish to modify the depth of a tension crack on the downhill side of isolated shear piles.

An active force, P_1, in addition to the design load, acts on the uphill side of the design pile (Figure 22.5). This force can be obtained from the same computer analysis that was used to back-analyze the landslide in its original condition, i.e., the interslice force measured at the location of the proposed pile stabilization (Figure 22.7).

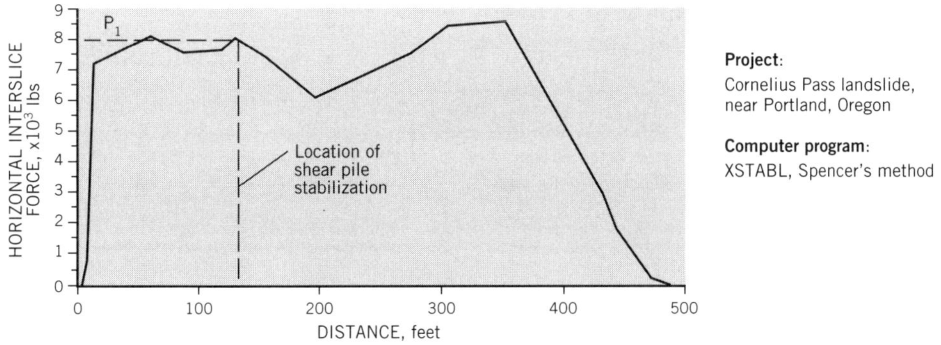

Figure 22.7 Determination of interslice force P_1 from back analysis of landslide.

The remainder of the calculations for soil-structure interaction can be performed by the computer program LPILE (Reese and Wang, 1997). For the pile anchorage *below* the slip surface of the landslide, the program is provided with the shear strength properties of the stable ground (normally obtained by selecting a soil type or rock in the computer program). However, unlike the calculated landslide force *above* the slip surface, a factor of safety (typically 3) is applied to the anchorage forces P_3 and P_4 in the analysis (Figure 22.5). The computer program calculates P_3, P_4, H_3, H_4 and the bending moment diagram, shear force diagram, and deflection of the pile.

The computer program derives p-y curves for the various soils in the analysis. However, if the program is told that there is no force in front of the pile (due to a crack as the landslide moves away from the pile), it assumes that there is no overburden pressure acting on the front of the pile. Using this incorrect assumption, the computer calculated p-y curves in the foundation in front of the pile are incorrect, corresponding to an excavation in front of the wall. However, the reality is that the full weight of overburden soil remains next to the wall. *Yamasaki's technique* is to have the computer calculate the p-y curves *for the full weight of the landslide in front of the wall,* then manually input the p-y foundation values into the computer analysis program. The same approach applies to an isolated shear pile analysis.

The Yamasaki procedure designs the laterally loaded pile as a cantilever above the slip surface of the landslide. However, as previously stated, this may be too conservative for deep-seated movements, soft clays, or cohesionless soils below the water table. Appropriate judgment can be exercised to reinstate lateral pressures at some level on the downslope side of the pile.

Assuming that the landslide is slow-moving or experiences seasonal movements (and pile construction is done at a condition of marginal stability during dry weather), the margin of safety added by the shear piles should fully stabilize the landslide.

Effect of Shear Pile Spacing on Passive Resistance

Reese, Wang, and Fouse (1992) provides a good summary of the research results concerning the effect of close spacing of piles on passive resistance development. As the pile spacing decreases, the interference from adjacent piles (Figure 22.8a) reduces the passive resistance compared to that of a single pile. For center-to-center spacing S and pile diameter B, the reduction factor is approximately 0.5 when the piles are contiguous (touching) in a wall (i.e., S/B = 1) as compared to a single isolated pile. The difference between group and single piles becomes insignificant when the S/B ratio reaches 3 to 4 (Figure 22.8b).

These experimental results are consistent with bearing capacity theories for clays. In a wall, the net passive resistance is the difference between the passive and active earth pressures; this is a maximum of 4c, where c is the undrained shear strength (see Figure 19.2, Chapter 19). For a single pile, the ultimate bearing capacity increases from about 6c at the surface to 9c at depths greater than H_1 = 10B. If the average resistance above the shear zone of a landslide is around 8c (part way between 6c and 9c), then the reduction from widely spaced piles to contiguous piles is 4c/8c = 0.5.

In summary, the group effect of contiguous piles in a wall is to reduce the bearing capacity of each pile to about 50% of its bearing capacity as a single isolated pile. Another way of thinking about this result is that single piles placed in a pattern that isolates them from other piles (i.e., S/B > 4) will provide approximately twice the resistance per pile than piles within a contiguous pile wall.

It is also interesting to note that a spacing S/B = 2 has a reduction factor of about 0.83. Thus piles with a gap between them of one pile diameter can provide almost as much resistance as if an extra pile was there in the gap.

A concern with piles embedded in very soft landslide debris is that the soil may pass between the piles. In such circumstances, the piles should be closely spaced irrespective of the loss of lateral bearing capacity due to group effect. However, isolated shear piles are not a good choice for soft

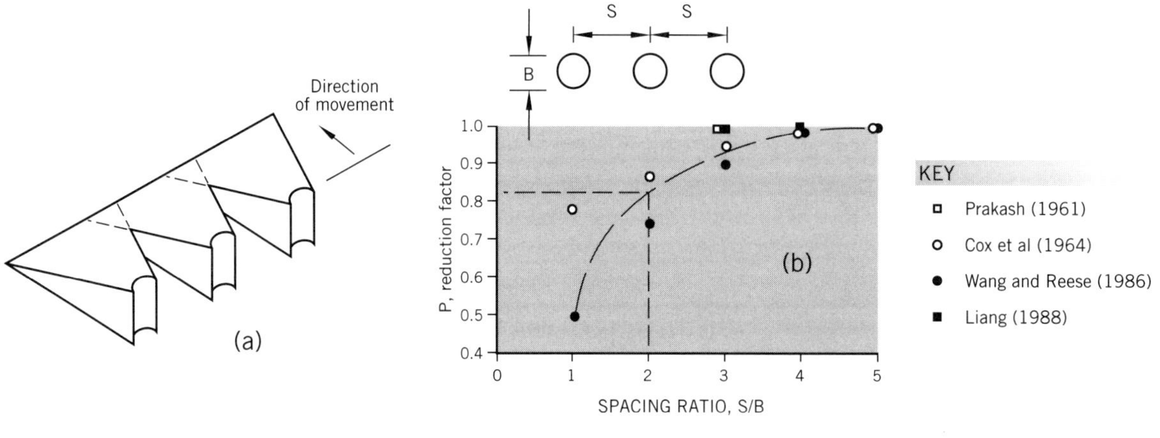

Figure 22.8 Spacing of shear piles:
(a) concept of interference in passive wedge development
(b) reduction factors for closely spaced piles
(after Reese, Wang and Fouse, 1992)

Table 22.1 Determination of Discrete Shear Zone Thickness

(1) Depth (feet)	(2) A readings A_0	(3) A_{180}	(4) Difference $(A_0 - A_{180})$	(5) Deflection upper to lower wheels (inches)	(6) Deflected shape of casing (inches)	(7) Subtraction correction (inches)	(8) Corrected shape of casing (inches)
62'-0"					6.427	5.60	0.83
62'-3"					6.267	5.46	0.81
62'-6"					6.104	5.32	0.78
62'-9"					5.939	5.18	0.76
63'-0"	1066	-1065	2151	1.2906	5.776	5.04	0.74
63'-3"	1070	-1058	2137	1.2822	5.616	4.90	0.72
63'-6"	1071	-1049	2120	1.2720	5.453	4.76	0.69
63'-9"	1064	-1041	2105	1.2630	5.301	4.62	0.68
64'-0"	1060	-1037	2097	1.2582	5.136	4.48	0.66
64'-3"	1057	-1035	2092	1.2552	4.985	4.34	0.65
64'-6"	1061	-1032	2093	1.2558	4.832	4.20	0.63
64'-9"	1068	-1045	2113	1.2678	4.676	4.06	0.62
65'-0"	1073	-1048	2121	1.2726	4.518	3.92	0.60
65'-3"	1079	-1054	2133	1.2798	4.361	3.78	0.58
65'-6"	1090	-1065	2155	1.2930	4.197	3.64	0.56
65'-9"	1123	-1102	2225	1.3350	4.033	3.50	0.53
66'-0"	1205	-1169	2374	1.4244	3.863	3.36	0.50
66'-3"	1249	-1222	2471	1.4826	3.705	3.22	0.49
66'-6"	1250	-1234	2484	1.4904	3.539	3.08	0.46
66'-9"	1249	-1231	2480	1.4880	3.341	2.94	0.40
67'-0"	1245	-1227	2472	1.4832	3.094	2.80	0.29
67'-3"	1242	-1223	2465	1.4790	2.878	2.66	0.22
67'-6"	1235	-1214	2449	1.4694	2.707	2.52	0.19
67'-9"	1208	-1174	2382	1.4292	2.545	2.38	0.17
68'-0"	1130	-1105	2234	1.3404	2.380	2.24	0.14
68'-3"	1074	-1061	2135	1.2810	2.226	2.10	0.12
68'-6"	1063	-1052	2105	1.2630	2.070	1.96	0.11
68'-9"	1054	-1035	2089	1.2534	1.911	1.82	0.09
69'-0"	1045	-1025	2070	1.2420	1.753	1.68	0.07
69'-3"	1035	-1015	2050	1.2300	1.597	1.54	0.06
69'-6"	1023	-1003	2026	1.2156	1.444	1.40	0.04
69'-9"	1009	-990	1999	1.1994	1.292	1.26	0.03
70'-0"	997	-977	1974	1.1844	1.138	1.12	0.02
70'-3"	987	-965	1952	1.1712	0.996	0.98	0.02
70'-6"	977	-956	1933	1.1598	0.854	0.84	0.01
70'9"	969	-947	1916	1.1496	0.712	0.70	0.01
71'-0"	959	-938	1897	1.1382	0.569	0.56	0.01
71'-3"					0.426	0.42	0.01
71'-6"					0.284	0.28	0
71'-9"					0.142	0.14	0
72'-0"					0		0

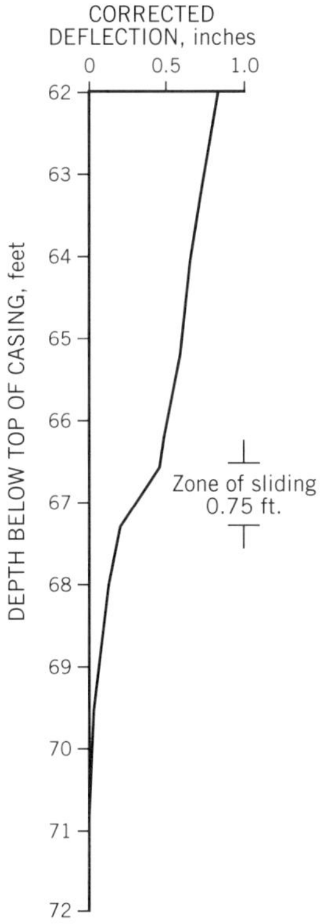

Figure 22.9 Measurement of discrete shear zone thickness using 3-inch interval inclinometer readings.

clays because the piles would need substantial reinforcement to counter the high bending moment.

Measurement of the Discrete Shear Zone Thickness

The thickness of the relatively weak discrete shear zone needs to be known as part of the shear pile design. This thickness can be measured by inclinometers passing through the shear zone using a modified reading procedure that more accurately measures the deformed shape of the casing in the shear zone. By measuring the thickness of the discrete shear zone on several inclinometers within a landslide, the variation in thickness can be assessed. The largest thickness should be used in the shear pile design.

Inclinometer probes in the United States have a 2-foot spacing between the wheels, and readings are usually taken at 2-foot intervals. Thus, in raising the probe from the bottom of the hole, the lower wheel is raised to the upper wheel location of the previous reading. This conventional reading procedure is too crude for measuring the thickness of the discrete shear zone. Instead, the readings should be taken at *3-inch* intervals; then, by tracking the position of the upper wheel, a more precise measurement of the discrete shear zone is obtained. Table 22.1 in Figure 22.9 demonstrates a typical set of computations.

The procedure is as follows:

1. Prepare a table (Figure 22.9) with the Depth column having an upper limit 1 foot higher and a lower limit 1 foot lower than the range of 3-inch field measurements. In the example, the shear zone is at a depth of about 67 feet, and the special 3-inch inclinometer readings were taken from 63 to 71 feet. Therefore, the depths in Table 22.1 range from 62 to 72 feet.
2. Calculate the deflection of the upper wheel relative to the lower wheel (column 5) to four places of decimals. For the Slope Indicator Digitilt inclinometer, deflection = 0.006 $(A_0 - A_{180})$ inch. Since the depths are measured to the mid-height of the inclinometer instrument, the lower wheel is 1 foot below the measured depth and the upper wheel is 1 foot above the measured depth. For the lowest reading, at 71 feet in Figure 22.9, first assume that the casing is linear and divide the 2-foot deflection value (1.1382 inch) into 8 equal increments (i.e., 0.142 inch) in column (6) from 72 feet to 70 feet.
3. To calculate the deflected shape of the casing 3 inches higher, at 69'-9", take the deflection, upper wheel to lower wheel, at 70'-9" and add the position of the lower wheel at 71'-9". In Table 22.1 of Figure 22.9:

Deflection, upper to lower wheel, at 70'-9"	=	1.1496"
Lower wheel at position 71'-9" reads		0.142"
∴ Upper wheel at position 69'-9" is at		1.2916"

This value is entered in column (6) at 69'-9" and shows the position of the upper wheel at this depth.

The procedure is repeated to track the position of the upper wheel at other levels up the casing, and column (6) provides the deflected shape of the casing.

After completing column (6), the deflected shape of the casing can be plotted to determine the location and thickness of the shear zone. However, it is easier to first make a correction for the out-of-plumb casing, as described below.

4. If the inclinometer casing is out-of-plumb, the data can be "corrected" by choosing a deflection for the lower 2-foot reading that brings it back to near-vertical.

   On Table 12.1, the out-of-plumb reading for the lowest measurement is 1.138 inches between the upper and lower wheels. A nearly equivalent value of 1.12 is a convenient value to break up into eighths, i.e., 0.14 inch per 3-inch interval. These changes (0.14 inch per 3 inches) are accumulated from bottom to top of the data set and are entered in column (7). The corrections of column (7) are subtracted from the calculated values in column (6) to obtain a "corrected" shape of casing in column (8). The "corrected" shape values have been rounded off to the nearest 0.01 inch and are plotted on Figure 22.9. It shows that the shear zone is 9 inches wide, from 66'-6" to 67'-3" based on the more precise inclinometer survey.

Applications

Isolated shear piles have been used effectively to stabilize landslides in several countries. The principal reasons for their infrequent use to date are probably the high cost and lack of consensus on design. At Landslide Technology, isolated shear piles for slip surface strengthening has been recommended at a conceptual level. The Washington Park Slide, Portland, Oregon, is a sporadically active lobe of ground within ancient landslide terrain. The lobe is about 2,700 feet long and about 1,000 feet wide. The Washington Park transit station is located near the center of the lobe (see Case History 4). The marginally stable slide area has a slope of about 5½° to the horizontal and much of the site is a paved parking area for three public structures. It would have been relatively easy to install deep isolated shear piles over a wide area and cover them with asphalt surfacing. However, the transit agency chose to implement extensive drainage to improve overall stability rather than shear piles.

22.2 OTHER TECHNIQUES

The principle of replacing weak soil on the discrete shear surface of a landslide with a stronger reinforcement has wider application. Although cast-in-place heavily reinforced concrete shear piles, as described in Section 22.1, are probably the optimum treatment where only a limited number of installations can be made, there are other alternatives for strengthening the slip zone incrementally. Most of these alternative techniques have been described elsewhere and an abbreviated summary will be provided below.

A key consideration is the difference in strength between the existing shear surface and the proposed remedy. The larger the difference between these two strengths the more improvement is achieved. Therefore, strengthening the slip surface is appropriate where the existing strength is relatively low (such as residual strength of a fat clay) and the introduced strength is high (crushed rockfill, concrete, or steel-reinforced).

Stone Columns

A column of compacted angular crushed rock installed in a clay landslide is shown on Figure 22.10. If the slip surface is passing just above a hard bedrock, the base of the column would be socketed into the rock 2 to 3 feet. If the clay stratum extends below the slip surface, a penetration of 6 to 10 feet past the slip surface would be appropriate. A strength of

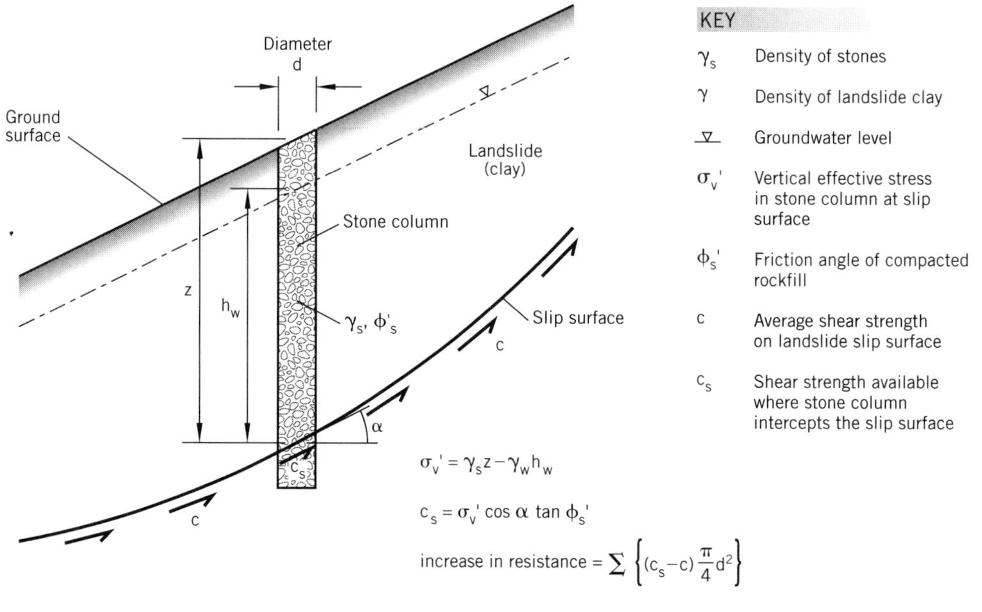

Figure 22.10 Stone column stabilization of a landslide in clay.

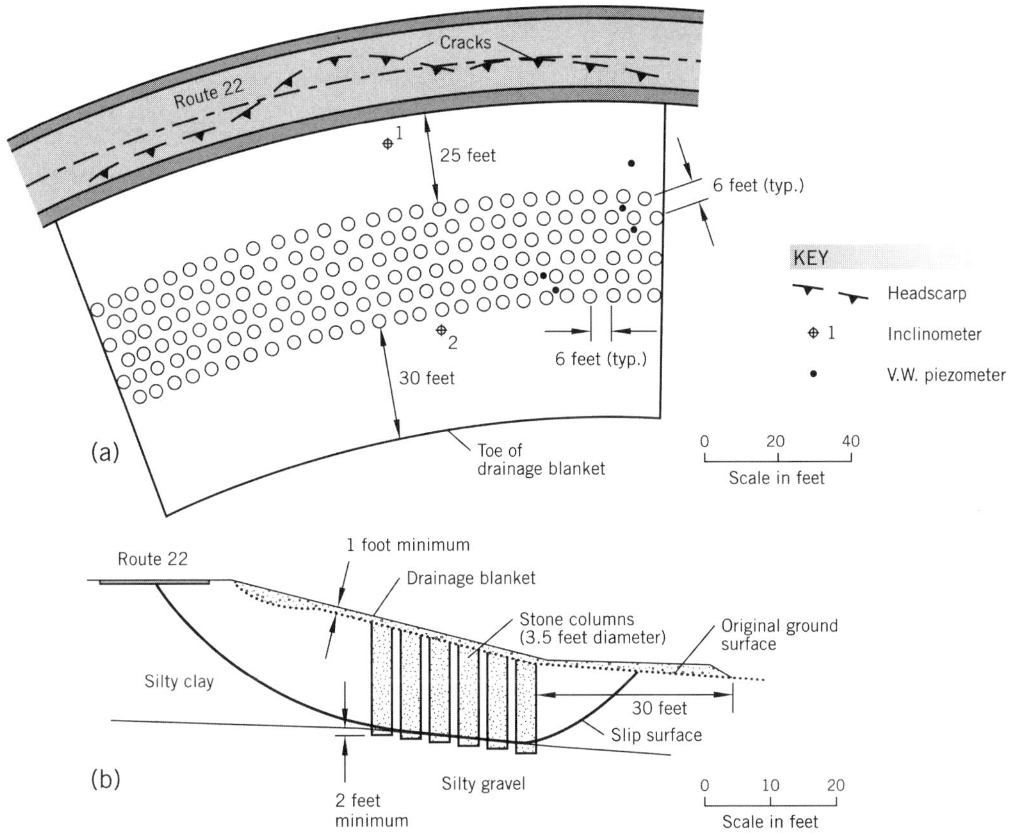

Figure 22.11 Stone column remediation of a landslide at Wadhams, New York State:
(a) plan
(b) section
(after Goughnour et al., 1991)

ϕ = 40°–43° can be assigned to the slip surface of the stone column, depending on depth below the surface and degree of compaction applied to the rockfill during construction.

The calculations for stability are relatively simple. A back analysis can provide both the average shear strength c acting on the slip surface of the landslide and the total resistance (in tons or similar force units) of the existing landslide. The *extra* resistance provided by the stone columns can be obtained by summing the calculated differences in available shear strength (c_s – c) at each column location. Additional details are shown on Figure 22.10.

Stone columns can also act as vertical drains and the tops of the columns can be connected to a system of near-surface trench drains. Such a system can prevent the build-up of groundwater levels during rainy seasons and provide rapid dissipation of excess groundwater pressures during storm events. Thus, the system simultaneously controls groundwater levels and increases shear resistance at the slip surface.

Example

Goughnour et al. (1991) describe three projects in which compacted stone columns were installed to stabilize landslides. One of these case histories was at Wadham, New York, where 156 stone columns, 3.5 feet diameter (within a triangular spacing of 6 feet on center) were installed to stabilize a slope below a highway (Figure 22.11a). The subsurface conditions consisted of 10 feet of firm silty clay overlying 10 to 20 feet of soft clay; below the clay was a gravel layer with an artesian head of up to 5 feet above ground. The slippage reached the clay/gravel boundary (Figure 22.11b).

The properties of the clays at this site are shown on Figure 22.12. The slip surface was mostly 15 to 16 feet below the surface. Field vane and laboratory consolidated-undrained triaxial tests at these depths gave shear strengths in the range of 280 to 440 lb./sq. ft. The back-calculated average shear strength along the slip surface was 270 lb./sq. ft.

Stone columns were installed by the bottom-feed dry installation method (see Chapter 21, Section 21.4). Of particular interest is that five vibrating wire piezometers were placed both outside the treated area and midway between selected stone columns (Figure 22.11a) to observe the excess pore pressures generated by column construction. The piezometer tips were either 12 or 18 feet below the ground surface. Within 4 feet of the stone columns the excess head ranged from 3.2 to 5.6 psi (7.4 to 12.9 feet). At a distance of 15 feet or greater, the excess heads were negligible. The raised pore pressures dissipated completely within 5 to 50 hours.

The slow-moving (0.9 inch in 6 months) landslide was stabilized by the stone columns. The stone columns served a dual function of relieving the pre-existing artesian head and providing additional resistance along the slip surface.

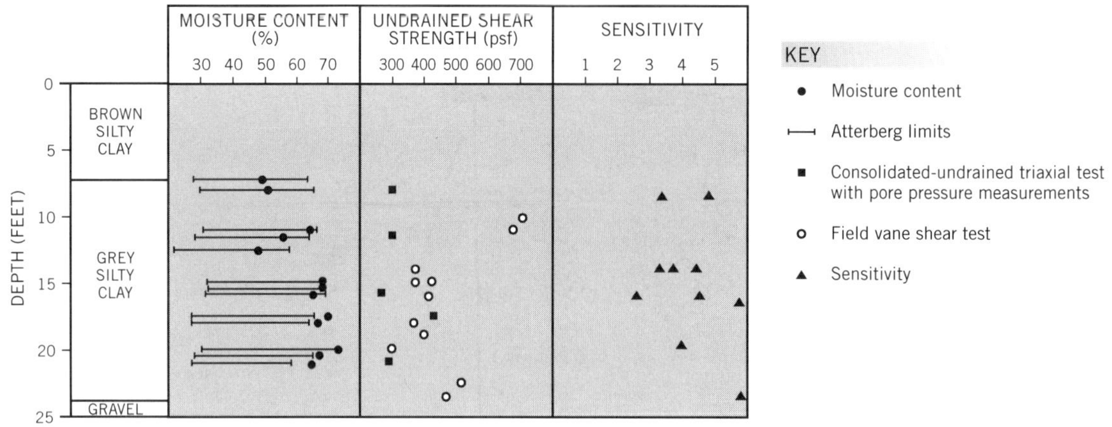

Figure 22.12 Properties of silty clay strata at Wadhams, New York, landslide (after Goughnour et al., 1991).

Deep Soil Mixing

Deep soil mixing with cement and/or lime can substantially increase the strength of a clay. Unconfined compressive strengths of up to 200 psi (28,800 psf) can be obtained from clay-cement admixtures. Allowing for some reduction in strength between laboratory and field conditions, an undrained shear strength in the field corresponding to hard clay (>6000 psf) is achievable. Furthermore, the quantity of cement grout injected into the clay can be targeted to a specified depth so that only the area around the slip surface needs to be subjected to the maximum cement content.

There are several disadvantages however. The equipment is large and expensive to mobilize and has limited mobility to move around a landslide site. More details of this technique are presented in Chapter 18, Section 18.4.

Lime-cement columns can improve the strength of soft clays, especially those with water contents above 60%. The columns are created by the dry mixing process. In Scandinavia the cement : lime ratio varies from 1:1 to 3:1. Advantages of the dry mixing process are the very low volume of spoils, efficient mixing, and relatively fast production rates. The techniques have received extensive research support in both Scandinavia and Japan, the two principal developers. Watn et al. (1999) discuss the use of lime-cement for slope stability.

Rubblized Slip Surface

The intent of this procedure is to set off a sufficient quantity of explosives within a drillhole to intermix soil and/or rock at and around the slip surface of a landslide. The weak soil on the slip surface is replaced by "rubble" having a much higher strength. By drilling numerous holes and rubblizing the slip surface at many places, the landslide should be stabilized. In principle, this technique should work satisfactorily for weak interbeds or gouge zones sandwiched between more competent rock, broken rock, or weathered rock.

A field experiment by a highway department to test this concept ended disastrously when the explosives were fired simultaneously in all holes. The resulting blast brought down the hillside onto a freeway. Although this event had an unsatisfactory outcome, the method may be feasible under more controlled conditions. For example, a slope with seasonal movements could be subjected to a progressive set of explosions during the dormant dry season without destabilizing the slope. Another option might be to perform a rubbling procedure for the slope that will be left in place and then make the

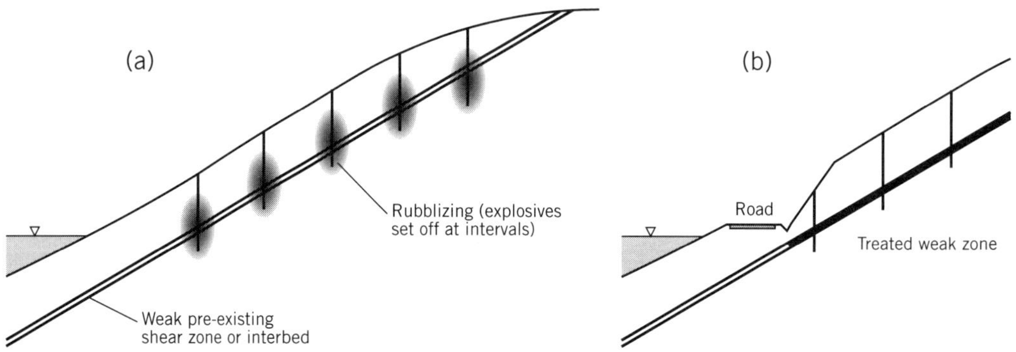

Figure 22.13 Rubblizing a slip surface (concept only):
(a) breaking up the shear zone
(b) completed road cut after ground treatment

cut slope (shown conceptually on Figure 22.13). Clearly, any such new techniques require extensive study of the likely slope behavior before it is put into practice.

Trench Drains

Trench drains that penetrate fully through the waterbearing stratum to denser soils or bedrock provide substantial resistance to landsliding. As more fully explained in Chapter 17, Section 17.3, the resistance developed at the slip surface due to a rockfilled trench may exceed the benefits attributed to groundwater drawdown.

Jet Grouting

A high pressure jet drill can selectively scour out a large radius in the soft interbed of a landslide and infill the void with a mixture of the native soil and cement (see Chapter 18, Section 18.3). By constructing a grid of treated ground over the landslide area, a landslide can be stabilized. One of the major benefits of jet grouting is that the equipment is relatively mobile. However, it would require a medium or larger landslide project for a grout mixing plant to be assembled at the site on a cost-competitive basis and the process generates large quantities of spoil. Cement usage is also high.

CHAPTER

23

Landslide Hazard

23.1 LANDSLIDE HAZARD MAPPING

The Case for Landslide Hazard Mapping

Landslide hazard mapping provides a service to society by producing maps that distinguish areas of different landslide risk. The maps are a useful tool for land-use planners, planning for lifeline facilities (e.g., power and water supply routes), local government building codes, and property owners.

Although landslide hazard maps have been produced for many years, interest in this aid to landslide prevention has been increasing. Landslide conferences often hold special sessions devoted to landslide risk.

Landslides account for considerable property damage and loss of life in regions with moderate to steep slopes and intense rainfall. In the United States, figures of $1 billion and 25 fatalities per year (on average) have been put forward. However, actual losses are difficult to quantify because data is not routinely collected after each occurrence. Moreover, there are consequential losses (loss of use, legal claims, sustained injuries, etc.) in addition to direct damage and loss of life. The potential benefit of landslide hazard mapping is to achieve a meaningful reduction in losses through awareness and avoidance of the hazard.

In addition to reducing the cost of landslide damage, a second reason for increased interest in landslide hazard mapping in the United States is that local authorities are being sued with increasing frequency for issuing permits for development on land that subsequently becomes involved in a landslide. Insurance against landslide risk is unavailable. Since home ownership is usually a family's largest single asset, the trauma of losing a substantial part of their wealth from an event over which they have no control inevitably causes homeowners to seek damages from anyone who may have contributed to the landslide. These groups include developers, builders, consulting engineers, local authorities, etc. who carry various types of liability insurance, general or professional. The connection between these design and construction groups and the cause of the landslide is often tenuous, to say the least, but multiple parties frequently become embroiled in a lawsuit. Furthermore, since juries are likely to be sympathetic toward homeowners, the parties being sued generally are loath to go to court unless they can prove complete innocence. Experienced geotechnical practitioners are often incensed to discover that their professional liability insurer prefers to settle a claim rather than defend the consultant's professional reputation.

The increasing use of lawsuits to provide retribution for landslides is a major problem in the United States, because the cost of defending such lawsuits can amount to millions of dollars per event. Legal costs frequently exceed remedial costs. It is, therefore, in a wide range of interests to prevent landslides wherever possible, and landslide hazard maps can be helpful in this endeavor.

Technical Issues of Mapping

Landslide hazard maps generally combine surface geology with slope gradient to determine the perceived level of hazard. The technical assumptions are that steep slopes and weak soils are more likely to experience instability than flatter slopes and stronger soils. Although this approach appears to be logical, a map combining slope and surface soils into a landslide hazard assessment is flawed for the following reasons:

1. *The surface soil may not represent the soil on the potential slip surface at depth.* Many examples of this difference are shown by the cross-sections of landslides included in this book. Therefore, knowledge of the soils at the ground surface may be irrelevant to the slope stability of the site.

2. *The groundwater levels in the slope cannot be seen and included as a factor in the slope stability.* Groundwater is a major influence on stability; in the infinite slope analysis, for example, the factor of safety of the slope with groundwater at the slip surface is approximately twice the factor of safety for groundwater at the ground surface. Therefore, a lack of knowledge of the groundwater levels within the slope introduces a huge potential error in the interpretation of slope stability.
3. *In many cohesive soils, the steeper slopes are stable and the flatter slopes are unstable.* This paradox arises from the natural development of the slope. Once an instability has been initiated (by erosion or other causes), the failed slope becomes a trough of ground between stable slopes. Surface runoff flows into the trough, raising groundwater levels within the sunken (unstable) soils. This perpetuates instability of the ground. Furthermore, continued movement lowers the shear strength of stiff clays to the residual strength whereas unfailed slopes retain a much higher strength. The clayey silt slopes of weathered siltstone around Hagg Lake (Case History 10) exhibit this behavior of the flatter slopes being unstable while the steeper slopes of the same soil are stable. It can be observed at many other sites with stiff clay slopes, including the well-studied London Clay.

Landslide hazard maps based solely on slope gradient and surface geology may cause unintended consequences. For example, someone may have owned a house for 30 years that has never experienced any distress associated with landslide movements. If a landslide map is produced that shows their property being located in a "moderate-to-high" risk area, any realtor or potential buyer of the property learning of this may either lose interest or reduce the value of any offer. The property's value has been lowered as a result of the landslide hazard map yet, for the three reasons stated earlier, the actual risk of a future landslide may be minimal.

A Suggested Approach

Given the need for landslide hazard maps, how can it be accomplished in a way that provides meaningful but not misleading information? The author suggests that the best approach is to provide only *factual* information on the maps and allow the users to make their own interpretation. If needed, interested parties can obtain professional advice from geotechnical practitioners for specific input on projects.

A factual landslide hazard map for an area can be produced by a two-step process:

1. ***Base Map.*** Put all pre-existing landslide conditions on a base map to a scale of 1:24,000 or larger. The map should show contours of ground elevations (for example, in the United States, U.S.G.S. maps typically provide contour intervals of 20 to 40 feet in hilly terrain). The mapped landslides would include: rockfall areas; active erosional slopes adjoining rivers, streams, springs; debris flow deposits; known shear failures; colluvium and other ancient landslide conditions; talus slopes, etc. Such maps can be compiled by experienced engineering geologists.
2. ***Manmade Landslides.*** Landslides would be reported by property owners as they occur via a required Landslide Report. These reports would allow the hazard map to be continually updated over time. Only slope failures above a minimum size (e.g., 50 cu. yd.) would have to be reported.

The Landslide Report could be a relatively simple one- or two-page document that provides: (i) date of occurrence; (ii) location by street address; (iii) plan sketch of the slide with approximate dimensions; (iv) whether a cut or fill is involved in the slide area; (v) weather conditions prior to, and at the time of, sliding; (vi) if the slide is a recurrence of previous movements or a new movement; (vii) type of soil in the landslide debris; and (viii) name of any professional engineer or geologist who has examined the slide.

The location of the slide would be plotted on the Base Map and thus would be available, together with the Landslide Report, as part of the public record. Clustering of landslide points on the map would provide a cautionary warning to persons interested in developing land close to such a cluster.

The city of Seattle, Washington, has a long history of slope instabilities within the Pleistocene deposits on which the city is built. A landslide hazard map was developed in the 1950s using point location of actual landslides reported by geotechnical consultants and others. It was kept by the public works department and was made available to the public.

An example of points representing past landslides in an area of north Seattle, Washington, is shown on Figure 23.1 (Coe et al., 2000; Laprade et al., 2000). This map includes more than 80 years of landslide records, although the record-keeping is acknowledged to be incomplete. The main benefit of the map is to show the clustering of the points in certain areas. For proposed new developments, a transparency can be made of these points so that it can be overlain on topographic, geologic, or street maps for additional interpretation of a specific site when combined with a field reconnaissance by a geotechnical professional.

23.2 ROCKFALL HAZARD RATING SYSTEM

Rocks falling onto a road can cause substantial damage, personal injuries, and occasionally deaths. They pose a similar hazard to railroads and other properties.

Since 1984 the Oregon Department of Transportation (ODOT) has been using a Rockfall Hazard Rating System (RHRS) to prioritize rock slopes for preventative remediation. This program has several objectives: (i) to promote knowledge about the condition of rock slopes adjacent to roads for cost-effective planning of available funds; (ii) to reassure the public that a protective program of dealing with slopes is in place; and (iii) provide the department with a legally defensible position

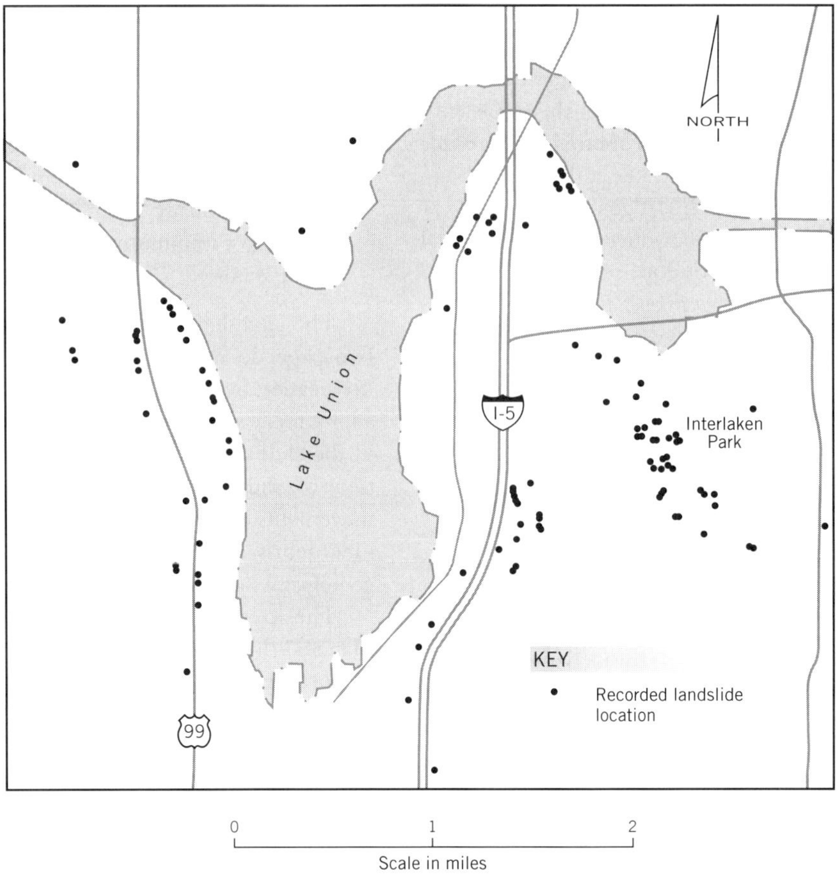

Figure 23.1 Map of recorded landslides in part of north Seattle, Washington; data compiled by Shannon & Wilson, Inc. (abstracted from USGS Open-File Report 00-303; Coe et al., 2000). Map simplified from the original.

that they are using a systematic approach to prioritize rockfall mitigation projects based on competent information regarding the relative hazard of each site.

The development of the Rockfall Hazard Rating System is an outgrowth of earlier work by Wyllie and Brawner (1975) and Wyllie (1987) for railways. The current system, intended for highways, was completed by Pierson and VanVickle (1993) and has been widely adopted by state highway departments in the United States, and by several other countries.

Slope Inventory and Preliminary Rating

The first step is for the rater, usually an engineering geologist, to visit the existing potential rockfall sites with a maintenance person who is knowledgeable about the highway's rockfall history and maintenance. The following information should be obtained:

1. Location of rockfall activity
2. Frequency of rockfall events
3. Time of year when activity is highest
4. Size/volume of rockfall per event
5. Physical characteristics of rockfall material
6. Where rockfalls have come to rest
7. Available accident history
8. Opinion of rockfall cause
9. Frequency of ditch cleaning/road patrol
10. Estimated annual cost of maintenance response

After inspecting the site to evaluate the potential for rockfall events, and considering the rockfall history, each slope is given a preliminary rating of A, B, or C to represent a high, moderate, or low potential for future rockfalls. The C rating, for example, means that it is unlikely that a rock will fall at the site or, if it does, it is unlikely to reach the roadway. Therefore, C rated sections are usually eliminated from further consideration. During development of the system several thousand slopes were examined by ODOT, of which 501 received an A rating and 839 received a B rating. Although B-rated slopes are included in the RHRS database, they are less likely to receive preventative treatment when the Agency has such a large A group. The A rated slopes advance to the detailed rating.

Detailed Rating

The Rockfall Hazard Rating System (RHRS) evaluates specific conditions that make the rock slope and highway dangerous to road users (Table 23.1). For each of 10 different categories, a score of up to 100 points is assigned. The points given for each category rise exponentially as the hazard increases, so that more hazardous sites stand out within a group of rated slopes.

For each category, a set of four "benchmarks" (based on description or values) are provided that represent a progressive increase in risk. At each benchmark, the score increases by a factor of 3, so the points progress from 3, 9, 27, to 81 points. A less-experienced rater can simply choose the more appropriate descriptive benchmark of the four choices; a more experienced rater can interpolate between the benchmark points. The benchmarks provide assistance to the rater and promote consistency between different raters.

A plot of the equation $y = 3^x$, Figure 23.2, can be used to determine the points for each rockfall risk category. Where numbers are calculated, there are separate scales on the abscissa; for descriptive variations, the worsening risk is shown by roman numerals I to IV.

The ten rockfall risk categories are described below:

1. ***Slope Height.*** The higher a loose rock is located on the slope, the higher is its potential energy, thus creating a greater hazard.

   The vertical slope height can be calculated by the methods described in Chapter 3, Section 3.2, using either a clinometer, Brunton compass, or surveying instrument. If rockfall is generated from a natural slope above the cut slope, the measurement should be taken to the maximum height on the rockfall source. The points for vertical slope height can be read directly from the graph, Figure 23.2(a).

2. ***Ditch Effectiveness.*** The risk associated with a rock slope is dependent on how well the roadside ditch captures the rockfall. If minor amounts of rock reach the roadway, the danger to the public is lower. The rating should consider the following: (i) slope height and angle; (ii) ditch width, depth, and shape; (iii) anticipated volume of rockfall per event; and (iv) impact of slope irregularities (launching features) on falling rock. Much of the information on ditch performance can be obtained from maintenance personnel.

   Points can be assessed from Figure 23.2(b) using the following guidelines:

   Good catchment. All or nearly all falling rocks are retained in the catch ditch.

   Moderate catchment. Falling rocks occasionally reach the roadway.

   Limited catchment. Falling rocks frequently reach the roadway.

   No catchment. No ditch, or ditch is totally ineffective. All or nearly all falling rocks reach the road.

Table 23.1 Summary Sheet of the Rockfall Hazard Rating System

		RATING CRITERIA AND SCORE		
ROCKFALL RISK CATEGORY	BENCHMARK I 3 POINTS	BENCHMARK II 9 POINTS	BENCHMARK III 27 POINTS	BENCHMARK IV 81 POINTS
1. Slope height	25 feet	50 feet	75 feet	100 feet
2. Ditch effectiveness	Good catchment	Moderate catchment	Limited catchment	No catchment
3. Average vehicle risk	25% of the time	50% of the time	75% of the time	100% of the time
4. Percent of decision sight distance	Adequate sight distance, 100% of low design value	Moderate sight distance, 80% of low design value	Limited sight distance, 60% of low design value	Very limited sight distance, 40% of low design value
5. Roadway width including paved shoulders	44 feet	36 feet	28 feet	20 feet
GEOLOGICAL CHARACTER — CASE 1: 6. Structural condition	Discontinuous joints, favorable orientation	Discontinuous joints, random orientation	Discontinuous joints, adverse orientation	Continuous joints, adverse orientation
GEOLOGICAL CHARACTER — CASE 1: 7. Rock friction	Rough, irregular	Undulating	Planar	Clay infilling, or slickensided
GEOLOGICAL CHARACTER — CASE 2: 6. Structural condition	Few differential erosion features	Occasional differential erosion features	Many differential erosion features	Major differential erosion features
GEOLOGICAL CHARACTER — CASE 2: 7. Difference in erosion rates	Small difference	Moderate difference	Large difference	Extreme difference
8. Block size / Volume of rockfall event	1 foot / 3 cubic yards	2 feet / 6 cubic yards	3 feet / 9 cubic yards	4 feet / 12 cubic yards
9. Climate and presence of water on slope	Low to moderate precipitation; no freezing periods; no water on slope	Moderate precipitation or short freezing periods or intermittent water on slope	High precipitation or long freezing periods or continual water on slope	High precipitation and long freezing periods or continual water on slope and long freezing periods
10. Rockfall history	Few falls	Occasional falls	Many falls	Constant falls

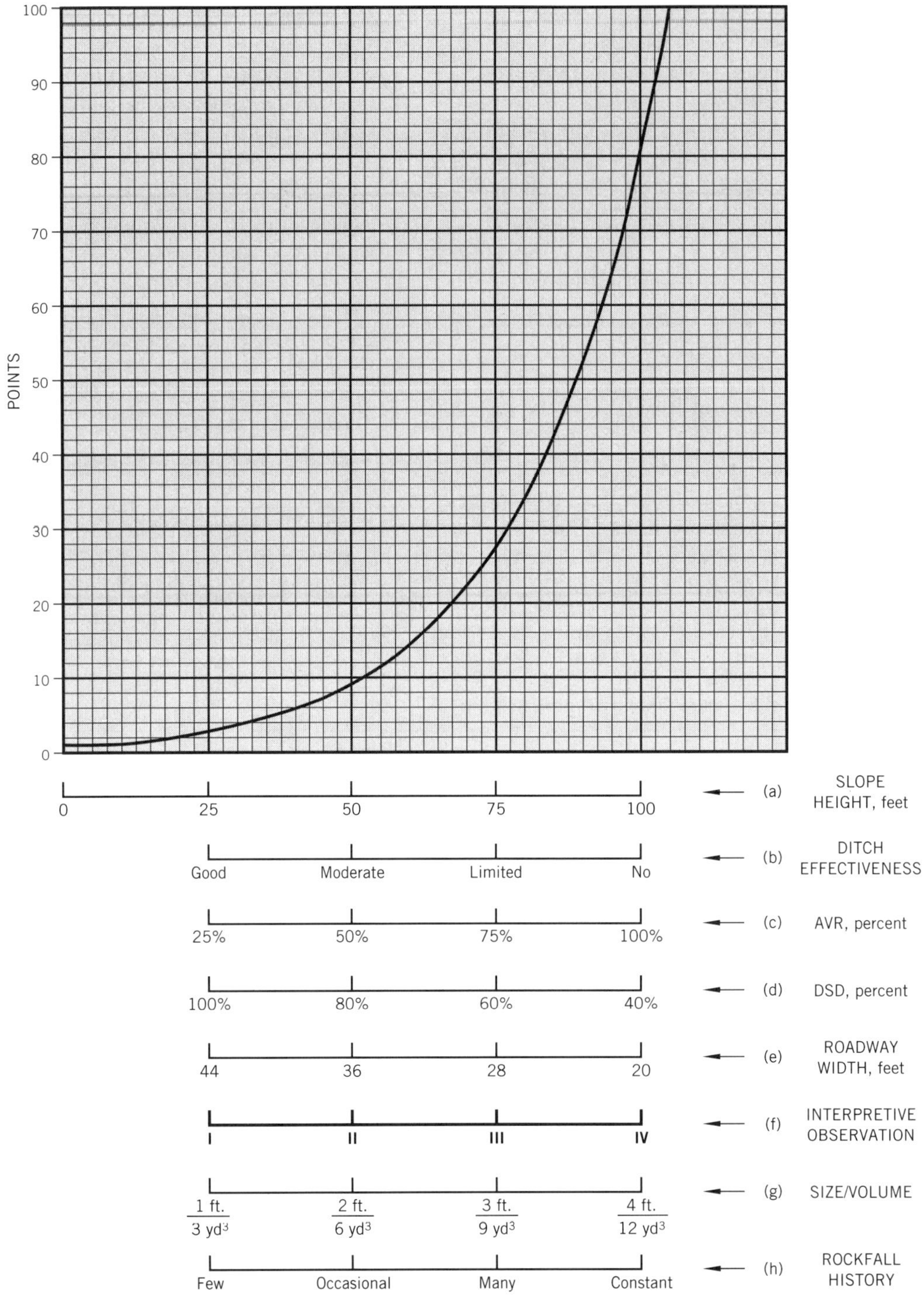

Figure 23.2 Rockfall Hazard Rating System (after Pierson and Van Vickle, 1993).

3. ***Average Vehicle Risk.*** This category represents the potential for a vehicle to be involved in a rockfall event. It shows how many vehicles are, on average, in the rockfall section at any time. The result also reflects the significance of the route.

$$\text{AVR} = \frac{[\text{Average Daily Traffic (cars/day)}][\text{Length of Cut Slope (miles)}]}{[\text{Posted Speed Limit (mph)}][24]} \times 100\%$$

Example: ADT = 3,000 cars per day
Length of cut = 0.42 mile
Speed limit = 40 miles per hour
AVR = 131%

For this example the cut is given 100 points for being above 105% AVR, as shown on Figure 23.2(c).

4. ***Percent of Decision Sight Distance.*** Highway curves and other visual obstructions (projecting rock outcrops, vegetation) can limit a driver's ability to see rockfall within the road and take evasive actions. The sight distance can vary considerably within a cut section.

   Sight distance is defined as the shortest distance along the road that a 6-inch high object can be continuously seen at an eye height of 3½ feet above the road surface. The most restricted sight distance is usually around the sharpest part of a road curve.

   "Decision Sight Distance" is the length of roadway required by a driver to perceive a problem and then bring a vehicle to a stop. The Decision Sight Distance is obtained from Table 23.2.

$$\text{Percent of Decision Site Distance} = \frac{\text{Actual Site Distance}}{\text{Decision Site Distance}} \times 100\%$$

   Points for percent of Decision Site Distance can be taken from Figure 23.2(d).

5. ***Road Width.*** The road width category relates to a driver's ability to avoid a rockfall event by providing room to maneuver. The width is measured at the narrowest paved width, including paved shoulders; on divided highways only the paved surface available to the driver is measured. The points assigned to roadway width can be taken from the graph, Figure 23.2(e).

6,7. ***Geologic Character.*** The Rockfall Hazard Rating System recognizes that slopes that produce rockfalls can be subdivided into two risk categories: Case 1—jointed rock slopes, and Case 2—slopes with rocks subject to erosion. Whichever case best fits the slope should be used for the rating. If both situations exist, scores are given for both cases and the worst case (highest score) is used in the rating.

   The Geologic Character is made up to two ratings, each given up to 100 points, for a total of up to 200 points. A person with a background in geology should make the ratings. If a localized area of the slope is causing a higher incidence of rockfalls it should be rated separately from the entire cut slope.

Table 23.2 Decision Sight Distance*

Posted Speed Limit (mph)	Decision Sight Distance (feet)	Posted Speed Limit (mph)	Decision Sight Distance (feet)
25	375	50	750
30	450	55	875
35	525	60	1,000
40	600	65	1,050
45	675		

**Source:* AASHTO (1984)

Case 1: Jointed Rocks

Where joints, bedding planes, or other discontinuities are oriented adversely to the slope, the potential for rockfall is greater. Adverse joints are those that (singularly or in combination with other joints) make planar, circular, block, wedge, or toppling failures possible. The benchmark descriptions are:

I (3 pts)	Discontinuous Joints, Favorable Orientation Slope contains jointed rock with no adversely oriented joints.
II (9 pts)	Discontinuous Joints, Random Orientation Slope contains randomly oriented joints creating a variable pattern. The slope is likely to have some scattered blocks with adversely oriented joints, but no dominant adverse pattern is present.
III (27 pts)	Discontinuous Joints, Adverse Orientation Rock slope exhibits a prominent joint pattern with an adverse orientation. These features have less than 10 feet of continuous length.
IV (81 pts)	Continuous Joints, Adverse Orientation Rock slope exhibits a dominant joint pattern with an adverse orientation and a length greater than 10 feet.

Points for joints patterns that are judged to be between the above benchmarks can be obtained from the Interpretative Observation scale on Figure 23.2(f). The same approach can be taken for all the other benchmark interpretations described below.

The second category for jointed rocks is to estimate the probable resistance along the joints. Macro roughness is the degree of undulation of the joint relative to the direction of possible movement. Micro roughness is the texture of the surface. Open joints, water-filled joints, joints with rock gouge, or weathered joints are weaker than tight unweathered joints and are assessed accordingly.

I (3 pts)	Rough, Irregular The surface of the joints are rough and the joint planes are irregular enough to cause interlocking.
II (9 pts)	Undulating Macro rough but without the interlocking ability
III (27 pts)	Planar Macro smooth and micro rough joint surfaces. Friction is derived strictly from the roughness of the rock surface.

IV (81 pts) Clay Infilling, or Slickensides Low friction materials separate the rock surfaces, negating any micro or macro roughness of the joint surfaces. Slickensided joints also have a lower friction angle, and belong in this category.

Case 2: Erodible Rocks

This case is used for slopes where differential erosion or oversteepening is the dominant condition that leads to a rockfall. This group includes layered rock containing erodible units that undermine more durable rock, rock/soil slopes that weather allowing materials to fall as the soil matrix is eroded, talus slopes, etc.

The structural condition analysis is based on the following benchmarks:

I (3 pts) Few Differential Erosion Features Minor differential erosion features that are not distributed throughout the slope.

II (9 pts) Occasional Erosion Features Minor differential erosion features that are widely distributed throughout the slope.

III (27 pts) Many Erosion Features Differential erosion features are large and numerous throughout the slope.

IV (81 pts) Major Erosion Features Severe cases, dangerous erosion-created overhangs, or significantly oversteepened soil/rock slopes or talus slopes.

The second category for erodible rocks is to estimate the rate of erosion in the slope. This includes the effect of both natural and manmade agents of erosion. This assessment should include consideration of the frequency of past rockfall events, amount of materials released and the size of rocks, blocks, or units being exposed in the slope. The degree of differential erosion can be related to the following benchmarks:

I (3 pts) Small Difference Erosion features take many years to develop. Slopes that are near equilibrium with their environment are covered by this category.

II (9 pts) Moderate Difference The difference in erosion rates allows erosion features to develop over a period of a few years.

III (27 pts) Large Difference The difference in erosion rates allows noticeable changes in the slope to develop annually.

IV (81 pts) Extreme Difference The difference in erosion rates allows rapid and continuous development of erosion features.

8. ***Block Size or Volume Per Event.*** Larger events obstruct more of the roadway and increase the possibility of accidents. The rater chooses block size or volume of rockfall, depending on the maintenance history. For new or recent cuts, the rater makes a judgment decision on the more likely type of rockfall.

The system benchmarks, Figure 23.2(g), are:

I (3 pts) 1 foot size block or 3 cu. yd. of materials.

II (9 pts) 2 foot size block or 6 cu. yd. of materials.

III (27 pts) 3 foot size block or 9 cu. yd. of materials.

IV (81 pts) 4 foot size block or 12 cu. yd. of materials.

The highest of the two scores is used in the rating.

9. ***Climate and Water on Slope.*** This factor takes account of weather conditions to rockfalls. The benchmarks, Figure 23.2(f), are:

I (3 pts) Low to moderate precipitation; no freezing; no water on slope.

II (9 pts) Moderate precipitation or short freezing periods or intermittent water on slope.

III (27 pts) High precipitation or long freezing periods or continual water on slope.

IV (81 pts) High precipitation and long freezing periods, or (ii) continual water on slope and long freezing periods.

High precipitation: More than 50 inches per year.
Low precipitation: Less than 20 inches per year.
Information on length of freezing periods can be obtained from the National Oceanic and Atmospheric Administration (NOAA) climate data in the United States. State agencies can provide data on precipitation.

10. ***Rockfall History.*** This is a direct measure of past rockfall experience and should be a prediction of future experience if no remediation is undertaken. It should reflect the results from the individual rockfall factors 1 through 9.

The benchmarks, Figure 23.2(h), are:

I (3 pts) Few Falls Rockfalls occur only a few times a year (or less), or only during severe storms. This category is also used if no rockfall history data is available.

II (9 pts) Occasional Falls Rockfall occurs regularly. Rockfall can be expected several times per year and during most storms.

III (27 pts) Many Falls Typically, rockfall occurs frequently during a certain season, such as the winter or spring wet period, or the winter freeze/thaw, etc. This category is for sites where frequent rockfalls occur during a certain season but are not a significant problem during the rest of the year. This category may also be used where severe rockfall events have occurred.

IV (81 pts) Constant Falls Rockfalls occur frequently throughout the year; also for sites where severe rockfall events are common.

Implementation of the Rating System

After completing the rockfall hazard rating, a conceptual construction cost estimate is made for remedial work for the more dangerous rock cuts (usually Category A slopes from the Preliminary Rating). The cost divided by the RHRS points can be used for comparison with other slopes for planning possible remediation of the site. In broad terms, the higher the ratio the more expensive is the remedial measure relative to the perceived risk of dangerous rockfall.

The developers of the RHRS suggest several ways in which the results can be implemented for rockfall hazard reduction:

1. *Total Score.* The most hazardous sites are at the top of this list.
2. *Ratio.* Projects are selected based on the construction cost/points ratio. The expectation with this selection method is that it will provide the greatest systemwide hazard reduction for the available money.
3. *Similar Remedial Designs.* Rockfall sites requiring similar construction features can be grouped into a single project. An example would be wire mesh screening.
4. *Proximity.* By grouping rockfall remediation into sections of roadway in close proximity to each other, a larger contract can be let with some overall savings. An example would be to select a 20-mile section of road and complete all A-rated sites at one time.

These four approaches can be combined in some fashion with treatment of some high cost sites and a broader group of other sites as an optimal solution to the use of available funds. Finally, it is essential that the RHRS, once established, be continually reviewed and adjusted as new experience occurs. The suggested review period is annually.

In summary, the Rockfall Hazard Rating System is a practical way of quantifying rockfall risk. It combines the scientific approach of observing the factors that lead to rockfalls with the experience of the maintenance personnel. The use of an exponential increase in points with increasing risk ensures that the more hazardous sites are differentiated from the mass of collected data. The system could be modified, as needed, to fit practices and geologic conditions in other parts of the world.

An example of the rating of an erosion slope (which failed) is given in Case History 12.

PART C Selected Case Histories

Part C describes 12 case histories of landslides that have been referenced earlier in the main text. These case histories make key points related to landslide studies; the key points are listed at the start of each case history. The earlier chapters of the book also contain case histories. These usually illustrate a specific issue or remedial technique.

1 Washington Park Reservoirs Slide

Figure CH-1.1 Washington Park Reservoirs Slide. The two reservoirs are at the lower left, and the outline of the 1700-foot long landslide is shown by the broken white line.

Key Points

- Reactivation of ancient landslide terrain
- Slide debris composition
- Drainage tunnel remediation
- Long-term monitoring
- Seismic-induced ground movement

Summary

When Portland, Oregon, developed two water supply reservoirs on the west side of the city, excavations were made into a slope of ancient landslide terrain. The ancient landslide reactivated, creating a large-to-very large landslide. Subsequently, extensive drainage tunnels installed near bedrock slowed the rate of movement to creep levels. Surface measurements of the slope movements have been taken for more than 100 years. The case history describes (i) the heterogeneous nature of ancient landslide terrain (based on the numerous vertical shafts dug from the surface to bedrock); (ii) observations of the slip surface; and (iii) intermittent penetration into waterbearing seams within the landslide debris. Of practical significance to similar sites, it can be noted that an excavation of only 3 percent of the landslide volume was sufficient to cause relatively high rates of movement in the previously marginally stable clay slope.

Background Information

In the 1890s, the city of Portland constructed two reservoirs on the west slopes of the city (Figures CH-1.1, CH-1.2). Reservoir sites were in a narrow ravine and required excavation of about 100,000 cu. yds., deepening the ravine to create the reservoirs. Unbeknown to the city engineers at that time, the long moderate slope above the ravine was a deep deposit of ancient landslide terrain.

The history of this landslide, and the efforts to stabilize it, has been chronicled in great detail by Clarke (1904). His technical paper, 91 pages long, must surely be one of the longest technical papers ever published. It describes the ground conditions within each of 22 shafts dug by hand through the deep landslide, the tunnels built to lower groundwater levels, and the court case brought by upslope landowners seeking compensation for damages to their properties from landslide movements.

Ancient Landslide Reactivation

In 1887, the slope above the reservoirs site was subdivided and laid out for residential development. A surveyor closed on the original stakes 6 years later, indicating that the slope was essentially stable between 1887 and 1893.

In the winter of 1891–92, the developers built a cable railway to link the base of the slope to the proposed residential

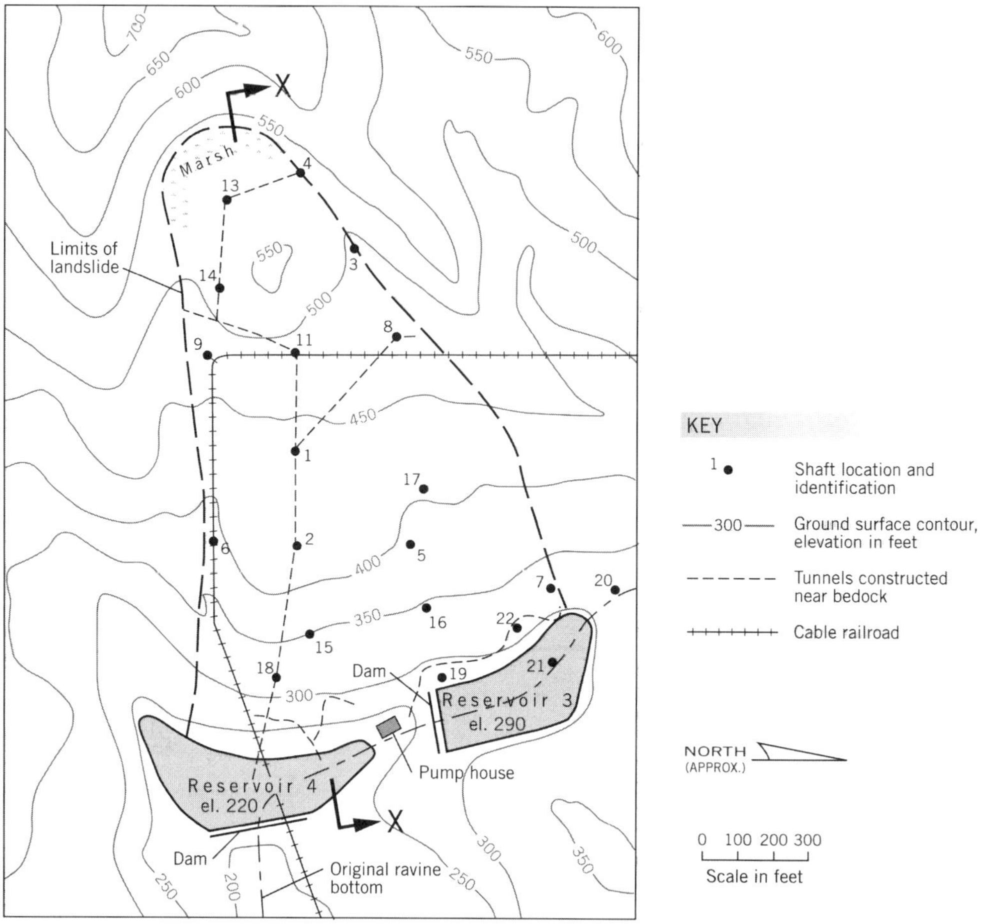

Figure CH-1.2 Washington Park Reservoirs Slide. Site plan, 1893–1900 period (after Clarke, 1904).

development (Figure CH-1.2). This required a sidehill embankment up to 25 feet high and two timber trestles, one of which crossed the site of Reservoir 4. Several other minor earthworks were needed for the railway; some local instabilities occurred. The railway was used for two summers but was abandoned when the reservoirs were built.

Construction of the two reservoirs began in October 1893 and was completed in September 1894. In August 1894, a crack 300 feet long was observed in the west wall of Reservoir 4 with outward movement of about ½ inch per day for a short time. A concrete wall was built to retain the slippage and the reservoir was filled on December 17, 1894. At Reservoir 3, higher up the ravine, the concrete lining in the base of the reservoir started to bulge after the excavation was completed. A concrete buttress and drainage tunnel were installed to correct what was thought to be a local condition.

New movements occurred during the winter of 1894–95 at both reservoirs. In 1895, the engineers found the head of the landslide in a marsh 1,700 feet upslope of the reservoirs. The graben feature was semicircular, 300 feet long, and 30–60 feet wide with about 20 feet depth of peat (Figure CH-1.2).

In 1895 and 1896, surveyors set out 29 survey lines across the slide and measured movements at 1 month intervals. The margins of the slide were easily defined, showing that the main body of the slide increased from 400 feet wide at the top to about 1,100 feet wide at the base, the latter bracketing the two reservoirs. Between December 31, 1894, and October 11, 1897 (2¾ years), the total downhill movement at Reservoir 4 was 3.24 feet. At Reservoir 3, where movement is partially buttressed by the masonry gravity dam (Figure CH-1.3) built across the ravine, the measured lateral movement was 1.69 feet.

It is apparent from the ground contours that substantial landslide movements had occurred historically long before any manmade construction. The contours change from concave in the stable ground above the top of the slide to convex on the slide (for example, compare the elev. 550 and elev. 500 contours on Figure CH-1.2).

LANDSLIDE DEBRIS

The landslide was investigated by 22 deep shafts and 9 borings. Some of the borings and shafts were at the same locations. The shafts were each 3.5 feet square, dug by hand, and dewatered by up to 4 pumps working together. On Figure CH-1.2, the shafts are shown but the boring locations are omitted. In 16 shafts, the base of the slide was reached and ranged from 48 to 111 feet below the ground surface (average

Figure CH-1.3 Washington Park Reservoirs Slide. The masonry dam at Reservoir 3 acts as a partial buttress to the landslide and has locally reduced downslope movement. The dam itself has suffered only minor distress.

78 feet). Contours of the slip surface are shown by broken lines on the plan, Figure CH-1.4. The narrower spacing of countours near the bottom of the slide indicates that the shear surface is more steeply inclined at the bottom than at the top.

This effect can be seen on the cross-section X-X taken near the center of the slide, Figure CH-1.5. The surface of the basalt changes its dip near the mid-length. On the upper half of the slide, clay soils are predominant and groundwater levels are near the ground surface. The lower half of the slide generally has more permeable soil (more broken rock) and groundwater levels are lower.

The graphic descriptions given by Clarke (1904) of the soils encountered in the shafts show the high degree of disturbance and intermixing of soils in the landslide debris. In Shaft 1 near the center of the slide, for example, the 75-foot depth of landslide soils included: clay; clay with small pieces of rotten

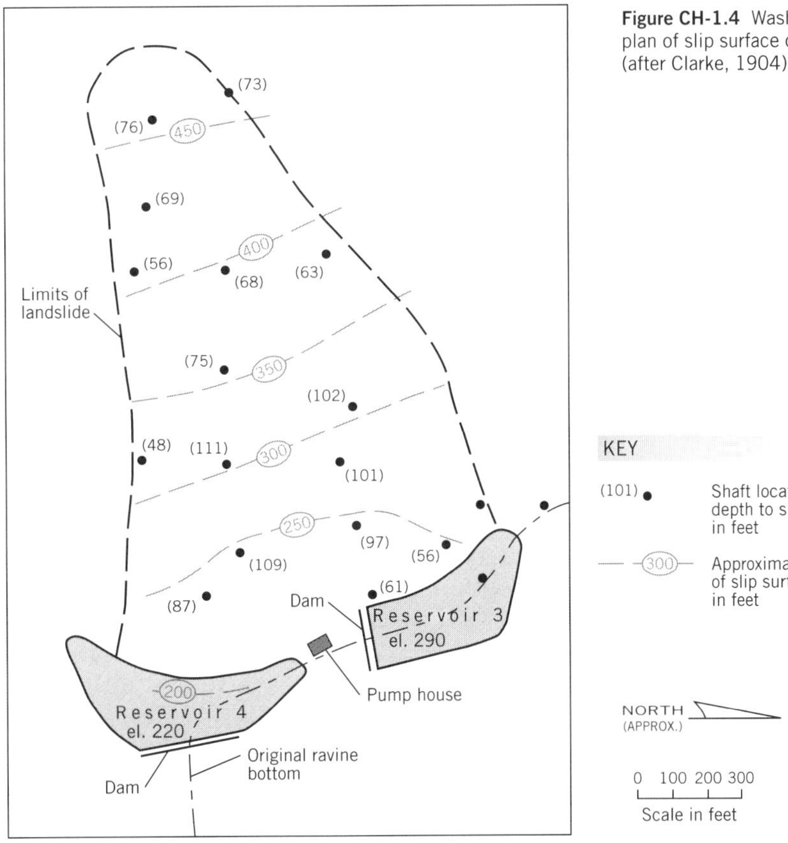

Figure CH-1.4 Washington Park Reservoirs Slide. Site plan of slip surface contours and depth of overburden (after Clarke, 1904).

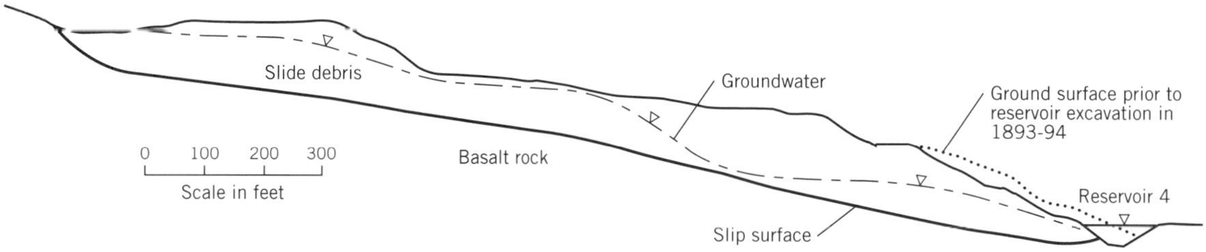

Figure CH-1.5 Washington Park Reservoirs Slide. Section X-X through center of slide (after Clarke, 1904).

rock; broken rock, clay, and gravel; sandy clay; blue clay with rock fragments; small fragments of wood; foul air; very black clay, tough and dry; clay mingled with broken rock and rounded gravel; heavy groundwater flows after rain. Other shafts encountered sand, highly fragmented hard basalt, water-bearing sand and loose rock; seamy rock, etc. However, the predominant soil was clay with more local (but sometimes extensive) pockets of broken rock.

At the slip surface, movement was invariably occurring within a seam of plastic clay, in several cases only a few inches thick. To quote Clarke (1904): the "line of cleavage" had "bright shining surfaces" (presumably slickensides). Where observed, the discrete shear zone varied from a few inches up to a maximum of about 18 inches thick.

These ground conditions reflect the heavy intermixing of soils and rocks within the slide area. It is a common occurrence in ancient landslide terrain, the most surprising aspects being (i) the degree of intermixing and (ii) the relatively high stiffness of clay debris at this and other sites.

Free groundwater varied considerably at the shafts. For example, 4 million gallons were pumped out of the 75-foot deep Shaft 1 during a 5½ month period from August to January, and the original water level was regained within one year after pumping was stopped. Other shafts were dry. Tapping into these porous zones via tunnels or horizontal drains can stabilize a landslide by eliminating the recharge of water into the slide debris and thus reducing the pressure head.

Two other items of interest can be noted from the studies and construction work. First, the excavation for Reservoir 4 passed through about 30 feet of landslide debris below the streambed. An older stream channel had apparently been overridden by the ancient landslide, moving it toward the east. The reservoir excavation took away this support, causing the reactivated movement to begin in August 1894 during construction. Second, a boring into the slide debris below Reservoir 3 encountered a bone fragment identified as that of a camel's molar from the Pleistocene era. The bone was embedded within the ancient landslide, at depths of 41 feet below ground and 17 feet above bedrock. It suggests that the ancient landslide occurred during the Pleistocene era.

The shafts were excavated between July 1897 and January 1899. At Shaft 1, where most groundwater was encountered, temporary dewatering during construction locally depressed the groundwater by 51 feet and the landslide stopped. When groundwater recovered over the following year, increased movements occurred during the winters of 1898–99 and 1899–1900. Further action had to be taken.

Drainage Tunnel Remediation

In June 1900, an ambitious underdrainage project was drawn up, requiring construction of drainage tunnels along the slip surface near bedrock. Work began immediately, starting near the center of the slide to the west of Reservoir 4. The tunnel section was 3.5 feet wide (on average) with a clear height of 5.3 feet. It used 6-inch by 8-inch timbers at 3-foot spacings with 2-inch by 8-inch timber lagging for support. The tunnel passed through the middle of the slide and branched in several directions near the head to intercept groundwater entering the slide area (Figure CH-1.2).

The contractor used small dump cars that were lowered down the vertical shafts. In the tunnel, cars with 9-inch diameter wheels traveled on an 18-inch gauge track. When filled with slide debris, they were pulled by hand to the foot of the shaft, where they were hoisted to the surface and dumped, then returned below.

Comparatively small amounts of water were encountered until they reached Shaft 1. There, a flow of 1.8 million gallons occurred over 20 days, but the rate of flow gradually diminished. Shaft 11 also had flows of around 165,000 gpd (115 gpm). Drainage work was completed in September 1901. Subsequently, the combined flow from all tunnels has ranged from 10,000 gpd (7 gpm) in summer to 50,000 gpd (35 gpm) in winter.

Since the major remedial works described by Clarke (1904), some additional tunnels were added in 1904–06 without encountering much groundwater (Clarke, 1918). In 1925–26, tunnels damaged by landslide movements were repaired or replaced Extensive surface drains were added. In 1928–29, most of the tunnels were backfilled with coarse gravel to prevent collapse of the deteriorating timber supports. There have been no significant changes since that date except that 9 inclinometers (Figure CH-1.6) were installed in 1987 to provide more accurate measurements of lateral movements.

Surface Monitoring

In 1895, additional survey lines were set up along and across the entire landslide in a grid of 100 feet spacing in the direction of slope and 50 feet spacing across the slope (260 survey points). Monthly readings were taken, and it was apparent that the tunnels were very effective in draining the pockets of water-bearing materials and bringing stability to the landslide mass. Three of the lines across the slope are shown on Figure CH-1.6.

This landslide has been monitored for more than 100 years. Many of the survey hubs are within the public park and have been damaged or destroyed.

Observations for two lines traversing laterally across the slide are shown on Figure CH-1.7. The locations of the survey lines, Line 80 near the top of the slide and Line 20 near the

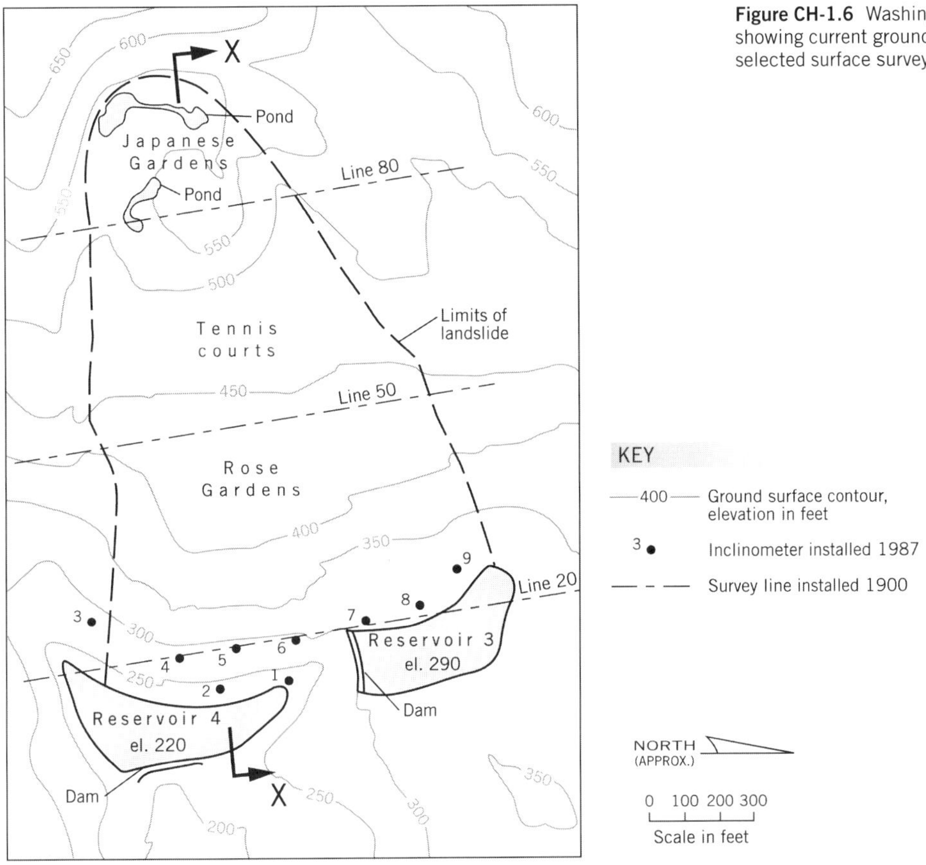

Figure CH-1.6 Washington Park Reservoirs Slide. Site plan, showing current ground surface contours, inclinometers, and selected surface survey lines.

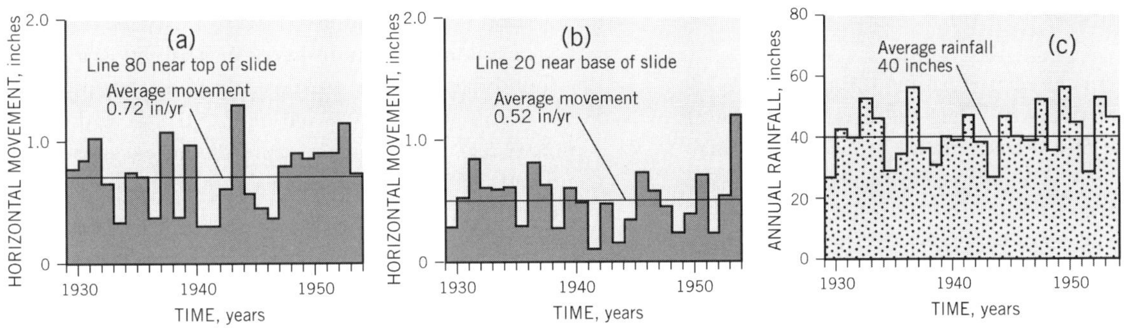

For location of lines, see Figure CH-1.8

(a) (b) Observed average annual horizontal movements of all points within the slide boundaries on the stated line

Figure CH-1.7 Washington Park Reservoirs Slide. Observations over a 25-year period, 1930–1954. Annual horizontal movements on

(a) line 80
(b) line 20
(c) annual rainfall

base of the slide, are shown on the plan, Figure CH-1.6. The 25-year time period of 1930–54 occurred after all remedial works had been completed. The average annual rainfall of 40 inches is approximately the long-term average for Portland.

The following comments are pertinent:

- Movements reduced to creep rates of about 0.6 inch/year.
- The survey line near the top of the landslide moved slightly faster than the line near the bottom.
- There is no discernible trend for the landslide to speed up or slow down over this time period.

The annual rates of movement probably are controlled by winter groundwater buildup plus major storm events (see Chapter 5, Section 5.1). Thus, there is no correlation between annual rainfall and ground movements. The random nature of the movements over so many years indicates that the slide, in a broad context, is moving at uniform average rates reflecting a residual condition in the discrete shear zone of clay. Since about 1975, the average rate of creep movement appears to have slowed down for unknown reasons. However, there are many facilities on the slide including residences and two tourist attractions: the Rose Gardens and Japanese Gardens. Their use of water may have been reduced in recent years.

DAMAGE TO STRUCTURES

The walls and floor of the pumphouse, located between the two reservoirs, have been extensively cracked by the slope movements from above. The pumphouse is at the base of the landslide where the movement meets stable ground on the opposite side of the original ravine (see Figure CH-1.2). Ground distress can be expected at such locations.

The masonry dam at Reservoir 3 (Figure CH-1.3) acts as a partial buttress to the landslide, as mentioned previously. The gravity dam has suffered surprisingly little distress. There are several mostly horizontal cracks, generally within the mortar joints on the west side (which abuts the slide); a minor amount of seepage is coming through the cracks.

STABILITY ANALYSIS

A stability analysis for the section passing through the middle of the slide (Figure CH-1.5), using the groundwater levels observed by Clarke (1904) and ground surface before reservoir excavation, gave a back-analyzed average strength on the slip surface of $c'=0$, $\phi'=13.3°$ for $F = 1.00$. This average value is a composite of the slickensided clay shear zones, the uneven slip surface, and probably some local seams of broken rock.

When the reservoir excavation is included, the *static* factor of safety F drops to 0.95, all other conditions remaining the same as in the back analysis. This small but significant decrease in F is due to the loss of resistance being greater than the decrease in driving force.

The estimated volume of excavation was 100,000 cu. yd. This is about 3% of the total landslide mass of 3,400,000 cu. yd. In the lawsuit brought by the uphill property owners against the city, this small percentage change was used by expert witnesses representing the city to argue that such a small change could not have caused the large observed movements. These engineers stated that the slope was moving prior to the reservoir excavations, citing the construction of the cable railway two years earlier when several local slips occurred during winter construction. The plaintiff's expert witnesses, who had less time to prepare and were not given access to the city's information, argued that the deep-seated slide had been quiescent and was reactivated by the excavation. As proof, they referred to the surveys of 1887 and 1893 on the upslope development that had closed on the original stakes, indicating no deep-seated movement in the six years before the reservoir excavation began. Their conclusion, as stated by one of the engineers: "When dealing with forces not far from the point of exact equilibrium, grave consequences may easily follow seemingly insignificant causes."

Since the Washington Park Reservoirs Slide, similar arguments have been used in lawsuits involving the reactivation of ancient landslide terrain. These cases have confirmed that relatively small changes in the slope configuration can cause accelerated ground movements in ancient landslide terrain (see also Case Histories 5 and 10). When a geotechnical engineer is faced with this situation, and accepting the difficulties of measuring the shear strength within the discrete shear zone of an ancient landslide, the recommended course of action is to assume that the slope is either marginally stable or even intermittently active in the wet season, and make appropriate design choices to leave the state of equilibrium unchanged (as a minimum) or preferably improved over the preexisting conditions (see Chapter 2, Section 2.14).

The directions of the judge in the Washington Park Reservoir lawsuit are of interest. In his charge to the jury the trial judge said, in part:

> If this land was sliding in recent years before the excavation of the reservoirs, and would have been sliding since the excavation of the reservoirs, it would make no difference in plaintiff's claim or defendant's liability if it should be found that the slide was not continuous but intermittent. For instance, if it should be found that in the years immediately preceding the time of the excavation the land had slid in the winter and stood still in the summer, or had slid one year and had stood still the next, the city would not be liable though the land may have been at complete rest at the time of the excavation of the reservoirs and have started to move immediately thereafter; provided, that this movement was only a continuation of the intermittent movement which had existed before and which would have existed thereafter whether the reservoirs were constructed or not.

Irrespective of the legal opinion on liability, the information in the Clarke (1904) technical paper suggests that the Washington Park slope was a dormant or possibly intermittently active ancient landslide before construction of the reservoirs. The excavations for the two reservoirs reduced the slope stability and caused the much increased activity in the three winters 1894–95, 1895–96, and 1896–97. The modest groundwater lowering achieved by the pumping at Shaft 1, and the permanent drainage tunnels built in 1900–01, have since slowed the rate of movement to a creep rate of less than 1 inch per year.

SEISMIC-INDUCED GROUND MOVEMENT

The effect of an earthquake on a preexisting landslide at marginal stability is an issue that is being studied with increasing frequency in earthquake-prone areas. At the time of writing, the analytical methods are of uncertain reliability, but empirical approaches, based on actual observations, are available (see Chapter 12, Section 12.3).

In 1987, nine inclinometers were installed at the Washington Park site to measure the rates of movement at the discrete shear zone more accurately than is possible from survey hubs. These inclinometers have been read at three-month intervals for many years. All are located on Water Bureau property near the base of the landslide where they are protected from public disturbance and vandalism.

On March 25, 1993, a crustal earthquake of magnitude 5.6 occurred with its epicenter about 35 miles south of the site. The focal depth was 10 miles. Significant levels of shaking were felt by residents of Portland, and the isoseismal map produced afterward put the intensity in Portland at Level V on the Modified Mercalli scale.

The inclinometer locations are shown on the plan, Figure CH-1.6. The effect of this earthquake, the largest ever felt in the Portland metropolitan area, was to cause an immediate jump in the rate of movement on some but not all the inclinometers. Two examples are given in Figure CH-1.8.

It should be noted that the rate of creep is low which makes it difficult to accurately measure small movements from periodic readings. Nevertheless, the typical jump movements shown on Figure CH-1.8 of 0.05 inch and 0.17 inch show that the landslide responded to the earthquake. Inclinometer SI-7 is close to the masonry dam at Reservoir 3 that partly buttresses the landslide and probably accounts for the slower creep rate.

COSTS

Clarke (1904) provides details of construction costs. The 2,507 lin. ft. of tunnel cost $14,161.14, at an average cost of $5.65 per lin. ft. This cost is for labor and materials but excludes survey and engineering superintendence. This work, performed in 1900–01, occurred before the first ENR Construction Cost Index. The first ENRCCI was 97 in 1908, and it rose to 6225 in mid-2000. Assuming a value of 90 for the years 1900–01, the multiplier is 69. Thus, the cost of the tunnel construction is approximately equivalent to a cost of about $1 million in the year 2000 ($400 per lin. ft.).

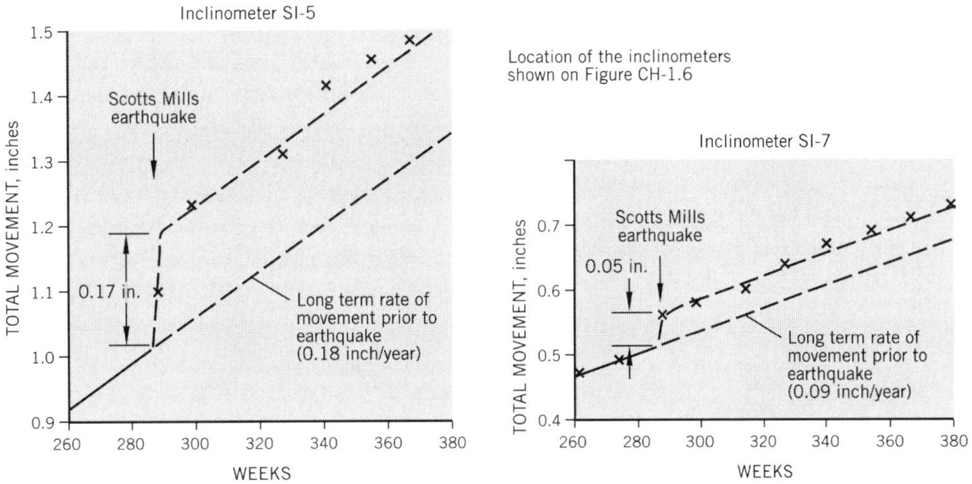

Figure CH-1.8 Washington Park Reservoirs Slide. Apparent movements of the landslide in response to the Scott Mills earthquake of 1993.

2 Beaver Shoreline Erosion

Figure CH-2.1 Beaver Shoreline Erosion. Aerial view of the erosion slope failure. Columbia River in the foreground.

Key Points

- River erosion
- Riverbank design and construction issues
- Filter design

Summary

The Beaver generating plant is built on the flood plain of the Columbia River. The slope originally was protected by a layer of riprap placed directly over sand fill and river alluvium (primarily silt). A section of riprap washed out twice in 1996.

The case history describes the repair of the small erosional slope failure using graded filters behind riprap. Similar designs can be used at other sites subject to erosion from river currents and waves.

Background Information

The Portland General Electric (PGE) oil- and gas-fired Beaver generating plant, near Clatskanie, Oregon, is built on the flood plain of the Columbia River about 40 miles upstream from the mouth. The site was protected from floods by a perimeter dike, and the area was later infilled with dredged sand. It has experienced river erosion, and parts of the shoreline are protected by riprap.

In early 1996, at the time of major storms and flooding in western Oregon, a section of riprap failed near the confluence of the Columbia River main channel and Bradbury Slough. It was quickly rebuilt, but failed again two months later. Subsequently, the eroded area continued to retrogress around the initial failure (Figure CH-2.1).

The riprap had been placed directly onto sand fill and river alluvium. The native soils have a typical profile of 2 to 3 feet of uniform fine to medium sand fill overlying medium stiff, slightly cohesive silt with thin sand lenses. The silt is river alluvium that has been subjected to desiccation hardening near the ground surface.

The erosion bowl (Figure CH-2.1) was 143 feet wide at the top of the river bank and extended back up to 38 feet from the original shoreline. Near the east side of the failure area, an erosion gully had cut through the sand fill. It carried runoff water during heavy rainfall (Figure CH-2.2). Around the perimeter of the erosion bowl, near-vertical sand faces had periodically collapsed onto the shelf of more erosion-resistant, slightly claycy silt below (Figure CH-2.3).

Causation

The erosion bowl developed because the sandy soils were pulled through the voids of the riprap by streamflow, undermining the riprap protection. Contributing factors include flood velocities, waves (3 to 4 feet high) made by passing ships, and runoff from the upslope gully.

Remediation

The remedial treatment was to reinstate the riverbank to its approximate pre-slide location by infilling the erosion bowl

Figure CH-2.2 Beaver Shoreline Erosion. Site plans:
(a) erosion bowl before remediation
(b) reconstructed slope

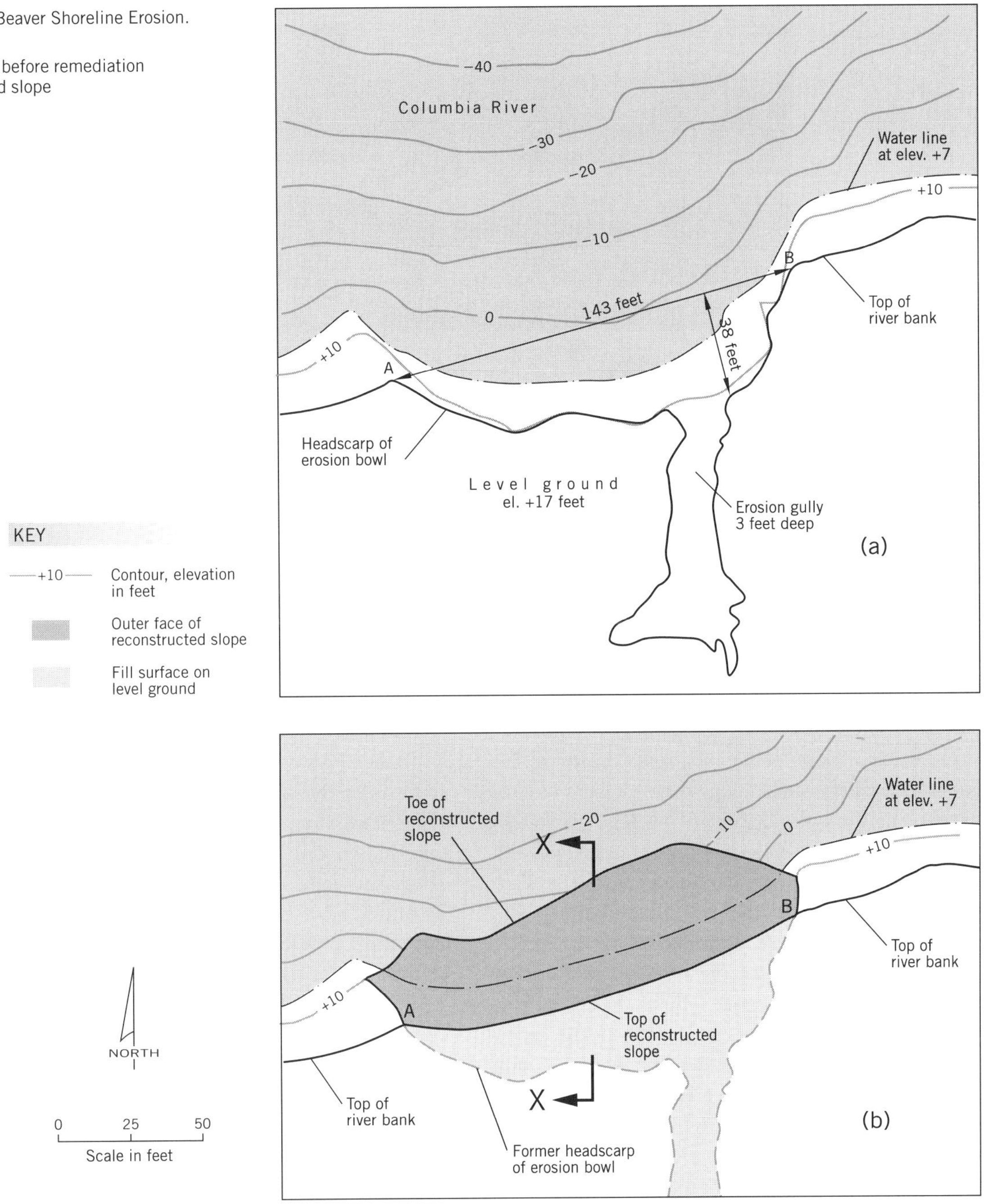

with free-draining rockfill behind an outer protection of riprap. The rockfill had to meet filter requirements as an intermediary between the riprap and fine-grained native soils. The free-draining rockfill prevents buildup of pore water pressures behind the riprap after heavy rainfall and at low tide.

This type of remediation is applicable to springs, seepage lines intercepted by cuts, uplift pressures, etc. The key requirement is to ensure that groundwater pressures are relieved behind the outer slope, using soils that meet filter requirements (Chapter 11). The most common mistake is to use rockfill that is too coarse, allowing the fines of the protected soil to migrate and ultimately fail the slope by internal erosion and undermining.

DESIGN

An underwater survey (bathymetry) was undertaken, which showed a somewhat bowl-shaped erosion surface offshore (Figure CH-2.2a). In some areas, the river bank below the waterline was as steep as 2:1 (horizontal: vertical).

Figure CH-2.3 Beaver Shoreline Erosion. Erosion bowl with near-vertical faces of sand fill and "beach" of silt alluvium with thin covering of sand.

The initial preference of the design team was to provide a reconstructed exterior slope of 2:1 (horizontal : vertical), but cross-sections showed that such a slope with a straight line top of slope from point A to point B (Figure CH-2.2b) would have required a large quantity of fill extending a considerable distance into the river. Instead, the top of the reconstructed slope was bowed inward (concave to the river), and the outer slope of the riprap was increased to 1½ : 1 (horizontal: vertical).

A typical design section near the center of the erosion bowl (Section X–X, Figure CH-2.4) shows the reconstruction of the slope failure. Environmental concerns (fish movements up and down the river) permitted a limited time frame of November 1 to February 28 for construction. Fortunately, this period coincides with relatively low water levels, and historical data indicated that the river, with tidal influences, could be expected to range from elevations +0.5 to +9.5 feet. Given this data and the need to minimize underwater construction, the bottom of the riprap was set at elev. −2 feet.

A bench was inserted into the relatively steep river bank to break up the linear contact between the rockfill and underlying native soils. This bench had a width of 10 feet at elev. −2 feet and required a temporary oversteeping to 1:1 of the native soil slope above it (Figure CH-2.4). Since this creates a potential safety hazard to the excavator working above the cut, the contract specified close sequence construction (Chapter 13) such that only a 20-foot length across the slide could be notched at one time. Each 20-foot long bench had to be backfilled before proceeding to the adjoining section of the repair.

The filter design required two filter layers to be used between the native soils and riprap. The fine filter layer forms a thin strip, 1 foot thick, covering the silt and fine sand fill. Of particular concern for protection was the sand fill; a particle size distribution is shown on Figure CH-2.5.

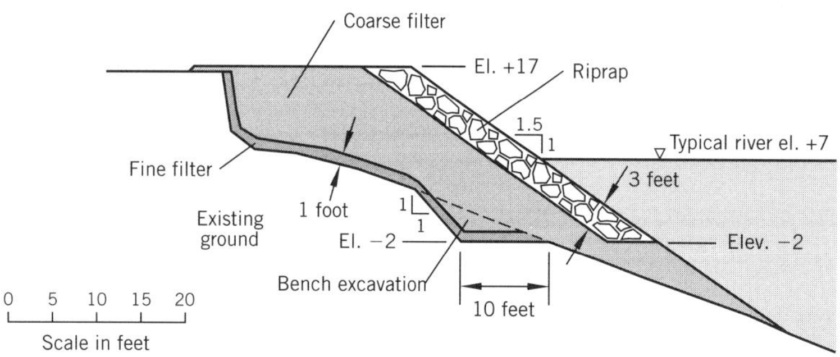

Figure CH-2.4 Beaver Shoreline Erosion. Section X-X showing the repair to the slope: riprap backed by filters.

The fine filter (¾ inch minus) was obtained from a hard rock quarry. It can be seen that it meets the filter requirements for retaining the sand fill, i.e., D_{15} of the filter is less than 5 times the d_{85} of the sand fill. The contract also required that the fines passing the No. 200 sieve of the fine filter should be less than 7% by weight; this requirement was also met.

The coarse filter gradation (8 inch minus) is estimated on Figure CH-2.5. It also meets filter requirements with respect to the fine filter and riprap.

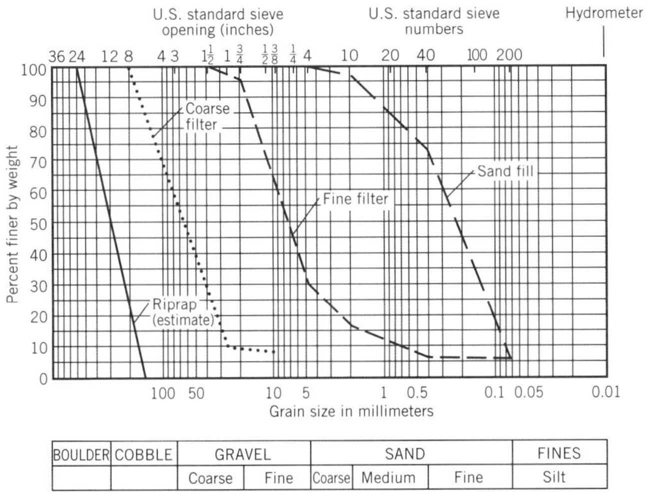

Figure CH-2.5 Beaver Shoreline Erosion. Gradations of soils used for repair of the eroded slope.

SAFETY

Safety was a major concern because of the marginal stability of the slope and the need to temporarily oversteepen the underwater slope for the bench. Several precautions were taken: (i) the key requirement was that the bench had to be cut in short segments and backfilled immediately (close sequence construction, already described); (ii) a D-7 dozer with winch was used as an anchor at the top of the slope to hold or slow down the trackhoe if the ground should start to move (Figure CH-2.6); (iii) the trackhoe was supported on timber spreader mats (Figure CH-2.6); (iv) a boat with outboard motor was kept at the site in case the operator fell or leapt into the water in an emergency; a lifebelt was provided at the water edge; (Figure CH-2.6); (v) all rockfill/riprap stockpiles had to be more than 50 feet away from the river edge. The contract was completed without incident.

CONSTRUCTION TECHNIQUE

The contractor's equipment was a Komatsu PC 650 trackhoe, Caterpillar 980C front end loader, D-7 dozer, and dump trucks. A project baseline was established close to the top of the riverbank, and parallel to the river, for survey control. First, the contractor infilled the erosion gully 80 feet long, 15 feet wide, and 3 feet deep, lining the sides and bottom with fine filter. A 8-inch diameter flexible plastic perforated pipe was installed along the entire length of the gully, the pipe being surrounded by fine filter. This drain was later extended as solid pipe through the main repair area to discharge surface runoff water into the river. The remainder of the gully was infilled with coarse filter.

Figure CH-2.6 Beaver Shoreline Erosion. Beach area infilled with coarse filter to elev. +9 feet. Headscarp cut to 1.5:1 with fine filter partially spread on it. Trackhoe on timber spreaders at east end of project. D-7 dozer with winch is above trackhoe. In the foreground is a boat and lifebelt for safety purposes.

An access road was built to the "beach" inside the main erosion bowl. After cleaning out trees and other debris, the near-vertical headscarp was laid back to 1½:1 (horizontal : vertical) to allow the fine filter layer to be placed on it at the approximate angle of repose. (*Note:* The design did not require slope flattening of the headscarp because a contractor could have set up a windrow at the base of the headscarp and pushed the fine filter up the slope face concurrent with coarse filter placement. This procedure is generally followed on steep slopes where slope flattening is not feasible.)

Next, the 1-foot layer of fine filter was installed on the beach at low water level and the whole beach area was raised to elev. +9 feet (i.e., high water) with coarse filter (Figure CH-2.6). The rockfill was placed in 12 inch lifts and compacted by the tracks of the trackhoe and the back of the trackhoe bucket.

Starting at the east end of the project, the contractor excavated the required 10-foot wide bench over a 20-foot length. The outer slope was backfilled with fine filter, coarse filter, and riprap per the design up to the elev. +9 feet level. After completing two segments, the trackhoe moved onto the completed platform and continued moving westward. This maneuver avoided the trackhoe being directly above the 1:1 temporary cut slope of the bench. After the work below water had been completed, the remainder of the construction above elev. +9 feet was performed. The final appearance of the repaired slope is shown on Figure CH-2.7.

On many projects in which fine filter is specified between rockfill and native soils, the contractor with the successful bid makes a request to substitute filter fabric in place of the fine filter layer. This option was considered in design, but the fine filter layer was selected at this site because a fabric is susceptible to damage during construction (which may not be detected) and provides a weaker shear strength on the steep cut slope. At this site with steep slopes, daily tides, and tight maneuvering space for construction equipment, a fine filter layer was considered preferable to filter fabric.

CONSTRUCTION COST

Quantities:	Excavation 600 cu. yd.; Fine filter 1,360 tons; coarse filter 3,465 tons; riprap 1,333 tons. Average haul distance: 8 miles.
Construction cost:	\$164,150 (1998); Cost/foot width of treatment: \$1,150
	ENR Construction Cost Index: 5850
Construction time:	3 weeks.

Figure CH-2.7 Beaver Shoreline Erosion. Finished project showing curved (in plan) outer surface of riprap.

3 Bonners Ferry Slide

Figure CH-3.1 Bonners Ferry Slide. Oblique aerial photograph of the major flow slide aftermath, showing rupture of the railroad embankment adjacent to the flood plain (in foreground) (Photo: David Kramer).

Key Points

- Artesian pressures
- Dewatering as a design requirement
- Retrogressive flow slide

Summary

Realignment of Highway 95 near Bonners Ferry, Idaho, required construction of a high embankment across a ravine. The ravine bottom had a covering of old landslide deposits: soft silts, sandy silt, and clay. A site investigation boring encountered an artesian pressure condition near the middle of the ravine. The contract required excavation of the soft, loose foundation soils before constructing the embankment. During excavation, a flow slide occurred on September 20, 1998, which was followed by a much larger flow slide on October 16–17 (Figure CH-3.1). The case history demonstrates the caution needed when an excavation is made where artesian conditions are present below a slope.

Background Information

U.S. Highway 95 formerly passed around a narrow ravine incised into glacial sediments near Bonners Ferry, Idaho. A road improvement project required the road to be realigned over a distance of 2.4 miles, including construction of an embankment fill up to 95 feet high across the ravine.

The site plan, Figure CH-3.2, shows the ravine at the road crossing. The embankment footprint is shown by broken lines. Also shown are original ground contours at 20-foot intervals.

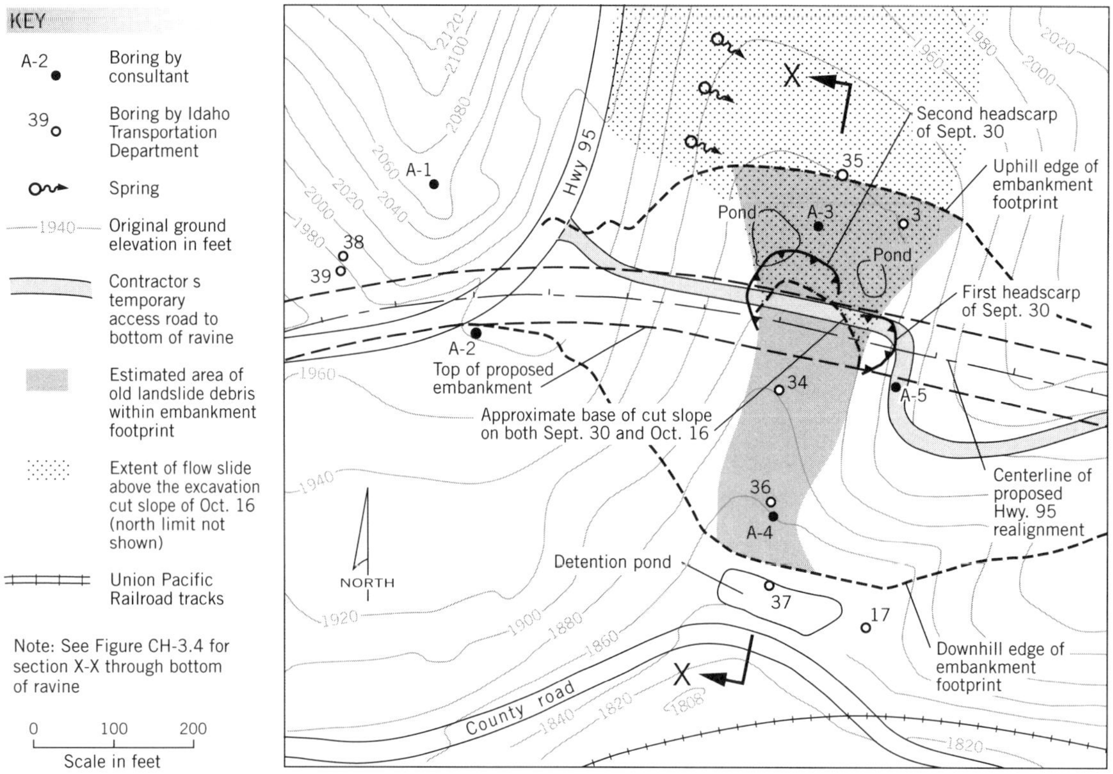

Figure CH-3.2 Bonners Ferry Slide. Site plan.

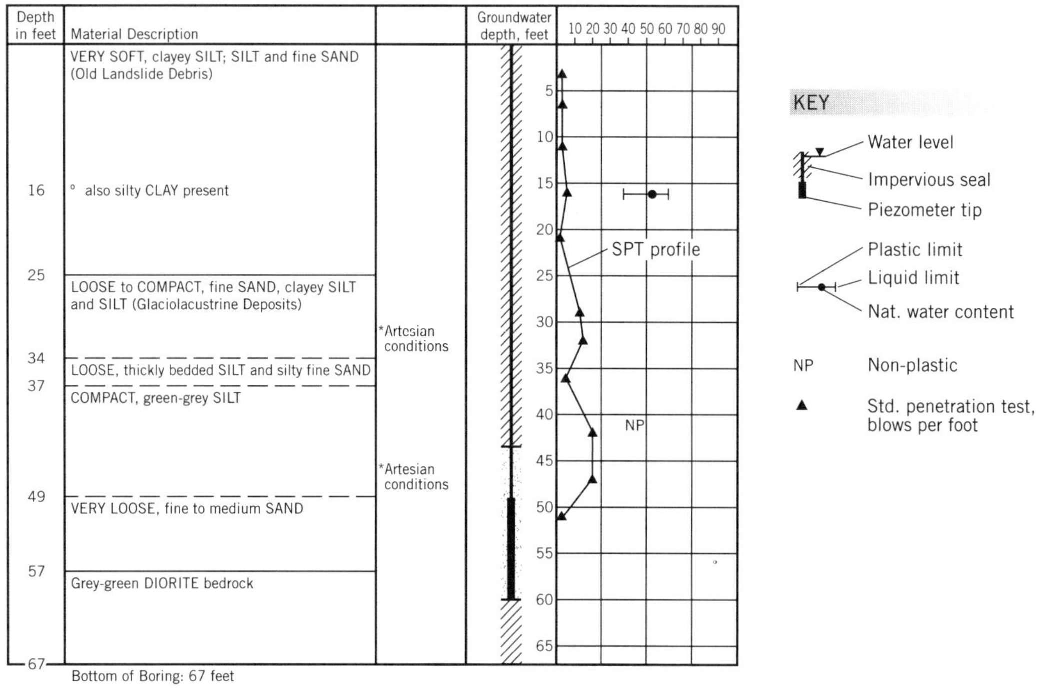

Figure CH-3.3 Bonners Ferry Slide. Summary log for boring A-3 (simplified from consultant's log).

Site investigations for the embankment were conducted by Idaho Transportation Department (ITD) assisted by an outside geotechnical consultant. It was evident that old landslide debris covered the valley floor. The soils on the valley floor were described as very soft and loose silt, fine sandy silt, and very silty clay. The boring log for hole A-3 is shown on Figure CH-3.3 (modified and abbreviated from the original). The maximum observed depth of the soft landslide debris was 25 feet in hole A-3. The estimated area of old landslide debris within the embankment footprint is shaded on Figure CH-3.2. Section X-X (Figure CH-3.4) is taken through the bottom of the ravine. The average slope is about 9° to the horizontal.

The old landslide debris was underlain by compact glaciolacustrine deposits of fine sand and clayey silt over most of the foundation; further downslope, very dense glacial till was below the landslide soils. There were insufficient borings to provide reliable contacts between these units, but both the glacio-lacustrine deposits and glacial till were denser and stronger than the landslide soils. Bedrock changed from igneous intrusives (upslope) to metamorphic rocks (downslope) in Section X-X, and this change may account for the irregular shape of the bedrock surface in the cross-section.

Artesian groundwater conditions were encountered in boring A-3 at 31 and 46 feet below the surface (Figure CH-3.3); the head was 9.6 feet above the ground surface in January 1997. There were also two ponds (Figure CH-3.2) that were fed year-round by natural springs. Other springs were observed near the base of the existing U.S. Highway 95. All the spring/pond data shown on Figure CH-3.2 are approximate and are based on photographs, anecdotes, etc.

DESIGN AND CONTRACT DOCUMENTS

The old slide deposits were considered dormant. The main concerns of the design team were differential settlements and seismically induced liquefaction if the soft materials were left in place.

The contract required all the slide debris in the bottom of the ravine to be excavated below the footprint of the embankment (50,000 cu. yds. for the shaded area, Figure CH-3.2). After excavating the slide debris, a 5-foot deep rockfill layer was to be placed over the excavated area. The maximum distance between the excavation cut face and the advancing edge of the rockfill was specified as 70 feet.

The contractor was warned that the slide debris was saturated and excavation would be needed below the water table. The contract also mentioned year-round springs at the site.

Special Provision SP-34 described Slide Debris Removal. It included: "Any dewatering necessary for the excavation operation shall be considered incidental to Slide Debris Removal."

LANDSLIDE EVENT

Construction began in August 1998 at the south end of the ravine and proceeded upslope. Initially, the track-mounted

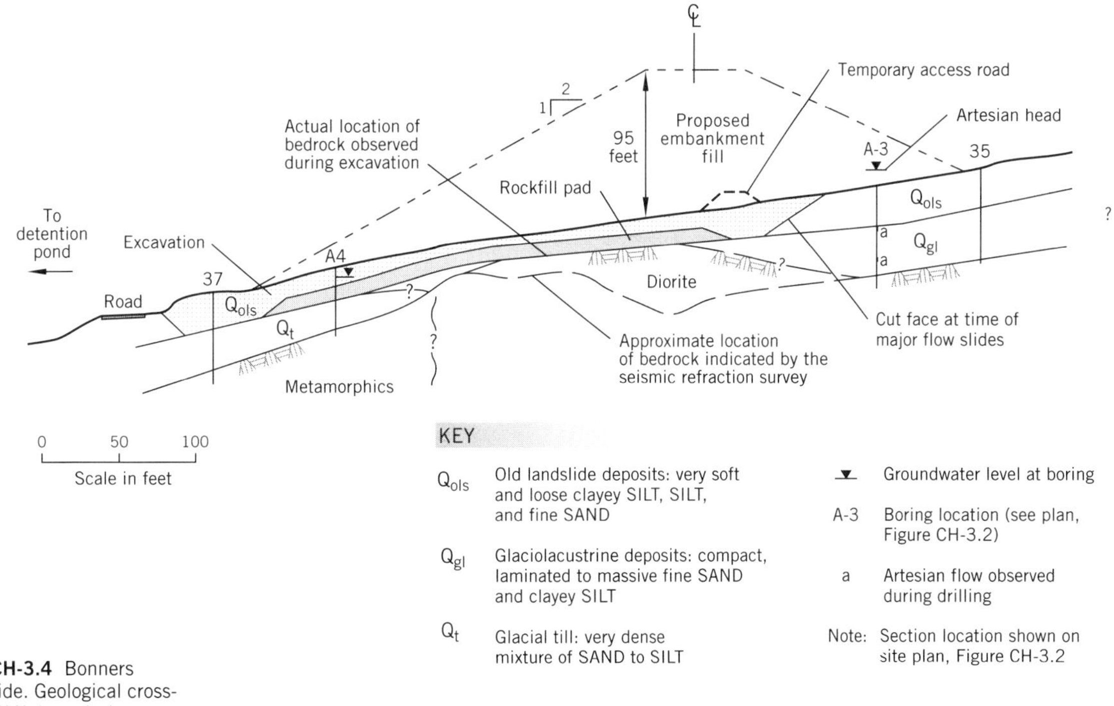

Figure CH-3.4 Bonners Ferry Slide. Geological cross-section X-X through the bottom of the ravine.

backhoe worked from the excavation floor, digging material from the toe of the cut face and loading directly into dump trucks. As the rockfill base advanced further into the excavation area, the backhoe occasionally was able to sit on the edge of the rockfill base as material was pushed off the slope toward the backhoe by a dozer. Seepage within the cut face caused the soft soils to slump, making it easier for the soil to be picked up.

On several occasions, fairly large collapses occurred due to the silt becoming liquid and flowing downslope toward the rockfill. A firmer crust rode on top of the flowing mud, eventually breaking up into the muddy mass. At times, a dozer would trim the open cut face to about 2 : 1 (horiz:vert), while being "anchored" by another dozer. Other techniques were used to channel the mud to the excavator and try to control the amount and rate of soil movement.

Diorite bedrock was encountered up to approximately the centerline, according to ITD field inspectors (see Figure CH-3.4). The surface was higher than was indicated by the seismic refraction survey line through the ravine bottom.

On September 30, a flow slide occurred in two "pulses" about 15 minutes apart. In the first flow, a dump truck was hit broadside and was pushed 100 feet downslope. The second flow pushed the truck into a detention pond (Figure CH-3.2) down the slope. Three other mud waves occurred on that day, filling the 22-foot deep detention pond and crossing the county road below.

The responses to this event were to: (i) install drainage to pick up the seepage from springs coming into the excavation from the base of the existing U.S. Highway 95 (Figure CH-3.2); (ii) build drains to the west pond; and (iii) raise the minimum height of rockfill from 5 feet to 8 feet. It took two weeks to clean up, add the extra rockfill, and return to the same position in the excavation (Figure CH-3.5).

On October 16, 1998, discussions began between the contractor and resident engineer about halting any further subgrade excavation on the north side of the embankment foundation. A small flow slide occurred at 3:00 p.m. At 5:50 p.m., major mudflows began, first filling up the ground behind the railroad embankment downslope. As the flow slide continued, it scoured through the railroad embankment, and mud moved onto the flood plain below (Figure CH-3.1). A front end loader, a dump truck and trailer, and a pickup truck were carried downslope from the site to the flood plain.

Numerous flow pulses continued until about 2:00 p.m. on the following day. Video films showed flowing soils, estimated to be up to 10 feet deep, coming down the ravine.

Figure CH-3.5 Bonners Ferry Slide. Extent of re-excavation into old landslide materials as of October 15, 1998, the day before the large flow slide. Rockfill layer is seen in lower left (Photo: Todd Williams, Idaho Transportation Department).

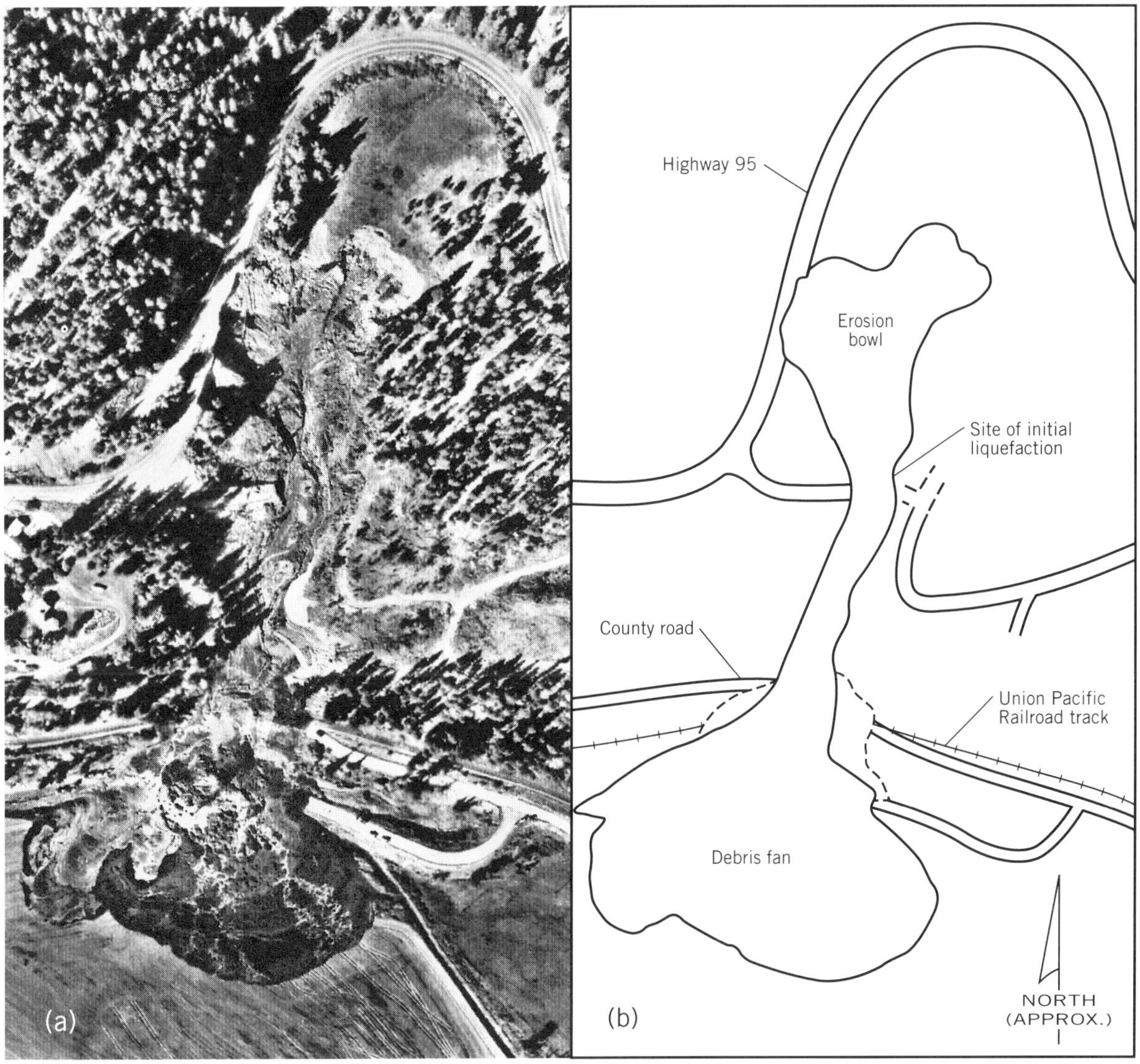

Figure CH-3.6 Bonners Ferry Slide.
(a) aerial photo of slide area (Photo: Idaho Transportation Department)
(b) same area as (a) showing the key features of the flow slide

When the movements stopped, the headscarp had moved 700 feet upslope from the position of the cut face prior to the flow slide (Figure CH-3.6). As seen on the same drawing, Figure CH-3.6(b), erosion also undermined the existing U.S. Highway 95, the headscarp extending behind the centerline of the road.

CAUSATION

The bottom of the ravine near the proposed highway centerline had a history of continuous springs. The two ponds, especially the west pond, had been built to provide a constant water supply for livestock. Boring A–3, in the same vicinity, had encountered almost 10 feet of flowing artesian head, the artesian strata being 31 and 46 feet below ground.

The temporary excavation to remove old slide debris was up to 25 feet deep. Therefore, the floor of the excavation was only about 6 feet above the upper artesian zone (Figure CH-3.3). The excavation probably caused a critical hydraulic gradient to develop between the base of the cut face and the artesian zone. The upwelling of groundwater loosened and liquefied the sand at the cut base to begin the flow slide. Once initiated, the loss of ground caused a chain reaction of: (i) erosion from flowing sand; (ii) undermining the slope face; and (iii) collapse of the face as the sand continued to liquefy. Thus, the landslide regressed upslope for many hours until flowing sand ceased.

PREFABRICATED VERTICAL DRAINS

A thick layer of granular fill was spread on the ravine floor up to the headscarp. In the footprint of the embankment, no additional excavation was made of old landslide debris, most of which had been scoured out by the landslide event. Instead, prefabricated vertical drains were installed in the foundation at about 7-foot centers to increase the rate of settlement during construction and reduce the risk of earthquake-induced liquefaction.

ECONOMIC EFFECTS OF THE FLOW SLIDE

The costs resulting from this landslide were substantial. The contractor was delayed, cleanup was required, and additional remediation was needed. The embankment failure on the existing Highway 95 had to be mitigated by shifting the highway alignment further into the hillside. In addition, residents and travelers were disrupted by a 2½-week closure of Highway 95 and the county road. During this period, large trucks had to make a 112-mile detour. Power was lost, schools were closed for several days, and the commerce of Bonners Ferry was adversely impacted. The Union Pacific Railroad had to rebuild their embankment, and there were consequential economic losses from the temporary closure of their tracks.

DEWATERING AS A DESIGN REQUIREMENT

The Bonners Ferry landslide of 1998 is a prime example of a slope failure occurring during temporary works for a project. It is fortunate that no construction workers or members of the public were killed or injured by the large flow slide.

It is a longstanding practice in civil engineering projects to make the contractor responsible for the construction and safety methods used to build a project. In this case, a very experienced contractor was following a risky method of excavation—allowing the springs in the cut face to cause a local failure that moved saturated soils toward the excavator. This rate of movement cannot be precisely controlled and there is evidence that some fairly significant flows and slumps occurred prior to the September 30 and October 16–17 mudflows. This method of excavation is a common practice in construction.

A preventative procedure would have been to install a line of deep wells or wellpoints across the old landslide deposits on the north and south sides of the excavation footprint. This would have allowed excavation to occur under drawndown groundwater conditions. The excavation/rockfill replacement could have taken place using closely sequenced construction methods (Chapter 13, Section 13.5).

Why was it not done? The author believes the main reason was the contract specification statement: "Any dewatering necessary for the excavation operation shall be considered incidental to Slide Debris Removal." Statements similar to this appear in many special provisions specifications for foundation excavations; therefore, the agency was using a common specification clause. The problem with the clause is that construction dewatering covers a wide range of costs. Since contractors have to keep their costs to a minimum in preparing a competitive bid, their bid usually allows for only minimal construction dewatering, such as sump pumping from within the excavation.

Claims for problems arising from inadequate site dewatering are fairly common on earthworks. One method of avoiding this type of claim is to specify that the contractor shall employ an experienced consultant to design a dewatering system for the site and verify that it is properly installed and working properly before excavation begins. It then becomes a required part of the *design* for the facility. The result would be that instability during construction is avoided, the contract proceeds smoothly, and the integrity of the foundation beneath and adjoining the structure has not been compromised by soil softening.

4 Washington Park Station Slide

Figure CH-4.1 Washington Park Station. Aerial view of East Pit excavation in progress on February 13, 1996. Looking northeast.

Key Points

- Landslide prevention
- Automatic data acquisition
- Shear pile design and performance

Summary

This case history concerns a deep excavation (Figure CH-4.1) made into ancient landslide terrain. As soon as the excavation work began, shear movements were observed at the ancient landslide slip plane deep below the ground surface. The movements were monitored continuously by an in-place inclinometer, which was part of an automatic data acquisition system (ADAS). A piezometer and rain gauge were also automated. Comparison of the shear movements with the rate of excavation and winter storms clearly showed that the excavations were the principal cause of the movements.

At about mid-depth, the excavation work was temporarily halted because of concerns about a potential landslide into the excavation. A shear pile wall was designed and constructed to support the temporary cut slope, the piles passing through the ancient slip plane to anchorage in hard rock. Subsequently, the excavation was continued to the full depth without incident. The shear pile wall was later incorporated into the pile foundation of the structure, thus making the pile wall a very cost-effective solution.

Background Information

The hillside to the north of Canyon Road in Portland, Oregon, is an extensive area of ancient landslide terrain extending from the road to the top of the hill, a distance of 2,700 feet. The Washington Park Light Rail Transit Station (transit station) is located near the middle of the gentle (5½° to the horizontal) slope. The foundation of the station overlies stiff silts and clays with the slip surface of the ancient landslide at a depth of 55 feet on the east side of the structure and 80 feet at the west side. The East Pit excavation (Figure CH-4.2) is 33 to 48 feet deep, and reaches to within 10 feet of the ancient slip surface in the northeast corner.

In the two years prior to construction, inclinometers near the station site measured an average shear movement of 0.07 inch per year at the slip surface, showing that the ancient landslide was marginally stable. Eight additional inclinometers were installed around the perimeter of the site before excavation began (Figure CH-4.2).

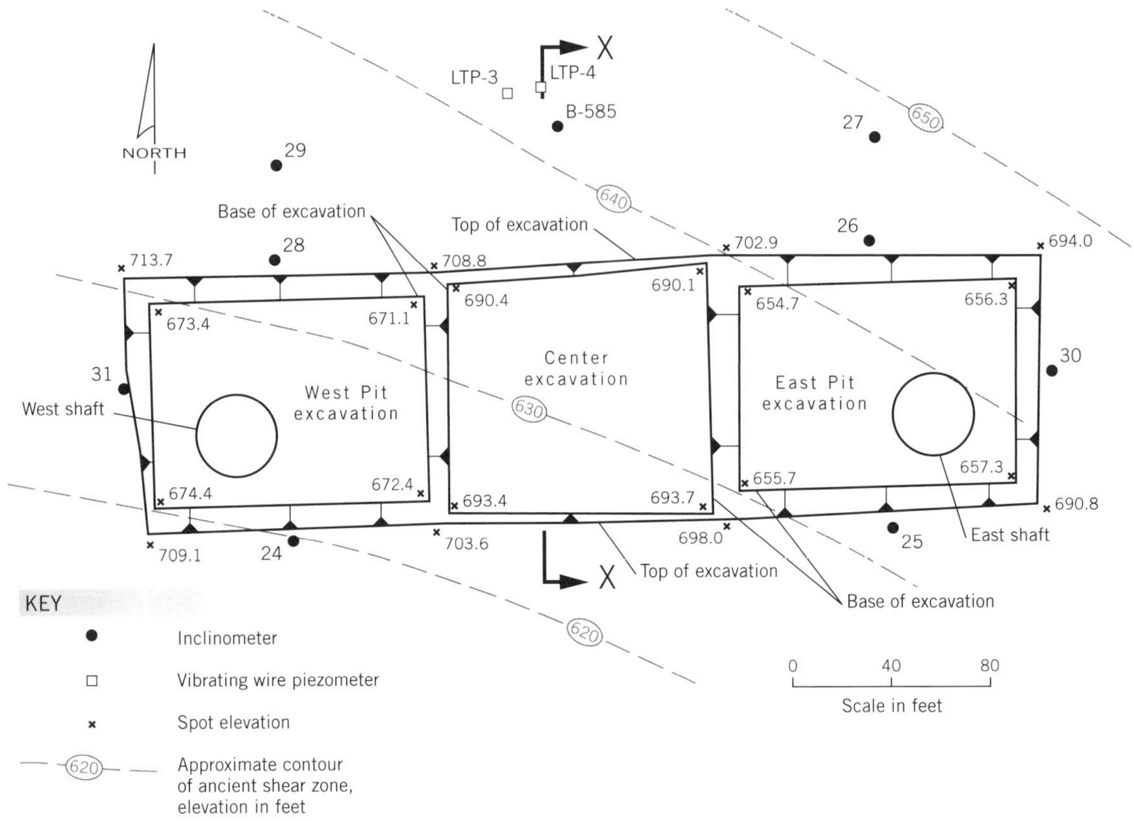

Figure CH-4.2 Washington Park Station. Plan of station excavations and field instrumentation.

AUTOMATIC DATA ACQUISITION SYSTEM

A decision was made to select two instrumentation sites for continuous readout monitoring: B-585 (Figure CH-4.2, CH-4.3), about 50 feet upslope from the center of the excavation, and B-583 (not shown on the figures), about 500 feet downslope of the excavation. At the location uphill from the excavation, one piezometer tip (LTP-3) was set just above the discrete shear zone of the ancient landslide as measured in the adjacent inclinometer casing, and one in-place inclinometer (B-585) was installed *across* the discrete shear zone. A rain gauge was installed to monitor rainfall.

Cross-section X-X (Figure CH-4.3) shows the position of the in-place inclinometer relative to the East Pit floor and the ancient slip surface. The summary boring for B-585 is also shown on the same figure.

In-Place Inclinometer

To determine a more precise thickness of the discrete shear zone, inclinometer readings were taken at 3 inch intervals as described in Chapter 22, Section 22.1. These measurements showed that the shear zone was less than 1 foot thick and a gauge length of 2 feet was selected for the in-place inclinometer. The wheels were placed above and below the discrete shear zone. The in-place inclinometer, supplied by Slope Indicator Co., is described in Chapter 4, Section 4.1.

Piezometers

The vibrating wire piezometers were supplied by Geokon. The sensors were installed within a saturated sand backfill that extended to the discrete shear zone. (e.g., LTP-3, shown on Figure CH-4.3). The sensor tips were located just above the shear zone to avoid damage from shear movements. About 2 feet depth of sand, sealed above by 2 feet of bentonite chips, were placed and saturated. The holes were grouted above the sensor zone with weak cement-bentonite.

Rain Gauge

A tipping bucket rain gauge from Texas Electronics was used to measure cumulative rainfall. The gauge was calibrated to tip for every 0.01 inch of precipitation, which is counted as one pulse by the data logger.

Data Loggers

The data logger was a Model CR10 from Campbell Scientific. The cable wires were terminated at a panel on top of the hermetically sealed, 12-channel data loggers.

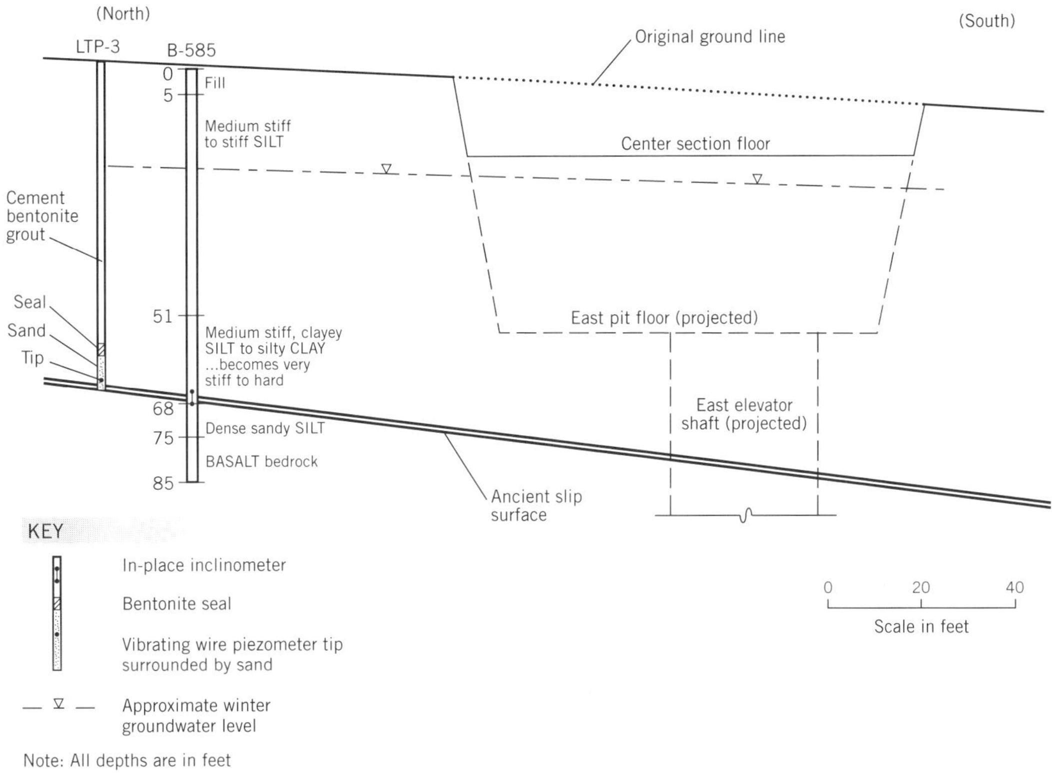

Figure CH-4.3 Washington Park Station. Geological section X-X through the center of the station.

A metal box housed the logger, a 12-volt battery pack, and vibrating wire interface (a similar unit is shown in Chapter 4, Figure 4.20). A 64 kilobyte RAM was used for program and data storage. Programming, logger communication, and data retrieval were accomplished by a laptop PC using software supplied by Campbell. The data loggers were placed in steel cabinets bolted to concrete pads for protection against vandalism.

Data Collection

The data loggers took readings every 10 minutes. After each hour, the data was averaged together with maximums and minimums. These were stored permanently in engineering units. Periodically (usually weekly) the data was retrieved using the menu-driven software into a laptop PC and put onto spreadsheets. The spreadsheets generated x-y plots of the data with time. The continuous monitoring equipment began collecting data on February 18, 1995 and performed flawlessly.

Reactivation of Shear Movements on the Ancient Landslide Slip Surface

Excavations for the East and West Pits at the transit station cut a "slot" into the ancient landslide and undermines the slope above it. Although the excavation depths never reached the full depth of the landslide, the continuous readout monitoring of in-place inclinometer B-585, uphill of the excavation, began to show shear movements on the ancient slip surface (plane) as soon as excavations started. The observed data, related to the West Pit excavation, is shown on Figure 4.12 of Chapter 4. It clearly demonstrated that the movements were directly related to the excavation activities and not to winter weather (Cornforth and Mikkelsen, 1996).

The ancient slip plane is much weaker than the soils above and below it. Triaxial tests on the overburden soils gave average effective stress strength parameters of $c' = 660$ psf, $\phi' = 33°$. A back analysis of the ancient landslide terrain gave an average strength along the slip plane of $c'_r = 0$, $\phi'_r = 7.9°$.

The deep-seated movements along the ancient slip plane were especially worrying at the East Pit excavation. At the northeast corner, the base of the excavation is estimated to be only 10 feet above the ancient slip plane (Figure CH-4.4). The design team, which included several geotechnical specialists, agreed that there was an unacceptable risk of a local landslide developing behind the 115-foot long north cut slope of the East Pit excavation. Construction was temporarily halted at about elevation 681, 40 feet above the ancient slip plane (Figure CH-4.5) on February 23, 1995. *Note:* Excavation for the West Pit, with a higher base level, was continued to completion.

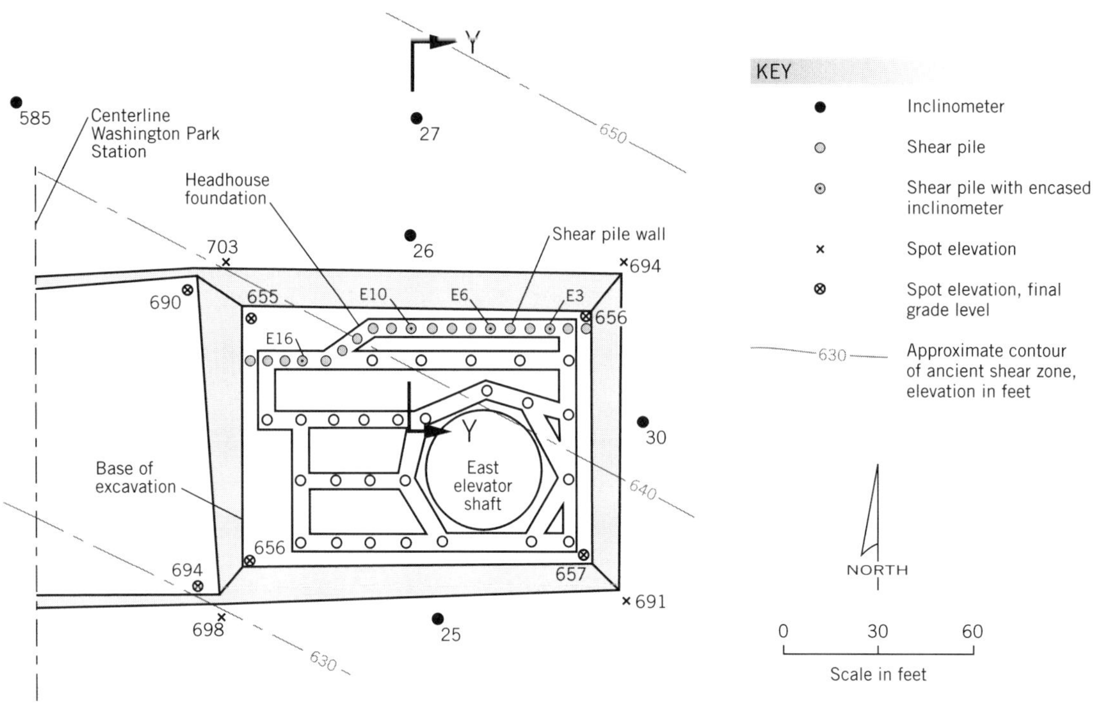

Figure CH-4.4 Washington Park Station. East Pit excavation showing shear pile wall and foundation.

TEMPORARY SUPPORT OF CUT SLOPE

The contractor had elected to support the vertical slopes of stiff silt and clay with soil nails. These comprised 8-inch diameter continuous flight auger holes 30 to 45 feet long in which No. 8 steel rebars were inserted and grouted in-place. The soil nails provide local slope stability for the cut slopes and were not designed to provide resistance against movements occurring below foundation level.

Several ideas were put forward to provide stabilization against a deep-seated failure, including a tied-back soldier pile wall and arrays of reticulated micropiles. However, Landslide Technology's proposal to install bored, cast-in-place shear piles was accepted by the owner as the better technical and economical solution to the problem. It involved construction of 19 vertical concrete piles along the north edge of the excavation (Figure CH-4.4). A major advantage of the line of shear piles is that they could be incorporated into the foundation of the building and thus serve a double purpose of temporary slope support and permanent foundation support. Using this concept, only one additional pile was needed for the slope stabilization project.

SHEAR PILE WALL DESIGN

Stability Analysis

The adopted model for the deep-seated slope instability analysis is shown on Section Y-Y, Figure CH-4.5. It was assumed to be a soil block moving down the ancient slip plane CD and entering the base of the excavation along a shear plane DE. The block failure would extend behind the slope reinforced by soil nails. A tension crack AB, water-filled in winter, could occur at this location. Shear surfaces BC and DE would be first-time sliding through the stiff clay, and shear surface CD would be at residual strength. It was assumed that the stabilizing shear force on the pile, P, acts immediately above the ancient slip plane (Figure CH-4.5).

Stability analyses indicated that the final excavation depth would be barely stable (calculated factor of safety F = 1.18) for a water-filled tension crack using the average shear strength parameters ($c' = 660$ psf, $\phi' = 33°$) measured in laboratory tests for BC and DE. However, it was anticipated that the cohesion intercept $c' = 660$ psf could decline to zero during winter as the stiff, overconsolidated clay softened in response to the excavation. Therefore, further calculations were made for $c' = 0$, and water levels within the tension crack being controlled by horizontal drains to 15 feet below the ground surface in winter. It was also assumed that shear strengths along both the ancient shear plane and the first-time sliding surfaces would be proportionately mobilized, i.e., $\tan \phi'_m = \dfrac{\tan \phi'}{F}$.

The stability analysis results for $c' = 0$ are plotted on the graph, Figure CH-4.6, and show that the calculated static factor of safety falls to 0.75 without shear piles. A minimum force of 28 kips per lin. ft. is required to maintain stability. It was recommended that the shear piles be capable of providing a resistance of 60 kips per lin. ft. Stability analyses on the east and

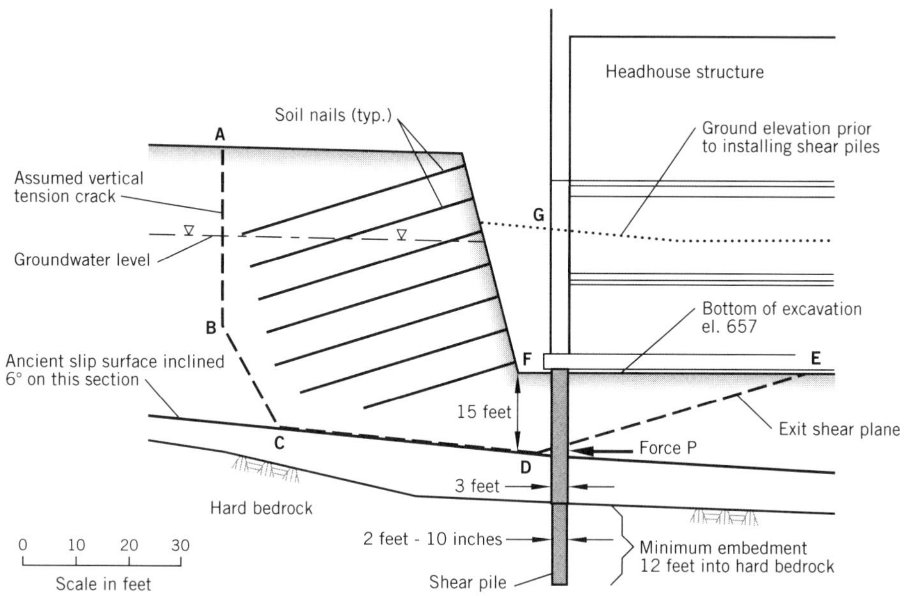

Figure CH-4.5 Washington Park Station. Section Y-Y through the middle of the East Pit excavation, showing the stability analysis model.

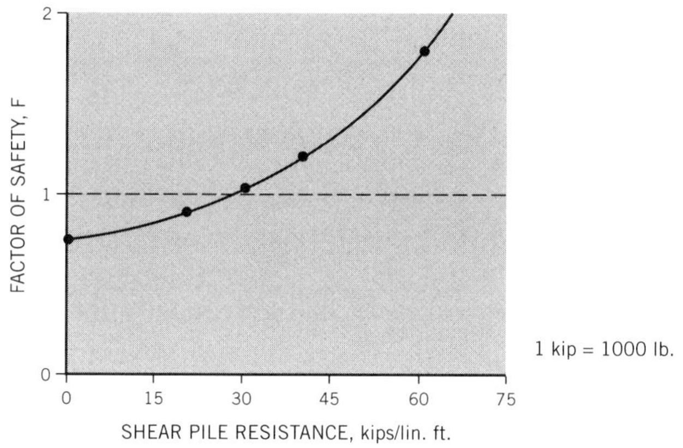

Figure CH-4.6 Washington Park Station. Graph showing the calculated shear pile resistance needed to maintain slope stability.

TABLE CH-4.1 Shear Pile Design Parameters

Soil Profile	Thickness	Parametric Range
Above shear zone	15 feet (approx.)	c = 1,500 to 2,000 psf
Shear zone	1.5 feet	$\phi_r' = 8°$
Decomposed basalt	9 feet	c = 6,000 psf **or** c′ = 0, $\phi' = 39°$
Weathered basalt	12 feet	c = 2,000 psi to 5,000 psi

south walls of the excavation, where the ancient slip plane does not dip toward the excavation, were found to be adequate.

Pile Design

As the cantilevered shear piles develop resistance to the landslide forces, they deflect laterally. Later, the vertical load of 145 tons per pile from the headhouse structure is applied to the deflected pile, creating additional bending moments that the pile must be designed to resist.

The lateral deflection of the pile wall depends on the *actual* force acting on it. For deflection calculations, a conservative value of 40 kips per lin. ft. was used in the design analysis.

The thickness of the ancient landslide shear zone was measured on three inclinometer casings by the method described in Chapter 22, Section 22.1. The values ranged from 0.55 to 0.85 feet and, for analysis purposes, the shear zone thickness was conservatively set at 1.5 feet. The soil parameters used in the shear pile design are listed in Table CH-4.1.

Structural engineers at Parsons Brinckerhoff performed the shear pile design. Soil-structure interaction was modeled by using modulus of subgrade reaction k from Reese, Wang, and Fouse (1992). The maximum deflection at the top of the pile was computed to be 0.7 to 1.0 inch, with base fixity about 14 feet below the slip plane. The maximum bending moment was about 1,000 kip-inch using procedures developed by Fukuoka (1977). The pile was conservatively designed for an eccentricity of 3.6 inches (10 percent diameter) for the axial load from the headhouse structure of 290 kips, giving a total design moment of 1,820 kip-inch. The final pile design required 32 No. 10 steel reinforcement bars in two concentric loops (Figure CH-4.7).

SHEAR PILE CONSTRUCTION

The shear piles were constructed by a large track-mounted drilling rig equipped with 36-inch and 42-inch diameter soil augers for the soil overburden, and a 34-inch rock auger for socketing into bedrock. The reinforcement cage extended from the base of the socket to the final grade level at the bottom of the excavation (F on Figure CH-4.5). A steel liner, 0.38-inch wall thickness, was installed

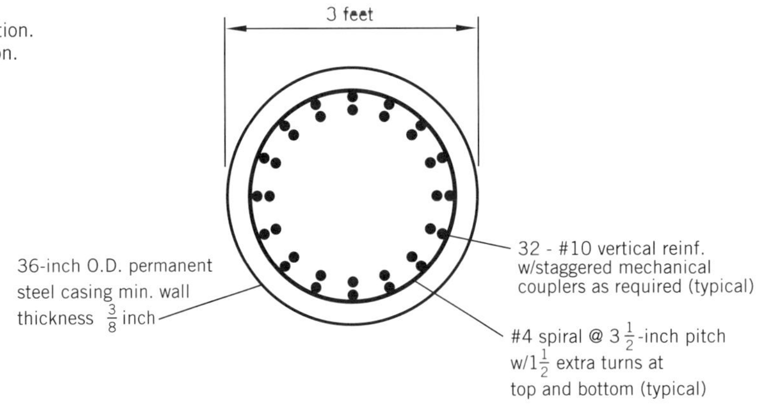

Figure CH-4.7 Washington Park Station. Reinforced concrete shear pile section.

through the overburden and left in-place after the concrete was poured. From foundation level F to the temporary excavation floor at G, the hole was filled with Controlled Density Fill (Lone Star Northwest Mix Design 564), which was later excavated.

The shear piles were constructed October 16–31, 1995. No pile drilling was permitted until the concrete in piles within 12 feet of it had cured for at least 48 hours. Excavation of the East Pit was resumed on December 4, 1995, and was completed on March 14, 1996.

PERFORMANCE OBSERVATIONS

Inclinometer casings were installed through the center of four selected shear piles (Figure CH-4.4). Readings were taken twice weekly as the excavation occurred. The upper casings were cut off as the excavation deepened.

The deflected profiles of the shear piles at the end of construction are shown on Figure CH-4.8. The estimated depths of the slip plane, based on inclinometers surrounding the transit station, are shown on the plots. Some observations of technical interest are:

- The maximum deflection of about 1.3 inch occurs in pile E10 near the mid-length of the wall and was within expectations.
- The piles near the ends of the shear pile wall (E3, E16) have smaller deflections, presumably due to end restraints.
- Deflections are essentially linear *above* the shear plane, suggesting that the assumption of the shear force acting directly above the discrete shear zone is valid.
- Deflections extend for 8–12 feet *below* the shear plane, i.e., down to the top of the hard bedrock (see Table CH-4.1).

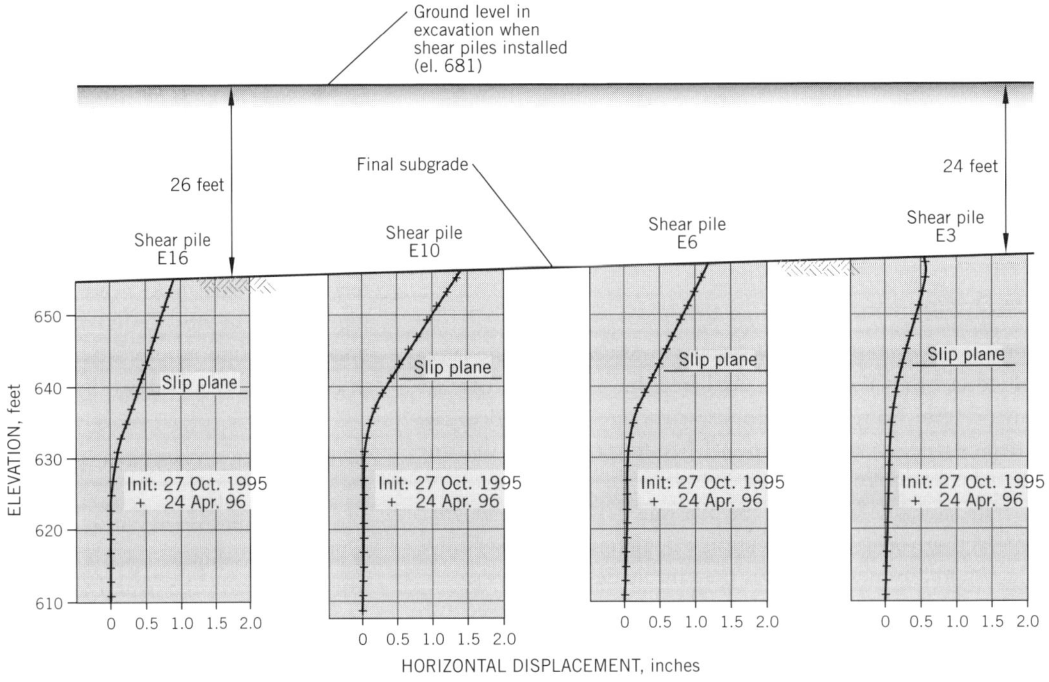

Figure CH-4.8 Washington Park Station. Measured deflection profiles of four shear piles at the end of site excavation in the East Pit.

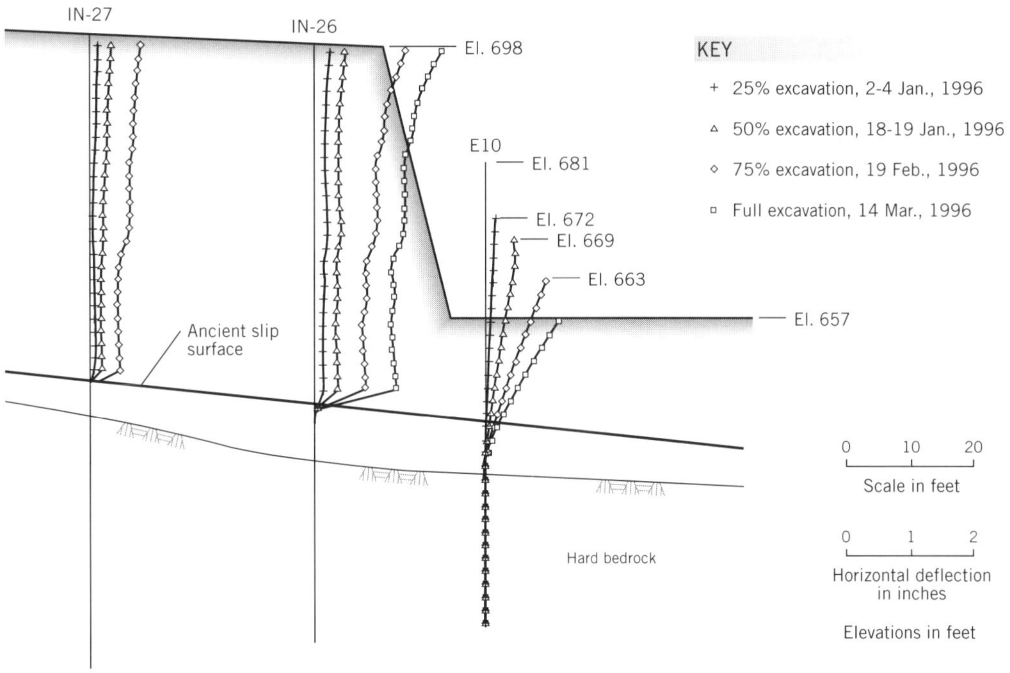

Figure CH-4.9 Washington Park Station. Movements of the ground on the north side of the East Pit during excavation (Section Y-Y).

A comparison of the deflection of pile E10 with inclinometers IN–26 and IN–27, upslope from E10, is shown on Figure CH-4.9. They cover the excavation period from December 4, 1995 to March 14, 1996. Cumulative movements corresponding approximately to 25%, 50%, 75% and full (100%) excavation are shown. The last set of readings were taken on April 24, 1996, approximately six weeks after the excavation had been completed, when the contractor was placing forms for the north wall. The movements within the shear piles had essentially stopped, having moved only an additional 0.04 inch during the final six-week period. Part of this very small movement may have been caused by construction of the adjacent foundation piles, which occurred March 22–April 2, 1996.

It is of interest that inclinometers IN-26 and IN-27 moved as a block during the excavation work. The shear is confined to the ancient shear plane; the soil immediately above the ancient shear plane retained a near-vertical face. This behavior contrasts with the bending of shear pile E10. Inclinometer IN-27 moved less than inclinometer IN-26 due to its greater distance from the excavation. Although the final movement of inclinometer IN-27 was not measured (it was replaced when the excavation was between 75% and 100%), the earlier movements are similar to inclinometer IN-26.

5 Pelton Park Slide

Figure CH-5.1 Pelton Park Slide. Pelton recreation park as seen from the air, looking northeast. The excavations at the Pelton Upper Slide and the debris heap, which caused the first set of movements, are in the upper right of the photo (Photo: May 25, 2990).

Key Points

- Reactivation of ancient slide by (i) fill and (ii) earthquake
- Marginal stabilization

Summary

This case history describes a large double wedge landslide, the long lower wedge having a near-horizontal slip surface. The ancient slide was barely stable and was reactivated in 1990 when a spoil heap was placed on the upper wedge. This movement was stabilized by removing the spoil heap. However, in March 1993, a crustal earthquake reactivated the slide.

The case history demonstrates the delicate balance between stability and instability on slopes with a factor of safety of approximately 1.00 under natural conditions. At this site, relatively small destabilizing factors caused significant movements. Due to the large size of the landslide, the only practical treatment was to excavate part of the upper wedge at the head of the slide, thereby improving stability by a marginal amount.

Background Information

Pelton Park is a recreation park on a wide bench about 2 miles upstream from Pelton Dam in central Oregon, Figure CH-5.1. The landslide tapers from narrow ledges at the two ends to a maximum length of about 550 feet near the center (Figure CH-5.2). The width of the landslide is approximately 2,300 feet; the volume is about 2,000,000 cu. yd. The bench is an eroded feature left behind when the Deschutes River cut its channel through beds of basalt and tuff.

Fill Loading Reactivation

The large ancient landslide was reactivated when a heap of soils and rocks was dumped at the head of the slide in May, 1990 (Figure CH-5.2). This material came from the Pelton Upper Slide (see Case History 6), where the soils were bulldozed over the edge of a cliff in an attempt to stabilize a smaller landslide further upslope. Although trucks removed materials from the heap, it grew faster than it was being

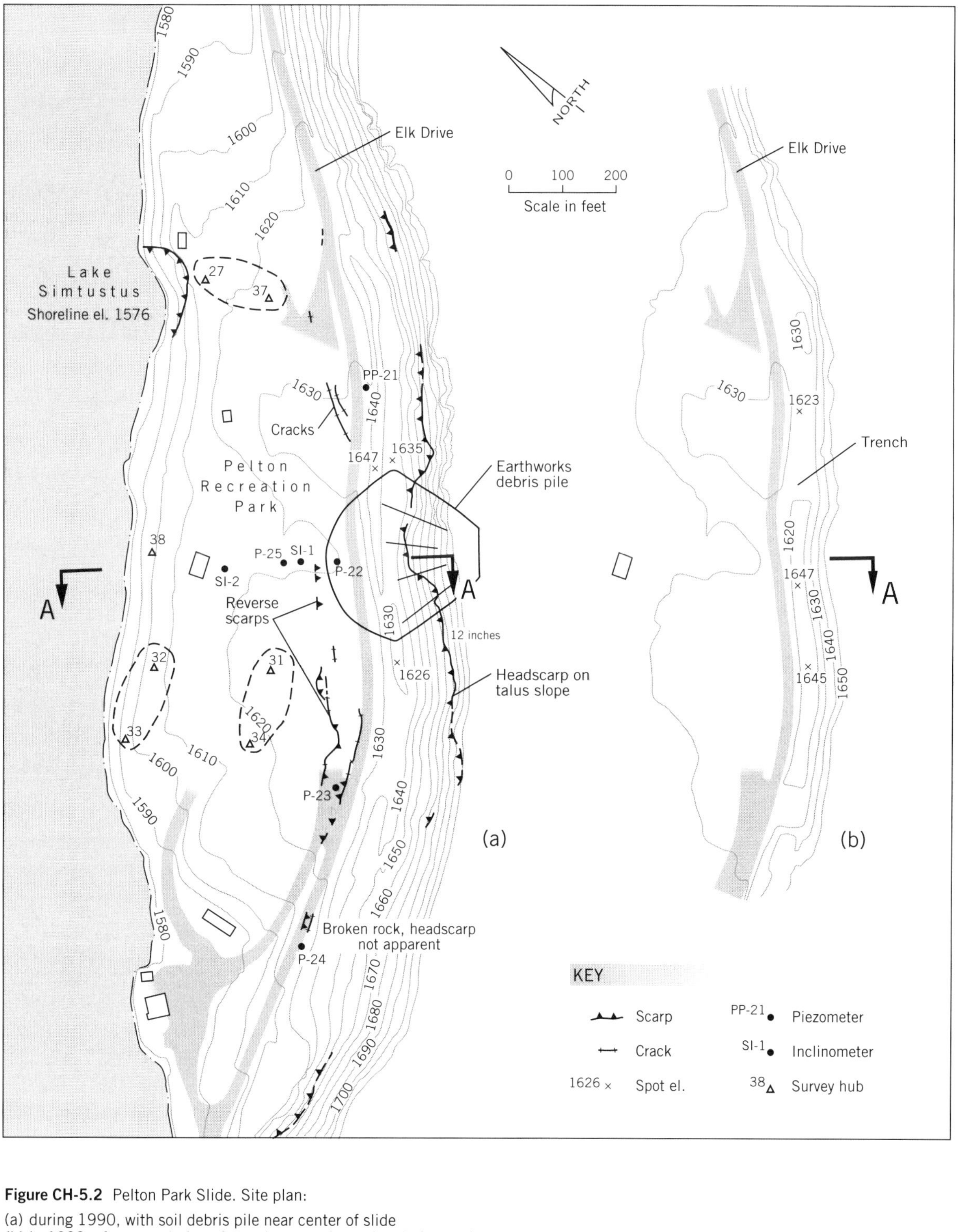

Figure CH-5.2 Pelton Park Slide. Site plan:
(a) during 1990, with soil debris pile near center of slide
(b) in 1993, after completion of trench excavation at head of landslide

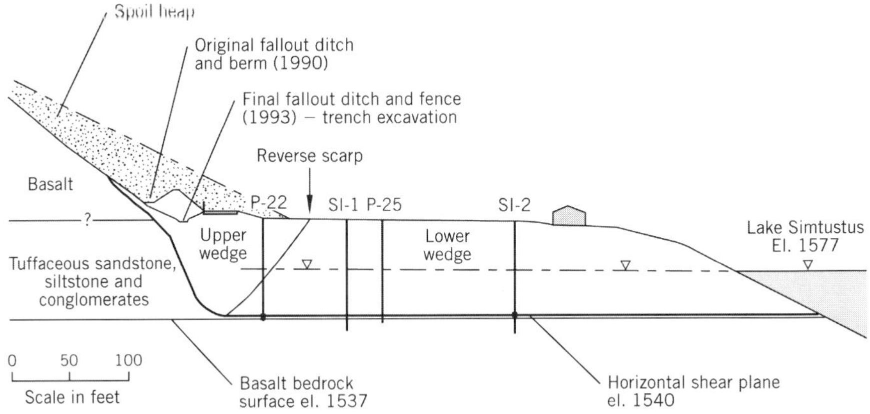

Figure CH-5.3 Pelton Park Slide. Cross-section A-A through the center of the landslide.

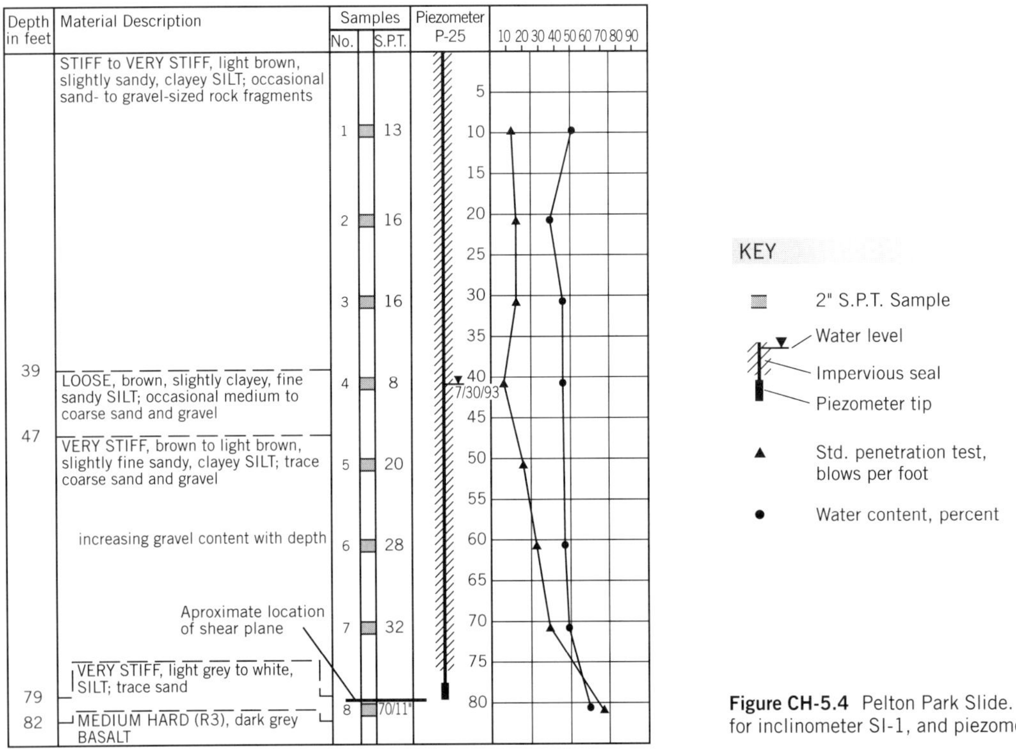

Figure CH-5.4 Pelton Park Slide. Summary boring log for inclinometer SI-1, and piezometer P-25 details.

removed. The debris fan reached its maximum size around June 5, when the earthmoving contract was halted. It remained in place for two months and was removed between July 25 and August 2, 1990.

Two inclinometers, SI-1 and SI-2, were installed on the bench in early 1990. When the soil piled up at the head of the park, the inclinometers showed that the ground was shearing horizontally at elev. 1540, approximately 75 to 85 feet below the ground (Figure CH-5.3). Shear was occurring within a stiff, tuffaceous silt (white volcanic ash) immediately above basalt bedrock, as shown on one of the boring logs reproduced in Figure CH-5.4. The toe of the landslide is below reservoir level. Subsequently, two piezometers were installed on the park slide, each with the tip sealed near the slide plane. Groundwater levels were only 1 to 2 feet above reservoir level (elev. 1577).

Several survey hubs monitored the amount and rate of movement in different areas of the park. Three areas have been selected for analysis and the survey hubs in each group are surrounded by a broken line on the plan (Figure CH-5.2). The pairs of hubs in these selected areas have been averaged. The movement-time plots (Figure CH-5.5) show a near-lin-

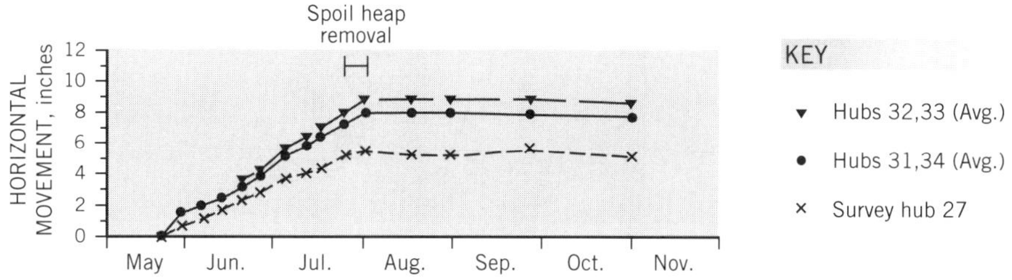

Figure CH-5.5 Pelton Park Slide. Horizontal movements vs. time for landslide reactivated by fill heap.

ear rate of movement from May 23 (when they were set) to August 2 (when the heap was removed). The rates of movement were as high as 1 inch per week. Movements stopped when the spoil heap was removed.

The tuffaceous silts in this part of Oregon are on essentially horizontal bedding planes with some crossbedding (Figure CH-5.6). Failure occurs as a double wedge with the upper wedge providing the push to the lower wedge (Figure CH-5.3). The lower wedge, typically much larger and having zero driving force, provides resistance along a very weak, slickensided shear zone. Residual strengths in these tuffaceous silts are around $\phi'_r = 8$ degrees.

The downslope edge of the upper wedge is defined by a reverse scarp, as shown on the plan and section. The headscarp is exposed on the talus slope above the park and is probably a thin soil slippage above bedrock. Once it reaches the tuff beds, the shear plane probably dips at about 60 degrees to the horizontal. This interpretation is shown on the cross-section (Figure CH-5.3). It can be seen that the large heap of soil sits above the upper wedge and increases the driving force of the landslide. Although limited in width compared to the entire slide, it was sufficient to reactivate the landslide.

After the soil heap had been removed, Portland General Electric (PGE) shaped the lower part of the slope to provide a fallout ditch and berm (Figure CH-5.3). These were intended to catch any rocks falling down the slope and protect the perimeter road (Elk Drive) which runs along the base of the basalt cliff. Numerous readings of the survey hubs showed no movements over the next two years.

Seismic Reactivation

On March 25, 1993, almost three years after the fill load reactivation had ceased, a crustal earthquake of magnitude 5.6

Figure CH-5.6 Pelton Park Slide. Exposure of tuffaceous soils in a road cut near the slide site. Upper beds: silty fine sandstone; lower beds: sandy, clayey silt (Photo: 1990).

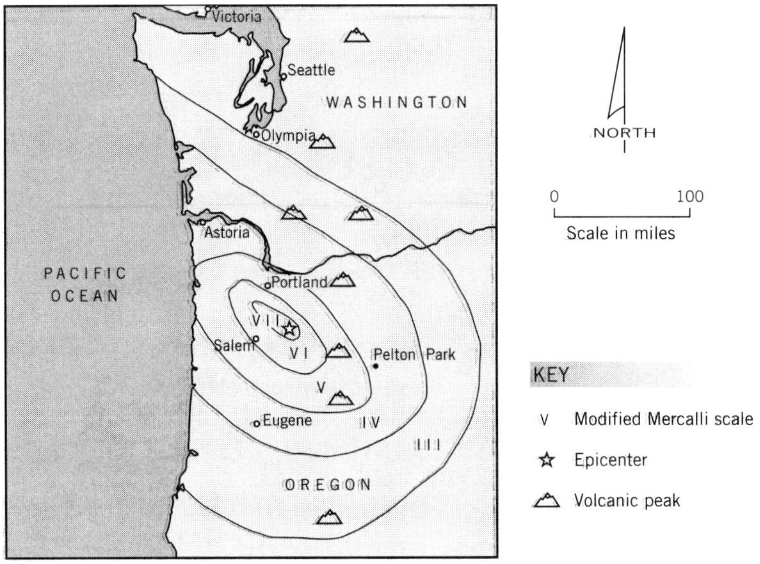

Figure CH-5.7 Pelton Park Slide. Intensity map for the Scott Mills earthquake of March 1993 (after Madin et al., 1993).

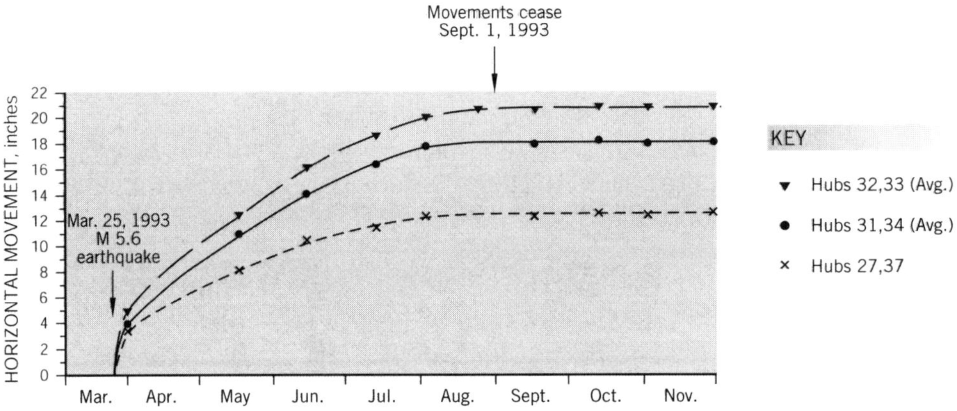

Figure CH-5.8 Pelton Park Slide. Horizontal movements vs. time for landslide reactivated by the 1993 earthquake.

occurred with its epicenter near Scotts Mills, about 72 miles northwest of the site. The focal depth was 10 miles. Although there are no seismometers installed on Pelton Dam, the maximum ground acceleration at this distance is estimated to be around 0.01g (Madin et al., 1993).

The intensity maps produced after the earthquake (Madin et al., 1993) indicate that the park landslide would have experienced Intensity level IV bordering on V (Figure CH-5.7). The Modified Mercalli scale Intensity IV is equivalent to: "Vibration like passing of heavy trucks; or sensation of a jolt like a heavy ball striking the walls." Intensity V: "Felt outdoors. Small unstable objects displaced or upset."

PGE surveyors were on the site within six days of the earthquake and measured renewed movements at the survey hubs. Subsequently, readings taken at monthly intervals showed substantial continuing movements (Figure CH-5.8). Total movements were as high as 21 inches.

The horizontal movements can be interpreted as follows:

1. An "instantaneous" movement, coincident with the shaking, of about 3 inches at these six survey points, followed by
2. A subsequent velocity of about 1 inch per week that slowly decreases over a three-month period, and
3. Cessation of movements around September 1, five months after the earthquake.

A surprising aspect of this case history is that such a modest earthquake shaking could cause this prolonged and significant amount of movement. The question arises: Could there have been some other cause?

Pelton Dam has year-round maintenance staff at the hydroelectric project. There was no change in dam operations that could account for the landslide reactivation.

Another possible cause is exceptional rainfall. The dam is located in one of the more arid climates of Oregon, with average annual precipitation of only 9 inches. March 1993 was wetter than normal, but at least 2 of the previous 10 March precipitations were higher. Since no movements occurred during these earlier high rainfall periods, it is highly unlikely that March rainfall was the causal factor, and rainfall does not account for the prolonged movements into summer; it is concluded that the movements were triggered by the mild earthquake shaking.

It is clear that the Pelton Park slide was finely balanced on the knife edge separating stability from instability. The horizontal slip plane and low residual strength could easily be disturbed by a horizontal inertia force. Two possible explanations for the mechanics of the failure are:

1. The lower wedge slip plane, being a preexisting ancient slide, is at residual strength. The shaking may have caused small excess pore water pressures to develop on the horizontal residual shear plane, thereby allowing movements to occur that lasted until the excess pore pressures dissipated (e.g., Seed and Chan, 1966).
2. The horizontal shaking from the earthquake may have increased the width of tension cracks at the headscarp. An increased separation would reduce the shearing resistance on the steeply inclined upper wedge and increase the force (push) of the upper wedge on the lower wedge.

MARGINAL REMEDIATION

PGE was interested in improving the stability of the Pelton Park to provide some small reserve of stability. Given the prevailing conditions at this site: deep slip plane, toe below water, infrastructure in place, there were few choices.

Landslide Technology recommended unloading the upper wedge to the maximum practicable extent to decrease the driving force. This solution required the fallout ditch to be excavated deeper, removing the protective berm and replacing it with a wire mesh fence at the roadside. These measures reduce the load on the upper wedge, as shown on Figure CH-5.3.

After regulatory approval, the reconstruction work occurred between early November 1993 and late January 1994. The owner excavated about 30,000 cu. yd. of soil. Calculations indicate that a 5 to 6 percent improvement of the factor of safety is achieved by the remediation work. This marginal level of remediation should help to reduce or eliminate movements that might occur in the future from rapid reservoir drawdown, exceptional precipitation, or another minor earthquake.

Since remediation, there have been two exceptionally wet winters in Oregon. Movements of the hubs over the past five years have averaged less than 0.2 inch per year. Crack healing could account for these small movements.

6 Pelton Upper Slide

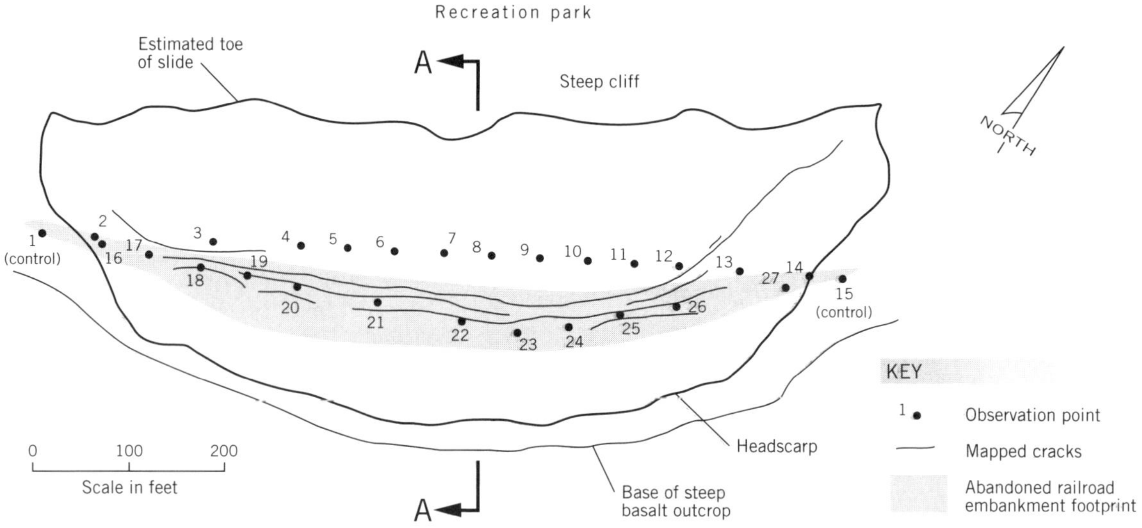

Figure CH-6.1 Pelton Upper Slide. Site plan showing old railroad embankment, ground cracks, and survey points of the 1975 landslide.

Key Points

- Double wedge failure
- Relationship between F and landslide velocity
- Delayed failure
- Three-dimensional stability analysis
- Regrading remediation

Summary

This double wedge failure became active without any apparent reason in 1975 and gradually slowed down to almost zero over the next 13 years. Acting on poor technical advice from a consultant, the owner attempted to improve stability by excavating soils from the lower wedge of the landslide in 1990; this caused the slide movements to reactivate. Subsequently, the slide was stabilized by a regrading contract that excavated soil from the upper wedge and placed it on the lower wedge.

The case history provides considerable insight into the nature of double wedge failures. The extensive data collection during movements also allows calculation of the "static" factor of safety in relation to the velocity of movements (see also Chapter 9, Section 9.4).

Background Information

Pelton Dam is part of a hydroelectric project on the Deschutes River near the town of Madras in central Oregon. A recreation park was created on a gentle slope adjoining the reservoir. This area, also a landslide, is described in Case History 5 (Pelton Park Slide).

On a bluff 150 to 200 feet above the park, a medium-to-large slide occurred rapidly in January 1975. The headscarp had an arcuate shape corresponding to the rock cliffs above it (Figure CH-6.1) and the cracks were mapped one month later as the slide continued to move. The toe of the slide outcropped onto a steep rock bluff. The landslide has a maximum width of 880 feet and maximum length (upslope/downslope) of 300 feet.

There was no readily apparent "triggering" event to explain the landslide cause. The only activity in this isolated area was the construction of a railroad up the side of the Deschutes canyon in 1909 that was never completed. The railroad company built an embankment across the slide area at the location shown shaded on the plan (Figure CH-6.1).

The facility owner, Portland General Electric Co. (PGE), set up 27 survey points near the head of the landslide in

February 1975, two weeks after the failure. These points were set at uniform intervals across the top of the abandoned railroad embankment and on the slope below it. Frequent measurements of horizontal and vertical movements were taken on the survey points over the next 13 years.

The data from two of these points near the center of the slide are typical of the observed movements. Point 22, located on top of the embankment and between the headscarp and ground cracks, experienced 6.0 feet of vertical movement and 6.9 feet of horizontal movement during 13 years of observations (Figure CH-6.2a). Point 8, further downslope, had a horizontal movement of 17.1 feet and no vertical movement over the same period (Figure CH-6.2b). It can be noted that all the other survey points below the cracks also had only horizontal movements, showing that the base of the landslide was a horizontal plane.

The graphs show that the rate of movement was fastest initially, and that movements gradually slowed down. The average initial rate of movement for the horizontally moving points 7 through 11 was 3.6 inches per week. By 1988, the movements had essentially stopped (0.02 inch per week).

The recreation park below the slide was closed to the public in 1984 because boulders rolling down the hillside had become an unacceptable hazard to park users. Although slide movements had almost stopped by 1990, PGE let a contract to excavate substantial quantities of the slide debris with the intent of improving stability so that the park could be reopened. Shortly after excavation began, renewed cracking and movements developed. After three weeks of excavation, horizontal movements had reached 9.6 inches per week, almost three times faster than the initial rate of movement in 1975.

INITIAL SITE VISIT

It was at that juncture that the author first visited the site to provide an opinion on the cause of the renewed slide activity. An aerial view of the site at that time (Figure CH-6.3) shows that the contractor had pushed landslide debris over the cliff edge onto the recreation park below. The fan-shaped pile of debris was being hauled away by trucks.

The site visit revealed that cracks were developing about 70 feet downslope from the headscarp and parallel to it. These cracks had reverse (uphill-facing) scarps (Figure CH-6.4), indicating that the soil between the headscarp and the crack had a greater vertical movement than the ground further downslope. It was evident that the landslide could be modeled as a double-wedge. A subsequent review of the survey point data quickly confirmed this field diagnosis.

Since the earthworks contractor was excavating soils almost entirely from the ground below the cracks (i.e., the lower wedge), the resistance to sliding was being reduced while the driving force remained relatively unchanged. Thus, the factor of safety was being progressively lowered, causing

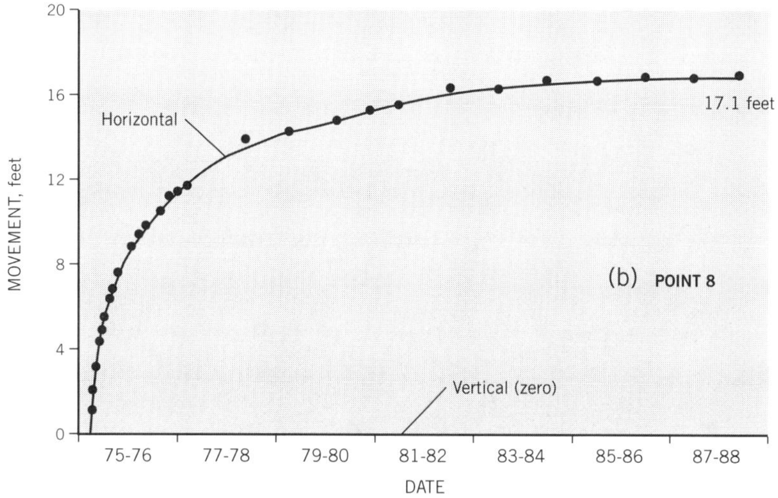

Figure CH-6.2 Pelton Upper Slide. Typical observed movements of survey points, 1975–1988
(a) on the upper wedge
(b) on the lower wedge

Figure CH-6.3 Pelton Upper Slide. Aerial view showing the excavation (on right) that reactivated the landslide (Photo: May 29, 1990).

Figure CH-6.4 Pelton Upper Slide. Reverse scarps separating the upper and lower wedges during the 1990 slide reactivation (Photo: June 11, 1990).

the landslide movements to accelerate. The client stopped the earthworks contract, and a site investigation was quickly undertaken.

SUBSURFACE INVESTIGATIONS

Two holes, LT-1 and LT-2, were drilled through the overburden soils into the underlying basalt bedrock (Figure CH-6.5). The drilling equipment was an air-rotary drill rig using an ODEX casing advancer system. Samples were taken at 5-foot intervals. Inclinometer casings were installed in each hole. The fast rates of landslide movements allowed the depth of sliding to be determined by sets of inclinometer readings taken on the same day.

Groundwater levels were measured in separate piezometer holes that were located about 20 feet away from the inclinometer holes and along the same contour line. The piezometer tips were set at the approximate failure zone. A third piezometer hole, P-3, was drilled into the graben area at the head of the landslide to complete the groundwater profile for the section under study. This hole was terminated just above the failure zone.

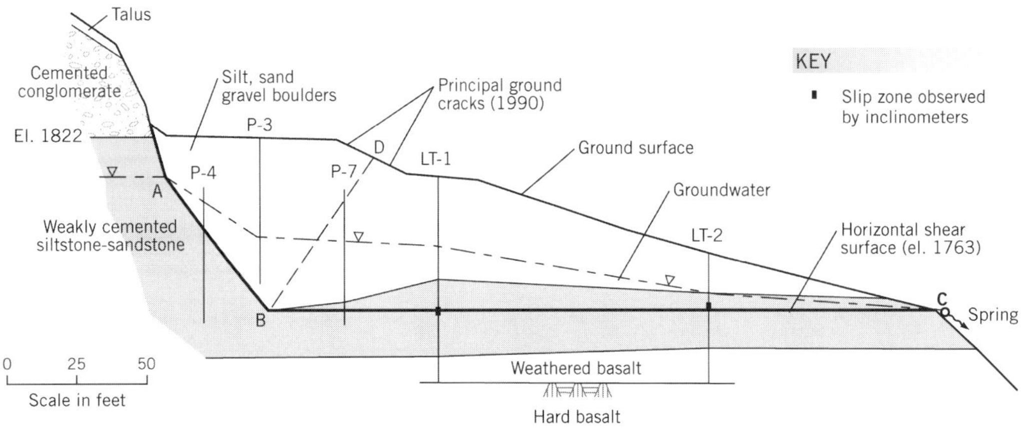

Figure CH-6.5 Pelton Upper Slide. Geological section A-A through center of landslide reactivated in 1990.

Figure CH-6.6 Pelton Upper Slide. Thin-wall tube sample taken at the slip surface in boring P-6 (3 inch dia. sample is supported on a cradle—crack on the extreme right of photo is the join between two samples, not the slip surface).

Geologic Section A-A (Figure CH-6.5), near the center of the landslide, shows that the landslide is sliding along a horizontal shear plane at about elevation 1763. Subsequent measurements by PGE surveyors of the ground break at the toe of the landslide were also at this elevation.

The landslide debris above the failure plane comprised two strata. The major portion of the debris was a heterogeneous mixture of sand, silt, volcanic ash, pebbles, and scattered large boulders. Below this soil mixture is a slightly cemented, tuffaceous sandstone-siltstone stratum. In soil mechanics terminology, the material can be described as hard gray-brown clayey silt, slightly cemented. Borings LT-1 and LT-2 encountered 12 feet and 3 feet of this material above the shear failure zone. The sandstone-siltstone stratum extended below the failure zone to basalt.

The probable explanation of the observed site conditions is that the landslide area is a bench left behind during the canyon development and the infilled soils have fallen onto it over time from higher up the slope. Below these infilled materials, a scabland of in-place tuffaceous siltstone/sandstone overlies basalt (Figure CH-6.5). The horizontal shear plane of the lower wedge passes through this tuffaceous siltstone layer.

Figure CH-6.7 Pelton Upper Slide. Block of clay taken from the slip surface of the landslide, showing striated and glossy (slickensided) surface.

Shear Strength Measurements

Undisturbed tube samples of the sandstone-siltstone showed considerable horizontal fracturing at the slip plane of the landslide (Figure CH-6.6). Test specimens for direct shear tests could not be obtained from these zones, but specimens were cut from the soil immediately above it. The direct shear tests required numerous stress reversals to reach residual strength. Results on one set of specimens gave $c'_r = 0$, $\phi'_r = 8\frac{1}{2}°$. Other samples gave higher values. One of the difficulties with the test is that the specimens continually shed silt into the bowl of the apparatus, making the specimen slightly thinner after each shearing cycle. After the shear test, the specimen easily separated into two parts at the shear plane; the shear surfaces had a glossy slickensided appearance.

Soils in the vicinity of the shear zone were described as hard, fissured, sandy, clayey silt. Average properties were LL 82, PL 57, PI 25, w 54, clay fraction (one test only) 11 percent, and wet density 105.4 pcf. Clay mineralogy tests conducted by Oregon State University showed that the predominant clay mineral was smectite (montmorillonite). A clay block taken from the slip surface (Figure CH-6.7) shows striations and slickensides.

Samples of the gouge soils from the upper headscarp were remolded at their natural water content into a direct shear (stress reversal) machine. Drained shear tests at a rate of 0.06 inch per hour gave a shear strength ϕ' of 33° with no significant drop-off past the peak strength. The soils were described as firm, sandy, clayey silt with scattered pebbles. Average index properties for the fines were: LL 56, PL 40, PI 16.

Stability Analyses

A two-dimensional stability analysis assumed $c'_r = 0$, $\phi'_r = 33°$ for the granular soils in the very steep headscarp of the slide above elevation 1806 and siltstone residual strength below this elevation. Using the 1988 Section A-A (Figure CH-6.8), representing a just-stable condition with a factor of safety F of

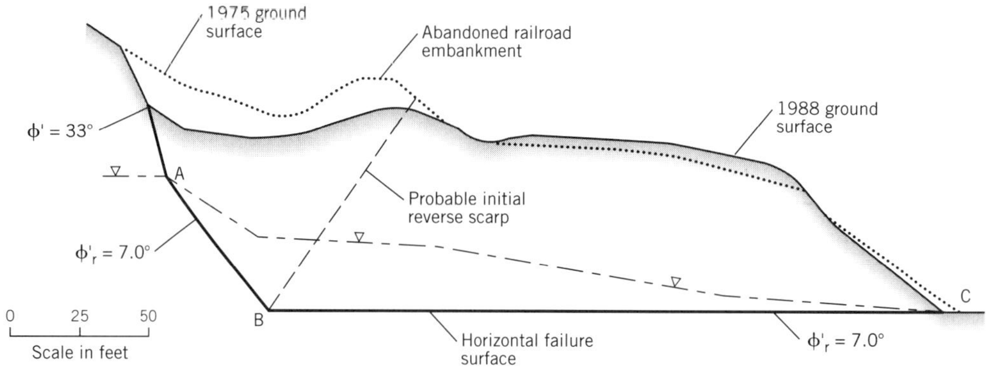

Figure CH-6.8 Pelton Upper Slide. Section A-A showing the ground profiles at the start (1975) and end (1988) of the original landslide.

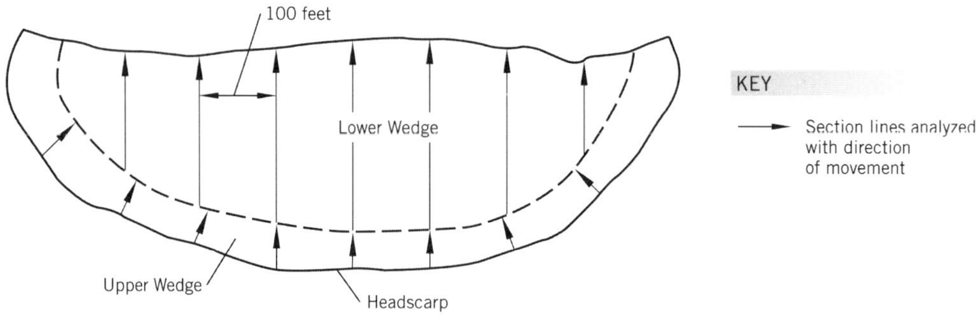

Figure CH-6.9 Pelton Upper Slide. Plan showing the seven sections taken for the three-dimensional stability analysis.

1.00, the back-analysis gave residual strength parameters of $c'_r = 0$, $\phi'_r = 7.0°$ for the siltstone.

The geometry of this landslide is conducive to a three-dimensional stability analysis (Figure CH-6.9). The extent of our knowledge of the subsurface conditions is limited to the borings through the center section and four other borings within the upper wedge at the top of the slide. However, the shear plane location at the toe of the lower wedge has been confirmed by survey over most of the width. The main purpose of a three-dimensional analysis is to attempt to take account of the varying angle of thrust of the upper wedge (Figure CH-6.9). The upper wedge puts the maximum pressure at right angles to the headscarp, which forms an arc around the landslide. The main body of the landslide is within the lower wedge, and the extensive survey measurements from 1975–1988 have shown that the lower wedge moves as a block in one direction. All survey points below the reverse crack of the graben (see Figure CH-6.1) moved by extremely consistent horizontal distances, averaging 17.1 feet.

Parallel sections were taken at 100-foot widths. The forces in the upper wedges were calculated normal to the headscarp and then were resolved in the direction of sliding of the lower wedge. Assumptions made in the analysis were:

- Groundwater falls uniformly in a pattern similar to the central section A-A; this assumption is supported by the approximately horizontal springline that was observed at the headscarp after the remedial excavation.
- The upper wedge shear zone below the exposed headscarp is represented by the conditions observed at the central section A-A.
- The thrust of the upper wedge on the lower wedge for the vertical plane BE (Figure CH-6.10) is at 10° to the horizontal.

For these assumptions, the back-calculated residual strength ϕ'_r is 8.6°. The back-calculated ϕ'_r in the three-dimensional analysis is in very good agreement with the residual strength $\phi'_r = 8\frac{1}{2}°$ measured in the laboratory.

REGRADING REMEDIATION

In a generalized double-wedge failure (Figure CH-6.10), the upper wedge moves down the shear surface of the headscarp AB while the lower wedge moves along the basal shear surface BC. As the upper wedge sinks down, its force on the lower wedge continuously decreases and the two wedges eventually reach equilibrium of driving and resisting forces. This progression explains the decreasing rate of movement from 1975 to 1990 (Figure CH-6.2).

To stabilize the landslide, it was reasoned that if additional soil is excavated from the upper wedge, it would further decrease the force on the lower wedge in much the same way that the force decreases under natural conditions, i.e., man-made removal of soils from the upper wedge would speed up what nature takes many years to accomplish. Furthermore, if the excavated materials are placed on the lower wedge, they will increase the resisting strength by increasing the effective normal stresses. Since the basal shear surface BC at Pelton Upper Slide is horizontal, the extra weight of soil does not add any driving force to the lower wedge.

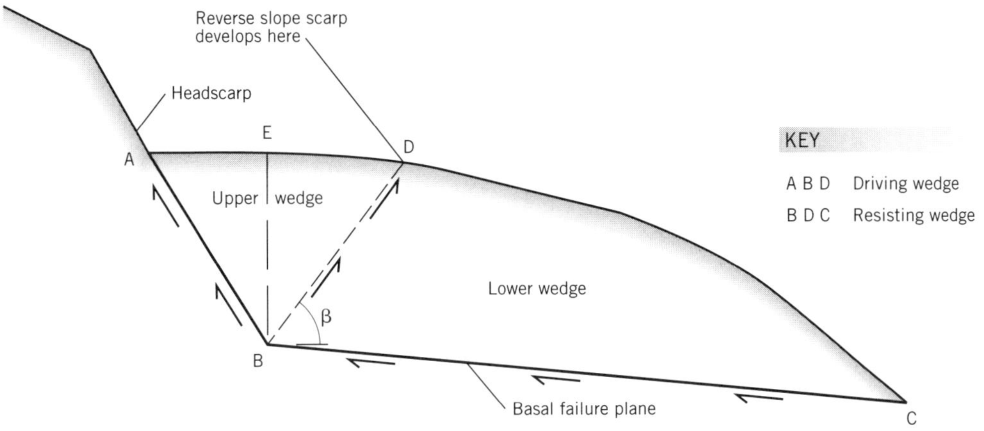

Figure CH-6.10 Generalized double wedge analysis.

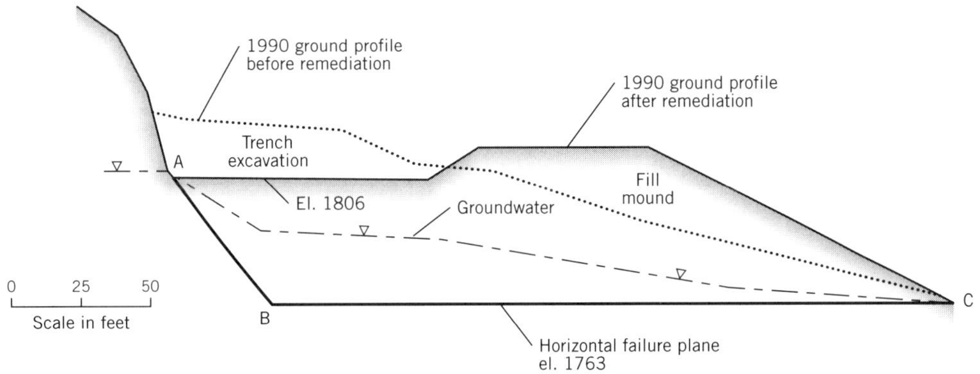

Figure CH-6.11 Pelton Upper Slide. Section A-A showing the ground profiles before and after the remediation of the slide.

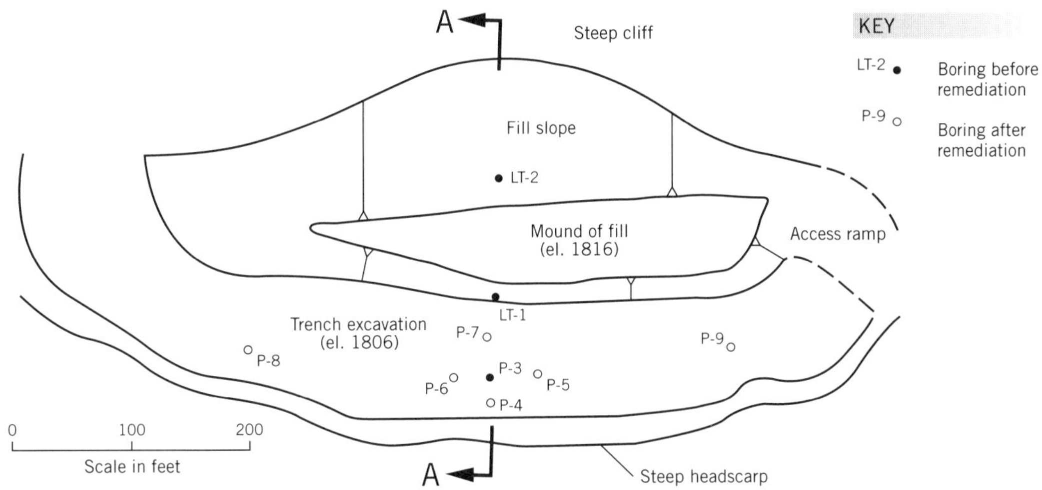

Figure CH-6.12 Pelton Upper Slide. Plan of site after remediation.

The remedial treatment followed these principles. A deep trench was excavated at the head of the landslide (Figures CH-6.11 and CH-6.12). The change from cut to fill was designed to occur where the reverse scarp cracks had been mapped (see Figure CH-6.1). Soils from the excavation were used to build a mound of fill on the lower wedge (Figure CH-6.12). The fill was spread by a dozer and compacted by the dozer tracks. Boulders were buried within the fill so that they would not be a future danger to the recreation park below.

The regrading work was a balanced cut and fill that was completed in only three weeks. It immediately stopped the landslide movements. A view of the nearly completed work is shown on Figure CH-6.13. The regrading treatment was extremely cost-effective compared to alternatives.

Figure CH-6.13 Pelton Upper Slide. Trench excavation (right) and fill mound (center) near the completion of regrading remediation (Photo: July 11, 1990).

The steep face of weakly cemented siltstone exposed by the trench excavation was expected to weaken with time through weathering. It was shotcreted as a preventative measure.

The Pelton Upper Slide has remained stable. Survey points were installed after remediation and have been monitored annually.

LANDSLIDE CAUSATION

The original landslide occurred unexpectedly in 1975 for "no apparent reason." It moved rapidly at first and gradually slowed down over the next 13 years (Figure CH-6.2) as the upper wedge sank and the lower wedge moved horizontally downslope (Figure CH-6.10).

The abandoned railroad embankment was built in 1909. Almost the entire railroad fill was between the headscarp and reverse cracks (Figure CH-6.1), which later became the upper wedge of the landslide. The failure occurred through very stiff clayey silt (weathered tuffaceous siltstone).

The landslide thus has: (i) stiff overconsolidated clayey soils; (ii) a manmade fill that increases the shear stresses in the foundation and decreases slope stability; and (iii) a large (17 feet) landslide movement occurring 66 years after construction. It is apparent that the landslide is a *delayed failure,* as discussed in Chapter 2, Section 2.15.

NATURE OF THE REVERSE SCARP

A double wedge landslide has a significant change in inclination of slip surfaces from the steeper upper wedge to the flatter lower wedge (Figure CH-6.10). The upper wedge usually has insufficient strength to be stable against movement along slip surface AB; thus, the upper wedge ABD pushes on the lower wedge BDC. The upper wedge is the *driving* wedge and the lower wedge is the *resisting* wedge.

The reverse scarp that occurs at D is a surface manifestation of the change in slope from AB to BC, and all translational slides typically develop a reverse scarp (sometimes referred to as the "antithetic" scarp).

In most analyses of double wedge failures, the analyst separates the two wedges by a vertical line BE and calculates the thrust of the wedge ABE rather than the wedge ABD. Although this practice is acceptable for stability calculations, it does not reflect the reality in which the upper wedge ABD concurrently shears along surfaces AB and BD during landslide movements and is the true driving wedge. For a level ground surface, the inclination of the surface BD should be close to the Rankine value of ($45° + \frac{1}{2}\phi'$) to the horizontal. At the Pelton Upper Slide, the strength of the landslide gouge material was measured as $\phi' = 33°$. Using this value, angle β for plane BD would be $61\frac{1}{2}°$ to the horizontal (Figure CH-6.10).

The numerous measurements taken on the survey points by PGE during the original landslide movements of 1975–1990 and the reactivation of 1990 provide a valuable insight into the nature of the reverse scarp BD. By comparing the slide movements for observation points on each side of the reverse scarp (for example, observation points 3 and 18 on Figure CH-6.1) the relative motion of the shear movements along the reverse scarp can be determined.

Analysis of the movements at four of the observation sets (20/4, 18/3, 19/3, 26/12) indicate that the direction of the slip zone in the reverse scarp is upslope at 51° to 59° (average 55°) to the horizontal. A line drawn at an angle of 55° to the horizontal from point B reaches the ground surface at D (Figure CH-6.5). It is noted that the cracks mapped in both 1975 (first slide) and 1990 (reactivation, after 17 feet of horizontal movement) were close to D determined in this manner. It suggests that the 1990 cracks originate from B and are not the reopening of the 1975 cracks, which have moved further downslope.

Nature of the Graben

The soils of the upper wedge sink down along the shear faces of the headscarp and reverse scarp (Figure CH-6.10). What may begin as an inverted triangle at the onset of sliding becomes a trapezoid as the lower wedge moves out and the upper wedge settles into the separation, producing a graben at the ground surface. Clearly, the zone at the base of the upper wedge undergoes intense local shearing.

Boring P-3 in the graben area (Figure CH-6.12) behaved differently from borings LT-1 and LT-2, both of which were on the lower wedge. Boring P-3 experienced considerable soil caving and loss of drill water circulation, especially in the 20 feet above the basal shear plane. A review of all the borings within the upper wedge area shows a noticeable drop in SPT blow count that starts to occur about 23 feet above the basal shear plane.

The practical significance of the high distress at the base of an upper wedge is that the graben at the head of translational landslides is often chosen for reservoirs, storm detention ponds, or building sites. The extra loads provided by these manmade structures are likely to induce differential movements within the distressed ground below, or may reactivate the ancient slide. Therefore, any structures built on top of the graben of a double-wedge landslide are likely to have a relatively weak foundation that is prone to differential settlement after construction.

Relationship between F and Rate of Movement

When a landslide stops moving, as the original landslide did by 1988, the static factor of safety is exactly 1.00. When movement is occurring, the "static" factor of safety falls *below* 1.00. This is an important consideration, because a back analysis of a landslide assumes that the factor of safety is 1.00, and the assumption is incorrect when some movement is still taking place. The implications are discussed in Chapter 9, Section 9.4.

The Pelton Upper Slide provided an opportunity to examine the relationship between the calculated "static" factor of safety F and the velocity of movement (in this case, the rate of horizontal movement of the lower wedge shearing along a horizontal plane). The PGE survey data provided the information needed for such stability analyses. Therefore, the factor of safety F can be calculated at any time during the initial landslide, based on F being 1.00 in 1988 after movements had essentially stopped. Furthermore, after the contractor started to excavate soil from the lower (resisting) wedge in 1990, the reactivation allows F to be calculated at the time the work was terminated.

The results are given in Table CH-6.1 and Figure CH-6.14. The approximate relationship $F = 1.00 - 0.1\sqrt{V}$ is a reasonable fit (Figure CH-6.14b) where V is the horizontal velocity of the lower wedge in inches per week. It can be speculated that this relationship may apply to other landslides

Table CH-6.1 Results of Stability Analyses

Situation		Velocity of Horizontal Movement	Stability Analyses: Two-dimensional	Stability Analyses: Three-dimensional
1975	February: Initial failure	3.6 inches/week	F = 0.80	F = 0.76
1975	December: 50% of final movement	1.0 inch/week	F = 0.90	(not calculated)
1988	Stability	0.02 inch/week	F = 1.00	F = 1.00
1990	Reactivation	9.6 inches/week	F = 0.63	F = 0.73

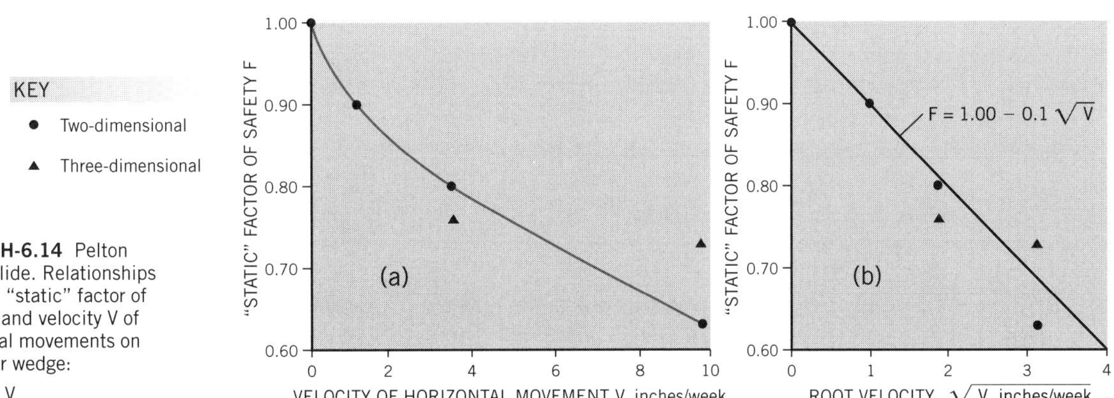

Figure CH-6.14 Pelton Upper Slide. Relationships between "static" factor of safety F and velocity V of horizontal movements on the lower wedge:
(a) F vs. V
(b) F vs. $\sqrt{V}$

that are similar to Pelton Upper Slide; i.e., a very thin discrete shear zone and PI of 25 (CI = 0.44, Chapter 8, Section 8.2).

The stability analyses show that the "static" factor of safety can fall well below 1.00 during active slide movements. Therefore, the amount of remedial improvement has to be raised accordingly in landslides where fairly rapid rates of movement are occurring. For example, at a horizontal velocity of 1 inch per week, F drops to 0.90 at Pelton Upper Slide. To achieve an *actual* margin of safety of 20 percent (F = 1.20), the factor of safety increase for the remedial works has to be 33 percent, not 20 percent.

As shown on Figure CH-6.14, the "static" factor of safety had dropped to about 0.63 for Section A-A when the toe unloading was stopped. Allowing for this initial condition, the calculated factor of safety after regrading remediation was 1.29.

There is obviously a misuse of terminology to discuss a factor of safety of less than 1.00 when the landslide is in a dynamic rather than a static condition. Thus $F < 1.00$ can be viewed as a measure of the deficiency of resistance that has to be made up in addition to the desired safety margin.

REFERENCES

This landslide information has been published earlier in the following two references that include some additional details.

Cornforth, D. H., and Vessely, D. A. 1992a. *Factor of Safety during Landslide Movements.* Proceedings, Sixth International Symposium on Landslides, Christchurch, New Zealand, A.A. Balkema, publisher, pp. 367–372.

Cornforth, D. H., and Vessely, D. A. 1992b. *Pelton Landslide: An Unusual Double Wedge Failure.* Stability and Performance of Slopes and Embankments II, GT Division, ASCE, Berkeley, California, pp. 310–324.

7 Skagway Marine Slide

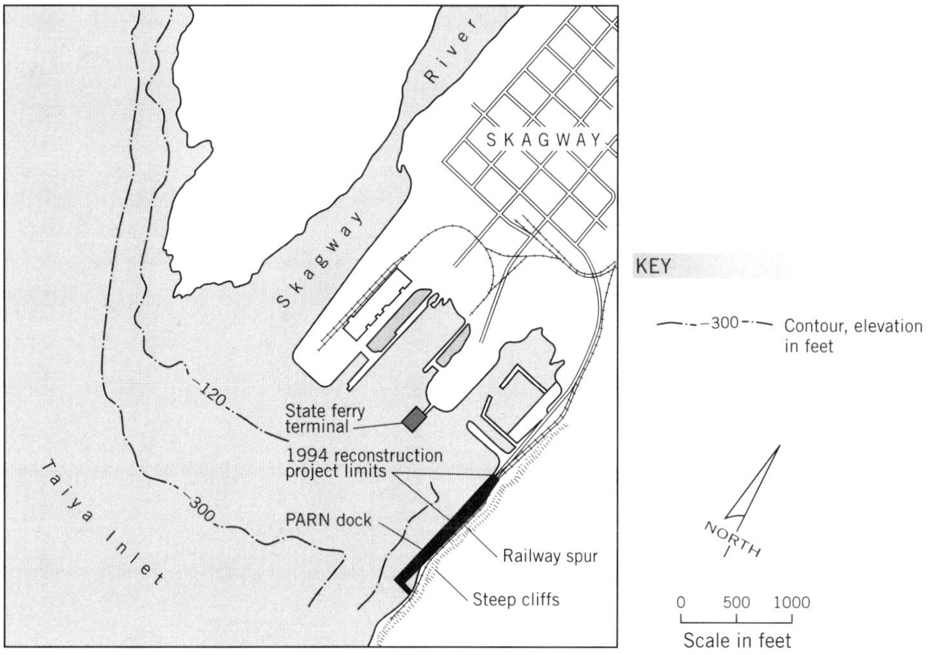

Figure CH-7.1 Skagway Marine Slide. Plan of Skagway harbor before the landslide.

Key Points

- Subaerial submarine flow slide
- Partly consolidated soil
- Total and effective stress stability analyses
- Artesian groundwater

Summary

On November 3, 1994, a very large flow slide destroyed 755 feet of the existing PARN wooden dock and 150 feet of partly built, open cell bulkhead of a dock improvement contract. A large wave crossed the harbor and pulled the state's floating ferry terminal from its moorings. One construction worker was killed. Three major lawsuits resulted from the landslide occurrence.

The case history describes the onset of a major subaerial submarine landslide as observed by construction workers. It also presents the stability analyses in terms of both total stresses and effective stresses. Parts of these analyses are described in detail in Chapter 9. The case history also includes the effect on slope stability of (i) buried piles; (ii) tide levels; (iii) partly consolidated foundation soils; and (iv) artesian pressures. Finally, the mechanism of slope failure is discussed.

Background Information

Skagway (Figure CH-7.1) is at the extreme northern end of Alaska's Inner Passage, and became prominent in 1896–1900 when gold was discovered on the Klondike River. Today, the small town is largely dependent on the tourist industry and many cruise ships visit Skagway during the summer months.

Dock Improvement Project

The PARN dock on the east shoreline of the harbor was built around 1898. It was supported on timber friction piles embedded in the underlying soft soils. A steep rock cliff forms the wall of the fjord adjoining the site. A narrow embankment between the dock and cliff carried a railway spur line (Figure CH-7.1).

A dock improvement project was undertaken in 1994 (Figure CH-7.2). It consisted of removing 545 feet of timber dock from the north end of the 1,300-foot long dock and replacing it with an open cellular structure constructed from driven sheetpiles infilled with dredge spoils (Figures CH-7.2b and CH-7.3). The harbor was to be dredged at the north end of the dock to create a deeper draft for ships, thereby providing the backfill needed for the cells.

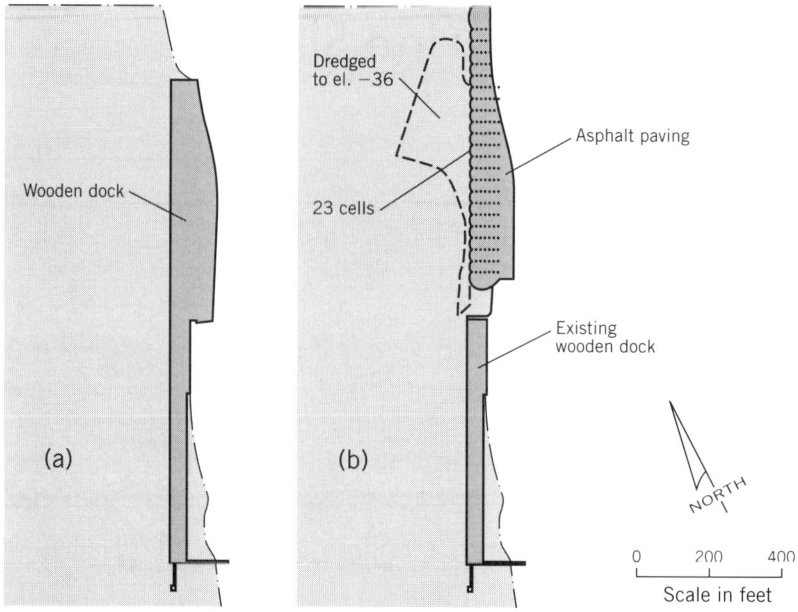

Figure CH-7.2 Skagway Marine Slide. Dock improvement project of 1994:
(a) existing structure
(b) proposed reconstruction at north end of dock

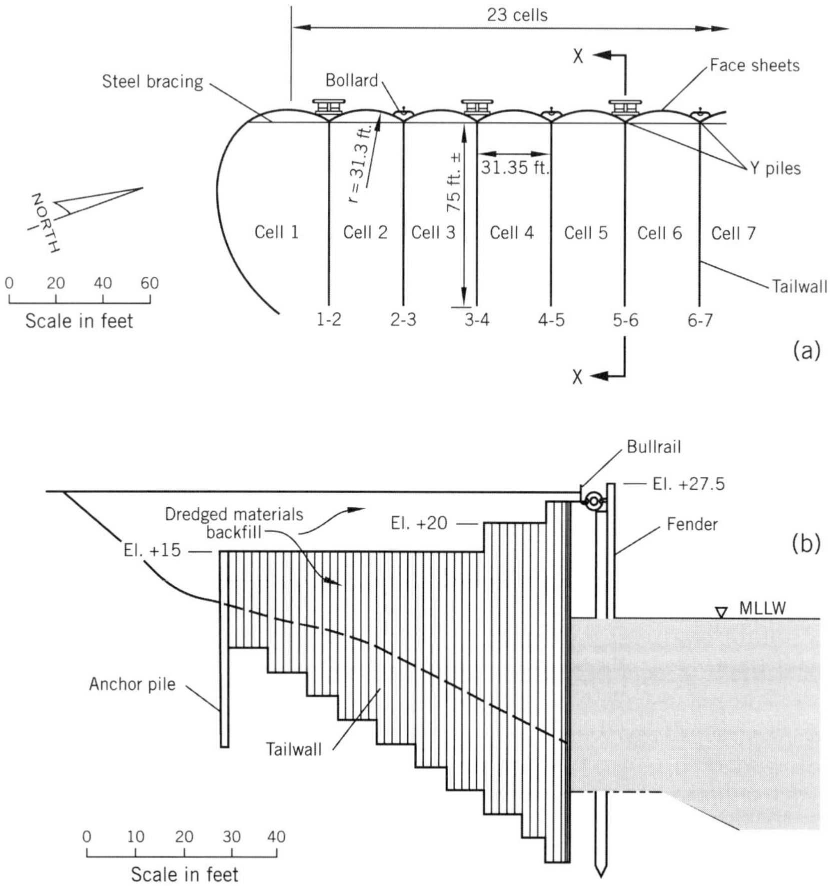

Figure CH-7.3 Skagway Marine Slide. Proposed dock reconstruction of 1994:
(a) plan of cellular structure
(b) cross-section X-X showing tailwall

Prior to the start of cell construction, the owner decided to stockpile riprap on the edge of the shoreline just south of the construction area (Figure CH-7.4). The 40-foot wide gap between the existing railroad embankment and dock was infilled with soil and rock to create a staging area (platform) for the riprap stockpile. The staging area fill began about September 27, 1994 (Figure CH-7.5). Riprap was brought to the site by railcars and stockpiled into a narrow heap, up to about 25 feet high, by October 13. From photographs and topography, it is estimated that the platform fill was about 345 feet long and had a volume of 6,700 cu. yds. The stockpile of riprap was estimated to be 6,000 cu. yds. Additional fill (estimate: 1,000 cu. yd.) was placed behind the newly constructed cells. Therefore, the total volume of fill was around 13,700 cu. yds.

The approximate rate of fill construction is shown on Figure CH-7.5. Below, and on the same time scale, is the graph of daily low tides. The landslide occurred at the lowest low tide level after the fill had been placed.

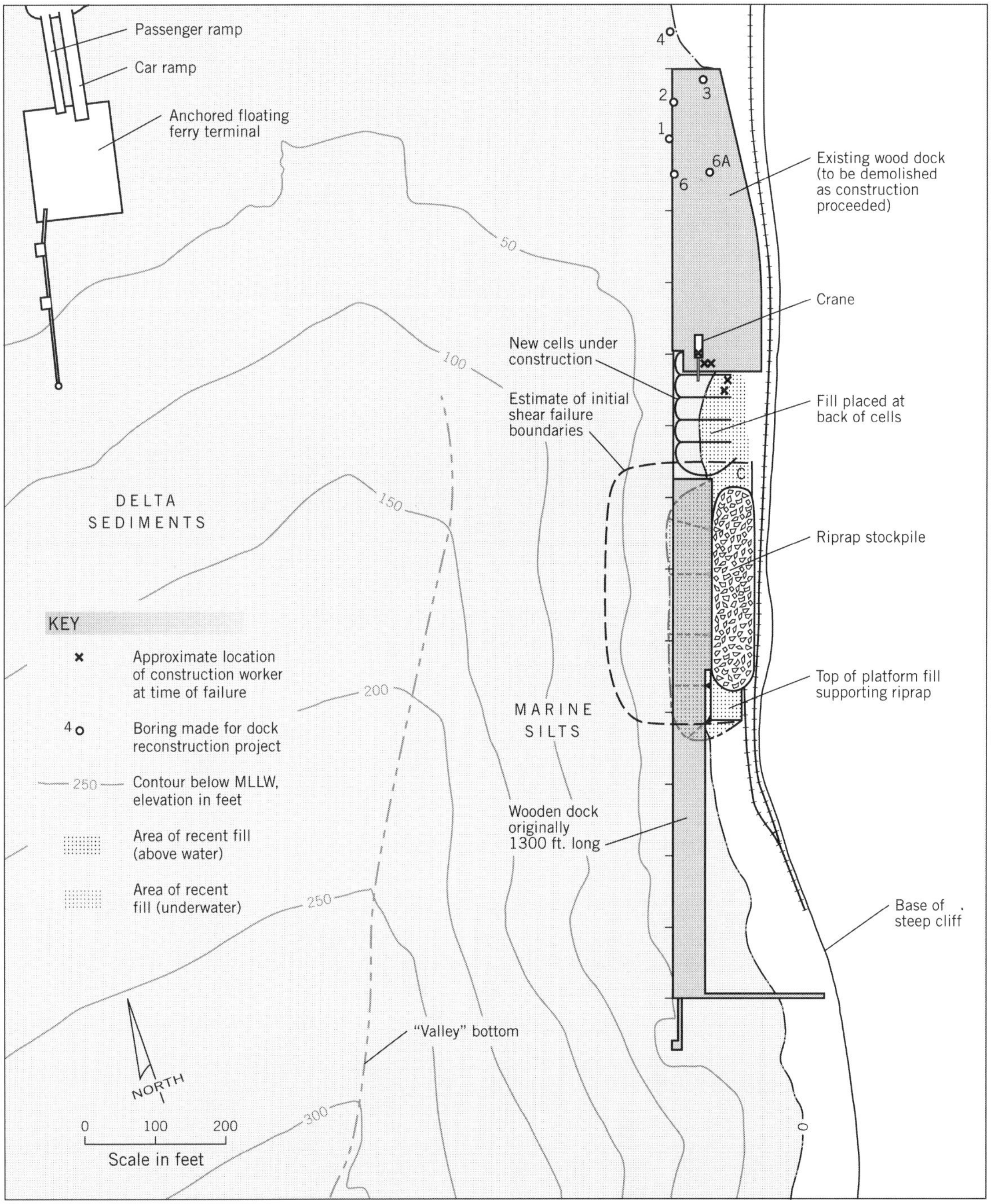

Figure CH-7.4 Skagway Marine Slide. Site plan just before the November 3, 1994, slide.

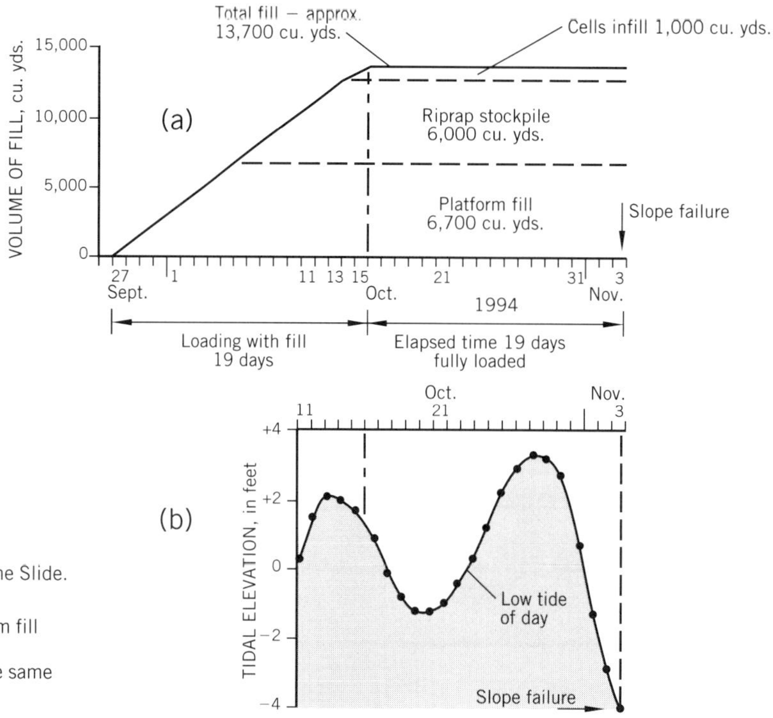

Figure CH-7.5 Skagway Marine Slide. Loading with fill:
(a) construction of the platform fill and riprap stockpile
(b) lowest daily tide during the same time period

The contractor first removed the decking and underlying timber piles before each cell was constructed. A large crane pulled out the longer timber piles and transferred them to a floating barge tied to the dock.

The facing sheets of the sheetpile cells (Figure CH-7.3a) were installed with the aid of a curved steel template. Sheet piles were driven into the mud below the site by a vibratory piledriver. After completing the convex outer face of a cell, a steel brace was installed across the back of the face to maintain the outward curvature.

Sheets for Cell 1, at the south end of the project, were driven on October 8–9. The other cells followed sequentially, and sheets for Cell 4 were driven on October 27–31. Interviews with the construction workers and site records show that most of the sheet piles were driven very easily and did not reach any hard stratum. Pile-driving records for tailwall 4–5 (between Cells 4 and 5) driven on November 2 and 3 (just before failure) showed that piles near the outer (west) side of the tailwall penetrated up to 22 feet below the mudline under the static weight of the hammer without any vibratory loading.

Possible Evidence of Slope Movements before Failure

Several events that occurred during the two to three days preceding the failure suggest that movement was beginning to occur in the underwater slope below the four cells, namely:

- On October 31, the construction workers spent two hours going back and forth over the outer face sheetpiles trying to hold them up to grade (by welding sheets together). The Y piles (see Figure CH-7.3a) were noted to have moved 6 inches toward the bay.
- On November 2, the day before the slide, the field inspector observed a long crack on top of the bank near the cells. During an interview with the OSHA area director about 2 months after the failure, he was asked to state the *exact* location of the crack and responded "between the new construction and stockpiled armor rock." (OSHA is the federal agency responsible for investigating construction-related accidents.) The line marked C on Figure CH-7.4 is the approximate location marked on the site plan by the field inspector during his OSHA interview.

  At least three construction workers saw the crack, and estimates of its length range from 30 to 60 feet. The crack had a small downdrop toward the ocean.
- On November 3, the day of the flow slide, the tops of the sheetpiles on Cells 1 and 2 were in disarray and had S-shaped configurations. The steel braces in Cells 1 and 2 had been destroyed. This damage could have been caused by loss of ground and downslope pulling action at the pile tips.

In summary, field observations around the sheetpile in the 1 to 2 days before the flow slide occurred suggest that some movements were occurring as the low tides became progressively lower.

FLOW SLIDE OF NOVEMBER 3, 1994

The site conditions just before the flow slide (based on depositions) is shown on the site plan (Figure CH-7.4) and a photograph of a scale model used in one of the trials (Figure CH-7.6). The scale model was prepared under the author's guidance, and the view shown is at an oblique angle looking northeast. The sea is not shown, and the steep underwater slope is seen as terraced steps.

The contractor was working two 10-hour shifts. The day shift on November 3 had pulled out about 50 wood piles from Cell 4. The two tailwalls for Cell 4 had been driven and the facing (outer) wall for Cell 5 was completed (Figures CH-7.4, CH-7.6).

The five members of the construction crew for the night shift had arrived on site at about 7 p.m. It was very dark and intermittently raining. Their illumination was a light on the front of the crane and a light tower with the lights directed toward the work area in Cell 4. The workers could only see as far as Cell 1, a distance of 130 feet from the crane. Their positions are shown on the plan, Figures CH-7.4.

Two workers entered Cell 4 and were standing on the shoreline of the cell awaiting instructions to begin lifting out piles with the help of the crane sitting on the dock above them. It was extreme low tide for Skagway (elev. −4 feet) (Figure CH-7.5b). Eyewitness accounts indicate that there was probably 4 to 8 feet of differential head between the water level in the cell and the low tide in the bay. At 7:12 p.m., failure occurred rapidly.

The first indication that movement was occurring was heavy and sustained rattling and twisting of the sheetpiles at the waterfront, and the sheetpiles started to settle. At about the same time, the two workers sitting on the slope within Cell 4 noticed that the surface pebbles were starting to move downslope. Alarmed, they both stood up. One scrambled upslope and managed to reach safety on the dock. He watched his coworker slide downslope and be lost with the slide debris as it moved into the bay.

As the slide gathered speed, the tops of both the timber piles and sheetpiles leaned toward the shore. The sheetpile cells moved down and outward, and a wall of water came into the cell area. The crane operator described the final loss of ground as "like flushing a toilet. Everything just went whoosh. . . ." The water was extremely turbulent, and the surface became covered with piles and parts of the wood dock from the south that had been destroyed by the flow slide.

Regarding the time for failure to occur, the four surviving workers gave slightly different accounts, but they all agreed that it was fast. An estimate of time based on their responses to questioning is that about 11 seconds ± 2 seconds elapsed from the time at which stones first began to roll down the slope to the total disappearance under water of the dock, cells, and riprap stockpile.

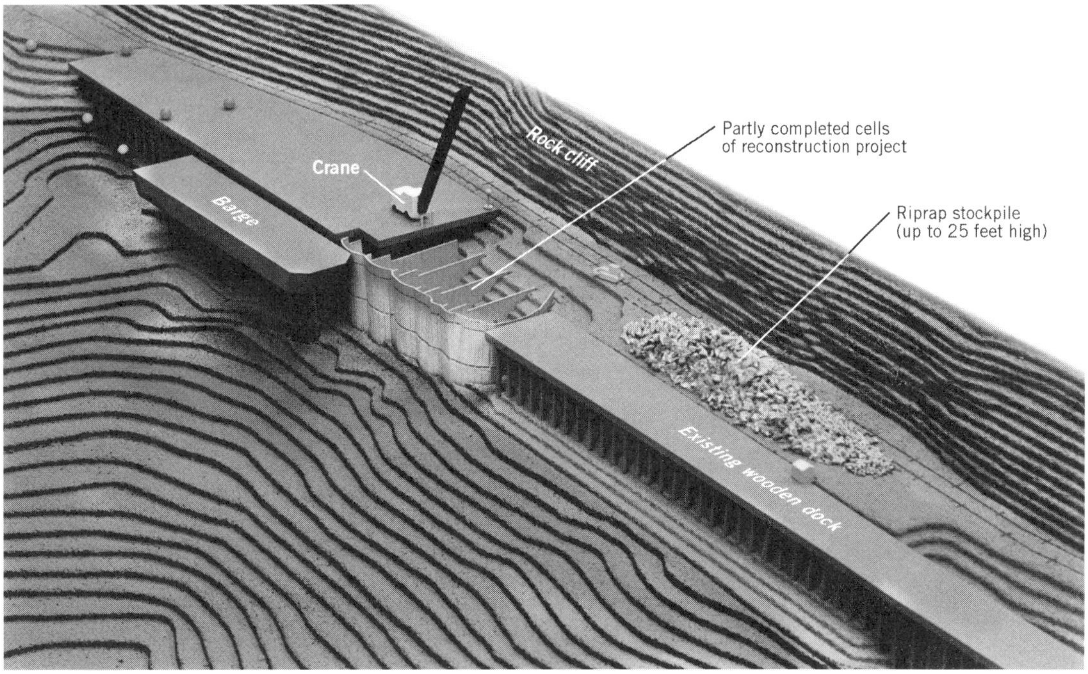

Figure CH-7.6 Skagway Marine Slide. Photograph of a scale model, used at trial, depicting the conditions at the site on November 3, 1994, just before the catastrophic flow slide. Four cells have been driven and partly backfilled. The heap of riprap, supported on a recently placed staging fill, is shown to the right (south) of the construction area.

A huge wave crossed the harbor and pulled out all anchors of the floating ferry terminal (Figure CH-7.4), sending the dock to the west side of the harbor. Damages to the floating terminal amounted to about $2 million.

The aftermath of the flow slide is shown on Figure CH-7.7. The landslide destroyed all (755 feet) of the wooden dock to the south of the work area, the partly completed cells, the riprap stockpile and platform fill, and part of the railway embankment. Comparison of bathymetric surveys before and after the slide show that the ground had been scoured to depths of up to 70 feet immediately downslope from the dock (Figure CH-7.7). Part of a scale model of the flow slide area after the failure is shown on Figure CH-7.8. The darker area in the center of the photograph represents scour depths exceeding 20 feet. The scour extended to the full length of the former wooden dock. Based on the scour contours, the volume of the flow slide immediately below the dock was around 1 million cu. yds. Bathymetric surveys showed that total scour, due to the flowing soil moving into the deeper water of the fjord, was several times larger than that at the site.

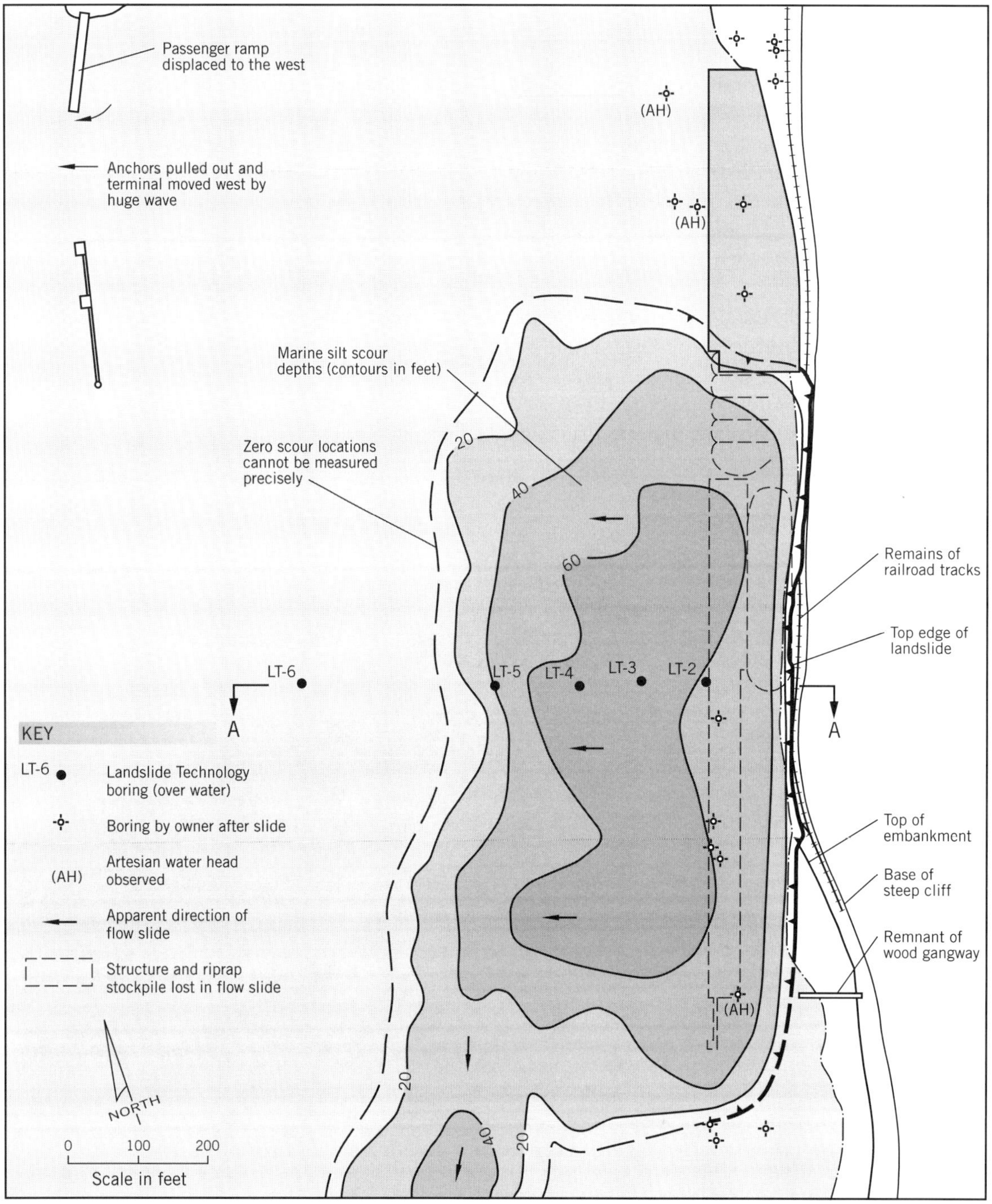

Figure CH-7.7 Skagway Marine Slide: Site plan after the November 3, 1994, slide.

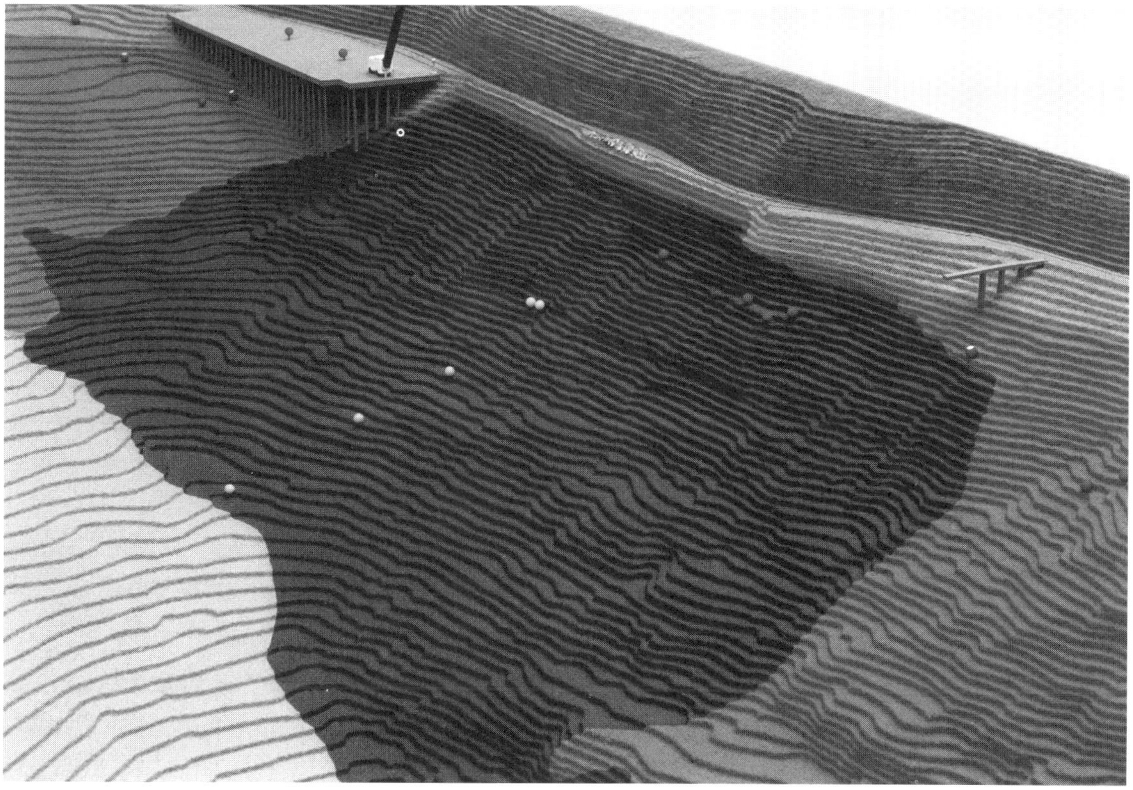

Figure CH-7.8 Skagway Marine Slide. Oblique view of a scale model showing "after failure" conditions. The dark area represents scour greater than 20 feet deep (scour is 70 feet deep near the center). Light-colored pinheads are Landslide Technology boring locations to investigate the failure. The crane is seen in the upper left.

The four surviving workers each gave eyewitness accounts of the event to the police within hours of the tragedy. The crane operator, who was in the best position to observe it, described the slide as follows:

> "I sat there in the crane (overlooking the site)... when out of my peripheral vision I saw or sensed movement... that the crane was starting to move toward the mountainside to my left. At that time I noticed all four cells and piling... everything in my field of view... begin to move and slide down and to the right into the bay. At this time I opened the door to the crane and jumped to the ground. I looked around and watched the cells disappear out of view and a wall of water rush in and envelop the area." He then ran for his life together with another worker. "When we stopped and looked back, I could see a wave moving across the bay toward the ferry terminal and heard a loud crash or boom at this time..."

Another worker wrote in his police statement:

> "... heard the sheet wall snapping, and saw it start to collapse... [referring to the lost worker, Paul Wallin] the ground was sliding out from under him, then the piling that was to be removed started falling over; their bottoms slid toward the water so the tops tipped toward the beach as they fell. At least one fell across Paul and then he and the piling disappeared as a huge wall of water and debris came crashing in over them."

One worker noted that the barge, used to load out the extracted piles, had risen to the level of the dock deck, indicating that the wave had reached a height of about 25 feet above low tide at the construction site. The wave, with an equal depth of trough, must have been initially 50 feet high.

In summary, it is clear from the witness statements that the slide began slowly with stones sliding down the surface of the slope and the tops of the sheet piles tipping backward. This may account for the first 10 seconds of above-water activity. When it became a rapid and widespread flow slide—the final second or so "like flushing a toilet"—the rapid ground displacement caused the 50-foot high wave that crossed the harbor, damaging the ferry terminal.

Landslide Investigations

Five overwater borings were made in February 1995 at Section A-A (Figures CH-7.7 and CH-7.8) near the center of the landslide. Drilling was performed from an anchored barge with twice-daily tidal variations of up to 25 feet during cold, wintry weather. Work continued around the clock, using mud-rotary techniques to bore through the soft overburden soils and HQ-coring within the bedrock or talus. Water depths ranged from 90 to 220 feet below mean sea level. Samples were recovered by thin-wall tubes and standard penetration test sampler.

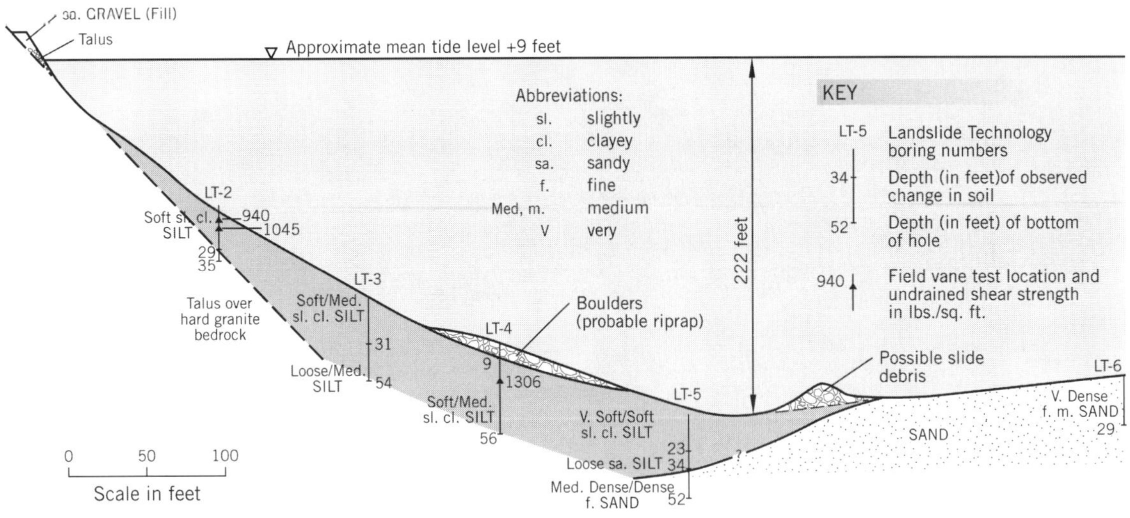

Figure CH-7.9 Skagway Marine Slide. Geological section A-A.

The geological section A-A (Figure CH-7.9) shows that the sloping ground on the east side of Skagway harbor comprises soft marine silt overlying talus or bedrock. Most borings did not reach bedrock, which plunges at a steep inclination below the harbor. To the west, LT-6 encountered very dense, fine to medium sand; this hole is located near the edge of the Skagway River delta on much flatter east-facing slopes. Thus, the "valley" bottom is where the marine silts on the east shoreline meet the alluvial fan of the river delta. The flow slide occurred in the marine silts. After moving downslope, the slide was turned south by dense sand (more resistant to erosion) and scoured a channel along the "valley" bottom to deeper water.

The marine silts are soft to medium stiff, sandy, slightly clayey silt. There are minor variations in soil composition from silt to clayey silt. The measured soil properties are summarized on Table CH-7.1.

Effective stress triaxial shear tests gave shear parameters of $c' = 0$, $\phi' = 38°$ for the marine silt. Only four in-place Geonor vane tests were attempted due to the difficult drilling conditions; one result was discarded due to barge movement during the test.

A concurrent site investigation was made by the dock owner to design the replacement dock. Most of these borings were located outside the landslide scour area. The main benefit of this information, apart from confirming the lateral continuity of the marine silt, was that three borings reached either bedrock or talus below the marine silt and measured artesian heads of around elev. +18.0 feet. The artesian borings are identified by symbol (AH) on the site plan, Figure CH-7.7. They encountered fresh water with flows of up to 60 gpm coming into the drill stem from the bedrock or talus. Therefore, the marine silt must extend up to about elev. +18 to block the release of groundwater and thus create the artesian conditions. This study, with mostly land-based borings, showed that talus deposits overlie bedrock.

Table CH-7.1 Summary of Marine Silt Properties

	Average	Range	Tests
Natural water content	31.3%	28.1 to 37.8%	14
Liquid limit LL	26.9%	22.4 to 31.7%	13
Plastic limit PL	21.9%	17.3 to 24.3%	13
Plasticity index PI	5.0%	Nonplastic to 12.5%	13
Wet density	125.0 lb./cu. ft.		4
In-place peak vane shear strength, c	1,100 lb./sq. ft.	940 to 1,310 lb./sq. ft.	3
In-place remolded shear strength	210 lb./sq. ft.	140 to 290 lb./sq. ft.	3
Coefficient of consolidation, c_v		790 to 29,700 sq. ft./year	5

SOIL PROPERTIES

Unit	γ pcf	γ' pcf	ϕ degrees
Riprap stockpile	120		45
Platform fill	125	61	40
Old fill	125	61	40
Marine silt	123	59	38

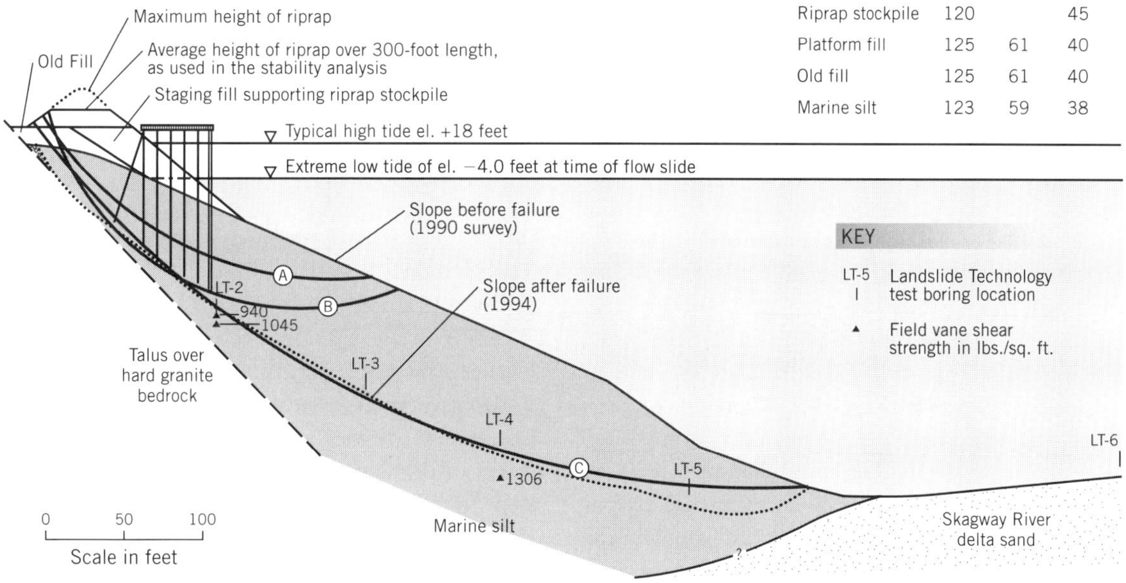

Figure CH-7.10 Skagway Marine Slide. Stability arcs used in the analysis of section A-A.

LANDSLIDE STABILITY ANALYSES

As presented in Chapter 2, Section 2.12, Bjerrum (1971) described similar submarine flow slides in Norway starting with an initial shear slide that can be predicted by a soil mechanics analysis. Bjerrum observed that if the initial slide involves saturated, loosely deposited, fine sand and silt, it will be followed by a flow slide. He noted that many submarine flow slides occur at extreme low tide, when stability is most critical due to loss of support from the water.

The purpose of the stability analysis, therefore, was to determine whether an initial shear slide would occur at Skagway under the observed conditions. The landslide can be analyzed in two ways: by total stress analysis or by effective stress analysis.

Three stability arcs (Figure CH-7.10) were chosen: Arc A is a shallow arc passing through the mid-depth of the marine silt stratum, where the least amount of consolidation would occur in the foundation silts after loading with fill; Arc B is similar to Arc A but reaches the full depth of scour at the top of the slope; and Arc C covers the entire scoured surface from headscarp to toe of the slope. Arc A intercepts the wood piles supporting the dock; Arcs B and C pass below the pile tips.

Total Stress Analysis

A shallow shear slide is a more probable initial failure, as described by Bjerrum (1971), and deduced from an inspection of the loaded cross-section of the slope. A total stress analysis was made using shallow arc A and the two field vane test results close to it. The average vane strength for the two tests is 992 psf. Using Bjerrum's chart (Figure 3.28, Chapter 3) to convert field vane shear test results to the mobilized shear resistance along a surface of sliding, the correction factor is 1.14. Therefore, the *available* field shear strength within the marine silt for arc A is 1131 psf. The calculated factor of safety for shallow arc A is:

$$F = \frac{1{,}131}{1{,}125} = 1.01$$

The total stress analysis shows that a shear slide is likely to occur at extreme low tide under the fill loads placed at the top of the slope.

Effective Stress Analysis

The Skagway landslide was selected to demonstrate several types of stability calculations in Chapter 9 and the details will not be reproduced again here. A sophisticated suite of computer software programs (Sigma/W, Seep/W and Slope/W from Geo-Slope, Calgary, Canada) was used to make the computations. The calculations were complex, involving a combination of: (i) partly consolidated soils (marine silt); (ii) artesian pressures; (iii) pile interference; and (iv) rapid drawdown.

(a) Partly consolidated soils

The excess pore pressures set up in the marine silts due to the weight of the platform fill and riprap were calculated as follows:

(i) The vertical stresses in the marine silt due to the fill loading were calculated (Figure 9.33, Chapter 9).

(ii) Assuming the initial excess pore pressures are equal to the applied fill stresses (realistic for soft, normally consolidated soil), the silt begins to consolidate as soon as the fill is placed. Drainage is immediate at the two drainage boundaries, i.e., the talus/bedrock boundary to the west and the sea to the east. The full height of fill was assumed to have been in place for 29 days when failure occurred (one-half of construction time plus 19 days under full load, Figure CH-7.5). The rate of dissipation was determined by calculating a "composite" value of the coefficient of consolidation (1,850 sq. ft./year) from the individual laboratory triaxial dissipation tests (see Chapter 9, Section 9.12). Knowing the initial excess pore pressures and the percent dissipation at various distances from the drainage boundaries, the contours of *undissipated* head can be drawn (Figure 9.35, Chapter 9). For each arc, (A,B,C), the excess pore pressure at the base of each slice was calculated and used in the stability analyses.

(iii) The calculated excess pore pressures were reduced by 50 percent due to dilatancy as the silt sheared to failure. The 50 percent drop was based on the results of laboratory consolidated-undrained triaxial tests, i.e., Skempton pore pressure parameter $A_f = 0.5$.

The calculation of the undissipated heads along trial arc A after 29 days of loading by the staging fill and riprap stockpile is given in column (8) of Table CH-7.2.

(b) Effect of artesian pressures

The measured artesian head at the talus/bedrock boundary is elev. +18.0 feet. Sea level varies from a high of about elev. +22 feet to an extreme low of elev. –4 feet. The mean sea level is elev. +8.7 feet. As explained in Chapter 9, Section 9.11(c), the marine silt is too impermeable for silt to drain in response to twice-daily tide changes except for the outer 3 feet of the slope; the latter has essentially no effect on the slope stability. As a result of many thousands of daily tide fluctuations, the groundwater within the slope will have reached a long-term steady state crossflow from elev. +18.0 feet (artesian level) to +8.7 feet (mean sea level), as shown on Figure 9.37 (Chapter 9). This remains unchanged during the daily tidal fluctuations. Based on the principle of effective stress, the shear strength within the slope also remains unchanged. This shear strength was determined for the three

Table CH-7.2 Calculations of Excess Pore Pressures Acting on Base of Slices for Trial Arc A

(1)	(2)	(3) = (2) – 8.7	(4)	(5)	(6) (4)(5)	(7) = 0.5(6)	(8) = (7)/64	(9) = (3)+(8)
	Artesian Head		Pore Pressure created by Fill					
Slice	Steady Crossflow (elev).	Head above Sea Level (feet)	Increase in Vertical Stress (psf)	Percent Undissipated at Base of Slice	Undissipated Pore Pressure (psf)	50% Undissipated Pore Pressure (psf)	Undissipated Head (feet)	Total Excess Head (feet)
3	14.4	5.7	1925	10%	193	96	1.5	7.2
4	14.7	6.0	2000	35%	700	350	5.5	11.5
5	14.4	5.7	1675	52%	871	435	6.8	12.5
6	14.2	5.5	1400	63%	882	441	6.9	12.4
7	13.8	5.1	1100	72%	792	396	6.2	11.3
8	12.8	4.1	700	75%	525	263	4.1	8.2
9	11.3	2.6	275	55%	151	75	1.2	3.8
10	9.7	1.0	100	10%	10	5	0.1	1.1

Explanations:

Col. (2): Steady state crossflow from elev. +18.0 feet (artesian head) to elev. +8.7 feet (mean sea level); $k_v = 10\ k_h$ (anisotropic)

Col. (4): Increase in vertical stress obtained for recently placed fill using Boussinesq distribution with depth

Col. (6): Undissipated pore pressure at base of slice assumes initial $\Delta u = \Delta\sigma_v$ and dissipation based on $c_v = 1{,}850$ sq. ft./year

Col. (7): 50% reduction of Column (6) is based on $A_f = 0.5$ (Skempton Pore Pressure Coefficient A at failure, based on testing of the marine silts at Skagway)

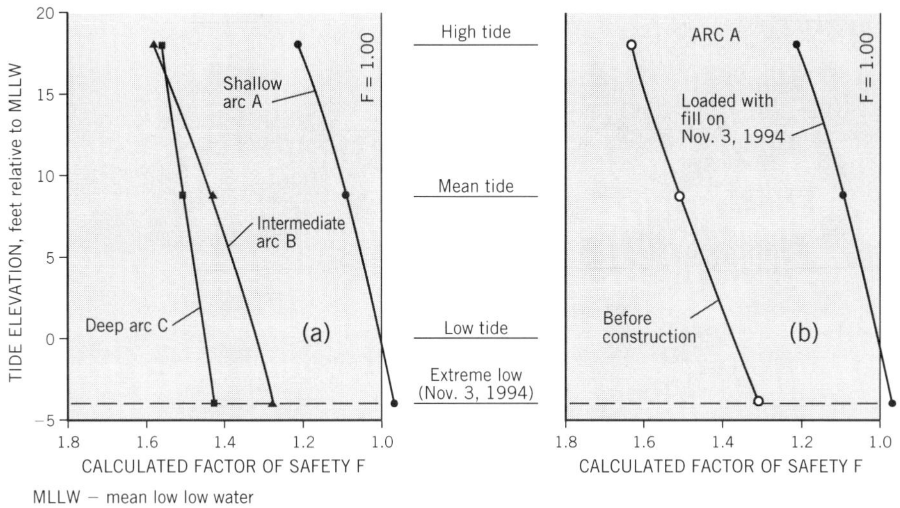

Figure CH-7.11 Skagway Marine Slide. Results of effective stress stability analyses on section A-A:
(a) three trial arcs (A, B, C) on November 3, 1994
(b) stability of the original slope before construction compared to the stability on November 3, 1994 (trial arc A)

trial arcs A, B, C, by performing a stability analysis with the sea at mean tide level. The analysis provides the shear strength available within the slope, and any tidal change is a $\phi = 0$ analysis. Therefore, changes in the factor of safety are the result of changes in the driving moment (WR sin α) of the analysis, which increases at low tide and decreases at high tide due to buoyancy.

The calculation of the excess head above mean sea level due to the artesian pressure is given in column (3) of Table CH-7.2. The table also gives the total excess head (column 9) obtained by combining the artesian pressure with the undissipated pressures resulting from the fill loads. The data are for shallow arc A.

(c) Effect of piles on shallow arc A

The contribution of the piles to stability is discussed in Chapter 9, Section 9.11(d). For stability arc A, the maximum shear resistance of the piles was calculated to be 8,600 lb./ft. This is about 3 percent of the total shear resistance along the arc and is, therefore, relatively insignificant.

(d) Results of stability analyses

The results of the stability analyses for arcs A, B, and C at the time of failure are shown on Figure CH-7.11(a). The trial arc A is the more critical (lowest factor of safety) at the extreme low tide of elev. –4.0 feet. This result (F = 0.97) is consistent with an initial shear slide causing the development of a much larger flow slide (Bjerrum, 1971).

Using arc A only, the stability of the slope before and after the construction of a fill at the top of the slope is shown on Figure CH-7.11(b). These calculations show that the factor of safety of this arc dropped by about 0.3 as a result of the fill construction.

(e) Conclusions

Engineers familiar with effective stress stability analyses will understand the difficulties of combining the factors present at Skagway into a single analysis. However, all the data have been obtained from factual field and laboratory sources. It is evident that the construction of the fill above the slow-draining marine silt produces a factor of safety of around 1.0 at extreme low tide.

In summary, both total stress and effective stress stability analyses show that a shear failure landslide probably would occur at the extreme low tide of November 3, 1994.

LANDSLIDE CAUSATION AND MECHANISM OF FAILURE

The landslide at the PARN dock in Skagway probably occurred in a similar manner to those described by Bjerrum (1971) for marine silt slope failures in the Norwegian fjords. The loading of the upper slope with 13,700 cu. yds. of fill and riprap (weight—approximately 22,500 tons) induced a shear failure at extreme low tide on November 3, 1994. The extent of the initial shear slide is approximately estimated by arc A and the broken line shown on the site plan, Figure CH-7.4. The ground crack at C, and the disarray of the sheet piles on the day preceding the failure, probably resulted from developing shear movements within the slope.

Within a few seconds of the onset of the initial shear failure, instability spread creating a major flow slide. The flow slide developed to the south because the initial shear slide caused a pulldown loading of the entire length of the dock to the south, thereby inducing liquefaction from the rapid thrust of the piles into the marine silt. The flow slide extended north below the construction cells because: (i) numerous wood piles had been extracted, causing weakening of the sensitive marine silts in this area; and (ii) there was a differential head of 4 to 8 feet of sea water between the inside and outside of the cells. Once initiated, the flow slide moved directly downslope to reach the compact sands of the river delta, and

Figure CH-7.12 Skagway Marine Slide. View of the landslide headscarp from the end of the remaining wooden dock (Photo: November 14, 1994).

then turned south along the "valley" bottom to scour a deep channel through the marine silts into the deeper water of the fjord (Figure CH-7.7). Since the shear failure below the stockpile occurred a few seconds before the general liquefaction of the marine silts, the slope immediately below the stockpile would slump first and sea water would enter the construction site at the south and southwest sides of Cell 1, as described by the workers.

AUTHOR INVOLVEMENT

The author visited Skagway about 10 days after the flow slide. A photograph taken by him from the end of the remaining dock (Figure CH-7.12) shows the loss of the wood dock, riprap stockpile, new construction, and much of the railway embankment. He participated in the two larger lawsuits as a geotechnical landslide expert regarding causation.

8 Faraday Slide

Figure CH-8.1 Faraday Slide. Rockfill buttress nearing completion. Outline of the landslide is shown by the dashed white line. Clackamas River in foreground; Faraday Canal at top of photograph (Photo: September 28, 1989).

Key Points

- Survey hubs monitor landslide movements
- Required depth of inclinometer casings
- Delayed failure
- Circular arc slip surface
- Buttress remediation

Summary

A canal built in 1907 conveys water diverted from the Clackamas River to the Faraday generating plant. The canal was widened and deepened in 1957 and fill spoil was placed on the adjoining slope. Slope movements were observed in 1976 and then accelerated, caused by a developing failure within the underlying hard clay (decomposed flow breccia). The slip surface was a circular arc failure. The slide was stabilized in 1989 by an external buttress built part way into the Clackamas River.

Measurements of surface movements were taken for almost 10 years before remediation and demonstrate the accelerating velocity of a delayed failure. The vertical and horizontal measurements of the survey hubs reflect the circular arc slip surface at depth.

Background Information

Portland General Electric's Faraday Project, near Estacada, Oregon, is a hydroelectric generating system on the Clackamas River. River water is diverted into a canal that runs parallel to the river before entering the forebay of the power plant (Figure CH-8.1).

Faraday Slide is a medium-size landslide on a hillside between the Faraday Canal and the Clackamas River. It is approximately 440 feet wide on average and 360 feet long. The vertical elevation change is 115 feet (Figure CH-8.2).

The hydroelectric project was built in 1907. Minor ground movements at the landslide site were noted in 1938 correspondence. In 1957, the canal was widened from 60 feet to 100 feet, and deepened from 12 to 16 feet. Spoil from the excavation was placed on the downhill slope below the canal. Some movements occurred, and the area was regraded in 1967 by flattening the outer slope from 1.4:1 to 2:1 (horizontal:vertical). Essentially, the fill was redistributed to cover the entire slope (Figure CH-8.3).

Surface Monitoring

Renewed movements of the slope began in 1976. Portland General Electric installed 14 surface survey hubs on the slope

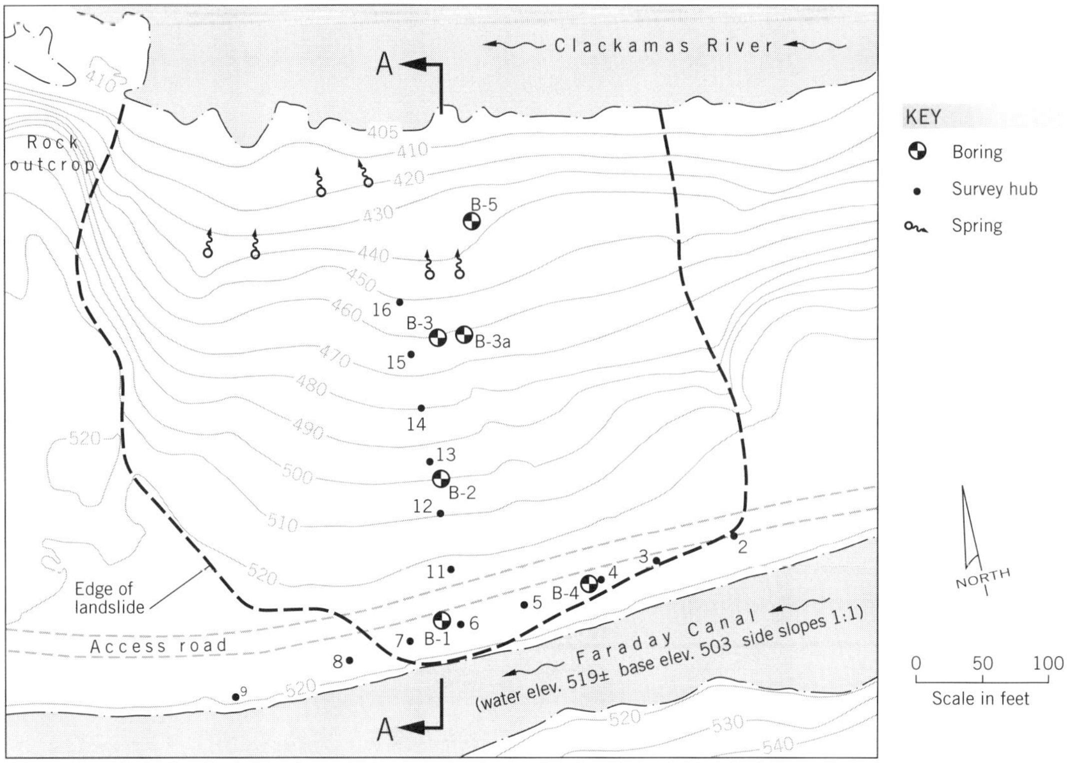

Figure CH-8.2 Faraday Slide. Site plan.

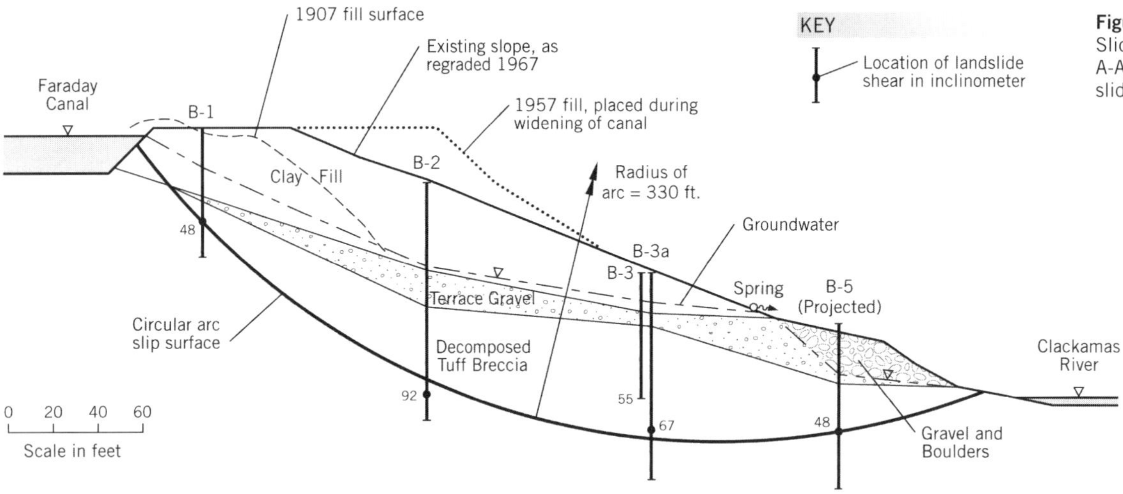

Figure CH-8.3 Faraday Slide. Geological section A-A through center of slide.

in December 1977 (Figure CH-8.2). Eight hubs were set along the crest of the slope alongside the canal to monitor the active and inactive areas and detect any lateral widening of the slide. Seven survey hubs, each 40 feet apart, were oriented in a line passing upslope to downslope through the center of the slide.

The initial velocities of movement were minor, but began to accelerate with time. A typical set of data for horizontal movements, at survey hub 12 (Figure CH-8.4), show accelerating movements, indicating that the stiff to hard clay was strain-softening (weakening) with time. This is the essence of a delayed failure (see Chapter 2, Section 2.15). If left untreated, a catastrophic landslide with release of canal water could have occurred within a few years.

CAUSATION

The landslide was caused by placing spoil from the canal excavation on the slope below the canal.

SITE GEOLOGY

At the landslide site, the Clackamas River has cut a narrow 700-foot deep canyon through the Sardine Formation of Miocene age. The formation is composed chiefly of tuff and tuff breccia with minor amounts of andesite and basalt flows. During the late canyon deepening, the river cut through old

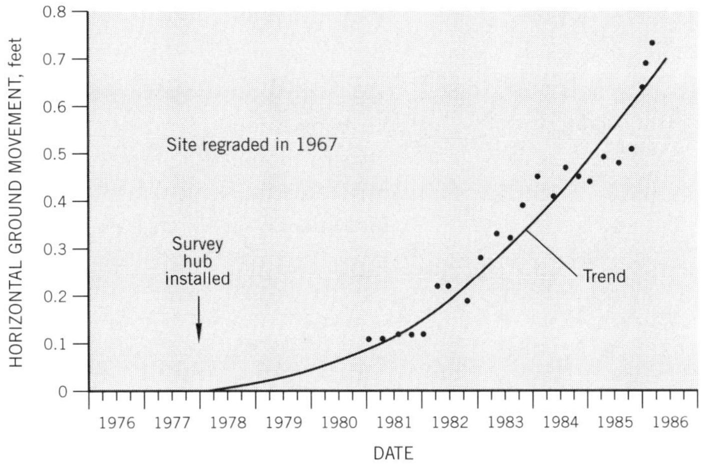

Figure CH-8.4 Faraday Slide. Horizontal movements of survey hub 12.

floodplains (Quaternary). Remnants of these terrace gravels were encountered below the recent fills (Figure CH-8.3).

Aerial photos taken around 1930 show that a large ravine drained the hillside. The ravine was crossed by the canal construction.

Outcrops of cemented tuff breccia with rock-like characteristics can be seen near river level. The presence of cemented tuff breccia on the river bottom indicates that the toe of the slide outcrops above the river.

Springs were observed on the slope as high as 40 to 50 feet above the river. They appeared to be emerging near the contact between the fill and terrace gravel.

SITE INVESTIGATION

The site was explored by five borings (Figures CH-8.2 and CH-8.3), which were advanced by the ODEX method (see Chapter 3, Section 3.7). Samples were mostly recovered from Standard Penetration Tests (SPT). Several thin-wall tube samples also were obtained for laboratory shear tests. The summary boring log of B-2 is shown on Figure CH-8.5.

The geological section through the center of the slide is shown on Figure CH-8.3. The subsurface conditions comprise three strata. Near the surface, fill was encountered to depths of up to 38 feet, and comprises old fill (placed during the original construction of the project) and more recent fill (placed during

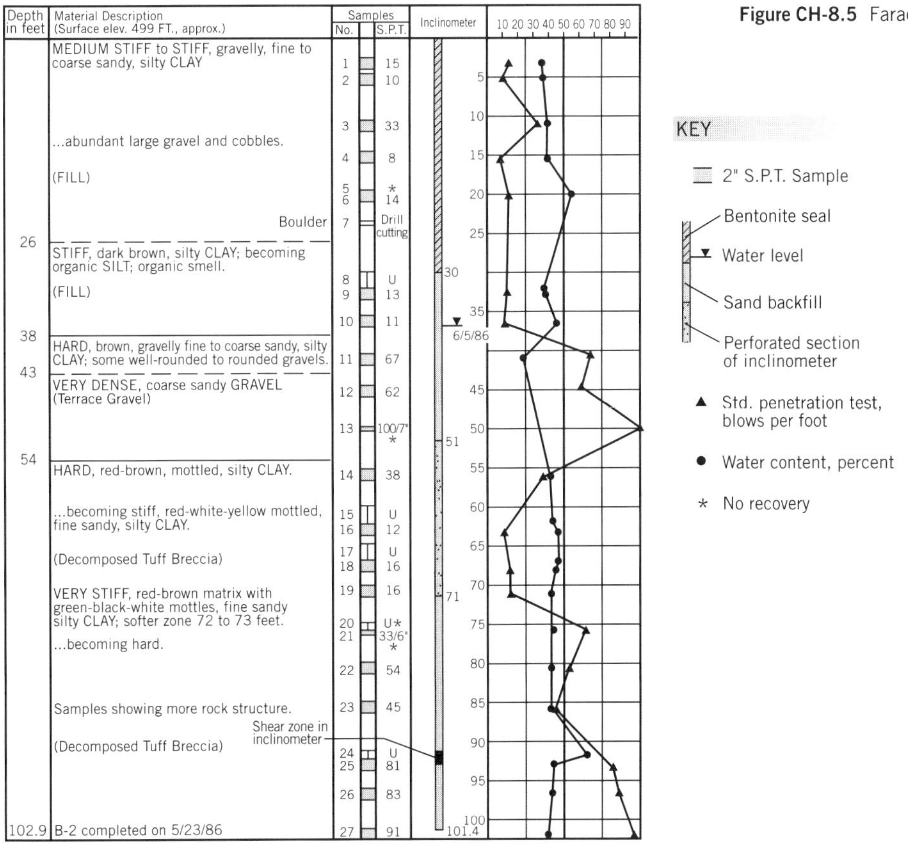

Figure CH-8.5 Faraday Slide. Boring log B-2.

the widening of the canal). Underneath the fill was rounded terrace gravel deposits of variable thickness. The lowest unit at the slide area was decomposed tuff breccia, generally described as very stiff to hard, multicolored, sandy silty clay.

INCLINOMETER DATA

At the completion of each boring, an inclinometer casing was installed to the bottom of the hole. The casing was perforated over a short length to allow water to enter the casing and thus "double" as a piezometer. (*Note:* This procedure of 1986 is no longer followed by Landslide Technology because of the slow response of an inclinometer casing to rapid rises in groundwater during storms.) The annular space between the inclinometer casing and the wall of the boring was backfilled with coarse sand and sealed over the upper 30 feet with bentonite.

The inclinometers showed that a circular arc slip surface was passing almost entirely through the hard clay of the decomposed flow breccia (Figure CH-8.3). The slip surface was confined to a very thin shear zone.

Movements were observed in the inclinometers of B-1, B-2, and B-5, but not in B-3. After reviewing the data, it became evident that the slip surface was passing below the bottom of B-3 (55 feet deep, with base at river level, see Figure CH-8.3). SPT blow counts were 63, 62, and 85 in the three samples from 45 to 55 feet. At that time, Landslide Technology had a policy of stopping drilling when three successive SPTs exceeded 50 blows per foot. Boring B-3 was deepened by drilling boring B-3a to a depth of 89 feet. There was a sharp drop in SPT (to 8 blows per foot) at 65 feet. Subsequently, shear movement was measured by the inclinometer casing at a depth of 67 feet.

It is not unusual to encounter relatively high SPT blow counts *above* the slip surface of a landslide. Therefore, the depth of inclinometer casings should be based on the deepest conceivable depth of shearing, and not on the SPT blow counts.

SURFACE MOVEMENTS ANALYSIS

The survey hubs monitored the surface movements of the landslide from 1977 to 1989. Hubs 6 and 11 to 16, which progress from head to toe along the center of the slide area, are plotted on the section, Figure CH-8.6. The average vector of movement is shown for each hub and the slopes (vertical: horizontal) are given in the table below the figure.

The circular arc for stability analysis is based on the best fit of the movement depths in the four inclinometer casings, the observed headscarp, and the toe at river level. The tangents to the circular arc at each survey hub are given in the table of Figure CH-8.6. There is very good agreement between the surface measurements at the hubs and the at-depth movements along the circular slip surface. Therefore, surface measurements can provide input concerning the type of landslide movement, i.e., planar or circular (see also Case History 6 for similar evidence). They can also provide continuing data on the amount and rate of movement long after inclinometer casings have sheared off.

LABORATORY TESTS

Average soil properties measured in the decomposed tuff breccia (fine sandy silty clay) are: density 114.5 lb/cu. ft; water content 41%; LL 70, PL 39; torvane strength > 3,000 lb./sq. ft.

Undisturbed samples taken in borings B-1 and B-2 coincided with the slip surface in the inclinometer casings. Two stress-reversal shear box tests were performed on each sample. The results for sample 24 of boring B-2, depth 91 to 92 feet, are shown on Figure CH-8.7. Note that this soil reached the residual strength within three cycles of shearing. The Mohr envelope, Figure CH-8.7(b), gives average residual strength parameters of c'_r = 435 psf, ϕ'_r = 17.5°. The low result at 4 tons/sq. ft. was ignored.

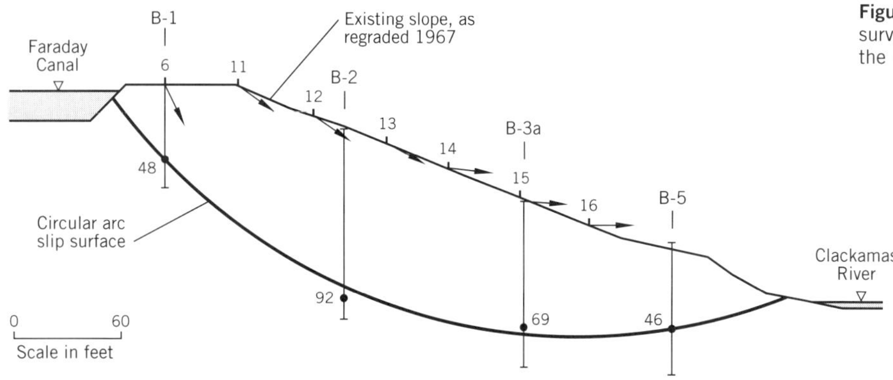

Survey pin	6	11	12	13	14	15	16
Measured slope	2.11	0.68	0.56	0.43	0.13	0.13	0.00
Circular arc slope	1.00	0.72	0.51	0.36	0.22	0.08	−0.01

slope = ratio of vertical:horizontal

Figure CH-8.6 Faraday Slide. Comparison of survey hub movements at the surface with the underlying landslide slip surface.

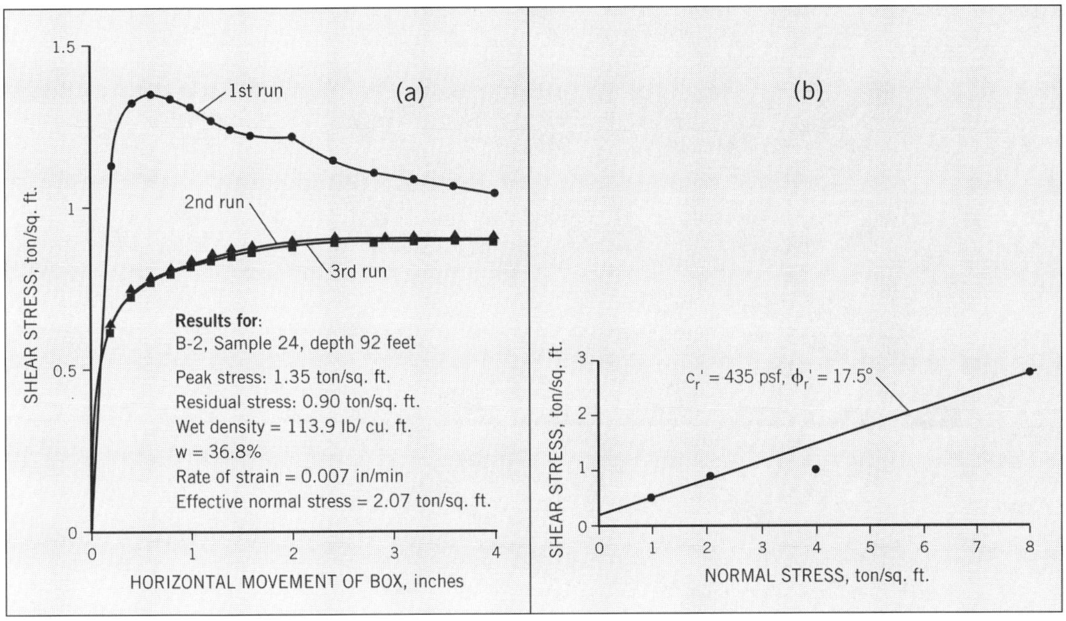

Figure CH-8.7 Faraday Slide. Direct shear tests to measure residual strength
(a) shear stress vs. horizontal displacement
(b) Mohr envelope

LANDSLIDE STABILITY ANALYSIS

The stability analysis for this landslide has been used as an example of a circular arc analysis in Chapter 9, Section 9.9. Using the residual strength parameters $c_r' = 435$ psf, $\phi_r' = 17.5°$ and the groundwater profile shown on the section, the stability analyses gave the results shown on Table CH-8.1.

In the unlikely event that the slip surface passes below the buttress, the calculated F values are 1.12 immediately after construction and 1.34 long term.

Table CH-8.1 Results of Effective Stress Stability Analyses

Conditions	Factor of Safety F
Existing slope	F = 0.97
Immediately after construction of buttress	F = 1.24
Long-term, clay foundation fully consolidated	F = 1.40

REMEDIATION

The landslide was remediated in late 1989 by construction of a 510-foot long rockfill buttress extending part way across the Clackamas River (which carries minor amounts of water at this location because of the upstream diversion of water into the Faraday Canal). Some difficulty was experienced in obtaining environmental permits to allow the encroachment into the riverbed. There was also a need to calculate the effect of the buttress on flood levels upstream from the project; fortunately, the affected areas were owned by the client. Another design detail was to carefully protect the upstream edge of the buttress to prevent erosion and eddies during flood.

The buttress design cross-section is shown on Figure CH-8.8. The buttress is about 50 feet wide, 34 feet high. The outer slope is a relatively steep 1½:1 (horizontal: vertical) slope

Figure CH-8.8 Faraday Slide. Rockfill buttress section.

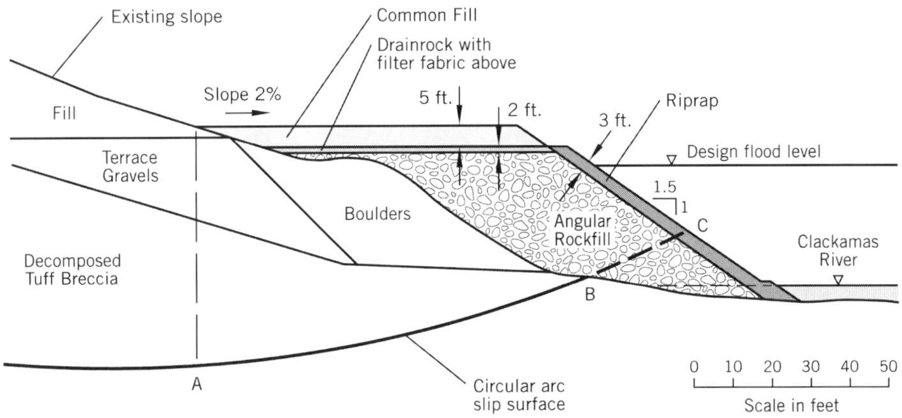

Figure CH-8.9 Faraday Slide. Constructing the riprap outer slope using a backhoe (Photo: September 5, 1989).

Figure CH-8.10 Faraday Slide. Filter fabric and common fill being placed above compacted drainrock (Photo: October 6, 1989).

to minimize encroachment into the river channel while obtaining a high benefit to stability. The riprap is 3 feet thick (measured normal to the slope) and is designed to resist a flood velocity of 23 feet/second (Figure CH-8.9).

The rockfill was a well-graded angular shot rock (basalt) with maximum stone size of 18 inches. The maximum lift thickness was 2 feet. The drainrock placed above it was a 3-inch minus shot rock. The rockfill was compacted by four passes of a D-8 dozer; the drainrock was compacted by a vibrating roller to give a smooth surface for the fabric. Fines were limited to 2% and 3% by weight, respectively, for the rockfill and drainrock, and were nonplastic. All rockfill and riprap placed in the river was pre-washed at the quarry to remove fines.

Some of the buttress weight was provided by common fill taken from the top of the slope, which helps slope stability and is inexpensive because of the short haul distance (Figure CH-8.10). Normally, a higher percentage of a buttress can be built from common fill (with significant cost savings), but at this site the need to prevent erosion during flood and provide a steep outer slope meant that rockfill was needed up to flood level.

CONSTRUCTION COST

Quantities:	Common excavation 8,400 cu. yd. (from top of slope)
Buttress Fill:	Rockfill 33,970 cu. yd.; Riprap 5,440 cu. yd.; Drainrock 3,175 cu. yd.
	Common fill 8,000 cu. yd.; Total fill 50,585 cu. yd.
Construction cost:	$790,000 (1989); Cost/lin. ft. of buttress: $1,550
Construction time:	3 months

Engineering News-Record Construction Cost Index: 4600

BUTTRESS BENEFITS

At this project, shear is occurring along a weak, well-defined circular slip surface within a much stronger stratum of stiff to hard clay. A rockfill buttress at the toe of the slope provides several advantages:

- It provides extra resistance to sliding because the discrete shear zone of the landslide must pass through the high strength rockfill of the buttress (line BC, Figure CH-8.8).
- It is free-draining, so pore water pressure will not build up within it.
- It provides a counter-rotational moment, point A being downslope from the center of rotation of the arc (Figure CH-8.8).
- It provides extra strength to the underlying clay stratum (line AB, Figure CH-8.8), but only in the longer term, after excess pore water pressures (in response to the buttress loading) have dissipated.

9 Goat Lick Slide

Figure CH-9.1 Goat Lick Slide. Bare ground between the road and the Flathead River is the slide area. The river makes a 180° bend, cutting into the bottom of the slope during flood (Photo: May 14, 1993).

Key Points

- Concrete shear pile wall
- Selective stabilization
- Tieback ground anchors
- Inclined inclinometers

Summary

Goat Lick Slide developed in a 275-foot high steep slope above the Flathead River in Glacier National Park, Montana. The river is actively eroding the base of the slope. Access for drilling equipment onto the slope was denied due to it being an environmentally sensitive site.

The case history describes the use of angled inclinometers to obtain information about the depth of slippage below the road. It also demonstrates the design of a tied-back shear pile wall remediation supporting a cantilevered outer lane on the highway. The case history is an example of selective stabilization (Chapter 14, Section 14.5) in which the highway is stabilized, but not the entire landslide.

Background Information

The landslide site is at the south boundary of Glacier National Park, Montana. Highway 2 passes through the park for a few miles on the east side of the Flathead River (Figure CH-9.1). Goat Lick is an environmentally sensitive area where mountain goats lick minerals from calcareous rocks exposed on the steep banks of the river.

In 1980, a highway bridge was destroyed by a snow avalanche; the bridge site is on the right in Figure CH-9.1. Highway 2 then was raised 12 to 16 feet by construction of a reinforced earth (R.E.) wall, one of the first to be built in the United States (Figure CH-9.2). The raised road extends from the reconstructed bridge to a new bridge that was built to allow goats to pass under the highway to reach the river banks.

Minor cracking and settlement of the new fill was first observed in the spring of 1985, about 4 years after construction. By 1989, distress had increased significantly. In June 1991, Montana Department of Transportation (MDT) retained Landslide Technology. Three borings (GL-1, -2, -3 with slope inclinometers installed) were drilled to supplement earlier MDT investigations. The site plan, Figure CH-9.3(a) shows the road cracks, which extended to the inner fogline, and boring locations. The outer face of the R.E. wall had settled about 1.4 feet at Section X-X (Figure CH-9.3c). Ground movements occurred in April coincident with snowmelt.

The borings encountered native fill underlain by highly weathered argillite (i.e., a wide range of particle sizes from clay to angular boulders) with fractured argillite bedrock below. The unconfined compressive strength of intact rock cores ranged from 2,300–6,900 lb./sq. in. Drilling contractors indicated that an auger with rock teeth could penetrate through rock with a strength of about 4,000 psi; the practical limit for a large auger is 6,000 psi compressive strength. Therefore, the use of a helical auger at this site was practical.

The road is 275 feet vertically above the Flathead River, which makes a 180° bend at the base of the slope (Figure CH-9.1). The river in flood (May–June) cuts into the soft argillite rocks of the slope causing slope toe erosion. The outer slope is at about 38° to the horizontal, which is the approximate angle of repose of the colluvium (Figure CH-9.2 and Figure CH-9.4). Bedding planes in the argillite bedrock dip 60°–70° to the south. A local thrust fault may have crushed the rock in the slide area, making it more susceptible to river erosion at this location.

The preliminary evaluation of the landslide causation was that it was a combination of: (i) erosion at the toe of the slope by the river, causing surficial movements and some loss of ground below the R.E. wall; (ii) elevated groundwater levels in the slope; and (iii) the R.E. wall locally overloading the marginally stable slope. Therefore, the cracks in the road initially were thought to be a circular arc failure passing behind and underneath the R.E. wall.

Figure CH-9.2 Goat Lick Slide. Highway 2 is supported by a reinforced earth wall. Below the wall, the slope is at 38° to the horizontal (Photo: May 14, 1993).

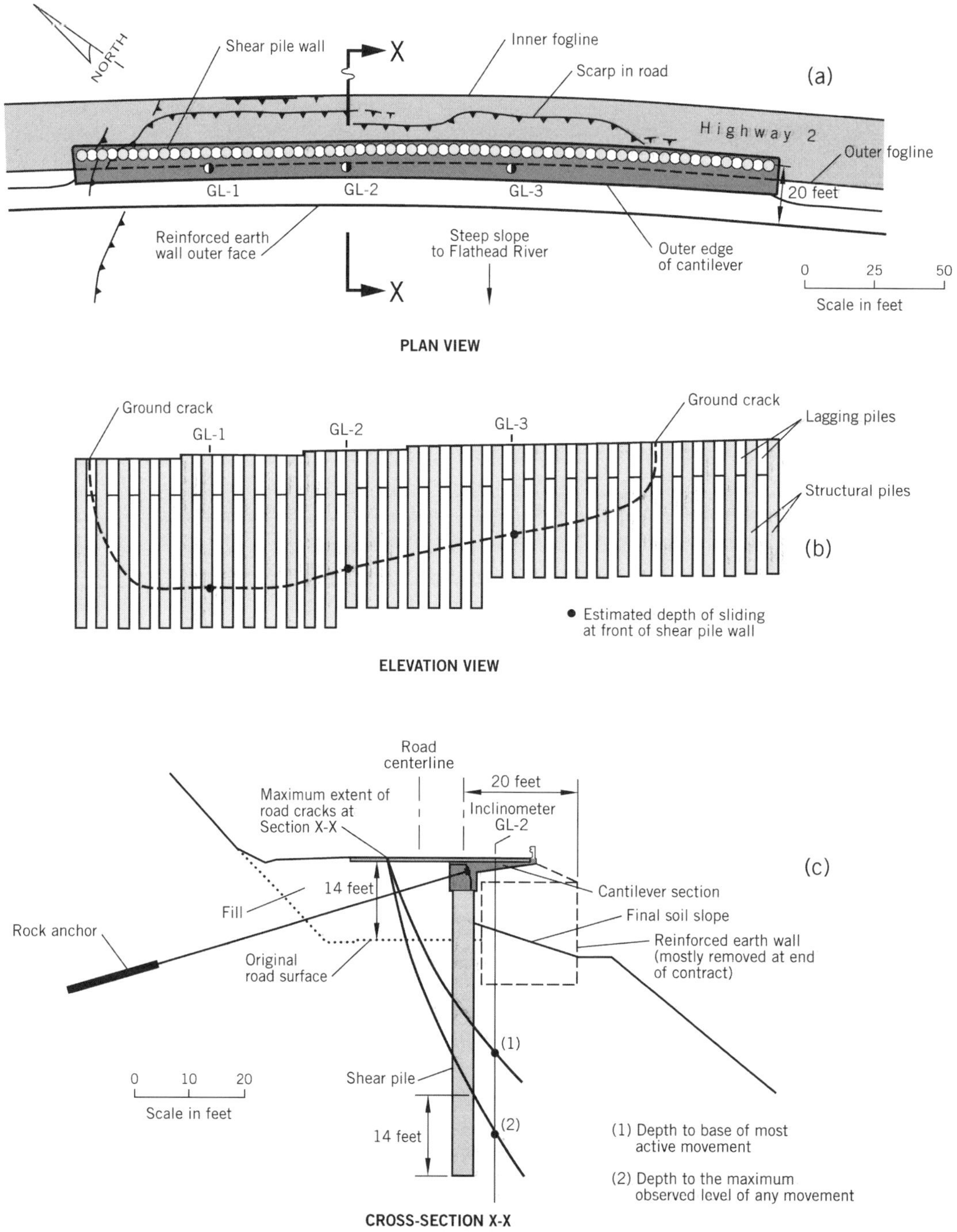

Figure CH-9.3 Goat Lick Slide:

(a) site plan
(b) elevation of shear pile wall
(c) section X-X

Note: Section X-X is drawn to twice the scale of the plan and elevation.

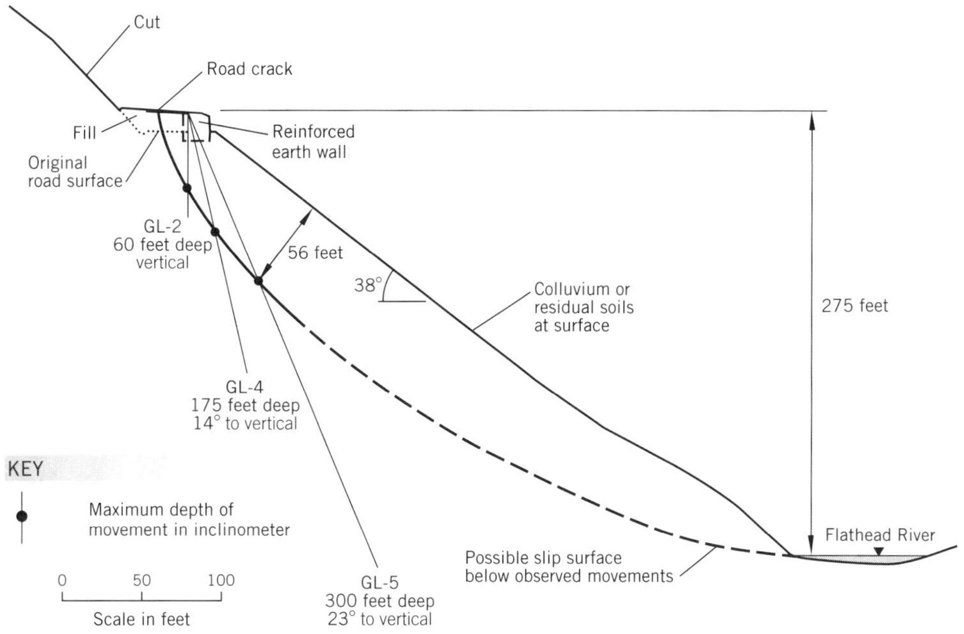

Figure CH-9.4 Goat Lick Slide. Section X-X near the center of landslide showing how the inclined inclinometers provided data on the landslide depth behind the slope.

Remedial Options

Some conceptual treatments considered for the distressed road included: (i) relocation of the road further into the slope to support the road on bedrock; (ii) bridge the slide area; and (iii) install a tiedback soldier pile wall immediately downhill from the face of the existing R.E. wall. MDT favored option (iii) because it would stabilize the road without major road reconstruction and would minimize environmental impacts.

Another recommended treatment was to install erosion control measures at the base of the slope to arrest river erosion. As a designated wild and scenic river, it is very difficult to obtain permits for construction in the river. However, such treatment has been recommended and may have to be implemented eventually.

Inclinometer Readings

Movements observed by the three inclinometers GL-1, -2, -3 during the spring runoff of 1992 showed that the landslide slip surface ranged from 53 to 38 feet deep (Figure CH-9.3b), well below the base of the R.E. wall at the inclinometer locations (Figure CH-9.3c). The headscarp between these slip surfaces and the road cracks was inclined at a steep angle to the horizontal (61° to 65°) showing that the slip surface was plunging deeply below the ground surface.

The Digitilt inclinometer (Slope Indicator Company) is capable of measuring shear movements in drillholes inclined at up to about 30° to the vertical. Accordingly, two additional deep inclinometers (GL-4, GL-5) were installed at 14.5° and 23° to vertical (Figure CH-9.4). Subsequent readings showed that deep-seated shear movements were occurring below the slope between the road and the river. Thus, it appears that the landslide is a translational slide that probably extends to the base of the slope.

When this information became available, preliminary design work on a tiedback retaining wall *downslope* of the R.E. wall was halted. Such a wall would have had to penetrate 60 feet vertically below ground before reaching the slip surface! The benefit of installing inclined inclinometers at an environmentally sensitive site was clearly demonstrated.

Geotechnical and Structural Design: Tied-back Concrete Shear Pile Wall

Landslide Technology brought Inca Engineers, Bellevue, Washington, into the project as structural engineering consultants. Since a shear pile wall could not be economically constructed downslope of the R.E. wall because of the depth to the slip surface, it was agreed that a shear pile wall should be designed further upslope below the outer highway lane.

The wall design (Figures CH-9.3 and CH-9.5) consisted of heavily reinforced concrete cylindrical piles, 4 feet diameter and spaced at 7½ feet center-to-center. These structural piles penetrate through the slip surface of the landslide and are

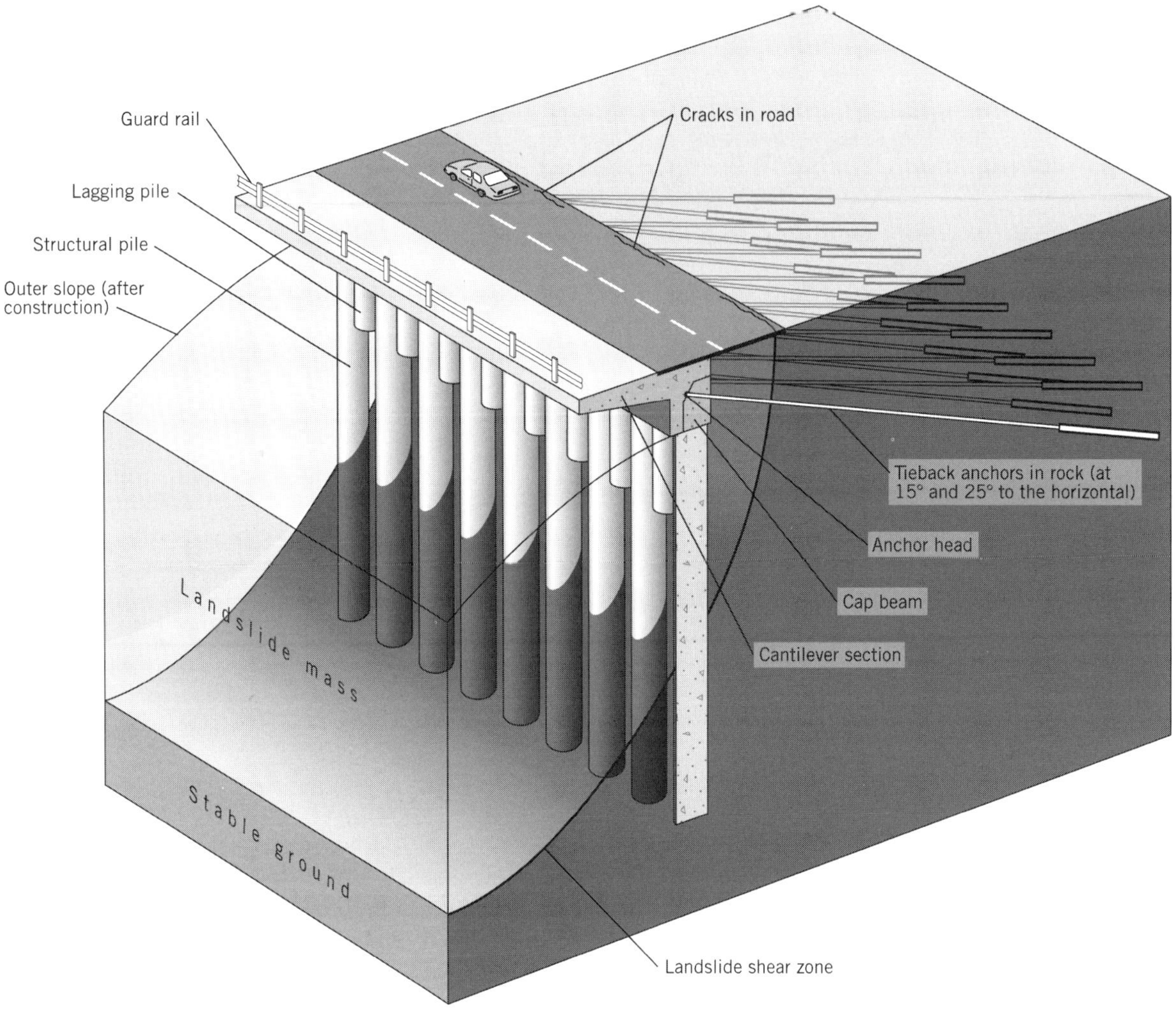

Figure CH-9.5 Goat Lick Slide. Three-dimensional block drawing showing the shear piles, lagging piles, and tieback anchors in relation to the landslide.

anchored at their lower end in stable bedrock. Between each structural pile is a lean concrete pile, also 4 feet diameter, such that the combination of structural and lean concrete piles provides a continuous secant pile wall.

The unreinforced lean concrete piles are installed ahead of the structural piles and extend 18 to 22 feet below the surface, slightly below the mid-depth of the overburden soils (Figures CH-9.3b and CH-9.5). Their purpose is to act as lagging to retain the backfill when the R.E. wall is removed on the downslope side. By partially penetrating through the overburden, the lagging piles are able to maintain the integrity of the wall should the ground downslope continue to move in the future, yet do not block groundwater that is able to pass below the pile tips.

The structural piles are heavily reinforced with steel bars around the perimeter of the pile, as shown on Figure 19.39 in Chapter 19. One benefit of steel reinforcing bars over, for example, a steel H beam, is that the amount of reinforcement can be varied over the length of the pile according to the bending and shear stresses of the design.

The location of the shear pile wall (Figure CH-9.3c) served three purposes: (i) it was placed immediately behind the horizontal steel strips of the existing R.E. wall so that the R.E. wall could continue to support the road throughout construction; (ii) it reduced the landslide load that the structural piles had to carry; in fact, the loading was a simple wedge loading, with strengths along the slip surface being determined by back analysis; and (iii) it left sufficient room for the ground anchors to be drilled from a level pad on the outside of the existing highway.

The concrete pile wall has a cap beam that ties together the pile heads. Tieback anchors were drilled at 15° and 25° to the horizontal and anchored into the cap beam. The design required the anchors to be installed, prestressed, and "locked off" before a cantilever section is built above the cap beam to carry the outer lane of the highway (Figure CH-9.6).

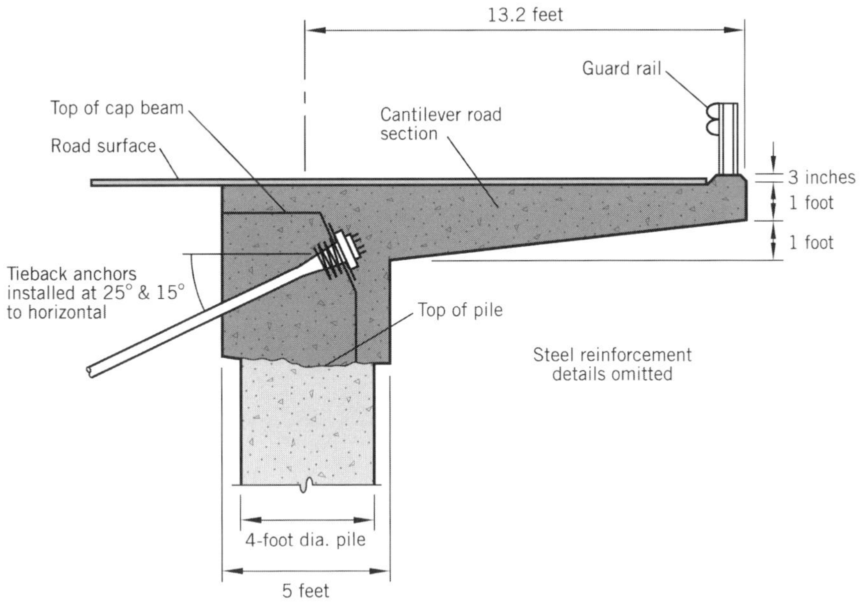

Figure CH-9.6 Goat Lick Slide. Section showing details of the pile top, cap beam, tieback anchor, and cantilever road section.

The pile wall was 254 feet wide and comprised 34 structural concrete piles and 33 lean concrete lagging piles. The lagging piles typically were 16 feet long below the cap beam. The structural piles had three groups of lengths ranging from 43 to 56 feet (Figure CH-9.3b) based on inclinometer observations of the slip surface depth. These piles were embedded an estimated 10 to 14 feet into relatively hard and competent bedrock below the slip surface. Because the inclinometers were located a few feet further downslope than the pile wall, the position of the slip surface was interpolated between the ground cracks in the road and the depths measured in the inclinometers (Figure CH-9.3c).

The wall design loads were based on three sections where ground movements had been measured by inclinometers (GL-1, GL-2, GL-3). At each section, the wall design load assumed that all support was lost on the downslope side of the pile due to continued landslide movement and formation of a deep crack (see Chapter 19, Section 19.5). The soil strength was conservatively estimated to be $\phi' = 25°$. Groundwater was assumed to be either (i) 5 feet higher than the highest level observed in the piezometers, or (ii) one-half of the wall height above the slip surface, whichever was higher. The calculated lateral loads ranged from 44 to 108 kips per foot width. However, the structural design used uniformly distributed lateral loading, ranging from 1,430 to 2,350 psf, acting from the ground surface to the depth of sliding. Using a pseudo-static seismic coefficient of 0.15, approximately equivalent to a $M_L = 8\frac{1}{2}$ earthquake (Seed 1979), the factors of safety for seismic loading exceed 1.30; the static factor of safety is greater than 1.50.

The ground anchor design is shown on Figure 19.9 of Chapter 19. It is a stranded cable using the Dwyidag system. The bonded length in the bedrock is 25 feet and the unbonded length between the ground anchor and the capping beam is 45 feet. The total length of 70 feet is the approximate limit for swinging leads where a small rig has to be held over a wall by a crane or put on a portable platform held by a crane. There were 34 ground anchors in the capping beam, one above each of the structural piles, installed at 15° to the horizontal (i.e., at 7½ foot centers). In addition, there were 19 anchors in the beam above the lean concrete piles on the northern half of the wall; these were installed at 25° to the horizontal. The ground anchors were designed to carry an allowable working load of 135 tons with a factor of safety of 3. This assumes that the argillite bedrock provides an ultimate pullout resistance of 130 psi. The cap beam detail (excluding steel reinforcement) and cantilevered road section is shown on Figure CH-9.6.

CONSTRUCTION

The lean concrete piles (minimum 28-day compressive strength : 500 psi) were constructed first, generally using an open flight auger and Kelly bar. The fill and overburden soils were easy to drill (Figure CH-9.7), but an occasional oversize boulder (up to 2 feet across) had to be pulled out of the hole. A temporary steel casing was put in the open hole so that lean concrete could be poured in from the top without intermixing with soils from the sides of the hole. After the lean concrete piles had been completed (Figure CH-9.8), the structural concrete piles (minimum 28-day compressive strength: 4,000 psi) were drilled between them by the same technique (plus occasional use of a bucket auger). The steel reinforcement cages for the structural piles were fabricated offsite.

Figure CH 9.7 Goat Lick Slide. Excavating overburden soil for a lagging pile (Photo: July 20, 1993).

Figure CH-9.8 Goat Lick Slide. Line of completed lean concrete (lagging) piles (Photo: July 21, 1993).

The cap beam was cast-in-place, allowing cut-outs for the tieback anchorage to the beam (Figures CH-9.6 and CH-9.9). After tiebacks had been installed (Figures CH-9.10), they were tested and locked off at about 70% of the design load. Most of the fill of the R.E. wall was removed from the front of the shear pile wall and the remaining face was shotcreted. Finally, the cantilever section was cast-in-place to complete the shear pile wall and road section.

The testing of ground anchors followed the procedure shown on Figure 19.11, Chapter 19. All the anchors were subjected to a Performance Test (cycled up to 125% of design load) and three anchors were checked by a Proof Test (up to 150% of design load).

CONSTRUCTION COSTS

The shear pile wall was constructed from early July to late September, 1993. The site is remote, and concrete had to be brought in from a batching plant 25 miles away. There were space and access limitations. A biologist was employed full-time to monitor the effect of construction on the goats and other wildlife.

The contractor's bid price was $1.44 million, close to the engineer's estimate. Extras were minimal. The bid price included resurfacing the affected road and dismantling the R.E. wall in front of the new retaining wall. The ENR construction cost index for June 1993 is 5260.

The cost of the two angled inclinometers, total 475 feet deep, including cased drillholes and installed inclinometer casings, were: traffic control $21,000; inclinometer installations $33,850; total $54,850.

Figure CH-9.9 Goat Lick Slide. Cap beam (on right) with cutout notches for ground anchors. The girders (left, center) provide support for the formwork of the roadway cantilever section.

Figure CH-9.10 Goat Lick Slide. Drilling for a tieback anchor.

10 Hagg Lake, Slides 4 and 3

Figure CH-10.1 Hagg Lake, Slide 4. Long, translational landslide extends from the road to below the reservoir on the right. Outline of the 600-foot long landslide is shown by the dashed white lines (Photo: 1981).

Key Points

- Residual strength of weathered siltstone
- Shear key stabilization
- Selective stabilization
- Buttress stabilization

Summary

Hagg Lake reservoir has a perimeter road built around the lake, most of the ground being geologically classified as ancient landslide terrain. Shortly after construction, a major storm caused numerous landslides affecting the perimeter road. Tests of residual strength performed on the stiff clayey silt overburden soils had relatively high residual strengths (ϕ'_r = 33°) that caused the consultants to arrive at incorrect remedial treatments. When landslide 4 (Figure CH-10.1) subsequently reactivated, it was found that soil samples taken at the ancient slip surface had much lower residual strengths than the overburden soils.

Two case histories, Slides 4 and 3, demonstrate the importance of collecting samples for residual testing from the discrete shear zone in ancient landslide terrain, and not from the overburden above it. It also shows that weathered bedrock may have radically different residual strengths at different sites despite the fact that the parent bedrock is the same.

Background Information

Scoggins Dam was built in 1972–73 and its reservoir (Hagg Lake) extends into a widespread area of ancient landslide terrain in the volcanic clays of the Yamhill Formation (Upper Eocene). The soils, resulting from weathering of the parent siltstones, are typically stiff clayey silts and silty clays with average properties: LL 59; PL 37; w 41%; clay fraction (<0.002 mm) 25%; wet density γ 115 lb./cu. ft. Undrained shear strengths averaged 1,500 lb./sq. ft. In December 1977, a storm caused numerous landslides along the perimeter road.

SLIDE 4

Site Investigation of Slide 4

The road fill was built over a shallow slope averaging 7° to the horizontal (Figure CH-10.2a). The fill side slopes were built at 1½ horizontal : 1 vertical (33° to the horizontal). Stress-reversal shear box tests performed on the overburden gave an average residual strength of 33° (Figure CH-10.3) and showed very little change in strength from peak to residual during three cycles of shear. At that time (1979), a three-reversal shear box test was the accepted procedure for measuring the residual strength of clays in commercial soils laboratories.

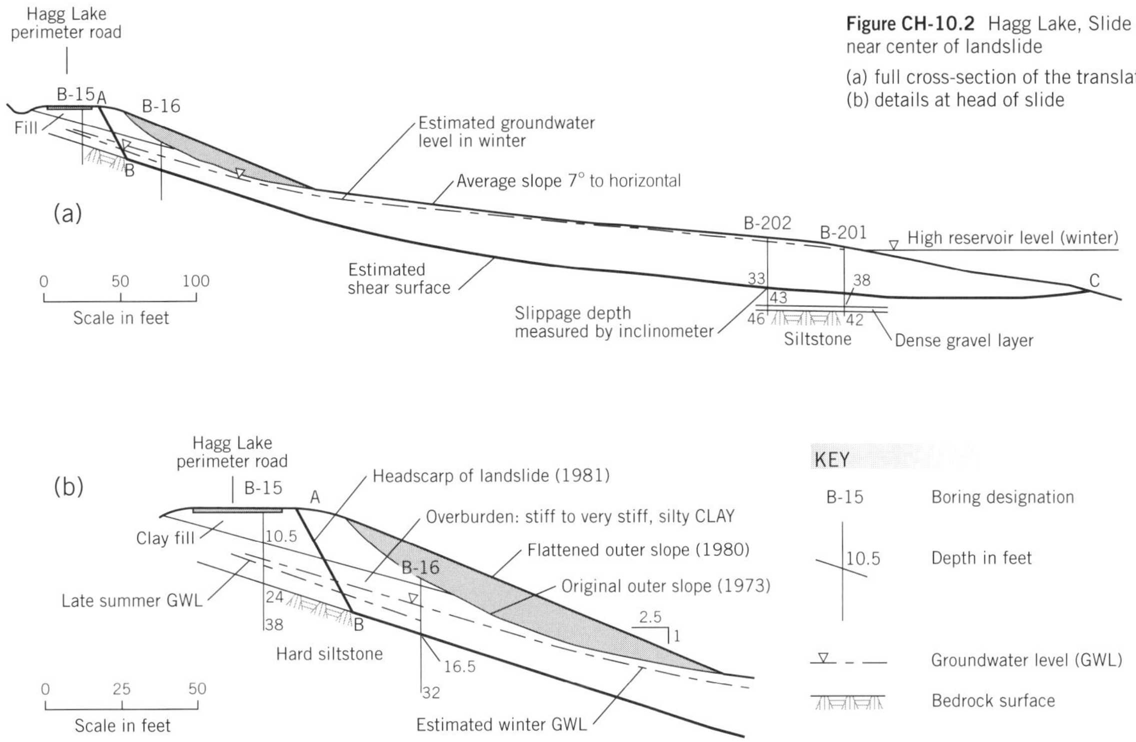

Figure CH-10.2 Hagg Lake, Slide 4. Section near center of landslide

(a) full cross-section of the translational slide
(b) details at head of slide

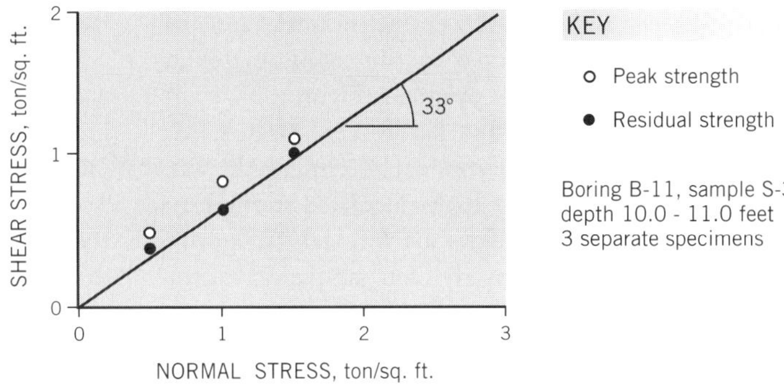

Figure CH-10.3 Hagg Lake, Slide 4. Residual strength measurements on clayey silt overburden using three stress reversals in a direct shear test.

Some minor slope failures occurred at this site in December 1977; essentially, these were small slumps along the outer edge of the fill. It was reasoned that the failures occurred because the outer fill slopes of 33° coincided with the measured shear strength of 33°, causing surficial instability. In the summer of 1980, the outer slope of the road fill was flattened from 1½:1 to 2½:1.

A medium-to-large landslide occurred in the following winter (Figure CH-10.1). It extended downslope more than 600 feet from the road centerline, the toe being below the reservoir (Figure CH-10.2a). The extra weight of the new fill had reactivated the ancient landslide terrain. The slide was 430 feet wide and, at the center, extended back across both lanes of the road.

An inclinometer casing (B-202) was installed 465 feet downslope from the road centerline (near the reservoir) and showed shear movements within discrete shear zones at 19–21 feet and 32–34 feet below the surface (Figure CH-10.2a). A 3-foot thick gravel layer, believed to be in-place, was a further 8 feet below the lower shear zone. The gravel was underlain by hard siltstone bedrock.

Since the slope had started to fail under the weight of the initial fill, a back analysis was performed using the original outer fill slope of 1½ : 1 (horizontal : vertical) and an average distance of 20 feet from the road centerline to the headscarp (Figure CH-10.2). The headscarp slope (AB) of 60° to the horizontal is based on observations of the 16-foot high headscarp, and was assigned a strength of $\phi' = 33°$ for first-time

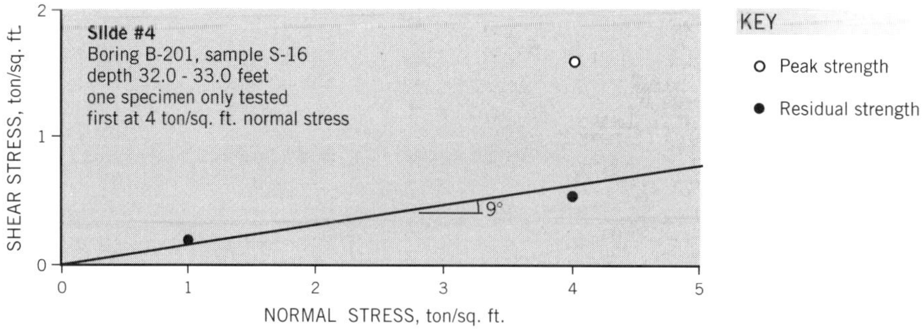

Figure CH-10.4 Hagg Lake, Slide 4. Residual strength measurements on sample taken from the discrete shear zone.

sliding. A non-circular stability analysis (Spencer's method) passing through the upper shear zone (which moved more than the lower shear zone) gave the back-calculated result: $c_r' = 0$, $\phi_r' = 12.2°$ for groundwater at the highest observed levels. For an infinite slope analysis, $\phi_r' = 10.9°$.

There is, of course, some uncertainty about the slip surface depth between the road and boring B-202, but the assumption that it is approximately parallel to the ground surface is reasonable for this site where the ancient landslide terrain is colluvium.

Residual Strength of the Discrete Shear Zone

Additional stress reversal direct shear tests were performed on a sample obtained from a depth of 32–33 feet in boring B-202. This sample is *within* the lower discrete shear zone of the ancient landslide, as observed by the inclinometer. A residual strength of $\phi_r' = 9°$ was obtained (Figure CH-10.4) after a shear displacement of 2.75 inches in the shear box, i.e., 11 reversals instead of the conventional 3. This result is in fairly good agreement with the back-calculated strength of 11° to 12°, and substantially below the residual strength of 33° measured on the general overburden sample above the shear zone.

The huge difference between the residual strength in the general overburden and the residual strength on the actual shear surface of the landslide may be the result of more advanced physiochemical deterioration of the clay on the ancient slip surface.

Shear Key Remedial Treatment

The road is near the top of the ancient landslide terrain at this site. A decision was made to provide selective stabilization of the road rather than to stabilize the entire landslide. Because the ancient landslide is relatively shallow, a rockfilled shear key was constructed that passes through the landslide debris into the hard siltstone below. The centerline of the road was moved further upslope to minimize the amount of unsupported ancient landslide above the temporary cut slope during construction. A typical section is shown on Figure CH-10.5. The photographs shown on Figures CH-10.6 and CH-10.7 were taken during the remedial construction. Excess cut soils were hauled off site.

Filter fabric separated the rockfills from the native soils and clay backfill. Rockfill was 6 inch minus in the road prism and 3 inch minus in the side fills. It was compacted in 6-inch to 9-inch lifts by several passes of a smooth-drum vibratory roller.

Shortly after construction, minor slippage developed in the cut slope above the reconstructed road section. Horizontal drains were installed at 15-foot centers in the cut slope. A subdrain, 30 inches wide, of crushed rock was constructed below each horizontal drain outlet. These subdrains adjoined a 15-inch thick slope protection blanket of crushed rock that extended along the base of the cut slope (Figure CH-10.5).

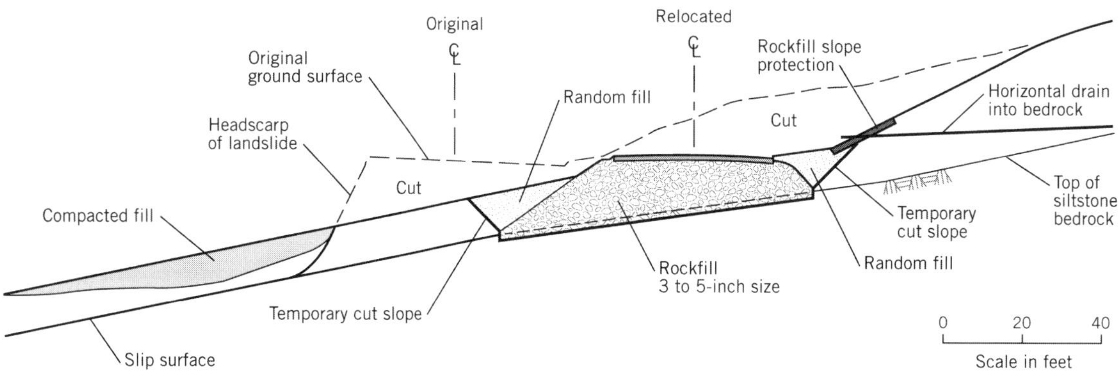

Figure CH-10.5 Hagg Lake, Slide 4. Shear key remediation.

Figure CH-10.6 Hagg Lake, Slide 4. Construction of shear key and roadbed. Overburden soils (ancient landslide) have been stripped off the siltstone bedrock. In foreground, rockfill for the shear key is being spread by a dozer. On left is the temporary 1:1 cut slope covered with white filter fabric (Photo: K. F. Fujitani, August 17, 1983).

Figure CH-10.7 Hagg Lake, Slide 4. Nearing completion of shear key. Smooth drum vibrating roller is compacting rockfill. Similar view to Figure CH-10.6 (Photo: September 1, 1983).

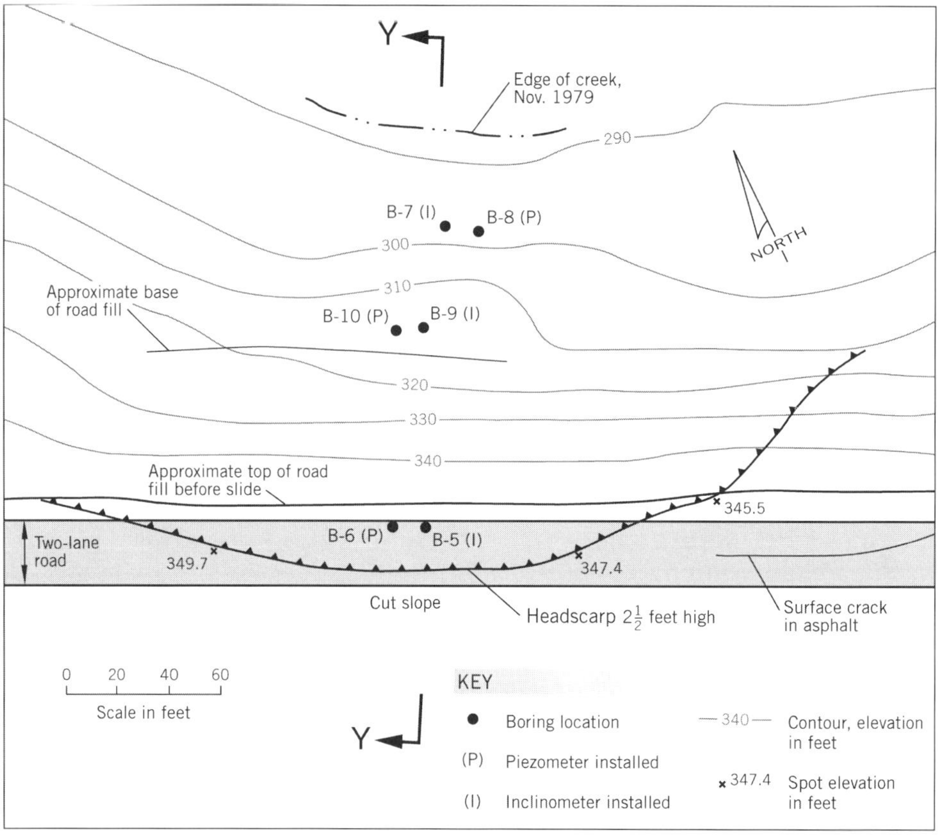

Figure CH-10.8 Hagg Lake, Slide 3. Plan of landslide.

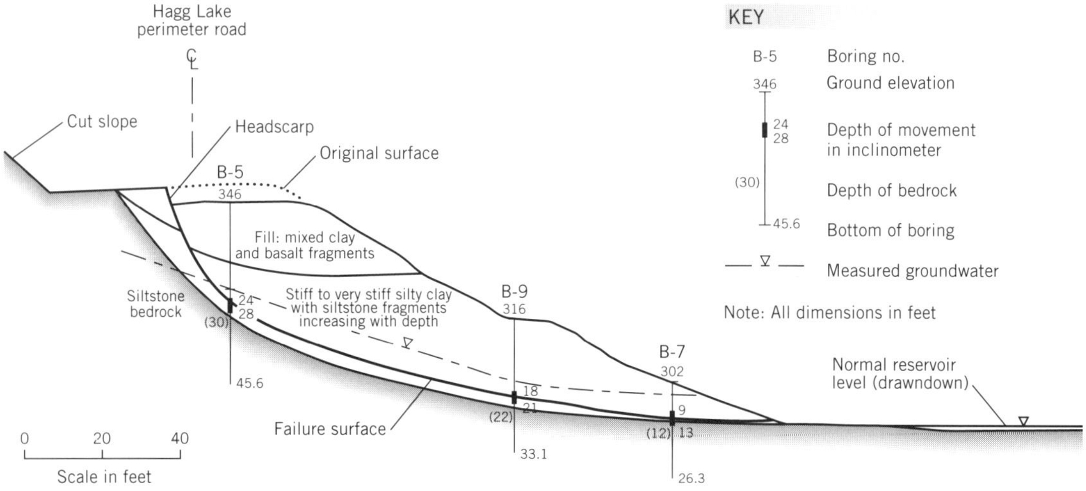

Figure CH-10.9 Hagg Lake, Slide 3. Geological section Y-Y.

SLIDE 3

SITE INVESTIGATION OF SLIDE 3

The landslide occurred at the site of a fairly steep cross-slope where a balanced cut and fill construction had been made. The headscarp of the landslide reached the inner lane of the two-lane road (Figure CH-10.8). The medium-size landslide is about 200 feet wide at the outer shoulder and gradually widens downslope. The maximum length, at the center of the slide, is 165 feet.

Inclinometers and pneumatic-type piezometers were installed in side-by-side borings. The inclinometers measured a shear zone 3 to 4 feet deep, the slippage occurring near the top of siltstone bedrock (Section Y-Y, Figure CH-10.9).

All the piezometer tips were placed within the shear zone and sealed with bentonite about 3 feet above the tips. Measured groundwater levels were only a few feet above the shear zone. This reservoir was drawn down (Figure CH-10.9) at the time of the landslide.

The subsurface soils generally were described as stiff to very stiff silty clay with siltstone fragments and nodules. At the

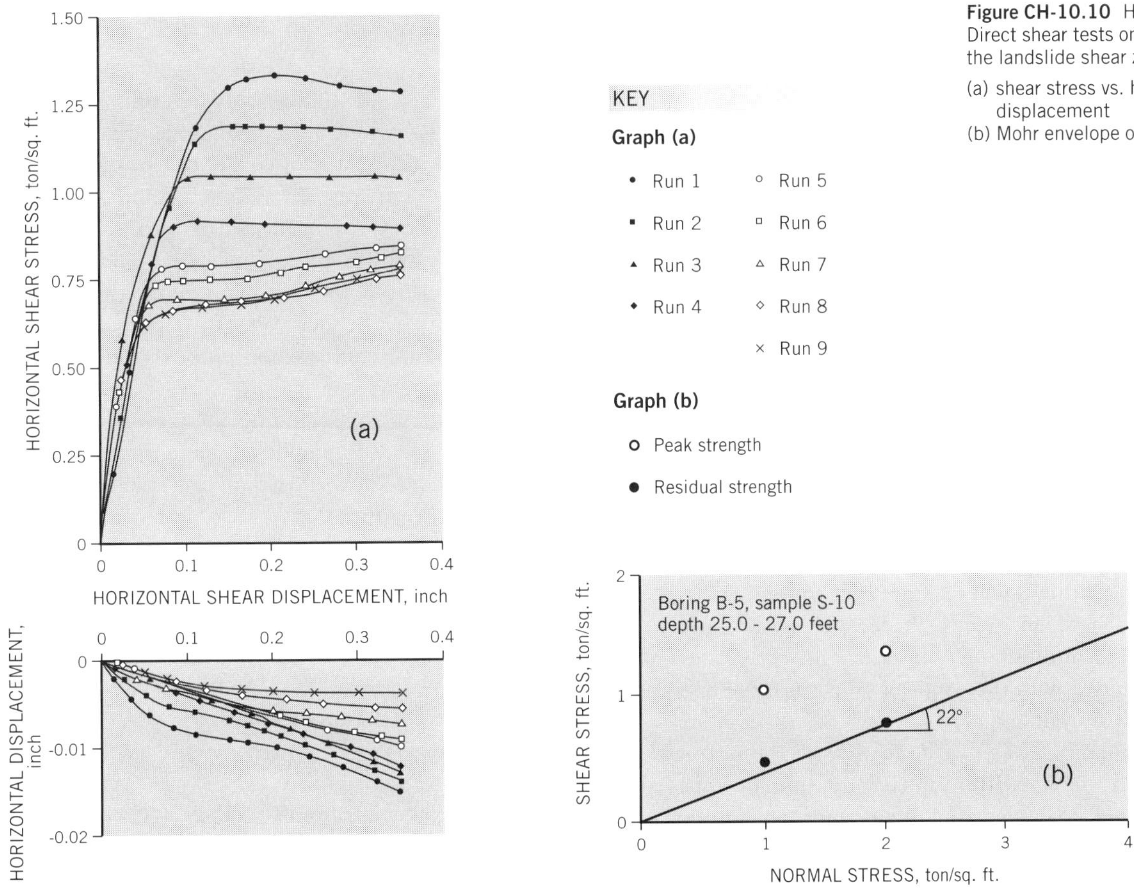

Figure CH-10.10 Hagg Lake, Slide 3. Direct shear tests on sample taken from the landslide shear zone
(a) shear stress vs. horizontal shear displacement
(b) Mohr envelope of residual strength

failure zone, small amounts of wood were noted at 24 feet depth in boring B-5, and basalt fragments were observed within clayey silt at 10 feet in boring B-7. At a depth of 16 feet in boring B-9, Atterberg limits were LL 60, PL 38. The amount of siltstone fragments present in the samples generally increased with depth and were more numerous than at the site of Slide 4.

A stress reversal direct shear test on a sample taken from the discrete shear zone shows a gradual drop in shear strength with increasing shear displacement (Figure CH-10.10). It required 7 runs to reach a condition approaching residual and the residual strength was $c'_r = 0$, $\phi'_r = 22°$.

The back analysis was performed on the failure profile using a low reservoir level but with groundwater levels in the slope arbitrarily raised by 5 feet above summer levels to simulate the conditions during the storm event that caused the slide. The back-calculated required strength of $\phi'_r = 20°$ (using Spencer's method) is in good agreement with the $\phi'_r = 22°$ measured in the laboratory.

Buttress Remedial Treatment

The landslide was stabilized by a rockfill buttress approximately 23 feet high (just above reservoir high water level), width of 300 feet, and a base length of 32 feet (Figure CH-10.11). The outer slope of the buttress is 2 : 1 (horizontal : vertical). Fabric was placed below the base and sides of the rockfill and the inner slope was cut at 1:1. The base was embedded 3 feet into hard siltstone bedrock. Compacted native fill, as needed, was used to restore the outer slope above the buttress to 2½:1 (horizontal : vertical). The calculated factor of safety with the buttress in place is F = 1.42 for the elevated (storm-simulated) groundwater with low water in the reservoir, and F = 1.52 for summer groundwater levels.

Because the toe excavation undermines the slope, the buttress was built by close sequence construction in which short lengths of the excavation were opened up and backfilling occurred as quickly as practicable (see Chapter 13, Section 13.5). The rockfill was a 6 inch minus crushed quarry rock and was compacted in about 9 inch lifts using a vibratory compactor. As requested by the owner, 27 horizontal drains were drilled 40 feet into the siltstone bedrock starting from the top of the buttress; the average length was 154 feet (Figure CH-10.11). The buttress was built in 1981.

Variable Residual Strength

From the work of Skempton and others, it is customary to think of residual strength as a basic property of an overconsolidated clay. Research has been performed on numerous overconsolidated clay strata, and the main issue in recent years has been the amount of shear displacement needed to reach residual strength (as discussed in Chapter 8, Section 8.8).

The Hagg Lake experience (Cornforth and Fujitani, 1991) is based on a moderately weak siltstone that weathers to a stiff clayey silt. At Slide 3, the laboratory-measured ϕ_r' of 22° was in close agreement with the back-calculated ϕ_r' of 20°. At Slide 4, the laboratory-measured ϕ_r' of 9° was in reasonable agreement with a back-calculated ϕ_r' of 11° to 12°. Both these results differ markedly from the ϕ_r' values 33° to 34° obtained on samples taken from the overburden soils *above* the shear zone. If such values are used in stability analyses, as they were initially at Slide 4, it would be concluded that a flat translation landslide will not occur. However, the reality is that the residual strengths within the discrete shear zone of the ancient landslide are very different from the rest of the overburden soils. Furthermore, these two case histories show that the value of ϕ_r' can differ between discrete shear zones at different locations. Yet all the soils at Hagg Lake are from the Yamhill Formation and are technically similar in classification properties.

The likely explanation is that the clayey silt within the discrete shear zone has been greatly altered by physical and chemical weathering. Groundwater movement may play a significant role combined with thousands of years since the origin of the landslide movement.

It would be improper to speculate further on the reason for the variations of ϕ_r' within apparently similar soils. However, from a practical standpoint, three points need to be emphasized for landslide investigations on ancient landslides derived from weathered residual soils:

1. Samples taken from the overburden for residual strength tests *must* be taken at the discrete shear zone of the ancient landslide.
2. Inclinometers are needed to determine the location of the discrete shear zone. At Slide 4, for example, the two shear zones are 6 feet and 18 feet above an in-place gravel layer. Collecting a residual strength test sample from just above the siltstone bedrock (or the gravel layer) may seem logical, but would be at the wrong place.
3. Sufficient shear displacement must occur to reach the residual condition (see Chapter 8, Section 8.8). A ring shear apparatus is preferable to a direct shear box for this purpose.

Since the Hagg Lake studies, similar results have been obtained at other sites where ancient landslide terrain overlies the parent bedrock. Since it is often difficult to obtain thin-wall samples from the discrete shear zone, a common practice at Landslide Technology is to collect data on the other elements of a landslide study (depth of slippage, groundwater levels, etc.) and determine the residual strength from back analysis of the landslide. In such cases, the laboratory tests, if performed, are used for confirmation purposes rather than as basic data.

The main purpose for including these two Hagg Lake landslides in this book is to emphasize these three points. To the best of the author's knowledge, the Hagg Lake experience in 1979–82 is the first time that (i) variable residual strength has been measured in the same soil formation at two nearby sites, probably due to variable weathering, and (ii) the residual strength on the slip surface can be very much lower than the residual strength measured in overburden samples above it.

Remedial Construction Costs

Slide 4

(i) Initial slope flattening from 1½:1 to 2½:1. Cost: $113,800
ENRCCI (1980) = 3250

(ii) Key trench, relocation of road further upslope, and local regrading. Cost: $455,000 (1983). (Rockfill was 18,400 cu. yd. @ $9.82/cu. yd. for 300 feet length of key trench.) Horizontal drains (27), total length: 2,227 feet. Cost: $76,700
ENRCCI (1983) = 4000.

Slide 3

Construction cost to build 250 feet wide buttress, provide 21 horizontal drains (154 feet avg. length) and some related slope treatment outside the buttress.

Cost: $370,000 (1981)

ENRCCI (1981) = 3550

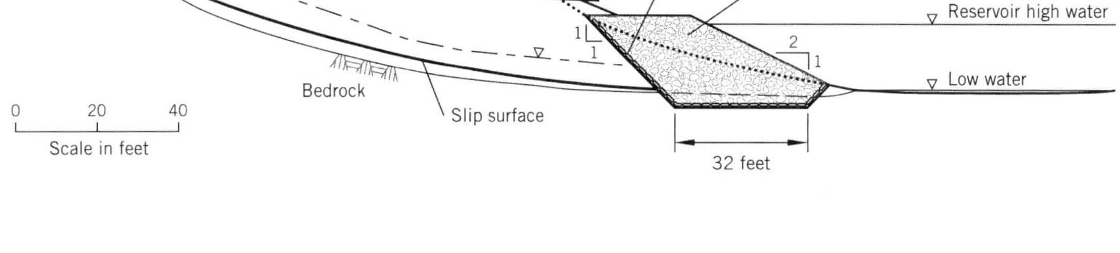

Figure CH-10.11 Hagg Lake, Slide 3. External buttress remediation.

11 Hagg Lake Slide 6

Figure CH-11.1 Hagg Lake, Slide 6. Headscarp crossing road as seen from the center of the slide. Cut slope is to the left of the photo and fill slope is on the right (Photo: 1979).

Key Points

- Trench drains
- Piezometer drawdown test
- Trench safety

Summary

Slide 6 on the Hagg Lake perimeter road involved a very high groundwater level within a natural bowl of ancient landslide terrain. The depth of the landslide was relatively shallow. This case history describes the effective use of trench drains to lower groundwater levels (and prevent them from rising again to unacceptable levels). This remediation was very successful, but the case history is not a good model of site construction practice. Recommendations are given for writing specifications to reduce stressful site relationships and construction claims, and to promote site safety.

Background Information

The 10 mile-long Hagg Lake perimeter road was constructed in 1973–74 after Scoggins Dam created the reservoir. Geologists estimated that more than 50 percent of the slopes around the reservoir were ancient landslide terrain. These slopes were colluvium of silts and clays derived from the siltstone bedrock.

In the fall of 1977, a heavy storm caused numerous landslides in the cut and fill slopes made through the uneven terrain. One section of the road, labeled Slide 6 after the major storm (Figure CH-11.1), had previously required annual maintenance since construction. At this site, the road was cut-and-fill construction and had been built at about the mid-length of a 750-foot long ancient slide. The reactivated part of the ancient slide had a width of 270 feet at the guardrail, and the headscarp extended back to the road centerline (see site plan, Figure CH-11.2).

Ground surveys and site investigations were performed in early 1980. Three borings in a line through the center of the landslide are shown on the site plan and cross-section (Figures CH-11.2 and CH-11.3). The borings showed consistent conditions as exemplified by a typical boring log (Figure CH-11.4). The ancient slide debris (colluvium) is about 18 feet thick and overlies a hard siltstone bedrock (Figure CH-11.3). At this section, the fill is about 10 feet deep at the outer shoulder and has a 2:1 (horizontal : vertical) outer slope.

The ancient landslide debris has low SPT blow counts and water contents around 45–50 percent. The average properties at this landslide are: LL 50, PL 31, γ = 115 lb./cu. ft. (CI = 0.64). It can be described as medium stiff clayey silt with siltstone fragments.

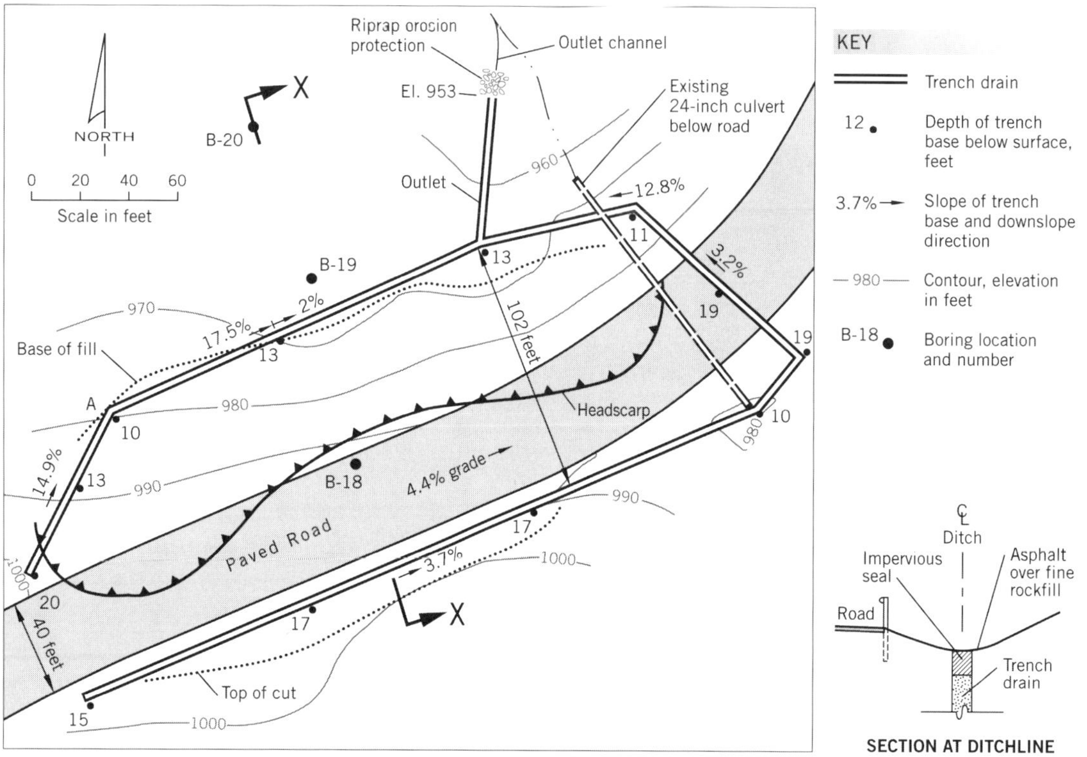

Figure CH-11.2 Hagg Lake, Slide 6. Plan of trench drains.

Figure CH-11.3 Hagg Lake, Slide 6. Geological section X-X.

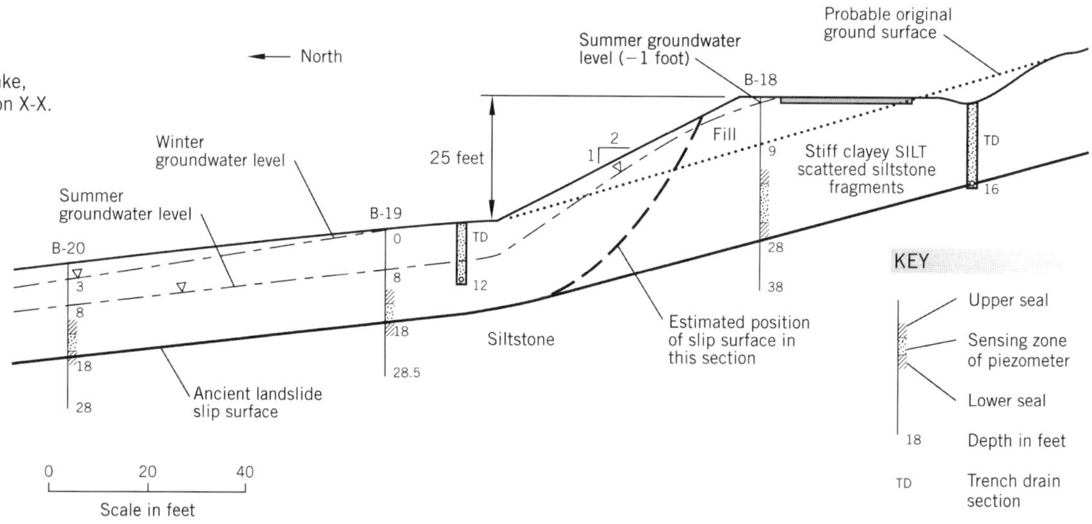

Figure CH-11.4 Hagg Lake, Slide 6. Summary log, boring B-19.

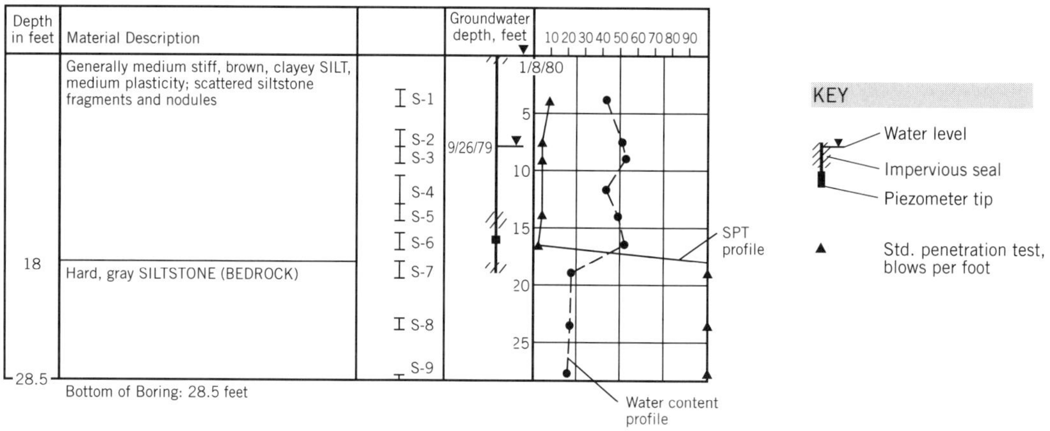

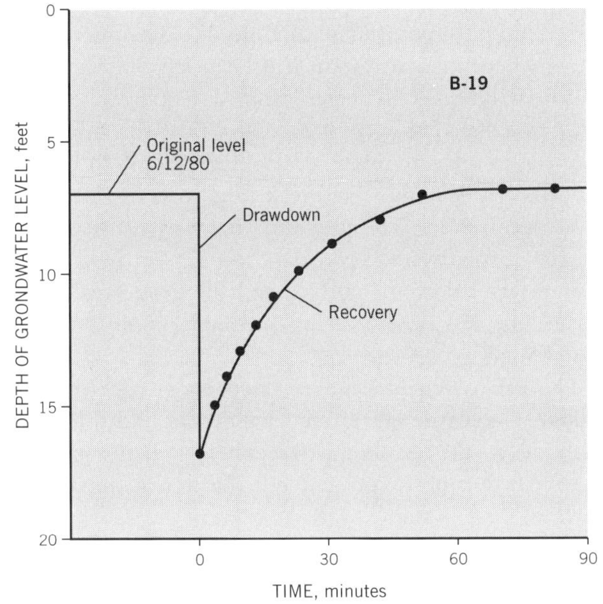

Figure CH-11.5 Hagg Lake, Slide 6. Drawdown test in the piezometer of boring B-19.

GROUNDWATER LEVELS

The groundwater levels are very high and close to the ground surface during wet periods of the winter (Figure CH-11.4). A surprising result was the high water table, winter and summer, in the compacted clayey silt road fill at boring B-18 (Figure CH-11.3). It is observed, however, that the ground on three sides of the landslide is higher, directing surface runoff toward the slide.

There was some concern that the piezometers may not be giving correct readings. The drilling contractor used rotary techniques with a proprietary mud fluid (Revertex) that is supposed to break down quickly after drilling and not affect groundwater measurements. To check this function, the groundwater in boring B–19 was drawn down by 10 feet. The result, Figure CH-11.5, shows a relatively quick recovery to the original summer level. It confirmed that the piezometer was performing satisfactorily.

COMPUTATION OF SLIDE MASS PERMEABILITY

A rising head permeability test in a piezometer can be used to measure the ground permeability around the sensing depth. If the permeability in the slide mass is assumed to be the same in all directions, the calculated coefficient of permeability for the slide mass is 3 x 10^{-6} cm/sec, as shown on Figure CH-11.6.

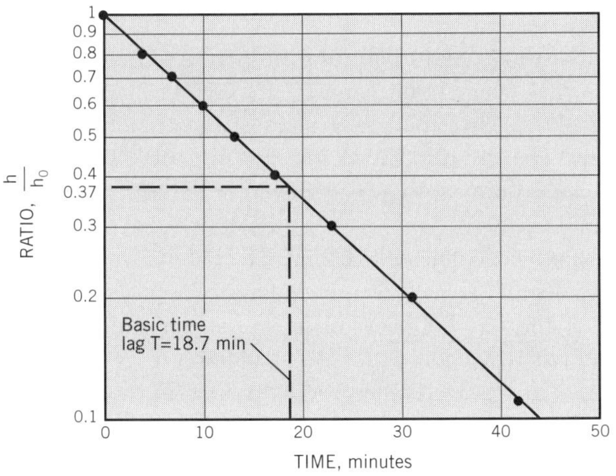

From graph, T = 18.7 mins

Using formula from Chapter 4: Piezometers and Inclinometers

$$k = \frac{d^2 \log_e \left(\frac{2\,m\,L}{D}\right)}{8\,L\,T}$$

d = 0.375 inch; $\log_e N = 2.3 \log_{10} N$; L = 36 inches; D = 5 inches; m = 1

$$k = \frac{(0.375)^2 (2.3)(\log_{10} 14.4)}{(8)(36)(18.7)} \times \frac{(2.54)}{(60)} \longleftarrow \text{Conversions}$$

$$k = 2.9 \times 10^{-6} \text{ cm/sec}$$

Figure CH-11.6 Hagg Lake, Slide 6. Measurement of soil permeability from the rising head test performed on standpipe piezometer of boring B-19.

LANDSLIDE CAUSATION

This landslide is the reactivation of part of an ancient landslide caused by the construction of a road fill. Prior to construction, this slope probably was marginally stable under winter conditions and may have moved periodically in storms.

TRENCH DRAIN DESIGN

The measured shear strength parameters for the clay overburden were: $c'=0$, $\phi' = 32\frac{1}{2}$ degrees. Allowing for groundwater at the ground surface during very wet winter weather (as measured by the three piezometers), the calculated factor of safety was 1.01 for a circular arc slip surface passing below the toe of the road fill and touching the top of the siltstone bedrock.

A more likely failure mode is a translational movement along the ancient landslide slip surface reactivated by the road fill. No inclinometers were installed at this site (owner decision) nor were any residual strength tests performed on samples taken at the ancient landslide slip surface, although tests at other landslide sites in the area (see Case History 10) showed that residual strengths were lower on the ancient slip surface than in the overburden. Therefore, it is probable that the actual slippage at Slide 6, between the road and Hagg Lake reservoir, was along the ancient landslide slip surface.

The high groundwater levels and relatively shallow depth to the siltstone made trench drains a good remedial choice at this site. A rock-filled shear key was also considered, but this would have required general site dewatering prior to construction and possible reactivation of sliding in the upslope area.

The decision was taken to increase the stability by constructing interceptor trench drains on both sides of the road, one drain to be built along the ditch line on the south (upslope) side and the other to be constructed along the toe of the fill slope (Figures CH-11.2 and CH-11.3). Although the upslope drain may have been sufficient, the high groundwater in the road fill piezometer (B-18) indicated a strong recharge and thus a probable need for the second drain. Estimates of drawdown (not reproduced) indicated that the factor of safety would rise to about 1.7. However, piezometers B-18 and B-19 were destroyed during construction so the actual drawdown was never measured. The trench drains were installed between August 20, 1980, and September 12, 1980. During the 24 years since construction, this section of road has remained in excellent condition.

The design was a 3-foot wide trench drain with a 6-inch diameter flexible drainpipe near the base. The backfill was an imported crushed basalt of 3 inch minus gradation with not more than 5 percent by weight of nonplastic fines passing the No. 200 U.S. sieve (in recent years, the author prefers to spec a limit of 3 percent fines). The upper 18 inches was to be constructed from locally available compacted impervious fill (see insert on Figure CH-11.2).

TRENCH DRAIN CONSTRUCTION

The contractor began work at the outlet drain on the north (downhill) side of the project. The depth of the trench varied from about 11 feet to 20 feet, the deepest sections being at the west end of the south leg, where it reaches into the road fill, and at the road crossing of the south leg (Figure CH-11.2).

The owner was unwilling to pay for full-time inspection of this project. Therefore, despite a contract requirement of closely sequenced construction, there were times during the early phases of construction when the inspector arrived on site to find up to 70 feet of open trench! Sloughing of the sides of the trench was a continual problem, made worse by the significant lengths of open trench.

The contractor was told to adhere to the specified closely sequenced method and generally speed up construction. As commonly occurs in these situations, the contractor's technique improved with time. In some parts of the trench, where the excavation had been left open too long, the top width had increased from 3 feet to as much as 10–12 feet. Records were kept, and a final cost adjustment was made to separate reasonable overbreak (due to site conditions) from excessive overbreak (due to contractor's poor work organization). The part-time field inspector observed that the groundwater level was seeping into the trench about 8 feet below the surface and the ground below the seepage line was slowly caving and undermining the drier ground above.

The open trench was excavated by a truck-mounted backhoe. Geotextile fabric was wrapped around the flexible drainpipe at the ground surface before being put into the trench. No personnel entered the trench during construction.

The geotextile fabric lining of the sides and bottom of the trench was accomplished by draping the lining across the trench. It should be noted that the fabric lining does not have to be laid perfectly against the sides of the trench, which compromises personnel safety. The gravel backfill presses the fabric against the trench sides. The fabric needs to extend well beyond the trench sides to avoid exposing workers to a possible sidewall collapse. Fabric is a low-cost item and a minimum length outside the trench on each side (for example: a length of H, where H is the trench depth) can be specified.

Extensive ground caving was occurring near the northwest corner (angle point A on Figure CH-11.2). A decision was made to raise the grade level of the trench drain in this area. The site plan shows the revised grades. The trench drain crossed the existing road culvert at the east side of the project.

The project comprised 765 linear feet of trench drain. The average depth was 14.8 feet, including the 1.5 feet clay soil depth at the top. For most of its length the drain reached bedrock, thereby adding strength in addition to groundwater lowering. In late December 1980, the flow from the trench drain was estimated to be 25–30 gpm.

CONSTRUCTION COST

The negotiated final cost: $66,957

Cost per lin. ft.: $ 87.53

ENR Construction Cost Index (August, 1980) = 3304

CONCLUDING COMMENTS

Trench drains provide an excellent form of remediation for relatively shallow landslides with high groundwater levels. However, the author strongly recommends a much tighter control on the construction site than occurred in the 1980 project described above. This can reduce stressful situations between the consultant and contractor and reduce the likelihood of construction claims for overbreakage and personal injuries.

All excavations of deep trenches with vertical sides carry some risk of caving. Although the soils at Hagg Lake Slide 6 were cohesive (cohesive index 0.64), significant caving of the trench sides occurred during construction, partly attributable to lack of supervision and poor construction practices. The general contractor is considered to be the party responsible for site safety, especially for temporary works. Furthermore, the owner sometimes refuses to pay for the full-time site presence of the design engineer's representatives, as was the situation in this 1980 case history. However, if a serious accident or death occurs, the design engineer will almost certainly become a defendant in a legal claim. Therefore, depending on specific site conditions and proposed excavation depths, some or all of the following requirements should be *specified* in trench drain contracts:

- Temporary dewatering of the excavation area before construction begins.
- Use of trench supports within the open trench (this is mandatory if personnel are to get into the trench for any reason).
- Limit the length of trench open at any time, and require partial or total backfill overnight and on non-workdays (closely sequenced construction, see Chapter 13).
- A full-time observer must be on site during trench excavation and backfill to monitor the ground conditions; such an observer should be provided at the expense of the owner and should be regarded as part of the project construction cost. Under normal circumstances, the observer should be a member of the design engineer's staff because the staff have the most knowledge of the subsurface conditions prior to construction. The consultant's observer is also readily available to provide technical advice should an unexpected condition be encountered.

These restrictions and requirements increase the construction cost. However, after allowing for such costs, trench drains are an economical remedial/preventative option for suitable sites.

12 Crown Point Highway Slide

Figure CH-12.1 Crown Point Highway Rock Slide. Rock face after the landslide. The slide was about 300 feet wide and 110 feet high. An overhang of cemented conglomerate and sandstone failed on September 24, 1990 (Photo: May 26, 1993).

Key Points

- Rockfall Hazard Rating System
- Differential erosion of rock

Summary

The Rockfall Hazard Rating System (RHRS) described in Chapter 23 helps to identify more hazardous rock slopes by assigning a geometrically higher score as the level of hazard increases. An example of the use of the RHRS scoring system, for a slope that failed shortly before remediation was to be implemented, is given in this case history.

Background Information

The site is east of Portland, Oregon, where a high steep cut was made through cemented conglomerate and sandstone in 1914 to provide a road alongside the Sandy River (Figure CH-12.1). The conglomerate consists of hard quartzite and igneous rock cobbles in a matrix of cemented sand. At road level, there is a 15- to 20-foot high stratum of weak siltstone/claystone beds underlying the conglomerate (Figure CH-12.2). Seepage passes over the siltstone/claystone layers.

Erosion of the siltstone/claystone created an overhang of 5 to 8 feet (an erosion rate of about 1 inch per year on average). Near-vertical cracks were observed underneath the overhang (Figure CH-12.3) and there were cracks at the top of the slope. In addition to the severe overhang, weathering (freeze/thaw, rainfall, wind, etc.) of the conglomerate face caused fairly frequent falls of cobbles onto the road—two to three times weekly during winter, according to maintenance crews. Large slab-type rockfalls occurred in February 1980, April 1984, and January 1990, the latter event being estimated as 300 cu. yds. of debris.

The Oregon Department of Transportation (ODOT) was very concerned with this section of road. Using the Rockfall

Figure CH-12.2 Crown Point Highway Rock Slide. Beds of claystone/siltstone below cemented conglomerate in slope remaining after the 1990 rock slide. Contact between the two strata occurs 19 feet above the road in the photograph. (Photo: May 26, 1993).

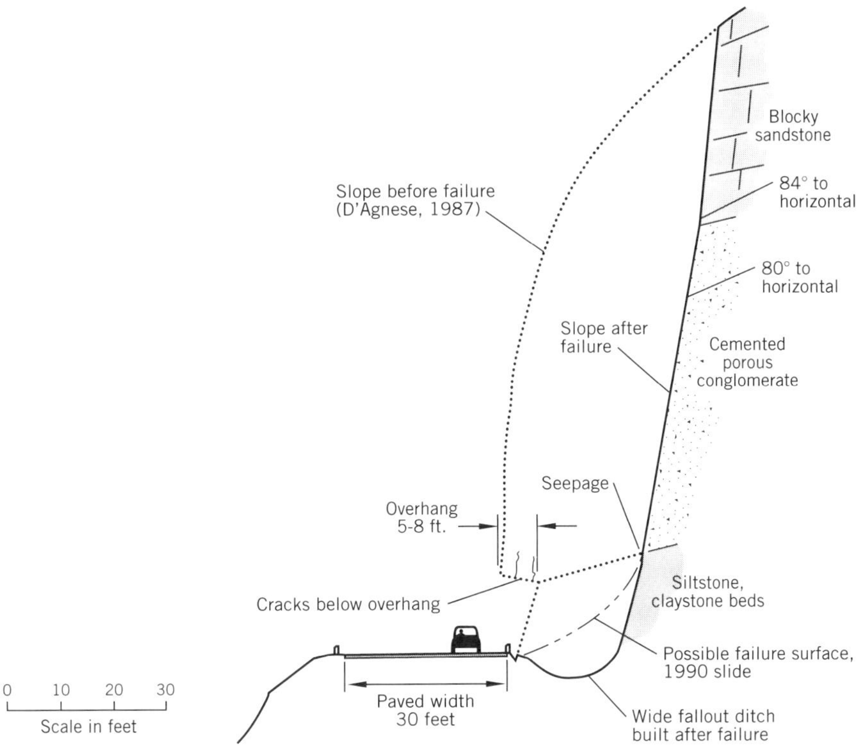

Figure CH-12.3 Crown Point Highway Rock Slide. Geological cross-section taken near the center of the 1990 landslide.

Hazard Rating System (described in Chapter 23, Section 23.2) the slope was rated as one of the potentially more dangerous rock slope hazards in the state and had been targeted for mitigation in 1993 or 1994. State geologists (D'Agnese, 1987) had studied various possible treatments.

LANDSLIDE EVENT

On September 24, 1990, a crew was replacing barriers between the road and the Sandy River. Minor rockfalls occurred at 8 a.m., 11 a.m., and 3 p.m., ranging in size from 1 to 3 cu. yds. At 4:30 p.m. a rockslide occurred that brought about 30,000 cu. yds. of rock onto the road.

All four members of the crew were injured, one suffering a significant back injury and broken leg (plus trauma). Amazingly, no one was killed; traffic had been temporarily stopped at the time.

Table CH-12.1 Slope Measurements at the Crown Point Slide

	Outer Slope of Rock*	
Section	Upper Sandstone	Conglomerate
100 feet east of bridge	84°	80°
200 feet east of bridge	76°	78$^1/_2$°
300 feet east of bridge	79°	85°
Average	80°	81°

*angles to the horizontal

The landslide was about 300 feet long, 110 feet high, and extended into the hillside up to 30 feet (Figure CH-12.3). The road was closed for six weeks to clear the debris and construct a fallout ditch, 15 to 19 feet wide and 5 feet deep, below the remaining cliff (Figures CH-12.1 and CH-12.3). Concrete barriers were installed to separate the traveling public from the ditch.

The slide extended east from Stark Street bridge. Slope measurements taken by clinometer at 100-foot intervals gave the information shown on Table CH-12.1.

These measurements show good consistency at around 80° to the horizontal. A commonly specified design slope for cemented rock slopes of 4 vertical : 1 horizontal corresponds to an angle of 76° to the horizontal.

CAUSATION

The landslide occurred as a result of progressive loss of support at the base of the slope due to erosion of weak sedimentary rocks.

REMEDIATION

After the failure, ODOT proposed to create a fallout ditch 22 feet wide throughout the length of road adjacent to the cliff of conglomerate and sandstone. The hard rock was to be pre-split, then blasted and excavated in benches 15 to 20 feet high. The final rock face would be 4V:1H (vertical:horizon-

Table CH-12.2 Rating of the Crown Point Slide Site Using the Rockfall Hazard Rating System*

Rockfall Risk Factors	Score (Points)
1. *Height of slope:* 110 feet.	100
2. *Ditch effectiveness:* Totally ineffective.	81
3. *Average vehicle risk:* Daily traffic 3,000 cars per day; length of cut slope 0.42 mile; posted speed limit 40 mph. AVR 131%.	100
4. *Percent of decision sight distance:* Site distance obstructed by horizontal curve at east end of site to 330 feet measured; decision site distance at posted speed limit is 600 feet. Percent of decision sight distance = 55%.	36
5. *Road width:* Shoulder 2 feet, 2–12 feet travel lanes, shoulder 4 feet. Total 30 feet. *Geologic Character:* Case 2: Erodible Rocks.	20
6. *Structural Condition:* Type IV major erosional features. Rater selected the maximum points.	100
7. *Rate of erosion:* Extreme, several inches per year due to year round seepage eroding the siltstone.	81
8. *Block size or Volume:* Block sizes of up to about 3 feet (27 points). Rockfall events can exceed 50 cu. yds. (maximum of 100 points). Volume is the key determinant for rating.	100
9. *Climate and water on slope:* High precipitation of 58 inches per year; short duration freezing; water seeps essentially year around. Category III rating.	27
10. *History:* Rockfalls are frequent; severe rockfalls occur on a regular basis	81
TOTAL SCORE	726

*See Chapter 23, Section 23.2

tal). The upper overburden, typically 8 feet thick, was to be reconstructed to 1½:1 (horizontal:vertical). The lower erodible siltstone/claystone stratum was to be shotcreted.

A contract let in 1997 encountered numerous difficulties in implementing the remediation. The successful bidder had concerns over safe operating conditions at the worksite. This contractor wanted minimum bench widths of 12 feet with local turnaround areas 24 feet wide to allow the backhoe to maneuver on the high benches. The landowner refused to give the contractor access to the top of the slope to remove trees. ODOT had difficulties in negotiating the purchase of the land outside the right-of-way. These issues had not been resolved at the time of writing this case history.

Rockfall Hazard Rating System

The rating system is described in Chapter 23, Section 23.2. The rating of the Crown Point Slide is shown on Table CH-12.2. (*Note:* These ratings are based on the site conditions *before* the 1990 landslide.)

References

AASHTO 1984. *A Policy on Geometric Design of Highways and Streets.* AASHTO, Washington, DC, 1087 pp.

AASHTO 1986. *Standard Specification for Highway Bridges.* 16th Edition AASHTO, Washington, D.C.

ASCE 1997. *Geotech Special Technical Publication 75*, Seattle, Washington

ASTM STP 1089. *Deep Foundation Improvements: Design, Construction and Testing.*

ASTM D 2573. *Standard Test Method for Field Vane Shear Test in Cohesive Soil.*

Allen, T.M., and Holtz, R.D. 1991. *Design of Retaining Walls Reinforced with Geosynthetics.* Proceedings of the ASCE Geotechnical Engineering Congress, Boulder, Colorado, pp. 970–987.

Anderson, L.R., Sharp, K.D., Bowles, D.S., and Canfield, R.V. 1984. *Application of Methods of Probabilistic Characterization of Soil Properties.* ASCE Symposium on Probabilistic Characterization of Soil Properties Atlanta, Georgia, pp. 90–105.

Andrus, R.D., and Stokoe, K.H. 1997. *Liquefaction Resistance Based on Shear Wave Velocity.* Proceedings: NCEER Workshop on Evaluation of Liquefaction Resistance of Soils, National Center for Earthquake Engineering Research, pp. 89–128.

Arcement, G.J., and Schneider, V.R. 1984. *Guide for Selecting Manning's Roughness Coefficient for Natural Channels and Flood Plains.* Report No. FHWA-TS-84-204.

Armour, T., Groneck, P., Keeley J., and Sharma, S. 2000. *Micropile Design and Construction Guidelines Implementation Manual.* Report FHWA-SA-97-070, Federal Highways Administration, 376 pp.

Austin, D.N., and Driver, T. 1995. *Classifying Rolled Erosion Control Products.* Erosion Control No. 2, pp. 48–53.

Azzouz, A.S., and Baligh, M.M. 1978. *Three Dimensional Stability of Slopes.* Research Report R78-8, No. 595, Massachusetts Institute of Technology, Cambridge, Massachusetts, 349 pp.

Azzouz, A.S., Baligh, M.M., and Ladd, C.C. 1981. *Three Dimensional Stability Analyses of Four Embankment Failures.* Proceedings 10th International Conference, SMFE, Stockholm, pp. 343–346.

Babasaki, R., Suzuki, K., Saitah, S., Suzuki, Y., and Tokitoh, K. 1991. *Construction and Testing of Deep Foundation Improvement Using the Deep Cement Mixing Method.* Deep Foundations Improvements: Design Construction and Testing. ASTM Special Technical Publication 1089, pp. 224–233.

Bahner, E.W., and Naguib, A.M. 1998. *Design and Construction of a Deep Soil Mix Retaining Wall for the Lake Parkway Freeway Extension.* Geo-Institute, ASCE. Geotechnical Special Publication No. 81, pp. 41–58.

Baker, W.H. 1982. *Planning and Performing Structural Chemical Grouting.* Proceedings of Conference on Grouting in Geotechnical Engineering, New Orleans ASCE, pp. 515–539.

Baker, W.H. 1985. *Embankment Foundation Densification by Compaction Grouting.* Issues in Dam Grouting. ASCE, Denver, Colorado, pp. 104–122.

Baligh, M.M., and Azzouz, A.S. 1975. *End Effects on Stability of Cohesive Slopes.* Journal Geotechnical Engineering Division, ASCE, pp. 1105–1117.

Ballard, R.F., Sjostrom, K.J., McGee, R.G., and Leist, R.L. 1993. *A Rapid Geophysical Technique for Subbottom Imaging.* Dredging Research Technical Notes DRP-2-07, U.S. Army Engineers Waterways Experiment Station, Vicksburg, MS.

Barksdale, R.D., and Bachus, R.C. 1983. *Design and Construction of Stone Columns.* FHWA Report SCEGIT-83-104 (E 20-686), Federal Highway Administration, Washington, DC.

Barley, A.D., and Woodward, M.A. 1992. *High Loading of Long Slender Minipiles*. Proceedings, ICE Conference on Piling European Practice and Worldwide Trends. Thomas Telford, London, pp. 131–136.

Barrett, R.K. 1979. *Arc of Horizontal Drains: Case Histories from the Colorado Division of Highways District III,* Grand Junction, Colorado.

Barrett, R.K. 1979. *Use of Horizontal Drains: Case Histories from the Colorado Division of Highways.* Colorado Division of Highways, District III, Grand Junction, Colorado.

Barron, R.A. 1948. *Consolidation of Fine-Grained Soils by Drain Wells.* ASCE Transactions, Paper 2346, Volume 113, pp. 718–724.

Baziar, M.H., and Dobry, R. 1995. *Residual Strength and Large-Deformation Potential of Loose Silty Sands.* Journal of Geotechnical Engineering, Vol. 121, No. 12, pp. 896–906.

Bell. A.L. 1993. *Jet Grouting.* Ground Improvement, Blackie. pp. 149–174.

Bishop, A.W. 1948. *A New Sampling Tool for Use in Cohesionless Sands Below Groundwater Level.* Geotechnique, Vol. 1. June, pp. 125–131.

Bishop, A.W. 1955. *The Use of the Slip Circle in the Stability Analysis of Slopes* Geotechnique, Vol. V, No. 1 March, p. 7–17.

Bishop, A.W., and Henkel, D.J. 1957. *The Measurement of Soil Properties in the Triaxial Test.* Edward Arnold Ltd., London, 190 pp.

Bishop, A.W., Green, G.E., Garga, V.K., Andresen, A., and Brown, J.D. 1971. *A New Ring Shear Apparatus and its Application to the Measurement of Residual Strength.* Geotechnique, December, pp. 273–328.

Bjerrum, L. 1954. *Geotechnical Properties of Norwegian Marine Clays.* Geotechnique Vol. 6, pp. 49–69.

Bjerrum, L., Casagrande, A., Peck, R.B., and Skempton, A.W. 1960. *From Theory to Practice in Soil Mechanics.* Selections from the writings of Karl Terzaghi, John Wiley, New York.

Bjerrum, L., Kringstad, S., and Kummeneje, O. 1961. *The Shear Strength of a Fine Loose Sand.* Proc. 5th International Conference on Soil Mechanics and Foundation Engineering, Paris, Vol. 1, pp. 29–37.

Bjerrum, L. 1971. *Subaqueous Slope Failures in Norwegian Fjords.* Norwegian Geotechnical Institute, Oslo, Publication 88, 8 pp.

Bjerrum, L. 1972. *Embankments on Soft Ground.* ASCE Conference on Performance of Earth and Earth-Supported Structures, Purdue University, Vol. 2, pp. 1–54.

Bjerrum, L. 1973. *Problems of Soil Mechanics and Construction on Soft Clays.* Proceedings 8th International Conference on Soil Mechanics and Foundation Engineering, Moscow, Vol. 3, pp. 111–159.

Blodgett, J.C. 1986. *Rock Riprap Design for Protection of Stream Channels Near Highway Structures: Volume 1—Hydraulic Characteristics of Open Channels.* U.S. Geological Survey Water Resources Investigations Report 86-4127.

Botto, G. 1985. *Developments in the Techniques of Jet-Grouting.* XII Ciclo di Conferenze di Geotecnica, Torino, Italy.

Brand, E.W. 1985. *Predicting the Performance of Residual Soil Slopes.* Proceedings International Conference on Soil Mechanics and Foundation Engineering, San Francisco, California, pp. 2541–2578.

Brand, E.W., Premchitt, J., and Phillipson, H.B. 1984. *Relationship Between Rainfall and Landslides in Hong Kong.* Proceedings, Fourth International Symposium on Landslides, Toronto, Vol. 1, pp. 377–384.

Brawner, C.O. 1994. *Rockfall Hazard Mitigation Methods.* National Highway Institute Course No. 13219 as published by Federal Highway Administration Publication No. FHWA SA-93-085, 436 pp.

British Standards Institution 1989. *Ground Anchorages.* BS 8081, British Standards Institution, London, England.

Bromhead, E.N. 1979. *A Simple Ring Shear Apparatus.* Ground Engineering, Vol. 12, No. 5, pp. 40–44.

Bromhead, E.N. 1986. *The Stability of Slopes.* Chapman & Hall, New York, 374 pp.

Broms, B.B. 1965. *Design of Laterally Loaded Piles.* Journal Soil Mechanics and Foundations Div., ASCE, Vol. 91, No. SM3, pp. 79–99.

Brooks, G.R., Aylsworth, J.M., Evans, S.G., and Lawrence, D.E. 1994. *The Lemieux Landslide of June 20, 1993, South Nation Valley, Southeastern Ontario—A Photographic Record.* Miscellaneous Report 56 Geological Society of Canada, Ottawa, Ontario, 18 pp.

Bruce, D.A. 1988. *Developments in Geotechnical Construction Processes for Urban Engineering.* Civil Engineering Practice, Vol. 3, No. 1, pp. 49–97.

Bruce, D.A., Granata, R., Mauro, M., and Cippo, A.P. 1993. *Some Recent Developments in Ground Treatment for Tunnelling.* Proceedings, Rapid Excavation and Tunneling Conference, Cushing-Malloy, Inc., Ann Arbor, Michigan.

Bruce, D.A. 2001. *An Introduction to the Deep Mixing Methods as Used in Geotechnical Applications.* Federal Highway Administration Report FHWA-RD-99-167, Volume III.

Bruce, D.A. 2002. *Micropiles: Think Small for Big Loads.* Geo-Strata, ASCE Geo Institute, July 2002, pp. 15–18.

Burmister, D.M. 1948. *The Importance and Practical use of Relative Density in Soil Mechanics.* Proceedings ASTM, Vol. 48, pp. 1249–1269.

Burns, S.F. 1998. *Landslides in the Portland Area Resulting from the Storm of February, 1996.* Environmental, Groundwater and Engineering Geology: Applications from Oregon. Star Publishing Co., Portland, Oregon, pp. 353–365.

Byle, M.J. 2000. *An Approach to the Design of LMD Grouting.* ASCE Geotechnical Special Publication No. 104, Advances in Grouting and Ground Modification ASCE Press, New York, pp. 94–110.

Byrne, R.J., Cotton, D.M., Porterfield, J., Wolschlag, C., and Ueblacker, G. 1996. *Manual for Design and Construction Monitoring of Soil Nail Walls.* Federal Highway Administration Report FHWA-SA-96-069R. 530 pp.

Casagrande, A. 1932. *Research on the Atterberg Limits of Soils.* Public Roads, Vol. 13, pp. 121–136.

Cavounidis, S. 1987. *On the Ratio of Factors of Safety in Slope Stability Analyses.* Geotechnique, Vol. 37. No. 2, pp. 207–210.

CCTG 1993. *Technical Rules for the Design and Calculation of Foundations for Civil Engineering Works.* Publication 62.

Cedergren, H.R. 1977. *Seepage, Drainage, and Flow Nets.* 2d ed., John Wiley, New York, 534 pp.

Chassie, R.G. 1992. *Soil Nailing Overview. Soil Nailing and Reinforced Soil Walls.* ASCE and University of Washington Seminar, March 28, 1992, Seattle, Washington, 58 pp.

Chen, R.H., and Chameau, J.L. 1983. *Three Dimensional Limit Equilibrium Analyses of Slopes.* Geotechnique, Vol. 32, pp. 31–40.

Cheney, R.S. 1984. *Permanent Ground Anchors.* Report FHWA-DP-68-IR, Federal Highway Administration.

Ching, R.K.H., Sweeney, D.J., and Fredlund, D.G. 1984. *Increase in Factor of Safety Due to Soil Suction for Two Hong Kong Slopes.* Proceedings 4th International Symposium of Landslides, Toronto, Canada, Vol. 1, pp. 617–624.

Chleborad, A.F. 1994. *Modeling and Analysis of the 1949 Narrows Landslides, Tacoma, Washington.* Bulletin of the Association of Engineering Geologists, Vol. XXXI, No. 3, pp. 305–327.

Chleborad, A.F., and Schuster, R.L. 1989. *Characteristics of Slope Failures Induced by the April 13, 1949 and April 29, 1969 Puget Sound Area Washington Earthquakes.* Third Annual Workshop on Earthquake Hazards in the Puget Sound-Portland Area, U.S. Geological Survey Open-File Report 89-465.

Christian, J.T., Ladd, C.C., and Baecher, G.B. 1994. *Reliability applied to Slope Stability Analysis.* Journal Geotechnical Engineering, Vol. 120, December, pp. 2180–2207.

Christopher, B.R., Gill, S.A., Giroud, J-P, Juran, I., Mitchell, J.K., Schlosser, F., and Dunnicliff, J. 1990. *Reinforced Soil Structures, Vol. 1: Design and Construction Guidelines.* FHWA Report FHWA-RD-89-043, 301 pp.

Clarke, D.D. 1904. *A Phenomenal Landslide.* Transactions ASCE, Vol. 53, Paper 984, pp. 321–411.

Clarke, D.D. 1918. *A Phenomenal Landslide: Supplement.* Transactions ASCE, Vol. 82, Paper 1415, pp. 767–796.

Clough, G.W., and O'Rourke, T.D. 1990. *Movements of InSitu Walls.* Proceedings, Design and Performance of Earth Retaining Structures, Cornell University, Ithaca, New York, pp. 439–470.

Clouterre 1991. *French Recommendations for Soil Nailing.* Projects Nationaux DeRecherche Developpement En Genie Civil Domaine De Saint-Paul, BP37-78470 Saint Remy Les Chev Reuse. (in French)

Clouterre Project, 1991. *French National Research Project on Soil Nailed Walls.* English translation by the Federal Highway Administration Report FHWA-SA-93-026.

Coe, J.A., Michael, J.A., Crovelli, R.A., and Savage, W.Z. 2000. *Preliminary Map Showing Landslide Densities, Mean Recurrence Intervals, and Exceedance Probabilities as Determined from Historic Records.* Seattle, Washington Open-File Report 00-303. U.S. Geological Survey.

Coppin, N.J., and Richards, I. 1990. *Use of Vegetation in Civil Engineering.* CIRIA Book 10, Butterworths, England.

Cornforth, D.H. 1961(a), *Plane Strain Failure Characteristics of Saturated Sand.* Ph.D. Thesis, Imperial College, University of London, 192 pp.

Cornforth, D.H. 1961(b), *Discussion on effectiveness of filter drains on laboratory triaxial tests.* Proceedings of 5th International Conference SMFE, Paris, France, Vol. 3, pp. 121–122.

Cornforth, D.H. 1964. *Some Experiments on the Influence of Strain Conditions on the Strength of Sand.* Geotechnique, Vol. 14, No. 2, pp. 145–167.

Cornforth, D.H. 1973. *Prediction of Drained Strength of Sands from Relative Density Measurements.* American Society for Testing and Materials STP 523, pp. 281–303.

Cornforth, D.H. 1974(a). *Performance Characteristics of the Slope Indicator Series 200-B Inclinometer.* Proceedings, Field Instrumentation in Geotechnical Engineering, Buttersworth, pp. 126–135.

Cornforth, D.H. 1974(b). *One-Dimensional Consolidation Curves of a Medium Sand.* Geotechnique, Vol. 24, No. 4, December, pp. 678–683.

Cornforth, D.H., Worth, E.G., and Wright, W. L. 1974. *Observations and Analysis of a Flow Slide in Sand Fill.* Proceedings, Field Instrumentation in Geotechnical Engineering, Buttersworth, pp. 136–151.

Cornforth, D.H. 1990. *Earthquake-Induced Landslides.* U.S. Dept. of Interior Geological Survey Proceedings, 4th Annual Workshop on Earthquake Hazards in the Puget Sound and Portland Areas, April 16–19, 1990, Seattle, Washington. Open-File Report 90–703, pp. 11–18.

Cornforth, D.H., and Fujitani, K.F. 1991. *Residual Strength of Volcanic Clay: Two Case Studies.* International Conference on Slope Stability Engineering, Isle of Wight, England, April, The Institution of Civil Engineers, London, England, pp.23–26.

Cornforth, D.H., and Vessely, D.A. 1992(a). *Pelton Landslide: An Unusual Double-Wedge Failure.* Stability and Performance of Slopes and Embankments-II, Geotechnical Engineering Division Specialty Conference, University of California. Vol. 1, pp. 310–324.

Cornforth, D.H., and Vessely, D.A. 1992(b). *Factors of Safety During Landslide Movements.* Proceedings of the Sixth International Symposium on Landslides, Christchurch, New Zealand, Vol. 1, pp. 367–372.

Cornforth, D. H. 1995. *Landslide Remediation Using Original Profile Analysis.* Ian Boyd Donald Symposium, Modern Developments in Geomechanics, Monash University, Melbourne, Australia, p. 115–126.

Cornforth, D.H., and Lowell, J.A. 1996. *The 1994 Submarine Slope Failure at Skagway, Alaska.* Proceedings of the 7th International Symposium on Landslides, Trondheim, Norway, Vol. 1, pp. 527–532.

Cornforth, D.H., and Mikkelsen, P.E. 1996. *Continuous Monitoring of the Slope Above an Excavation Within a Marginally Stable Landslide.* Proceedings of the 7th International Symposium on Landslides, Trondheim, Norway, Vol.3, pp. 1539–1544.

Cornforth, D.H., Arango, I., Vessely, D. A., and Hill, R.J. 1996. *Seismic Design of the Barney Reservoir Dam, Oregon.* 16th Annual USCOLD Lecture Series, U.S. Committee on Large Dams, Los Angeles, California, July, pp. 75–90.

Cornforth, D.H., Schroeder, W.L., Hill, R.J., and Vessely, D.A. 1996. *Slope Stabilization Using Reinforced Concrete Piles.* Proceedings, Oregon ASCE Geotechnical Seminar on Ground Stabilization and Seismic Mitigation: Theory and Practice, Portland, Oregon, November 6–7, 30 pp.

Cornforth, D.H. 2000. *Seismic Reactivation of Pelton Park Landslide.* Proceedings of the 8th International Symposium on Landslides, Cardiff, Wales, Landslides in Research, Theory, and Practice Vol. 1, pp. 317–322.

Cornforth, D.H., and Mikkelsen, P.E. 2000. *Continuous Monitoring of Groundwater Response to Winter Rainfall.* Proceedings of the 8th International Symposium on Landslides, Cardiff, Wales, Landslides in Research, Theory and Practice, Vol. 1, pp. 323–328.

Cornforth, D.H., Duncan, J.M., and Morgenstern, N.R. 2001. *Stability Review of the Oahe Dam, Pierre, South Dakota.* Golder Associates, Inc. report to U.S. Army Corps of Engineers, Omaha District, 38 pp.

Cotton, D.M. 1992. *Development of the Top-Down Permanent Wall Method and Stability Problems in Seattle Native Outwash Glacial Till and Lacustrine Deposits*. ASCE and University of Washington Seminar, March 28, 1992. Seattle, Washington.

Coulter, H.W., and Migliaccio, R.R. 1966. *Effects of the Earthquake of March 27, 1964 at Valdez, Alaska*. Geological Survey Professional Paper 542-C, U.S. Department of the Interior.

Croce, P., and Flora, A. 2000. *Analysis of Single-Fluid Jet Grouting*. Geotechnique, Vol. 50, No. 6., pp. 739–748.

Crom, T.R. 1966. *Dry-Mix Shotcrete Practice*. Shotcreting SP-14, American Concrete Institute, Detroit, Michigan, pp. 15–32.

Cruden, D.M. 1991. *A Simple Definition of a Landslide*. Bulletin of the International Association of Engineering Geology, No. 43, pp. 27–29.

Cruden, D.M., and Varnes, D.J. 1996. *Landslide Types and Processes*. Landslides: Investigation and Mitigation. Special Report 247, National Academy Press, Washington D.C., pp. 36–75.

D'Agnese, S.L. 1987. *Rockfall Hazard at the Stark Street Intersection, Crown Point Highway*. Geology Report, Oregon Department of Transportation, Region 1, Portland, Oregon, 14 pp.

Dai, S.H., and Wang, M.O. 1992. *Reliability Analysis in Engineering Applications*. Van Nostrand, Reinhold, New York.

Dames and Moore. 1989. *The October 17, 1989 Loma Prieta Earthquake*. Special Report by Dames and Moore, Los Angeles, California, 29 pp.

Daromola, O. 1980. *Effect of Consolidation Age on Stiffness of Sand*. Geotechnique Vol. 30, No. 2, pp. 213–216.

Das, B.M. 1994. *Principles of Geotechnical Engineering*. 3d ed. PWS Publishing, Boston, 672 pp.

Davidson, R.R., Levallois, J., and Graybeal, K. 1992. *Seepage Cutoff Wall for Mud Mountain Dam*. Slurry Walls: Design Construction and Quality Control ASTM STP 1126, pp. 309–323.

Davis, S., and Karzulovic, J.K. 1961. *Deslizamientos en el valle del rio San Pedro Provincia de Valdivia Chile, Anales de la Faculted de Ciencieas Fisical y Matematicos, Santiago, Chile*. University of Chile, Institute of Geology, Publication No. 20.

Day, S.R., and Ryan, C.R. 1992. *State of the Art in Bio-Polymer Drain Construction*. Slurry Walls: Design Construction and Quality Control, ASTM STP 1129, pp. 333–343.

Day, S.R., and Ryan, C.B. 1995. *Containment, Stabilization and Treatment of Contaminated Soils using In-situ Soil Mixing*. Proceedings Specialty Conference, New Orleans, Louisiana, February 1969, ASCE Geotechnical Special Publication No. 46, Volume 2, pp. 1349–1365.

Decker, M.D. 1983. *Dissertation*, Purdue University.

Degen, W.S. 1997. *56m Deep Vibrocompaction at German Lignite Mining Area*. Ground Improvement Geosystems—Densification and Reinforcement, Thomas Telford Ltd., London, pp. 127–133.

Dise, K., Stevens, M.G., and Von Thun, J.L. 1994. *Dynamic Compaction to Remediate Liquefiable Embankment Foundation Soils*. ASCE Geotechnical Special Publication No. 45, pp. 1–25.

DJM Association of Japan 1993. DJM Manual published by the Dry Jet Mixing Association of Japan (in Japanese), 75 pp.

Dobbels, D., and Lyman, T. 2002. *Microtunneling—What Is It?* Geo-Strata, ASCE Geo Institute, July 2002, pp. 10–12.

Domenico, P.A., and Schwartz, F.W. 1990. *Physical and Chemical Hydrogeology*. John Wiley & Sons, New York, 824 pp.

Douglas, B.J., and Strutznsky, A. 1984. *SPT Hammer Energy Measurements*. Report to U.S. Geological Survey.

Driscoll, F.G. 1986. *Groundwater and Wells* 2d ed. Johnson Filtration Systems, Inc., St. Paul, Minnesota, 1089 pp.

Duncan, J.M., and Buchignani, A.L. 1973. *Failure of Underwater Slope in San Francisco Bay*. Journal of Soil Mechanics and Foundation Engineering Div., ASCE, pp. 687–703.

Duncan, J.M. 1992. *State-of-the-Art: Static Stability and Deformation Analysis*. Proceedings ASCE Conference on Stability and Performance of Slope and Embankments II. Geotechnical Special Publication 31.

Duncan, J.M. 1996. *Soil Slope Stability Analysis*. Landslides: Investigation and Mitigation, Transportation Research Board National Academy Press, Washington, D.C., pp. 337–371.

Duncan, J.M. 2000. *Factors of Safety and Reliability in Geotechnical Engineering*, Journal Geotech and Geo Env. Engineering Vol. 126, No. 4, April 2000, pp. 307–316.

Duncan, J.M., and Seed, R.B. 1986. *Compaction-induced Earth Pressures under K_0 Conditions*. ASCE, Journal of Geotechnical Division, Vol. 112, GT1, pp. 1–22.

Duncan, J.M., Buchignani, A.L., and De Wet, M. 1987. *An Engineering Manual for Slope Stability Studies*. Virginia Polytechnic Institute and State University, Blacksburg, Virginia, 80 pp.

Duncan, J.M., Seed, R.B., Wong, K.S., and Ozawa, Y. 1984. *FEADAM84: A Computer Program for Finite Element Analysis of Dams*. Geotechnical Research Department, Report SU/GT/84-03, Stanford University.

Duncan, J.M., Clough, G.W., and Ebeling, R.M. 1990. *Behavior and Design of Gravity Earth Retaining Structures*. Proceedings, ASCE Conference on Design and Performance of Earth Retaining Structures, Ithaca, New York, pp. 251–277.

Duncan, J.M., Wright, S.G., and Wong, K.S. 1990. *Slope Stability During Rapid Drawdown*. H. Bolton Seed Memorial Symposium, Vol. 2, pp. 253–272.

Duncan, J.M., Williams, G.W., Sehn, A.L., and Seed, R.B. 1991. *Estimation of Earth Pressures Due to Compaction*. ASCE Journal of Geotechnical Division, Vol. 117, No. 12, pp. 1833–1847.

Duncan, J.M., Williams, G.W., Sehn, A.L., and Seed, R.B. 1993. *Estimation of Earth Pressures Due to Compaction*. Closure of discussion, ASCE Journal of Geotechnical Engineering, Vol. 119, No. 7, pp. 1172–1177.

Dunnicliff, J. 1988. *Geotechnical Instrumentation for Monitoring Field Performance*. John Wiley & Sons, New York, 577 pp.

EERI 1993. *The March 25, 1993 Scotts Mills Earthquake—Western Oregon Wake-Up Call*. Earthquake Engineering Research Institute Special Earthquake Report, May, pp. 1–8.

Eckel, E.B. 1958. *Introduction Landslides and Engineering Practice*. Special Report 29, National Research Council Highway Research Board, pp. 1–5.

Elias, V., Christopher, B.R., and Berg, R.R. 2001. *Mechanically Stabilized Earth Walls and Reinforced Slopes: Design and Construction Guidelines*. Publication FHWA-NHI-00-043, National Highway Institute, FHWA, Washington, D.C., 393 pp.

Eldin, A.K.G. 1951. *Some Fundamental Factors Controlling the Shear Properties of Sand*. Ph.D. thesis, University of London.

Erwin, E.D., and Glenn, J.M. 1991. *Plastic Concrete Slurry Wall for Wister Dam*. Slurry Walls: Design Construction and Quality Control, ASTM STP 1129, pp. 251–267.

Essler, R.D., Drooff, E.R., and Falk, E. 2000. *Compensation Grouting: Concept, Theory, and Practice.* ASCE Geotechnical Special Publication No. 104, pp. 1–15.

Fannin, R.J. 2000. *Basic Geosynthetics: A Guide to Best Practices.* BiTech Publishers, Vancouver, B.C., 86 pp.

Felio, B.Y., Vucetic, M., Hudson, M., Barar, P., and Chapman, R. 1990. *Performance of Soil Nailed Walls During the October 17, 1989 Loma Prieta Earthquake.* Proceedings, 43rd Canadian Geotechnical Conference, Quebec, Canada, pp. 165–173.

Fetter, C.W. 1988. *Applied Hydrogeology.* 2d ed. Merrill Publishing Company, Columbus, Ohio, 592 pp.

Field, C.R., and Stone, S.D. 1995. *Flexible Membrane Liners.* Civil Engineering magazine, April, pp. 60–61.

FLAC. *Fast Lagrangian Analysis of Continua.* Itasca Software.

Ford, W.H. 1974. *Low Pressure Jet Cleaning of Plastic Drains in Sandy Soils.* Transactions, American Society of Agricultural Engineers, Vol. 17, pp. 895–897.

Forrester, K., and Nyland G. 1980. *Two Landslides on New South Wales Highways.* Proceedings, 3rd International Symposium on Landslides, New Delhi, Vol. 1, pp. 181–184.

Forrester, K. 2001. *Subsurface Drainage for Slope Stabilizaton.* ASCE Press, Reston, Virginia, 208 pp.

Fraser, A.M. 1957. *The Influence of Stress Ratio on Compressibility and Pore Pressure Coefficients in Compacted Soils.* Ph.D. thesis, University of London.

Fredlund, D.G., and Krahn, J. 1977. *Comparison of Slope Stability Methods of Analysis.* Canadian Geotechnical Journal 14, pp. 429–439.

Fredlund, D.G., and Morgenstern, N.R. 1977. *Stress State Variables for Unsaturated Soils.* ASCE Journal, Geotechnical Engineering Division, Vol. 103, pp. 447–466.

Fredlund, D.G. Morgenstern, N.R., and Widger, R.A. 1978. *Shear Strength of Unsaturated Soils.* Canadian Geotechnical Journal, Vol. 15, pp. 313–321.

Fredlund, D.G., Krahn, J., and Pufahl, D.E. 1981. *The Relation Between Limit Equilibrium Slope Stability Methods.* Proceedings of the International Conference on Soil Mechanics and Foundation Engineering, Stockholm, Vol. 3, pp. 409–416.

Fredlund, D.G. 1984. *Analytical Methods for Slope Stability Analysis.* Proceedings 4th International Symposium Landslides, Toronto, Vol. 1, pp. 229–250.

Fredlund, D.G. 1987. *Slope Stability Analysis Incorporating the Effect of Soil Suction.* Slope Stability: Geotechnical Engineering and Geomorphology, John Wiley & Sons, pp. 113–144.

Fukuoka, M. 1977. *The Effects of Horizontal Loads on Piles Due to Landslide.* Proceedings of the Ninth International Conference on Soil Mechanics and Foundation Engineering, Tokyo, Japan, pp. 27–42.

Geotechnical Control Office 1984. *Geotechnical Manual for Slopes.* 2d ed. Geotechnical Control Office, Hong Kong, 295 pp.

Gibbs, H.J., and Holtz, W.G. 1957. *Research on Determining the Density of Sands by Spoon Penetration Testing.* Proceedings of the 4th International Conference on Soil Mechanics and Foundation Engineering, London, Vol. 1, pp. 35–39.

Gibson, R.E., and Henkel, D.J. 1954. *Influence of Duration of Tests at Constant Rate of Strain on Measured "Drained" Strength.* Geotechnique Vol. 4, pp. 6–15.

Gifford, D.G., and Wheeler, J.R. 1992. *Concrete Slurry Wall for Temporary and Permanent Foundation Wall at Gallery at Harborplace—Baltimore, Maryland.* Slurry Walls: Design Construction and Quality Control ASTM STP 1126, pp. 151–163

Goble, G.G. 1995. *SPT Improvements.* GRL Newsletter, December.

Goodman, R.E., and Seed, H.B. 1966. *Earthquake-Induced Displacements in Sand Embankments.* Journal of the Soil Mechanics and Foundation Division, ASCE, Vol 92, SM2, March 1966, pp. 125–146.

Goughnour, R.R., Teh Sung, J., and Ramsey, J.S. 1991. *Slide Correction by Stone Columns.* Deep Foundation Improvement, ASTM Special Technical Publication 1089, pp. 131–147.

Gray, D.H., and Sotir, R.B. 1996. *Biotechnical and Soil Bioengineering Slope Stabilization.* John Wiley & Sons, Inc., New York, 378 pp.

Greenway, D.R. 1987. *Vegetation and Slope Stability.* Chapter 6, Slope Stability: Geotechnical Engineering and Geomorphology, John Wiley & Sons, pp. 187–230.

Guarino, L. 1985. *Pore Pressure Changes Due to Bentonite Pellet Seals.* M.S. Thesis, University of Massachusetts, Amherst, MA.

Habib, P. 1953. *Influence de la Variation de la Contrainte Principale Moyenne sur la Resistance au Cisaillement des Sols.* Proceedings 3rd International Conference on Soil Mechanics, Vol. 1, pp. 131–136.

Hafiz, M.A.A. 1950. *Strength Characteristics of Sands and Gravels in Direct Shear.* Ph.D. thesis, University of London.

Haley and Aldrich, Inc. 1986. *Prefabricated Vertical Drains: Volumes 1 and 2.* Federal Highway Administration Report FHWA/RD-86/168.

Hampton, M.A., Lee, H.J., and Locat, J. 1996. *Submarine Landslides.* Reviews of Geophysics 34 February American Geophysical Union, pp. 33–59.

Hannan, R., and Devine, H. 1998. *Ruckel Landslide, Columbia River Gorge, Cascade Locks, Oregon: A Case Study of Successful Landslide Remediation.* Environmental, Groundwater and Engineering Geology. Applications from Oregon. Edited by Scott Burns, Star Publishing Company, Belmont, California, pp. 367–372.

Hansbo, S. 1979. *Consolidation of Clay by Band-Shaped Prefabricated Drains.* Ground Engineering Vol. 12, No. 5, pp. 21–25.

Harder, L.F. 1997. *Application of the Becker Penetration Test for Evaluating the Liquefaction Potential of Gravelly Soils.* Proceeding, NCEER Workshop on Evaluation of Liquefaction Resistance of Soils, National Center for Engineering Research, Buffalo, pp. 129–148.

Harder, L.F., and Seed, H.B. 1986. *Determination of Penetration Resistance for Coarse-grained Soils Using the Becker Hammer Drill.* Report UCB/EERC-86/06, National Science Foundation Research Report, University of California, Berkeley, 126 pp.

Harp, E.L., Wilson, R.C., and Wieczorek, G.F. 1981. *Landslides from the February 4, 1976 Guatemala Earthquake.* U.S. Geological Survey Professional Paper, 1204-A.

Hassan, A.M., and Wolff, T.F. 1999. *Search Algorithm for Minimum Reliability Index of Earth Slopes.* Journal GGE, ASCE, Vol. 125, pp. 301–308.

Hazen, A. 1930. *Water Supply.* American Civil Engineers Handbook, John Wiley, New York.

Head, K.H. 1982. *Manual of Soil Laboratory Testing, Volume 2: Permeability, Shear Strength and Compressibility Tests.* Pentech Press Ltd., London, England, pp. 335–747.

Head, K.H. 1986. *Manual of Soil Laboratory Testing, Volume 3: Effective Stress Tests.* Pentech Press Ltd., London, England, pp. 743–1238.

Head, K.H. 1992. *Manual of Soil Laboratory Testing, Volume 1: Soil Classification and Compaction Tests.* 2d ed. Halsted Press (John Wiley & Sons), 384 pp.

Helland, A. 1894. *Opdyrkning of Lerfaldet Vaerdalen.* Norges Geologiske Undersøkelse, Vol. 14.

Hencher, S.R., and Martin, R.P. 1982. *The Description and Classification of Weathered Rocks in Hong Kong for Engineering Purposes.* Proceedings, 7th S.E. Asian Geotechnical Conference, Hong Kong, Vol. 1, pp. 125–142.

Henkel, D.J., and Wade, N.H. 1966. *Plane Strain Tests on a Saturated Remolded Clay.* Journal SMFE Division, ASCE, Vol. 92, SM6, pp. 67–80.

Henkel, D.J. 1971. *The Calculation of Earth Pressures in Open Cuts in Soft Clays.* The Arup Journal, Vol. 6, No. 4, pp. 14–15.

Henn, R.W. 1996. *Practical Guide to Grouting of Underground Structures.* ASCE Press, New York, 191 pp.

Hetenyi, M. 1946. *Beams on Elastic Foundation.* University of Michigan Press, Ann Arbor.

Higgins, J.D., Fragaszy, R.J., and Beard, L.D. 1983. *Engineering Geology of Loess in Southeastern Washington.* Washington Division of Geology and Earth Resources. Bulletin 78 (Engineering Geology in Washington), Vol. 2, pp. 887–898.

Higgins, J.D., and Fragaszy, R.J. 1988. *Design Guide for Cut Slopes in Loess of Southeastern Washington.* Report WA-RD 145-2, Washington State Department of Transportation, Olympia, WA, 65 pp.

Higgins, J.D., and Modeer, V.A. 1996. *Loess.* In Landslide Investigation and Mitigation. Transportation Research Board Special Report 247, National Academy Press, Washington D.C., pp. 585–606.

Ho, D.Y.F., and Fredlund, D.G. 1982(a). *A Multistage Triaxial Test for Unsaturated Soils.* Geotechnical Testing Journal, ASTM, June, pp. 18–25.

Ho, D.Y.F., and Fredlund, D.G. 1982(b). *Increase in Strength Due to Suction for Two Hong Kong Soils.* Proceedings, Conference on Engineering and Construction in Tropical Residual Soils, ASCE, Honolulu, Hawaii, pp. 263–295.

Holmsen, P. 1953. *Landslides in Norwegian Quick-Clays.* Geotechnique, Vol. 3, pp. 187–194.

Holtz, W.G., and Gibbs, H.J. 1951. *Consolidation and Related Properties of Loessial Soils.* ASTM Special Technical Publication 126, pp. 9–26.

Holzer, T.L. 1948. *The Loma Prieta California Earthquake of October 17, 1989—Liquefaction.* U.S. Geological Survey Professional Paper 1551, U.S. Government Printing Office.

Houlsby, A.C. 1990. *Construction and Design of Cement Grouting.* John Wiley & Sons, New York, 442 pp.

Hryciw, R.D., Raschke, S.A., Ghalib, A.M., and Shin, S. 2002. *A Cone with a View: The VisCPT.* Geo-Strata, ASCE Geo Institute, July 2002, pp. 25–29.

Hudson, R.V. 1961. *Laboratory Investigation of Rubble-mound Breakwaters.* Transactions, ASCE Paper 3213, Vol. 126, Part IV, p. 492.

Hungr, O. 1987. *An Extension of Bishop's Simplified Method of Slope Stability Analysis to Three Dimensions.* Geotechnique Vol. 37, No. 1, pp. 113–117.

Hungr, O., Salgado, F.M., and Byrne, P.M. 1989. *Evaluation of a Three-Dimensional Method of Slope Stability Analysis.* Canadian Geotechnical Journal, Vol. 26, pp. 679–686.

Hunter, J.H., and Schuster, R.L. 1968. *Stability of Simple Cuttings in Normally Consolidated Clays.* Geotechnique, Vol. 18, No. 3, pp. 372–378.

Hutchinson, J.N. 1977. *Assessment of the Effectiveness of Corrective Measures in Relation to Geological Conditions and Types of Slope Movement.* Bulletin of International Association of Engineering Geologists, Vol. 16, pp. 131–155.

Hutchinson, J.N. 1978. *Assessment of the Effectiveness of Corrective Measures in Relation to Geological Conditions and Types of Movement.* Norwegian Geotechnical Institute Publication, No. 124, pp. 1–25.

Hvorslev, M.J. 1939. *Torsion Shear Tests and Their Place in the Determination of the Shearing Resistance of Soils.* Proceedings of ASTM, Vol. 39, pp. 999–1022.

Hvorslev, M.J. 1949. *Subsurface Exploration and Sampling of Soils for Civil Engineering Purposes.* Waterways Experiment Station, Vicksburg, Mississippi, 521 pp.

Hvorslev, M.J. 1949. *Time Lag in the Observation of Groundwater Levels and Pressures.* U.S. Army Waterways Experimental Station, Vicksburg, Mississippi.

Hvorslev, M.J. 1951. *Time Lag and Soil Permeability in Groundwater Observations.* Bulletin No. 36, Waterways Experiment Station, U.S. Corps of Engineers, Vicksburg, Miss.

Hynes, M.E., and Olsen, R.S. 1999. *Influence of Confining Stress on Liquefaction Resistance.* Proceedings International Workshop on Physics and Mechanics of Soil Liquefaction, Balkema Publishers, Rotterdam, pp. 145–152.

Iai, S., Noda, S., and Tsuchida, H. 1988. *Basic Consideration for Designing the Area of the Ground Compaction as a Remedial Measure Against Liquefaction.* U.S.-Japan Joint Workshop on Remedial Measures for Liquefiable Soils, Jackson, Wyoming.

Idriss, I.M. 1991. *Earthquake Ground Motions at Soft Soil Sites.* Proceedings: 2nd International Conference on Recent Advances in Geotechnical Earthquake Engineering and Soil Dynamics, Vol. 3, pp. 2265–2271.

Idriss, I.M., and Sun, J.I. 1992. *Users Manual for SHAKE91.* Center for Geotechnical Modeling, Dept. of Civil Engineering, University of California, Davis.

Ingold, T.S., and Myles, B. 1996. *Ballistic Soil Nailing: Earth Reinforcement.* Proceedings, International Symposium on Earth Reinforcement, Japan.

Institution of Civil Engineers, 1999. *Specification for the Construction of Slurry Trench Cut-off Walls.* Thomas Telford Books, London, England, 70 pp.

Ivanetich, K., Gularte, F., and Dees, B. 2000. *Compaction Grout: A Case History of Seismic Retrofit.* ASCE Geotechnical Special Publication No. 104, pp. 83–93.

Jaky, J. 1948. *Pressure in Silos.* Proceedings of 2nd International Conference on Soil Mechanics, Rotterdam, Vol. 1, pp. 103–107.

Janbu, N. 1954. *Stability Analyses of Slopes with Dimensionless Parameters.* Harvard Soil Mechanics Series No. 46, Cambridge, Massachusetts, 81 pp.

Janbu, N. 1957. *Stability Analysis of Slopes with Dimensionless Parameters.* Harvard University Soil Mechanics Series, No. 46.

Janbu, N. 1968. *Slope Stability Computations.* Soil Mechanics and Foundation Engineering Report, Technical University of Norway, Trondheim.

Janbu, N. 1996. *Slope Stability Evaluations in Engineering Practice Special Lecture,* Proceedings of the 7th International Symposium on Landslides, Trondheim, Norway, Vol. 1, pp. 17–34.

Japanese Geotechnical Society 1996. *Remedial Measures Against Soil Liquefaction*. Balkema Publishers, 443 pp.

Jefferies, M.G. 1993. *Nor Sand: A Simple Critical State Model for Sand*. Geotechnique 43, pp. 91–103.

Jibson, R.W. 1993. *Predicting Earthquake-Induced Landslide Displacements Using Newmark's Sliding Block Analysis.* Transportation Research Record, No. 1411, pp. 9–17.

Jibson, R.W., Prentice, C.S., Langer, C.J., Rogozhin, E.A., and Borissoff, B.A. 1994. *Unusual Landslides and Other Strong-Shaking Effects of the 29 April 1991 Racha Earthquake in the Republic of Georgia*. Bulletin of the Seismological Society of America, Vol. 84, No. 4, pp. 963–973.

Kane, H. 1968. *A Mechanistic Explanation of the Physical Properties of Undisturbed Loess.* University of Iowa Research Project HR-126, 113 pp.

Karol, R.H. 1983. *Chemical Grouting.* Marcel Dekker, Inc., New York.

Keefer, D.S. 1984. *Landslides Caused by Earthquakes.* Geological Society of America Bulletin, Vol. 95, April, pp. 406–421.

Kenney, T.C., Pazin, M., and Choi, W.S. 1977. *Design of Horizontal Drains for Soil Slopes.* Journal Geotechnical Engineering Div., ASCE, Vol. 103, GT 11, November, pp.1311–1323.

Kenney, T.C., Lau, D., and Ofoegbu, G.I. 1984. *Permeability of Compacted Granular Materials.* Canadian Geotechnical Journal, Vol. 21, No. 4, pp. 726–729.

Kirkpatrick, W.M. 1957. *The Condition of Failure for Sands.* Proceedings 4th International Conference on Soil Mechanics, London, Vol. 1, pp. 172–178.

Koerner, J., Soong, T-Y., and Koerner, R.M. 1998. *Earth Retaining Wall Costs in the U.S.A.* GRI Report #20, Geosynthetics Institute, Folsom, Pennsylvania.

Kolbuszewski, J.J. 1948a. *An Experimental Study of the Maximum and Minimum Porosities of Sand.* Proceedings, 2nd International Conference on Soil Mechanics, Rotterdam, Vol. 1, pp. 158–165.

Kolbuszewski, J.J. 1948b. *General Investigation of the Fundamental Factors Controlling Loose Packing of Sands.* Proceedings, 2nd International Conference on Soil Mechanics, Rotterdam, Vol. VII, pp. 47–49.

Kovacs, W.D., Salomone, L.A., and Yokel, F.Y. 1983. *Comparison of Energy Measurements in the Standard Penetration Test Using the Cathead and Rope Method.* National Bureau of Standards Report to the U.S. Nuclear Regulatory Commission.

Krizek, R.J., and Spino, M.J. 2000. *Spatial and Directional Variations in Engineering Properties of an In Situ Silicate-Grouted Sand.* ASCE Geotechnical Special Publication No. 104, Advances in Grouting and Ground Modification, pp. 139–154.

Kulhawy, F.H., and Mayne, P.W. 1990. *Manual on Estimating Soil Properties for Foundation Design.* Report prepared by Cornell University for the Electric Power Research Institute.

Ladd, C.C., and Foott, R. 1974. *New Design Procedure for Stability of Soft Clays.* Journal of Geotechnical Engineering Div., ASCE, Vol. 100, No. 7, pp. 763–786.

Ladd, C.C., Foote, R., Ishihara, K., Schlosser, F., and Poulos, H.G. 1977. *Stress-deformation and Strength Characteristics.* Proceedings of 9th International Conference on Soil Mechanics and Foundation Engineering, Tokyo, 2: pp. 421–494.

Lambe, T.W., and Whitman, R.V. 1969. *Soil Mechanics.* John Wiley & Sons, New York, 553 pp.

Lander, J.F. 1996. *Tsunamis Affecting Alaska, 1737–1996.* Geophysical Research Documentation No. 31, U.S. Department of Commerce, N.O.A.A., Boulder, CO, 195 pp.

Landslide Technology 1986. *Washington Park Geotechnical Study.* Report to City of Portland, Bureau of Waterworks, 61 pp.

Laprade, W.T., Kirkland, T.E., Nashem, W.D., and Robertson, C.A. 2000. *Seattle Landslide Study.* Shannon and Wilson Report W-7992-01, 164 pp.

La Rochelle, P., Roy, M., and Tavenas, F. 1973. *Field Measurements of Cohesion in Champlain Clays.* Proceedings Eighth International Conference on Soil Mechanics and Foundation Engineering, Moscow, Vol. 1, pp. 229–236.

Lee M.K.W., and Finn, W.D.L. 1978. *DESRA-2: Dynamic Effective Stress Response Analysis of Soil Deposits with Energy Transmitting Boundary Including Assessment of Liquefaction Potential.* Soil Mechanics Series Report 38, Dept. of Civil Engineering, University of British Columbia, Vancouver, Canada.

Lefebvre, G. 1996. *Soft Sensitive Clays.* Landslides Investigation and Mitigation. Chapter 24, National Academy Press, Washington, D.C., pp. 607–619.

Lemos, L., Skempton, A.W., and Vaughan, P.R. 1985. *Earthquake Loading of Shear Surfaces in Slopes.* Proceedings: Eleventh International Conference on Soil Mechanics and Foundation Engineering, August 12–16, San Francisco, California, Vol. 4.

Lenzini, P.A., and Hendron, A.J. 1998. *Report on Adequacy of Design PARN Dock, Skagway, Alaska,* 20 pp.

Leschinsky, D., Baker, R., and Silver, M.L. 1985. *Three Dimensional Analysis of Slope Stability.* International Journal Numerical and Analytical Methods in Geomechanics, Vol. 9, pp. 199–223.

Leshchinsky, D. 2000. *Basics of Mechanically Stabilized Earth Wall Systems.* Geo Denver 2000, ASCE.

Liang, N. 1983. *An Examination of the SPT.* Thesis, University of British Columbia.

Lo, K., Ooi, P.L., and Lee, S.L. 1990. *Unified Approach to Ground Improvement by Heavy Tamping.* Journal of Geotechnical Engineering Division, ASCE Vol. 116, No. 3, pp. 514–527.

Long, E., and George, W. 1967. *Buttress Design Earthquake-Induced Slides.* Journal of the Soil Mechanics and Foundations Division, ASCE, Vol. 93, SM4, July 1967, pp. 595–610.

Long, M.T. (1986). *Camp Five Slide—Exploration, Design and Construction of a Horizontal Drain Solution.* Proceedings, 22nd Symposium on Engineering Geology and Soils Engineering, Boise, Idaho, pp. 246–265.

Long, J.H., Weatherby, D.E., and Cording, E.J. 1998. *Summary Report of Research on Permanent Ground Anchor Walls.* Volume 1: Current Practice and Limiting Equilibrium Analyses. Report FHWA-RD-98-065. Federal Highway Administration, McLean, Virginia.

Louis, C. 1986. *Theory and Practice in Soil Nailing Temporary or Permanent Works.* ASCE Annual Conference, Boston, Massachusetts.

Lovell, C.W. 1984. *Three Dimensional Analysis of Landslides.* Proceedings 4th International Symposium Landslides, Toronto, Vol. 2, pp. 451–455.

Lowe, J., and Karafiath, L. 1960. *Stability of Earth Dams upon Drawdown.* Proceedings of First Pan-American Conference on Soil Mechanics and Foundation Engineering, Mexico City, Vol. 2, pp. 537–560.

Lukas, R.G. 1986. *Dynamic Compaction for Highway Construction: Volume 1, Design and Construction Guidelines.* FHWA Report, FHWA-RD-86-133.

Lumb, P. 1975. *Slope Failures in Hong Kong.* Quarterly Journal of Engineering Geology, pp. 31–65.

Lunne, T., Robertson, P.K., and Powell, J.J.M. 1997. *Cone Penetration Testing in Geotechnical Practice.* Blackie Academic Publishers, London, 312 pp.

Lupini, J.F., Skinner, A.E., and Vaughan, P.R. 1981. *The Drained Residual Strength of Cohesive Soils.* Geotechnique Vol. 31, No. 2, pp. 181–213.

Madin, I.P., Priest, G.R., Mabey, M.A., Malone, S., Yelin, T.S., and Meier, D. 1993. *March 25, 1993 Scott Mills Earthquake—Western Oregon's Wake-up Call.* Oregon Geology Vol. 55, No. 3, May, pp. 51–57.

Mahar, J.W., Parker H.W., and Wuellner, W.W. 1975. *Shotcrete Practice in Underground Construction.* Report FRA-OR and D 75-90, U.S. Department of Transportation, Washington D.C.

Makdisi, F.I., and Seed, H.B. 1978. *Simplified Procedure for Estimating Dam and Embankment Earthquake-Induced Deformations.* Journal of the Geotechnical Engineering Division, ASCE, Vol. 104, July 1978, pp. 849–868.

Manson, M.W., Keefer, D.K., and McKittrick, M.A. 1992. *Landslides and Other Geologic Features in the Santa Cruz Mountains, California, Resulting from the Loma Prieta Earthquake of October 17, 1989.* DMG Open-File Report 91-05, California Department of Conservation, Division of Mines and Geology.

Mansur, C.I., Postol, G., and Salley, J.R. 2000. *Performance of Relief Well Systems along Mississippi River Levees.* Journal of Geotechnical and Geoenvironmental Engineering, Vol. 126, No. 8, pp. 727–738.

Marachi, N.D., Chan, C.K., Seed, H.B., and Duncan, J.M. 1969. *Strength and Deformation Characteristics of Rockfill Materials.* University of California Report TE-69-5, Berkeley, California.

Marcuson, W.F., and Franklin, A.G. 1979. *State of the Art of Undisturbed Sampling of Cohesionless Soils.* Paper GL-79-16, U.S. Waterways Experiment Station, Vicksburg, Mississippi.

McCulloch, D.S., and Bonilla, M.G. 1967. *Railroad Damage in the Alaska Earthquake.* Journal of the Soil Mechanics and Foundations Division, ASCE, Vol. 93, SM5, September 1967, pp. 89–100.

McKenna, G.T. 1995. *Grouted-In Installation of Piezometers in Boreholes.* Canadian Geotechnical Journal, Vol. 32, pp. 355–363.

McKibben, J.R. 1984. *Using Alternative Retaining Walls in Rebuilding the Freeway System in Atlanta.* TR News, No. 114, Transportation Research Board, Washington, D.C.

Meyerhof, G.G. 1953. *The Bearing Capacity of Foundations under Eccentric and Inclined Loads.* Proceedings of 3rd International Conference on Soil Mechanics and Foundation Engineering, Zurich, Vol. 1, pp. 440–445.

Meyerhof, G.G. 1957. *Discussion on Research on Determining the Density of Sands by Spoon Penetration Testing.* Proc. 4th International Conference on Soil Mechanics and Foundation Engineering, London, Vol. 3, p 110.

Michalowski, R.L. 2002. *Stability Charts for Uniform Slopes.* Journal of Geotechnical and Geoenvironmental Engineering, ASCE Vol. 128, No. 4, pp. 351–355.

Mikkelsen, P.E. 1996. *Field Instrumentation.* Chapter 11: Landslides: Investigation and Mitigation. Special Report 247 Transportation Research Board, National Academy of Sciences, pp. 278–316.

Mikkelsen, P.E. 2002. *Cement-Bentonite Grout Backfill for Borehole Instruments.* Geotechnical News, BiTech Publications, Vancouver, Canada, December pp. 38–42.

Mikkelsen, P.E. 2003. *Advances in Inclinometer Data Analysis.* Symposium on Field Measurements in Geomechanics, Oslo, Norway, 13 pp.

Mikkelsen, P.E., and Green, G.E. 2003. *Piezometers in Fully Grouted Boreholes.* Symposium on Field Measurements in Geomechanics, Oslo, Norway, 10 pp.

Millett, R.A., Perez, J-Y, and Davidson, R.R. 1992. *U.S.A. Practice Slurry Wall Specifications 10 Years Later.* Slurry Walls: Design Construction and Quality Control. ASTM STP 1129, pp. 42–66.

Milligan, V. 2003. *Some Uncertainties in Embankment Dam Engineering.* Terzaghi Lecture. Journal of Geotechnical and Geoenvironmental Engineering, Volume 129, No. 9, September, pp. 785–797.

Mitchell, J.K. 1970. *In-place Treatment of Foundation Soils.* Journal of SMFE Div., ASCE, Vol. 96, No. 1, pp. 73–110.

Mitchell, J.K., and Solymar, Z.V. 1984. *Time Dependent Strength Gain in Freshly Deposited or Densified Sand.* Journal of Geotechnical Engineering, ASCE, Vol. 110, No. 11, pp. 1559–1576.

Mitchell, J.K., and Villet, W.C.B. 1987. *Reinforcement of Earth Slopes and Embankments.* National Cooperative Highway Research Program Report 290, Transportation Research Board, 323 pp.

Mitchell, J.K., and Christopher, B.R. 1990. *North American Practice in Reinforced Soil Systems.* Proceedings Design and Performance of Earth Retaining Structures. ASCE Geotechnical Special Publication No. 25, pp. 322–346.

Morgenstern, N.R., and Price, V.E. 1965. *The Analysis of the Stability of General Slip Surfaces.* Geotechnique, Vol. 15, pp. 79–93.

Munfakh, G.A., Samtani, N.D., Castelli, R.J., and Wang, J-N. 1999. *Earth Retaining Structures Reference Manual.* NHI Course No. 13236-Module 6, publication No. FHWA-NHI-99-025, Federal Highway Administration, Washington D.C., 444 pp.

Nash, D. 1987. *A Comparative Review of Limit Equilibrium Methods of Stability Analysis.* Chapter 2 Slope Stability, edited by M.G. Anderson and K.S. Richards, John Wiley & Sons, pp. 11–75.

NAVFAC 1982. *Design Manual 7.2: Foundations and Earth Structures.* Department of the Navy, U.S. Government Printing Office, Washington, D.C.

Newmark, N.M. 1965. *Effects of Earthquakes on Dams and Embankments.* Geotechnique, Vol. 15, No. 2, pp. 139–160.

Nishida, K., and Yoga, T. 1996. *Grouting and Deep Mixing.* Balkema, Rotterdam.

Nixon, J.K. 1954. *Some Investigations on Granular Soils with Particular Reference to the Compressed Air Sand Sampler.* Geotechnique, March, pp. 16–31.

Norrish, N.I., and Wyllie, D.C. 1996. *Rock Slope Stability Analysis.* Chapter 15, Landslide Investigation and Mitigation. Special Report 247, Transportation Research Board, pp. 391–428.

Noson, L.L., Quamar A., and Thorsen, G.W. 1988. *Washington State Earthquake Hazards.* Washington Division of Geology and Earth Resources, Information Circular 85, 77 pp.

O'Brien, A.S. 1997. *Vibrocompaction of Loose Estuarine Sands.* Ground Improvement Geosystems—Densification and Reinforcement, Thomas Telford Ltd., London, pp. 120–126.

O'Rourke, T.D., and Jones, C.J.F.P. 1992. *Overview of Earth Retention Systems: 1970–1990.* Proceedings ASCE Conference on the Design and Performance of Earth Retaining Structures, Ithaca, New York, pp. 22–51.

Palmerton, J.B. 1984. *Stabilization of Moving Land Masses by Cast-in-Place Piles.* Report GL-84-4, U.S. Corps of Engineers, Waterways Experiment Station, Vicksburg, Mississippi, Report GL-84-4, 134 pp.

Peck, R.B. 1969. *Advantages and Limitations of the Observational Method in Applied Soil Mechanics.* Geotechnique Vol. 19, No. 2, pp. 171–187.

Peck, R.B., and Bazaraa, A.R.S. 1969. *Discussion on Settlement of Spread Footings on Sand.* Journal Soil Mechanics and Foundation Engineering Division, ASCE, Vol. 95, pp. 905–909.

Peck, R.B., Hanson, W.E., and Thornburn, T.H. 1974. *Foundation Engineering.* 2d ed. John Wiley, New York.

Peck, R.B., Ireland, H.O., and Teng, C.Y. 1948. *A Study of Retaining Wall Failures.* Proceedings of 2nd International Conference on Soil Mechanics, Rotterdam, Vol. 3, pp. 296–299.

Peltier, M.R. 1957. *Recherches Expérimentales sur la Courbe Intrinsèque de Rupture des Sols Pulvérulents.* Proceedings 4th International Conference on Soil Mechanics, Vol. 1, pp. 179–182.

Penman, A.D.M. 1960. *A Study of the Response Time of Various Types of Piezometers.* Conference on Pore Pressure and Suction in Soils, British Geotechnical Society, Butterworths, London, pp. 53–58.

Peterson, G.L., Scofield, D.H., Squier, L.R., and Toor, F. 1998. *Engineering Geology and Drainage of the Arizona Inn Landslide, U.S. Highway 101, Southern Oregon Coast.* Environmental, Groundwater and Engineering Geology. Applications from Oregon. Edited by Scott Burns. Star Publishing Company, Belmont, California, pp. 231–248.

Pierson, L.A., Davis, S.A., and VanVickle, R. 1990. *The Rockfall Hazard Rating System Implementation Manual.* Oregon Department of Transportation, Report No. FHWA-OR-EG-90-01.

Pierson, L.A., and Van Vickle, R. 1993. *Rockfall Hazard Rating System—Manual.* Federal Highway Administration, Washington, D.C. Report, FHWA-SA-93-057, 104 pp.

Plumelle, C., Schlosser, R., Delage, P., and Knochenmus, G. 1990. *French National Research Project on Soil Nailing: Clouterre.* ASCE Conference on Design and Performance of Earth Retaining Structures, Ithaca, New York, pp. 660–675.

Post Tensioning Institute 1996. *Recommendations for Prestressed Rock and Soil Anchors.* Post Tensioning Manual, Fifth Edition, Phoenix, Arizona, 405 pp.

Powers, J.P. 1981. *Construction Dewatering: A Guide to Theory and Practice.* John Wiley & Sons, New York, 484 pp.

Powers, J.P. 1992. *Construction Dewatering.* 2d ed. John Wiley & Sons, New York, 492 pp.

Prellwitz, R.W. 1978. *Analysis of Parallel Drains for Highway Cut-Slope Stabilization.* Proceedings, 16th Annual Engineering Geology and Soils Engineering Symposium, Boise, Idaho, pp. 153–180.

Pujol-Rius, A., Griffin, P., Neal, J., and Taki, O. 1989. *Foundation Stabilization of Jackson Lake Dam.* Proceedings, 12th International Conference on Soil Mechanics and Foundation Engineering, San Paulo, Brazil.

Quigley, R.M. 1980. *Geology, Mineralogy, and Geochemistry of Canadian Soft Soils: A Geotechnical Perspective.* Canadian Geotechnical Journal, Vol. 17, No. 2, pp. 261–285.

Reese, L.C. 1958. *Discussion of Soil Modulus for Laterally Loaded Piles* by McClelland, B., and Focht, J.A., Jr., Transactions, ASCE Vol. 123.

Reese, L.C., Cox. W.R., and Koop, F.D. 1974. *Analysis of Laterally Loaded Piles in Sand.* Proceedings, Fifth Offshore Technology Conference, Vol. II, Paper 2080, Houston, Texas, pp. 473–485.

Reese, L.C., Cox, W.R., and Koop, F.D. 1975. *Field Testing and Analysis of Laterally Loaded Piles in Stiff Clay.* Proceedings, 7th Offshore Technology Conference, Vol. 2, Houston, Texas, pp. 672–690.

Reese, L.C. 1984. *Handbook on Design of Piles and Drilled Shafts under Lateral Load.* Federal Highway Administration Report, FHWA-IP-84-11, 360 pp.

Reese, L.C., and O'Neill, M.W. 1988. *Drilled Shafts: Construction Procedures and Design Methods.* Federal Highway Administration Publication FHWA-HI-88-042, McLean, Virginia, 564 pp.

Reese, L.C., and Wang, S.T. 1989. *Documentation of Computer Program Lpile+.* Published by Ensoft, Inc., Austin, Texas.

Reese, L.C., Wang, S.T., and Fouse, J.L. 1992. *Use of Drilled Shafts in Stabilizing a Slope.* Conference on Stability of Slopes and Embankments II, ASCE, New York, pp. 1318–1332.

Reese, L.C., and Wang S.T. 1997. *Computer Program LPILE Plus, Version 3.0. A Program for the Analysis of Piles and Drilled Shafts under Lateral Loads.* Written for Ensoft, Inc., Austin, Texas, 232 pp.

Reusch, H. 1901. *Nogle Optegnelser Fra Vaerdalen (with summary in English).* Norges Geologiske Undersøkelse, Vol. 32, pp. 218–223.

Riemer, M.F. 1995. *Cyclic and Post-cyclic Testing of Saturated, Alluvial Silts.* Report to Cornforth Consultants, Inc., Geotechnical Engineering Report, UCB/GT/95-02, May, 16 pp.

Reinforced Earth Company 1995. *Minimum Embankment Requirements for MSE Structures.* Technical Bulletin MSE-7.

Riggs, C.O., Schmidt, N.O., and Rassieur, C.L. 1983. *Reproducible SPT Hammer Impact Force with an Automatic Free Fall SPT Hammer System.* ASTM Geotechnical Testing Journal, Vol. 6, No. 3, December, pp. 201–209.

Ripley, C.F. 1986. *Discussion of Internal Stability of Granular Filters* by T.C. Kenney and D. Lau. Canadian Geotechnical Journal, Vol. 23 (2), pp. 255–258.

Rixner, J.J., Kraemer, S.R., and Smith, A.D. 1986. *Prefabricated Vertical Drains. Vol. 1 Engineering Guidelines* Report FHWA/RD-86/168, Federal Highway Administration, McLean, Virginia, 117 pp.

Rowe, P.W. 1969. *The Relation Between the Shear Strength of Sands in Triaxial Compression, Plane Strain and Direct Shear.* Geotechnique, Vol. 19, No. 1, pp. 75–86.

Rowe, P.W. 1972. *The Relevance of Soil Fabric to Site Investigation Practice.* Geotechnique, Vol. 22, No. 2, June, pp. 193–300.

Royster, D.L. 1977. *Some Observations on the Use of Horizontal Drains in the Correction and Prevention of Landslides.* 28th Annual Highway Geology Symposium, Rapid City, South Dakota, 55 pp.

Royster, D.L. 1980. *Horizontal Drains and Horizontal Drilling: An Overview.* 59th Annual Meeting of Transportation Research Board, Washington, D.C., 21 pp.

Ruxton, B.P., and Berry, L. 1957. *Weathering of Granite and Associated Erosional Features in Hong Kong.* Bulletin, Geological Society of America, Vol. 68, pp. 1263–1291.

Sabatini, P.J., Elias, V., Schmertmann, G.R., and Bonaparte, R. 1997. *Earth Retaining Systems.* Geotechnical Engineering Circular No. 2. Report FHWA-SA-96-038 Office of Technology Application, FHWA, Washington, D.C., 161 pp.

Sabatini, P.J., Pass, D.G., and Bachus, R.C. 1999. *Ground Anchors and Anchored Systems.* Geotechnical Engineering Circular No. 4, FHWA-IF. 99-015, 176 pp.

Sarma, S.K. 1973. *Stability Analyses of Embankments and Slopes.* Geotechnique, Vol. 23, pp. 423–433.

Sarma, S.K. 1979. *Stability Analysis of Embankments and Slopes.* Journal Geotechnical Eng. Division, ASCE, Vol. 105, pp. 1511–1524.

Salah-Mars, S., Green, R.K., Kanakari, H., Mejia, L.H., Weaver, K.D., and Boddie, P.J. 1995. *Evaluation of Earthquake-Induced Slope Displacements.* Proceedings: Third International Conference on Recent Advances in Geotechnical Earthquake Engineering and Soil Dynamics, April 2–7, Vol. 1, St. Louis, Missouri.

Scherer, S.D., and Gay, R.L. 2000. *Compaction Grouting: Three Midwest Case Histories.* ASCE Geotechnical Special Publication No. 104, pp. 65–82.

Schlosser, F., and Delage, P. 1987. *Reinforced Soil Retaining Structures.* The Application of Polymeric Reinforcement in Soil Retaining Structures. pp. 3–65.

Schlosser, F. 1990. *Mechanically Stabilized Earth Retaining Structures in Europe.* Proceedings ASCE Conference on Design and Performance of Earth Retaining Structures. Ithaca, N.Y. pp. 347–378.

Schmertmann, J.H., and Smith, T.V. 1977. *A Summary of SPT Energy Calibration Services Performed for the Florida Department of Transportation.* Final Research Report 245 * D73 College of Engineering, University of Florida, Gainesville, Florida, 21 pp.

Schmertmann, J.H., and Palacios, A. 1979. *Energy Dynamics of SPT.* Journal of Geotechnical Engineering ASCE, Vol. 105, No. GT8, pp. 899–926.

Schnabel, P.B., Lysmer, J., and Seed, H.B. 1972. *SHAKE: A Computer Program for Earthquake Response Analysis of Horizontally Layered Sites.* Report UCB/EERC-72/12 University of California, Berkeley.

Schoenwolf, D.A., and Dobbels, D.J. 1992. *Underreamed Drilled Shafts Installed Using Slurry Methods.* Slurry Walls: Design Construction and Quality Control. ASTM STP 1126, pp. 194–206.

Schor, H. 1992. *Hills Like Nature Makes Them.* Urban Land, pp. 40–43.

Seed, H.B., and Goodman, R.E. 1964. *Earthquake Stability of Slopes of Cohesionless Soils.* Journal of the Soil Mechanics and Foundation Division, ASCE, Vol. 90, SM6, November 1964, pp. 43–73.

Seed, H.B., and Chan, C.K. 1966. *Clay Strength under Earthquake Loading.* Journal, Soil Mechanics and Foundation Div., ASCE, Vol. 92, SM2, March, pp. 53–78.

Seed, H.B., and Lee, K.L. 1966. *Liquefaction of Saturated Sands During Cyclic Loading.* Journal of Soil Mechanics and Foundations Division, ASCE, Vol. 92, No. SM6, Paper 4972, pp. 105–134.

Seed, H.B. 1967. *Slope Stability During Earthquakes.* Journal of the Soil Mechanics and Foundations Division, ASCE, Vol. 93, No. SM4, July 1967, pp. 299–323.

Seed, H.B., and Wilson, S.D. 1967. *The Turnagain Heights Landslide, Anchorage, Alaska.* Journal of Soil Mechanics and Foundations Division, ASCE, Vol. 93, SM 4, pp. 325–353.

Seed, H.B. 1970. *Earth Slope Stability During Earthquakes.* Chapter 15, Earthquake Engineering, edited by Robert Wiegel, Prentice-Hall, pp. 383–401.

Seed, H.B. 1970. *Soil Problems and Soil Behavior.* Chapter 10, Earthquake Engineering, edited by Robert Wiegel, Prentice-Hall, pp. 227–251.

Seed, H.B., and Whitman, R.V. 1970. *Design of Earth Retaining Structures for Dynamic Loads.* Proceedings, ASCE Specialty Conference on Lateral Stresses and Earth Retaining Structures. Cornell University, Ithaca, New York, pp. 103–147.

Seed, H.B., and Idriss, I.M. 1971. *Simplified Procedure for Evaluating Soil Liquefaction Potential.* Proceedings Journal of Soil Mechanics and Foundations Division, ASCE, Vol. 97, No. SM9, September, pp. 1249–1273.

Seed, H.B., Arango, I., and Chan, C.K. 1975. *Evaluation of Soil Liquefaction Potential During Earthquakes.* Report EERC 75-28, Earthquake Engineering Research Center, University of California, Berkeley.

Seed, H.B., Idriss, I.M., Makdisi, F., and Banerjee, N. 1975. *Representation of Irregular Stress Time Histories by Equivalent Uniform Stress Series in Liquefaction Analyses.* Report UCB/EERC-75/29, University of California, Berkeley.

Seed, H.B. 1979. *Considerations in the Earthquake-Resistant Design of Earth and Rockfill Dams.* Geotechnique Vol. 29, No. 3, pp. 215–263.

Seed, H.B., and Idriss, I.M. 1982. *Ground Motions and Soil Liquefaction During Earthquakes.* Earthquake Engineering Research Institute monograph.

Seed, H.B. 1983. *Stability of Port Fills and Coastal Deposits.* Seventh Asian Regional Conference SMFE, Israel Institute of Technology, Haifa, Israel.

Seed, H.B., Idriss, I.M., and Arango, I. 1983. *Evaluation of Liquefaction Potential using Field Performance Data.* Journal Geotechnical Engineering Division, ASCE, Vol. 109, No. 3, pp. 458–482.

Seed, H.B., Tokimatsu, K., Harder, L.F., and Chung, R.M. 1985. *Influence of SPT Procedures in Soil Liquefaction Resistance Evaluations.* Journal of Geotechnical Engineering Division, ASCE, Vol. 111, pp. 1425–1445.

Seed, H.B. 1986. *Design Problems in Soil Liquefaction.* Journal Geotechnical Engineering Division, ASCE, Vol. 113, No. 8, pp. 827–845.

Seed, H.B. 1987. *Design Problems in Soil Liquefaction.* Journal Geotechnical Engineering Division, ASCE, Vol. 113, No. 8, pp. 827–845.

Seed, R.B., and Harder, L.F. 1990. *SPT-Based Analysis of Cyclic Pore Pressure Generation and Undrained Residual Strength.*

Proceedings, H. Bolton Seed Memorial Synposium, Vol. 2, Bi Tech Publishers, pp. 351–376.

Seed, R.B., Mitchell, J.K., and Seed, H.B. 1990. *Kettleman Hills Waste Landfill Slope Failure II; Stability Analyses.* Journal of Geotechnical Engineering, ASCE, Vol. 116, No. 4, pp. 669–691.

Sellers, J.B. 2000. *Vibrating Wire Sensors and Their Applications.* Geotechnical Field Instrumentation Seminar, ASCE, Seattle Section, April, 9 pp.

Serota, S., and Jangle, A. 1972. *A Direct-reading Pocket Shear Vane.* Civil Engineering Magazine, ASCE, January.

Shannon, W.L. 1966. *Slope Failures at Seward, Alaska.* ASCE Conference on Stability and Performance of Slopes and Embankments, Berkeley, California.

Sherard, J.L., Woodward, R.J., Gizienski, S.F., and Clevenger, W.A. 1963. *Earth and Earth-Rock Dams.* John Wiley & Sons, New York.

Sherard, J.L., Dunnigan, L.P., and Talbot, J.R. 1984(a). *Basic Properties of Sand and Gravel Filters.* Journal of Geotechnical Engineering, ASCE, Vol. 110, No. 6, pp. 684–700.

Sherard, J.L., Dunnigan, L.P., and Talbot, J.R. 1984(b). *Filters for Silts and Clays.* Journal of Geotechnical Engineering, ASCE, Vol. 110, No. 6, pp. 701–718.

Shuttle, D.A., and Jefferies, M.G. 1998. *Dimensionless and Unbiased CPT Interpretation in Sand.* International Journal for Numerical and Analytical Methods in Geomechanics, Vol. 22, pp. 351–391.

Shuttle, D., and Jefferies, M. 2000. *Prediction and Validation of Compaction Grout Effectiveness.* ASCE Geotechnical Special Publication No. 104, pp. 48–64.

Skempton, A.W. 1954. *The Pore-Pressure Coefficients A and B.* Geotechnique, pp. 143–147.

Skempton, A.W. 1957. *Discussion on The Planning and Design of the New Hong Kong Airport.* Proceedings, Institution of Civil Engineers, pp. 305–307.

Skempton, A.W. 1964. *Long-term Stability of Clay Slopes.* Fourth Rankine Lecture. Geotechnique Vol. 14, No. 2, June, pp. 75–102.

Skempton, A.W., and Hutchinson, J.N. 1969. *Stability of Natural Slopes and Embankment Foundations: State-of-the-Art Report.* Proceedings 7th International Conference on SMFE, Mexico City, pp. 291–340.

Skempton, A.W. 1985. *Residual Strength of Clays in Landslides, Folded Strata and the Laboratory.* Geotechnique Vol. 35, No. 1, pp. 3–18.

Skempton, A.W. 1986. *Standard Penetration Test Procedures and the Effects in Sands of Overburden Pressure, Relative Density, Particle Size, Ageing and Overconsolidation.* Geotechnique. Vol. 36, No. 3, pp. 425–448.

Slocombe, B.C., Bell, A.L., and Baez, J.I. 2000. *The Densification of Granular Soils Using Vibro Methods.* Geotechnique Vol. 50, No. 6, pp. 715–725.

Slope Indicator Company 1994. *Applications Guide.* 2d ed. Slope Indicator Co., Seattle, Washington, 170 pp.

Smith, D.D. 1980. *The Effectiveness of Horizontal Drains.* California DOT Report, FHWA/CA/TL-80/16, 79 pp.

Solymar, Z.V. 1984. *Compaction of Alluvial Sands by Deep Blasting.* Canadian Geotechnical Journal Vol. 21, pp. 305–321.

Soydemir, C., Swekosky, F.J., Baez, J.I., and Mooney, J.S. 1997. *Ground Improvement at Albany Airport, New York.* ASCE Geotechnical Special Publication No. 69, pp. 506–524.

Spencer, E. 1967. *A Method of Analysis of the Stability of Embankments Assuming Parallel Inter-Slice Forces.* Geotechnique, Vol. 17, pp. 11–26.

Spiker, E.C., and Gori, P.L. 2000. *National Landslide Hazards Mitigation Strategy—A Framework for Loss Reduction.* U.S.G.S. Open File Report 00-450, USGS National Center, Reston, Virginia, 58 pp.

Squier, L.R., and Harvey, A.F. 2000. *Two Debris Flows: Logging Impacts in Coast Range, Oregon, USA.* Second International Conference on Debris-Flow Hazards Mitigation: Mechanics, Prediction and Assessment, August 16–20, Taipei, Taiwan, 31 pp.

Stanic, B. 1984. *Influence of Drainage Trenches on Slope Stability.* Journal of Geotechnical Engineering, ASCE, Vol. 110, No. 11, November, pp. 1624–1636.

Stark, T.D., and Vettel 1992. *Bromhead Ring Shear Test Procedure.* ASTM Geotechnical Testing Journal, Vol. 15, No. 1, March, pp. 24–32.

Stark, T.D., and Mesri, G. 1992. *Undrained Shear Strength of Liquefied Sands for Stability Analysis.* Journal of Geotechnical Engineering, ASCE, Vol. 118, No. 11, pp. 1727–1747.

Stark, T.D., and Eid, H.T. 1993. *Modified Bromhead Ring Shear Apparatus.* Geotechnical Testing Journal, ASTM, March, pp. 100–107.

Steinberg, S. 1982. *Energy Calibration and Hammer Influence on SPT.* Conference on Updating Subsurface Sampling of Soils and Rocks and Their In Situ Testing, Santa Barbara, California.

Swihart, J., and Haynes, J. 2002. *Canal Lining Demonstration Project: Year 10 Final Report.* Report R-02-03 Bureau of Reclamation, Boise, Idaho and Denver, Colorado, 230 pp.

Sy, A., and Campanella, R.G. 1993. *Dynamic Performance of the Becker Hammer Drill and Penetration Test.* Canadian Geotechnical Journal, Vol. 30, pp. 607–619.

Sy. A., and Campanella, R.G. 1994. *Becker and Standard Penetration Tests (BPT-SPT) Correlations with Consideration of Casing Friction.* Canadian Geotechnical Journal, Vol. 31, No. 3, pp. 343–356.

Taki, O., and Yang, D.S. 1991. *Soil-cement Mixed Wall Technique.* Proceedings, Geotechnical Engineering Congress ASCE, Denver, Colorado, pp. 298–309.

Tallard, G.R. 1992. *New Trenching Method Using Synthetic Bio-Polymers.* Slurry Walls: Design, Construction and Quality Control ASTM STP 1129, pp. 86–102.

Tamaro, G.J., and Poletto, R.J. 1992. *Slurry Walls—Construction Quality Control.* Slurry Walls: Design, Construction and Quality Control, ASTM STP 1129, pp. 26–41.

Tang, W.H., Stark, T.D., and Angulo, M. 1999. *Reliability in Back Analysis of Slope Failures.* Journal Soil Mechanics and Foundations, Tokyo, October.

Tavenas, F., and Leroueil. S. 1987. *Laboratory and In Situ Stress-Strain-Time Behavior of Soft Clays.* Proc. International Symposium on Geotechnical Engineering of Soft Soils, Mexico City, Vol. 2, pp. 3–48.

Taylor, D.W. 1937. *Stability of Earth Slopes.* Journal of Boston Society of Civil Engineers, Vol. 24, No. 3, p. 197–246.

Taylor, D.W. 1941. *Cylindrical Compression Research Program on Stress-Deformation and Strength Characteristics of Soils.* 7th Progress Report to U.S. Army Corps of Engineers.

Taylor, D.W. 1948. *Fundamentals of Soil Mechanics.* John Wiley & Sons, New York, 700 pp.

Terzaghi, K. 1943. *Theoretical Soil Mechanics.* John Wiley & Sons, New York, 500 pp.

Terzaghi, K., and Peck, R.B. 1948. *Soil Mechanics in Engineering Practice.* 1st ed. John Wiley & Sons, New York, 729 pp.

Terzaghi, K. 1956. *Varieties of Submarine Slope Failures.* Harvard University Soil Mechanics Series No. 52. Also published as Norwegian Geotechnical Institute Publication 25.

Terzaghi, K. 1957. *Opening Address.* Proceedings, 4th International Conference on Soil Mechanics and Foundation Engineering, London. Vol. 3, p. 58.

Terzghi, K., Peck, R.B., and Mesri, G. 1996. *Soil Mechanics in Engineering Practice.* John Wiley & Sons, Inc., 549 pp.

Thompson S.N., and Miller, I.R. 1990. *Design, Construction, and Performance of a Soil Nailed Wall in Seattle, Washington.* Proceedings ASCE Conference on Design and Performance of Earth Retaining Structures, Ithaca, New York, pp. 629–643.

Tofani, G. 2000. *Grout-In-Place Installation Procedures for Combination Slope Inclinometer/Multi-Stage Piezometer Installations.* Seminar on Geotechnical Field Instrumentation ASCE, Seattle, 12 pp.

Tunne, T., Robertson, P.K., and Powell, J.J.M. 1997. *Cone Penetration Testing in Geotechnical Practices.* Blackie Academic Publishers, London, England, 312 pp.

Turnbull, W.J. 1968. *Construction Problems Experienced with Loess Soils.* Highway Research Record 212, Highway Research Board, National Research Council, Washington, D.C., pp. 10–27.

Ueblacker, G. 1996. *Portland Westside Lightrail Corridor Project Micropile Retaining Wall.* Foundation Drilling, November, pp. 8–12.

UNESCO Working Party for World Landslide Inventory. 1993. *Multilingual Landslide Glossary.* BiTech Publishers Ltd., Richmond, B.C., Canada.

Updike, R.G. 1999. *Man in a Changing Landscape.* Proposed National Agenda for Landslide Hazard Mitigation, U.S. Geological Survey, Denver, Colorado, 79 pp.

U.S. Bureau of Reclamation. 1960. *Earth Manual.* U.S. Government Printing Office, Washington, D.C., 751 pp.

U.S. Corps of Engineers 1960. *Severe Windstorms of Record.* Technical Bulletin No. 2 (CW-178), January.

U.S. Corps of Engineers. 1962. *Waves in Inland Reservoirs.* Technical memorandum No. 132 (CW-164 and CW-165), November.

U.S. Corps of Engineers. 1970. *Engineering and Design—Stability of Earth and Rockfill Dams.* Engineering Manual EM 1110-2-1902.

U.S. Corps of Engineers. 1973. *Shore Protection Manual.* Coastal Engineering Research Center, Fort Belvoir, Virginia.

U.S. Corps of Engineers. 1975. *Large Wave Tank Tests of Riprap Stability.* Technical Memorandum No. 51, May.

U.S. Corps of Engineers. 1976. *Wave Runup and Wind Setup on Reservoir Embankments.* Report ETL 1110-2-221.

U.S. Corps of Engineers. 1978. *Slope Protection Design for Embankments in Reservoirs.* Engineer Technical Letter ETL 1110-2-222, July, 11 pp.

U.S. Corps of Engineers. 1986. *Seepage Analysis and Control for Dams.* Manual EM 1110-2-1901, September.

U.S. Corps of Engineers. 1992. *Design, Construction, and Maintenance of Relief Wells.* Engineers Manual EM 1110-2-1914, 86 pp.

U.S. Corps of Engineers. 1993. *Standard Practice for Shotcrete.* Publication EM 1110-2-2005, Washington, D.C.

U.S. Corps of Engineers. 1994. *Hydraulic Design of Flood Control Channels.* Manual EM 1110-2-1601, June.

U.S. Corps of Engineers. 1996. *Soil Sampling.* Manual EM 1110-1-1906, September, 1996, NTIS, Springfield, Virginia.

U.S. Corps of Engineers. 1997. *Chemical Grouting.* Adapted from Corps of Engineers Manual 24, ASCE Press, New York, 30 pp.

U.S. Corps of Engineers. 1998. *Geophysical Exploration for Engineering and Environmental Investigations.* USACE Technical Engineering and Design Guide, No. 23, 204 pp.

U.S. Corps of Engineers. 1998. *Risk-based Analysis in Geotechnical Engineering for Support of Planning Studies.* Engineering Circular 1110-2-554.

U.S. Federal Highway Administration. 1980. *Highway Subdrainage Design.* Report No. FHWA-TS-80-224, 113 pp.

U.S. Federal Highway Administration. 1989. *Design of Riprap Revetment.* Hydraulic Engineering Circular No. 11, Publication FHWA-1P-89-016, 156 pp.

U.S. Federal Highway Administration. 1991. *Soil Nailing for Stabilization of Highway Slopes and Excavations.* Federal Highway Administration Publication No. FHWA-RD-89-193, 209 pp.

U.S. Forest Service 1994. *Application Guide for Launched Soil Nails, Vols. 1 and 2.* FHWA-FPL-93-003, Washington D.C., 60 pp.

U.S. Soil Conservation Service. 1978. *Riprap for Slope Protection Against Wave Action.* Water Resources Publications, Littleton, Colorado, 53 pp.

U.S. Soil Conservation Service. 1986. *Guide for Determining the Gradation of Sand and Gravel Filters.* Soil Mechanics Note No. 1, 210-VI. U.S. Dept. of Agriculture.

Vanel, P. 1992. *Making Diaphragm Wall Joints Watertight with the CWS System.* Slurry Walls: Design Construction and Quality Control ASTM STP 1126, pp. 16–25.

Varnes, D.J. 1978. *Slope Movement Types and Processes.* Chapter 2, Landslides: Analysis and Control, Special Report 176, Transportation Research Board, National Academy of Sciences, Washington, D.C., 234 pp.

Vaughan, P.R. 1969. *A Note on Sealing Piezometers in Boreholes.* Geotechnique Vol. 19, No. 3, September, pp. 405–413.

Vessely, D.A., and Cornforth, D.H. 1998. *Estimating Seismic Displacements of Marginally Stable Landslides Using Newmark Approach.* Geotechnical Earthquake Engineering and Soil Dynamics, Seattle, Washington, American Society of Civil Engineers, Geotechnical Special Publication No. 75, Vol. 1, pp. 800–811.

Walker, A.D. 1994. *A Deep Soil Mix Cutoff Wall at Lockington Dam, Ohio.* Proceedings, In Situ Deep Soil Improvement ASCE Convention, Atlanta, Georgia, pp. 133–145.

Walkinshaw, J.L. 1975. *Reinforced Earth Construction.* Report FHWA-DP-18, Federal Highway Administration, Washington, D.C., 70 pp.

Wang, S.T., and Reese, L.C. 1986. *Study of Design Methods for Vertical Drilled Shaft Retaining Walls.* Texas Department of Highways and Public Transportation, Austin, Texas.

Watn, A., Christensen, S., Emdal, A., and Nordal, S. 1999. *Lime-Cement Stabilization of Slopes—Experiences and a Design Approach.* Proceedings, International Conference on Dry Deep Mix Methods for Deep Soil Stabilization, Stockholm, Sweden, pp. 169–176.

Weatherby, D.E. 1998. *Design Manual for Permanent Ground Anchor Walls.* Report FHWA-RD-97-130, Federal Highway Administration, McLean, Virginia, 242 pp.

Weaver, K.D. 1991. *Dam Foundation Grouting.* ASCE Press, New York, 178 pp.

Weaver, K.D. 2000. *A Critical Look at Use of "Rules of Thumb" for Selection of Grout Injection Pressures.* ASCE Geotechnical Special Publication No. 104, Advances in Grouting and Ground Modification, pp. 173–180.

Welsh, J. 1986. *In Situ Testing for Ground Modification Techniques.* ASCE Geotechnical Special Publication No. 6, pp. 322–335.

Wilde, S.A. 1958. *Forest Soils: Their Protection and Relation to Silviculture.* Ronald Press, New York, 537 pp.

Winterkorn, H.F., and Fang, H-Y. 1975. *Foundation Engineering Handbook.* Van Nostrand Reinhold Co., 751 pp.

Wolff, T.F. 1994. *Evaluating the Reliability of Existing Levees.* Report for U.S. Corps of Engineers, Waterways Experiment Station, Vicksburg, Mississippi.

Wright, S.G. 1991. *UTEXAS3: A Computer Program for Slope Stability Calculations.* Shinoak Software, Austin, Texas, 138 pp.

Wu, T.H., Tang, W.H., and Einstein, H.H. 1996. *Landslide Hazard and Risk Assessment.* Landslides Investigation and Mitigation Special Report 247. Transportation Research Board, National Research Council. National Academy Press, Washington DC, pp. 106–118.

Wu, T.H., and Sangrey, D.A. 1978. *Strength Properties and Their Measurement.* Chapter 6, Landslides: Analysis and Control Special Report 176, Transportation Research Board, National Academy of Sciences, Washington, D.C., 234 pp.

Wyllie, D.C., and Brawner, C.O. 1975. *Rock Slope Stability on Railway Projects.* Proceedings, American Railway Engineering Association meeting, Vancouver, British Columbia, Canada.

Wyllie, D.C. 1987. *Rock Slope Inventory System.* Proceedings, FHWA Rockfall Mitigation Seminar, Region 10, Portland, Oregon.

Wyllie, D.C., and Norrish, N.I. 1996. *Rock Strength Properties and Their Measurement.* Chapter 14, Landslides Investigation and Mitigation Special Report 247, Transportation Research Board, pp. 372–390.

Yang, D.S., and Takeshima, S. 1994. *Soil Mix Walls in Difficult Ground.* Proceedings In Situ Deep Soil Improvement ASCE Convention, Atlanta, Georgia, pp. 106–120.

Yang, D.S., Yagihashi, J.N., and Yashizawa, S.S. 1998. *SWING Method for Deep Mixing.* Geo-Institute ASCE Soil Improvement for Big Digs. Geotechnical Special Publication No. 81, pp. 111–121.

Yoshimi, Y., and Tokimatsu, K. 1983. *SPT Practice: Survey and Comparative Tests.* Soils and Foundations, Vol. 23, No. 3, pp. 106–111.

Yoshimi, Y., Hatanaka, M., and Oh-Oka, H. 1978. *Undisturbed Sampling of Saturated Sands by Freezing.* Soils and Foundations, Japanese Society SMFC, Tokyo, Japan, Vol. 18, No. 3, pp. 59–73.

Youd, T.L., and Idriss, I.M. 2001. *Liquefaction Resistance of Soils.* Summary Report from the 1996 NCEER and 1998 NCEER/NSF Workshop on Evaluation of Liquefaction Resistance of Soils, Journal of Geotechnical and Geoenvironmental Engineering, Vol. 127, No. 4, pp. 297–313.

Zaruba, Q., and Mencl, V. 1969. *Landslides and Their Control.* Elsevier Publishers, Amsterdam.

Zen, K., Yamazaki, H., Watanabe, A., Yoshizawa, H., and Tamai, A. 1987. *Study on a Reclamation Method with Cement-Mixed Sandy Soils.* Technical Note of the Port and Harbour Research Institute, No. 579.

Credits

Figures 2.16, 8.1(a)(b), 8.3: © John Wiley & Sons Limited; reprinted with permission
Figures 2.30, 18.2, 18.4, 18.7, 18.9, 19.44, 19.45, 19.46, 19.47, 21.19, 22.11, 22.12: copyright ASTM International; reprinted with permission
Figure 20.18: reproduced with permission of the Institution of Civil Engineers, London
Figure 21.1: copyright Thomas Telford Ltd., London
Figure 20.32: © 2000 reprinted in "Design Manual for Reinforced Earth Walls"

Case History Cross-References

It may be of interest to some readers to be provided with the cross-references to the twelve case histories within the main text (Parts A,B).* The pages are listed below.

*In total, the book refers to about 100 actual examples of landslides or ground treatments.

Index